AVERAGE RATE OF CHANGE

Average Rate of Change $= \dfrac{y_2 - y_1}{x_2 - x_1}$

GRAPHING QUADRANTS

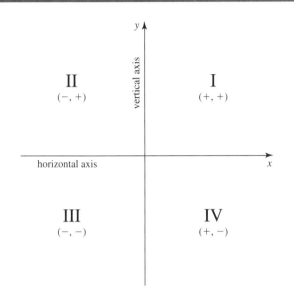

PYTHAGOREAN THEOREM

$a^2 + b^2 = c^2$

CONVERTING ANGLES

Multiply by $\dfrac{\pi}{180°}$ to convert degrees to radians.

Multiply by $\dfrac{180°}{\pi}$ to convert radians to degrees.

$y = $

$y = \cos x$

$y = \tan x = \dfrac{\sin x}{\cos x}$

$y = \csc x = \dfrac{1}{\sin x}$

$y = \sec x = \dfrac{1}{\cos x}$

$y = \cot x = \dfrac{\cos x}{\sin x}$

UNIT CIRCLE

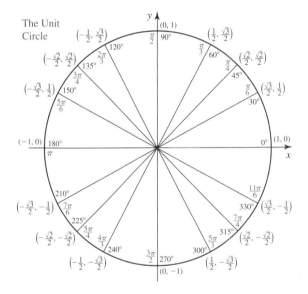

The coordinates of any point on the unit circle are $(\cos \theta, \sin \theta)$.

LAW OF COSINES

$$c^2 = a^2 + b^2 - 2ab\cos(C)$$

LAW OF SINES

$$\frac{\sin(A)}{a} = \frac{\sin(B)}{b} = \frac{\sin(C)}{c}$$

POLAR COORDINATES

(r, θ) where $r = \sqrt{x^2 + y^2}$ and $\theta = \tan^{-1}\left(\frac{y}{x}\right)$

TRIGONOMETRIC IDENTITIES

Pythagorean Identities:

$$\cos^2(\theta) + \sin^2(\theta) = 1$$

$$1 + \tan^2(\theta) = \sec^2(\theta)$$

$$\cot^2(\theta) + 1 = \csc^2(\theta)$$

Product to Sum Identities:

$$\cos(\theta)\cos(\phi) = \frac{\cos(\theta + \phi) + \cos(\theta - \phi)}{2}$$

$$\sin(\theta)\sin(\phi) = \frac{\cos(\theta - \phi) - \cos(\theta + \phi)}{2}$$

$$\sin(\theta)\cos(\phi) = \frac{\sin(\theta + \phi) + \sin(\theta - \phi)}{2}$$

Sum to Product Identities:

$$\cos(\theta) + \cos(\phi) = 2\cos\left(\frac{\theta + \phi}{2}\right)\cos\left(\frac{\theta - \phi}{2}\right)$$

$$\cos(\theta) - \cos(\phi) = -2\sin\left(\frac{\theta + \phi}{2}\right)\sin\left(\frac{\theta - \phi}{2}\right)$$

$$\sin(\theta) + \sin(\phi) = 2\sin\left(\frac{\theta + \phi}{2}\right)\cos\left(\frac{\theta - \phi}{2}\right)$$

$$\sin(\theta) - \sin(\phi) = 2\sin\left(\frac{\theta - \phi}{2}\right)\cos\left(\frac{\theta + \phi}{2}\right)$$

Sum and Difference Identities:

$$\cos(\theta + \phi) = \cos(\theta)\cos(\phi) - \sin(\theta)\sin(\phi)$$

$$\cos(\theta - \phi) = \cos(\theta)\cos(\phi) + \sin(\theta)\sin(\phi)$$

$$\sin(\theta + \phi) = \sin(\theta)\cos(\phi) + \cos(\theta)\sin(\phi)$$

$$\sin(\theta - \phi) = \sin(\theta)\cos(\phi) - \cos(\theta)\sin(\phi)$$

$$\tan(\theta + \phi) = \frac{\tan(\theta) + \tan(\phi)}{1 - \tan(\theta)\tan(\phi)}$$

$$\tan(\theta - \phi) = \frac{\tan(\theta) - \tan(\phi)}{1 + \tan(\theta)\tan(\phi)}$$

Double-Angle Identities:

$$\cos(2\theta) = \cos^2(\theta) - \sin^2(\theta)$$
$$= 2\cos^2\theta - 1$$
$$= 1 - 2\sin^2\theta$$

$$\sin(2\theta) = 2\sin(\theta)\cos(\theta)$$

$$\tan(2\theta) = \frac{2\tan(\theta)}{1 - \tan^2(\theta)}$$

Half-Angle Identities:

$$\cos\left(\frac{\theta}{2}\right) = \pm\sqrt{\frac{1 + \cos(\theta)}{2}}$$

$$\sin\left(\frac{\theta}{2}\right) = \pm\sqrt{\frac{1 - \cos(\theta)}{2}}$$

$$\tan\left(\frac{\theta}{2}\right) = \frac{1 - \cos(\theta)}{\sin(\theta)}$$

or

$$\tan\left(\frac{\theta}{2}\right) = \frac{\sin(\theta)}{1 + \cos(\theta)}$$

Precalculus: A Make It Real Approach
Frank C. Wilson, Scott L. Adamson, Trey Cox, Alan E. O'Bryan

Acquisitions Editor: Gary Whalen
Developmental Editor: Carolyn Crockett
Assistant Editor: Cynthia Ashton
Editorial Assistant: Sabrina Black
Media Editor: Lynh Pham
Senior Marketing Manager: Danae April
Marketing Communications Manager: Mary Anne Payumo
Content Project Manager: Jennifer Risden
Design Director: Rob Hugel
Art Director: Vernon Boes
Print Buyer: Karen Hunt
Rights Acquisitions Specialist: Tom McDonough
Production Service: Lachina Publishing Services
Text Designer: Terri Wright
Art Editor: Amy Mayfield
Photo Researcher: Bill Smith Group
Text Researcher: Karyn Morrison
Copy Editor: Amy Mayfield
Illustrator: Matrix Art Services; Lachina Publishing Services
Cover Designer: Terri Wright
Cover Image: Robert George Young/Masterfile
Compositor: Lachina Publishing Services

© 2013 Brooks/Cole, Cengage Learning

ALL RIGHTS RESERVED. No part of this work covered by the copyright herein may be reproduced, transmitted, stored, or used in any form or by any means graphic, electronic, or mechanical, including but not limited to photocopying, recording, scanning, digitizing, taping, Web distribution, information networks, or information storage and retrieval systems, except as permitted under Section 107 or 108 of the 1976 United States Copyright Act, without the prior written permission of the publisher.

> For product information and technology assistance, contact us at
> **Cengage Learning Customer & Sales Support, 1-800-354-9706.**
> For permission to use material from this text or product,
> submit all requests online at **www.cengage.com/permissions**.
> Further permissions questions can be e-mailed to
> **permissionrequest@cengage.com**.

Library of Congress Control Number: 2011939007

ISBN-13: 978-0-618-94991-5

ISBN-10: 0-618-94991-7

Brooks/Cole
20 Davis Drive
Belmont, CA 94002-3098
USA

Cengage Learning is a leading provider of customized learning solutions with office locations around the globe, including Singapore, the United Kingdom, Australia, Mexico, Brazil, and Japan. Locate your local office at **www.cengage.com/global**.

Cengage Learning products are represented in Canada by Nelson Education, Ltd.

To learn more about Brooks/Cole, visit **www.cengage.com/brookscole**.

Purchase any of our products at your local college store or at our preferred online store **www.cengagebrain.com**.

Printed in the United States of America
1 2 3 4 5 6 7 15 14 13 12 11

Precalculus
A Make It Real Approach

Frank C. Wilson
Chandler-Gilbert Community College

Scott L. Adamson
Chandler-Gilbert Community College

Trey Cox
Chandler-Gilbert Community College

Alan E. O'Bryan
Arizona State University

BROOKS/COLE
CENGAGE Learning™

Australia • Brazil • Japan • Korea • Mexico • Singapore • Spain • United Kingdom • United States

This book is dedicated to every learner who has asked the question, "When am I ever going to use this?" In this book, we answer the question repeatedly through the use of interesting and engaging real-world contexts.

BRIEF CONTENTS

Chapter 1
Mathematical Modeling, Functions, and Change   1

Chapter 2
Linear Functions and Matrices   67

Chapter 3
Transformations of Functions   161

Chapter 4
Quadratic Functions   227

Chapter 5
Polynomial, Power, and Rational Functions   281

Chapter 6
Exponential and Logarithmic Functions   339

Chapter 7
Modeling with Other Types of Functions   409

Chapter 8
Trigonometric Functions   493

Chapter 9
Triangle Trigonometry and Applications 617

Chapter 10
Trigonometric Identities 679

Chapter 11
Conic Sections 723

Chapter 12
Sequences and Series 779

CONTENTS

Preface xiii
About the Authors xxi

1 Mathematical Modeling, Functions, and Change 1

1.1 Mathematical Modeling 2
EXERCISES 6
1.2 Functions and Function Notation 11
EXERCISES 18
1.3 Functions Represented by Tables and Formulas 22
EXERCISES 29
1.4 Functions Represented by Graphs 34
EXERCISES 39
1.5 Functions Represented by Words 42
EXERCISES 46
1.6 Preview to Inverse Functions 51
EXERCISES 56
STUDY SHEET 59
REVIEW EXERCISES 60
MAKE IT REAL PROJECT 65

2 Linear Functions and Matrices 67

2.1 Functions with a Constant Rate of Change 68
EXERCISES 81
2.2 Modeling with Linear Functions 85
EXERCISES 92
2.3 Linear Regression 94
EXERCISES 103
2.4 Systems of Linear Equations 109
EXERCISES 124
2.5 Using Matrices to Solve Linear Systems 128
EXERCISES 138
2.6 Matrix Operations and Applications 142
EXERCISES 149
STUDY SHEET 153
REVIEW EXERCISES 154
MAKE IT REAL PROJECT 159

viii   CONTENTS

3 Transformations of Functions   161

3.1 Vertical and Horizontal Shifts   162
EXERCISES   170
3.2 Vertical and Horizontal Reflections   175
EXERCISES   187
3.3 Vertical Stretches and Compressions   192
EXERCISES   198
3.4 Horizontal Stretches and Compressions   203
EXERCISES   212
STUDY SHEET   219
REVIEW EXERCISES   220
MAKE IT REAL PROJECT   226

4 Quadratic Functions   227

4.1 Variable Rates of Change   228
EXERCISES   237
4.2 Modeling with Quadratic Functions   246
EXERCISES   254
4.3 Forms and Graphs of Quadratic Functions   257
EXERCISES   271
STUDY SHEET   275
REVIEW EXERCISES   276
MAKE IT REAL PROJECT   279

5 Polynomial, Power, and Rational Functions   281

5.1 Higher-Order Polynomial Functions   282
EXERCISES   294
5.2 Power Functions   298
EXERCISES   308
5.3 Rational Functions   314
EXERCISES   325
STUDY SHEET   332
REVIEW EXERCISES   333
MAKE IT REAL PROJECT   337

6 Exponential and Logarithmic Functions 339

6.1 Percentage Change 340
EXERCISES 349
6.2 Exponential Function Modeling and Graphs 355
EXERCISES 364
6.3 Compound Interest and Continuous Growth 368
EXERCISES 377
6.4 Solving Exponential and Logarithmic Equations 379
EXERCISES 390
6.5 Logarithmic Function Modeling 392
EXERCISES 398
STUDY SHEET 403
REVIEW EXERCISES 404
MAKE IT REAL PROJECT 408

7 Modeling with Other Types of Functions 409

7.1 Combinations of Functions 410
EXERCISES 418
7.2 Piecewise Functions 424
EXERCISES 431
7.3 Composition of Functions 436
EXERCISES 443
7.4 Logistic Functions 449
EXERCISES 456
7.5 Choosing a Mathematical Model 464
EXERCISES 475
STUDY SHEET 483
REVIEW EXERCISES 484
MAKE IT REAL PROJECT 491

x CONTENTS

8 Trigonometric Functions 493

8.1 Periodic Functions 494
EXERCISES 502
8.2 Angle Measure 509
EXERCISES 521
8.3 Unit Circle and Trigonometric Functions 523
EXERCISES 541
8.4 Graphing Cosine and Sine Functions 544
EXERCISES 558
8.5 Modeling with Trigonometric Functions 562
EXERCISES 567
8.6 Other Trigonometric Functions 574
EXERCISES 585
8.7 Inverse Trigonometric Functions 588
EXERCISES 605
STUDY SHEET 609
REVIEW EXERCISES 610
MAKE IT REAL PROJECT 615

9 Triangle Trigonometry and Applications 617

9.1 Right Triangle Trigonometry 618
EXERCISES 623
9.2 Law of Cosines 627
EXERCISES 634
9.3 Law of Sines 637
EXERCISES 645
9.4 Polar Coordinates 648
EXERCISES 658
9.5 Vectors 660
EXERCISES 670
STUDY SHEET 674
REVIEW EXERCISES 675
MAKE IT REAL PROJECT 678

10 Trigonometric Identities 679

10.1 Basic Identities 680
EXERCISES 690
10.2 Verifying Identities 692
EXERCISES 701
10.3 Other Trigonometric Identities 703
EXERCISES 716
STUDY SHEET 719
REVIEW EXERCISES 720
MAKE IT REAL PROJECT 721

CONTENTS xi

11 Conic Sections 723

11.1 Slices of a Cone 724
11.2 Ellipses and Circles 727
EXERCISES 741
11.3 Parabolas 745
EXERCISES 754
11.4 Hyperbolas 757
EXERCISES 769
STUDY SHEET 773
REVIEW EXERCISES 774
MAKE IT REAL PROJECT 777

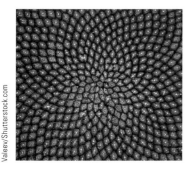

12 Sequences and Series 779

12.1 Sequences 780
EXERCISES 787
12.2 Arithmetic and Geometric Sequences 789
EXERCISES 794
12.3 Series 796
EXERCISES 806
STUDY SHEET 809
REVIEW EXERCISES 810
MAKE IT REAL PROJECT 811

ANSWERS 813
INDEX I1

PREFACE

Mathematics is a lens through which we can see and better understand the world. Yet, too often, teachers and textbooks fail to help students connect the mathematics they are learning to their personal lives. In *Precalculus: A Make It Real Approach* we use numerous and varied real-world contexts to introduce mathematical concepts. Users of this book answer for themselves the question, "When am I ever going to use this?" Readers learn the mathematical concepts behind the procedures and discover how to apply quantitative thinking in their daily lives. The textbook includes engaging features and invaluable support materials that help teachers and learners make sense of mathematical concepts.

■ Textbook Features

Explanations and Examples

The conversational tone and detailed explanations in the text make the math come alive for students. One student commented, "It's like having the teacher standing over my shoulder explaining things to me." The intriguing, relevant photographs and carefully illustrated graphics pique the interest and capture the attention of students. The meaningful real-world contexts inform students about the world and motivate learners to investigate mathematical concepts. The idea of "function" and "rate of change" are major themes that connect the content throughout the text. Students will find helpful *Just in Time* boxes throughout the textbook as reminders for previously learned skills.

Homework Exercises

The overarching philosophy of developing students' mathematical reasoning and persistent problem solving is evident throughout the homework exercises. These exercises are divided into four sections: *Skills and Concepts*, *Show You Know*, *Make It Real*, and *Stretch Your Mind*. *Skills and Concepts* exercises allow students to practice procedures and demonstrate a basic understanding of key concepts. *Show You Know* exercises require learners to demonstrate a conceptual understanding of the big ideas through written explanations. *Make It Real* exercises integrate a broad variety of real-world contexts from golf course ratings to Egyptian pyramids. Through these exercises, learners increase their understanding of the world while making connections between mathematics and their daily lives. *Stretch Your Mind* exercises allow students to tackle challenging problems that require deep thought, creativity, and persistence.

Study Sheets

The end-of-chapter study sheets present a series of questions focused on the big ideas of the chapter. As students reflect upon these questions, they solidify their understanding of the key mathematical concepts. The study sheets are an ideal chapter review activity to help students prepare for the end-of-chapter exam.

---

17. **Salary Adjustments** A person's salary is reduced by 5% one year due to necessary budget cuts. The next year (after business has improved) the person is given a 5% raise. Is the employee's income back to where it was originally? Explain your reasoning and support your argument with at least one of the following: equations, graphs, or tables.

MAKE IT REAL

18. **U.S. Immigrants** The number of immigrants coming into the United States increased from 385,000 in 1975 to 1,122,000 in 2005. (*Source: Statistical Abstract of the United States, 2007, Table 5*) Calculate the average rate of change, the 30-year growth factor, the annual growth factor, the 30-year percentage change, and the average annual percentage change.

19. **Stock Market** According to the 2004 Andex Chart, the average return of a $1 investment made in 1925—with no acquisition costs or taxes, and all income reinvested into the S&P 500—would have grown to $2641 by 2005. (*Source: www.andexcharts.com*)
    a. What is the 80-year growth factor?
    b. What is the 80-year percentage growth rate?
    c. What is the average annual growth factor?
    d. What is the average annual percentage change?
    e. Write an equation for the function $I(y)$, which would model the value of the initial $1 investment, I, as a function of the number of years since 1925.
    f. Evaluate $I(90)$ and explain the meaning of the numerical value in the context of the problem.

20. **Golf Course Management** Managers of golf pro shops anticipate the total revenue for an upcoming year based on the predicted total number of rounds that will be played at their course. They do this by looking at the total number of rounds played in prior years. Assume there were 33,048 rounds played 3 years ago and 38,183 rounds played this year. Calculate the average rate of change and the annual percentage change, and use each to predict the number of rounds to be played 2 years from now.

21. **Internet Usage** The graph displays the incredible growth in worldwide Internet usage from 1996 to 2006. Using successive differences and ratios, determine whether a linear, quadratic, or exponential function would be the best mathematical model.

Source: www.internetworldstats.com/emarketing.htm

22. **Professional Baseball Salaries** Professional athletes are some of the highest paid people in the world. The average major league baseball player's salary climbed from $1,998,000 in 2000 to $2,866,500 in 2006. (*Source: www.sportsline.com*)

    a. Assuming *linear* growth in the average players' salaries, find a formula $L(y)$ for the average salary L as a function of the year since 2000, y. Explain what the slope and vertical intercept mean in the real-world context. Use $L(y)$ to predict this year's average salary.
    b. Assuming *exponential* growth in the average players' salaries, find a formula $E(y)$ as a function of the year since 2000, y. Explain what the initial value and growth factor mean in the real-world context. Use $E(y)$ to predict this year's average salary.

23. **Purchasing Power of the Dollar** The purchasing power of the dollar as measured by consumer prices from 1980 to 2005 is given in the table. (Assume that in 1982, $1 was worth $1.)

Years Since 1980 y	Value of the Dollar V
0	1.215
5	0.928
10	0.766
15	0.656
20	0.581
25	0.512

Source: Statistical Abstract of the United States, 2007, Table 705

CHAPTER 8 Study Sheet

As a result of your work in this chapter, you should be able to answer the following questions, which are focused on the "big ideas" of this chapter.

SECTION 8.1
1. What does it mean for something to be "periodic"?
2. What is the relationship between maximum value, minimum value, midline, and amplitude?

SECTION 8.2
3. How are arc length and radius related?
4. What is a radian?

SECTION 8.3
5. What is the relationship between the sine and cosine functions and the unit circle?
6. Why will the values of $f(\theta) = r\sin(\theta)$ and $g(\theta) = r\cos(\theta)$ never exceed r?

SECTION 8.4
7. How do you solve a trigonometric equation graphically?
8. How do you solve a trigonometric equation with a data table?

SECTION 8.5
9. How can you tell if a sinusoidal function will model a data set well?
10. How do you construct a sinusoidal model?

SECTION 8.6
11. How are tangent, cotangent, secant, and cosecant related to sine and cosine?
12. Why do tangent, cotangent, cosecant, and secant have restricted domains?

SECTION 8.7
13. What is the relationship between a trigonometric function and its inverse?
14. What are the benefits of using inverse trigonometric functions in solving trigonometric equations?
15. What are the drawbacks of using inverse trigonometric functions in solving trigonometric equations?

Review Exercises

End-of-chapter review exercises allow students to refine their conceptual understanding, further develop their ability to make sense of real-world situations, and sharpen their mathematical expertise. Grouped by section number, these exercises make it easy to focus on particular areas of study.

Make It Real Projects

End-of-chapter projects allow students to apply the concepts learned in the chapter to data collected from an area of personal interest. By allowing students to focus on a real-world context of their choosing, their motivation and interest in doing the mathematics increases. Additionally, the *Make It Real Projects* provide learners another avenue for developing mathematical reasoning and persistent problem solving as they make sense of the mathematical ideas.

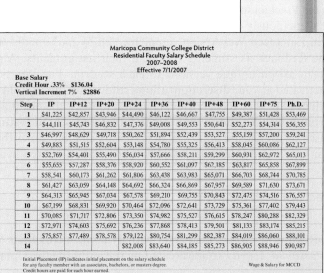

Instructor Supplements

Instructor's Resource Manual
ISBN-10: 1-111-98841-2; ISBN-13: 978-1-111-98841-8

The *Instructor's Resource Manual* is the new teacher's best friend and the veteran teacher's trusted colleague. The manual illustrates the best teaching practices of the award-winning author team and highlights common areas where students struggle. Detailed lesson plans help teachers new to the course create highly effective teaching and learning environments. The manual offers teaching tips, team-building ideas, and active learning strategies that make great teachers even better at developing student understanding. The *Instructor's Resource Manual* is written by the author team.

Classroom Activity Guide

The *Classroom Activity Guide* is available on the PowerLecture CD and offers two classroom activities for each section of the text. These in-class activities help teachers implement active learning and collaborative group work in the classroom. Each activity stimulates student discussions focused on the section's learning objectives. The *Classroom Activity Guide* is written by the author team.

Test Bank
ISBN-10: 1-111-98852-8; ISBN-13: 978-1-111-98852-4

The *Test Bank* aligns with the philosophy and approach of the text and includes multiple choice, short answer, and free response questions that focus on concepts and skills. Many of the problems integrate interesting and engaging real-world contexts. The *Test Bank* is written by the author team.

Instructor Solutions Manual
ISBN-10: 1-111-58086-3; ISBN-13: 978-1-111-58086-5

The *Instructor Solutions Manual* details how to solve all of the exercises in the text. The problem-solving approach demonstrated in the solutions is consistent with the strategies taught in the text. The *Instructor Solutions Manual* is written by the author team.

Enhanced WebAssign
ISBN-10: 0-538-73810-3; ISBN-13: 978-0-538-73810-1

Exclusively from Cengage Learning, Enhanced WebAssign® offers an extensive online program for Precalculus to encourage the practice that's so critical for concept mastery. The meticulously crafted pedagogy and exercises in this text become even more effective in Enhanced WebAssign, supplemented by multimedia tutorial support and immediate feedback as students complete their assignments. Algorithmic problems allow you to assign unique versions to each student. The Practice Another Version feature (activated at your discretion) allows students to attempt the questions with new sets of values until they feel confident enough to work the original problem. Students benefit from a new YouBook with highlighting and search features; Personal Study Plans (based on diagnostic quizzing) that identify chapter topics they still need to master; and links to video solutions, interactive tutorials, and even live online help.

PowerLecture with ExamView
ISBN-10: 1-111-98836-6; ISBN-13: 978-1-111-98836-4

This CD-ROM provides the instructor with dynamic media tools for teaching, including the *Classroom Activity Guide* created by the author team. Create, deliver, and customize tests (both print and online) in minutes with ExamView® Computerized Testing Featuring Algorithmic Equations. Easily build solution sets for homework or exams using Solution Builder's online solutions manual. Microsoft® PowerPoint® lecture slides and figures from the book are also included on this CD-ROM.

ExamView Computerized Testing

ExamView® testing software allows instructors to quickly create, deliver, and customize tests for class in print and online formats, and features automatic grading. It includes a test bank with hundreds of questions customized directly to the text. ExamView is available within the PowerLecture CD-ROM.

Solution Builder
www.cengage.com/solutionbuilder

This online instructor database offers complete worked solutions to all exercises in the text, allowing you to create customized, secure solutions printouts (in PDF format) matched exactly to the problems you assign in class.

■ Student Supplements

Student Solutions Manual
ISBN-10: 1-111-98838-2; ISBN-13: 978-1-111-98838-8

The *Student Solutions Manual* details how to solve all of the odd exercises in the text. The problem-solving approach demonstrated in the solutions is consistent with the strategies in the text because the *Student Solutions Manual* is written by the author team.

Student Study Guide
ISBN-10: 1-111-98834-X; ISBN-13: 978-1-111-98834-0

The *Student Study Guide* helps students focus on the key concepts in each section of the text. Extra examples, key terms, and additional explanations help students succeed. The *Student Study Guide* is written by the author team.

Enhanced WebAssign
ISBN-10: 0-538-73810-3; ISBN-13: 978-0-538-73810-1

Exclusively from Cengage Learning, Enhanced WebAssign® offers an extensive online program for Precalculus to encourage the practice that's so critical for concept mastery. You'll receive multimedia tutorial support as you complete your assignments. You'll also benefit from a new YouBook with highlighting and search features; Personal Study Plans (based on diagnostic quizzing) that identify chapter topics you still need to master; and links to video solutions, interactive tutorials, and even live online help.

CengageBrain.com

To access additional course materials, please visit **www.cengagebrain.com**. At the CengageBrain.com home page, search for the ISBN of your title (from the back cover of your book) using the search box at the top of the page. This will take you to the product page where these resources can be found.

Reviewers

Om P. Ahuja, *Kent State University*; Angela Angeleska, *University of Tampa*; Chris Bendixen, *Lake Michigan College*; Jeffery Berg, *Arapahoe Community College*; Kevin Bolan, *Everett Community College*; Nathan Borchelt, *Clayton State University*; Dwayne Brown, *University of South Carolina*; Annette M. Burden, *Youngstown State University*; Chapin P. Carnes, *Central New Mexico Community College*; Chapin Carnes, *College of Santa Fe, Albuquerque*; Phillip Clark, *Scottsdale Community College*; Yong S. Colen, *Indiana University of Pennsylvania*; Schery Collins, *Virginia Highlands Community College*; Ana-Maria Croicu, *Kennesaw State University*; Ann Darke, *Bowling Green State University*; Alicia Serfaty de Markus, *Miami-Dade College, Kendall Campus*; Lara K. Dick, *Meredith College*; Letitia Downen, *Southern Illinois University, Edwardsville*; Jerrett Dumouchel, *Florida Community College at Jacksonville*; Joanne Duvall, *Coastal Carolina University*; Donna Fatheree, *University of Louisiana at Lafayette*; Cathy Ferrer, *Valencia Community College, East*; Joe Fox, *Salem State College*; Cynthia Francisco, *Oklahoma State University*; Bill Gallegos, *Truckee Meadows Community College*; Lidia Gonzalez, *CUNY–York College*; Jacqueline Grace, *State University of New York at New Paltz*; Zdenka Guadarrama, *Rockhurst University*; Judy Hayes, *Lake-Sumter Community College, South Lake Campus*; Celeste Hernandez, *Richland College*; Jeffrey Hood, *Midwestern State University*; Kevin Hopkins, *Southwest Baptist University*; R. Michael Howe, *University of Wisconsin, Eau Claire*; Susan Howell, *University of Southern Mississippi*; Rhonda Hull, *Clackamas Community College*; Joseph Jordan, *John Tyler Community College*; Brian Jue, *California State University, Stanislaus*; Susan Kellicut, *Seminole Community College*; Roseanne Killion, *Missouri State University*; William Krant, *Palo Alto College*; Richard Larson, *Jackson Community College*; Pam Littleton, *Tarleton State University*; Habib Maagoul, *Northern Essex Community College*; Antoinette Marquard, *Cleveland State University*; Christian Mason, *Virginia Commonwealth University*; Meagan McNamee, *Central Piedmont Community College, North Campus*; Grzegorz Michalski, *Georgia Southern University*; Brian Milleville, *Erie Community College*; Mitsue Nakamura, *University of Houston, Downtown*; Bette Nelson, *Alvin Community College*; Daniel Nearing, *Scottsdale Community College*; Dennis Nickelson, *William Woods University*; Stephen Nicoloff, *Paradise Valley Community College*; Lilia Orlova, *Nassau Community College*; Martha Pate, *Lamar State College, Port Arthur*; Faith Peters, *Miami-Dade College, Wolfson Campus*; Sandra Poinsett, *College of Southern Maryland*; Arturo Presa, *Miami-Dade College, Wolfson Campus*; David Price, *Tarrant County College, Southeast*; Michael Price, *University of Oregon*; Adelaida Quesada, *Miami-Dade College, Kendall Campus*; Michelle Ragle, *Northwest Florida State College*; Leela Rakesh, *Central Michigan University*; Kevin Reeves, *East Texas Baptist University*; Czarina Reyes, *Brookhaven College*; Behnaz Rouhani, *Georgia Perimeter College*; Susan Sabrio, *Texas A&M University, Kingsville*; Bjorn Schellenberg, *College of Count Saint Vincent*; Rachel Schwell, *Connecticut State University*; Keith Sinkhorn, *Peru State College*; Diana Staats, *Dutchess Community College*; James H. Stewart, *Jefferson Community and Technical Colleges*; Sharon Stewart, *Hazard Community College*; Dennis Stramiello, *Nassau Community College*; Debra Swedberg, *Casper College*; Sara Taylor, *Dutchess Community College*; Mary Ann Teel, *University of North Texas*; Abby Train, *New Mexico State University, Main Campus*; Lynn Trimpe, *Linn-Benton Community College*; Michael Waters, *Northern Kentucky University*; Yongjun Yang, *University of Wisconsin, Sheboygan*; Diane Zych, *Erie Community College, State University of New York*

Acknowledgements

We express gratitude to Howard Speier, Linda Meng, Dave Quadlin, and the administration and faculty at Chandler-Gilbert Community College for their willingness to conduct a comprehensive class test of this book. The suggestions from our colleagues were invaluable in helping us refine the manuscript.

We thank Linda Meng for researching the data and creating many of the mathematical models used in the text. We thank Jo-Ann Williams for her contributions to an early version of the text.

We thank the accuracy checkers, Sue Steele and Milos Podmanik, and the copy editor, Amy Mayfield, for the countless hours invested in verifying the mathematical accuracy and clarity of the text. Their thorough review enhanced the quality of the final text.

We are grateful for Gary Whalen and Carolyn Crockett, our editors at Brooks/Cole, for their commitment to this project. With the help of Liz Covello, Jennifer Risden, Sabrina Black, and Cynthia Ashton they put the pieces in place that helped our vision become a reality.

We are thankful for all of the reviewers who provided candid feedback and suggestions for improvement. As a result, we were able to modify the text to better meet the needs of teachers and students.

We acknowledge the significant contributions of Dr. Pat Thompson and Dr. Marilyn Carlson in helping shape our mathematical thinking and focus on sense-making. Although they did not contribute directly to this text, their influence has made us better mathematicians, authors, and educators.

Most of all, we thank our families for the tireless support during the four years we spent working on this project. Their love and encouragement kept us moving forward even when things were difficult. For this, we are grateful.

ABOUT THE AUTHORS

Frank C. Wilson

Frank is a popular professor and award-winning textbook author whose passion is helping students see how to apply mathematics in their personal lives. Frank shares his passion with colleagues worldwide through engaging workshops, keynote presentations, journal articles, and curricular materials. Frank's classes are highly interactive with students actively participating in the learning process. Student presentations, group discussions, and humor create the rich and rewarding learning environment of his classes. Frank holds B.S. and M.S. degrees in Mathematics from Brigham Young University and will complete his Ph.D. in Business Administration (Business Quantitative Methods) in 2012.

Scott Adamson

Scott is an award-winning professor who strives to help students develop mathematical reasoning and persistant problem solving as they make sense of big mathematical ideas. He structures the classroom environment so that students are afforded the opportunity to make sense of mathematics and strives to develop enthusiastic learners in the classroom. Scott holds a B.S. and M.A.T. in Mathematics Education from Northern Arizona University and a Ph.D. in Curriculum and Instruction–Mathematics Education from Arizona State University.

Trey Cox

Trey's goal for mathematics education is to help students see the relevance of mathematics in their lives and to learn the meaning and purpose of the "big" mathematical ideas. His courses are taught in a fun and interesting environment where students have the opportunity to enjoy learning and create a strong, personal sense of motivation and responsibility for their own learning. Trey holds a B.A. in Secondary Education–Mathematics and a M.S. degree in Educational Administration from Concordia University, Wisconsin, and a Ph.D. in Curriculum and Instruction–Mathematics Education from Arizona State University.

Alan O'Bryan

Alan has worked as a high school mathematics instructor with Gilbert Public Schools in Gilbert, Arizona, and as a faculty associate with Arizona State University assisting with research projects that seek to improve teachers' and students' understanding of key ideas in algebra and precalculus. He believes strongly in the benefits of having students explain their thinking and understanding of concepts, both orally and in writing, and in helping students make connections among the many topics taught within a mathematics course. Alan holds a B.A. in Secondary Education from Arizona State University and a M.A. in Teaching from Grand Canyon University.

CHAPTER 1

Mathematical Modeling, Functions, and Change

Home ownership is a hallmark of the American dream. By looking at housing trends, homebuyers may forecast home values. Mathematical modeling is one primary tool used to forecast home values.

- **1.1** Mathematical Modeling
- **1.2** Functions and Function Notation
- **1.3** Functions Represented by Tables and Formulas
- **1.4** Functions Represented by Graphs
- **1.5** Functions Represented by Words
- **1.6** Preview to Inverse Functions

STUDY SHEET
REVIEW EXERCISES
MAKE IT REAL PROJECT

SECTION 1.1 Mathematical Modeling

LEARNING OBJECTIVES

- Identify real uses for symbolic, numeric, and graphical forms of mathematical models
- Analyze mathematical models and use them to create and answer real-world questions

GETTING STARTED

The word *model* has many everyday meanings. A model car is a replica of a real car. A role model is a person who represents admirable qualities such as honesty and integrity that others try to copy. A supermodel exhibits the physical qualities many people strive to reproduce. Thus, the word *model* is synonymous with *replica, representation, copy,* or *reproduction*.

In this section we give a broad overview of how *mathematical models* are used to represent real-world problems. In later chapters, we demonstrate the process for finding such a model for a given situation.

> **MATHEMATICAL MODEL**
>
> A **mathematical model** is a symbolic, numerical, graphical, or verbal representation of a problem situation.

■ Decision-Factor Equation

Mathematical models help us understand the nature of problem situations. They are often helpful in making predictions or solving problems in real-world situations.

One type of mathematical model is the *decision-factor equation*. We will use a decision-factor equation to model the decision process for purchasing a preowned vehicle.

When buying a vehicle, we carefully consider important features of a car such as price, manufacturer, engine type, fuel economy, color, model year, and body style, just to name a few. Suppose we are planning to purchase a preowned Toyota Corolla. We have decided that the three features most important to us are mileage, price, and color.

Table 1.1 shows Corollas meeting our criteria offered by Autotrader.com on July 6, 2007 in the vicinity of Indian Orchard, Massachusetts. We can use the data in this table to create a decision-factor equation that will produce a number called the *decision factor* that will help us decide which car to buy. The car that best fits our criteria will have the smallest decision factor.

Table 1.1

Car	Mileage	Price	Color
1	61,671	$11,495	Grey
2	23,258	$15,997	Red
3	14,865	$15,995	Silver
4	5,295	$16,495	Silver
5	35,671	$11,995	Red
6	3,446	$16,495	Grey

EXAMPLE 1 ■ **A Decision-Factor Equation**

Create a decision-factor equation with mileage, price, and color as the criteria. Assume that mileage is most important followed by price then color.

Solution *Mileage* and *price* have numeric values; *color* does not. We need to assign numeric values to the color options *red, grey,* and *silver* to use in our equation. We choose red = 1, grey = 2, and silver = 3, making *red* our first choice, *grey* our second choice, and *silver* our third choice.

One decision-factor equation for this situation is

$$\text{decision factor} = \text{mileage} + \text{price} + \text{color}$$

Using Car 1 from Table 1.1, we get

$$\text{decision factor} = 61{,}671 + 11{,}495 + 2$$
$$= 73{,}168$$

Notice that the color number has a negligible effect on the decision factor. If we want the color to have a greater effect, we can modify the decision-factor equation. For example,

$$\text{decision factor} = \text{mileage} + \text{price} + (1000 \cdot \text{color})$$

We use this modified equation to calculate the decision factor for each of the six cars on the list, as shown in Table 1.2.

Table 1.2

Car	Mileage	Price	Color	Decision Factor	Rank
1	61,671	$11,495	Grey = 2	75,166	6
2	23,258	$15,997	Red = 1	40,255	4
3	14,865	$15,995	Silver = 3	33,860	3
4	5,295	$16,495	Silver = 3	24,790	2
5	35,671	$11,995	Red = 1	48,666	5
6	3,446	$16,495	Grey = 2	21,941	1

Using this model, we find that Car 6 has the lowest decision factor. Notice that it has the lowest mileage, the highest price, and the second-choice color.

To quadruple the effect of the price on the decision factor, we can modify the equation as shown.

$$\text{decision factor} = \text{mileage} + (4 \cdot \text{price}) + (1000 \cdot \text{color})$$

This gives the results shown in Table 1.3.

Table 1.3

Car	Mileage	Price	Color	Decision Factor	Rank
1	61,671	$11,495	Grey = 2	109,651	6
2	23,258	$15,997	Red = 1	88,246	5
3	14,865	$15,995	Silver = 3	81,845	3
4	5,295	$16,495	Silver = 3	74,275	2
5	35,671	$11,995	Red = 1	84,651	4
6	3,446	$16,495	Grey = 2	71,426	1

Although the numerical value of the decision factors changed, Car 6 still has the lowest decision factor. We decide to buy Car 6.

■ Mathematical Models Presented Numerically

Just as we may use equations to model a situation, we may also use a table of values. For example, one representation of a mathematical model for the increasing number of registered vehicles in the United States is a table of data such as Table 1.4.

Table 1.4

Year	Number of Registered Vehicles
1980	155,796,000
1990	188,798,000
1995	201,530,000
2000	221,475,000
2001	230,428,000
2002	229,620,000
2003	231,390,000

Source: Statistical Abstract of the United States, 2006, Table 1078

Whether in symbolic form (like a decision-factor equation) or numerical form (like a table of data), one of the purposes of a mathematical model is to make sense of the world around us. As we examine the data in Table 1.4, we may ask a variety of questions about the situation: *How is the number of registered vehicles in the United States changing? To whom is this information important?* or *Why is the number of registered vehicles in the United States changing the way it is? Is the number of registered vehicles keeping pace with the increase in population?* We explore these questions in the following examples.

EXAMPLE 2 ■ Analyzing a Mathematical Model in Numerical Form

Describe the change in the number of registered vehicles in the United States. Then identify to whom this analysis may be important and why.

Solution The number of registered vehicles tends to be increasing. We can examine the amount of increase by subtracting one value from the next. The results are shown in Table 1.5.

Table 1.5

Year	Number of Registered Vehicles	Difference
1980	155,796,000	33,002,000
1990	188,798,000	12,732,000
1995	201,530,000	19,945,000
2000	221,475,000	8,953,000
2001	230,428,000	−808,000
2002	229,620,000	1,770,000
2003	231,390,000	

By examining these differences, we can describe how the number of registered vehicles is changing. From 1980 to 1990, the number increased by more than 33 million vehicles (about 3.3 million vehicles per year). From 1990 to 1995, the number increased by nearly 13 million vehicles (about 2.6 million vehicles per year). In the next 5-year interval (1995 to 2000), there was an increase of about 20 million vehicles (4 million vehicles per year). From 2000 to 2001, the number increased by nearly 9 million vehicles. From 2001 to 2002, the number dropped by about 808,000 vehicles. Finally, from 2002 to 2003, the number of registered vehicles increased by nearly 2 million vehicles.

Although most drivers may not care about these statistics, many government agencies do. For example, state motor vehicle divisions that recognize these trends may be able to better forecast tax revenue from licensing fees. State and county officials may use these data to help shape plans for upgrades to transportation infrastructure such as roads, bridges, and highways.

Table 1.6

Year	Population of United States
1980	227,726,000
1990	250,132,000
1995	266,557,000
2000	282,402,000
2001	285,329,000
2002	288,172,000
2003	291,028,000

Source: Statistical Abstract of the United States, 2006, Table 2

In Example 3, we continue to explore the data about the number of registered vehicles in the United States by comparing it to the U.S. population data shown in Table 1.6.

EXAMPLE 3 ■ Thinking about Trends in Data

Offer a possible reason the number of registered vehicles in the United States is changing the way it is. Is the number of registered vehicles keeping pace with the increase in population?

Solution One reason for the increase in the number of registered vehicles in the United States is that the population is increasing. We expect that the number of registered vehicles will keep up with the population. In Table 1.7 we compare the number of registered vehicles data with the U.S. population data to check our assumptions. We see that, indeed, as the population increases, the number of registered vehicles tends to increase. This is true everywhere except for the change from 2001 to 2002, where the number of registered vehicles decreased even though the U.S. population increased.

Table 1.7

Year	Number of Registered Vehicles	Population of United States
1980	155,796,000	227,726,000
1990	188,798,000	250,132,000
1995	201,530,000	266,557,000
2000	221,475,000	282,402,000
2001	230,428,000	285,329,000
2002	229,620,000	288,172,000
2003	231,390,000	291,028,000

We can better understand these trends by computing the number of cars per capita (per person) for each of the given years. This value is found by dividing the number of registered vehicles for the given year by the population in that year. As shown in Table 1.8, the number of registered vehicles per capita initially increased but leveled off in the early 2000s.

Table 1.8

Year	Number of Registered Vehicles	Population of United States	Number of Vehicles Per Capita
1980	155,796,000	227,726,000	0.68
1990	188,798,000	250,132,000	0.75
1995	201,530,000	266,557,000	0.76
2000	221,475,000	282,402,000	0.78
2001	230,428,000	285,329,000	0.81
2002	229,620,000	288,172,000	0.80
2003	231,390,000	291,028,000	0.80

■ **Mathematical Models Presented Graphically**

Newspapers and magazines present information in graphical form every day. Let's use the context of median home prices in the metropolitan Phoenix area to investigate a mathematical model presented graphically.

The metropolitan Phoenix area was one of the fastest growing areas in the United States in the early 2000s. As a result, home prices skyrocketed in 2005. The graph in

Figure 1.1 shows that the median price of homes in the Phoenix area increased from $155,800 in the fourth quarter of 2003 to $260,190 in the first quarter of 2006.

In the graph, the fourth quarter of 2003 (October–December) is represented by $t = 0$. The first quarter of 2004 (January–March) is $t = 1$, the second quarter of 2004 (April–June) is $t = 2$, and so on. The vertical axis of the graph shows the median price in dollars.

Figure 1.1

EXAMPLE 4 ■ Analyzing a Mathematical Model Presented Graphically

Describe the trend seen in the graph (Figure 1.1) of the median home price in the metropolitan Phoenix area.

Solution The median home prices increased from the fourth quarter of 2003 ($t = 0$) until the third quarter of 2005 ($t = 7$). Starting in quarter 7 (third quarter 2005), median home prices stabilized at approximately $260,000. Median home prices increased very quickly between quarter 4 (fourth quarter 2004) and quarter 7 (third quarter 2005).

EXAMPLE 5 ■ Using a Mathematical Model

Based on the graph (Figure 1.1) of the median home price in the metropolitan Phoenix area, determine the best time to have sold a home between the fourth quarter of 2003 and the first quarter of 2006.

Solution Assuming that circumstances were such that a homeowner could decide to either wait to sell or sell immediately, it would have been wise for the homeowner to wait to sell until quarter 7 (third quarter 2005). This is when the median price peaked.

SUMMARY

In this section you learned that a mathematical model is a representation of a real-world situation. You discovered that this representation may be symbolic (as in the decision-factor equation), numerical (as in the population data), or graphical (as in the median home price). We will use mathematical models throughout the text to solve problems and make predictions.

1.1 EXERCISES

■ SKILLS AND CONCEPTS

In Exercises 1–5, use the following decision-factor equation.

$$\text{decision factor} = 100(\text{gas mileage}) + 1000(\text{style}) - \frac{1}{10}(\text{miles})$$

where the style variable is as follows

$$\text{style} = \begin{cases} 2 & \text{if economy car} \\ 1 & \text{if midsize sedan} \\ 0 & \text{if other} \end{cases}$$

1. What kind of a car does this decision-factor equation suggest is desired? Why?

2. Decide which of the following cars is the best to buy. Explain why the results make sense.

   Car A: 2002 Toyota Echo (economy car) with 18,298 miles that gets 35 miles per gallon

   Car B: 2002 Chevrolet Suburban (sport utility vehicle) with 16,000 miles that gets 16 miles per gallon

   Car C: 2002 Dodge Stratus (midsize sedan) with 19,845 miles that gets 28 miles per gallon

3. Suppose the decision factor is 2500 on a car that gets 25 miles per gallon and has 20,000 miles on the odometer. What style car is this?

4. If the decision factor is −1500 on a midsize sedan with 45,000 miles on the odometer, what is the car's gas mileage?

5. Based on the three cars in Exercise 2, the highest decision factor is 3670.2. Now consider another car that gets 30 miles per gallon and is a midsize sedan. What is the maximum mileage it can have to generate a decision factor greater than 3670.2?

For Exercises 6–10, write a decision-factor equation that will accurately score each type of car desired. Use the variables style, color, mileage, cost, year. *Be sure to define any scales used for any of these variables, as necessary.*

6. A relatively new red sports car. Money is not an object.

7. An economical, older-model vehicle for a family of five. Anything will do.

8. A vintage model car that could be restored and shown at car shows.

9. An off-road vehicle that hides dirt well and is very reliable.

10. A vehicle that could be used in the construction field. The company colors are blue and white. Price is not a concern.

■ SHOW YOU KNOW

11. Write an explanation of what a mathematical model is and how the concept is similar and dissimilar from the non-mathematical use of the word *model*.

12. What are some reasons that mathematical models are created and studied?

13. Describe the different ways that mathematical models may be represented.

■ MAKE IT REAL

For Exercises 14–18, analyze the mathematical model given in numerical form. Then answer the questions that follow each numerical model.

14. **Number of Homes Sold** The number of resale homes sold in the Metropolitan Phoenix area for selected quarters is given in the table.

Quarter	Number of Homes Sold
Fourth Quarter 2003	18,350
First Quarter 2004	19,460
Second Quarter 2004	28,760
Third Quarter 2004	27,580
Fourth Quarter 2004	26,315
First Quarter 2005	27,325
Second Quarter 2005	30,705
Third Quarter 2005	30,715
Fourth Quarter 2005	22,090

Source: www.poly.asu.edu

a. Describe when the number of homes sold is increasing and when the number of homes sold is decreasing.

b. Compare the difference in the number of homes sold from one quarter to the next. What patterns do you notice?

15. **McDonald's Revenue per Year** The total annual revenue for McDonald's Corporation for the 6-year period beginning in 2000 is given in the table.

Year	Total Revenue ($ millions)
2000	14,243
2001	14,870
2002	15,406
2003	17,140
2004	19,065
2005	20,460

Source: www.mcdonalds.com

a. Describe the trend seen in these data. Are revenues increasing or decreasing over time?

b. Compare the difference in revenue from one year to the next. What patterns do you notice?

16. **McDonald's Revenue and Number of Locations** The total annual revenue for McDonald's Corporation compared to the number of store locations is given in the table.

Number of Locations	Total Revenue ($ millions)
28,707 in 2000	14,243
30,093 in 2001	14,870
31,108 in 2002	15,406
31,129 in 2003	17,140
31,561 in 2004	19,065
31,886 in 2005	20,460

Source: www.mcdonalds.com

a. Describe the trend seen in these data. What connections are there between the increasing revenues and the number of McDonald's locations?

b. Predict the revenue when there are 32,000 locations. Justify your answer.

17. **Value of a Car** According to www.bankrate.com, vehicles depreciate by about 15% each year. The table projects the value of a 2006 Ford Mustang with a sticker price of $19,439.

Year	Value
2006	$19,439
2007	$16,523
2008	$14,045
2009	$11,938
2010	$10,147

Source: www.kbb.com

a. What will be the value of a 2006 Ford Mustang in the year 2011?

b. Find the difference in the value of the 2006 Ford Mustang from one year to the next. What patterns do you notice?

c. Will the value of the 2006 Ford Mustang ever be $0? If so, when? If not, why not?

18. **Teacher Salary Comparison** Over 60% of men not in the teaching profession earn a higher salary than men who are teachers. The table shows how much more money the average college-educated male non-teacher made as compared to the average male teacher.

Year	Percent More Earned by Non-Teachers as Compared to Teachers
1940	−3.6%
1950	2.1%
1960	19.7%
1970	33.1%
1980	36.1%
1990	37.5%
2000	60.4%

Source: www.nea.org

For example, in 1990 male non-teachers made 37.5% more than male teachers on average.

a. Describe the trend observed in these data.

b. Why was there such a big jump in the percentage of non-teachers who earn a higher salary than teachers from 1990 to 2000?

c. What does the −3.6% in 1940 indicate about salaries of male teachers?

For Exercises 19–23, analyze the mathematical model given in numerical form. For each model,

a. Describe any trends you notice.

b. Write and answer at least two questions related to the situation.

19. **Teacher Salary Comparison** Over 16% of women not in the teaching profession earn a higher salary than women who are teachers. The table shows how much more money the average college-educated female non-teacher makes as compared to the average female teacher.

Year	Percent More Earned by Non-Teachers as Compared to Teachers
1940	−15.8%
1950	−11.2%
1960	−12.7%
1970	−3.1%
1980	−3.7%
1990	4.5%
2000	16.4%

Source: www.nea.org

For example, female non-teachers made 3.7% less than female teachers in 1980 but by 2000 female non-teachers made 16.4% more than female teachers.

20. **NBA Minimum Salary** The table gives the minimum salary paid in 2005–2006 to NBA players with the given years of service to the league.

Years of Service	Minimum Salary
0	$398,762
1	$641,748
2	$719,373
3	$745,248
4	$771,123
5	$835,810
6	$900,498
7	$965,185
8	$1,029,873
9	$1,035,000
10+	$1,138,500

Source: www.insidehoops.com

21. **Super Bowl Ticket Prices** The table shows the price of a Super Bowl ticket for selected Super Bowls.

Super Bowl	Ticket Face Value
I (1)	$10
V (5)	$15
X (10)	$20
XV (15)	$40
XX (20)	$75
XXV (25)	$150
XXX (30)	$300
XXXV (35)	$325
XL (40)	$600

Source: www.superbowl.com

22. **Price of Gasoline** The table shows the average price per gallon for unleaded, regular gasoline for selected years.

Year	Average Price
1990	$1.16
1995	$1.15
1997	$1.23
1998	$1.06
1999	$1.17
2000	$1.51
2001	$1.46
2002	$1.36
2003	$1.59
2004	$1.88

Source: Statistical Abstract of the United States, 2006, Table 722

23. **Comparing Price of Gasoline and Annual Fuel Consumption** The table shows the average price of unleaded, regular gasoline

for selected years and the fuel consumption at that price. (Note: The years are not shown.)

Price of Gasoline	Fuel Consumption (billions of gallons)
$1.06	155.4
$1.15	143.8
$1.16	130.8
$1.17	161.4
$1.23	150.4
$1.36	168.7
$1.46	163.5
$1.51	162.5
$1.59	169.6

Source: Statistical Abstract of the United States, 2006, Table 1085

For Exercises 24–28, analyze the mathematical model given in graphical form. Then answer the questions that follow each graph.

24. Apple iPod Sales The graph shows iPod unit sales per quarter.

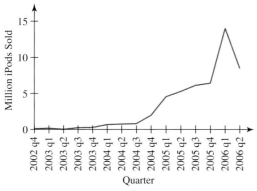

Source: www.applematters.com

a. Describe the amount of iPod sales from quarter to quarter. Are sales increasing or decreasing? Are they increasing or decreasing quickly or slowly?

b. Provide a possible explanation for the slow sales initially followed by a rapid increase in sales.

c. Provide a possible explanation for the drastic drop in sales in the beginning of 2006.

25. Number of Hurricanes by Decade The graph shows the number of hurricanes that have hit the mainland United States each decade, beginning in 1900–1909.

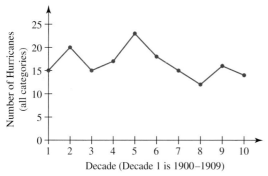

Source: www.aoml.noaa.gov

a. Over what time period did the number of hurricanes decrease the longest?

b. In which decade did the greatest number of hurricanes strike the mainland United States?

c. In which decade did the least number of hurricanes strike the mainland United States?

26. Alternative-Fuel Vehicles The graph shows the total number of alternative-fuel vehicles in use in the United States for selected years.

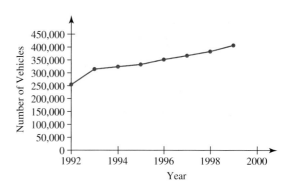

Source: Quantitative Environmental Learning Project at www.seattlecentral.org

a. What is the overall trend in the number of alternative-fuel vehicles in use in the United States?

b. Approximately how many more alternative-fuel vehicles are in use in 1999 compared to 1992?

c. Predict how many alternative-fuel vehicles will be in use in 2000.

27. Doctoral Degrees in Mathematics The graph shows the number of doctoral degrees in mathematics awarded in selected years.

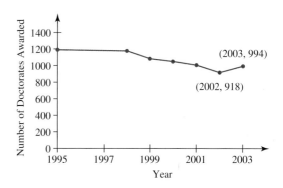

Source: Statistical Abstract of the United States, 2006, Table 784

a. Describe the trend in the number of doctoral degrees in mathematics awarded from 1995–2002.

b. How many more doctoral degrees in mathematics were awarded in 2003 than in 2002?

c. If the trend from 2002 to 2003 continued, how many doctoral degrees in mathematics were awarded in 2006?

28. Successful Space Launches in the United States The following data and graph show the number of successful space launches in the United States for selected groups of years.

Year	Year Group	Number of Successful Launches
1957–1964	0	207
1965–1969	1	279
1970–1974	2	139
1975–1979	3	126
1980–1984	4	93
1985–1989	5	61
1990–1994	6	122
1995–2002	7	227

Source: Statistical Abstract of the United States, 2006, Table 784

a. What factors may have caused the dramatic decrease in the number of successful space launches from the 1960s to the early 1970s?

b. What factors may have caused the dramatic increase in the number of successful space launches from the late 1980s and into the 1990s?

c. How many successful launches each year, on average, were there between 1957 and 1964?

d. How many successful launches each year, on average, were there between 1985 and 1989?

For Exercises 29–33, analyze the mathematical model given in graphical form. For each model,

　a. Describe any trends you notice.

　b. Write and answer at least two questions related to the situation.

29. The Labor Force and Head Start Enrollment The graph shows the number of children enrolled in the Head Start program as the number of women in the workforce increased.

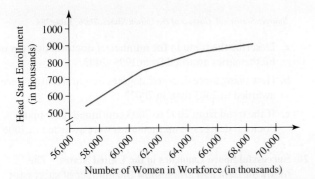

Source: Modeled from Statistical Abstract of the United States, 2006, Tables 564 and 579

30. Arizona Diamondbacks Average Salary The graph shows the average salary for Arizona Diamondback players from 1998–2005.

Source: sports.espn.go.com

31. New York Yankees and Boston Red Sox Salaries The graph shows the average salary (in millions of dollars) for New York Yankee players compared to Boston Red Sox players from 2001–2009.

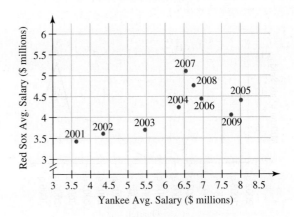

Source: Boston Red Sox, New York Yankees

32. Student–Teacher Ratio The graph shows the number of students per teacher in Texas public schools for the specific years.

Source: www.sbec.state.tx.us

33. **Average SAT scores** The graph shows the average SAT scores for students over the 10-year period from 1992–2002.

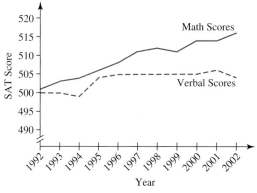

Source: www.collegeboard.com

■ **STRETCH YOUR MIND**

Exercises 34–35 are intended to challenge your understanding of mathematical models.

34. Explain the advantages and disadvantages of mathematical models.

35. A particular mathematical model in graphical form passes through all of the points in the given data set. A classmate claims that since the model fits the data perfectly, all predictions made using the model will be accurate. Explain why your classmate may be wrong.

SECTION 1.2 Functions and Function Notation

LEARNING OBJECTIVES

■ Write and interpret functions using function notation

■ Explain how a function is a process or a correspondence

■ Solve function equations for a given variable using an equation, table, and graph

GETTING STARTED

One of the credit scores most widely used by financial institutions is the FICO® score formulated by the Fair Isaac Corporation. Engineer Bill Fair and mathematician Earl Isaac founded the Fair Isaac Corporation in 1958 to provide a way for lenders to quantify their investment risk. The loan interest rates that a consumer is offered depends largely upon the FICO score.

In this section we introduce the concept of function and show how to apply functions in real-world situations such as getting a loan to buy a car.

■ **Functions**

For a car loan, two primary factors come into play: the applicant's *credit score* and the *interest rate* a lender is willing to offer. Since both of these factors can change, we call them **variables** and denote them with letters such as c for *credit score* and r for *interest rate*. A numeric value that does not change is called a **constant**. For instance, a constant in this situation is the price of the car, assuming that the price will not change during the negotiation process.

> **VARIABLE**
>
> A **variable** is a quantity that changes value.

> **CONSTANT**
>
> A **constant** is a numeric value that remains the same.

Functions are typically expressed as a combination of variables and constants.

> **FUNCTION (SINGLE VARIABLE)**
>
> A **single-variable function** is a process or correspondence relating two quantities in which each input value generates exactly one output value.

PEER INTO THE PAST

LEIBNIZ

Gottfried Wilhelm von Leibniz was born on July 1, 1646, in what is now Germany. Leibniz was a philosopher, theologian, and poet. He created much of the mathematical notation that we use today. He was the first to use the terms *coordinates* and *axes of coordinates*. The terms *function* and *variable* are also credited to Leibniz.

Source: www.mathforum.org

For example, consider the household freezer. If we put water into a freezer, what will it become? Ice! Since the input (water) put into the freezer will become exactly one thing (ice), we can say that the freezer is a function. We represent this situation in Figure 1.2.

A Function as a Process

Input   Function (process)   Output

Figure 1.2

Alternatively, a function may be thought of as a *correspondence* between two sets of values. That is, each item from a set of inputs is matched with a single item from a set of outputs, as shown in Figure 1.3. Although a function process may be implied in the correspondence, the correspondence doesn't explicitly state what the process is.

Function correspondences do not have to include numbers. Figure 1.4 shows a correspondence relating tennis tournaments with the locations of the tournaments.

A Function as a Correspondence

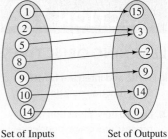

Set of Inputs   Set of Outputs

Figure 1.3

Tennis Tournament Correspondence

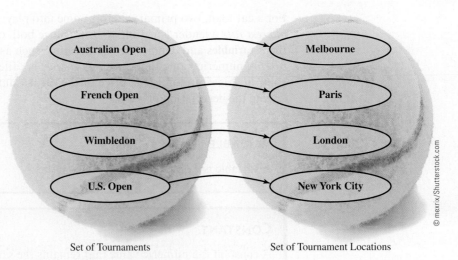

Set of Tournaments   Set of Tournament Locations

Figure 1.4

■ Determining If a Relationship Is a Function

Let's look at a function as a correspondence in a relationship between credit score and loan interest rate. A credit score can range from 300 to 850. According to Fair Isaac Corporation, most people have scores between 600 and 800. Table 1.9 shows how the FICO score impacts the interest rate a person is able to obtain on a loan.

Table 1.9

Credit Score c	Annual Interest Rate (percent) r
625	7.89
642	7.34
668	6.91
685	6.70
744	6.52
793	6.30

Source: www.fico.com
(Data is accurate as of 2007.)

The relationship between the person's credit score, c, and the annual interest rate, r, is a function because each credit score (input) produces only one interest rate (output). For example, a person who has a credit score (input) of 744 can secure an annual interest rate (output) of 6.52%.

In this example of a function as a correspondence, we examined two different sets (*credit score* and *annual interest rate*) and noted a link between them. Because the interest rate is affected by the credit score, we assume there is an implicit function process that relates the two sets. However, we do not know and thus cannot state explicitly what that process is.

We also need to be aware that there are relationships between variables that are not functions. For example, consider the relationship that exists between the US Airways' Italian destination cities from Tucson, AZ, and Reno, NV, shown in Figure 1.5.

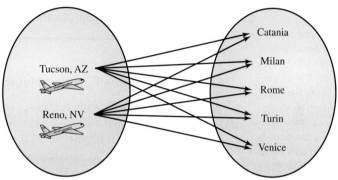

Figure 1.5

Source: Adapted from www.usairways.com

We can see that both Reno and Tucson have multiple destination cities in Italy. The fact that each input (origination city) has more than one output (destination city) makes this relationship not a function. To be a function, each input must only have one output—Tucson and Reno would need to have only one destination city each.

■ Vertical Line Test

If a relationship between two variables exists and is represented as a graph in the rectangular coordinate system (see Section 1.4), the **vertical line test** may be used to determine if the relationship is a function.

> **VERTICAL LINE TEST**
>
> If each vertical line drawn on a graph intersects the graph in at most one place, the graph is the graph of a function.

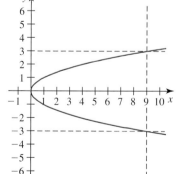

Figure 1.6

Using the vertical line test, we determine that the graph in Figure 1.6 is not a function because the vertical line crosses the graph more than once. For example, the vertical line shows that the input of 9 has two outputs, 3 and −3.

■ Function Notation

Functions may also be defined using **function notation**, a formal mathematical notation developed to communicate mathematical concepts on a universal scale. Consider the car loan scenario we discussed earlier. To express the relationship between the two variables in verbal terms we say that the interest rate, r, depends on the person's credit score, c. In mathematical terms we say that "r is a function of c." We are expressing the

PEER INTO THE PAST

EULER

Leonhard Euler (1707–1783) was the first to use the notation $f(x)$ as the symbol for a function. On page 268 of his work titled *Commentarii Academiae Scientiarum Petropolitanae* (Cajori, vol. 2), Euler introduces the notation as a way of writing "the function f of x."

Euler was from Basel, Switzerland, and was born to Margaret and Reverend Paul Euler. His father, a mathematician and a minister, groomed Leonhard to become a minister from birth. However, Leonhard chose not to be in the ministry and went on to study mathematics, physics, and engineering. He wrote over 866 books relating to these topics. It was once said that *"Euler calculated without apparent effort, as men breathe..."*

Euler had partial blindness when he was 20 and then lost almost all of his sight as he got older. Despite this difficult handicap, he persevered and discovered very important mathematical concepts.

Source: www.andrews.edu/~calkins/math/biograph/bioeuler.htm

---

credit score c as an input variable whose value determines the interest rate r through a function correspondence. In function notation we denote this as $f(c) = r$. The input variable c is the **independent variable** and the output variable r is the **dependent variable** because the value of r depends on the value of c.

> ### FUNCTION NOTATION
>
> The component parts of function notation include an *input*, an *output*, and a *function name* as detailed in the following diagram.
>
> function name → $f(x) = y$ ← output
> (input: x)
>
> The input x is called the **independent variable** and the output y is called the **dependent variable**. This equation may equivalently be written $y = f(x)$.

EXAMPLE 1 ■ Using Function Notation

Write each of the following sentences in function notation by choosing meaningful letters to represent each variable. Then identify the independent variable (input) and the dependent variable (output).

a. A person's weight is a function of the person's height.
b. The current gas price is a function of the amount of available crude oil.

Solution

a. Selecting w for the person's weight and h for the height, we represent this function as $f(h) = w$. The function f takes a value for the independent variable h and generates a value for the dependent variable w.

b. Selecting g for the gas price and c for the available crude oil, we represent this function as $f(c) = g$. The function f will take a value for the independent variable c and generates a value for the dependent variable g.

■ Evaluating Functions and Solving Function Equations

We can *evaluate* functions using function notation.

> ### EVALUATING A FUNCTION
>
> The process of finding the *output* of a function that corresponds with a given *input* is called **evaluating a function**.

For example, if we say "evaluate $f(642)$," we mean "find the output value that corresponds with an input value of 642."

EXAMPLE 2 ■ Evaluating a Function from a Formula

Suppose that your current job pays $11.50 per hour. Your salary, S, is calculated by multiplying $11.50 by the number of hours you work, h. That is, *Salary* = $11.50 · *hours* or, in function notation, $S = f(h) = 11.5h$. Evaluate $f(80)$ and explain what the numerical answer means.

Solution To evaluate $f(80)$ means to find the numerical value of S that results from "plugging in" 80 for the number of hours. We have

$$f(h) = 11.5h$$
$$f(80) = 11.5(80)$$
$$= 920$$

In other words, when $h = 80$, $S = 920$. This means that if you work 80 hours your salary will be $920.

Function notation is extremely versatile. Suppose we are given the function $f(x) = x^2 - 2x + 1$. We may evaluate the function at numerical values as well as non-numerical values. For example,

$$f(2) = (2)^2 - 2(2) + 1 \qquad f(\Delta) = (\Delta)^2 - 2(\Delta) + 1$$
$$= 4 - 4 + 1$$
$$= 1$$

In each case, we replaced the value of x in the function $f(x) = x^2 - 2x + 1$ with the quantity in the parentheses. Whether the independent variable value was 2 or Δ, the process was the same.

EXAMPLE 3 ■ Evaluating a Function at Nonnumeric Values

Given $f(x) = 5x^2 - 9x + 4$, find $f(\square)$ and $f(\nabla + \diamond)$.

Solution Admittedly, it feels a bit strange to evaluate functions at nonnumeric values. Nevertheless, we use the exact same strategy as if we were using numeric values. We replace all x values on the right-hand side of the equation with the quantity in the parentheses.

$$f(\square) = 5(\square)^2 - 9(\square) + 4 \qquad f(\nabla + \diamond) = 5(\nabla + \diamond)^2 - 9(\nabla + \diamond) + 4$$

To find the input value that corresponds with a given output, we must solve a function equation.

SOLVING A FUNCTION EQUATION

The process of finding the *input* of a function that corresponds with a given *output* is called **solving a function equation**.

For example, if we say "solve $f(c) = 6.91$," we mean "find the input value that corresponds with an output value of 6.91."

EXAMPLE 4 ■ Solving an Equation from a Formula

As stated earlier, the salary from a job that pays $11.50 per hour is given by the function $S = f(h) = 11.5h$. Solve $f(h) = 805$ for h. Explain what the numerical answer means in its real-world context.

Solution To solve $f(h) = 805$ means to find the number of hours h that must be put into function f to generate $805 in salary. That is, we need to find the number of hours that must be worked to earn $805. To find this, we set the salary, $f(h)$, equal to 805 and solve for h.

$$f(h) = 11.5h$$
$$805 = 11.5h$$
$$\frac{805}{11.5} = \frac{11.5h}{11.5}$$
$$h = 70$$

This means that if you want to earn $805 at a job that pays $11.50 per hour, you must work 70 hours.

■ Condensed Function Notation

Mathematicians constantly look for ways to more efficiently communicate mathematical concepts. For example, we can represent a function that converts h hours into D dollars using the function notation $D = f(h)$. However, this notation becomes cumbersome when calculating particular values. If a person earns $8 per hour, we have $D = f(h) = 8h$. Given that the person has worked 25 hours, we use the equation to calculate the worker's earnings as

$$D = f(25) = 8(25)$$
$$= 200$$

The multiple equal signs makes this difficult to read and understand quickly. To make things easier, we often rewrite the equation $D = f(h) = 8h$ as $D(h) = 8h$. This *condensed notation* is less cumbersome to work with and avoids the series of equal signs.

> **CONDENSED FUNCTION NOTATION**
>
> When calculating particular function values, it is customary to condense the function notation $D = f(h)$ to $D(h)$.
>
> $$\underset{\text{output}}{D}(\underset{\text{input}}{h})$$
>
> The equation $D(a) = b$ means that when $h = a$, $D = b$. This equation is written as $D = f(a) = b$ in standard function notation.

When we discuss inverse functions, we will see there are some mathematical concepts that are better understood using the standard function notation. So, although we frequently use condensed function notation, we will return to the standard function notation when discussing inverses.

Table 1.10

Golf Courses

Years (Since 1980) t	Golf Facilities G
0	12,005
5	12,346
10	12,846
15	14,074
20	15,489

Source: Statistical Abstract of the United States, 2004–2005, Table 1240

EXAMPLE 5 ■ Evaluating a Function and Solving an Equation from a Table

The number of golf courses in the United States has increased over the years. As seen in Table 1.10, the number of golf facilities, G, is a function of the years since 1980, t. We write this in condensed function notation as $G(t)$. Use the table to do each of the following.

a. Solve $G(t) = 12,846$ for t and explain what the numerical value means in this context.

b. Evaluate $G(20)$ and explain what the numerical value means in this context.

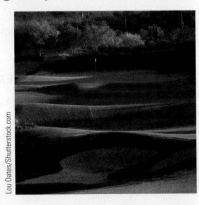

Solution

a. To "solve $G(t) = 12{,}846$ for t" means to determine the year when the number of golf courses was 12,846. From the table we can see that there were 12,846 golf courses 10 years after 1980. That is, in 1990.

b. To evaluate $G(20)$ means to determine the number of golf courses in the year 20. When we locate 20 in the Years column of the table, we see that there were 15,489 golf courses in that year. In other words, in 2000 there were 15,489 golf courses in the United States.

EXAMPLE 6 ■ Evaluating and Solving a Function Equation from a Graph

The sales of blank audio cassettes in the United States have declined since 1990. As shown in Figure 1.7, the sales of blank audio cassettes, C, is a function of the years since 1990, t. We write this in function notation as $f(t) = C$. Use the graph to do each of the following.

a. Estimate $f(12)$ and explain what the solution means in this context. Then write the result in function notation.

b. Find a value of t such that $f(t) = 334$ and explain what the solution means in this context. Then write the result in function notation.

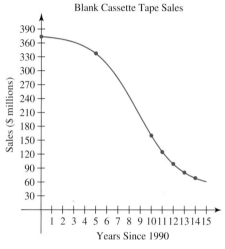

Figure 1.7

Solution

a. To estimate $f(12)$ means to find the value of the cassette tape sales for the year 12 years after 1990. That is, 2002. Using Figure 1.8, we find the t value of 12 on the horizontal axis, go "up" to the function graph, and then "over" to the vertical axis and find a C value of approximately 98.

   Therefore, according to the mathematical model, in the year 2002 there were approximately $98 million in blank cassette sales. In function notation, we write $f(12) = 98$.

b. To find a value of t such that $f(t) = 334$, we estimate the year in which there were $334 million in audio cassette sales. Using Figure 1.9, we locate 334 on the vertical axis, go "over" to the function graph, and then "down" to the horizontal axis to find a t value of approximately 5.

Figure 1.8

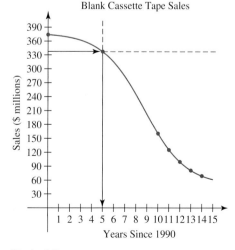

Figure 1.9

Therefore, audio cassette sales were $334 million around 1995, according to the mathematical model. In function notation, we write $f(5) = 334$.

■ Multivariable Functions

Sometimes functions have more than one dependent variable. For example, a credit report contains information such as the number and type of accounts, bill-paying history, collection actions, outstanding debt, and the age of the accounts. The credit scoring system predicts who is most likely to repay a debt by awarding points for these creditworthiness factors. Thus, the credit scoring system is a mathematical model represented by a multivariable function.

Other common multivariable functions include the compound interest formula $A = P\left(1 + \dfrac{r}{n}\right)^{nt}$ and the formula for volume of a box $V = lwh$. We'll look into these in more detail in future sections.

SUMMARY

In this section you learned that mathematical models are represented by functions. You discovered that a function is a process or correspondence relating two quantities in which each input value generates exactly one output value. You learned that functions are expressed in function notation $f(x) = y$, where y is the output value, x is the input value, and f is the function name. You discovered that the output value is called the dependent variable and the input value is called the independent variable because the output value depends on the input value. You learned that to evaluate a function means to determine the output value given an input value and that to solve a function means to find the input value that yields a given output value. Finally, you saw that functions may be either single variable or multivariable.

1.2 EXERCISES

■ SKILLS AND CONCEPTS

For Exercises 1–5, write each of the following expressions in function notation by choosing meaningful letters to represent each variable. Also identify the independent variable (input) and the dependent variable (output).

1. The amount of property tax you owe is a function of the assessed value of your home in dollars.

2. The length of your fingernails is a function of the amount of time that has passed since your last manicure.

3. The cost of mailing a letter is a function of the weight of the package in ounces.

4. The amount of water required for your lawn (in gallons) is a function of the temperature (in degrees).

5. A person's blood alcohol level is a function of the number of alcoholic drinks consumed in a 2-hour period.

In Exercises 6–15, use the vertical line test to determine whether the graph represents a function in the rectangular coordinate system.

6.

7.

8.

9.

10.

11.

12.

13.

14.

15.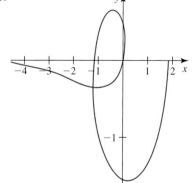

In Exercises 16–25, determine whether the table represents a function.

16.
x	y
9	7
10	7
11	7
12	7
15	7
18	7

17.
x	y
1	6
2	7
3	8
4	7
5	6
6	7

18.
x	y
9	10
10	9
11	8
10	7
9	6
8	5

19.
x	y
9	17
18	12
27	13
36	15
36	7
45	9

20.
x	y
−2	4
−1	1
0	0
1	1
2	4

21.
x	y
1.1	12
2.2	9
3.3	3
2.2	4
4.4	5

22.
x	y
1	7
1	7
1	7
1	7
1	7
1	7

23.
x	y
2	1
5	2
4	3
7	4
6	5
8	6

24.
x	y
19	17
19	7
19	−7
19	7
19	17
19	7

25.
x	y
3	17
4	13
5	9
3	17
4	13
5	19

In Exercises 26–35, evaluate the function for the specified value. Do not simplify. Note that the special symbols (such as #, Δ, and Θ) do not have any special mathematical meaning. They are just symbols.

26. $f(x) = 2x^2 - 3x$; $f(4)$

27. $v(t) = \dfrac{\sqrt{0.3t}}{10}$; $v(3)$

28. $r(s) = |9s^3 - 2s + 18|$; $r(-2)$

29. $t(v) = -v^2 + 3v - \dfrac{4}{v}$; $t(-4)$

30. $h(x) = 3^x - 17x + x^2$; $h(b)$

31. $m(x) = \sqrt{x^2 - 4x}$; $m(\# + 3)$

32. $n(d) = -2d + d^3$; $n(\square)$

33. $r(s) = |9s^3 - 2s + 18|$; $r(\Theta + \Delta)$

34. $t(v) = -v^2 + 3v - \dfrac{4}{v}$; $t(\Delta - \nabla)$

35. $h(x) = 3^x - 17x + x^2$; $h(b^2 + 7)$

36. **Weight Loss** Suppose that you go on a diet and the following graph displays your progress in losing weight. As seen in the graph, your weight (w) is a function of the time since your diet started ($t = 0$) because each week results in only one weight (in pounds).

a. Evaluate $w(3)$ and explain in a complete sentence what your solution means in its real-world context.

b. Find a value of t such that $w(t) = 160$ and explain in a complete sentence what your solution means in its real-world context.

c. Estimate $w(2.5)$ and discuss the accuracy of your prediction.

d. Estimate the solution to $w(t) = 100$ and discuss the accuracy of your approximation.

37. **Per-Gallon Fuel Cost** The fuel cost, C, of operating a car that goes 25 miles per gallon of gasoline is a function of the price of gasoline, g (in dollars per gallon), and the distance driven, D (in miles). The formula is

$$C(g, d) = \dfrac{gd}{25} \text{ dollars}$$

Use function notation to express the fuel cost of operation if gasoline costs $2.76 per gallon and the car travels 310 miles. Then calculate the cost.

38. **Grocery Cost** A friend wants to purchase cereal, milk, and bananas on her next trip to the grocery store. The

grocery bill in dollars, g, is a function of the number of boxes of cereal, c, gallons of milk, m, and bunches of bananas, b, she buys. Assume cereal is $3.75 a box, milk is $2.79 a gallon, and a bunch of bananas is $1.15. The grocery bill is given by

$$g(c, m, b) = 3.75c + 2.79m + 1.15b$$

Use function notation to express the grocery bill if she buys two boxes of cereal, three gallons of milk, and one bunch of bananas.

SHOW YOU KNOW

39. Choose five exercises from this section. For each exercise, describe whether the function represented a process or a correspondence. Explain your reasoning.

40. Explain why the vertical line test is a valid way to check to see if a graph represents a function.

41. What is the practical advantage of requiring a function to have exactly one output for each input?

42. Give an example of a nonmathematical process that represents the function concept.

43. Give an example of a nonmathematical correspondence that represents the function concept.

MAKE IT REAL

44. **Children in Preschool and Kindergarten** Based on data provided by the Census Bureau, the number of American children age 3 to 5 years enrolled in preprimary school (preschool and kindergarten) can be modeled by the function

$$p(y) = 121.46y + 5231.31$$

where p represents the number of children enrolled (in thousands) and y is the number of years since 1980. (*Source: www.census.gov*)

Evaluate $p(27)$ and explain what the numerical answer represents in its real-world context. Then write the solution in function notation.

45. **Consumer Expenditures** The average annual expenditures of all U.S. consumers can be modeled by the formula

$$E(t) = 1019.65t + 27{,}861.97$$

where E represents the annual expenditures in dollars and t is the years since 1990. (*Source: www.census.gov*)

Evaluate $E(20)$ and explain what the numerical answer represents in its real-world context. Then write the solution in function notation.

46. **College Enrollment** The number of students enrolled for the Spring semester at Chandler-Gilbert Community College has been growing in recent years. The number of students enrolled can be modeled by the function

$$C(y) = 168.9y + 6741$$

where C is the number of students enrolled and y is the years since 2003. (*Source: Modeled from data at www.azcentral.com*)

Solve $C(y) = 9000$ for y and explain what the numerical answer represents in its real-world context. Then write the solution in function notation.

47. **Homes Sales Price** The median sales price of new homes since 1980 can be modeled by the formula

$$m(t) = 5844.95t + 56{,}589.91 \text{ dollars}$$

where m is the median sales price and t is the year since 1980. (*Source: www.census.gov*)

Solve $m(t) = 300{,}000$ for t and explain what the numerical answer represents in its real-world context. Then write the solution in function notation.

48. **U.S. National Parks Visits** The number of recreational visits to the National Parks of the United States is displayed in the table. The number of visits to the national parks, p, is a function of the year, t.

Year	Recreational Visits to U.S. National Parks (millions of people)
1990	258.7
1995	269.6
1999	287.1
2000	285.9
2001	279.9
2002	277.3
2003	266.1
2004	276.4

Source: www.census.gov

a. Solve $p(t) = 277.3$ for t and explain the meaning of the solution.

b. Evaluate $p(2000)$ and write a sentence explaining what the numerical value you find means in its real-world context.

c. Estimate $p(2010)$ and discuss the accuracy of your prediction.

d. Estimate the solution to $p(t) = 300$ and discuss the accuracy of your approximation.

49. **Most Visited States by Foreigners** The twelve most frequently visited American states by overseas travelers in 2004 are displayed in the following table. The number of visits, v, is a function of the U.S. state, s.

U.S. State	Number of Visits (in thousands)
New York	5426
Florida	4430
California	4207
Hawaii	2215
Nevada	1626
Illinois	975
Massachusetts	935
Texas	874
New Jersey	833
Pennsylvania	691
Arizona	630
Georgia	427

Source: www.census.gov

a. Solve $v(s) = 630$ for s and explain the meaning of the solution.

b. Evaluate $v(Hawaii)$ and explain the meaning of the result.

50. Home Value Zillow.com is a website that approximates the value of a home based on its address. Based on data from the website, we created the following graph showing the value of one Arizona home, h, as a function of the month, m.

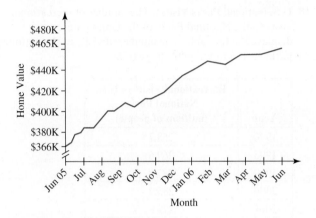

a. Evaluate $h(Aug05)$ and explain what the solution means in its real-world context.

b. Find a value of m such that $h(m) = 440,000$ and explain what the solution means in its real-world context.

c. Estimate the solution to $h(m) = 480,000$ and discuss factors that may affect the accuracy of your estimate.

■ STRETCH YOUR MIND

Exercises 51–52 are intended to challenge your understanding of functions and function notation.

51. A classmate solves the equation $f(x) = 4x - 3$ for x as follows.

$$f(x) = 4x - 3$$
$$f(x) - 4x = -3$$
$$x(f - 4) = -3$$
$$x = \frac{-3}{f - 4}$$

Explain what is wrong with this problem-solving process.

52. In the equation $x(f) = 9f + 6$, which variable is independent and which is dependent?

SECTION 1.3 Functions Represented by Tables and Formulas

LEARNING OBJECTIVES

- Determine if a data table represents a function
- Calculate and interpret the meaning of an average rate of change from a table
- Create and use basic function formulas to model real-world situations

GETTING STARTED

Both consumer debt and fuel prices for Americans have been increasing at an ever-increasing rate while total savings have decreased at an increasingly rapid rate. Although no one knows what the future will bring, we can forecast what might happen by analyzing tables of data related to these issues.

In this section we explain how to analyze functions in tables and formulas. We'll also demonstrate how to find average rates of change and use them to help forecast unknown results.

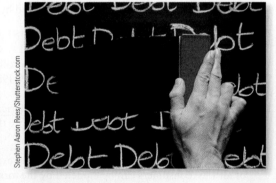

Before going on, let's step back and review what we know about mathematical models and functions. We've shown that a function is a process or correspondence relating a set of inputs with a set of outputs in such a way that each input is paired with a single output. We have also seen that many real-world situations and data sets can be effectively modeled using functions. In the remaining sections of this chapter, we look in greater detail at functions and function models of real-world data sets in four different ways: tables, formulas, graphs, and words (Figure 1.10).

Figure 1.10

SECTION 1.3 ■ FUNCTIONS REPRESENTED BY TABLES AND FORMULAS

Table 1.11

Year d	Total Consumer Debt ($ billions) C
1985	593.00
1990	789.10
1995	1095.80
2000	1556.25
2005	2175.25

Source: Federal Reserve Board

■ Determining If a Table Represents a Function

The Federal Reserve Board monitors total consumer debt (excluding loans secured by real estate). Table 1.11 shows several years of consumer debt with the year as the input and the total consumer debt as the output.

To determine if the table represents a function, we only need to verify that each input value corresponds with a single output value. That is, for each year, there can only be one value for the total consumer debt. Intuitively, we know that this must be the case, and a quick inspection of the table verifies this conclusion. Therefore, total consumer debt is a function of the calendar year.

■ Average Rate of Change

One way to analyze a table of data is to calculate an *average rate of change*.

HOW TO: ■ CALCULATE AN AVERAGE RATE OF CHANGE

The average rate of change of a function f over an interval $[a, b]$ is calculated by dividing the difference of two outputs by the difference in the corresponding inputs. That is,

$$\text{average rate of change} = \frac{f(b) - f(a)}{b - a}$$

where $(a, f(a))$ and $(b, f(b))$ are any two data points in the table.

EXAMPLE 1 ■ Calculating and Interpreting an Average Rate of Change

Table 1.12 shows the age-adjusted death rate due to heart disease. (Age-adjusting is a statistical technique used by analysts to help avoid distortions in data interpretation when comparing disease rates over time.)

Table 1.12

Years since 1990 t	Age-Adjusted Death Rate Due to Heart Disease (deaths/100,000 people) r
0	321.8
1	313.8
2	306.1
3	309.9
4	299.7
5	296.3
6	288.3
7	280.4
8	272.4
9	267.8
10	257.6

Source: Statistical Abstract of the United States, 2006, Table 106

a. Calculate the average rate of change in the death rate between 1990 and 2000.
b. Interpret the real-world meaning of the average rate of change from part (a).
c. Use the average rate of change from part (a) and the death rate in 2000 to forecast the 2003 death rate. Explain why the forecasted 2003 death rate differs from the actual death rate reported by the Census Bureau (232.1 deaths per 100,000 people).

Solution

a. In 1990, the death rate was 321.8 deaths per 100,000 people. In 2000, the death rate was 257.6 deaths per 100,000 people. We calculate the average rate of change over the 10-year period.

$$\text{average rate of change} = \frac{257.6 - 321.8}{10 - 0} \frac{\text{deaths per 100,000 people}}{\text{year}}$$

$$= -6.42 \text{ deaths per 100,000 people per year}$$

b. The average rate of change finds the rate at which the death rate changed each year *if we assume that the death rate changed by the same amount each year.* The rate of change is negative since the death rate is decreasing. If the death rate had decreased by the same amount each year over the 10-year period, it would have decreased by 6.42 deaths per 100,000 people each year.

c. To forecast the 2003 death rate, we assume that the average rate of change between 2000 and 2003 will be the same as that between 1990 and 2000. In part (a), we showed that the average rate of change was -6.42 deaths per 100,000 people each year. Because there are 3 years between 2000 and 2003, the 3-year change will be 3 times the average rate of change. We add this change to the 2000 death rate to forecast the 2003 death rate.

$$2003 \text{ death rate} = 2000 \text{ death rate} + 3(\text{average rate of change})$$
$$2003 \text{ death rate} = 257.6 + 3(-6.42)$$
$$\approx 238.3$$

We predict that the death rate due to heart disease in 2003 was 238.3 deaths per 100,000 people. Our projection differs from the death rate of 232.1 reported by the Census Bureau because we used the average rate of change between 1990 and 2000. The actual average rate of change between 2000 and 2003 was

$$\frac{232.1 - 257.6}{3 - 0} = -8.5 \text{ deaths per 100,000 people per year}$$

As Example 1 shows, we should exercise caution when using an average rate of change to forecast function values. The rate at which the function is changing could vary widely.

EXAMPLE 2 ■ **Identifying Limitations in Using an Average Rate of Change**

Refer to Table 1.12 and calculate the average rate of change in the death rate due to heart disease over the following time intervals: 1990 to 1991, 1992 to 1993, 1995 to 1996, and 1999 to 2000. Does the average rate of change between 1990 and 2000 (-6.42 deaths per 100,000 people per year) accurately represent those values?

Solution We calculate the average rate of change in the death rate due to heart disease over the given time intervals.

1990 to 1991: $\dfrac{313.8 - 321.8}{1 - 0} = -8.0$ \qquad 1992 to 1993: $\dfrac{309.9 - 306.1}{3 - 2} = 3.8$

1995 to 1996: $\dfrac{296.3 - 288.3}{5 - 4} = -8.0$ \qquad 1999 to 2000: $\dfrac{257.6 - 267.8}{10 - 9} = -10.2$

We see that the average rate of change varied widely, ranging from a 10.2 decrease to a 3.8 increase. The -6.42 average rate of change from 1990 to 2000 we calculated in Example 1 is a *constant* annual change that does not accurately predict wide variations in the actual annual rates of change for intermediate years.

SECTION 1.3 ■ FUNCTIONS REPRESENTED BY TABLES AND FORMULAS

EXAMPLE 3 ■ Calculating and Interpreting an Average Rate of Change

The movie *Ice Age: The Meltdown* was introduced on March 31, 2006. Table 1.13 shows the gross receipts for selected weekends within the first 9 weeks that the movie was shown in U.S. theaters.

Table 1.13

Weekend w	Gross Receipts (in millions) G
1	$68.0
3	$20.0
6	$4.2
9	$0.8

Source: www.the-numbers.com

a. Calculate the average rate of change for each pair of consecutive data points. Include units with each average rate of change.

b. Describe what is happening to the average rate of change for the first nine weekends.

Solution

a. The average rates of change are shown in Table 1.14.

Table 1.14

Weekend w	Gross Receipts (in millions) G	Average Rate of Change (in millions per weekend)
1	$68.0	
		$\dfrac{20.0 - 68.0}{3 - 1} = -\$24.0$
3	$20.0	
		$\dfrac{4.2 - 20.0}{6 - 3} \approx -\$5.3$
6	$4.2	
		$\dfrac{0.8 - 4.2}{9 - 6} \approx -\$1.1$
9	$0.8	

b. Between the first and third weekend, the average gross receipts were decreasing at an average rate of $24.0 million per weekend. The theaters were bringing in an average of $5.3 million less per weekend between the third and sixth weekends. Between the sixth and ninth weekend, gross receipts were decreasing by a lesser average amount of $1.1 million per weekend. Notice that although the rates of change of the gross receipts are all negative, they are becoming less and less negative as time goes on. The *scatter plot* of the data shown in Figure 1.11 validates our conclusion. (A **scatter plot** is a graphical representation of a data table.)

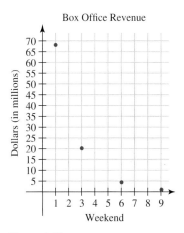

Figure 1.11

Predicting Unknown Data Values Using a Table

Despite the limitations of using the average rate of change to predict unknown data values, we often use it in forecasting when additional data are unavailable. Using the average rate of change is frequently more accurate than guessing.

EXAMPLE 4 ■ **Estimating the Unknown Value of a Function at a Specified Input Value**

Monthly life insurance premiums are based on the age and the gender of the insured. Table 1.15 records the monthly premium for a $500,000 life insurance policy for a male. The table works well for the ages shown but if a man's age falls between those ages, we must estimate the monthly premium value. Estimate the monthly premium for a 53-year-old male.

Table 1.15

Life Insurance for Males	
Age a	Monthly Premiums for $500,000 Worth of Coverage P
35	$14
40	$18
45	$27
50	$40
55	$65
60	$103
65	$180
70	$322

Source: www.Insure.com

Solution We first calculate the average rate of change for the data points that encompass the age 53: (50, $40) and (55, $65).

$$m = \frac{P(55) - P(50)}{55 - 50} \quad \frac{\text{dollars}}{\text{year of age}}$$

$$= \frac{65 - 40}{55 - 50}$$

$$= \$5.00 \text{ per year of age}$$

For each year of age between 50 and 55, we estimate there is a $5 increase in premium. Since 53 is 3 years more than 50, the premium for a 53-year-old male will be $15 more than the premium for a 50-year-old male. The premium is given by

$$\text{age 53 premium} = \text{age 50 premium} + (\text{average rate of change})(\text{difference in age})$$

$$= 40 + (5)(3)$$

$$= 55 \text{ dollars}$$

We estimate a 53-year-old male will pay a monthly premium of $55 for $500,000 worth of life insurance.

Functions Represented by Formulas

Although tables are useful, their value is limited because they can only show a finite number of values. In contrast, a function formula (equation) may be used to calculate any number of values. A **formula** is a succinct mathematical statement expressing a relationship between quantities.

For example, when a 154-pound woman exercises on a stair-step machine, the number of calories burned is related to the number of minutes she spends exercising on the machine. For every minute she exercises, she burns around 9 calories. (*Source:* www.fns.usda.gov) In other words, the total number of calories burned, C, is equal to nine times the number of minutes, m, spent operating the stair-step machine. Symbolically, we write

$$C(m) = 9m$$

To determine how many calories she burns in 5 minutes, we *evaluate* the function at $m = 5$.

$$C(m) = 9m$$
$$C(5) = 9(5)$$
$$= 45$$

When she exercises on the stair-step machine for 5 minutes, she burns 45 calories.

To determine how long it will take her to burn 150 calories, we set $C(m) = 150$ and solve the function for m. In other words, we substitute 150 for $C(m)$ and solve the resultant equation.

$$C(m) = 9m$$
$$150 = 9m$$
$$m \approx 16.67$$

If she wants to burn 150 calories, she must exercise on the stair-step machine for 16.67 minutes. Since most people don't end their timed exercise on a fraction of a minute, we'll round the number up to 17 minutes.

Is this calorie-burning formula, $C(m) = 9m$, a function? That is, does each value of the independent variable m result in exactly one dependent variable value C? Yes. A woman exercising for any number of minutes m would burn about the same number of calories C each time she exercised for m minutes in the same activity at the same level of intensity.

EXAMPLE 5 ■ Finding a Formula for a Cell Phone Plan

The Sprint Fair & Flexible cell phone plan costs one professor $35.10 a month (including fees and taxes), if she does not go over her allotted 200 anytime minutes. However, she must pay $0.28 a minute (including fees and taxes) for every minute over her limit. (*Source:* Sprint cell phone bill)

a. If she does not go over 200 minutes, the professor's monthly bill is $35.10. Write a formula that she can use to calculate her monthly cell phone bill if she exceeds 200 minutes.

b. If she talks a half-hour over her monthly limit, what will be the total amount of her monthly cell phone bill?

Solution

a. The total cost, C, depends on the extra minutes, m, she uses. Since each extra minute costs $0.28, the total cost will be the product of 0.28 and the number of extra minutes plus the cost of the basic plan. Symbolically, we write

$$C(m) = 0.28m + 35.10$$

b. A half-hour is 30 minutes, so $m = 30$. To calculate her bill, we evaluate $C(m)$ at $m = 30$.

$$C(m) = 0.28m + 35.10$$
$$C(30) = 0.28 \cdot 30 + 35.10$$
$$= 8.40 + 35.10$$
$$= 43.50$$

The professor's monthly phone bill is $43.50.

For real-world situations that are complex and rely on multiple inputs, mathematicians have developed standardized formulas to help ordinary people make important life decisions. One such formula, the monthly loan payment formula, is illustrated in Example 6.

EXAMPLE 6 ■ Using a Multivariable Formula to Find the Amount of a Car Payment

The manufacturer's suggested retail price (MSRP) for a new 2006 Toyota Corolla was approximately $15,000. (*Source:* www.toyota.com) As of May 30, 2006, Pentagon Federal Credit Union offered a 5-year new car loan with a 7.5% interest rate compounded monthly. The monthly payment formula is

$$M(p, i, n) = p\left[\frac{i(1 + i)^n}{(1 + i)^n - 1}\right]$$

where
- M = monthly payment in dollars
- p = amount borrowed in dollars
- n = number of monthly payments
- i = interest rate per payment period as a decimal

What is the monthly payment on the loan if a person finances the full retail price of the car?

Solution To use the formula, we must first determine the value of each of the input variables: p, n, and i. Since $15,000 is to be borrowed, $p = 15,000$. To determine the number of monthly payments, n, we must determine the number of months in 5 years.

$$\frac{5 \text{ years}}{1} \cdot \frac{12 \text{ months}}{1 \text{ year}} = 60 \text{ months}$$

So $n = 60$. To obtain the monthly periodic rate, i, we divide the decimal form of the annual rate, 7.5%, by the number of months in a year.

$$i = \frac{0.075}{12}$$
$$= 0.00625$$

The monthly periodic rate is 0.625%, but we use the decimal form in the formula.

Now that we know the values of the three input variables, we can calculate the payment amount.

$$M(p, i, n) = p\left[\frac{i(1 + i)^n}{(1 + i)^n - 1}\right]$$

$$M(15000, 0.00625, 60) = 15,000\left[\frac{(0.00625)(1 + 0.00625)^{60}}{(1 + 0.00625)^{60} - 1}\right]$$

$$= 15,000\left[\frac{(0.00625)(1.4533)}{1.4533 - 1}\right]$$

$$= 15,000\left(\frac{0.009083}{0.4533}\right)$$

$$\approx 300.57 \text{ rounded to the nearest cent}$$

For a $15,000 car financed with a 7.5% loan for a period of 5 years, the monthly payment is $300.57.

EXAMPLE 7 ■ Using a Multivariable Formula to Determine a Loan Amount

On May 31, 2006, the national average rate on a 48-month car loan was 9.3%. (*Source:* www.bankrate.com) If we can afford a $250 monthly car payment, what price car can we afford at this rate and loan length?

Solution We use the same formula as we did in Example 6. We know the monthly payment is $250, so $M = 250$. Since the annual rate is 9.3%, the monthly periodic rate is

$$i = \frac{0.093}{12} = 0.00775$$

Finally, the loan period is 48 months, so $n = 48$. To find the affordable car price, which is the loan amount p, we substitute the values we know into the formula and solve for p.

$$M(p, i, n) = p\left[\frac{i(1+i)^n}{(1+i)^n - 1}\right]$$

$$250 = p\left[\frac{0.00775(1 + 0.00775)^{48}}{(1 + 0.00775)^{48} - 1}\right]$$

$$250 \approx p\left[\frac{0.00775(1.4486)}{1.4486 - 1}\right]$$

$$250 \approx p\left(\frac{0.01123}{0.4486}\right)$$

$$250 \approx p(0.02503)$$

$$p \approx 9988.92$$

Given a 9.3%, 48-month car loan with a monthly payment of $250, we can afford to buy a vehicle that costs $9,988.92.

You may have noticed in Example 7 that $\frac{250}{0.02503} \approx 9988.01$ instead of the 9988.92 shown. The difference in the two values occurs because of *round-off error*. To minimize the error introduced by rounding intermediate computations, we kept the calculated values in our calculator and used them to calculate our final answer instead of the rounded values we recorded in intermediate steps of the example. We deferred all rounding to the end of the problem. This approach helps reduce the error that occurs due to rounding and allows us to write the intermediate values with fewer significant digits.

SUMMARY

In this section you learned how to analyze functions in tables and formulas. Additionally, you discovered how to find average rates of change and use them to help forecast unknown results.

1.3 EXERCISES

SKILLS AND CONCEPTS

In Exercises 1–6 use the given formula to do each of the following.

  a. Determine the indicated output value.
  b. Interpret the meaning of the result.

1. The formula for calculating the perimeter of a rectangle, P, is $P(w, l) = 2w + 2l$, where w is the width in inches and l is the length in inches. Calculate the value of $P(5, 8)$.

2. Use the formula in Exercise 1 to calculate the value of $P(103, 808)$.

3. $V(r) = \frac{4}{3}\pi r^3$ is the formula for computing the volume of a sphere, V (in cubic centimeters), where r is the radius of the sphere (in centimeters). Compute the value of $V(100)$.

4. Use the formula in Exercise 3 to compute the value of $V(63)$.

5. The formula for calculating the area of a trapezoid, A, is $A(h, b_1, b_2) = \frac{h(b_1 + b_2)}{2}$, where h is the height, b_1 is one base, and b_2 is the second base. Each of the lengths is measured in meters. Calculate the value of $A(9, 14, 24)$.

6. Use the formula in Exercise 5 to calculate the value of $A(17, 25, 30)$.

SHOW YOU KNOW

7. Explain what is meant by "average rate of change." Particularly address the word *average*.
8. Find a situation where the average rate of change can be calculated. What does this calculation mean in the context of the situation?
9. Explain how someone's car insurance premium can be computed using a multivariable function. That is, describe which variables could be the independent variables and which could be the dependent variable.

MAKE IT REAL

For Exercises 10–15, do each of the following.

a. Determine if the table of data represents a function.
b. Calculate the average rates of change for consecutive pairs of data values.
c. Explain the meaning of the average rates of change in the context of the data.

10. **Life Insurance Premiums**

Life Insurance for Males		
Age a	Monthly Premiums for $1,000,000 of Coverage P	Average Rate of Change
35	$21	
40	$30	
45	$47	
50	$75	
55	$124	
60	$198	
65	$348	
70	$628	

Source: www.Insure.com

11. **Life Insurance Premiums**

Life Insurance for Females		
Age a	Monthly Premiums for $1,000,000 of Coverage P	Average Rate of Change
35	$21	
40	$25	
45	$41	
50	$57	
55	$88	
60	$130	
65	$209	
70	$361	

Source: www.Insure.com

12. **Percent of 16-Year-Old Drivers**

Year t	Percent of 16-Year-Olds with Driver's Licenses L	Average Rate of Change
1998	44.0%	
1999	37.0%	
2000	34.0%	
2001	32.5%	
2002	31.5%	
2003	30.5%	
2004	30.0%	

Source: Federal Highway Administration

13. **Baby Girls' Average Weight**

Age (in months) m	Weight (in pounds) W	Average Rate of Change
Birth	7.5	
6	16.0	
12	21.0	
18	24.0	
24	26.5	
30	28.5	
36	30.5	

Source: National Center for Health

14. **Baby Boys' Average Weight**

Age (in months) m	Weight (in pounds) W	Average Rate of Change
Birth	7.9	
6	17.4	
12	22.8	
18	25.9	
24	28.0	
30	29.8	
36	31.5	

Source: National Center for Health

15. **Divorce Rates—Sixties and Seventies**

Year t	Number of Divorces per 1000 Married Women, Age 15 and Older D	Average Rate of Change
1960	9.2	
1965	10.1	
1970	15.0	
1975	20.1	

Source: Rutgers National Marriage Project

In Exercises 16–29, use the data in the table to determine each solution.

16. Divorce Rates—From the Eighties

Year t	Number of Divorces per 1000 Married Women, Age 15 and Older D	Average Rate of Change
1980	22.7	
1985	21.7	
1990	20.7	
1995	19.7	
2000	18.7	
2005	17.7	

Source: **Rutgers National Marriage Project**

Estimate $D(1982)$ and interpret what it means in a real-world context.

17. China's Oil Demand

Year t	Demand (in millions of barrels per day) D	Average Rate of Change
2002	5.0	
2003	5.7	
2004	6.1	
2005	6.3	

Source: **International Energy Agency**

Estimate $D(2006)$ and interpret what it means in a real-world context.

18. Personal Bankruptcy in Japan

Year d	Number of Filings (in thousands) N	Average Rate of Change
1994	45.0	
1998	100.0	
2000	149.0	
2002	225.0	

Source: **Financial Services Agency**

Estimate the number of bankruptcy filings in 2001 and write your answer in function notation.

19. U.S. Lawmakers' Private Trips in 2006

Month m	Number of Trips N	Average Rate of Change
January	158	
February	61	
March	29	

Source: **Political Money Line**

Estimate the number of private trips that were given to U.S. Lawmakers in April 2006.

20. Amazon's Net Income

Year Y	Net Income ($ billions) R	Average Rate of Change
2002	3.9	
2003	5.3	
2004	6.9	

Source: **www.Amazon.com**

Estimate $R(2001)$ and interpret what it means in a real-world context.

21. Amazon's Net Income

Year Y	Net Income ($ billions) R	Average Rate of Change
2005	8.5	
2006	10.7	
2007	14.8	

Source: **www.Amazon.com**

Estimate the net income of Amazon for 2008. Write your answer in function notation.

22. Large U.S. Churches

Estimated Number of U.S. Churches with an Average Weekly Attendance of at Least 2000 People		
Year d	Churches C	Average Rate of Change
1960	0	
1970	10	
1980	115	
1990	300	
2000	600	
2003	800	
2005	1200	

Source: **Hartford Institute for Religion Research**

Estimate and explain the meaning of $C(1965)$.

23. Transportation Security Administration Fines

Year t	Fines F	Average Rate of Change
2002	279	
2003	3426	
2004	9741	

Source: **TSA**

a. What is the value of $F(2003)$ and what does it mean?

b. Estimate the number of fines issued by the TSA in 2005 and write your answer in function notation.

24. Daytime Emmy Award Viewers

Year t	Viewers (in millions) V	Average Rate of Change
2001	10.30	
2002	9.62	
2003	8.94	
2004	8.26	
2005	7.58	

Source: Nielsen Media Research

a. What is the value of $V(2003)$ and what does it mean?

b. How many viewers do you estimate watched the daytime Emmy awards in 2006? Explain your reasoning.

25. Exxon Mobil 2005 Net Income

Quarter q	Net Income (in billions) N	Average Rate of Change
2	6.9	
3	10.0	
4	10.5	

Source: Exxon Mobil Corporation

a. What is the value of $N(2)$ and what does it mean?

b. Estimate the net income for Exxon Mobil Corporation in the first quarter of 2006. Explain your reasoning.

26. Hispanic Marines

Year t	Percentage of Hispanic Marine Corps Recruits H	Average Rate of Change
2002	13.5%	
2003	14.5%	1.00
2004	16.0%	1.50
2005	16.5%	0.50

Source: CNA Corporation

a. What is the value of $H(2005)$ and what does it mean?

b. A newspaper stated, "The Marines are increasingly turning to Hispanics to fill their ranks." Do you agree with this conclusion?

27. Foreign Vehicle Demand in the U.S.

New Vehicles Sold in the U.S. (in thousands)			
Year	Nissan	Honda	Toyota
2000	700	1000	1400
2001	600	1050	1550
2002	620	1100	1550
2003	690	1190	1600
2004	830	1200	1800
2005	920	1250	1999

Source: Autodata

a. In each pair of consecutive years, which car had the largest average rate of change?

b. Which car would you predict had the largest average rate of change in 2006? Explain your reasoning.

28. Rates of Home Ownership

Year	United States	Arizona	Utah
1930	47.8%	44.8%	60.9%
1940	43.6%	47.9%	61.1%
1950	55.0%	56.4%	65.3%
1960	61.0%	63.9%	71.7%
1970	62.9%	65.3%	69.3%
1980	64.4%	68.3%	70.7%
1990	64.2%	64.2%	68.1%
2000	66.2%	68.0%	71.5%

Source: U.S. Census Bureau, Housing and Household Economic Statistics Division Revised: 2004

a. Which state has had the greatest increase over one decade? Include statistics to support your answer.

b. Which state has had the greatest decrease over one decade? Include statistics to support your answer.

29. U.S. Home Sales

2005 U.S. Home Sales ($ millions)		
Months m	New Home Sales N	Existing Home Sales E
February	1.24	6.87
March	1.30	6.88
April	1.27	7.08
May	1.28	7.07
June	1.29	7.22
July	1.36	7.06
August	1.27	7.20
September	1.24	7.19
October	1.34	7.02
November	1.23	7.01
December	1.27	6.68

Source: Department of Commerce, National Association of Realtors

a. Determine the months between which the function $N(m)$ is *increasing*. Support your decision with average rates of change.

b. Determine the months between which the function $N(m)$ is *decreasing*. Support your decision with average rates of change.

c. Determine the months between which the function $E(m)$ is decreasing. Support your decision with average rates of change.

30. Investment Account $B(p, r, n, t) = p\left(1 + \frac{r}{n}\right)^{nt}$ is

the formula used to calculate the balance, B, in an investment account with a lump sum invested and interest compounded a fixed number of times a year for a certain number of years. The independent variables represent the following:

p is the amount invested.
r is the nominal interest rate as a decimal.
n is the number of times the compounding occurs in a year.
t is the years the money is invested.

Compute the value of $B(500, 0.05, 12, 2)$.

31. **Investment Account** Use the formula in Exercise 30 to compute the value of $B(350, 0.06, 12, 5)$.

32. **Skid Distance** The formula for calculating the minimum speed of a car in miles per hour at the beginning of a skid, S, is $S(d, f, n) = \sqrt{30\,dfn}$, where d is the skid distance in feet, f is the drag factor for the road surface, and n is the braking efficiency as a percent (100% = 1.00). (*Source:* www.harristechnical.com) Calculate $S(60, 0.75, 1)$.

33. **Skid Distance** Use the formula in Exercise 32 to calculate $S(155, 0.2, 0.9)$.

In Exercises 34–35, use the formula to
  a. Determine the value of the indicated input.
  b. Interpret the meaning of the calculated value.

34. **Commission-Based Income** People who work in sales often earn a base salary plus a commission based on their sales. If the annual salary for a certain real estate agent is $12,000 plus 6% of total sales, then

$$T(s) = 0.06s + 12{,}000$$

is the formula used to compute the total salary, T, as a function of the amount of sales, s, the person brings in. What is the value of s if $T = 30{,}000$?

35. **Commission-Based Income** Use the formula in Exercise 34 to find s given $T = 60{,}000$.

The formulas for Exercises 36–41 are standard measurement conversion formulas. (*Source:* www.wikipedia.org)

36. **Temperature Conversion** The formula for converting temperatures on the Kelvin scale, K, to temperatures on the Fahrenheit scale, F, is

$$F(K) = \frac{9}{5}K - 459.67$$

What is the value of K if $F = 98.33$?

37. **Temperature Conversion** Use the formula in Exercise 36 to find K given $F = 70$.

38. **Temperature Conversion** The formula

$$F(C) = \frac{9}{5}C + 32$$

converts the temperature on the Celsius, C, scale to an equivalent temperature on the Fahrenheit, F, scale. What is C if $F = 70$?

39. **Temperature Conversion** Use the formula in Exercise 38 to find F given $C = 70$.

40. **Temperature Conversion** The formula

$$C(F) = \frac{5}{9}(F - 32)$$

converts the temperature on the Fahrenheit scale, F, to the temperature on the Celsius scale, C. What is F if $C = 30$?

41. **Temperature Conversion** Use the formula in Exercise 40 to find F given $C = 10$.

42. **Area of a Trapezoid** The formula for the area, A, of a trapezoid is

$$A(h, b_1, b_2) = \frac{h(b_1 + b_2)}{2}$$

where h is the height, b_1 is one base, and b_2 is the second base. If $A = 50$, $b_1 = 7$, $b_2 = 13$, what is h?

43. **Area of a Trapezoid** Using the formula in Exercise 42, what is h if $A = 120$, $b_1 = 8$, $b_2 = 14$?

44. **Investment Account** Using the formula

$$B(p, r, n, t) = p\left(1 + \frac{r}{n}\right)^{nt}$$

what is p, if $B = 5395.40$, $r = 0.06$, $n = 12$, $t = 5$?

45. **Investment Account** Using the formula in Exercise 44, what is p, if $B = 5395.40$, $r = 0.03$, $n = 12$, $t = 5$?

46. **Vehicle Cost** A vehicle owner wants to calculate the total cost of his 2007 Jeep Compass with a MSRP of $18,366. His monthly loan payment is $317.54 for 5 years after he puts down a $2000 down payment. (*Source:* Car price at www.edmunds.com)
  a. Write a formula for the total amount he has paid toward the cost of the car (including down payment), T, as a function of the number of months he has made payments on the loan, m.
  b. What is the total cost of the Jeep, after he has made all of the payments?
  c. How much money has he paid in interest for his Jeep?

47. **Vehicle Cost** A student has a monthly car payment of $172.55 a month for 6 years on a 2007 Blazing Blue two-door hatchback Toyota Yaris valued at $13,210 in June 2006. (*Source:* www.edmunds.com) Her down payment was $3000.
  a. Write a formula for the total cost of the car, T, as a function of the number of months she will have to pay on the loan, m.
  b. What is the total cost of the Yaris after she has paid all of the payments?
  c. How much money has she paid in interest for her Toyota?

48. **Car Payments** The formula for monthly payments on a vehicle loan is

$$M(p, i, n) = p\left[\frac{i(1 + i)^n}{(1 + i)^n - 1}\right]$$

where
  p = amount borrowed in dollars
  n = number of monthly payments
  i = interest rate per payment period as a decimal

  a. Compute $M(2000, 0.005, 24)$.
  b. Interpret the meaning of the value of the output, M.

49. Car Payments Use the formula in Exercise 48 to

a. Compute $M\left(600, \dfrac{0.098}{12}, 12\right)$.

b. Interpret the meaning of the value of the output.

50. Investment Future Value The formula for calculating the future value of an investment with constant periodic payments at a fixed rate for an established number of months is

$$FV(p, i, n) = p\left[\dfrac{(1+i)^n - 1}{i}\right]$$

where
  p = constant monthly payment
  i = monthly interest rate
  n = number of monthly payments

Compute $FV(10, 0.00425, 60)$ and explain the meaning of the value of the output.

51. Investment Future Value Use the formula in Exercise 50 to compute $FV(100, 0.00425, 60)$ and explain the meaning of the value of the output.

52. Investment Future Value An investor plans to contribute $50 a month for 4 years in an ING Direct account with a 4.16% annual interest rate compounded monthly. (*Source: Rate quote at www.ingdirect.com*)

a. Write a formula for her investment account value using the formula found in Exercise 50.

b. Calculate the future value of the investment at 4 years.

53. Investment Future Value An investor plans to save $75 a month for 5 years in a Citibank account earning 4.64% compounded monthly as advertised at www.direct.citibank.com in June 2006.

a. Write a formula for his investment account value using the formula found in Exercise 50.

b. Calculate the future value of the investment at 5 years.

■ STRETCH YOUR MIND

Exercises 54–55 are intended to challenge your understanding of functions.

54. Suppose that between 2000 and 2005, a person's salary increases at an average rate of $2200 per year. Then, between 2005 and 2009, the person's salary increases at an average rate of $1600 per year. What is the average rate of change in salary between 2000 and 2009?

55. The price of a particular item *increases* at an average rate of $0.45 per month for the first 5 months of the year. For the last 7 months of the year, the price *decreases* by $0.30 per month. What is the average rate of change in the price over the year?

SECTION 1.4 Functions Represented by Graphs

LEARNING OBJECTIVES

- Graph functions on the rectangular coordinate system
- Find a function's practical domain and practical range values from its graph
- Determine the vertical intercept (initial value) and horizontal intercepts of a function from a graph and interpret their real-world meanings

GETTING STARTED

Medical professionals often use growth charts like the one shown in Figure 1.12. By knowing how to read the graph, they are able to determine if the head circumference and weight of a child fall within an expected range. They may also predict the future growth of the child.

In this section we see how functions are represented graphically. We also discuss how to interpret a function graph.

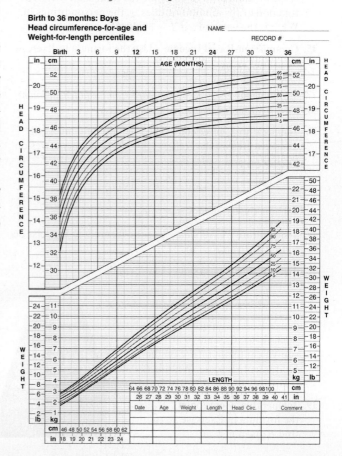

Figure 1.12

Source: www.cdc.gov

PEER INTO THE PAST

DESCARTES

René Descartes, born in France on March 31, 1596, was considered a "jack of all trades," contributing to mathematics, philosophy, anatomy, psychology, and optics. At the very young age of eight, he entered college. A sickly and feeble youth, Descartes was allowed by his parents to stay in bed until 11 o'clock in the morning—a habit he continued to do until the year of his death. While in bed, he tried to bridge the gap between the seeming certainty of mathematics and the controversial nature of philosophy.

In mathematics, Descartes unified algebra and geometry through Cartesian coordinates (rectangular or x-y coordinates). In his other work, Descartes explored the concepts of self, God, and mind. He asked "if our senses are all an illusion created by a malicious deceiver, what can we trust?" His answer was that we can doubt, and that the deceiver cannot cause us to doubt our own existence. Thus, the famous "cogito ergo sum"—"I think, therefore I am."

Source: http://www.math.psu.edu/tseng/class/descartes.html

■ Graphs of Functions

When given the equation of a function, we can generate a table of values and then plot the corresponding points on the **rectangular (Cartesian) coordinate system**. On the graph, y is frequently used in place of the function notation $f(x)$. That is, $y = f(x)$. Then the horizontal axis shows the value of the independent variable, x, and the vertical axis shows the value of the dependent variable, y.

Once we have drawn a sufficient number of points to determine the basic shape of the graph, we typically connect the points with a smooth curve. For example, the function $y = x^3 - 9x$ has the corresponding table of values (Table 1.16) and graph (Figure 1.13). (Note: Use caution when connecting the points with a smooth curve. Be sure that the graph transitions smoothly from one point to the next.)

Table 1.16

x	y
−3	0
−2	10
−1	8
0	0
1	−8
2	−10
3	0

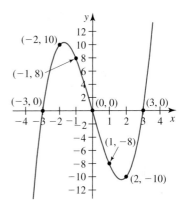

Figure 1.13

EXAMPLE 1 ■ Estimating Function Values from a Graph

Estimate $f(-3)$ and $f(2)$ using the graph of $f(x) = x^3 - 16x$ shown in Figure 1.14.

Solution

It appears from the graph in Figure 1.15 that $f(-3) \approx 20$ and $f(2) \approx -25$. Calculating these values with the algebraic equation, we see that our estimates are very close.

$$f(-3) = (-3)^3 - 16(-3) \qquad f(2) = (2)^3 - 16(2)$$
$$= -27 + 48 \quad \text{and} \quad = 8 - 32$$
$$= 21 \qquad \qquad = -24$$

Figure 1.14

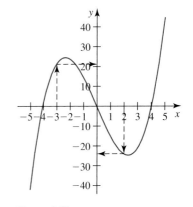

Figure 1.15

EXAMPLE 2 ■ Determining Input and Output Values from a Graph

The graph in Figure 1.16 is a function $f(x)$ with labeled points A, B, C, D, E, and F; input values g, h, j, k, m, n, p, and q; and output values a, b, c, d, and e.

a. Evaluate $f(h)$.
b. Evaluate $f(n)$.
c. Evaluate $f(0)$.
d. Solve $f(x) = d$ for x.

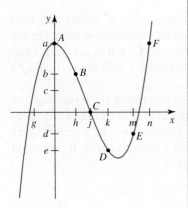

Figure 1.16

Solution

a. To evaluate $f(h)$ means to find an output value for the input value h. In other words, we need to find a y-value for the x-value h. We locate h on the horizontal axis, go up to the function graph f, and then over to find the output value of b. Therefore, b is the answer, written in function notation as $f(h) = b$ or as the ordered pair (h, b). (Note that B is *not* the answer because B is simply the name of the point and not the output value of the point.)

b. Again, to evaluate $f(n)$ means to find an output value for the input value n. We locate n on the horizontal axis, go up to the function graph f, and then over to find the output value of a. Therefore, a is the answer, written in function notation as $f(n) = a$ or as the ordered pair (n, a).

c. To evaluate $f(0)$, we locate 0 on the horizontal axis, go up to the function graph f, and encounter the output value a. Therefore, a is the answer, written in function notation as $f(0) = a$ or as the ordered pair $(0, a)$.

d. This problem is different from parts (a)–(c). To "solve $f(x) = d$ for x" means that we know the output is d and we are to find the input value for x. To do this, we locate d on the *vertical axis*, go over to the function graph f to the point labeled E, and then up to the horizontal axis. When we do this we see that we are near the value m so our estimate is m. We can write the solution in function notation as $f(m) = d$ or as the ordered pair (m, d). We also observe that $f(x) = d$ for an unlabeled value of x between j and k and an unlabeled value of x to the left of g.

■ Domain and Range

When considering the graph of a function, we need to consider what values are reasonable for the input and output. The concepts of *domain* and *range* address this issue.

> **DOMAIN AND RANGE**
> - The set of all possible values of the *independent* variable (input) of a function is called its **domain**. When we limit the domain to values that make sense in the real-world context of a problem, we get the **practical domain**.
> - The set of all possible values of the *dependent* variable (output) of a function is called its **range**. When we limit the range to values that make sense in the real-world context of a problem, we get the **practical range**.

(One way to keep the terms straight is to observe that the terms *in*put, *in*dependent variable, and doma*in* all contain the word *in*.) When a value from the domain is substituted for the independent variable and the corresponding output is evaluated, the result is a value of the dependent variable.

EXAMPLE 3 ■ Determining Practical Domain and Range

Determine (a) the practical domain and (b) the practical range for the head circumference of a boy who is in the 50th percentile as shown in Figure 1.17. Use the child's age as the independent variable, t, and the head circumference as the dependent variable, H.

SECTION 1.4 ■ FUNCTIONS REPRESENTED BY GRAPHS

Figure 1.17
Source: www.cdc.gov

Solution

a. The practical domain for the head circumference function is $0 \leq t \leq 36$ because the graph goes from age 0 months to 36 months in the horizontal direction.

b. We find the curve on the graph marked with the 50. This is the graph representing the 50th percentile. The practical range for the function is $36 \leq H \leq 49.6$ because the head circumference ranges from 36 cm at birth to 49.6 cm at age 36 months.

■ Interpreting Graphs in Context

When we use a graph to model a real-world situation, it is important to keep track of the independent and dependent variables, and to understand the contextual meaning of the coordinates of a point. For example, the graphs of functions often intersect the vertical and horizontal axes at what we call the *vertical* and *horizontal intercepts*. A vertical intercept may also be referred to as the **initial value**. We investigate this in the following example.

EXAMPLE 4 ■ Interpreting the Real-World Meaning of a Graph

The graph in Figure 1.18 is a mathematical model of the temperature of an oven over time. The independent variable is time in minutes and the dependent variable is temperature in degrees Fahrenheit (°F). Describe what the graph indicates and include an explanation of any intercepts.

Figure 1.18

Solution The vertical intercept (where the graph intersects the vertical axis) shows that the oven started at approximately 75° (room temperature). We represent this by the ordered pair (0, 75). This vertical intercept may also be described as the initial value because at time 0 minutes (when we started keeping time) the initial oven temperature was 75°. The oven then heated up slightly past 350°. The oven then repeatedly cooled off a few degrees and heated up a few degrees above 350°, possibly trying to maintain a temperature of 350°. Near the 30-minute mark the oven may have been turned off, because it began to cool off. The graph does not cross the horizontal axis so there is no horizontal intercept. A horizontal intercept would show us the time in minutes when the oven temperature was 0°, and it is reasonable to assume that the oven temperature would never reach that low (unless it was in a landfill in the Arctic).

At first the relationship between data from a table and its associated graph may seem trivial. In reality, it is fundamental to developing a deep understanding of graphical representations of functions. Let's investigate this in Examples 5 and 6.

EXAMPLE 5 ■ Drawing the Graph of a Real-World Situation

Due to excessive air traffic, the air traffic control tower informs a pilot that she must go into a holding pattern and wait for her turn to land. Given that she has been told to circle the airport at a radius of 3 miles for 10 minutes, construct a graph to represent the distance (in miles) of the plane from the center (the airport) of its circular flight path as a function of the time (in minutes) during those 10 minutes.

Solution Initially many of us think that the graph of the distance of the plane from the airport is a circle because the plane is circling the airport. However, if we pay close attention to what the coordinates of the points mean we can create an accurate model. "A radius of 3 miles from the airport" means 3 miles from *the center* of the circular path. Thus the independent variable is time, t, (in minutes) and the dependent variable is distance, d, (in miles) *from the center of the circular flight path*. We begin by plotting points for 0 minutes, 1 minute, and 2 minutes, as shown in Figure 1.19: (0, 3), (1, 3), and (2, 3).

We can see that at each point the plane is 3 miles from the center. This pattern continues for the 10 minutes that the plane circles, forming the horizontal line at $d = 3$ shown in Figure 1.20.

Figure 1.19

Figure 1.20

EXAMPLE 6 ■ Interpreting a Graph of a Real-World Situation

A child tosses a ball straight up and allows it to fall to the ground. He tosses the ball from 2 feet above the ground with an initial velocity of $3 \frac{\text{feet}}{\text{sec}}$. We can model this by the formula $h(t) = -16t^2 + 3t + 2$. (Note: We demonstrate how to generate this formula in a later section.) Use the graph of the height function found in Figure 1.21 to answer the following questions.

a. Explain what the graph means in its real-world context.
b. Is the graph of the height function the actual path of the ball after the child tosses it? Explain.

Figure 1.21

SUMMARY

In this section you learned that functions may be represented by graphs in addition to formulas and tables. You learned that graphs are made up of individual points, often connected with a smooth curve, that represent data. To understand what a graph means in context, you learned to focus on what the points represent in terms of the data. You discovered how to find the input and output of a function from its graph by locating values on the independent and dependent axes. Finally, using a graph and its real-world context you learned how to interpret the meaning of graphs, find the practical domain and range values, and determine and interpret the vertical (initial value) intercept.

1.4 EXERCISES

SKILLS AND CONCEPTS

In Exercises 1–5, refer to the graph of $f(x)$.

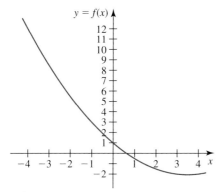

1. Evaluate $f(-3)$.
2. Solve $f(x) = 2$ for x.
3. Solve $f(x) = 0$ for x.
4. Evaluate $f(0)$.
5. Evaluate $f(4)$.

In Exercises 6–10, refer to the graph of $g(x)$.

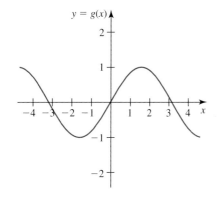

6. Evaluate $g(-1)$.
7. Solve $g(x) = 1.5$ for x.
8. Solve $g(x) = 0$ for x.
9. Evaluate $g(0)$.
10. Evaluate $g(1.5)$.

For Exercises 11–15, refer to the graph of $f(x)$ with x-values g, h, j, k, m, n and y-values a, b, c, d, e. Determine a possible solution to each equation and a function value for each expression.

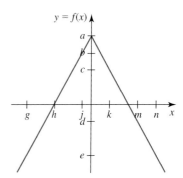

11. Evaluate $f(g)$.
12. Solve $f(x) = a$ for x.
13. Solve $f(x) = e$ for x.
14. Evaluate $f(k)$.
15. Evaluate $f(j)$.

For Exercises 16–20, match the five written scenarios with the most appropriate graph given in A–E. As you look at each graph, remember that time is advancing from left to right.

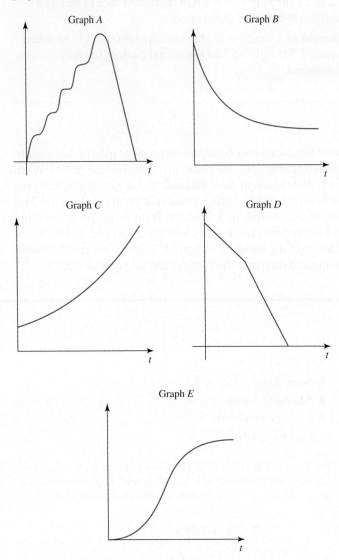

16. It took me several breaths to inflate a balloon. As I started to tie it off, it slipped from my hand and flew around the room. The amount of air in the balloon is the dependent variable.

17. At the beginning of spring, the grass grew slowly and I seldom had to mow the lawn. By midsummer it was growing very fast, so I mowed twice a week. In fall, I only had to mow once in a while. When winter came I didn't mow at all. The cumulative number of times the lawn has been mowed to date is the dependent variable.

18. The amount of money in my savings account started out growing slowly because I didn't have much money in it. However, now that I have a larger amount in the account, interest is helping my balance grow much faster. The balance in my savings account is the dependent variable.

19. I put water in the ice-cube tray and placed it in the freezer to make ice. The temperature of the water in the ice-cube tray is the dependent variable.

20. I started walking to class at a constant pace but realized that I was going to be late so I started walking faster and maintained my fastest rate until I reached class. My distance from the classroom is the dependent variable.

SHOW YOU KNOW

For Exercises 21–25, use the graphs A–E (given in the instructions for Exercises 16–20) to answer each question.

21. Describe what the vertical and horizontal intercepts for graph A mean in its real-world context.

22. Describe what the vertical and horizontal intercepts for graph D mean in its real-world context.

23. Describe what the vertical and horizontal intercepts for graph E mean in its real-world context.

24. Determine the practical domain and range for the function represented by graph C.

25. Determine the practical domain and range for the function represented by graph B.

MAKE IT REAL

26. **Cell Phone Subscribers** Industry statistics reveal that the number of cellular subscribers in the United States totaled nearly 195 million at the end of 2005. A recently completed study revealed that the majority of consumer households are now acquiring multiple cellular phones. The following graph shows the percentage of households with multiple cellular phones, m, as a function of the number of years since the second quarter of 2002, y.

Source: www.icrsurvey.com

a. Determine the practical domain and range for $m = f(y)$.

b. Explain what the graph means in its real-world context.

27. **Cigarette Consumption** The number of cigarettes consumed by adults over the age of 18 in the United States, c, is a function of the years since 1900, t, and is given in the following graph.

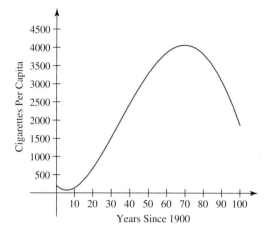

Source: www.infoplease.com

a. Determine the practical domain and range for $c = f(t)$.

b. What conclusions about cigarette consumption can you draw from the graph?

c. What does the vertical intercept mean in its real-world context?

28. **Golf Pro Drives** The following scatter plot shows data collected during a study at the driving range at Bay Hill Country Club. The data show the average drive in yards of 35 Professional Golf Association tour pros versus the height at which they tee the ball up (in inches). Use the plot of the data to answer the following questions.

   a. Do these data represent a function? Explain.

   b. Why do you think the data are spread out as they are? Provide a reasonable explanation.

   c. If an additional professional golfer joins the study and chooses to tee up the ball 1.55 inches, do you feel confident in your ability to estimate how far of a drive he will hit?

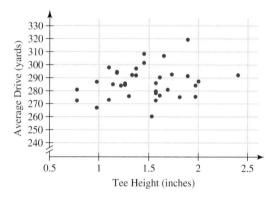

Source: Golf Magazine, June 2006

29. **Jefferson Memorial** The Jefferson Memorial (also commonly known as the St. Louis Gateway Arch) is an elegant monument to westward expansion in America. Located on the banks of the Mississippi River in St. Louis, Missouri, the 630-foot-tall stainless steel arch dominates the city skyline.

A graphical representation of the arch based on the equation used by the architect is shown in the following figure. Use it to answer the following questions. (Note: The graphical representation does not take into account the thickness of the arch.)

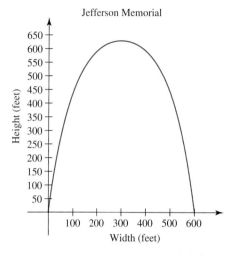

a. Inside the arch is a tram that takes visitors to the top of the arch. Approximately how high off the ground will the tram be when it is 100 horizontal feet away from its original position (the origin)?

b. When the tram is 500 vertical feet above the ground, what is its horizontal distance from its original position (the origin)? (Note: There are two possible solutions.)

c. The *Titanic* was a famous cruise ship that ran into an iceberg and sank in 1912 even though it was touted as being virtually unsinkable. The ship had a length of 882 feet. If the *Titanic* had been placed lengthwise on the ground below the St. Louis Arch, would it fit between the two legs of the arch?

d. On July 2, 1982, the *St. Louis Post-Dispatch* posed the following hypothetical problem: If the *Spirit of St. Louis* airplane (wingspan of 46 feet) was flown directly through the middle of the arch, 200 feet above the base, how far would the wingtips be from the inside edge of the arch? Give an answer to this question and provide a justification for your result.

■ STRETCH YOUR MIND

Exercises 30–31 are intended to challenge your ability to work with graphs.

30. **Cyclist Position** The following diagram is a side view of an individual cycling up and over a hill. Considering the information provided on the diagram, draw a sketch of the

speed of the cyclist as a function of the horizontal position of the cyclist from the start.

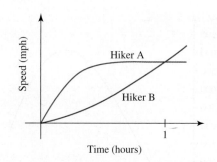

31. **Hiker Position** The given graph represents speed vs. time for two hikers. (Assume the hikers start from the same position and are traveling in the same direction.)

   a. What is the relationship between the position of Hiker A and Hiker B 1 hour into the hike? In other words, is Hiker A or Hiker B ahead or are they at the same position? Justify your conclusion.
   b. Which hiker is going faster 30 minutes into the hike? Explain how you know.
   c. Between 45 and 60 minutes into the hike, is one hiker pulling away from the other? Justify your response.

SECTION 1.5 Functions Represented by Words

LEARNING OBJECTIVES

- Convert words representing function relationships into symbolic and graphical representations
- Translate functions given in equations, tables, and graphs into words

GETTING STARTED

Environmentalists and concerned citizens have actively promoted Earth-friendly policies over the past several decades. In 1970, Congress created the Environmental Protection Agency (EPA) "to protect human health and the environment." (*Source:* www.epa.gov) One of the roles of the EPA is to monitor air quality and crack down on polluters. By analyzing air quality trends, the EPA is able to measure its own effectiveness.

In this section we discuss how to convert words representing function relationships into symbolic and graphical representations. We also detail how to translate functions given in equations, tables, and graphs into words. These skills may be used by all of us to interpret public reports such as those issued by the EPA.

■ Recognizing a Function in Words

One of the EPA goals is to reduce smog-forming nitrogen oxides (NO_x). In a 2007 report, the EPA stated, "In 2007, . . . sources emitted approximately 506,000 tons of NO_x, an overall decrease of about 1,300 tons from 2006." (*Source: NO_x Budget Trading Program: Compliance and Environmental Results 2007*, p. 17).

We can represent the relationship between the 2006 and 2007 nitrogen oxide emission levels using functions. We let the variable x represent the 2006 emissions level and the variable y represent the 2007 emissions level. Since the 2007 level is 1300 tons less than the 2006 level, the relationship between x and y is

$$y(x) = x - 1300$$

What was the 2006 emissions level? Since we know $y = 506,000$ tons, we can find x.

$$506,000 = x - 1300$$
$$x = 507,300$$

The 2006 level was 507,300 tons.

The process for converting a function given in words to symbolic notation is given in the box below. With a little practice, you will become proficient in translating words into symbols for simple mathematical relationships. For more complex relationships, you will need to develop additional mathematical skills to accomplish Step 2 of the process.

> **HOW TO:** ■ **TRANSLATE A FUNCTION GIVEN IN WORDS INTO SYMBOLIC NOTATION**
>
> 1. Identify the things that are being related to each other.
> 2. Express the mathematical relationship between the things using words.
> 3. Select variables to represent the things being related to each other.
> 4. Rewrite the mathematical relationship from Step 2 using the variables from Step 3 and appropriate mathematical notation.

EXAMPLE 1 ■ **Translating a Function in Words into Symbolic Notation**

In 2008, the Bureau for Economic Analysis (BEA) issued a press release detailing per capita income rankings for 2007. An excerpt from the report is provided here.

> Connecticut led the nation with a per capita income of $54,117. . . . Mississippi had the lowest per capita income of all states, $28,845. . . . [The national average was $38,611.] (*Source:* www.bea.gov)

a. How many dollars above (or below) the national average is the per capita income of Connecticut and Mississippi?

b. Let n represent the national average per capita income. Use function notation to represent the per capita income of each state as a function of the national average per capita income.

Solution

a. We calculate the difference between each state's average per capita income and the national average per capita income.

$$54{,}117 - 38{,}611 = 15{,}506 \qquad 28{,}845 - 38{,}611 = -9766$$

The per capita income of Connecticut was $15,506 above the national average in 2007. The per capita income of Mississippi was $9766 below the national average in 2007.

b. Let c represent the per capita income of Connecticut and m represent the per capita income of Mississippi. We have

$$c(n) = n + 15{,}506 \quad \text{and} \quad m(n) = n - 9766$$

We can often represent functions given in words graphically, as demonstrated in Example 2. Graphical representations are especially useful in depicting rates of change.

EXAMPLE 2 ■ **Translating a Function in Words into a Graph**

Apple Computer Corporation has captured the attention of millions of consumers with its innovative iPod digital music player. The company published the following in its 2005 Annual Report.

> Net sales of iPods rose $3.2 billion . . . during 2005 compared to 2004. Unit sales of iPods totaled 22.5 million in 2005 . . . [up] from the 4.4 million iPod units sold in 2004. (*Source:* Apple Computer Corporation 2005 Annual Report, p. 32)

The report further indicated that iPod sales revenue was $1.3 billion in 2004 and that 0.9 million iPod units were sold in 2003, generating $0.3 billion in revenue. Use a graph to model the revenue from iPod sales as a function of iPod units sold. Then use the graph to predict sales revenue when 15 million iPods are sold. (Note: The terms *sales*, *revenue*, and *sales revenue* are often used interchangeably.)

Solution We are asked to draw a graph representing iPod sales revenue as a function of iPod units sold. We first construct a data table to record what we know (Table 1.17). We make a scatter plot for the data and then draw a continuous, smooth curve through the data points, as shown in Figure 1.22.

From the graph, it appears that when 15 million iPod units are sold the sales revenue will be roughly 3.5 billion dollars.

Table 1.17

Year	Units Sold (in millions)	Revenue ($ billions)
2003	0.9	0.3
2004	4.4	1.3
2005	22.5	1.3 + 3.2 = 4.5

Figure 1.22

As stated earlier, graphs are especially useful in depicting rates of change. The graph in Example 2 was increasing, indicating that an increase in units sold corresponded with an increase in sales revenue. The graph is also curved downward, indicating that the rate at which sales revenue is increasing is decreasing. In other words, as the number of iPods sold increases, the revenue per unit sold decreases.

■ Interpreting the Meaning of a Function Using Words

Although it is important to be able to translate words into function notation and graphs, it is equally important to be able to reverse the process. A mathematical model is of no use if we are unable to interpret the meaning of the results in a real-world context. As we show in Example 3, keeping track of the units, or quantities represented by the variables, is one of the best ways to make sense out of function results.

EXAMPLE 3 ■ Interpreting in Words the Meaning of a Function Equation

Based on data from 1990 to 2003, the number of golf facilities in the United States can be modeled by

$$g(t) = \frac{4633}{1 + 59.97e^{-0.2567t}} + 12{,}000 \text{ facilities}$$

where t is the number of years since the end of 1990. (*Source:* Modeled from *Statistical Abstract of the United States, 2004–2005*, Table 1240) What does $g(23) = 15{,}982$ mean?

Solution The function g looks very complex and intimidating. In fact, we might initially think we will not be able to answer the question. However, let's apply a little mathematical reasoning.

What is the meaning of each of the variables? The independent variable t is the number of years since the end of 1990. Since $t = 23$, we are evaluating the function at the point in time 23 years after the end of 1990.

$$1990 + 23 = 2013$$

That is, we are evaluating the function in 2013.

What is the meaning of the dependent variable *g*? From the function equation we see that *g* is the number of golf facilities in year *t*. Therefore, *g*(23) is the number of golf facilities in year 2013. Thus *g*(23) = 15,982 means that at the end of 2013 there will be 15,982 golf facilities in the United States (according to the model).

What about the variable *e*? Actually, *e* is not a variable. It is an irrational number so commonly used that it has its own special name: *e*. The number $e \approx 2.71828$. In later chapters, we will discuss the origin of *e* and its many uses.

Interpreting the real-world meaning of a table of values also requires that we keep track of the meaning of the variables. This is especially true when we are asked to use computations to explain the meaning of table results.

EXAMPLE 4 ■ Interpreting in Words the Meaning of a Function Table

Describe in words the meaning and real-world significance of the data in Table 1.18.

Solution There are multiple ways to interpret the data from a table in words. Let's calculate the average rate of change and then write statements to describe the significance of the data in the table.

$$\text{average rate of change} = \frac{605 - 158}{13 - 0} \frac{\text{dollars}}{\text{year}}$$

$$= 34.38 \text{ dollars per year}$$

The table shows that per capita spending on prescription drugs has increased rapidly. In 2003, this spending reached a new high ($605 per person). Although per capita prescription drug spending increased an average of $34.38 per year between 1990 and 2003, it jumped by $53 between 2002 and 2003. The table may indicate that future increases in spending are likely to be even more extreme.

Table 1.18

Years Since 1990 *t*	Per Capita Spending on Prescription Drugs (dollars) *P*
0	158
5	224
8	311
9	368
10	423
11	485
12	552
13	605

Source: Statistical Abstract of the United States, 2006, Table 121

In Example 4, we gave one interpretation of the data given in the table and gave our opinion on what we expected for the future. A different person summarizing the data may write factual statements very different from those we presented. Because there could be several ways to interpret the same data set correctly, another person's written analysis should not be rejected simply because it is not identical to one's own conclusions.

As was the case with symbolic and tabular representations of functions, keeping track of the meaning of the independent and dependent variables is essential to understanding the graphical results.

EXAMPLE 5 ■ Interpreting in Words the Meaning of a Function Graph

Malaria is a major killer of children under 5 years old in Africa. The World Health Organization and a variety of humanitarian organizations are working together with the people of Africa to implement innovative strategies to save lives. One such strategy includes the distribution of insecticide-treated nets (ITNs). The nets purportedly can reduce malaria transmission by more than half. Families who choose to sleep under the nets are protected during the times of the day when mosquitoes, the transmitters of the malaria infection, are most active.

Based on data from 1999–2003, the number of insecticide-treated nets distributed in Africa may be modeled as a function of the year. The raw data and a function model of the data are shown in Figure 1.23.

Figure 1.23

Source: Modeled from data at www.afro.who.net

Given that the independent variable t represents the years since 1999 and the dependent variable n represents the number of insecticide-treated nets distributed in Africa (in thousands), analyze and interpret in words the meaning of the graph.

Solution We first observe that the graph is increasing. That is, the number of insecticide-treated nets distributed is increasing as time goes on. In terms of malaria control and prevention, this is a good thing.

We next observe that the graph curves upward. This means that the rate at which the graph is increasing is also increasing. Notice that between roughly (0, 500) and (1, 1000), the average rate of increase in the model was about 500 thousand nets per year. However, between roughly (3, 4400) and (4, 9500), the average rate of increase in the model was about 5100 thousand nets per year. We summarize these results in words as follows.

> The distribution of insecticide-treated nets in Africa increased from about 500 thousand in 1999 to about 9500 thousand in 2003. The rate at which the nets were being distributed was also increasing. Between 1999 and 2000, the annual distribution rate increased by about 500 thousand nets per year. However, between 2002 and 2003, the annual distribution rate increased by about 5100 thousand nets per year. Since the model graph is curved upward, we anticipate that future rates of increase will be even more dramatic, at least in the short term.

Readers who want to become engaged in this humanitarian issue on a personal level may visit www.nothingbutnets.net to learn about easy ways to get involved.

SUMMARY

In this section you learned how to represent verbal descriptions of relationships using function concepts. You also discovered how to interpret in words the meaning of functions in equations, tables, and graphs. These skills will be used extensively throughout the rest of this book and the rest of your life.

1.5 EXERCISES

SKILLS AND CONCEPTS

In Exercises 1–7, be sure to identify the meaning of any variables you use in your function equation, table, or graph.

1. **Gasoline Prices** On July 10, 2008, the average price of gas in Utah was $4.109 per gallon. (*Source:* www.utahgasprices.com) Use function notation to represent the cost of buying g gallons of gas at this price.

2. **Disposable Personal Income** The following is an excerpt from a May 1, 2006, press release issued by the Bureau of Economic Analysis.

    [monthly] disposable personal income (DPI) increased $78.4 billion . . . in March [2006] (*Source:* U.S. Bureau of Economic Analysis)

    Use function notation to represent the disposable personal income at the end of March 2006 as a function of disposable personal income at the end of February 2006.

3. **Wheat Production** The following is an excerpt from May 12, 2006, press release issued by the National Agricultural Statistics Service of the U.S. Department of Agriculture.

    Winter wheat production is forecast at 1.32 billion bushels, down [0.18 billion] from 2005. Based on May 1 conditions, the U.S. yield is forecast at 42.4 bushels per acre, 2.0 bushels [per acre] less than last year. (*Source:* www.usda.gov)

    Create a table that gives total winter wheat production as a function of the number of bushels produced per acre.

4. **Orange Juice Concentrate** The following is an excerpt from a May 12, 2006, press release issued by the National Agricultural Statistics Service of the USDA.

    Florida frozen concentrated orange juice (FCOJ) yield for the 2005–2006 season, at 1.62 gallons per box at 42.0 degrees Brix, is increased from the 1.58 gallons last season. (*Source:* www.usda.gov)

(Note: 42 degrees Brix means the juice has 42 grams of sucrose sugar per 100 grams of liquid.)

Draw a function graph that gives the Florida frozen concentrated orange juice yield as a function of the number of years since the 2004–2005 season. Then use the graph to forecast the yield for the 2007–2008 season.

5. **Arizona University Enrollment** The following is an excerpt from the Arizona University System FY 2003–2004 Annual Report published by the Arizona Board of Regents.

- In Fall 2003, the universities' headcount enrollment increased by 1,595 students, from 113,869 to 115,464.
- Arizona's universities are expecting tremendous growth in the next two decades. Student enrollments are projected to increase to more than 170,000 students by the year 2020. (*Source:* www.abor.asu.edu)

Assuming that annual enrollment will continue to increase by 1595 students per year, draw a function graph that models the enrollment at Arizona universities between Fall 2002 and Fall 2020. Then use the graph to estimate the enrollment in Fall 2010.

6. **Apple Macintosh Sales** Apple Computer Corporation published the following in its 2005 Annual Report.

Total Macintosh net sales increased $1.4 billion . . . during 2005 compared to 2004. Unit sales of Macintosh systems increased 1.2 million units . . . during 2005 compared to 2004. (*Source:* Apple Computer Corporation 2005 Annual Report, p. 32)

The report further indicated that 3.3 million Macintosh computers were sold in 2004, generating sales revenue of $4.9 billion. In 2003, 3.0 million Macintosh units were sold, generating $4.5 billion in revenue. Create a graph to model the revenue from Macintosh sales as a function of Macintosh units sold.

7. **Community College Enrollment** In her May 2006 speech to faculty and staff, the president of Chandler-Gilbert Community College indicated that although fall semester enrollments continued to increase, they were not increasing as rapidly as they had in previous years. (*Source:* author's notes) Draw a graph to represent fall enrollments as a function of time.

SHOW YOU KNOW

8. A classmate claims that one of the most important things to focus on when translating a function given in words into symbolic notation is to determine what the variables and constants are that are involved in the problem. Do you agree or disagree? Explain your reasoning.

9. Explain how to determine which variables should be assigned to the horizontal and vertical axes when translating a function given in words to a graphical representation.

10. For the function $f(x) = y$, explain what each part of the symbolic notation means. In other words, what does the f, x, and y represent in terms of inputs and outputs of functions?

11. A classmate states, "My weekly income is a function of the number of hours I work at $9.00 per hour." In creating the equation of the mathematical model, what are the advantages of choosing the variable w to represent *weekly income* and the variable h to represent *hours worked*?

12. When interpreting the meaning of a function graph using words, what information do the labels on the horizontal and vertical axes reveal?

MAKE IT REAL

In Exercises 13–19, explain in words the real-world meaning of the indicated function value.

13. **College Attendance** Based on Census Bureau data from 2000–2002 and projections for 2003–2013, *private* college enrollment can be modeled by

$$P(x) = 0.340x - 457 \text{ thousand students}$$

where x is the number of students (in thousands) enrolled in *public* colleges. (*Source:* Modeled from *Statistical Abstract of the United States, 2006*, Table 204) What does $P(14,000) = 4303$ mean?

14. **Greenhouse Gas Emissions** Based on data from 1990–2002, carbon dioxide emissions can be modeled by

$$C(g) = 1.095g - 1743 \text{ million metric tons}$$

where g is the total amount of greenhouse gas emissions (in millions of metric tons). (*Source:* Modeled from *Statistical Abstract of the United States, 2006*, Table 362) What does $C(6400) = 5265$ mean?

15. **Apple Computer Net Sales** Based on data from 2001–2005, the net sales of Apple Computer Corporation can be modeled by

$$S(c) = 1.45c - 408 \text{ million dollars}$$

where c is the cost of sales (in millions of dollars). (*Source:* www.apple.com) For what value of c does $S(c) = 0$ and what does that mean for Apple Computer?

16. **Life Expectancy** Based on data from 1980–2003 and projections for 2005 and 2010, the average life expectancy of a woman can be modeled by

$$f(m) = 0.045645m^3 - 9.9292m^2 + 720.34m - 17350 \text{ years}$$

where m is the average life expectancy of a man (in years) born in the same year as the woman. (*Source:* Modeled from *Statistical Abstract of the United States, 2006*, Table 96) What does $f(73) = 79$ mean?

17. **Per Capita Pineapple Consumption** Based on data from 1999–2003, the per capita consumption of pineapple can be modeled by

$$p(f) = 1.461f^3 - 35.40f^2 + 285.9f - 766.4 \text{ pounds}$$

where f is the combined per capita consumption of apricots, avocados, cherries, cranberries, kiwis, mangoes, papayas, and honeydew melons (in pounds) consumed in the same year. (*Source:* Modeled from *Statistical Abstract of the United States, 2006,* Table 203) What does $p(9) = 4.4$ mean?

18. **Per Capita Pineapple Consumption** According to the model in Exercise 17, $p(19) = 1907$. What does this mean? Does it make sense in the real-world context of the problem? Explain.

19. **Per Capita Pineapple Consumption** According to the model in Exercise 17, $p(3) = -188$. What does this mean? Does it make sense in the real-world context of the problem? Explain.

Exercises 20–29 focus on interpretations of real-world data sets.

20. **Marketing Costs for Farm Foods**

Years Since 1990 t	Marketing Labor Cost ($ billion) L
0	154.0
4	186.1
5	196.6
6	204.6
7	216.9
8	229.9
9	241.5
10	252.9
11	263.8
12	273.1
13	285.9

Source: Statistical Abstract of the United States, 2006, Table 842

Analyze and interpret in words the meaning and real-world significance of the data in the table.

21. **Spending on Medical Services**

Years Since 1990 t	Per Capita Spending on Physician and Clinical Services (dollars) P
0	619
5	813
8	914
9	954
10	1010
11	1085
12	1162
13	1249

Source: Statistical Abstract of the United States, 2006, Table 121

Analyze and interpret in words the meaning and real-world significance of the data in the table.

22. **Chicken Egg Production**

Years Since 1990 t	Egg Production (billions) E
0	68.1
5	74.8
7	77.5
8	79.8
9	82.9
10	84.7
11	86.1
12	87.3
13	87.5
14	89.1

Source: Statistical Abstract of the United States, 2006, Table 842

Analyze and interpret in words the meaning and real-world significance of the data in the table.

23. **Bowling** The table shows the variable number of bowling establishments and bowling membership based on data from 1990, 1995, and 2000–2004.

Tenpin Bowling Establishments b	Bowling Membership (thousands) M
7611	6588
7049	4925
6247	3756
6022	3553
5973	3382
5811	3246
5761	3112

Source: Statistical Abstract of the United States, 2006, Table 1234

Analyze and interpret in words the meaning and real-world significance of the data in the table.

24. **Movie Theaters** The table shows the movie attendance and number of motion picture screens based on data from 2000–2004.

Movie Attendance (millions) m	Motion Picture Screens S
1421	38,000
1487	37,000
1639	36,000
1574	37,000
1536	37,000

Source: Statistical Abstract of the United States, 2006, Table 1234

Analyze and interpret in words the meaning and real-world significance of the data in the table.

25. DVD Player Sales Based on data from 1997–2004, DVD player sales can be modeled by the function shown in the graph.

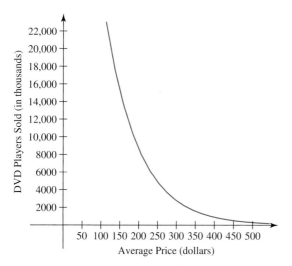

Source: **Modeled from Consumer Electronics Association data (www.ce.org)**

Describe in words the meaning and real-world significance of the function graph.

26. Carbon Monoxide Pollution Based on data from 1990–2003, carbon monoxide pollutant concentration can be modeled by the linear function shown in the graph.

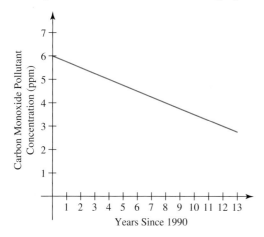

Source: **Modeled from *Statistical Abstract of the United States, 2006*, Table 359**

Describe in words the meaning and real-world significance of the function graph.

27. Cassette Tape Shipment Value Based on data from 1990–2004, the value of music cassette tapes shipped can be modeled by the function shown in the graph.

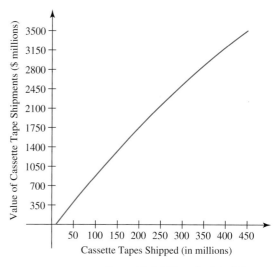

Source: **Modeled from *Statistical Abstract of the United States, 2006*, Table 1131**

Describe in words the meaning and real-world significance of the function graph.

28. Cell Phones and Pagers Based on data from 2000–2004, cell phone sales can be modeled by the cubic function shown in the graph.

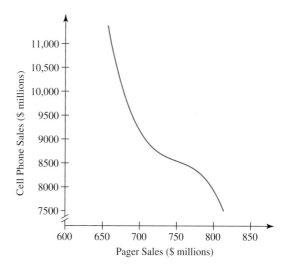

Source: **Modeled from *Statistical Abstract of the United States, 2006*, Table 1003**

Describe in words the meaning and real-world significance of the function graph.

29. Blank Audio Cassette Tape Sales Based on data from 1990–2004, blank audio cassette tape sales can be modeled by the logistic function shown in the graph.

Source: Modeled from *Statistical Abstract of the United States, 2006*, Table 1003

Describe in words the meaning and real-world significance of the function graph.

■ STRETCH YOUR MIND

Exercises 30–34 are intended to challenge your ability to interpret the meaning of functions in words.

30. Income Ranking In March 2006, the Bureau for Economic Analysis (BEA) issued a press release detailing per capita income rankings for the states. An excerpt from the report is provided here.

> Connecticut led the nation with a per capita income . . . 38 percent above the national average. Louisiana's per capita income was . . . 28 percent below the national average . . . (*Source:* www.bea.gov)

Write the equation of a function that relates the per capita income of Connecticut to the per capita income of Louisiana.

31. Autism Study In June 2006, a son of one of the authors was diagnosed with autism, a condition commonly characterized by repetitive motions, limited communication skills, and impaired social skills. In the ensuing quest to better understand the condition, the author discovered that a popular theory was that the MMR vaccine was one of the causes of autism.

In March 2001, the *Journal of the American Medical Association* published the results of a study investigating this theory. The study indicated that the number of autism cases in the study population increased from 44 cases per 100,000 live births in 1980 to 208 cases per 100,000 live births in 1994. Over the same time period, the percentage of children receiving immunizations by the age of 24 months increased from 72% to 82%. (*Source: JAMA*, Vol. 285, No. 9; March 2001)

Based on these results, explain whether or not the study supports the hypothesis that the MMR vaccine is responsible for the increase in autism.

32. Darden Restaurants Financials Darden Restaurants, Inc., is the parent company of a number of popular restaurants including The Olive Garden, Red Lobster, Bahama Breeze, and Smokey Bones Barbeque & Grill. According to the company's 2005 annual report, the cost of sales increased from $3.1 billion in 2001 to $4.1 billion in 2005. During the same time period, revenue from sales increased from $4.0 billion to $5.3 billion. (*Source:* **Darden Restaurants, Inc., 2005 Annual Report, p. 60**) Answer the following questions using mathematical computations to substantiate your results.

a. Between 2001 and 2005, was the cost of sales increasing more rapidly than the revenue from sales? Explain.

b. What is an equation of a linear function that models revenue from sales as a function of cost from sales?

c. Given the model you found in part (b), what do you estimate the revenue from sales was when the cost of sales was $3.9 billion?

d. According to the company's annual report, the *cost of sales* was $3.9 billion in 2004 and the revenue from sales was $5.0 billion. How well did your estimate in part (c) fit the actual data in the company's annual report?

33. Chipotle Mexican Grill Income In 2001, Chipotle Mexican Grill, Inc., earned $131,598 thousand in revenue and had a net income of −$24,000 thousand. Between 2001 and 2005, revenue and net income increased dramatically for the restaurant. In 2005, the company earned $627,695 thousand in revenue and $37,696 thousand in net income. (*Source:* Chipotle Mexican Grill, Inc., 2005 Annual Report, p. 24)

Write a function equation with a constant rate of change to model the net income of the company as a function of its revenue.

34. Chipotle Mexican Grill Revenue According to the company's 2005 annual report, Chipotle Mexican Grill, Inc., classifies its revenue as *restaurant sales* or *franchise royalties and fees.* A following graph relates franchise royalties and fees to restaurant sales, based on data from 2001–2005.

Source: Chipotle Mexican Grill, Inc., 2005 Annual Report, p. 24

Describe what the graph communicates regarding the relationship between sales and franchise revenue.

SECTION 1.6 Preview to Inverse Functions

LEARNING OBJECTIVES

- Explain the relationship between a function and its inverse
- Explain and use inverse function notation to solve real-world problems
- Find the inverse of a function from a table or graph and interpret its practical meaning

GETTING STARTED

Automobile engineers focused on improving a car's braking system need to be able to predict the car's braking distance (output) given the car's speed (input). Conversely, police officers need to be able to calculate a car's speed (output) based on its braking distance (input) as indicated by skid marks on the road. The mathematical concept of *inverse functions* allows the engineers and the police officers to get the information they need.

In this section we introduce the concept of inverse functions. Since we study inverse functions in depth in future sections, the purpose here is to develop a basic understanding of the inverse function concept.

■ The Inverse Function

Recall that a function is a process or correspondence relating two quantities in which each input value generates exactly one output value. Some functions have the additional characteristic that each output value has exactly one input value. Functions with this property are *reversible*. The inverse of a function reverses the process or correspondence of the original function. The function reversing the process or correspondence, if it exists, is called an **inverse function**.

INVERSE FUNCTION

A function f^{-1} (read "f inverse") is the function whose inputs are the outputs of f and whose outputs are the inputs of f.

Figure 1.24 is a graphical representation of the concept that an inverse function reverses the correspondence between two data sets. The arrows in the figure point to the outputs of the function. Notice that in the inverse function correspondence, the set of inputs of the original function becomes the set of outputs for the inverse function. Similarly, the set of outputs of the original function becomes the set of inputs for the inverse function. Observe that the inverse is indeed a function since each input of the inverse function corresponds with a single output.

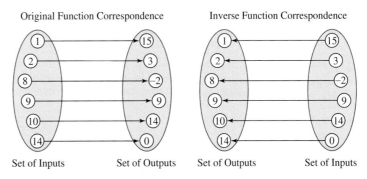

Figure 1.24

We can use a similar approach when looking at a function as a process, as shown in the following diagram.

$$\text{input } x \to \boxed{f} \to y \text{ output}$$
$$\text{output } x \leftarrow \boxed{f^{-1}} \leftarrow y \text{ input}$$

Notice that $y = f(x)$ and $x = f^{-1}(y)$. At this point in the text, the goal is not to compute inverse function models but rather to interpret their meaning in the context of different real-world situations. We revisit the inverse function concept repeatedly throughout the text as we introduce different types of functions. When we do, we will demonstrate how to find inverses algebraically. Let's now return to the braking distance scenario.

■ Modeling with Inverse Functions

Suppose you are in a car, traveling at a safe speed, and the traffic light ahead turns red. You press the brake pedal, and the car slows down and eventually stops. We call the distance that the car travels while the driver is braking the **braking distance**. We used the website www.phy.ntnu.edu.tw/ntnujava/viewtopic.php?t=224 to simulate this situation and record the data shown in Table 1.19.

Table 1.19

Braking Distance	
Speed (mph)	Distance (feet)
0	0.00
20	16.70
25	26.12
30	37.63
35	51.21
40	66.90
45	84.65
50	104.53

Using a technique called *quadratic regression*, we found the function in Example 1 to model the data set. We will explain this technique in detail in Chapter 4 as a part of our in-depth discussion of quadratic functions.

EXAMPLE 1 ■ Using a Braking Distance Model

The braking distance as a function of the speed of the car can be modeled by

$$T = f(S) = 0.042(S)^2 - 0.00077(S) - 0.0026$$

where T is the braking distance in feet and S is the speed of the car in miles per hour. Find $f(100)$ and explain what it means.

Solution The notation $f(100)$ means the braking distance when the car is traveling 100 mph. We compute the value of $f(100)$ by substituting 100 for the variable S in the function.

$$f(100) = 0.042(100)^2 - 0.00077(100) - 0.0026$$
$$= 419.920 \text{ feet}$$

According to the model, the braking distance for a car traveling 100 mph is 419.920 feet.

We have created a model to predict the braking distance (dependent variable) given the speed of the car (independent variable). But what if the braking distance is known and the speed is unknown? Can we determine the original speed of the car? Police department accident reconstruction officers often measure the skid marks left by a vehicle prior to an accident to answer this very question.

To simplify the mathematical model we will use in Example 2, we interpret the length of the skid marks left by the car to be exactly the braking distance. We also ignore the fact that variables other than braking distance and speed enter into this application. (Ignored variables include driving surface, driver alertness, and so on.)

Table 1.20

Braking Distance	
Speed (mph)	Distance (feet)
0	0.00
20	16.70
25	26.12
30	37.63
35	51.21
40	66.90
45	84.65
50	104.53

EXAMPLE 2 ■ Using Skid Marks to Determine the Speed of a Car

Table 1.20 shows the braking distance as a function of speed.

Assuming that the skid marks represent the braking distance of a car, describe how a police officer could use this data to determine the pre-accident speed of a vehicle.

Solution In Table 1.20, *speed* is the independent variable and *braking distance* is the dependent variable. However, for the police officer *braking distance* is the independent variable and *speed* is the dependent variable. To represent this, we switch the columns as shown in Table 1.21.

To estimate the speed, the police officer must find the braking distance (length of skid marks) and predict the corresponding speed. For example, if the braking distance was about 67 feet, he could predict that the car was traveling approximately 40 mph. The function used by the police officer is the *inverse* of the braking distance function.

Table 1.21

Braking Distance	
Distance (feet)	Speed (mph)
0.00	0
16.70	20
26.12	25
37.63	30
51.21	35
66.90	40
84.65	45
104.53	50

We say that two functions have an **inverse relationship** if the *independent* variable of the first function is the *dependent* variable of the second function and the *dependent* variable of the first function is the *independent* variable of the second function. That is, interchanging the inputs and outputs of one function yields the other function. For example, the functions $f(s) = t$ and $g(t) = s$ (shown in Tables 1.22 and 1.23) have an inverse relationship.

Table 1.22

Braking Distance as a Function of Speed $f(s) = t$	
Speed (mph) s	Braking Distance (feet) t
0	0.00
20	16.70
25	26.12
30	37.63
35	51.21
40	66.90
45	84.65
50	104.53

Table 1.23

Speed as a Function of Braking Distance $g(t) = s$	
Braking Distance (feet) t	Speed (mph) s
0.00	0
16.70	20
26.12	25
37.63	30
51.21	35
66.90	40
84.65	45
104.53	50

We say that f is the inverse function of g and that g is the inverse function of f. Symbolically, we write $f = g^{-1}$ and $g = f^{-1}$. Note that the independent variable of f (speed) is the dependent variable of g. Similarly, the dependent variable of f (braking distance) is the independent variable of g.

We may also represent the inverse relationship between speed and braking distance graphically. We demonstrate this in Example 3.

EXAMPLE 3 ■ Graphing the Braking Distance Function

Using the data in Table 1.22, draw a scatter plot of the braking distance data as a function of speed. Then graph the inverse of the braking distance function.

Solution For the braking distance function, the speed of the car is the independent variable and is measured along the horizontal axis. The braking distance is the dependent variable and is measured along the vertical axis. See Figure 1.25.

Figure 1.25

Figure 1.26

The inverse function is determined by interchanging the independent and dependent variables as shown in Figure 1.26. In this case, the braking distance (length of skid marks) becomes the independent variable and is measured along the horizontal axis. The speed of the car becomes the dependent variable and is measured along the vertical axis.

EXAMPLE 4 ■ Interpreting an Inverse Function Model

Super Bowl ticket prices have increased in the 40-year history of the game (see Table 1.24).

a. Interpret the meaning of $P = f(G)$ and $G = f^{-1}(P)$, where P represents the value of a ticket and G represents a particular Super Bowl game.

b. Calculate and explain the meaning of $P = f(20)$ and $G = f^{-1}(20)$.

Table 1.24

Super Bowl Game Number G	Ticket Face Value P
I (1)	$10
V (5)	$15
X (10)	$20
XV (15)	$40
XX (20)	$75
XXV (25)	$150
XXX (30)	$300
XXXV (35)	$325
XL (40)	$600

Source: www.superbowl.com

Solution

a. The notation $P = f(G)$ indicates that the price of a Super Bowl ticket, P, depends on the Super Bowl game, G, for which it was used. That is, if we know the Super Bowl game number, we can determine the price of a ticket. The function $G = f^{-1}(P)$ is the inverse of $P = f(G)$. That is, if we know how much a ticket cost, we can tell which Super Bowl game that ticket represents.

b. $P = f(20)$ is the price of a ticket in Super Bowl 20 (XX), so $P = \$75$. $G = f^{-1}(20)$ is the game in which the price of the ticket was $20, so $G = 10$: In Super Bowl X (10), the price of a ticket was $20.

EXAMPLE 5 ■ Evaluating an Inverse Function Given a Graph

The graph in Figure 1.27 shows the function model $H = f(N)$ where H represents Head Start enrollment (in thousands) and N represents the number of women in the workforce (in thousands). Evaluate $f(66000)$ and $f^{-1}(600)$ and explain what each means in this context.

Solution The notation $f(66000)$ tells us to find the Head Start enrollment, H, when the number of women in the workforce, N, is 66,000,000 (remember that the numbers are in thousands). From the graph in Figure 1.28, we estimate that $H(66000) \approx 850$. That is, when there are 66,000,000 women in the workforce, Head Start enrollment is approximately 850,000.

The notation $f^{-1}(600)$ tells us to find the number of women in the workforce when Head Start enrollment is 600,000. Although we don't have a graph of f^{-1}, we can use the graph of f to find the desired result by reversing the process.

From the graph of f in Figure 1.29, we estimate that $f^{-1}(600) \approx 58,000$. That is, when Head Start enrollment is 600,000, the number of women in the workforce is about 58,000,000.

The relationship between Head Start enrollment and women in the workforce is a function correspondence. Although we related the variables to each other with a function, an increase in women in the workforce does not necessarily *cause* an increase in Head Start enrollment.

Figure 1.27

Source: Statistical Abstract of the United States, 2006

Figure 1.28

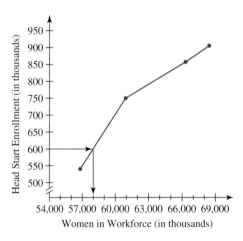

Figure 1.29

■ Inverse Relationships That Are Not Inverse Functions

Not all functions have an inverse function. Some processes cannot be reversed and some reversible processes do not yield exactly one output for each input to the inverse function. Example 6 shows one such function.

EXAMPLE 6 ■ Determining If an Inverse Relationship Is a Function

Projectile motion refers to the motion seen when an object, such as a baseball hit by a bat, is propelled into the air before returning to the ground due to the effects of gravity. Using the website www.exploratorium.edu/baseball/scientificslugger.html, we determined that if a hit ball has an initial velocity of 242 feet per second and is hit at an angle of 20°, the ball will travel 417 feet. The function $h = f(t)$ represents a model for the vertical height, h, of the baseball (in feet) as a function of the time, t seconds. This function is shown in the graph in Figure 1.30. Does the inverse relationship represent a function? Why or why not?

Solution: The inverse function, $f^{-1}(h) = t$, would allow us to determine at what time the ball was at a particular height. That is, if we wanted to know at what time the height of the ball was 90 feet, we could evaluate $f^{-1}(90)$. Using the graph of $f(t)$ shown in Figure 1.31, we find that there are two times where the height is 90 feet: at about 1.5 seconds into its flight and at about 3.6 seconds into its flight.

Although an inverse relationship exists, the inverse is not a function since there are two outputs for a single input.

Figure 1.30   Figure 1.31

In Example 6, we saw that the baseball reaches a height of 90 feet two times. In cases such as this, it is not appropriate to use the function notation $f^{-1}(h) = t$ since the inverse does not represent a function. However, we can still discuss the idea of the inverse relationship by finding the two times as output values for a particular input value.

SUMMARY

In this section you discovered that if a situation may be explained as a function process or correspondence, the inverse function reverses the process or correspondence of the original function. You also learned how to evaluate inverse functions and interpret their meaning in the context of real-world situations. You also found that an inverse relationship can exist even though an inverse function does not.

1.6 EXERCISES

SKILLS AND CONCEPTS

For Exercises 1–10, do each of the following.
  a. Identify the independent and dependent variables of the function.
  b. Identify the independent and dependent variables for the inverse of the function.
  c. Determine whether or not the inverse relationship represents a function. Justify your conclusion.

1. The height of water in a bathtub (in inches) that is being filled after the water has been running for t minutes.

2. Your distance from home t minutes after leaving home.

3. The length of your hair t days after getting a haircut.

4. The total amount of rain that falls on a person's lawn in Phoenix as a function of the days since June 1.

5. An electric bill total as a function of the temperature at which the thermostat is set.

6. The full name of each student in your class and the number of hours the student works during a given week.

7. The speed of a car as a function of the time since entering a freeway from the on-ramp.

8. The height of a child as a function of her age between birth and 10 years.

9. The weight of an adult as a function of his age between 18 and 40 years.

10. The number of points scored in each quarter of a basketball game.

Exercises 11–15 are based on the following data table, which shows the speed of a car in the car chase scene in the movie The Blues Brothers.

Time (seconds)	Speed (mph)
4	45
10	36
12	45
15	72
17	72
20	72
22	54
26	75
31	80
36	54
41	36

Source: www.hypertextbook.com

11. If S represents the speed of the car at a given time t, explain what the notation $S(t)$ represents.
12. Evaluate $S(22)$ and explain what it means.
13. Create a scatter plot of the function $S(t)$.
14. What does the inverse relationship represent in this context?
15. Create a scatter plot of the inverse relationship. Does the inverse relationship represent a function? Explain your reasoning.

■ SHOW YOU KNOW

16. Your friend missed class today and asks you what is meant by "inverse function." Write an explanation of the concept for your friend.
17. Construct two tables of values that would represent inverse functions of each other. Explain how you know that they are inverse functions.
18. What does the notation $x = f^{-1}(y)$ mean?

■ MAKE IT REAL

19. **Percentage of Cremations** The percentage of cremations as a function of years since 1980 can be modeled by $C = f(t) = 0.8071t + 9.5811$, where C is the percent of people cremated at death and t is the number of years since 1980. (*Source: Modeled from data at www.cremationassociation.org*)
    a. Describe what the notation $f^{-1}(C) = t$ means in this context.
    b. Solve the equation $40 = 0.8071t + 9.5811$ and explain what the result means.
    c. The inverse function for this situation is $f^{-1}(C) = 1.239C - 11.871$. Evaluate $f^{-1}(40)$ and explain the practical meaning of the solution.
    d. Explain the relationship between the results in parts (b) and (c).

20. **Female Participation in Athletics** The number of females participating in high school athletics has been increasing steadily since 1990. (*Source: Statistical Abstract of the United States, 2001, Table 1241*) The number of females who participated in high school athletics can be modeled by
$$F = f(t) = 0.104t + 1.83$$
where F is the number of female participants (in millions) and t is the number of years since the 1990–1991 school year.
    a. Describe what the notation $f^{-1}(F) = t$ means in this context.
    b. Solve the equation $3 = 0.104t + 1.83$ and explain the practical meaning of the result.
    c. The inverse function for this situation can be modeled by $f^{-1}(F) = 9.615F - 17.596$. Evaluate $f^{-1}(3)$ and explain the practical meaning of the result.
    d. Explain the relationship between the results in parts (b) and (c).

21. **Fiddler Crab Claws** Male fiddler crabs have one large claw and one small claw. Scientists have found that there is a relationship between the weight of the large claw and the weight of the crab's body, as shown in the table.

Body Weight (grams)	Large Claw Weight (grams)
199.7	38.3
238.3	52.5
270.0	59.0
300.2	78.1
355.2	104.5
420.1	135.0
470.1	164.9
535.7	195.6
617.9	243.0

Source: Problems of Relative Growth, J. S. Huxley, Dover, 1972, p. 12, Table 1

   a. Assume that the claw weight, C, is a function of the body weight, B. Describe what the notation $C = f(B)$ means in this context.
   b. Create a scatter plot of the function $C = f(B)$. Then evaluate and interpret $C = f(300.2)$.
   c. Create a scatter plot of the inverse function $f^{-1}(C) = B$. Explain why this is a function.
   d. Evaluate $f^{-1}(52.5) = B$ using the table and interpret the practical meaning of the result.

22. **Buying Power of the Dollar** Due to inflation, the buying power of the dollar decreases over time. Based on data from 1990–2004, the buying power of the current dollar can be modeled by
$$V(t) = -0.0107t + 0.836 \text{ of a 1980 dollar}$$
where t is the number of years after 1990. Assuming $V = f(t)$, what does $f^{-1}(0.44) = 37$ mean?

23. Apple Computer International Sales Based on data from 2001–2005, the international net sales of Apple Computer Corporation can be modeled by

$$I(d) = \frac{3217}{1 + 9446e^{-0.001704d}} + 2400 \text{ million dollars}$$

where d is the domestic net sales (in millions of dollars). (*Source:* www.apple.com) Assuming $I = f(d)$, what does $f^{-1}(5597) = 8338$ mean?

24. Pager and Cell Phone Sales Based on data from 2000–2004, the amount of revenue generated by cell phone sales (in millions of dollars) can be modeled by

$$C = f(p) = -0.002745p^3 + 6.173p^2 - 4633p + 1{,}169{,}000$$

where p is the amount of revenue generated by pager sales (in millions of dollars). What does $f^{-1}(8077) = 811$ mean?

■ STRETCH YOUR MIND

Exercises 25–28 are intended to challenge your understanding of inverse functions. For these exercises, refer to the functions $g(y) = 2y + 4$ *and* $f(x) = \dfrac{x - 4}{2}$.

25. Explain why it makes sense that the function $g(y) = 2y + 4$ is the inverse of the function $f(x) = \dfrac{x - 4}{2}$.

26. Explain the meaning of the following diagram.

$$\text{input } x \rightarrow \boxed{f} \rightarrow y \text{ output}$$
$$\text{output } x \leftarrow \boxed{f^{-1}} \leftarrow y \text{ input}$$

27. Given $f(10) = 3$, find $f^{-1}(3)$. How can you use the function $g(y)$ to confirm your answer?

28. If $g(2) = 8$, find $g^{-1}(8)$. How can you use the function $f(x)$ to confirm your answer?

CHAPTER 1 Study Sheet

As a result of your work in this chapter, you should be able to answer the following questions, which are focused on the "big ideas" of this chapter.

SECTION 1.1
1. What is the purpose of a mathematical model?
2. What are the different representations of a mathematical model and what are the advantages of each?

SECTION 1.2
3. What is the relationship between the terms *independent variable, dependent variable, input, output, domain,* and *range*?
4. What is *function notation* and how does it work?
5. In mathematical terms, what do the words *solve* and *evaluate* mean?

SECTION 1.3
6. What is an average rate of change and what does it represent?

SECTION 1.4
7. What do we mean when we say *practical domain* and *practical range*?

SECTION 1.5
8. Why is it important to keep track of variables when working with functions in any real-world context?

SECTION 1.6
9. What does it mean for one function to be an inverse of another? Use tables and graphs as a part of your explanation.
10. What does the notation $x = f^{-1}(y)$ mean?

ENTIRE CHAPTER
11. How do you determine if a data set, graph, equation, or verbal expression represents a function? Your explanation should apply to each of the function representations listed.
12. How do you evaluate and solve a function in all of its representations?

REVIEW EXERCISES

■ SECTION 1.1 ■

1. A high school student is trying to decide what college to attend after graduation. The student has identified cost, distance from home (closer is better), and reputation (on a scale of 0 to 10 with 0 being very bad and 10 being very good) as the most important characteristics in deciding which college to attend. Create an example of a decision-factor equation for this situation and explain why you created it in the way you did.

In Exercises 2–3, analyze the given mathematical model given in numerical form. Then, answer the questions that follow each numerical model.

2. **Attendance at NFL Football Games**   The table shows the total attendance (in thousands) at all National Football League games for select years.

Year	Total Attendance (in thousands)
1985	14,058
1990	17,666
1995	19,203
2000	20,954
2001	20,590
2002	21,505
2003	21,639
2004	21,709

Source: Statistical Abstract of the United States, 2006, Table 1233

   a. Describe the trend observed in these data.
   b. Based on the data, predict what the total attendance at NFL games will be in 2006. Explain how you arrived at your prediction.

3. **Tennis Participation by Age Group**   The table shows the number of people who participate in the sport of tennis in the United States (in thousands) for the given age groups.

Age Group	Number of Participants (in thousands)
7–11	997
12–17	2054
18–24	1161
25–34	2312
35–44	1609
45–54	683
55–64	429
65 and over	327

Source: Statistical Abstract of the United States, 2006, Table 1238

   a. Describe the trend observed in these data.
   b. Why do you think the number of participants increases and decreases in the way that it does?

In Exercises 4–5, analyze the mathematical model given in graphical form. Then, answer the questions that follow each graph.

4. **U.S. Mortgage Interest Rates**   The average interest rate for a home mortgage in the United States between 1970 and 2005 is shown in the graph.

Source: Modeled from www.federalreserve.gov data

   a. In what year was the interest rate the highest? What was the interest rate in this year?
   b. Between what two consecutive years did the interest rate increase the most?
   c. Between what two consecutive years did the interest rate decrease the most?

5. **Travel Trends**   The average number of persons per vehicle on the road for any purpose between 1977 and 2002 is shown in the graph.

Source: Modeled from Statistical Abstract of the United States, 2006, Table 1254

   a. Describe the trend observed in this graph.
   b. What does it mean to say the average number of persons per vehicle is 1.6?
   c. From 1977 to 1990, the average vehicle occupancy declined. Give some possible reasons for this decline.

SECTION 1.2

6. Write the following expression in function notation by choosing meaningful letters to represent each variable. Also identify the independent variable (input) and the dependent variable (output).

   The number of calories a person takes into his body is a function of the amount of food consumed by that person.

7. Evaluate the function $f(x) = -2x^2 + 3x$ at $x = -1$.

8. **Income of Texans** Based on data provided by the Bureau of Economic Analysis, the per capita income of Texans from 2000 to 2004 can be modeled by the formula

   $$T = f(y)$$
   $$= 133.75y^3 - 666.71y^2 + 1131.61y + 28339.67$$

   where T represents the per capita income of Texans (in dollars) and y is the years since 2000. (*Source:* www.bea.gov) Evaluate $f(10)$ and explain what the numerical answer represents in its real-world context. Finally, write your solution in function notation.

9. **Tuition and Fees of Four-Year Colleges** The College Board, a nonprofit membership association of 4500 schools, colleges, and universities has consistently shown that the average tuition and fees for undergraduate students attending 4-year public universities are rising. The average cost of tuition and fees for 2004 in various regions of the country is shown in the table. The cost of tuition and fees, t, is a function of the U.S. region, r, and therefore, $f(r) = t$.

U.S. Region	2004 Average Cost of Tuition and Fees (in dollars)
New England	6839
Middle States	6300
Midwest	6085
Southwest	4569
South	4143
West	4130

   Source: www.collegeboard.com

   a. Solve $f(r) = 4569$ for r and write a sentence explaining in words what your solution means in its real-world context.

   b. Evaluate $f(\text{New England})$ and write a sentence explaining in words what the numerical value you find means in its real-world context.

10. **New Business Profits** Suppose that you have started a new company called Gus's Goat Grooming Service. The following graph of $p = f(g)$ displays your profit, p, in dollars for the number of goats you have groomed.

    a. Estimate the maximum amount of profit made and how many goats are groomed to make that profit.

    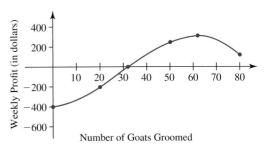

    b. When are the profits negative? Why could this happen?

    c. Evaluate $f(20)$ and explain what your solution means in its real-world context.

    d. Find a value of g such that $f(g) = 300$ and explain what the solution means in its real-world context. Are there any other solutions for g such that $f(g) = 300$? If so, give them.

11. **The Harris–Benedict Formula for Caloric Intake** When designing a personal nutrition plan for yourself you should first calculate the number of calories your body burns in one day. This number is known as your total daily energy expenditure, E, or "maintenance level." The average value of E for women in the United States is between 2000 and 2100 calories per day and for men it is between 2700 and 2900 per day. (*Source:* www.weightlossforall.com) These are only averages, however, as caloric expenditure can vary widely depending on one's level of fitness and activity level.

    To calculate the total daily energy expenditure, E, we consider E a multivariable function dependent on variables including height in inches, h, weight in pounds, w, and age in years, a. E is also dependent on how active a person is so a level of activity factor is used to adjust the caloric intake. A factor of 1.55 is used for moderately active people.

    Therefore, E is written in function notation as $E = f(h, w, a)$; this is known as the Harris–Benedict equation. The multivariable functions for both men and women who are moderately active are given below.

    Men:
    $$E(w, h, a) = 1.55(66 + 6.23w + 12.7h - 6.8a)$$

    Women:
    $$E(w, h, a) = 1.55(655 + 4.35w + 4.7h - 4.7a)$$

    where w is in pounds, h is in inches, and a is in years.

    a. Use function notation to express the total daily energy expenditure, E, for a moderately active male, who weighs 210 pounds, is 6 foot 2 inches tall (74 inches), and is 42 years of age. Calculate the total daily energy expenditure, E, for this person.

    b. Write your total daily energy expenditure, E, in function notation and then calculate its value assuming that you are moderately active.

SECTION 1.3

12. Baby Girls' Average Height

Age (months) m	Height (inches) H	Average Rate of Change
Birth	19.4	
6	25.6	
12	29.0	
18	31.6	
24	33.8	
30	35.8	
36	37.4	

Source: National Center for Health

a. Calculate the average rates of change for consecutive pairs of data values.

b. Describe how the height changes as age increases.

13. Baby Boys' Average Height

Age (months) m	Height (inches) H	Average Rate of Change
Birth	19.6	
6	26.5	
12	29.8	
18	32.3	
24	34.5	
30	36.2	
36	37.8	

Source: National Center for Health

a. Calculate the average rates of change for consecutive pairs of data values.

b. Explain the meaning of the rates of change in the context of the data.

c. Estimate $H = f(27)$ and interpret what it means in a real-world context.

14. Perception Reaction Time Distance *Perception reaction time distance* is the distance a vehicle will travel during the time when a hazard first becomes visible to a driver and the driver takes action to avoid the hazard.

Perception Time of 1.4 Seconds for Nominal Hazard

Miles Per Hour m	Distance (in feet) D
30	61.60
40	82.13
50	102.67
60	123.20

Source: www.harristechnical.com/articles/skidmarks.pdf

Given that $D = f(m)$, estimate and explain the meaning of $f(35)$, $f(57)$, and $f(64)$.

15. Gross Movie Receipts The movie *Mission Impossible III* made its debut on May 5, 2006.

Weekend w	Receipts in Millions of Dollars R
1	47.743
2	25.009
3	11.350
4	7.002
5	4.685
6	3.021
7	1.343

Source: www.the-numbers.com

Estimate the receipts for the eighth weekend.

16. Skid Mark Distances on Snow

Speed (mph) s	Distance (feet) in Snow D
30	100
40	178
50	278
60	400

Source: www.harristechnical.com/articles/skidmarks.pdf

Given that $D = f(m)$, use the data in the table to estimate and explain the meaning of $f(47)$, $f(55)$, and $f(68)$.

17. Investment Account $B(p, r, n, t) = p\left(1 + \dfrac{r}{n}\right)^{nt}$ is the formula used to calculate the balance, B, in an investment account into which a lump sum is invested and interest is compounded a fixed number of times a year for a certain number of years. The independent variables represent the following:

p is the amount invested.
r is the nominal interest rate as a decimal.
n is the number of times the compounding occurs in a year.
t is the years the money is invested.

Compute the value of $B(5000, 0.067, 12, 3)$.

18. Vehicle Loan Payments The formula for monthly payments on a vehicle loan is

$$M(p, i, n) = p\left[\dfrac{i(1 + i)^n}{(1 + i)^n - 1}\right]$$

where

p = amount borrowed in dollars
n = number of monthly payments
i = interest rate per monthly payment period as a decimal

a. Calculate the monthly payment for a 2006 Chrysler Sebring Convertible with a MSRP of $26,115 in June 2006. (*Source:* price at www.edmunds.com) E-Loans has offered 5.89% APR for a 5-year loan.

b. How much interest will be paid over the life of the 5-year loan?

19. **Internet Plans** Nextel offered a BlackBerry plan that includes unlimited email and Internet with pay-as-you-go voice service for $49.99 a month plus 20 cents per minute for voice service. (*Source:* nextelonline.nextel.com)

   a. Write an equation that models the monthly bill T as a function of the number of voice service minutes that have been used, m.

   b. What is the total monthly bill (minus taxes and fees) when 100 voice service minutes are used?

20. The formula $C = f(F) = \frac{5}{9}(F - 32)$ converts the temperature on the Fahrenheit scale F to the temperature on the Celsius scale, C. What is F if $f(F) = 35$?

21. The formula for calculating the minimum speed of a car at the beginning of a skid S is $S(d, f, n) = \sqrt{30dfn}$

   where

   d = skid distance in decimal feet
   f = drag factor for the road surface
   n = percent braking efficiency written as a decimal (i.e., 100% = 1.00)

   How fast was a car traveling if it left a 64.3-foot skid on a road with a drag factor of 0.7 (asphalt) in a car with a braking efficiency of 100%?

■ SECTION 1.4 ■

22. Estimate the function values using the following graph of $h(x)$ below.

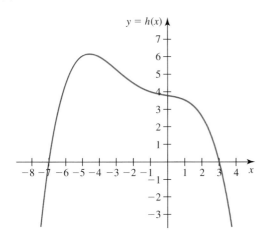

   a. Evaluate $h(-4)$.
   b. Solve $h(x) = 4$ for x.
   c. Solve $h(x) = 0$ for x.
   d. Evaluate $h(0)$.
   e. Evaluate $h(3)$.

23. **Charitable Giving by Americans** In 2005, Americans gave an estimated $260 billion to charity, marking a 5-year high in American giving. The rise came partly in response to a wave of natural disasters, including Hurricane Katrina, the South Asian tsunami, and the earthquake in Kashmir. The following graph shows the gift amount as a percentage of assets, g, as a function of the 2003 income, m.

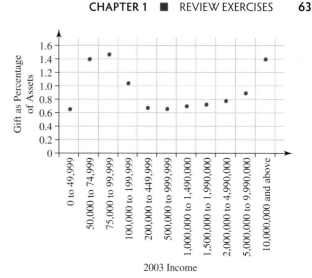

   a. Determine the practical domain and range for $g = f(m)$.
   b. Explain what the graph means in its real-world context.

24. Match the three written scenarios with the most appropriate graph given in D–F.

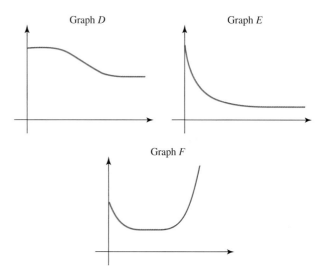

   a. With the temperature more than 100° outside, I got in my car to cool off. I started the car and turned on the air conditioner. At first, the air conditioner did not blow cool air; however, after the car ran for a few minutes, the air conditioner began to cool down the car. Once the temperature inside the car was comfortable, I adjusted the air conditioner to maintain that temperature. The temperature inside the car is the dependent variable.

   b. In 1964 I purchased a brand new car and as soon as I drove it off the lot the value of the car depreciated—dramatically for the first three years, then at a slower rate. I have kept the car and it is in great condition so the value has risen over the last 15 years. The value of the car in dollars is the dependent variable.

   c. It has been a long time since I have worked out. I decided recently to begin walking on the treadmill to lower my heart rate. I walk three miles four times a week. When I started my workout regimen my heart rate was high, it stayed that way for six months and then slowly started to lower. Now it has reached a plateau. My heart rate in beats per minute is the dependent variable.

25. For graph D in Exercise 24, do each of the following.
   a. Describe what the vertical intercept means in the real-world context.
   b. Determine the practical domain and range for the function represented by the graph.

26. **Gunsight Butte** A picture of Gunsight Butte in Lake Powell near Page, Arizona, has been transposed onto a coordinate system that displays the elevation of the butte as a function of the horizontal distance. Assume that the horizontal and vertical scales of the graph are the same.

   a. Explain what the graph means in its real-world context.
   b. If the maximum height of Gunsight Butte is 290 feet, what can we say about the horizontal length of the butte as shown in the figure?
   c. What do the vertical and horizontal intercepts mean in this context?
   d. Suppose you were able to hike along the rim (top) of Gunsight Butte starting at the point labeled a and ending where the crown begins (labeled b). Estimate your change in horizontal position and in vertical position. Justify your numerical estimates.

■ SECTION 1.5 ■

27. **Gunsight Butte** The graph of the vertical height of Gunsight Butte in Exercise 26 is given by a function $v = f(h)$, where v is the vertical distance (in feet) and h is the horizontal distance (in feet) measured from the origin of the coordinate system. Explain in words what $v(415) = 290$ means in its real-world context.

28. **Chipotle Mexican Grill Sales** Restaurant sales for Chipotle Mexican Grill increased every year between 2001 and 2005. In 2001, annual sales were $131,331 thousand but they increased to $625,077 thousand in 2005. Additionally, the *annual change* in restaurant sales was increasing between 2001 and 2005. (*Source:* Chipotle Mexican Grill, Inc, 2005 Annual Report, p. 24) Draw a graph to model sales as a function of years since 2001.

■ SECTION 1.6 ■

For situations in Exercises 29–30, do each of the following.
   a. Use function notation to represent the scenario, if possible. (Be sure to identify the meaning of any variables used.)
   b. Describe the advantages that knowing the inverse relationship to the situation would provide.
   c. Determine whether or not the inverse relationship represents a function and justify your conclusion.

29. The price of a new electronics item (e.g., DVD player, MP3 player, plasma television) over time.

30. The temperature in your backyard from midnight one night to midnight the next.

31. **Number of Children in Head Start** The number of children enrolled in the Head Start program has increased as the number of women in the workforce has increased. (*Source: Statistical Abstract of the United States, 2006,* Tables 564 and 579) The number of children enrolled in Head Start may be modeled by the function

$$H = f(N) = 0.0301N - 1134.9$$

where h is the number of children in the Head Start program (in thousands) and N is the number of women in the workforce (in thousands).

   a. Describe what information the notation $f^{-1}(H) = N$ provides in this situation.
   b. Solve the equation
$$700 = 0.0301N - 1134.9$$
and explain what the result means in the context of this situation.
   c. The inverse function for this situation can be modeled by
$$f^{-1}(H) = N = 33.22H + 37{,}704.32$$
Evaluate $f^{-1}(700)$ and explain what it means in the context of this situation.
   d. Explain the relationship between parts (b) and (c).

32. **Daily Newspaper Circulation vs. Cable TV Subscribers** Data reveal that as the number of cable television subscribers increases, the daily newspaper circulation decreases. (*Source:* www.census.gov) Daily newspaper circulation as a function of cable TV subscribers can be modeled by the function

$$N = f(C) = -0.3824C + 81.574$$

where N is the daily newspaper circulation (in millions) and C is the number of cable television subscribers (in millions).

   a. Describe what information the notation $f^{-1}(N) = C$ provides in this situation.
   b. Solve the equation $60 = -0.3824C + 81.574$ and explain what the result means in the context of this situation.

c. The inverse function for this situation may be modeled by $f^{-1}(N) = C = -2.62N + 213.32$. Evaluate $f^{-1}(60)$ and explain what it means in the context of this situation.

d. Explain the relationship between parts (b) and (c).

33. **Minimum Wage Salary** In 2006, workers earning minimum wage earned $5.15 per hour. (*Source:* www.dol.gov) The weekly gross income of a worker earning minimum wage can be modeled by the function $D = f(h) = 5.15h$, where D is the gross income (in dollars) and h is the number of hours worked.

a. Describe what information the notation $f^{-1}(D) = h$ provides in this situation.

b. Solve the equation $206 = 5.15h$ and explain what the result means in the context of this situation.

c. The inverse function for this situation may be modeled by $f^{-1}(D) = h = \dfrac{D}{5.15}$. Evaluate $f^{-1}(206)$ and explain what it means in the context of this situation.

d. Explain the relationship between parts (b) and (c).

Make It Real Project

What to Do

1. Obtain a recent copy of a local, regional, or national newspaper or magazine.

2. Read through the paper looking for functions represented in words, tables, and graphs. Select a minimum of two samples for each function representation type (words, tables, and graphs).

3. For each sample, represent the data in all four function representations: formulas, tables, graphs, and words.

4. Select one of the function models and describe how you could use it to enhance your quality of life.

Where to Find Data

The following newspapers are typically accessible in public libraries:

- The Wall Street Journal
- USA Today
- The New York Times

The following magazines are typically accessible in public libraries:

- Time
- Newsweek
- People Magazine

Since many periodicals publish digital versions of their graphics and data online, you may find it helpful to access online versions for your report.

Rubberball/Nicole Hill/Getty Images

CHAPTER 2

Linear Functions and Matrices

Among all cancers, lung cancer is the number-one killer. Smoking is the leading cause of lung cancer. Fortunately, smoking rates in the United States have declined dramatically since 1950. The smoking rate in the United States can be represented by a linear function.

2.1 Functions with a Constant Rate of Change
2.2 Modeling with Linear Functions
2.3 Linear Regression
2.4 Systems of Linear Equations
2.5 Using Matrices to Solve Linear Systems
2.6 Matrix Operations and Applications

STUDY SHEET
REVIEW EXERCISES
MAKE IT REAL PROJECT

SECTION 2.1 Functions with a Constant Rate of Change

LEARNING OBJECTIVES

- Explain the real-world meaning of slope, horizontal intercept, and vertical intercept
- Write linear functions in slope-intercept form, point-slope, and standard form
- Graph linear functions

GETTING STARTED

In April 1896, the first modern Olympics were held in Athens, Greece. The U.S. Olympic Team Manager for those games, John Graham, was inspired by the marathon race he witnessed there. With the help of a Boston businessman, Graham planned the first annual Boston Marathon in April 1897. The Boston Marathon is now considered one of the premier distance races in the world, pitting the best runners against each other in a grueling run of over 26.2 miles. In 2009, the fastest runner was Ethiopian Deriba Merga, who completed the race in just over 2 hours and 8 minutes. His average speed, which was less than 5 minutes per mile, is an example of a constant rate of change. (*Source:* www.bostonmarathon.org)

In this section we investigate the concept of a constant rate of change, classify functions with constant rates of change as linear functions, and explore such functions with words, graphs, tables, and formulas. We develop strategies for graphing linear functions given in the three common forms—slope-intercept, point-slope, and standard.

■ Constant Speed

If you have traveled in a car with the cruise control on, you have some sense of the concept of a constant speed, or constant rate of change in distance with respect to time. This mathematically powerful concept is closely linked to the linear function concept. Let's look at this relationship in the context of a marathon race.

Marathon courses have mile markers to help runners monitor their progress. Suppose that between two such mile markers, Deriba Merga ran at a *constant speed* of 17.6 feet per second (completing the mile in 5 minutes). This means that *for every second that elapsed since the time he passed the mile marker, he moved exactly 17.6 feet*. If Merga runs at this constant speed for 2 seconds, he will travel 35.2 feet past the mile marker: $17.6 \frac{\text{feet}}{\text{second}}(2 \text{ seconds}) = 35.2$ feet. If he runs at this constant speed for 14.9 seconds, he will travel 262.24 feet: $17.6 \frac{\text{feet}}{\text{second}}(14.9 \text{ seconds}) = 262.24$ feet. To generalize, if Merga runs at a constant speed of $17.6 \frac{\text{feet}}{\text{second}}$ for t seconds after passing the mile marker, he will travel $17.6t$ feet past the mile marker.

or

More broadly stated, if an object is traveling at a constant speed, then

change in distance = (constant speed)(time elapsed)

Calling the constant speed m, we say "the change in distance will always be m times the amount of time elapsed." For example, if the constant speed is 2.9 miles per hour, we say "the change in distance (in miles) will always be 2.9 miles per hour times the number of hours elapsed."

EXAMPLE 1 ■ Interpreting Constant Speed

Interpret each of the following constant speeds first with an example and then in general terms.

a. 3 miles per hour **b.** 6.14 meters per day

Solution

a. If an object travels at a constant speed of 3 miles per hour for 1 hour, the object travels 3 miles: $\left(3 \frac{\text{miles}}{\text{hour}}\right)(1 \text{ hour}) = 3$ miles. In general, if the object travels for h hours at a constant speed of 3 miles per hour it will travel $3h$ miles.

b. If an object travels at a constant speed of 6.14 meters per day for 1 day, the object travels 6.14 meters: $\left(6.14 \frac{\text{meters}}{\text{day}}\right)(1 \text{ day}) = 6.14$ meters. In general, if the object travels for d days at a constant speed of 6.14 meters per day, it will travel $6.14d$ meters.

■ Constant Rate of Change

The idea of constant speed can be generalized to make sense of a *constant rate of change*. A situation has a **constant rate of change** m if, whenever the input changes, the output changes by m times as much.

$$\text{change in output value} = m(\text{change in input value})$$

For example, let x be the input and y the output for some function. If that function has a constant rate of change of 5, then

$$\text{change in } y = 5(\text{change in } x)$$

Now suppose x changes by 8. By how much will y change?

$$\text{change in } y = 5(\text{change in } x)$$
$$\text{change in } y = 5(8)$$
$$\text{change in } y = 40$$

The value of y will change by 40. By how much will y change if x changes by, say, 0.0004?

$$\text{change in } y = 5(\text{change in } x)$$
$$\text{change in } y = 5(0.0004)$$
$$\text{change in } y = 0.002$$

If x changes by 0.0004, then the value of y will increase by 0.002.

EXAMPLE 2 ■ Interpreting Constant Rates of Change

Each of the following values of m is a constant rate of change for some function. Explain what each value tells about the relationship between the input and output variables of the function.

a. $m = -2.7$; input variable is x, output variable is y.
b. $m = 113.44$; input variable is t, output is $f(t)$.

Solution

a. change in $y = -2.7$(change in x)
b. change in $f(t) = 113.44$(change in t)

In mathematical notation, we use the Greek letter delta (Δ) to represent the phrase "change in." Thus, for example, "change in $f(t) = 113.44$(change in t)" is written "$\Delta f(t) = 113.44 \Delta t$."

EXAMPLE 3 ■ Interpreting Constant Rates of Change

Explain what each of the following constant rates of change mean in the context of the given situation.

a. From 1990 to 2003, the concentration of carbon monoxide in the atmosphere had a near constant rate of change of −0.248 parts per million per year. (*Source: Statistical Abstract of the United States, 2006*, Table 359)

b. Based on data from 1974–2003, the death rate due to heart disease as a function of the percentage of people who smoke has a constant rate of change of 14.08 deaths per 100,000 people per percentage point. (*Source:* Modeled from CDC and Census Bureau data)

Solution

a. In this context, the dependent (output) variable is the concentration of carbon monoxide in parts per million, C, and the independent (input) variable is the year, t. A constant rate of change of -0.248 parts per million per year means

$$\text{change in carbon monoxide concentration} = -0.248(\text{number of years that pass})$$

$$\Delta C = -0.248 \Delta t$$

Note that the rate of change is negative. This means that as time passes, the carbon monoxide concentration decreases. Also, since this problem has a practical domain of years from 1990 to 2003, the change in years is restricted to this domain.

b. In this context, the dependent (output) variable is the death rate due to heart disease in deaths per 100,000 people, R, while the independent (input) variable is the percentage of people who smoke, p. A constant rate of change of 14.08 deaths per 100,000 people per percentage point means

$$\text{change in death rate} = 14.08 \begin{pmatrix} \text{change in percentage} \\ \text{of people who smoke} \end{pmatrix}$$

$$\Delta R = 14.08 \Delta p$$

Note that the rate of change is positive. This means that as the percentage of people who smoke increases, the death rate due to heart disease increases.

In general, we have the following definition for the concept of the constant rate of change.

> **CONSTANT RATE OF CHANGE**
>
> A function has a **constant rate of change** m if, for any change in the independent variable, the dependent variable always changes by exactly m times as much.

EXAMPLE 4 ■ Calculating a Constant Rate of Change in Context

The number of Medicare enrollees between 1980 and 2004 can be modeled by a function with a constant rate of change. In 1980 there were 28.4 million Medicare enrollees. By 2004 the number of Medicare enrollees had increased to 41.7 million. (*Source: Statistical Abstract of the United States, 2006*, Table 132) Determine the constant rate of change by which the number of Medicare enrollees increased over time.

Solution Calling the constant rate of change m, we have

$$\begin{pmatrix} \text{change in Medicare} \\ \text{enrollees (in millions)} \end{pmatrix} = m \begin{pmatrix} \text{number of} \\ \text{years that pass} \end{pmatrix}$$

From 1980 to 2004, 24 years elapsed (2004 − 1980 = 24). During this time period, the number of Medicare enrollees increased by 13.3 million (41.7 − 28.4 = 13.3). Thus,

$$13.3 \text{ million enrollees} = m(24 \text{ years})$$

$$\frac{13.3 \text{ million enrollees}}{24 \text{ years}} = \frac{m(24 \text{ years})}{24 \text{ years}}$$

$$\left(\frac{13.3}{24}\right)\frac{\text{million enrollees}}{\text{year}} = m$$

$$m \approx 0.554 \text{ million enrollees per year}$$

This means that for each year that passed, the number of Medicare enrollees increased by about 0.554 million, or about 554,000. More generally, from 1980 to 2004 a change of t years means a change of approximately $0.554t$ million Medicare enrollees.

Example 4 illuminates two very important ideas. First, the units of the constant rate of change come from the units of the dependent and independent variables.

Units of a Rate of Change

The units of a rate of change are the units of the dependent variable divided by the units of the independent variable.

$$\text{units of rate of change} = \frac{\text{units of dependent variable}}{\text{unit of independent variable}}$$

We commonly write this as "units of output per unit of input."

The second idea from Example 4 is that we can create a formula to calculate a constant rate of change. Recall that there were 28.4 million Medicare enrollees in 1980 and 41.7 million Medicare enrollees in 2004. We can represent this information with the ordered pairs (1980, 28.4) and (2004, 41.7). Then

$$\begin{pmatrix}\text{change in Medicare}\\ \text{enrollees (in millions)}\end{pmatrix} = m\begin{pmatrix}\text{number of years}\\ \text{since 1980}\end{pmatrix}$$

$$(41.7 - 28.4) = m(2004 - 1980)$$

$$\frac{(41.7 - 28.4)}{(2004 - 1980)} = \frac{m(2004 - 1980)}{(2004 - 1980)}$$

$$m = \frac{41.7 - 28.4}{2004 - 1980}$$

Because the rate of change is constant (i.e., for any change in the input, the output always changes by exactly m times as much), m will be the same when comparing *any* ordered pairs for the function. Thus, if we say a function has a constant rate of change and that (x_1, y_1) and (x_2, y_2) are ordered pairs for the function, then

$$\text{change in } y = m(\text{change in } x)$$

$$\frac{\text{change in } y}{\text{change in } x} = \frac{m(\text{change in } x)}{\text{change in } x}$$

$$m = \frac{\text{change in } y}{\text{change in } x}$$

$$m = \frac{y_2 - y_1}{x_2 - x_1} \quad \begin{array}{l} y_2 - y_1 \text{ tells us how much } y \text{ changes}\\ x_2 - x_1 \text{ tells us how much } x \text{ changes}\end{array}$$

HOW TO: ■ CALCULATE A CONSTANT RATE OF CHANGE

To calculate a constant rate of change m, divide the change in the dependent variable by the change in the independent variable.

$$m = \frac{\text{change in the dependent variable}}{\text{change in the independent variable}}$$

If given ordered pairs (x_1, y_1) and (x_2, y_2),

$$m = \frac{y_2 - y_1}{x_2 - x_1}$$

In Chapter 1 we calculated an average rate of change by finding the difference of two outputs divided by the difference in the corresponding inputs. We used the formula $m = \frac{y_2 - y_1}{x_2 - x_1}$, which we now know is the method for calculating a constant rate of change. This is not a coincidence—an average rate of change *is* a constant rate of change. Specifically, it is the constant rate of change necessary for the function values to change by the same amount for the same change in the input. Thus, when using an average rate of change, we are reasoning about a function *as if* it has a constant rate of change over a given interval.

To illustrate, we consider the graphs of two functions shown in Figures 2.1 and 2.2. We calculate the average rate of change in the functions over the intervals [3, 5], [3, 6], [3, 7], and [3, 8]. (The red line segments between each pair of points indicate which pairs of points are being used to calculate the average rate of change.)

Figure 2.1

Figure 2.2

For the function in Figure 2.1, the average rate of change varies as soon as we vary the second point. For the intervals given, the average rates of change (r) are 2, 1, 0, and −1, respectively. The function in Figure 2.2 has a constant rate of change, so the average rate of change (r) over *any* interval is always 0.5. In fact, for any function with a constant rate of change, the average rate of change of the function over any interval is the same as the constant rate of change in the function.

EXAMPLE 5 ■ Using a Constant Rate of Change

Table 2.1 gives data points for a function with a constant rate of change. Find the value of a.

Solution To find a, we need to know the constant rate of change for this function. Since for any change in x, the value of y always changes by exactly m times as much, it does not matter which ordered pairs we choose. But for the sake of comparison, we show the calculations for several ordered pairs.

Table 2.1

x	y
−4	5.8
1	−0.2
2.5	a
3	−2.6

(−4, 5.8) and (3, −2.6)	(−4, 5.8) and (1, −0.2)	(1, −0.2) and (3, −2.6)
$m = \dfrac{\text{change in } y}{\text{change in } x}$	$m = \dfrac{\text{change in } y}{\text{change in } x}$	$m = \dfrac{\text{change in } y}{\text{change in } x}$
$m = \dfrac{-2.6 - 5.8}{3 - (-4)}$	$m = \dfrac{-0.2 - 5.8}{1 - (-4)}$	$m = \dfrac{-2.6 - (-0.2)}{3 - 1}$
$m = \dfrac{-2.6 - 5.8}{3 + 4}$	$m = \dfrac{-0.2 - 5.8}{1 + 4}$	$m = \dfrac{-2.6 + 0.2}{3 - 1}$
$m = \dfrac{-8.4}{7}$	$m = \dfrac{-6}{5}$	$m = \dfrac{-2.4}{2}$
$m = -1.2$	$m = -1.2$	$m = -1.2$

Thus the constant rate of change for this function is −1.2. We now find a using the ordered pairs (1, −0.2) and (2.5, a) and the constant rate of change, $m = -1.2$.

$$m = \frac{\text{change in } y}{\text{change in } x}$$
$$-1.2 = \frac{-0.2 - a}{1 - (2.5)}$$
$$-1.2 = \frac{-0.2 - a}{-1.5}$$
$$1.8 = -0.2 - a$$
$$a + 1.8 = -0.2$$
$$a = -2.0$$

■ Linear Functions

A **linear function** is any function that has a constant rate of change. The constant rate of change is also called the **slope**. Linear functions are used extensively to model many real-world situations.

> **LINEAR FUNCTION**
>
> Any function with a constant rate of change is called a **linear function**.

> **SLOPE OF A LINEAR FUNCTION**
>
> The constant rate of change of a linear function is called the **slope** of the function. The term *slope* is also used to refer to the steepness of the graph of a linear function.

EXAMPLE 6 ■ Creating a Linear Function from a Verbal Description

Based on data from 1990–2003, carbon monoxide pollutant concentrations in the United States have been decreasing at a rate of about 0.248 parts per million per year. In 1990, carbon monoxide pollutant concentrations were 6 parts per million. (*Source: Statistical Abstract of the United States, 2006*, Table 359) Create a function to model the carbon monoxide pollutant concentration in the United States as a function of years since 1990. Then estimate the carbon monoxide pollutant concentration for 2006.

Solution We are to find a pollutant concentration function $C(t)$ with the dependent variable, C, representing the carbon monoxide pollutant concentration (in parts per million) and the independent variable, t, representing the number of years since 1990.

We know $C(0) = 6$ because $t = 0$ corresponds to 1990, and in 1990 the pollutant concentration was 6 parts per million. We also know the pollutant concentration is decreasing by 0.248 parts per million each year. There are t years beyond 1990 so the expression $-0.248t$ will calculate the amount of change in the pollutant level over a period of t years. Therefore, we will add the change in pollutant level, $-0.248t$, to the pollutant level of 6 parts per million. The function is then

$$C(t) = 6 + (-0.248t)$$
$$C(t) = 6 - 0.248t$$

To estimate the pollutant concentration in 2006, we evaluate $C(t)$ at $t = 16$.

$$C(16) = 6 - 0.248(16)$$
$$= 2.032$$

We estimate the carbon monoxide pollutant concentration in 2006 to be 2.032 parts per million.

■ Slope-Intercept Form of a Linear Function

Linear functions may be represented in slope-intercept, point-slope, or standard form. Each form has its benefits and we will use them all.

> **SLOPE-INTERCEPT FORM OF A LINEAR FUNCTION**
>
> A linear function with slope m and vertical intercept $(0, b)$ is written in **slope-intercept form** as
>
> $$y = mx + b$$
>
> The value b is commonly referred to as the **initial value**.

EXAMPLE 7 ■ **Graphing a Function in Slope-Intercept Form**

Graph the function $C(t) = -0.248t + 6$ from Example 6. Recall that C represents the carbon monoxide concentration (in parts per million) and t represents the number of years since 1990.

Solution We have $C(t) = -0.248t + 6$. We note the initial value is 6 so the vertical intercept is $(0, 6)$. Starting at $(0, 6)$, we increase the number of years since 1990. Since the slope is -0.248, each 1-year increase will result in a 0.248 parts per million *decrease* in the pollutant concentration. In other words, for each increase of t years, C changes by $-0.248t$. We calculate the concentration (C) values for the arbitrary values of $t = 4$, 10, and 13, then plot and connect the resultant ordered pairs. The graph is shown in Figure 2.3.

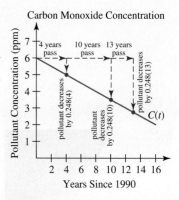

Figure 2.3

EXAMPLE 8 ■ Interpreting the Slope and Vertical and Horizontal Intercepts

The annual U.S. lumber imports from Canada from 1998 to 2003 can be approximated with the linear function $L(t) = 598.74t + 18{,}895$, where L is measured in millions of board feet and t is the number of years since 1998. (*Source:* Modeled from *Statistical Abstract of the United States, 2006*, Table 852) Find and interpret the slope, vertical intercept, and horizontal intercept of the linear function.

Solution The slope (constant rate of change) of the function is 598.74 million board feet per year since 1998. Thus, the United States increased its lumber imports from Canada by 598.74 million board feet per year between 1998 and 2003.

The vertical intercept is (0, 18,895). This means that in 1998 ($t = 0$), the United States imported 18,895 million board feet of lumber from Canada.

We find the horizontal intercept by letting $L(t) = 0$ and solving for t.

$$L(t) = 598.74t + 18{,}895$$
$$0 = 598.74t + 18{,}895$$
$$-18{,}895 = 598.74t$$
$$\frac{-18{,}895}{598.74} = \frac{598.74t}{598.74}$$
$$-31.6 \approx t$$

The horizontal intercept is approximately (−31.6, 0). This tells us that 31.6 years before the end of 1998 (that is, mid-1957) the United States did not import any lumber from Canada. We are skeptical of this prediction because it lies well outside the domain used to create the model.

We saw that the graph of the function in Example 7 is a line. In fact, since every linear function has a constant rate of change, the graph of every linear function will be a line. This means that we only need two points to graph a linear function. Once we have plotted the two points, we can connect them with a line and know that every ordered pair that satisfies the function equation will lie on that line.

EXAMPLE 9 ■ Graphing a Function in Slope-Intercept Form

Graph $f(x) = -\dfrac{3}{2}x + 4$.

Solution To graph the line that represents this linear function, we use the vertical intercept (0, 4) as our first point. To find a second point, we evaluate the function at any other value of x. We choose $x = 6$.

$$f(x) = -\frac{3}{2}x + 4$$
$$f(6) = -\frac{3}{2}(6) + 4$$
$$= -9 + 4$$
$$= -5$$

Thus a second point is (6, −5). The graph is shown in Figure 2.4.

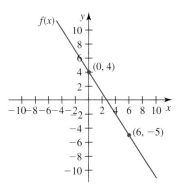

Figure 2.4

EXAMPLE 10 ■ Horizontal and Vertical Lines

Find the slope and equation for each line shown in Figure 2.5.

Solution We begin with A, the horizontal line. We pick two points on the line and determine the slope. We choose $(0, -5)$ and $(6, -5)$.

$$m = \frac{-5 - (-5)}{6 - 0}$$
$$= \frac{-5 + 5}{6 - 0}$$
$$= \frac{0}{6}$$
$$= 0$$

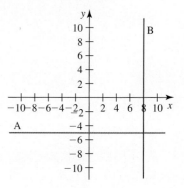

Figure 2.5

The slope is zero. What does this mean? Let's refer back to our understanding of constant rate of change.

$$\text{change in } y = m(\text{change in } x)$$
$$\text{change in } y = 0(\text{change in } x)$$
$$\text{change in } y = 0$$

Thus a slope of zero means that no matter what the change in x is, the change in y is zero. That is, y does not change. When x varies, y remains constant. Since the value of y is constant, the equation of the line does not depend on x but is defined entirely by that constant y value. So the equation of Line A in Figure 2.5 is $y = -5$.

We now consider B, the vertical line. We pick two points and determine the slope. We choose $(8, 0)$ and $(8, 2)$.

$$m = \frac{2 - 0}{8 - 8}$$
$$= \frac{2}{0} \rightarrow \text{undefined}$$

In this case we do not get a numeric value and we say that the slope is *undefined*. In other words, the slope does not exist. Again, let's return to our understanding of a constant rate of change.

$$\text{change in } y = m(\text{change in } x)$$
$$\text{change in } y = m0$$
$$\text{change in } y = 0$$

This does not make sense because we know the change in y is not 0. In fact, no value of m can make the statement true. Thus, there is no defined slope for a vertical line.

For a vertical line, the x-value is constant but y can be any number. Consequently, the equation for a vertical line is of the form $x = a$ or, in this case, $x = 8$. Note that this is not a function—it fails the vertical line test (the same input has multiple outputs).

EQUATIONS FOR HORIZONTAL AND VERTICAL LINES

- A horizontal line with vertical intercept $(0, b)$ has equation $y = b$.
- A vertical line with horizontal intercept $(a, 0)$ has equation $x = a$.

Interpreting the Graphical Meaning of the Slope of a Line

The slope of a line tells us much about the graph of the linear function and, if the linear function represents a real-world situation, much about the underlying context.

> **RELATIONSHIP BETWEEN THE SLOPE OF A LINE AND ITS GRAPH**
>
> - A line with *positive* slope is *increasing* (going up) from left to right.
> - A line with *negative* slope is *decreasing* (going down) from left to right.
> - A line with *zero* slope is *horizontal*.
> - A line with *undefined* slope is *vertical*.
> - The greater the *magnitude* (absolute value) of the slope, the steeper the line.

■ Point-Slope Form of a Linear Function

The point-slope form of a line is most useful when we are given a point and a slope for an underlying context.

EXAMPLE 11 ■ Writing a Linear Function in Point-Slope Form

Between 1980 and 2004, the number of Medicare enrollees increased by approximately 0.554 million enrollees per year. By 2004 the number of Medicare enrollees had increased to 41.7 million. (*Source:* Modeled from *Statistical Abstract of the United States, 2006*, Table 132) Create a linear model for the number of Medicare enrollees as a function of the number of years since 2000.

Solution We first need to define our variables. We let E represent the number of Medicare enrollees (in millions) and t represent the number of years since 2000. Using these variables, we know the linear function has slope $m = 0.554$ and passes through the point (4, 41.7) since $t = 4$ in 2004.

Observe that $E - 41.7$ represents the change in the Medicare enrollees since 2004 (in millions) and that $t - 4$ represents the change in years since 2004. From our earlier discussion of constant rate of change we know that

$$\text{change in output} = m(\text{change in input})$$

Applying that knowledge to this context, we have

$$E - 41.7 = 0.554(t - 4)$$

This form is referred to as *point-slope form*.

Given any constant rate of change m and ordered pair (x_1, y_1), we can write a formula to find any ordered pair (x, y) as follows.

$$\text{change in } y = m(\text{change in } x)$$
$$y - y_1 = m(x - x_1)$$

> **POINT-SLOPE FORM OF A LINEAR FUNCTION**
>
> A linear function with slope m and a point (x_1, y_1) is written in **point-slope form** as $y - y_1 = m(x - x_1)$.

When using technology to graph the line, we typically convert the point-slope form into the modified form $y = m(x - x_1) + y_1$, which is a format a graphing calculator accepts.

EXAMPLE 12 ■ Writing and Graphing Linear Functions in Point-Slope Form

The slope of a line is -2 and the line passes through the point $(3, -4)$. Write an equation of the line in point-slope form. Then graph the function.

Solution We know that the change in $y = m$(change in x). Applying that knowledge to this problem, we have

$$\begin{pmatrix} \text{change from} \\ y = -4 \text{ to some} \\ \text{other } y\text{-value} \end{pmatrix} = -2 \times \begin{pmatrix} \text{change from} \\ x = 3 \text{ to some} \\ \text{other } x\text{-value} \end{pmatrix}$$

$$\underbrace{y - (-4)}_{\substack{\text{the change from} \\ y = -4 \text{ to some} \\ \text{other } y\text{-value}}} \underbrace{=}_{\text{is equal to}} \underbrace{-2 \times (x - 3)}_{\substack{\text{the rate of change} \\ \text{times the change in } x}}$$

$$y + 4 = -2(x - 3)$$

To graph the function, we plot the known point $(3, -4)$. Since the slope is -2, a 1-unit increase in x corresponds with a 2-unit decrease in y. So as x increases from 3 to 4, y decreases from -4 to -6. Thus, the point $(4, -6)$ is also on the line. We plot this point and draw a line passing through both points, as shown in Figure 2.6.

Figure 2.6

■ Standard Form of a Linear Function

In some situations we do not have a designated independent and dependent variable. Instead, there are two variable quantities that relate to each other. In these circumstances, slope-intercept and point-slope forms are not natural representations. Instead, we use the *standard form* of a linear function.

STANDARD FORM OF A LINEAR FUNCTION

A linear function can be written in **standard form** as

$$Ax + By = C$$

where A, B, and C are real numbers.

EXAMPLE 13 ■ Writing a Formula for a Linear Function in Standard Form

The American Heart Association recommends that adults eat 25 to 30 grams of fiber each day. (*Source:* www.americanheart.org) Metamucil Orange Coarse Milled Fiber Supplement provides 3 grams of fiber per tablespoon and a cup of Kashi GOLEAN Crunch! Cereal provides 8 grams of fiber. (*Source:* www.metamucil.com and www.kashi.com) Suppose an adult male wants to use these products to get 15 grams of fiber each day. (The rest of the suggested fiber amount will come from other food sources.) Construct a linear function to model how much Metamucil supplement and Kashi cereal the man needs to consume each day to meet this goal.

SECTION 2.1 ■ FUNCTIONS WITH A CONSTANT RATE OF CHANGE

Solution Let c be the tablespoons of Metamucil the man takes in a day and k be the cups of Kashi the man eats in a day. Since Metamucil has 3 grams of fiber per tablespoon and Kashi has 8 grams of fiber per cup, we have

$$3 \frac{\text{grams}}{\text{tablespoon}} \cdot c \text{ tablespoons} + 8 \frac{\text{grams}}{\text{cup}} \cdot k \text{ cups} = 15 \text{ grams}$$

$$3 \frac{\text{grams}}{\cancel{\text{tablespoon}}} \cdot c \cancel{\text{tablespoons}} + 8 \frac{\text{grams}}{\cancel{\text{cup}}} \cdot k \cancel{\text{cups}} = 15 \text{ grams}$$

$$3c \text{ grams} + 8k \text{ grams} = 15 \text{ grams}$$

$$3c + 8k = 15$$

To show that the function $3c + 8k = 15$ is a linear function, we can solve for one of the variables.

$$3c + 8k = 15$$
$$8k = -3c + 15$$
$$k = \frac{-3c + 15}{8}$$
$$k = -\frac{3}{8}c + \frac{15}{8}$$

Since there is a constant rate of change ($-\frac{3}{8}$ cups of Kashi per tablespoon of Metamucil), the function is linear.

EXAMPLE 14 ■ Graphing a Linear Function in Standard Form

We choose to plot c on the horizontal axis and k on the vertical axis for the function $3c + 8k = 15$ from Example 13. Find the vertical and horizontal intercepts of the function and interpret their meanings in the given context. Then graph the function.

Solution To find the vertical intercept, we set $c = 0$ and solve for k.

$$3c + 8k = 15$$
$$3(0) + 8k = 15$$
$$0 + 8k = 15$$
$$8k = 15$$
$$k = \frac{15}{8}$$

The vertical intercept $\left(0, \frac{15}{8}\right)$ tells us that if the man only eats Kashi, he will need to eat $\frac{15}{8}$ cups to obtain 15 grams of fiber.

To find the horizontal intercept, we set $k = 0$ and find the value of c.

$$3c + 8k = 15$$
$$3c + 8(0) = 15$$
$$3c + 0 = 15$$
$$3c = 15$$
$$c = 5$$

The horizontal intercept $(5, 0)$ tells us that if the man only takes Metamucil, he must take 5 tablespoons to obtain 15 grams of fiber.

Using the two intercept points, we construct the graph shown in Figure 2.7. Each point on the line shows combinations of Metamucil and Kashi the man can take to obtain 15 grams of fiber.

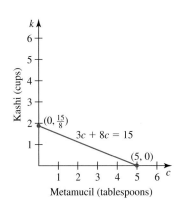

Figure 2.7

Since the intercepts for a linear function in standard form are so convenient for graphing the function, let's streamline the process of finding them.

$$Ax + By = C \qquad\qquad Ax + By = C$$
$$Ax + B(0) = C \qquad\qquad A(0) + By = C$$
$$Ax = C \qquad\qquad By = C$$
$$x = \frac{C}{A} \qquad\qquad y = \frac{C}{B}$$

Thus the horizontal intercept is $\left(\frac{C}{A}, 0\right)$ and the vertical intercept is $\left(0, \frac{C}{B}\right)$.

HOW TO: ■ GRAPH A LINEAR FUNCTION IN STANDARD FORM

To graph the linear function $Ax + By = C$,

1. Plot the horizontal intercept $\left(\frac{C}{A}, 0\right)$.
2. Plot the vertical intercept $\left(0, \frac{C}{B}\right)$.
3. Connect the two intercepts with a straight line.

EXAMPLE 15 ■ Graphing a Linear Equation in Standard Form

Graph $2x - 6y = 10$.

Solution We begin by finding the intercepts.

$$x = \frac{10}{2} \qquad\qquad y = \frac{10}{-6}$$
$$x = 5 \qquad\qquad y = -\frac{5}{3}$$

The horizontal intercept is $(5, 0)$ and the vertical intercept is $\left(0, -\frac{5}{3}\right)$. The graph is shown in Figure 2.8.

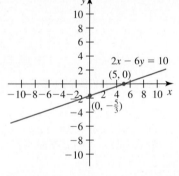

Figure 2.8

SUMMARY

In this section, you learned the meaning of constant rate of change and discovered that any function with a constant rate of change is a linear function. You learned how to create and graph linear functions given in words, graphs, tables, and formulas. Additionally, you learned how to use the three common forms of linear functions: slope-intercept, point-slope, and standard.

TECHNOLOGY TIP — GRAPHING A FUNCTION

1. Bring up the graphing list by pressing the $\boxed{Y=}$ button.

2. Type in the function(s) using the $\boxed{X,T,\theta,n}$ button for the variable.

3. Specify the size of the viewing window by pressing the $\boxed{WINDOW}$ button and editing the parameters. The Xmin is the minimum x-value, Xmax is the maximum x-value, Ymin in the minimum y-value, and Ymax is the maximum y-value. The Xscl and Yscl are used to specify the spacing of the tick marks on the graph. Xres refers to the resolution of the graph. Changing this value will change the speed at which the graph is drawn.

4. Draw the graph by pressing the $\boxed{GRAPH}$ button.

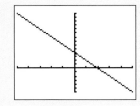

2.1 EXERCISES

SKILLS AND CONCEPTS

In Exercises 1–10, you are given information about a linear function. Determine the slope of the linear function and its vertical and horizontal intercepts.

1. $y = -4x + 10$
2. $y = 0.5x - 19$
3. $y = -\dfrac{4}{5}(x - 1) - \dfrac{1}{2}$
4. $y = (x + 6) + 7$
5. The line passing through (4, 0), (0, 8)
6. The line passing through (4, 4), (9, 9)
7. The line passing through (2, 9), (5, 4)
8. The line passing through (−15, 4), (21, 4)
9. $4x + 2y = -10$
10. $3x - 8y = 14$

In Exercises 11–16, determine the constant rate of change (slope) of the linear function and explain what it means in each context.

11. Annual fees, which are presently $55, are projected to increase by $5 per year for the next decade.
12. The car, originally valued at $12,800, has been decreasing in value at a constant rate over the past eight years. It is now worth $8200.
13. At a local community college, a student's tuition bill was $1008 for 9 credit hours. The student adds 4 credit hours and his tuition increases to $1456.
14. A high school basketball team notices that attendance at its games changes at a constant rate based on the number of losses the team has suffered. When the team had lost six games, 275 people attended the next game. When the team had lost 11 games, 180 people attended the next game.
15. Due to a salary freeze, my salary will remain the same for the next 3 years.

16. The store starts its retail workers at $7.00 per hour but guarantees fixed-value raises every 6 months. The manager says I will be making $10.00 per hour after working for the company for 3 years. (*Hint:* The wage is a linear function of the number of 6-month periods worked.)

In Exercises 17–26, you are given information about a linear function. For each exercise,

    a. Graph the linear function by hand.

    b. Determine the vertical and horizontal intercepts of the graph algebraically and plot them on the graph.

17. $y = -4x + 12$
18. $y = x + 3$
19. $y = 2x$
20. $y = 6$
21. $y = 1.5(x - 2) + 7$
22. $y = -3(x + 5) - 1$
23. $9x - 4y = 36$
24. $11x + 2y = 10$
25. The line passing through (3, 1) and (5, −1)
26. The line passing through (4, 2) and (6, 5)

In Exercises 27–33, you are given a pair of coordinates, a graph, or a verbal description of a linear function. For each exercise, find a formula that defines the function.

27. (0, 3), (4, 7)
28. (−5, 13), (5, 17)
29. (4, 27), (21, 27)

30.

31.

32.

33. **Vehicle Speed** A driver is presently driving 50 miles per hour and begins accelerating at a constant rate of 1 mile per hour per second. Model speed as a function of seconds elapsed since acceleration began.

In Exercises 34–37, you are given a table of values for a linear function. For each table, find the missing value, a.

34.

x	y
−2	7
10	a
14	55

35.

x	y
3	−2
6	−12
8	a

36.

x	y
6	87
a	83
11	77

37.

x	y
a	10.5
−9	15
−7	18

SHOW YOU KNOW

38. Explain why a horizontal line has zero slope and a vertical line has undefined slope.

39. Describe the connection between the often-used interpretation of slope as "rise over run" and the interpretation of slope as a number relating the relative changes in the output to the relative changes in the input.

40. When graphing linear functions, why does a greater magnitude slope create a steeper line? (Assume the scale of the graph remains unchanged.)

41. A classmate states that the slope-intercept form of a linear function can be thought of as a special case of the point-slope form of a linear function. Explain what your classmate means.

42. A classmate claims all linear functions have a vertical intercept. Is your classmate correct? Explain.

43. A classmate erroneously claims the terms *no slope*, *zero slope*, and *undefined slope* all mean the same thing. Explain the difference between each of the terms.

MAKE IT REAL

In Exercises 44–49,

a. Determine the vertical and horizontal intercepts of the graph of the implied linear function algebraically.

b. Using the intercepts from part (a), graph each of the linear functions by hand.

c. Explain the meaning of the intercepts in part (a) within the given context.

44. **Orange Prices** A devastating freeze in January 2007 destroyed roughly 75% of California Central Valley's orange crop. Market analysts predicted that as a result of the freeze, an orange that cost $0.50 before the freeze would cost $1.50 after the freeze. Model the price of an orange as a linear function of the percentage of the crop that was destroyed. (*Hint:* When 0% of the crop was destroyed, the price was $0.50.)

45. **Death Rate from Heart Disease** In 1980, the age-adjusted death rate due to heart disease was 412.1 deaths per 100,000 people. Between 1980 and 2003, the death rate decreased at a near-constant rate. In 2003, the death rate was 232.1 deaths per 100,000 people. (*Source: Statistical Abstract of the United States, 2006,* **Table 106**) Model the death rate due to heart disease as a linear function of years since 1980.

46. **McDonald's Dividends** Between 2000 and 2002, McDonald's Corporation dividends per share increased at a constant rate of $0.01 per year. In 2000, the dividend per share was $0.215 (*Source:* **McDonald's Investor Fact Sheet, January 2006**) Model the dividend per share as a linear function of years since 2000.

47. **Highway Signs** Many cross-country highway travelers are accustomed to seeing road signs warning of a steep downward slope such as the 8% decline shown in the figure. An 8% decline means that the road descends 8 feet for each 100 feet of horizontal distance traveled.

Given that the elevation of the road adjacent to the sign is 8240 feet, model the road elevation as a function of the horizontal distance from the sign. (Assume we are measuring the elevation as the road descends away from the sign.)

48. **Book Prices** In 2009, bestselling fiction books in mass-market paperback editions sold for about $8. At the same time, bestselling fiction hardcover editions sold for about $27. A reader wanted to spend about $100 on new books in some combination of paperback and hardcover editions. (*Source:* www.publishersweekly.com)

49. **Fiber Intake** One cup of Total Whole Grain cereal contains 4 grams of fiber. One scoop of ProFiber fiber supplement contains 6 grams of fiber. A person wants to consume 10 grams of fiber at breakfast using some combination of these products. (*Source:* product packaging)

For Exercises 50–56, define the variables and then write the linear function for each real-world situation described.

50. **Voting Age Population** Between 1980 and 2000, the number of U.S. residents of voting age increased at a nearly constant rate of 2.1 million people per year. In 1992, there were 189.5 million U.S. residents of voting age. Model the number of U.S. residents of voting age as a function of the number of years since 1980. (*Source: Statistical Abstract of the United States, 2006,* **Table 407**)

51. **Civil Service Retirement System** Between 1980 and 2000, the number of participants in the Federal Civil Servant Retirement System changed at a nearly constant rate. In 1995, there were 3.73 million participants. In 1999, there were 3.36 million participants. Model the number of Federal Civil Servant Retirement System participants as a function of years since 1980. (*Source: Statistical Abstract of the United States, 2006,* **Table 539**)

52. **Fiber Intake** Experts recommend eating several servings of fresh fruit each day to receive tremendous health benefits. Suppose you want to increase your daily fiber intake by 10 grams by consuming fresh blueberries and blackberries. Blueberries contain about 2 grams of fiber per serving and blackberries contain about 5 grams of fiber per serving. Model the combinations of blueberry and blackberry servings you could consume to meet your goal. (*Source:* www.uhs.wisc.edu)

53. **Income Tax Returns** Between 1990 and 2000, the number of personal income tax returns filed in the United States increased at a nearly constant rate of 1,555,000 returns per year. In 1990, the number of personal income tax returns that were filed was 109,868,000. Model the number of personal income tax returns filed as a function of years since 1990. (*Source: Statistical Abstract of the United States, 2006,* **Table 471**)

54. **Protein Intake** You are planning a meal of oven-roasted chicken breast with macaroni and cheese, and you want the meal to contain 8 grams of protein. Oven-roasted chicken breast contains about 0.17 grams of protein per gram of chicken and macaroni and cheese contains 0.03 grams of protein per gram. Model the combinations of grams of chicken and macaroni and cheese you could consume to meet your goal. (*Source:* www.highproteinfoods.com)

55. **Retirement Systems** During a 13-year period, the number of beneficiaries for state and local retirement systems changed at a nearly

constant rate compared to the amount of employee contributions. When there were $13.9 billion in employee contributions, there were 4.03 million beneficiaries. When there were $28.8 billion in employee contributions, there were 6.49 million beneficiaries. Model the number of beneficiaries as a function of the employee contributions.
(*Source: Statistical Abstract of the United States, 2006*, Table 541)

56. **Orthodontist Bill** One of the authors set up the following payment plan with his daughter's orthodontist.

    - Overall Treatment fee: $4950
    - Down payment, insurance payment, and records fee paid by author: $1980
    - Balance financed at 0% interest: $2970
    - Monthly payment: $135

    Write a formula for the loan balance as a function of the number of monthly payments that have been made.

■ **STRETCH YOUR MIND**

Exercises 57–61 are intended to challenge your understanding of linear functions.

57. Two lines are said to be *parallel* if they do not intersect at any point. If $f(x) = mx + b$ and $g(x) = nx + c$ are parallel lines, what conclusions can you draw about m, n, b, and c?

58. Two lines are said to be *perpendicular* if they intersect at a 90° angle (see the figure).

    a. What is the relationship between the slopes of the two perpendicular lines in the figure?
    b. Based on your result in part (a), what do you predict is the relationship between the slopes of any two perpendicular lines?
    c. Using a ruler and graph paper, draw several examples of perpendicular lines to test your prediction in part (b). Did your prediction hold true? Explain.

59. Explain why the equation of a vertical line cannot be written in slope-intercept form.

60. Do all lines represent functions? Explain.

61. **Slope Rating in Golf** *Slope rating* is a measurement of the difficulty of a golf course that makes it possible for poorer golfers to compete against better golfers in head-to-head matchups. The *handicap* of a golfer is a measurement of how good the golfer is compared to par (the number of strokes a round is expected to take without any mistakes).

    Imagine two golfers:

    - Golfer A is a very good golfer with a handicap of 3. This means that Golfer A would expect to earn a score of 3 over par. For example, on a par 75 course, she would expect to score 78.
    - Golfer B is a recreational golfer who only occasionally breaks a score of 100 so he has a handicap of 25 for a par 75 course.

    On a course of average difficulty (slope rating of 113), both golfers would be expected to shoot near their handicaps. But as the course difficulty increases, Golfer B's scores will rise faster than Golfer A's. On the very challenging Pebble Beach–Spyglass Hill Golf Course (slope of 147), Golfer B will need some help if the two golfers are to have a competitive match against each other. Otherwise Golfer A will most likely win easily.

    The number of shots a golfer is expected to shoot over par, E, is equal to the ratio of the course slope, s, to the slope of an average course (113) multiplied by the person's handicap, h,

    $$E(s, h) = \left(\frac{s}{113}\right)(h)$$

    For each of the following questions, assume that a golfer has a handicap of 25 ($h = 25$).

    a. What score will the golfer most likely shoot at Pebble Beach–Spyglass Hill Golf Course (par 72 and slope of 147)?
    b. What score will the golfer most likely shoot if he plays the relatively easy Greenfield Lakes Golf Club in Gilbert, Arizona (par 62 and slope of 91)?
    c. Solve $E(s, 25) = 30$ for s and then explain what your answer means in the context of golf courses.
    d. Explain how a golfer playing on a course with slope 110 could use the idea of constant rate of change to anticipate how his score will change on this course in comparison to playing an average course with slope 113.

    (*Hint:* Make use of the $\frac{s}{113}$ part in the $E(s, h)$ formula.)

SECTION 2.2 Modeling with Linear Functions

LEARNING OBJECTIVES

- Determine if two quantities are directly proportional
- Construct linear models of real-world data sets and use them to predict results
- Find the inverse of a linear function and interpret its meaning in a real-world context

GETTING STARTED

One challenge for international travelers is having to convert back and forth between metric and English units of measurement. For example, if the posted speed limit is 50 kilometers per hour, what is the speed in miles per hour? Knowing how to use a linear function to convert to miles per hour could save you the cost of a speeding ticket.

In this section, we continue our discussion of linear functions, refining the process of constructing linear functions from real-world data sets and scenarios. We also revisit the concept of inverse functions and demonstrate how to find the inverse of a linear function algebraically.

■ Recognizing When to Use a Linear Model

Several key phrases alert us to the fact that a linear model may be used to model a data set. Table 2.2 details how to interpret the mathematical meaning of some commonly occurring phrases.

Table 2.2

Phrase	Mathematical Meaning
Increasing at a rate of 20 people per year	The constant rate of change of the linear function is $20 \frac{\text{people}}{\text{year}}$.
Tickets cost $37 per person	The constant rate of change is $37 \frac{\text{dollars}}{\text{person}}$.
Sales decrease by 100 tickets for every $1 increase in price	The constant rate of change is $m = \frac{-100 \text{ tickets}}{1 \text{ dollar}} = -100$ tickets per dollar.
There are 350 students today. The number of students is increasing by 10 students per month.	The initial value is 350 students and the constant rate of change is $10 \frac{\text{students}}{\text{month}}$.
The price is $12.25 and is decreasing at a constant rate of $0.02 per day.	The initial value is $12.25 and the constant rate of change is $-0.02 \frac{\text{dollars}}{\text{per day}}$.

Some of the simplest linear models to construct are those that model direct proportionalities, where one quantity is a constant multiple of another quantity.

DIRECT PROPORTIONALITY

Two quantities are **directly proportional** when one quantity is a constant multiple of the other. That is,

$$y = kx \quad \text{for a constant } k$$

k is called the **constant of proportionality**.

Solving the equation $y = kx$ for k yields $k = \dfrac{y}{x}$. Thus another way to define direct proportionality is to say that two quantities are directly proportional if the output divided by the input is a constant.

EXAMPLE 1 ■ Constructing a Linear Model for Directly Proportional Quantities

The 1-month average retail price for gasoline in Texas was $3.219 per gallon on February 26, 2011. (*Source:* www.TexasGasPrices.com) Write a function that will give the total cost of a gasoline purchase as a function of the number of gallons purchased at this average retail price. Then calculate the cost of 20 gallons of gasoline.

Solution Let C be the total cost (in dollars) and g be the number of gallons purchased. Since the price per gallon is constant, the total cost will be the product of the price per gallon and the number of gallons purchased. That is,

$$C(g) = 3.219g$$

To determine the cost of 20 gallons of gasoline, we evaluate this function at $g = 20$.

$$C(20) = 3.219(20)$$
$$= \$64.38$$

Twenty gallons of gasoline cost $64.38.

One of the most well-known examples of direct proportionality is the relationship between the diameter of a circle, d, and its circumference, C. Scholars and builders in antiquity knew that the circumference of a circle was directly proportional to its diameter, that is $C = kd$, yet the value of the constant of proportionality, k, eluded them for centuries. Ultimately, the number $\pi \approx 3.1416$ was discovered to be the constant of proportionality k, giving us the familiar formula $C = \pi d$. (See the Peer into the Past for details on the history of π.)

Direct proportions can also be used to convert from English to metric units of measure and vice versa. One such conversion is demonstrated in Example 2.

EXAMPLE 2 ■ Using Direct Proportionality to Convert Units of Measure

The speed of 100 kilometers per hour is roughly equivalent to 62.14 miles per hour. Additionally, 0 kilometers per hour is equivalent to 0 miles per hour. Find a function that converts kilometers per hour into miles per hour. Then calculate the speed in miles per hour that is equivalent to 50 kilometers per hour.

Solution We let the variable m represent miles per hour and the variable x represent kilometers per hour. Since x is the independent variable and m is the dependent variable, ordered pairs will be of the form (x, m). We have two points: (100, 62.14) and (0, 0). We calculate the slope of the line between the two points.

PEER INTO THE PAST

THE NUMBER π

From ancient times, it has been known that the ratio of the circumference of a circle to its diameter, π, is a very special value. Estimating the value of π has quite a history and has even had a competitive flair. As far back as 2000 B.C. the Babylonians and Egyptians had estimates for π of $\dfrac{25}{8} \approx 3.125$ and $\dfrac{256}{81} \approx 3.160$, respectively. The Chinese ($\approx$1200 B.C.) as well as the Israelites ($\approx$550 B.C.) estimated π as 3 (see 1 Kings 7:23 in the Bible). Archimedes ($\approx$300 B.C.) estimated an average of 3.1463, and Fibonacci (A.D. 1220) arrived at 3.141818. In more modern times yet prior to computer technology, Ludolph van Ceulen (A.D. 1596) calculated π to 35 correct decimal places, Machin (A.D. 1766) to 100, Richter (A.D. 1855) to 500, and Ferguson (A.D. 1947) to 808 correct decimal places. Since the advent of supercomputers, we now know π to more than 10 billion decimal places. (*Source:* http://mathforum.org/isaac/problems/pi2.html)

On a humorous note, March 14 (the date 3/14 and Albert Einstein's birthday) has been dubbed Pi Day. Mathematics educators, students, and enthusiasts around the country take some time to celebrate π. Fun festivities include pie-eating and pi-memorization contests. For information on Pi Day go to http://en.wikipedia.org/wiki/Pi_day.

$$\text{slope} = \frac{62.14 - 0}{100 - 0} \frac{\text{miles per hour}}{\text{kilometers per hour}}$$

$$= 0.6214 \text{ miles per kilometer} \quad \text{The "per hour" terms cancel out.}$$

Since the vertical intercept is (0, 0), the linear model in condensed function notation is

$$m(x) = 0.6214x$$

Consequently, kilometers per hour and miles per hour are directly proportional.

To determine the speed in miles per hour that is equivalent to 50 kilometers per hour, we evaluate this function at $x = 50$.

$$m(50) = 0.6214(50)$$
$$\approx 31$$

Therefore, 50 kilometers per hour is approximately equal to 31 miles per hour.

As pointed out earlier, identifying key phrases in a verbal expression helps us recognize when linear relationships exist. The next example focuses on recognizing this.

EXAMPLE 3 ■ Constructing a Linear Model from a Verbal Description

T-Mobile's Even More™ 500 Talk plan cost $39.99 per month in 2011. The plan included free evening and weekend minutes and 500 whenever minutes. Additional minutes used cost $0.45 per minute. (*Source:* www.t-mobile.com)

a. Write a function that will give the monthly cell phone cost as a function of the number of additional minutes used over 500 minutes. (T-Mobile defines *additional minutes* to be the minutes used beyond a subscriber's monthly allotment.)

b. Calculate the cost of using a total of 600 non-weekend minutes.

Solution In solving any word problem, identifying the meaning of all variables is critical. We let c be the monthly cell phone cost (in dollars) and m be the number of additional minutes used. Since the monthly cost depends on the number of additional minutes used, c is the dependent variable and m is the independent variable.

a. The first indicator that this is a linear function is the constant rate of change: $0.45 per minute. This will be the slope of the linear function. The $39.99 fixed value is the initial value of our function. We have

$$c(m) = 0.45m + 39.99$$

b. We are asked to find the cost of using 600 non-weekend minutes. (Since weekend minutes are free, they will not affect the monthly cell phone cost.) Since 500 minutes are included with the plan, 100 additional minutes will be used. We evaluate the function at $m = 100$.

$$c(100) = 0.45(100) + 39.99$$
$$= 45 + 39.99$$
$$= 84.99$$

It costs $84.99 to use 600 non-weekend minutes. Note that although only 20% more minutes were used than the plan allowed, the monthly cost more than doubled.

EXAMPLE 4 ■ Constructing a Linear Model from a Table of Data

Table 2.3 shows the total undergraduate nonresident tuition and fees cost per semester for students enrolled in the Golf Management Program at Arizona State University—Polytechnic in 2006–2007.

Table 2.3

Nonresident	Enrolled Hours				
	7	8	9	10	11
Nonresident Undergraduate Tuition	$4592	$5248	$5904	$6560	$7216
Program Tuition	600	600	600	600	600
Financial Aid Trust	22	22	22	22	22
Association of Students of AZ	1	1	1	1	1
Total Undergraduate Nonresident Tuition & Fees	$5215	$5871	$6527	$7183	$7839

Source: www.asu.edu

a. Find a function that models the total undergraduate nonresident tuition and fees as a function of the number of enrolled hours for students taking 7–11 credits.

b. Predict the cost of enrolling in 10.5 credit hours.

Solution We first identify the variables, letting t represent the total tuition and fees cost and h represent the enrolled hours. Since the total tuition and fees cost depends on the number of enrolled hours, t is the dependent variable and h is the independent variable.

a. Observe that the program tuition, financial aid trust, and Association of Students of Arizona fees do not change as the number of enrolled hours increases. These fees total $623 regardless of whether a student enrolls in 7 or 11 credit hours. Note also that the nonresident undergraduate tuition increases. If the tuition increases at a constant rate, this rate will be the slope of a linear function. We construct Table 2.4 to determine the rate of change.

Table 2.4

Enrolled Hours	Nonresident Undergraduate Tuition	Change in Nonresident Undergraduate Tuition
7	$4592	
8	$5248	$5248 − $4592 = $656
9	$5904	$5904 − $5248 = $656
10	$6560	$6560 − $5904 = $656
11	$7216	$7216 − $6560 = $656

Table 2.4 shows that the tuition increases at a constant rate of $656 per enrolled credit hour. Consequently, the total nonresident undergraduate tuition and fees may be modeled by a linear function.

When we add the $623 fee to the $4592 tuition cost for 7 credit hours, we get $5215. Thus, the point (7, 5215) is on the line. With this point and the slope of $656 per credit hour, we can construct the linear model.

$$\underbrace{t - 5215}_{\text{tuition and fees above \$5215}} = \underbrace{656}_{\substack{\text{tuition} \\ \text{increase} \\ \text{per credit} \\ \text{hour}}} \underbrace{(h - 7)}_{\substack{\text{credit} \\ \text{hours} \\ \text{above 7}}}$$

Since we are going to use this model to forecast a value of t, we solve for t.

$$t = 656(h - 7) + 5215$$

Note that the practical domain for this model is $7 \leq h \leq 11$ since the data only included values of h between 7 and 11. Forecasting the cost of tuition and fees for credit hours outside of this range may not be accurate.

b. To determine the cost of 10.5 credit hours, we evaluate this function at $h = 10.5$.

$$t(h) = 656(h - 7) + 5215$$

$$t(10.5) = 656(10.5 - 7) + 5215$$

$$= 7511$$

We predict it costs $7511 to enroll in 10.5 credit hours of classes. (This assumes that tuition is prorated for partial credits.)

■ Inverses of Linear Functions

In Chapter 1 we introduced the concept of inverse functions. Recall the domain of a function f was the range of its inverse function f^{-1} and the range of the function f was the domain of its inverse function f^{-1}. This notion is represented as follows.

$$x \rightarrow \boxed{f} \rightarrow y = f(x)$$

$$f^{-1}(y) = x \leftarrow \boxed{f^{-1}} \leftarrow y$$

The phrase "y is the function of x" corresponds with the phrase "x is the inverse function of y." Symbolically, $y = f(x)$ is related to $x = f^{-1}(y)$. Recognizing this relationship between a function and its inverse is critical for a deep understanding of inverse functions, especially in a real-world context.

Let's investigate inverse functions in the context of the tuition and fees equation we created in Example 4. If we let $t = f(h)$, we can write the tuition and fees equation in slope-intercept form as

$$t = 656h + 623$$

which we can represent as

$$\textit{enrolled hours} \rightarrow \boxed{f} \rightarrow \textit{tuition and fees}$$

We would like to find a function that reverses the process. That is, we want a function that converts *tuition and fees* into *enrolled hours*. This function will be the inverse function f^{-1}, as represented here:

$$\textit{enrolled hours} \leftarrow \boxed{f^{-1}} \leftarrow \textit{tuition and fees}$$

Fortunately, we can easily find the inverse function f^{-1} by solving the equation $t = 656h + 623$ for h.

$$t = 656h + 623$$

$$t - 623 = 656h$$

$$h = \frac{t - 623}{656}$$

$$h = \frac{1}{656}t - \frac{623}{656}$$

Notice the independent variable of this new function is t and the dependent variable is h. In inverse function notation, we write the inverse as

$$f^{-1}(t) = \frac{1}{656}t - \frac{623}{656} \quad \text{\small since } h = f^{-1}(t)$$

Let's check to see if this function does indeed convert *tuition and fees* into *enrolled hours*. From Table 2.3, we know that 8 enrolled hours cost $5871 so we expect our inverse function to give us the same result. We write the equation in inverse function notation and evaluate the function at $t = 5871$.

$$f^{-1}(5871) = \frac{1}{656}(5871) - \frac{623}{656}$$

$$f^{-1}(5871) = \frac{5871}{656} - \frac{623}{656}$$

$$f^{-1}(5871) = \frac{5248}{656}$$

$$f^{-1}(5871) = 8$$

$$t = 8 \quad \text{since } f^{-1}(h) = t$$

Thus, $5871 will cover the cost of nonresident undergraduate tuition and fees for 8 credit hours. This agrees with the information presented in Table 2.3.

Many students struggle with the concept of inverse functions. To help you get a better grasp on this concept, we will work a few straightforward examples before summarizing the process of finding an inverse of a linear function.

EXAMPLE 5 ■ Finding the Inverse of a Linear Function

Find the inverse of the function $y = 4x - 2$.

Solution In the function we are given, x is the independent variable and y is the dependent variable. We solve this equation for x.

$$y = 4x - 2$$

$$y + 2 = 4x$$

$$\frac{y + 2}{4} = x$$

$$x = \frac{1}{4}y + \frac{1}{2}$$

In this new equation, y is the independent variable and x is the dependent variable. We can write the original function as

$$f(x) = 4x - 2 \quad \text{since } y = f(x)$$

Similarly, we can write the inverse function as

$$f^{-1}(y) = \frac{1}{4}y + \frac{1}{2} \quad \text{since } x = f^{-1}(y)$$

The function $f^{-1}(y) = \frac{1}{4}y + \frac{1}{2}$ is the inverse of $f(x) = 4x - 2$.

EXAMPLE 6 ■ Finding the Inverse of a Linear Function

We can convert degrees Celsius into degrees Fahrenheit using the equation

$$F = \frac{9}{5}C + 32$$

where F is degrees Fahrenheit and C is degrees Celsius. Given that $F = f(C)$, find the function f^{-1}. Then calculate $f^{-1}(50)$ and interpret what it means in its real-world context.

Solution We solve the equation $F = \frac{9}{5}C + 32$ for C.

$$F = \frac{9}{5}C + 32$$

$$F - 32 = \frac{9}{5}C$$

$$C = \frac{F - 32}{\frac{9}{5}}$$

$$C = \frac{5}{9}F - \frac{160}{9}$$

$$f^{-1}(F) = \frac{5}{9}F - \frac{160}{9} \quad \text{since } f^{-1}(F) = C$$

The input to f^{-1} is degrees Fahrenheit and the output is degrees Celsius. $f^{-1}(50)$ will convert 50 degrees Fahrenheit into degrees Celsius.

$$f^{-1}(50) = \frac{5}{9}(50 - 32)$$

$$= \frac{5}{9}(18)$$

$$= 10$$

Thus 50 degrees Fahrenheit is the same as 10 degrees Celsius.

We now summarize the process of finding the inverse of a linear function. Observe that since dividing by 0 is undefined, this process only works for linear functions with nonzero slopes.

HOW TO: ■ FIND THE INVERSE OF A LINEAR FUNCTION

1. Write the function $f(x) = mx + b$ as $y = mx + b$.
2. Solve the function equation for x.

$$y = mx + b$$

$$y - b = mx$$

$$x = \frac{y - b}{m}$$

$$x = \frac{1}{m}y - \frac{b}{m}$$

3. Write the inverse function using the notation

$$f^{-1}(y) = \frac{1}{m}y - \frac{b}{m}$$

Recall that horizontal lines can be written in the form $y = b$. The inverse of a horizontal line is a vertical line $x = a$; however, a vertical line is not a function. Therefore, only linear functions with nonzero slopes have an inverse *function*.

It is customary in many textbooks to write the inverse of a function $y = f(x)$ as $y = f^{-1}(x)$. Some teach that to find the inverse function we interchange the x and y variables and solve the resultant equation for y. But this strategy does not make sense when working with inverses in a real-world context. For example, we stated earlier that the function $f(C) = \frac{9}{5}C + 32$ converts degrees Celsius into degrees Fahrenheit and the function $f^{-1}(F) = \frac{5}{9}F - \frac{160}{9}$ converts degrees Fahrenheit into degrees Celsius. What does $f^{-1}(C) = \frac{5}{9}C - \frac{160}{9}$ do? Nothing except confuse us! It has no meaning in a real-world context because the input to the inverse function must be degrees Fahrenheit, not degrees Celsius. By replacing the variable F with the variable C in the inverse function, we completely obscured the real-world meaning of the functions. For this reason, we have chosen to use a less-traditional, more meaningful approach to working with inverses.

SUMMARY

In this section you refined your ability to construct linear functions from real-world data sets and scenarios. You also learned how to calculate and interpret average rates of change. Finally, you learned how to find the inverse of a linear function algebraically.

2.2 EXERCISES

SKILLS AND CONCEPTS

In Exercises 1–5, find the inverse of the function $y = f(x)$. Write your solution in inverse function notation, $f^{-1}(y)$.

1. $y = -4x + 10$
2. $y = 0.5x - 19$
3. $y = 9x - 18$
4. $y = 6x - 15$
5. $y = \frac{1}{3}x + 4$

SHOW YOU KNOW

6. What are some key phrases that indicate that a verbal description can be modeled by a linear function?
7. What does it mean to say two quantities are *directly proportional*?
8. There is one type of linear function that does not have an inverse function. What type of linear function is it and why does it not have an inverse function?
9. Explain what it means for one function to be the inverse of another function. Use the terms *domain* and *range* as a part of your explanation. Use diagrams as appropriate.
10. Why don't we find the inverse function by interchanging the x and the y and solving for y when working with functions in a real-world context?

MAKE IT REAL

In Exercises 11–18, determine if a linear function will be a good model for the data set or verbal description. If the data appears to be linear or nearly linear, state why and find a linear function that models the data set. Otherwise, explain why a linear model will not be a good fit.

11. **Gas Prices** On January 20, 2007, gasoline in Ozark, Alabama, was priced at $1.889 per gallon. (*Source:* www.alabamagasprices.com)
    Model the total cost of the gasoline as a function of the number of gallons purchased.

12. **Gas Prices** On January 20, 2007, gasoline in Queens, New York, was priced at $2.339 per gallon. (*Source:* www.newyorkgasprices.com)
    Model the total cost of the gasoline as a function of the number of gallons purchased.

13. **Retail vs. Wholesale** A six-pack of Stylish Plaid Capri Pants featured at wholesaleclothingmart.com cost $103.50 in 2007. In the retail clothing industry, it is customary to mark up the wholesale price of an item by 100%. (That is, the retail price is double the wholesale price.)
    Model the total retail revenue in 2007 as a function of the number of pairs of pants sold at retail price.

Jason Stitt/Shutterstock.com

14. **Retail vs. Wholesale** A six-pack of Stylish Pleated Maternity Tops with Lace Trim featured at wholesaleclothingmart.com cost $93 in 2007. In the retail clothing industry, it is customary to mark up the wholesale price of an item by 100%. (That is, the retail price is double the wholesale price.)
    Model the total retail revenue from tops in 2007 as a function of the number of tops sold at retail price.

15. **Restaurant Expenses** The total company restaurant expenses incurred by Burger King Holdings, Inc., between 2004 and 2006 are given in the table. Model the expenses as a function of the fiscal year.

Fiscal Year	Total Company Restaurant Expenses ($ millions)
2004	1087
2005	1195
2006	1296

Source: **Burger King Annual Report, 2006**

16. McDonald's Restaurants The total revenues earned by McDonald's restaurants between 2000 and 2005 are given in the table. Model total revenues as a function of the fiscal year.

Fiscal Year	Total Revenues ($ millions)
2000	14,243
2001	14,870
2002	15,406
2003	17,140
2004	19,065
2005	20,460

Source: **McDonald's Investor Fact Sheet, Jan. 2006**

17. Children in Madagascar Model the number of children under 5 years of age as a function of years since 1990.

Years Since 1990 t	Children under 5 (thousands) C
9:0	2120
7	2630
8	2707
9	2787
10	2859
11	2946
12	3036

Source: **World Health Organization**

18. Prescription Drug Spending Model the per capita prescription drug spending as a function of years since 1990.

Years Since 1990 t	Per Capita Spending on Prescription Drugs (dollars) P
0	158
5	224
8	311
9	368
10	423
11	485
12	552
13	605

Source: Statistical Abstract of the United States, 2006, **Table 121**

In Exercises 19–20, assume that the real-world context can be accurately modeled with a linear function.

19. Aging U.S. Population In 1950, the number of people age 65 and older who lived in the United States was 12 million. By 2005, that number had grown to 37 million people. (*Source: Health United States 2006*, p. 16)

Model the number of people who are age 65 and older as a linear function of the number of years since 1950.

20. Life Expectancy in the United States Between 1900 and 2003, the life expectancy for men and women in the United States increased dramatically. For men, life expectancy increased from 48 to 75 years. For women, life expectancy increased from 51 to 80 years. (*Source: Health United States 2006*, p. 30)

Model male life expectancy as a function of female life expectancy.

21. Frozen Oranges A devastating freeze in California's Central Valley in January 2007 wiped out approximately 75% of the state's citrus crop. According to an Associated Press news report, 40-pound boxes of oranges that were selling for $6 before the freeze were selling for $22 after the freeze.

A linear model for the price of a box of oranges as a function of the percentage of the citrus crop that was frozen is given by

$$P = f(F) = \frac{16}{75} F + 6$$

where P is the price of a 40-pound box of oranges and F is the percentage of the citrus crop that was frozen.

Find the inverse function. Then determine the value of $f^{-1}(14)$ and interpret the real-world meaning of the result.

22. Leaning Tower of Pisa Construction of the Tower of Pisa was completed in 1360. By 1990, the tilt of the tower was so severe that it was closed for renovation. Renovators were able to reduce the tower's 1990 tilt by 17 inches. The resultant tower leans 13.5 feet (162 inches) off the perpendicular. When the tower was reopened in 2001, officials forecast that it would take 300 years for the tower to return to its 1990 tilt (*Source: TIME Magazine*, June 25, 2001)

a. Construct a linear formula that models the lean of the renovated tower, where l is the number of inches from the perpendicular and t is the number of years since 2001.

b. Use the formula from part (a) to predict the lean of the tower in 2100.

23. **Smoking and Heart Disease** Based on data from 1974 to 2003, the death rate due to heart disease in the United States (in deaths per 100,000 people) can be modeled by

$$D = f(p) = 14.08p - 53.87$$

where p is the percentage of people who smoke (*Source: Modeled from CDC and Census Bureau data*)

Find the inverse function. Then determine the value of $f^{-1}(101)$ and interpret the real-world meaning of the result.

24. **Snow Runoff** Based on data from June 2006, the forecast for maximum 5-day snow runoff volumes for the American River at Folsom, CA, can be modeled by

$$v = f(t) = -2.606t + 131.8$$

thousand acre-feet, where t is the number of days since the end of May 2006. (*Source: Modeled from National Weather Service data*) That is, the model forecasts the snow runoff for the 5-day period beginning on the selected day of June 2006.

Find the inverse function. Then determine the value of $f^{-1}(61.4)$ and interpret the real-world meaning of the result.

25. **Marketing Labor Costs** Based on data from 1990 to 2003, the marketing labor cost for farm foods may be modeled by

$$L = f(t) = 10.47t + 146.8 \text{ billion dollars}$$

where t is the number of years since 1990. (*Source: Modeled from Statistical Abstract of the United States, 2006, Table 842*)

Find the inverse function. Then determine the value of $f^{-1}(251.5)$ and interpret the real-world meaning of the result.

■ **STRETCH YOUR MIND**

Exercises 26–29 are intended to challenge your understanding of linear function models and inverse functions.

26. A linear function model passes through every point of a data set. Will the model perfectly forecast unknown function values? Explain.

27. A classmate claims function modeling is a waste of time since models do not always accurately predict the future. Provide a convincing argument to refute this claim.

28. A classmate claims $f^{-1}(x)$ is equivalent to $\dfrac{1}{f(x)}$. Do you agree? Explain.

29. What are some techniques you can use to quickly determine whether or not a data table represents a linear or nearly linear function?

SECTION 2.3 Linear Regression

LEARNING OBJECTIVES

- Use linear regression to find the equation of the line of best fit
- Use a linear regression model to make predictions
- Explain the meaning of the correlation coefficient (r) and the coefficient of determination (r^2)

GETTING STARTED

In 1938, the U.S. Congress passed the Fair Labor Standards Act, which was signed into law by President Franklin Roosevelt. The intent of this act was to eliminate "labor conditions detrimental to the maintenance of the minimum standards of living necessary for health, efficiency and well-being of workers." (*Source:* www.dol.gov) The federal minimum wage, begun as part of this act, required that employers pay their workers $0.25 per hour in 1938. Over time, the minimum wage has increased in an attempt to keep up with the rising costs of goods and services. In 2011, the federal minimum wage was $7.25 per hour. (*Source:* www.dol.gov) Information such as this can be modeled using linear regression.

In this section we use linear regression to determine the equation of the linear function that best fits a corresponding data set. We then use linear regression models to make predictions. We also discuss ways to determine how well the linear model represents a data set.

Linear Regression

The federal minimum wage has increased since its inception in 1938. Table 2.5 gives the hourly minimum wage each decade beginning in 1950.

Table 2.5

Year	Years Since 1950 t	Federal Minimum Wage (dollars) W
1950	0	0.75
1960	10	1.00
1970	20	1.60
1980	30	3.10
1990	40	3.80
2000	50	5.15

Source: www.dol.gov

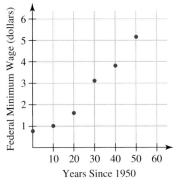

Figure 2.9

Let's investigate the relationship between the quantities given in the table—namely, the value of t (years since 1950) and W (federal minimum wage). We begin by examining the scatter plot of these data shown in Figure 2.9.

We see that a relationship between the two quantities does exist: As the number of years since 1950 increases, the federal minimum wage increases. We can also describe this relationship quantitatively using a linear model, as in Example 1.

EXAMPLE 1 ■ Selecting a Linear Function to Model a Data Set

Determine which of the three linear function graphs in Figure 2.10 best models the federal minimum wage data. Then use the graph that best models the data to predict the minimum wage in 2010.

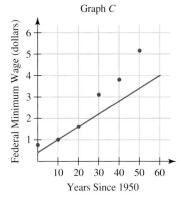

Figure 2.10

Solution We are looking for a linear function that "goes through" the middle of the scatter plot as much as possible. Graph A connects the first and last data points but does not represent the other data points well. Graph C connects the second and third data points but also does not represent the other data points well. It appears as though Graph B is the linear function that best models the federal minimum wage data.

To use the model to predict the minimum wage in 2010, we must find the value of W when $t = 60$ since 2010 is 60 years after 1950. Using Graph B as shown in Figure 2.11, we estimate the minimum wage to be $5.20.

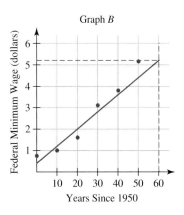

Figure 2.11

Least Squares

Our choice of Graph B in Example 1 was based on a visual analysis only. To mathematically determine which is truly the line of best fit, we compute the *sum of squares*. The **sum of squares** is calculated by finding the difference between each output of the data set and the corresponding "predicted" value from the linear model, squaring each of these differences, and then adding them up. (We must square the differences since some of the differences will be positive and some will be negative. Adding them without squaring them would result in cancellation of positive and negative values.) Each of these individual differences is called a **residual** because it measures the "residue" in our prediction (the function value) from the actual data value of the given data. The line whose sum of squares is closest to 0 is called the **line of best fit**.

To illustrate, we calculate the sum of squares for each of the three graphs of Example 1 by using the raw data and the equation of each line. The results are shown in Tables 2.6, 2.7, and 2.8.

Table 2.6 Graph A

Years Since 1950 t	Federal Minimum Wage (dollars) W	Linear Model (dollars) $A(t) = 0.088t + 0.75$	Residual $(W - A)$	Square of Residual $(W - A)^2$
0	0.75	0.75	0	0
10	1.00	1.63	−0.63	0.3969
20	1.60	2.51	−0.91	0.8281
30	3.10	3.39	−0.29	0.0841
40	3.80	4.27	−0.47	0.2209
50	5.15	5.15	0	0
			Total Sum of Squares	**1.5300**

Table 2.7 Graph B

Years Since 1950 t	Federal Minimum Wage (dollars) W	Linear Model (dollars) $C(t) = 0.0783t + 0.50$	Residual $(W - B)$	Square of Residual $(W - B)^2$
0	0.75	0.50	0.25	0.0625
10	1.00	1.28	−0.28	0.0784
20	1.60	2.07	−0.47	0.2209
30	3.10	2.85	0.25	0.0625
40	3.80	3.63	0.17	0.0289
50	5.15	4.42	0.73	0.5329
			Total Sum of Squares	**0.9861**

Table 2.8 Graph C

Years Since 1950 t	Federal Minimum Wage (dollars) W	Linear Model (dollars) $C(t) = 0.06t + 0.40$	Residual $(W - C)$	Square of Residual $(W - C)^2$
0	0.75	0.40	0.35	0.1225
10	1.00	1.00	0	0.0000
20	1.60	1.60	0	0.0000
30	3.10	2.20	0.90	0.8100
40	3.80	2.80	1.00	1.0000
50	5.15	3.40	1.75	3.0625
			Total Sum of Squares	4.9950

Of the three sums of the squares of the residuals, Graph B has the smallest sum (0.9861). Thus Graph B is the most accurate of the three graphs.

Interpolation and Extrapolation

One reason we want to find the linear model that most accurately represents the actual data is so that our predictions will be as accurate as possible.

> **INTERPOLATION AND EXTRAPOLATION**
>
> - **Interpolation** is the process of predicting the output value for an input value that lies between the maximum and minimum input values of the data set.
> - **Extrapolation** is the process of predicting the output value for an input value that comes before the minimum input value or after the maximum input value of a data set.

To highlight these concepts, we use the three models from Graph A, Graph B, and Graph C from Example 1 to predict the federal minimum wage in 1966 (interpolation) and in 2010 (extrapolation). In comparing the values generated by the three models, we will see the importance of determining the most accurate model.

EXAMPLE 2 Predicting Values Using a Linear Model

Using each of the linear models for the federal minimum wage given in Tables 2.6–2.8, predict the minimum wage in 1966 and in 2010. Then compare the values produced by each model.

Solution Since 1966 is 16 years after 1950, we substitute $t = 16$ into each function to predict the federal minimum wage in 1966. Since 2010 is 60 years after 1950, we substitute $t = 60$ into each function to predict the federal minimum wage in 2010. The differing results are shown in Table 2.9.

Table 2.9

Years Since 1950 t	Model A $A(t) = 0.088t + 0.75$	Model B $B(t) = 0.0783t + 0.50$	Model C $C(t) = 0.06t + 0.40$
16	2.16	1.75	1.36
60	6.03	5.20	4.00

Model A projects the highest minimum wage and Model C predicts the lowest minimum wage in 1966 and 2010. Since Model B was the line of best fit, we conclude that Model B is the most accurate predictor of the three models.

■ Linear Regression Model

To find the line of best fit, we could do as we have done so far in this section: draw several lines that we think best fit the scatter plot of the data, compute the total sum of the squares of the differences, and choose the model that produces the smallest sum. But how would we know we had found the best line? We do not want to go through the least squares process for every possible model! Fortunately, we can use technology to easily find the line of best fit, which is also called the **linear regression model**. The Technology Tip at the end of this section details the steps needed to find this linear model.

> ### LINEAR REGRESSION MODEL
> The equation of the line that best fits a data set, as determined by the least value of the total sum of the squares of the residuals, is known as the **linear regression model** or the **least squares regression line**.

PEER INTO THE PAST
REGRESSION

Why is the process of finding the line of best fit called "linear regression"? During the 1870s, Sir Francis Galton investigated the relationship between the average height of parents and that of their offspring. What Galton observed and recorded was that the offspring of particularly tall parents were also tall—but not as tall as their parents. The offspring of particularly short parents were also short—but not as short as their parents. That is, the offspring of these parents tended to be less tall or less short; they *regressed* toward the mean height of the population.

We see this "regression toward the mean" in many real-world situations. For example, if a basketball player scores an extraordinarily high number of points in one game, he most likely will not score as many points in the next game. The number of points will "regress toward the mean" or be closer to the player's average number of points per game.

Sir Francis Galton
1822–1911

EXAMPLE 3 ■ Determining a Linear Regression Model

Use linear regression to determine the equation of the line of best fit for the federal minimum wage data. Describe the relationship between the quantities t (years since 1950) and W (federal minimum wage, in dollars) using this regression model.

Solution   First, we use a graphing calculator to determine the equation of the line of best fit. (Refer to the Technology Tip at the end of this section for detailed instructions on computing the linear regression model.) Screen shots from the steps are shown in Figures 2.12 and 2.13.

Figure 2.12

Figure 2.13

We find $W(t) = 0.091t + 0.288$ is the equation of the line of best fit. Using this model, we predict that in 1950 the minimum wage was $0.288 and that it increases by $0.091 each year.

In Table 2.10 we confirm that the line in Example 3 has the least squares value when compared to those found in Tables 2.6–2.8 for Models A–C.

Table 2.10

Years Since 1950 t	Federal Minimum Wage (dollars) W	Linear Model (dollars) $F(t) = 0.091t + 0.288$	Residual $(W - F)$	Square of Residuals $(W - F)^2$
0	0.75	0.29	0.46	0.2116
10	1.00	1.20	−0.20	0.0400
20	1.60	2.11	−0.51	0.2601
30	3.10	3.02	0.08	0.0064
40	3.80	3.93	−0.13	0.0169
50	5.15	4.84	0.31	0.0961
			Total Sum of Squares of Residuals	0.6311

We see this linear regression model produce the smallest least squares value when compared to those computed for the linear models representing Graph *A* (1.5300), Graph *B* (0.9861), and Graph *C* (4.995).

Coefficient of Determination

You may have noticed that the graphing calculator outputs a value, r^2, when computing a linear regression model, as shown in Figure 2.14.

This value, known as the **coefficient of determination**, describes the strength of the fit of a linear regression model to a set of data. The stronger the fit, the closer this value, r^2, is to 1.

Figure 2.14

COEFFICIENT OF DETERMINATION

The **coefficient of determination**, r^2, is a value that describes the strength of fit of a linear regression model to a set of data. The closer the value of r^2 is to 1, the stronger the fit.

EXAMPLE 4 ■ Computing a Linear Regression Model and Coefficient of Determination

Private philanthropy is the act of donating money to support a charitable cause. For example, many colleges and universities accept private philanthropy to fund scholarships for financially needy students.

Table 2.11 shows the amount of money donated by U.S. residents, corporations, and foundations for philanthropic purposes between 2000 and 2003.

a. Find the linear regression model for the data.
b. Interpret the meaning of the slope and the vertical intercept of the model in terms of private philanthropy funds, *F*, and years since 2000, *t*.
c. Use the model to predict how much money in private philanthropy funds will be given in 2008.
d. State the coefficient of determination.

Table 2.11

Years Since 2000 t	Funds ($ billion) F
0	227.7
1	229.0
2	234.1
3	240.7

Source: Statistical Abstract of the United States, 2006, Table 570

Solution

a. We use a graphing calculator to find the linear regression model, as shown in Figure 2.15.

Figure 2.15

The linear regression model is $P(t) = 4.41t + 226.26$.

b. The model suggests that in 2000 ($t = 0$), $226.26 billion in private philanthropy funds was given and the amount increases each year at a rate of $4.41 billion per year.

c. Since 2008 is 8 years after 2000, we substitute $t = 8$ into the regression model.

$$P(t) = 4.41t + 226.26$$
$$P(8) = 4.41(8) + 226.26$$
$$= 261.54$$

We predict that $261.54 billion in private philanthropy funds will be given in 2008.

d. The coefficient of determination is approximately 0.93.

Correlation Coefficient

Another way to measure the strength of fit of a linear regression model to a data set involves the **correlation coefficient**, r. When using a graphing calculator for linear regression, the value of r is shown below the value of r^2 (see Figure 2.14). The computation of this number requires statistical knowledge and will not be addressed here, but we will explain how to interpret it.

The correlation coefficient, r, is a value such that $-1 \leq r \leq 1$. If $r = 1$, we have a perfectly linear data set with a positive slope. If $r = -1$, we have a perfectly linear data set with a negative slope. If $r = 0$, we say there is no correlation between the input and output.

> **CORRELATION COEFFICIENT**
>
> The **correlation coefficient**, r, is a value that describes the strength of fit of a linear regression model to a set of data. The closer the value of $|r|$ is to 1, the better the fit. When $r > 0$, the regression line is increasing. When $r < 0$, the regression line is decreasing.

EXAMPLE 5 ■ **Interpreting the Correlation Coefficient**

Match each graph with the correct correlation coefficient value and explain your reasoning.

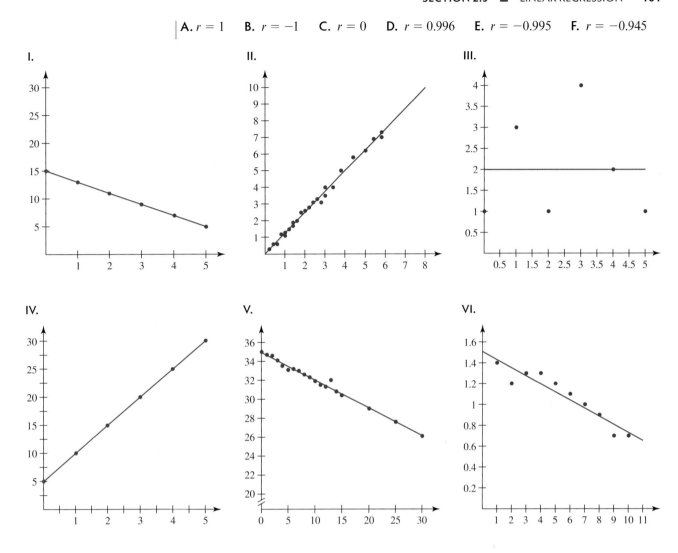

Solution I—B, II—D, III—C, IV—A, V—E, VI—F

Graph I is perfectly linear with a negative slope so the correlation coefficient is $r = -1$. Graph II, although not perfectly linear, seems to have a strong linear fit with a positive slope. Therefore, the correlation coefficient must be $r = 0.996$. Graph III seems to have no correlation. As the dependent variable value increases, the independent variable value neither increases (indicating a positive correlation) nor decreases (indicating a negative correlation). Therefore, the correlation coefficient must be $r = 0$. Graph IV is perfectly linear with a positive slope so the correlation coefficient is $r = 1$. Graph V shows a negative correlation, as does Graph VI. However, Graph V seems to be more strongly fit to the linear model shown than Graph VI. Therefore, Graph V must have a correlation coefficient of $r = -0.995$ and Graph VI must have a correlation coefficient of $r = -0.945$.

SUMMARY

In this section you learned how to use linear regression to find the line of best fit, or linear regression model, for a data set. You also learned that the process of predicting an unknown data value using a linear model is called interpolation or extrapolation, depending on the value of the input variable. And you discovered that both the coefficient of determination and the correlation coefficient indicate the strength of fit of the line of best fit to the data set.

TECHNOLOGY TIP ■ CREATING LISTS OF VALUES FOR A DATA SET

1. Bring up the Statistics Menu by pressing the [STAT] button.

2. Bring up the List Editor by selecting **1:EDIT** and pressing [ENTER].

3. Clear the lists. If there exists data in the list, use the arrows to move the cursor to the list heading, **L1**. Press the [CLEAR] button and press [ENTER]. This clears all of the list data. Repeat for each list with data. (Warning: Be sure to use [CLEAR] instead of [DELETE]. [DELETE] removes the entire column.)

4. Enter the numeric values of the *inputs* in list **L1** and press [ENTER] after each entry.

5. Enter the numeric values of the *outputs* in list **L2** and press [ENTER] after each entry. When complete, press [2ND] then [MODE] to exit out of the List Editor.

TECHNOLOGY TIP ■ DRAWING A SCATTER PLOT

1. Bring up the Statistics Plot Menu by pressing the [2ND] button then the [Y=] button.

2. Open **Plot1** by pressing [ENTER].

3. Turn on **Plot1** by moving the cursor to **On** and pressing [ENTER]. Confirm that other menu entries are as shown.

4. Graph the scatter plot by pressing [ZOOM] and scrolling to **9:ZoomStat**. Press [ENTER]. This will graph the entire scatter plot along with any functions in the Graphing List. The ZoomStat feature automatically adjusts the viewing window so that all of the data points are visible.

TECHNOLOGY TIP — LINEAR REGRESSION

1. Enter the numeric values of the *inputs* and *outputs* of the data set using the List Editor.

2. Return to the Statistics Menu by pressing the [STAT] button.

3. Bring up the Calculate Menu by using the arrows to select [CALC].

4. Select 4:LinReg(ax+b). To automatically paste the regression equation into the Y= Editor, press [VARS], Y-VARS, 1:Function, 1:Y1.

5. Press [ENTER]. The line of best fit is $y = 0.6008x + 2.693$ and has coefficient of determination $r^2 = 0.9999$.

Optional Step:
If the correlation coefficient and coefficient of determination do not appear, do the following:
   Press [2ND] then [0], scroll to DiagnosticOn and press [ENTER] twice. This will ensure that the correlation coefficient r and the coefficient of determination r^2 will appear the next time you do a regression.

2.3 EXERCISES

SKILLS AND CONCEPTS

In Exercises 1–5, compute the total sum of the squares of the differences between the data points and each of the linear models given. Then use the result to determine which linear model, y_1 or y_2, best fits the data.

1. $y_1 = \dfrac{7}{4}x + 1$ or $y_2 = \dfrac{3}{2}x + 2$

x	y
0	1
2	5
4	6
6	12
8	16
10	19
12	20
14	25
16	29

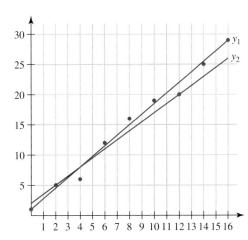

2. $y_1 = -\dfrac{5}{2}x + 10$ or $y_2 = -\dfrac{7}{3}x + 10$

x	y
0	10
1	8
2	4
3	3
4	0

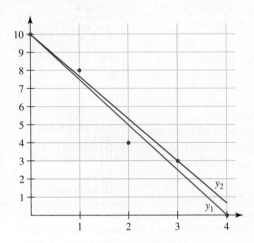

4. $y_1 = \dfrac{13}{8}x + \dfrac{75}{4}$ or $y_2 = 2x$

x	y
10	35
20	40
30	75
40	70
50	100

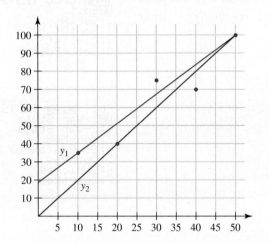

3. $y_1 = 17.5x + 12.5$ or $y_2 = 30.5x - 180$

x	y
5	100
10	125
15	320
20	430
25	450

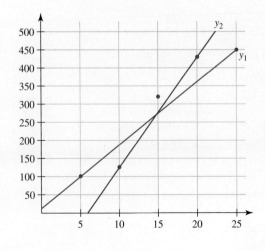

5. $y_1 = -3x + 15$ or $y_2 = -3x + 13.9$

x	y
0	15
1	10
2	9
3	4
4	2

■ SHOW YOU KNOW

6. What is linear regression?
7. Explain what the value of the coefficient of determination indicates.
8. Explain the difference between interpolation and extrapolation.
9. Why is it important to look at the coefficient of determination or correlation coefficient before using a linear model to interpolate or extrapolate?
10. If the coefficient of determination of a model is 1, what is the relationship between the model and the original data set?

■ MAKE IT REAL

11. **Taxable Income** The data in the table show the amount of taxes to be paid by married couples filing a joint tax return for the given taxable income amounts.

Taxable Income ($ thousands) I	Tax Paid (dollars) T
20	2249
30	3749
40	5249
50	6749
60	8249
70	10,621
80	13,121
90	15,621

Source: 2006 IRS Tax Table, www.irs.gov

   a. Create a linear regression model for these data.
   b. Using the model found in part (a), interpret the practical meaning of the slope and vertical intercept of the model.
   c. Use the linear regression model to predict the amount of tax paid if a married couple's income is $72,000. Is this an example of interpolation or extrapolation? Explain.
   d. Use the linear regression model to predict the amount of tax paid if a married couple's income is $110,000. Is this an example of interpolation or extrapolation? Explain.
   e. Based on the coefficient of determination, do you think the interpolation or extrapolation values you found are accurate?

12. **Households with Televisions** The data in the table show the number of households in the United States with televisions.
   a. Create a linear regression model for these data.
   b. Using the model found in part (a), interpret the practical meaning of the slope and vertical intercept of the model.
   c. Use the linear regression model to predict the number of households with televisions in 1985. Is this an example of interpolation or extrapolation? Explain.
   d. Use the linear regression model to predict the number of households with televisions in 2010. Is this an example of interpolation or extrapolation? Explain.

Years Since 1980 t	Number of Households (in millions) H
0	76
10	92
15	95
17	97
18	98
20	101
21	102
22	106
23	107

Source: Statistical Abstract of the United States, 2006, Table 1117

   e. Based on the coefficient of determination, do you think the interpolation or extrapolation values you found are accurate?

13. **Carbon Monoxide Pollutant Concentrations** The data in the table show the carbon monoxide pollutant concentration, in parts per million, for years since 1990. The data are based on 359 monitoring stations from various locations in the United States.

Years Since 1990 t	Carbon Monoxide Pollutant Concentration (parts per million) P
0	6
5	4.7
9	3.9
10	3.4
11	3.2
12	3
13	2.8

Source: Statistical Abstract of the United States, 2006, Table 359

   a. Create a linear regression model for these data.
   b. Using the model found in part (a), interpret the practical meaning of the slope and vertical intercept of the model.
   c. Use the linear regression model to predict the carbon monoxide pollutant concentration in 1992. Is this an example of interpolation or extrapolation? Explain.
   d. Use the linear regression model to predict the carbon monoxide pollutant concentration in 2010. Is this an example of interpolation or extrapolation? Explain.
   e. Based on the coefficient of determination, do you think the interpolation or extrapolation values you found are accurate?

14. Home Run Hitting

The table shows the cumulative number of home runs hit by Luis Gonzalez up to the All-Star break in 2006.

Game Number g	Cumulative Number of Home Runs Hit H
4	1
5	2
12	3
13	4
16	5
76	6
87	7

Source: www.mlb.com

a. Create a linear regression model for these data.

b. Using the model found in part (a), interpret the practical meaning of the slope and vertical intercept of the model.

c. Use the linear regression model to predict the number of home runs hit by Luis Gonzales by the 100th game of the season. Is this an example of interpolation or extrapolation? Explain.

d. Use the linear regression model to predict the number of home runs hit by Luis Gonzales by the final game of the season, game 162. Is this an example of interpolation or extrapolation? Explain.

e. Based on the coefficient of determination, do you think the interpolation or extrapolation values you found are accurate?

15. Famous Skyscrapers

The data in the table show the number of stories and the height of the 10 tallest skyscrapers in the world.

Skyscraper	Number of Stories n	Height (feet) h
Taipei 101	101	1667
Petronas Tower 1	88	1483
Willis Tower	110	1451
Jin Mao Building	88	1381
Two International Finance Centre	88	1362
CITIC Plaza	80	1283
Shun Hing Square	69	1260
Empire State Building	102	1250
Central Plaza	78	1227

Source: www.infoplease.org

a. Create a linear regression model for these data.

b. Using the model found in part (a), interpret the practical meaning of the slope of the model.

c. Use the linear regression model to predict the height of a new building that has 50 stories. Is this an example of interpolation or extrapolation? Explain.

d. Use the linear regression model to predict the height of a new building that has 120 stories. Is this an example of interpolation or extrapolation? Explain.

e. Based on the coefficient of determination, do you think the interpolation or extrapolation values you found are accurate?

16. Manatee Population

Because the Florida manatee population is threatened, the Florida Manatee Sanctuary Act of 1978 was enacted to protect the species. Scientists interested in the relationship between the number of manatee deaths and time collected the data shown in the table.

a. Create a scatter plot of these data.

b. Find the linear regression model for these data.

c. Referring to the coefficient of determination, r^2, and the correlation coefficient, r, explain whether or not the linear model represents the situation well.

d. Use the linear regression model to predict the year in which the number of manatee deaths will be 450.

Year	Manatee Deaths	Year	Manatee Deaths
1974	7	1991	174
1975	25	1992	163
1976	62	1993	146
1977	114	1994	192
1978	84	1995	201
1979	77	1996	416
1980	63	1997	242
1981	116	1998	232
1982	114	1999	269
1983	81	2000	272
1984	128	2001	325
1985	119	2002	305
1986	122	2003	380
1987	114	2004	276
1988	133	2005	396
1989	168	2006	416
1990	206		

Source: research.myfwc.com

17. Manatee Population

Because many manatee deaths can be attributed to a boating incident, scientists are interested in the relationship between the number of manatee deaths and the number of registered boats in Florida. Use the data set to respond to the items that follow.

Year	Registered Boats (in thousands)	Manatee Deaths
1990	716	206
1991	715	174
1992	710	163
1993	730	146
1994	748	192
1995	766	201
1996	789	416
1997	803	242
1998	824	232

Source: research.myfwc.com

a. Create a scatter plot of the number of manatee deaths as a function of the number of registered boats (in thousands).

b. Find the linear regression model for these data.

c. Explain the practical meaning of the slope and vertical intercept of the model.

d. Referring to the coefficient of determination, r^2, and the correlation coefficient, r, explain whether or not the linear model represents the situation well.

e. Suppose the number of registered boats increased to 3 million boats. Use the linear regression model to predict the number of manatee deaths. Discuss the accuracy of your result.

18. **Chipotle Mexican Grill** The table shows Chipotle Mexican Grill's costs for food, beverage, and packaging along with labor costs for the indicated years.

Years Since 2000 t	Food, Beverage, and Packaging Costs ($ thousands) f	Labor Costs ($ thousands) L
1	45,236	46,048
2	67,681	66,515
3	104,921	94,023
4	154,148	139,494
5	202,288	178,721

Source: Chipotle Mexican Grill, Inc., 2005 Annual Report, p. 24

a. Create a scatter plot of the labor costs as a function of food, beverage, and packaging costs.

b. Find the linear regression model for these data. Use the model to describe the labor costs and the food, beverage, and packaging costs.

c. Referring to the coefficient of determination, r^2, and the correlation coefficient, r, explain whether or not the linear model represents the situation well.

d. Suppose the labor costs for Chipotle Mexican Grill were 200,000 ($ thousands). Predict the food, beverage, and packaging costs associated with this labor cost. Discuss the accuracy of this result.

19. **Cigarette Smoking** The data in the table give the percentage of people who smoke cigarettes for selected years.

a. Create a scatter plot of the percentage of people who smoke as a function of the year.

b. Find the linear regression model for these data.

c. Referring to the coefficient of determination, r^2, and the correlation coefficient, r, explain whether or not the linear model represents the situation well.

d. In what year does the model predict that the percentage of people who smoke will be 0%? Discuss the accuracy of this result.

Year t	People Who Smoke Cigarettes (percent) P
1974	36.9
1979	33.1
1983	31.6
1985	30
1987	28.8
1988	28.1
1990	25.4
1991	25.8
1992	26.3
1993	24.7
1994	24.9
1995	24.5
1997	24
1998	23.4
1999	22.7
2000	22.6
2001	22
2002	21.4
2003	21.1

Source: www.cdc.gov

20. **Heart Disease and Cigarette Smoking** The data in the table give the percentage of people who smoke cigarettes and the death rate (deaths per 100,000 people) caused by heart disease.

Year t	People Who Smoke Cigarettes (percent) P	Heart Disease Death Rate (per 100,000) D
1974	36.9	458.8
1979	33.1	401.6
1983	31.6	388.9
1985	30	375
1987	28.8	355.9
1988	28.1	352.5
1990	25.4	321.8
1991	25.8	313.8
1992	26.3	306.1
1993	24.7	309.9
1994	24.9	299.7
1995	24.5	296.3
1997	24	280.4
1998	23.4	272.4
1999	22.7	267.8
2000	22.6	257.6
2001	22	247.8
2002	21.4	240.8
2003	21.1	232.1

Source: www.cdc.gov and *Statistical Abstract of the United States, 2006,* Table 106)

a. Create a scatter plot of heart disease death rate as a function of the percentage of people who smoke.

b. Find the linear regression model for these data.

c. Explain the practical meaning of the slope of the model.

d. Use the linear regression model to predict the death rate due to heart disease if 10% of people smoke cigarettes.

21. **Golf Ball Collecting** An Arizona man living in a home alongside the fairway of a golf course collected the balls he found in his yard from January 2008 to January 2009 and created the following table of data.

Months Since Jan. 2008 m	Average Number of Golf Balls Found per Day G
0	5.80
1	5.25
2	7.93
3	12.33
4	13.32
5	12.53
6	13.75
7	14.94
8	16.28
9	15.38
10	13.33
11	13.07
12	12.77

Source: Data collected by Jim Simpson, Gilbert, Arizona

a. Create a scatter plot of average number of golf balls found as a function of the months since January 2008.

b. Find the linear regression model for these data.

c. Explain the practical meaning of the slope of the model.

d. Use the linear regression model to predict the average number of golf balls the man will find in March 2010.

■ **STRETCH YOUR MIND**

For Exercises 22–26, the residual plot for a linear model is shown. A residual plot has the same independent variable as the scatter plot of the original data set; however, the dependent variable is the residual (the difference in the actual output and the predicted output). Based on the graph of the residual plot, explain whether or not the linear model fits the original data set well.

22.

23.

24.

25.

26.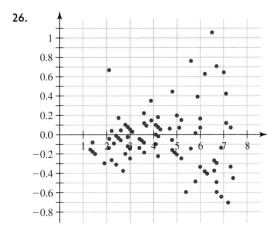

SECTION 2.4 Systems of Linear Equations

LEARNING OBJECTIVES

- Determine the solution to a system of equations algebraically, graphically, and using technology and interpret the real-world meaning of the results

- Use the substitution and elimination methods to solve linear systems that model real-world scenarios

- Determine if systems of linear equations are dependent or inconsistent and explain the real-world meaning of the results

GETTING STARTED

In many real-world contexts, it is common to work with more than one equation at the same time. For example, a snack company combines several ingredients to create a batch of a trail mix of a specific volume. The volume of the mix is equal to the sum of the volumes of the individual ingredients, and the relationship between the volumes can be represented by an equation. The company may also have nutritional targets related to the fat, protein, and sugar content of the mix. The company can create equations that relate the nutritional value of the ingredients to the nutritional value of the mix. To determine how much of each ingredient to put in the mix, the company would solve a volume equation and a nutritional-value equation as a system of equations.

In this section we solve systems of linear equations using a variety of methods. We also discuss how to determine if a system is dependent or inconsistent and discover the real-world meaning of those terms.

■ Systems of Linear Equations

A **system of equations** is a group of two or more equations. To solve a system of equations means to find values for the variables that satisfy all of the equations in the system. Systems of equations can involve any number of equations and variables; however, we will limit ourselves to situations containing two variables in this section.

SYSTEM OF LINEAR EQUATIONS IN TWO VARIABLES

A **system of linear equations** in two variables is a group of two or more linear equations that use the same variables.

To develop an understanding of systems of linear equations, let's look at some variables that affect the cost of a taxi ride. Taxis play a key role in cities around the world, ferrying tourists, taking people to and from airports, and even saving lives by keeping intoxicated people from driving. The cost of hiring a taxi varies from city to city and usually depends on how far a person travels and how much time the taxi spends waiting (in traffic or for a client).

Table 2.12 compares the costs for hiring a taxi in Seattle, Washington, and Dallas, Texas, in July 2009.

Table 2.12

City	First Mile	Cost per Additional Mile	Cost per Minute Waiting
Seattle	$2.50	$2.00	$0.50
Dallas	$4.00	$1.80	$0.30

Source: www.taxifarefinder.com

We will simplify our discussion by assuming that a typical taxi ride involves 5 minutes of wait time. Using this assumption, the wait-time cost in Seattle is $2.50 and in Dallas is $1.50.

EXAMPLE 1 ■ Exploring a System of Equations Using a Table of Values

Suppose we hired a taxi in Seattle and in Dallas. Using the fare data in Table 2.12, create a table of values and estimate the number of additional miles that must be driven for the trip costs to be equal.

Solution We begin with the fixed costs.

Seattle: first mile cost + wait-time cost = $2.50 + $2.50 = $5.00

Dallas: first mile cost + wait-time cost = $4.00 + $1.50 = $5.50

Each additional mile costs $2.00 in Seattle and $1.80 in Dallas. So the variable costs are $2.00 × additional miles in Seattle and $1.80 × additional miles in Dallas. To create a table of values, we note the calculations necessary to determine the total cost of a trip. In Seattle, we begin with a $5.00 charge and add an additional $2.00 for each mile traveled. For example, suppose we traveled 3 additional miles. Then

$$\text{Seattle taxi cost} = \$5.00 + \frac{\$2.00}{1 \text{ mi}} \cdot 3 \text{ mi}$$

$$= \$5.00 + \frac{\$2.00}{1 \text{ mi}} \cdot 3 \text{ mi}$$

$$= \$5.00 + \$2.00(3)$$

$$= \$11.00$$

Using similar calculations for Dallas, we create Table 2.13.

Table 2.13

Additional Miles Traveled	Cost in Seattle	Cost in Dallas
0	$5.00	$5.50
1	$7.00	$7.30
2	$9.00	$9.10
3	$11.00	$10.90
4	$13.00	$12.70
5	$15.00	$14.50

We notice the cost for hiring a taxi in Seattle, which is less expensive for a trip of less than 2 additional miles, becomes more expensive by the time we have traveled 3 additional miles. However, it is not clear exactly how far we must travel for the cost in Seattle to equal the cost in Dallas. We can get a better idea by creating a table for values between 2 and 3 additional miles, as shown in Table 2.14. To create the table, we use the cost equation for Seattle, Cost = 5.00 + 2(additional miles), and the cost equation for Dallas, Cost = 5.50 + 1.80(additional miles).

Table 2.14

Additional Miles Traveled	Cost in Seattle	Cost in Dallas
2	$9.00	$9.10
2.2	$9.40	$9.46
2.4	$9.80	$9.82
2.5	$10.00	$10.00
2.6	$10.20	$10.18
2.8	$10.60	$10.54
3.0	$11.00	$10.90

Now we can see that at 2.5 additional miles the cost in Seattle and the cost in Dallas are both $10.00. The *solution to the system of linear equations* is (2.5, 10.00).

SOLUTION TO A SYSTEM OF EQUATIONS IN TWO VARIABLES

The **solution to a system of linear equations** in two variables (if it exists) is a pair of values (x, y) such that when x is used as the input, each equation returns the same y-value. Graphically this will appear as a point (x, y) where the lines of all the equations intersect.

■ Solving a System Using Graphs

One common approach to solving a system of equations is to graph all of the equations simultaneously and find the point of intersection. We do this in Example 2.

EXAMPLE 2 ■ Solving a System of Equations Graphically

Write a system of linear equations to represent the cost of hiring a taxi in Seattle and the cost of hiring a taxi in Dallas as functions of the number of additional miles traveled. Then graph the functions on the same set of axes and discuss what information the graphs reveal.

Solution We let S represent the cost in dollars of hiring a taxi in Seattle, D represent the cost in dollars of hiring a taxi in Dallas, and x represent the number of additional miles traveled. Combining the fixed and variable costs for each city, we get

$$\text{Seattle: } S(x) = 2.00x + 5.00$$
$$\text{Dallas: } D(x) = 1.80x + 5.50$$

We now graph these functions as shown in Figure 2.16. The graphs have been restricted to domains of $x \geq 0$ because the number of additional miles traveled must be nonnegative to make sense.

The two lines appear to intersect at (2.5, 10). We estimate this to be the solution of the system; however, we can use algebra to verify if this is an exact solution. We will do this in Example 3.

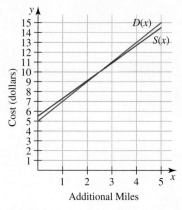

Figure 2.16

■ Solving a System Using the Substitution Method

For many systems of equations, the solution includes "messy" numbers that are difficult to determine from a table or graph. Fortunately, most graphing calculators have an *intersect* command that allows us to quickly determine the point of intersection of two graphs. We detail this process in the Technology Tip at the end of the section.

EXAMPLE 3 ■ **Solving a System Using the Substitution Method**

Use algebra to solve the system of equations in Example 2.

Solution From Example 2, we have the system

$$S(x) = 2.00x + 5.00$$
$$D(x) = 1.80x + 5.50$$

We also know a solution to this system means that hiring a taxi in either city costs exactly the same for some number of additional miles traveled. This tells us that we want to find the value for x such that $S(x) = D(x)$. In other words, we want

$$\underbrace{S(x)}_{\substack{\text{the cost of hiring a} \\ \text{taxi in Seattle for} \\ x \text{ miles beyond the} \\ \text{first mile}}} \underbrace{=}_{\text{is the same as}} \underbrace{D(x)}_{\substack{\text{the cost of hiring a} \\ \text{taxi in Dallas for} \\ x \text{ miles beyond the} \\ \text{first mile}}}$$

$$S(x) = D(x)$$
$$2.00x + 5.00 = 1.80x + 5.50$$

This method of solving a system of equations is known as the **substitution method** because we are substituting the expressions $2.00x + 5.00$ and $1.80x + 5.50$ for S and D in the relationship $S(x) = D(x)$. We solve the equation.

$$2.00x + 5.00 = 1.80x + 5.50$$
$$0.20x + 5.00 = 5.50$$
$$0.20x = 0.50$$
$$x = 2.5$$

A hired taxi traveling 2.5 additional miles will cost the same in either city. We verify the answer by substituting this value into each equation of the original system.

$$S(x) = 2.00x + 5 \qquad\qquad D(x) = 1.80x + 5.50$$
$$S(2.5) = 2.00(2.5) + 5 \qquad\qquad D(2.5) = 1.80(2.5) + 5.50$$
$$= 10.00 \qquad\qquad\qquad\qquad = 10.00$$

The solution to the system is (2.5, 10.00). When the taxi travels 2.5 additional miles the cost is $10.00 in both cities.

HOW TO: ■ USE THE SUBSTITUTION METHOD

To solve a system of linear equations of the form

$$y = ax + b$$
$$y = cx + d$$

using the substitution method,

1. Replace the value of y in the first equation with $cx + d$. This creates a new equation $cx + d = ax + b$.
2. Solve the new equation $cx + d = ax + b$ for x. The solution will be $x = \dfrac{c - a}{b - d}$.
3. Substitute the value of x back into either the first or second equation and solve for y.

If the system of equations models real-world data, you should also complete Steps 4 and 5.

4. Ask yourself if the mathematical solution makes sense in its real-world context.
5. Reevaluate the functions, as necessary, and verbally express the real-world meaning of the result.

We have demonstrated how to solve a system of equations using a table, a graph, and algebraic methods. There is another clever approach worth considering. We observe there is a $0.50 difference ($5.50 − $5.00 = $0.50) in the cost for the first mile in each city. We also note the cost for each additional mile differs by $0.20 per mile ($2.00 − $1.80 = $0.20). For the total cost to be the same, the taxi must travel enough additional distance that the difference in fares will make up for the difference in initial costs. Therefore, we divide $0.50 by $0.20 per mile.

$$\frac{\$0.50}{\$0.20/\text{mile}} = 2.5 \text{ miles}$$

Note this approach does not require a table, graph, or formal equations. However, it does require that we have a solid understanding of the situation.

We have just seen how each of the four representations of a function—table, graph, formula, and verbal description (with a few easy calculations)—can be used to find the solution to a system. Each method has its own advantages and disadvantages. Being flexible enough to work with all four representations gives us more ways to approach a problem and provides us with a deeper understanding of what the solution means in a given context. Note also that in many real-world scenarios, an exact solution to a system of equations is not necessary; an estimate from a table of values or a graph is often good enough.

EXAMPLE 4 ■ Solving a System of Equations

Solve the following system of equations using the substitution method. Then check your answer by graphing.

$$y = -3x + 2$$
$$y = 0.5x - 34.75$$

Solution We know we want the outputs of both functions to be the same for a given input. For sake of clarity, we label the equations, making them $y_1 = -3x + 2$ and $y_2 = 0.5x - 34.75$. We want to know the value of x that makes $y_1 = y_2$. Then

$$y_1 = y_2$$
$$-3x + 2 = 0.5x - 34.75$$
$$2 = 3.5x - 34.75$$
$$36.75 = 3.5x$$
$$x = 10.5$$

We substitute 10.5 for x in either original equation to get the y-value that completes the solution. It doesn't matter which equation we use. To demonstrate, we use both.

$y = -3x + 2$	$y = 0.5x - 34.75$
$y = -3(10.5) + 2$	$y = 0.5(10.5) - 34.75$
$y = -31.5 + 2$	$y = 5.25 - 34.75$
$y = -29.5$	$y = -29.5$

Thus, the solution to this system of equations is $(10.5, -29.5)$. We graph the system of equations to check our work, as shown in Figure 2.17.

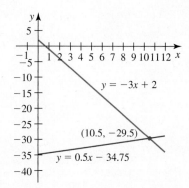

Figure 2.17

The graph confirms that our solution is correct. (Note: If you are using a calculator and the intersection point does not appear, adjust your viewing window to make sure that the solution is included in the domain and range of the viewing window.)

Business Applications of Linear Systems

Business analysts are often interested in profit, revenue, and costs of a company. In evaluating a business plan, it is important to know at what sales level revenue and costs are expected to be equal. This sales level is referred to as the **break-even point** and is the point at which the company begins to turn a profit.

Any business has two types of costs: *fixed costs* and *variable costs*. **Fixed costs** are those that remain constant regardless of production levels. For example, building rent, product research and development, and advertising are usually fixed costs. **Variable costs** are those that vary with the level of production. For example, raw materials and production-line worker wages are variable costs.

Break-even analysis may be applied to large companies or small home-based businesses. An example of one such business is given in Example 5.

EXAMPLE 5 ■ Determining a Break-Even Point

An artisan wants to sell her handmade craft angels online. She estimates her material cost for each angel to be \$3.50. As of January 2007, the online merchant craftmall.com charged a \$14.95 per month fee for a Premier account featuring up to 25 products. Comparing her craft to similar crafts on the market, the artisan estimates she can sell the craft angel for \$9.95. How many angels will she have to sell each month to break even? At that production level, what will be her production cost, revenue, and profit?

Solution The cost equation for the craft angels is the sum of the variable cost, \$3.50 per angel, and the fixed cost, \$14.95. Let a be the number of angels sold in a month. The cost equation is

$$C(a) = 3.50a + 14.95 \text{ dollars}$$

The revenue equation is

$$R(a) = 9.95a \text{ dollars}$$

We want to determine when her revenue will equal her cost. In other words, we want to find the value of a such that $R(a) = C(a)$. Graphing the two functions simultaneously results in the graphs shown in Figure 2.18.

It appears the graphs intersect near (2.3, 23) at the intersection point, $R(a) = C(a)$. We can find the exact point of intersection by using the substitution method.

$$R(a) = C(a)$$
$$9.95a = 3.50a + 14.95$$
$$6.45a = 14.95$$
$$a = \frac{14.95}{6.45}$$
$$a \approx 2.318$$

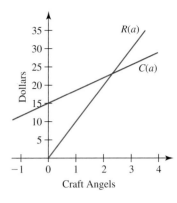

Figure 2.18

We evaluate $R(a)$ at $a = 2.318$ and determine that

$$R(2.318) = 9.95(2.318)$$
$$\approx 23.06$$

Thus the break-even point is roughly (2.318, 23.06). In the context of the problem, though, it does not make sense to talk about 2.318 angels. So we conclude she must sell 3 angels per month to cover her costs.

The cost to produce and advertise 3 angels is

$$C(3) = 3.50(3) + 14.95$$
$$= \$25.45$$

The revenue from the sale of 3 angels is

$$R(3) = 9.95(3)$$
$$= \$29.85$$

She will profit \$4.40 if she sells 3 angels (\$29.85 − \$25.45 = \$4.40).

It is essential when solving a real-world problem (such as the ones shown in Examples 1–3 and 5) to make sure the solution makes sense in the context of the problem. A common error among students is to accept a mathematically correct answer (e.g., (2.318, 23.06)) without verifying that it makes sense in the context of the problem.

EXAMPLE 6 ■ Solving a System in Standard Form Using Substitution

Suppose in January 2009 a local community college ordered replacement computers for employees in two departments, math and history. For the math department, a purchase of \$17,427.89 was made that consisted of 6 Dell OptiPlex 360 desktops and 5 Dell Latitude E4200 laptops. For the history department, a purchase of \$13,211.91 was made that consisted of 6 Dell OptiPlex 360 desktops and 3 Dell Latitude E4200 laptops. (*Source:* www.dell.com)

a. Write a system of equations for this situation and explain what each term in the equations represents.

b. Solve the system algebraically and explain what the answer means in its real-world context.

Solution

a. The unknowns are the prices per computer for the desktop and laptop models purchased. Let d be the price in dollars for a Dell OptiPlex 360 desktop computer and let t be the price in dollars for a Dell Latitude E4200 laptop. We use the information given to create the following equations.

$$\underbrace{6d}_{\text{the total cost of 6 desktops}} + \underbrace{5t}_{\text{the total cost of 5 laptops}} \underbrace{=}_{\text{must equal}} \underbrace{17{,}427.89}_{\text{the total cost of the order for the math department}}$$

$$\underbrace{6d}_{\text{the total cost of 6 desktops}} + \underbrace{3t}_{\text{the total cost of 3 laptops}} \underbrace{=}_{\text{must equal}} \underbrace{13{,}211.91}_{\text{the total cost of the order for the history department}}$$

b. Solving this system will tell us the prices the community college paid for each type of computer. To use the substitution method, we will solve the first equation for t and substitute this expression into the second equation.

$$6d + 5t = 17{,}427.89$$
$$5t = 17{,}427.89 - 6d$$
$$t = \frac{17{,}427.89 - 6d}{5}$$

Now we use the second equation and substitute the above expression for t.

$$6d + 3t = 13{,}211.91$$

$$6d + 3\left(\frac{17{,}427.89 - 6d}{5}\right) = 13{,}211.91 \qquad \text{Substitute } \frac{17{,}427.89 - 6d}{5} \text{ for } t.$$

$$6d + \frac{3(17{,}427.89) - 3(6d)}{5} = 13{,}211.91$$

$$6d + \frac{52{,}283.67 - 18d}{5} = 13{,}211.91$$

$$5(6d) + 5\left(\frac{52{,}283.67 - 18d}{5}\right) = 5(13{,}211.91) \qquad \text{Multiply through by 5 to eliminate the fraction.}$$

$$30d + \cancel{5}\left(\frac{52{,}283.67 - 18d}{\cancel{5}}\right) = 66{,}059.55$$

$$30d + 52{,}283.67 - 18d = 66{,}059.55$$
$$12d + 52{,}283.67 = 66{,}059.55$$
$$12d = 13{,}775.88$$
$$d = 1147.99$$

This tells us the college paid $1147.99 for each desktop. We substitute this value for d in either original equation to determine the price for a laptop.

$$6d + 5t = 17{,}427.89$$
$$6(1147.99) + 5t = 17{,}427.89$$
$$6887.94 + 5t = 17{,}427.89$$
$$5t = 10{,}539.95$$
$$t = 2107.99$$

Therefore, the solution is $d = 1147.99$ and $t = 2107.99$. This means each Dell OptiPlex 360 Desktop cost $1147.99 and each Dell Latitude E4200 laptop cost $2107.99.

■ Solving a System Using the Elimination Method

Example 6 demonstrates the challenges of using substitution to solve a system of equations in standard form. However, there is another method that works much better for such systems.

To illustrate, let's return to the situation in Example 6. The math department and history department both ordered 6 desktops. We assume they paid the same price per computer, so the total price for each order was different only because the number of laptops purchased was different. The difference in order totals was $4215.98, and there was a difference of 2 laptops purchased. This means that 2 laptops must cost $4215.98. Therefore,

$$2t = 4215.98$$
$$t = 2107.99$$

Plugging this answer back into one of the original equations will tell us the price for a desktop, d, and we arrive at the same solution found in Example 6 much more efficiently.

Let's examine this approach in a more formal algebraic way. By taking the difference of the number of laptops ordered and the difference of the order totals we were able to create the equation $2t = 4215.98$, which was easy to solve.

$$\begin{aligned} 6d + 5t = 17{,}427.89 &\rightarrow 6d + 5t = 17{,}427.89 \\ -(6d + 3t = 13{,}211.91) &\rightarrow -6d - 3t = -13{,}211.91 \\ \hline & 0d + 2t = 4215.98 \end{aligned}$$

There is no difference in the number of desktops purchased, but there is a difference of 2 laptops.

There was a $4215.98 difference in price between the two orders.

$$0d + 2t = 4215.98$$
$$2t = 4215.98 \quad \text{The difference of 2 laptops is responsible for the \$4215.98 difference in price.}$$
$$t = 2107.99 \quad \text{Each laptop costs \$2107.99 (then use this to find } d\text{).}$$

This approach is called the **elimination method** because it eliminates one of the variables to get an equation that is easier to solve. In the example, we found the difference between the two equations, thereby eliminating the variable d and leaving us with the equation $2t = 4215.98$, which was very easy to solve.

HOW TO: ■ USE THE ELIMINATION METHOD WITH TWO EQUATIONS AND TWO VARIABLES

To solve a system of two equations using the elimination method,

1. Write both equations in standard form.
2. Vertically align the variables in one equation with the corresponding variables in the other equation.
3. If one of the variables has the same coefficient in both equations (or the same magnitude coefficient with the opposite sign), move on to Step 4. If not, multiply one (or both) equations by a constant to create this situation.
4. Combine the equations using addition or subtraction. This will eliminate one of the variables.
5. Solve the resulting equation for the remaining variable.
6. Substitute the value from Step 5 back into one of the original equations and solve for the other variable.

EXAMPLE 7 ■ Using the Elimination Method

Suppose a business is purchasing new cell phones from Verizon Wireless for some of the company's executives and office staff. The company executives asked for 4 Blackberry 8830 Smartphones and 2 Samsung Knack cell phones, which totaled $518.82. The office staff requested 2 Blackberry 8830 Smartphones and 7 Samsung Knack cell phones, which totaled $518.79. (Prices accurate as of 2009.)

a. Write a system of equations to represent this situation and explain what each term in your equations represents.

b. Solve the system using the elimination method.

Solution

a. The unknowns in this situation are the prices for the individual cell phones. Let b be the price in dollars of a Blackberry 8830 Smartphone and let k be the price in dollars of a Samsung Knack.

$$\underbrace{4b}_{\substack{\text{The total}\\\text{cost of 4}\\\text{Blackberry}\\\text{phones}}} + \underbrace{2k}_{\substack{\text{the total}\\\text{cost of 2}\\\text{Samsung}\\\text{phones}}} \underbrace{=}_{\substack{\text{must}\\\text{equal}}} \underbrace{518.82}_{\substack{\text{the total cost}\\\text{of the order for}\\\text{the company}\\\text{executives}}}$$

$$\underbrace{2b}_{\substack{\text{The total}\\\text{cost of 2}\\\text{Blackberry}\\\text{phones}}} + \underbrace{7k}_{\substack{\text{the total}\\\text{cost of 7}\\\text{Samsung}\\\text{phones}}} \underbrace{=}_{\substack{\text{must}\\\text{equal}}} \underbrace{518.79}_{\substack{\text{the total cost of}\\\text{the order for the}\\\text{office staff}}}$$

b. Applying the elimination method to this system creates an immediate problem. If we find the difference between the orders, we will see that there is a difference of two Blackberry phones and five Samsung phones. This means that subtracting the two equations will not eliminate one of the variables. Thus, it appears that the elimination method will not help us here.

But let's play a game of "what if?" Suppose instead of ordering 2 Blackberry phones and 7 Samsung phones for the office staff, we double the order. This means the total price for the order will also double: 4 Blackberry phones, 14 Samsung phones, and a total cost of $1037.58.

$2b + 7k = 518.79$	Start with the original equation.
$2(2b + 7k = 518.79)$	Multiply all of the terms by 2.
$4b + 14k = 1037.58$	Get a new equation without affecting the values of b and k.

We now have a system with the equations $4b + 2k = 518.79$ and $4b + 14k = 1037.58$. Although this no longer exactly represents the original situation, we have done nothing to alter the prices of each phone (the values of b and k are unaffected by this change). We have created an **equivalent system of equations** that has the same solution as the original system but is easier to solve.

$$4b + 2k = 518.82$$
$$4b + 14k = 1037.58$$

We now apply the elimination method to find the solution.

$$\begin{array}{rl} 4b + 2k = 518.82 & \rightarrow \quad 4b + 2k = 518.82 \\ -(4b + 14k = 1037.58) & \rightarrow \quad \underline{-4b - 14k = -1037.58} \\ & \qquad\quad\; 0b - 12k = -518.76 \end{array}$$

There is no difference in the number of Blackberry phones purchased, but there is a difference of 12 Samsung phones (the negative means that the second order had more phones).

There was a $518.76 difference in price between the two orders (the negative means that the second order was more expensive).

$$0b - 12k = -518.76$$
$$-12k = -518.76$$

The difference of 12 Samsung phones is responsible for the $518.76 difference in price.

$$k = 43.23$$

Each Samsung phone costs $43.23.

We now use the value of k to find b.

$$4b + 2k = 518.82$$
$$4b + 2(43.23) = 518.82$$
$$4b + 86.26 = 518.82$$
$$4b = 432.36$$
$$b = 108.09$$

Thus, the solution is $b = 108.09$ and $k = 43.23$. The company paid $108.09 for each Blackberry 8830 Smartphone and $43.23 for each Samsung Knack.

Equivalent Systems of Equations

A system of equations can be modified to create an **equivalent system of equations** that has the same solution as the original system. The following operations on a system will not affect the solution.

1. Interchange (change the position of) two equations.
2. Multiply an equation by a nonzero number.
3. Add (or subtract) a nonzero multiple of one equation to a nonzero multiple of another equation.

EXAMPLE 8 ■ Using the Elimination Method to Solve a System of Equations

A 1-cup serving of oil-roasted, salted peanuts contains 37.94 grams of protein and 27.26 grams of carbohydrates. A 1-cup serving of seedless raisins (not packed) contains 4.67 grams of protein and 114.74 grams of carbohydrates. (*Source:* www.nutri-facts.com) GORP (Good Ol' Raisins and Peanuts) is a popular snack food that provides short-term energy from carbohydrates and long-term energy from protein. How many cups of peanuts and how many cups of raisins are needed to create a 70-cup mix of GORP containing 1325 grams of protein?

Solution We let p be the number of cups of peanuts and r be the number of cups of raisins in the mix. Since the mix contains 70 cups of GORP, we create the following equation.

$$\underbrace{P}_{\substack{\text{the number} \\ \text{of cups of} \\ \text{peanuts}}} + \underbrace{}_{\text{plus}} \underbrace{r}_{\substack{\text{the number} \\ \text{of cups of} \\ \text{raisins}}} \underbrace{=}_{\substack{\text{must} \\ \text{equal}}} \underbrace{70}_{\substack{70 \text{ cups} \\ \text{total}}}$$

Although both protein and carbohydrate information is provided, only the protein information is needed to solve this problem. Since each cup of peanuts contains 37.94 grams of protein, the peanuts contribute $37.94p$ grams of protein to the mix. Similarly, since each cup of raisins contains 4.67 grams of protein, the raisins contribute $4.67r$ grams of protein to the mix. We want the total amount of protein to be 1325 grams, so we create the following equation.

$$\underbrace{37.94p}_{\substack{\text{the total grams} \\ \text{of protein in } p \\ \text{cups of peanuts}}} + \underbrace{}_{\text{plus}} \underbrace{4.67r}_{\substack{\text{the total grams} \\ \text{of protein in } r \\ \text{cups of raisins}}} \underbrace{=}_{\substack{\text{must} \\ \text{equal}}} \underbrace{1325}_{\substack{1325 \text{ total} \\ \text{grams of protein} \\ \text{in the mix}}}$$

This yields the following system of equations.

$$p + r = 70$$
$$37.94p + 4.67r = 1325$$

We first create an equivalent system of equations and then solve the system using the elimination method.

$37.94(p + r = 70) \rightarrow 37.94p + 37.94r = 2655.8$ Create an equivalent equation.

$$\begin{array}{r} 37.94p + 37.94r = 2655.8 \\ -(37.94p + 4.67r = 1325) \\ \hline 0p + 33.27r = 1330.8 \end{array}$$ Solve the equivalent system of equations.

$$0p + 33.27r = 1330.8$$
$$33.27r = 1{,}330.8$$
$$r = 40$$

The mixture contains 40 cups of raisins. Since the mixture contains a total of 70 cups, there must be 30 cups of peanuts in the mixture. Therefore, 30 cups of peanuts and 40 cups of raisins are needed for a 70-cup mixture of GORP to have 1325 grams of protein.

EXAMPLE 9 ■ Determining When a System of Equations Has No Solution

Find the solution to the following system of linear equations.

$$4x - 5y = 20$$
$$8x - 10y = 30$$

Solution We use the elimination method. We multiply the first equation by -2 with the aim of eliminating x.

$$\begin{array}{rl} -2(4x - 5y = 20) \rightarrow & -8x + 10y = -40 \\ 8x - 10y = 30 \rightarrow & \underline{8x - 10y = 30} \\ & 0x + 0y = -10 \end{array}$$

$$0x + 0y = -10$$
$$0 = -10$$

But $0 \neq -10$. Since the system of equations led to a contradiction, the system does not have a solution. This can readily be seen by writing both equations in slope-intercept form.

$$y = 0.8x - 4$$
$$y = 0.8x - 3$$

The lines have the same slope but different y-intercepts, so they are parallel. Therefore, the two lines do not intersect and the system of equations does not have a solution. Graphing both lines as shown in Figure 2.19 validates our conclusion.

Figure 2.19

EXAMPLE 10 ■ Solving a System of Linear Equations with Infinitely Many Solutions

One of the authors is a fan of Nature Valley's Sweet and Salty Nut Granola Bars. According to the package labeling, the Peanut and the Almond bars have the nutritional values shown in Table 2.15.

Table 2.15

Variety	Calories from Fat	Protein (in grams)	Fat (in grams)	Carbohydrates (in grams)
Peanut	80	4	9	19
Almond	60	3	7	22

If the author wants to consume exactly 26 grams of protein and 520 calories from fat, how many bars of each type should he eat?

Solution We begin by identifying our variables. We let p be the number of Peanut bars eaten and a be the number of Almond bars eaten. Since the question does not ask about grams of fat or carbohydrates, we ignore those table values.

We need to set up an equation for *protein* and an equation for *calories from fat*. The amount of protein consumed by eating p Peanut bars is $4p$ since each bar contains 4 grams of protein. Similarly, the amount of protein from eating a Almond bars is $3a$ since each bar contains 3 grams of protein. The total amount of protein we want to consume is 26 grams. Therefore,

$$\text{protein: } 4p + 3a = 26$$

The calories from fat from eating p Peanut bars is $80p$ since each peanut bar contains 80 calories from fat. Similarly, the calories from fat from eating a Almond bars is $60a$ since each almond bar contains 60 calories from fat. The total number of calories from fat we want to consume is 520. Therefore,

$$\text{calories from fat: } 80p + 60a = 520$$

Since we want both equations to be simultaneously true, we combine them into a system of equations.

$$4p + 3a = 26$$
$$80p + 60a = 520$$

Since we have focused on the elimination method in recent examples, let's use the substitution method for more practice. We choose to solve the first equation $4p + 3a = 26$ for p.

$$4p + 3a = 26$$
$$4p = -3a + 26$$
$$p = -0.75a + 6.5$$

Substituting this value of p into the second equation yields

$$80p + 60a = 520$$
$$80(-0.75a + 6.5) + 60a = 520$$
$$(-60a + 520) + 60a = 520$$
$$520 = 520$$

What happened? The variable a dropped out and the equation resulted in a true statement, $520 = 520$. This indicates that the system of equations is *dependent*. Looking again at the original equations, we can see the second equation is equivalent to 20 times the first equation. In other words, multiplying the first equation by 20 will yield an equivalent system involving two identical equations. Therefore, any solution that satisfies the first equation will satisfy the second equation. We generate a table of a few values for the first equation, Table 2.16.

Whole-number solutions include $a = 2$, $p = 5$ and $a = 6$, $p = 2$. That is, eating two Almond bars and five

Table 2.16

a	$p = -0.75a + 6.5$
0	6.5
1	5.75
2	5
3	4.25
4	3.5
5	2.75
6	2

Peanut bars or eating six Almond bars and two Peanut bars will result in the desired amounts of protein and calories from fat.

In addition to these whole-number solutions, there are infinitely many real-number solutions, some of which are shown in the table. In fact, any point on this line is a solution to the system of equations. However, the only solutions that make sense in the context of this real-world application are solutions for which both *a* and *p* are nonnegative.

SOLUTIONS TO SYSTEMS OF EQUATIONS

Systems of linear equations will have 0, 1, or infinitely many solutions.

- A system of equations without a solution is said to be **inconsistent** (Figure 2.20a).
- A system of equations with infinitely many solutions is said to be **dependent** (Figure 2.20b).

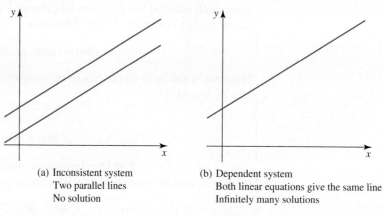

(a) Inconsistent system
Two parallel lines
No solution

(b) Dependent system
Both linear equations give the same line
Infinitely many solutions

Figure 2.20

■ Solving a System of Three or More Equations

For a system of three or more linear equations to have a solution, all of the lines must intersect at the same point. The fact that two lines intersect at a point (a, b) does not ensure that the third line will intersect the first two lines at the same point.

EXAMPLE 11 ■ Solving a System of Three Equations

Solve the following system of equations.

$$2x - y = 5$$
$$3x + 2y = 11$$
$$4x - 4y = 8$$

Solution We will find the point of intersection of the first two equations and then check to see if the point is a solution to the third equation. The first equation may be written as $y = 2x - 5$. Substituting this value in for y in the second equation and solving for x yields

$$3x + 2y = 11$$
$$3x + 2(2x - 5) = 11$$
$$3x + 4x - 10 = 11$$
$$7x - 10 = 11$$
$$7x = 21$$
$$x = 3$$

We substitute this value of x into $y = 2x - 5$ and solve.

$$y = 2(3) - 5$$
$$= 1$$

The point of intersection of the first two lines is (3, 1). We will check to see if this point satisfies the third equation.

$$4x - 4y = 8$$
$$4(3) - 4(1) = 8$$
$$12 - 4 = 8$$
$$8 = 8$$

Since the resultant statement is true, the solution to the system of equations is $x = 3$ and $y = 1$. (A false statement would have shown that the system was inconsistent.) We confirm the result in Figure 2.21.

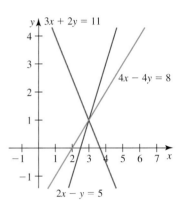

Figure 2.21

SUMMARY

In this section you learned how to find the solution to a system of linear equations using tables, graphs, and algebraic approaches. You learned two algebraic methods for solving linear systems of equations: substitution and elimination. You also learned systems may have no solution, one solution, or infinitely many solutions.

TECHNOLOGY TIP ■ FINDING A POINT OF INTERSECTION

1. Simultaneously graph both functions in the system of linear equations. Adjust the window as necessary to make the point of intersection visible.

2. Press [2ND] [TRACE] to bring up the Calculate Menu.

3. Select **5: intersect**. The calculator asks **"First curve?"** Use the blue arrow buttons to select either curve then press [ENTER]. The calculator asks **"Second curve?"** Again, use the arrow buttons to select the second line and press [ENTER].

4. The calculator asks **"Guess?"** If there is more than one point of intersection, move the cursor near the desired point of intersection and press [ENTER]. Otherwise, just press [ENTER]. The point of intersection is highlighted on the graph and its coordinate is displayed.

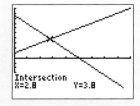

2.4 EXERCISES

SKILLS AND CONCEPTS

In Exercises 1–10,

a. Solve the system of equations algebraically. If a solution does not exist, so state.

b. Use a graphing calculator to find the point(s) of intersection of the lines. Compare your solutions to those obtained in part (a). (*Hint:* Remember that to graph a linear equation given in standard form on your graphing calculator, you will need to solve the equation for *y*.)

1. $y = 5x - 9$
   $y = 2x - 3$

2. $-5x + 2y = 8$
   $y = -3x + 26$

3. $y = 5$
   $y = -3.21x + 4.32$

4. $-26x + 10y = 112$
   $21x + 100y = 1000$

5. $-98x + 10y = -76$
   $-98x + 10y = -28$

6. $y = 0.5x - 9$
   $y = 0.5x + 10$

7. $y = 2x - 9$
   $y = -3x - 9$
   $y = \frac{1}{2}x - 9$

8. $y = 0.25x + 1$
   $4x + y = 18$
   $y = 2$

9. $y = 4.8x + 6$
   $y = 1.9x - 2.1$
   $y = 2x + 1.6$

10. $y = 11x - 12$
    $y = 15x - 21$
    $y = 15x - 12$

In Exercises 11–16, solve the system of equations. Verify your solution by graphing the system of equations or by substituting your solution back into the original equations. If the system of equations is inconsistent or dependent, so state.

11. $2x + 2y = 20$
    $3x + 2y = 27$

12. $4x + 5y = 61$
    $-4x - 6y = -74$

13. $3x + 2y = -50$
    $9x - y = -80$

14. $4x - y = 5$
    $12x - 3y = 0$

15. $x + y = 0.9$
    $2x + 3y = 4.2$

16. $9x - 2y = 648$
    $7x + 5y = 504$

In Exercises 17–22, you are given incomplete tables of values for two linear functions. For each exercise,

a. Determine the constant rate of change for each function and use it to complete the table of values.

b. Write a formula for each linear function.

c. Solve the system, then use the table of values to justify that your solution makes sense.

d. Graph the system to verify your solution in part (c).

17.
x	f(x)	g(x)
-2	1	
-1		
0	7	2
1		0
2		

18.
x	f(x)	g(x)
-2		2
0	34	4
2		
4	18	
6		

19.
x	f(x)	g(x)
-5	4	
0		-13
1		
4		7
5	-6	

20.
x	f(x)	g(x)
-10		
-5	-3	
0	7	
3		-2
7		0

21.

x	f(x)	g(x)
−2	3.36	11.15
−0.5		
0		
2.5		4.4
3	5.36	

22.

x	f(x)	g(x)
−3		
−1.5	−0.16	
2.5		12.04
3.5	−14.16	
4		14.59

SHOW YOU KNOW

23. What is the relationship between the solution to a system of equations and the graph of the system?

24. While solving a system of equations algebraically, we get part of the solution first, such as $x = 3.5$, then substitute this value back into one of the original equations to find the other part of the solution. Why doesn't it matter which original equation we use for this final step?

25. Can a system of equations be simultaneously inconsistent and dependent? Explain.

26. The graph of a system of linear equations consists of a single line. How many solutions are there to the system? Explain.

27. Summarize the basic ideas behind the substitution method.

28. Summarize the basic ideas behind the elimination method.

MAKE IT REAL

29. **Diabetes** The incidence of diabetes in men and women aged 45–64 years increased between 1994 and 1999. The incidence rate for women can be modeled by

$$F(t) = 3.75t + 131 \text{ people per thousand}$$

and the incidence rate for men can be modeled by

$$M(t) = 20t + 109 \text{ people per thousand}$$

where t is the number of years after 1994. (*Source: Modeled from Statistical Abstract of the United States, 2001, Table 109*)

a. Solve the system of equations algebraically and explain what the solution means.

b. Check your solution by graphing the system or substituting your solution back into the original equations.

30. **Gasoline Prices** Based on data from 1990 to 2000, the retail cost of unleaded regular gasoline can be modeled by

$$R(t) = 0.035t + 1.16 \text{ dollars per gallon}$$

and the retail cost of unleaded premium gasoline can be modeled by

$$P(t) = 0.034t + 1.35 \text{ dollars per gallon}$$

where t is the number of years after 1990. (*Source: Modeled from Statistical Abstract of the United States, 2001, Table 704*)

a. Solve the system of equations algebraically and explain what your solution means.

b. Based on your experience of purchasing gasoline, does your solution seem reasonable? Explain.

31. **Weekly Food Cost** Based on data from 1990 and 2000, the weekly food cost for a 15- to 19-year-old male in a four-person family t years after 1990 can be modeled by one of the following functions, depending on the family's food spending behavior. (*Source: Modeled from Statistical Abstract of the United States, 2001, Table 705*)

Thrifty Plan
$$T(t) = 0.56t + 21.40 \text{ dollars}$$

Moderate-Cost Plan
$$M(t) = 1.06t + 36.80 \text{ dollars}$$

Liberal Plan
$$L(t) = 1.21t + 42.60 \text{ dollars}$$

According to the models, will the food cost of all three plans ever be the same? Explain.

32. **Weekly Food Cost** Based on data from 1990 and 2000, the weekly food cost for a 12- to 19-year-old female in a four-person family t years after 1990 can be modeled by one of the follow functions, depending on the family's food spending behavior. (*Source: Modeled from Statistical Abstract of the United States, 2001, Table 705*)

Thrifty Plan
$$T(t) = 0.55t + 20.80 \text{ dollars}$$

Moderate-Cost Plan
$$M(t) = 0.85t + 30.10 \text{ dollars}$$

Liberal Plan
$$L(t) = 1.04t + 36.30 \text{ dollars}$$

According to the models, will the food cost of all three plans ever be the same? Explain.

33. **Bread and Gasoline Prices** Based on data from 1990 to 2000, the retail cost of unleaded regular gasoline can be modeled by

$$R(t) = 0.035t + 1.16 \text{ dollars per gallon}$$

where t is the number of years since 1990. Based on data from 1993 to 2000, the price of a loaf of whole wheat bread may be modeled by

$$C(t) = 0.0404t + 0.991$$

where t is the number of years since 1990. (*Source: Modeled from Statistical Abstract of the United States, 2001, Tables 704, 706*)

Rounded up to the nearest year, when will a loaf of whole wheat bread cost more than a gallon of unleaded regular gasoline?

34. Entertainment Revenues Based on data from 1998 and 1999, the amount of revenue brought in by amusement and theme parks can be modeled by

$$A(t) = 177t + 7335 \text{ million dollars}$$

where t is the number of years since the end of 1998. The amount of revenue brought in by racetracks may be modeled by

$$R(t) = 507t + 4599 \text{ million dollars}$$

(*Source:* Modeled from *Statistical Abstract of the United States, 2001,* Table 1231)

a. Solve the system of equations algebraically and explain what the solution means.

b. Check your solution by graphing the system or by substituting your solution back into the original equations.

35. High School Athletics The table shows the number of men and women participating in high school sports from 1990 to 2002 (the given year is when the school year began).

Years Since 1990	Men (in millions)	Women (in millions)
0	3.41	1.89
2	3.42	2.00
4	3.54	2.24
6	3.71	2.47
8	3.83	2.65
10	3.92	2.78
12	3.99	2.86

Source: Statistical Abstract of the United States, 2006, Table 1237

a. Use linear regression to create a system of equations that models the high school athletics participation of men and women. Round the initial value and slope of each model to 2 decimal places.

b. What are the constant rates of change used in the models and what do they represent?

c. Solve the system algebraically and explain what the solution represents. Check your answer by graphing the system or by substituting your solution back into the original equations.

d. What assumption(s) do we make when solving this system of equations?

36. Chicken and Fish Consumption The table shows the per capita chicken and fish consumption in pounds per person in the United States from 1980 to 2003.

Years Since 1980	Per Capita Chicken Consumption (pounds per person)	Per Capita Fish Consumption (pounds per person)
0	32.7	12.4
5	36.4	15
10	42.4	15
15	48.2	14.8
20	54.2	15.2
22	56.8	15.6
23	57.5	16.3

Source: Statistical Abstract of the United States, 2006, Table 202

a. Use linear regression to create a system of equations that models the per capita consumption of chicken and fish. Round your models off to 2 decimal places.

b. What are the constant rates of change in your models? What do they represent?

c. Solve the system algebraically and explain what the solution represents. Check your answer by graphing the system or by substituting your solution back into the original equations.

d. Explain whether or not you think the solution in part (c) is reasonable. How could you check whether this solution matches what really happened?

37. GORP Mix In Example 8 we learned that a cup of oil-roasted, salted peanuts has 37.94 grams of protein and 27.26 grams of carbohydrates and a cup of seedless raisins contains 4.67 grams of protein and 114.74 grams of carbohydrates. (*Source:* www.nutri-facts.com) If you created a 30-cup mixture with 2130 grams of carbohydrates, how many cups of peanuts and how many cups of raisins are in the mixture?

For Exercises 38–41, refer to the following table of values for taxi fares in various cities. The term flag drop *refers to the cost of the first portion of a mile that is included in the initial fare. Depending on the city, the included portion may be as little as 0.10 mile or as much as 1 mile.*

City	Flag Drop	Cost per Additional Mile	Cost per Minute Waiting
New York City	$2.50	$2.00	$0.40
Baltimore	$1.80	$2.20	$0.40
Atlanta	$2.50	$2.00	$0.35
Seattle	$2.50	$2.00	$0.50

Source: www.taxifarefinder.com

For each exercise,

a. Write a system of equations to represent the total costs of hiring a taxi in each indicated city as functions of the number of additional miles traveled beyond the flag drop. Be sure to identify what your variables represent.

b. Solve the system of equations using algebra. Verify your solution by graphing the system or by substituting your solution back into the original equations. If no solution exists, so state.

c. Explain what your answer in part (b) means and explain whether it makes sense in the context of the problem.

(*continued*)

38. Taxi Fares Two passengers, one in New York City and the other in Baltimore, each hire a taxi. Assume each encounters 4 minutes of wait time during their rides.

39. Taxi Fares Two passengers, one in Baltimore and the other in Seattle, each hire a taxi. Assume each encounters 2 minutes of wait time during their rides.

40. Taxi Fares Two passengers, one in New York City and the other in Atlanta, each hire a taxi. Assume each encounters 6 minutes of wait time during their rides.

41. Taxi Fares Two passengers, one in Seattle and the other in Atlanta, each hire a taxi. Assume neither passenger encounters any wait time during their rides.

42. Computer Prices A graphic design firm placed two orders for computers from Apple Inc. For their web design department, they purchased 2 15-inch MacBook Pro laptops and 3 Mac Pro Quad-core desktops for a total order price of $10,895. For their graphic arts department they purchased 4 15-inch MacBook Pro laptops and 3 Mac Pro Quad-core desktops for a total order price of $14,293. (*Source:* www.apple.com, June 2009) Assume the prices of the computers did not change between orders.
   a. Create a system of equations to model this situation. Be sure to identify what your variables represent.
   b. Solve the system of equations. How much did each type of computer cost?

43. Computer Prices A university placed two orders for computers from Hewlett-Packard. The order for the science department totaled $11,449.87, and consisted of 7 HDX 16t Premium Series laptops and 6 Pavilion dv7t Series laptops. The order for the English department totaled $9,699.89 and consisted of 6 HDX 16t Premium Series laptops and 5 HP Pavilion dv7t Series laptops. (*Source:* www.hp.com, June 2009) Assume the prices of the computers did not change between orders.
   a. Create a system of equations to model this situation. Be sure to identify what your variables represent.
   b. Solve the system of equations. How much did each type of computer cost?

44. Mixing Cereals One of the authors enjoys mixing his cereal, eating 2 cups of a combination of Cheerios and Kashi GOLEAN! Crunch each morning. Cheerios contain 3 grams of fiber per cup while Kashi GOLEAN! Crunch contains 8 grams of fiber per cup. (*Source:* product labels) The author wants to eat 2 cups of cereal, consisting of a combination of Cheerios and Kashi, and wants his bowl of cereal to contain 13 grams of fiber.
   a. What are the unknowns in this situation?
   b. Create a system of equations to model this situation.
   c. Solve the system of equations and explain what your solution means.
   d. Verify your solution by graphing the system or by substituting the solution into the original equations.

45. Trail Mix Planters Honey Nut Medley Trail Mix contains 4 grams of protein per 1-ounce serving. Planters

Dry Roasted Peanuts contain 8 grams of protein per 1-ounce serving. (*Source:* product labels) Suppose you want to create an 8-ounce mixture of trail mix and peanuts that has a combined total of 42 grams of protein.
   a. What are the unknowns in this situation?
   b. Create a system of equations to model this situation.
   c. Solve the system of equations and explain what your solution means.
   d. Verify your solution by graphing the system or by substituting the solution into the original equations.

For Exercises 46–49, refer to the following table of values for taxi fares in various cities.

City	Flag Drop	Cost per Additional Mile	Cost per Minute Waiting
Chicago	$2.25	$1.80	$0.33
St. Louis	$2.50	$1.70	$0.37
Phoenix	$2.50	$1.80	$0.33
San Francisco	$3.10	$2.25	$0.45

Source: www.taxifarefinder.com, updated June 2008

For each exercise,
   a. Write a system of equations to represent the situation described. Be sure to explain what your variables represent.
   b. Solve the system of equations. Verify your solution by graphing the system or by substituting your solution back into the original formulas. If the system is inconsistent or dependent, so state.
   c. Explain what your answer in part (b) means and explain whether it makes sense in the context of the problem.

46. Taxi Fares Two passengers hired taxis, one in San Francisco and the other in Chicago, and each rode the same additional distance beyond the flag drop and had the same wait time during their rides. The passenger in San Francisco paid a total of $17.05, while the passenger in Chicago paid a total of $13.38.

47. Taxi Fares Two passengers hired taxis, one in St. Louis and the other in Phoenix, and each rode the same additional distance beyond the flag drop and had the same wait time during their rides. The passenger in St. Louis paid a total of $6.53, while the passenger in Phoenix paid a total of $6.52.

48. Taxi Fares Two passengers hired taxis, one in San Francisco and the other in St. Louis, and each rode the same additional distance beyond the flag drop and had the same wait time during their rides. The passenger in San Francisco paid a total of $11.65, while the passenger in St. Louis paid a total of $9.02.

49. Taxi Fares Two passengers hired taxis, one in Chicago and the other in Phoenix, and each rode the same additional distance beyond the flag drop and had the same wait time during their rides. The passenger in Chicago paid a total of $10.44, while the passenger in Phoenix paid a total of $10.69.

■ STRETCH YOUR MIND

Exercises 50–53 are intended to challenge your understanding of systems of linear equations.

50. **Part-Time Employment** A full-time student works two jobs. The first job pays $6.75 per hour and the second job pays $8.00 per hour. The student plans to work 20 hours per week. The first job offers 4-hour shifts and is near the student's apartment so she uses a total of 0.25 gallon of gas for a round trip to that job. The second job also offers 4-hour shifts; however, it is several miles from her home so she uses 1 gallon of gas for a round trip to the job. How many hours should she work in each job if she wants to work 20 hours, earn $155, and use exactly 2 gallons of gas?

51. A system of linear equations consists of the functions $f(x) = mx + b$ and $g(x) = \frac{1}{m}x - \frac{b}{m}$. For what values of m and b will the system be inconsistent or dependent? If the system has a unique solution, what will it be?

52. **Solving Systems with Three or More Variables** A system of equations with more than two variables can be solved using the elimination method. For example, a system with three equations and three variables can be solved by taking a pair of equations and using the elimination method to eliminate one of the variables, then taking a different pair of equations and eliminating the same variable. The resulting two equations with two variables can then be solved as we have seen in this section. Using this approach, solve each of the following systems.

a. $x - y + z = 6$
$x + y + z = 8$
$x - y - z = -4$

b. $3x - 3y + z = 5$
$x + y + z = 11$
$2x + 2z = 10$

c. $-2x - y + 4z = 8$
$2x + y - 4z = -8$
$x + y + z = 6$

d. $x + y + z + w = 1$
$x - y + z - w = 2$
$x - y - z + w = 3$
$x - y - z - w = 4$

53. A system of equations has more variables than it has equations. Is it possible for such a system to have exactly one solution? Explain.

SECTION 2.5 Using Matrices to Solve Linear Systems

LEARNING OBJECTIVES

■ Write systems of linear equations in augmented matrices

■ Use technology to reduce augmented matrices to reduced row echelon form

■ Solve linear systems using matrices and interpret the meaning of the results in a real-world situation

GETTING STARTED

Financial advisors typically encourage investors to diversify their investment portfolios. By investing in a variety of different types of investments, investors can reduce their risk while increasing the probability of a good financial return on their investment. Determining how much money to place in each type of investment account is not an exact science; however, a system of linear equations can be used to help decide which mix of investments is best.

In this section we introduce matrix notation and demonstrate how matrices are used to represent and solve systems of linear equations. We also show how to use technology to simplify an augmented matrix to reduced row echelon form.

Leah-Anne Thompson/Shutterstock.com

■ Modeling with a System of Three Equations

An investor with $30,000 in an existing mutual fund IRA wants to roll over her investment into a new retirement plan with TIAA-CREF. The company offers a variety of accounts with varying returns, as shown in Table 2.17.

Table 2.17

CREF Variable Annuity Accounts	Unit Value	Average Annual Return			
		1-year	5-year	10-year	Since Inception
Bond Market	$79.67	5.80%	4.39%	5.86%	6.95%
Equity Index	$103.79	19.53%	11.05%	7.29%	11.22%
Global Equities	$112.52	23.70%	13.41%	6.54%	10.01%
Growth	$72.53	18.33%	8.23%	3.02%	8.19%
Inflation-Linked Bond	$46.72	3.53%	5.58%	6.33%	6.24%
Money Market	$24.33	5.02%	2.50%	3.68%	4.72%
Social Choice	$131.54	13.08%	8.71%	6.97%	9.99%
Stock	$269.02	22.19%	12.71%	7.55%	10.67%

Source: www.tiaa-cref.com; as of 6/30/07

The investor looks at the long-term performance of each account as an indicator of a likely long-term return. She decides to invest her money in the Bond Market, Equity Index, and Social Choice accounts. After analyzing her *risk tolerance*, she decides to invest twice as much money in the Bond Market account as in the Equity Index account because it appears to be more stable. (A person with a high risk tolerance is more comfortable with the possibility of large fluctuations in the annual percentage return than a person with a low risk tolerance.) The investor wants to earn a 7% return on her investments. She assumes she will be able to earn the 10-year average annual return.

To determine how much money she should invest in each account, we begin by defining variables to represent the amount invested in each account. We let x be the amount of money invested in the Bond Market account, y the amount invested in the Equity Index account, and z the amount invested in the Social Choice account. Since she has $30,000 to invest, we have

$$x + y + z = 30{,}000$$

Since she plans to invest twice as much money in the Bond Market account as in the Equity account, we have

$$x = 2y$$
$$x - 2y = 0$$

The Bond Market account has a 5.86% return rate, the Equity Index account has a 7.29% return rate, and the Social Choice account has a 7.55% return rate. Since she wants to earn a 7% return on her total investment, we have

$$0.0586x + 0.0729y + 0.0755z = 0.07(30{,}000)$$
$$0.0586x + 0.0729y + 0.0755z = 2100$$

Thus we need to solve the following system of equations.

$$x + y + z = 30{,}000$$
$$x - 2y = 0$$
$$0.0586x + 0.0729y + 0.0755z = 2100$$

■ Matrix Notation

Solving a system of three equations with three unknowns is somewhat complex. However, with the mathematical machinery of matrices, we will be able to solve such systems efficiently. We begin by introducing the necessary notation.

An $m \times n$ Matrix

An $m \times n$ matrix A is an array of numbers with m rows and n columns. The plural of matrix is *matrices*.

Capital letters are typically used to represent matrices. The following examples show matrices of various dimensions.

$$A = \begin{bmatrix} 2 & 4 & 6 \\ 5 & 3 & 1 \end{bmatrix}, \quad B = \begin{bmatrix} 3 \\ -1 \\ 6 \end{bmatrix}, \quad C = \begin{bmatrix} 2.1 & 1.9 \\ 0.1 & 0.8 \end{bmatrix}, \quad D = \begin{bmatrix} 2 & 7 \end{bmatrix}$$

Matrix A is a 2×3 matrix because it has two rows and three columns. Similarly, B is a 3×1 matrix, C is a 2×2 matrix, and D is a 1×2 matrix.

A matrix that consists of a single row of numbers, such as matrix D, is called a **row matrix**. A matrix that consists of a single column of numbers, such as matrix B, is called a **column matrix**. A matrix that has the same number of rows as columns, such as matrix C, is called a **square matrix**.

The numbers (or variables) inside the matrix are called the **entries** of the matrix. The entries of an $m \times n$ matrix A may be represented as

$$A = \begin{bmatrix} a_{11} & a_{12} & \cdots & a_{1n} \\ a_{21} & a_{22} & \cdots & a_{2n} \\ \cdots & \cdots & \cdots & \cdots \\ a_{m1} & a_{m2} & \cdots & a_{mn} \end{bmatrix}$$

The subscript of each entry indicates its row and column position. For example a_{21} refers to the entry in the second row and first column. In general, an entry a_{ij} is the term in row i and column j.

EXAMPLE 1 ■ Determining the Dimensions and Entry Values of a Matrix

Determine the dimensions of the matrix A and the value of the entries a_{12}, a_{21}, and a_{24}.

$$A = \begin{bmatrix} 1 & 2 & 3 & 4 \\ 5 & 6 & 7 & 8 \\ 9 & 10 & 11 & 12 \end{bmatrix}$$

Solution The matrix has three rows and four columns so it is a 3×4 matrix. For this matrix $a_{12} = 2$, $a_{21} = 5$, and $a_{24} = 8$.

■ Representing a System of Linear Equations with an Augmented Matrix

To represent the system of linear equations

$$2x + 3y = 8$$
$$4x - y = 2$$

with a matrix, we write

$$\begin{bmatrix} 2 & 3 & | & 8 \\ 4 & -1 & | & 2 \end{bmatrix}$$

The matrix is called an **augmented matrix** because the *coefficient matrix*, $\begin{bmatrix} 2 & 3 \\ 4 & -1 \end{bmatrix}$, is augmented (added on to) with the *column matrix*, $\begin{bmatrix} 8 \\ 2 \end{bmatrix}$. The vertical bar between the last two columns of numbers indicates that the matrix is an augmented matrix. The first column of the matrix contains the coefficients of the *x*-terms. In this case, the *x*-terms $2x$ and $4x$ have coefficients 2 and 4, respectively. The second column contains the coefficients of the *y*-terms. In this case, the *y*-terms $3y$ and $-y$ have coefficients 3 and -1, respectively.

To solve a system of equations with a matrix, we must first write the matrix in *reduced row echelon form*, using matrix row operations or technology.

■ Row Operations

When we discussed the elimination method in Section 2.4, we identified three operations that yielded an equivalent system of equations.

1. Interchange (change the position of) two equations.
2. Multiply an equation by a nonzero number.
3. Add a nonzero multiple of one equation to a nonzero multiple of another equation.

We can apply similar operations, called **row operations**, to augmented matrices.

Row Operations

For any augmented matrix of a system of equations, the following row operations yield an augmented matrix of an equivalent system of equations.

1. Interchange (change the position of) two rows.
2. Multiply a row by a nonzero number.
3. Add a nonzero multiple of one row to a nonzero multiple of another row and replace either row with the result.

Before the technology that is available today, mathematicians had to repeatedly apply row operations to convert a matrix into *reduced row echelon form*. A matrix is said to be in **reduced row echelon form** if it meets the following criteria.

Reduced Row Echelon Form

An augmented matrix is said to be in reduced row echelon form if it satisfies each of the following conditions.

1. The leading entry (first nonzero entry) in each row is a 1.
2. The leading entry of each row is the only nonzero entry in its corresponding column.
3. The leading entry in each row is to the right of the leading entry in the row above it.
4. All rows of zeros are at the bottom of the matrix.

A matrix in reduced row echelon form is also referred to as a **reduced matrix**.

The Technology Tip at the end of the section shows how to use a graphing calculator to write an augmented matrix in reduced row echelon form. In subsequent examples we will use this technology, but first we will demonstrate the manual process to help you grasp the concepts.

Suppose we are given the system of equations

$$2x + y = 11$$
$$4x + 3y = 27$$

The corresponding matrix is $\begin{bmatrix} 2 & 1 & | & 11 \\ 4 & 3 & | & 27 \end{bmatrix}$ and the corresponding graph is shown in Figure 2.22. Note that the two lines intersect at (3, 5).

For the first step of the row reduction process, we multiply the first row by 2, subtract the second row, and place the result in the second row. This row operation eliminates the 4 in the first column. The resultant matrix is $\begin{bmatrix} 2 & 1 & | & 11 \\ 0 & -1 & | & -5 \end{bmatrix}$ and the corresponding graph is shown in Figure 2.23. Notice that these two lines intersect at the same point, (3, 5), as the first pair of lines.

Figure 2.22

We next add the first two rows and place the result in the first row. This eliminates the 1 in the first row and second column. The resultant matrix is $\begin{bmatrix} 2 & 0 & | & 6 \\ 0 & -1 & | & -5 \end{bmatrix}$ and the corresponding graph is shown in Figure 2.24. This pair of lines also intersects at (3, 5).

Figure 2.23

Figure 2.24

The equations $2x = 6$ and $-y = -5$ simplify to $x = 3$ and $y = 5$, respectively. The corresponding final matrix is $\begin{bmatrix} 1 & 0 & | & 3 \\ 0 & 1 & | & 5 \end{bmatrix}$. Now we can read the intersection point directly from the matrix.

Let's now return to the investment scenario introduced at the start of the section. Recall that we had created the following system of equations.

$$x + y + z = 30{,}000$$
$$x - 2y = 0$$
$$0.0586x + 0.0729y + 0.0755z = 2100$$

In the equations, x is the amount of money invested in the Bond Market account, y is the amount invested in the Equity Index account, and z is the amount invested in the Social Choice account. The corresponding augmented matrix is

$$\begin{bmatrix} 1 & 1 & 1 & | & 30{,}000 \\ 1 & -2 & 0 & | & 0 \\ 0.0586 & 0.0729 & 0.0755 & | & 2100 \end{bmatrix}$$

Using technology, we reduce the matrix to

$$\begin{bmatrix} 1 & 0 & 0 & | & 9065.93 \\ 0 & 1 & 0 & | & 4532.97 \\ 0 & 0 & 1 & | & 16{,}401.10 \end{bmatrix}$$

The reduced matrix corresponds with the system of equations

$$1x + 0y + 0z = 9065.93$$
$$0x + 1y + 0z = 4532.97$$
$$0x + 0y + 1z = 16{,}401.10$$

which simplifies to

$$x = 9065.93$$
$$y = 4532.97$$
$$z = 16{,}401.10$$

These equations show that the investor should invest \$9065.93 in the Bond Market account, \$4532.97 in the Equity Index account, and \$16,401.10 in the Social Choice account.

EXAMPLE 2 ■ **Creating a System of Equations and Solving It with Matrices**

A college student works three jobs. The first job pays \$7 per hour, the second job pays \$8 per hour, and the last job pays \$10 per hour. Between the three jobs, the student works 30 hours a week. The student enjoys the first job (since it is related to his field of study) and hates the third job even though it pays the most. The student needs to earn \$250 a week. How many hours should the student work in each job?

Solution We begin by defining variables to represent the number of hours worked in each job. We let f represent the number of hours spent working in the first job, s represent the number of hours spent working the second job, and t represent the number of hours working the third job.

Since the student works 30 hours a week, we know that

$$f + s + t = 30$$

The total amount of pay for a job is determined by multiplying the pay rate by the number of hours worked. Since the student must earn \$250, we have

$$7f + 8s + 10t = 250$$

Using each equation as a row in the augmented matrix, we rewrite the system of equations as

$$\begin{bmatrix} 1 & 1 & 1 & | & 30 \\ 7 & 8 & 10 & | & 250 \end{bmatrix}$$

Using technology, we reduce the augmented matrix to

$$\begin{bmatrix} 1 & 0 & -2 & | & -10 \\ 0 & 1 & 3 & | & 40 \end{bmatrix}$$

We then rewrite the matrix as a system of equations.

$$1f + 0s - 2t = -10$$
$$0f + 1s + 3t = 40$$

This is not too meaningful as written, but if we rewrite each equation, we can make sense of it.

$$f = 2t - 10$$
$$s = 40 - 3t$$

The first equation tells us the number of hours spent in the first job is 10 hours less than twice the number of hours worked in the third job. The second equation tells us the number of hours spent in the second job is 40 hours minus three times the number of hours worked in the third job.

Since the student hates the third job, we want t to be as small as possible while keeping f and s nonnegative. Let's try $t = 5$ hours.

$$f = 2(5) - 10 = 0$$
$$s = -3(5) + 40 = 25$$

One solution to the system of equations is to work 0 hours at the first job, 25 hours at the second job, and 5 hours at the third job. (In fact, the student cannot work fewer than 5 hours at the third job because it would make f a negative number of hours.) Let's double-check to make sure that this adds up to 30 hours and will earn the required $250.

$$0 + 25 + 5 = 30$$
$$7(0) + 8(25) + 10(5) = 250$$

This solution allows the student to make the needed $250 while working the least amount of hours possible at the third job. It also shows the student doesn't need the first job—the one he likes most.

■ Dependent Systems of Equations

Although we chose exactly one solution in Example 2, the system of equations actually has infinitely many solutions. Recall from Section 2.4 that a system of equations with infinitely many solutions is said to be **dependent**. To find a few more solutions for the system from Example 2,

$$f = 2t - 10$$
$$s = 40 - 3t$$

we construct Table 2.18 with different values of t and calculate f and s.

Table 2.18

Hours in Third Job t	Hours in First Job $f = 2t - 10$	Hours in Second Job $s = 40 - 3t$
6	2	22
8	6	16
10	10	10
12	14	4

Any of these combinations of hours will result in 30 hours of work and $250 in earnings.

To ensure all variables are nonnegative, the following conditions must be met.

$$2t - 10 \geq 0 \qquad 40 - 3t \geq 0$$
$$2t \geq 10 \qquad 40 \geq 3t$$
$$t \geq 5 \qquad 13\tfrac{1}{3} \geq t$$

So any value of t between 5 and $13\tfrac{1}{3}$ hours can be used to calculate a solution that will make sense in the context of our problem.

EXAMPLE 3 ■ Using Matrices to Solve a Dependent System of Equations

According to the product packaging, Trader Joe's Semi Sweet Chocolate Chips contain 4 grams of fat, 1 gram of protein, and 1 gram of fiber in each 15-gram serving. (One tablespoon of chocolate chips weighs about 15 grams.) A 30-gram serving of shelled walnuts contains 20 grams of fat, 5 grams of protein, and 2 grams of fiber. A 30-gram serving of Diamond Premium Almonds contains 15 grams of fat, 6 grams of protein, and 4 grams of fiber. (One-fourth cup of walnuts or almonds weighs 30 grams.)

A hiker plans to make a trail mix with chocolate chips, walnuts, and almonds. She wants 150 grams of the mix and wants the mix to contain 16 grams of fiber. How many servings of chocolate chips, walnuts, and almonds should she use to make the mix?

Solution We first define variables, letting c be the number of servings of chocolate chips, w be the number of servings of walnuts, and a be the number of servings of almonds.

The first equation of the system will be related to the weight of the mix. Recall that a serving of chocolate weighs 15 grams and a serving of nuts weighs 30 grams. We need 150 grams total in the mix.

$$\text{Grams equation: } 15c + 30w + 30a = 150$$

We readily notice each coefficient is a multiple of 15, so we simplify the equation by dividing each side by 15.

$$\text{Grams equation: } c + 2w + 2a = 10$$

The second equation will be related to the fiber content in the mix. Recall that a serving of chocolate has 1 gram of fiber, a serving of walnuts has 2 grams of fiber, and a serving of almonds has 4 grams of fiber. We need 16 grams of fiber in the mix.

$$\text{Fiber equation: } 1c + 2w + 4a = 16$$

We consolidate these equations into a system of equations and write the augmented matrix.

$$\begin{array}{c} c + 2w + 2a = 10 \\ 1c + 2w + 4a = 16 \end{array} \qquad \begin{bmatrix} 1 & 2 & 2 & | & 10 \\ 1 & 2 & 4 & | & 16 \end{bmatrix}$$

Using technology, we reduce the matrix to

$$\begin{bmatrix} 1 & 2 & 0 & | & 4 \\ 0 & 0 & 1 & | & 3 \end{bmatrix}$$

We then write the resultant system of equations

$$c + 2w + 0a = 4$$
$$0c + 0w + 1a = 3$$

which simplifies to

$$c + 2w = 4$$
$$a = 3$$

The mix requires 3 servings of almonds but we have flexibility in the amount of chocolate chips and walnuts to include. We can rewrite the equation relating these two quantities as $c = -2w + 4$ and construct Table 2.19 to show some of the possible options for chocolate and walnuts.

Table 2.19

Servings of Walnuts w	Servings of Chocolate Chips $c = -2w + 4$
0	4
0.5	3
1	2
2	0

■ Inconsistent Systems of Equations

Not every system of equations has a solution. Recall from Section 2.4 that a system of equations with no solution is said to be **inconsistent**. When a system of equations has no solution, a contradiction will occur in the augmented matrix. This is illustrated in Example 4.

EXAMPLE 4 ■ Using Matrices with an Inconsistent System of Equations

Continuing with the trail mix scenario in Example 3, we add an additional constraint. We want the trail mix to contain exactly 2 servings of almonds. How many servings of chocolate chips, walnuts, and almonds should we include in the mix?

Solution The new constraint is given by

$$\text{Almond equation: } 0c + 0w + 1a = 2$$

We add this equation to the other two equations from Example 3.

$$\begin{array}{c} c + 2w + 2a = 10 \\ 1c + 2w + 4a = 16 \\ 0c + 0w + 1a = 2 \end{array} \quad \begin{bmatrix} 1 & 2 & 2 & | & 10 \\ 1 & 2 & 4 & | & 16 \\ 0 & 0 & 1 & | & 2 \end{bmatrix}$$

Using technology, we reduce the matrix to

$$\begin{bmatrix} 1 & 2 & 0 & | & 0 \\ 0 & 0 & 1 & | & 0 \\ 0 & 0 & 0 & | & 1 \end{bmatrix}$$

This is equivalent to the following system of equations.

$$c + 2w = 0$$
$$a = 0$$
$$0 = 1$$

Notice the bottom equation states that 0 is the same as 1, which is a false statement. As a result, we know the system of equations does not have a solution. In other words, if we use exactly 2 servings of almonds, there is no way to create a 150-gram trail mix that contains exactly 16 grams of fiber.

EXAMPLE 5 ■ Using a System of Equations in a Real-World Context

A chef has been commissioned to create a party mix containing pretzels, bagel chips, and Chex® cereal for a corporate gathering of 180 people. He estimates that on average each person will consume 1/2 cup of the party mix. The Original Chex Party Mix recipe calls for 9 cups of Chex cereal, 1 cup of pretzels, and 1 cup of bagel chips. (*Source:* www.chex.com) However, the chef plans to modify the recipe so that there are half as many bagel chips as pretzels and three times as much cereal as pretzels. How many cups of each ingredient will be needed to make the party mix?

Solution Let p be the number of cups of pretzels, b be the number of cups of bagel chips, and c be the number of cups of Chex cereal. Since each of the 180 people is expected to consume 1/2 cup of party mix, a total of 90 cups of the party mix is needed. Therefore,

$$p + b + c = 90$$

Since there are half as many bagel chips as pretzels, we have

$$b = 0.5p$$
$$2b = p$$
$$-p + 2b = 0$$

Since there is three times as much cereal as pretzels, we obtain

$$c = 3p$$
$$-3p + c = 0$$

No other constraints are given, so the system of equations and corresponding matrix are

$$\begin{aligned} p + b + c &= 90 \\ -p + 2b &= 0 \\ -3p + c &= 0 \end{aligned} \qquad \begin{bmatrix} 1 & 1 & 1 & | & 90 \\ -1 & 2 & 0 & | & 0 \\ -3 & 0 & 1 & | & 0 \end{bmatrix}$$

Using technology, we reduce the matrix to

$$\begin{bmatrix} 1 & 0 & 0 & | & 20 \\ 0 & 1 & 0 & | & 10 \\ 0 & 0 & 1 & | & 60 \end{bmatrix}$$

This corresponds with the system

$$p = 20$$
$$b = 10$$
$$c = 60$$

The chef should use 20 cups of pretzels, 10 cups of bagel chips, and 60 cups of cereal.

SUMMARY

In this section you learned basic matrix notation and saw how to use matrices to represent and solve systems of linear equations. You learned how to formulate linear systems from real-world data and use the solutions to these systems to make decisions. You also learned how to use technology to simplify an augmented matrix to reduced row echelon form.

TECHNOLOGY TIP ■ ENTERING A MATRIX

1. Activate the Matrix Menu by pressing [2nd] [x⁻¹]. You may or may not have some matrices displayed in the name list.

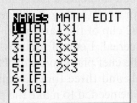

2. Create the augmented matrix *A* by moving the cursor to **EDIT** and pressing [ENTER].

3. Enter the dimensions of the matrix. Type the number of rows and press [ENTER]. Then type the number of columns and press [ENTER]. The example matrix is a 3 × 4 augmented matrix.

4. Enter the individual values of the matrix. Use the arrow keys to move from one entry to another. When you have entered all of the values, press [2nd] [MODE] to exit the Matrix Editor.

TECHNOLOGY TIP ■ FINDING THE REDUCED ROW ECHELON FORM OF A MATRIX

1. Activate the Matrix Math Menu by pressing [2nd] [x⁻¹] and moving the cursor to the **MATH** menu item.

2. Select the **rref** operation by scrolling down the list to item **B:**. Press [ENTER] to place the operation on the home screen. (The **rref(** function will convert a matrix to reduced row echelon form.)

3. Select the matrix *A* from the Matrix Names Menu by pressing [2nd] [x⁻¹], moving the cursor to matrix *A*, and pressing [ENTER]. This places the matrix *A* on the home screen.

4. Calculate the reduced row echelon form of the matrix by pressing [ENTER].

2.5 EXERCISES

SKILLS AND CONCEPTS

In Exercises 1–5, determine the dimensions of the matrix A.

1. $A = \begin{bmatrix} 19 & 5 & 1 \\ 2 & 8 & 18 \end{bmatrix}$

2. $A = \begin{bmatrix} -10 \\ 42 \\ 17 \end{bmatrix}$

3. $A = \begin{bmatrix} 14 & 0 & -5 \end{bmatrix}$

4. $A = \begin{bmatrix} -4 & 5 & 8 & 11 \\ 11 & 5 & 4 & -6 \end{bmatrix}$

5. $A = \begin{bmatrix} 17 \end{bmatrix}$

In Exercises 6–10, write the augmented matrix as a system of linear equations using the variables x, y, and z as appropriate.

6. $\begin{bmatrix} 2 & 1 & 0 & | & 11 \\ 1 & 0 & 4 & | & 0 \end{bmatrix}$

SECTION 2.5 ■ EXERCISES

7. $\begin{bmatrix} 1 & 3 & | & 12 \\ 4 & 1 & | & 7 \\ -1 & 6 & | & 9 \end{bmatrix}$

8. $\begin{bmatrix} 1 & 0 & 4 & | & 0 \\ 0 & 1 & 2 & | & 19 \\ 6 & 0 & 3 & | & 22 \end{bmatrix}$

9. $\begin{bmatrix} 3 & -1 & 9 & | & 13 \\ -3 & 1 & 0 & | & -31 \end{bmatrix}$

10. $\begin{bmatrix} 5 & 2 & 4 & | & 11 \\ -1 & 2 & 2 & | & 3 \\ 5 & 10 & 3 & | & 18 \end{bmatrix}$

In Exercises 11–20,

a. Rewrite the system of equations as an augmented matrix.

b. Use technology to simplify the matrix to reduced row echelon form.

c. Identify the solution(s) to the system of equations. If the system is inconsistent or dependent, so state.

11. $2x + 5y = 2$
    $3x - 5y = 3$

12. $6x + 2y = 10$
    $-x - 2y = -5$

13. $-2x + 6y + 4z = 10$
    $4x - 12y + 2z = -20$
    $3x + 4y - z = 11$

14. $10x - y + z = -7$
    $9x - 2y + 4z = 7$
    $x + 2y - 4z = -17$

15. $3x - 2y + z = 6$
    $11x - 20y - z = 0$
    $y + z = 3$

16. $9x - 6y = 0$
    $4x + 5y = 23$
    $x - z = 6$

17. $x - y = -5$
    $9x + y = 25$
    $29x + y = 65$

18. $x - 2y = -7$
    $6x + 5y = 94$
    $10x + 3y = 114$

19. $2x - 4y = 16$
    $9x + y = -4$
    $-3x + 6y = -24$

20. $x - y = 3$
    $6x + 7y = 44$
    $6x - 7y = 16$

■ SHOW YOU KNOW

21. Explain the conceptual meaning of the solution to a system of equations.

22. How are systems of equations and augmented matrices related?

23. What row operations result in an equivalent system of equations?

24. Explain what is meant by an inconsistent system of equations.

25. Suppose a friend missed class and asks you what is meant by "reduced row echelon form." How do you respond?

■ MAKE IT REAL

In Exercises 26–35, set up and solve the system of linear equations using matrices.

26. **Investment Choices** The table shows the average annual rate of return of a variety of TIAA-CREF investment accounts over a 10-year period.

CREF Variable Annuity Accounts	10-Year Average
Bond Market	5.86%
Equity Index	7.29%
Global Equities	6.54%
Growth	3.02%
Inflation-Linked Bond	6.33%
Money Market	3.68%
Social Choice	6.97%
Stock	7.55%

Source: www.tiaa-cref.com; as of 6/30/07

An investor chooses to invest $3000 in the Inflation-Linked Bond, Global Equities, and Stock accounts. He assumes he will be able to get a return equal to the 10-year average and wants the total return on his investment to be 7%. He decides to invest three times as much money in the Inflation-Linked Bond account as in the Global Equities account. How much money should he invest in each account? (For ease of computation, round each percentage to the nearest whole number percent, e.g., 7.55% ≈ 8%.)

27. **Investment Choices** An investor chooses to invest $5000 in the Global Equities, Money Market, and Social Choice accounts shown in Exercise 26. She wants to put five times as much money in the Money Market account as in the Global Equities account. She assumes she will be able to get a return equal to the 10-year average and wants the total return on her investment to be 6%.

How much money should she invest in each account? (For ease of computation, round each percentage to the nearest whole number percent.)

28. **Resource Allocation: Sandwiches** A plain hamburger requires one ground beef patty and a bun. A cheeseburger requires one ground beef patty, one slice of cheese, and a bun. A double cheeseburger requires two ground beef patties, two slices of cheese, and a bun.

Frozen hamburger patties are typically sold in packs of 12; hamburger buns, in packs of 8; and cheese slices, in packs of 24.

A family is in charge of providing burgers for a neighborhood block party. They have purchased 13 packs of buns, 11 packs of hamburger patties, and 3 packs of cheese slices. How many of each type of sandwich should they make if they want to use up all of the buns, patties, and cheese slices?

29. **Pet Nutrition: Food Cost** PETsMART sold the following varieties of dog food in June 2003. The price shown is for an 8-pound bag.

Pro Plan Adult Chicken & Rice Formula
25% protein, 3% fiber, $7.99

Pro Plan Adult Lamb & Rice Formula
28% protein, 3% fiber, $7.99

Pro Plan Adult Turkey & Barley Formula
26% protein, 3% fiber, $8.49

Source: www.petsmart.com

A dog breeder wants to make 120 pounds of a mix containing 27% protein and 3% fiber. Give two possible answers for how many 8-pound bags of each dog food variety the breeder should buy. For each answer, calculate the total cost of the dog food.

30. **Pet Nutrition: Food Cost** PETsMART sold the following varieties of dog food in June 2003.

Nature's Recipe Venison Meal & Rice Canine
20% protein, $21.99 per 20-pound bag

Nutro Max Natural Dog Food
27% protein, $12.99 per 17.5-pound bag

PETsMART Premier Oven Baked Lamb Recipe
25% protein, $22.99 per 30-pound bag

Source: www.petsmart.com

A dog breeder wants to make 300 pounds of a mix containing 22% protein. Give two possible answers for how many bags of each dog food variety the breeder should buy. For each answer, calculate the total cost of the dog food. (*Hint:* Note each bag is a different weight. Fractions of bags may not be purchased.)

31. **Utilization of Ingredients** A custard recipe calls for 3 eggs and 2.5 cups of milk. A vanilla pudding recipe calls for 2 eggs and 2 cups of milk. A bread pudding recipe calls for 2 eggs, 2 cups of milk, and 8 slices of bread.

A stocked kitchen contains 18 eggs, 1 gallon of milk, and 24 slices of bread. How many batches of each recipe should a chef make to use up all of the ingredients?

32. **First Aid Kit Supplies** Safetymax.com sells first aid and emergency preparedness supplies to businesses. A company that assembles first aid kits for consumers purchases 3500 1-inch by 3-inch plastic adhesive bandages, 1800 alcohol wipes, and 220 tubes of antibiotic ointment from Safetymax.com.

The company assembles compact, standard, and deluxe first aid kits for sale to consumers. A *compact* first aid kit contains 20 plastic adhesive bandages, 8 alcohol wipes, and 1 tube of antibiotic ointment. A *standard* first aid kit contains 40 plastic adhesive bandages, 20 alcohol wipes, and 2 tubes of antibiotic ointment. A *deluxe* first aid kit contains 50 plastic adhesive bandages, 28 alcohol wipes, and 4 tubes of antibiotic ointment.

How many of each type of kit should the company assemble to use up all of the bandages, wipes, and antibiotics ordered?

33. **Concert Ticket Sales** On the weekend of July 23–25, 2004, the House of Blues Sunset Strip in Hollywood, California, hosted three concerts: Saves the Day, Jet, and BoDeans. Saves the Day tickets cost $15, Jet tickets cost $20, and BoDeans tickets cost $22. (*Source:* www.ticketmaster.com)

If concert planners expected that a total of 1200 tickets would be sold over the weekend, how many tickets needed to be sold for each concert to bring in $23,500? Find three different solutions.

34. **Concert Ticket Sales** On July 7, 2004, Shania Twain was scheduled to perform at the TD Waterhouse Centre in Orlando, Florida. The center offered 18,039 seats for the concert in three seating classifications: floor, lower, and upper. Based upon their location, tickets were offered at three different prices: $80, $65, and $45. (*Source:* www.ticketmaster.com) Suppose the average price of a floor ticket was $80, the average price of a lower ticket was $65, and the average price of an upper ticket was $45. If concert planners expected a total of 12,000 tickets would be sold, including 5 times as many tickets on the lower level as on the floor, how many of each type of ticket would have to be sold to reach $675,000 in ticket revenue?

35. **Concert Ticket Sales** On June 24, 2004, Madonna was scheduled to perform at Madison Square Gardens in New York, New York. Tickets were offered at four different prices: $49.50, $94.50, $154.50, and $304.50. (*Source:* www.ticketmaster.com) If concert planners expected a total of 25,000 tickets would be sold, including 10 times as many of the least expensive ticket as of the most expensive ticket and one-third as many $154.50 tickets as of $94.50 tickets, how many of each type of ticket would need to be sold to earn $2,130,000 in revenue?

STRETCH YOUR MIND

Exercises 36–40 are intended to challenge your understanding of linear system applications.

36. **Amusement Park Rides**   Rides at an amusement park require 3, 4, 5, or 6 tickets. A family purchases 75 tickets and receives 2 tickets from a park guest who had leftover tickets. The family wants to use all their tickets and wants the sum of 5- and 6-ticket rides to be twice as much as the sum of 3- and 4-ticket rides. What is the largest possible number of 6-ticket rides that the family can go on subject to these constraints?

37. **Assortment of Coins**   A coin purse contains 64 coins including pennies, nickels, dimes, and quarters. The total value of the coins is $4.94. There is the same number of nickels as dimes. How many of each type of coin is in the bag?

38. **Investments**   The following table shows the average annual rate of return of a variety of TIAA-CREF investment accounts over a 10-year period.

CREF Variable Annuity Accounts	10-Year Average
Bond Market	5.86%
Equity Index	7.29%
Global Equities	6.54%
Growth	3.02%
Inflation-Linked Bond	6.33%
Money Market	3.68%
Social Choice	6.97%
Stock	7.55%

Source: www.tiaa-cref.com; as of 6/30/07

An investor wants to earn a 7% annual return on a $68,500 investment. The investor expects the annual return on each account will be equal to the 10-year rate, rounded to the nearest whole number percentage.

The investor wants to invest the same amount of money in the Stock account as in the Global Equities account and twice as much in the Social Choice account as in the Growth account. How much money should the investor place in each account?

39. **Traffic Flow**

The figure shows the flow of traffic at four city intersections.

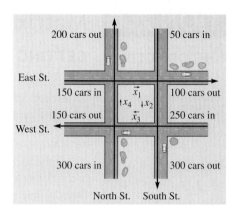

For each intersection, the number of cars entering the intersection must equal the number of cars leaving the intersection. For example, the number of cars entering the intersection of North and West is $x_3 + 300$. The number of cars leaving the intersection is $x_4 + 150$. Therefore, $x_3 + 300 = x_4 + 150$.

Find two separate sets of values x_1, x_2, x_3, and x_4 that work in the traffic-flow system.

40. **Traffic Flow**   Repeat Exercise 39 for the following traffic-flow diagram. Then explain who could benefit from this type of traffic-flow analysis.

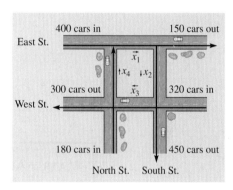

SECTION 2.6 Matrix Operations and Applications

LEARNING OBJECTIVES

- Identify properties of matrix addition and scalar multiplication
- Use matrix addition and scalar multiplication in real-world situations

GETTING STARTED

Many schools use a salary schedule to determine how much to pay their teachers. Typically, as teachers gain experience and participate in professional development activities, they are advanced on the salary schedule. However, due to inflation (rising prices), a teacher's buying power may be reduced even though his or her salary is increased. A cost of living allowance (COLA) is designed to offset the effects of inflation on a person's salary.

In this section we further develop the concept of matrices by demonstrating matrix addition and scalar multiplication. We also demonstrate how to use these techniques to analyze real-world financial situations such as COLAs for teachers.

■ Matrix Addition

We add two matrices of the same dimension by adding their corresponding entries. For example, to add the 2×2 matrices $A = \begin{bmatrix} 1 & 2 \\ 3 & 4 \end{bmatrix}$ and $B = \begin{bmatrix} 4 & -2 \\ -3 & 1 \end{bmatrix}$ to get matrix C, we have

$$C = A + B$$
$$= \begin{bmatrix} 1 & 2 \\ 3 & 4 \end{bmatrix} + \begin{bmatrix} 4 & -2 \\ -3 & 1 \end{bmatrix}$$
$$= \begin{bmatrix} 1+4 & 2+(-2) \\ 3+(-3) & 4+1 \end{bmatrix} \quad \text{Add corresponding entries.}$$
$$= \begin{bmatrix} 5 & 0 \\ 0 & 5 \end{bmatrix}$$

MATRIX ADDITION

An $m \times n$ matrix A and an $m \times n$ matrix B can be added together to form a new $m \times n$ matrix, C. The value of the entry in the ith row and jth column of C is $c_{ij} = a_{ij} + b_{ij}$.

If A and B are not of the same dimension, matrix addition is undefined.

EXAMPLE 1 ■ Calculating the Sum of Two Matrices

Calculate $C = A + B$ given $A = \begin{bmatrix} 2 & 0 \\ -1 & 6 \\ 4 & 5 \end{bmatrix}$ and $B = \begin{bmatrix} 9 & 2 \\ -7 & -6 \\ 8 & -1 \end{bmatrix}$.

SECTION 2.6 ■ MATRIX OPERATIONS AND APPLICATIONS

Solution

$$C = A + B$$

$$= \begin{bmatrix} 2 & 0 \\ -1 & 6 \\ 4 & 5 \end{bmatrix} + \begin{bmatrix} 9 & 2 \\ -7 & -6 \\ 8 & -1 \end{bmatrix}$$

$$= \begin{bmatrix} 2+9 & 0+2 \\ -1+(-7) & 6+(-6) \\ 4+8 & 5+(-1) \end{bmatrix}$$

$$= \begin{bmatrix} 11 & 2 \\ -8 & 0 \\ 12 & 4 \end{bmatrix}$$

Mikael Damkier/Shutterstock.com

EXAMPLE 2 ■ Using Matrix Addition in a Real-World Context

According to the package labeling, a 0.5-ounce serving of M&Ms Mini Baking Bits contains 3.5 grams of fat, 9 grams of carbohydrates, and 1 gram of protein. A 1-ounce serving of Walmart Party Size Party Peanuts contains 14 grams of fat, 5 grams of carbohydrates, and 8 grams of protein. A 1/2-cup serving of Diamond Sliced Almonds contains 15 grams of fat, 6 grams of carbohydrates, and 6 grams of protein. Create a matrix for each ingredient showing the fat, carbohydrate, and protein content of the product. Then add the three matrices together and interpret the meaning of the result.

Solution The matrix for each ingredient will be of the form $B = \begin{bmatrix} fat \\ carbohydrates \\ protein \end{bmatrix}$. We have

$$\text{M\&Ms: } M = \begin{bmatrix} 3.5 \\ 9 \\ 1 \end{bmatrix}$$

$$\text{Peanuts: } P = \begin{bmatrix} 14 \\ 5 \\ 8 \end{bmatrix}$$

$$\text{Almonds: } A = \begin{bmatrix} 15 \\ 6 \\ 6 \end{bmatrix}$$

We then add the three matrices.

$$M + P + A = \begin{bmatrix} 3.5 \\ 9 \\ 1 \end{bmatrix} + \begin{bmatrix} 14 \\ 5 \\ 6 \end{bmatrix} + \begin{bmatrix} 15 \\ 6 \\ 6 \end{bmatrix}$$

$$= \begin{bmatrix} 3.5 + 14 + 15 \\ 9 + 5 + 6 \\ 1 + 6 + 6 \end{bmatrix}$$

$$= \begin{bmatrix} 32.5 \\ 20 \\ 13 \end{bmatrix}$$

This means that if we create a mix from one serving of each ingredient, the mix will contain 32.5 grams of fat, 20 grams of carbohydrates, and 13 grams of protein.

■ Scalar Multiplication

There are two types of multiplication used with matrices: **scalar multiplication** and **matrix multiplication**. We will address scalar multiplication here. The term **scalar** means *constant* or *number*. Scalar multiplication "scales" the entries of a matrix by making them larger or smaller by a given factor.

For example, if $A = \begin{bmatrix} 1 & 3 & 0 \\ -2 & 4 & 6 \\ 5 & 7 & 9 \end{bmatrix}$ then

$$5A = 5 \begin{bmatrix} 1 & 3 & 0 \\ -2 & 4 & 6 \\ 5 & 7 & 9 \end{bmatrix}$$

$$= \begin{bmatrix} 5 \cdot 1 & 5 \cdot 3 & 5 \cdot 0 \\ 5 \cdot (-2) & 5 \cdot 4 & 5 \cdot 6 \\ 5 \cdot 5 & 5 \cdot 7 & 5 \cdot 9 \end{bmatrix} \quad \text{Multiply each entry by the scalar 5.}$$

$$= \begin{bmatrix} 5 & 15 & 0 \\ -10 & 20 & 30 \\ 25 & 35 & 45 \end{bmatrix}$$

In this example, the scalar 5 increased the magnitude of each entry by a factor of 5.

SCALAR MULTIPLICATION

An $m \times n$ matrix A and a real number k can be multiplied together to form a new $m \times n$ matrix, C. The value of the entry in the ith row and jth column of C is $c_{ij} = k \cdot a_{ij}$.

SECTION 2.6 ■ MATRIX OPERATIONS AND APPLICATIONS

EXAMPLE 3 ■ **Multiplying by a Scalar**

Calculate $C = -\frac{1}{2}A$ given $A = \begin{bmatrix} 2 & 0 \\ -4 & 6 \\ 1 & 8 \end{bmatrix}$.

Solution

$$C = -\frac{1}{2}A$$

$$= -\frac{1}{2}\begin{bmatrix} 2 & 0 \\ -4 & 6 \\ 1 & 8 \end{bmatrix}$$

$$= \begin{bmatrix} -\frac{1}{2} \cdot 2 & -\frac{1}{2} \cdot 0 \\ -\frac{1}{2} \cdot (-4) & -\frac{1}{2} \cdot 6 \\ -\frac{1}{2} \cdot 1 & -\frac{1}{2} \cdot 8 \end{bmatrix}$$

$$= \begin{bmatrix} -1 & 0 \\ 2 & -3 \\ -\frac{1}{2} & -4 \end{bmatrix}$$

EXAMPLE 4 ■ **Using Scalar Multiplication in a Real-World Context**

Table 2.20 shows the base salary schedule for educators at University of Nevada–Las Vegas (UNLV) for 2005–2006.

Table 2.20

Rank	Minimum	Maximum
IV: Professor	59,683	123,958
III: Associate	43,681	90,722
II: Assistant	37,103	77,061
I: Instructor	30,584	63,520

Source: hr.unlv.edu

For the 2006–2007 school year, faculty received a 4% cost of living adjustment. Create a new salary schedule for the 2006–2007 school year for UNLV faculty.

Solution We can solve the problem by writing the table as a matrix and multiplying by a scalar. Since salaries are to increase by 4%, the 2006–2007 salaries will be 104% of the 2005–2006 salaries.

$$104\% = \frac{104}{100}$$

$$= 1.04$$

Thus we will multiply the matrix by 1.04 to increase each entry by 4%.

$$S = \begin{bmatrix} 59{,}683 & 123{,}958 \\ 43{,}681 & 90{,}722 \\ 37{,}103 & 77{,}061 \\ 30{,}584 & 63{,}520 \end{bmatrix}$$

$$1.04S = 1.04 \begin{bmatrix} 59{,}683 & 123{,}958 \\ 43{,}681 & 90{,}722 \\ 37{,}103 & 77{,}061 \\ 30{,}584 & 63{,}520 \end{bmatrix}$$

$$\approx \begin{bmatrix} 62{,}070 & 128{,}916 \\ 45{,}428 & 94{,}351 \\ 38{,}587 & 80{,}143 \\ 31{,}807 & 66{,}061 \end{bmatrix}$$

The new matrix represents the calculated 2006–2007 pay scale for UNLV faculty, as shown in Table 2.21.

Table 2.21

Rank	Minimum	Maximum
IV: Professor	62,070	128,916
III: Associate	45,428	94,351
II: Assistant	38,587	80,143
I: Instructor	31,807	66,061

To aid in the application of the concepts we have discussed in this section, we provide a summary of key properties for the processes.

MATRIX ADDITION AND SCALAR MULTIPLICATION PROPERTIES

Let A, B, and C be $m \times n$ matrices and let c and k be real numbers. Let O be the $m \times n$ **zero matrix** (a matrix with entries of all zeros). The following properties hold.

Matrix Addition Properties

1. Additive Associative $\quad A + (B + C) = (A + B) + C$
2. Additive Commutative $\quad A + B = B + A$
3. Additive Identity $\quad A + O = O + A = A$
4. Additive Inverse $\quad (-A) + A = A + (-A) = 0$

Scalar Multiplication Properties

5. Distributive $\quad c(A + B) = cA + cB$
6. Distributive $\quad (c + k)A = cA + kA$
7. Multiplicative Associative $\quad c(kA) = (ck)A$
8. Scalar Unit $\quad 1A = A$
9. Scalar Zero $\quad 0A = O$

No matrix subtraction properties are listed because we view matrix subtraction as a combination of scalar multiplication and matrix addition. In other words, the matrix expression $A - B$ is equivalent to $A + (-1)B$. As shown in Example 5, in practice we typically simplify the matrix expression $A - B$ by subtracting the entries of B from the corresponding entries of A.

EXAMPLE 5 ■ Calculating the Difference of Two Matrices

Let $A = \begin{bmatrix} 4 & -1 & 2 \\ 3 & -5 & 0 \\ 9 & 11 & 20 \end{bmatrix}$ and $B = \begin{bmatrix} -4 & 7 & 3 \\ 10 & -9 & 10 \\ -2 & 4 & 8 \end{bmatrix}$. Show $A - B$ is equivalent to $A + (-1)B$.

Solution

$$A - B = \begin{bmatrix} 4 & -1 & 2 \\ 3 & -5 & 0 \\ 9 & 11 & 20 \end{bmatrix} - \begin{bmatrix} -4 & 7 & 3 \\ 10 & -9 & 10 \\ -2 & 4 & 8 \end{bmatrix}$$

$$= \begin{bmatrix} 4-(-4) & -1-7 & 2-3 \\ 3-10 & -5-(-9) & 0-10 \\ 9-(-2) & 11-4 & 20-8 \end{bmatrix}$$ Subtract the entries of B from the corresponding entries of A.

$$= \begin{bmatrix} 8 & -8 & -1 \\ -7 & 4 & -10 \\ 11 & 7 & 12 \end{bmatrix}$$

$$A + (-1)B = \begin{bmatrix} 4 & -1 & 2 \\ 3 & -5 & 0 \\ 9 & 11 & 20 \end{bmatrix} + (-1)\begin{bmatrix} -4 & 7 & 3 \\ 10 & -9 & 10 \\ -2 & 4 & 8 \end{bmatrix}$$

$$= \begin{bmatrix} 4 & -1 & 2 \\ 3 & -5 & 0 \\ 9 & 11 & 20 \end{bmatrix} + \begin{bmatrix} (-1)(-4) & (-1)7 & (-1)3 \\ (-1)10 & (-1)(-9) & (-1)10 \\ (-1)(-2) & (-1)4 & (-1)8 \end{bmatrix}$$ Multiply the entries of B by the scalar -1.

$$= \begin{bmatrix} 4 & -1 & 2 \\ 3 & -5 & 0 \\ 9 & 11 & 20 \end{bmatrix} + \begin{bmatrix} 4 & -7 & -3 \\ -10 & 9 & -10 \\ 2 & -4 & -8 \end{bmatrix}$$

$$= \begin{bmatrix} 8 & -8 & -1 \\ -7 & 4 & -10 \\ 11 & 7 & 12 \end{bmatrix}$$

Matrix expressions may combine one or more matrix operations. Example 6 includes matrix addition and scalar multiplication with three different matrices including the zero matrix.

EXAMPLE 6 ■ Solving a Matrix Algebra Problem Involving the Zero Matrix

Let $A = \begin{bmatrix} 1 & 0 & 4 \\ -1 & 2 & -5 \end{bmatrix}$, $B = \begin{bmatrix} 5 & 7 & 0 \\ 3 & -2 & -4 \end{bmatrix}$, and $O = \begin{bmatrix} 0 & 0 & 0 \\ 0 & 0 & 0 \end{bmatrix}$. Calculate $A - (2B + O)$.

Solution

$$A - (2B + O) = \begin{bmatrix} 1 & 0 & 4 \\ -1 & 2 & -5 \end{bmatrix} - \left(2 \cdot \begin{bmatrix} 5 & 7 & 0 \\ 3 & -2 & -4 \end{bmatrix} + \begin{bmatrix} 0 & 0 & 0 \\ 0 & 0 & 0 \end{bmatrix}\right)$$

$$= \begin{bmatrix} 1 & 0 & 4 \\ -1 & 2 & -5 \end{bmatrix} + (-1)\left(\begin{bmatrix} 10 & 14 & 0 \\ 6 & -4 & -8 \end{bmatrix} + \begin{bmatrix} 0 & 0 & 0 \\ 0 & 0 & 0 \end{bmatrix}\right) \quad \text{Multiply } B \text{ by 2.} \\ \text{Rewrite "–" as "+(–1)."}$$

$$= \begin{bmatrix} 1 & 0 & 4 \\ -1 & 2 & -5 \end{bmatrix} + (-1)\begin{bmatrix} 10 & 14 & 0 \\ 6 & -4 & -8 \end{bmatrix}$$

$$= \begin{bmatrix} 1 & 0 & 4 \\ -1 & 2 & -5 \end{bmatrix} + \begin{bmatrix} -10 & -14 & 0 \\ -6 & 4 & 8 \end{bmatrix} \quad \text{Multiply } 2B \text{ by } -1.$$

$$= \begin{bmatrix} -9 & -14 & 4 \\ -7 & 6 & 3 \end{bmatrix}$$

SUMMARY

In this section you learned how to do matrix addition and scalar multiplication. You saw how these techniques could be used in life and business to analyze real-world situations.

TECHNOLOGY TIP ■ ADDING TWO MATRICES

1. Enter matrix A and matrix B into the calculator using the Matrix Menu.

3. Press ENTER to display the sum of the two matrices.

2. Use the Matrix Names Menu to place matrix A on the home screen. Press the $+$ key. Then place matrix B on the home screen.

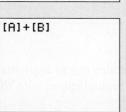

Error Alert:
If the DIM MISMATCH error message appears, double-check that you have entered the matrices correctly. If you have, this message tells you that the matrix addition is not possible because the two matrices are of different dimensions.

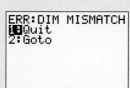

TECHNOLOGY TIP — SCALAR MULTIPLICATION

1. Enter matrix A into the calculator using the Matrix Menu.

2. Type in the scalar on the home screen, press the $\boxed{\times}$ key, and then use the Matrix Names Menu to place matrix A on the home screen.

3. Press $\boxed{\text{ENTER}}$ to display the product of the scalar and the matrix.

4. If you would like to convert decimal entries to fractions, press $\boxed{\text{MATH}}$, select FRAC, then press $\boxed{\text{ENTER}}$.

2.6 EXERCISES

SKILLS AND CONCEPTS

In Exercise 1–10, perform the indicated matrix operation, if possible, given the following matrices. Solve these problems without technology.

$$A = \begin{bmatrix} 1 & 6 \\ 3 & 4 \\ 5 & 2 \end{bmatrix}, B = \begin{bmatrix} -5 & 0 \\ 7 & 8 \\ -9 & 1 \end{bmatrix},$$

$$C = \begin{bmatrix} 2 & 3 & 0 \\ -1 & 4 & 1 \\ 5 & -2 & 0 \end{bmatrix}, D = \begin{bmatrix} 1 & -3 & 4 \\ 0 & 5 & 7 \\ 9 & 8 & 2 \end{bmatrix}$$

1. $A + B$
2. $A - B$
3. $B + C$
4. $C - D$
5. $D + C$
6. $2A$
7. $3B$
8. $2A + 3B$
9. $-2C + 2D$
10. $2A + 2C$

In Exercises 11–20, use technology to simplify the matrix expressions given the following matrices.

$$A = \begin{bmatrix} 1.2 & 6.3 & 0.4 \\ -9.1 & 4.2 & 1.7 \\ 0.9 & -2.0 & 0.3 \end{bmatrix},$$

$$B = \begin{bmatrix} 1.4 & -0.3 & 0.4 \\ -2.8 & 5.5 & 7.1 \\ 9.2 & 8.6 & 2.0 \end{bmatrix}$$

11. $1.2A$
12. $-2.3B$
13. $1.2A - 2.3B$
14. $4.1A + 0.1B$
15. $-1.1A + 2.9B$
16. $2.9A + 0.1B$
17. $-8.7A + 8.7B$
18. $-A - 9.2B$
19. $7.8A + 9.9B$
20. $-1.7A - 2.1B$

21. Personal Debt A couple is planning to get married. He has the following debts: a $2700 consumer loan at Loan Shark Larry's, a $26,500 car loan, and an $8200 credit card balance at Risky Bank, plus a $2700 consumer loan at Mastercraft Jewelers (for the ring). She has the following debts: a $1200 car loan, an $82,500 home mortgage, and a $200 credit card balance at Risky Bank, plus a $250 consumer loan at Mastercraft Jewelers. Determine their combined debts of each type at the various financial institutions.

22. Food Supply Bed and breakfasts attract visitors with their intimate ambiance and delicious cuisine. Many establishments serve wedding brunches in addition to providing lodging services.

A bed and breakfast hostess is planning an upcoming wedding brunch. She plans to make 4 dozen muffins (2 dozen apple and 2 dozen blueberry) and 6 fruit crisps (3 apple and 3 blueberry). The amount of flour, sugar, and fruit required to make a single fruit crisp is shown in Table A and the amount required to make a dozen muffins is shown in Table B.

Table A

Crisp	Flour	Sugar	Fruit
apple	1/2 cup	3/4 cup	3 cups
blueberry	1/2 cup	1 cup	4 cups

Table B

Muffins	Flour	Sugar	Fruit
apple	2 cups	1/4 cup	3/4 cup
blueberry	2 cups	1/3 cup	1 cup

Use matrices to create a table that shows how much flour, sugar, and fruit will be required for the apple desserts and the blueberry desserts.

■ SHOW YOU KNOW

23. What is scalar multiplication?

24. Explain why two matrices must have the same dimension for them to be added together.

25. Describe the conditions under which the sum of two matrices results in a matrix consisting only of zeros.

26. Create an argument to justify the commutative property, $A + B = B + A$, for the sum of two matrices.

27. Suppose the difference, $A - B$, of two matrices is calculated. Describe the relationship between $A - B$ and the difference $B - A$.

■ MAKE IT REAL

In Exercises 28–41, use matrix addition and/or scalar multiplication to find the solution.

28. Faculty Salaries The 2001–2002 pay scale for the Green River Community College faculty is shown in the table. For the 2002–2003 academic year, the faculty received a 3.432% cost of living increase. Determine the 2002–2003 pay scale.

	240 Credits (dollars)	300 Credits (dollars)	360 Credits (dollars)
Level 1	35,131	38,742	42,354
Level 2	37,016	40,627	44,238
Level 3	38,900	42,511	46,122
Level 4	40,784	44,395	48,006

Source: Green River Community College

29. Faculty Salaries If the Green River Community College faculty in Exercise 28 gets a 3.432% raise annually, determine the 2005–2006 pay scale.

30. Condiment Prices On July 20, 2002, Albertsons.com advertised the following items at the indicated prices.

	Albertsons	Hunt's	Kraft
24-ounce ketchup	$0.69	$1.89	
18-ounce barbecue sauce		$0.99	$0.99
32-ounce mayonnaise	$2.19		$3.19

If the prices are only affected by inflation and the annual rate of inflation is 3%, determine the price of each of the items on July 20, 2006.

31. Soda Prices On July 20, 2002, Albertsons.com advertised the following items at the indicated prices.

	Albertsons	A&W	Henry Weinhard's
6-pack root beer	$1.59	$1.67	$4.50
12-pack cream soda	$3.18	$3.34	$9.00
2-liter club soda	$0.99		

If the prices are only affected by inflation and the annual rate of inflation is 3%, determine the price of each of the items on July 20, 2006.

32. Auto Prices The average trade-in values of a Volkswagen New Beetle and Volkswagen Golf in July 2002 are shown in Table C. The average retail values of the two vehicles are shown in Table D.

Ruben Enger/Shutterstock.com

Table C

Average Trade-In		
	Golf	New Beetle
2000 Model	$11,000	$11,850
2001 Model	$11,875	$13,175

Source: www.nada.com

Table D

Average Retail		
	Golf	New Beetle
2000 Model	$13,050	$14,000
2001 Model	$14,025	$15,475

Source: www.nada.com

Use matrices to create a table that shows the average dealer markup for each of the vehicles.

33. **Auto Prices** The average trade-in values of a Honda Civic and Honda Accord in July 2002 are shown in Table E. The average retail values of the two vehicles are shown in Table F.

Table E

Average Trade-In		
	Accord	Civic
2000 Model	$14,800	$8,925
2001 Model	$16,575	$9,850

Source: www.nada.com

Table F

Average Retail		
	Accord	Civic
2000 Model	$17,100	$10,800
2001 Model	$18,975	$11,825

Source: www.nada.com

Use matrices to create a table that shows the average dealer markup for each of the vehicles.

34. **Energy Usage** The amount of natural gas and coal energy produced and consumed in the United States is shown in the tables. Use matrix operations to create a table that shows the difference between energy production and consumption.

Energy Production		
Years Since 1960	Natural Gas (quadrillion BTUs)	Coal (quadrillion BTUs)
0	12.66	10.82
10	21.67	14.61
20	19.91	18.60
30	18.36	22.46
40	19.74	22.66

Energy Consumption		
Years Since 1960	Natural Gas (quadrillion BTUs)	Coal (quadrillion BTUs)
0	12.39	9.84
10	21.80	12.27
20	20.39	15.42
30	19.30	19.25
40	23.33	22.41

Source: Statistical Abstract of the United States, 2001, **Table 891**

35. **Energy Policy** If you were a lobbyist for the natural gas industry, how would you use the results of Exercise 34 to persuade legislators to support further natural gas exploration?

36. **Energy Usage** The amount of nuclear electric power and coal energy produced in the United States is shown in the tables. Use matrix operations to create a table that shows the difference between energy production and consumption.

Energy Production		
Years Since 1960	Nuclear Electric Power (quadrillion BTUs)	Coal (quadrillion BTUs)
0	0.01	10.82
10	0.24	14.61
20	2.74	18.60
30	6.16	22.46
40	8.01	22.66

Energy Consumption		
Years Since 1960	Nuclear Electric Power (quadrillion BTUs)	Coal (quadrillion BTUs)
0	0.01	9.84
10	0.24	12.27
20	2.74	15.42
30	6.16	19.25
40	8.01	22.41

Source: Statistical Abstract of the United States, 2001, **Table 891**

37. **Energy Policy** An environmentalist believes the United States should only produce as much energy as it consumes. Based on the results of Exercise 36, which energy technology (coal or nuclear electric power) seems to best support the environmentalist's position? Defend your conclusion.

38. **Renewable Energy Consumption** The amount of renewable energy consumed in the United States in various years is shown in the tables. Use matrices to make a table showing the total amount of energy consumed between the beginning of 1997 and the end of 1999.

1997 Energy Consumption	
Renewable Energy Type	Energy Consumed (quadrillion BTUs)
conventional hydroelectric power	3.94
geothermal energy	0.33
biomass	2.98
solar energy	0.07
wind energy	0.03

1998 Energy Consumption	
Renewable Energy Type	Energy Consumed (quadrillion BTUs)
conventional hydroelectric power	3.55
geothermal energy	0.34
biomass	2.99
solar energy	0.07
wind energy	0.03

1999 Energy Consumption	
Renewable Energy Type	Energy Consumed (quadrillion BTUs)
conventional hydroelectric power	3.42
geothermal energy	0.33
biomass	3.51
solar energy	0.08
wind energy	0.04

Source: Statistical Abstract of the United States, 2001, Table 896

39. Average Energy Consumption Using the results from Exercise 38, use matrix operations to construct a table showing the average annual amount of energy consumed between the start of 1997 and the end of 1999.

40. Energy Usage Trends Using the data tables from Exercise 38 and the results from Exercise 39, construct a table showing the difference between the 1999 energy consumption and the average annual energy consumption. What conclusions can you draw from your result?

41. Organized Physical Activity The tables show the percentage of high school students involved in physical education classes. Based on the information given, can you determine the percentage of ninth-graders who exercise 20 or more minutes per class? Justify your answer.

Males Enrolled in a Physical Education Class		
Grade	Enrolled in a P.E. Class (percent)	Exercised 20 Minutes or More per Class (percent)
9	60.7	82.1
10	82.3	84.4
11	65.3	79.4
12	44.6	82.0

Females Enrolled in a Physical Education Class		
Grade	Enrolled in a P.E. Class (percent)	Exercised 20 Minutes or More per Class (percent)
9	51.5	69.6
10	75.6	72.5
11	56.6	70.2
12	36.8	68.0

Source: Statistical Abstract of the United States, 2001, Table 1246

STRETCH YOUR MIND

Exercises 42–45 are intended to challenge your ability to apply matrix addition and scalar multiplication properties. For each exercise, use

$$A = \begin{bmatrix} 2 & 15 & 0 \\ 6 & -4 & -9 \\ -8 & \frac{1}{2} & \frac{2}{3} \end{bmatrix}, C = \begin{bmatrix} 12 & 1 & 4 \\ 3 & 0 & 5 \\ -2 & \frac{3}{2} & -\frac{1}{3} \end{bmatrix}$$

42. Solve the matrix equation for B.

$$5A - B = C$$

43. Solve the matrix equation for B.

$$-2A + 3B = C$$

44. Solve the matrix equation for B.

$$\tfrac{1}{2}A + \tfrac{2}{3}B = \tfrac{1}{6}C$$

45. Solve the matrix equation for B.

$$-100A - 200B = 330C$$

CHAPTER 2 Study Sheet

As a result of your work in this chapter, you should be able to answer the following questions, which are focused on the "big ideas" of this chapter.

SECTION 2.1
1. What distinguishes linear functions from other types of functions?
2. What does it mean to say a function has a constant rate of change?
3. How do you determine if a data set, graph, equation, or verbal expression represents a linear function?
4. What is an efficient way to graph a linear function whose equation is given in standard form? In slope-intercept form? In point-slope form?

SECTION 2.2
5. What key phrases indicate that a linear function can model a phenomenon that is being described verbally?

SECTION 2.3
6. What does it mean for a line to be the line of best fit for a data set?
7. What is the relationship between the phrases *least squares* and *line of best fit*?
8. What does the coefficient of determination represent?
9. How do you interpret the value of the correlation coefficient?
10. What is linear regression used for?

SECTION 2.4
11. Describe the different ways to solve a system of equations and describe the benefits of each.
12. What is the relationship between the algebraic solution to a system of linear equations and the point of intersection of the graphs of the equations in the system?
13. Explain, using examples, the meaning of the terms *inconsistent* and *dependent* system of equations.

SECTION 2.5
14. How are systems of equations and augmented matrices related?
15. What row operations result in an equivalent system of equations?

SECTION 2.6
16. What is scalar multiplication?

REVIEW EXERCISES

SECTION 2.1

In Exercises 1–3, determine the slope of the linear function that passes through each point and then give its vertical and horizontal intercept.

1. $(1, 7)$ and $(6, 2)$
2. $(2, 9)$ and $(4, 3)$
3. $(-3, -4)$ and $(0, -2)$

In Exercises 4–6, determine the slope and any horizontal or vertical intercepts of each linear function.

4. $y = -5x + 10$
5. $y = -3x + 18$
6. $y = 2x - 12$

In Exercises 7–12, linear functions are represented numerically, symbolically, or verbally.

    a. Graph each linear function by hand.

    b. Determine the slope and any vertical or horizontal intercepts of the graph algebraically and plot them on your graph.

    c. For real-world contexts, explain the meaning of the slope and horizontal and vertical intercepts. If any of the values do not make sense in the given context, so state.

7. $y = 5x + 4$
8. $y = -9$
9. The line passing through $(5, 6)$, $(-3, -9)$
10. The line that passes through $(-8, 2)$, $(0, 5)$
11. **Medical Research** Medical researchers have found that there is a linear relationship between a person's blood pressure and their weight. In males 35 years of age, for every 5-pound increase in the person's weight there is generally an increase in the systolic blood pressure of 2 millimeters of mercury (mmHg). Moreover, for a male 35 years of age and 190 pounds the preferred systolic blood pressure is 125 mmHg.
12. **Business Costs** The weekly payroll cost C (in dollars) of running a cell phone company is related to the number n of salespersons. Suppose there are fixed costs of $4800 per week, and each salesperson costs the company $1100 per week. For example, if there are 10 salespersons, then the weekly cost is $11,000.
13. **Car Sales** A car company has found a linear relationship between the amount of money it spends on advertising and the number of cars it sells. Suppose when it spent $50,000 on advertising, it sold 500 cars. Moreover, assume for each additional $5000 spent, it will sell 20 more cars.

    a. Find a formula for c, the number of cars sold, as a linear function of the amount spent on advertising, a.

    b. What is the slope of the linear function and its meaning in this context?

    c. What is the vertical intercept of the function and its meaning in this context?

SECTION 2.2

In Exercises 14–16, find a model for the data and answer the given questions.

14. **Nutrition** The Recommended Daily Allowance (RDA) for fat for a person on a 2000 calorie per day diet is less than 65 grams. A McDonald's Big 'N' Tasty™ sandwich contains 32 grams of fat. A super size order of French fries contains 29 grams of fat. (*Source:* www.mcdonalds.com)

    a. Write the equation for fat grams consumed as a function of super size orders of French fries eaten.

    b. Write the equation for fat grams consumed as a function of Big 'N' Tasty sandwiches eaten.

    c. How many combination meals (Big 'N' Tasty sandwich and super size order of French fries) can a person eat without exceeding the RDA for fat?

15. **Used Car Value** In 2011, the average price of a 2007 Toyota Prius was $14,800. The average retail price of a 2010 Toyota Prius was $21,950. (*Source:* www.nadaguides.com) Find a linear model for the value of a Prius in 2011 as a function of its production year.

16. **Used Car Value** In 2001, the average retail price of a 1998 Mercedes-Benz Roadster two-door SL500 was $51,400. The average retail price of a 2000 Mercedes-Benz Roadster two-door SL500 was $66,025. (*Source:* www.nadaguides.com)

    a. Find a linear model for the value of a Mercedes-Benz Roadster two-door SL500 in 2001 as a function of its production year.

    b. Use your linear model to predict the 2001 value of a 1999 Mercedes-Benz Roadster two-door SL500.

    c. A 1999 Mercedes-Benz Roadster two-door SL500 had an average retail price of $58,500 in 2001. How good was your linear model at predicting the value of the vehicle?

In Exercises 17–18, you are given an incomplete table of values for a real-world function. Determine the average rate of change over the given interval, then use it to estimate the remaining values in the table.

17. Admissions to Spectator Amusements

Year	Admissions to Spectator Amusements ($ billions)
2003	$36.0
2004	
2005	
2006	$39.9

Source: Statistical Abstract of the United States, 2008, Table 1193

18. Total Attendance at National Football League Games

Year	Total Attendance at NFL Games (thousands of people)
2003	21,709
2004	
2005	
2006	22,256

Source: Statistical Abstract of the United States, 2008, Table 1204

■ SECTION 2.3 ■

19. Consider the number of hours of sleep per day for a baby relative to the age of the baby (starting with newborn). Would the correlation coefficient of a linear model of this situation be positive, negative, or zero? Explain.

20. Consider the average winning speed of cars competing in the Indianapolis 500 relative to the year in which the race took place (starting in 1912). Would the correlation coefficient in this situation be positive, negative, or zero? Explain.

21. Researchers collected data on 100 individuals' grade point averages and the number of hours per week spent watching television. In the linear regression done on this data, the correlation coefficient, r, was -0.95. What does this tell about the relationship between grade point averages and television watching in this study? Explain what the negative sign as well as the 0.95 numerical value mean in terms of this study.

22. Which line (a or b) is the line of best fit in the following figure? Write a verbal explanation of why the line you chose is the line of best fit.

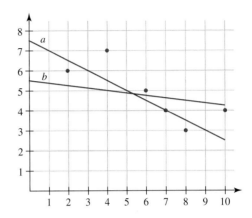

23. Sleep Hours The table shows the recommended number of hours of sleep each day for the typical child as given by Lucile Packard Children's Hospital at Stanford.

Age (months) a	Total Sleep Hours per Day H
0	16
1	15.5
3	15
6	14
9	14
12	14
18	13.5
24	13

Source: www.lpch.org

a. Create a scatter plot of the total sleep hours per day as a function of the age.

b. Find the linear regression model for these data.

c. Does the linear model fit the data well? Explain. (Refer to the correlation coefficient or coefficient of determination in your explanation.)

d. Use the model to predict the total sleep hours per day for a 3-year-old child.

24. **Net Sales vs. Cost of Sales** The cost of sales and net sales for Apple Computer, Inc. between 2000 and 2005 is given in the table.

Years Since 2000	Cost of Sales ($ millions)	Net Sales ($ millions)
1	4128	5363
2	4139	5742
3	4499	6207
4	6020	8279
5	9888	13,931

Source: www.apple.com

a. Which of the following is best modeled by a linear function?
   i. cost of sales as a function of year
   ii. net sales as a function of year
   iii. net sales as a function of cost of sales
b. Find the equation of the linear model that best fits the data set you identified in part (a).
c. According to the model, what will be the cost of sales when Apple Computer, Inc. generates 20,000 million dollars in net sales?

25. **Height and Weight of a Professional Basketball Team** The table shows the height and weight of the players on the Phoenix Suns 2006–2007 roster.

Height (inches) H	Weight (pounds) W	Height (inches) H	Weight (pounds) W
74	200	82	250
75	188	75	195
77	210	79	215
83	250	80	215
80	230	82	245
80	220	81	235
80	230	74	200
79	228		

Source: www.nba.com

a. For the data provided, find the equation for the line of best fit and label it $w(h)$.
b. Interpret the meaning of the slope and vertical intercept in the context of this problem.
c. Find $w(70)$ and interpret the meaning of the answer in the context of this problem.
d. Solve $w(h) = 200$ and interpret the meaning of the answer in the context of this problem.

■ SECTION 2.4 ■

In Exercises 26–29,

a. Solve the system of equations algebraically. If a solution does not exist, so state.
b. Use a graphing calculator to find the point(s) of intersection of the lines. Compare your solutions to those obtained in part (a). (*Hint:* Remember that to graph a standard form equation using your graphing calculator, you will need to solve the equation for *y*.)

26. $y = 3x + 4$
    $y = 1.7x + 5.3$

27. $y = 4.6x + 10$
    $y = -10x + 44$

28. $y = 7.1x - 41.5$
    $y = -1.9x + 3.5$

29. $y = 2.5x - 1$
    $y = -0.25x + 10$

30. **Artist and Athlete Earnings** In 1999 the number of paid employees in the independent artist, writer, and performer industry surpassed the number of paid employees in the sports team industry. However, the annual payroll for the artist industry still trailed the sports team industry payroll by $4.1 billion. Based on data from 1998 and 1999, the payroll for the sports team industry can be modeled by

$$S(t) = 990t + 5718 \text{ million dollars}$$

and the payroll for the independent artist industry can be modeled by

$$A(t) = -35t + 3494 \text{ million dollars}$$

where *t* is the number of years since 1998. (*Source:* Modeled from *Statistical Abstract of the United States, 2001*, Table 1232)

a. According to the models, when were the payrolls for the two industries equal? Check your solution by graphing the system or by substituting the solution back into the original equations.
b. Does this solution seem reasonable? Explain.

31. Pork and Beef Consumption The table shows the per capita retail consumption for pork and beef in pounds per person in the United States from 1981 to 2001.

Years Since 1980	Retail Pork Consumption (pounds per person)	Retail Beef Consumption (pounds per person)
1	54.7	78.3
8	52.5	72.7
11	50.2	66.7
16	48.5	67.3
21	50.3	66.3

Source: www.agmanager.info

a. Use linear regression to create a system of equations that models the per capita retail pork and beef consumption. Round off to 2 decimal places.

b. What are the constant rates of change in your models? What do they represent?

c. Solve the system algebraically and explain what the solution represents. Check your answer by graphing the system or by substituting your solution back into the original equations.

d. What assumption(s) do we make when solving this system of equations?

In Exercises 32–33, solve the system of equations. Verify your solution by graphing the system or by substituting your solution back into the original equations. If the system of equations is inconsistent or dependent, so state.

32. $2x + 2y = 20$
$3x + 2y = 27$

33. $4x + 5y = 61$
$-4x - 6y = -74$

34. Computer Prices A small business placed two orders for computers from Hewlett-Packard. The order for the human resources department totaled $12,576.24 and consisted of 8 HDX 16t Premium Series laptops and 5 Pavilion dv7t Series laptops. The order for the marketing department totaled $10,524.30 and consisted of 7 HDX 16t Premium Series laptops and 4 HP Pavilion dv7t Series laptops. (*Source:* www.hp.com, June 2009) Assume that the prices of the computers did not change between orders.

a. Create a system of equations to model this situation. Be sure to identify what your variables represent.

b. Solve the system of equations. How much did each type of computer cost?

35. Trail Mix Planters Honey Nut Medley Trail Mix contains 8 grams of protein per 2-ounce serving. Planters Dry Roasted Peanuts contain 16 grams of protein per 2-ounce serving. (*Source:* product labels) Suppose you want to create a 16-ounce mixture of trail mix and peanuts that has a combined total of 84 grams of protein.

a. What are the unknowns in this situation?

b. Create a system of equations to model this situation.

c. Solve the system of equations and explain what your solution means.

d. Verify your solution by graphing the system or by substituting the solution into the original equations.

■ **SECTION 2.5** ■

In Exercises 36–40, write the system of equations as an augmented matrix then simplify the matrix to reduced row echelon form. Identify the solution(s) to the system of equations.

36. $2.1x - y = -8$
$3.4x + y = 8$

37. $5.2x - 1.3y = 12$
$-10.4x + 2.6y = 24$

38. $4x - 8y = -16$
$x + y = 5$

39. $2.9x - 8.1y = 4$
$3.7x + 16.2y = 1.5$

40. $x + y + z = 6$
$2x - y + z = 3$
$3x + 2z = 9$

In Exercises 41–44, use technology to write the matrix in reduced row echelon form.

41. $\begin{bmatrix} 8.1 & 4.3 & 6.2 & | & 3.1 \\ 2.7 & -2.8 & 7.5 & | & 7.2 \\ 8.2 & -5.3 & -1.1 & | & 5.9 \end{bmatrix}$

42. $\begin{bmatrix} 2.6 & 3.0 & 0.2 & | & 9.6 \\ 6.1 & 4.0 & 6.5 & | & 0.8 \\ 9.1 & -5.0 & -0.8 & | & 9.9 \end{bmatrix}$

43. $\begin{bmatrix} 1 & 5 & -1 & | & -11 \\ 0 & 0 & 3 & | & 12 \\ 2 & 4 & -2 & | & 8 \end{bmatrix}$

44. $\begin{bmatrix} 1 & 2 & 3 & | & 9 \\ 2 & -1 & 1 & | & 8 \\ 3 & 0 & -1 & | & 3 \end{bmatrix}$

SECTION 2.6

In Exercises 45–49, perform the indicated matrix operation, if possible, given the following matrices. Solve these problems without technology.

$$A = \begin{bmatrix} 2 & 3 \\ -3 & -4 \\ 0 & 7 \end{bmatrix}, B = \begin{bmatrix} 5 & 6 \\ 3 & -4 \\ 9 & 2 \end{bmatrix}$$

45. $A + B$
46. $A - B$
47. $2A$
48. $3B$
49. $2A + 3B$

In Exercises 50–53 use technology to simplify the matrix expressions given

$$A = \begin{bmatrix} 1.2 & 6.1 & -0.4 \\ -9.1 & 4.2 & 1.7 \\ 0.9 & 3.0 & 3.3 \end{bmatrix}, B = \begin{bmatrix} 3.4 & -0.3 & -0.4 \\ -3.8 & 5.6 & 7.2 \\ 2.2 & 2.6 & 2.0 \end{bmatrix}$$

50. $3.2A$
51. $5.3B$
52. $3.2A - 5.3B$
53. $4.1A + 0.1B$
53. $-1.1A + 2.9B$

55. Auto Prices The average trade-in values of a Toyota Celica and Toyota MR2 Spyder in July 2002 are shown in Table G. The average retail values of the two vehicles are shown in Table H.

Table G

Average Trade-In		
	Celica	MR2 Spyder
2000 Model	$13,025	$18,750
2001 Model	$14,850	$20,100

Source: www.nada.com

Table H

Average Retail		
	Celica	MR2 Spyder
2000 Model	$15,300	$21,300
2001 Model	$17,250	$22,725

Source: www.nada.com

Use matrices to create a table that shows the average dealer markup for each of the vehicles.

Make It Real Project

What to Do

1. Find a set of at least six data points from an area of personal interest.
2. Draw a scatter plot of the data and explain why you think a linear model would or would not fit the data well.
3. Find the equation of the line of best fit for the data.
4. Interpret the physical meaning of the slope and *y*-intercept of the model.
5. Use the model to predict the value of the function at an unknown point. Do you think the prediction is accurate? Explain.
6. Explain how a consumer and/or a businessperson could benefit from the model.

Where to Find Data

Box Office Guru

www.boxofficeguru.com

Look at historical data on movie revenues.

Nutri-Facts

www.nutri-facts.com

Compare nutritional content of common foods based on serving size.

Quantitative Environmental Learning Project

www.seattlecentral.org/qelp

Look at environmental information in easy-to-access charts and tables.

U.S. Census Bureau

www.census.gov

Look at data on U.S. residents ranging from Internet usage to family size.

Local Gas Station or Supermarket

Track an item's price daily for a week.

School Registrar

Ask for historical tuition data.

Utility Bills

Look at electricity, water, or gas usage.

Employee Pay Statements

Look at take-home pay or taxes.

CHAPTER 3

Transformations of Functions

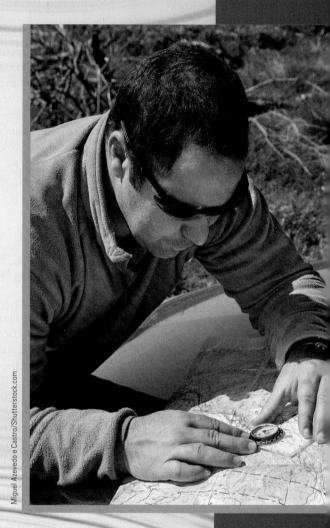

Getting lost in the back country is frightening and potentially life threatening. For most avid hikers, a topographic map and a compass are as important as an adequate supply of water.

Topographic maps use contour lines to show the variability of the terrain. Hikers are typically less interested in the elevation of a particular geographical feature (ridge, peak, saddle) than in the elevation change between their current position and the geographical feature. The concept of function transformations can help hikers transform map data into practical information and determine their best hiking route.

3.1 Vertical and Horizontal Shifts
3.2 Vertical and Horizontal Reflections
3.3 Vertical Stretches and Compressions
3.4 Horizontal Stretches and Compressions
STUDY SHEET
REVIEW EXERCISES
MAKE IT REAL PROJECT

SECTION 3.1 Vertical and Horizontal Shifts

LEARNING OBJECTIVES

- Identify what change in a function equation results in a vertical shift
- Identify what change in a function equation results in a horizontal shift

GETTING STARTED

With an elevation of 8850 meters (29,035 feet) at its highest point, Mount Everest is the tallest mountain in the world. Every year people gather in Nepal in organized expeditions to "summit"—climb to the peak of—Mount Everest. As the expeditions make their way up the mountain, climbers typically ascend to a certain elevation, spend some time resting, and then hike partway back down the mountain. In this way, their bodies can acclimatize to the extreme conditions of high elevation, low oxygen levels, physical strain, and cold.

In this section we explore the relationship between a function and a transformation of the function. Specifically, we investigate vertical and horizontal shifts and see how to model real-world situations such as the elevation of mountain climbers with such transformations. We also look at the symmetry of some function graphs.

■ Vertical Shifts

We know that functions may be represented by equations, graphs, tables, or words, but **transformations** (changes) in functions are often easiest to see using a graph. We will use the Mount Everest example to illustrate one of the most common transformations, the *vertical shift*.

On April 10, 2005, a Mount Everest expedition arrived at its base camp (elevation: 5165 meters). Between April 10 and May 30, 2005, the group followed the recommended pattern for acclimatization—spending periods of time resting and hiking up and down the mountain to ensure a successful summit. Table 3.1 and Figure 3.1 show the climbers' elevation, E (in meters), as a function of the days, d, since their arrival at base camp.

Table 3.1

Days Since April 10, 2005 d	Elevation (meters) $E(d)$
0	5165
4	5600
5	6363
6	5782
7	6400
15	7040
20	7680
21	5165
22	7040
24	7680
26	5165
33	6400
36	7040
39	7700
40	7040
41	6400
48	7040
49	7800
50	8850

Figure 3.1

Source: www.project-himalaya.com

EXAMPLE 1 ■ Shifting a Graph Vertically

Based on the Mount Everest expedition data in Table 3.1 and Figure 3.1, create a table of values and draw the associated graph that represents the climbers' elevation relative to their base camp (elevation: 5165 meters) d days after April 10. How do your table and graph compare to the model E?

Solution To express the elevation relative to base camp, the elevation for any given day must reflect the difference between the actual elevation, $E(d)$, and the elevation at base camp, $B(d)$. For example, on April 25 ($d = 15$) the actual elevation is 7040 meters, which is 1875 meters higher than the elevation at base camp. Partial data showing this relationship is given in Table 3.2.

SECTION 3.1 ■ VERTICAL AND HORIZONTAL SHIFTS

Table 3.2

Days Since April 10, 2005 d	Elevation (meters above sea level) $E(d)$	Elevation (meters above base camp) $B(d)$
0	5165	$5165 - 5165 = 0$
5	6363	$6363 - 5165 = 1198$
15	7040	$7040 - 5165 = 1875$
20	7680	$7680 - 5165 = 2515$
40	7040	$7040 - 5165 = 1875$
50	8850	$8850 - 5165 = 3685$

For every value of d, $B(d)$ is 5165 meters less than $E(d)$. This means that the graph of B is formed by shifting E downward 5165 units, as shown in Figure 3.2a. This vertical shift occurs because we have changed the vertical point of reference for the function. The graph of the new function is shown in Figure 3.2b.

Figure 3.2

Note that the shifted graph in Figure 3.2b still possesses the basic properties of the original function. The relative changes in altitude from day to day remain the same, as do the number of days it took the expedition to summit the mountain. The only change is the vertical point of reference that results in the **vertical shift**.

VERTICAL SHIFTS

The graph of $g(x) = f(x) + k$ is the graph of f shifted vertically by $|k|$ units. If k is positive, the shift is upward. If k is negative, the shift is downward.

EXAMPLE 2 ■ Determining the Relationship between Two Functions That Differ by a Vertical Shift

Refer to the functions introduced in Example 1 and write an equation that relates $B(t)$ and $E(t)$.

Solution Recall that $B(t)$ was obtained by subtracting 5165 from $E(t)$. Therefore, $B(t) = E(t) - 5165$. The graph of B is the graph of E shifted downward by 5165.

■ Generalizing Transformations: Vertical Shifts

The key to understanding vertical shifts, including how the graphs of two functions are related, is to realize that we are defining the outputs of one function using the outputs of another function. In other words, we have a function defined such that we know its outputs for some set of inputs. This can be provided as a graph, a table, a description, or a formula. We then want to construct a new function, called the **image function**, based on the **parent function** (the function we already know).

For example, consider $g(x) = f(x) + 5$. We are implying there exists a function f whose outputs can be used to define the outputs of function g. For all values of x in the domain of f, we think of $g(x) = f(x) + 5$ as follows:

$$\underbrace{g(x)}_{\text{the output of } g \text{ at } x} \underbrace{=}_{\text{is equal to}} \underbrace{f(x)}_{\text{the output of } f \text{ at } x} \underbrace{+ \ 5}_{\text{increased by 5}}$$

Because the basic idea is true no matter how the function f (the parent function) is given (description, table, graph, or formula), we can understand the relationship between the parent and image functions in all settings. With vertical shifts, the *pattern* of the outputs does not change. That is, the relationship between the outputs does not change even though their vertical placement on the graph varies.

EXAMPLE 3 ■ Using Vertical Shifts in a Real-World Context

Based on 2007 ticket prices, the cost of an adult Disney Park Hopper® Bonus Ticket can be modeled by

$$T(d) = -5.357d^2 + 59.04d + 28.20 \text{ dollars}$$

where d is the number of days the ticket authorizes entrance into Disneyland and Disney California Adventure. If Disney executives authorize a $10 across-the-board increase in ticket prices for 2012, what will be the model for 2012 ticket prices?

Solution Since each ticket price is increased by $10, the 2012 ticket price model will be the 2007 model plus $10.

$$\begin{aligned} N(d) &= T(d) + 10 \\ &= (-5.357d^2 + 59.04d + 28.20) + 10 \\ &= -5.357d^2 + 59.04d + 38.20 \text{ dollars} \end{aligned}$$

■ Horizontal Shifts

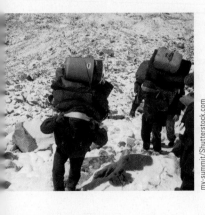

Two functions may also be related by a shift to the left or right. In a **horizontal shift**, two functions will have identical output values but these output values will occur for different input values. We will illustrate this concept by returning to the Mount Everest expedition data.

EXAMPLE 4 ■ Shifting a Graph Horizontally

According to the electronic journal kept by members of the April 2005 Mount Everest expedition, the group was concerned their excursion might be delayed due to political instability in Nepal. To avoid potential delays, they sent their guides to base camp five days earlier than originally scheduled. (*Source:* www.project-himalaya.com)

a. Suppose the entire expedition had arrived at base camp five days early with their guides and then followed the pattern of climbing shown in Table 3.1 and Figure 3.1. How would we need to change the graph of E to model the elevation of the climbers?

b. Suppose political instability had delayed the expedition and the climbers reached their base camp a week later than planned but then followed the same pattern of climbing. How would we need to change the graph of E to model the elevation of the climbers?

Solution

a. If the climbers had arrived five days earlier with their guides, they would have reached base camp on April 5. Considering our original graph (Figure 3.1) and definition of d as "days since April 10," this corresponds to $d = -5$.

If we assume the climbers followed the same pattern of climbing from this date, then they would have reached the summit on May 25 instead of May 30. Likewise, the climbers would have reached *all* of the same elevations five days earlier, as shown in Table 3.3.

Table 3.3

d Value When Climbers Reached This Elevation if They Arrived 5 Days Early	Days Since April 10, 2005 d	Elevation (meters) $E(d)$
$0 - 5 = -5$	0	5165
$5 - 5 = 0$	5	6363
$15 - 5 = 10$	15	7040
$20 - 5 = 15$	20	7680
$40 - 5 = 35$	40	7040
$50 - 5 = 45$	50	8850

This change will cause the graph to shift to the left 5 units, as shown in Figures 3.3a and 3.3b.

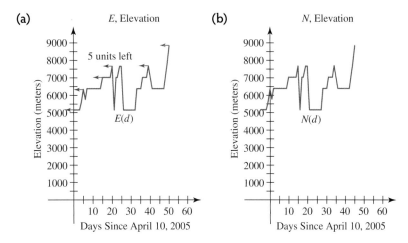

Figure 3.3

Observe that $N(d) = E(d + 5)$. We can verify this by checking a couple of points.

$$N(0) = E(0 + 5) \qquad N(35) = E(35 + 5)$$
$$= E(5) \qquad\qquad\quad = E(40)$$
$$= 6363 \qquad\qquad\;\, = 7040$$

The results are consistent with the graphs of N and E.

b. If the expedition began seven days late, the group would have reached base camp on April 17 instead of April 10. Again, the climbers would have reached all of the same elevations after the same number of days climbing but seven days later for each, as shown in Table 3.4.

Table 3.4

d Value When Climbers Reached This Elevation if They Arrived 7 Days Late	Days Since April 10, 2005 d	Elevation (meters) E(d)
0 + 7 = 7	0	5165
5 + 7 = 12	5	6363
15 + 7 = 22	15	7040
20 + 7 = 27	20	7680
40 + 7 = 47	40	7040
50 + 7 = 57	50	8850

We can model this new situation by shifting the original graph to the right 7 units, as shown in Figures 3.4a and 3.4b.

Figure 3.4

Observe that $V(d) = E(d - 7)$. We can verify this by checking a couple of points.

$$V(7) = E(7 - 7) \qquad V(50) = E(50 - 7)$$
$$= E(0) \qquad \qquad = E(43)$$
$$= 5165 \qquad \qquad = 6400$$

The results are consistent with the graphs of V and E.

We summarize our observations regarding horizontal shifts as follows.

HORIZONTAL SHIFTS

The graph of $g(x) = f(x - h)$ is the graph of f shifted horizontally by $|h|$ units. If h is positive, the shift is to the right. If h is negative, the shift is to the left.

■ Generalizing Transformations: Horizontal Shifts

Again, the key to transformations is to remember that we are using the outputs of a parent function to define the outputs of an image function. Consider the function $g(x) = f(x - 2)$. We observe that g and f have identical *outputs*.

$$\underbrace{g(\ \ )}_{\text{the outputs of } g} = \underbrace{f(\ \ )}_{\text{the outputs of } f}$$

If $g(x) = f(x)$, then the functions are identical. If $g(x) = f(x − 2)$, the outputs are identical but they occur at *different inputs*. The expression $x − 2$ indicates that if we want to evaluate g for some input x, we need to look to the output of f when the input is two units less than x. For example, to find $g(−1)$ and $g(7)$, we do the following.

$$g(x) = f(x − 2) \qquad g(x) = f(x − 2)$$
$$g(−1) = f(−1 − 2) \qquad g(7) = f(7 − 2)$$
$$g(−1) = f(−3) \qquad g(7) = f(5)$$

The output of g at -1 is defined by the output of f at -3.

The output of g at 7 is defined by the output of f at 5.

If f is given by the graph in Figure 3.5a, then g is defined by the graph in Figure 3.5b.

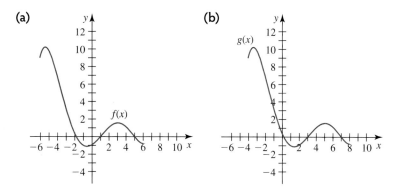

Figure 3.5

Note we can verify statements such as $g(7) = f(5)$ by using the graphs, where we can see that $g(7) = 0$ and $f(5) = 0$. If there is any confusion as to whether one function's graph is to the right or left of another function's graph, we simply look at the inputs in the definition. In the case of $g(x) = f(x − 2)$, x will always be larger than $x − 2$. Thus the same outputs occur in g for larger inputs than in f. This means the graph of g is to the right of the graph of f by 2 units (or the graph of f is to the left of the graph of g by 2 units).

The opposite is true for a relationship defined by $h(x) = f(x + 2)$. In this case the inputs in f are larger for the same output (since $x + 2$ is always larger than x). As shown in Figure 3.6, the graph of f is to the right of h (or the graph of h is to the left of f) by 2 units.

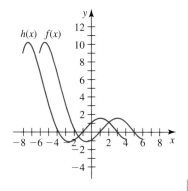

Figure 3.6

EXAMPLE 5 ■ Graphing Function Transformations

Use the graph of f shown in Figure 3.7 to draw the graph of $g(x) = f(x − 2) + 3$.

Solution As shown in Figure 3.8, the graph of g will be the graph of f shifted right 2 units and shifted upward 3 units since the outputs of g occur for larger inputs (x is larger than $x − 2$), and those outputs are then increased by 3. When completing multiple transformations like this we typically follow the order of operations.

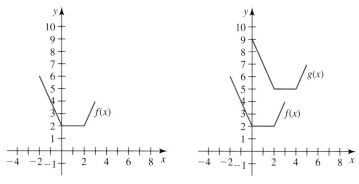

Figure 3.7

Figure 3.8

EXAMPLE 6 ■ Performing Shifts on a Table

The function f is defined by Table 3.5. Using this table,

a. Evaluate $f(x) + 6$ when $x = 2$.
b. Evaluate $f(x - 3)$ when $x = 1$.
c. Evaluate $f(0) - f(-3)$.
d. Create a table of values for $g(x)$ if $g(x) = f(x + 1) - 4$.

Table 3.5

x	$f(x)$
−3	4
−2	5
−1	7
0	11
1	19
2	28
3	46

Solution

a. $f(x) + 6 = f(2) + 6$ since $x = 2$
            $= 28 + 6$ since $f(2) = 28$ from the table
            $= 34$

b. $f(x - 3) = f(1 - 3)$ since $x = 1$
             $= f(-2)$
             $= 5$ from the table

c. $f(0) - f(-3) = 11 - 4$
                  $= 7$

d. To create the table of values for g, we could evaluate g for different x-values. For example,

$$g(-4) = f(-4 + 1) - 4$$
$$= f(-3) - 4$$
$$= 4 - 4$$
$$= 0$$

This shows $(-4, 0)$ is a coordinate point for g. We could also find this point by taking the coordinate point $(-3, 4)$ from f and shifting it left 1 unit and downward 4 units. Using either of these procedures, we obtain the table of values for g shown in Table 3.6.

Table 3.6

x	$g(x)$
−4	0
−3	1
−2	3
−1	7
0	15
1	24
2	42

EXAMPLE 7 ■ Analyzing Shifts

The graph of $f(x) = x^3$ is given in Figure 3.9. Function g is a transformation of f and is shown in Figure 3.10. Describe how f was transformed to create g and write the formula for g in terms of f. Then simplify the result.

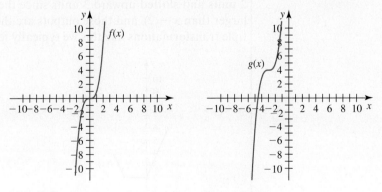

Figure 3.9   Figure 3.10

Solution As shown in Figure 3.11, function f has been shifted left 3 units and upward 4 units to create g.

The function notation for this relationship is

$$g(x) = f(x + 3) + 4$$

This shows that we find the outputs of g by using an input value 3 units less and then increasing the output of f by 4 units. We can now use this relationship to find the formula for g.

$$g(x) = f(x + 3) + 4$$
$$= (x + 3)^3 + 4 \quad \text{since } f(x) = x^3$$

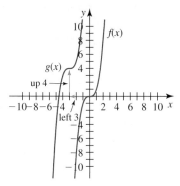

Figure 3.11

Aligning Data Horizontally

We often use a horizontal shift to *align data* before creating a mathematical model to reduce the size of the numbers used in the computation of model coefficient values. To align data, we determine the input value we want to use as a starting point. For example, if we want to model the population of the United States in the 21st century, we might choose to align the data to the year 2000, meaning $t = 0$ represents 2000, $t = 1$ represents 2001, and so on.

EXAMPLE 8 ■ Aligning Data Horizontally

The population of the United States increased early in the 21st century, as shown in Table 3.7. (*Source: Statistical Abstract of the United States, 2006*, Table 2)

a. Rewrite the population table as a function of t, where t is defined to be the number of years since 2000.

b. Describe the relationship between y and t, then write $P(t)$ in terms of y.

Table 3.7

Year y	U.S. Population (in thousands) $P(y)$
2000	282,402
2001	285,329
2002	288,173
2003	291,028
2004	293,907

Solution

a. The rewritten data is shown in Table 3.8.

Table 3.8

Years Since 2000 t	U.S. Population (in thousands) $P(t)$
0	282,402
1	285,329
2	288,173
3	291,028
4	293,907

b. Since y represents the year and t represents the number of years since 2000, $t = y - 2000$. Therefore, $P(t) = P(y - 2000)$.

SUMMARY

In this section you learned how to use vertical and horizontal shifts to transform a function equation and its associated graph. You also learned how to determine the relationship between the equations of graphs that differ by a horizontal or vertical shift. Additionally, you learned how to align data to simplify the modeling process.

3.1 EXERCISES

SKILLS AND CONCEPTS

In Exercises 1–4, graph each transformation of f and write the formula for the transformed function in the form $y = mx + b$.

1. $f(x) - 2$
2. $f(x - 3)$
3. $f(x + 1) + 2.5$
4. $f(x - 3) + 2$

In Exercises 5–8,
  a. Create a table of values for f and g.
  b. Graph f and g.
  c. Describe the relationship between the graph of f and the graph of g.

5. $f(x) = x^2$, $g(x) = f(x + 2) + 7$
6. $f(x) = \sqrt{x}$, $g(x) = f(x - 4) + 2$
7. $f(x) = |x| + 5$, $g(x) = f(x - 1) - 3$
8. $f(x) = \dfrac{1}{x}$, $g(x) = f(x + 6) + 1$

In Exercises 9–14, use Table A to evaluate each expression.

Table A

x	f(x)	x	f(x)
−4	12	1	−3
−3	10	2	4
−2	7	3	5
−1	2	4	6
0	−1		

9. $f(x) + 2$ when $x = -3$
10. $f(x + 4)$ when $x = -1$
11. $f(x + 1) - 9$ when $x = -2$
12. $f(x - 5) - 4$ when $x = 4$
13. $f(-3) + f(1)$
14. $2f(0) - f(-2)$

In Exercises 15–18, use Table A to solve each equation for x.

15. $f(x) = -3$
16. $f(x + 5) = 2$
17. $f(x) + 10 = 15$
18. $f(x - 1) + 2.5 = 14.5$

In Exercises 19–20, refer to Table A to answer each question.

19. Create a table of values for g given $g(x) = f(x - 4)$.
20. Create a table of values for k given $k(x) = f(x + 1) + 5$.

In Exercises 21–24, match the formula for each transformation of $f(x) = x^2$ with graph A, B, C, or D. Do not use a calculator.

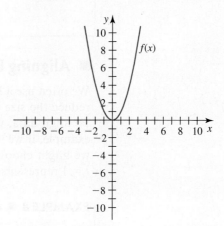

21. $g(x) = (x + 4)^2$
22. $h(x) = (x - 4)^2$
23. $j(x) = x^2 - 4$
24. $k(x) = x^2 + 4$

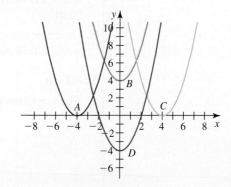

In Exercises 25–30, draw the graph of each function as a transformation of the given function f.

25. $A(x) = f(x + 1)$
26. $B(x) = f(x) - 3$
27. $C(x) = f(x - 3) + 2$
28. $D(x) = f(x + 4) - 1$
29. $E(x) = f(x) + \pi$
30. $G(x) = f(x - \sqrt{8})$

In Exercises 31–34, you are given the graph of a transformation of $f(x) = \sqrt{x}$.

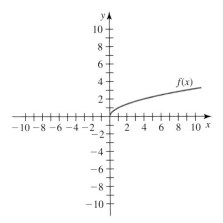

For each exercise,
  a. Describe the transformations required on f to create the new function.
  b. Use function notation to write each new function in terms of f.
  c. Write the formula for the new function.

31.

32.

33.

34.

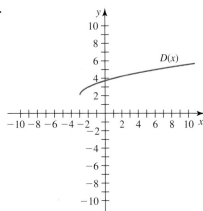

In Exercises 35–37, use the table to identify the shift(s) required to create the indicated function. Then write a formula for each transformed function in terms of f.

x	$f(x)$	$g(x)$	$h(x)$	$j(x)$
-1	-0.25	0.25	2.75	-1
0	0	2	3	0.75
1	0.25	6.75	3.25	1
2	2	16	5	1.25
3	6.75	31.25	9.75	3
4	16	54	19	7.75
5	31.25	85.75	34.25	17
6	54	128	57	32.25

35. $g(x)$ 36. $h(x)$ 37. $j(x)$

SHOW YOU KNOW

38. Explain why $f(x) + k$ shifts the graph of f vertically instead of horizontally and why $f(x - h)$ shifts the graph of f horizontally instead of vertically.

39. Sketch a graph of a nonlinear function on graph paper, then pick two points on your graph and calculate the average rate of change between these points. Shift your function vertically and note where the two points you chose previously are now located. Use this example to

explain why vertical shifts do or do not affect the average rate of change of a function.

40. After completing Exercise 39, one of your classmates repeats the exercise using a horizontal shift and claims that horizontal shifts change the average rate of change of a function. A second classmate disagrees and says the average rate of change remains the same. Explain how each of these students could reach his or her conclusion.

41. One of your classmates tells you that horizontal shifts seem backward. He does not understand why $f(x - 4)$ would shift f to the right while $f(x + 4)$ would shift f to the left. Write an explanation for your classmate.

■ MAKE IT REAL

42. **Satellite Television** In January 2007, satellite television service through DirecTV was offered at a monthly rate of $55.97 for 140 channels, high-definition access, and DVR service. The high definition DVR unit cost an additional $299. (*Source:* DirecTV)

    a. Write a function to model the total cost for this satellite television service for m months of subscription.

    b. A company salesman offered a $100 rebate on the high definition DVR. Show how to transform the original function to accommodate this rebate. Write the formula for the new function.

43. **Professional Football** From 1985 to 2006 the average weight of professional football players increased from about 225 to 248 pounds. (*Source:* www.palmbeachpost.com) The average weight of a professional football player can be modeled by

    $W(t) = 225(1.00465)^t$ pounds

    where t is the years since 1985. Assuming football pads weigh about 13 pounds, what is the function that gives the average weight of a professional football player in full pads?

44. **Baseball Salaries** Based on data from 1990 to 2006, the average annual salary of a major league baseball player can be modeled by

    $A(t) = 0.1474t + 0.4970$ million dollars

    where t is the number of years since 1990. (*Source:* Modeled from CBSSportsline.com, MLB Baseball 2006 Salaries)

    a. According to the model, what was the average salary in 1990?

    b. Create a new function $I(t)$ that gives the *increase* in the average salary over the 1990 salary level.

    c. If the meaning of t was changed from *years since 1990* to *years since 1900*, how would the function $A(t) = 0.1474t + 0.4970$ need to change for the results to still make sense?

In Exercises 45–48,

a. Create a table of values of aligned data.

b. Describe the horizontal shift required to align the data.

c. If the original function is f and the aligned function is g, use function notation to write g in terms of f.

45. Align the data to 1980.

Year	Number of Public Airports in the U.S.	Year	Number of Public Airports in the U.S.
1980	4814	1995	5415
1985	5858	2000	5317
1990	5589	2002	5286

Source: Statistical Abstract of the United States, 2006, **Table 1062**

46. Align the data to 2000.

Year	Retail Sales of Indoor Houseplants (in millions)
2000	1332
2001	1784
2002	2128
2003	1571
2004	1495

Source: Statistical Abstract of the United States, 2006, **Table 1231**

47. Align the data to 1998.

Year	Number of Doctorate Degrees Awarded in Astronomy
1995	173
1998	206
1999	159
2000	185
2001	186
2002	144
2003	167

Source: Statistical Abstract of the United States, 2006, **Table 784**

48. Align the data to 1995.

Year	Reported Burglaries (in thousands)	Year	Reported Burglaries (in thousands)
1989	3168	1997	2641
1991	3157	1999	2101
1993	2835	2001	2117
1995	2594	2003	2153

Source: Statistical Abstract of the United States, 2006, **Table 293**

49. Golf Courses The number of golf facilities in the United States as a function of years since 1980 can be modeled by function G.

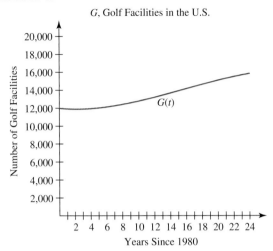

G, Golf Facilities in the U.S.

a. Suppose you shifted this function to the right 5 units. What would your new graph represent? Use function notation to write your new function in terms of G.

b. Suppose you shifted this function to the left 10 units. What would your new graph represent? Use function notation to write your new function in terms of G.

c. Suppose you shifted this function downward 12,000 units. What would your new graph represent? Use function notation to write your new function in terms of G.

50. Oxygen Levels Function H shows the approximate percent of hemoglobin saturation in the blood at an elevation of x thousand meters above sea level. Hemoglobin is the protein in red blood cells that transports oxygen throughout the body. Thus, this function shows how efficiently the human body can absorb and distribute oxygen at various elevations. (*Source:* Modeled from www.dr-amy.com/rich/oxygen)

H, Hemoglobin Saturation

a. The human body requires a minimum of 87% hemoglobin saturation to function normally. At approximately what elevation is this saturation reached? Use this information to discuss the health dangers of climbing to Mount Everest's summit.

b. Research your city's elevation. About what percent of hemoglobin saturation would you expect your city's citizens to have? (If you can only find your city's elevation in feet, the conversion is 1000 feet = 304.8 meters.)

c. Use a vertical shift to create a new graph with the vertical axis representing the hemoglobin saturation percentage above the safe level of 87%.

d. Write the formula for the graph from part (c) in terms of H.

51. Atmospheric Pressure The first table shows the common atmospheric pressure for different altitudes.

Altitude (meters) a	Atmospheric Pressure (psi) $P(a)$
0	14.70
1000	13.04
2000	11.53
3000	10.17
4000	8.94
5000	7.83
6000	6.84
7000	5.96
8000	5.16

a. Use a graphing calculator to plot the points.

b. Find a linear regression model for this data. Is it a reasonable fit? Explain.

c. The next table represents a transformation of the function P. Describe the transformation, then explain what the new function shows and how the table should be labeled. Then write the formula for the new function.

−4000	14.70
−3000	13.04
−2000	11.53
−1000	10.17
0	8.94
1000	7.83
2000	6.84
3000	5.96
4000	5.16

d. The next table represents a different transformation of the function *P*. Describe the transformation, then explain what the new function shows and how the table should be labeled. Then write the formula for the new function.

0	0
1000	−1.66
2000	−3.17
3000	−4.53
4000	−5.76
5000	−6.87
6000	−7.86
7000	−8.74
8000	−9.54

52. Sprinter's Time and Distance The function *D* models the distance a 100-meter sprinter has traveled *t* seconds after the beginning of a race.

a. Sketch a graph that could represent this situation. Be sure to label your axes appropriately.

b. In some races, such as junior high track meets, volunteers time runners with a handheld stopwatch. However, the volunteers may not all begin their timers precisely with the start of the race. Suppose it takes a volunteer 0.75 seconds after the race begins to start the stopwatch. Draw a graph for the sprinter's distance traveled as a function of the time since the volunteer started the stopwatch.

c. If possible, use transformations to describe the relationship between *D* and your new function, then discuss how this relates to the domain and range of your new function.

d. Repeat parts (b) and (c) for the following situations.

  i. The volunteer accidentally started the stopwatch 0.25 seconds before the race actually began.

  ii. The volunteer accidentally stopped the stopwatch halfway through the race and then took 2 full seconds to start it again.

53. Mount Everest Summit Temperatures The graph shows the average temperature, *f*, on the summit of Mount Everest during the mth month of the year.

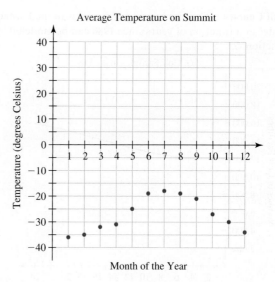

Source: www.mounteverest.net

Interpret the transformations $f(m + 12)$ and $f(m - 12)$ in this context.

■ STRETCH YOUR MIND

Exercises 54–57 are intended to challenge your understanding of vertical and horizontal shifts.

54. Use the table to evaluate the expression

$$f(x - f(x)) + f(f(x + 4) - x)$$

when $x = -1$.

x	f(x)
−4	12
−3	10
−2	7
−1	2
0	−1
1	−3
2	4
3	5
4	6

55. The function $P = f(n)$ describes the price you pay for buying *n* items. Suppose you have a coupon for 15% off the total price of your purchase. Can you model this using only shifts of the original function? If so, show how. If not, explain why it is impossible.

56. You are told the function *f* has the following property: $f(x) = f(x + 5n)$ for any integer values of *n*. Describe any characteristics this function will possess.

57. Create the graphs of a function and its inverse. Explain how and why a vertical shift and horizontal shift of the original function affect the inverse function.

SECTION 3.2 Vertical and Horizontal Reflections

LEARNING OBJECTIVES

- Identify what change in a function equation results in a horizontal reflection
- Identify what change in a function equation results in a vertical reflection
- Understand the concept of symmetry and determine if a function is even, odd, or neither

GETTING STARTED

One person's debt payment is another person's income. Although taking on excessive debt as a consumer is dangerous financially, it is beneficial financially for a consumer to receive debt payments from others. Whether debt is viewed as financially positive or negative depends upon whether you owe money or are owed money. This idea can be explained mathematically using the concept of reflections.

In this section we explore mathematical reflections in equations, graphs, and tables.

■ Vertical Reflections

In January 2003, one of the authors took out a $13,460 loan to buy a new car. Over the next few years, he monitored the loan balance and total mileage. The function A, graphed in Figure 3.12, models the situation where the amount owed, A, is in thousands of dollars and the miles driven, m, are in thousands of miles.

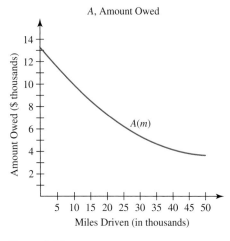

Figure 3.12

The equation for the model is $A(m) = 0.003569m^2 - 0.3715m + 13.29$. This formula was created from the actual data by using *quadratic regression*, a technique we will address in Chapter 4.

EXAMPLE 1 ■ Reflecting a Function Vertically

Net worth is the difference between a person's assets and liabilities. In short, *assets* are the things you own that have a dollar value and *liabilities* are the debts you owe to others. Assuming your assets remain constant, a reduction in your liabilities increases your net worth. Using this definition of net worth and the graph shown in Figure 3.12,

a. Draw a graph for the function N, the effect the car loan has on the author's net worth as a function of the miles he has driven the car. Explain how the graphs of A and N are related.

b. Use function notation to write N in terms of the function A.

c. Write a formula for $N(m)$.

176 CHAPTER 3 ■ TRANSFORMATIONS OF FUNCTIONS

Solution

a. If the car loan amount is $13,460, the loan *reduces* the author's net worth by $13,460. We represent this idea of reduction with −$13,460. The car loan has a negative effect on the net worth. In other words, when the loan function A is positive, the effect on net worth function N is negative. Table 3.9 shows the relationship between A and N.

Table 3.9

Miles Driven (thousands) m	Amount Owed ($ thousands) $A(m)$	Effect on Net Worth ($ thousands) $N(m)$
4	11.86	−11.86
7.5	10.70	−10.70
15	8.52	−8.52
22.5	6.74	−6.74
0	13.46	−13.46

The graph for $N(m)$ is shown in Figure 3.13. This change demonstrates a **vertical reflection** (also called a **reflection about the horizontal axis**). As shown in Figure 3.14, each of the output values is the same number of units from the horizontal axis, but the values are positive for A and negative for N.

Figure 3.13

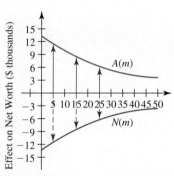
Figure 3.14

b. The function $N(m) = -A(m)$ describes the relationship between N and A. For example, to find the car loan's effect on the net worth after the author drove the car 15,000 miles, we can use the function relationship.

$$N(m) = -A(m)$$
$$N(15) = -A(15)$$
$$= -(8.57)$$
$$= -8.57$$

After he drove the car 15,000 miles, the car loan reduced the author's net worth by $8570.

c. Since $N(m) = -A(m)$ and $A(m) = 0.002994m^2 - 0.3421m + 13.03$, we have

$$N(m) = -A(m)$$
$$N(m) = -(0.002994m^2 - 0.3421m + 13.03)$$
$$= -0.002994m^2 + 0.3421m - 13.03$$

VERTICAL REFLECTIONS

The graph of $g(x) = -f(x)$ is the graph of f reflected vertically about the horizontal axis.

■ Generalizing Transformations: Vertical Reflections

Returning to the idea that in a transformation we use the outputs of one function to define the outputs of a new function, we generalize vertical reflections as follows.

$$\underbrace{g(x)}_{\text{the output of } g \text{ at } x} \underbrace{=}_{\text{is equal to}} \underbrace{-}_{\text{the opposite of}} \underbrace{f(x)}_{\text{the output of } f \text{ at } x}$$

EXAMPLE 2 ■ Graphing a Vertical Reflection

Given the graph of f in Figure 3.15, graph $g(x) = -f(x)$.

Solution As shown in Figure 3.16, to obtain the function g, we make all positive values of f negative and all negative values of f positive.

Figure 3.15   Figure 3.16

EXAMPLE 3 ■ Determining the Equation of a Vertically Reflected Graph

The graph of $f(x) = x^4 - 4x^2$ is shown in Figure 3.17 together with the graph of a function g. What is the function equation for g?

Solution From the graph, we see g is a vertical reflection of f. Therefore,

$$g(x) = -f(x)$$
$$= -(x^4 - 4x^2) \quad \text{since } f(x) = x^4 - 4x^2$$
$$= -x^4 + 4x^2$$

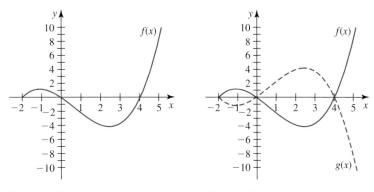

Figure 3.17

■ Horizontal Reflections

When a person travels underwater, the person's body bears the weight of the water above it. The pressure of the water on the person's body increases rapidly as she descends below the surface. For each 33 feet she descends underwater, the pressure on her body increases by 14.7 pounds per square inch (psi). (*Source:* www.onr.navy.mil)

EXAMPLE 4 ■ Reflecting a Function Horizontally

Based on the pressure information just given,

a. Write a function for P, the increased pressure on a human body d feet under the surface of the ocean. Then create a table of values and graph the function.

b. Let x represent a person's elevation (feet above sea level). Then graph a new function I to model the increased pressure at elevation x.

c. Use function notation to write function I in terms of function P.

Solution

a. Since the rate of change is constant, P will be a linear function. For each 33-foot increase in depth, the pressure increases by 14.7 pounds. The rate of change is given by

$$\frac{14.7 \text{ psi}}{33 \text{ feet}} \approx 0.45 \text{ psi per foot}$$

So $P(d) = 0.45d$. The table and graph of the model are shown in Table 3.10 and Figure 3.18.

Table 3.10

Depth (feet) d	Increased Pressure (psi)
20	9
40	18
60	27
80	36

Figure 3.18

b. Because the surface of the ocean has an elevation of 0 feet, negative numbers represent elevations below sea level and positive numbers represent elevations above sea level. When the depth below the surface is d feet, the elevation is $x = -d$. Table 3.11 is a partial table of values.

If we plot the points from the table using elevation as the independent variable instead of depth, we get the graph in Figure 3.19.

Table 3.11

Depth (feet) d	Elevation (feet) $x = -d$	Increased Pressure (psi)
20	−20	9
40	−40	18
60	−60	27
80	−80	36

Figure 3.19

c. Figures 3.18 and 3.19 show that the basic relationship between the distance below the surface and the increased pressure is the same whether we determine it using depth or elevation. Because the pressure increases as *depth increases*, when we use depth as the independent variable the pressure change function has a positive rate of change. Because the pressure increases as the *elevation decreases*, when we use elevation as the independent variable the pressure change function has a negative rate of change. Thus, the new function I that models the increased pressure as a function of elevation is $I(x) = -0.45x$.

The key to writing the relationship between these functions is to see that $x = -d$ and $d = -x$. The same depth will be represented by numbers of the same magnitude but with opposite signs in each function. Thus

$$P(d) = P(-x) \quad \text{and} \quad I(x) = I(-d) \quad \text{where } x = -d$$

In addition, both functions will output the same values for increased pressure at the same depths as long as the appropriate input values are used. We can say that

$$P(d) = I(-d) \quad \text{and} \quad I(x) = P(-x)$$

In other words, using the opposite input value in each function will return the same output value.

To see this, consider the increased pressure 40 feet below the surface of the ocean. This is a d value of 40 or an x value of -40.

$$\begin{array}{ll}
P(d) = I(-d) & I(x) = P(-x) \\
P(40) = I(-40) & I(-40) = P(-(-40)) \\
\quad = -0.45(-40) & \quad = P(40) \\
\quad = 18 \text{ psi} & \quad = 0.45(40) \\
& \quad = 18 \text{ psi}
\end{array}$$

The pressure at an elevation of -40 feet is identical to the pressure 40 feet below the surface of the ocean.

We can also use this relationship to verify the formula we wrote for $I(x)$.

$$\begin{aligned}
I(x) &= P(-x) \\
&= 0.45(-x) \\
&= -0.45x
\end{aligned}$$

We summarize our observations on horizontal reflections as follows.

Horizontal Reflections

The graph of $g(x) = f(-x)$ is the graph of f reflected horizontally about the vertical axis.

■ Generalizing Transformations: Horizontal Reflections

In general, we think of horizontal reflections as follows.

$$\underbrace{g(x)}_{\substack{\text{the output} \\ \text{of } g \text{ at } x}} \underbrace{=}_{\text{is equal to}} \underbrace{f(-x)}_{\substack{\text{the output of } f \text{ at} \\ \text{the opposite of } x}}$$

EXAMPLE 5 ■ Reflecting a Graph Horizontally

Use the graph of f shown in Figure 3.20 to draw the graph of $g(x) = f(-x)$.

Solution As shown in Figure 3.21, the graph of g will be the graph of f reflected horizontally about the vertical axis. If the point (a, b) is on the graph of f, the point $(-a, b)$ will be on the graph of g.

Figure 3.20

Figure 3.21

■ Combining Shifts and Reflections

Sometimes changes in a situation will require that we combine shifts and reflections to modify an original function to create our new model.

EXAMPLE 6 ■ Combining Shifts and Reflections

The value 14.7 psi is the pressure exerted by Earth's atmosphere on the human body at sea level. Recall that when a person descends 33 feet underwater, the pressure increases by 14.7 psi of pressure. Thus the pressure on a person 33 feet underwater is twice that of the pressure on a person at sea level.

a. Use a table and graph to model the total amount of pressure, T, exerted on a person underwater at elevation t.

b. In Example 4, we found the function showing increased pressure as a function of depth in feet, $P(d) = 0.45d$. How can this function be transformed to create the function in part (a)?

c. Use function notation to write the function in part (b) in terms of the original function P and the variable x.

d. Write the formula for $T(x)$.

Table 3.12

Elevation (feet above sea level) x	Total Pressure (psi) $T(x)$
0	14.7
−10	14.7 + 0.45(10) = 19.2
−20	14.7 + 0.45(20) = 23.7
−33	14.7 + 14.7 = 29.4
−40	14.7 + 0.45(40) = 32.7
−50	14.7 + 0.45(50) = 37.2
−66	14.7 + 14.7(2) = 44.1
−99	14.7 + 14.7(3) = 58.8

Solution

a. The pressure at sea level is 14.7 psi, and the pressure increases by 14.7 psi for each additional 33 feet of underwater descent. Using the rate of pressure change we determined in Example 4, about 0.45 psi per foot, we can fill in values that are not multiples of 33 in Table 3.12. The graph is shown in Figure 3.22.

Figure 3.22

b. As shown in Figure 3.23, the function *T* is a result of two transformations from the original function *P*. First, *T* is a horizontal reflection of *P* since the input values are *elevation* instead of *depth*. The resultant function has also been shifted upward 14.7 units. This shift is required since the output values are for *total pressure* instead of *increased pressure* because the pressure at sea level is equal to 14.7 psi.

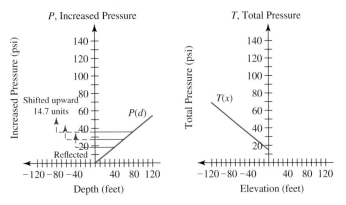

Figure 3.23

c. To write the function *T* in terms of *P* we need to incorporate the two transformations just described. Since $d = -x$, $P(d) = P(-x)$. The transformation $P(-x)$ reflects the graph of $P(x)$ horizontally. The total pressure is 14.7 psi greater than *P*, so $T(x)$ may be written as

$$T(x) = P(-x) + 14.7$$

d. From Example 4 we know that $P(d) = 0.45d$, so $P(-x) = 0.45(-x)$. Thus

$$T(x) = P(-x) + 14.7$$
$$= 0.45(-x) + 14.7$$
$$= -0.45x + 14.7$$

EXAMPLE 7 ■ Identifying and Defining Multiple Transformations

The graph of $y = f(x)$ is given in Figure 3.24.

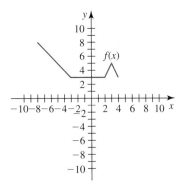

Figure 3.24

For each of the following functions, describe the transformation(s) used on the function $y = f(x)$ to create the new function, then use function notation to write each new function in terms of *f*.

182 CHAPTER 3 ■ TRANSFORMATIONS OF FUNCTIONS

a. Graph A

b. Graph B

c. Graph C

Solution

a. In Graph A, the original function has been shifted downward 6 units. We express r in terms of f as

$$r(x) = f(x) - 6$$

b. In Graph B, the original function has been reflected in both the vertical and horizontal directions. We express t in terms of f as

$$t(x) = -f(-x)$$

c. In Graph C, the original function has been reflected vertically. We express u in terms of f as

$$u(x) = -f(x)$$

EXAMPLE 8 ■ Writing Formulas for Transformed Functions

The graph of $f(x) = x^2 - 3$ is shown in Figure 3.25.

Figure 3.25

Match each of the following formulas to the appropriate transformed graph, then write the formula of the transformed function.

a. $g(x) = f(x + 4) - 3$ b. $g(x) = -f(x)$ c. $g(x) = -f(x - 5) + 4$

Graph A

Graph B

Graph C

Solution

a. $g(x) = f(x + 4) - 3$. This relationship shows that shifting the original function f to the left 4 units and downward 3 units will yield the graph of g, which is displayed in Graph C. The formula for $y = g(x)$ is

$$\begin{aligned} g(x) &= f(x + 4) - 3 \\ &= [(x + 4)^2 - 3] - 3 \quad \text{since } f(x) = x^2 - 3 \\ &= [(x^2 + 8x + 16) - 3] - 3 \\ &= (x^2 + 8x + 13) - 3 \\ &= x^2 + 8x + 10 \end{aligned}$$

b. $g(x) = -f(x)$. From this relationship we can see g is the vertical reflection of the original function f. This is shown in Graph A. The formula for $y = g(x)$ is

$$\begin{aligned} g(x) &= -f(x) \\ &= -(x^2 - 3) \quad \text{since } f(x) = x^2 - 3 \\ &= -x^2 + 3 \end{aligned}$$

c. $g(x) = -f(x - 5) + 4$. This relationship shows g to be the result of a vertical reflection of f that was then shifted to the right 5 units and upward 4 units, which we can see in Graph B. The formula for $y = g(x)$ is

$$\begin{aligned} g(x) &= -f(x - 5) + 4 \\ &= -[(x - 5)^2 - 3] + 4 \quad \text{since } f(x) = x^2 - 3 \\ &= -[(x^2 - 10x + 25) - 3] + 4 \\ &= -(x^2 - 10x + 22) + 4 \\ &= -x^2 + 10x - 22 + 4 \\ &= -x^2 + 10x - 18 \end{aligned}$$

184 CHAPTER 3 ■ TRANSFORMATIONS OF FUNCTIONS

■ Symmetry

The concept of symmetry is closely related to the concept of reflections. Artists, architects, fashion designers, and many others use symmetry—a characteristic of objects that have two identical halves—to create products that appeal to our sense of beauty and interest. In fact, human beings are creatures of symmetry. Except for slight imperfections and the placement of some internal organs, our bodies are essentially symmetrical.

Functions can display symmetry as well. In mathematics we most often use symmetry to describe shapes or graphs that are unchanged after being reflected across a straight line, called a **line of symmetry**. Some examples of figures and graphs with lines of symmetry are shown in Figure 3.26.

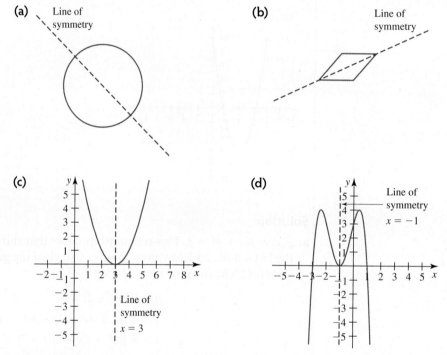

Figure 3.26

Functions that are symmetric with respect to the vertical axis are called **even functions** and those that are symmetric with respect to the origin are called **odd functions**.

EVEN FUNCTIONS

A function f is an **even function** if $f(-x) = f(x)$. Even functions are symmetric with respect to the vertical axis, meaning the vertical axis is the line of symmetry.

ODD FUNCTIONS

A function f is an **odd function** if $f(-x) = -f(x)$. Odd functions are symmetric with respect to the origin.

In practical terms, if reflecting a graph f about the vertical axis results in the same graph as that obtained by reflecting the graph f about the horizontal axis, then the function is an odd function.

EXAMPLE 9 ■ Testing for Even and Odd Symmetry

Determine if each of the following functions display even symmetry, odd symmetry, or neither.

a. $f(x) = \dfrac{1}{3}x$

b. $f(x) = 3x^2 + 3x$

c. $f(x) = x^4 - 2x^2$

Solution

a. First, let's examine the graph shown in Figure 3.27. It appears this function will have odd symmetry. If the function is reflected horizontally (Figure 3.28) or vertically (Figure 3.29) it will yield the same function.

Figure 3.27

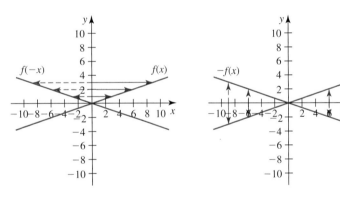

Figure 3.28   Figure 3.29

We can verify this by finding the formula for the function $f(-x)$ and showing that this is the same as the function for $-f(x)$:

$$f(x) = \dfrac{1}{3}x \qquad\qquad f(x) = \dfrac{1}{3}x$$

$$f(-x) = \dfrac{1}{3}(-x) \qquad -f(x) = -\left(\dfrac{1}{3}x\right)$$

$$= -\dfrac{1}{3}x \qquad\qquad = -\dfrac{1}{3}x$$

$$f(-x) = -f(x)$$

Therefore, this function has odd symmetry (is symmetrical with respect to the origin).

b. First, let's examine the graph shown in Figure 3.30. The function does not have a line of symmetry at $x = 0$, so it will not have even symmetry. Furthermore, it will not generate the same function when reflected vertically (Figure 3.31) and horizontally (Figure 3.32) so it does not have odd symmetry.

Figure 3.30

Figure 3.31

Figure 3.32

We can verify this observation algebraically as well.

$$f(x) = 3x^2 + 3x \qquad\qquad f(x) = 3x^2 + 3x$$
$$f(-x) = 3(-x)^2 + 3(-x) \qquad -f(x) = -(3x^2 + 3x)$$
$$= 3(x^2) - 3x \qquad\qquad = -3x^2 - 3x$$
$$= 3x^2 - 3x$$

Since $f(x) \ne f(-x)$, the function does not have even symmetry. Also, since $f(-x) \ne -f(x)$, the function does not have odd symmetry.

c. Again, let's begin by looking at the graph in Figure 3.33. This function appears to have even symmetry. We can verify this by comparing $f(x)$ and $f(-x)$.

$$f(x) = x^4 - 2x^2$$
$$f(-x) = (-x)^4 - 2(-x)^2$$
$$= x^4 - 2(x^2)$$
$$= x^4 - 2x^2$$
$$f(x) = f(-x)$$

This function has even symmetry (is symmetrical with respect to the vertical axis).

Figure 3.33

SUMMARY

In this section you learned how to transform functions with vertical and horizontal reflections. You also discovered how to determine if a function has even or odd symmetry.

3.2 EXERCISES

SKILLS AND CONCEPTS

In Exercises 1–4, graph each transformation of f and write the formula for the transformed function in the form $y = mx + b$.

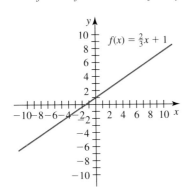

1. $-f(x)$
2. $f(-x)$
3. $-f(x - 3)$
4. $f(-x) - 6$

In Exercises 5–10, use the values shown in Table B to evaluate each expression.

Table B

x	f(x)
−8	26
−6	20
−4	13
−2	5
0	−4
2	−14
4	−25
6	−37
8	−50

5. $-f(x)$ when $x = -6$
6. $f(-x)$ when $x = 2$
7. $f(-x - 2) + 1$ when $x = 4$
8. $-f(-x) - 3.3$ when $x = -6$
9. $-f(-2) + f(6)$
10. $f(-8) - 3f(2)$

In Exercises 11–14, solve each equation for x using Table B.

11. $f(-x) = 13$
12. $-f(x) = 25$
13. $-f(x + 6) - 12 = 13$
14. $f(-x + 4) + 7 = 3$

In Exercises 15–16, refer to Table B to answer each question.

15. Create a table of values for j if $j(x) = -f(-x)$.
16. Create a table of values for k if $k(x) = -f(x + 6) - 4$.

In Exercises 17–20, match the formula for the transformation of $f(x) = 2^x$ with its Graph A, B, C, or D. Do not use a calculator.

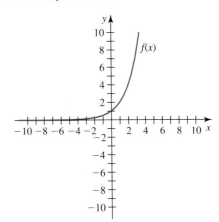

17. $g(x) = 2^{-x} + 3$
18. $h(x) = 2^{-x} - 3$
19. $j(x) = -2^x + 3$
20. $k(x) = -2^x - 3$

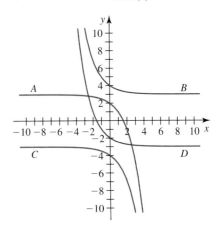

In Exercises 21–24, draw the graph of each function as a transformation of f.

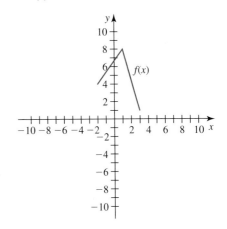

21. $A(x) = -f(x)$
22. $C(x) = f(-x) + 3$
23. $D(x) = -f(x - 5)$
24. $E(x) = -f(-x) + 1$

In Exercises 25–28, you are given the graph of a transformation of $f(x) = \sqrt{x}$.

For each exercise,

a. Describe the transformations required on f to create the new function.

b. Use function notation to write each new function in terms of f.

c. Write the formula for the new function.

25.

26.

27.

28.

In Exercises 29–31, use the table to identify the transformation(s) on f required to create the indicated function. Then write a formula for each transformed function in terms of f.

x	f(x)	g(x)	h(x)	j(x)
−4	0	3	0	3
−3	3	1	2	1
−2	5	0	1	2
−1	6	−3	−1	4
0	4	−5	−4	7
1	1	−6	−6	9
2	−1	−4	−5	8
3	−2	−1	−3	6
4	0	1	0	3

29. $g(x)$

30. $h(x)$

31. $j(x)$

For Exercises 32–35,

a. Complete the graph to make it an odd function.

b. Complete the graph to make it an even function.

c. Complete the graph to make it neither even nor odd.

Use only continuous functions. If it is not possible to create a graph of the indicated type, say not possible *and explain why.*

32.

33.

34.

35.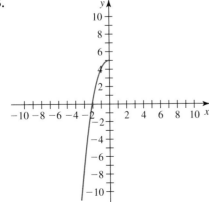

In Exercises 36–40, determine whether each of the following functions is even, odd, or neither.

36. $a(x) = \dfrac{1}{x^2}$

37. $b(x) = \sqrt[3]{x}$

38. $c(x) = 4x^3 - x$

39. $d(x) = x^2 - 2x + 1$

40. Create a table that shows a function with
    a. Even symmetry.
    b. Odd symmetry.
    c. No symmetry.

SHOW YOU KNOW

41. Refer to your work on Exercises 32–35. What appears to be a condition for a function to have odd symmetry? Use the definition of odd function to explain why this condition must exist, then explain how to use this fact to quickly identify functions that could not be odd.

42. How do each of the following transformations affect the rate of change, m, and vertical intercept, b, of a linear function $y = mx + b$?
    a. Horizontal shift
    b. Vertical shift
    c. Horizontal reflection
    d. Vertical reflection

43. Given the function $f(x) = mx + b$, determine what values of m and b are required for the function to have the indicated symmetry. If it is not possible for the function to have the indicated symmetry, say *not possible* and explain why.
    a. Odd
    b. Even
    c. Both odd and even
    d. Neither odd nor even

44. Function f is even and function g is odd. Which transformations have the ability to change these symmetries and which do not?

45. Given $g(x) = f(-x)$, we know that g will be a horizontal reflection of f. Explain why this change to the function causes this type of transformation.

46. Explain why there are no functions (other than $y = 0$) that possess symmetry with respect to the horizontal axis.

MAKE IT REAL

47. **Mortgage Payments** The table shows mortgage payments as a function of the total amount borrowed on a 30-year

fixed loan at 7% annual interest. (Note: We do not factor in additional costs such as taxes, mortgage insurance, etc.)

Amount Borrowed ($ thousands) b	Monthly Payment (dollars) $P(b)$
100	665.30
200	1330.60
225	1496.93
280	1862.85
330	2195.50

a. Is this a linear relationship? Defend your answer.

b. Write a formula for this situation.

c. Reflect your function horizontally. Draw and label the new function and explain how you could interpret your new graph.

d. Reflect your function vertically. Draw and label the new function and explain how you could interpret your new graph.

e. Would a horizontal shift have any practical meaning in this situation? Explain.

f. Would a vertical shift have any practical meaning in this situation? Explain.

For Exercises 48–49, use the graph of P, which shows the increased pressure in psi as a function of the underwater depth in feet.

48. Pressure Underwater Maroon Lake, located in Colorado's Rocky Mountains, is at an elevation of about 2920 m (9580 ft). At this elevation the atmospheric pressure is about 10.3 psi.

a. Create a table, graph, and formula for L, the total pressure on the human body r feet below the surface of Maroon Lake.

b. Explain how to transform the function P to create L.

49. Pressure Underwater Many experts recommend that recreational divers avoid underwater depths of greater than 100 feet.

a. Construct a table and graph showing B, the increased pressure on the human body z feet less than the maximum recommended safe diving depth.

b. Explain the transformations on the function P that we use to create B.

c. Use an equation to show the relationship between d and z.

d. Use function notation to write B in terms of P.

50. Car Loans For this exercise, recall the function A, which modeled the amount owed on the author's car loan in thousands of dollars after driving m thousand miles:

$$A(m) = 0.003569m^2 - 0.3715m + 13.29$$

a. Explain the transformations on A that can be used to create function P.

b. What does the new function represent and how should we label the vertical axis?

c. Write the formula for the function P.

d. P can also be thought of as a transformation of N, the effect on the author's net worth in thousands of dollars after driving m thousand miles.

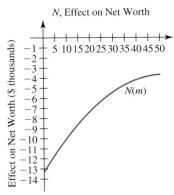

Use function notation to write N in terms of P.

51. **Atmospheric Temperatures** As altitude increases, temperature tends to decrease according to the model $T(A) = -6.505A + 15.01$, where T is the temperature in degrees Celsius and A is the altitude above sea level in kilometers.

   a. Construct the table of values and graph for R, the difference between room temperature (about 22°C) and the actual temperature at a specific elevation.

   b. Describe how the function T can be transformed to generate the graph of R.

   c. Use function notation to write R in terms of T, then write the formula for R.

52. **Mount Everest** For this exercise, use the graph of E, the climbers' elevation d days after April 10, 2005.

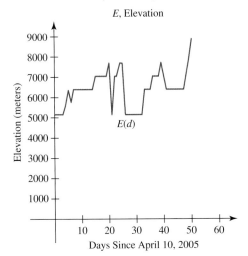

a. Construct the graph for S, the difference in elevation between the summit and the climbers' elevation d days after April 10.

b. Describe how the function E, which modeled the climbers' elevation d days after April 10, can be transformed to generate the graph of S.

c. Use function notation to write S in terms of E.

■ STRETCH YOUR MIND

Exercises 53–56 are intended to challenge your understanding of reflections and symmetry.

53. Is it possible for a function to have "translational symmetry," meaning that a horizontal or vertical shift of the function could yield an identical function? If so, give an example. If not, explain why it is not possible.

54. Explain any differences and similarities in the following functions: $p(x) = -f(x) + k$, $q(x) = -[f(x) + k]$, and $r(x) = -f(x) - k$.

55. Is it possible for transformations of a function to create a relationship that is no longer a function? Explain.

56. Create the graphs of a function and its inverse. Explain how and why horizontal and vertical reflections in the original function affect the inverse function.

SECTION 3.3 Vertical Stretches and Compressions

LEARNING OBJECTIVES

- Identify what change in a function equation results in a vertical stretch
- Identify what change in a function equation results in a vertical compression

GETTING STARTED

According to www.caloriecontrol.org, about 65% of adults in the United States are overweight. Extra weight can increase the risk of heart disease and diabetes. Overweight Americans spend billions of dollars a year on diet books and diet programs, yet most diet counselors give the same advice: to lose weight you must burn more calories than you consume. A 3500-calorie difference between calories consumed and calories burned will result in 1 pound of weight lost. If you decide to burn calories by running, how far would you have to run to burn 200 extra calories each day? We will answer questions such as these in this section.

In this section we expand our understanding of function transformations by introducing vertical and horizontal stretches and compressions. We also see how such transformations can be used to model real-world problems such as exercise plans for weight loss.

■ Vertical Stretches and Compressions

The number of calories you burn depends on a number of factors including the type of exercise you do, how strenuously you exercise, and your body weight. Table 3.13 shows the average number of calories a 155-pound person burns while running.

Table 3.13

Distance (miles) d	Calories Burned (155-pound person) $C(d)$
5	563
6	704
7	809
8	950
9	1056
10	1126

Source: www.nutristrategy.com

EXAMPLE 1 ■ Stretching and Compressing a Function Vertically

Using Table 3.13,

a. Use linear regression to model C, the calories burned by a 155-pound person who runs d miles. Then explain what the slope of the model represents.

b. A 190-pound person will burn about 22.6% more calories per mile than a 155-pound person. If H represents the calories burned for a 190-pound person running d miles, create a table, formula, and graph for H and use function notation to write H in terms of C. (Note: A 22.6% increase means the number of calories burned is larger by a factor of 1.226.)

c. A 130-pound person will burn about 83.8% as many calories per mile as a 155-pound person. If L represents the calories burned for a 130-pound person running d miles, create a table, formula, and graph for L and use function notation to write L in terms of C. (Note: If a number is 83.8% of another number, then it is 0.838 times as large.)

Solution

a. Using linear regression on our graphing calculator, we find $C(d) = 114.63d + 8.2857$, which is graphed in Figure 3.34. The slope is 114.63 calories per mile. That is, for each mile a 155-pound person runs, the total calories burned increases by almost 115 calories.

b. To represent "22.6% more calories," we need to multiply the output values from C by a factor of 1.226, as shown in Table 3.14. The points are plotted in Figure 3.35.

Figure 3.34

Table 3.14

Distance (miles) d	Calories Burned (155-pound person) $C(d)$	Calories Burned (190-pound person) $H(d)$
5	563	$1.226(563) \approx 690$
6	704	$1.226(704) \approx 863$
7	809	$1.226(809) \approx 992$
8	950	$1.226(950) \approx 1165$
9	1056	$1.226(1056) \approx 1295$
10	1126	$1.226(1126) \approx 1380$

The table shows that the outputs for each value of d are greater for H than they are for C. However, as indicated on Figure 3.35, the difference between each output is not constant: as the output values in C get larger, the amount of increase to the next output value of H also gets larger. The outputs do not change by a constant *amount* (as in a vertical shift) but are related by a constant *factor*: each output for H is 1.226 times the corresponding output for C. This is very clear when we compare the graphs. The calculations used to fill in the table for $H(d)$ yield the following relationship:

$$H(d) = 1.226C(d)$$

Now we can use this relationship to find the formula for H and graph the function, as shown in Figure 3.36.

$$H(d) = 1.226C(d)$$
$$= 1.226(114.63d + 8.2857)$$
$$= 140.5d + 10.16$$

Figure 3.35

Figure 3.36

The difference in the slopes of the two functions solidifies our earlier point. The 190-pound person is burning about 22.6% more calories per hour than the 155-pound person, and thus the difference between the total calories they burn will continue to increase as the distance run increases.

c. We can model the calories burned by a 130-pound person (83.8% of those burned by a 155-pound person) by multiplying all of the outputs of C by 0.838, as shown in Table 3.15.

Table 3.15

Distance (miles) d	Calories Burned (155-pound person) $C(d)$	Calories Burned (130-pound person) $L(d)$
5	563	$0.838(563) \approx 472$
6	704	$0.838(704) \approx 590$
7	809	$0.838(809) \approx 678$
8	950	$0.838(950) \approx 796$
9	1056	$0.838(1056) \approx 885$
10	1126	$0.838(1126) \approx 944$

Again, although the difference between the outputs is not constant, the values of L and C are related: each output of L is 0.838 times the corresponding output in C. Table 3.15 yields the following relationship between L and C:

$$L(d) = 0.838 C(d)$$

Now we can use this relationship to find the formula for L and graph the function, as shown in Figure 3.37.

$$L(d) = 0.838 \, C(d)$$
$$= 0.838(114.63d + 8.2857)$$
$$= 96.06d + 6.943$$

In Figure 3.37 we can again see how the percentage difference in the number of calories burned per hour yields different slopes in the two functions. This means the difference between the calories burned will continue to grow larger as the number of miles run increases.

Figure 3.37

The transformations in Example 1 represent a *vertical stretch* (part b) and a *vertical compression* (part c). This type of transformation occurs any time the outputs of two functions differ by a constant *factor*. When the magnitude (absolute value) of the constant factor is greater than 1 as in part (b), the result is a **vertical stretch**. We call it a stretch because the graph appears distorted as if someone physically stretched the graph to be farther from the horizontal axis. When the magnitude (absolute value) of the constant factor is less than 1 as in part (c), the result is a **vertical compression**. We call it a compression because the graph appears distorted as if someone physically pushed the graph closer to the horizontal axis. We summarize our observations of vertical stretches and compressions as follows.

VERTICAL STRETCHES AND COMPRESSIONS

The graph of $g(x) = af(x)$ is the graph of f stretched or compressed vertically by a factor of $|a|$ units. If $|a| > 1$, the transformation is a **vertical stretch**. If $0 < |a| < 1$, the transformation is a **vertical compression**.

Generalizing Transformations: Vertical Stretches and Compressions

As in previous sections, we can generalize vertical stretches and compressions by remembering that transformations use the outputs of the parent function to define the outputs of the image function. In the case of vertical stretches and compressions, we multiply the outputs of the parent function by some constant factor for each input to define the image function. For example, consider $g(x) = 4f(x)$ and $g(x) = \frac{1}{3}f(x)$.

$$\underbrace{g(x)}_{\text{the output of } g \text{ at } x} \underbrace{=}_{\text{is equal to}} \underbrace{4}_{\text{four times}} \underbrace{f(x)}_{\text{the output of } f \text{ at } x}$$

$$\underbrace{h(x)}_{\text{the output of } h \text{ at } x} \underbrace{=}_{\text{is equal to}} \underbrace{\frac{1}{3}}_{\text{one-third}} \underbrace{f(x)}_{\text{the output of } f \text{ at } x}$$

EXAMPLE 2 ■ Interpreting and Graphing Function Transformations

Determine the relationship between the function $f(x) = x^3 - 3x^2$ and $g(x) = 0.5(x^3 - 3x^2)$. Then graph both functions.

Solution We observe $g(x) = 0.5f(x)$. That is, the image function g has all of the same input values as the parent function f, but the output values of g are one-half (0.5) as big as those of f. Thus the graph of g is the graph of f compressed vertically by a factor of 0.5. The graphs of both functions are shown in Figure 3.38.

Notice that $(1, -2)$ on f becomes $(1, -1)$ on g, $(2, -4)$ on f becomes $(2, -2)$ on g, and so on. The horizontal intercepts of each function are the same because half of 0 is still 0.

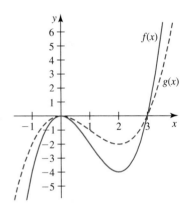

Figure 3.38

EXAMPLE 3 ■ Determining the Effect of Stretches and Compressions on Rates of Change

The British rail system, called BritRail, has a special offer for tourists called a BritRail Pass that provides unlimited train travel for a specified number of days. The pass must be purchased prior to arriving in the country, and the price is the same regardless of how many days it is actually used (up to the maximum).

In 2004, one of the authors went to Scotland on his honeymoon and traveled in the country via passenger train. The BritRail Freedom of Scotland Pass, which cost $189 in 2004, gave the author unlimited travel on trains in Scotland for up to 8 days. The same pass was priced at $292 in 2007. (*Source:* www.acprailnet.com)

a. Create a formula and graph for the average cost per day of each pass given that the pass is used for exactly u days.

b. Write the 2007 average daily cost as a function of the 2004 average daily cost. Then explain how the graphs of the two functions are related.

c. Write the 2004 average daily cost as a function of the 2007 average daily cost. Then explain how the graphs of the two functions are related.

Solution

a. We obtain the average cost per day by dividing the ticket cost by the number of days the ticket is used, u. Thus we have

$$C(u) = \frac{189}{u} \quad \text{and} \quad T(u) = \frac{292}{u}$$

where $C(u)$ is the average daily cost of the $189 pass (in dollars per day) and $T(u)$ is the average daily cost of the $292 pass (in dollars per day).

Since the passes are valid for 8 days, we graph the functions on the interval [0, 8]. Furthermore, since the passes may only be used for whole numbers of days, we plot only the points that correspond with whole days. See Figure 3.39. This type of graph is called a *discrete* graph since it is defined for a finite number of values.

Figure 3.39

b. From part (a), we know $C(u) = \frac{189}{u}$ gives the 2004 average cost and $T(u) = \frac{292}{u}$ gives the 2007 average cost. Since both functions are a multiple of $\frac{1}{u}$, we can write one function in terms of the other.

$$T(u) = \frac{292}{u} \cdot \frac{189}{189} \qquad \text{Multiply by 1.}$$

$$= \frac{292}{189} \cdot \frac{189}{u} \qquad \text{Regroup terms.}$$

$$= \frac{292}{189}(C(u)) \qquad \text{since } C(u) = \frac{189}{u}$$

$$\approx 1.54 C(u)$$

The graph of T is the graph of C *stretched* vertically by a factor of 1.54. That is, the output values of T are 1.54 times as big as the output values of C for any value of u.

c. From part (a), we know $C(u) = \frac{189}{u}$ gives the 2004 average cost and $T(u) = \frac{292}{u}$ gives the 2007 average cost. Since both functions are a multiple of $\frac{1}{u}$, we can write one function in terms of the other.

$$C(u) = \frac{189}{u} \cdot \frac{292}{292} \qquad \text{Multiply by 1.}$$

$$= \frac{189}{292} \cdot \frac{292}{u} \qquad \text{Regroup terms.}$$

$$= \frac{189}{292}(T(u)) \qquad \text{since } T(u) = \frac{292}{u}$$

$$\approx 0.65 T(u)$$

The graph of C is the graph of T *compressed* vertically by a factor of 0.65. The output values of C are 0.65 times as big as the output values of T for any value of u.

JUST IN TIME ■ DISCRETE VERSUS CONTINUOUS FUNCTIONS

Functions may be classified as either *discrete* or *continuous*.

A **discretely defined function** has a finite number of input values. For example, if a movie costs $5.75 per person, then the linear function $P(x) = 5.75x$ models the total cost for x people to attend the movie. This function is defined only for nonnegative integer values of x (such as 0, 1, 2, 3, 4, etc.). The graph is a series of disconnected points.

A **continuously defined function** is defined for every value within the interval of its domain. For example, if $d(t)$ models the depth of water in a bathtub t minutes after turning the water on and if it takes 3 minutes to completely fill the tub, we can evaluate $d(t)$ for any value of t between 0 and 3. All values of t (such as 1.3, 2.0459, 2.9999, and so on) exist and provide us with a meaningful output.

■ Using Stretches and Compressions to Change Units

To make numbers less cumbersome, we often change the units used in a data set. For example, consider Table 3.16, which shows the same relationship between values using different units.

Table 3.16

Year t	Total Consumer Debt (in billions of dollars) $C(t)$	Total Consumer Debt (in dollars) $D(t)$
1985	593.00	593,000,000,000
1990	789.10	789,100,000,000
1995	1095.80	1,095,800,000,000
2000	1556.25	1,556,250,000,000
2005	2175.25	2,175,250,000,000

Source: Federal Reserve Board

The numbers under *billions of dollars* are much easier to work with. Changing output units in this way is an example of a *vertical compression* or *stretch* since, for the same input value, the output values differ by a constant factor. We can show the relationship between D and C as follows:

$$D(t) = 1{,}000{,}000{,}000\, C(t)$$

or

$$C(t) = 0.000000001\, D(t)$$

■ Combining Transformations

We can use vertical stretches and compressions with other types of transformations. We just need to reason through the transformation by considering the relative inputs and outputs of the related functions. In doing so, we will follow the order of operations.

EXAMPLE 4 ■ Using Multiple Transformations to Change a Function

The graph of g is the graph of f reflected vertically, stretched vertically by a factor of 2, and shifted downward 3 units. Write the equation of g in terms of f.

Solution The equation is $g(x) = -2f(x) - 3$. The -2 stretches the graph of f vertically by a factor of 2 and reflects the resultant graph vertically. The -3 reduces the resultant output values by 3 units, resulting in a vertical shift downward by 3 units.

EXAMPLE 5 ■ Combining Transformations

Explain the relationship between the graph of g and the graph of f given that the function $f(x) = \sqrt{x}$ and the function $g(x) = 3f(x + 2)$. Then graph both functions to verify the result.

Solution Since $x + 2$ is larger than x, the related outputs occur at larger inputs in g than they do in f, telling us that g is 2 units to the left of f (see the green curve in Figure 3.40). The outputs we get from f are then multiplied by 3 to get the final output for g (the red curve). Therefore, the graph of g is the graph of f shifted left by 2 units and then stretched vertically by a factor of 3.

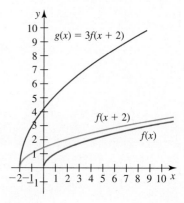

Figure 3.40

SUMMARY

In this section you learned how to transform functions using vertical stretches and compressions. You also learned how to use vertical stretches and compressions in modeling real-world data sets.

3.3 EXERCISES

SKILLS AND CONCEPTS

In Exercises 1–6, graph each transformation of f or g and find the formula for the transformed function.

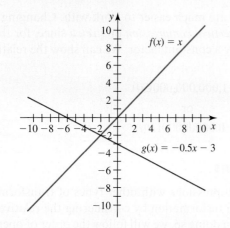

1. $y = 4f(x) + 1$
2. $y = \dfrac{3}{4} f(-x)$
3. $y = 3g(x + 2)$
4. $y = -\dfrac{2}{3} g(x)$
5. $y = -0.5f(x - 5)$
6. $y = 6g(-x) - 4.5$

In Exercises 7–10, explain the transformations on f required to create g. It is not necessary to graph the functions.

7. $f(x) = x^3$, $g(x) = 8(x - 2)^3$
8. $f(x) = x$, $g(x) = \dfrac{4}{3} x + 3$
9. $f(x) = 3x^5$, $g(x) = 3(-x)^5$
10. $f(x) = (x + 5)^3$, $g(x) = -\dfrac{1}{7}(x + 5)^3$

In Exercises 11–16, use the values shown in Table C to evaluate each expression.

Table C

x	f(x)
−4	7
−3	13
−2	29
−1	20
0	6
1	−1
2	−4
3	−13
4	−28

11. $-\frac{2}{3}f(x)$ when $x = -3$
12. $4f(x - 2)$ when $x = 1$
13. $-3f(-x) + 7$ when $x = -2$
14. $3.5f(-x) - 2.1$ when $x = 3$
15. $4f(-3) + \frac{1}{2}f(0)$
16. $1.7f(4) - 3.25f(0)$

In Exercises 17–20, solve each equation for x using Table C.

17. $-2f(x) = -40$
18. $-\frac{1}{4}f(x) - 6 = 1$
19. $5f(x - 2) + 3 = -62$
20. $\frac{1}{2}f(x - 4) + 10 = 13.5$

In Exercises 21–22, refer to Table C to answer each question.

21. Create a table of values for h if $h(x) = -\frac{1}{2}f(x)$.
22. Create a table of values for j if $j(x) = 2f(-x)$.

For Exercises 23–26, draw the graph of each transformation of the function $f(x) = x^2$, then write the formula for the transformed function.

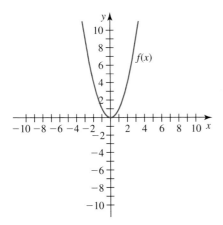

23. $g(x) = -f(x) + 1$
24. $h(x) = 3f(x + 1)$
25. $j(x) = \frac{1}{3}f(x - 2) + 5$
26. $k(x) = -\frac{3}{4}f(x) - 5$

For Exercises 27–30, draw the graph of each transformation of the given function.

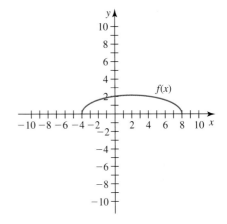

27. $g(x) = f(-x) - 3$
28. $h(x) = \frac{1}{4}f(x - 2) + 1$
29. $j(x) = -6f(-x) - 2$
30. $k(x) = \frac{5}{3}f(x - 3) + 4$

In Exercises 31–33, use the table to identify the transformation(s) on f required to create the indicated function. Then use function notation to write each transformed function in terms of f.

x	f(x)	g(x)	h(x)	j(x)
−1	−12	−48	−6	36
0	−2	−8	−1	6
1	5.5	22	2.75	−16.5
2	4	16	2	−12
3	−1.5	−6	−0.75	4.5
4	3	12	1.5	−9
5	7	28	3.5	−21
6	−1	−4	−0.5	3

31. $g(x)$
32. $h(x)$
33. $j(x)$

For Exercises 34–35, list at least three coordinate points that lie on each transformation of f.

34.

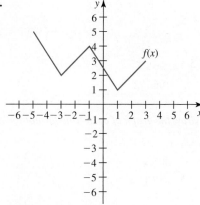

a. $g(x) = -3f(x) + 4$
b. $h(x) = 2f(x - 3)$
c. $k(x) = -4f(x - 6) + 5$

35. $f(x) = 6x^2$
a. $g(x) = 2f(x + 3)$
b. $h(x) = -3.5f(x) - 2$
c. $k(x) = 4.25f(-x)$

SHOW YOU KNOW

36. The function D models the cost of admission for n adults at Disneyland. Explain how vertical stretches or compressions can be used to model the cost of admission if a group has a coupon giving them 15% off each ticket.

37. The function C models the average cost of electricity for a single-family home in Chicago during month m of the year. Explain how vertical stretches or compressions can be used to model the average cost after a 3% increase in the cost of electricity.

38. When you perform a vertical stretch or compression, what happens to the vertical intercept? Why?

39. When you perform a vertical stretch or compression, what happens to the horizontal intercept(s)? Why?

40. How do vertical stretches and compressions affect the rate of change, m, and vertical intercept, b, of a linear function $y = mx + b$?

41. A classmate says any linear function can be created by performing transformations on the function $f(x) = x$. Is your classmate correct? Explain.

42. Sketch a graph of a nonlinear function on graph paper, then pick two points on your graph and calculate the average rate of change between these points. Stretch or compress your function vertically and note where the two points you chose previously are now located. Use this example to explain the effect a vertical stretch or compression has on the average rate of change of a function.

MAKE IT REAL

43. Orange Prices A devastating freeze in January 2007 destroyed roughly 75% of California Central Valley's orange crop. Market analysts predicted this freeze would cause the price of an orange in supermarkets to triple. The graph shows the function C, the cost in dollars of purchasing x oranges before the 2007 freeze. (In this model, we assume no discounts are given for large orders of oranges.)

a. What is the slope of C and what does it represent in the situation?
b. Find a function F that models the cost of oranges after the freeze. Then write F in terms of C.
c. What is the slope of F and how is it related to the slope of C?

44. Value of the Dollar Based on data from 1980 to 2005, the value of the dollar based on producer prices can be modeled by
$V(t) = -0.00004785t^3 + 0.02314t^2 - 0.04774t + 1.137$
where t is the number of years since 1980. (*Source: Modeled from Statistical Abstract of the United States, 2007, Table 705*)
  Write the formula for $P(t)$ given $P(t) = 100V(t)$. What does the function P represent in this situation?

45. Travel Costs In February 2007, a BritRail 15-day Flexipass sold for $644. This pass allowed a traveler unlimited travel on British trains for 15 days during a 60-day period. (*Source: www.raileurope.com*)

a. Write a formula to model the per-day cost if a traveler purchased one of these passes and rode the train d days. Then draw a graph of this function.
b. Travelers can also purchase a BritRail 15-day Consecutive Pass, which allows unlimited travel over a 15-day period. In February 2007 this pass sold for $499. Explain how this situation can be modeled by

transforming the function created in part (a). Then graph the transformed function.

c. Which option is cheaper for the same number of days of travel? Why do you think BritRail sells this pass for less?

46. **Travel Costs** A Eurail Select Pass allows travelers to customize their rail passes by choosing the number of countries they want to visit and the number of travel days they want to use. In February 2007, a first-class 3-country, 10-day unlimited-use pass sold for $609. (*Source:* www.raileurope.com)

a. Write a formula to model the per-day cost of using this pass if the traveler rides the train on d days. Then graph the function.

b. In the same month, a 5-country, 10-day unlimited-use pass sold for $705. Explain how the per-day cost for the 5-country ticket can be found by transforming the function created in part (a). Then draw the transformed graph.

c. In February 2007, Eurail offered a special promotion. If travelers purchased a 10-day ticket they would receive an additional day of travel for free. Can you use transformations to model this special promotion? Explain.

Exercises 47–52 focus on vertical stretches and compressions used to change the units for functions.

47. **Registered Vehicles** The table shows the total number of registered vehicles in the United States as a function of the year.

Year t	Number of Registered Vehicles $N(t)$
1980	155,796,000
1990	188,798,000
1995	201,530,000
2000	221,475,000
2001	230,428,000
2002	229,620,000
2003	231,390,000

Source: Statistical Abstract of the United States, 2006, Table 1078

a. Function M represents the approximate number of registered vehicles in the United States in *millions* as a function of the year. Describe how N can be transformed to create M, then use function notation to write M in terms of N.

b. The function L can be created by shifting M to the left 1970 units. Explain what L models, then use function notation to write L in terms of N.

48. **Speed of Sound** The table shows the speed of sound at sea level in feet per second for different Celsius temperatures.

Temperature (°C) t	Speed of Sound (feet per second) $S(t)$
0	1087.003
5	1096.907
10	1106.722
15	1116.450
20	1126.095
25	1135.658
30	1145.141

Source: www.digitaldutch.com/atmoscalc

a. Without calculating or graphing, does it appear that a linear model will fit this data well? Explain.

b. What is the average rate of change of S between 0°C and 30°C?

c. The conversion between feet and meters is 1 foot = 0.3048 meters. Explain the transformation needed on S so that it models the speed of sound in meters per second.

d. Celsius and kelvin are temperature measurements such that a change of one degree Celsius is the same change in temperature as one kelvin, but they have a different point of reference: 0°C = 273.15 K. Use function notation to represent the relationship between S and P, the speed of sound at sea level in meters per second for a temperature of m kelvin.

49. **Organic Growers** Based on data from 2000 to 2003, the number of certified growers of organic crops and animals can be modeled by N, as shown in the graph.

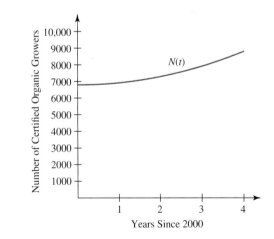

Source: **Modeled from *Statistical Abstract of the United States, 2006*, Table 803**

a. A function R models the number of certified organic growers in *thousands*, where t is the number of years since 2000. Explain the transformation of N required to create R, then use function notation to write N in terms of R.

b. Draw the graph of R.

50. Pressure and Depth The graph shows the function P, the total pressure in pounds per square inch at a depth of d feet.

An *atmosphere* is another unit for measuring pressure, where 1 atmosphere is equal to the pressure of the atmosphere at sea level (14.7 pounds per square inch).

  a. Explain how P can be transformed to create R, the total pressure in atmospheres at a depth of d feet.

  b. Draw the graph of R.

51. Salaries Based on data from 1993 to 2003, the median annual salary in *thousands of dollars* for individuals in the science and engineering field can be modeled by $S(t) = 1.899t + 41.75$, where t is the number of years since 1990. (*Source: Modeled from Science and Engineering Indicators 2006, National Science Foundation, Table 3-8*)

  a. A function $L(t)$ models the median annual salary in *dollars* for individuals in the science and engineering field t years since 1990. Use function notation to write L in terms of S, then write the formula for L.

  b. Function Y is defined in terms of S as $Y(t) = 0.001\ S(t)$. Describe what Y represents, then write the formula for Y.

52. Nursing Home Care Based on data from 1960 to 2004, the cost of nursing home care can be modeled by
$$N(t) = 0.06759t^2 - 0.2958t + 0.0624$$
where t is years since 1960 and N is in *billions of dollars*. (*Source: Modeled from Statistical Abstract of the United States, 2007, Table 120*)

  a. A function H models the cost of nursing home care in *dollars*, where t is years since 1960. Use function notation to write H in terms of N, then write the formula for H.

  b. Function C in terms of N is defined as $C(t) = 1000N(t)$. Describe what C represents, then write the formula for C.

■ STRETCH YOUR MIND

Exercises 53–57 are intended to challenge your understanding of vertical stretches and compressions.

53. Gas Prices The EPA estimates a 2007 PT Cruiser has a fuel efficiency of about 19 miles per gallon for city driving. (*Source: www.fueleconomy.gov*)

  a. Write a formula to show $T(c)$, the cost of driving c miles in the city in a 2007 PT Cruiser if the driver filled up in Fairbanks for $2.19 per gallon.

  b. Explain how you could transform T to determine the cost of driving c miles in the city if the driver filled up in Amarillo for $2.44 per gallon, then use function notation to write your new function in terms of T.

  c. The EPA estimates a 2007 PT Cruiser has a fuel efficiency of about 26 miles per gallon on the highway. (*Source: www.fueleconomy.gov*) Explain how you could transform T to determine the cost of driving h miles on the highway if the driver filled up in Fairbanks, then use function notation to write your new function in terms of T.

  d. Repeat part (c) given that the driver filled up in Amarillo.

54. Racing Speeds The following racetrack is being used for a professional race car event.

  a. Assuming racers will slow down for a turn and speed up for a straightaway, sketch a graph that could model a racer's speed versus time for one lap of the race. Label this function as $s(t)$.

  b. Using your answer to part (a), sketch a graph for $d(t)$, the total distance traveled throughout one lap of the race as a function of time.

  c. To encourage families to come to the main event, race organizers constructed a scale model of the racetrack that kids can race around in go-carts. The racetrack is a one-tenth scale replica of the original, and the go-carts have been designed to have a maximum speed of one-tenth that of a real race car. Explain the relationship between the time it takes the professional drivers to finish one lap versus the time it takes for a go-cart to finish one lap. Do not take into consideration the skill level of the drivers.

  d. Use the transformations of the functions in parts (a) and (b) to create models for the speed and distance traveled of a go-cart over one lap.

55. Examine the functions f and g in the figure.

Is it possible to create g by transforming f using a vertical compression? Explain.

56. **Average Temperatures** The average temperature, A, in Spokane, Washington, in degrees Fahrenheit as a function of m, the month of the year, is shown in the graph.

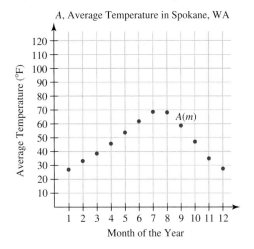

Source: www.cityrating.com

a. The following formula can be used to convert Fahrenheit temperatures to Celsius temperatures: $C = \frac{5}{9}(F - 32)$. Use this information to describe how to transform function A to represent the average temperature in Spokane in degrees Celsius. (*Hint: Be careful!*)

b. Draw the graph of the transformed function.

57. The function P gives the U.S. per capita spending on prescription drugs in dollars t years after 1970. A classmate says a vertical stretch factor equal to the U.S. population will create a function that shows the total amount spent on prescription drugs in the United States. Explain the error in your classmate's thinking.

SECTION 3.4 Horizontal Stretches and Compressions

LEARNING OBJECTIVES

- Identify what change in a function equation results in a horizontal stretch
- Identify what change in a function equation results in a horizontal compression

GETTING STARTED

When a person takes medicine, the body immediately begins to break down the drug. Consequently, medicine has a limited time of peak effectiveness before an additional dose is required. An important measurement in determining dosage size and schedule is the *half-life* of the drug, or the amount of time it takes the body to eliminate half of the medicine in its system.

In this section we demonstrate how horizontal stretches and compressions can be used to model real-world applications such as half-lives. We also look at the effect of horizontal stretches and compressions on graphs, tables, and function equations.

■ Horizontal Stretches and Compressions

The half-life of a drug may vary from person to person, depending on a number of factors. Let's see how horizontal transformations can be used to determine times between doses for various half-lives.

EXAMPLE 1 ■ Stretching a Function Horizontally

Cefotetan (SEF oh tee tan), a prescription antibiotic, has a normal half-life of about 4 hours and is usually prescribed in 2-gram doses. (*Source:* www.merck.com)

a. Create a table of values and a graph for A, the amount of Cefotetan present in the body t hours after taking a 2-gram dose (assuming no further doses are taken).

b. Some people process the drug more slowly. In their bodies, Cefotetan's half-life may be up to 5 hours. Create a table of values and a graph for S, the amount of Cefotetan present in the body t hours after taking a 2-gram dose for these people (assuming no further doses are taken).

c. Explain the connection between S and A. Then use function notation to write S in terms of A.

Solution

a. A half-life of 4 hours tells us that for every 4 hours that passes the amount of Cefotetan will be reduced by half. This is shown in Table 3.17.

Table 3.17

Time Since Taking One Dose (hours) t	Amount of Cefotetan Remaining with a 4-Hour Half-Life (grams) $A(t)$
0	2
4	1
8	0.5
12	0.25
16	0.125

The graph of these values is shown in Figure 3.41. Each of the points is connected by a smooth curve since we may calculate the amount of Cefotetan at any time.

Figure 3.41

b. Using the same idea with a 5-hour half-life gives us the values shown in Table 3.18 and the graph in Figure 3.42.

Table 3.18

Time Since Taking One Dose (hours) t	Amount of Cefotetan Remaining with a 5-Hour Half-Life (grams) $S(t)$
0	2
5	1
10	0.5
15	0.25
20	0.125

Figure 3.42

c. For any $t > 0$, we can see that $S(t) > A(t)$. With a longer half-life, the drug is not eliminated as quickly. Thus there will always be more Cefotetan remaining in S than in A for the same value of t (when $t > 0$). We see the following relationships:

SECTION 3.4 ■ HORIZONTAL STRETCHES AND COMPRESSIONS

i. $S(5) = A(4)$. In both scenarios the patient has 1 gram remaining after one half-life period, but this is 5 hours in S and 4 hours in A.
ii. $S(10) = A(8)$. In both scenarios the patient has 0.5 gram remaining after two half-life periods, but this is 10 hours in S and 8 hours in A.
iii. $S(15) = A(12)$. In both scenarios the patient has 0.25 gram remaining after three half-life periods, but this is 15 hours in S and 12 hours in A.

Notice that the difference between corresponding inputs is not constant; however, they are related. We can see this relationship when the inputs are compared as a ratio. Notice that each ratio is equivalent to $\frac{4}{5}$.

$$\frac{4}{5} \quad \frac{8}{10} = \frac{4}{5} \quad \frac{12}{15} = \frac{4}{5}$$

To find $S(t)$, we have to use an input in A that is $\frac{4}{5}$ of t. That is,

$$S(t) = A\left(\frac{4}{5}t\right)$$

To verify this relationship, let's use it to find $S(20)$

$$S(t) = A\left(\frac{4}{5}t\right)$$

$$S(20) = A\left(\frac{4}{5}(20)\right)$$

$$= A\left(\frac{80}{5}\right)$$

$$= A(16)$$

Using Tables 3.17 and 3.18, we see that $A(16) = 0.125$ gram and $S(20) = 0.125$ gram.

The relationship $S(t) = A\left(\frac{4}{5}t\right)$ is an example of a *horizontal stretch* of A; the graph appears to have been *stretched* so that it is farther from the vertical axis. In a **horizontal stretch**, the same output values occur for input values that are different by a constant *factor*. (Recall that to be a horizontal *shift*, the input values must change by a constant *amount*.) A common mistake is to think of this relationship as a *compression* because the fraction $\frac{4}{5}$ is less than 1. However, if we recall that $S(5) = A(4)$, $S(10) = A(8)$, and $S(15) = A(12)$, we can see the same output values are occurring *later* in S than in A. So this is a stretch. To determine the stretch factor, let's take a close-up look at the graphs of these functions in Figure 3.43.

The input values of S are $\frac{5}{4}$ of the corresponding inputs in A (the inputs giving the same amount of medicine remaining). So, $S(t) = A\left(\frac{4}{5}t\right)$ shows a horizontal stretch of A by a factor of $\frac{5}{4}$.

Figure 3.43

EXAMPLE 2 ■ Compressing a Function Horizontally

As noted in Example 1, Cefotetan has a normal half-life of 4 hours. However, some people process the drug more quickly. In their bodies Cefotetan's half-life may be as short as 3 hours.

a. Create a table of values and a graph for F, the amount of Cefotetan present in the body t hours after taking a 2-gram dose for a person who processes the drug more quickly (assuming no further doses are taken).

b. Explain the connection between F and A, then use function notation to write F in terms of A.

Solution

a. A half-life of 3 hours tells us that for every 3 hours that passes the amount of Cefotetan will be reduced by half. This is shown in Table 3.19 and Figure 3.44.

Table 3.19

Time Since Taking One Dose (hours) t	Amount of Cefotetan Remaining with a 3-Hour Half-Life (grams) $F(t)$
0	2
3	1
6	0.5
9	0.25
12	0.125

Figure 3.44

b. For any $t > 0$, we can see $F(t) < A(t)$. With a shorter half-life, the drug is eliminated more quickly. Thus there will always be less Cefotetan remaining in F than in A for the same value of t (when $t > 0$). We see the following relationships between F and A:

  i. $F(3) = A(4)$. In both scenarios the patient has 1 gram remaining after one half-life period, but this is 3 hours in F and 4 hours in A.

  ii. $F(6) = A(8)$. In both scenarios the patient has 0.5 gram remaining after two half-life periods, but this is 6 hours in F and 8 hours in A.

  iii. $F(9) = A(12)$. In both scenarios the patient has 0.25 gram remaining after three half-life periods, but this is 8 hours in F and 12 hours in A.

The relationship between the inputs in A and F may be compared as a ratio. Notice that each ratio is equivalent to $\frac{4}{3}$.

$$\frac{4}{3} \qquad \frac{8}{6} = \frac{4}{3} \qquad \frac{12}{9} = \frac{4}{3}$$

To find $F(t)$, we must use an input in A that is $\frac{4}{3}$ of t. That is,

$$F(t) = A\left(\frac{4}{3}t\right)$$

To verify this relationship, let's use it to find $F(12)$:

$$F(t) = A\left(\frac{4}{3}t\right)$$

$$F(12) = A\left(\frac{4}{3}(12)\right)$$

$$= A\left(\frac{48}{3}\right)$$

$$= A(16)$$

Using Tables 3.17 and 3.19, we see that $A(16) = 0.125$ gram and $F(12) = 0.125$ gram.

The relationship $F(t) = A\left(\dfrac{4}{3}t\right)$ discussed in Example 2 is a *horizontal compression* of A because the graph appears to have been squeezed closer to the vertical axis. In a **horizontal compression**, the same output values occur for input values that are different by a constant *factor*. A common mistake is to think of this relationship as a *stretch* because the fraction $\dfrac{4}{3}$ is greater than 1. However, if we recall that $F(3) = A(4)$, $F(6) = A(8)$, and $F(9) = A(12)$, we can see the same output values are occurring *earlier* in F than in A. So this is a compression. To determine the compression factor, let's take a close-up look at the graphs of these functions in Figure 3.45.

Figure 3.45

The input values of F are $\dfrac{3}{4}$ the magnitude of the corresponding inputs in A (the inputs giving the same amount of medicine remaining). Thus, $F(t) = A\left(\dfrac{4}{3}t\right)$ shows a horizontal compression of A by a factor of $\dfrac{3}{4}$.

Notice that the stretch or compression factor is always the reciprocal of the coefficient of the input variable. If the reciprocal is less than 1, the transformation is a horizontal compression. If the reciprocal is greater than 1, the transformation is a horizontal stretch. We summarize our observations as follows.

HORIZONTAL STRETCHES AND COMPRESSIONS

The graph of $g(x) = f(bx)$ is the graph of f stretched or compressed horizontally by a factor of $\left|\dfrac{1}{b}\right|$.

- If $\left|\dfrac{1}{b}\right| < 1$, the transformation is a **horizontal compression**.
- If $\left|\dfrac{1}{b}\right| > 1$, the transformation is a **horizontal stretch**.

■ Generalizing Transformations: Horizontal Stretches and Compressions

Horizontal compressions are the perhaps the trickiest of the transformations even when we remember that transformations use the outputs of a parent function to define the outputs of an image function. In the case of horizontal stretches and compressions, we look for a relationship between inputs of the two functions that will result in the same output values. Consider the relationship $g(x) = f(3x)$. This tells us the two functions will have *identical* outputs.

$$\underbrace{g(\ \ )}_{\text{the outputs of } g} = \underbrace{f(\ \ )}_{\text{the outputs of } f}$$

However, the expression $3x$ tells us these outputs result from different inputs. If we want to evaluate g for some input x, we need to look to the output of function f when its input is 3 times as great as x (a larger magnitude input).

$$\underbrace{g(x)}_{\text{the outputs of } g \text{ at } x} = \underbrace{f(3x)}_{\text{the outputs of } f \text{ at 3 times } x}$$

To illustrate, let's find $g(-2)$ and $g(1)$.

$$\underbrace{\begin{aligned} g(x) &= f(3x) \\ g(-2) &= f(3 \cdot -2) \\ g(-2) &= f(-6) \end{aligned}}_{\substack{\text{The output of } g \text{ at } -2 \text{ is} \\ \text{defined by the output} \\ \text{of } f \text{ at } -6.}} \qquad \underbrace{\begin{aligned} g(x) &= f(3x) \\ g(1) &= f(3 \cdot 1) \\ g(1) &= f(3) \end{aligned}}_{\substack{\text{The output of } g \text{ at } 1 \text{ is} \\ \text{defined by the output} \\ \text{of } f \text{ at } 3.}}$$

If f is given by the values in Table 3.20, then g has the values in Table 3.21.

Table 3.20

x	$f(x)$
-12	20
-6	15
-1	12
0	11
3	10
5	9
9	6

Table 3.21

x	$g(x)$
-4	20
-2	15
$-1/3$	12
0	11
1	10
5/3	9
3	6

If f is defined by the graph in Figure 3.46a, then g is defined by the graph in Figure 3.46b.

Figure 3.46

We can verify statements such as $g(1) = f(3)$ using either the tables or the graphs. We see that this is a *horizontal compression* by a factor of $\frac{1}{3}$.

To determine which function has outputs that occur for inputs closer to zero (and thus determine whether the image function is a compression or a stretch), we simply need to look at the inputs in the definition. In the case of $g(x) = f(3x)$, x will always be closer to zero than $3x$, so the same outputs occur in g for inputs closer to zero than in f. The opposite is true for a relationship defined by a statement such as $h(x) = f(0.5x)$. In this case the inputs in f are closer to zero for the same output.

$$\underbrace{h(x)}_{\substack{\text{the outputs} \\ \text{of } h \text{ at } x}} = \underbrace{f(0.5x)}_{\substack{\text{the outputs of } f \\ \text{at half of } x}}$$

This is a *horizontal stretch* by a factor of 2 (because $\frac{1}{0.5} = 2$).

EXAMPLE 3 ■ Determining a Compression from an Equation

Describe the graphical relationship between $f(x) = 5x^2$ and $g(x) = 5(2x)^2$. Then graph both functions to confirm the accuracy of your conclusion.

Solution Notice $g(x) = f(2x)$. Since the reciprocal of the coefficient of the input variable is $\frac{1}{2}$, the graph of g is the graph of f compressed horizontally by a factor of $\frac{1}{2}$. The graphs are shown in Figure 3.47.

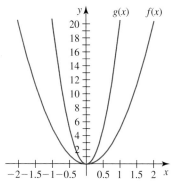

Figure 3.47

■ Using Stretches and Compressions to Change Units

Horizontal stretches and compressions are often used to change the units on the input variable. For example, consider Table 3.22 showing the amount of taxes to be paid by married couples filing a joint tax return for the given income amounts in two different functions.

Table 3.22

Income ($ thousands) n	Tax Paid (dollars) $T(n)$	Income (dollars) $d = 1000n$	Tax Paid (dollars) $A(d)$
20	2,249	20,000	2,249
30	3,749	30,000	3,749
40	5,249	40,000	5,249
50	6,749	50,000	6,749
60	8,249	60,000	8,249
70	10,621	70,000	10,621
80	13,121	80,000	13,121
90	15,621	90,000	15,621

Source: 2006 IRS Tax Table, www.irs.gov

Function A is a *horizontal stretch* of T because the same output values now occur for input values 1000 times as great. We can also say T is a *horizontal compression* of A because the same output values occur for input values 0.001 times as great. Since $d = 1000n$ and $n = 0.001d$, we can relate A and $T(n)$ as follows:

$$A(d) = T(0.001d)$$
or
$$T(n) = A(1000n)$$

■ Combining Transformations

Horizontal stretches and compressions may be combined with the other transformations we have studied: horizontal and vertical shifts, horizontal and vertical reflections, and vertical stretches and compressions.

EXAMPLE 4 ■ Combining Transformations

For each of the following, draw the graph of g(x).

a. The function f is graphed in Figure 3.48. Draw $g(x) = f(-2x) - 6$.

b. Given $f(x) = x^2$, graph $g(x) = -\dfrac{1}{4}f\left(\dfrac{1}{3}x\right) + 2$.

Solution

a. The function $g(x) = f(-2x) - 6$ is the combination of three transformations on f: a horizontal compression by $\dfrac{1}{2}$, a horizontal reflection, and a vertical shift downward 6 units. Using the graph of f to find the points, we do the stretches, compressions, and reflections first and the shifts last, as shown in Table 3.23 and Figure 3.49.

Figure 3.48

Table 3.23

x	$f(x)$ (Figure 3.49a)	Horizontal Compression by $\dfrac{1}{2}$ $f(2x)$ (Figure 3.49b)	Horizontal Reflection $f(-2x)$ (Figure 3.49c)	Vertical Shift Downward 6 $f(-2x) - 6$ (Figure 3.49c)	Coordinate Points for $g(x)$ (Figure 3.49c)
−4	1	undefined	undefined	undefined	undefined
−3	1	undefined	$f(-2(-3)) = f(6) = 5$	$5 - 6 = -1$	$(-3, -1)$
−2	1	$f(2(-2)) = f(-4) = 1$	$f(-2(-2)) = f(4) = 4$	$4 - 6 = -2$	$(-2, -2)$
−1	1	$f(2(-1)) = f(-2) = 1$	$f(-2(-1)) = f(2) = 0$	$0 - 6 = -6$	$(-1, -6)$
0	1	$f(2(0)) = f(0) = 1$	$f(-2(0)) = f(0) = 1$	$1 - 6 = -5$	$(0, -5)$
1	1	$f(2(1)) = f(2) = 0$	$f(-2(1)) = f(-2) = 1$	$1 - 6 = -5$	$(1, -5)$
2	0	$f(2(2)) = f(4) = 4$	$f(-2(2)) = f(-4) = 1$	$1 - 6 = -5$	$(2, -5)$
3	1	$f(2(3)) = f(6) = 5$	undefined	undefined	undefined
4	4	undefined	undefined	undefined	undefined
5	4.5	undefined	undefined	undefined	undefined
6	5	undefined	undefined	undefined	undefined

Figure 3.49

SECTION 3.4 ■ HORIZONTAL STRETCHES AND COMPRESSIONS

b. We are given $f(x) = x^2$ and asked to graph $g(x) = -\frac{1}{4}f\left(\frac{1}{3}x\right) + 2$. Function g is the combination of four transformations on f: a horizontal stretch by a factor of 3, a vertical compression by $\frac{1}{4}$, a vertical reflection, and a vertical shift upward 2 units. We can again examine how these transformations affect the coordinates of the original function, as shown in Table 3.24. To do this we will take a slightly different approach than in the previous tables to give us another way of approaching transformations. This method keeps track of what happens to individual points as the function transforms instead of looking at what happens at fixed values of x. The graphical progression of the transformations is shown in Figure 3.50.

Table 3.24

Coordinate Points for $f(x) = x^2$ (Fig. 3.50a; blue)	Horizontal Stretch by 3 $f\left(\frac{1}{3}x\right)$ (Fig. 3.50b; purple)	Vertical Compression by $\frac{1}{4}$ and Vertical Reflection $-\frac{1}{4}f\left(\frac{1}{3}x\right)$ (Fig. 3.50a; green)	Vertical Shift Upward 2 $-\frac{1}{4}f\left(\frac{1}{3}x\right) + 2$ (Fig. 3.50b)	Coordinate Points for $g(x)$ (Fig. 3.50b)
$(-3, 9) \to$	$(-9, 9) \to$	$\left(-9, -\frac{9}{4}\right) = (-9, -2.25) \to$	$(-9, -0.25)$	$(-9, -0.25)$
$(-2, 4) \to$	$(-6, 4) \to$	$\left(-6, -\frac{4}{4}\right) = (-6, -1) \to$	$(-6, 1)$	$(-6, 1)$
$(-1, 1) \to$	$(-3, 1) \to$	$\left(-3, -\frac{1}{4}\right) = (-3, -0.25) \to$	$(-3, 1.75)$	$(-3, 1.75)$
$(0, 0) \to$	$(0, 0) \to$	$(0, 0) \to$	$(0, 2)$	$(0, 2)$
$(1, 1) \to$	$(3, 1) \to$	$\left(3, -\frac{1}{4}\right) = (3, -0.25) \to$	$(3, 1.75)$	$(3, 1.75)$
$(2, 4) \to$	$(6, 4) \to$	$\left(6, -\frac{4}{4}\right) = (6, -1) \to$	$(6, 1)$	$(6, 1)$
$(3, 9) \to$	$(9, 9) \to$	$\left(9, -\frac{9}{4}\right) = (9, -2.25) \to$	$(9, -0.25)$	$(9, -0.25)$

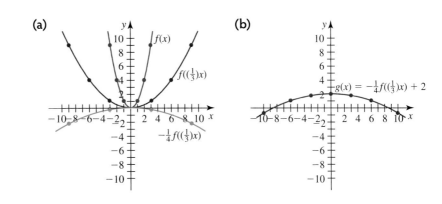

Figure 3.50

212 CHAPTER 3 ■ TRANSFORMATIONS OF FUNCTIONS

EXAMPLE 5 ■ Solving a Transformed Function Equation

Given the Table 3.25 for $f(x)$, solve the transformed function equation $f(2x) = 1$.

Solution Since $f(2x) = 1$ and $f(4) = 1$, $2x = 4$. Dividing each side by 2 yields the solution $x = 2$.

Table 3.25

x	$f(x)$
0	8
1	12
2	10
3	6
4	1

SUMMARY

In this section you learned how to transform functions using horizontal stretches and compressions. You also learned how to use horizontal stretches and compressions in real-world contexts.

3.4 EXERCISES

SKILLS AND CONCEPTS

In Exercises 1–4, graph each transformation of f or g.

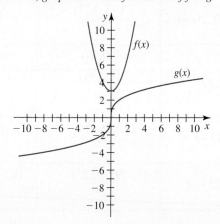

1. $y = f\left(\dfrac{1}{4}x\right)$
2. $y = g(-2x)$
3. $y = -g(0.5x) - 5$
4. $y = -\dfrac{1}{2}f(3x)$

In Exercises 5–10, use the values shown in Table D to evaluate each expression.

Table D

x	$f(x)$	x	$f(x)$	x	$f(x)$
−4	3	−1	10	2	−5
−3	6	0	−2	3	24
−2	1	1	15	4	−9

5. $f(3x)$ when $x = 1$
6. $f\left(-\dfrac{2}{3}x\right)$ when $x = 3$
7. $f(0.5x) + 2$ when $x = 4$
8. $2f(2x)$ when $x = -2$
9. $-10f\left(-\dfrac{4}{3}x\right) - 2.5$ when $x = -3$
10. $7f\left(\dfrac{1}{6}(x - 5)\right) - 1$ when $x = 11$

In Exercises 11–14, solve each equation for x using Table D.

11. $f\left(\dfrac{1}{3}x\right) = 24$
12. $3f(4x) = -15$
13. $2f(2.5x) - 7 = 5$
14. $\dfrac{6}{5}f(4(x + 2)) = -\dfrac{12}{5}$

In Exercises 15–18, refer to Table D to answer each question.

15. Create a table of values for g if $g(x) = f(4x)$.
16. Create a table of values for h if $h(x) = f\left(\dfrac{1}{3}x\right)$.
17. Create a table of values for j if $j(x) = -2f(3x)$.
18. Create a table of values for k if $k(x) = f\left(-\dfrac{1}{2}x\right) + 3$.

In Exercises 19–24, match each graph with the appropriate transformation of the given function.

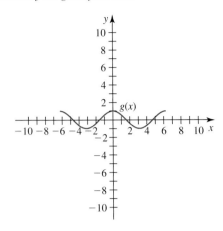

a. $A(x) = g(2x) + 3$

b. $B(x) = g\left(\dfrac{1}{4}x\right) + 3$

c. $C(x) = g\left(\dfrac{1}{2}x\right) + 3$

d. $D(x) = g(4x) + 3$

e. $E(x) = g(2(x - 1)) + 3$

f. $F(x) = g\left(\dfrac{1}{2}(x - 1)\right) + 3$

19.

20.

21.

22.

23.

24.

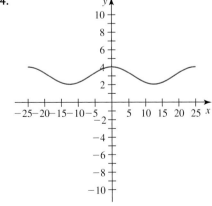

For Exercises 25–28, draw the graph of each transformation of the function f.

25. $g(x) = f(2x)$
26. $h(x) = f(-3x)$
27. $j(x) = f\left(\dfrac{1}{4}x\right) + 2$
28. $k(x) = f\left(-\dfrac{3}{4}x\right) - 2$

For Exercises 29–31,

a. Describe the horizontal transformations required on f to create the new function.
b. Use function notation to write each new function in terms of f.
c. Write the formula for the new function. (You may wish to check your work using a graphing calculator.)

29.

30.

31.

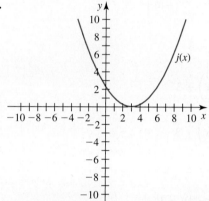

For Exercises 32–35, draw the graph of each transformation of the given function f.

32. $g(x) = 4f(-0.5x)$

33. $h(x) = \dfrac{1}{2}f\left(\dfrac{1}{2}x\right)$

34. $j(x) = -2f(-1.75x)$

35. $k(x) = \dfrac{3}{5}f\left(-\dfrac{2}{3}x\right)$

For Exercises 36–37, list three coordinate points that lie on each transformation of the given function f.

36.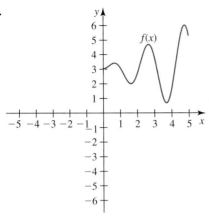

   a. $g(x) = f(3x)$
   b. $h(x) = -f\left(\dfrac{1}{5}x\right) - 3$
   c. $k(x) = -6f\left(\dfrac{2}{3}x\right)$

37. $f(x) = \dfrac{1}{x}$

   a. $g(x) = f(2x)$
   b. $h(x) = f\left(-\dfrac{1}{4}x\right) + 7$
   c. $k(x) = -2f(1.75x)$

38. **Swimming Competition** During a race a swimmer swims at a constant speed from A to B and back in lane 3 of the swimming pool shown in the figure. Judges standing at positions A, B, and C are observing the race.

   a. Sketch a graph for D_A, the distance between the swimmer and Judge A as a function of the time since the race began. Assume the race involves two laps and that the swimmer begins the race on Judge A's side of the pool.
   b. Sketch a graph of D_B, the distance between the swimmer and Judge B as a function of the time since the race began. Is the graph of D_B a transformation of D_A? Explain.
   c. Sketch a graph of D_C, the distance between the swimmer and Judge C as a function of the time since the race began.
   d. In the next race, the swimmer in lane 3 swims twice as fast as the previous swimmer. How will this affect the graphs of the functions in parts (a) through (c)? Explain.

In Exercises 39–40, the graphs of functions f and g are given.

   a. Explain how f may be transformed vertically to create g, then use function notation to write g in terms of f.
   b. Explain how f may be transformed horizontally to create g, then use function notation to write g in terms of f.

39.

40.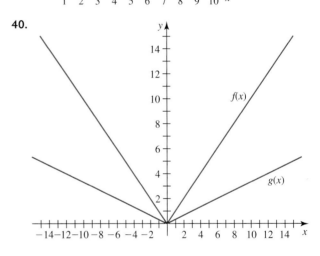

SHOW YOU KNOW

41. When you perform a horizontal stretch or compression, what happens to the vertical intercept?

42. When you perform a horizontal stretch or compression, what happens to the horizontal intercept(s)?

43. In the figure, function g is a transformation of $f(x) = x^2$.

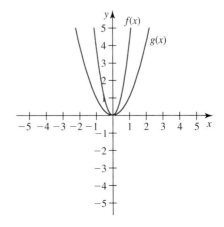

One of your classmates says *g* is created by vertically compressing *f* by a factor of $\frac{1}{4}$, while another classmate says it is a horizontal stretch by a factor of 2. Which classmate is correct? Explain.

44. Is it always true that a function defined by using a vertical stretch or compression can also be defined by using a horizontal stretch or compression instead, and vice versa? Explain.

45. Function *g* is a transformation of *f* such that $g(x) = af(bx)$. When transforming *f* to create *g*, which transformation should be done first? Explain.

46. We have looked at six types of transformations: horizontal and vertical shifts, horizontal and vertical reflections, and horizontal and vertical stretches/compressions. Which of these transformations may impact the following function characteristics? List all that apply.
    a. Vertical intercept
    b. Horizontal intercept(s)
    c. Average rate of change from $x = 1$ to $x = 3$

47. Sketch a graph of a nonlinear function on graph paper, then pick two points on the graph and calculate the average rate of change between the points. Stretch or compress the function vertically and note where the two points previously chosen are now located. Use this example as an aid to explain the effect a horizontal stretch or compression has on the average rate of change of a function.

■ MAKE IT REAL

48. **Medicine** Rifampin (rif AM pin) is a drug used to help manage active tuberculosis and has a half-life of about 4 hours in many people. (*Source:* www.merck.com)
    a. Create a table of values to model the amount of Rifampin present in a person's body as a function of the time since taking a 300-milligram dose. (Assume no further doses are taken.)
    b. In some patients the half-life can be as short as 3 hours. Explain how to transform the function created in part (a) to model the amount of the drug present in a person's body if the half-life is 3 hours.

49. **Medicine** Gentamicin (jen ta MYE sin) is an antibiotic with a half-life of about 3 hours in many people. (*Source:* www.merck.com)
    a. Create a table of values to model the amount of Gentamicin present in a person's body as a function of the time since taking a 150-milligram dose. (Assume no further doses are taken.)
    b. In some patients the half-life can be as short as 2 hours. Explain how to transform the function created in part (a) to model the amount of the drug present in a person's body if the half-life is 3 hours.
    c. In people with advanced kidney disease, the half-life of the drug can be as long as 70 hours. Explain how to transform the function created in part (a) to model the amount of the drug present in a person's body if the half-life is 70 hours.

50. **Scale Models** *E* approximates the shape of the Eiffel Tower. Its output is the height in meters of the tower *x* meters from its center, as shown in the graph. (A "negative" distance is interpreted to mean a distance measured from the center to the left side of the tower.)

a. Kings Island Amusement Park in Mason, Ohio, has a one-third scale model of the Eiffel Tower. Explain the transformations on *E* necessary to create a graph showing *K*, the height of Kings Island's replica in meters as a function of the distance from its center.

b. Paris Las Vegas Hotel in Las Vegas, Nevada, originally planned to create a full-scale replica of the Eiffel Tower, but had to create a half-scale replica because of the hotel's proximity to the Las Vegas airport. Explain the transformations on *K* needed to create the graph for the Paris Las Vegas's replica.

51. **Scale Models** The Gateway Arch (Jefferson National Expansion Memorial) in St. Louis, Missouri was built in the 1960s to celebrate the westward expansion of the United States. Its shape can be modeled by the function *A*, the height of the arch in feet *x* feet from the base of its left leg.

Paperlandmarks sells paper kits that hobbyists assemble to create scale replicas of famous landmarks. For the Gateway Arch, the company sells 1/600th scale replicas and 1/1000th scale replicas. (*Source:* www.paperlandmarks.com)

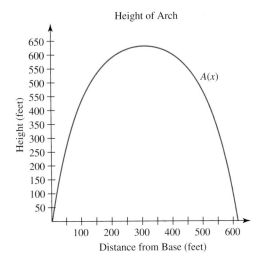

Height of Arch

a. Approximately how tall will each scale model be (in feet)?
b. For each kit, describe how A must be transformed to create a graph of the height of each model as a function of the distance from the base of its left leg.
c. Use function notation to write each of the functions in part (b) in terms of A.
d. Since it is uncommon to measure models of this size in feet, rewrite the functions in part (c) so that the dimensions of the models are measured in inches.

Exercises 52–55 focus on stretches and compressions used to change the units for functions.

52. Mortgage Payments The following tables show principal and interest payments as a function of the total amount borrowed for a 30-year fixed loan at 6% annual interest.

Amount Borrowed (dollars) w	Monthly Payment (dollars) $Y(w)$
100,000	599.55
200,000	1199.10
225,000	1348.99
280,000	1678.74
330,000	1978.52

Amount Borrowed ($ thousands) b	Monthly Payment (dollars) $P(b)$
100	599.55
200	1199.10
225	1348.99
280	1678.74
330	1978.52

a. Explain the transformation required on Y to create the function P, then use function notation to demonstrate this relationship.
b. Given that $T(b) = 0.001P(b)$, explain what the function T models.
c. Explain the similarities and differences in the rates of change for functions Y and P.

53. Minimum Wage The first table shows the minimum wage mandated by federal law t years after 1950.

Years Since 1950 t	Federal Minimum Wage (dollars) W
0	0.75
10	1.00
20	1.60
30	3.10
40	3.80
50	5.15

Source: www.dol.gov

The next table shows a transformation of the minimum wage table.

0	0.75
1	1.00
2	1.60
3	3.10
4	3.80
5	5.15

a. Explain the transformation on W required to create the new function.
b. Use function notation to write the new function in terms of W.
c. Place appropriate labels on the second table to describe what it models.
d. Should these functions be graphed on the same set of axes? Explain.

54. Home Prices The graph below shows M, the median price of homes in the Boston area. The value $x = 0$ corresponds to the third quarter of 2005 (Jul–Sep 2005), and x is in quarters of a year (3-month periods).

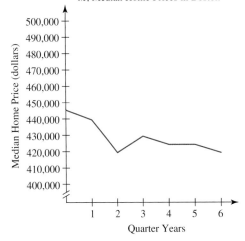

M, Median Home Prices in Boston

Source: www.housingtracker.com

a. Create an estimated table of values for a new function that models the median home price in *thousands of dollars* as a function of the number of quarter years after the third quarter of 2005.

b. Use function notation to write the new function in terms of *M*.

c. Create an estimated table of values for a new function that models the median home price in *thousands of dollars* as a function of the *number of years* after the third quarter of 2005.

d. Use function notation to show the relationship between the new function in part (c) and *M*.

55. **Atmospheric Pressure** Based on data from sea level to 8500 meters, the atmospheric pressure can be modeled by $P(a) = 14.96(0.9998)^a$ pounds per square inch, where *a* is the altitude in meters. (*Source: Modeled from Digital Dutch 1976 Standard Atmosphere Calculator*)

a. Function $R(k)$ models the atmospheric pressure in pounds per square inch where *k* is the altitude in *kilometers* (1000 meters = 1 kilometer). Explain the transformation on *P* required to create *R*, then use function notation to write *R* in terms of *P*.

b. Write the formula for *R* as a function of *k*.

c. Function *S* models the atmospheric pressure in pounds per square inch where *c* is the altitude in *centimeters* (1 meter = 100 centimeters). Explain the transformation on *P* required to create *S*, then use function notation to write *S* in terms of *P*.

d. Write the formula for *S* as a function of *c*.

■ **STRETCH YOUR MIND**

Exercises 56–60 are intended to challenge your understanding of vertical and horizontal stretches and compressions.

56. When performing transformations, not every point of a function has to change. Explain the characteristics of a point if it did not change after performing the indicated transformation.

a. Horizontal stretch/compression
b. Vertical stretch/compression
c. Horizontal reflection
d. Vertical reflection

57. Water is poured into two vases at a constant rate.

a. Sketch graphs showing the height of the water in each vase as a function of time.

b. If the width of each vase was increased, what would happen to the graphs drawn in part (a)? Explain.

c. If the rate at which the water was poured into the vases increased, what would happen to the graph drawn in part (a)? Explain.

58. Examine the following functions *f* and *g*.

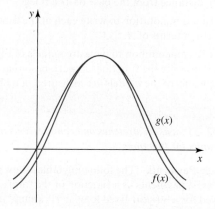

Is it possible to create *g* by transforming *f* using only a horizontal stretch or compression? Explain.

59. Create the graphs of a function and its inverse. Explain how a vertical stretch or compression of the original function will affect the inverse function.

60. Create the graphs of a function and its inverse. Explain how a horizontal stretch or compression of the original function will affect the inverse function.

CHAPTER 3 Study Sheet

As a result of your work in this chapter, you should be able to answer the following questions, which are focused on the "big ideas" of this chapter.

SECTION 3.1
1. What happens numerically, graphically, and symbolically when a function is shifted vertically? Why?
2. What happens numerically, graphically, and symbolically when a function is shifted horizontally? Why?

SECTION 3.2
3. What happens numerically, graphically, and symbolically when a function is reflected either horizontally or vertically? Why?
4. What is meant by even and odd symmetry? How can we test for even or odd symmetry?

SECTION 3.3
5. What is a vertical stretch or compression?
6. What impact do vertical stretches and compressions have on the rate of change?

SECTION 3.4
7. What is a horizontal stretch or compression?
8. What impact do horizontal stretches and compressions have on rate of change?
9. What effect does a negative stretch or compression have on a function?

REVIEW EXERCISES

■ SECTION 3.1 ■

For each given function in Exercises 1–3,
  a. Graph the function.
  b. Describe the transformations necessary on f to create g.
  c. Draw the graph for g by performing the necessary transformations.
  d. List at least three coordinate points on the graph of g.
  e. Write the formula for g.

1. $f(x) = x^3$, $g(x) = f(x - 1) + 6$
2. $f(x) = 4\sqrt{x}$, $g(x) = f(x + 2) - 5$
3. $f(x) = \dfrac{1}{2}x + 3$, $g(x) = f(x - 4) - 2$

In Exercises 4–6, use the table to determine the transformations on f required to create the indicated function. Then use function notation to write each transformed function in terms of f.

x	$f(x)$	$g(x)$	$h(x)$	$j(x)$
−4	2	1	11	10
−3	4	1	5	12
−2	7	2	1	15
−1	11	4	0	19
0	5	7	−1	13
1	1	11	−3	9
2	0	5	−7	8
3	−1	1	−12	7
4	−3	0	−20	5

4. $g(x)$
5. $h(x)$
6. $j(x)$

In Exercises 7–8, describe the transformation on f required to create each function. Then use function notation to write each transformed function in terms of f.

7.

8.

In Exercises 9–10,
  a. Create a table of values for the aligned data.
  b. Describe the horizontal translation required to align the data.
  c. If the original function is f and the aligned function is g, use function notation to write g in terms of f.

9. Align the data to 1992.

Year	Median Family Income in Constant (2003) Dollars
1980	44,452
1985	45,223
1990	48,248
1992	46,992
1994	47,615
1996	49,378
1998	52,675
2000	54,191

Source: Statistical Abstract of the United States, 2006, Table 679

10. Align the data to 1999.

Year	Individual Income Tax Returns Filed (in millions)
1996	116.1
1997	118.4
1998	120.3
1999	122.5
2000	124.9
2001	127.1
2002	129.4
2003	130.3
2004	130.1

Source: Statistical Abstract of the United States, 2006, **Table 471**

11. Refer to the table in Exercise 9.
    a. Create a table that shows the increase in median family income since 1980.
    b. Describe the transformation required to create the table in part (a).
    c. If the original function is f and the transformed function is g, use function notation to write g in terms of f.

12. Refer to the table in Exercise 10.
    a. Create a table that shows the increase in individual tax returns filed since 1998.
    b. Describe the transformation required to create the table in part (a).
    c. If the original function is f and the transformed function is g, use function notation to write g in terms of f.

In Exercises 13–16, use the values shown in Table E to evaluate each expression.

Table E

x	$f(x)$
−3	−11
−2	−8
−1	−1
0	3
1	5
2	4
3	2

13. $f(x) - 6$ when $x = 2$
14. $f(x) + 2.5$ when $x = 0$
15. $f(x + 1)$ when $x = -3$
16. $f(x - 5)$ when $x = 3$

In Exercises 17–18, solve each equation for x using Table E.

17. $f(x) + 1 = 6$
18. $f(x - 4) = -1$

In Exercises 19–20, refer to Table E to answer each question.

19. Create a table of values for g if $g(x) = f(x + 2)$.
20. Create a table of values for h if $h(x) = f(x) - 4.5$.

SECTION 3.2

For each given function in Exercises 21–23,
  a. Graph the function.
  b. Describe the transformations necessary on f to create g.
  c. Draw the graph for by performing the necessary transformations.
  d. List at least three coordinate points on the graph of g.
  e. Write the formula for g.

21. $f(x) = x^3$, $g(x) = f(-x) + 2$
22. $f(x) = 2|x|$, $g(x) = -f(x + 1)$
23. $f(x) = \dfrac{2}{5}x - 7$, $g(x) = -f(-x) - 6$

24. **Falling Objects** The function $d(t) = 16t^2$ can be used to model the distance an object travels in feet t seconds after it has been dropped.
    a. Draw a graph for d.
    b. Suppose you dropped an object and it fell for 1.6 seconds before hitting the ground. How high did it fall from?
    c. Suppose you drop an object from your hand 4 feet above ground. How long will it take to hit the ground? How long will it take to reach a height of 2 feet off the ground?
    d. Imagine dropping an object off the roof of a 100-foot-tall building. Graph the function h, the height of the object above ground t seconds after being dropped.
    e. Describe how you could have created the graph for function h by performing transformations on d. Then use function notation to demonstrate the relationship between the functions.

25. **Business Expenses** The table shows the labor expenses for Chipotle Mexican Grill in years after 2000.

Year Since 2000 t	Labor Costs ($ thousands) $L(t)$
1	46,048
2	66,515
3	94,023
4	139,494
5	178,721

Source: **Chipotle Mexican Grill, Inc., 2005 Annual Report, p. 24**

  a. Why are labor costs considered negative numbers when determining their effect on profit? Explain how L can be transformed to represent this.
  b. Use function notation to write N, the effect on profit of the labor costs, in terms of L.

c. Describe what function S represents if $S(y) = N(y - 2000)$.
d. Create a table of values for S. Would a linear function be a good model for S? Explain why or why not.

In Exercises 26–27, complete each of the following for the given graph.

a. Explain what it means for a function to display even symmetry. Then complete the graph so that the function displays even symmetry.

b. Explain what it means for a function to display odd symmetry. Then complete the graph so that the function displays odd symmetry.

26.

27.

In Exercises 28–30, determine whether each of the following functions has even symmetry, odd symmetry, or neither type of symmetry.

28. $f(x) = 4x + 3$
29. $f(x) = \dfrac{2}{x^2 + 1}$
30. $f(x) = 2.3x^5 + x^3$

SECTION 3.3

For each given function in Exercises 31–33,
a. Graph the function.
b. Describe the transformations necessary on f to create g.
c. Draw the graph for g by performing the necessary transformations.

d. List at least three coordinate points on the graph of g.
e. Write the formula for g.

31. $f(x) = 2x$, $g(x) = 5f(x - 4)$
32. $f(x) = \dfrac{1}{x^2}$, $g(x) = -0.75f(x)$
33. $f(x) = \sqrt{x} - 1$, $g(x) = 1.5f(-x) + 8$
34. Complete the table of values as much as possible. You will not have enough information to fill in every missing number.

x	$f(x)$	$\dfrac{1}{2}f(x)$	$4f(-x)$	$-2f(x-3)$
−9	2			
−6	4			
−3	7			
0	11			
3	5			
6	1			
9	0			

In Exercises 35–40, use the values shown in Table F to evaluate each expression.

Table F

x	$f(x)$
−3	14
−2	11
−1	10
0	12
1	16
2	22
3	31

35. $4f(x)$ when $x = -1$
36. $-f(x + 4)$ when $x = -2$
37. $\dfrac{4}{3}f(x - 2) + 1$ when $x = 0$
38. $3.25 f(-x) + 7$ when $x = 3$
39. $9f(-2) + 0.5f(0)$
40. $\dfrac{5}{2}f(1) - 3f(3)$

In Exercises 41–42, solve each equation for x using Table F.

41. $3f(-x) = 30$
42. $\dfrac{1}{3}f(x - 5) + 6 = 10$

In Exercises 43–44, refer to Table F to answer each question.

43. Create a table of values for g if $g(x) = -f(-x)$.
44. Create a table of values for h if $h(x) = 2.5f(x) - 4$.
45. **Member Benefits** In January 2007, one of the authors visited a Barnes & Noble bookstore with a list of hardcover

books he was considering for purchase. Let *R* represent the retail cost of purchasing of these books.

a. The author is a Barnes & Noble member, which entitles him to 20% off the cost of hardcover books. Use function notation to show the relationship between the author's cost, *A*, of purchasing *x* of the books on his list and the retail cost.

b. When he visited the store, the author had a $50 gift card that he intended to use toward the purchase of these books. Use function notation to write *G*, the author's cost to purchase *x* books from his list as a member with the gift card, in terms of *R*.

c. In performing the transformations on *R* to create *G*, which transformation must occur first? Explain why this order makes sense in the given situation.

d. What do you know about this situation if function *G*'s output is negative?

46. Baseball Salaries The average salary of a major league baseball player in millions of dollars can be modeled by $A(t) = 0.1474t + 0.4970$, where *t* is the years since 1990.
(*Source:* http://sportsline.com/mlb/salaries/avgsalaries)

a. A function *S* models the average salary of a major league baseball player in *dollars*. Explain how to transform *A* to create *S*. Then use function notation to write *S* in terms of *A*.

b. Write the formula for *S*.

c. *V* models the average salary in dollars of a major league baseball player in year *y*. Use function notation to write *V* in terms of *S*.

d. Write the formula for *V*.

47. Military Spending The table shows the total spent for military payroll in thousands (i.e., $15,375 represents $15,375,000) between 1990 and 2003.

Year *y*	Total Military Payroll Expenses ($ thousands) *P(y)*
1990	88,650
1995	98,396
2000	103,447
2001	106,013
2002	114,950
2003	122,270

Source: Statistical Abstract of the United States, 2006, Table 496.

a. *S* represents the total military payroll expenses in *dollars*. Explain how to transform *P* to create *S*. Then use function notation to write *S* in terms of *P*.

b. *M* represents the total military payroll expenses in *millions of dollars*. Explain how to transform *P* to create *M*. Then use function notation to write *M* in terms of *P*.

48. The graph shows *f* and three functions created by performing vertical stretches or compressions on *f*. Estimate the stretch or compression factor necessary to create each function.

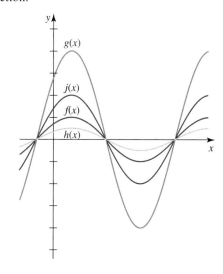

■ SECTION 3.4 ■

For each given function in Exercises 49–51,

a. Graph the function.

b. Describe the transformations necessary on *f* to create *g*.

c. Draw the graph for *g* by performing the necessary transformations.

d. List at least three coordinate points on the graph of *g*.

e. Write the formula for *g*.

49. $f(x) = 5x^2$, $\quad g(x) = f(0.25x) - 10$

50. $f(x) = x + 4$, $\quad g(x) = -f(2x) + 6$

51. $f(x) = \sqrt{x}$, $\quad g(x) = f(-3x)$

In Exercises 52–55, describe the transformation on f required to create each function. Then graph each transformed function.

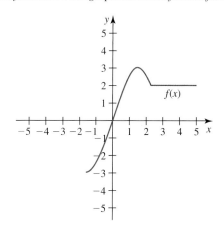

52. $g(x) = f\left(\dfrac{1}{3}x\right) - 2$

53. $g(x) = -\dfrac{1}{2}f(x - 2) + 4$

54. $g(x) = 2f(-x)$

55. $g(x) = f(2(x + 4))$

56. U.S. Population The following graph of P is the projected population of the United States (in millions) throughout the 21st century, assuming that the growth rate from 1995–2004 continues.

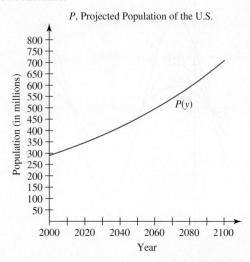

Source: Modeled from World Health Organization

a. The function A models the U.S. population in millions t years after 2000. Describe how to transform P to create A. Then use function notation to write A in terms of P.

b. The function N represents the population of the United States in millions x decades after 2000. Use function notation to write N in terms of P and describe the transformations involved.

c. Explain what M models if
$$M(x) = 1{,}000{,}000 P(10x + 2000)$$

57. Real Estate The function $V(T) = 0.096345T + 15.099$ approximates the number of vacant housing units (in thousands) in a state that contains T thousand total housing units. (This function model is valid for up to 5900 thousand total housing units.)

a. Explain what f models if
$$f(x) = 1000 V(0.001x)$$

b. Write the formula for f.

In Exercises 58–60, describe the transformation required on the given function f to create each function. Then use function notation to write each transformed function in terms of f.

58.

59.

60.

61. The function $g(x) = \dfrac{1}{3}x$ is a transformation of the function $f(x) = x$. Is $g(x)$ the result of a vertical compression by a factor of $\dfrac{1}{3}$ or a horizontal stretch by a factor of 3? Explain.

62. Master's Degrees Awarded
The table shows the number of master's degrees awarded (in thousands) in the United States between 1960 and 2000.

Decades Since 1960 t	Number of Master's Degrees Awarded (in thousands) $M(t)$
0	75
0.5	121
1	209
1.5	293
2	298
2.5	286
3	325
3.5	398
4	457

Source: Statistical Abstract of the United States, 2006, Table 286.

a. $D(y)$ models the number of master's degrees awarded y years after 1960. Explain how to transform M to create D. Then use function notation to write d in terms of M.

b. Explain what N represents if
$$N(y) = 1000D(y)$$

63. U.S. Exports
Based on data from 1990 to 2005, the value of U.S. goods exported to Colombia may be modeled by
$$C(t) = 122.81(1.22)^t \text{ million dollars}$$
where t is the number of years since 1990. (*Source:* Modeled from *Statistical Abstract of the United States, 2006*, Table 1293)

a. E models the U.S. exports to Colombia d decades after 1990. Explain how to transform C to create E. Then use function notation to write E in terms of C.

b. Write the formula for E.

c. Explain what P represents if
$$P(d) = 1{,}000{,}000E(d)$$

In Exercises 64–67, explain the effect, if any, of each type of transformation on the given function characteristic.

    a. Vertical translation
    b. Horizontal translation
    c. Vertical reflection
    d. Horizontal reflection
    e. Vertical stretch/compression
    f. Horizontal stretch/compression

64. Horizontal intercept(s)

65. Vertical intercept

66. Sign of the rate of change (positive vs. negative)

67. Magnitude of the rate of change

Make It Real Project

What to Do

1. From an area of personal interest, find a set of data that is a function of year.

2. Pick a year and align the data to that year. Describe the transformation of the data needed to achieve this alignment.

3. Perform a vertical shift on the data. Describe what the vertical shift represents.

4. Starting over with the original data set, use a vertical stretch or compression to change the units. Explain what the transformed data set represents.

5. Use a horizontal stretch or compression to organize the data by decades instead of years. Explain what the transformed data set represents.

6. Would performing a vertical or horizontal reflection on the data set have any real-world meaning? Explain.

Where to Find Data

The following websites may contain data that may be helpful for this project.

InfoPlease

- www.infoplease.com/almanacs.html

Information from sports to education to science and health

U.S. Census Bureau

- www.census.gov

A variety of information on various characteristics of the U.S. population

BBall Sports

- www.bballsports.com/

A database of statistics from professional baseball, basketball, hockey, and football

United Nations Statistics Division

- unstats.un.org/unsd/databases.htm

Demographics and business statistics from countries around the world

CHAPTER 4

Quadratic Functions

"The only thing constant in life is change."
François de la Rochefoucauld

Nothing demonstrates this maxim better than the world of technology. The 8-track tape, which was cutting-edge music storage and distribution technology in the mid-1960s, gave way to the cassette tape by the mid-1970s. The cassette tape in turn gave way to the compact disc. The emergence of digital audio players, such as the Apple iPod, has, in turn, diminished the role of the compact disc. The rise and decline of different types of music storage and distribution media can be modeled by mathematical functions with variable rates of change.

4.1 Variable Rates of Change
4.2 Modeling with Quadratic Functions
4.3 Forms and Graphs of Quadratic Functions
STUDY SHEET
REVIEW EXERCISES
MAKE IT REAL PROJECT

SECTION 4.1 Variable Rates of Change

LEARNING OBJECTIVES

- Understand rates of change in a function model
- Calculate first and second differences of a table of data
- Determine the concavity and increasing/decreasing behavior of a function from a table or graph
- Interpret the meaning of inflection points in real-world contexts

GETTING STARTED

The average cost of a movie ticket in the United States has been continually rising. According to the website "Box Office Mojo" (www.boxofficemojo.com), there was an estimated 18-cent increase in the average price per ticket from 2006 to 2007. The increase from 1999 to 2000 was 31 cents per ticket. Notice that the increase in price each year has not been *constant* but has *varied*.

In this section we discuss the difference between constant and variable rates of change and see how to apply this knowledge to real-world data such as movie ticket prices. We use average rates of change to estimate unknown data values, to estimate the rate of change of a function at a single data point, and to analyze the *concavity* of a graph. We also use first and second differences and inflection points to describe the change in a function.

■ Variable Rates of Change

Table 4.1 gives the average movie ticket price in the United States at 5-year intervals, beginning in 1975. The difference in the successive values shown in Table 4.2 shows the increase in the average ticket price over each 5-year interval change.

Table 4.1

Year	Years Since 1975 t	Average Cost of a Movie Ticket (dollars) M
1975	0	2.05
1980	5	2.69
1985	10	3.55
1990	15	4.23
1995	20	4.35
2000	25	5.39
2005	30	6.40

Source: www.boxofficemojo.com

Table 4.2

Years Since 1975 t	Average Cost of a Movie Ticket (dollars) M	Change over 5 Years
0	2.05	
5	2.69	$2.69 - $2.05 = $0.64
10	3.55	$3.55 - $2.69 = $0.86
15	4.23	$4.23 - $3.55 = $0.68
20	4.35	$4.35 - $4.23 = $0.12
25	5.39	$5.39 - $4.35 = $1.04
30	6.40	$6.40 - $5.39 = $1.01

From Table 4.2 we first observe that over each 5-year interval, the average cost of a movie ticket always increases. A function such as this is known as an **increasing function** because the output values continually increase as the input values increase. Conversely, a function whose output values decrease as the input values increase is known as a **decreasing function**.

INCREASING AND DECREASING FUNCTIONS

- An **increasing function** is a function whose output values *increase* as its input values increase.
- A **decreasing function** is a function whose output values *decrease* as its input values increase.

A function may or may not be increasing or decreasing over its entire domain. In fact, many functions increase on some intervals and decrease on others.

Next we observe that even though the average cost of a movie ticket is an increasing function, the amount of increase in the average ticket price over each 5-year period is not constant. Figure 4.1 shows the scatter plot of these data. Let's focus on two particular 5-year intervals: 1980–1985 and 1990–1995.

Over the 5-year period between 1980 and 1985, the ticket price increased from $2.69 to $3.55, an $0.86 change. Between 1990 and 1995, the ticket price increased from $4.23 to $4.35, a $0.12 change. Because the rate at which the ticket price is changing is not constant, we say the ticket price has a *variable rate of change*.

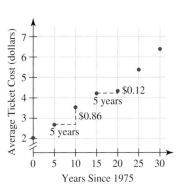

Figure 4.1

VARIABLE RATE OF CHANGE

Any function whose rate of change varies (is not constant) is said to have a **variable rate of change**. All nonlinear functions have variable rates of change.

EXAMPLE 1 ■ Interpreting Increasing and Decreasing Functions

Figure 4.2 displays the median age of the first marriage for American women for every decade from 1900 to 2000. (*Source:* www.census.gov)

a. Determine between which years the median marriage age is increasing most rapidly. Then calculate the average annual change over that time period.

b. Determine between which years the median marriage age is decreasing most rapidly. Then calculate the average annual change over that time period.

Figure 4.2

Solution Since the points of the scatter plot do not form a straight line, we know the function is nonlinear and has a variable rate of change.

a. We are looking for the two consecutive data points between which there is the greatest vertical increase. It appears the most pronounced vertical increase occurred between 1980 and 1990. Over that 10-year period, the median marriage age rose from 22 years to about 23.9 years.

$$\frac{23.9 - 22}{90 - 80} \frac{\text{years of age}}{\text{years since 1900}} = \frac{1.9}{10} \text{ year of age per year}$$

$$= 0.19 \text{ year of age per year}$$

Between 1980 and 1990, the average rate of change in the median marriage age of women was 0.19 year of age per year.

b. We are looking for the two consecutive data points between which there is the greatest vertical decrease. It appears the greatest vertical decrease occurred between 1940 and 1950. Over that 10-year period, the median marriage age appears to drop from about 21.5 years to roughly 20.3 years.

$$\frac{20.3 - 21.5}{50 - 40} \frac{\text{years of age}}{\text{years since 1900}} = -\frac{1.2}{10} \text{ year of age per year}$$

$$= -0.12 \text{ year of age per year}$$

Between 1980 and 1990, the average rate of change in the median marriage age of women was -0.12 year of age per year.

■ Average Rate of Change in Nonlinear Functions

When a function has a variable rather than a constant rate of change over a given interval, it is often helpful to determine the average rate of change of the function over the interval, as we did in Example 1. The average rate of change then can be used to fill in (interpolate) missing data over an interval or predict (extrapolate) unknown values outside of the domain of the function.

To see how this is done, consider the *line graph* (Figure 4.3) of the data from Example 1, along with the table of values (Table 4.3) that was used to produce the graph. (A **line graph** is a scatter plot with the data points connected by lines.)

Table 4.3

Years Since 1900 t	Median Age of First Marriage M
0	21.9
10	21.6
20	21.2
30	21.3
40	21.5
50	20.3
60	20.3
70	20.8
80	22.0
90	23.9
100	25.1

Figure 4.3

Figure 4.4

We can see that the rate of change over each 10-year interval of time varies. Focusing more closely on the 1940–1970 interval (boxed in Figure 4.3 and enlarged in Figure 4.4), we calculate the change over three 10-year intervals.

$$\frac{21.5 - 20.3}{40 - 50} = \frac{1.2}{-10}$$
$$= -0.12$$

$$\frac{20.3 - 20.3}{50 - 60} = \frac{0}{-10}$$
$$= 0$$

$$\frac{20.3 - 20.8}{60 - 70} = \frac{-0.5}{-10}$$
$$= 0.05$$

Between 1940 and 1950, the median marriage age decreased by 0.12 years of age per year. Between 1950 and 1960, the median marriage age was unchanged. Between 1960 and 1970, the median marriage age increased by 0.05 year of age per year.

What was the median marriage age in, for example, 1945, 1952, and 1968? We do not have data for these years but we can use the average rates of change to estimate. For example, in 1940 the median marriage age was 21.5 and decreased at an average rate of 0.12 year of age per year between 1940 and 1950. Since there are 5 years between 1940 and 1945, we have

$$21.5 + 5(-0.12) = 21.5 - 0.6$$
$$= 20.9$$

So we estimate that the median marriage age in 1945 was 20.9. Similarly, we can estimate the median marriage ages in 1952 and 1968.

1952	*1968*
$20.3 + 2(0) = 20.3$	$20.3 + 8(0.05) = 20.3 + 0.4$
	$= 20.7$

We estimate the median marriage age was 20.3 years in 1952 and was 20.7 in 1968.

■ Rates of Change at an Instant

Let's now investigate how we can estimate the rate of change at a single instant using the average rate of change.

As a basketball passes through a hoop, the ball's height in relation to the basketball court decreases. As the ball falls, its speed increases due to the effects of gravity until the conflicting forces of friction and gravity cause the ball to fall at a constant rate (known as the *terminal velocity*).

To model this phenomenon, we dropped a basketball repeatedly through a hoop from a height of approximately 10 feet. We measured the height of the ball over time using a motion detector. Table 4.4 shows the average of the height readings collected every 0.2 seconds in four trials.

Table 4.4

Time (seconds) t	Height of Basketball (feet) H
0	9.98
0.2	9.34
0.4	7.49
0.6	4.42
0.8	0.13

We can estimate the velocity of the ball at 0.6 second by calculating the average rate of change in the height between 0.4 second and 0.6 second.

$$\frac{\Delta H}{\Delta t} = \frac{4.42 - 7.49}{0.6 - 0.4} \frac{\text{feet}}{\text{second}}$$

$$= \frac{-3.07}{0.2} \text{ feet per second} \quad \text{The ball falls 3.07 feet over the 0.2 second}$$

$$= -15.35 \text{ feet per second}$$

The ball falls at an average velocity of 15.35 feet per second *between* the 0.4 and 0.6 second marks.

How fast was the ball falling *right at* the 0.6 second mark? In other words, what was the velocity of the ball *at that instant in time*? Determining an answer to this question is problematic because two points are needed to find an average rate of change. Nevertheless, we can use the average rate of change concept to answer this question by making the time interval extremely small.

Table 4.5 provides additional information about the height of the ball near (before and after) the 0.6 second mark. If we use the time just before 0.6 second in the table—0.59 second—we can estimate the instantaneous rate of change at 0.60 second, which is the velocity of the ball at 0.6 second.

$$\frac{\Delta H}{\Delta t} = \frac{4.42 - 4.60}{0.60 - 0.59} \frac{\text{feet}}{\text{second}}$$

$$= \frac{-0.18}{0.01} \text{ feet per second} \quad \text{The ball falls 0.18 feet over the 0.01 second.}$$

$$= -18 \text{ feet per second}$$

Thus we estimate that the ball is falling at a velocity of 18 feet per second at 0.60 second.

By picking a small Δt, we arrived at a reasonable estimate for how fast the ball was falling at a single point in time. The process of calculating an average rate of change with a small Δt is referred to as **estimating the instantaneous rate of change**.

Table 4.5

Time (seconds) t	Height of Basketball (feet) H
0.57	4.96
0.58	4.78
0.59	4.60
0.60	4.42
0.61	4.23
0.62	4.04
0.63	3.85

HOW TO: ■ ESTIMATE THE INSTANTANEOUS RATE OF CHANGE

To estimate the instantaneous rate of change of a function at a point, determine the average rate of change of the function over a very small interval containing the point. In other words, find $\frac{\Delta y}{\Delta x}$ for Δx close to 0.

Note that decreasing the value of Δx increased the accuracy of the estimate. It is customary to use the phrase "Δx approaches 0" and the notation $\Delta x \to 0$ to represent the idea of using ever decreasing positive values for Δx.

■ Successive Differences of Functions

Another way to analyze the behavior of a function with a variable rate of change is to use *successive differences*. When using successive differences, we look for patterns in the rates of change. To calculate **first differences** of a data table with equally spaced inputs, we calculate the difference in consecutive output values. The differences in consecutive first differences are referred to as **second differences**.

EXAMPLE 2 ■ Interpreting First and Second Differences

The per capita amount of money spent on prescription drugs, P, as a function of the years since 1990, t, can be modeled by Table 4.6, which was generated from the function $P(t) = 2.9t^2 - 2.6t + 158.7$. The first differences, ΔP, are shown in the table.

a. Explain the practical meaning of the first differences in this context.
b. Calculate the second differences. Then explain what the second differences tell us about the relationship between the per capita spending on prescription drugs and the years since 1990.
c. Using first and second differences, predict the per capita prescription drug spending in 1996.

Table 4.6

Years Since 1990 t	Per Capita Spending on Prescription Drugs (dollars) P	First Differences ΔP
0	158.7	
1	159.0	$159.0 - 158.7 = 0.3$
2	165.1	$165.1 - 159.0 = 6.1$
3	177.0	$177.0 - 165.1 = 11.9$
4	194.7	$194.7 - 177.0 = 17.7$
5	218.2	$218.2 - 194.7 = 23.5$

Source: Modeled from *Statistical Abstract of the United States, 2006*, Table 121

Solution

a. The first differences show us the annual rate of change in the per capita prescription drug spending in dollars per year. We observe that the first differences are increasing. That is, the annual rate of change in spending is increasing.
b. In Table 4.7, we calculate the differences of the first differences and see that the second differences are all 5.8.

Table 4.7

Years Since 1990 t	Per Capita Spending on Prescription Drugs (dollars) P	First Differences ΔP	Second Differences $\Delta(\Delta P)$
0	158.7		
1	159.0	0.3	
2	165.1	6.1	$6.1 - 0.3 = 5.8$
3	177.0	11.9	$11.9 - 6.1 = 5.8$
4	194.7	17.7	$17.7 - 11.9 = 5.8$
5	218.2	23.5	$23.5 - 17.7 = 5.8$

The second differences tell us that the annual rates of change in spending are increasing at a constant rate of 5.8 dollars per year each year.

c. Since the function has a constant second difference of 5.8, we can calculate the first difference between $t = 5$ and $t = 6$ by adding 5.8 to the first difference between $t = 4$ and $t = 5$ ($\Delta P = 23.5$).

$$23.5 + 5.8 = 29.3 \text{ dollars per year}$$

Between 1995 and 1996, the per capita spending on prescription drugs increased by \$29.30. Therefore, the function value at $t = 6$ will be 29.3 dollars more than the function value at $t = 5$ ($P = 218.2$).

$$218.2 + 29.3 = 247.5 \text{ dollars}$$

In 1996, the per capita prescription drug spending was 247.50 dollars.

Concavity and Second Differences

In general, second differences tell us about the concavity (curvature) of a nonlinear graph. When the second differences are positive, the function is *concave up* (curved upward). When the second differences are negative, the function is *concave down* (curved downward). Let's examine this concept within a real-world scenario.

In times of increasing gas prices, many drivers become concerned about the fuel efficiency of their vehicles. The U.S. Department of Energy reports that although each vehicle reaches its optimal fuel economy at a different speed, gas mileage usually increases up to speeds near 45 miles per hour and then decreases rapidly at speeds above 60 miles per hour. (*Source:* www.fueleconomy.gov)

Consider the fuel economy function F shown in Table 4.8 and Figure 4.5. Notice that the first differences decrease as speed increases. Between 10 and 20 miles per hour, gas mileage is increasing at a rate of 6.5 miles per gallon per additional 10 miles per hour of speed. Between 20 and 30 miles per gallon, gas mileage is still increasing but at the lesser rate of 4.5 miles per gallon per additional 10 miles per hour of speed. Between 40 and 50 miles per hour, the gas mileage is increasing at the much smaller rate of 0.6 miles per gallon per additional 10 miles per hour of speed.

Figure 4.5

Table 4.8

Speed (mph) s	Fuel Economy (mpg) F	First Differences ΔF	Second Differences $\Delta(\Delta F)$
10	17.6		
		6.5	
20	24.1		−2.0
		4.5	
30	28.6		−2.0
		2.5	
40	31.1		−1.9
		0.6	
50	31.7		−2.1
		−1.5	
60	30.2		−2.0
		−3.5	
70	26.7		−1.9
		−5.4	
80	21.3		

Let's look more closely at the first part of the graph, where the fuel economy is increasing as speed increases. See Figure 4.6. We can see although the fuel economy is increasing, it does not increase as much between 30 and 40 miles per hour as it does between 10 and 20 miles per hour. This is an example of a function that is *increasing* and *concave down*.

Table 4.8 shows that between 50 and 60 miles per hour, the gas mileage is decreasing by 1.5 miles per gallon per additional 10 miles per hour of speed. Between 70 and 80 miles per hour, the gas mileage is decreasing by 5.4 miles per gallon per 10 miles per hour of speed.

The magnitude of the decrease becomes greater as the speed increases, as shown in Figure 4.7. Since the first differences are decreasing (the second differences are negative), this part of the function is *decreasing* and *concave down*.

Figure 4.6

Figure 4.7

Notice that the second differences in Table 4.8 are all negative. This tells us that the first differences are decreasing. This means that the graph will be concave down on the entire domain of the function.

CONCAVITY

- The graph of a function *f* is said to be **concave up** if its rate of change *increases* as the input values increase. Concave up functions curve upward.

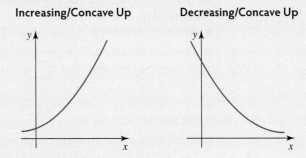

- The graph of a function *f* is said to be **concave down** if its rate of change *decreases* as the input values increase. Concave down functions curve downward.

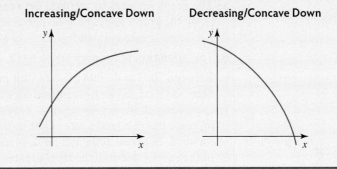

■ Inflection Points

Many graphs are concave up on portions of their domain and concave down on others. We refer to points on a graph where the concavity changes as **inflection points**. To show where the concavity changes on the graph in Figure 4.8, we mark the inflection points.

The first inflection point occurs where the graph changes from concave up to concave down. The second inflection point occurs where the graph changes from concave down to concave up.

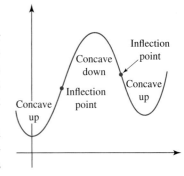

Figure 4.8

INFLECTION POINT

The point on a graph where the function changes concavity is called an **inflection point**. The inflection point is the point where the instantaneous rate of change is locally maximized or minimized.

EXAMPLE 3 ■ Interpreting Inflection Points in a Real-World Context

Savvy investors in multifamily properties (apartment buildings) closely monitor the markets in which they invest. Marcus and Millichap Real Estate Investment Services helps investors by providing in-depth reports on various sectors of the U.S. rental market. Based on data from 2003 to 2007, the average price per unit (apartment) for multifamily properties in Columbus, Ohio, can be modeled by

$$p(t) = -2.417t^3 + 14.14t^2 - 20.65t + 48.99 \text{ thousand dollars}$$

where t is the number of years since 2003. (*Source:* Modeled from data in Marcus and Millichap's 2008 National Apartment Report) A graph of the model is shown in Figure 4.9.

a. Estimate the intervals over which the function is increasing, decreasing, concave up, and concave down.

b. Determine if there are any inflection points on the graph.

c. Explain what the answers in parts (a) and (b) tell us about multifamily housing prices in Columbus, Ohio, between 2003 and 2007.

Figure 4.9

Solution

a. The graph appears to increase between $t = 1$ and $t = 3$. On the intervals [0, 1] and [3, 4], the graph appears to be decreasing. The graph is concave up from $t = 0$ to $t = 2$ and concave down from $t = 2$ to $t = 4$.

b. The inflection point appears to be (2, 45).

c. Between 2003 and 2004 median unit prices were decreasing; however, the rate of decrease lessened as time moved forward. Between 2004 and 2005, prices were increasing at an increasing rate. Between 2005 and 2006 prices continued to increase but at a lesser rate. That is, the rate of increase lessened as time moved forward. Between 2006 and 2007, prices again decreased; however, the magnitude of the rate of decrease became greater as time moved forward.

The inflection point indicates that in 2005 the median price was about $45,000 per unit. This was the time when the instantaneous rate of change was locally maximized; that is, when prices were increasing most rapidly.

SUMMARY

In this section you learned the difference between constant and variable rates of change. You used average rates of change to estimate unknown data values, to estimate the rate of change of a function at a single data point, and to analyze the concavity of a graph. You also used first and second differences to help describe the change in a function.

4.1 EXERCISES

SKILLS AND CONCEPTS

For Exercises 1–10, create a table of values and an associated graph that

1. Are increasing and concave up.
2. Are decreasing and concave down.
3. Are increasing and concave down.
4. Are decreasing and concave up.
5. Are increasing at a constant rate.
6. Are decreasing at a constant rate.
7. Are constant.
8. Are increasing, with a point of inflection.
9. Have two points of inflection.
10. Are concave up twice and concave down once.
11. Each of the functions f, g, and h in Table A are increasing, but each increases in a different way. Which of the graphs in Figure A best fits each function?

Table A

x	$f(x)$	$g(x)$	$h(x)$
1	23	10	2.2
2	24	20	2.5
3	26	29	2.8
4	29	37	3.1
5	33	44	3.4
6	38	50	3.7

Figure A
(a) (b)

(c) (d)

12. Each of the functions f, g, and h in Table B are decreasing, but each decreases in a different way. Which of the graphs in Figure B best fits each function?

Table B

x	$f(x)$	$g(x)$	$h(x)$
2	−4	−2	−8
5	−6	−12.5	−17
7	−12	−24.5	−26
11	−27	−60.5	−35
12	−32	−72	−44
18	−38	−162	−53

Figure B
(a) (b)

(c) 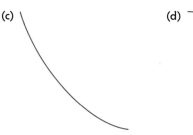 (d)

13. Determine whether each function f, g, and h in the table has a constant rate of change or a variable rate of change.

x	$f(x)$	$g(x)$	$h(x)$
−4	14	18	39
−2	19	26	36
0	22	38	33
2	23	56	30
4	20	83	27

14. In the following table, there are missing data values. Use the information you have been given to fill in the missing data.

x	y	First Differences Δy	Second Differences $\Delta(\Delta y)$
−3			
		17	
−2	−16		−6
−1			−6
		5	
0	0		−6
1			−6
		−7	
2	−8		−6
3	−21		

15. Label points A, B, C, D, E, and F on the graph shown.

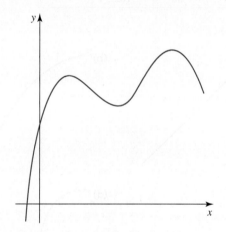

a. Point A is a point on the curve where the instantaneous rate of change is negative.
b. Point B is a point on the curve where the value of the function is positive.
c. Point C is a point on the curve where the instantaneous rate of change is the most positive.
d. Point D is a point on the curve where the instantaneous rate of change is 0.
e. Points E and F are different points on the curve where the instantaneous rate of change is about the same.

For Exercises 16–19, use the table to answer each question.

x	f(x)	g(x)
−2	0	5
−1	3	3
0	4	2
1	−1	1
2	6	−1
3	−2	0

16. Compute the average rate of change of f from $x = -2$ to $x = 3$.
17. Compute the average rate of change of g from $x = -1$ to $x = 2$.
18. Estimate the instantaneous rate of change of g at $x = -1$.
19. Estimate the instantaneous rate of change of f at $x = 0$.
20. Using the following graph, explain the behavior of function f on the interval from $x = 0$ to $x = 15$. State over which intervals f is increasing or decreasing and concave up or concave down. What does the point **P** indicate?

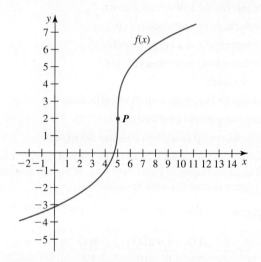

21. **Fertilizer** The following figure shows the yield, Y, of an apple orchard (in bushels) as a function of the amount of fertilizer, f (in pounds), used on the orchard.

a. Over what interval(s) is the function increasing? Decreasing?
b. Discuss the concavity of the function and then explain what that tells about the apples and fertilizer.

22. **Sprinter** A poorly conditioned sprinter starts a 400-meter race at a rapid pace; however, as the race progresses his speed decreases. By the time he reaches the finish line, he is walking. Sketch a graph of the sprinter's distance traveled as a function of time since the start of the race.

23. **Racing** A race car is being driven around an oval racetrack at a constant speed.

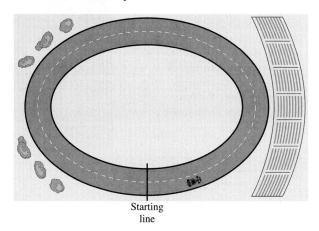

Construct a rough sketch of a function that represents the *shortest* distance between the car and the starting line while imagining the car moving around the track at a constant rate. Identify any important aspects of the graph such as increasing, decreasing, concave up, concave down, and points of inflection.

■ SHOW YOU KNOW

For Exercises 24–27, determine if each function has a constant or a variable rate of change over the interval provided. Explain or show how you know.

24.

x	y
−8	80
−7	63
−6	48
−5	35
−4	24
−3	15
−2	8
−1	3

25.

x	y
0	1
2	5
4	6
6	12
8	16
10	19
12	20
14	25
16	29

26.

27.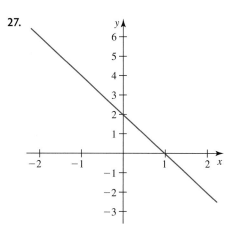

28. The figure shows a portion of the graph of a function. If it is possible, explain whether the function has a constant or variable rate of change. If it is impossible to determine, explain why.

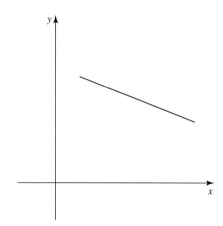

29. Given the following table and graph of the function f, a classmate claims the average rate of change of f between $x = 3$ and $x = 5$ is 18. You know that this response is incorrect. What does the classmate's response tell you about his thinking? Explain.

x	0	1	2	3	4	5	6
$f(x)$	0.75	1.5	3	6	12	24	48

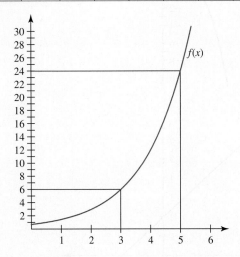

MAKE IT REAL

30. **Marathon Runner** After recovering from an injury, an athlete begins training again for a marathon. As she gets stronger, she is able to run faster. Although her run times improve each week, they improve by a lesser amount as time goes on. She runs a full marathon once a week. Sketch a graph of the number of minutes it takes her to run a marathon as a function of the number of weeks since her recovery from the injury. Label the independent and dependent axes.

31. **Traffic Fatalities** A model F representing the percentage of traffic accidents that are fatal, F, as a function of the posted speed limit, s, is shown in the figure.

Source: Modeled from www-nrd.nhtsa.dot.gov

a. For what speeds is F increasing and for what speeds is it decreasing?

b. Is F primarily concave up or concave down? What does this tell us about the fatality rates?

32. **Phoenix High Temperatures** In the summer of 1990, the temperatures in Arizona reached an all-time high (so high, in fact, that some airlines decided it might be unsafe to land their planes there). The daily temperatures in Phoenix for June 19–29, 1990 are given in the table.

Date: June 1990 (day of month)	Temperature (°F)
19	109
20	113
21	114
22	113
23	113
24	113
25	120
26	122
27	118
28	118
29	108

Source: www.weather.com

a. Identify over what time intervals the function is constant, increasing, and decreasing.

b. Between June 24 and June 26, is the graph of the data concave up or concave down? Explain.

c. In practical terms, what does the concavity of the data set tell about the Phoenix temperature between June 24 and June 26?

33. **The Yard House** The Yard House restaurant opened in 1996 in Long Beach, California. It derives its name from a 3-foot-tall glass (called a yard-of-ale glass) originally designed for stagecoach drivers. The elongated neck made it possible to hand the stagecoach driver his drink without climbing up onto or down from the coach. The basic shape of the glass is displayed in the figure. (*Source:* www.yardhouse.com)

As a beverage is poured into the glass at a constant rate until the glass is full, the height of the liquid will rise in the glass over time.

a. Sketch a graph of the height of the liquid in the glass as a function of time.

b. As you sketch the graph of the function, is it important to consider that the liquid is being poured into the glass at a constant rate? Explain.

34. **Identity Theft** A January 26, 2006, headline in *The Wall Street Journal* read, "ID Theft Complaints Still Rising, but Rate of Increase Slows."

a. Write a description of what this headline means.

b. Sketch a possible graph of the number of ID theft complaints as a function of time. Be sure the shape of the graph matches what the headline states.

c. Imagine you are a newspaper reporter analyzing the following graphs. Write a headline that would represent accurately each graph. Make sure your headline clearly describes the special characteristics of the graph.

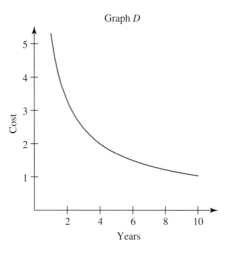

35. **Home-Field Advantage** A common belief in athletics is that the home team has an advantage due to hometown fans, familiar surroundings, and short travel times. To test this hypothesis, two Georgia Southern University professors analyzed data collected from Major League Baseball. As expected, the results indicated that a home-field advantage does exist in the major leagues, but only under certain circumstances. Specifically, the strength of the home-field advantage varies with the number of runs scored by the home team. A claim made in their published research article stated, "The probability of a home team winning a game increases as it scores more runs, but it increases at a decreasing rate." (*Source:* "An Analysis of the Home-Field Advantage in Major League Baseball Using Logit Models: Evidence from the 2004 and 2005 Seasons," Levernier & Barilla, J. Quant. Analysis in Sports, www.bepress.com/jqas/vol3/iss1/1/)

a. Sketch a graph of a function that could reasonably approximate the researchers' claim. Make sure to properly label the independent and dependent variables.

b. Describe in your own words what their claim, "The probability of a home team winning a game increases as it scores more runs, but it increases at a decreasing rate," means in terms of the likelihood of the home team winning baseball games.

c. Explain why you think the claim the researchers make could be true in terms of a baseball contest.

36. **Home Foreclosures** A headline in the *East Valley Tribune* (February 10, 2007) stated, "Numbers are Rising in the Valley, but Not as Bad as 2002: Foreclosure Fears." The accompanying article stated "foreclosures shot up in some East Valley communities in 2006 but remain far below the number of foreclosures four years ago." Using the information in the newspaper's article, complete the following table with reasonable values that could model how the

number of foreclosures may have changed as described from 2002 to 2006.

Years Since 2002 d	Number of Foreclosures F
0	1563
1	
2	
3	
4	314

37. Per Capita Income The per capita personal income of each resident of the United States from 1960 to 2000 is given in the table. The dependent variable P represents the per capita income (in dollars) and t represents the number of years since 1960.

Years Since 1960 t	Per Capita Income (dollars) P
0	1,832
10	4,334
20	9,865
30	18,425
40	30,013

Source: U.S. Department of Commerce

a. Using *averages*, estimate $P(5)$ and explain what the value of the answer means in the real-world context.

b. Using the *average rate of change* from $t = 0$ to $t = 10$, estimate $P(5)$ and explain what the value of the answer means in the real-world context.

c. If you were to estimate $P(13)$, would it be possible to use *either* averages *or* the average rate of change to arrive at the estimate? Justify your answer.

d. Using *successive differences*, estimate $P(50)$ and explain what the value of the answer means in the real-world context.

38. Fast-Food Sales Actual and predicted fast-food restaurant sales from 1990 to 2010 are given in the table.

Years Since 1990 t	Fast-Food Restaurant Sales ($ millions) S
0	69,840
2	82,433
4	96,341
6	112,882
8	133,372
10	159,126
12	191,461
14	231,693
16	281,138
18	341,112
20	412,932

Source: Statistical Abstract of the United States, 2006, **Table 1269**

a. Using *averages*, estimate $S(11)$ and explain what the value of the answer means in the real-world context.

b. Using the *average rate of change* from $t = 0$ to $t = 10$, estimate $S(11)$ and explain what the value of the answer means in the real-world context.

c. If you were to estimate $S(13)$, would it be possible to use *either* averages *or* the average rate of change to arrive at the estimate? Justify your answer.

d. Using *successive differences*, estimate $S(22)$ and explain what the value of the answer means in the real-world context.

39. Cable Television The following table gives the number of basic cable television subscribers (in thousands).

Years Since 1970 d	Number of Basic Cable TV Subscribers (in thousands) F
5	9,800
10	17,500
15	35,440
20	50,520
25	60,550
30	66,250
31	66,732
32	66,472
33	66,050

Source: Statistical Abstract of the United States, 2007, **Table 1134**

a. Use the table to estimate the point of inflection and interpret the meaning of the result.

b. From the table, estimate the instantaneous rate of change in 2002 and interpret the practical meaning of the result (including units).

40. Abortions The table gives the number of reported abortions in South Carolina from 1984 to 2004.

Years Since 1984 t	Number of Reported Abortions A
0	11,704
2	12,174
4	14,133
6	13,285
8	11,008
10	10,992
12	9,326
14	8,801
16	7,527
18	6,657
20	6,565

Source: Citizen magazine, January 2007

a. Find the average rate of change in the number of abortions between 1984 and 1990. Then interpret the meaning of the result, including units.

b. Find the average rate of change in the number of abortions between 1990 and 2004. Then interpret the meaning of the result, including units.

c. In 1990, South Carolina passed a law requiring parental consent for minors to have an abortion and a 1-hour waiting period after abortion counseling. Does it appear from the data that the law had an effect on the number of reported abortions? Refer to the first differences and average rates of change in your response.

41. **World Records** The year of a world record set in the men's 100-meter freestyle swim, y, and the record time in seconds, T, is given in the scatter plot along with a function model $T(y)$.

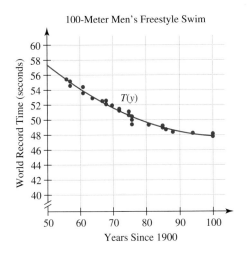

Source: www.duelinthepool.com/USASWeb

a. Is the graph of T increasing or decreasing? Explain what this tells you about the world record times.

b. Is the graph of T concave up or concave down? Explain what the concavity tells you about the decline in the world record times.

42. In a race for political office, an incumbent politician claims, "As long as I have been in office, the crime rate has dropped!" In the same campaign, the incumbent's opponent claims, "We need to vote Mrs. X out because crime continues to rise!" Is it possible that both political candidates are telling the truth? Justify your answer using a graph.

43. **World Record Time for the Mile** Since 1913, the world record time for the runners of the mile has decreased, as shown in the table.

Year	Record Time (minutes)	Year	Record Time (minutes)
1913	4.24	1965	3.89
1923	4.17	1975	3.82
1933	4.13	1985	3.77
1943	4.04	1993	3.74
1954	3.99	1999	3.72

Source: International Association of Athletics Federation

a. Between 1913 and 1999, what was the average annual decrease in the world record time?

b. Draw a scatter plot of the data set. Then describe the practical meaning of the concavity of the graph.

44. **Women Earning Ph.D.s** The function shown in the figure estimates the number of women who earned a Ph.D. in the 20th century.

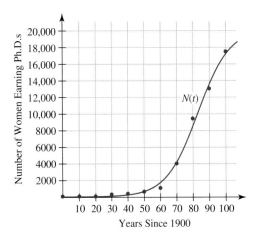

a. Over what interval(s) is the function increasing and concave up? Over what interval(s) is it increasing and concave down?

b. What does the concavity indicate about the rate of increase in the number of women earning Ph.D.s?

c. Estimate the average rate of change from $t = 1950$ to $t = 1960$, from $t = 1960$ to $t = 1970$, and from $t = 1970$ to $t = 1980$.

45. **Immigration** The percentage of the total U.S. population made up of immigrants from 1850 to 2006 is shown in the graph.

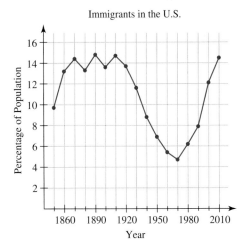

Source: www.census.gov

a. Describe what the graph indicates about the percentage of immigrants in the United States. Refer to the concepts of *increasing*, *decreasing*, and *concavity* as appropriate.

b. Give possible historical reasons for why there have been such drastic changes in the percentage of immigrants in the United States.

46. Growth Charts The U.S. Centers for Disease Control produces growth charts to help pediatricians and parents assess the health and growth patterns of children. (*Source:* www.cdc.gov) For each chart shown, write a description of what the chart indicates about boys and girls of various ages. Include a description of the nature of the function in terms of whether it is increasing or decreasing, its concavity, and inflection points.

SECTION 4.1 ■ EXERCISES 245

For Exercises 48–49, the graph of f is provided. Sketch the graph of a function g in which the output values of g(x) represent the instantaneous rate of change for f(x) at each value of x.

48.

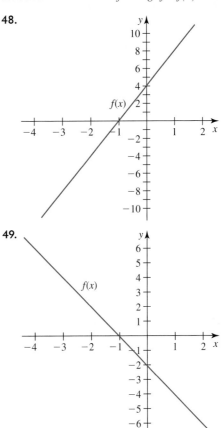

49.

50. Water is poured into a spherical fish bowl at a constant rate.

a. Sketch a graph of the function that best models the height of water in the spherical bowl as a function of the amount of water (volume) in the bowl.
b. Discuss what the concavity of the graph represents in this context.

■ **STRETCH YOUR MIND**

Exercises 47–50 are intended to challenge your understanding of rates of change and concavity of graphs.

47. Sketch the graph of a function f with all of the following properties:

$f(0) = 3$.
$f(x)$ is decreasing for $0 \leq x \leq 4$.
$f(x)$ is increasing for $4 \leq x \leq 6$.
$f(x)$ is decreasing for $x > 6$.
$f(x) \to 11$ as $x \to \infty$.

SECTION 4.2 Modeling with Quadratic Functions

LEARNING OBJECTIVES

- Recognize the relationship between a quadratic equation and its graph
- Use second differences to determine if a quadratic equation represents a data set
- Construct and use quadratic models to predict unknown results and interpret these findings in a real-world context

GETTING STARTED

Successful business executives know how to make money in an ever-changing marketplace. For companies in the retail business, customer loyalty is one of the hallmarks of financial success. The *quantity discount*—where an item's price is reduced for customers who buy large quantities of the item—is one marketing strategy used by successful retailers to attract and retain customers.

In this section we investigate quadratic functions and their applications. We look at equations, data tables, and graphs of these functions. We determine how to find a quadratic function from a data table by using quadratic regression. We apply these concepts to a number of real-world situations, including the business strategy of the quantity discount.

■ **The Quadratic Equation in Standard Form**

In 2007, National Pen Company offered the promotion shown in Table 4.9 for the Dynagrip Pen in their online catalog.

Table 4.9

Paid Order Size	Price per Pen
50	$0.79
100	$0.77
150	$0.75
250	$0.73
500	$0.71

Source: www.pens.com

Notice that as the order size increases, the company reduces the price per pen, employing the quantity discount strategy. When using such pricing strategies, the company must be aware of the impact the price reductions will have on its revenue.

From the table we see that the first $0.02 price reduction is offered for an order of 50 additional pens (100 pens), and the next for an order of 50 more (150 pens). We would expect each following discount to be given for orders of 50 more each time, but the table shows this is not the case. The next discount is given instead for an order of 100 more than the previous order (250 pens) and the final discount for an order of 250 more (500 pens).

To see why the company does not offer the $0.02 discount for each additional 50 pens ordered, we begin by considering the hypothetical pricing structure shown in Table 4.10, which assumes that for every increase of 50 pens in the order, the price per pen decreases by $0.02. To make our calculations simpler, we rewrite the pricing structure in terms of 50-pen sets and price per 50-pen set. See Table 4.11.

Table 4.10

Paid Order Size	Price per Pen
50	$0.79
100	$0.77
150	$0.75
250	$0.71
500	$0.61

Table 4.11

Number of 50-Pen Sets x	Price per Set p
1	$39.50
2	$38.50
3	$37.50
5	$35.50
10	$30.50

The revenue from pen sales is the product of the price per 50-pen set and the number of sets sold. That is,

$$R(x) = px$$

where R is the revenue (in dollars), p is the price (in dollars per set), and x is the number of sets sold. We update our data table to show the revenue generated (Table 4.12) and plot revenue as a function of the number of 50-pen sets in Figure 4.10.

Table 4.12

Number of 50-Pen Sets x	Price per Set p	Revenue R
1	$39.50	$39.50
2	$38.50	$77.00
3	$37.50	$112.50
5	$35.50	$177.50
10	$30.50	$305.00

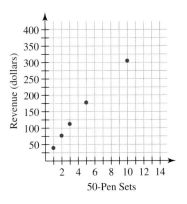

Figure 4.10

So far, it appears that this pricing strategy will continue to increase the company's revenue. But this will not always be the case. Since the price per set decreases by $1 for each additional set sold, 15 sets will sell for $25.50 each, 20 sets will sell for $20.50 each, 30 sets will sell for $10.50 each, and 40 sets will sell for $0.50 each. As shown in Figure 4.11, the corresponding revenues are $382.50, $410.00, $315.00, and $20.00.

What happened? Between 0 and 20 sets revenue is increasing although at a decreasing rate. Beyond 20 sets, revenue is decreasing and at an increasingly rapid rate. So at what point should the company stop offering the additional $0.02 per pen discount? It appears from the graph that the additional discount should not be offered for orders consisting of more than 20 50-pen sets. Since revenue does not take into account the cost of producing the pens, the company may need to further adjust the price reduction limit when additional factors are taken into consideration.

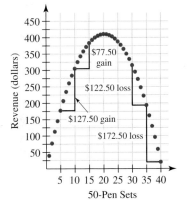

Figure 4.11

Notice that the revenue function depends on price and quantity sold. Is there a way to write revenue as a function of quantity sold only? Yes. We return to the pen-set hypothetical pricing data table, repeated here as Table 4.13.

Recall that each 1-set increase corresponds with a $1 decrease in price per set. Since the rate of change in price is constant, price as a function of sets is a linear function with slope $m = -1$. We substitute in (1, 39.50) to determine the initial value.

Table 4.13

Number of 50-Pen Sets x	Price per Set P
1	$39.50
2	$38.50
3	$37.50
5	$35.50
10	$30.50

$$p(x) = -1x + b$$
$$39.50 = -1(1) + b$$
$$b = 40.50$$

Therefore,

$$p(x) = -x + 40.50$$

We can now write the revenue function $R(x) = px$ exclusively in terms of x.

$$R(x) = px$$
$$= (-x + 40.50)x$$
$$= -x^2 + 40.50x$$

The graph of the revenue function is the parabola shown in Figure 4.12. A function equation of this form is called a *quadratic equation in standard form*.

Figure 4.12

QUADRATIC EQUATION IN STANDARD FORM

A function equation of the form

$$y = ax^2 + bx + c$$

with constants a, b, and c and with $a \neq 0$ is called a **quadratic equation in standard form**. The graph of a quadratic equation is a **parabola**.

To discover the meaning of the *parameters a, b,* and *c* in the quadratic equation $y = ax^2 + bx + c$, let's use the revenue function $R(x) = -x^2 + 40.50x$ and first determine the units of the parameters. In this equation, $a = -1$, $b = 40.50$, and $c = 0$. The output of the revenue function is *dollars*, so the units of each of the terms of the quadratic equation must be *dollars*. That is, the units of $-1x^2$, $40.50x$, and 0 must all be *dollars*. But the input of the function is *50-pen sets*, so the units of x are pen sets, not dollars. Thus the units of the coefficients of each term must compensate for this. We have

(units of a)x^2 = dollars   (units of b)x = dollars

(units of a)(pen sets)^2 = dollars   (units of b)(pen sets) = dollars

$$\text{units of } a = \frac{\text{dollars}}{(\text{pen sets})^2} \qquad \text{units of } b = \frac{\text{dollars}}{\text{pen set}}$$

So the units of *a* are *dollars per pen set squared*, the units of *b* are *dollars per pen set*, and the units of *c* are *dollars*. We now use this information to help define the meanings of the parameters.

The *c* is the initial value of the function. That is, $f(0) = c$. In this case, $f(0) = 0$, so the vertical intercept of the function's parabola is $(0, 0)$. In other words, when 0 pens have been sold, 0 dollars of revenue have been earned.

From its units (dollars per pen set), we know that *b* is a rate of change. But what does it represent? Let's evaluate $R(x) = -x^2 + 40.50x$ at $x = 0$ and $x = h$, where h is some value of x "close" to 0. We have $R(0) = 0$ and $R(h) = -h^2 + 40.50h$. Let's calculate the average rate of change between these two values.

$$\text{average rate of change} = \frac{R(h) - R(0)}{h - 0} \frac{\text{dollars}}{\text{pen sets}}$$

$$= \frac{(-h^2 + 40.50h) - 0}{h}$$

$$= -h + 40.50 \text{ dollars per pen set}$$

Observe as the value of h gets close to 0, the average rate of change approaches 40.50. Thus the *instantaneous* rate of change at $x = 0$ is 40.50 dollars per pen set. On the graph, the slope of the parabola at the vertical intercept is 40.50. In other words, when 0 pen sets have been sold, revenue is increasing at a rate of 40.50 dollars per pen set.

The value of *a* relates to how much the rate of change itself is changing. As is clear from the graph and our previous discussion, the rate of change is not constant and thus has its own rate of change. The value for *a* is half of the rate of change in the rate of change. (We'll show why later.) In this case, $a = -1$ so the rate of change in the rate of change is -2 dollars per pen set for each pen set. That is, the revenue per pen set decreases by 2 dollars for each additional pen set sold. Since the revenue per pen set is itself decreasing, the parabola is concave down.

We summarize our conclusions as follows.

THE MEANING OF *a*, *b*, AND *c* IN A QUADRATIC EQUATION

In the quadratic equation $y = ax^2 + bx + c$, the parameters *a*, *b*, and *c* represent the following:

a = one half of the rate of change in the rate of change

b = the instantaneous rate of change at $x = 0$ (initial rate of change)

c = the value of y at $x = 0$ (initial value)

EXAMPLE 1 ■ Interpreting the Meaning of the Parameters in a Quadratic Equation

In 1962, Sam Walton opened the first Walmart store. The chain grew rapidly to 24 stores by 1967. In that year, the company generated $12.6 million in sales. Today Walmart is one of the world's premier retailers, generating $312 billion in net sales in 2006. (*Source:* walmartstores.com) Based on data from 1996 to 2006, the net sales of Walmart can be modeled by the quadratic function

$$s(t) = 0.8636t^2 + 14.39t + 84.72 \text{ billion dollars}$$

where *t* is the number of years since the end of 1996. Explain the meaning of the parameters in the model in their real-world context. Then explain the graphical meaning of the parameters.

Solution Since $c = 84.72$ billion dollars, the model estimates that Walmart earned 84.72 billion dollars in revenue in 1996. Since $b = 14.39$ billion dollars per year, the model estimates that at the end of 1996, Walmart sales revenue was increasing at a rate of 14.39 billion dollars per year. Since $a = 0.8636$, the model estimates that the increase in revenue is increasing at a rate of 1.73 ($2 \cdot 0.8636 \approx 1.73$) billion dollars per year each year. For example, since revenue was increasing at a rate of 14.39 billion dollars per year in 1996, we expect that in 1997 revenue will be increasing at a rate of about 16.12 billion dollars per year ($14.39 + 1.73 = 16.12$).

Since $c = 84.72$, the vertical intercept of the parabola is (0, 84.72). Since $b = 14.39$, the slope of the parabola at the vertical intercept is 14.39. Since $a = 0.8636$ is a positive number, the rate of change is itself increasing, which means the parabola is concave up. See Figure 4.13.

Figure 4.13

Although the model in Example 1 is quadratic, only a portion of the parabola is used to model the data set. This is often the case with quadratic models of real-world data sets.

Determining If a Data Set Represents a Quadratic Function

In Section 4.1 we defined first differences to be the set of differences in outputs for equally spaced inputs, and second differences to be the set of differences in the first differences. We also saw that since linear functions have a constant rate of change, they have constant first differences.

To see how we can use successive differences to determine if a data set represents a quadratic function, let's return to the pen-set revenue function. First we reconstruct the table of values for the function and then calculate the first differences. See Table 4.14.

Table 4.14

x	$R(x) = -x^2 + 40.50x$	First Differences
0	0	
		39.50
1	39.50	
		37.50
2	77.00	
		35.50
3	112.50	

From Table 4.14 we note that the first differences in the quadratic function are not constant. But we see from our calculations in Table 4.15 that the *second* differences of the function *are* constant.

Table 4.15

x	$R(x) = -x^2 + 40.50x$	First Differences	Second Differences
0	0		
		39.50	
1	39.50		-2
		37.50	
2	77.00		-2
		35.50	
3	112.50		

We can also see this by looking at Figure 4.14, which shows that the first differences decrease by 2 for each 1-unit increase in the number of 50-pen sets.

Another way to observe the constant second differences is to graph the rate-of-change function, as shown in Figure 4.15. That function is a line with a slope of -2 and a vertical intercept of 40.50, the initial rate of change in the revenue function.

Figure 4.14

Figure 4.15

Will the second differences always be constant for a quadratic function? That is, will the rate of change always be changing at a constant rate? We investigate this idea using the quadratic function $y = ax^2 + bx + c$. We evaluate this function at five input values, each spaced 1 unit apart. (Note: Each value of y in Table 4.16 has been simplified algebraically.)

Table 4.16

x	$y = ax^2 + bx + c$	First Differences	Second Differences
x_1	$y = ax_1^2 + bx_1 + c$		
		$2ax_1 + (a + b)$	
$x_1 + 1$	$y = ax_1^2 + (2a + b)x_1 + (a + b + c)$		$2a$
		$2ax_1 + (3a + b)$	
$x_1 + 2$	$y = ax_1^2 + (4a + b)x_1 + (4a + 2b + c)$		$2a$
		$2ax_1 + (5a + b)$	
$x_1 + 3$	$y = ax_1^2 + (6a + b)x_1 + (9a + 3b + c)$		$2a$
		$2ax_1 + (7a + b)$	
$x_1 + 4$	$y = ax_1^2 + (8a + b)x_1 + (16a + 4b + c)$		

So yes, the second differences are always constant in a quadratic function and do not depend on the value of x_1. Thus, as stated earlier, we can use second differences to determine if a data set represents a quadratic function.

DIFFERENCE PROPERTIES OF QUADRATIC FUNCTIONS GIVEN IN TABLES

For equally spaced input values, quadratic functions have *linear* first differences and *constant* second differences.

EXAMPLE 2 ■ Determining If a Table Represents a Quadratic Function

A golden rectangle is said to be the most aesthetically pleasing of all rectangles. Artists and architects have incorporated the shape into drawings, buildings, and works of art such as the canvas of Salvador Dali's *The Sacrament of the Last Supper* shown here. Table 4.17 shows the width and area in centimeters (cm) of various golden rectangles.

Determine if the area of a golden rectangle is a quadratic function of its width.

Table 4.17

Width (cm)	Area (cm²)
10	161.80
20	647.21
30	1456.23
40	2588.85
50	4045.08

Solution We construct Table 4.18 to calculate first and second differences.

Table 4.18

Width (cm)	Area (cm²)	First Differences	Second Differences
10	161.80		
		485.41	
20	647.21		323.61
		809.02	
30	1456.23		323.61
		1132.62	
40	2588.85		323.61
		1456.23	
50	4045.08		

Since the second differences are constant, the area of a golden rectangle is a quadratic function of its width. (In the exercises at the end of the chapter, we will further investigate the properties of golden rectangles.)

Using Quadratic Regression to Find a Quadratic Function of Best Fit

Many real-world data sets have second differences that are not constant but are nearly so. Such data sets can still be modeled with a quadratic function. Just as we used linear regression to find a line of best fit in Chapter 2, we can use quadratic regression to find the quadratic function that best fits a data set. Although the model of best fit will not pass through every point of the data set, it is often sufficiently accurate to describe the relationship between the values of the data set.

EXAMPLE 3 ■ Using Quadratic Regression to Find a Model of Best Fit

The number of hospital beds in the United States has been *decreasing* since 1990 despite the fact that the population of the United States has been increasing. Table 4.19 shows the number of hospital beds in the United States by year.

Use quadratic regression to find the quadratic function model for the data set. Then explain the meaning of the function parameters.

Table 4.19

Years Since 1990 t	Hospital Beds (in thousands) h
0	1213
4	1128
8	1013
10	984
12	976
14	956

Source: Health, United States, 2006

Solution The full development of quadratic regression is beyond the scope of this text. Instead, we use a graphing calculator and the Technology Tip at the end of this section. We determine that the quadratic equation that best models the number of hospital beds in the United States is $H(t) = 0.9002t^2 - 31.67t + 1220$ thousand beds, where t is the number of years since 1990.

According to the model, there were 1220 thousand beds in 1990 and at that time the number of beds was decreasing at a rate of 31.57 thousand beds per year. However, the rate of change will increase (become less negative) by 1.8004 thousand beds per year each year ($2(0.9002) = 1.8004$).

EXAMPLE 4 ■ Using Quadratic Regression to Find a Model of Best Fit

Use quadratic regression to find the quadratic function that best models the data set in Table 4.20, the per capita spending on prescription drugs in the United States. Interpret the meaning of the parameters of the model. Then predict the per capita spending for 2006.

Table 4.20

Years Since 1990 t	Per Capita Spending on Prescription Drugs (dollars) P
5	224
9	368
10	423
11	485
12	552
13	605

Source: Statistical Abstract of the United States, 2006, Table 121

Solution Using a graphing calculator and the Technology Tip at the end of this section, we determine that the quadratic model is $P(t) = 2.76t^2 - 1.11t + 159$.

According to the model, the per capita prescription drug spending was $159 in 1990 and was decreasing at a rate of $1.11 per year. The rate of change itself was increasing at a rate of $5.52 per year each year.

Since the year 2006 corresponds with $t = 16$, we have

$$P(16) = 2.76(16)^2 - 1.11(16) + 159$$
$$\approx 848$$

According to the model, per capita prescription drug spending in the United States was about $848 in 2006.

EXAMPLE 5 ■ Using Quadratic Regression to Model a Data Set

Chipotle Mexican Grill achieved remarkable financial results between 2001 and 2005, as shown in Table 4.21.

Table 4.21

Year t	Franchise Royalties and Fees ($1000s) f	Restaurant Sales ($1000s) s
2001	267	131,331
2002	753	203,892
2003	1493	314,027
2004	2142	468,579
2005	2618	625,077

Source: Chipotle Mexican Grill, Inc., 2005 Annual Report, p. 24

Draw a scatter plot of restaurant sales as a function of franchise royalties and fees. If the data appears to be concave up or concave down, use quadratic regression to find the quadratic model that best fits the data. Then use the model to estimate restaurant sales when franchise royalties and fees reach $3 million.

Solution The scatter plot shown in Figure 4.16 appears to be concave up. Using a graphing calculator and the Technology Tip at the end of this section, we determine that the quadratic model for sales (accurate to 4 significant digits) is

$$s(f) = 0.05196f^2 + 54.89f + 120{,}000 \text{ thousand dollars}$$

where f is the amount of franchise royalties and fees (in thousand dollars). To verify the accuracy of our work, we graph the model and the scatter plot together in Figure 4.17.

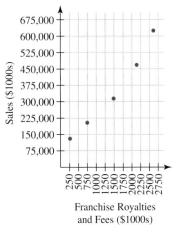

Figure 4.16 Figure 4.17

We are to forecast sales when franchise royalties and fees reach $3 million, which is equivalent to 3000 thousand dollars. Since the franchise royalties and fees are given in thousands, we must evaluate $s(3000)$.

$$s(f) = 0.05196f^2 + 54.89f + 120{,}000$$
$$s(3000) = 0.05196(3000)^2 + 54.89(3000) + 120{,}000$$
$$= 752{,}310$$
$$= 752{,}300 \text{ (accurate to 4 significant digits)}$$

We estimate that when franchise royalties and fees reach $3 million, sales will reach $752.3 million.

SUMMARY

In this section you learned the standard form of the quadratic equation and discovered the meaning of its parameters. You also learned how to use first and second differences to determine if a data set represents a quadratic function. Additionally, you learned how to use quadratic regression to find the quadratic function of best fit for a data set and how to interpret the model in its real-world context.

TECHNOLOGY TIP ■ QUADRATIC REGRESSION

1. Enter the data using the Statistics Menu List Editor.

2. Bring up the Statistics Menu Calculate feature by pressing STAT and using the blue arrows to move to the CALC menu. Then select item 5:QuadReg, and press ENTER.

3. If you want to automatically paste the regression equation into the Y= Editor so that you can easily graph the model, press the key sequence VARS: Y-VARS: Function: Y₁ and press ENTER. Otherwise press ENTER.

4.2 EXERCISES

SKILLS AND CONCEPTS

In Exercises 1–5, calculate the first and second differences of the data table. Then indicate whether the data represents a linear or quadratic function or neither.

1.

x	y
0	1
1	3
2	9
3	19
4	33

2.

x	y
2	−4
4	−16
6	−36
8	−64
10	−100

3.

x	y
5	2
10	4
15	6
20	8
25	10

4.

x	y
4	5
8	10
12	20
16	40
20	80

5.

x	y
−3	−1
0	−10
3	−1
6	26
9	71

SHOW YOU KNOW

6. Given that the quadratic function $s(t) = at^2 + bt + c$ represents the distance of a car from Orlando (in miles) after t hours of travel, explain what a, b, and c represent including units.

7. How can you tell if a data set with equally spaced inputs represents a quadratic function?

8. What does it mean to say that a data set has linear first differences?

9. What does a second difference indicate?

10. What is the relationship between the second difference of a quadratic equation and its associated graph?

11. In terms of a rate of change, what does it mean for a parabola to be concave down?

12. In terms of a rate of change, what does it mean for a parabola to be concave up?

13. A concave up parabola passes through the point (0, 3). From this information, what do you know about the parameters of $y = ax^2 + bx + c$?

14. In the equation $y = ax^2 + bx + c$, what is the *graphical* significance of the value of b?

15. In the equation $y = ax^2 + bx + c$, what is the *graphical* significance of the value of c?

Exercises 16–20 focus on the relationship between a quadratic model equation and the situation being modeled.

16. What do we mean when we say the "initial value" of a quadratic function model?

17. What do we mean when we say the "initial rate of change" of a quadratic function model?

18. If $a < 0$ in the quadratic model $y = ax^2 + bx + c$, what do we know about the rate of change of the model?

19. If $a > 0$ in the quadratic model $y = ax^2 + bx + c$, what do we know about the rate of change of the model?

20. Why can't $a = 0$ for a quadratic function model $y = ax^2 + bx + c$? Use the concept of rate of change in your explanation.

MAKE IT REAL

In Exercises 21–30, explain the real-world meaning of the parameters a, b, and c of the quadratic function model.

21. **Medicare Enrollees** Based on data from 1980 to 2004, the number of Medicare enrollees (in millions) can be modeled by

$$M(t) = -0.00472t^2 + 0.663t + 28.4$$

where t is the number of years since 1980. (*Source:* Modeled from *Statistical Abstract of the United States, 2006*, Table 132)

22. **Prescriptions** Based on data from 1990 to 2003, the amount of money spent on prescription drugs (per capita) can be modeled by

$$P(t) = 2.889t^2 - 2.613t + 158.7$$

dollars, where t is the number of years since 1990. (*Source: Statistical Abstract of the United States, 2006*, Table 121)

23. **Children in Madagascar** Based on data from 1990 to 2002, the number of children under 5 in Madagascar can be modeled by

$$C(t) = 1.046t^2 + 60.82t + 2152$$

thousand children, where t is the number of years since 1990. (*Source:* Modeled from World Health Organization data)

24. **Malaria Cases** Based on data from 1998 to 2002, the number of clinical malaria cases reported in children under 5 years of age in Ghana can be modeled by

$$C(t) = -140{,}281t^2 + 658{,}186t + 583{,}452$$

cases, where t is the number of years since 1998. (*Source:* www.afro.who.int)

25. **USAA Membership** Based on data from 2003 to 2007, the number of members of the USAA (an insurance and financial services company) can be modeled by

$$M(t) = 0.05t^2 + 0.15t + 5.0$$

million members, where t is the number of years since 2003. (*Source:* Modeled from USAA 2007 Report to Members, p. 21)

26. **U.S. Population** Based on data from 1990 to 2004, the population of the United States can be modeled by

$$P(t) = -19.56t^2 + 3407t + 250{,}100$$

thousand people, where t is the number of years since 1990. (*Source:* Modeled from *Statistical Abstract of the United States, 2006*, Table 2)

27. **Online School Enrollment** Based on data from 2003–2004 through 2006–2007, the number of students enrolled in the Arizona Virtual Academy can be modeled by

$$s(x) = 141.25x^2 + 358.75x + 318.75$$

students, where x is the number of years since 2003–2004. (*Source:* Modeled from Arizona Virtual Academy Fact Sheet)

28. **Consumer Spending on Books** Based on data from 2004 to 2005 and projections for 2006 to 2009, the amount of money spent by consumers on books classified as *adult trade* can be modeled by

$$b(x) = -31.15x^2 + 556.1x + 14970$$

million dollars, where x is the number of years since 2004.
(*Source:* Modeled from *Statistical Abstract of the United States, 2007,* Table 1119)

29. **U.S. Oil Production vs. Imports** Based on data from 1985 to 2004, the difference between U.S. oil field production and net oil imports can be modeled by

$$b(t) = 4.294t^2 - 278.3t + 2251$$

million barrels, where t is the number of years since 1985.
(*Source:* Modeled from *Statistical Abstract of the United States, 2007,* Table 881)

30. **Yogurt Production** Based on data from 1997 to 2005, the amount of yogurt produced in the United States annually can be modeled by

$$y(x) = 14.99x^2 + 62.14x + 1555$$

million pounds, where x is the number of years since 1997.
(*Source:* Modeled from *Statistical Abstract of the United States, 2007,* Table 846)

In Exercises 31–35, use quadratic regression and a graphing calculator to find the quadratic function that best fits the data set. Then use the model to forecast the value of the function at the indicated point.

31. **Live Births by Race**

Years Since 1990 t	Total Live Births (in thousands) b	Live Births to Women Racially Classified as *White* (in thousands) w
0	1165	670
5	1254	785
9	1308	840
10	1347	866
11	1349	880
12	1366	904
13	1416	947
14	1470	983

Source: Statistical Abstract of the United States, 2007, Table 83

Model *white* births as a function of total live births. How many *white* births will there be when live births reach 1500 thousand?

32. **Abortions**

Years Since 1985 x	Abortions (per 1000 live births) a
0	422
5	389
10	350
15	324
16	325
17	319

Source: Statistical Abstract of the United States, 2007, Table 96

What was the abortion rate in 2005?

33. **Manufacturing Employees**

Years Since 2000 x	Computer and Electronic Products Industry Employees (in thousands) e
0	1820
2	1507
3	1355
4	1323
5	1320

Source: Statistical Abstract of the United States, 2007, Table 980

How many computer and electronic products industry employees were there in 2009?

34. **Manufacturing Employees**

Years Since 1990 x	Aerospace Products and Parts Industry Employees (in thousands) e
0	841
10	517
12	470
13	442
14	442
15	456

Source: Statistical Abstract of the United States, 2007, Table 980

How many aerospace products and parts industry employees were there in 2009?

35. **NFL Player Salaries**

Years Since 2000 x	NFL Player Average Salary ($1000s) s
0	787
1	986
2	1180
3	1259
4	1331
5	1400

Source: Statistical Abstract of the United States, 2007, Table 1228

What was the NFL player average salary in 2008?

■ **STRETCH YOUR MIND**

Exercises 36–38 are intended to challenge your understanding of quadratic functions.

36. Show there does not exist a quadratic function that passes through all of the following points: (0, 4), (3, 13), (10, 34).

37. A classmate claims that the following table has a constant second difference of 8. Do you agree? Explain.

x	y
0	−1
2	9
4	27
8	53
12	87

38. The vertex of a certain concave down parabola is (0, 5). Explain as completely as possible what you know about the parameters of its equation $y = ax^2 + bx + c$.

SECTION 4.3 Forms and Graphs of Quadratic Functions

LEARNING OBJECTIVES

- Recognize and use the vertex, standard, and factored forms of quadratic functions
- Determine the vertex, horizontal intercepts, and vertical intercept of a quadratic function from its equation, data table, or graph
- Use the quadratic formula to solve real-world problems

GETTING STARTED

The rectangle is said to be the most common geometric shape we encounter in our daily lives. Whether in art or architecture, the properties of rectangles are fascinating.

Numerous applications of mathematics are represented using the concept of rectangles. The fact that the area of a rectangle may be written as a quadratic function of the length of its shortest side has relevance in settings seemingly unrelated to geometry.

In this section we introduce several forms of the quadratic equation. We also discuss the parabolic graphs of quadratic functions. By knowing the concavity, horizontal and vertical intercepts, and the vertex of a parabola, we are able to better understand the meaning of quadratic function models such as that for a rectangle.

■ Vertex Form of a Quadratic Function

In Section 4.2 we learned the standard form of a quadratic function, $y = ax^2 + bx + c$, with initial value c and initial rate of change b. We also noted that a was half of the rate of change in the rate of change. Quadratic functions may also be written in *vertex form*. This form is especially useful for graphing and relies heavily on the concept of function transformations.

EXAMPLE 1 ■ Constructing a Quadratic Equation in Vertex Form

A square is to be cut out of the middle of a 12-inch by 12-inch matting board to make a frame. The mat frame will be placed over a square picture and should overlap the picture by 0.5 inch on each side, as shown in Figure 4.18.

Write a quadratic equation in vertex form that represents the area of the mat frame after the square in the middle is removed. Then use the equation to calculate the mat frame area for a 6-inch by 6-inch picture and an 8-inch by 8-inch picture. Finally, graph the parabolas representing the area of any x-inch by x-inch picture and the area of the corresponding mat frame.

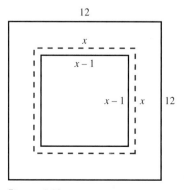

Figure 4.18

Solution Since both sides of the matting board are 12 inches long, the area of the mat before the center is removed is given by

$$M = 12 \times 12$$
$$= 144 \text{ square inches}$$

The area of the picture is $P = x^2$. The square opening of the mat is to overlap a picture by 0.5 inch on each side. That is, the length of each side of the square opening must be reduced by 1 inch (0.5 inch from each side). The area of the square that is removed from the center of the mat is given by

$$C = (x - 1)^2$$

The area of the mat frame will be the difference in the area of the original mat and the area of the square that is removed. That is,

$$A = M - C$$
$$= 144 - (x - 1)^2$$
$$= -(x - 1)^2 + 144$$

So the area of the mat frame is $A(x) = -(x - 1)^2 + 144$. A quadratic function written this way is said to be in **vertex form**.

Using the function, we find that the area of the mat frame for a 6-inch by 6-inch picture is

$$A(6) = -(6 - 1)^2 + 144$$
$$= -25 + 144$$
$$= 119 \text{ square inches}$$

The area of the mat frame for an 8-inch by 8-inch picture is

$$A(8) = -(8 - 1)^2 + 144$$
$$= -49 + 144$$
$$= 95 \text{ square inches}$$

The graphs representing the area of any x-inch by x-inch square picture and its corresponding mat frame are shown in Figure 4.19.

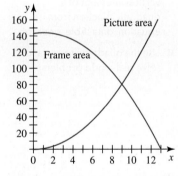

Figure 4.19

Vertex Form of a Quadratic Function

The equation of a parabola written in the form

$$y = a(x - h)^2 + k$$

with $a \neq 0$ is said to be in **vertex form**. The point (h, k) is called the **vertex** of the parabola.

The **vertex** of a parabola is the point on the graph where the rate of change is equal to zero. For the basic quadratic function, $y = x^2$ (which is equivalent to $y = x^2 + 0x + 0$), we know the rate of change is equal to zero at $(0, 0)$ since the initial value is $c = 0$ and the initial rate of change is $b = 0$. From the graph of the function shown in Figure 4.20, we see the vertex is the "turning point" of the graph. To the left of the vertex, the graph is decreasing. To the right of the vertex, the graph is increasing. In terms of a rate of change, the rate of change is negative to the left of the vertex, 0 at the vertex, and positive to the right of the vertex.

The minimum or maximum value of the function occurs at the vertex of the parabola.

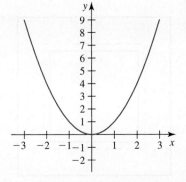

Figure 4.20

SECTION 4.3 ■ FORMS AND GRAPHS OF QUADRATIC FUNCTIONS

MAXIMUM AND MINIMUM VALUES OF A QUADRATIC FUNCTION

- The *maximum* value of a *concave down* parabola occurs at the vertex. A concave down parabola does not have a minimum value.
- The *minimum* value of a *concave up* parabola occurs at the vertex. A concave up parabola does not have a maximum value.

Thus when working with quadratic functions, the statement "find the maximum (minimum) value of the function" is equivalent to "find the y-coordinate of the vertex."

How does the graph of $y = a(x - h)^2 + k$ compare to $y = x^2$? From our understanding of transformations, we know that a vertically stretches or compresses the graph of $y = x^2$ by a factor of a. If $a < 0$, the graph will also be reflected vertically about the horizontal axis. Recall also that h shifts the graph horizontally $|h|$ units and k shifts the graph vertically $|k|$ units. Table 4.22 gives several quadratic functions in vertex form and the corresponding effects of a, h, and k.

Table 4.22

Equation	Effect of a	Effect of h	Effect of k
$y = x^2$	none	none	none
$y = (x - 1)^2$	none	shift right 1 unit	none
$y = 3(x + 2)^2 - 8$	vertically stretch by a factor of 3	shift left 2 units	shift downward 8 units
$y = -x^2$	reflect about horizontal axis	none	none
$y = -x^2 + 4$	reflect about horizontal axis	none	shift upward 4 units
$y = -3(x - 2)^2 + 6$	reflect about horizontal axis, vertically stretch by a factor of 3	shift right 2 units	shift upward 6 units

Note: When graphing a function from its equation, *reflections, stretches, and compressions must be done **before** horizontal and vertical shifts.*

EXAMPLE 2 ■ Graphing a Quadratic Equation in Vertex Form by Hand

Graph the function $y = -0.5(x + 2)^2 + 4$ by plotting points.

Solution The vertex is $(-2, 4)$. We create Table 4.23, a table of values for the quadratic equation $y = -0.5(x + 2)^2 + 4$ by selecting x values near $x = -2$. We then plot the points and connect them with a smooth curve, as shown in Figure 4.21.

Table 4.23

x	y
-5	-0.5
-4	2
-3	3.5
-2	4
-1	3.5
0	2
1	-0.5

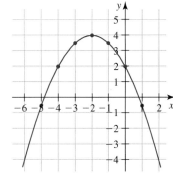

Figure 4.21

JUST IN TIME ■ THE ABSOLUTE VALUE FUNCTION, |x|

The absolute value function of a number is the distance between the number and 0. For example, $|-5| = 5$ and $|5| = 5$. The absolute value function is formally defined as a *piecewise* function:

$$|x| = \begin{cases} -x & \text{if } x < 0 \\ x & \text{if } x \geq 0 \end{cases}$$

This notation says if $x \geq 0$, the function value is equal to x but if $x < 0$, the function value is equal to $-x$. We'll use this formal definition to evaluate $|-5|$.

$$|-5| = -(-5) \text{ since } -5 < 0$$
$$|-5| = 5$$

As a sort of shorthand, we sometimes write $|x| = \pm x$.

JUST IN TIME ■ $\sqrt{x^2}$

Is $\sqrt{x^2} = x$? Consider the following table of values

x	x^2	$\sqrt{x^2}$
-2	4	2
-1	1	1
0	0	0
1	1	1
2	4	2

Observe that $\sqrt{x^2} = -x$ when $x < 0$ and $\sqrt{x^2} = x$ when $x \geq 0$. That is,
$\sqrt{x^2} = \begin{cases} -x & \text{if } x < 0 \\ x & \text{if } x \geq 0 \end{cases}$, which is equivalent to $\sqrt{x^2} = |x|$.

■ Finding the Horizontal Intercepts of a Quadratic Function in Vertex Form

We can determine the location of the horizontal intercepts of a quadratic function by setting the vertex form equal to zero and solving.

$$0 = a(x - h)^2 + k$$
$$-a(x - h)^2 = k$$
$$(x - h)^2 = -\frac{k}{a}$$
$$\sqrt{(x - h)^2} = \sqrt{-\frac{k}{a}}$$
$$|x - h| = \sqrt{-\frac{k}{a}}$$
$$\pm(x - h) = \sqrt{-\frac{k}{a}}$$

The plus-or-minus sign indicates that this breaks out into two separate cases as follows:

$$x - h = \sqrt{-\frac{k}{a}} \qquad -(x - h) = \sqrt{-\frac{k}{a}}$$

$$x = h + \sqrt{-\frac{k}{a}} \qquad x - h = -\sqrt{-\frac{k}{a}}$$

$$x = h - \sqrt{-\frac{k}{a}}$$

Combining both results, we have $x = h \pm \sqrt{-\frac{k}{a}}$. The horizontal intercepts lie $\sqrt{-\frac{k}{a}}$ units to the left and to the right of the x-coordinate of the vertex, h.

> **HORIZONTAL INTERCEPTS OF A QUADRATIC FUNCTION IN VERTEX FORM**
>
> The horizontal intercepts of a quadratic function
>
> $$y = a(x - h)^2 + k$$
>
> with $a \neq 0$ occur at $x = h \pm \sqrt{-\frac{k}{a}}$ provided $-\frac{k}{a} > 0$. If $-\frac{k}{a} < 0$, the parabola does not have any horizontal intercepts.

EXAMPLE 3 ■ **Modeling a Real-World Situation with a Quadratic Function**

As of 2011, the highest jump (68 inches) by a dog was achieved by Cinderella May (a greyhound) at the Purina Dog Chow Incredible Dog Challenge show in 2006. (*Source:* www.guinessrecords.com)

Write a quadratic equation in vertex form to model the vertical height of the dog t seconds after leaping into the air. Determine the horizontal intercepts of the function and interpret what they mean in this context. Then graph the height function to verify your conclusions. (*Hint:* The rate of change in the vertical velocity is referred to as *acceleration due to gravity* and is approximately equal to -32 feet per second per second on Earth.)

Solution We are to write the function in the form $y = a(t - h)^2 + k$. Recall a is half of the rate of change in the rate of change. In this case, a is half of the acceleration due to gravity (-32 feet per second per second), or -16 feet per second per second. Since the maximum height occurs at the y-coordinate of the vertex, k is the maximum height obtained, 68 inches. However, before substituting the value for k into the equation, we need to convert the value from inches to feet.

$$68 \text{ inches} \cdot \frac{1 \text{ foot}}{12 \text{ inches}} = 5.67 \text{ feet}$$

Now we can write

$$y = -16(t - h)^2 + 5.67$$

We also know that at time $t = 0$, the dog was on the ground ($y = 0$). Substituting this point into the equation, we solve to find the value of h.

$$0 = -16(0 - h)^2 + 5.67$$
$$0 = -16h^2 + 5.67$$
$$16h^2 = 5.67$$
$$h^2 = 0.3542$$
$$h \approx 0.595$$

Thus the model for the height of the dog t seconds after leaping into the air is
$$y = -16(t - 0.595)^2 + 5.67$$

The graph of this function, shown in Figure 4.22, is a concave down parabola with vertex (0.595, 5.67). The horizontal intercepts are determined by

$$x = h \pm \sqrt{-\frac{k}{a}}$$

$$t = h \pm \sqrt{-\frac{k}{a}}$$

$$\approx 0.595 \pm \sqrt{-\frac{5.67}{(-16)}}$$

$$\approx 0.595 \pm 0.595$$

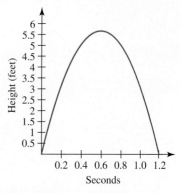

Figure 4.22

So $t = 0.595 - 0.595 = 0$ seconds or $t = 0.595 + 0.595 = 1.190$ seconds. In this context, the first horizontal intercept indicates that at 0 seconds the dog had not left the ground. The second horizontal intercept indicates that at 1.190 seconds the dog had returned to the ground after her leap.

EXAMPLE 4 ■ Determining the Vertex Form of a Quadratic Equation from a Graph

The graph in Figure 4.23 models the height of a ball that is propelled into the air at a rate of 32 feet per second from a height of 3 feet. Determine the vertex form of the quadratic equation represented by the graph. Then describe the practical meaning of the graph.

Solution From the graph, we can see the vertex occurs at (1, 19). Consequently, the vertex form of the quadratic equation will be $h(t) = a(t - 1)^2 + 19$. To determine the value of a, we must substitute in another point from the graph. For ease of computation, we choose (0, 3).

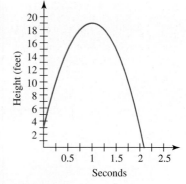

Figure 4.23

$$3 = a(0 - 1)^2 + 19$$
$$3 = a + 19$$
$$a = -16$$

So the vertex form of the quadratic equation is $h(t) = -16(t - 1)^2 + 19$.

From the graph, we can see the initial height of the ball is 3 feet above the ground. Its height increases for the first second although at a decreasing rate. At the end of the first second, the ball reaches its maximum height (19 feet). At that instant, its velocity (rate of change) is 0 feet per second. The ball then begins to fall toward the ground as a result of gravity. It strikes the ground shortly after 2 seconds of flight time.

■ Comparing the Vertex and Standard Forms

So far in this section we have focused exclusively on the *vertex form* of the equation of a parabola. We can readily see the relationship between the vertex form and standard form discussed in Section 4.2 by expanding the vertex form.

$$y = a(x - h)^2 + k$$
$$= a(x - h)(x - h) + k$$
$$= a(x^2 - 2hx + h^2) + k$$
$$= ax^2 - 2ahx + ah^2 + k$$
$$= ax^2 + (-2ah)x + (ah^2 + k)$$
$$= ax^2 + bx + c \quad \text{Let } b = -2ah \text{ and } c = ah^2 + k.$$

This conversion between forms yields a significant result:

$$b = -2ah$$
$$h = -\frac{b}{2a}$$

That is, the x-coordinate of the vertex is equal to $-\frac{b}{2a}$.

x-COORDINATE OF THE VERTEX

The x-coordinate of the vertex of a quadratic function in standard form, $y = ax^2 + bx + c$, is $x = -\frac{b}{2a}$.

EXAMPLE 5 ■ Finding the Vertex of a Parabola from an Equation

Although virtually nonexistent in the United States, malaria is one of the primary killers of children in Africa, claiming over a million lives annually. International efforts to help curb the spread of the disease have made some inroads yet there remains a great need for a long-term solution.

Based on data from 1998 to 2002, the number of clinical malaria cases in children under 5 years of age in Ghana can be modeled by

$$C(t) = -140{,}281t^2 + 658{,}186t + 583{,}452 \text{ cases}$$

where t is the number of years since the end of 1998.

According to the model, when will the number of cases reach a maximum? At that time, how many cases will there be?

Solution Since $-140{,}281 < 0$, this parabola will be concave down. The maximum value of the function will occur at the vertex.

$$t = -\frac{b}{2a}$$
$$= -\frac{658{,}186}{2(-140{,}281)}$$
$$\approx 2.346$$

The number of cases reached a maximum roughly 2.346 years after the end of 1998, or 4 months into 2001. (The 2 gets us to the end of 2000. The 0.346 is a portion of the year 2001. We convert it to months as shown here: $0.346 \text{ year} \cdot \frac{12 \text{ months}}{1 \text{ year}} \approx 4 \text{ months.}$)

To determine the number of cases at that time, we evaluate the quadratic function at $t = 2.346$.

$$C(2.346) = -140{,}281(2.346)^2 + 658{,}186(2.346) + 583{,}452$$
$$\approx 1{,}355{,}490$$

According to the model, in the 1-year period ending at the end of April 2001, there were 1,355,490 clinical malaria cases in Ghana in children under the age of 5 years.

Recall that a quadratic function in *vertex form* has horizontal intercepts at $x = h \pm \sqrt{-\dfrac{k}{a}}$. To find the horizontal intercepts for a quadratic function in *standard form* we write h and $-\dfrac{k}{a}$ in terms of a, b, and c. Recall $h = -\dfrac{b}{2a}$ and $c = ah^2 + k$. Thus,

$$c = ah^2 + k$$

$$k = c - ah^2$$

$$\frac{k}{-a} = \frac{c - ah^2}{-a}$$

$$-\frac{k}{a} = -\frac{c - ah^2}{a}$$

$$-\frac{k}{a} = \frac{ah^2 - c}{a} \qquad \text{Distribute the negative sign and reorder terms in numerator.}$$

$$-\frac{k}{a} = h^2 - \frac{c}{a} \qquad \text{Divide both terms in numerator by } a.$$

$$-\frac{k}{a} = \left(-\frac{b}{2a}\right)^2 - \frac{c}{a} \qquad \text{since } h = -\frac{b}{2a}$$

$$-\frac{k}{a} = \left(\frac{b^2}{4a^2}\right) - \frac{c}{a}$$

$$-\frac{k}{a} = \frac{b^2}{4a^2} - \frac{c}{a} \cdot \frac{4a}{4a} \qquad \text{Create a common denominator.}$$

$$-\frac{k}{a} = \frac{b^2 - 4ac}{4a^2} \qquad \text{Write as a single fraction.}$$

Therefore,

$$x = h \pm \sqrt{-\frac{k}{a}}$$

$$= -\frac{b}{2a} \pm \sqrt{\frac{b^2 - 4ac}{4a^2}} \qquad \text{since } h = -\frac{b}{2a} \text{ and } -\frac{k}{a} = \frac{b^2 - 4ac}{4a^2}$$

$$= -\frac{b}{2a} \pm \frac{\sqrt{b^2 - 4ac}}{\sqrt{4a^2}}$$

$$= -\frac{b}{2a} \pm \frac{\sqrt{b^2 - 4ac}}{2a}$$

This result is referred to as the **quadratic formula**.

QUADRATIC FORMULA

The horizontal intercepts of $y = ax^2 + bx + c$ with $a \neq 0$ occur at

$$x = -\frac{b}{2a} \pm \frac{\sqrt{b^2 - 4ac}}{2a}$$

If the *discriminant* $b^2 - 4ac$ is negative, the function does not have any horizontal intercepts.

It is customary in many textbooks to write the quadratic formula in the equivalent form of

$$x = \frac{-b \pm \sqrt{b^2 - 4ac}}{2a}$$

One drawback of this equivalent form is that it obscures the fact that the horizontal intercepts are equidistant from the *x*-coordinate of the vertex, $-\frac{b}{2a}$.

EXAMPLE 6 ■ Finding the Horizontal Intercepts of a Quadratic Function

The Space Needle is a popular tourist attraction in Seattle, Washington. The observation deck stands 520 feet above the ground. If a water droplet falls from the observation deck, how long will it take to reach the ground? The height of the water droplet above the ground *t* seconds after it is dropped is given by $h(t) = -16t^2 + 520$. (This model neglects air resistance.) Graph the function to verify your results.

Solution The droplet will be on the ground when $h = 0$, so we need to find the horizontal intercepts of the function. Using the quadratic formula, we have

$$t = -\frac{b}{2a} \pm \frac{\sqrt{b^2 - 4ac}}{2a}$$

$$= -\frac{0}{2(-16)} \pm \frac{\sqrt{(0)^2 - 4(-16)520}}{2(-16)}$$

$$= 0 \pm \frac{\sqrt{33280}}{-32}$$

$$\approx \pm 5.7 \text{ seconds}$$

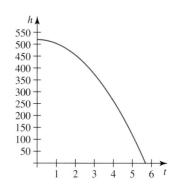

Figure 4.24

In this context, it only makes sense to talk about a positive value of *t*. According to the model, the water droplet will strike the ground 5.7 seconds after it is dropped. The graph in Figure 4.24 verifies this result.

■ Factored Form of a Quadratic Function

When working with graphs, the *factored form* of a quadratic function is especially useful for determining the location of horizontal intercepts.

> **FACTORED FORM OF A QUADRATIC FUNCTION**
>
> The equation of a parabola written in the form
>
> $$y = a(x - x_1)(x - x_2)$$
>
> with $a \neq 0$ is said to be in **factored form**.
>
> - The horizontal intercepts of the parabola are $(x_1, 0)$ and $(x_2, 0)$.
> - The vertex of the parabola lies halfway between the horizontal intercepts at $x = \frac{x_1 + x_2}{2}$.
> - The vertical intercept is $(0, ax_1x_2)$.

We saw earlier how the vertex and standard forms of a quadratic function were related. To see how the factored form is related to the standard form, we multiply out the factored form.

$$y = a(x - x_1)(x - x_2)$$
$$= a(x^2 - x_1 x - x_2 x + x_1 x_2)$$
$$= a[x^2 - (x_1 + x_2)x + x_1 x_2]$$
$$= ax^2 - a(x_1 + x_2)x + a x_1 x_2$$
$$= ax^2 + bx + c \qquad \text{Let } b = -a(x_1 + x_2) \text{ and } c = a x_1 x_2$$

We see that the vertical intercept c is the product of the horizontal intercepts and a, whereas b is the sum of the horizontal intercepts times the opposite of a.

EXAMPLE 7 ■ Working with a Quadratic Function in Factored Form

By summing the lengths of the sides, we can see that the rectangle in Figure 4.25 has a perimeter of 8 units. The equation for the area of this rectangle is

$$A = (8 - 2x)(2x - 4)$$
$$= -2(-4 + x)(2(x - 2))$$
$$= -4(x - 4)(x - 2)$$

Figure 4.25

a. Describe the appearance of the graph based on its equation.
b. Graph the function to verify your answer to part (a).
c. Determine the practical domain for the area function.
d. Explain the practical meaning of the vertex.

Solution

a. Since $-4 < 0$, the parabola is concave down. It has horizontal intercepts at $x = 4$ and at $x = 2$. The vertex lies halfway between these horizontal intercepts, at $x = \dfrac{2 + 4}{2} = 3$. Evaluating the function at $x = 3$ yields

$$A(3) = -4(3 - 4)(3 - 2)$$
$$= 4$$

The vertex is $(3, 4)$.

b. The graph in Figure 4.26 verifies the results.

c. We know the area of a rectangle cannot be negative. We see from the graph that $A \geq 0$ when $2 \leq x \leq 4$. So the practical domain of the function is $2 \leq x \leq 4$.

d. The vertex shows that the area is maximized at 4 square units when $x = 3$. When $x = 3$, the rectangle has sides of length 2 since $8 - 2(3) = 2$ and $2(3) - 4 = 2$.

Figure 4.26

■ Factoring Quadratic Functions

Once you have become skilled at reading and interpreting quadratic function graphs, you can quickly determine if a quadratic equation is factorable and how it factors. The horizontal intercepts of a quadratic function graph correspond directly with the factors of the quadratic function, as shown in Figure 4.27. Every factorable quadratic function can be written in one of the two forms shown in the figure.

SECTION 4.3 ■ FORMS AND GRAPHS OF QUADRATIC FUNCTIONS   267

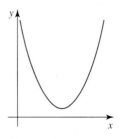

(a) two *x*-intercepts
$y = a(x - p)(x - q)$

(b) one *x*-intercept
$y = a(x - p)^2$

(c) no *x*-intercepts
not factorable

Figure 4.27

EXAMPLE 8 ■ **Determining the Factored Form of the Quadratic Equation from a Graph**

Determine the factored form of the quadratic equation whose graph is shown in Figure 4.28.

Solution We can see from the graph the horizontal intercepts are (1, 0) and (10, 0). So the factored form of the equation will be $y = a(x - 1)(x - 10)$. It also appears the graph passes through $(0, -15)$. We use this point to determine the value of a.

$$-15 = a(0 - 1)(0 - 10)$$
$$-15 = a(-1)(-10)$$
$$-15 = 10a$$
$$a = -1.5$$

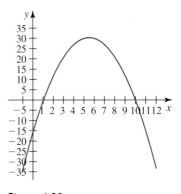

Figure 4.28

The factored form of the quadratic equation is $y = -1.5(x - 1)(x - 10)$.

To determine how to factor $y = x^2 + bx + c$ algebraically, we observe $(x + p)(x + q) = x^2 + (p + q)x + pq$. If we let $b = p + q$ and $c = pq$, we have determined the factored form for every factorable quadratic function of the form $y = x^2 + bx + c$.

EXAMPLE 9 ■ **Factoring a Quadratic Function Algebraically**

Factor $y = x^2 + 8x + 12$ algebraically.

Solution We are looking for values for p and q that add together to equal 8 and multiply together to equal 12. We begin by creating a list of number pairs that multiply together to be 12. We then add together each number in the pair. See Table 4.24.

Table 4.24

p	q	Product	Sum
1	12	12	13
2	6	12	8
3	4	12	7
−1	−12	12	−13
−2	−6	12	−8
−3	−4	12	−7

Notice the factors that add to be 8 are 2 and 6. Therefore, $y = x^2 + 8x + 12$ in factored form is $y = (x + 2)(x + 6)$.

As we have seen throughout this section, the vertex, standard, and factored forms all have their advantages. Some situations are better suited for one form than another.

EXAMPLE 10 ■ Using Quadratic Function Graphs in a Real-World Context

A homeowner plans to enclose a portion of her backyard for a garden as depicted in Figure 4.29. The area of her yard is 2700 square feet. She has 70 feet of fencing and wants to enclose an area as large as possible for the garden. She will use the existing block wall for two sides of her garden but will need to fence the remaining two sides. What is the maximum area that she will be able to enclose with the 70 feet of fencing?

Figure 4.29

Solution We let l represent the length of the garden and w represent the width. The amount of fencing that will be used is $l + w = 70$. That is, $l = -w + 70$. The area of the garden is

$$\begin{aligned} A &= lw \\ &= (-w + 70)w \\ &= -(w - 70)w \quad \text{factored form} \\ &= -w^2 + 70w \quad \text{standard form} \end{aligned}$$

This is a concave down parabola. From the factored form, we can tell the vertex will be at $w = 35$ since 35 lies halfway between the horizontal intercepts of $w = 0$ and $w = 70$. We get the same result if we calculate $-\dfrac{b}{2a}$ in standard form.

$$\begin{aligned} w &= -\frac{b}{2a} \\ &= -\frac{70}{2(-1)} \\ &= 35 \end{aligned}$$

If the width is 35 feet, then the length of the garden is $l = -35 + 70 = 35$ feet. At first glance, it appears a 35-foot by 35-foot garden will maximize the area. However, if we look closely, we recognize that we cannot have a garden 35 feet wide because the width of the yard itself is 30 feet. The graph of the function in Figure 4.30 will help us see what is going on.

The red dashed line indicates the maximum width of the garden that is possible in the yard: 30 feet. From the graph, we see a width of 30 feet will yield the maximum area for gardens of width 0 feet to 30 feet. We calculate the corresponding length and area.

Figure 4.30

$$\begin{aligned} l &= -w + 70 & A &= lw \\ &= -30 + 70 & &= 40(30) \\ &= 40 & &= 1200 \end{aligned}$$

For this backyard, a garden of width of 30 feet and length of 40 feet will yield the maximum area that can be enclosed with 70 feet of fencing. The maximum area is 1200 square feet.

SECTION 4.3 ■ FORMS AND GRAPHS OF QUADRATIC FUNCTIONS

EXAMPLE 11 ■ **Using Quadratic Function Graphs in a Real-World Context**

A rancher plans to build three adjacent corrals for his flock of sheep. The adult rams (males) will be placed in the first corral, the lambs 4 months old and older will be placed in the second corral, and the adult ewes (females) and their lambs younger than four months old will be placed in the third corral. See Figure 4.31.

Figure 4.31

The perimeter of the ram corral must be 200 feet. The total enclosed area of all three corrals must be at least 6000 square feet. What is the minimum amount of fencing that will be needed?

Solution Since the perimeter of the ram corral is 200, we have $2x + 2y = 200$.

$$2x + 2y = 200$$
$$x + y = 100$$
$$y = 100 - x$$

The combined width of the corrals is $x + 2x + 3x = 6x$. The combined area is

$$A = (\text{width})(\text{length})$$
$$= (6x)(y)$$
$$= 6x(100 - x) \quad \text{since } y = 100 - x$$
$$= -6x(x - 100)$$

From the factored form, we can see the area function has horizontal intercepts $(0, 0)$ and $(100, 0)$. The vertex will be halfway between, at $x = 50$. The graph of the parabola is shown in Figure 4.32.

The red dashed line is drawn on the graph at 6000 square feet, the minimum enclosed area desired. The width of the corral must be of sufficient size as to generate an area value at or above that horizontal line. To find where the graph of the parabola and the line intersect, we set the equation of the horizontal line equal to the equation of the parabola.

$$-6x^2 + 600x = 6000$$
$$-x^2 + 100x = 1000$$
$$x^2 - 100x + 1000 = 0$$

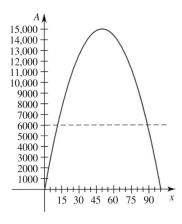

Figure 4.32

We determine the intersection points by using the quadratic formula.

$$x = -\frac{-100}{2(1)} \pm \frac{\sqrt{(-100)^2 - 4(1)(1000)}}{2(1)}$$

$$= 50 \pm \frac{\sqrt{6000}}{2}$$

$$= 50 \pm \frac{\sqrt{400(15)}}{2}$$

$$= 50 \pm \frac{20\sqrt{15}}{2}$$

$$= 50 \pm 10\sqrt{15}$$

So $x \approx 11.3$ or $x \approx 88.7$. If the width of the ram corral is between 11.3 feet and 88.7 feet, the enclosed area will be at least 6000 square feet.

The total amount of fencing needed is found by adding the lengths of all of the sides.

$$\text{total fencing} = 12x + 4y$$
$$= 12x + 4(100 - x)$$
$$= 12x + 400 - 4x$$
$$= 8x + 400$$

We want this number to be as small as possible. We can see the smaller the value of x, the smaller the amount of fencing will be needed. Consequently, the value of x that minimizes the amount of fence needed while still meeting the square footage requirement is $x = 11.3$. The corresponding value of y is 88.7 since $y = 100 - 11.3$. The minimum amount of fencing needed is $8(11.3) + 400 = 490.4$ feet.

■ Inverses of Quadratic Functions

Do quadratic functions have inverse functions? Let's consider the function $y = f(x) = x^2$ and solve this function for x.

$$y = x^2$$
$$\sqrt{y} = \sqrt{x^2}$$
$$\sqrt{y} = |x|$$
$$\sqrt{y} = \pm x$$
$$x = \pm\sqrt{y}$$
$$f^{-1}(y) = \pm\sqrt{y}$$

We know that in a function, each input value must correspond with exactly one output value. In this case, we see that each positive input value will correspond with two output values instead of one. For example, $f^{-1}(4)$ equals 2 and -2. Therefore, this quadratic function does not have an inverse function. In general, *quadratic functions do not have inverse functions*.

SUMMARY

In this section you learned how to work with the graphs of quadratic functions with equations given in vertex, standard, and factored forms and saw how these forms are related. You also learned strategies for finding the horizontal intercepts and the vertex of a parabola using these forms. Additionally, you learned how to use and interpret these forms and graphs in real-world situations.

4.3 EXERCISES

SKILLS AND CONCEPTS

In Exercises 1–15, determine the coordinates of the vertex and horizontal intercepts of the parabola. If no horizontal intercepts exist, so state.

1. $f(x) = 4(x - 3)^2 + 2$
2. $f(x) = -9(x + 12)^2 + 18$
3. $f(x) = 0.25(x - 2)^2 - 1$
4. $g(x) = -2(x - 5)^2 + 10$
5. $f(x) = 8(x - 0.5)^2 - 2$
6. $y = -2x^2 + 4x - 8$
7. $y = 0.1x^2 + 0.4x - 1.2$
8. $y = x^2 + 4x + 5$
9. $y = 9x^2 - 4$
10. $y = 3x^2 + 4x + 1$
11. $f(x) = 6(x - 4)(x + 2)$
12. $g(x) = -5x(x - 10)$
13. $f(x) = -3(x + 7)(x + 11)$
14. $g(x) = 0.2(x - 4)(x + 12)$
15. $f(x) = -7(x + 1)(x - 1)$

In Exercises 16–20, sketch the graph of the quadratic function by hand.

16. $g(x) = -2(x - 5)^2 + 10$
17. $y = -2x^2 + 4x - 8$
18. $y = 9x^2 - 4$
19. $g(x) = -5x(x - 10)$
20. $g(x) = 0.2(x - 4)(x + 12)$

In Exercises 21–25, determine the equation of the quadratic function from its graph.

21.

22.

23.

24.

25.

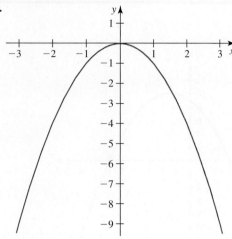

In Exercises 26–30, factor the quadratic function.

26. $y = x^2 - 11x + 24$
27. $y = x^2 + 2x + 1$
28. $y = x^2 - 16$
29. $y = x^2 - 2x - 24$
30. $y = x^2 - 5x - 6$

In Exercises 31–34, create a quadratic function to model the situation. Then answer all given questions.

31. Picture Frame A square is to be cut out of the middle of an 18-inch by 18-inch matting board to make a frame. The mat frame will be placed over a square picture and should overlap the picture by 1 inch on each side, as shown in the figure.

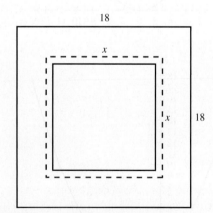

What will be the mat frame area for a 12-inch by 12-inch picture and a 15-inch by 15-inch picture?

32. Picture Frame A square is to be cut out of the middle of a 16-inch by 16-inch matting board to make a frame. The mat frame will be placed over a square picture and should overlap the picture by 1 inch on each side, as shown in the picture.

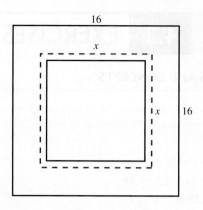

What will be the mat frame area for a 10-inch by 10-inch picture and a 14-inch by 14-inch picture?

33. Livestock Pens A rancher has 500 feet of fencing to construct two adjacent rectangular pens (see figure).

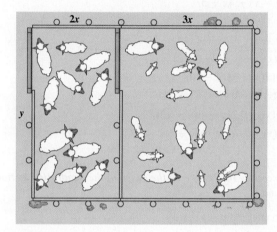

What are the dimensions of the pens with maximum combined area?

34. Livestock Pens A rancher has 600 feet of fencing to construct two adjacent rectangular pens (see figure).

What are the dimensions of the pens with maximum combined area?

SHOW YOU KNOW

35. Describe the advantages and disadvantages of each form of a quadratic function: standard, vertex, factored.

36. How is the *vertex* form of a quadratic function related to the *standard* form?

37. How is the *factored* form of a quadratic function related to the *standard* form?

38. A quadratic function cannot have an inverse function. Why not?

39. Describe how you can quickly find the second horizontal intercept of a quadratic function if you know the first horizontal intercept and the vertex.

MAKE IT REAL

In Exercises 40–45, create a quadratic function to model the situation. Then answer all given questions.

40. **Men's High Jump Record** As of 2011, the men's outdoor high jump record was held by Javier Sotomayor of Cuba. On July 27, 1993, he jumped 2.45 meters (8 feet $1/2$ inch) into the air. Assuming air resistance was negligible and that he landed on a cushion 3 feet above the height from which he jumped, how long was he airborne? (*Hint:* The height above the ground of a person t seconds after he jumps into the air can be modeled by $s(t) = -16(t - h)^2 + k$ feet, where h is the time he reaches his maximum height and k is the maximum height (in feet).)

41. **Women's High Jump Record** As of 2011, the women's outdoor high jump record was held by Stefka Kostadinova of Bulgaria. On August 30, 1987, she jumped 2.09 meters (6 feet 10 inches) into the air. Assuming air resistance was negligible and that she landed on a cushion 3 feet above the height from which she jumped, how long was she airborne? (*Hint:* The height above the ground of a person t seconds after she jumps into the air can be modeled by $s(t) = -16(t - h)^2 + k$ feet, where h is the time she reaches her maximum height and k is the maximum height (in feet).)

42. **Longest Vertical Drop on Earth** Mount Thor in Canada is said to have the longest vertical drop (4100 feet) on Earth. (*Source:* www.wikipedia.com) How long would it take for a pebble dropped off the cliff to reach the ground? (For the purpose of this model, neglect air resistance.) (*Hint:* The height of a falling object above the ground t seconds after it is dropped can be modeled by $s(t) = -16t^2 + s_0$, where s_0 is the initial height.)

43. **Swimming Pool Design** Many counties require in-ground swimming pools to be set back at least 5 feet from the fence line and from all buildings.
    A family plans to install a rectangular pool into their 30-foot by 60-foot backyard. They want the pool to have a 110-foot perimeter and contain as much area as possible. One pool shape they are considering is a rectangle.

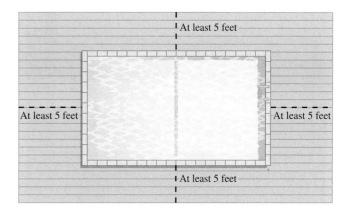

What will be the dimensions of the rectangular pool with maximum surface area?

44. **Swimming Pool Design** Many counties require in-ground swimming pools to be set back at least 5 feet from any fence line and from all buildings.
    A family plans to install a pool into their 30-foot by 60-foot backyard. They want the pool to have a 110-foot perimeter and contain as much area as possible. One pool shape they are considering is a rectangle with two semicircles attached at each end.

What will be the dimensions of the pool of this shape with maximum surface area? (*Hint:* Recall that the area of a circle is $A = \pi r^2$ and the circumference of a circle is $C = 2\pi r$.)

45. **Pool Decking** Single-family, in-ground swimming pools are typically bordered with pool decking. It is customary to have a limited amount of decking on the far side of the pool and a larger amount of decking on the near side of the pool (to accommodate deck furniture). A rectangular pool with a 100-foot perimeter is bordered on three sides by 2 feet of decking and on one side by 8 feet of decking. The pool surface area is to be at least 500 square feet. (See figure on next page.)

What pool dimensions will require the least amount of decking?

■ **STRETCH YOUR MIND**

Exercises 46–50 are intended to challenge your understanding of quadratic functions.

46. **Water Rockets** As a youth, the favorite toy of one of the authors was a red water rocket, similar to that shown in the photo.

    Assuming air resistance is negligible, the rocket is launched from a height of 2 feet, and the maximum height attained by the water rocket is 66 feet, determine the initial velocity of the rocket. That is, find the velocity of the rocket at the moment it is launched.

47. **Water Rockets** As of March 2007, the world record for the greatest height reached by a water rocket was reported to be 2088 feet. (*Source:* www.wikipedia.com) Assuming the rocket was launched from a height of 3 feet and air resistance was negligible, determine how long it took for the rocket to reach its maximum height and its total flight time from launch to landing.

48. Show how to convert a quadratic function in factored form, $y = a(x - x_1)(x - x_2)$, into a quadratic function in vertex form, $y = a(x - h)^2 + k$. Specify the relationship between x_1 and x_2 and h and k.

49. Explain why some quadratic functions *cannot* be written in factored form.

50. How can you tell if a quadratic function in vertex form can be rewritten in factored form?

CHAPTER 4 Study Sheet

As a result of your work in this chapter, you should be able to answer the following questions, which are focused on the "big ideas" of this chapter.

SECTION 4.1
1. What is meant by *increasing* and *decreasing* functions?
2. What does the term *variable rate of change* mean?
3. How can average rates of change be used to fill in gaps in a table of data?
4. What does it mean to find the rate of change at an instant? How is the rate of change at an instant calculated? How can the rate of change at an instant be visualized on a graph?
5. How are rates of change and successive differences used to describe the concavity of a graphical model?
6. How can rates of change be used to describe inflection points?

SECTION 4.2
7. What distinguishes a quadratic function from a linear function? Use the language of rate of change in your response.
8. Interpret the parameters of a quadratic model, $y = ax^2 + bx + c$. What does each mean in a modeling situation?

SECTION 4.3
9. What do the parameters a, h, and k represent in a quadratic equation in vertex form, $y = a(x - h)^2 + k$?
10. How can you find the horizontal intercepts of a quadratic equation in vertex form?
11. How can you find the vertex of a quadratic model in standard form? What does the vertex represent?
12. What is the quadratic formula used for? What is a discriminant?

REVIEW EXERCISES

■ SECTION 4.1 ■

For Exercises 1–3, create a table of values and an associated graph that

1. Is decreasing and concave down.
2. Has a point of inflection.
3. Is concave down twice and concave up once.
4. Each of the functions f, g, h, and j shown in the table are decreasing, but each decreases in a different way. Which of the graphs best fits each function? Explain how you made your choice.

x	$f(x)$	$g(x)$	$h(x)$	$j(x)$
−6	34.3	9.4	30	17.49
−4	32.7	7.9	10	8.55
−2	30.8	6.4	−2	4.18
0	28.5	4.9	−4	2.05
2	25.7	3.4	−5	1
4	22.2	1.9	−6	0.49
6	17.2	0.4	−8	0.24
8	8.5	−0.9	−20	0.12

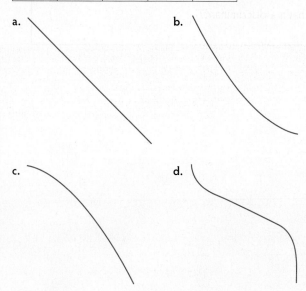

a.   b.
c.   d.

5. Identify any inflection points in the functions of Exercise 4. Then explain how to find an inflection point from a table and how to find an inflection point from a graph.

6. Label points A, B, C, D, E, and F on the graph of f according to the following descriptions of each point.
   a. Point A is a point on the curve where the instantaneous rate of change is positive.
   b. Point B is a point on the curve where the value of the function is positive.
   c. Point C is a point on the curve where the instantaneous rate of change is the most negative.
   d. Point D is a point on the curve where the instantaneous rate of change is 0.
   e. Points E and F are different points on the curve where the instantaneous rates of change have the same magnitude but opposite signs.

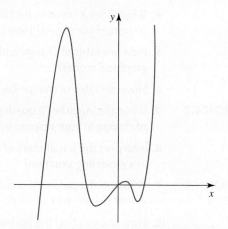

7. The table shows the number of fighter aircraft produced by three of the major powers in World War II from 1940 to 1944. For each country,
   a. Describe the concavity of the function, including identifying any possible inflection points. Then interpret your findings in the context of the situation.
   b. Sketch a graph that could model each function and label the independent and dependent axes.

Year y	Germany $G(y)$	USSR $U(y)$	United Kingdom $K(y)$
1940	2,746	4,574	4,283
1941	3,744	7,086	7,064
1942	5,515	9,924	9,849
1943	10,898	14,590	10,727
1944	26,326	17,913	10,730

Source: *World War II: The Encyclopedia of Facts and Figures*, John Ellis, Table 93

For Exercises 8–12, refer to the fighter aircraft production data given in Exercise 7.

8. Compute the average rate of change of G from 1942 to 1944, give the units on this rate, and interpret what this number means.

9. Compute the average rate of change of U from 1940 to 1943, give the units on this rate, and interpret what this number means.

10. Estimate the instantaneous rate of change of K at $y = 1941$, give the units on this rate, and interpret what this number means.

11. Estimate the instantaneous rate of change of G at y = 1943, give the units on this rate, and interpret what this number means.

12. Using *successive differences*, predict what each country's production would have been in 1945 (assuming the war continued through the entire year).

13. The table shows the temperatures in Seattle, Washington, from 11 P.M. on March 13, 2007 to 3 P.M. on March 14, 2007.

Hours Since 11 P.M. March 13, 2007 t	Temperature in Seattle, WA (°F) $F(t)$
0	43
2	42
4	41
6	41
8	39.9
10	37
12	39
14	42
16	42.1

Source: weather.noaa.gov

a. Plot the points and connect them with a smooth curve to model the temperature over this time period.

b. Describe the intervals where the function is increasing, decreasing, and constant. Then explain what this tells us about temperatures in Seattle over these intervals.

c. Describe the intervals where the function is concave up and concave down. Then explain what this tells us about temperatures in Seattle over these intervals.

d. Use *averages* to find $F(3)$, $F(9)$, and $F(13)$.

e. Use *average rates of change* to find $F(3)$, $F(9)$, and $F(13)$. Then compare these results to part (d).

f. What would $t = 6.75$ represent? Could you use either averages or the average rate of change to find $F(6.75)$? Explain.

14. **Blank Cassette Tape Sales** Over the past couple of decades, sales of blank cassette tapes have declined, as shown in the table and graph.

Years Since 1990 t	Blank Cassette Tape Sales ($ millions) $S(t)$
0	376
5	334
10	162
12	98
14	66

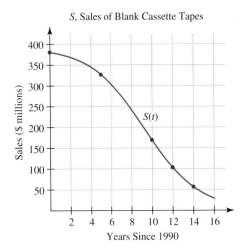

S, Sales of Blank Cassette Tapes

a. Describe the intervals over which the function is concave up and concave down. Then explain what this tells about the sales of blank cassette tapes.

b. Use *averages* to estimate $S(11)$ and $S(13)$.

c. Use *average rates of change* to estimate $S(11)$ and $S(13)$. Then compare these results to part (b).

d. Use *successive differences* to predict $S(15)$ and $S(16)$.

15. Five years ago Coach Anderson took over as head coach of the worst basketball team in the league, at which time the team began to improve. After this season, however, some of the fans want Coach Anderson fired. They claim his early gains have not continued, while the coach's supporters say the team's record has improved each year under Coach Anderson. Using a graph and a discussion of concavity, explain how both groups can be correct.

16. Over the last three years, a company has begun to lose money. The CEO says the losses are not too severe and that the company should be able to keep doing business, but analysts have predicted big trouble for this company in the near future. Is the graph of the company's profits concave up or concave down? Explain.

■ **SECTION 4.2** ■

Exercises 17–19 focus on a conceptual understanding of quadratic functions.

17. Describe how to determine if a table of data, a graph, or an equation represents a quadratic function.

18. If the second differences of a table of data are negative, what do you know about its graph?

19. Describe two different methods that can be used to find a quadratic model for a table of data.

In Exercises 20–21, explain the meaning of the parameters of the quadratic function.

20. **Nuclear Power Consumption** Based on data from 1970 to 2004, the consumption of nuclear power in the United States can be modeled by

$$N(t) = -0.003084t^2 + 0.3512t + 0.04516$$

quadrillion BTUs, where t is the number of years since 1970. (**Source: Modeled from** *Statistical Abstract of the United States, 2007*, **Table 895**)

21. **Consumer Spending on Farm Foods** Based on data from 1990 to 2004, consumer expenditure for farm foods can be modeled by

$$E(t) = 1.008t^2 + 9.951t + 450.5$$

billion dollars, where t is the number of years since 1990. (**Source: Modeled from** *Statistical Abstract of the United States, 2007*, **Table 818**)

In Exercises 22–23, use a graphing calculator and quadratic regression to find the quadratic model that best fits the data set.

22. **U.S. Wine Imports**

Years Since 1990 t	Wine Imports (in thousands of hectoliters) W
0	2510
5	2781
10	4584
12	5655
13	6214
14	6549
15	7262

Source: Statistical Abstract of the United States, 2007, **Table 819**

23. **U.S. Exports to Canada**

Years Since 1990 t	Exports to Canada ($ millions) C
0	4217
10	7640
11	8121
12	8660
13	9313
14	9741
15	10570

Source: Statistical Abstract of the United States, 2007, **Table 827**

■ SECTION 4.3 ■

In Exercises 24–26, find the vertex and horizontal intercepts of the quadratic function.

24. $y = -10(x + 1)^2 - 20$
25. $y = 4x^2 + 16x + 30$
26. $y = -6(x + 20)(x - 8)$

In Exercises 27–29, graph the quadratic function by hand. (Hint: Refer to your answers in Exercises 24–26.)

27. $y = -10(x + 1)^2 - 20$
28. $y = 4x^2 + 16x + 30$
29. $y = -6(x + 20)(x - 8)$

In Exercises 30–31, create a quadratic function to model the situation. Then answer all given questions.

30. **Golden Rectangle** A golden rectangle is a rectangle with the property that the ratio of its length to its width is equal to the ratio of the sum of its length and width to its length. That is, $\dfrac{l}{w} = \dfrac{l + w}{l}$. The width of a particular golden rectangle is 2 meters. What is the length and area of the rectangle?

31. **Orange Prices** On March 15, 2007, Safeway.com offered individual large Navel oranges for $1.61 per pound. A 4-pound bag of oranges was offered for $4.99. (**Source:** www.safeway.com) Assuming that the total cost of purchasing n pounds of oranges can be modeled by a quadratic function, calculate the price the store should charge for a 10-pound bag of oranges and the corresponding price per pound.

In Exercises 32–33, determine the equation of the quadratic function from the graph.

32.

33.

Make It Real Project

What to Do

1. Find a set of at least six data points from an area of personal interest. Choose data that appear to be concave up or concave down.

2. Draw a scatter plot of the data and explain why you do or do not believe a quadratic model would fit the data well.

3. Find a quadratic regression model for your data.

4. Interpret the physical meaning of the parameters, a, b, and c.

5. Use the model to predict the value of the function at an unknown point and explain why you do or do not think the prediction is accurate.

6. Describe the transformations (shifts, stretches, compressions, reflections) that your regression model shows compared to a simple quadratic equation, $y = x^2$.

7. Write your quadratic model in vertex form and interpret the meaning of the parameters.

8. Describe your function using the language of rate of change (increasing, decreasing, concave up, concave down). Be sure to explain what these terms mean in the context of your data set.

CHAPTER 5

Polynomial, Power, and Rational Functions

Photofish/Shutterstock.com

On March 24, 1989, the Exxon Valdez oil tanker spilled more than 10.8 million gallons of oil, polluting more than 1300 miles of Alaskan shoreline. The cleanup effort cost approximately $2.1 billion. (*Source:* www.evostc.state.ak.us)

People overseeing cleanup efforts must determine when the expected gains from spending additional money on cleanup are not worth the additional cost. Cost-benefit decisions such as this can sometimes be modeled with rational functions.

5.1 Higher-Order Polynomial Functions
5.2 Power Functions
5.3 Rational Functions
STUDY SHEET
REVIEW EXERCISES
MAKE IT REAL PROJECT

SECTION 5.1 Higher-Order Polynomial Functions

LEARNING OBJECTIVES

- Use higher-order polynomial functions to model real-world situations
- Use the language of rate of change to describe the behavior of a higher-polynomial function
- Find the inverse of a polynomial function

GETTING STARTED

The U.S. Postal Service limits the size of the packages it will send. To determine if a package qualifies to be sent Parcel Post®, USPS employees add the length of the package to the distance around the thickest part (girth). The sum of the length and girth cannot exceed 130 inches. For a rectangular package with equal height and width, the relationship between the length of the package and its volume can be described by a cubic polynomial function.

In this section we look at higher-order *polynomial* functions, beginning with cubic functions. We investigate the rates of change of these functions and discuss concavity, inflection points, and end behavior of their graphs. We also see how polynomial functions are applied to a variety of real-world contexts, including the size of shipping boxes.

■ Cubic Functions

Figure 5.1 is a representation of six boxes whose length plus girth equal 130 inches. The length, width, height, and volume of each box are recorded in Table 5.1. Recall that volume is the product of length, width, and height. That is, $V = lwh$.

Figure 5.1

Table 5.1

Box	Length (inches)	Width (inches)	Height (inches)	Volume (cubic inches)
A	110	5	5	2750
B	90	10	10	9000
C	70	15	15	15,750
D	50	20	20	20,000
E	30	25	25	18,750
F	10	30	30	9000

To determine the relationship between the width and volume of the box we calculate successive differences. That is, we find the first differences (ΔV), second differences ($\Delta(\Delta V)$), and third differences ($\Delta(\Delta(\Delta V))$), as shown in Table 5.2. This strategy requires that each of the widths in the table be equally spaced. In this case, the widths are equally spaced 5 inches apart.

Table 5.2

Box	Width (inches)	Volume (cubic inches)	First Difference ΔV	Second Difference $\Delta(\Delta V)$	Third Difference $\Delta(\Delta(\Delta V))$
A	5	2750	6250	500	−3000
B	10	9000	6750	−2500	−3000
C	15	15,750	4250	−5500	−3000
D	20	20,000	−1250	−8500	
E	25	18,750	−9750		
F	30	9000			

We know from Chapter 4 that functions with constant first differences are linear and functions with constant second differences are quadratic. Functions with constant third differences, as in Table 5.2, are **cubic functions**.

PROPERTY OF CUBIC FUNCTIONS

Any function with constant third differences is a **cubic function**.

EXAMPLE 1 ■ Finding the Equation of a Cubic Function Algebraically

A rectangular package has a square end (see Figure 5.2). The sum of the length and girth of the package is equal to 130 inches. Find an equation that relates the width of the package to the volume of the package.

Figure 5.2

Solution The girth of the package is $4w$ because the distance across the top of the package, the distance down the front of the package, the distance across the bottom of the package, and the distance up the back of the package are each w inches. Since the sum of the length and girth is 130 inches, we have

$$l + 4w = 130$$
$$l = 130 - 4w$$

The volume of a rectangular box is the product of its length, width, and height. Thus,

$$V = lwh$$
$$V = lw^2 \quad \text{since } h = w$$
$$V = (130 - 4w)w^2 \quad \text{since } l = 130 - 4w$$
$$V = -4w^3 + 130w^2$$

The equation that relates the volume of the package to its width is $V = -4w^3 + 130w^2$.

STANDARD FORM OF A CUBIC FUNCTION

A cubic function has an equation of the form

$$y = ax^3 + bx^2 + cx + d$$

with constants a, b, c, and d and $a \neq 0$.

Graphs of Cubic Functions

In Example 1 we saw that the width and the volume of the package were related by $V = -4w^3 + 130w^2$. We can verify the accuracy of the equation by graphing it together with a scatter plot of the data presented in Table 5.1, as shown in Figure 5.3.

Observe that the graph is initially concave up but changes to concave down around $w = 10$. The point where the function changes concavity is called an **inflection point**.

Figure 5.3

INFLECTION POINT OF A CUBIC FUNCTION

The point on a graph where the concavity changes is called an **inflection point**. All cubic functions have exactly one inflection point.

Figure 5.4 shows the graphs with inflection points marked for several cubic functions.

(a) $y = x^3$

(b) $y = -x^3 + 6x^2 - 4x + 11$

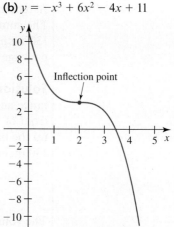

(c) $y = x^3 - x^2 - 2x$

(d) $y = -x^3 + 6x^2 - 11x + 8$

Figure 5.4

Relationship between Inflection Points and Rates of Change

The concavity of the graph and the rates of change of the corresponding function are closely related. When the graph is concave up, the rates of change (and first differences) are increasing. When the graph is concave down, the rates of change (and first differences) are decreasing. Thus inflection points, which occur where the concavity changes, also indicate where the rates of change switch from increasing to decreasing or vice versa.

For example, the graph in Figure 5.5 shows the rates of change of a cubic function at various points. Observe that when the graph is concave down, the rates of change are decreasing (12.9, 6.7, 2.1, 0, −1.9, −3.0). When the graph is concave up, the rates of change are increasing (−3.0, −2.5, 0, 2.1, 4.7, 9.0). At the inflection point (3, 6), the rates of change stop decreasing and begin to increase.

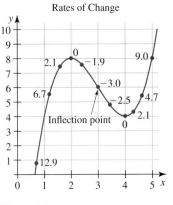

Figure 5.5

■ Modeling with Cubic Functions

Scatter plots that appear to change concavity exactly one time can be modeled by cubic functions. The resultant model will have constant third differences and either increasing or decreasing second differences.

EXAMPLE 2 ■ Using a Cubic Function in a Real-World Context

The per capita consumption of breakfast cereals (ready to eat and ready to cook) since 1980 is given in Table 5.3.

Table 5.3

Years Since 1980 t	Consumption (pounds) C	Years Since 1980 t	Consumption (pounds) C
0	12	10	15.4
1	12	11	16.1
2	11.9	12	16.6
3	12.2	13	17.3
4	12.5	14	17.4
5	12.8	15	17.1
6	13.1	16	16.6
7	13.3	17	16.3
8	14.2	18	15.6
9	14.9	19	15.5

Source: Statistical Abstract of the United States, 2001, Table 202

a. Create a scatter plot of these data and explain what type of function might best model the data.

b. Find the cubic regression model for the situation and graph the model together with the scatter plot.

c. Use the model from part (b) to predict the per capita consumption of breakfast cereal in 2000.

Solution

a. The scatter plot is shown in Figure 5.6. It appears that the per capita consumption (after a brief decline early on) increases at an increasing rate (concave up) until about 1988. Then, the per capita consumption increases at a decreasing rate (concave down) until 1994, where it begins to decrease but remains concave down. Because the graph appears to change concavity once, a cubic model may be appropriate. We see a possible inflection point (where the per capita consumption of cereal is increasing at the greatest rate) at approximately 1988 ($t = 8$).

Figure 5.6

b. We use the graphing calculator to find the cubic regression model.

$$c(t) = -0.004718t^3 + 0.1165t^2 - 0.3585t + 12.17$$

where c is the per capita consumption of cereal (in pounds) and t is the number of years since 1980. (The cubic regression process is identical to that used for linear regression except that CubicReg is selected instead of LinReg.) The graph of the model is shown in Figure 5.7.

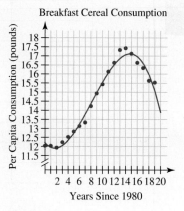

Figure 5.7

c. Since $t = 20$ represents the year 2000, we substitute this value into the function to predict the per capita consumption of cereal in 2000.

$$c(t) = -0.004718t^3 + 0.1165t^2 - 0.3585t + 12.17$$
$$c(20) = -0.004718(20)^3 + 0.1165(20)^2 - 0.3585(20) + 12.17$$
$$= 13.86$$

In 2000, each person in the United States consumed nearly 14 pounds of cereal, on average, according to the model.

■ Polynomial Functions

Linear, quadratic, and cubic functions are all **polynomial functions**. Polynomial functions are formally defined as follows.

POLYNOMIAL FUNCTION

For whole number n, a function of the form

$$y = a_n x^n + a_{n-1} x^{n-1} + a_{n-2} x^{n-2} + \cdots + a_1 x + a_0$$

with $a_n \neq 0$ is called a **polynomial function of degree n**. Each $a_i x^i$ is called a **term**. The a_i are real-number values called the **coefficients** of the terms.

Note that the terms of polynomial functions have coefficients labeled a_n, a_{n-1}, and so on. Using this indexing system, we refer to the coefficient on the x^3-term as a_3, the coefficient of the x^2-term as a_2, and so on. The exponent of each term of a polynomial must be a nonnegative whole number. Table 5.4 shows examples of polynomial functions and Table 5.5 shows examples of nonpolynomial functions.

Table 5.4

Polynomial
$y = 2x - 5$
$y = -x^2 + 3x + 2$
$y = -\dfrac{2}{3}x^3 - 7x + \dfrac{2}{7}$
$y = 0.5x^5 - 2.5x^2 + 1.75x + 6.3$
$y = -5$

Table 5.5

Nonpolynomial	Reason
$y = \dfrac{2x - 5}{x^2 + 1}$	Not a sum of terms of the form $a_i x^i$.
$y = x^{0.5} + x^{-3.4}$	The exponents are not whole numbers.
$y = \sqrt{x} + x^2$	Since $\sqrt{x} = x^{1/2}$, one exponent is not a whole number.
$y = 2^x$	The exponent is a variable.
$y = 2x^{1/3} + 6x^{-1}$	The exponents are not whole numbers.

The graphs of polynomial functions are fairly predictable. We summarize the characteristics and appearance of the graphs of polynomial functions of the first through fifth degree in Table 5.6.

Table 5.6

Function Name and Degree	Concavity, Inflection Points, and End Behavior	Constant Difference	Sample Graph
linear first degree	no concavity no inflection points one end $\to \infty$ one end $\to -\infty$	first	
quadratic second degree	concave up only or concave down only no inflection points both ends $\to \infty$ or both ends $\to -\infty$	second	
cubic third degree	concave up and concave down one inflection point one end $\to \infty$ one end $\to -\infty$	third	
quartic fourth degree	concavity changes zero or two times zero or two inflection points both ends $\to \infty$ or both ends $\to -\infty$	fourth	

Table 5.6 (continued)

Function Name and Degree	Concavity, Inflection Points, and End Behavior	Constant Difference	Sample Graph
quintic fifth degree	concavity changes one or three times one or three inflection points one end → ∞ one end → −∞	fifth	(graph shown)

EXAMPLE 3 Selecting a Higher-Order Polynomial Model

The per capita consumption of chicken in the United States has continued to increase since 1985; however, the rate of increase has varied. Create a scatter plot of the data in Table 5.7 and determine which polynomial function best models the situation. Then describe the relationship between the per capita consumption of chicken and time (in years) using the language of rate of change.

Table 5.7

Per Capita Consumption of Chicken (Boneless, Trimmed Weight)		Per Capita Consumption of Chicken (Boneless, Trimmed Weight)	
Years Since 1985 t	Consumption (pounds) C	Years Since 1985 t	Consumption (pounds) C
0	36.4	8	48.5
1	37.2	9	49.3
2	39.4	10	48.8
3	39.6	11	49.5
4	40.9	12	50.3
5	42.4	13	50.8
6	44.2	14	54.2
7	46.7		

Source: Statistical Abstract of the United States, 2001, Table 202

Solution We create the scatter plot of the data shown in Figure 5.8 and look for changes in concavity.

Although it is not always totally clear where the changes in concavity occur when looking at real-world data, we can look for trends and approximate. If we sketch a rough line graph as in Figure 5.9, we can better see that this data is initially concave up (roughly) but changes to concave down around 1990. Then, it changes to concave up again around 1996.

From Table 5.6 we see that the concave up, concave down, concave up pattern is best modeled using a quartic (fourth-degree polynomial) function. Using quartic regression, we determine that the function

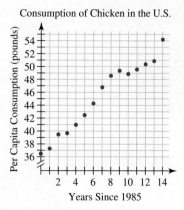

Figure 5.8

$$C(t) = 0.002623t^4 - 0.07545t^3 + 0.6606t^2 - 0.4688t + 36.92$$

best fits the data. We draw a graph of the model together with the scatter plot, as shown in Figure 5.10.

Figure 5.9

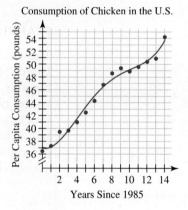

Figure 5.10

According to the model, per capita chicken consumption increased at an increasing rate from 1985 until approximately 1990. Then, from 1990 to approximately 1996, per capita chicken consumption continued to increase but at a decreasing rate. From 1996 to 1999, per capita chicken consumption again increased at an increasing rate.

■ End Behavior of Polynomial Functions

Polynomial functions have predictable long-run behavior, known as a function's **end behavior**.

> **END BEHAVIOR OF A POLYNOMIAL FUNCTION**
>
> For any polynomial function, as x approaches $\pm\infty$, $f(x)$ approaches $\pm\infty$. That is, as the magnitude (absolute value) of x gets larger and larger, the magnitude of the function values will also get larger and larger. Symbolically we write
>
> $$\text{as } x \to \pm\infty, \; f(x) \to \pm\infty$$

EXAMPLE 4 ■ Determining the End Behavior of a Polynomial Function

Use a table to determine the end behavior of $y = x^3 - x^2$.

Solution We create Table 5.8 and observe that as x approaches $-\infty$, the values of y approach $-\infty$. As x approaches ∞, the values of y approach ∞.

Table 5.8

x	y
−10,000	−1,000,100,000,000
−1000	−1,001,000,000
−100	−1,010,000
−10	−1100
−1	−2
1	0
10	900
100	990,000
1000	999,000,000
10,000	99,990,000,000,000

Symbolically, we write

$$\text{as } x \to -\infty, y \to -\infty \text{ and as } x \to \infty, y \to \infty$$

It is important to keep end behavior in mind when extrapolating with polynomial function models. We illustrate this idea in the following example, where we see that extrapolation can give inaccurate predictions.

EXAMPLE 5 ■ Extrapolating Using a Cubic Regression Model

The data in Table 5.9 show the projected Internet usage (in hours per person per year) in the United States.

Table 5.9

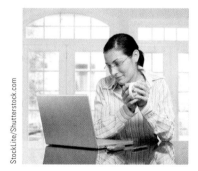

Projected Internet Usage: Hours per Person per Year (Based on 1995–1999 Data)	
Years Since 1995 t	Usage per Person (hours per year) H
0	5
1	10
2	34
3	61
4	99
5	135
6	162
7	187
8	208
9	228

Source: Statistical Abstract of the United States, 2001, Table 1125

a. Referring to a scatter plot of the data, determine if a cubic regression model fits the data. Then find and graph the cubic regression model that shows the relationship between Internet usage and time.

b. Predict the Internet usage (hours per person per year) in the year 2012. Discuss the accuracy of this prediction.

c. In light of the results from part (b), determine a practical domain for the cubic regression model.

Solution

a. We analyze the scatter plot of the data shown in Figure 5.11. The data is concave up initially but changes to concave down, so a cubic polynomial function is appropriate.

Using a graphing calculator, we find the cubic regression model

$$L(t) = -0.4514t^3 + 6.125t^2 + 6.171t + 1.994$$

where t is measured in years since 1995 and L is Internet usage per person per year (in hours). The model has a coefficient of determination of $r^2 = 0.9982$ so it is a great fit for the data. The graph of the model is shown in Figure 5.12.

Figure 5.11

Figure 5.12

b. We use the cubic regression model to predict the amount of Internet usage in 2012. Since 2012 is 17 years since 1995, we evaluate L when $t = 17$.

$$L(17) = -0.4514(17)^3 + 6.125(17)^2 + 6.171(17) + 1.994$$
$$= -341.2122$$

According to the model, people will spend a negative amount of time using the Internet, which is, of course, impossible. Observe from the graph in Figure 5.13 that the cubic model decreases quickly after 2005. Based on our life experience, we have no reason to anticipate that Internet usage will decrease.

Figure 5.13

c. It is clear that when we extrapolate using the cubic model, we may obtain results that do not make sense. For this regression model, we estimate the practical domain to be [0, 10].

EXAMPLE 6 ■ Graphing Polynomial Functions to Match a Verbal Description

Graph a polynomial function f that meets the following criteria.

- Fourth degree.
- As $x \to \infty$, $f(x) \to \infty$.
- The graph of f has two inflection points.
- $f(x) = 0$ exactly twice.

Solution A function that is concave up, then concave down, then concave up will have exactly two inflection points. A function with exactly two x-intercepts will meet the condition $f(x) = 0$ exactly twice. A fourth-degree polynomial has the property that both ends approach the same value. Since we know that as $x \to \infty$, $f(x) \to \infty$, it must be true that as $x \to -\infty$, $f(x) \to \infty$. The graph in Figure 5.14 meets the criteria. (Note: There are other graphs that also meet this criteria.)

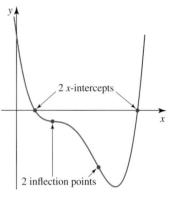

Figure 5.14

■ Relative Extrema of Polynomial Functions

A **relative maximum** occurs at the point where a graph changes from increasing to decreasing. A **relative minimum** occurs at the point where a graph changes from decreasing to increasing. The term **relative extrema** is used to refer to maxima and minima simultaneously. The graph of a polynomial function of degree n will have at most $n - 1$ relative extrema but it may have fewer.

EXAMPLE 7 ■ Identifying Relative Extrema

Graph the function $y = x^5 - 4x^4 + 4x^3$ and identify the points where the relative extrema occur.

Solution The function will have at most 4 relative extrema (since the degree is 5). Graphing the function as shown in Figure 5.15, we see that there is one point where a relative maximum occurs and one point where a relative minimum occurs.

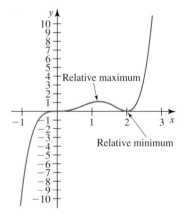

Figure 5.15

■ Inverses of Polynomial Functions

Not all polynomial functions have inverse functions; however, some do. Any polynomial function whose graph is strictly increasing or strictly decreasing will have an inverse function. Any polynomial function whose graph changes from increasing to decreasing or vice versa at any point in its domain will not have an inverse function.

EXAMPLE 8 ■ Finding the Inverse of a Polynomial Function

Find the inverse of $f(x) = x^5$.

Solution

$$y = x^5$$
$$(y)^{1/5} = (x^5)^{1/5}$$
$$y^{1/5} = x$$
$$f^{-1}(y) = y^{1/5}$$

The inverse of $f(x) = x^5$ is $f^{-1}(y) = y^{1/5}$.

SUMMARY

In this section you learned how to determine which higher-polynomial function can best model a real-world situation. You also learned how to describe the graph of any polynomial function by using the language of rate of change.

5.1 EXERCISES

SKILLS AND CONCEPTS

In Exercises 1–4, numerical representations of three functions are shown in a table. Each function is either linear, quadratic, cubic, or none of these. Use successive differences to identify each function appropriately.

1.

x	$f(x)$	$g(x)$	$h(x)$
0	−1.00	0.00	2.50
2	−0.92	−0.88	5.00
4	−0.68	−3.04	7.50
6	−0.28	−5.76	10.0
8	0.28	−8.32	12.5
10	1.00	−10.00	15.0
12	1.88	−10.08	17.5
14	2.92	−7.84	20.0
16	4.12	−2.56	22.5
18	5.48	6.48	25.0

2.

x	$f(x)$	$g(x)$	$h(x)$
0	5.00	0.76	1.00
2	5.51	1.32	53.28
4	6.08	2.00	111.24
6	6.70	2.80	176.56
8	7.39	3.72	250.92
10	8.14	4.76	336.00
12	8.98	5.92	433.48
14	9.90	7.20	545.04
16	10.91	8.60	672.36
18	12.03	10.12	817.12

3.

x	$f(x)$	$g(x)$	$h(x)$
−5	−105	−5	11.27
−4	−48	5	9.58
−3	−15	15	8.14
−2	0	25	6.92
−1	3	35	5.88
0	0	45	5.00
1	−3	55	4.25
2	0	65	3.61
3	15	75	3.07
4	48	85	2.61

4.

x	$f(x)$	$g(x)$	$h(x)$
0	0	0	0.09
1	4	9	0.10
2	0	16	0.11
3	−6	21	0.13
4	−8	24	0.14
5	0	25	0.17
6	24	24	0.20
7	70	21	0.25
8	144	16	0.33
9	252	9	0.50

In Exercises 5–8, examine the given graph and indicate the number of times the concavity changes. Use this result to determine which type of polynomial function is represented by the graph.

5.

6.

7.

8.
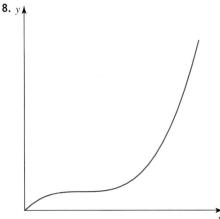

In Exercises 9–16,

a. Describe the end behavior of the given polynomial function.

b. Make a table of values that confirms the end behavior you described. Create your table in such a way that it shows what happens to function values as $x \to \infty$ and $x \to -\infty$.

9. $y = -2x^2 - 12x - 10$
10. $y = x^3 - 6x + 1$
11. $y = -3x^3 + 2x^2 - 2x - 3$
12. $y = x^4 - 2x^2 - 5$
13. $y = x^5 - 3x^3 + 2$
14. $y = 0.5x^2 - 10x - 5$
15. $y = -x^5 + x^4 + 2x^3 + 3x^2$
16. $y = x^6 - 10x^5 - x - 5$

In Exercises 17–22, sketch the graph of a function with the given description. Then classify the function by its degree.

17. The function is always increasing. It is initially concave up, then concave down, then concave up.

18. The function is initially increasing and concave up. It continues to increase but changes to be concave down. Finally, the function decreases.

19. The function is always increasing at a constant rate.

20. The function is initially decreasing and concave up. The function then changes to increasing but remains concave up. Finally, the function changes to concave down but remains increasing.

21. The function is always increasing and concave down.

22. The function is always increasing. It is initially concave down but changes to concave up before returning to concave down.

In Exercises 23–27, sketch the graph of a polynomial function that has the given characteristics.

23. • Third degree
    • As $x \to \infty, f(x) \to -\infty$
    • 1 maximum and 1 minimum

24. • Fourth degree
    • As $x \to \infty, f(x) \to \infty$
    • 2 minimums and 1 maximum
    • $f(x) \neq 0$

25. • Fourth degree
    • As $x \to \infty, f(x) \to -\infty$
    • 2 maximums and 1 minimum
    • $f(x) = 0$ exactly twice

26. • Fifth degree
    • As $x \to \infty$, $f(x) \to \infty$
    • As $x \to -\infty$, $f(x) \to -\infty$
    • Always increasing

27. • Second degree
    • As $x \to \infty$, $f(x) \to -\infty$
    • $f(x) = 0$ exactly once

SHOW YOU KNOW

28. Explain how to use successive differences to determine if a numerical representation of a function is linear, quadratic, cubic, quartic, etc.

29. Explain how to use rates of change and concavity to determine which polynomial type would best model a scatter plot of data.

30. How can we tell if a given function is a polynomial function?

31. Explain why the following function is or is not a polynomial function:
$$f(x) = 3x^2 - \sqrt{x} + 8$$

32. Explain why the following function is or is not a polynomial function:
$$f(x) = 2x^5 - x\sqrt{2} + \frac{1}{3}$$

33. How do we determine the end behavior of a polynomial function? Why is it important to understand this end behavior when modeling real-world data?

MAKE IT REAL

34. **Drug Use Among Eighth Graders** The data in the table show the percentage of eighth graders who admit to using the drug Ecstasy.

Years Since 1990 t	Drug Use (percent) D
6	3.4
7	3.2
8	2.7
9	2.7
10	5.2
11	4.3
12	3.2

Source: www.ojp.usdoj.gov

a. Make a scatter plot of these data.

b. Which type of polynomial function (quadratic, cubic, quartic) does the scatter plot best represent?

35. **Adjustable Rate Mortgage** In 2001, adjustable rate mortgages (ARMs) were offered at different rates at different times of the year. The table shows the average rate for each month in 2001.

Month t	Interest Rate (percent) P
1	6.69
2	6.49
3	6.35
4	6.24
5	6.17
6	6.08
7	6.07
8	5.89
9	5.74
10	5.48
11	5.38
12	5.55

Source: www.hsh.com

a. Make a scatter plot of these data and, using the idea of rate of change, explain why a quartic function best models the data.

b. Use regression to find the quartic model equation.

36. **U.S. Homicide Rate** The U.S. homicide rate in the 1990s is shown in the table.

Years Since 1990 t	Homicides (per 100,000 people) H
0	9.4
1	9.8
2	9.3
3	9.5
4	9
5	8.2
6	7.4
7	6.8
8	6.3
9	5.7
10	5.5

Source: www.infoplease.lycos.com

a. Make a scatter plot of these data and, using the idea of rate of change, explain why a cubic function best models the data.

b. Use regression to find the cubic model equation.

37. Average Hotel Room Rate The average hotel room rate is shown in the table.

Years Since 1990 t	Room Rate (dollars) R	Years Since 1990 t	Room Rate (dollars) R
0	57.96	5	66.65
1	58.08	6	70.93
2	58.91	7	75.31
3	60.53	8	78.62
4	62.86	9	81.33

Source: Statistical Abstract of the United States, 2001, Table 1266

a. Make a scatter plot of these data and, using the idea of rate of change, explain why a cubic function best models the data.

b. Use regression to find the cubic model equation.

38. Ford Employee Earnings The average hourly earnings for a Ford Motor Company employee increased throughout the 1990s. However, the rate of increase was not constant, as shown in the table.

Years Since 1990 t	Hourly Wage (dollars) W	Years Since 1990 t	Hourly Wage (dollars) W
1	19.10	7	22.95
2	19.92	8	24.30
3	20.94	9	25.58
4	21.81	10	26.73
5	21.79	11	27.38
6	22.30		

Source: Ford Motor Company 2001 Annual Report, p. 71

a. Make a scatter plot of these data and, using the idea of rate of change, explain why a quartic function best models the data.

b. Use regression to find the quartic model equation.

c. Describe the end behavior for the function model and discuss whether or not you think this end behavior will accurately predict future earnings.

39. Nonbusiness Bankruptcy Filings The number of nonbusiness Chapter 11 bankruptcies filed in the United States is given in the table.

Years Since 1998 t	Bankruptcies B
0	981
1	731
2	722
3	745
4	894
5	966
6	935

Source: Statistical Abstract of the United States, 2006, Table 749

a. Make a scatter plot of these data and, using the idea of rate of change, explain why a cubic function best models the data.

b. Use regression to determine the cubic model equation.

40. Mail Order and Supermarket Prescriptions The data show the number of mail order prescriptions filled as a function of the number of supermarket prescriptions filled for selected years.

Years Since 1995 t	Prescriptions Filled at Supermarkets (millions) s	Prescriptions Filled by Mail Order (millions) m
0	221	86
2	269	109
3	306	123
4	357	134
5	394	146
6	418	161
7	444	174
8	462	189
9	470	214

Source: Statistical Abstract of the United States, 2006, Table 126

a. Make a scatter plot of these data and, using the idea of rate of change, explain why a cubic function best models the data.

b. Use regression to find the cubic model equation.

c. Describe the relationship between the two quantities. Specifically, describe what the concave up and concave down portions of the graph tell us about the relationship between the quantities.

■ **STRETCH YOUR MIND**

Exercises 41–47 are intended to challenge your understanding of higher-order polynomial functions. In Exercises 41–45, describe the end behavior of the function when $a > 0$ and again for when $a < 0$.

41. $y = ax + b$

42. $y = ax^2 + bx + c$

43. $y = ax^3 + bx^2 + cx + d$

44. $y = ax^4 + bx^3 + cx^2 + dx + e$

45. $y = ax^5 + bx^4 + cx^3 + dx^2 + ex + f$

46. Write a general statement that describes the maximum number of x-intercepts for the general polynomial

$$y = a_n x^n + a_{n-1} x^{n-1} + a_{n-2} x^{n-2} + \cdots + a_1 x + a_0$$

47. Write a general statement that describes the end behavior of the general polynomial

$$y = a_n x^n + a_{n-1} x^{n-1} + a_{n-2} x^{n-2} + \cdots + a_1 x + a_0$$

SECTION 5.2 Power Functions

LEARNING OBJECTIVES

- Use power functions to model real-world situations
- Use the language of rate of change to describe the behavior of power functions
- Determine whether a power function represents direct or inverse variation

GETTING STARTED

Many species of animals migrate from one area to another for the purpose of feeding, giving birth, or escaping seasonal climate changes. For some species, migration occurs every season. For example, certain species of birds migrate south for winter and return north in the summer. For other species, migration happens just once in a lifetime. For example, Pacific salmon are born in fresh water, migrate to ocean waters, and then return to their freshwater birthplace to breed before dying. An article published in the *Journal of Zoology* revealed that the migration speed of animals that run, swim, and fly depends on the mass of the animal and can be modeled using *power functions*. (*Source:* Hedenstrom, A. (2003), "Scaling Migration Speed in Animals That Run, Swim and Fly," *Journal of Zoology 259*, 155–160)

In this section we use the real-world context of animal migration speed to learn about power functions. In addition to looking at the equations and graphs of power functions, we investigate the rates of changes of these functions. We also discuss direct and inverse variation.

■ Power Functions

Migrating land animals, such as caribou, are called *runners*. Runners migrate seasonally from one region to another. The migration speed of such animals, in kilometers per day, is dependent on the mass of the animal, given in kilograms. In Table 5.10, we see the migration speed for several different runners. The mass given is the median mass for each animal.

Table 5.10

Animal	Mass (kg) m	Migration Speed (km/day) V
Grey wolf	40	14.5
Gazelle	55	15.1
Dall sheep	140	16.8
Zebra	270	18.2
Wildebeest	275	18.3
Polar bear	500	19.6
Caribou	600	20.1
American buffalo	900	21.1
African elephant	5500	26.2

Source: Hedenstrom (2003)

To understand the relationship between the mass of the migrating land animal and its migration speed, let's consider the average rate of change between data points, shown in Table 5.11.

Table 5.11

Mass (kg) m	Migration Speed (km/day) V	Average Rate of Change (km/day/kg)
40	14.5	
		$\dfrac{0.6 \text{ km/day}}{15 \text{ kg}} = 0.04 \dfrac{\text{km per day}}{\text{kg}}$
55	15.1	
		$\dfrac{1.7 \text{ km/day}}{85 \text{ kg}} = 0.02 \dfrac{\text{km per day}}{\text{kg}}$
140	16.8	
		$\dfrac{1.4 \text{ km/day}}{130 \text{ kg}} = 0.01 \dfrac{\text{km per day}}{\text{kg}}$
270	18.2	
		$\dfrac{0.1 \text{ km/day}}{5 \text{ kg}} = 0.02 \dfrac{\text{km per day}}{\text{kg}}$
275	18.3	
		$\dfrac{1.3 \text{ km/day}}{225 \text{ kg}} = 0.006 \dfrac{\text{km per day}}{\text{kg}}$
500	19.6	
		$\dfrac{0.5 \text{ km/day}}{100 \text{ kg}} = 0.005 \dfrac{\text{km per day}}{\text{kg}}$
600	20.1	
		$\dfrac{1.0 \text{ km/day}}{300 \text{ kg}} = 0.003 \dfrac{\text{km per day}}{\text{km}}$
900	21.1	
		$\dfrac{5.1 \text{ km/day}}{4600 \text{ kg}} = 0.001 \dfrac{\text{km per day}}{\text{kg}}$
5500	26.2	

Observe that the rate of change is approaching zero. That is, as the mass of the animal increases, the number of additional kilometers per day that it travels when 1 additional kilogram is added to its mass approaches 0 kilometers. In other words, the migration speed initially increases quickly for every 1-kilogram increase in mass, but as the mass increases, the migration speed increases at an ever-decreasing rate.

The scatter plot in Figure 5.16 confirms our findings from examining the average rates of change. When the animal mass is small, even small changes in the mass of the animal result in relatively large increases in the migration speed. As the animal mass increases, the migration speed increases at a smaller rate.

Functions that demonstrate this behavior can often be modeled with a **power function**.

Figure 5.16

POWER FUNCTION

A function with equation of the form

$$y = ax^b$$

where a and b are constants, is called a **power function**.

One main difference between a power function and a polynomial function is that in a power function the exponent, b, can take on any real-number value rather than just positive integer values. Also, a power function is a single-term function whereas a polynomial function may have multiple terms.

EXAMPLE 1 ■ Using a Power Function in a Real-World Context

Create a power function to describe the relationship between the mass of a runner and its migration speed as shown in Table 5.12. Use this model to determine the migration speed of the bull moose, which typically has a mass of around 650 kilograms.

Table 5.12

Migration Speed of Runners		
Animal	Mass (kg) m	Migration Speed (km/day) V
Grey wolf	40	14.5
Gazelle	55	15.1
Dall sheep	140	16.8
Zebra	270	18.2
Wildebeest	275	18.3
Polar bear	500	19.6
Caribou	600	20.1
American buffalo	900	21.1
African elephant	5500	26.2

Source: Hedenstrom (2003)

Solution Using the Technology Tip at the end of this section, we find a power regression model

$$V(m) = 9.31m^{0.12}$$

where m is the mass of the animal in kilograms and V is the migration speed in kilometers per day. We use this model to find the migration speed of a bull moose with a mass of 650 kilograms.

$$V(m) = 9.31m^{0.12}$$
$$V(650) = 9.31(650^{0.12})$$
$$= 20.25$$

According to the model, the migration speed of a bull moose is 20.25 kilometers per day.

Let's check to see if the function $V(m) = 9.31m^{0.12}$ makes sense for the data showing the migration speed of runners. Recall that the scatter plot of the data was concave down, indicating that as the mass of the animal increases, the speed of the animal increases by a lesser amount. Table 5.13 shows that the function $V(m) = 9.31m^{0.12}$ is also concave down: As m increases, the value of $m^{0.12}$ increases by a lesser amount.

Table 5.13

m	$m^{0.12}$	Increase	m	$m^{0.12}$	Increase
0	0	1.213	30	1.504	0.028
5	1.213	0.105	35	1.532	0.025
10	1.318	0.066	40	1.557	0.022
15	1.384	0.049	45	1.579	0.020
20	1.433	0.039	50	1.599	
25	1.472	0.032			

Power Functions with $x \geq 0$ and $0 < b < 1$

In Example 1, we looked at the power function $V(m) = 9.31m^{0.12}$. Notice that the exponent b had the property that $0 < b < 1$ and that $m \geq 0$. Figure 5.17 illustrates the behavior of several power functions $y = ax^b$ with the characteristic that $x \geq 0$ and $0 < b < 1$.

We see that each power function contains the point $(1, a)$ since $1^b = 1$ for all b. Further, as b gets closer and closer to 1, the power function approaches the linear power function $y = ax^1$.

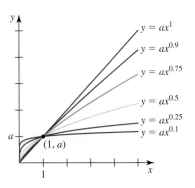

Figure 5.17

JUST IN TIME ■ SOLVING POWER FUNCTION EQUATIONS

To solve a power function equation $c = ax^b$ for x, apply the following steps.

General Procedure for $c = ax^b$

1. Divide each side by a.
$$\frac{c}{a} = x^b$$

2. Raise each side to $\frac{1}{b}$.
$$\left(\frac{c}{a}\right)^{1/b} = (x^b)^{1/b}$$

3. Simplify.
$$x = \left(\frac{c}{a}\right)^{1/b}$$

Specific Procedure for $58 = 27x^{0.75}$

1. $\dfrac{58}{27} = x^{0.75}$

2. $\left(\dfrac{58}{27}\right)^{1/0.75} = (x^{0.75})^{1/0.75}$

3. $x = \left(\dfrac{58}{27}\right)^{1/0.75} \approx 2.77$

When solving power function equations using the streamlined method shown here, it is important to remember the meaning of the rational exponent. For example, since $0.75 = \dfrac{3}{4}$, we have

$$x = \left(\frac{58}{27}\right)^{1/0.75}$$
$$= \left(\frac{58}{27}\right)^{\frac{1}{3/4}}$$
$$= \left(\frac{58}{27}\right)^{4/3}$$
$$= \sqrt[3]{\left(\frac{58}{27}\right)^4}$$

EXAMPLE 2 ■ Using a Power Function in a Real-World Context

Using the Technology Tip at the end of this section, we determine a model for the migration speed of a swimmer (such as a whale, salmon, or shark): $V(m) = 5.41m^{0.16}$, where m is the mass of the swimmer in kilograms. Given that the bottlenose dolphin has a migration speed of 13 kilometers per day, use the model to determine the mass of the dolphin.

Solution In this case, we know the value of V, the migration speed, and need to find the mass, m.

$$V(m) = 5.41m^{0.16}$$
$$13 = 5.41m^{0.16}$$
$$\frac{13}{5.41} = m^{0.16}$$
$$m = \left(\frac{13}{5.41}\right)^{1/0.16}$$
$$m \approx 239.7$$

According to the model, the bottlenose dolphin that has a migration speed of 13 kilometers per day has a mass of 239.7 kilograms (about 530 pounds).

JUST IN TIME ■ RATIONAL EXPONENTS AND RADICALS

Rational exponents can be expressed using radical notation. For example,

$$x^{1/2} = \sqrt{x}$$
$$x^{1/3} = \sqrt[3]{x}$$
$$x^{1/4} = \sqrt[4]{x}$$

If the exponent is not a unit fraction like these, we use properties of exponents to first rewrite the expression.

$$x^{2/3} = (x^2)^{1/3} = \sqrt[3]{x^2}$$
$$x^{0.35} = x^{35/100} = x^{7/20} = (x^7)^{1/20} = \sqrt[20]{x^7}$$

These radical expressions can also be written as

$$\sqrt[3]{x^2} = (\sqrt[3]{x})^2$$
$$\sqrt[20]{x^7} = (\sqrt[20]{x})^7$$

In general,

$$x^{m/n} = \sqrt[n]{x^m} = (\sqrt[n]{x})^m$$

EXAMPLE 3 ■ Using a Power Function in a Real-World Context

Birds that migrate are classified by how they fly: flapping flight or thermal-soaring flight. In this example, we focus on birds that use flapping flight.

a. Use the data in Table 5.14 to create a scatter plot of migration speed versus mass.
b. Describe the relationship between the variables using the language of rate of change.
c. Find a power function model for the data.

Table 5.14

Animal	Mass (kg) m	Migration Speed (km/day) V
Passerine	0.025	107.63
Arctic tern	0.110	81.22
Red knot shorebird	0.150	76.57
Duck	0.5	60.92
Canada goose	5	39.33
Tundra swan	7.26	36.64
Trumpeter swan	12.7	32.95

Source: Hedenstrom (2003)

Solution

a. Figure 5.18 shows the scatter plot of the data.

b. The migration speed decreases as the mass of the bird increases. The graph shows that when the mass is small, a small increase in mass results in a large decrease in migration speed. When the mass is larger, a similar increase in mass results in a significantly smaller decrease in migration speed.

c. We use the Technology Tip at the end of this section to find the model $s(m) = 53.4m^{-0.190}$ for the migration speed of flapping-flight migrators.

Figure 5.18

■ Power Functions with $x > 0$ and $b < 0$

Note that the power function model in Example 3 has a negative exponent. Figure 5.19 illustrates the behavior of such power functions with the characteristic that $x > 0$ and $b < 0$.

Each power function contains the point $(1, a)$. Also, the more negative the value of b, the more quickly the function values approach 0. Furthermore, as b approaches 0, the power function gets closer and closer to the linear power function $y = a$.

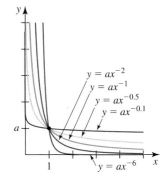

Figure 5.19

■ Direct and Inverse Variation

The power function $y = ax^b$ with $a > 0$ is an *increasing* function if $b > 0$ and a *decreasing* function if $b < 0$. For increasing power functions, the quantities x^b and y are said to vary *directly*. For decreasing power functions, the quantities x^b and y are said to vary *inversely*.

DIRECT VARIATION

A power function $y = ax^b$ with $b > 0$ represents direct variation. We say that "y varies directly with x^b" or "y is directly proportional to x^b." a is called the **constant of proportionality**.

A power function $y = ax^b$ with $b < 0$ may be rewritten as $y = ax^{-c}$ where c is a positive number equal to $|b|$. This alternate form is useful for interpreting inverse variation.

INVERSE VARIATION

A power function $y = ax^{-c} = \dfrac{a}{x^c}$ with $c > 0$ represents inverse variation. We say that "y varies inversely with x^c" or "y is inversely proportional to x^c." a is called the **constant of proportionality**.

Many real-world quantities are directly proportional to each other, as shown in Table 5.15. Some quantities that are inversely proportional to each other are shown in Table 5.16.

Table 5.15

Related Quantities	Formula	In Words
Cost of a fuel purchase and the amount of fuel bought	$C = kg$	The cost of a fuel purchase is directly proportional to the amount of fuel purchased (in gallons). k is the constant of proportionality and represents the fuel price per gallon.
Area of a circle and its radius	$A = \pi r^2$	The area of a circle is directly proportional to the square of its radius. π is the constant of proportionality.
Blood flow in an artery and the radius of the artery	$F = kr^4$	The rate at which blood flows in an artery (in mL per minute) is directly proportional to the fourth power of the radius of the artery. k is the constant of proportionality.

Table 5.16

Related Quantities	Formula	In Words
Average earnings per hour when paid a fixed amount of money to complete a task and hours worked	$A = \dfrac{k}{x}$	The average earnings per hour, A, is inversely proportional to the amount of hours worked, x. k is the constant of proportionality and represents the fixed amount of money paid for the job.
Length of a 4-cubic-foot box with equal height and width and box width	$L = \dfrac{4}{w^2}$	The length of a box, L, with equal height and width, w, is inversely proportional to the square of the width with constant of proportionality 4.

JUST IN TIME ■ NEGATIVE EXPONENTS

Recall the following property of negative exponents:

$$x^{-p} = \frac{1}{x^p}$$

We can make sense of this rule by investigating patterns.

$$x^4 = x \cdot x \cdot x \cdot x$$

$$x^3 = \frac{x \cdot x \cdot x \cdot x}{x} = x \cdot x \cdot x$$

$$x^2 = \frac{x \cdot x \cdot x}{x} = x \cdot x$$

$$x^1 = \frac{x \cdot x}{x} = x$$

For each decrease in 1 of the exponent, we remove one factor of x.

$$x^0 = \frac{x}{x} = 1$$

Continuing to decrease the exponent by 1 and continuing to divide by x produces the following pattern.

$$x^{-1} = \frac{1}{x}$$

$$x^{-2} = \frac{\frac{1}{x}}{x} = \frac{1}{x^2}$$

$$x^{-3} = \frac{\frac{1}{x^2}}{x} = \frac{1}{x^3}$$

EXAMPLE 4 ■ Modeling an Inverse Variation Relationship

Table 5.17 shows the top five countries to which the United States exports cotton.

a. By analyzing a scatter plot of the data and investigating rates of change, explain why a power function may best model the data.
b. Determine a power regression model to represent the data and use it to describe the relationship between the amount of cotton exports and the ranking of the country.
c. According to the model, how much cotton would the United States export to a country whose rank is 6?

Table 5.17

Ranking of Country r	Cotton Exports (in thousand metric tons) C
1. China	1234
2. Turkey	478
3. Mexico	329
4. Indonesia	231
5. Thailand	160

Source: Statistical Abstract of the United States, 2007, Table 825

Solution

a. Using Figure 5.20, we analyze the scatter plot of the data to determine if a power function might model the situation. We observe the rate of change is initially very dramatic but lessens as the country rank increases. This is characteristic of a power function describing an inverse variation relationship between the quantities.

b. We use the Technology Tip at the end of this section to calculate the power regression model for this situation, and graph the model in Figure 5.21. We find the power function $C(r) = 1208r^{-1.229}$ models the situation well. We can also express this model as $C(r) = \dfrac{1208}{r^{1.229}}$ and say that the amount of cotton exports, C, varies inversely (with a constant of proportionality of 1208) with the rank of the country raised to the power of 1.229.

Figure 5.20

Figure 5.21

c. To determine the amount of cotton exported to the sixth-ranked country, we substitute $r = 6$ into the model.

$$C(r) = \frac{1208}{r^{1.229}}$$

$$C(6) = \frac{1208}{6^{1.229}}$$

$$C(6) \approx 133.6$$

According to the model, the sixth-ranked country receives about 133.6 metric tons of cotton from the United States.

Inverses of Power Functions

Power functions that are strictly increasing or strictly decreasing will have an inverse function. The process for finding the inverse is identical for all such functions. We calculate the inverse of $f(x) = ax^n$:

$$y = ax^n$$

$$\frac{y}{a} = x^n$$

$$\left(\frac{y}{a}\right)^{1/n} = (x^n)^{1/n}$$

$$x = \left(\frac{y}{a}\right)^{1/n}$$

$$f^{-1}(y) = \left(\frac{y}{a}\right)^{1/n}$$

Although the term *inverse* is used when discussing *inverse variation* and *inverse function*, the two concepts are not synonymous.

SUMMARY

In this section you discovered how power functions behave and saw how to use a power function to model real-world situations. You also learned how to use the language of rate of change to describe a power function model. Finally, you learned that power functions can represent directly proportional or inversely proportional relationships.

TECHNOLOGY TIP ■ POWER REGRESSION

1. Press [2nd] then [0], scroll to **DiagnosticOn** and press [ENTER] twice. This will ensure that the correlation coefficient *r* and the coefficient of determination r^2 will appear.

2. Bring up the Statistics Menu by pressing the [STAT] button.

3. Bring up the List Editor by selecting **EDIT** and pressing [ENTER].

4. If there are data in the lists, clear the lists. Use the arrows to move the cursor to the list heading, **L1**, then press the [CLEAR] button and press [ENTER]. This clears all of the list data. Repeat for each list with data. (Warning: Be sure to use [CLEAR] instead of [DELETE]. [DELETE] removes the entire column.)

5. Enter the numeric values of the *inputs* in list **L1**, pressing [ENTER] after each entry.

6. Enter the numeric values of the *outputs* in list **L2**, pressing [ENTER] after each entry.

7. Return to the Statistics Menu by pressing the [STAT] button.

8. Bring up the Calculate Menu by using the arrows to select **CALC**. Use arrows to move down to **A:PwrReg**.

9. Calculate the power equation of the model by selecting **A:PwrReg** and pressing [ENTER] twice. The power regression model is $y = 1208.86x^{-1.22943}$ and has correlation coefficient $r = -0.9974$.

5.2 EXERCISES

SKILLS AND CONCEPTS

In Exercises 1–10, a power function is given in tabular or graphical form. Determine whether the power function represents direct or inverse variation and explain how you know.

1.

x	y
0	0
2	3.138
4	4.925
6	6.410
8	7.728
10	8.934
12	10.06
14	11.12

2.

3.

x	y
1	75.0
2	24.7
3	12.9
4	8.16
5	5.71
6	4.27

4.

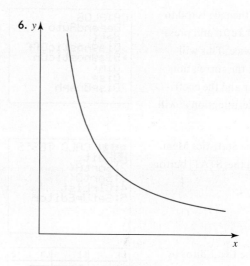

5.

x	y
0	0
1	0.25
2	1.6245
3	4.85476
4	10.5561
5	19.2823
6	31.5463
7	47.8305

6.

7.

x	y
5	2.55402
10	1.19149
15	0.762765
20	0.555851
25	0.434868
30	0.355843
35	0.300342
40	0.259314

8.

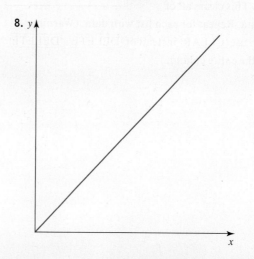

9.

x	y
2	4
4	16
6	36
8	64
10	100
12	144

10.

In Exercises 11–15, write a power function representing the verbal statement.

11. The area, A, of a square is directly proportional to the square of the length of one of its sides, s.
12. A quantity W is inversely proportional to the square root of a quantity n.
13. Your weekly earnings, W, are directly proportional to the number of hours you work, h, with a constant of proportionality of 7.95.
14. The volume of a cylinder with a height of 4 centimeters is directly proportional to the square of the radius of the base of the cylinder with a constant of proportionality of 4π.
15. The time, t, in hours, needed to drive a distance of 400 miles is inversely proportional to the average speed, s, in miles per hour.

In Exercises 16–25, solve the given equation.

16. $2x^{3/2} = 10$
17. $5g^{-2} = 25$
18. $4\sqrt[3]{t^5} = 50$
19. $1.5m^{0.65} = 25$
20. $\dfrac{5}{w^{2.5}} = 0.5$
21. $2\sqrt[4]{t^7} = 25$
22. $10x^{-2.5} = 80$
23. $4.25v^{0.38} = 15$
24. $\dfrac{1}{f^{1/2}} = 0.75$
25. $x^{2/3} = 16$

In Exercises 26–30, complete each table so that it accurately represents the given verbal description of a power function. Then write the equation of the function.

26. The value of y is inversely proportional to the square of the value of x with a constant of proportionality of 3.

x	1	2	3	4
y				

27. The value of y is inversely proportional to the square root of the value of x with a constant of proportionality of 1.

x	1	4	16	25
y				

28. The value of y is directly proportional to the cube of the value of x with a constant of proportionality of -2.

x	0	1	2	3
y				

29. The value of y is directly proportional to the value of x raised to the 0.25 power with a constant of proportionality of 5.

x	1	4	16	25
y				

30. The value of y is inversely proportional to the value of x raised to the 1.5 power with a constant of proportionality of 4.

x	0	1	2	3
y				

SHOW YOU KNOW

31. The graphs of two power functions are shown. One of the graphs represents the function $y = ax^n$ and the other represents the function $y = ax^{1/n}$ with n being an integer greater than 1. Explain which is which and how you can tell. Then find the coordinates of their point of intersection.

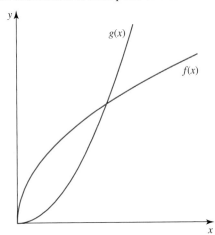

32. The graphs of two power functions are shown. One of the graphs represents the function $y = ax^{-n}$ and the

other represents the function $y = ax^{1/n}$ with n being an integer greater than 1. Explain which is which and how you can tell. Then find the coordinates of their point of intersection.

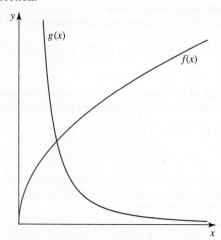

33. Based on the information given in the following graph of a power function, can you determine the equation of the power function? Why or why not? If you can, find the equation of the power function. If not, describe the equation of the power function as much as you can.

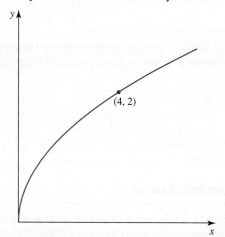

34. Based on the information given in the following graph of a power function, can you determine the equation of the power function? Why or why not? If you can, find the equation of the power function. If not, describe the equation of the power function as much as you can.

35. Match each function with its graph. Explain how you were able to make each choice.

a. $y = x^{-2}$
b. $y = x^2$
c. $y = x^{1/2}$
d. $y = x^{-1/2}$

Graph A

Graph B

Graph C

Graph D

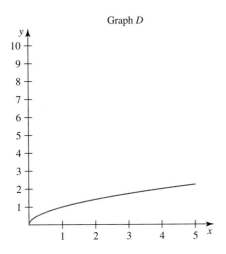

MAKE IT REAL

36. Running Speed The table shows the maximum relative running speed, in meters per second, of several different animals compared to their body mass in kilograms.

  a. Examine the data and explain why an inverse variation relationship might best model the situation.

  b. Use a graphing calculator to find a power function model. Then use the model to describe the relationship between maximum relative running speed and body mass of these animals.

  c. Use the power regression model to determine the mass of a zebra with a relative running speed of 8.16 meters per second.

Animal	Mass (kg) m	Relative Speed (m/s) V
Fallow deer	55	11.51
Desert warthog	85	12.38
Mountain goat	114	6.36
Antelope	227	6.65
Black wildebeest	300	10.97
Camel	550	3.03
European bison	865	5.4
Giraffe	1075	3.8
Black rhinoceros	1200	3.6
White rhinoceros	2000	1.79
Hippopotamus	3800	1.7
Asian elephant	4000	1.18
African elephant	6000	1.4

Source: Journal of Experimental Biology 205, 2897–2908

37. Nail Sizes The common nail is used for many home improvement and construction projects. Nails are made in a variety of lengths to meet the specific needs of different projects. According to www.sizes.com, the rule of thumb is to use nails that are three times the thickness of the board being fastened with the nail.

  a. This relationship between the length of the nail and the thickness of the board being fastened can be described using a power function. Will the power function be a direct or inverse variation? Explain.

  b. What is the constant of proportionality?

  c. Without using a calculator, write the power function that represents the rule of thumb described.

38. Nail Sizes The common nail comes in different lengths. In general, the longer the nail, the greater the diameter of the head of the nail. The table gives the length and diameter of a series of common nails.

Nail Length (inches) L	Nail Head Diameter (inches) D
1	0.0700
2	0.1130
3	0.1483
4	0.1920
5	0.2253
6	0.2625

Source: www.sizes.com

  a. Assuming the data will be modeled by a power function $D = aL^b$, which of the following inequalities will the parameter b satisfy: $b < 0$, $0 < b < 1$, or $b > 1$? Explain.

  b. Use a graphing calculator to find a power function to model the data. Then use the model to predict the nail head diameter of a nail with length 3.5 inches.

  c. If you found a nail with head diameter 0.2492 inches, what is the length of the nail, according to the model?

39. Blood Velocity with Stenosis A stenosis is an abnormal narrowing in a blood vessel. As the blood vessel narrows, blood flow is often affected. The following graph models blood velocity relative to blood vessel radius.

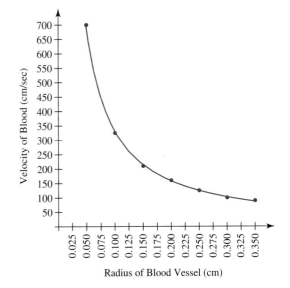

Source: Adapted from data at www.personal.engin.umich.edu

a. Explain why a power function model may be appropriate for this situation.

b. Will the power function model $y = ax^b$ have a value for b such that $b < 0$, $0 < b < 1$, or $b > 1$? Explain.

c. Use a graphing calculator to find a power function to model the relationship between the blood vessel radius and the velocity of blood.

d. What does the model suggest concerning blood flow as the blood vessel becomes increasingly large (less and less narrow)?

e. What does the model suggest concerning blood flow as the blood vessel becomes increasingly small (more and more narrow)?

40. **Wind Power** According to www.virtualsciencefair.org, the power generated by a windmill, in watts, is proportional (with a constant of proportionality of $\frac{1}{2}$) to the density of the air, the area swept out by the windmill rotor blade, and the cube of the velocity of the wind.

a. Write the given verbal description as a function where P is the power (W), D is the density of the air (kg/m³), A is the area swept out by the rotor blade (m²), and V is the velocity of the wind (m/s).

b. Suppose D and A are held constant. Describe the impact on P if the wind velocity, V, is doubled.

c. Suppose D and A are held constant. Describe the impact on P if the wind velocity, V, is tripled.

d. Suppose D and A are held constant. Describe the impact on P if the wind velocity, V, is cut in half.

41. **Terminal Velocity** Terminal velocity may be thought of as the maximum velocity obtained by a free-falling object. For example, when a skydiver first falls from the plane, she will free-fall faster and faster until terminal velocity is reached (unless the parachute is deployed first). Holding all other variables constant, terminal velocity is directly proportional to the square root of the mass (in kilograms) of the object falling.

a. Without using a calculator, write a power function for the terminal velocity, V, of a skydiver relative to the mass of the skydiver, m.

b. If the mass of the skydiver doubles, what will be the impact on the terminal velocity for that skydiver?

c. If the mass of the skydiver quadruples, what will be the impact on the terminal velocity for that skydiver?

d. If the mass of the skydiver is cut in half, what will be the impact on the terminal velocity for that skydiver?

42. **Hydroplaning Speeds** When airplanes land on a wet runway, *hydroplaning* may occur—the tires on the airplane slide across the top of the water on the runway rather than roll along the pavement. The pilot does not have complete control of the airplane while it is hydroplaning. The minimum speeds at which a landing airplane will hydroplane depend on the air pressure in the tires, as shown in the table.

Tire Pressure (psi) P	Minimum Speed (mph) S
25	45
50	63.64
75	77.942
100	90
125	100.62
150	110.23

Source: www.flightsafety.org

a. Calculate the rate of change between each pair of consecutive data points.

b. Do the rates of change indicate a power function may be appropriate? Explain.

c. Use a graphing calculator to find a power function that models the minimum speed as a function of the tire pressure.

d. If the air pressure of the tires is 110 psi, what is the minimum speed at which hydroplaning is predicted to occur?

e. If the pilot will be landing at 95 mph, what must be the air pressure in the tires so the plane will not hydroplane?

43. **Pendulum Swing** The time it takes for a pendulum to complete one period (swing forward then back to its initial position) depends on the length of the pendulum. Data collected from a pendulum experiment by students, showing the period (in seconds) and the length of the pendulum (in meters), are given in the table.

Length of Pendulum (meters) L	Period (seconds) P
0.145	0.7601
0.2305	0.9583
0.3056	1.1116
0.3815	1.2347
0.434	1.3339
0.483	1.3639
0.6	1.5532

Source: phoenix.phys.clemson.edu

a. Using rates of change, show that the function $P(L)$, the period of the pendulum relative to the length of the pendulum, can be modeled using a power function.

b. Use a graphing calculator to find a power function that models the data.

c. In the short story "The Pit and the Pendulum" by Edgar Allen Poe, a 30-foot (about 9.1-meter) pendulum swings, ultimately to the demise of the main character. What does the model predict as the period of this pendulum?

d. If a pendulum's period is 2 seconds, what is the length of the pendulum according to the model?

44. **Galileo Galilei** Galileo Galilei is attributed with discovering that the distance traveled by a falling object is directly proportional to the square of the time that the object has been in the air. Prior to this, people thought that the time depended also on the mass of the object. Galileo showed that objects of different masses dropped at the same moment hit the ground at the same time. More specifically, the distance traveled by a falling object is proportional to the square of the time elapsed with a constant of proportionality of 16.

a. Write the equation that models the distance traveled d (in feet) as a function of time t (in seconds).

b. Use a graphing calculator to graph $d(t)$.

c. Use the model to find the average rate of change of a falling object over the first 2 seconds, the next 2 seconds, and the next 2 seconds. What can you say about the average rate of change of a falling object?

45. **Light Intensity** The intensity of a beam of light can provide valuable information about the source of the light. For example, astronomers sometimes measure the distance to a star by measuring the brightness of the star. In a classroom activity, students collected data showing the luminous intensity of light (in milliwatts per square centimeter) as a function of the distance (in meters) away from a light sensor, using a flashlight as the source of light.

Distance (meters) d	Light Intensity (mw/cm²) L
0.5	0.462
0.6	0.341
0.7	0.280
0.8	0.203
0.9	0.161
1.0	0.164
1.1	0.111
1.2	0.097
1.3	0.091
1.4	0.091
1.5	0.080

a. Calculate the rates of change between consecutive data values. Do the rates of change indicate that a power function model may be appropriate for the data set?

b. Use a graphing calculator to find a power function that models the data.

c. For a light source 2 meters from the light sensor, use the model to predict the luminous intensity of the light.

d. If the luminous intensity of a light is known to be 0.25 mw/cm², how far from the light sensor is the light source?

46. **Cell Phone Plans** In 2007, T-Mobile offered the following My Faves™ for Individuals plan.

Minutes	Cost (dollars)
300	39.99
600	49.99
1000	59.99
1500	69.99

Source: www.t-mobile.com

a. Use a graphing calculator to find a power function to model the cost of the plan as a function of minutes.

b. According to the model, how many minutes would one get for $79.99?

■ **STRETCH YOUR MIND**

Exercises 47–51 are intended to challenge your understanding of power functions.

47. Graph several power functions of the form $y = ax^b$ and $b < 0$. Consider all values of x including $x < 0$. Write a description of the behavior of the resulting functions.

48. Graph several power functions of the form $y = ax^b$ and $0 < b < 1$. Consider all values of x including $x < 0$. Write a description of the behavior of the resulting functions.

49. Graph several power functions of the form $y = ax^b$ and $b > 1$. Consider all values of x including $x < 0$. Write a description of the behavior of the resulting functions.

50. Consider the following power functions: $f(x) = x^{0.25}$, $g(x) = x^{0.5}$, and $h(x) = x^{0.75}$. Write a convincing argument that shows that when $0 < x < 1$, $f(x) > g(x) > h(x)$ but when $x > 1$, $h(x) > g(x) > f(x)$.

51. Consider the following power functions: $f(x) = x^{-1}$, $g(x) = x^{-2}$, and $h(x) = x^{-4}$. Write a convincing argument that shows that when $0 < x < 1$, $f(x) < g(x) < h(x)$ but when $x > 1$, $h(x) < g(x) < f(x)$.

SECTION 5.3 Rational Functions

LEARNING OBJECTIVES

- Find and interpret the meaning of asymptotes in real-world applications
- Analyze rational function graphs and identify vertical and horizontal asymptotes as well as removable discontinuities
- Find the domain of rational functions
- Find the inverse of a rational function

GETTING STARTED

In 1989, the Exxon Valdez oil tanker struck a reef and spilled at least 11 million gallons of oil, polluting approximately 1180 miles of Alaskan coastline in Prince William Sound. This disastrous event has had far-reaching and long-lasting effects. As recently as 2006, an appeal in the 17-year-old federal court case was being convened to reconsider the punitive damages that ExxonMobil Corporation would be required to pay to the thousands of Alaskans whose lives were affected. (*Source:* www.answers.com/topic/exxon-valdez-oil-spill)

In this section, we investigate *rational functions* and see how real-world situations such as the cleanup costs of the Exxon Valdez oil spill can be modeled by such functions. Using the notion of rate of change, we look at vertical and horizontal asymptotes, removable discontinuities, vertical and horizontal intercepts, and domain.

■ Rational Functions

Recall that polynomial functions are defined as functions that have the form $f(x) = a_n x^n + a_{n-1} x^{n-1} + \cdots + a_1 x + a_0$, where n is a nonnegative integer. When one polynomial function is divided by another, a **rational function** is created.

> ### RATIONAL FUNCTION
> A **rational function** is a function of the form
> $$f(x) = \frac{p(x)}{q(x)}$$
> where $p(x)$ and $q(x)$ are polynomial functions with $q(x) \neq 0$.

Cleanup efforts initially included relatively simple and inexpensive techniques, but the techniques became more complex and costly as time progressed. Rational functions can be used to model a variety of real-world phenomena, such as the cleanup cost for the Exxon Valdez oil spill in 1989. Assume that by employing a variety of cleanup techniques, pollutants from an oil spill can be removed. The initial removal of oil at sea is far easier and less expensive than the final stages of cleanup. In fact, it may be quite costly, or even impossible, to remove all of the pollutants. Explore North estimates the total cleanup cost for the Valdez spill was over 2.1 billion dollars. (*Source:* www.explorenorth.com) Based on this estimate and the cleanup processes involved, we created a rational function model for the Valdez oil spill cleanup cost: $C(p) = \dfrac{0.08p}{100 - p}$, which models the cleanup cost, C (in billions of dollars), to remove p percent of the pollutants. (Note: Because this model is based on a single data point and an awareness of the underlying behavior of cleanup costs, its accuracy may be limited.)

SECTION 5.3 ■ RATIONAL FUNCTIONS

■ Vertical Asymptotes

Let's investigate the pollution cleanup function C by graphing it over a reasonable domain. Since p is the percentage of pollutants removed, anywhere from 0% to near 100% of the pollution could be removed. Therefore, we choose $0 \leq p < 100$ as the domain. The graph is shown in Figure 5.22.

From the graph we see that this function appears to approach the red dashed vertical line, which is called a **vertical asymptote**. This line is not part of the graph of the function, but defines a kind of *boundary* for the function.

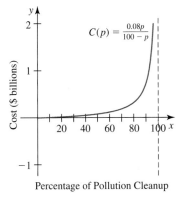

Figure 5.22

VERTICAL ASYMPTOTE

A **vertical asymptote** of a function $f(x)$ is a vertical line, $x = a$, that the graph of $f(x)$ approaches but does not cross. More formally, as x approaches a, $f(x)$ approaches $\pm\infty$. Symbolically, we write this as $x \to a, f(x) \to \pm\infty$.

Even though vertical asymptotes are not part of the graph of a rational function, they are often drawn because they are helpful in describing how the function behaves.

EXAMPLE 1 ■ Exploring Vertical Asymptotes in a Real-World Context

The cost of cleaning up p percent of the pollutants from an oil spill can be modeled by $C(p) = \dfrac{0.08p}{100 - p}$, where C is the cost in billions of dollars.

a. Create a table of values for C and ΔC in terms of p from $p = 0$ to $p = 100$ in increments of 25 percent.

b. Referring to the values of ΔC from part (a), describe what is happening to the cleanup cost as each additional 25 percent of pollutants is removed.

c. Explain why it is impossible to use the model to calculate the cost of removing 100 percent of the pollution.

d. Determine the equation of the vertical asymptote for $C(p)$ and justify why it is a vertical asymptote.

Solution

a. To create Table 5.18, we evaluate $C(p)$ for each of the values of p from 25 up to 100 percent. For instance, $C(25) = \dfrac{(0.08 \cdot 25)}{(100 - 25)} = \dfrac{2}{75} = 0.02667$. We cannot evaluate $C(100)$ because

$$C(100) = \dfrac{(0.08 \cdot 100)}{(100 - 100)}$$
$$= \dfrac{8}{0} \qquad \text{\textbf{Division by 0 is undefined.}}$$
$$= \text{undefined}$$

To fill in the ΔC column, we simply calculate the first differences.

Table 5.18

Percent of Pollution Removed p	Cost ($ billions) C	Change in the Cleanup Cost ($ billions) ΔC
0	0	0.02667
25	0.02667	0.05333
50	0.08000	0.16000
75	0.24000	unknown
100	undefined	

b. The rate of change shown in Table 5.18 tells us that the costs are continually increasing and at an increasing rate. We can see this as well on the graph shown in Figure 5.23. Notice the slope keeps getting steeper. The slopes of the tangent lines (red dashed lines) show us that the rate of change becomes extremely large near $p = 100$.

c. Recall that when we tried to calculate the cost of removing 100 percent of the pollutants, we got $C(100) =$ undefined. Consequently, we cannot determine the cost of removing 100 percent of the pollutants with the model.

d. Since $C(p)$ is undefined at $p = 100$, $p = 100$ may be a vertical asymptote. We simply need to confirm that as $p \to 100$, $C(p) \to \infty$. To do this, we create a table of values for $C(p)$ and for the rate of change of $C(p)$ near $p = 100$. See Table 5.19.

$$C(p) = \frac{0.08p}{100 - p}$$

Percentage of Pollution Cleanup

Figure 5.23

Table 5.19

Percent of Pollution Removed P	Cost ($ billions) C	Rate of Change ($ billions per percent) $\dfrac{\Delta C}{\Delta p}$
93	1.06	0.19
94	1.25	0.27
95	1.52	0.40
96	1.92	0.67
97	2.59	1.33
98	3.92	4.00
99	7.92	cannot calculate
100	undefined	

Notice how rapidly the cost is increasing as p nears 100. Increasing the amount of pollution removed from 97 percent to 98 percent would cost 1.33 billion dollars.

Increasing the amount of pollution removed from 98 percent to 99 percent would cost an additional 4.00 billion dollars. We conclude that as $p \to 100$, $C(p) \to \infty$. Therefore, $p = 100$ is the vertical asymptote, shown as the red dashed vertical line in Figure 5.24. In the real-world context of oil cleanup, the cost to clean up a spill is relatively low at first but skyrockets as we get closer and closer to cleaning up 100 percent of the spill.

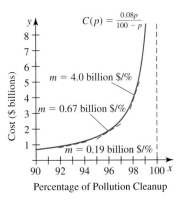

Figure 5.24

HOW TO: ■ FIND VERTICAL ASYMPTOTES

To find all vertical asymptotes of $f(x) = \dfrac{p(x)}{q(x)}$,

1. Factor the numerator $p(x)$ and denominator $q(x)$.
2. Cancel out any factors that the numerator and denominator have in common. (This puts the rational function in *simplified form*.)
3. Set the resultant denominator equal to zero and solve. The solutions to this equation are the vertical asymptotes of $f(x) = \dfrac{p(x)}{q(x)}$.

EXAMPLE 2 ■ Finding Vertical Asymptotes from a Function Equation

Determine the vertical asymptotes of the graph of $y = \dfrac{2x(x + 4)(x - 2)}{(x - 1)(x + 5)(x - 5)}$. Then explain the behavior of the function near its vertical asymptotes.

Solution Since the numerator and denominator do not have any factors in common, we simply need to determine the x-values that make the denominator equal 0: $x = 1$, $x = -5$, and $x = 5$. Consequently, vertical asymptotes occur at $x = 1$, $x = -5$, and $x = 5$. We confirm our conclusions with the graph in Figure 5.25. The asymptotes are indicated by the red dashed vertical lines.

Observe that as $x \to -5$ from the left, $y \to \infty$ and as $x \to -5$ from the right, $y \to -\infty$. As $x \to 1$ from the left, $y \to -\infty$ and as $x \to 1$ from the right, $y \to \infty$. As $x \to 5$ from the left, $y \to -\infty$ and as $x \to 5$ from the right, $y \to \infty$.

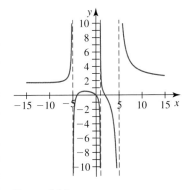

Figure 5.25

Removable Discontinuities (Holes)

To see what happens graphically if the numerator and denominator have a nonconstant factor in common, let's consider the rational function $f(x) = \dfrac{x^2 - x - 2}{x - 2}$. This function is undefined at $x = 2$ because the denominator equals zero when $x = 2$:

$$f(2) = \frac{(2)^2 - 2 - 2}{2 - 2}$$
$$= \frac{0}{0}$$
$$= \text{undefined}$$

However, the graph of f shown in Figure 5.26 appears linear, even at $x = 2$. To check this, we factor the numerator of f and simplify the function:

$$f(x) = \frac{(x - 2)(x + 1)}{x - 2}$$
$$= \left(\frac{x - 2}{x - 2}\right)(x + 1)$$
$$= 1 \cdot (x + 1) \text{ for } x \neq 2$$
$$= x + 1$$

Thus the graph of f is the line $y = x + 1$ except at $x = 2$, where f is undefined. That is, although the line $y = x + 1$ contains the point $(2, 3)$, the graph of f does not. We say the graph of f has a **removable discontinuity** (a hole) in it at the point $(2, 3)$. We redraw the graph (Figure 5.27) to show this.

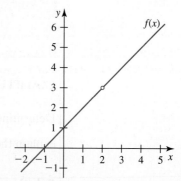

Figure 5.26   Figure 5.27

HOW TO: ■ FIND A REMOVABLE DISCONTINUITY (HOLE)

To find all removable discontinuities (holes) of $f(x) = \dfrac{p(x)}{q(x)}$,

1. Factor the numerator $p(x)$ and denominator $q(x)$.
2. Determine which factors the numerator and denominator have in common.
3. Set each of the common factors equal to zero and solve. The solutions to these equations are the removable discontinuities of $f(x) = \dfrac{p(x)}{q(x)}$.

■ Horizontal Asymptotes

A rational function may have a *horizontal asymptote* as well as vertical asymptotes. The *horizontal asymptote* of a rational function is a horizontal line, $y = b$, that the function approaches as the independent variable approaches $-\infty$ or ∞. To find horizontal asymptotes, we need to know how the output values of the function behave as the input values approach $\pm\infty$.

EXAMPLE 3 ■ Exploring Horizontal Asymptotes in a Real-World Context

When a new drug is first marketed, it is usually under a patent that restricts drug sales to the pharmaceutical company that developed it. This gives the company a chance to recoup its development costs, which average around $800,000,000 for research, testing, and equipment. (*Source:* www.wikipedia.org)

Suppose the total cost C, in dollars, of producing g grams of a new drug is given by the linear function $C(g) = 800{,}000{,}000 + 10g$, where $10g$ represents the additional costs associated with manufacturing each gram of the drug.

a. Evaluate $C(0)$ and explain what the numerical answer means in its real-world context.

b. Find the slope of $C(g)$ and explain what the numerical answer means in its real-world context.

c. Form the rational function $A(g) = \dfrac{C(g)}{g}$ and explain what $A(g)$ represents in its real-world context.

d. Discuss the rate of change of $A(g)$ as g increases and explain what this tells about the cost of the new drug.

e. Complete Table 5.20 for the given values of g. Then describe what happens to $A(g)$ as g approaches ∞. Explain what this means in terms of the real-world context.

Table 5.20

Grams of New Drug g	Average Cost (dollars per gram) $A(g) = \dfrac{C(g)}{g}$
0	
100	
1000	
10,000	
100,000	
1,000,000	
10,000,000	
100,000,000	
1,000,000,000	
10,000,000,000	
100,000,000,000	

f. Determine the horizontal asymptote for $A(g)$.

Solution

a. $C(0) = 800{,}000{,}000$. This means that the cost of producing 0 grams of the new drug is \$800,000,000. The \$800,000,000 is the initial or fixed cost of development that must be recouped (or lost) by the company.

b. The slope of $C(g)$ is 10 dollars per gram. That is, the drug costs 10 dollars per gram to produce.

c. We have

$$A(g) = \frac{C(g)}{g} \frac{\text{dollars}}{\text{grams}}$$

$$= \frac{800{,}000{,}000 + 10g}{g} \text{ dollars per gram}$$

The function $A(g)$ gives the average cost per gram of producing a total of g grams of the drug.

d. We create Table 5.21 and the graph in Figure 5.28 to show the rate of change of $A(g)$. As the number of grams produced increases, the average cost decreases but at a lesser and lesser rate.

Table 5.21

Grams of New Drug g	Average Cost (dollars per gram) $A(g) = \dfrac{C(g)}{g}$	Rate of Change $\dfrac{\Delta A}{\Delta g}$
0	undefined	cannot calculate
10	80,000,010	$\dfrac{-\$4{,}000{,}000}{1 \text{ gram}}$
20	40,000,010	$\dfrac{-\$1{,}333{,}333.33}{1 \text{ gram}}$
30	26,666,676.67	$\dfrac{-\$666{,}666.67}{1 \text{ gram}}$
40	20,000,010	$\dfrac{-\$400{,}000}{1 \text{ gram}}$
50	16,000,010	

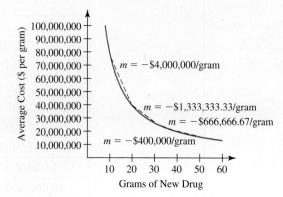

Figure 5.28

e. The completed table is shown in Table 5.22. As $g \to \infty$, $A(g) \to 10$. This means as the number of grams produced becomes very large, the average cost per gram approaches \$10.

SECTION 5.3 ■ RATIONAL FUNCTIONS

Table 5.22

Grams of New Drug g	Average Cost (dollars per gram) $A(g) = \dfrac{C(g)}{g}$
0	undefined
100	$8,000,010
1000	$800,010
10,000	$80,010
100,000	$8010
1,000,000	$810
10,000,000	$90
100,000,000	$18
1,000,000,000	$10.80
10,000,000,000	$10.08
100,000,000,000	≈ $10.00

f. Since $A(g) \to 10$ as $g \to \infty$, $y = 10$ is the horizontal asymptote for $A(g)$. We indicate this with a red dashed horizontal line on the graph shown in Figure 5.29.

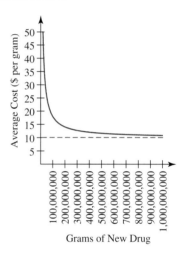

Figure 5.29

We now formally define the term **horizontal asymptote**.

HORIZONTAL ASYMPTOTE

A **horizontal asymptote** of a function f is a horizontal line that the graph of f approaches as x approaches positive or negative infinity. More formally, a horizontal asymptote occurs at $y = b$ if and only if the graph of f approaches the line $y = b$ as x approaches either ∞ or $-\infty$.

The graph of a rational function never crosses a vertical asymptote. However, the graphs of some rational functions do cross their horizontal asymptotes. The difference is that a vertical asymptote occurs where the function is undefined, whereas a horizontal asymptote represents a *limiting value* of the function as $x \to \pm\infty$. There is no reason the function cannot take on this limiting value for some finite x-value. For example, the graph of

$$h(x) = \frac{2x^2 + 3x - 2}{x^2}$$

(Figure 5.30) crosses the line $y = 2$, its horizontal asymptote; however, the graph does not cross its vertical asymptote, the y-axis.

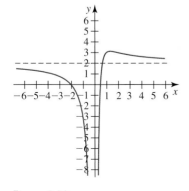

Figure 5.30

Another example is the graph shown in Figure 5.31. Note the graph moves back and forth (oscillates) over the horizontal line $y = 3$ so f crosses this line many times, but as $x \to \infty$, $f(x) \to 3$. We can see this from Table 5.23 as well. A similar type of situation can also occur for a function that approaches a limiting value as $x \to -\infty$, as shown in Figure 5.32.

Figure 5.31

Table 5.23

x	y	x	y
0	4	3.5	2.969
0.5	2.393	4	3.018
1	3.367	4.5	2.988
1.5	2.776	5	3.006
2	3.135	5.5	2.995
2.5	2.917	6	3.002
3	3.049		

Figure 5.32

■ Finding Horizontal Asymptotes

Recall that the *leading coefficient* of a polynomial is the coefficient on the term with the largest exponent. Leading coefficients and the degrees of the polynomials in the numerator and denominator play a key role in determining the location of horizontal asymptotes on a rational function graph.

Due to the end behavior of polynominals, as the magnitude of x becomes increasingly large, the graph of $f(x) = \dfrac{ax^n + \cdots}{bx^m + \cdots}$ will be more and more influenced by the leading terms. In fact, for values of x near $\pm\infty$, $f(x) \approx \dfrac{ax^n}{bx^m}$. Consider the following three functions:

$$f(x) = \frac{3x^2 + 2x + 1}{4x^3 - 5}$$

$$g(x) = \frac{3x^2 + 2x + 1}{4x^2 - 5}$$

$$h(x) = \frac{3x^2 + 2x + 1}{4x - 5}$$

For large values of x, the value of each function may be approximated by the ratio of the leading terms.

$$f(x) \approx \frac{3x^2}{4x^3} \qquad g(x) \approx \frac{3x^2}{4x^2} \qquad h(x) \approx \frac{3x^2}{4x}$$
$$\approx \frac{3}{4x} \qquad \approx \frac{3}{4} \qquad \approx \frac{3x}{4}$$

Observe that as $x \to \infty$, $f(x) \to 0$, $g(x) \to \dfrac{3}{4}$, and $h(x) \to \infty$. Consequently, f has a horizontal asymptote at $y = 0$, g has a horizontal asymptote at $y = \dfrac{3}{4}$, and h does not have a horizontal asymptote.

HOW TO: ■ FIND HORIZONTAL ASYMPTOTES

For a rational function $f(x) = \dfrac{p(x)}{q(x)} = \dfrac{ax^n + \cdots}{bx^m + \cdots}$, where a is the leading coefficient of the numerator and b is the leading coefficient of the denominator,

- If $n < m$ (i.e., if the degree of the numerator is less than that of the denominator), a horizontal asymptote occurs at $y = 0$.
- If $n = m$ (i.e., if the degree of the numerator is equal to that of the denominator), a horizontal asymptote occurs at $y = \dfrac{a}{b}$.
- If $n > m$ (i.e., if the degree of the numerator is greater than that of the denominator), the function does not have a horizontal asymptote.

EXAMPLE 4 ■ Finding a Horizontal Asymptote from an Equation

The ratio of the surface area of any cube to its volume is given by $r(s) = \dfrac{6s^2}{s^3}$, where s is the length of any side. Determine the horizontal asymptote of the function and interpret what it means in its real-world context.

Solution The degree of the numerator is 2 and the degree of the denominator is 3. Since $2 < 3$, this function has a horizontal asymptote at $r = 0$. Increasing the length of the side of any cube decreases its area to volume ratio. For large values of s, this ratio is near 0.

Recall that the x-intercepts of a function occur where $y = 0$ and the y-intercepts occur where $x = 0$. If we know the x-intercepts, y-intercepts, horizontal asymptote, and vertical asymptotes of a rational function, it is relatively easy to determine the basic shape of its graph by calculating just a few additional points.

EXAMPLE 5 ■ Drawing a Rational Function Graph by Hand

Draw the graph of $f(x) = \dfrac{3x^2 - 15x + 18}{2x^2 - 18}$.

Solution

- horizontal asymptote:

    Since the degrees of the numerator and denominator are equal ($m = n = 2$), the graph has a horizontal asymptote at $y = \dfrac{3}{2}$.

- y-intercept:

    $f(0) = \dfrac{18}{-18} = -1$ so there is a y-intercept at $(0, -1)$.

To find the vertical asymptotes and x-intercept, it is helpful to write the function in factored form.

$$f(x) = \frac{3x^2 - 15x + 18}{2x^2 - 18}$$

$$= \frac{3(x-2)(x-3)}{2(x+3)(x-3)}$$

Observe that the factor $x - 3$ occurs in both the numerator and denominator. As long as $x \neq 3$, $\frac{x-3}{x-3} = 1$. Therefore, we can rewrite the function in simplified form as

$$f(x) = \frac{3(x-2)}{2(x+3)}, \quad x \neq 3$$

We see that a removable discontinuity (hole) in the graph occurs at $x = 3$.

We now proceed to finding the x-intercept and vertical asymptotes.

- x-intercepts:

    x-intercepts occur at values of x that make the numerator of the simplified rational function equal to zero. So an x-intercept occurs at $(2, 0)$.

- vertical asymptotes:

    Vertical asymptotes occur at values of x that make the denominator of the simplified rational function equal to zero. So a vertical asymptote occurs at $x = -3$.

In Table 5.24 we evaluate the function at a few other key points. We plot the points and then connect them with a smooth curve. At $x = 3$, we place an open dot on the curve to indicate that the function is discontinuous there, as shown in Figure 5.33.

Table 5.24

x	$f(x) = \dfrac{3(x-2)}{2(x+3)},\ x \neq 3$
-10	2.57
-4	9.00
-2	-6.00
10	0.92

Figure 5.33

EXAMPLE 6 ■ Finding the Domain of a Rational Function

Find the domain of each of the following rational functions.

a. $f(x) = \dfrac{3x^2 - 5}{x + 4}$

b. $g(x) = \dfrac{2}{x^2 - 25}$

c. $h(x) = \dfrac{2x^4}{x^2 + 1}$

Solution

a. The domain of the rational function $f(x) = \dfrac{3x^2 - 5}{x + 4}$ consists of all real values of x except $x = -4$ since this value makes the denominator equal to zero.

b. The domain of the rational function $g(x) = \dfrac{2}{x^2 - 25}$ consists of all real numbers x except $x = 5$ and $x = -5$ since these values make the denominator equal to zero.

c. The domain of the rational function $h(x) = \dfrac{2x^4}{x^2 + 1}$ consists of all real numbers x. There is no value of x that makes the denominator equal to zero.

■ Inverses of Rational Functions

A rational function that is strictly increasing or strictly decreasing will have an inverse function. One that changes from increasing to decreasing or vice versa will not. Often the process of finding the inverse of a rational function is algebraically complex.

EXAMPLE 7 ■ Finding the Inverse of a Rational Function

Find the inverse of $f(x) = \dfrac{x + 2}{3x}$.

Solution

$$y = \frac{x + 2}{3x}$$
$$3xy = x + 2$$
$$3xy - x = 2$$
$$x(3y - 1) = 2$$
$$x = \frac{2}{3y - 1}$$
$$f^{-1}(y) = \frac{2}{3y - 1}$$

SUMMARY

In this section you learned about rational functions and their graphs. You discovered how to determine the location of vertical and horizontal asymptotes as well as removable discontinuities (holes). You also modeled real-world contexts with rational functions. Finally, you learned how to determine the domain and the inverse of a rational function.

5.3 EXERCISES

■ SKILLS AND CONCEPTS

In Exercises 1–4, find the horizontal and vertical asymptotes, if any, for each function.

1. $f(x) = \dfrac{2x}{x + 4}$

2. $h(x) = \dfrac{3}{2x - 6}$

3. $g(x) = \dfrac{(1 - x)(2 + 3x)}{2x^2 + 1}$

4. $k(x) = \dfrac{2x - 2}{2x^2 + x - 3}$

In Exercises 5–8, what are the x-intercepts, y-intercepts, and horizontal and vertical asymptotes (if any)?

5. $f(x) = \dfrac{x - 3}{x - 5}$

6. $g(x) = \dfrac{x^2 - 16}{x^2 + 16}$

7. $k(x) = \dfrac{x^2 - 9}{x^3 + 9x^2}$

8. $m(x) = \dfrac{x(2 - x)}{(x^2 - 10x + 12)}$

9. Let $f(x) = \dfrac{1}{x-2}$.

   a. Complete the table for x-values close to 2. What happens to the values of $f(x)$ as $x \to 2$ from the right and from the left?

x	f(x)
1.00	
1.90	
1.99	
2.00	
2.01	
2.10	
3.00	

   b. Complete the following tables. Use the tables of values to determine what happens to the values of $f(x)$ as $x \to \infty$ and as $x \to -\infty$.

x	f(x)
5	
50	
500	
5000	
50,000	

x	f(x)
−5	
−50	
−500	
−5000	
−50,000	

   c. Without using a calculator, graph $f(x)$. Give the equations for the vertical and horizontal asymptotes.

10. Let $h(x) = \dfrac{1}{(x+3)^2}$.

    a. Complete the table for x-values close to −3. What happens to the values of $h(x)$ as x approaches 3 from the left and from the right?

x	h(x)
−4.00	
−3.10	
−3.01	
−3.00	
−2.99	
−2.90	
−2.00	

    b. Complete the following tables. Use the tables of values to determine what happens to the values of $h(x)$ as $x \to \infty$ and as $x \to -\infty$.

x	h(x)
5	
50	
500	
5000	
50,000	

x	h(x)
−5	
−50	
−500	
−5000	
−50,000	

    c. Without using a calculator, graph $h(x)$. Give the equations for the vertical and horizontal asymptotes.

11. Use the graph of $h(x)$ to describe the following transformations.

    a. $y = -h(x) - 3$
    b. $y = h(x-2) + 4$
    c. $y = -2h(x+3) - 5$

Exercises 12–13 show a transformation of $y = \dfrac{2}{x}$.

   a. Find a possible formula for the graph.
   b. Give the coordinates of any x- and y-intercepts.
   c. Find the equations for the vertical and horizontal asymptotes.

12.

14.

13.

15.

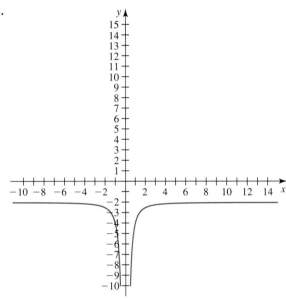

Exercises 14–15 show a transformation of $y = \dfrac{2}{x^2}$.

  a. Find a possible formula for the graph.
  b. Give the coordinates of any x- and y-intercepts.
  c. Find the equations for the vertical and horizontal asymptotes.

16. For the following table, give the coordinates of the x-intercept and the equation of the vertical asymptote.

x	y
3.0	3.0
3.1	2.8
3.2	2.5
3.3	2.1
3.4	1.7
3.5	1.0
3.6	0
3.7	−1.7
3.8	−5.0
3.9	−15.0
4.0	undefined
4.1	25
4.2	15

17. For the following table, give the coordinates of the y-intercept and the equation of the horizontal asymptote.

x	y
−6.0	−3.000
0.0	0.000
6.0	0.600
12.0	0.750
16.0	0.800
21.0	0.840
36.0	0.900
46.0	0.920
96.0	0.960
196.0	0.980
396.0	0.990
1096.0	0.996
4096.0	0.999

In Exercises 18–19, sketch a possible graph for each function described.

18. The function f has one vertical asymptote at $x = -2$ and a horizontal asymptote at $y = 2$. The graph has a horizontal intercept at $(4, 0)$ and a vertical intercept at $(0, -6)$. The graph is concave down for $-2 < x < 4$ and has an inflection point at $(6, 1)$.

19. The function g has two vertical asymptotes, one at $x = -1$ and one at $x = 5$. There is a horizontal asymptote at $y = 4$ as $x \to -\infty$ and at $y = -6$ as $x \to \infty$. The graph is concave down for $-1 < x < 5$ and has an x-intercept at $(1, 0)$. Finally, the graph is concave up for $x < -1$ and concave down for $x > 5$.

For Exercises 20–21, create a table of values that satisfies each condition for the functions described.

20. The function f has an x-intercept at 4, a y-intercept at -2, is concave down for $x > 0$, and has a horizontal asymptote at $y = 1$.

21. The function g has a vertical asymptote at $x = -10$, is concave up for $x > -10$, and has an x-intercept at $x = -8$.

In Exercises 22–23, write the formula for a rational function that satisfies each condition for the functions described.

22. The function g has two vertical asymptotes, the line $x = 6$ and the line $x = -2$.

23. A rational function has only one vertical asymptote and no positive function values.

In Exercises 24–27, graph the function and label all of the important features including any x- and y-intercepts, vertical and horizontal asymptotes, and removable discontinuities (holes).

24. $f(x) = \dfrac{2x + 1}{x - 3}$

25. $f(x) = \dfrac{4 - 3x}{2x + 1}$

26. $f(x) = \dfrac{3x^2 - 3x - 6}{x^2 + 8x + 16}$

27. $f(x) = \dfrac{x^2 - 2x - 3}{2x^2 - x - 10}$

28. A function f is defined by the following graph. Which of the following describes the behavior of f?

   A. As the value of x approaches 0, the value of f increases.

   B. As the value of x increases, the value of f approaches 0.

   C. As the value of x approaches 0, the value of f approaches 0.

29. Which of the following best describes the behavior of the function f defined by $f(x) = \dfrac{x^2}{x - 2}$?

   A. As the value of x gets very large, the value of f approaches 2.

   B. As the value of x gets very large, the value of f increases.

   C. As the value of x approaches 2, the value of f approaches 0.

■ SHOW YOU KNOW

30. What is the definition of a rational function?

31. Use tables, symbols, graphs, and words to explain the behavior of a function near a vertical asymptote.

32. Use tables, symbols, graphs, and words to explain the behavior of a function near a horizontal asymptote.

33. What role do asymptotes play in rational functions? How do we determine if and where they exist?

34. What is meant by the term *removable discontinuity*? Under what conditions will one exist?

35. Describe a real-world interpretation of a vertical and horizontal asymptote.

■ MAKE IT REAL

36. **Life Expectancies in the United States** The graphs of M and W illustrate the life expectancies for men, M, and women, W, in the United States based on their current age a.

[Graph: W(a) and M(a) showing Expected Years to Live vs Current Age in Years]

Source: Annuity Advantage Life Expectancy Tables, www.annuityadvantage.com/lifeexpectancy.htm

a. Estimate $M(0)$ and $W(0)$ and explain what each answer means in this context.

b. Estimate $M(20)$ and $W(20)$ and explain what each answer means in this context.

c. Estimate the limiting value for each function.

d. Using the language of rate of change, explain what the horizontal asymptote of each function means in terms of men's and women's life expectancy.

37. **Travel Time** The speed limit on most interstate highways in Arizona through rural areas is 75 mph. (*Source:* www.dot.state.az.us/highways/Traffic/Speed.asp)

    a. If a driver maintains an average speed of 75 mph, how long will it take to make a 150-mile journey?

    b. Write an equation that defines travel time, t, as a function of the average speed, s.

    c. As the average speed increases, what happens to the travel time? Explain what this means in this real-world context.

    d. When there are fatal accidents, highway patrol officers slow or stop the traffic on the interstate. Create a table for the time it takes to travel 150 miles when traveling at $s = 20, 15, 10, 5, 4, 3, 2,$ and 1 miles per hour.

    e. Taking into account your answers to parts (a)–(d), sketch a graph of $t(s)$ on the interval $0 < s < 100$.

    f. Find the equations for the horizontal and vertical asymptotes and explain what these asymptotes mean in this context.

38. **Depreciation** A new car depreciates in value very quickly right after it is driven off the lot. Suppose a new car has the initial value of C dollars, and the value of the car as scrap metal is T dollars. If the life of the car is N years, then the average amount, D, by which the car depreciates in value each year is given by the multivariable function $D(C, T, N) = \dfrac{C - T}{N}$.

a. If the purchase price of the car is $34,000 and the value of the scrap metal is $2000, write a formula for D as a function of N.

b. Using your answer to part (a), complete the following table of values for D.

N	D
1	
3	
5	
7	
9	
11	
13	
15	

c. Give the practical domain of $D(N)$.

d. Sketch a graph of $D(N)$.

e. Explain what the vertical and horizontal asymptotes of $D(N)$ mean in terms of the car's value.

39. **Sound Intensity** The loudness (or intensity) of any sound is a function of the listener's distance from the source of the sound. In general, the relationship between the intensity, I, in decibels (dB) and the distance, d, in feet can be modeled by the function $I(d) = \dfrac{k}{d^2}$. The constant k is determined by the source of the sound and the surroundings. A reasonable value for the human voice is $k = 1495$. (*Source:* www.glenbrook.k12.il.us)

a. Complete the following table for $I(d)$.

d	I
0.1	
0.5	
1	
2	
5	
10	
20	
30	

b. Give a practical domain for $I(d)$.

c. Sketch a graph of $I(d)$.

d. As you move closer to the person speaking, what happens to the intensity of the sound? Be sure to include the idea of rate of change in your explanation.

40. Body Mass Index Doctors, physical therapists, and fitness specialists often use a person's body mass index to determine whether the person should lose weight. The formula for the body mass index, B, is the multivariable function $B(w, h) = \dfrac{705w}{h^2}$, where w is weight (in pounds) and h is height (in inches). (*Source:* Centers for Disease Control; www.cdc.gov/nccdphp/dnpa/bmi/index.htm)

a. Give the formula for a 190-pound person's body mass index.

b. Complete the table for $B(h)$ for a 190-pound person.

h	B
60	
62	
64	
66	
68	
70	
72	
74	
76	

c. Sketch a graph of the function $B(h)$ for a 190-pound person.

d. Explain what happens to B as h increases. Include the notion of rate of change in your explanation. Why does this make sense in the context of this situation?

41. Blood Alcohol Levels In 1992, the National Highway Traffic Safety Administration recommended that states adopt 0.08% blood alcohol concentration as the legal measure of drunk driving. (*Source:* www.nhtsa.dot.gov) A regular 12-ounce beer is 5% alcohol by volume and the normal bloodstream contains 5 liters (169 ounces) of fluid. A person's maximum blood alcohol concentration, C, can be approximately modeled by the multivariable function

$C(w, n) = \dfrac{600n}{w(169 + 0.6n)}$, where n is the number of beers consumed in one hour and w is a person's body weight in pounds.

a. How many beers can a person have and legally drive if the person weighs 200 pounds? 110 pounds?

b. Does a person's weight have any impact on the upper limit of beers that can be consumed legally?

42. Walking Speed A heart patient recovering from surgery is on a rehabilitation program that includes the use of a treadmill to increase his cardiovascular fitness. His pace of walking/running is defined by the function $r(t) = \dfrac{5280}{t}$, where r is measured in feet per minute and t is the time (in minutes) it takes to complete the workout.

a. Graph the function $r(t)$ and label the independent and dependent axes.

b. Give the domain of the function $r(t)$.

c. Why would some of the domain values not make sense in a real-world context?

d. Explain what happens to $r(t)$ as t gets larger and larger. What does this mean in terms of the heart patient?

43. Google Share Price On November 22, 2006, the share price of Google reached a new high of $513 per share. This share price reflected an increase of $181.45 over its 52-week low price. Investors who purchased at the 52-week low price and sold at the new high earned an astounding 54.7% return on their investment. One question many investors ask before they buy shares in a company is "If I buy today, what is the lowest share price at which I can sell and still break even?" The answer takes into account the commissions that are paid when shares are bought or sold. USAA Investment Management Company charges its occasional investors $19.95 per trade whether buying or selling shares. Suppose a new investor buys a number of Google shares at $513 per share through this company.

a. Determine the total amount of money needed (including the transaction fees) to buy and sell 1 share, 10 shares, 100 shares, and 1000 shares.

b. Determine the selling price needed for the investor to make exactly enough money to cover the transaction cost of buying and selling the Google shares (the break-even share price) for 1 share, 10 shares, 100 shares, and 1000 shares.

c. As the number of shares initially purchased increases, what happens to the break-even share price?

44. Video Rentals and Sales The table provides information about video rentals (in millions of dollars) and video sales (in millions of dollars) for select years.

Years Since 1990 t	Video Rentals ($ millions) r	Video Sales ($ millions) s
1	8400	3600
2	9100	4000
4	9500	5500
6	9300	7300
10	8250	10800

The dollar value of video rentals can be modeled by the function $r(t) = -54.3t^2 + 562t + 8030$ million dollars. The dollar value of video sales can be modeled by $s(t) = 19.5t^2 + 603t + 2860$ million dollars. The population of the United States over the same time period can be modeled by $P(t) = 0.0067t^2 + 2.56t + 250$, where t is years since 1990.

a. Write a formula for the total revenue from video rentals and video sales.

b. Write a formula for the average annual cost per person in the United States for videos. Graph this formula.

c. Use the formula from part (b) to estimate the average annual video cost per person in the year 2004.

d. When will the average annual cost per person be $65?

e. Describe the concavity of this annual cost per person graph. Explain what this concavity means in practical terms.

45. Wait Time An area of mathematics known as *queuing theory* addresses the issue of how a business can provide adequate customer service by analyzing the wait time of customers in line. (*Source:* http://en.wikipedia.org/wiki/Queueing_theory) When the ticket window opens for a concert of a very popular music group, there are already 150 people in line. People arrive and join the line at a rate of 50 people per hour. Customer service representatives start serving at a rate of 90 people per hour but, due to fatigue, slow down by 2 people per hour as time passes. The number of people in line at time t hours is

$$P(t) = 150 + (50 - (90 - 2t))t$$
$$= 150 - 40t + 2t^2$$

The representatives are serving people at a rate of $90 - 2t$ people per hour. To determine the wait time, we divide the number of people in line by the rate at which they are being served. That is,

$$R(t) = \frac{150 - 40t + 2t^2}{90 - 2t} \text{ hours}$$

a. What is a practical domain for the model $P(t)$?

b. What will be the wait time for someone who arrives 2 hours after the ticket window opens?

c. How long after the ticket window opens should a person arrive if they want to wait only 0.5 hours?

d. How long will it take for the length of the line to reach 0 people?

■ STRETCH YOUR MIND

Exercises 46–49 are designed to challenge your understanding of rational functions.

46. The function $f(x) = \dfrac{-3x^2 + 2}{x - 1}$ has a special kind of asymptote known as "slant" or "oblique." Graph $f(x)$. Then determine the equation of the asymptote for $f(x)$ algebraically.

47. Are the following statements true or false? Explain your answer.

  a. "An asymptote of a function $f(x)$ is a straight line."

  b. "An asymptote of $f(x)$ can be approached but never reached or crossed by $f(x)$."

48. The function $f(x) = \dfrac{15x(x + 3)}{(x^2 + 1)(x - 4)(x + 3)}$ has a vertical asymptote, a horizontal asymptote, and a removable discontinuity. Find each. Also, by inspecting the graph of $f(x)$ explain over what intervals $f(x)$ is concave up and concave down. Finally, estimate its point(s) of inflection.

49. Create a table of values that could demonstrate what the values of the function $f(x)$ are doing in the following graph. Explain how the table and the graph demonstrate the same thing.

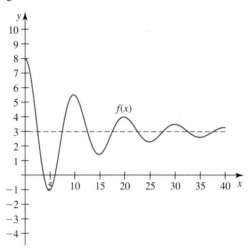

CHAPTER 5 Study Sheet

As a result of your work in this chapter, you should be able to answer the following questions, which are focused on the "big ideas" of this chapter.

SECTION 5.1
1. How can we use the idea of successive differences to determine if a numerical representation of a function is linear, quadratic, cubic, quartic, etc.?
2. How can we use the idea of rate of change and concavity to determine which polynomial type might best model a scatter plot of data?
3. What are the characteristics of a polynomial function? That is, what makes a polynomial a polynomial?
4. How can we determine the end behavior of a polynomial function? Why is it important, when modeling, to understand this end behavior?

SECTION 5.2
5. How can we use the idea of rate of change to describe the behavior of a power function?
6. What are the differences between direct and inverse variation?
7. Generally, what is the behavior of any power function with $b > 1$? $0 < b < 1$? $b < 0$?

SECTION 5.3
8. What is the definition of a rational function?
9. What role do asymptotes play in rational functions? What different types are there? How do we determine if and where they exist?
10. What is meant by a removable discontinuity? Under what conditions will one exist?
11. Describe a real-world interpretation of a vertical and horizontal asymptote.

REVIEW EXERCISES

■ SECTION 5.1 ■

In Exercises 1–2, numerical representations of three functions are shown in a table. The functions are either linear, quadratic, cubic, or none of these. Using the idea of rate of change, identify each function appropriately.

1.

x	f(x)	g(x)	h(x)
−9	170	14	−34.65
−7	104	12	−18.90
−5	54	10	−8.75
−3	20	8	−3.00
−1	2	6	−0.45
1	0	4	0.10
3	14	2	−0.15
5	44	0	0.00
7	90	−2	1.75
9	152	−4	6.30

2.

x	f(x)	g(x)	h(x)
−5	−9	200	50
−4	−7	108	36
−3	−5	48	24
−2	−3	14	14
−1	−1	0	6
0	1	0	0
1	3	8	−4
2	5	18	−6
3	7	24	−6
4	9	20	−4

In Exercises 3–7,
  a. Describe the end behavior of the given polynomial function.
  b. Make a table of values that confirms the end behavior you described. Create your table in such a way that it shows what happens to function values as $x \to \infty$ and $x \to -\infty$.

3. $y = x^2 + 2x - 8$
4. $y = x^3 - 2x^2 + x - 2$
5. $y = -x^5 + 3x^2 + 2x - 7$
6. $y = -x^4 - 2x^3 + 3x^2 - x + 2$
7. $y = x^6$

In Exercises 8–10, determine the minimum degree of the polynomial function by observing the number of changes in concavity in the graph.

8.

9.

10.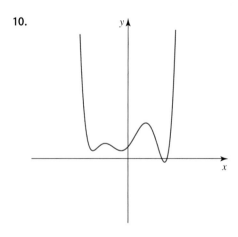

In Exercises 11–12, apply your knowledge of higher-order polynomial functions to model real-world situations.

11. **Cigarette Use among Eighth Graders** The data in the table show the percentage of eighth graders who admit to using cigarettes.

Years Since 1990 t	Cigarette Use (percent) C
1	44.0
2	45.2
3	45.3
4	46.1
5	49.2
6	47.3
7	45.7
8	44.1
9	40.5
10	36.6
11	31.4
12	28.4

Source: www.ojp.usdoj.gov

a. Make a scatter plot of these data.

b. Analyze the scatter plot and explain which type of polynomial function (based on degree) might best model the situation. Base your decision on the rate of change observed in the scatter plot.

c. Use regression to find a function model equation of the polynomial type you identified in (b). Then graph the function.

d. Use your model to predict when the percentage of eighth graders who use cigarettes will be 0. Do you think this result is reasonable? Why or why not?

12. **Egg Production in the United States** The data in the table show the egg production in the United States for selected years.

Years Since 1990 t	Egg Production (billions) E
0	68.1
5	74.8
7	77.5
8	79.8
9	82.9
10	84.7
11	86.1
12	87.3
13	87.5
14	89.1

Source: Statistical Abstract of the United States, 2006, Table 842

a. Make a scatter plot of these data and, using the idea of rate of change, explain why a cubic function best models the data. Use regression to find the model.

b. Describe the end behavior for the function model and discuss whether or not you think this end behavior will accurately predict future egg production amounts.

SECTION 5.2

In Exercises 13–16, a power function is given in tabular or graphical form. Determine whether the power function represents direct or inverse variation and explain how you know.

13.

x	y
0	0.00
2	4.76
4	11.31
6	18.78
8	26.91
10	35.57
12	44.67

14.

15.

x	y
1	5.00
2	2.50
3	1.67
4	1.25
5	1.00
6	0.83

16.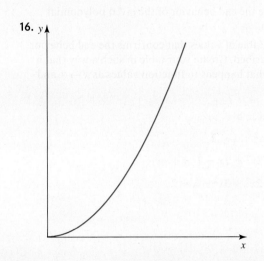

In Exercises 17–21, write a power function representing the verbal statement.

17. The luminosity, L, of a star is directly proportional to the fourth power of its surface temperature, T. (*Source:* www.astronomynotes.com)

18. The cost, C, of putting 10 gallons of gas into a gas tank is directly proportional to the price per gallon, p.

19. The amount of force, F, acting on a certain object from the gravity of Earth at sea level is directly proportional to the object's mass, m. The gravitational constant, -9.8 m/s^2, is the constant of proportionality.

20. The area, A, of a rectangle is directly proportional to its width if its length is fixed to be 7 centimeters.

21. The average speed, s (in miles per hour), needed to drive a distance of 1200 miles is inversely proportional to the time, t (in hours), spent traveling.

In Exercises 22–23, use your knowledge of power functions to answer the questions about real-world situations.

22. **Top Oil Exporters** The table shows the 10 countries with the largest daily export rates of oil, measured in million barrels per day.

   a. Examine the data and determine why an inverse variation relationship might best model the situation.

   b. Find a power regression and use it to describe the relationship between the rank of the country and its daily export of oil.

   c. Use the power regression model to determine the amount of oil exported daily by the country that ranks 11th in oil exports.

Top 10 Countries c	Oil Exports (in million bbl/day) E
1. Saudi Arabia	7.920
2. Russia	7.000
3. Norway	3.466
4. United Arab Emirates & Iran	2.500
5. Venezuela	2.100
6. Kuwait	1.970
7. Mexico	1.863
8. Canada	1.600
9. Iraq	1.500
10. United Kingdom	1.498

Source: CIA—The World Factbook, 2006

23. **Public Debt of Nations** The table shows the 10 countries with the largest public debt, measured as the percent of the gross domestic product.

   a. Examine the data and determine why an inverse variation relationship might best model the situation.

   b. Find a power regression and use it to describe the relationship between the rank of the country and the amount of its public debt.

   c. Use the power regression model to determine the public debt of a country that ranks 11th.

Top 10 Countries c	Public Debt (% of GDP) P
1. Lebanon	209.0
2. Japan	176.2
3. Seychelles	166.1
4. Jamaica	133.3
5. Zimbabwe	108.4
6. Italy	107.8
7. Greece	104.6
8. Egypt	102.9
9. Singapore	100.6
10. Belgium	90.3

Source: CIA—The World Factbook, 2006

■ SECTION 5.3 ■

In Exercises 24–25, find the horizontal and vertical asymptotes, if any, of each function.

24. $f(x) = \dfrac{3x}{4x - 1}$

25. $h(x) = \dfrac{3}{(x - 6)(2x - 2)}$

In Exercises 26–27, what are the x-intercepts, y-intercepts, and horizontal and vertical intercepts (if any)?

26. $f(x) = \dfrac{x - 2}{x + 4}$

27. $m(x) = \dfrac{x(x + 4)}{(x^2 + 2x - 8)}$

28. Let $f(x) = \dfrac{5}{x - 6}$.

   a. Complete the table for x-values close to 6. What happens to the values of $f(x)$ as $x \to 6$ from the right and from the left?

x	5	5.9	5.99	6	6.01	6.1	7
$f(x)$							

   b. Complete the following tables. Use the tables of values to determine what happens to the values of $f(x)$ as $x \to \infty$ and as $x \to -\infty$?

x	1	10	100	1000	10,000
$f(x)$					

x	-1	-10	-100	-1000	$-10,000$
$f(x)$					

   c. Without using a calculator, graph $f(x)$. Give the equations for the vertical and horizontal asymptotes.

29. For the following data, there is one vertical asymptote. Give its equation and explain.

x	y
6.6	−16
6.7	−22
6.8	−34
6.9	−68
7	undefined
7.1	71
7.2	36
7.3	24
7.4	19

30. For the following data, there is a horizontal asymptote. Give its equation and explain.

x	y
−20	−4.091
−15	−4.118
−10	−4.167
−5	−4.286
0	−5
5	−15

31. A function f has one vertical asymptote at $x = 4$ and a horizontal asymptote at $y = 0$. The graph has no horizontal intercepts and a vertical intercept at $(0, -10)$. The graph is concave down for $-6 < x < 4$ and concave down for $4 < x < 14$. Sketch a possible graph for the function described.

32. **Electrical Appliances** Every electrical appliance has two costs associated with its use: the purchase price and the operating cost. The following table shows a typical appliance's annual cost based on national averages.

Appliance	Average Cost per Year in Electricity C
Home computer	$10
Television	$16
Microwave	$17
Dishwasher	$51
Clothes dryer	$75
Washing machine	$79
Refrigerator	$92

Source: www1.eere.energy.gov

a. For a new refrigerator that costs $1150, determine the total annual cost if the refrigerator lasts for 15 years. Assume the only costs associated with the refrigerator are its purchase price and the cost for electricity.

b. Develop a function, $C(y)$, that gives the annual cost, C, of a refrigerator as a function of the number of years, y, the refrigerator lasts.

c. Use a calculator to graph $C(y)$.

d. Determine the equation for the vertical and horizontal asymptote of this function and explain the significance of each with respect to the refrigerator and its cost.

e. If a company offers a refrigerator that costs $1700, but says that it will last at least 25 years, is the refrigerator worth the difference in cost?

Make It Real Project

What to Do

1. Find a set of at least six data points from an area of personal interest. Choose data that appear to change concavity. Also try to choose data that does not involve time as the independent variable.

2. Draw a scatter plot of the data and explain which polynomial or power function might best model the situation.

3. Find a regression model for your data.

4. Using the idea of rate of change, describe the relationship between the two quantities.

5. Use the model to predict the value of the function at an unknown point and explain why you think the prediction is or is not accurate.

6. Determine if your regression model has an inverse function. If not, explain why and determine a meaningful way to restrict the domain of your function so that the inverse would be a function.

7. Interpret the meaning of the function in the context of your situation. If possible, compute the inverse function (if it is a quadratic, power, or rational function).

8. Explain how a consumer and/or a businessperson could benefit from the model.

CHAPTER 6

Exponential and Logarithmic Functions

The term "exponential" is used widely in the media. Consider the following excerpts from actual news articles:

"The **exponential** amount of digitized content..."

"...demand continues to increase at an **exponential** pace..."

"...economic indicators continue their **exponential** rise..."

"...[they] were given **exponential** worth in advertising dollars..."

As sophisticated as these phrases may sound, they are vague at best and senseless at worst. There is a clear need for an increased understanding of exponential relationships.

© Beathan/Corbis

- **6.1** Percentage Change
- **6.2** Exponential Function Modeling and Graphs
- **6.3** Compound Interest and Continuous Growth
- **6.4** Solving Exponential and Logarithmic Equations
- **6.5** Logarithmic Function Modeling

STUDY SHEET
REVIEW EXERCISES
MAKE IT REAL PROJECT

SECTION 6.1 Percentage Change

LEARNING OBJECTIVES

- Calculate change factors from tables and graphs
- Calculate percentage rates of change from tables, graphs, and change factors
- Recognize that functions with a constant percentage change are exponential functions

GETTING STARTED

Consistent advances in computer technology have produced vastly increased processing power since the early 1970s. In 1965, one of Intel's cofounders, Gordon Moore, made the empirical observation now known as Moore's Law: the number of transistors on an integrated circuit doubles every 2 years. (*Source:* www.intel.com/technology/mooreslaw) One way to analyze the doubling behavior of a situation such as this is to consider the growth percentage change that occurs over time.

In this section we distinguish between the rate of change we studied in previous chapters and percentage change. We also derive the standard equation for an exponential function (any function with a constant percentage change). Additionally, we show how to calculate growth and decay factors as well as percentage change. We use these concepts to "fill in the gaps" in data.

■ Patterns of Growth

The phrase "the population of the city of Gilbert, Arizona, is *increasing at a rate of 1000 people per month*" describes growth using an average rate of change, whereas the phrase "the population of Tempe, Arizona, is *increasing at 0.019% per year*" describes growth as a percentage change. Although these two familiar ways of describing growth sound similar and may have similar patterns in the short term, they are fundamentally different over the long term. Recall any function with a constant rate of change, such as the Gilbert population function, is a *linear function*. On the other hand, any function with a constant *percentage* rate of change, such as the population growth of Tempe, is an *exponential function*.

For example, consider the growth in the number of transistors on an integrated circuit. Table 6.1, generated using Moore's law, displays the predicted number of transistors over a 10-year period beginning in 1971.

It is apparent from these data that the number of transistors is increasing over time. Thus, the function is an increasing function. From our prior study of polynomial functions, we know that to determine if a data set is linear or quadratic we need to find the successive differences over equal intervals of time. We see from Table 6.1 that the change in the years is consistently in equal intervals, so let's begin by looking at the first and second differences in Table 6.2.

Table 6.1

Years Since 1971 x	Number of Transistors T
0	2300
2	4600
4	9200
6	18,400
8	36,800
10	73,600

Table 6.2

Years Since 1971 x	Number of Transistors T	First Differences ΔT	Second Differences $\Delta(\Delta T)$
0	2300	2300	2300
2	4600	4600	4600
4	9200	9200	9200
6	18,400	18,400	18,400
8	36,800	36,800	
10	73,600		

Since neither the first nor the second differences are constant, the function is neither linear nor quadratic.

A different way of quantifying the change in the number of transistors is to calculate the ratio (quotient) of successive output values for the equally spaced input values, as shown in Table 6.3. Notice that all of the successive ratios are equal. We say the data has a *2-year growth factor* of 2 since the output values are multiplied by the factor of 2 for each 2-year increase.

Table 6.3

Years Since 1971 x	Number of Transistors T_x	Ratio $\dfrac{T_{x+2}}{T_x}$
0	2300	$\dfrac{4600}{2300} = 2$
2	4600	$\dfrac{9200}{4600} = 2$
4	9200	$\dfrac{18{,}400}{9200} = 2$
6	18,400	$\dfrac{36{,}800}{18{,}400} = 2$
8	36,800	$\dfrac{73{,}600}{36{,}800} = 2$
10	73,600	

If the same pattern of growth continues, we can estimate the number of transistors in 2-year increments beyond year 10 by using the 2-year growth factor to expand the data as shown in the shaded portion of Table 6.4.

Table 6.4

Years Since 1971 x	Number of Transistors T_x	Tax-Year Growth Factor (Ratio) $\dfrac{T_{x+2}}{T_x}$
0	2300	2
2	4600	2
4	9200	2
6	18,400	2
8	36,800	2
10	73,600	2
12	147,200	2
14	294,400	2
16	588,800	

■ Exponential Functions

Finding and applying the growth factor works well if we continue to estimate the number of transistors at 2-year intervals. But to fill in the data (interpolate) for missing years or forecast the data (extrapolate) for future years, we need a function that calculates the number of transistors for any year.

From the preceding tables we see that the transistor function will have an initial value of 2300. That is, $f(0) = 2300$. Furthermore, by observing patterns we find that

$f(2) = 4600$	$f(4) = 9200$	$f(6) = 18{,}400$	$f(8) = 36{,}800$
$= 2300(2)$	$= 2300(4)$	$= 2300(8)$	$= 2300(16)$
$= 2300(2^1)$	$= 2300(2^2)$	$= 2300(2^3)$	$= 2300(2^4)$
$= 2300(2^{\frac{1}{2}(2)})$	$= 2300(2^{\frac{1}{2}(4)})$	$= 2300(2^{\frac{1}{2}(6)})$	$= 2300(2^{\frac{1}{2}(8)})$

Thus it appears that $f(x) = 2300(2^{\frac{1}{2}x})$, where x is the number of years since 1971. With this function, we can calculate the number of transistors in 1972 (1 year after 1971) or 2012 (41 years after 1971).

Interpolation

$$f(1) = 2300(2^{\frac{1}{2}(1)})$$
$$= 2300(2^{0.5})$$
$$\approx 2300(1.414)$$
$$\approx 3252$$

Extrapolation

$$f(41) = 2300(2^{\frac{1}{2}(41)})$$
$$= 2300(2^{20.5})$$
$$\approx 2300(1482910.4)$$
$$\approx 3{,}410{,}693{,}921$$

We estimate there were 3252 transistors on an integrated circuit in 1972 and that there will be 3.4 billion transistors on an integrated circuit in 2012.

The **exponential function** $f(x) = 2300(2^{\frac{1}{2}x})$ can be rewritten as

$$f(x) = 2300(2^{\frac{1}{2}x})$$
$$= 2300(2^{\frac{1}{2}})^x \qquad \text{since } b^{mn} = (b^m)^n$$
$$= 2300(\sqrt{2})^x \qquad \text{since } b^{\frac{1}{2}} = \sqrt{b}$$
$$\approx 2300(1.414)^x \qquad \text{since } \sqrt{2} \approx 1.414$$

As stated earlier, 2300 is the initial value of the function. The value 1.414 is the annual growth factor or, more commonly, the **growth factor**.

Exponential Function

A function of the form

$$y = ab^x$$

with $a \neq 0$, $b > 0$, and $b \neq 1$ is an **exponential function**.

a is the **initial value** of the function.

b is the **growth factor** if $b > 1$ and the **decay factor** if $b < 1$. (For convenience, we will refer to growth and decay factors as **change factors** when discussing strategies or concepts that apply to both types.)

EXAMPLE 1 ■ **Writing the Equation for an Exponential Function**

The annual growth factor for the population of the United States is approximately 1.09. In 2005, the population of the United States was estimated to be 298.2 million. (*Source:* World Health Organization) Write the equation of the exponential function that represents this situation.

Solution Let t represent the number of years since 2005 and P be the population of the United States (in millions). Since 2005 is 0 years after 2005, the initial value of the population function is 298.2 million. We have

$$P(t) = 298.2(b)^t$$

Since the annual growth factor is $b = 1.09$, the exponential function model equation is $P(t) = 298.2(1.09)^t$.

Finding Change Factors

If the input values of an exponential function are 1 unit apart, the *change factor* is simply equal to the ratio of consecutive output values. However, finding the change factor when the input values are more or less than 1 unit apart requires a bit more work.

For example, in the transistor scenario the input values were 2 units apart, so the ratio we calculated was the 2-year growth factor. Let's see how we can accurately find the annual (1-year) growth factor (which we found earlier by trial-and-error). Calling the annual growth factor b, we get

$$b \cdot b = 2 \text{ (2-year growth factor from Table 6.3)}$$
$$b^2 = 2$$
$$b = \sqrt{2} \approx 1.414$$

This agrees with our earlier result when we found the function $f(x) = 2300(1.414)^x$. We can now use the annual growth factor 1.414 to interpolate for years 1, 3, and 5, as shaded in Table 6.5.

Table 6.5

Years Since 1971 x	Number of Transistors T	Annual Growth Factor $\dfrac{T_{x+1}}{T_x}$	Two-Year Growth Factor $\dfrac{T_{x+2}}{T_x}$
0	2300	$\sqrt{2} \approx 1.414$	
1	3252	1.414	2
2	4600	1.414	
3	6502	1.414	2
4	9200	1.414	
5	13,001	1.414	2
6	18,400	1.414	

To generalize, consider the exponential function $y = ab^x$ that passes through the points (x_1, y_1) and (x_2, y_2). Calculating the ratio of the output values, we get

$$\frac{y_2}{y_1} = \frac{ab^{x_2}}{ab^{x_1}}$$

$$\frac{y_2}{y_1} = \frac{b^{x_2}}{b^{x_1}}$$

$$\frac{y_2}{y_1} = b^{x_2 - x_1}$$

Recall from the rules of rational exponents (see Just in Time in Section 6.4) that $(x^n)^{1/n} = x$. Therefore, we raise each side of the equation to the power $\dfrac{1}{x_2 - x_1}$ to solve for b.

$$b^{x_2 - x_1} = \frac{y_2}{y_1}$$

$$(b^{x_2 - x_1})^{1/(x_2 - x_1)} = \left(\frac{y_2}{y_1}\right)^{1/(x_2 - x_1)}$$

$$b = \left(\frac{y_2}{y_1}\right)^{1/(x_2 - x_1)}$$

In words, the change factor is the ratio of the outputs raised to 1 over the difference in the inputs.

HOW TO: CALCULATE CHANGE FACTORS

To calculate the change factor of an exponential function with points (x_1, y_1) and (x_2, y_2), raise the ratio of the output values to 1 over the difference in the input values. That is,

$$b = \left(\frac{y_2}{y_1}\right)^{1/(x_2 - x_1)}$$

Table 6.6

Years Since 2000 y	Number of Cremations C
0	629,362
5	778,025

Source: www.cremationassociation.org

EXAMPLE 2 ■ Finding Change Factors

According to the Cremation Association of North America (CANA), the number of people choosing to be cremated is increasing dramatically. Table 6.6 displays the number of people choosing cremation in the United States for the years 2000 and 2005.

a. Assuming these data have a common growth factor, determine the *5-year* growth factor and the *annual* growth factor.

b. Use the annual growth factor to interpolate the number of cremations in 2001, 2002, 2003, and 2004.

Solution

a. To find the 5-year growth factor, we find the ratio by dividing the number of cremations for the year 2005 by the number of cremations for the year 2000.

$$\frac{C(5)}{C(0)} = \frac{778,025}{629,362} \quad \begin{array}{l}\text{number of cremations in 2005}\\ \text{number of cremations in 2000}\end{array}$$

$$\approx 1.236 \quad \text{5-year growth factor}$$

The 5-year growth factor is approximately 1.236.
   The annual growth factor is given by

$$b = \left(\frac{778,025}{629,362}\right)^{1/(5-0)}$$

$$\approx (1.236)^{1/5}$$

$$\approx 1.043$$

The annual growth factor is approximately 1.043.
   To check this, recall that the annual growth factor is the number we would have to multiply by itself 5 times to get the 5-year growth factor, as shown in Table 6.7.

Table 6.7

Years Since 2000 y	Number of Cremations C	Annual Growth Factor	Five-Year Growth Factor
0	629,362	b	
1	?	b	
2	?	b	
3	?	b	
4	?	b	1.236
5	778,025		

In other words, multiplying by 5 annual growth factors is equivalent to multiplying by the 5-year growth factor once. Letting b represent the annual growth factor, we can write this as

$$b \cdot b \cdot b \cdot b \cdot b = 1.236$$
$$b^5 = 1.236$$

Solving for b, we get

$$b^5 = 1.236$$
$$\sqrt[5]{b^5} = \sqrt[5]{1.236}$$
$$b = 1.043$$

Therefore, the annual growth factor is approximately 1.043, which confirms our earlier result.

b. We can now fill in Table 6.8 for 2001 to 2004. (Note that although we rounded b to 1.043 when we wrote it down, we used the more accurate estimate $b = 1.043322581$ to generate the table. All numbers have been rounded to the nearest whole number.)

Table 6.8

Years Since 2000 y	Number of Cremations C	Annual Growth Factor	Five-Year Growth Factor
0	629,362	1.043	
1	656,628	1.043	
2	685,074	1.043	
3	714,754	1.043	
4	745,719	1.043	1.236
5	778,025		

■ Percentage Change

So far in this text, we have described change in terms of the average rate of change or the change factor. However, change is often discussed in terms of a percentage change.

EXAMPLE 3 ■ Calculating a Percentage Change

One avenue people use to help generate adequate funds for retirement is the stock market. The Standard and Poor's 500 Index (S&P 500), considered by many to be the best indicator for the U.S. stock market as a whole, provides the benchmark by which other investments and portfolio managers are measured. (*Source:* www.zealllc.com) Table 6.9 shows how the S&P 500 Index changed between January 3, 1997, and January 3, 2007. Find the 10-year percentage change and use it to extrapolate the S&P 500 Index for the year 2017.

Table 6.9

Years Since January 3, 1997 y	S&P 500 Index S
0	748.03
10	1416.60

Source: finance.yahoo.com

Solution To find the 10-year percentage change we first calculate the 10-year growth factor, which is the ratio of the two indices for 1997 and 2007.

$$\text{10-year growth factor} = \frac{1416.60}{748.03}$$
$$\approx 1.89$$

To convert the 10-year growth factor (1.89) to a percentage we multiply by 100%.

$$1.89(100\%) = 189\%$$

The 2007 index value is 189% of the 1997 value. If it were 100% of the 1997 index value, the index values would be the same. Therefore, the additional 89% represents the 10-year growth in the index value. Assuming the 2017 index value will be 189% of the 2007 index value, we predict the S&P 500 for the year 2017 to be

$$S(20) = 1416.60(1.89)$$
$$\approx 2677.37$$

As we saw in Example 3, change factors and percentage rates of change are closely related.

CHANGE FACTORS AND PERCENTAGE RATES OF CHANGE

The **change factor**, b, of an exponential function is given by $b = 1 + r$, where r is the **percentage rate of change** (as a decimal).

- If $r > 0$, b is called a **growth factor** and r is called the **percentage growth rate**.
- If $r < 0$, b is called a **decay factor** and r is called the **percentage decay rate**.

EXAMPLE 4 ■ **Finding an Exponential Function from a Percentage Rate**

On March 31, 2007, USAA Federal Savings Bank advertised a 5-year certificate of deposit (CD) with an *annual percentage yield* of 5.0%. (**Annual percentage yield**, discussed in Section 6.3, is the percentage increase in the value of an investment over a 1-year period.) (*Source:* www.usaa.com)

Find an exponential function that models the value of a $1000 investment in the 5-year CD as a function of the number of years the money has been invested. Then calculate the value of the CD when it matures 5 years later.

Solution We convert the percentage growth rate into a growth factor.

$$b = 1 + r$$
$$= 1 + 0.05 \quad \text{since 5\% = 0.05}$$
$$= 1.05$$

Since the initial investment is $1000, the value of the investment after t years is given by $V(t) = 1000(1.05)^t$. To determine the value of the investment at maturity, we evaluate the function at $t = 5$.

$$V(5) = 1000(1.05)^5$$
$$= 1276.28$$

A $1000 investment into a CD with an annual percentage yield of 5% will be valued at $1276.28 when it matures 5 years later.

Not only do exponential functions have constant change factors, they also have constant percentage rates of change. In fact, any function that is changing at a constant percentage rate is an exponential function.

> ## Exponential Growth and Decay
>
> Any function that *increases* at a constant percentage rate is said to demonstrate **exponential growth**. This growth may be very rapid (e.g., 44% per year) or very slow (e.g., 0.01% per year).
>
> Any function that *decreases* at a constant percentage rate is said to demonstrate **exponential decay**. This decay may be very rapid (e.g., −50% per year) or very slow (e.g., −0.02% per year).

EXAMPLE 5 ■ Determining Decay Factors and Percentage Decay Rates

Although depreciation rates vary among vehicles, a typical car will lose about 15% to 20% of its value each year. (*Source:* www.kbb.com) Suppose you purchase a new Toyota Camry for $21,500 and want to estimate its worth over the next 5 years. Assuming it will lose 15% of its value each year,

a. Determine the annual percentage decay rate, the annual decay factor, the 5-year decay factor, and the 5-year percentage decay rate.

b. Create a table of values for the car's value over this period.

Solution

a. Since the car's value is depreciating, the annual percentage decay rate will be a *negative* value (−15%) and is written as the decimal $r = -0.15$.

Since the decay factor is given by $b = 1 + r$, we have

$$b = 1.00 - 0.15 \quad \text{decay factor = 1 − percentage decay}$$
$$b = 0.85 \quad \text{annual decay factor}$$

The annual decay factor is 0.85.

To determine the 5-year decay factor, we raise the annual decay factor to the fifth power.

$$(0.85)^5 \approx 0.4437$$

The 5-year decay factor is 0.4437. That is, after 5 years, the car will be worth 44.37% of its original value.

To determine the 5-year decay rate, we subtract 1 from the 5-year decay factor since $r = b - 1$.

$$0.4437 - 1 = -0.5563$$

The car is depreciating at a rate of 55.63% every 5 years.

b. The values for the car over the time period are shown in Table 6.10.

Table 6.10

Age of Car a	Value Computation	Value of the Car V
0	$21{,}500(0.85)^0$	21,500.00
1	$21{,}500(0.85)^1$	18,275.00
2	$21{,}500(0.85)^2$	15,533.75
3	$21{,}500(0.85)^3$	13,203.69
4	$21{,}500(0.85)^4$	11,223.13
5	$21{,}500(0.85)^5$	9539.66

EXAMPLE 6 ■ **Finding an Exponential Model**

A news article titled "Study Indicates Volunteerism Rises Among Collegians" made the claim that "The number of college students volunteering grew more than 20 percent, from 2.7 million to 3.3 million, between 2002 and 2005." (*Source: East Valley Tribune*: October 16, 2006)

Find the annual growth factor, the annual percentage growth rate, and an exponential model for the number of college-aged volunteers. Use this to predict the number of volunteers in the year 2012.

Solution Since the percentage of college-aged volunteers grew by 20% over a 3-year interval, 20% is the 3-year percentage growth rate. We need to calculate the *annual* growth factor and the *annual* percentage growth rate. To do this, we first need to find the 3-year growth factor by using the fact that $b = 1 + r$.

$$b = 1 + 0.20$$
$$b = 1.20$$

Therefore, the 3-year growth factor is 1.20. We take the third root of the 3-year growth factor to find the annual growth factor.

$$b = 1.20^{1/3}$$
$$= \sqrt[3]{1.20}$$
$$\approx 1.063$$

Since the annual growth factor is related to the annual percentage growth rate by the formula $b = 1 + r$, we have

$$1.063 = 1 + r$$
$$1.063 - 1 = r$$
$$0.063 = r$$
$$0.063(100\%) = r$$
$$6.3\% = r$$

Therefore, the annual percentage growth rate is 6.3%.

Since the initial number of volunteers was 2.7 million and the annual growth factor is 1.063, the exponential model that represents the number of college-aged volunteers is $V(y) = 2.7(1.063)^y$, where V is the number of volunteers (in millions) and y is the number of years since 2002. We can use this to predict the number of college-aged volunteers in the year 2011 ($y = 9$).

$$V(9) = 2.7(1.063)^9$$
$$\approx 4.7 \text{ million student volunteers}$$

SUMMARY

In this section you learned how to distinguish between the rate of change and the percentage rate of change. You also learned that an exponential function has a constant percentage rate of change and the standard form $y = ab^x$. Additionally, you discovered how to calculate change factors and percentage rates of change and how to use these values to fill in the gaps in data sets.

6.1 EXERCISES

SKILLS AND CONCEPTS

In Exercises 1–4, determine whether the following tables are linear, quadratic, exponential, or none of these by calculating successive differences and/or change factors.

1.
x	y
0	80
1	40
2	20
3	10
4	5
5	2.5

2.
x	y
0	3
1	10
2	21
3	36
4	55
5	78

3.
x	y
0	0
1	3
2	16
3	45
4	96
5	175

4.
x	y
0	−5
1	−15
2	−45
3	−135
4	−405
5	−1215

5. Match the data sets in the following tables with the functions shown in A, B, and C.

a.
x	y
0	4
1	6
2	9
3	13.5
4	20.25

b.
x	y
1	80
2	64
3	51.2
4	40.96
5	32.768

c.
x	y
−5	0.69632
−4	1.7408
−3	4.352
−2	10.88
−1	27.2

A. $k(x) = 68(2.5)^x$
B. $f(x) = 4(1.5)^x$
C. $h(x) = 100(0.8)^x$

6. The table shows some values of an exponential function, f, and a linear function, g. Find the equation for $f(x)$ and $g(x)$ and use the functions to complete the table.

x	f(x)	g(x)
0	?	5
1	0.36	?
2	0.216	8.2
3	?	?
4	0.07776	?
5	?	13

7. After examining the following graph you determine that distance traveled increases at an increasing rate. Which of the following best describes the reasoning that can be used to determine this response?

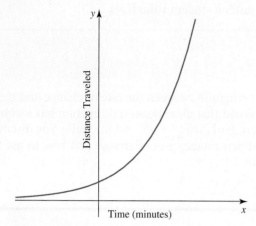

A. The amount of distance traveled is greater for each successive second.

B. As you move to the right on the graph the slope gets steeper.

C. The rate is greater for each successive second.

D. All of the above.

E. None of the above.

8. **Allowance Riddle** Suppose a child is offered two choices to earn an increasing weekly allowance: the first option he can choose begins at 1 cent and doubles each week, while the second option begins at $1 and increases by $1 each week. How much allowance would the child earn in 4 weeks, 8 weeks, and a year? Which option is the best choice for the child?

In Exercises 9–12, use the graph of f(x) below to answer each question.

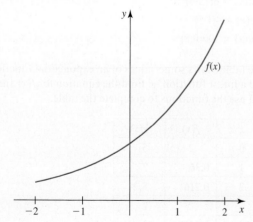

9. Considering the information provided by the graph of $f(x)$, is it possible to determine if the function is quadratic or exponential? Why or why not?

10. In the following table, find values of $f(x)$ that will make the function quadratic.

x	f(x)
-2	
-1	
0	
1	
2	

11. In the following table, find values of $f(x)$ that will make the function exponential.

x	f(x)
-2	
-1	
0	
1	
2	

12. Give a possible equation for $f(x)$.

SHOW YOU KNOW

13. Explain what we mean when we talk about *percentage growth* or *percentage decay*.

14. Write an explanation comparing *constant percentage change* to *constant rate of change*.

15. **Microorganism Growth** A microbiologist generated a model that describes the number of bacteria in a culture after t days but has just updated the model from $P(t) = 7(2)^t$ to $P(t) = 7(3)^t$. Which of the following implications can you draw from this information? Defend your choice(s).

A. The final number of bacteria is three times as much instead of two times as much as the original amount.

B. The initial number of bacteria is three instead of two.

C. The number of bacteria triples every day instead of doubling every day.

D. The growth rate of the bacteria in the culture is 30 percent per day instead of 20 percent per day.

16. **Jack and the Beanstalk** Jack plants a 5-centimeter-tall beanstalk in his backyard. When the beanstalk is 5-centimeters tall it grows by about 15% per day for the next month. What formula represents the height of the beanstalk as a function of the number of days since it was transplanted? Explain your choice.

A. $H(t) = 5 + 0.15t$

B. $H(t) = 5.15t$

C. $H(t) = 5(1.15)^t$

D. $H(t) = 5.15^t$

E. $H(t) = 5 + 1.15t$

17. Salary Adjustments A person's salary is reduced by 5% one year due to necessary budget cuts. The next year (after business has improved) the person is given a 5% raise. Is the employee's income back to where it was originally? Explain your reasoning and support your argument with at least one of the following: equations, graphs, or tables.

MAKE IT REAL

18. U.S. Immigrants The number of immigrants coming into the United States increased from 385,000 in 1975 to 1,122,000 in 2005. (*Source: Statistical Abstract of the United States, 2007*, **Table 5**) Calculate the average rate of change, the 30-year growth factor, the annual growth factor, the 30-year percentage change, and the average annual percentage change.

19. Stock Market According to the 2004 Andex Chart, the average return of a $1 investment made in 1925—with no acquisition costs or taxes, and all income reinvested into the S&P 500—would have grown to $2641 by 2005. (*Source: www.andexcharts.com*)

a. What is the 80-year growth factor?

b. What is the 80-year percentage growth rate?

c. What is the average annual growth factor?

d. What is the average annual percentage change?

e. Write an equation for the function $I(y)$, which would model the value of the initial $1 investment, I, as a function of the number of years since 1925.

f. Evaluate $I(90)$ and explain the meaning of the numerical value in the context of the problem.

20. Golf Course Management Managers of golf pro shops anticipate the total revenue for an upcoming year based on the predicted total number of rounds that will be played at their course. They do this by looking at the total number of rounds played in prior years. Assume there were 33,048 rounds played 3 years ago and 38,183 rounds played this year. Calculate the average rate of change and the annual percentage change, and use each to predict the number of rounds to be played 2 years from now.

21. Internet Usage The graph displays the incredible growth in worldwide Internet usage from 1996 to 2006. Using successive differences and ratios, determine whether a linear, quadratic, or exponential function would be the best mathematical model.

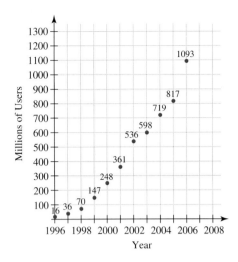

Source: www.internetworldstats.com/emarketing.htm

22. Professional Baseball Salaries Professional athletes are some of the highest paid people in the world. The average major league baseball player's salary climbed from $1,998,000 in 2000 to $2,866,500 in 2006. (*Source: www.sportsline.com*)

a. Assuming *linear* growth in the average players' salaries, find a formula $L(y)$ for the average salary L as a function of the year since 2000, y. Explain what the slope and vertical intercept mean in the real-world context. Use $L(y)$ to predict this year's average salary.

b. Assuming *exponential* growth in the average players' salaries, find a formula $E(y)$ as a function of the year since 2000, y. Explain what the initial value and growth factor mean in the real-world context. Use $E(y)$ to predict this year's average salary.

23. Purchasing Power of the Dollar The purchasing power of the dollar as measured by consumer prices from 1980 to 2005 is given in the table. (Assume that in 1982, $1 was worth $1.)

Years Since 1980 y	Value of the Dollar V
0	1.215
5	0.928
10	0.766
15	0.656
20	0.581
25	0.512

Source: Statistical Abstract of the United States, 2007, **Table 705**

a. Use the data to calculate the change factor for the purchasing power of the dollar for each of the following time intervals: (i) 1980 through 2005, (ii) 1990 through 1995, and (iii) 2000 through 2005.

b. Express the changes in part (a) as percentage changes. Write a sentence interpreting each answer in the real-world context.

c. Calculate the average rate of change and the percentage change in purchasing power of the dollar from 1985 through 2005.

d. What is the practical meaning to consumers if the percentage change in the purchasing power of the dollar is negative? Positive? Zero?

24. **Tuition at American Private Universities** The table shows the average yearly tuition and required fees charged by American private universities in the school year beginning in 2000.

Years Since 2000 y	Average Tuition (dollars) T
0	16,072
1	17,377
2	18,060
3	18,950
4	20,045
5	21,235

Source: Annual Survey of Colleges, The College Board, New York

a. Using a rate of change and percentage change, show why these data can be modeled by a linear function or an exponential function.

b. Plot the data points and add the graph of the linear function that best models $T(y)$.

c. What is the slope for the linear function modeling tuition and fees for private universities? What does this mean in the real-world context?

d. What prediction does this formula give for average tuition and fees at American private universities for the academic year beginning in 2011?

25. **Tuition at American Public Universities** The table shows the average yearly in-state tuition and required fees charged by American public universities in the school year beginning in 2000.

Years Since 2000 y	Average Tuition (dollars) T	Years Since 2000 y	Average Tuition (dollars) T
0	3508	3	4645
1	3766	4	5126
2	4098	5	5491

Source: Annual Survey of Colleges, The College Board, New York

a. Using a rate of change and percentage change, show why these data can be modeled by a linear function or an exponential function.

b. What is the slope for the linear function modeling tuition and fees for public universities? What does this mean in the real-world context?

c. Explain what the slope of the linear function model tells you about the rate of increase in tuition at public versus private institutions.

26. **Celebrity Spending Power**
Johnny Depp is estimated to have earned $18 million in 2006. In contrast, the average American citizen made roughly $30,000. To get a sense of what it feels like to make $18 million a year, the prices of a number of common products are listed together with the "feels like" price.

Product	At This Price (dollars)	Feels Like (dollars)
house	275,000.00	458.33
car	20,000.00	33.33
laptop	2000.00	3.33
boom box	300.00	0.50
hotel room	100.00	0.17
nice meal	40.00	0.07
hamburger	2.29	0.00
soda	1.00	0.00

Source: www.bankrate.com/brm/news/financial_literacy/movies_salary.asp?web=brm&athlete=163&submit=submit

Using percentages, explain how the table converts the actual product price for the average American citizen to the "feels like" price for Johnny Depp.

27. **U.S. Metro Populations** According to the U.S. Census Bureau, Atlanta, Georgia, added more people than any other metropolitan area from 2000 to 2006. The population of Atlanta experienced a 6-year percentage growth rate of 21%. The New Orleans area, still recovering from Hurricane Katrina, suffered a 6-year percentage decay rate of 22.2% over the same time period. (*Source: Arizona Republic, April 5, 2007*) Give the annual percentage growth rate for Atlanta and the annual percentage decay rate for New Orleans from 2000 to 2006.

28. **Computer Transistors** In the Getting Started feature of this section, we stated that in 1965 Moore's Law predicted that the number of transistors on a circuit would double every 2 years. Assuming this is true, we can model the growth of the number of transistors as $T(y) = 2300(\sqrt{2})^y$, where y is the number of years since 1971. The table shows the actual number of transistors for different Intel computers. Do these data provide evidence for or against using the mathematical model $T(y)$ and Moore's Law?

Intel Product Name	Years Since 1971 y	Number of Transistors on Integrated Circuit T
4004 Microprocessor	0	2300
8008 Microprocessor	1	3500
8080 Microprocessor	3	6000
8086 Microprocessor	7	29,000
286 Microprocessor	11	134,000
386 Microprocessor	14	275,000
486 Microprocessor	18	1,200,000
Pentium	22	3,100,000
Pentium II	26	7,500,000
Pentium III	28	9,500,000
Pentium IV	29	42,000,000
Itanium 2	31	220,000,000
Dual Core Itanium 2	35	1,700,000,000

Source: Intel: www.intel.com

29. **Jeans Sale** In 2007, Lucky Brand Jeans advertised their Socialite Jean for $118. (*Source:* www.luckybrandjeans.com) If a pair of jeans costs $118 today, how much will a pair cost in w weeks if the price is reduced by

   a. $10 per week?

   b. 10% per week?

30. **Photocopies** Most photocopy machines allow for enlarging or reducing the size of the original. Suppose you have a chart you wish to photocopy for a report. The chart is 8 inches by 5 inches. To reduce the size of the copy by 20%, you set the machine to reduce the chart to 80% of its original dimensions.

   a. What will be the dimensions of the photocopy?

   b. What is the percentage reduction in the *area* of the photocopy?

   c. If the chart must fit into a space 3 inches high in your report, how many reductions must you make with the photocopier set at 80% to reduce the height of the image from the original 5 inches to 3 inches?

   d. If you were to actually perform ten 80% reductions (photocopies of photocopies), what will the dimensions of your chart be?

   e. Find formulas for the width, w, and height, h, of the chart in terms of the number of 80% reductions, r.

31. **In the News** Find a newspaper article that claims something is "growing exponentially" and either support or refute that claim. Do you think the reporter simply meant the data is increasing rapidly or was the reference truly to a constant percentage change?

32. **Hotel Revenue** In 2006, hotels in Mesa and Chandler, Arizona, raised room rates by 8.7% over the prior year's rates and still filled more rooms than the year before. The average daily rate of a room in 2005 was $81.77. (*Source: East Valley Tribune*, April 12, 2007)

   a. To what price did the room rates rise in 2006?

   b. If the room rates continue to rise at 8.7% a year for the next few years, write a function, $R(y)$, to determine the room rate, R, in dollars for a given year, y, since 2005.

   c. How long do you think the room rates could continue to rise at 8.7% a year? Why?

33. **Radioactive Fallout** During the late 1950s and early 1960s, atmospheric tests of nuclear weapons became a global political concern because of the radioactive substances they released into the air (fallout). The most problematic of these substances was iodine-131, a radioactive isotope of iodine. Iodine-131 can settle on grass, be consumed by cows, become concentrated in milk, and ultimately end up in the thyroid glands of human beings who drink the milk. (*Source:* www.answers.com) The half-life of iodine-131 is about eight days, meaning half of the original quantity is left after eight days. What fraction of the iodine-131 released in an atmospheric nuclear test would be left after 30 days?

34. **Coca-Cola Production** The 2001 annual report of the Coca-Cola Company stated:

   "Our worldwide unit case volume increased 4 percent in 2001, on top of a 4 percent increase in 2000. The increase in unit case volume reflects consistent performance across certain key operations despite difficult global economic conditions. Our business system sold 17.8 billion unit cases in 2001." (*Source:* Coca-Cola Company 2001 Annual Report, p. 46)

   A unit case is equivalent to 24 8-ounce servings of finished beverage.

   a. Would a linear or exponential function best model the unit case volume of Coca-Cola? Explain.

   b. Based on the given data, it appears unit case volume has an annual percentage growth of 4%. If the unit case volume is modeled with an exponential function, what will be the annual growth factor?

   c. Find the exponential function that best models the unit case volume of Coca-Cola.

   d. Use the exponential model from part (c) to forecast the unit case volume sold in 2003 and 2004.

   e. According to Coca-Cola's 2004 Annual Report, the company sold "approximately 19.8 billion unit cases of [their] products in 2004 and approximately 19.4 billion cases in 2003." (*Source:* Coca-Cola Company 2004 Annual Report, p. 47) How accurate was the exponential model from part (c) at forecasting the 2003 and 2004 results? Explain.

35. **Family Trees** Have you ever constructed, or looked at, your family tree? Ignoring divorces, second marriages, and adoptions, answer each of the following questions, assuming the people would be alive.

   a. How many parents do you have?

   b. How many direct ancestors two generations before your generation (grandparents) do you have?

c. How many direct ancestors three generations before your generation (great-grandparents) do you have?

d. How many direct ancestors four generations before your generation do you have?

e. Find a formula for the exponential function that would give you the number of ancestors, A, based on the number of prior generations, g, you go back.

36. **Airplane Takeoffs** There are many factors that can require aircraft to allow for longer takeoff distances, called takeoff rolls. To ensure a safe takeoff, it is the responsibility of the pilot to assess all factors and conditions to calculate the cumulative total distance required for takeoff. The table indicates the possible cumulative effect of some takeoff conditions on takeoff rolls for a certain light airplane.

Additional Takeoff Condition	Cumulative Takeoff Roll (meters)
normal	400
tail wind (+20%)	400(1.20) = 480
weight (+20%)	480(1.20) = 576
temperature (+20%)	
grass landing strip (+25%)	
altitude (+10%)	
upslope (+10%)	

Source: www.auf.asn.au/index.html

a. Given that the normal run for the airplane is 400 meters, complete the table of takeoff values, accounting for the additional percentages as each additional takeoff condition is added.

b. What is the overall percentage increase in takeoff roll when all six takeoff conditions exist?

37. **Atmospheric Pressure** The physics of atmospheric pressure is well known—air resistance is lower at higher altitudes. Therefore, baseballs hit at Denver's Coors Field carry farther than in any other stadium in the country. Research confirms that a homerun that travels 400 feet in Miami would travel 420 feet in Denver. Atmospheric pressure decreases approximately exponentially with increasing height above sea level, at a rate of about 0.4% every 100 feet. (*Source:* www.aip.org/dbis/stories/2006/15254.html) By what percentage is air pressure reduced by moving from sea level to Denver?

38. **Home Values** The median price of a home in Las Vegas, Nevada dropped from $312,346 in 2006 to $306,100 in 2007. Let t be the number of years since 2006.

a. Assume the decrease in housing prices has been linear. Give an equation for the line representing price, P, in terms of t. What would the value of the home be in years 2 and 3?

b. If instead the housing prices have been falling exponentially, find an equation, $V(t)$, of the form $y = ab^x$ to represent housing prices. What would the value of the home be in years 2 and 3?

c. On the same set of axes, sketch the functions $P(t)$ and $V(t)$.

d. Which model for the price growth do you think is more realistic? Explain.

39. **Rumor Spread** The spread of a rumor can be modeled mathematically using an exponential function. Assume one person knows something about another person and tells two people and this process continues, with each person telling two more. Find the equation for a function that gives the number of people, n, who have heard the rumor after a number of iterations, x.

40. **Chain Letters** A once-common but now-illegal money-making scheme is the *chain letter*. Each new recipient is to send a small sum of money (typically ranging from $1 to $5) to the first person on the list. The new recipient must then remove the first name on the list, move the remaining names up, and add his name at the bottom of the list. The recipient must then copy the letter and mail it out to five or ten more people. The hope is this procedure will continue indefinitely with each recipient receiving a large sum of money. The success of such ventures rests solely on the notion of exponential growth in new recipients and the increasing number at each successive layer.

  Show that within a few mailings the entire global population would need to participate in order for those listed on the letter to earn any income, and therefore the majority of people participating will lose their invested money.

41. **Swimming Pool** Imagine a swimming pool that is 32 feet by 16 feet, or 512 square feet. Now suppose the owner has neglected maintaining the pool and algae begins to grow on the bottom surface. The first day the algae has covered about 4 square inches of the bottom of the pool, or $\frac{1}{16}$ square foot.

a. Show that if the area covered by the algae doubles each day, then before the end of the 16th day it will have covered the entire bottom of the pool.

b. Suppose you visited the pool when it was half covered by algae. How much time do you have to act before the pool is completely covered? Provide the reasoning behind your answer.

■ STRETCH YOUR MIND

The following exercise is intended to challenge your understanding of percentage change.

42. **Population Growth** According to the *East Valley Tribune*, Gilbert, AZ, had an estimated population of 161,059 resi-

dents through June 2004, while Tempe, AZ, had an estimated 162,652 residents. Gilbert adds 1000 new residents a month, while Tempe calculates its population by using an annual percentage growth rate of 0.019%.

a. According to these models, will Gilbert's population ever exceed Tempe's? Explain using a table, equation, or graph of two functions that model each city's growth rates.

b. Explain why even though Tempe's population is described as having a percentage growth rate, it is not growing very rapidly.

SECTION 6.2 Exponential Function Modeling and Graphs

LEARNING OBJECTIVES

- Construct exponential models algebraically from tables or words
- Use exponential regression to model real-world data sets
- Graph exponential functions given in equations, tables, or words

GETTING STARTED

Although virtually nonexistent in the United States, mosquito-transmitted malaria is a major killer of children in Africa. There currently is no vaccine, but a number of strategies can reduce the transmission of the disease, including the use of insecticide treated nets (ITNs). The nets can reduce malaria transmission by more than half. In recent years, the distribution of nets in Africa has increased exponentially.

In this section we focus on creating exponential models of real-world situations such as the distribution of ITNs. We use both algebraic and technological methods to create the models from data given in tables or words. We also discover how the growth factor and initial value affect the shape of the exponential function graph.

■ Modeling with Constant Percentage Rates of Change

As discussed in the previous section, anything that grows or decays at a constant percentage rate can be modeled by an exponential function. Consequently, a reference to a percentage of growth in a verbal description often signals underlying exponential behavior.

EXAMPLE 1 ■ Creating an Exponential Model from a Verbal Description

According to the World Health Organization, 538 thousand ITNs were distributed in the African region in 1999. In 2003, 9485 thousand nets were distributed. Between 1999 and 2003, net distribution increased at a nearly constant percentage rate. Assuming net distribution will increase at a constant percentage rate, find the function that models the distribution and forecast the number of nets that will be distributed in 2011. Then explain whether or not the estimate is realistic.

Solution Let t be the number of years since 1999 and let n be the number of nets distributed (in thousands). Since the distribution of nets is anticipated to increase at a constant percentage rate, we can use an exponential model. Since $t = 0$ corresponds with 1999, the initial value is 538. So far we have $n(t) = 538(b)^t$. Although the growth factor is not readily apparent, we can calculate it by substituting the second data point into the equation. Since 2003 corresponds with $t = 4$, we have

$$9485 = 538(b)^4$$
$$17.63 = b^4$$
$$(17.63)^{1/4} = (b^4)^{1/4}$$
$$2.049 = b$$

Thus the exponential model is $n(t) = 538(2.049)^t$.

To determine the net distribution level in 2011, we evaluate this function at $t = 12$.

$$n(12) = 538(2.049)^{12}$$
$$\approx 2{,}946{,}000 \text{ (accurate to 4 significant digits)}$$

According to the model, 2,946,000 thousand (2.946 billion) nets will be distributed in 2011.

Although the model gives a good estimate for years near the original data set, the further we move away from 2003, the less confident we are in the prediction because few things can sustain exponential growth indefinitely. Since 2011 is relatively far away from the last year in the data set (2003), we question the accuracy of the forecast.

EXAMPLE 2 ■ Comparing a Rate of Change to a Percentage Change Rate

The Netto Extra Treated Net was launched by Netto Manufacturing Co., Ltd., in April 2005 to meet the high demand for insecticide treated nets. According to a certificate of analysis issued with the net, the initial deltamethrin (insecticide) content in the net is 50.40 milligrams per square meter. After six washes, the deltamethrin residue was measured to be 34.72 milligrams per square meter. (*Source:* www.nettogroup.com)

Calculate the average rate of change and percentage rate of change. Which one more accurately represents this situation?

Solution To calculate the average rate of change, we use the average rate of change formula from Chapter 4.

$$\frac{50.40 - 34.72}{0 - 6} \frac{\text{mg per square meter}}{\text{washes}} \approx -2.613 \text{ mg per square meter per wash}$$

On average, each wash removes 2.613 milligrams of deltamethrin per square meter.

To determine the percentage change, we must first determine the decay factor.

$$\left(\frac{y_2}{y_1}\right)^{\frac{1}{x_2 - x_1}} = \left(\frac{50.40}{34.72}\right)^{\frac{1}{0-6}}$$
$$\approx 0.9398$$

We subtract 1 from the decay factor to determine the percentage rate of change.

$$0.9398 - 1 = -0.0602$$

Approximately 6.02% of the deltamethrin residue is removed with each wash.

We expect that as the quantity of deltamethrin residue available in the net decreases, the amount that is removed in each subsequent wash will also decrease. For this reason, the percentage rate of change seems to more accurately represent what is going on in this situation.

■ Modeling Half-Life and Doubling Time

In November 2006, Alexander Litvinenko, a former Russian spy living in Britain, was killed by poisoning while investigating the death of a Russian journalist. The poison used was polonium-210, a radioactive substance. The ensuing investigation led to international finger-pointing; however, as of April 2011, the case remained unsolved.

Radioactive substances such as polonium-210 decay exponentially. The *half-life* of polonium-210 is 138.376 days (138 days, 9 hours, 1 minute, and 26 seconds). The **half-life** of a substance is the amount of time it takes for half of the initial amount of the substance to remain. Half-lives are used widely in chemistry when comparing various radioactive elements.

EXAMPLE 3 Determining a Percentage Rate of Decay from a Half-Life

Polonium-210 has a half-life of 138.376 days. What percentage of the substance decays each day?

Solution Since the substance is decaying exponentially, we can model the amount remaining by $y = ab^t$. Since half of the initial value remains after 138.376 days, we have

$$\frac{1}{2}a = ab^{138.376}$$

$$\frac{1}{2} = b^{138.376}$$

$$\left(\frac{1}{2}\right)^{1/138.376} = (b^{138.376})^{1/138.376}$$

$$0.9950 = b$$

Since $b = 1 + r$, $r = -0.005$. The amount of polonium-210 remaining is decreasing at a rate of 0.5% per day.

Doubling time is the amount of time it takes for something that is growing to double. As was the case with half-life, doubling time is independent of the initial value of the exponential function.

EXAMPLE 4 Determining the Percentage Rate of Growth from a Doubling Time

The median price of a home in the United States was about $250,000 in March 2007. (*Source:* www.zillow.com) At what annual percentage rate would property values have to increase for the median price to double by March 2017?

Solution Since we are assuming a constant percentage growth, we can use an exponential function to model the value of the investment.

$$2(250{,}000) = 250{,}000(1 + r)^{10}$$
$$2 = (1 + r)^{10}$$
$$(2)^{1/10} = [(1 + r)^{10}]^{1/10}$$
$$1.072 \approx 1 + r$$
$$0.072 \approx r$$

A doubling time of 10 years corresponds with an annual percentage rate of about 7.2%.

EXAMPLE 5 Determining If Data Can Be Represented by an Exponential Model

Table 6.11 shows the average price for brand-name prescription drugs from 2000 to 2004. Determine if the data set has a constant or nearly constant percentage rate of change. If it does, model the data set with an exponential function and estimate the average drug price in 2006.

Solution Since the input values are equally spaced, we need only determine if consecutive output values have a constant ratio. See Table 6.12.

Table 6.11

Years Since 2000	Average Brand-Name Drug Price (dollars)
0	65.29
1	69.75
2	77.49
3	85.57
4	95.86

Source: Statistical Abstract of the United States, 2006, Table 126

Table 6.12

Years Since 2000	Average Brand-Name Drug Price (dollars)	Ratio of Consecutive Output Values
0	65.29	1.06831061
1	69.75	1.11096774
2	77.49	1.10427152
3	85.57	1.12025242
4	95.86	

The ratios are all approximately equal to 1.1, so they are nearly constant. Thus an exponential model is appropriate for this data set. Using 65.29 as the initial value, we construct an exponential model

$$p(t) = 65.29(1.1)^t \text{ dollars}$$

where t is the number of years since 2000.

To determine the average drug price in 2006, we evaluate the function at $t = 6$.

$$p(6) = 65.29(1.1)^6$$
$$\approx 115.7$$

According to the model, the average brand-name drug price in 2006 was $115.70.

■ Using Regression to Find an Exponential Model

In Example 5, we constructed an exponential model for a data set with nearly constant ratios. By using exponential regression, we can find the exponential model that best fits the data set. Using the Technology Tip at the end of this section, we determine the exponential model of best fit for the brand-name prescription drug price is

$$p(t) = 64.25(1.102)^t \text{ dollars}$$

where t is the number of years since 2000. Using the model of best fit, we get a slightly different estimate for the average brand-name prescription drug price in 2006.

$$p(6) = 64.25(1.102)^6$$
$$\approx 115.10$$

Table 6.13

Years Since 1990 t	U.S. Exports to Colombia ($ millions) C
0	119
10	415
11	452
12	520
13	512
14	593
15	677

Source: Statistical Abstract of the United States, 2007, Table 827

EXAMPLE 6 ■ Using Exponential Regression to Model a Data Set

Find the equation of the exponential function that best fits the data set shown in Table 6.13. Then forecast the U.S. exports to Colombia in 2010.

Solution Using the Technology Tip at the end of this section, we obtain

$$C(t) = 122.8(1.122)^t \text{ million dollars}$$

where t is the number of years since the end of 1990.

To forecast exports in 2010, we evaluate the function at $t = 20$.

$$C(20) = 122.8(1.122)^{20}$$
$$= 1228 \text{ million dollars}$$

We estimate that U.S. exports to Colombia will be $1228 million in 2010.

Graphing Exponential Functions

The growth factor plays a significant role in determining the shape of the graph of an exponential function. Let's investigate its effect in the context of economic forecasting.

The Congressional Budget Office (CBO) is given the responsibility of forecasting the economic future of the U.S. government. In a March 2007 report entitled *The Uncertainty of Budget Projections: A Discussion of Data and Methods*, the CBO explained that "uncertainty increases as the projections extend into the future" (Preface).

For example, the CBO estimated the *nominal gross domestic product* (GDP) was $13,235 billion in 2006 and that it would increase by 4.3% in 2007. (*Source:* www.cbo.gov) (The GDP of a country is the market value of its final goods and services produced within a year.) Assuming growth at a constant percentage rate in the future, we can construct an exponential function model for the GDP. Since the initial value is $13,235 billion and the annual growth rate is 4.3%, we have

$$G(t) = 13{,}235(1.043)^t \text{ billion dollars}$$

where t is the number of years since 2006. The graph of the function is shown in Figure 6.1.

From the graph, we see that the projected GDP in 2016 ($t = 10$) is just over $20,000 billion. However, with any projection there is a level of uncertainty. To address this concern, let's look at the additional projections shown in Figure 6.1 for rates somewhat near 4.3%—2.3%, 3.3%, 5.3%, and 6.3%. This will give us an idea of possible ranges of values. We see that the larger the percentage rate, the steeper the graph. That is, the higher the percentage rate, the higher the rate of change in the GDP. We also see that although the difference between the projections for 2016 is substantial, the difference in the projections for 2007 is relatively small.

Thus, the change factor tells us much about the graph. Recall that the change factor b is equivalent to 1 plus the percentage change rate. That is, $b = 1 + r$.

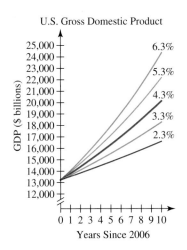

Figure 6.1

THE GRAPHICAL SIGNIFICANCE OF THE CHANGE FACTOR

The change factor, b, controls the steepness and increasing/decreasing behavior of the exponential function $y = ab^x$. For positive a,

- if $b > 1$, the graph is increasing, and increasing the value of b will make the graph increase more rapidly.
- if $0 < b < 1$, the graph is decreasing, and decreasing the value of b will make the graph decrease more rapidly.

EXAMPLE 7 ■ **Identifying an Exponential Function Based on Its Graph**

Figure 6.2 shows population models for Cambodia, Kazakhstan, Rwanda, and Hungary. Based on data from 1995 to 2004, the population percentage rates of change for the countries are 2.0%, −0.7%, 5.0%, and −0.2%, respectively. (*Source:* World Health Statistics, 2006) Identify which graph corresponds to each country. Then estimate the 2005 population for each country.

Solution Both Cambodia and Rwanda have positive growth rates, so their graphs will be increasing. However, Rwanda has a higher percentage rate of growth so its graph will be steeper. Graph *B* corresponds with Rwanda and Graph *A* with Cambodia.

Both Kazakhstan and Hungary have a negative growth rate so their populations are decreasing. Since Hungary is decreasing at a less negative percentage rate, its graph will be less steep. So Graph *D* corresponds with Hungary and Graph *C* with Kazakhstan.

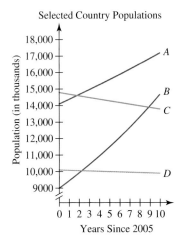

Figure 6.2

The vertical intercept of each graph is the initial population of the corresponding country. The initial populations (in thousands) are approximately

Kazakhstan: 14,800

Cambodia: 14,100

Hungary: 10,100

Rwanda: 9000

In Example 7, we stated that the vertical intercept of the graph corresponded with the initial value of the exponential function. To see why, let's consider a generic exponential function, $y = ab^x$. To find the vertical intercept, we set x equal to zero.

$$y = ab^0$$
$$= a(1) \quad \text{since } b^0 = 1 \text{ for } b \neq 0$$
$$= a$$

So the vertical intercept is $(0, a)$ and the initial value of the exponential function corresponds with the vertical intercept.

GRAPHICAL MEANING OF THE INITIAL VALUE

The exponential function $y = ab^x$ has vertical intercept $(0, a)$, where a is the initial value of the function.

EXAMPLE 8 ■ Interpreting Exponential Function Graphs

Inflation refers to the increase in prices that occurs over time. For example, if the annual inflation rate is 3% then an item that costs $100 today will cost $100 + 100(0.03) = 103 a year from now.

Consider the price of the chic leather boots for women shown in the ad below. In 1911, they cost $3.79. A comparable boot, shown next to the ad, retailed for $57.00 in 2011.

a. Determine the average annual inflation rate for the boots.
b. Graph an exponential function model for the price of the boots.
c. Estimate the price of the boots in 1963 and 2003 from the graph.

Solution
a. We first need to determine the ratio of the prices.

$$\frac{57}{3.79} = 15.04$$

The 2011 price was about 15 times more than the 1911 price. To determine the annual growth factor, we raise this number to 1 over the length of the period between 1911 and 2011 (100 years).

$$(15.04)^{(1/100)} \approx 1.0275$$

We subtract 1 from the growth factor to obtain the annual percentage change rate.

$$1.0275 - 1 = 0.0275$$
$$= 2.75\%$$

The average annual rate of inflation on the boots was about 2.75%.

b. A model for the price of the boots t years after 1911 is $p(t) = 3.79(1.0275)^t$. The graph is shown in Figure 6.3. Although the percentage change rate is constant (2.75%), the annual increase in price increases as the price itself increases, so the graph is concave up.

c. In 1963, $t = 52$. From the graph it appears as if the boot price was approximately $15 in 1963. In 2003, $t = 92$. From the graph it appears as if the boot price was approximately $46 in 2003.

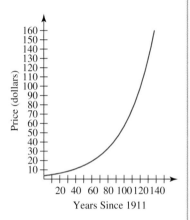

Figure 6.3

All exponential functions have a *horizontal asymptote* at the horizontal axis. To see why, suppose we have a decreasing exponential function with initial value 100 and constant percentage change -10%. This means for each 1 unit increase in the input, the output decreases by 10%. Table 6.14 shows the values of the function for the first four nonnegative integer values of the input.

Table 6.14

x	y	10% of y
0	100	$0.1(100) = 10$
1	$100 - 10 = 90$	$0.1(90) = 9$
2	$90 - 9 = 81$	$0.1(81) = 8.1$
3	$81 - 8.1 = 72.9$	$0.1(72.9) = 7.29$

Notice as the value of y becomes smaller, 10% of y also becomes smaller. That is, the amount by which y is decreasing is getting smaller as x increases. Also notice each function value is 90% of the value before it (e.g., 90 is 90% of 100, 81 is 90% of 90, and so on). This pattern will continue with the y-values becoming smaller while remaining positive. In other words, as $x \to \infty$, $y \to 0$. Thus a horizontal asymptote occurs at the horizontal axis. A similar argument applies for increasing exponential functions. In that case, as $x \to -\infty$, $y \to 0$.

In general, the graph of an exponential function will take on one of the four basic shapes shown in Figure 6.4. Notice in each case the line $y = 0$ (the horizontal axis) is a horizontal asymptote. Notice also that, as was the case with quadratic functions, the value of a controls the concavity of the graph. If $a > 0$, the graph is concave up. If $a < 0$, the graph is concave down.

362 CHAPTER 6 ■ EXPONENTIAL AND LOGARITHMIC FUNCTIONS

(a) $a > 0, b > 1$
concave up, increasing

(b) $a > 0, 0 < b < 1$
concave up, decreasing

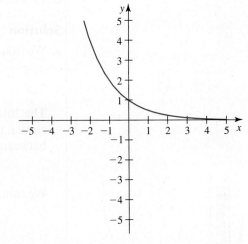

(c) $a < 0, b > 1$
concave down, decreasing

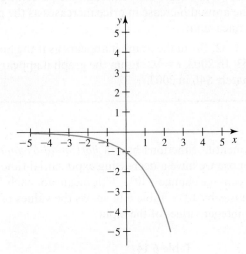

(d) $a < 0, 0 < b < 1$
concave down, increasing

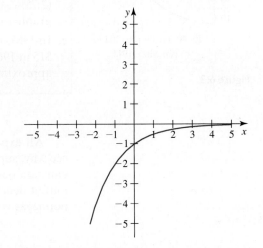

Figure 6.4

As was the case with other functions we have investigated, exponential function graphs may be altered using transformations, as shown in the next example.

EXAMPLE 9 ■ **Using Exponential Regression to Model an Aligned Data Set**

Table 6.15 shows the amount of land in farms in the United States between 1978 and 1997.

Table 6.15

Years Since 1978 t	Land in Farms (million acres) F	Years Since 1978 t	Land in Farms (million acres) F
0	1014.8	14	945.5
4	986.8	19	931.8
9	964.5		

Source: Statistical Abstract of the United States, 2001, Table 796

Find a function model for the data set. Then use the model to forecast the land in farms in 2007.

Solution We draw the scatter plot of the data, shown in Figure 6.5, to get an idea of what type of function may fit the data set. The data set appears to be concave up and decreasing so an exponential model may fit the data set well. We use the Technology Tip at the end of this section to find the exponential model of best fit, $F(t) = 1008(0.9956)^t$, and graph the resultant function in Figure 6.6.

Figure 6.5

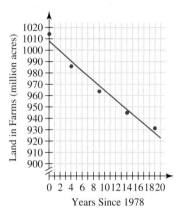

Figure 6.6

The graph, which appears nearly linear, does not fit the data set. The coefficient of determination from the graphing calculator, $r^2 = 0.976$, is not as close to 1 as we had expected. What happened? Recall the calculator assumes that an exponential function has a horizontal asymptote at $y = 0$, but this particular data set is "bending" too quickly to have a horizontal asymptote at $y = 0$. We can visually estimate that this data set will have a horizontal asymptote at $y = 900$ and create an aligned set of data by subtracting 900 from each of the farm size values as shown in Table 6.16.

Table 6.16

Years Since 1978 t	Land in Farms (million acres) F	Aligned Data $F - 900$
0	1014.8	114.8
4	986.8	86.8
9	964.5	64.5
14	945.5	45.5
19	931.8	31.8

Now we repeat the regression using the aligned data as the dependent variable. The resultant model, $A(t) = 115.1(0.9352)^t$, has a coefficient of determination much closer to 1 ($r^2 = 0.9990$). Because this value is close to 1, we see that our guess that the horizontal asymptote is $y = 900$ was a good one. To vertically shift this model to the position of the farm land data, we add back 900 to get $L(t) = 115.1(0.9352)^t + 900$. We graph the new model along with the scatter plot and the original model in Figure 6.7. We see that the new model fits the data much better than did the original.

To forecast the 2007 land in farms, we evaluate $L(t)$ at $t = 29$.

$$L(t) = 115.1(0.9352)^t + 900$$
$$L(29) = 115.1(0.9352)^{29} + 900$$
$$\approx 916.5$$

We estimate in 2007 there were 916.5 million acres of farm land in the United States.

Figure 6.7

364 CHAPTER 6 ■ EXPONENTIAL AND LOGARITHMIC FUNCTIONS

SUMMARY

In this section you learned how to construct an exponential function model from tables and words. You discovered how to use exponential regression to find the exponential model that best fits a data set. You also studied the various characteristics of an exponential function's graph, including the effect of the initial value and growth factor on the graph. Finally, you learned that sometimes it is necessary to align a data set to find the best model.

TECHNOLOGY TIP ■ EXPONENTIAL REGRESSION

1. Enter the data using the Statistics Menu List Editor.

2. Bring up the Statistics Menu Calculate feature by pressing [STAT] and using the blue arrows to move to CALC. Then select item 0: ExpReg and press [ENTER].

3. If you want to automatically paste the regression equation into the Y = Editor, press the key sequence [VARS]; Y-VARS; Function; Y1 and press [ENTER]. Otherwise press [ENTER].

6.2 EXERCISES

■ SKILLS AND CONCEPTS

In Exercises 1–5, identify which of the following exponential graphs corresponds with each given equation.

1. $y = 350(1.3)^x$
2. $y = 305(1.2)^x$
3. $y = 600(0.82)^x$
4. $y = 300(0.82)^x$
5. $y = 600(0.45)^x$

In Exercises 6–10, describe the graph of the function without drawing the graph. Use the terms increasing, decreasing, concave up, concave down, vertical intercept, and horizontal asymptote, as appropriate.

6. $f(x) = 21(0.9)^x$
7. $g(x) = -4(2.9)^x$
8. $h(x) = -0.8(1.9)^x$
9. $f(t) = 4(1.8)^t + 17$
10. $f(t) = 0.25(4)^t - 2$

■ SHOW YOU KNOW

11. How can you tell if a table of data represents an exponential function?

12. A classmate claims that any two points define a unique exponential function. Do you agree? Explain.
13. What are some key words in a verbal description that indicate an exponential function can be used to model the situation? What is meant by such words?
14. What does the term *half-life* mean?
15. What does the term *doubling time* mean?

■ **MAKE IT REAL**

In Exercises 16–20, determine which of the following graphs corresponds with a population model for each of the given countries. Then estimate the 2005 population of the country.

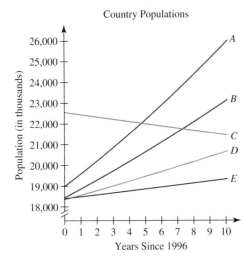

Source: **Modeled from World Health Organization data**

16. **Australia Population**   Between 1996 and 2006, the population of Australia grew at a rate of 1.2% annually.
17. **Ghana Population**   Between 1996 and 2006, the population of Ghana grew at a rate of 2.3% annually.
18. **Romania Population**   Between 1996 and 2006, the population of Romania *decreased* at a rate of 0.5% annually.
19. **Sri Lanka Population**   Between 1996 and 2006, the population of Sri Lanka grew at a rate of 0.5% annually.
20. **Afghanistan**   Between 1996 and 2006, the population of Afghanistan grew at a rate of 3.2% annually.

In Exercises 21–30, model the data with an exponential function, if appropriate, and answer the given questions. If an exponential function model is not appropriate for the situation, explain why. Do not use regression for these exercises.

21. **Population of the United States**   According to the World Health Organization, the population of the United States was 298,213 thousand in 2005. Between 1995 and 2004, the population grew at an average rate of 0.9% annually. (*Source:* **World Health Statistics, 2006**) Assuming the percentage growth rate will remain the same in the future, model the U.S. population as a function of years since 2005.

22. **Population of China**   According to the World Health Organization, the population of China was 1,323,345 thousand in 2005. Between 1995 and 2004, the population grew at an average rate of 0.7% annually. (*Source:* **World Health Statistics, 2006**) Assuming the population will continue to grow at this percentage rate, model the population of China as a function of years since 2005. What is the projected population of China in 2011?

23. **Crude Oil Production**   Based on data from 1970 to 2004, the production of crude oil in the United States has decreased at a rate of roughly 0.27 quadrillion BTU per year. In 1970, the production level was 20.4 quadrillion BTU. (*Source: Statistical Abstract of the United States, 2007,* Table 895)

24. **National Health Expenditures**   Between 1960 and 2004, the total national expenditure on health costs increased by roughly 10.1% annually. In 1960, national health expenditures were $28 billion. (*Source: Statistical Abstract of the United States, 2007,* Table 120)

25. **Research and Development Spending**   In 2003, Microsoft Corporation was ranked number 1 in research and development spending. That year, the company spent $7779 million on research and development. The 10th-ranked company was Glaxo Smith Kline, which spent $4910 million on research and development. According to the list of the top 10 companies, for each increase in rank number, the amount of spending decreased at a nearly constant percentage rate. (*Source: Science and Engineering Indicators 2006,* National Science Foundation, Table 4-6).

    Use the model to predict how much money was spent on research and development by the 5th-ranked company (Toyota Motor). Then compare your model result to the actual amount spent ($6210 million).

26. **Insurance Expenditures for Health Care**   Between 1960 and 2004, insurance company expenditures for health care increased at an ever-increasing rate. In 1960, $6 billion was spent on health care. In 2004, $659 billion was spent on health care. (*Source: Statistical Abstract of the United States, 2007,* Table 120)

27. **Atmospheric Pressure**   At 0 meters elevation, atmospheric pressure is 14.696 pounds per square inch. At 8500 meters, atmospheric pressure is 4.801 pounds per square inch. As elevation increases, atmospheric pressure decreases less rapidly. (*Source:* Modeled from Digital Dutch 1976 Standard Atmosphere Calculator) According to the model, what is the atmospheric pressure at 6000 meters?

28. **Salary for Engineers and Scientists**   Between 1993 and 2003, annual median salaries for science and engineering occupations increased from $48,000 to $66,000. Over that time period, median salaries increased at a rate of roughly $1800 per year. (*Source: Science and Engineering Indicators 2006,* National Science Foundation, Table 3-8)

29. **Adults with Only an Elementary Education**   In 1910, 23.8% of the adults in the United States had only an elementary education. By 1950 that number had fallen to 11.1%, and by 1990 it had dropped to 2.4%. (*Source: Digest of Education Statistics, 2005,* National Center for Education Statistics)

30. **Coca-Cola Production**   In its 2001 Annual Report, the Coca-Cola Company reported

    "Our worldwide unit case volume increased 4 percent in 2001, on top of a 4 percent increase in 2000. The increase in unit case volume reflects consistent

performance across certain key operations despite difficult global economic conditions. Our business system sold 17.8 billion unit cases in 2001." (**Source: Coca-Cola Company 2001 Annual Report, p. 46**)

A unit case is equivalent to 24 8-ounce servings of finished beverage.

In Exercises 31 to 40, use exponential regression to find a model for the data set. Then use the model to answer the given question. (As appropriate, align the data before finding the model equation.)

31. School Expenditures

School Year Since 1990–1991 t	Expenditure per Pupil (dollars) E
0	4902
2	5160
4	5529
6	5923
8	6508
10	7380
12	8044

Source: **National Center for Education Statistics**

What are the projected school expenditures per pupil in 2010–2011?

32. Atmospheric Pressure

Altitude (meters) a	Atmospheric Pressure (psi) p
0	14.696
1000	13.035
2000	11.530
3000	10.168
4000	8.940
5000	7.835
6000	6.843
7000	5.955
8000	5.163

Source: **Digital Dutch 1976 Standard Atmosphere Calculator**

What is the atmospheric pressure at 4500 meters?

33. National Health Spending

Years Since 1995 t	Health Expenditures ($ billions) H	Years Since 1995 t	Health Expenditures ($ billions) H
0	1020	5	1359
1	1073	6	1474
2	1130	7	1608
3	1196	8	1741
4	1270	9	1878

Source: Statistical Abstract of the United States, 2007, **Table 120**

At what percentage rate are health expenditures increasing?

34. Brand-Name Drug Prices

Years Since 1995 t	Average Brand-Name Drug Price (dollars) d
0	40.22
2	49.55
3	53.51
4	60.66
5	65.29
6	69.75
7	77.49
8	85.57
9	95.86

Source: Statistical Abstract of the United States, 2006, **Table 126**

What do the initial value and growth factor of the model represent in this real-world context?

35. Highway Accidents

Years Since 2000 t	Accidents Resulting in Injuries (percent) p
0	49.9
1	48.0
2	46.3
3	45.6
4	45.1

Source: Statistical Abstract of the United States 2007, **Table 1047**

What does the change factor tell about highway accident injuries?

36. Municipal Governments

Number of Local Municipal Governments (in thousands) m	Years Since 1965 Y
18.048	2
18.517	7
18.862	12
19.076	17
19.200	22
19.279	27
19.372	32
19.429	37

Source: Statistical Abstract of the United States, 2007, **Table 415**

In what year is the number of municipal governments projected to reach 20 thousand?

37. Government Expenditures

Years Since 1990 y	Expenditures ($ billions) E
0	1872.6
5	2397.6
10	2886.5
11	3061.6
12	3240.8
13	3424.7
14	3620.6
15	3877.2

Source: Statistical Abstract of the United States, 2007, Table 418

What are the projected government expenditures for 2010?

38. Bottled Water

Years Since 1980 t	Per Capita Bottled Water Consumption (gallons) w
0	2.4
5	4.5
10	8.0
14	10.7
15	11.6
16	12.5
17	13.1
18	16.0
19	18.1

Source: Statistical Abstract of the United States, 2001, Table 204

Who could benefit from the per capita bottled water consumption model?

39. Professional Basketball Salaries

Years Since 1980 t	NBA Average Salary ($1000s) b
0	170
5	325
10	750
15	1900
16	2000
17	2200
18	2600

Source: Statistical Abstract of the United States, 2001, Table 1324

Will the average NBA salary be more or less than $3 million in 2010?

40. U.S. Gross Domestic Product

Years Since 1930 t	GDP ($ billions) g
0	91.3
10	101.3
20	294.3
30	527.4
40	1039.7
50	2795.6
60	5803.2
70	9872.9

Source: www.lycos.com

What will be the gross domestic product of the United States in 2010?

■ **STRETCH YOUR MIND**

Exercises 41–45 are intended to challenge your understanding of exponential functions.

41. Given that an exponential function $y = ab^x$ passes through the points (c, ab^c) and $(c + h, ab^{c+h})$,

   a. Create a formula that calculates the average rate of change of the exponential function over the interval $[c, c + h]$. (*Hint:* $[c, c + h]$ means $c \leq x \leq c + h$.)

   b. Describe the relationship between the formula for the average rate of change of an exponential function and the equation of the exponential function.

42. Explain why a function with a constant percentage change cannot be linear.

43. Explain why an exponential function $y = ab^x$ does not have a horizontal intercept.

44. A classmate claims that two exponential functions, $f(x) = ab^x$ and $g(x) = cd^x$, will never intersect if $a > c$ and $b > d$. Do you agree? Justify your conclusion.

45. A classmate incorrectly claims any function that is concave up and has a horizontal asymptote at the *x*-axis is an exponential function. Give the equation of a function that could be used to persuade your classmate that he is incorrect.

SECTION 6.3 Compound Interest and Continuous Growth

LEARNING OBJECTIVES

- Use the compound interest formula to calculate the future value of an investment
- Construct and use continuous growth models
- Use exponential models to predict and interpret unknown results

GETTING STARTED

Saving for retirement is an important goal for many people, but Americans are saving less money than they should. When planning for retirement, the earlier you can start saving and investing the better off you are, thanks to the benefits of compound interest.

In this section we investigate the compound interest formula and its importance in financial planning. In addition, we learn about continuous growth and the number e. These concepts are used in real-world situations as diverse as saving for retirement, population growth, and radioactive half-life.

■ Compound Interest

Banks make a profit by loaning money to people and charging interest for this service. Some of the money that banks lend comes from customers who deposit money with the bank. To encourage customers to deposit their money with the bank, a bank pays its customers *interest*, calculated as a percentage of the amount of money the customer has in the bank during each interest period.

Consider the following situation. In April 2007, Heritage Bank offered a 1-year certificate of deposit (CD) with a minimum investment of $1000 at a *nominal interest rate* of 4.42% compounded quarterly. (*Nominal* means "in name only.") (*Source:* www.bankrate.com) We will use this interest rate to calculate the future value of a $1000 investment in the certificate of deposit after 1 year. The future value relies on the interest rate and the **compounding frequency**, the number of times per year that interest is paid. This particular CD is compounded quarterly (four times a year). To find the impact this will have on the future value of the CD after 1 year, we need to calculate the quarterly **periodic rate**, or the interest rate that will be applied to the CD every 3 months. We do this by dividing the nominal interest rate by the compounding frequency.

$$\frac{4.42\%}{4} = 1.105\%$$

Thus, every 3 months Heritage Bank will pay interest equivalent to 1.105% of the money in the account.

PERIODIC RATE

The **periodic rate** is calculated as

$$\text{periodic rate} = \frac{\text{nominal interest rate}}{\text{compounding frequency}}$$

We can now find the future value of the CD after 1 year by using a quarterly growth factor of 1.01105 (an increase of 1.105%) as shown in Table 6.17.

Table 6.17

End of Quarter	Value Calculation	End-of-Quarter Value
1 (3 months)	1000(1.01105)	$1011.05
2 (6 months)	1011.05(1.01105)	$1022.22
3 (9 months)	1022.22(1.01105)	$1033.52
4 (12 months)	1033.52(1.01105)	$1044.94

Since the nominal rate was 4.42%, we may have expected the final value to be $1044.20 instead of $1044.94. Why the $0.74 difference? To find out, let's look more closely at the values of the CD at the end of each of the four quarters of the year. See Table 6.18.

Table 6.18

Quarter	End-of-Quarter Value	Interest Payment
0 (initial deposit)	$1000.00	
1	$1011.05	$11.05
2	$1022.22	$11.17
3	$1033.52	$11.30
4	$1044.94	$11.42

Note that the interest payments continue to increase throughout the year. This is due to **compound interest**, which occurs when interest is paid on the initial amount invested as well as all previously earned interest.

■ The Compound Interest Formula

The preceding method to find a future value is not very efficient. Let's use our knowledge of exponential functions to determine a formula that will allow us to quickly and easily calculate future value. Since the initial value of the CD was $1000 and the quarterly growth factor was 1.0115, the value of the CD after q quarters is given by

$$V(q) = 1000(1.0115)^q \text{ dollars}$$

In general, the equation will be of the form

future value = (initial amount) · (periodic growth factor)$^{\text{(total number of compoundings)}}$

= (initial amount) · (1 + periodic rate)$^{\text{(total number of compoundings)}}$

The final value is usually described in terms of the number of years that have passed:

future value = (initial amount) · (1 + periodic rate)$^{\text{(total number of compounding)}}$

$$= \text{(initial amount)} \cdot \left(1 + \frac{\text{nominal rate}}{\text{compoundings per year}}\right)^{\left(\frac{\text{compoundings}}{\text{year}} \cdot \text{years}\right)}$$

To simplify the notation, we represent each of the quantities with a variable when we write the compound interest formula.

COMPOUND INTEREST FORMULA

The future value A of an initial investment P is given by

$$A = P\left(1 + \frac{r}{n}\right)^{nt}$$

where n is the compounding frequency (the number of times interest is paid per year), t is the number of years the money is invested, and r is the nominal interest rate in decimal form $\left(r = \dfrac{\text{nominal rate}}{100\%}\right)$.

Table 6.19 shows compounding frequencies and associated formulas for compound interest.

Table 6.19

Compounding Frequency	Number of Compoundings per Year	Compound Interest Formula
annually	1	$A = P\left(1 + \dfrac{r}{1}\right)^{1t}$
semiannually	2	$A = P\left(1 + \dfrac{r}{2}\right)^{2t}$
quarterly	4	$A = P\left(1 + \dfrac{r}{4}\right)^{4t}$
monthly	12	$A = P\left(1 + \dfrac{r}{12}\right)^{12t}$
daily	365	$A = P\left(1 + \dfrac{r}{365}\right)^{365t}$

EXAMPLE 1 ■ **Comparing Future Values**

In March 2007, MetLife Bank advertised a 5-year CD with a 4.64% nominal interest rate compounded daily. (*Source:* www.bankrate.com) Assuming someone could continue to invest their money at this rate and compounding frequency up until retirement at age 65, compare the future value of a $5000 investment of a 25-year-old to the future value of a $10,000 investment for a 45-year-old. Assume no additional deposits or withdrawals are made.

Solution The 25-year-old will be accumulating interest 365 days a year for 40 years (the time until she retires), while the 45-year-old will be accumulating interest 365 days a year for 20 years (the time until he retires). Using the compound interest formula, we have

The 25-year-old

$$A = P\left(1 + \frac{r}{n}\right)^{nt}$$
$$= 5000\left(1 + \frac{0.0464}{365}\right)^{365 \cdot 40}$$
$$= 5000(1 + 0.00013)^{14{,}600}$$
$$= 5000(1.00013)^{14{,}600}$$
$$= \$31{,}986.69$$

The 45-year-old

$$A = P\left(1 + \frac{r}{n}\right)^{nt}$$
$$= 10{,}000\left(1 + \frac{0.0464}{365}\right)^{365 \cdot 20}$$
$$= 10{,}000(1 + 0.00013)^{7300}$$
$$= 10{,}000(1.00013)^{7300}$$
$$= \$25{,}292.96$$

Deducting the amount of the initial investment, we see that the 25-year-old earned $26,986.69 in interest while the 45-year-old earned $15,292.96 in interest.

EXAMPLE 2 ■ **Comparing the Effects of Different Interest Rates**

In March 2007, Third Federal Savings and Loan offered a 5-year CD at a 5.15% nominal interest rate compounded quarterly. At the same time, Chase Bank offered a 5-year CD at a 3.68% nominal interest rate compounded daily. (*Source:* www.bankrate.com) If you want the future value of the CD to be $5000 in 5 years, determine which account would require you to invest the least amount.

Solution In this case we know the future value A and need to find the initial investment P.

Third Federal S&L

$$A = P\left(1 + \frac{r}{n}\right)^{nt}$$

$$5000 = P\left(1 + \frac{0.0515}{4}\right)^{4(5)}$$

$$5000 = P(1 + 0.013)^{20}$$

$$5000 = P\,(1.013)^{20}$$

$$5000 = P(1.292)$$

$$\frac{5000}{1.292} = P$$

$$\$3871.27 = P$$

Chase Bank

$$A = P\left(1 + \frac{r}{n}\right)^{nt}$$

$$5000 = P\left(1 + \frac{0.0368}{365}\right)^{365(5)}$$

$$5000 = P(1 + 0.0001)^{1825}$$

$$5000 = P(1.0001)^{1825}$$

$$5000 = P(1.202)$$

$$\frac{5000}{1.202} = P$$

$$\$4159.72 = P$$

The Third Federal S&L CD requires an initial deposit $288.45 less ($4159.72 − $3871.27) than the Chase Bank CD.

Compound Interest as an Exponential Function

Since compound interest causes an account to grow by a fixed percent each year, we may categorize the compound interest formula as an exponential function. First, recall the following property of exponents.

$$(x^a)^b = x^{ab} \quad \text{or} \quad x^{ab} = (x^a)^b$$

From this property we can rewrite the compound interest formula as follows.

$$A = P\left(1 + \frac{r}{n}\right)^{nt}$$

$$= P\left[\left(1 + \frac{r}{n}\right)^n\right]^t$$

If we then let $b = \left(1 + \frac{r}{n}\right)^n$, we can rewrite the compound interest formula in the form $y = ab^x$.

$$A = Pb^t$$

For example, a 4.4% CD compounded monthly with an initial deposit of $1000 could be written as

$$A = P\left[\left(1 + \frac{r}{n}\right)^n\right]^t$$

$$= 1000\left[\left(1 + \frac{0.044}{12}\right)^{12}\right]^t$$

$$= 1000[(1 + 0.00367)^{12}]^t$$

$$= 1000[(1.00367)^{12}]^t$$

$$= 1000(1.0449)^t \quad \text{Evaluate } (1.00367)^{12}.$$

Written in this form, the 1.0449 is the annual growth factor and the corresponding rate (4.49%) is the *annual percentage yield*. The **annual percentage yield (APY)** is the *actual* amount of interest earned during the year and takes into account not only the nominal rate but also the compounding frequency. The Truth in Savings Act of 1991 requires banks to publish the APY for all savings accounts and CDs so consumers can more easily compare offers from different banks and make informed financial decisions.

ANNUAL PERCENTAGE YIELD (APY)

The **annual percentage yield** of an investment earning a nominal interest rate r compounded n times per year is given by

$$\text{APY} = \left(1 + \frac{r}{n}\right)^n - 1$$

$$= \text{annual growth factor} - 1$$

EXAMPLE 3 ■ Modeling Compound Interest with Exponential Functions

An investor makes a $3000 initial deposit in a CD with a 3.68% interest rate compounded weekly.

a. Use the compound interest formula to find an exponential function to model the future value of the investment.

b. Determine the annual percentage yield for the investment.

Solution

a.
$$A = P\left(1 + \frac{r}{n}\right)^{nt}$$

$$= 3000\left(1 + \frac{0.0368}{52}\right)^{52t} \quad \text{1 year = 52 weeks}$$

$$= 3000(1 + 0.0007)^{52t}$$

$$= 3000(1.0007)^{52t}$$

$$= 3000[(1.0007)^{52}]^t$$

$$= 3000(1.0375)^t$$

So $A(t) = 3{,}000(1.0375)^t$ models the future value of the CD.

b. From the exponential model, we can see the annual growth factor is 1.0375 and the corresponding annual percentage yield is 3.75% (1.0375 − 1).

Does increasing the compounding frequency always increase the future value of an investment? In Example 4, we explore this question.

SECTION 6.3 ■ COMPOUND INTEREST AND CONTINUOUS GROWTH

EXAMPLE 4 ■ Finding the Continuous Growth Factor

Using the compound interest formula, examine what happens to $1 invested in an account with a 100% interest rate if interest is compounded an increasing number of times per year. Discuss what you discover.

Solution Based on the examples we have seen thus far, it appears that the value of the account will increase as we increase the compounding frequency. To see whether compounding every hour, minute, or second will dramatically increase the account value, consider Table 6.20.

Table 6.20

Compounding Method	Compounding Frequency n	Growth Factor $\left(1 + \frac{1}{n}\right)^n$	Annual Percentage Yield (APY)	Account Value after 1 Year
annually	1	2	100%	$2
semiannually	2	2.25	125%	$2.25
quarterly	4	2.441	144.1%	$2.44
monthly	12	2.613	161.3%	$2.61
weekly	52	2.693	169.3%	$2.69
daily	365	2.715	171.5%	$2.71
hourly	8760	2.718	171.8%	$2.72
every minute	525,600	2.718	171.8%	$2.72
every second	31,536,000	2.718	171.8%	$2.72
ten times per second	315,360,000	2.718	171.8%	$2.72

The table reveals an interesting result. While it is initially true that the account value increases as the compounding frequency increases, the account value then levels off. In other words, there is a *limiting value* to the future value of the account as the compounding frequency increases. It appears the account will never be worth more than $2.72 after 1 year, and the maximum annual growth (APY) will never be more than 171.8% (a growth factor of approximately 2.718). The graph in Figure 6.8 demonstrates the impact of the compounding frequency on the future value of the account and clearly shows this limiting value.

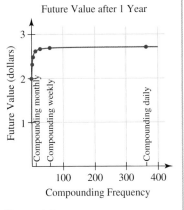

Figure 6.8

From the table and graph in Example 4, we observe $\left(1 + \frac{1}{n}\right)^n \to \approx 2.718$ as the compounding frequency $n \to \infty$. This number (≈ 2.718) is so important in mathematics that it has its own name: e.

THE NUMBER e

e is the irrational number 2.718281828459 . . . .

Since e is reserved to represent this special number, using e to represent a variable quantity in a function should be avoided. See the *Peer into the Past* sidebar on page 375 for more information about the number e.

Let's see how the number e emerges from the compound interest formula as we let the compounding frequency become infinitely large. Recall $\left(1 + \dfrac{1}{x}\right)^x \to e$ as $x \to \infty$. We let $x = \dfrac{n}{r}$. Observe that if r is any positive constant, then as $n \to \infty$, $x \to \infty$. Since $x = \dfrac{n}{r}$, $n = rx$. Let's return to the compound interest formula, $A = P\left(1 + \dfrac{r}{n}\right)^{nt}$. We can rewrite this as

$$A = P\left[\left(1 + \dfrac{r}{n}\right)^n\right]^t$$

$$= P\left[\left(1 + \dfrac{1}{x}\right)^{rx}\right]^t \quad \text{since } \dfrac{r}{n} = \dfrac{1}{x} \text{ and } n = rx$$

$$= P\left[\left(1 + \dfrac{1}{x}\right)^x\right]^{rt}$$

But $\left(1 + \dfrac{1}{x}\right)^x \to e$ as $x \to \infty$. So, as the compounding frequency, n, gets infinitely large, x also gets infinitely large and the compound interest formula becomes $A = Pe^{rt}$. We represent the notion of infinitely large n by using the term **continuous compounding**.

Continuous Compound Interest

The future value A of an initial investment P earning a **continuous compound interest** rate r is given by

$$A = Pe^{rt}$$

where t is the number of years after the initial investment is made. The annual percentage yield is APY $= e^r - 1$.

To understand how we determined the APY, observe how the continuous compound interest formula can be converted to the compound interest formula:

$$A = Pe^{rt}$$
$$= P(e^r)^t$$
$$= Pb^t \quad \text{Let } b = e^r.$$
$$= P(1 + \text{APY})^t \quad \text{since } b = 1 + \text{APY}$$

Notice $e^r = 1 + \text{APY}$ or APY $= e^r - 1$.

Thus, the growth factors for continuous compound interest can be found using e as follows:

100% interest rate, compounded continuously: $\quad e^1 \approx 2.7183$

5% interest rate, compounded continuously: $\quad e^{0.05} \approx 1.0513$

10% interest rate, compounded continuously: $\quad e^{0.10} \approx 1.1052$

Notice the annual growth factor for continuous compound interest is e^r.

EXAMPLE 5 ■ Continuous Compound Interest

Find the future value after 4 years of $2000 invested in an account with a 5.3% nominal interest rate compounded continuously.

Solution Using the formula for continuous compound interest, we have

$$A = Pe^{rt}$$
$$= 2000e^{0.053(4)}$$
$$= 2000e^{0.212}$$
$$= 2000(1.236)$$
$$= \$2472.30$$

After 4 years, the account will be worth $2472.30.

■ e: The Natural Number

Is e only helpful for continuous compound interest situations? The answer is a resounding *no*! In fact, e has so many amazing uses and appears in such a wide variety of contexts that it is often called the *natural number*. When working with exponential functions, it is very common to use e^k as the growth factor when an exponential function models something that is growing or decaying continuously instead of periodically.

Consider the following situation. The total annual health-related costs in the United States in billions of dollars can be modeled by the function

$$H(t) = 30.917(1.1013)^t$$

where t is the number of years since 1960. (*Source: Statistical Abstract of the United States, 2007,* Table 120). The annual growth rate is 10.13%, but is that growth *periodic* or *continuous*? In other words, do health-related costs only increase once a year, twice a year, every month, and so on, or are the costs increasing all of the time? Since the function models total costs for all people in the United States, every time health-related costs increase for *any person*, the annual costs for the country as a whole will increase. Given that there are so many products, medicines, medical researchers, and medical facility construction projects that are considered to be health-related costs, we can assume that the growth is continuous. Based on this explanation, we see that this exponential growth is more closely related to continuous compound interest than it is to periodically compounded interest.

The function H is similar to continuous compound interest, so we could use an exponential function involving the number e to model the annual health-related expenditures. Since we are not dealing with an interest-bearing account, we will use the more generalized form of $f(x) = ae^{kx}$ instead of the more specific $A = Pe^{rt}$. By finding the value of k such that $e^k = b$, we can rewrite the exponential function using the number e.

$$e^k = b$$
$$e^k = 1.1013$$

We use a system of equations and a calculator to find the value of k, as shown in Figure 6.9.

$$y_1 = e^k$$
$$y_2 = 1.1013$$

> **PEER INTO THE PAST**
>
> **HISTORY OF *e***
>
> The number *e* indirectly appears in mathematics literature in the 1600s without being given a special name. Despite its vital importance to logarithms, its connections to graphs of hyperbolas, and its powerful implications in calculus, the formalized discovery of *e* took place much as you have seen it in this section—from the compound interest formula.
>
> The renowned mathematician Wilhelm Liebnitz was the first to use *e* as a number in writing in 1690, although he called it *b* at that time. It was Euler, the man who is responsible for much of modern algebraic notation, who named it *e*.
>
> (*Sources:* http://www-groups.dcs.st-and.ac.uk/~history/PrintHT/e.html; Eli Maor, *e: The Story of a Number*, Princeton University Press, 1994.)

Figure 6.9

We see that $k = 0.09649$, so $e^{0.09649} = 1.1013$. Thus, we rewrite H as

$$H(t) = 30.917(1.1013)^t$$
$$= 30.917(e^{0.09649})^t$$
$$= 30.917e^{0.09649t}$$

The functions $H(t) = 30.917(1.1013)^t$ and $H(t) = 30.917e^{0.09649t}$ are equivalent, provided we disregard round-off error.

> **GENERAL FORM OF AN EXPONENTIAL FUNCTION USING e**
>
> Exponential functions of the form $f(x) = ab^x$ can be written in the form
>
> $$f(x) = ae^{kx}$$
>
> where a is the initial value, k is the continuous growth rate, and $b = e^k$.

EXAMPLE 6 ■ Converting $f(x) = ae^{kx}$ to $f(x) = ab^x$

Rewrite the function $f(x) = 1000e^{0.07x}$ in the form $f(x) = ab^x$.

Solution Since $b = e^k$, we can find the growth factor b by evaluating $e^{0.07}$.

$$f(x) = 1000e^{0.07x}$$
$$= 1000(e^{0.07})^x$$
$$= 1000(1.0725)^x$$

■ Continuous Growth (Decay) Rate

We know that functions of the form $f(x) = ab^x$ exhibit exponential growth when $b > 1$ and exponential decay when $b < 1$. Since $b = e^k$, what effect does k have on the value of b? Let's consider the results when $k = 0.1$ and when $k = -0.1$.

$$b = e^k \qquad\qquad b = e^k$$
$$= e^{0.1} \qquad\qquad = e^{-0.1}$$
$$\approx 1.105 \qquad\qquad \approx 0.905$$

It appears that $f(x) = ae^{kx}$ has exponential growth when $k > 0$ and exponential decay when $k < 0$.

> **CONTINUOUS GROWTH (DECAY) RATE**
>
> The **continuous growth (decay) rate** of an exponential function is the value of k that makes the following relationship true.
>
> $$e^k = b$$
>
> where b is the growth (decay) factor of an exponential function.
>
> If $k > 0$, then $b > 1$ and the function is growing.
>
> If $k < 0$, then $0 < b < 1$ and the function is decaying.

SUMMARY

In this section you learned how to calculate the future value of a compound interest account. You also learned that continuous compound interest creates a continuous exponential function and relies on the natural number *e*. Finally, you learned that *e* may be used as part of the base of an exponential function using the relationship $e^k = b$.

6.3 EXERCISES

SKILLS AND CONCEPTS

In Exercises 1–6, choose the approximate value of each expression from the following list: {1.4, 1.7, 2.7, 5.4, 5.7, 7.4}. Do not use a calculator.

1. e
2. $e - 1$
3. $e + 3$
4. $2e$
5. $\dfrac{e}{2}$
6. e^2

In Exercises 7–10, use <, >, or = to accurately describe the relationship between each number, then explain how you know this to be true. Do not use a calculator.

7. e^3 ___ 27
8. e^0 ___ 1
9. e^{-2} ___ 0.25
10. $\sqrt{e}$ ___ 2

11. Match each function with its graph.
    a. $y = e^{2x}$
    b. $y = e^{-0.5x}$
    c. $y = 2e^{-x}$
    d. $y = 0.5e^x$

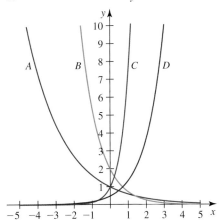

In Exercises 12–15, you are given a continuous growth rate for an account that compounds interest continuously.

a. What is the APY of the account?
b. What nominal rate would be necessary for an account with *quarterly* compound interest to have the same APY?

12. Continuous rate of 3%
13. Continuous rate of 0.9%
14. Continuous rate of 2.75%
15. Continuous rate of 5.55%

In Exercises 16–19, you are given a nominal rate for an account. For each, find the continuous growth rate that has the same APY.

16. Nominal rate of 4% compounded annually
17. Nominal rate of 5.5% compounded monthly
18. Nominal rate of 2.75% compounded semiannually
19. Nominal rate of 3.61% compounded quarterly

SHOW YOU KNOW

20. Explain what $\dfrac{r}{12}$ and $12t$ represent in the compound interest formula $A = P\left(1 + \dfrac{r}{12}\right)^{12t}$.

21. Is there a limit to the future value of $4000 invested at a nominal rate of 3.5% if interest is compounded an increasing number of times? Explain.

22. Is the APY of an account always higher than the nominal rate? Explain.

23. Suppose you invest $2000 evenly in two different company stocks. Over 3 years, the first stock increases in value with an average of 3% APY while the second stock decreases in value by an average of 3% APY. At the end of 3 years have you gained money, lost money, or broken even? Explain.

24. Consider the functions $y = 2^x$, $y = 3^x$, and $y = e^x$.
    a. For what values of x is $2^x < e^x < 3^x$? Explain how you know this.
    b. For what values of x is $3^x < e^x < 2^x$? Explain how you know this.

MAKE IT REAL

Certificates of Deposit

In Exercises 25–30, use the given information to determine the future value of each CD when it matures, assuming the minimum amount is invested.

	Institution	Maturity	Nominal Rate	Compounding Method	Minimum Investment
25.	Bank of America	5 years	3.64%	monthly	$1000
26.	First Arizona Savings	3 years	4.91%	quarterly	$500
27.	Discover Bank	4 years	5.07%	daily	$2500
28.	World Savings Bank	2.5 years	4.31%	daily	$1000
29.	Queens County Savings Bank	6 months	3.96%	semi-annually	$2500
30.	Guarantee Bank	6 months	4.67%	quarterly	$1000

Quoted rates are from www.bankrate.com and were accurate as of April 2007.

Certificates of Deposit

In Exercises 31–34, use the given information to answer the questions for each CD.

a. Each time interest is compounded, what is the percentage gain in the CD's value?

b. What is the future value of the CD when it matures, assuming the minimum amount is invested?

c. What is the average rate of change of the investment from the time when the CD is purchased until the time it matures?

d. What is the APY for the CD?

e. What initial investment would be required for the CD to be worth $5000 when it matures?

f. How much more money would the CD be worth if the interest was compounded continuously instead of periodically?

	Institution	Maturity	Nominal Rate	Compounding Method	Minimum Investment
31.	State Bank & Trust	2 years	4.75%	annually	$500
32.	Bank of America	1 year	2.86%	monthly	$1000
33.	Bank of Albuquerque	3 years	4.18%	quarterly	$1000
34.	Crescent State Bank	6 months	3.05%	daily	$500

Quoted rates are from www.bankrate.com and were accurate as of April 2007.

In Exercises 35–40, use the given information to determine which CD will have a higher annual percentage yield (APY). Quoted rates are from www.bankrate.com and were accurate as of April 2007.

35. Certificates of Deposit ING Direct offers a nominal rate of 5.01% compounded annually and BankDirect offers a nominal rate of 4.88% compounded daily.

36. Certificates of Deposit Fireside Bank offers a nominal rate of 5.10% compounded monthly and Discover Bank offers a nominal rate of 5.07% compounded daily.

37. Certificates of Deposit Eastern Savings Bank offers a nominal rate of 5.03% compounded monthly and Intervest National Bank offers a nominal rate of 5.02% compounded daily.

38. Certificates of Deposit National Bank of Kansas City offers a nominal rate of 5.07% compounded daily and Flagstar Bank offers a nominal rate of 5.11% compounded quarterly.

39. Certificates of Deposit Nova Savings Bank offers a nominal rate of 4.93% compounded daily and NBC Bank offers a nominal rate of 5.00% compounded semiannually.

40. Certificates of Deposit Flagstar Bank offers a nominal rate of 5.10% compounded quarterly and California First National Bank offers a nominal rate of 5.05% compounded monthly.

In Exercises 41–42, population growth models are given for countries based on the World Health Organization's 2006 World Health Statistics. In each model, the population is in millions t years after 2006.

a. Find the annual growth/decay rate of the population.

b. Rewrite the model in the form $f(t) = ab^t$.

c. State whether the population is increasing or decreasing and explain how you know.

41. Cambodia: $C(t) = 9.038e^{0.0198t}$

42. Kazakhstan: $K(t) = 14.825e^{-0.007t}$

In Exercises 43–44, population growth models are given for countries based on the World Health Organization's 2006 World Health Statistics. In each model the population is in millions t years after 2006.

a. Find the continuous growth/decay rate of the population.

b. Rewrite the model in the form $f(t) = ae^{kt}$.

c. State whether the population is increasing or decreasing and explain how you know.

43. Rwanda: $R(t) = 9.038(1.05)^t$

44. Hungary: $H(t) = 10.098(0.998)^t$

In Exercises 45–46, use the following information. Oxaprozin is a medicine prescribed to manage osteoarthritis and rheumatoid arthritis. After the drug reaches peak concentration in the bloodstream, the body begins to reduce the amount of medicine present according to the formula $A(t) = 600e^{-0.0173t}$, where A is the amount of medicine (in milligrams) left in the body t hours after peak concentration. (*Source:* Modeled from www.merck.com data)

45. **Prescription Medication** What is the continuous decay rate? What does this tell you about the situation?

46. **Prescription Medication** Rewrite the formula in the form $A(t) = ab^t$. What does b represent in this situation?

47. **Available Light Underwater** When light hits water, like the surface of the ocean, some of the light is reflected, some is absorbed, and some is scattered, resulting in a decreasing amount of surface light available as depth increases. In clear ocean water, the percent of surface light P available at a depth of d meters can be modeled by $P(d) = 100e^{-0.0307d}$. (*Source:* www.oceanexplorer.noaa.gov) Also, in clear water, photosynthesis can occur at depths of up to 200 meters. (*Source:* www.waterencyclopedia.com)

a. What percent of surface light is required for photosynthesis to occur?

b. At what depth is 50% of surface light available? 1% of surface light?

c. The model can also be used for murkier water by changing the value of k. Explain how you think k would change to represent water that is not very clear.

d. Write P in the form $P(d) = ab^d$ and explain what b represents.

■ **STRETCH YOUR MIND**

Exercises 48–52 are intended to challenge your understanding of compound interest and continuous growth.

48. How long will it take for $12,000 invested at a nominal rate of 5.12% compounded every second to have the same value as $14,000 invested at a nominal rate of 5.01% compounded every minute?

49. $2000 is invested at a nominal rate of 3% compounded monthly. $1000 is invested in an account compounding interest monthly. If after 8 years the two investments have the same value, how many times greater is the nominal rate of the second account than that of the first account?

50. An investment doubles over 13 years. How often was interest compounded if the nominal rate was 5.3374%?

51. **Population Growth** In 1869, Argentina's population was about 1,800,000. By 1947, its population was about 16,000,000. Write an exponential model for Argentina's population growth in the form $P(t) = ae^{kt}$. (*Source:* www.yale.edu/ynhti)

52. Explain difficulties you may face in finding the inverse of an exponential function.

SECTION 6.4 Solving Exponential and Logarithmic Equations

LEARNING OBJECTIVES

■ State and use the rules of logarithms

■ Solve exponential equations using logarithms and interpret the real-world meaning of the results

■ Solve logarithmic equations using exponentiation and interpret the real-world meaning of the results

GETTING STARTED

According to the World Health Organization's 2006 World Health Statistics, Bangladesh's population was 141.8 million in 2005 and had been growing at an average annual rate of 1.8% over the previous decade. We can model this population (in millions) with $p(t) = 141.8(1.018)^t$, where t is the number of years since 2005. We can use this model to predict the population for any value of t in the practical domain. But what if we want to algebraically determine the year the model predicts the population to reach 250 million?

In this section we discuss logarithmic functions, which are the inverses of exponential functions, and see how they are used in real-world contexts such as population predictions. We also explore the rules of logarithms and learn how they can be used to solve exponential and logarithmic equations.

Logarithms

Suppose we want to know the year in which the population of Bangladesh should reach 250 million, assuming the current growth rate continues. Using the model $P(t) = 141.8(1.018)^t$, we predict the population over the next 50 years, as shown in Table 6.21 and graphed in Figure 6.10.

Table 6.21

Years Since 2005 t	Population (in millions) $p(t)$
0	141.8
10	169.5
20	202.6
30	242.2
40	289.5
50	346.0

Figure 6.10

In the original function, t is the input and p is the output. But since we know the population, p, (250 million) and want to find the value of t, we need to work with the *inverse* of the exponential function to input a p-value and receive t as the output. Let's first look at the table of values (Table 6.22) and graph (Figure 6.11) of the inverse of the original relationship.

Table 6.22

Population (in millions) p	Years Since 2005 $t(p)$
141.8	0
169.5	10
202.6	20
242.2	30
289.5	40
346.0	50

Figure 6.11

From the table, it appears Bangladesh will have 250 million people sometime between 2035 and 2045 ($30 < t < 40$). The graph gives a little better estimate, showing that the population should reach 250 million in about 2037 ($t = 32$).

To algebraically verify this value, we return to the original exponential function $p(t) = 141.8(1.018)^t$ and solve for t when $p = 250$ (which represents 250,000,000):

$$250 = 141.8(1.018)^t$$

$$\frac{250}{141.8} = \frac{141.8(1.018)^t}{141.8}$$

$$1.763 = (1.018)^t$$

Now, to solve for t in the exponent, we must use the inverse of the exponential function. Observe that t is the exponent we place on 1.018 to get 1.763. We use the symbol *log* (which stands for *logarithm*) to represent the phrase "the exponent we place on."

t is	the exponent we place on	1.018	to get 1.763
$t =$	log	1.018	1.763

$$t = \log_{1.018}(1.763)$$

We read this as "t equals log base 1.018 of 1.763." Using properties and methods we will soon discuss, we can find $t \approx 31.785$, or a little less than 32 years after 2005. Thus, $t \approx 31.785$ is the answer to the question "What exponent on 1.018 is required to get 1.763?" In other words, $1.018^{31.785} \approx 1.763$.

■ Logarithmic Functions

We can generalize the preceding discussion to apply to any exponential situation. The symbol **log** is short for **logarithm**, and the two are used interchangeably. The equation $y = \log_b(x)$, which we read "y equals log base b of x," means "y is the exponent we place on b to get x." We can also think of it as answering the question "What exponent on b is necessary to get x?" That is, y is the number that makes the equation $b^y = x$ true.

> ### LOGARITHMIC FUNCTIONS
>
> Let b and x be real numbers with $b > 0$, $b \neq 1$, and $x > 0$. The function
>
> $$y = \log_b(x)$$
>
> is called a **logarithmic function**. The value b is called the **base** of the logarithmic function. We read the expression $\log_b(x)$ as "log base b of x."

A logarithmic function is the inverse of an exponential function. So if x is the independent variable and y is the dependent variable for the *logarithmic* function, then y is the independent variable and x is the dependent variable for the corresponding *exponential* function.

> ### INVERSE RELATIONSHIP BETWEEN LOGARITHMIC AND EXPONENTIAL FUNCTIONS
>
> $y = \log_b(x)$ is equivalent to $b^y = x$

For example, $y = \log_6(36)$ answers the question "What exponent do we place on 6 to get 36?" In other words, what value of y makes $6^y = 36$ true? Since $6^2 = 36$, $y = 2$. Symbolically, we write $\log_6(36) = 2$. Thus, "the exponent we place on 6 to get 36" is 2 and "log base 6 of 36 is 2."

> ### JUST IN TIME ■ PROPERTIES OF EXPONENTS
>
> If b, m, and n are real numbers with $b > 0$, then the following properties hold.
>
> **Property**
>
> 1. $b^{-n} = \dfrac{1}{b^n}$
> 2. $b^m \cdot b^n = b^{m+n}$
> 3. $\dfrac{b^m}{b^n} = b^{m-n}$
> 4. $b^{mn} = (b^m)^n = (b^n)^m$
> 5. $\sqrt[m]{b} = b^{1/m}$
>
> **Example**
>
> 1. $2^{-3} = \dfrac{1}{2^3}$
> 2. $4^3 \cdot 4^2 = (4 \cdot 4 \cdot 4)(4 \cdot 4) = 4^{3+2} = 4^5$
> 3. $\dfrac{7^5}{7^3} = \dfrac{7 \cdot 7 \cdot 7 \cdot 7 \cdot 7}{7 \cdot 7 \cdot 7} = 7^{5-3} = 7^2$
> 4. $5^6 = (5 \cdot 5 \cdot 5)(5 \cdot 5 \cdot 5) = (5^3)^2$ or $5^6 = (5 \cdot 5)(5 \cdot 5)(5 \cdot 5) = (5^2)^3$
> 5. $\sqrt[3]{6} = 6^{1/3}$

EXAMPLE 1 ■ Evaluating a Logarithm

a. Find the value of y given that $y = \log_3(81)$.
b. Estimate the value of y given that $y = \log_4(50)$.

Solution

a. $y = \log_3(81)$ answers the question: "What exponent do we place on 3 to get 81?" That is, what value of y makes the equation $3^y = 81$ true? Since $3^4 = 81$, $y = 4$. Symbolically, we write $\log_3(81) = 4$ and say the "log base 3 of 81 is 4."

b. $y = \log_4(50)$ answers the question: "What exponent do we place on 4 to get 50?" That is, what value of y makes the equation $4^y = 50$ true? The answer to this question is not a whole number. Since $4^2 = 16$ and $4^3 = 64$, we know y is a number between 2 and 3.

■ Rules of Logarithms

When equations involving logarithms become more complex, we use established rules of logarithms to manipulate or simplify the equations. There are two keys to understanding these rules. First, *logarithms are exponents*. Thus, the rules of logarithms come from the properties of exponents. Second, any number may be written as an exponential of any base.

Consider the number 81. Using exponents, we can say $3^4 = 81$ or $9^2 = 81$. Thus, we can substitute either 3^4 or 9^2 in any equation or formula that contains the number 81. When we apply this concept to logarithms, we see the benefit of this insight.

EXAMPLE 2 ■ Solving Logarithmic Equations

Solve each of the following equations for y.

a. $y = \log_3(81)$ b. $y = \log_9(81)$

Solution

a. We know that $y = \log_3(81)$ means $3^y = 81$. Since $3^4 = 81$, $y = 4$. Another way to approach this problem is to rewrite 81 as 3^4 in the equation.

$$y = \log_3(81)$$
$$y = \log_3(3^4)$$

Now the equation says "What exponent on 3 gives you 3^4?" The answer to this question is more obvious: 4. Thus,

$$y = \log_3(3^4)$$
$$= 4$$

b. We translate $y = \log_9(81)$ into the equation $9^y = 81$. Since $9^2 = 81$, $y = 2$. Using the alternative approach, we have

$$y = \log_9(81)$$
$$= \log_9(9^2) \quad \text{Substitute } 9^2 \text{ for 81.}$$
$$= 2 \quad \text{2 is the exponent on 9 that gives } 9^2.$$

We generalize our results from Example 2 in the following rule.

LOGARITHM RULE 1

$$\log_b(b^m) = m$$

There are two immediate results of Log Rule 1 that will prove extremely useful. They are $\log_b 1 = 0$ and $\log_b b = 1$. Recall that $b^0 = 1$ for all nonzero values of b and $b^1 = b$ for all b. Therefore,

$$\log_b(b^m) = m \quad \text{Log Rule 1}$$
$$\log_b(b^0) = 0 \quad \text{Set } m = 0.$$
$$\log_b(1) = 0 \quad \text{since } b^0 = 1$$

So log base b of 1 is always 0 no matter the value of b. Similarly,

$$\log_b(b^m) = m \quad \text{Log Rule 1}$$
$$\log_b(b^1) = 1 \quad \text{Set } m = 1.$$
$$\log_b(b) = 1 \quad \text{since } b^1 = b$$

So log base b of b is always 1 no matter the value of b.

EXAMPLE 3 ■ Using the Rules of Logarithms

Calculate the value of y in each of the following equations.

a. $y = \log_7(49)$
b. $y = \log_{100}(0.01)$
c. $y = \log_2(0.25)$

Solution

a.
$$y = \log_7(49)$$
$$= \log_7(7^2) \quad \text{since } 49 = 7^2$$
$$= 2 \quad \text{Log Rule 1}$$

So 2 is the exponent we place on 7 to get 49.

b.
$$y = \log_{100}(0.01)$$
$$= \log_{100}\left(\frac{1}{100}\right) \quad \text{since } 0.01 = \frac{1}{100}$$
$$= \log_{100}(100^{-1}) \quad \text{since } \frac{1}{x^n} = x^{-n}$$
$$= -1 \quad \text{Log Rule 1}$$

So -1 is the exponent we place on 100 to get 0.01.

c.
$$y = \log_2(0.25)$$
$$= \log_2\left(\tfrac{1}{4}\right) \quad \text{since } 0.25 = \tfrac{1}{4}$$
$$= \log_2\left(\tfrac{1}{2^2}\right) \quad \text{since } 4 = 2^2$$
$$= \log_2(2^{-2}) \quad \text{since } \frac{1}{x^n} = x^{-n}$$
$$= -2 \quad \text{Log Rule 1}$$

So -2 is the exponent we place on 2 in order to get 0.25.

Using what we just practiced, we can develop other rules involving logarithms. For Log Rule 2, recall that $b^y = m$ means the same thing as $\log_b(m) = y$. Thus we have

$$b^y = m$$
$$b^{\log_b(m)} = m \quad \text{since } y = \log_b(m)$$

LOGARITHM RULE 2

$$b^{\log_b(m)} = m$$

For Log Rule 3, let's begin with the following equation: $y = \log_2(16 \cdot 64)$. One way to approach this problem is to multiply 16 and 64 first.

$$y = \log_2(16 \cdot 64)$$
$$= \log_2(1024) \qquad \text{Multiply } 16 \cdot 64.$$
$$= \log_2(2^{10}) \qquad \text{Substitute } 2^{10} \text{ for 1024.}$$
$$= 10 \qquad \text{Log Rule 1}$$

Another way to approach this relies on the properties of exponents. Observe that the equation $y = \log_2(16 \cdot 64)$ is a base 2 logarithm. This tells us that we want to be sure to get a base 2 exponential inside of the parentheses.

$$y = \log_2(16 \cdot 64)$$
$$= \log_2(2^4 \cdot 2^6) \qquad \text{Substitute } 2^4 \text{ for 16 and } 2^6 \text{ for 64.}$$
$$= \log_2(2^{4+6}) \qquad \text{Exponent Property 2: } x^a \cdot x^b = x^{a+b}$$
$$= 4 + 6 \qquad \text{Log Rule 1}$$

Before adding the 4 and 6 together to get 10, let's consider what these numbers represent. The 4 came from the fact that $2^4 = 16$, so $4 = \log_2 16$. The 6 came from the fact that $2^6 = 64$, so $6 = \log_2 64$. Thus, $4 + 6$ may be rewritten as $\log_2(16) + \log_2(64)$. In other words,

$$\log_2(16 \cdot 64) = \log_2 16 + \log_2 64$$

and both expressions are equal to 10.

Let's repeat this process to generalize the rule. We start with $y = \log_b(m \cdot n)$. We have

$$y = \log_b(m \cdot n)$$
$$b^y = m \cdot n \qquad \text{relationship between logs and exponentials}$$
$$b^y = (b^{\log_b m})(b^{\log_b n}) \qquad \text{Log Rule 2}$$
$$b^y = (b^{\log_b m + \log_b n}) \qquad \text{Exponent Property 2: } x^n \cdot x^m = x^{n+m}$$
$$y = \log_b m + \log_b n \qquad \text{Equal exponentials with same bases have equal exponents.}$$

But $y = \log_b(mn)$, so $\log_b(mn) = \log_b m + \log_b n$.

Logarithm Rule 3

$$\log_b(mn) = \log_b(m) + \log_b(n)$$

For the next rule of logarithms, let's consider the equation $y = \log_2\left(\dfrac{64}{16}\right)$. One way to approach this problem is to first divide 64 by 16.

$$y = \log_2\left(\dfrac{64}{16}\right)$$
$$= \log_2(4) \qquad \text{Divide 64 by 16.}$$
$$= \log_2(2^2) \qquad \text{Substitute } 2^2 \text{ for 4.}$$
$$= 2 \qquad \text{Log Rule 1}$$

Alternatively, we can solve the problem using a more cumbersome approach that will lead us to Log Rule 4.

$$y = \log_2\left(\frac{64}{16}\right)$$

$$= \log_2\left(\frac{2^6}{2^4}\right)$$

$$= \log_2(2^6 \cdot 2^{-4})$$

$$= \log_2(2^6) + \log_2(2^{-4}) \quad \text{Log Rule 3}$$

$$= \log_2(2^6) + (-4) \quad \text{Log Rule 1}$$

$$= \log_2(2^6) - (4)$$

$$= \log_2(2^6) - \log_2(2^4) \quad \text{Log Rule 1}$$

The key relationship we want to recognize is that $\log_2\left(\frac{2^6}{2^4}\right)$ is equal to $\log_2(2^6) - \log_2(2^4)$.

LOGARITHM RULE 4

$$\log_b\left(\frac{m}{n}\right) = \log_b(m) - \log_b(n)$$

The final rule of logarithms is one of the most useful in solving exponential equations. Consider $\log_3(8^5)$. We have

$$\log_3(8^5) = \log_3(8 \cdot 8 \cdot 8 \cdot 8 \cdot 8)$$

$$= \log_3(8) + \log_3(8) + \log_3(8) + \log_3(8) + \log_3(8) \quad \text{Log Rule 3}$$

$$= 5\log_3(8) \quad \text{There are 5 } \log_3 8 \text{ terms.}$$

This observation yields the fifth rule of logarithms.

LOGARITHM RULE 5

$$\log_b(m^n) = n\log_b(m)$$

Although not explicitly a logarithm rule, "taking the log" of both sides is a common mathematical technique used in solving exponential equations. Assume $a = b$. Let x be the exponent we place on 10 to get a and let y be the exponent we place on 10 to get b. That is, $10^x = a$ and $10^y = b$. Since $a = b$, $10^x = 10^y$. But if $10^x = 10^y$ then $x = y$. We represent the verbal descriptions of x and y with logarithmic notation: $x = \log_{10}(a)$ and $y = \log_{10}(b)$. Since $x = y$, $\log_{10}(a) = \log_{10}(b)$. Therefore, if $a = b$, $\log_{10}(a) = \log_{10}(b)$. When we "take the log of both sides" we are simply applying this observation in a problem-solving situation. We use this technique in Example 4.

EXAMPLE 4 ■ Solving an Exponential Equation Using Logarithms

The total annual health-related costs in the United States in billions of dollars may be modeled by the function $H(t) = 30.917(1.1013)^t$, where t is the number of years since 1960. (*Source: Statistical Abstract of the United States, 2007*, Table 120) According to the model, when will health-related costs in the United States reach 250 billion dollars?

Solution

$$250 = 30.917(1.1013)^t$$
$$8.086 \approx 1.1013^t$$
$$\log_{10}(8.086) = \log_{10}(1.1013)^t \quad \text{Take the log of both sides.}$$
$$\log_{10}(8.086) = t \log_{10}(1.1013) \quad \text{Log Rule 5}$$
$$\frac{\log_{10}(8.086)}{\log_{10}(1.1013)} = t$$
$$21.66 = t \quad \text{Evaluate with a calculator.}$$

According to the model, 21.66 years after 1960 (8 months into 1982), the health-related costs in the United States reached 250 billion dollars.

We summarize the rules of logarithms as follows.

Rules of Logarithms

Let b, m, and n be real numbers with $b > 0$, $b \neq 1$, $m > 0$, and $n > 0$. Under these constraints, the following rules are always true.

Rule 1: $\log_b(b^m) = m$

Rule 2: $b^{\log_b(m)} = m$

Rule 3: $\log_b(m \cdot n) = \log_b(m) + \log_b(n)$

Rule 4: $\log_b\left(\dfrac{m}{n}\right) = \log_b(m) - \log_b(n)$

Rule 5: $\log_b(m^n) = n \log_b(m)$

■ Common and Natural Logarithms

Although any positive number other than 1 may be used as a base for a logarithm, there are two bases that are used so frequently that they have special names. A base-10 logarithm, $\log_{10}(x)$, is called the **common log**. When writing the common log, it is customary to omit the "10" and simply write $\log(x)$. Thus, $y = \log(x)$ means "What exponent on 10 gives us x?" The answer is "y is the exponent on 10 that gives us x ($x = 10^y$)."

A base-e logarithm, $\log_e(x)$, is called the **natural log** (so named because $e \approx 2.71828$ is called the *natural number*). When writing the natural log we write $\ln(x)$ instead of $\log_e(x)$. Thus, $y = \ln(x)$ asks "What exponent on e gives us x?" and the answer is "y is the exponent on e that gives us x ($x = e^y$)."

Common and Natural Logarithms

The **common logarithm**, $y = \log_{10}(x)$, is typically written as

$$y = \log(x)$$

This is equivalent to $x = 10^y$.

The **natural logarithm**, $y = \log_e(x)$, is typically written as

$$y = \ln(x)$$

This is equivalent to $x = e^y$.

SECTION 6.4 ■ SOLVING EXPONENTIAL AND LOGARITHMIC EQUATIONS

EXAMPLE 5 ■ **Using Common and Natural Logs**

Evaluate each of the following expressions:

a. $\log(1000)$

b. $\log(0.1)$

c. $\ln(e^5)$

d. $\ln\left(\dfrac{1}{\sqrt[4]{e}}\right)$

Solution

a.
$\log(1000)$
$\log(10^3)$ — $10^3 = 1000$
3 — 3 is the exponent on 10 that gives us 10^3.

b.
$\log(0.1)$
$\log\left(\dfrac{1}{10}\right)$ — $0.1 = \dfrac{1}{10}$
$\log(10^{-1})$ — $\dfrac{1}{10} = 10^{-1}$
-1 — -1 is the exponent on 10 that gives us 10^{-1}.

c.
$\ln(e^5)$
5 — 5 is the exponent on e that gives us e^5.

d.
$\ln\left(\dfrac{1}{\sqrt[4]{e}}\right)$
$\ln\left(\dfrac{1}{e^{1/4}}\right)$ — $e^{1/4} = \sqrt[4]{e}$
$\ln(e^{-1/4})$ — $e^{-1/4} = \dfrac{1}{e^{1/4}}$
$-\dfrac{1}{4}$ — $-\dfrac{1}{4}$ is the exponent on e that gives us $e^{-1/4}$.

■ Finding Logarithms Using a Calculator

When logarithms do not evaluate to integers, a calculator can help us find accurate decimal approximations. We use the $\boxed{\text{LOG}}$ and $\boxed{\text{LN}}$ buttons on the calculator to find the common and natural logs of any number.

EXAMPLE 6 ■ **Calculating Exact Logarithms**

Use a calculator to find $\log(600)$ and $\ln(100)$. Then interpret and check your answers.

Solution Using a calculator, we get the results shown in Figure 6.12. This tells us that $10^{2.778} \approx 600$ and $e^{4.605} \approx 100$. Since we used rounded numbers in these statements, when we check our answers we see they are slightly off. Our accuracy may be improved by using the exact values stored in the calculator, as shown in Figure 6.13.

```
log(600)
       2.77815125
ln(100)
       4.605170186
```

Figure 6.12

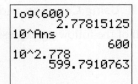

Figure 6.13

Unfortunately, some calculators are not programmed to calculate logarithms of other bases, such as $\log_3(17)$. However, using the rules of logarithms we can *change the base* of any logarithm and make it a common or natural logarithm.

CHANGE OF BASE FORMULA

For all $x > 0$, $y = \log_b(x)$ may be written as

$$y = \frac{\log(x)}{\log(b)} \text{ or } y = \frac{\ln(x)}{\ln(b)}$$

In either of these forms, the logarithm may be evaluated or graphed with a calculator.

The origin of this formula is not entirely obvious. However, with a little creativity we can use the existing rules of logarithms to show it holds true.

Suppose that $y = \log_b(x)$. Recall that $y = \log_b(x)$ is equivalent to $b^y = x$. Now consider $\dfrac{\log(x)}{\log(b)}$.

$$\frac{\log(x)}{\log(b)} = \frac{\log(b^y)}{\log(b)} \quad \text{since } b^y = x$$

$$= \frac{y \log(b)}{\log(b)} \quad \text{Log Rule 5}$$

$$= y \quad \text{since } \frac{\log(b)}{\log(b)} = 1$$

$$= \log_b(x) \quad \text{since } y = \log_b(x)$$

Thus $\log_b(x) = \dfrac{\log(x)}{\log(b)}$.

EXAMPLE 7 ■ Changing the Base of a Logarithm

Using the rules of logarithms, solve the logarithmic equation $y = \log_3(17)$.

Solution We know from the definition of a logarithm $y = \log_3(17)$ means $3^y = 17$. We also know $2 < y < 3$ since $3^2 = 9$, $3^3 = 27$, and $9 < 17 < 27$. To determine the exact value for y we apply the change of base formula.

$$y = \log_3(17)$$

$$y = \frac{\log(17)}{\log(3)} \quad \text{change of base formula}$$

$$y \approx 2.579 \quad \text{Evaluate with a calculator.}$$

This tells us that $\log_3(17) \approx 2.579$, so $3^{2.579} \approx 17$. The same formula can be used with natural logs, yielding the same result.

$$y = \log_3(17)$$

$$y = \frac{\ln(17)}{\ln(3)} \quad \text{change of base formula}$$

$$y \approx 2.579 \quad \text{Evaluate with a calculator.}$$

■ Solving Exponential and Logarithmic Equations

Now let's use some of the rules and techniques we have learned in this section to solve equations.

EXAMPLE 8 ■ Solving an Exponential Equation

The populations of India, I, and China, C (in thousands), can be modeled using the functions $I(t) = 1{,}103{,}371(1.015)^t$ and $C(t) = 1{,}323{,}345(1.007)^t$, where t is in years since 2005. According to these models, when will the populations of the two countries be equal?

Solution We set the model equations equal to each other and solve for t.

$$1{,}103{,}371(1.015)^t = 1{,}323{,}345(1.007)^t$$

$$\frac{1{,}103{,}371(1.015)^t}{1{,}103{,}371} = \frac{1{,}323{,}345(1.007)^t}{1{,}103{,}371}$$

$$(1.015)^t = 1.199(1.007)^t$$

$$\frac{(1.015)^t}{(1.007)^t} = \frac{1.199(1.007)^t}{(1.007)^t}$$

$$\left(\frac{1.015}{1.007}\right)^t = 1.199$$

$$(1.008)^t = 1.199$$

$$\ln(1.008)^t = \ln(1.199)$$

$$t\ln(1.008) = \ln(1.199)$$

$$t = \frac{\ln(1.199)}{\ln(1.008)}$$

$$t \approx 22.97$$

Almost 23 years after 2005, or just before the end of 2028, we expect China and India to have the same population. (*Note:* As in other examples and exercises in this text, we are keeping the actual values in our calculator even though we round the values when we write them down. By waiting to round until we have our final answer, we obtain a more accurate result.)

EXAMPLE 9 ■ Solving a Logarithmic Equation

Solve the equation $\log_5(2x + 3) = 3$ for x.

Solution

$$\log_5(2x + 3) = 3$$
$$5^3 = 2x + 3$$
$$125 = 2x + 3$$
$$122 = 2x$$
$$x = 61$$

For some students, logarithms are confusing. The following box gives a list of common student errors when using logarithms.

COMMON ERRORS WITH LOGARITHMS

Each of the following represent common errors when calculating with logarithms:

$$\log_b(m + n) \neq \log_b(m) + \log_b(n)$$

$$\log_b(m - n) \neq \log_b(m) - \log_b(n)$$

$$\log_b\left(\frac{m}{n}\right) \neq \frac{\log_b(m)}{\log_b(n)}$$

$$\log_b(m \cdot n) \neq (\log_b(m))(\log_b(n))$$

$$\log_b(m \cdot n^p) \neq p \log_b(m \cdot n)$$

SUMMARY

In this section you discovered the inverse relationship of logarithmic functions and exponential functions and learned that a logarithm is an exponent. You also developed and used the rules of logarithms to simplify logs and to solve exponential and logarithmic equations.

6.4 EXERCISES

SKILLS AND CONCEPTS

In Exercises 1–4, rewrite each statement in exponential form.

1. $\log_7(343) = 3$
2. $\log_5\left(\dfrac{1}{15,625}\right) = -6$
3. $\log(600) \approx 2.778$
4. $\ln(x) = y$

In Exercises 5–8, rewrite each statement in logarithmic form.

5. $4^7 = 16,384$
6. $5.5^{-2} = \dfrac{1}{30.25}$
7. $e^{3.2} \approx 24.53$
8. $10^y = x$

In Exercises 9–20, evaluate each expression without using a calculator.

9. $\log_4(64)$
10. $\log_5(25)$
11. $\log_2(64)$
12. $\log_8\left(\dfrac{1}{64}\right)$

13. $\log(10^4)$
14. $\log(1000)$
15. $\log\left(\dfrac{1}{100}\right)$
16. $\log(0.0001)$
17. $\ln(e^{0.2})$
18. $\ln(1)$
19. $\ln\left(\dfrac{1}{e^{1.3}}\right)$
20. $\ln(\sqrt{e})$

In Exercises 21–24, state the two integer values between which each expression falls. Then use the change of base formula to evaluate the expression exactly.

21. $\log_2(20)$
22. $\log_4\left(\dfrac{1}{50}\right)$
23. $\ln(10)$
24. $\log(0.003)$

In Exercises 25–28 you will examine common mistakes students make when simplifying logarithms. Use $m = 10$ and $n = 100$ to explore each of the following.

25. Show $\log(m + n) \neq \log(m) + \log(n)$.
26. Show $\log\left(\dfrac{m}{n}\right) \neq \dfrac{\log(m)}{\log(n)}$.
27. Show $\log(m \cdot n) \neq (\log(m))(\log(n))$.
28. Show $\log(m \cdot n^2) \neq 2\log(m \cdot n)$.

In Exercises 29–41, use the inverse relationship between logarithmic and exponential functions to solve each equation for x. Simplify your answers using the rules of logarithms and the change of base formula.

29. $5^x = 625$
30. $2^t = 0.25$
31. $3^x = 13$
32. $15^x = 2$
33. $e^x = 3$
34. $13^{5x} - 4 = 39$
35. $\log_3(x) = 5$
36. $\log_4(x) = 0$
37. $\log(x) = 3$
38. $\ln(x) = -5$
39. $\log_6(24x) = 3$
40. $\ln\left(\dfrac{5}{x}\right) = -2$
41. $ab^x = y$

■ SHOW YOU KNOW

42. The domain of the function $y = \log_b(x)$ is $x > 0$. Explain why the negative real numbers and zero are not in the domain of the function.

43. One of the rules of logarithms is $b^{\log_b(x)} = x$. Use the example $10^{\log(x)} = x$ to explain why this is true.

44. Can you combine $\log(8) + \ln(2)$ into a single logarithm? Defend your answer.

45. Explain how you can use the information $\log_3(10) \approx 2.1$ and $\log_3(4) \approx 1.26$ to approximate $\log_3(400)$.

■ MAKE IT REAL

In Exercises 46–47, use the following information. The population of India is currently growing according to the formula $P(t) = 1{,}103.4e^{0.0149t}$, where P is the population in millions and t is the number of years since 2005.

46. **Population** According to the model, in what year should the population of India reach 1,500,000,000 people (be careful with the units!)?

47. **Population** Find the inverse function of $P(t)$, then explain what it models.

In Exercises 48–49, use the following information. The antibiotic clarithromycin is eliminated from the body according to the formula $A(t) = 500e^{-0.1386t}$, where A is the amount remaining in the body (in milligrams) t hours after the drug reaches peak concentration.

48. **Medicine** How much time will pass before the amount of drug in the body is reduced to 100 milligrams?

49. **Medicine** Find the inverse of $A(t)$ and explain what the inverse function models.

In Exercises 50–51 you are given models for the population (in millions) of different countries t years after 2005. For each exercise, determine the year in which the models predict the populations will be equal. (Source: World Health Organization's 2006 World Health Statistics)

50. **Population**
    Rwanda: $R(t) = 9.04(1.05)^t$
    Hungary: $H(t) = 10.1(0.98)^t$

51. **Population**
    Cambodia: $C(t) = 14.07(1.02)^t$
    Kazakhstan: $K(t) = 14.83(0.93)^t$

In Exercises 52–53 you are given information on different CDs. For each exercise, determine how long it would take for the values of each CD to be equal. (Source: www.bankrate.com, April 2007)

52. **Certificates of Deposit**
    - $4000 invested at Charter One Bank with a nominal rate of 3.31% compounded quarterly
    - $3000 invested at Dearborn Federal Savings Bank with a nominal rate of 4.95% compounded annually

53. **Certificates of Deposit**
    - $2000 invested at Wachovia Bank with a nominal rate of 3.73% compounded daily
    - $2500 invested at Bank of America with a nominal rate of 3.252% compounded monthly

In Exercises 54–55, use the following information. A pH reading is used to measure the relative acidity of a substance. A pH of 7 is considered neutral (the pH of distilled water), a pH of less than 7 is acidic, and a pH above 7 is alkaline (basic). The pH, P, is measured by finding the concentration of hydrogen ions x in moles per liter in the substance, using the formula $P(x) = -\log(x)$. (Source: wateronttheweb.org)

54. **Chemistry** Healthy human blood should have hydrogen ion concentrations between 4.467×10^{-8} and 3.548×10^{-8} moles per liter. (Source: www.trans4mind.com) What is the pH range of healthy human blood? Is this acidic or alkaline?

55. **Chemistry** The pH values of some common substances are given in the table on the next page. For each, find the hydrogen ion concentration.

392 CHAPTER 6 ■ EXPONENTIAL AND LOGARITHMIC FUNCTIONS

Substance	pH P
battery acid	0.3
orange juice	4.3
sea water	8.0
bleach	12.6

Source: watertontheweb.org

■ STRETCH YOUR MIND

Exercises 56–60 are intended to challenge your understanding of logarithms and logarithmic functions.

56. If $b > c > 1$, for what values of x is $\log_b(x) > \log_c(x)$?

57. Solve each of the following equations. *Hint: After finding potential solutions, make sure they work in the original equation.*

    a. $\log(x) + \log(x + 21) = 2$
    b. $\log_6(x + 4) + \log_6(x - 1) = 1$
    c. $\log_2(x - 2) + \log_2(3x + 1) = 3$

58. For $\log_b(x) = 2 \log_c(x)$, what must be the relationship between b and c?

59. Health Insurance The formula
$$E(t) = \frac{3328.1008}{1 + 247.4153e^{-0.0929t}}$$
can be used to model insurance companies' total annual expenditures on health care costs in billions of dollars t years after 1960. (*Source: Statistical Abstract of the United States, 2007,* **Table 120**) During what year did the total annual costs reach $330 billion? (Solve algebraically without graphing.)

60. Without using the change of base formula and a calculator, sketch a graph of $y = \log_x(4)$. Explain why the function behaves as it does.

SECTION 6.5 Logarithmic Function Modeling

LEARNING OBJECTIVES

- Graph logarithmic functions from equations and tables
- Use logarithmic regression to model real-world data sets
- Use logarithms to linearize exponential data to find an exponential model

GETTING STARTED

Excessive government spending increases the deficit and may increase the tax burden on a country's citizens. In 1990, the United States government spent $1872.6 billion. In 2005, the government spent $3877.2 billion—more than twice the 1990 level of spending. (*Source: Statistical Abstract of the United States, 2007*) Economists monitoring government spending are interested not only in projecting the spending for future years but also in predicting when government spending will reach certain levels.

    In this section we look at logarithmic function modeling and see how it can be used to analyze issues such as government spending. We also investigate logarithmic function graphs and use logarithms to linearize a data set to find an exponential model.

■ Graphing Logarithmic Functions

Using the techniques addressed earlier in the chapter, we model United States government spending with the exponential function

$$s(t) = 1872.6(1.0497)^t$$

where s represents the spending (in billion dollars) and t represents years since 1990. Using the model, we can determine the amount of government spending in a particular year. By solving the equation for t, we can create a model that will give us the year in which a particular level of spending is projected to occur.

$$s = 1872.6(1.0497)^t$$

$$\frac{s}{1872.6} = 1.0497^t$$

$$t = \log_{1.0497}\left(\frac{s}{1872.6}\right) \quad\quad \text{Rewrite in logarithmic form.}$$

$$t = \frac{\ln\left(\dfrac{s}{1872.6}\right)}{\ln(1.0497)} \quad\quad \text{change of base formula}$$

$$t = \frac{\ln(s) - \ln(1872.6)}{\ln(1.0497)} \quad \text{Log Rule 4}$$

$$t = \frac{\ln(s)}{\ln(1.0497)} - \frac{\ln(1872.6)}{\ln(1.0497)}$$

$$t = 20.62 \ln(s) - 155.3$$

The function $t(s) = 20.62 \ln(s) - 155.3$ models the number of years since 1990, t, in which government spending will be s billion dollars. This logarithmic function is the inverse of the exponential function $s(t) = 1872.6(1.0497)^t$.

HOW TO: ■ FIND THE INVERSE OF AN EXPONENTIAL FUNCTION

The inverse of the exponential function $p = ab^t$ is the logarithmic function

$$t = \log_b\left(\frac{p}{a}\right)$$

$$= \frac{\ln(p) - \ln(a)}{\ln(b)}$$

$$= \frac{\ln(p)}{\ln(b)} - \frac{\ln(a)}{\ln(b)}$$

To further understand the logarithmic model $t(s) = 20.62 \ln(s) - 155.3$, we graph the equation in Figure 6.14. The graph is increasing and concave down with a horizontal intercept at the initial 1990 level of government spending ($1872.6 billion). When did government spending reach $3000 billion? From the graph, we see that approximately 10 years after 1990 government spending reached $3000 billion.

By learning the basic shapes of logarithmic function graphs, we can quickly determine from a scatter plot if a logarithmic model is appropriate for a particular real-world situation. The shape of a logarithmic function graph depends on the base of the logarithm. However, regardless of the base, the graph will have a vertical asymptote at the vertical axis, as shown in Figure 6.15.

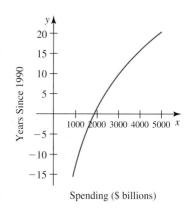

Figure 6.14

(a) $y = \log_b(x)$ with $b > 1$
concave down and increasing

(b) $y = \log_b(x)$ with $0 < b < 1$
concave up and decreasing

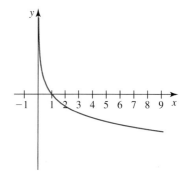

Figure 6.15

EXAMPLE 1 ■ Using Regrecvssion to Find a Logarithmic Model for a Data Set

The data set in Table 6.23 and scatter plot in Figure 6.16 show the inflation rates of the top 10 countries with the lowest rates of inflation. Determine if a logarithmic function model is appropriate for this situation. If a logarithmic function is appropriate, use regression to find the logarithmic model.

Table 6.23

Top 10 Countries C	Percent Inflation P
1. Nauru	−3.6
2. Vanautu	−1.6
3. San Marino	−1.5
4. N. Mariana Islands	−0.8
5. Barbados	−0.5
6. Dominica	−0.1
7. Israel	−0.1
8. Niger	0.2
9. Japan	0.3
10. Kiribati	0.5

Figure 6.16

Source: CIA–The World Factbook, 2006

Solution The data set and scatter plot appear to be more or less increasing and concave down. Since the countries are listed in rank order, we know as the rank number increases the inflation rate will also increase (or remain the same). A logarithmic model is appropriate for this situation.

Using the Technology Tip at the end of this section, we determine the logarithmic equation of best fit is $P(c) = -3.237 + 1.666 \ln(c)$. A graph of the model and the data is shown in Figure 6.17.

Figure 6.17

■ Finding an Exponential Model Using Logarithms

Recall that an exponential data set is characterized by a constant ratio for equally spaced values. Another way to detect if a data set is exponential is to take the logarithm of the output values, as shown in Example 2.

SECTION 6.5 ■ LOGARITHMIC FUNCTION MODELING

EXAMPLE 2 ■ Using Logarithms to Linearize Data

Complete Table 6.24 by calculating the logarithm of each of the output values. Then identify the mathematical relationship between the resultant values of $y = \log(3(2^t))$.

Table 6.24

t	$f(t) = 3(2^t)$	$y = \log(3(2^t))$	t	$f(t) = 3(2^t)$	$y = \log(3(2^t))$
−2	0.75		1	6	
−1	1.5		2	12	
0	3				

Solution We complete the table as shown in Table 6.25 and then look for a pattern by calculating the average rates of change.

Table 6.25

t	$f(t) = 3(2^t)$	$y = \log(3(2^t))$	Average Rate of Change
−2	0.75	$\log(0.75) \approx -0.1249$	$\dfrac{0.1761 - (-0.1249)}{-1 - (-2)} \approx 0.30$
−1	1.5	$\log(1.5) \approx 0.1761$	$\dfrac{0.4771 - 0.1761}{0 - (-1)} \approx 0.30$
0	3	$\log(3) \approx 0.4771$	$\dfrac{0.7782 - 0.4771}{1 - 0} \approx 0.30$
1	6	$\log(6) \approx 0.7782$	$\dfrac{1.079 - 0.7782}{2 - (1)} \approx 0.30$
2	12	$\log(12) \approx 1.079$	

Since $y = \log(3(2^t))$ has a constant rate of change, it must be a linear function. To write $y = \log(3(2^t))$ as a linear function, observe that

$$\log(3(2^t)) = \log(3) + \log(2^t) \quad \text{since } \log(ab) = \log(a) + \log(b)$$
$$= \log(3) + t\log(2) \quad \text{since } \log(b)^n = n\log(b)$$

Since $\log(3) \approx 0.4771$ and $\log(2) \approx 0.3010$, $\log(3(2^t)) \approx 0.3010t + 0.471$. If we let $y = \log(3(2^t))$, then $y \approx 0.3010t + 0.4771$. We readily recognize that y is an increasing linear function with slope 0.3010 and initial value 0.4771. We complete the table again (Table 6.26), this time using the alternate form, $y = \log(3) + t\log(2)$.

Table 6.26

t	$f(t) = 3(2^t)$	$y + \log(3) + t\log(2)$	Average Rate of Change
−2	0.75	$\log(3) - 2\log(2) \approx -0.1249$	$\log(2) \approx 0.30$
−1	1.5	$\log(3) - 1\log(2) \approx 0.1761$	$\log(2) \approx 0.30$
0	3	$\log(3) - 0\log(2) \approx 0.4771$	$\log(2) \approx 0.30$
1	6	$\log(3) + 1\log(2) \approx 0.7782$	$\log(2) \approx 0.30$
2	12	$\log(3) + 2\log(2) \approx 1.079$	

The results agree with our earlier computations.

As shown in Example 2, we can use logarithms to linearize an exponential data set. This approach provides yet another way to find an exponential model, as demonstrated in Example 3.

EXAMPLE 3 ■ **Finding an Exponential Model from a Linearized Data Set**

Table 6.27 shows the number of insecticide treated nets (ITNs) sold or distributed in the African region in the fight against malaria.

Table 6.27

Years Since 1999 t	Total Number ITNs Sold or Distributed (in thousands) N
0	538
1	886
2	2228
3	4346
4	9485

Source: www.afro.who.int

a. Calculate $\log(N)$ at each data point.
b. Use regression to find the linear equation that relates t and $\log(N)$.
c. Use the result from part (b) to find the exponential equation that relates t and N.

Solution

a. We create Table 6.28 to calculate $\log(N)$.

Table 6.28

Years Since 1999 t	Total Number of ITNs Sold or Distributed (in thousands) N	$\log(N)$
0	538	2.731
1	886	2.947
2	2228	3.348
3	4346	3.638
4	9485	3.977

b. Using linear regression on the data in columns t and $\log(N)$, we obtain $\log(N) = 0.3183t + 2.692$.

c. Rewriting $\log(N) = 0.3183t + 2.692$ in exponential form yields

$$N = 10^{0.3183t+2.692}$$
$$= 10^{0.3183t} 10^{2.692} \quad \text{since } b^{n+m} = b^n b^m$$
$$= (2.081)^t (492.0) \quad \text{since } b^{nt} = (b^n)^t$$
$$= 492.0(2.081)^t$$

So the exponential function model is $N(t) = 492.0(2.081)^t$ thousand ITNs, where t is the number of years since 1999.

The steps to find an exponential model by linearizing a data set are summarized below.

HOW TO: ■ FIND AN EXPONENTIAL MODEL USING LOGARITHMS

To find an exponential model for t as a function of p,
1. Calculate $\log(p)$ for each value of p.
2. Use linear regression on the data set with input value t and output value $\log(p)$.
3. Write the linear regression equation in the form $\log(p) = mt + b$.
4. Rewrite the logarithmic equation in exponential form:

$$p = 10^{mt+b}$$
$$= (10^{mt})(10^{b})$$
$$= (10^{b})(10^{m})^{t}$$

SUMMARY

In this section you learned how to use logarithmic functions to model real-world data sets. You also learned how to use logarithms to linearize a data set and find an exponential function model.

TECHNOLOGY TIP ■ LOGARITHMIC REGRESSION

1. Enter the data using the Statistic Menu List Editor.

2. Bring up the Statistics Menu Calculate feature by pressing STAT and using the blue arrows to move to **CALC**. Then select item **9:LnReg**, and press ENTER.

3. If you want to automatically paste the regression equation into the Y = Editor so that you can easily graph the model, press the key sequence VARS; **Y-VARS**; **Function**; **Y1** and press ENTER. Otherwise press ENTER.

6.5 EXERCISES

SKILLS AND CONCEPTS

In Exercises 1–5, determine algebraically the equation of the linear function that passes through points of the form $(t, \log(p))$.

1.
t	p
1	1
2	2
3	4
4	8

2.
t	p
1	6
2	18
3	54
4	162

3.
t	p
0	1000
1	1500
2	2250
3	3375

4.
t	p
−2	4000
−1	2000
1	500
3	125

5.
t	p
2	200
3	20
4	2
5	0.2

6. Given a logarithmic function, $f(x) = \ln(x)$, write a formula that calculates the average rate of change in the function between a point $(a, f(a))$ and $(a + 1, f(a) + 1))$.

7. Using the formula from Exercise 6, calculate the average rate of change for the values of a given in the table.

a	Average Rate of Change
1	
10	
100	
1000	

SHOW YOU KNOW

8. Explain how exponential and logarithmic functions are related.

9. What is the relationship between the initial value of an exponential function and the graph of its logarithmic function inverse?

10. Describe the appearance of the graph of the logarithmic function $y = \log_b(x)$ with $0 < b < 1$.

11. Describe the appearance of the graph of the logarithmic function $y = \log_b(x)$ with $1 < b$.

12. Describe the appearance of a scatter plot that may be effectively modeled with a logarithmic function.

13. Describe in words what happens to the average rate of change of $f(x) = \ln(x)$ as $x \to \infty$.

14. In calculus, we learn the *instantaneous* rate of change of $f(x) = \ln(x)$ is $\frac{1}{x}$. Describe what happens to the instantaneous rate of change of $f(x) = \ln(x)$ as $x \to \infty$ and as $x \to 0$.

15. Describe what the graph of the function $f(x) = \ln(x)$ tells you about the instantaneous rate of change of the function.

MAKE IT REAL

In Exercises 16–20, determine from the scatter plot if a logarithmic function model is a good fit for the data. Explain your reasoning.

16. **Americans Who Go Online**

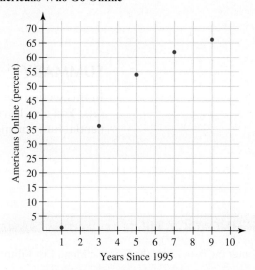

Source: Science and Engineering Indicators 2006, National Science Foundation, Table 7-8

17. **Insurance Expenditures**

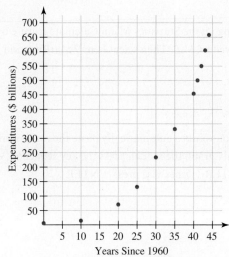

Source: Statistical Abstract of the United States, 2007, Table 120

18. Drug Prescriptions

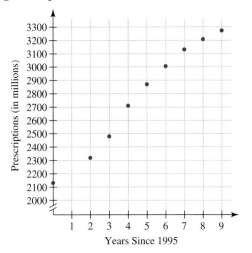

Source: Statistical Abstract of the United States, 2006, Table 126

19. DVD Player Sales

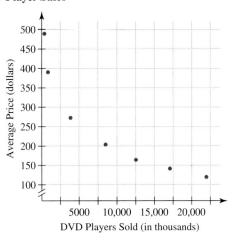

Source: Consumer Electronics Association

20. Elected Officials

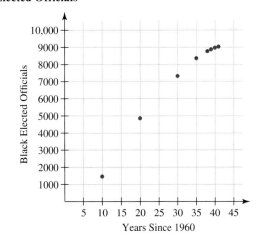

Source: Statistical Abstract of the United States, 2007, Table 403

In Exercises 21–28, find a logarithmic model for the data set algebraically.

21. Municipal Governments

Years Since 1965 Y	Number of Local Municipal Governments (in thousands) m
2	18.048
7	18.517
12	18.862
17	19.076
22	19.200
27	19.279
32	19.372
37	19.429

Source: Statistical Abstract of the United States, 2007, Table 415

22. U.S. Government Spending

Expenditures ($ billions) E	Year y	Expenditures ($ billions) E	Year y
1872.6	1990	3240.8	2002
2397.6	1995	3424.7	2003
2886.5	2000	3620.6	2004
3061.6	2001	3877.2	2005

Source: Statistical Abstract of the United States, 2007, Table 418

23. Elected Officials

Years Since 1960 Y	Number of Black Elected Officials N
10	1469
20	4890
30	7335
35	8385
38	8830
39	8896
40	9001
41	9061

Source: Statistical Abstract of the United States, 2007, Table 403

24. Pressure and Altitude

Atmospheric Pressure (psi) P	Altitude (meters) a	Atmospheric Pressure (psi) P	Altitude (meters) a
14.696	0	7.835	5000
13.035	1000	6.843	6000
11.530	2000	5.955	7000
10.168	3000	5.163	8000
8.940	4000		

Source: Digital Dutch 1976 Standard Atmosphere Calculator

400 CHAPTER 6 ■ EXPONENTIAL AND LOGARITHMIC FUNCTIONS

25. Purchasing Power of the Dollar

Years Since 1975 t	Value of the Dollar (using Consumer Price Index, 1982 = \$1) v
5	1.215
10	0.928
15	0.766
20	0.656
25	0.581
30	0.512

Source: Statistical Abstract of the United States, 2007, Table 705

26. Births to White Mothers

Year t	Total Live Births (in thousands) b	Live Births to Women Racially Classified as *White* (in thousands) w
1990	1165	670
1995	1254	785
1999	1308	840
2000	1347	866
2001	1349	880
2002	1366	904
2003	1416	947
2004	1470	983

Source: Statistical Abstract of the United States, 2007, Table 83

Model live births to white mothers as a function of total live births.

27. Disneyland Tickets—Adult

Days in Park d	2007 Park Hopper® Bonus Ticket Cost (dollars) T
1	83
2	122
3	159
4	179
5	189

Source: www.disneyland.com

28. Disneyland Tickets—Child

Days in Park d	2007 Park Hopper® Bonus Ticket Cost (dollars) T
1	73
2	102
3	129
4	149
5	159

Source: www.disneyland.com

Frank C. Wilson

In Exercises 29–36, use regression to find the logarithmic model of best fit for the data set. Compare your results to the corresponding exercise in Exercises 21–28.

29. Municipal Governments

Years Since 1965 Y	Number of Local Municipal Governments (in thousands) m
2	18.048
7	18.517
12	18.862
17	19.076
22	19.200
27	19.279
32	19.372
37	19.429

Source: Statistical Abstract of the United States, 2007, Table 415

30. U.S. Government Spending

Expenditures (\$ billions) E	Year y	Expenditures (\$ billions) E	Year y
1872.6	1990	3240.8	2002
2397.6	1995	3424.7	2003
2886.5	2000	3620.6	2004
3061.6	2001	3877.2	2005

Source: Statistical Abstract of the United States, 2007, Table 418

31. Elected Officials

Years Since 1960 Y	Number of Black Elected Officials N
10	1469
20	4890
30	7335
35	8385
38	8830
39	8896
40	9001
41	9061

Source: Statistical Abstract of the United States, 2007, Table 403

32. Pressure and Altitude

Atmospheric Pressure (psi) P	Altitude (meters) a	Atmospheric Pressure (psi) P	Altitude (meters) a
14.696	0	7.835	5000
13.035	1000	6.843	6000
11.530	2000	5.955	7000
10.168	3000	5.163	8000
8.940	4000		

Source: Digital Dutch 1976 Standard Atmosphere Calculator

33. Purchasing Power of the Dollar

Years Since 1975 t	Value of the Dollar (using Consumer Price Index, 1982 = $1) v
5	1.215
10	0.928
15	0.766
20	0.656
25	0.581
30	0.512

Source: Statistical Abstract of the United States, 2007, **Table 705**

34. Births to White Mothers

Year t	Total Live Births (in thousands) b	Live Births to Women Racially Classified as *White* (in thousands) w
1990	1165	670
1995	1254	785
1999	1308	840
2000	1347	866
2001	1349	880
2002	1366	904
2003	1416	947
2004	1470	983

Source: Statistical Abstract of the United States, 2007, **Table 83**

Model live births to white mothers as a function of total live births.

35. Disneyland Tickets—Adult

Days in Park d	2007 Park Hopper® Bonus Ticket Cost (dollars) T
1	83
2	122
3	159
4	179
5	189

Source: www.disneyland.com

36. Disneyland Tickets—Child

Days in Park d	2007 Park Hopper® Bonus Ticket Cost (dollars) T
1	73
2	102
3	129
4	149
5	159

Source: www.disneyland.com

In Exercises 37–42,

a. Linearize the data.
b. Use linear regression to find a model of the form $\log(p) = mt + b$.
c. Solve the equation in part (b) for p and rewrite the result in the standard form of an exponential function.

37. School Expenditures

School Year Since 1990–1991 t	Expenditure per Pupil (dollars) E
0	4902
2	5160
4	5529
6	5923
8	6508
10	7380
12	8044

Source: National Center for Education Statistics

38. Atmospheric Pressure

Altitude (meters) a	Atmospheric Pressure (psi) P
0	14.696
1000	13.035
2000	11.530
3000	10.168
4000	8.940
5000	7.835
6000	6.843
7000	5.955
8000	5.163

Source: Digital Dutch 1976 Standard Atmosphere Calculator

39. National Health Spending

Years Since 1995 t	Health Expenditures ($ billions) H
0	1020
1	1073
2	1130
3	1196
4	1270
5	1359
6	1474
7	1608
8	1741
9	1878

Source: Statistical Abstract of the United States, 2007, **Table 120**

40. Brand-Name Drug Prices

Years Since 1995 t	Average Brand-Name Drug Price (dollars) d
0	40.22
2	49.55
3	53.51
4	60.66
5	65.29
6	69.75
7	77.49
8	85.57
9	95.86

Source: Statistical Abstract of the United States, 2006, Table 126

41. Highway Accidents

Years Since 2000 t	Accidents Resulting in Injuries (percent) p
0	49.9
1	48.0
2	46.3
3	45.6
4	45.1

Source: Statistical Abstract of the United States, 2007, Table 1047

42. Municipal Governments

Number of Local Municipal Governments (in thousands) m	Years Since 1965 Y
18.048	2
18.517	7
18.862	12
19.076	17
19.200	22
19.279	27
19.372	32
19.429	37

Source: Statistical Abstract of the United States, 2007, Table 415

■ STRETCH YOUR MIND

Exercises 43–46 are intended to challenge your understanding of logarithmic function modeling.

43. A data set contains positive and negative domain (input) values. A classmate claims that it is impossible to model the data with a logarithmic function. Do you agree or disagree? Explain.

44. As with other functions, logarithmic functions may be transformed using shifts, reflections, stretches, and compressions. Describe how the graph of $h(x) = -2 \log(4(x + 1)) - 3$ looks in comparison to $g(x) = \log(x)$.

45. What types of asymptotes do logarithmic functions have? Explain.

46. Where will the graph of $f(x) = 3((0.5)^x)$ and the graph of $f^{-1}(x) = \dfrac{\ln(x)}{\ln(0.5)} - \dfrac{\ln(3)}{\ln(0.5)}$ intersect?

CHAPTER 6 Study Sheet

As a result of your work in this chapter, you should be able to answer the following questions, which focus on the "big ideas" of this chapter.

SECTION 6.1
1. What do we mean when we talk about percentage growth or percentage decay?
2. What is a growth factor and what is a decay factor?
3. What is the relationship between a change factor and a change rate?
4. How do you determine from an equation whether the exponential function is growing or decaying?

SECTION 6.2
5. What key verbal indicators suggest an exponential model may be appropriate for a particular real-world situation?
6. How can you tell if a data set represents an exponential function?
7. What effect does changing the initial value and change factor of an exponential function have on the graph of the function?
8. In terms of a rate of change, what does the concavity of an exponential function tell about the function?

SECTION 6.3
9. What is compound interest?
10. What is a periodic growth rate?
11. As the compounding frequency increases, what happens to the growth factor in the compound interest formula?
12. What is the difference between a nominal interest rate and an annual percentage yield?
13. What is the difference between periodic growth and continuous growth?

SECTION 6.4
14. What is a logarithm?
15. How are exponential and logarithmic functions related?
16. How are logarithms used in solving exponential equations?
17. How are exponential functions used in solving logarithmic equations?

SECTION 6.5
18. In terms of a rate of change, what does the concavity of a logarithmic function tell us about the function?
19. How are logarithmic function graphs distinguished from other function graphs?
20. What features of a scatter plot indicate a logarithmic function model may be appropriate?

REVIEW EXERCISES

■ SECTION 6.1 ■

In Exercises 1–5, determine if the table of values is linear, quadratic, exponential, or none of these by calculating successive differences and/or change factors.

1.
x	y
10	25
20	55
30	85
40	115
50	145
60	175

2.
x	y
−3	−0.125
−2	−0.25
−1	−0.5
0	−1
1	−2
2	−4

3.
x	y
−8	−81
−7	−64
−6	−49
−5	−36
−4	−25
−3	−16

4.
x	y
11	5.55
12	5.60
13	5.65
14	5.70
15	5.75
16	5.80

5.
x	y
−2	81
−1	9
0	1
1	0.11111
2	0.12345
3	0.00137

6. **Professional Baseball Tickets** Team Marketing Report released its Fan Cost Index, an annual survey of how expensive it is to attend a game at each of the 30 major league baseball parks. (The index takes into account the price of four tickets, food, drink, parking, and souvenirs.) As usual, Boston's Fenway Park topped the list: On average it cost a family of four $313.83 to take in a game there in the 2007 season. Ticket prices from 2000 to 2007 for the Boston Red Sox are shown in the following scatter plot. Calculate the growth factor and the percentage increase for each year from 2000 to 2007.

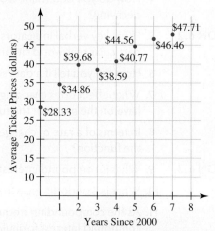

Source: www.teammarketing.com

7. **Salary of Professors** The average annual percentage increase in salaries for full professors from universities reporting to the American Association of University Professors (AAUP) was 3.55% from 2000 to 2007. The average salary for a full professor with a doctorate was $115,475 in 2007. Estimate the average salaries for the years 2008 and 2009 by assuming the percentage change of 3.55% continues.

8. **Disinfecting Swimming Pools** Chlorine is used to disinfect swimming pools. The chlorine concentration should be between 1.5 and 2.5 parts per million (ppm). On sunny, hot days, up to 30% of the chlorine dissipates into the air or combines with other chemicals. Therefore, the chlorine concentration, C, in parts per million in a particular pool after s sunny days can be modeled by $C(s) = 2.5(0.7)^s$.

   a. What is the initial concentration of chlorine in the swimming pool?

   b. Sketch a graph of $C(s)$.

   c. Evaluate $C(4)$ and explain what the value means in the real-world context.

   d. Estimate when more chlorine should be added.

9. **Investment Returns** From January 3, 1957 to January 3, 2007 (50 years), the percentage increase for the S&P 500 was 2940%. (*Source:* finance.yahoo.com)

   a. What is the annual percentage increase and what is the annual growth factor?

   b. Using the 50-year percentage increase of 2940%, determine the 2007 value of a $10,000 initial investment in 1957 that grew at the same rate as the S&P 500.

■ SECTION 6.2 ■

In Exercises 10–12, construct an exponential function model for the situation and answer any given questions.

10. **Investment Value** Based on data from August 1, 1952 through May 31, 2007, the average annual return on an investment in the CREF Stock Account was 10.72%. (*Source:* www.tiaa-cref.com)
    According to the model, how much would a $1000 investment made in the account on August 1, 1952 be worth on August 1, 2007?

11. **Investment Value** Based on data from March 1, 1990 through May 31, 2007, the average annual return on an investment in the CREF Social Choice Account was 10.14%. (*Source:* www.tiaa-cref.com)
    According to the model, how much would a $5000 investment made in the account on March 1, 1990 be worth on March 1, 2010?

12. **Population Growth** In 2005, Qatar had a population of 813,000 and was growing at an average annual rate of 4.0%. (*Source:* World Health Organization)
    According to the model, what will be the population of Qatar in 2015?

In Exercises 13–15, describe in words the appearance of the graph from the equation, using the terms increasing, decreasing, concave up, concave down, vertical intercept, and horizontal asymptote, as appropriate. Then check your work by using a calculator to graph the function.

13. $y = 2(1.5)^x$
14. $y = 0.5(2.3)^x$
15. $y = 17(0.8)^x$

In Exercises 16–18, use exponential regression to find the model of best fit for the Consumer Price Index data. (The Consumer Price Index is used to measure the increase in prices over time. In each of the following tables, the index is assumed to have the value 100 in the year 1984.)
Then use the model to predict the value of the function when $t = 25$ and interpret the real-world meaning of the result.

16. **Price of Dental Services**

Years Since 1980 t	Price Index I
0	78.9
5	114.2
10	155.8
15	206.8
20	258.5

Source: Statistical Abstract of the United States, 2001, Table 694

17. **Price of a Television Set**

Years Since 1980 t	Price Index I
0	104.6
5	88.7
10	74.6
15	68.1
20	49.9

Source: Statistical Abstract of the United States, 2001, Table 694

18. **Price of Admission to Entertainment Venues**

Years Since 1980 t	Price Index I
0	83.8
5	112.8
10	151.2
15	181.5
20	230.5

Source: Statistical Abstract of the United States, 2001, Table 694

■ SECTION 6.3 ■

Certificates of Deposit

In Exercises 19–21, use the given information to answer the questions for each CD.

a. Each time interest is compounded, what is the percentage gain in the CD's value?

b. What is the future value of the CD when it matures, assuming the minimum amount is invested?

c. What is the average rate of change of the investment from when the CD is purchased until the CD matures?

d. What is the APY for the CD?

e. What initial investment would be required for the CD to be worth $5000 when it matures?

f. How much more money would the CD be worth if the interest was compounded continuously instead of periodically?

	Institution	Maturity	Nominal Rate	Compounding Method	Minimum Investment
19.	National City Bank of Kentucky	5 years	4.40%	monthly	$2500
20.	Regions Bank	3 years	3.55%	quarterly	$500

Institution	Maturity	Nominal Rate	Compounding Method	Minimum Investment
21. Countrywide Bank, FSB	3 months	6.16%	daily	$10,000

Source: Quoted rates are from www.bankrate.com and were accurate as of April 2007.

In Exercises 22–23, you are given a continuous growth rate for an account that compounds interest continuously.

a. What is the APY of the account?

b. What nominal rate would be necessary for an account with quarterly compound interest to have the same APY?

22. Continuous rate of 2%

23. Continuous rate of 1.83%

In Exercises 24–25, you are given a periodic growth rate for an account. For each, find the continuous growth rate that has the same APY.

24. Nominal rate of 6% compounded monthly

25. Nominal rate of 3.25% compounded daily

In Exercises 26–27, you are given population growth models based on information from www.census.gov. In each model the population is in millions t years after 2006.

a. Find the annual growth/decay rate of the population.

b. Rewrite the model in the form $f(t) = ab^t$.

26. **Population** Saudi Arabia: $S(t) = 26.42e^{0.0227t}$

27. **Population** Bulgaria: $B(t) = 7.45e^{-0.009t}$

In Exercises 28–29, you are given population growth models based on information from www.census.gov. In each model the population is in millions t years after 2006.

a. Find the continuous growth/decay rate of the population.

b. Rewrite the model in the form $f(t) = ae^{kt}$.

28. **Population** Egypt: $E(t) = 77.5(1.018)^t$

29. **Population** Lithuania: $L(t) = 3.6(0.997)^t$

In Exercises 30–31, use the following information. Celecoxib is a medicine prescribed to manage osteoarthritis and rheumatoid arthritis. After the drug reaches peak concentration in the blood stream, the body begins to reduce the amount of medicine present according to the model $A(t) = 200(0.9389)^t$, where A is the amount of medicine (in milligrams) left in the body t hours after peak concentration. (*Source:* Modeled using information from www.merck.com)

30. **Medicine** Find the continuous decay factor, explain what it represents, and rewrite the model in the form $A(t) = ae^{kt}$.

31. **Medicine** Estimate the half-life of celecoxib, then explain how you found your answer.

SECTION 6.4

In Exercises 32–33, rewrite each statement in logarithmic form.

32. $8^3 = 512$

33. $10^{-4} = \dfrac{1}{10,000}$

In Exercises 34–35, rewrite each statement in exponential form.

34. $\log_2(128) = 7$

35. $\ln(19) \approx 2.944$

In Exercises 36–39, evaluate each expression without using a calculator.

36. $\log_3(81)$ 37. $\log_2\left(\dfrac{1}{32}\right)$

38. $\log(100{,}000)$ 39. $4^{\log_4(6)}$

In Exercises 40–41, state the two integer values between which each expression falls. Then check your estimate using the change of base formula.

40. $\log(49)$ 41. $\log_5(600)$

In Exercises 42–43,

a. Determine the year in which the model predicts each country's population will be 18 million.

b. Find the inverse of the population model and explain what the inverse function is used to find.

42. **Population** Ecuador's population in millions can be modeled by $E(t) = 13.36(1.0124)^t$, where t is in years since 2006. (*Source:* www.census.gov)

43. **Population** Guatemala's population in millions can be modeled by $G(t) = 12.18e^{0.02205t}$, where t is in years since 2006. (*Source:* www.census.gov)

Misha shiyanov/Shutterstock.com

44. **Astronomy** The function $A(m) = 0.6735e^{0.423m}$ models the aperture size A (in millimeters) required to see an object with a magnitude of m. (*Source:* Modeled using data from www.ayton.id.au) Find the inverse of this function and explain what it models.

45. **Certificates of Deposit** Use the compound interest formula to write an exponential function to model the value of $8000 invested in a 3-year CD from UFBDirect.com with a nominal rate of 4.75% compounded monthly. (*Source:* www.bankrate.com in April 2007) Then find the inverse of your function and explain what the inverse models.

46. **Population** Sudan's population (in millions) can be modeled by $S(t) = 40.19(1.026)^t$ and Ukraine's population by $U(t) = 46.96(0.9925)^t$, where t is in years since 2005. (*Source:* www.census.gov) In what year do these models predict the two countries will have the same population?

47. **Certificates of Deposit** Determine how long it would take for the following investments to have the same value: $5000 invested in a Capital One CD with a nominal rate of 4.88% compounded daily and $5500 invested in an Integra Bank CD with a nominal rate of 3.96% compounded semiannually. (*Source:* www.bankrate.com in April 2007)

In Exercises 48–50, use the following information. The brightest celestial body seen from Earth appears to be the sun, but this is because the sun is relatively close to Earth, not because the sun is a very bright star in the universe. For this reason, scientists sometimes refer to an object's absolute magnitude, which measures how bright an object would appear if it was located 10 parsecs (or 32.6 light years) away. The formula $A(d) = 11 - 5 \log(d)$ models the absolute magnitude of an object with an apparent magnitude of 6 (barely visible with the naked eye) if that object is d parsecs from Earth. (**Source:** www.astro.northwestern.edu)

48. **Astronomy** Find $A(12)$ and explain what your answer represents.

49. **Astronomy** Find d if $A(d) = 2.3$ and explain what your answer represents.

50. **Astronomy** Find the inverse of $A(d)$ and explain what the inverse function models.

■ SECTION 6.5 ■

In Exercises 51–53, describe the appearance of the graph based on its equation.

51. $y = \log_2(x)$
52. $y = \log_{0.4}(x)$
53. $y = 3 \log_4(x)$

In Exercises 54–56, find the logarithmic function that best fits the data. Then answer the given question.

54. **Percentage of TV Homes with a VCR**

Years Since 1984 t	Homes with a VCR (percent) V
1	20.8
3	48.7
5	64.6
7	71.9
8	75.0
9	77.1
10	79.0
11	81.0
12	82.2
13	84.2
14	84.6

Source: Statistical Abstract of the United States, 2001, Table 1126

Evaluate $V(4)$ and explain what the solution means in this real-world context.

55. **Price of Wine Consumed at Home**

Consumer Price Index i	Years Since 1980 t
89.5	0
100.2	5
114.4	10
133.6	15
151.6	20

Source: Statistical Abstract of the United States, 2001, Table 694

Evaluate $t(140)$ and explain what the solution means in this real-world context.

56. **Price of a Television Set**

Consumer Price Index i	Years Since 1980 t
104.6	0
88.7	5
74.6	10
68.1	15
49.9	20

Source: Statistical Abstract of the United States, 2001, Table 694

Evaluate $t(90)$ and explain what the solution means in this real-world context.

Make It Real Project

What to Do

1. Find a set of at least six data points from an area of personal interest. Choose data that appear to increase or decrease by a constant percentage rate.

2. Draw a scatter plot of the data and explain why an exponential growth or exponential decay function might best model the situation.

3. Find a regression model for your data.

4. Using rate of change ideas, describe the relationship between the two quantities.

5. Explain the practical meaning of the initial value and change factor of the model.

6. Use the model to predict the value of the function at an unknown point and explain why you think the prediction is accurate or not accurate.

7. Determine the inverse of the exponential model.

8. Interpret the meaning of the inverse function in the context of the situation.

9. Explain how a consumer and/or a businessperson could benefit from the model and from the inverse of the model.

CHAPTER 7

Modeling with Other Types of Functions

Adam Gryko/Shutterstock.com

The Maricopa Community College District typically offers annual salary increases for faculty for the first several years of employment. A *salary increase* is designed to increase employees' standard of living. Additionally, the Faculty Association seeks to negotiate a COLA each year. A COLA, or *cost-of-living allowance*, is designed to keep pace with inflation and thus allows employees to maintain their standard of living. Modeling the projected earnings of a faculty member working in the district requires the use of a combination of functions including linear, exponential, and piecewise as well as a product of functions.

- **7.1** Combinations of Functions
- **7.2** Piecewise Functions
- **7.3** Composition of Functions
- **7.4** Logistic Functions
- **7.5** Choosing a Mathematical Model

STUDY SHEET
REVIEW EXERCISES
MAKE IT REAL PROJECT

SECTION 7.1 Combinations of Functions

LEARNING OBJECTIVES

- Compute the sum, difference, product, or quotient of functions to model a real-world situation
- Determine the practical and theoretical domain of the combination of functions

GETTING STARTED

Advances in technology and medicine as well as a focus on healthy living have resulted in an increase in the life expectancy of men and women in the United States. The life expectancy of women is greater than that for men, but the gap is closing.

In addition to modeling life expectancy as a function of year, we can model the gap between the life expectancies of men and women. We can then use that model to predict if the life expectancy of men can ever be equal to that of women.

In this section we learn how to combine functions by addition, subtraction, multiplication, and division. This knowledge helps us better understand the relationships between quantities and make predictions such as for life expectancies.

■ Combining Functions Graphically and Numerically

The graph in Figure 7.1 shows the life expectancy of men and women born between 1980 and 2004. (*Source:* Statistical Abstract of the United States, 2007, Table 98) Looking at the overall trend in life expectancies in Figure 7.2, we see that the difference in life expectancy between women and men (D) decreases as time increases.

Figure 7.1

Figure 7.2

To see this more clearly, we calculate the differences throughout the data set, as shown in Table 7.1.

Table 7.1

Birth Year t	Male Life Expectancy (years) M	Female Life Expectancy (years) F	Difference in Life Expectancy $D = F - M$
1980	70.0	77.4	7.4
1982	70.8	78.1	7.3
1984	71.1	78.2	7.1
1986	71.2	78.2	7.0
1988	71.4	78.3	6.9

Table 7.1 (continued)

Birth Year t	Male Life Expectancy (years) M	Female Life Expectancy (years) F	Difference in Life Expectancy $D = F - M$
1990	71.8	78.8	7.0
1992	72.3	79.1	6.8
1994	72.4	79.0	6.6
1996	73.1	79.1	6.0
1998	73.8	79.5	5.7
2000	74.3	79.7	5.4
2002	74.5	79.9	5.4
2004	75.2	80.4	5.2

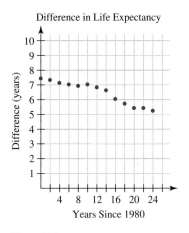

Figure 7.3

The table confirms the overall trend that the life expectancy gap is decreasing as time increases. The difference in life expectancy generally decreases from 7.4 years in 1980 to 5.2 years in 2004. We create a scatter plot of the differences in Figure 7.3 and observe the same trend.

■ Combining Functions Symbolically

We have just seen how to combine functions using graphs and tables. In the next example we show how to combine functions represented symbolically.

EXAMPLE 1 ■ Combining Functions Symbolically

Using linear regression, we determine that the life expectancy for women can be modeled by $F(t) = 0.106t + 77.6$ and the life expectancy for men by $M(t) = 0.204t + 70.0$, where t is the number of years since 1980.

a. Determine a function that models the difference in life expectancy between men and women. Name this function $D(t)$.

b. Interpret the meaning of the slope and vertical intercept of $D(t)$ in the context of this situation.

c. Graph $D(t)$ along with the scatter plot shown in Figure 7.3. Comment on the accuracy of the model.

d. Describe the practical domain and range of $D(t)$.

Solution

a. As we did graphically in Figure 7.2 and numerically in Table 7.1, we determine the model for the differences in life expectancies by subtracting the function $M(t) = 0.204t + 70.0$ from $F(t) = 0.106t + 77.6$. We combine the like terms to simplify the function.

$$D(t) = F(t) - M(t)$$
$$D(t) = 0.106t + 77.6 - (0.204t + 70.0)$$
$$D(t) = -0.098t + 7.6$$

b. The slope of -0.098 means the difference in life expectancy between men and women is decreasing by 0.098 years of age per year. The vertical intercept of $(0, 7.6)$ means that the difference in 1980 is 7.6 years. (The actual difference is 7.4 but the model predicts 7.6.)

c. Figure 7.4 shows the graph of $D(t) = -0.098t + 7.6$ along with the scatter plot of the data. We see that this function models the data reasonably well.

d. The practical domain for the function is the years for which the model can reasonably represent the situation. Determining the interval of values over which the model is valid is somewhat subjective. We choose the interval $0 \leq t \leq 30$, which means we reserve the use of our model for years between 1980 ($t = 0$) and 2010 ($t = 30$).

The practical range for the function is the differences in life expectancy for which the model can reasonably represent the situation. Based on the practical domain, we determine that the interval $4.66 \leq D \leq 7.60$ is the practical range. This corresponds to the difference in life expectancy in 2010 (4.66) and in 1980 (7.60).

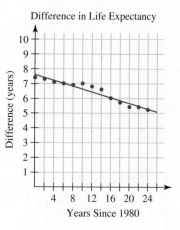

Figure 7.4

EXAMPLE 2 ■ Making a Prediction Using a Combination of Functions

Use the function model $D(t) = -0.098t + 7.6$ to predict the difference in life expectancy for men and women born in 2010. Then confirm the result using $M(t) = 0.204t + 70.0$ and $F(t) = 0.106t + 77.6$.

Solution Since 2010 is 30 years since 1980, we evaluate the function at $t = 30$.

$$D(30) = -0.098(30) + 7.6$$
$$= 4.66$$

The difference between the life expectancy of men and women born in 2010 is predicted to be 4.66 years.

We confirm this result and determine the life expectancies by evaluating both $M(t) = 0.204t + 70.0$ and $F(t) = 0.106t + 77.6$ at $t = 30$ and subtracting the results.

$$M(30) = 0.204(30) + 70.0 \qquad F(30) = 0.106(30) + 77.6$$
$$= 76.12 \qquad\qquad\qquad = 80.78$$

Women born in 2010 are expected to live 80.78 years and men born in 2010 are expected to live 76.12 years. The difference between these two values is 4.66 years, which confirms our result.

■ Dividing Functions

The U.S. federal government brings in money (revenue) primarily through taxes paid by its citizens. It spends money (outlays) on military, social programs, and education, among other programs. Currently, the government spends more money than it receives. To make up for this deficit, it borrows money by selling securities like Treasury bills, notes, bonds, and savings bonds to the public. According to the U.S. Treasury Department, the government's national debt is an accumulation of the deficit spending experienced each year. (*Source:* www.treasurydirect.gov)

Based on data from 1940 to 2004, the national debt can be modeled by the exponential function $D(t) = 0.086(1.07^t)$, where t is measured in years since 1940 and D is measured in trillions of dollars. (Modeled from www.whitehouse.gov data) The population of the United States can be modeled by the linear function $P(t) = 2.55t + 127$, where t is years since 1940 and P is millions of people. (*Source:* Modeled from www.census.gov data) We use these models in the next two examples as we combine functions using division.

SECTION 7.1 ■ COMBINATIONS OF FUNCTIONS

EXAMPLE 3 ■ **Combining Functions Using Division**

Suppose every person (including children) in the United States was to pay an equal amount of money to eliminate the national debt. Using the functions D and P just defined, create a function A that would give the amount of money each person in the United States would need to pay. Then evaluate and interpret $A(70)$.

Solution The function D gives the total national debt (in trillions of dollars) for a given year (since 1940). If we divide this value by the population for that same year, we get an average amount of debt per person in the United States.

$$A(t) = \frac{D(t)}{P(t)} = \frac{0.086(1.07^t)}{2.55t + 127} \frac{\text{trillion dollars}}{\text{million people}}$$

Note the units used in this combination of functions is trillion dollars per million people.

Evaluating $A(70)$ will provide us the amount of money required per person to completely pay for the national debt in 2010, since 2010 is 70 years after 1940.

$$A(70) = \frac{0.086(1.07)^{70}}{2.55(70) + 127}$$

$$\approx 0.0321 \text{ trillion dollars per million people.}$$

We express this result as a fraction and simplify.

$$\frac{0.0321 \text{ trillion dollars}}{1 \text{ million people}} = \frac{0.0321(1{,}000{,}000{,}000{,}000)\text{dollars}}{1{,}000{,}000 \text{ people}}$$

$$= \frac{32{,}100{,}000{,}000 \text{ dollars}}{1{,}000{,}000 \text{ people}}$$

$$= \$32{,}100 \text{ per person}$$

If everyone paid an equal amount, every man, woman, and child in the United States would have to pay $32,100 in 2010 in order to pay off the national debt.

EXAMPLE 4 ■ **Determining the Domain and Range**

The function $A(t) = \dfrac{0.086(1.07^t)}{2.55t + 127}$ gives the amount of money (in trillions of dollars per million people) that each person in the United States would need to pay in year t (where t is the number of years since 1940) to eliminate the national debt.

a. Determine the theoretical domain of this function.
b. Determine the practical domain of this function. Then graph the function.

Solution

a. The theoretical domain refers to the set of input and output values of this function independent of the real-world context. In this case, the theoretical domain is all values of t except the value that makes the denominator zero. To find this value, we set the denominator equal to zero and solve for t.

$$2.55t + 127 = 0$$
$$2.55t = -127$$
$$t = \frac{-127}{2.55}$$
$$t \approx -49.8$$

The theoretical domain is all values of t except $t = -49.8$.

b. The practical domain refers to input values that seem reasonable for the real-world context. We know it is not good practice to extrapolate too far outside of the original data set. Since the original data set included years 1940–2004, we choose domain values that represent the years 1940–2010. Thus we describe the practical domain to be values of t such that $0 \leq t \leq 70$. (Note: This choice of domain values is subjective and a different choice of values may also be acceptable.) Figure 7.5 shows the graph of

$$A(t) = \frac{0.086 \cdot 1.07^t}{2.55t + 127}$$ using the practical domain.

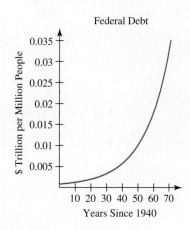

Figure 7.5

Multiplying Functions

The National Automobile Dealers Association (NADA) reports statistics on the sales of new vehicles in the United States. The number of new vehicles sold per dealership in the United States is given in Table 7.2 and Figure 7.6. We model these data using the quartic function $V(t) = 0.216t^4 - 4.50t^3 + 27.6t^2 - 28.6t + 653$, as seen in Figure 7.7.

Table 7.2

	Number of New Vehicles Sold
Years Since 1995 t	New Vehicles Sold per Dealership V
0	648
1	664
2	668
3	694
4	759
5	783
6	785
7	774
8	769
9	779
10	788

Source: www.autoexcemag.com

Figure 7.6

Figure 7.7

The NADA also reports the average retail selling price of these new vehicles. Table 7.3 and Figure 7.8 provide this information. We model these data using the linear function $P(t) = 777t + 21{,}000$, as seen in Figure 7.9.

SECTION 7.1 ■ COMBINATIONS OF FUNCTIONS

Table 7.3

Years Since 1995 t	New Vehicle Sales Price — Average Retail Selling Price (dollars) P
0	20,450
1	21,900
2	22,650
3	23,600
4	24,450
5	24,900
6	25,800
7	26,150
8	27,550
9	28,050
10	28,400

Source: www.autoexcemag.com

Figure 7.8

Figure 7.9

In the next example, we use these models to investigate the product of two functions.

EXAMPLE 5 ■ Interpreting the Product of Two Functions

The function $V(t) = 0.216t^4 - 4.50t^3 + 27.6t^2 - 28.6t + 653$ models the number of new vehicles sold, where t is measured in years since 1995 and V is the number of new vehicles per dealership in the United States. The function $P(t) = 777t + 21,000$ models the average retail selling price, P, of a new vehicle t years after 1995.

a. Find and interpret $f(15)$ given $f(t) = V(t) \cdot P(t)$.
b. Express the function $f(t) = V(t) \cdot P(t)$ symbolically.
c. Interpret the meaning of the function $f(t) = V(t) \cdot P(t)$.

Solution

a. We first evaluate and interpret $V(15)$.

$$V(15) = 0.216(15)^4 - 4.50(15)^3 + 27.6(15)^2 - 28.6(15) + 653$$
$$\approx 2180 \text{ (accurate to 3 significant digits)}$$

This tells us that 15 years after 1995 (2010), the number of new vehicles sold per dealership will be 2180 vehicles.

Next we evaluate and interpret $P(15)$.

$$P(15) = 777(15) + 21,000$$
$$= 32,655$$
$$\approx 32,700 \text{ (accurate to 3 significant digits)}$$

This tells us that 15 years after 1995 (2010), the average retail selling price of a new vehicle will be approximately $32,700.

To evaluate $f(15)$, we calculate $f(15) = V(15) \cdot P(15)$.

$$f(15) = V(15) \cdot P(15)$$
$$= 2180 \cdot 32{,}700$$
$$\approx 71{,}300{,}000 \text{ (accurate to 3 significant digits)}$$

This tells us that 15 years after 1995 (2010), a dealership will generate approximately $71 million in income from the sales of new vehicles. That is, if 2180 new vehicles are sold at an average cost of $32,700 each, roughly $71 million in income is generated.

b. We now express $f(t) = V(t) \cdot P(t)$ symbolically.

$$f(t) = V(t) \cdot P(t)$$
$$= (0.216t^4 - 4.50t^3 + 27.6t^2 - 28.6t + 653) \cdot (777t + 21{,}000)$$

c. Since $V(t)$ is the number of new vehicles sold per dealership and $P(t)$ is the average retail selling price per vehicle (in dollars), we interpret $f(t) = V(t) \cdot P(t)$ to be the total income from new vehicle sales per dealership t years after 1995. That is, if each dealership sold $V(t)$ vehicles at a price of $P(t)$ each, the income for the dealership would be $f(t)$ dollars.

The following box summarizes our work so far in this section and introduces formal notation for the combination of functions.

COMBINATIONS OF FUNCTIONS

Sum of two functions	$(f + g)(x) = f(x) + g(x)$
Difference of two functions	$(f - g)(x) = f(x) - g(x)$
Product of two functions	$(f \cdot g)(x) = f(x) \cdot g(x)$
Quotient of two functions	$\left(\dfrac{f}{g}\right)(x) = \dfrac{f(x)}{g(x)},\ g(x) \neq 0$

EXAMPLE 6 ■ **Understanding Combinations of Functions**

Now let's consider the conceptual aspects of combining functions when a real-world context is not provided. Given the graphs of two functions f and g shown in Figure 7.10, sketch the graph of the given combination of functions.

a. $(f + g)(x)$

b. $\left(\dfrac{f}{g}\right)(x)$

Figure 7.10

Solution

a. We know $(f + g)(x) = f(x) + g(x)$. In terms of the graphs, we are adding together the y-values of each function while leaving the x-values unchanged. Although there are many ways to approach this problem, one of the easiest is to create a table of values such as Table 7.4 for f and g and then sum the function values. To ensure as much accuracy as possible, we select the *x*-values of the intercepts, intersection points, and extrema for each graph. We use our best judgment to estimate the coordinates.

Then we connect the plotted points of $f(x) + g(x)$ with a smooth curve, as in Figure 7.11.

Table 7.4

x	f(x)	g(x)	f(x) + g(x)
0	10	−4	6
3	0	−2.5	−2.5
4	−2	−2	−4
8	−6	0	−6
13	0	2.5	2.5
14	3	3	6

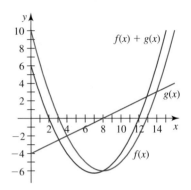

Figure 7.11

b. We use the same points used in part (a) and calculate the quotient of the two function values, as shown in Table 7.5.

A vertical asymptote occurs where $g(x) = 0$ provided $f(x) \neq 0$. We connect the plotted points of $\dfrac{f(x)}{g(x)}$ with a smooth curve, as in Figure 7.12.

Table 7.5

x	f(x)	g(x)	$\dfrac{f(x)}{g(x)}$
0	10	−4	−2.5
3	0	−2.5	0
4	−2	−2	1
8	−6	0	undefined
13	0	2.5	0
14	3	3	1

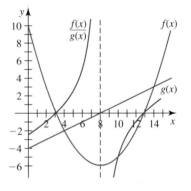

Figure 7.12

In practice, we typically use technology such as a graphing calculator to graph functions. However, graphing functions by hand helps our understanding of the underlying function concepts.

SUMMARY

In this section you learned how to combine functions by adding, subtracting, multiplying, and dividing them. You also learned how to interpret combinations of functions in real-world contexts and use the combinations to define additional function relationships and make predictions. Additionally, you examined graphs and tables to better understand the conceptual meanings of combinations of functions.

7.1 EXERCISES

SKILLS AND CONCEPTS

In Exercises 1–5, use the graphs of functions f and g to evaluate each given function. If it is not possible to evaluate the function, explain why not.

1. a. $(f + g)(2)$ b. $(f \cdot g)(2)$

2. a. $(g - f)(0)$ b. $\left(\dfrac{g}{f}\right)(0)$

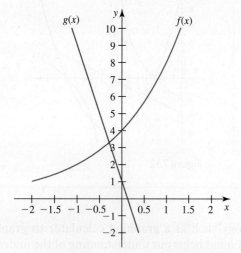

3. a. $(g \cdot f)(-2)$ b. $\left(\dfrac{f}{g}\right)(-2)$

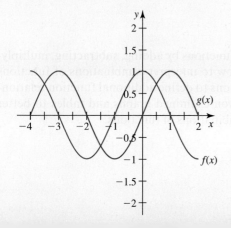

4. a. $(g + f)(1)$ b. $(f \cdot g)(1)$

5. a. $(g + f)(1)$ b. $(g \cdot f)(2)$

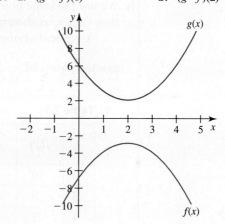

In Exercises 6–10, use the given table to evaluate each given function. If it is not possible to evaluate the function, explain why not.

6.

x	-2	0	2	4	6
$f(x)$	1	5	11	19	29
$g(x)$	5	1	5	17	37

a. $(f + g)(4)$
b. $(f - g)(4)$
c. $\left(\dfrac{f}{g}\right)(4)$
d. $(f \cdot g)(4)$

7.

x	-2	0	2	4	6
$f(x)$	-1	1	3	5	7
$g(x)$	-6	-4	-2	0	2

a. $(f + g)(4)$
b. $(f - g)(4)$
c. $\left(\dfrac{f}{g}\right)(4)$
d. $(f \cdot g)(4)$

8.

x	0	1	2	3	4
f(x)	0	1	8	27	64
g(x)	0	-1	-8	-27	-64

a. $(f + g)(4)$
b. $(f - g)(4)$
c. $\left(\dfrac{f}{g}\right)(4)$
d. $(f \cdot g)(4)$

9.

x	-4	0	4	8	16
f(x)	2	4	0	4	8
g(x)	-2	-6	-5	-7	-9

a. $(f + g)(4)$
b. $(f - g)(4)$
c. $\left(\dfrac{f}{g}\right)(4)$
d. $(f \cdot g)(4)$

10.

x	4	8	12	16	20
f(x)	0	2	4	6	8
g(x)	1	1	2	3	5

a. $(f + g)(4)$
b. $(f - g)(4)$
c. $\left(\dfrac{f}{g}\right)(4)$
d. $(f \cdot g)(4)$

In Exercises 11–20, combine the functions f and g symbolically as indicated. Then state the domain of the combination.

11. $f(x) = 3x - 5 \quad g(x) = -4x + 7$
    a. $(f + g)(x)$
    b. $(f - g)(x)$

12. $f(x) = -3x + 6 \quad g(x) = 5x - 10$
    a. $(f - g)(x)$
    b. $\left(\dfrac{f}{g}\right)(x)$

13. $f(x) = x - 1 \quad g(x) = x + 3$
    a. $(f \cdot g)(x)$
    b. $\left(\dfrac{g}{f}\right)(x)$

14. $f(x) = x^2 + 2x - 1 \quad g(x) = -2x^2 - 5$
    a. $(f + g)(x)$
    b. $(f \cdot g)(x)$

15. $f(x) = x^2 - 4 \quad g(x) = 2x + 8$
    a. $(g - f)(x)$
    b. $\left(\dfrac{f}{g}\right)(x)$

16. $f(x) = \sqrt{x - 2} \quad g(x) = 5$
    a. $(f + g)(x)$
    b. $(f \cdot g)(x)$

17. $f(x) = 2^x \quad g(x) = x$
    a. $(g \cdot f)(x)$
    b. $\left(\dfrac{g}{f}\right)(x)$

18. $f(x) = \ln(x) \quad g(x) = 4x$
    a. $(f + g)(x)$
    b. $(f \cdot g)(x)$

19. $f(x) = 2x^3 + x^2 - 4x + 7$
    $g(x) = -x^2 + 7x - 2$
    a. $(f + g)(x)$
    b. $(g \cdot f)(x)$

20. $f(x) = x^3 - x^2 + 3x - 1$
    $g(x) = 2x^3 - 2x^2 - 5x + 6$
    a. $(g \cdot f)(x)$
    b. $(f + g)(x)$

21. **Operating a Business** Suppose two large warehouse stores are operating in the same medium-sized city. Corporate headquarters has asked them to estimate the number of employees they will need for the upcoming year. Each store manager produced a graph (based on historical data) of the anticipated number of customers per day at that store. Their two graphs are shown on the same coordinate system. The horizontal axis represents the number of months in the upcoming year. The vertical axis represents the number of people expected at a store each day.

a. What does the point (6, 700) represent on Store 2's graph?

b. The function f represents the number of customers per day at Store 1 and g represents the number of customers at Store 2, both as a function of the number of months, m. Interpret the meaning of $(f + g)(m)$.

c. Since the corporate office hires for both stores, it needs an estimate of the stores' *total* number of customers. Sketch a graph of the daily number of customers at both stores.

22. High School Prom

Students on a prom committee need to set the price for prom tickets. From past experience, they know that if tickets cost $200 per couple, 100 couples buy tickets for the prom. They also predict from past experience that for every $5 decrease in price, an additional 2 couples will buy tickets.

a. Create a function representing the price, P, of a prom ticket per couple as a function of the number of $5 price decreases, n.

b. Create a function representing the number of couples attending prom, A, as a function of the number of $5 price decreases, n.

c. Compute and interpret the meaning of the function $(P \cdot A)(n)$.

d. Graph $(P \cdot A)(n)$ and describe the behavior of the function in the context of the situation.

e. Solve $(P \cdot A)(n) = 0$ for n and explain what these values mean in the context of the situation.

In Exercises 23–32, use the given graphs of the functions f and g to sketch a graph of each combination of functions.

23. Sketch $(f + g)(x)$.

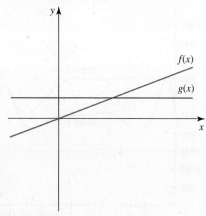

24. Sketch $(f - g)(x)$.

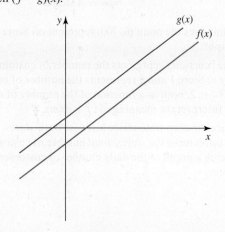

25. Sketch $(f \cdot g)(x)$.

26. Sketch $\left(\dfrac{f}{g}\right)(x)$.

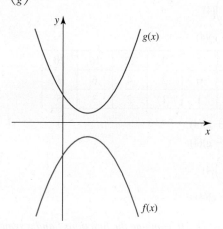

27. Sketch $(f + g)(x)$.

28. Sketch $(f \cdot g)(x)$.

29. Sketch $\left(\dfrac{f}{g}\right)(x)$.

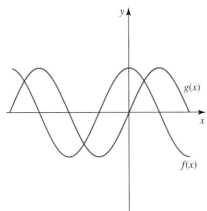

30. Sketch $(f + g)(x)$.

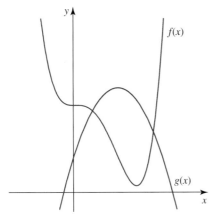

31. Sketch $(f - g)(x)$.

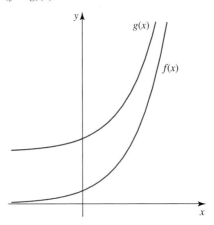

32. Sketch $(f \cdot g)(x)$.

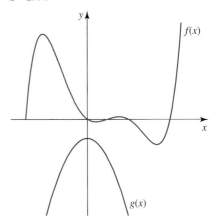

SHOW YOU KNOW

33. Suppose the function f has x-intercept $(a, 0)$ and a function g passes through the point (a, b). What are the coordinates of the function $h(x) = f(x) + g(x)$ at $x = a$?

34. A function $h(x) = f(x) + g(x)$ has an x-intercept at $x = a$. What is the relationship between $f(a)$ and $g(a)$?

35. Under what conditions will $h(x) = \dfrac{f(x)}{g(x)}$ have a vertical asymptote?

36. How are the horizontal intercepts of $h(x) = \dfrac{f(x)}{g(x)}$ related to the horizontal intercepts of f?

37. The units of a function s are million dollars and the units of a function q are dollars per hour. What are the units of $r(t) = \dfrac{s(t)}{q(t)}$?

38. The points of intersection of the graphs of a function f and a function g have what graphical significance for the function $j(x) = f(x) - g(x)$?

39. The graph of a function g is a horizontal line and the graph of a function h is a concave up parabola that lies above the graph of g. Describe the graph of $j(x) = g(x) - h(x)$.

40. The graph of a function f is a line with slope m. The graph of a function g is a line with slope $-\dfrac{1}{m}$. Both functions pass through the origin. Describe the graph of $f(x) = j(x) \cdot h(x)$.

41. If the units of f are dollars per day and the units of g are days per week, what are the units of $h(t) = f(t) \cdot g(t)$?

42. A function f is positive for all values of x and a function g is negative for all values of t. Does $h(t) = f(t) \cdot g(t)$ have any horizontal intercepts? Explain.

■ MAKE IT REAL

43. Stopping Distance The total stopping distance required to stop a moving car is typically based on reaction distance and braking distance. The reaction distance is the distance the car travels from the moment the driver decides to apply the brake to the moment the brake is applied. Braking distance is the distance traveled from the moment the driver applies the brake to the moment the car comes to a complete stop. Use the data in the table to respond to the following items.

Speed (mph) S	Reaction Distance (feet) R	Braking Distance (feet) B
20	22.0	22.2
25	27.5	34.7
30	33.0	50.0
35	38.5	68.0
40	44.0	88.8
45	49.5	112.4
50	55.0	138.8

Source: Minnesota Driving Manual

a. Determine a linear regression model representing $R(S)$, the reaction distance as a function of the speed of the car.

b. Determine a quadratic regression model representing $B(S)$, the braking distance as a function of the speed of the car.

c. Compute the function $(R + B)(S)$ and explain what it means in the context of this situation.

d. Evaluate and interpret $(R + B)(75)$.

44. Buying and Selling Homes The graph shows the median price of new and resale homes in the Greater Phoenix, Arizona, area. (*Source:* www.poly.asu.edu)

The median price of resale homes as a function of years since 1988 is given by $R(t)$, in thousands of dollars. The median price of new homes as a function of years since 1988 is given by $N(t)$, in thousands of dollars.

a. Interpret the meaning of $(N - R)(t)$.

b. Estimate and interpret $(N - R)(6)$.

c. Sketch a graph of $(N - R)(t)$.

d. Interpret the meaning of $\left(\dfrac{N}{R}\right)(t)$.

e. Estimate and interpret $\left(\dfrac{N}{R}\right)(17)$.

f. Sketch a graph of $\left(\dfrac{N}{R}\right)(t)$.

45. Registered Vehicles The number of registered vehicles in the United States, in millions, can be modeled by $V(t) = 3.383t + 154.8$, where t is the number of years since 1980. (*Source: Statistical Abstract of the United States, 2007*, Table 1074)

The population of the United States, in millions, can be modeled by $P(t) = 2.821t + 225.5$, where t is the number of years since 1980. (*Source: Statistical Abstract of the United States, 2006*, Table 2)

a. Compute and interpret $\left(\dfrac{V}{P}\right)(t)$.

b. Determine the value of t such that $\left(\dfrac{V}{P}\right)(t) = 1$. Explain what this means in this context.

46. SAT Scores The function M represents the mean SAT score for males relative to the number of years, t, since 1987. The function F represents the mean SAT score for females relative to the number of years, t, since 1987. Use the graph of these functions to respond to the following items.

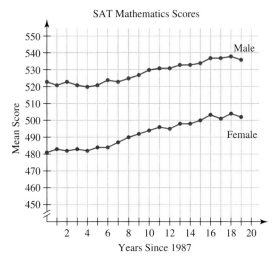

Source: www.collegeboard.com

a. Explain what information $M(15)$ provides.
b. Explain what information $F(15)$ provides.
c. Interpret the meaning of $(M - F)(t)$.
d. Sketch a graph of $(M - F)(t)$. Describe what information can be gleaned from this graph. That is, what trend do you notice and what does it mean in the context of the situation?

47. Soda Consumption The function $S(t) = -0.22t + 53.18$ represents the per capita consumption of soda pop in the United States, where S is measured in gallons of soda and t represents years since 2000. (*Source: Statistical Abstract of the United States, 2007*, Table 201) The function $P(t) = 2870.9t + 282{,}426$ represents the population, P, of the United States in thousands as a function of t, the number of years since 2000. (*Source: Statistical Abstract of the United States, 2006*, Table 2)

a. Compute and interpret $(S \cdot P)(t)$.
b. Graph $(S \cdot P)(t)$ and describe the behavior of this function.
c. Evaluate and interpret $(S \cdot P)(10)$.

48. Violent Crime The second column in the table shows the total number of violent crimes committed in the United States in the given years. The third column shows the number of violent crimes that are classified as murders. Use this table to respond to the following items.

Years Since 1990 t	Total Number of Violent Crimes (thousands) V	Total Number of Murders (thousands) M
0	1820	23
1	1912	25
2	1932	24
3	1926	25
4	1858	23
5	1799	22
6	1689	20

(*continued*)

Years Since 1990 t	Total Number of Violent Crimes (thousands) V	Total Number of Murders (thousands) M
7	1636	18
8	1534	17
9	1426	16
10	1425	16
11	1439	16
12	1424	16
13	1384	17
14	1367	16

Source: Statistical Abstract of the United States, 2007, Table 295

a. Interpret the meaning of $\left(\dfrac{M}{V}\right)(t)$.
b. Create a fourth column in the table showing the values of $\left(\dfrac{M}{V}\right)(t)$.
c. Use a graphing calculator to create a scatter plot of $\left(\dfrac{M}{V}\right)$ as a function of t.
d. Determine and create a polynomial regression model for $\left(\dfrac{M}{V}\right)(t)$. Explain why you decided on the polynomial you used to model these data.

49. Football Game Attendance and Player Salaries Attendance at National Football League games has consistently increased over the past decade. The total attendance, in thousands of people, at all NFL games during the year can be modeled by $A(t) = 323.3t + 17{,}284$, where t is the number of years since 1990. (*Source: Statistical Abstract of the United States, 2007*, Table 1228) Player salaries have also increased over this same time period. The average salary, in thousands of dollars, of an NFL player can be modeled by

$$S(t) = -0.1812t^4 + 5.758t^3 - 54.05t^2 + 199.6t + 311.5$$

(*Source: Statistical Abstract of the United States, 2007*, Table 1228)

a. Calculate and interpret the meaning of $A(18)$.
b. Calculate and interpret the meaning of $S(18)$.
c. Determine and interpret the meaning of $\left(\dfrac{S}{A}\right)(t)$.
d. Determine the practical domain and range of $\left(\dfrac{S}{A}\right)(t)$.

50. Cheese Consumption The function $C(t) = 0.5161t + 18.89$ represents the per capita consumption of cheese in the United States, where C is measured in pounds of cheese and t represents years since 1980. (*Source: Statistical Abstract of the United States, 2007*, Table 202) The function $P(t) = 2.943t + 222.7$ represents the population of the

United States, in millions of people, as a function of the number of years since 1980. (*Source:* Modeled from www.census.gov)

a. Compute and interpret $(C \cdot P)(t)$.
b. Graph $(C \cdot P)(t)$ and describe the behavior of this function.
c. Evaluate and interpret $(C \cdot P)(30)$.

■ STRETCH YOUR MIND

Exercises 51–55 are intended to challenge your understanding of the combination of functions. For each, suppose f(x) is always increasing and g(x) is always decreasing. Determine if each of the following combinations of functions is always increasing, always decreasing, or is it impossible to tell. Fully explain your rationale for your response.

51. $(f + g)(x)$
52. $(f - g)(x)$
53. $(f \cdot g)(x)$
54. $\left(\dfrac{f}{g}\right)(x)$
55. $f^{-1}(x) + g^{-1}(x)$

SECTION 7.2 Piecewise Functions

LEARNING OBJECTIVES

- Define piecewise functions using equations, tables, graphs, and words
- Determine function values of piecewise functions from a graph, equation, and table

GETTING STARTED

Many parking facilities assess fees based on the length of time a vehicle is left in the care of the attendants. Often the amount charged increases over set time intervals. When the parking fee is best defined by different equations over distinct time intervals, a piecewise function can be used to describe the fee assessment structure.

In this section we investigate piecewise functions such as this and discuss whether a given piecewise function is continuous or discontinuous. We demonstrate how to create piecewise functions in table, graph, and equation form and use the functions to determine specific values.

■ Piecewise Functions

Throughout this text we have seen that many different types of functions can be used to model real-world situations. We now look at scenarios that are best modeled by a combination of functions over distinct intervals of the domain.

Consider Table 7.6, which shows the fees to park in the East Economy Garage at Sky Harbor International Airport in Phoenix, Arizona for a single day. We see that for any time over 0 minutes through 60 minutes, the fee is $4.00; for time over 60 through 120 minutes, the fee is $8.00; and for any time over 120 minutes (for one day), the fee is $10.00. Graphing this information as in Figure 7.13, we see that $F(m)$ is a discontinuous combination of three linear functions. Note that we use an open circle to denote that a value is not included in the function and an arrow to denote that the function continues beyond the graph.

A combination of functions such as the parking fee function is called a **piecewise function**.

Table 7.6

Parking Time (minutes) m	Parking Fee (dollars) F
Over 0 through 60	4.00
Over 60 through 120	8.00
Over 120	10.00

Source: www.phxskyharbor.com

Figure 7.13

Piecewise Functions

A **piecewise function** is defined using two or more expressions over given intervals of the domain. Piecewise functions are written in the form

$$f(x) = \begin{cases} \text{Rule 1} & \text{if Condition 1} \\ \text{Rule 2} & \text{if Condition 2} \\ \text{Rule 3} & \text{if Condition 3} \\ \vdots & \vdots \end{cases}$$

The conditions define the input values for which each rule applies. The graphs of piecewise functions may be *continuous* or *discontinuous*. Intuitively, a **discontinuous** function is one with a "break," "hole," or "jump" in its graph and a **continuous** function is one whose graph can be drawn without lifting one's pencil.

■ Defining a Piecewise Function with an Equation

The rules of a piecewise function may be an algebraic expression such as $2x + 9$ or a verbal expression such as "pays $260 per credit," as the following examples show.

$$|x| = \begin{cases} x & \text{if } x \geq 0 \\ -x & \text{if } x < 0 \end{cases} \qquad f(\text{age}) = \begin{cases} \text{not eligible for driver's license} & \text{if age} < 16 \\ \text{eligible for driver's license} & \text{if age} \geq 16 \end{cases}$$

$$h(t) = \begin{cases} 5t - 6 & \text{if } t \leq 1 \\ t^2 & \text{if } 1 < t < 5 \\ -3t & \text{if } t \geq 5 \end{cases} \qquad y = \begin{cases} x^2 - 2x + 1 & \text{if } x < 0 \\ \text{undefined} & \text{if } x = 0 \\ -x + 4 & \text{if } x > 0 \end{cases}$$

To model the parking-fee schedule using an algebraic equation, let's look again at the different pieces that define $F(m)$. For example, for any time up to and including 60 minutes, the parking fee is $4.00. We write

$$F(m) = 4 \quad \text{if } 0 < m \leq 60$$

Combining the separate equations for each level of parking fees, we have

$$F(m) = \begin{cases} 4 & \text{if } 0 < m \leq 60 \\ 8 & \text{if } 60 < m \leq 120 \\ 10 & \text{if } 120 < m \leq 1440 \end{cases} \quad \text{since there are 1440 minutes in 1 day}$$

It is important to note that $F(m)$ is a single function defined in many *pieces*, not many functions.

EXAMPLE 1 ■ Evaluating Piecewise Functions

Use the parking-fee function, $F(m)$, to evaluate each of the following. Then explain what each result means in its real-world context.

a. $F(40)$
b. $F(76)$
c. $F(480)$
d. $F(1800)$

Solution

a. To evaluate $F(40)$ means to find the fee for parking 40 minutes in the garage. Since 40 is greater than 0 but less than or equal to 60, it satisfies the condition $0 < m \leq 60$. The corresponding "rule" is $4.00. So $F(40) = 4$, which means the fee for parking 40 minutes is $4.00.

b. To evaluate $F(76)$ means to find the fee for parking 76 minutes in the garage. Since 76 is greater than 60 but less than or equal to 120, it satisfies the condition $60 < m \leq 120$. The corresponding "rule" is $8.00. So $F(76) = 8$, which means the fee for parking 76 minutes is $8.00.

c. To evaluate $F(480)$ means to find the fee for parking 480 minutes in the garage. Since 480 is greater than 120, it satisfies the condition $120 < m \leq 1440$. The corresponding "rule" is $10.00. So $F(480) = 10$, which means the fee for parking 480 minutes is $10.00.

d. To evaluate $F(1800)$ means to find the fee for parking 1800 minutes in the garage *on a single day*. Since 1800 is greater than 1440, this value of m does not satisfy any of the conditions for the piecewise function. Therefore, $F(1800)$ is undefined.

EXAMPLE 2 ■ Solving Piecewise Functions

Use the parking-fee function, $F(m)$, to solve each of the following equations for m. Then explain what each result means in its real-world context.

a. $F(m) = 4$

b. $F(m) = 9$

Solution

a. To solve $F(m) = 4$ for m means to determine how many minutes a person can park in the parking garage for $4.00. From the function equation, we find that $4.00 is associated with the condition $0 < m \leq 60$. Thus a person who pays the $4.00 fee may park for more than 0 minutes but no more than 60 minutes.

b. To solve $F(m) = 9$ for m means to determine how many minutes a person can park in the parking garage for $9.00. From the function equation, we see there is no $9.00 parking-fee rule. Since there is no parking fee of exactly $9.00, we are unable to solve the equation. There is no way to spend exactly $9.00 on parking.

EXAMPLE 3 ■ Creating and Using a Piecewise Function

Simply Fresh Designs is a small business that designs and prints creative greeting cards featuring client photos. The 2006 pricing structure for ordering personalized 4-inch by 6-inch holiday greeting cards from Simply Fresh Designs is shown in Table 7.7.

Table 7.7

Design, Print, & Ship		
4 by 6	$1.00/card $0.75/card for orders of 50 or more	white 4¾ by 6½ envelope included
(25 minimum order)		
$6 shipping fee on all orders		

Source: www.simplyfreshdesigns.com

a. Create a piecewise function that can be used to calculate the total cost of purchasing n cards.
b. Use the piecewise function to calculate the cost of purchasing 40 cards and 50 cards.
c. How many cards could we buy if we want to spend at most $45.00?

Solution

a. We let n represent the number of cards and $C(n)$ represent the total cost (in dollars) of purchasing n cards. To determine the function conditions, we note that the card pricing changes from $1.00 per card to $0.75 per card when the order size reaches 50 cards. We also note the minimum order size is 25 cards. We have

$$C(n) = \begin{cases} \text{Rule 1} & \text{if } 25 \leq n < 50 \\ \text{Rule 2} & \text{if } 50 \leq n \end{cases}$$

To determine the function rules, we note that the total cost of the cards is a function of the individual card price plus the shipping cost ($6.00). If fewer than 50 cards are ordered the cost is $C(n) = 1.00n + 6.00$. If 50 or more cards are ordered, the cost is $C(n) = 0.75n + 6.00$. We now have

$$C(n) = \begin{cases} 1.00n + 6.00 & \text{if } 25 \leq n < 50 \\ 0.75n + 6.00 & \text{if } 50 \leq n \end{cases}$$

b. We are asked to use the piecewise function to calculate the cost of purchasing 40 cards and 50 cards.

$$C(40) = 1.00(40) + 6.00 \qquad\qquad C(50) = 0.75(50) + 6.00$$
$$= 46.00 \qquad\qquad\qquad\qquad\quad = 43.50$$

It costs $46.00 to order 40 cards.   It costs $43.50 to order 50 cards.

Due to the volume discount, the cost of 50 cards is actually $2.50 less than the 40-card cost.

c. To determine how many cards we could buy if we wanted to spend at most $45.00, we first use the first rule of the function.

$$1.00n + 6.00 = 45.00$$
$$n = 39$$

Since $25 \leq 39 < 50$, the solution meets the required condition for the rule. We could order 39 cards for $45.00.

Now we consider the second rule of the function.

$$0.75n + 6.00 = 45.00$$
$$0.75n = 39$$
$$n = 52$$

Since $50 \leq 52$, the solution meets the required condition for the rule. We could order 52 cards for $45.00.

For a price of $45.00, we could either order 39 cards or 52 cards. In this case, we are able to get an additional 13 cards for no additional cost by understanding piecewise functions.

When a scatter plot of a data set appears to have distinct pieces, we can use regression to find a model equation for each piece. The resulting piecewise function is often the best fit to the data.

EXAMPLE 4 ■ Using Regression to Create a Piecewise Function

From 1981 to 1995, the annual number of adult and adolescent AIDS deaths in the United States increased dramatically. However, from 1995 to 2001, the annual death rate plummeted, as shown in Table 7.8.

Table 7.8

Adult and Adolescent AIDS Deaths in the U.S.	
Years Since 1980 t	Number of Deaths during Year D
1	122
2	453
3	1481
4	3474
5	6877
6	12,016
7	16,194
8	20,922
9	27,680
10	31,436
11	36,708
12	41,424
13	45,187
14	50,071
15	50,876
16	37,646
17	21,630
18	18,028
19	16,648
20	14,433
21	8963

Source: *HIV/AIDS Surveillance Report*, Dec. 2001; Centers for Disease Control and Prevention, p. 30

a. Draw a scatter plot of the data.
b. Based on the scatter plot, what types of functions will best model each piece? Explain.
c. Use regression to find a piecewise function, $D(t)$, to model the AIDS death data.

Solution

a. The scatter plot is shown in Figure 7.14.
b. From 1981 to 1995, the data is increasing. From 1995 until 2001, the data is decreasing. From 1981 until about 1989, the data appear concave up. A quadratic function may fit this piece well. Between 1989 and 1994, the data appear somewhat linear. A linear function may fit this piece well. Between 1995 and 2001, the scatter plot looks somewhat like a cubic function.

Figure 7.14

c. We use the data for each piece to calculate the corresponding regression models. For the quadratic model, we used data from $t = 1$ through $t = 9$. For the linear model, we used data from $t = 9$ through $t = 14$. For the cubic model, we used data from $t = 15$ through $t = 21$. The resultant function is

$$D(t) = \begin{cases} 417.5t^2 - 682.0t + 101.0 & \text{if } 1 \le t \le 9 \\ 4512.1t - 13{,}138 & \text{if } 9 < t < 15 \\ -381.1t^3 + 21{,}913.3t^2 - 422{,}152.0t + 2{,}739{,}830.5 & \text{if } 15 \le t \le 21 \end{cases}$$

EXAMPLE 5 ■ Graphing a Piecewise Function from an Equation

In 1998, Sammy Sosa of the Chicago Cubs and Mark McGuire of the St. Louis Cardinals were engaged in a race to break the all-time Major League Baseball single-season home run record set in 1962 by Roger Maris of the New York Yankees. As is shown in Figure 7.15, early in the season Sosa was hitting home runs at a relatively slow pace but in June he hit a record number of home runs. In the latter part of the season, his home run pace slowed from his scorching June pace but still remained quite brisk.

The number of home runs Sosa hit in 1998 can be modeled by the piecewise function

$$H(g) = \begin{cases} 0.22g - 0.22 & \text{if } 0 \le g < 55 \\ 2.23(1.035^g) & \text{if } 55 \le g < 82 \\ 0.0021g^2 - 0.07g + 23.95 & \text{if } 82 \le g \le 163 \end{cases}$$

Figure 7.15

Source: www.amstat.org

where H is the cumulative number of home runs hit and g is the number of games played. (Note: Recall we can choose the "best" function for a data set by considering the type of change in the output values [e.g., first differences, second differences, and percentage change], the shape of the graph, and the r and r^2 values. The difference with piecewise functions is that we consider distinct intervals of the domain that seem to be modeled better by different function types rather than trying to model the entire domain with a single equation.)

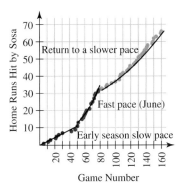

a. Sketch a graph of $H(g)$. Be sure to label the axes appropriately.

b. Use the formula of $H(g)$ to evaluate $H(30)$ and $H(60)$. Explain what each solution means in its real-world context.

c. Solve $H(g) = 30$ for g.

d. Use $H(g)$ to predict the number of home runs Sosa hit in 1998. Does the mathematical model show he would break Maris's record of 62 home runs? Explain. (Note: The 1998 Major League Baseball season was 162 games, but the Cubs played 163 games in that season due to a one-game playoff with the San Francisco Giants.)

Solution

a. The graph is shown in Figure 7.16.

b. To evaluate $H(30)$ means to find the cumulative number of home runs Sosa had hit after 30 games. Since $g = 30$ satisfies the condition $0 \le g < 55$, we use the corresponding rule. Therefore,

$$H(30) = 0.22(30) - 0.22$$
$$= 6.38$$
$$\approx 6$$

We estimate Sosa had hit 6 home runs after 30 games.

Figure 7.16

To evaluate $H(60)$ means to find the number of home runs Sosa had hit after 60 games. Since $g = 60$ satisfies the condition $55 \leq g \leq 82$, we use the corresponding rule. Therefore,

$$H(60) = 2.23(1.035^{60})$$
$$= 17.57$$
$$\approx 18$$

We estimate Sosa had hit 18 home runs after 60 games.

c. To solve $H(g) = 30$ for g, we need to use the graph in Figure 7.17 to find the game in which Sosa hit his 30th home run. We locate 30 home runs on the vertical axis and then go over to the graph and then down to the horizontal axis to find the game that is associated with 30 home runs. It appears that at about game 75 Sosa hit his 30th home run of the season.

d. Since the Cubs played 163 games in 1998, we need to evaluate $H(163)$. Since $g = 163$ satisfies the condition $82 \leq g \leq 163$, we apply the corresponding rule.

Figure 7.17

$$H(163) = 0.0021(163)^2 - 0.07(163) + 23.95$$
$$= 68.33$$
$$\approx 68$$

According to our model, Sosa hit 68 home runs in 1998 and broke Maris's record of 62. (Actually, he hit 66 and broke the record set by Maris.)

EXAMPLE 6 ■ Creating a Table and Graph for the Absolute Value Function

The absolute value function, $f(x) = |x|$ is formally defined as the piecewise function

$$f(x) = \begin{cases} -x & \text{if } x < 0 \\ x & \text{if } x \geq 0 \end{cases}$$

Create a table of values and graph of $f(x) = |x|$ for $-4 \leq x \leq 4$. Then find $f(-2.5)$.

Solution We create Table 7.9 by using selected x-values between -4 and 4. To create the graph, we plot these points to see the pattern and then connect the points. The graph of the absolute value function, shown in Figure 7.18, includes the points in the table as well as the value of the function for all values $-4 \leq x \leq 4$. Note that this is a piecewise continuous function.

Table 7.9

x	y
-4	$-(-4) = 4$
-3	$-(-3) = 3$
-2	$-(-2) = 2$
-1	$-(-1) = 1$
0	0
1	1
2	2
3	3
4	4

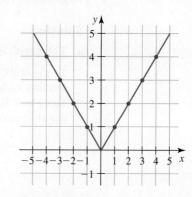

Figure 7.18

Since $x = -2.5$ satisfies the condition $x < 0$, we apply the corresponding rule, $f(x) = -x$, to find $f(-2.5)$.

$$f(-2.5) = -(-2.5)$$
$$= 2.5$$

We can easily verify our result by looking at the graph.

SUMMARY

In this section you learned that the mathematical modeling of data can be represented by piecewise functions and that piecewise functions can be represented by tables, equations, and graphs. You also learned some piecewise functions are discontinuous, characterized by breaks, holes, and jumps. Additionally, you discovered that piecewise functions lend flexibility to mathematical modeling by using a variety of functions over the domain of the function.

7.2 EXERCISES

SKILLS AND CONCEPTS

In Exercises 1–4, graph each piecewise function.

1. $f(x) = \begin{cases} 5 & \text{if } -3 \leq x < 1 \\ -4 & \text{if } x \geq 1 \end{cases}$

2. $f(x) = \begin{cases} -2x & \text{if } x < 0 \\ -3x^2 & \text{if } x \geq 0 \end{cases}$

3. $f(x) = \begin{cases} -x^2 & \text{if } x \leq 2 \\ \sqrt{x} & \text{if } 2 < x < 5 \\ 7 + 15x & \text{if } x \geq 5 \end{cases}$

4. $f(x) = \begin{cases} x - 1 & \text{if } -2 \leq x < 0 \\ 2^x & \text{if } 1 < x < 6 \\ 10x & \text{if } x \geq 6 \end{cases}$

In Exercises 5–8, write the equation for each piecewise function, $f(x)$.

5.

6.

7.

8.

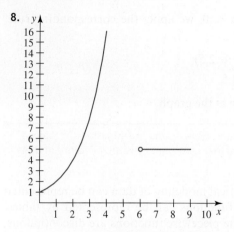

9. The following data table is best defined by a piecewise function, $f(x)$, that can be modeled by an exponential, quadratic, and linear function over three distinct intervals of the domain. Specify over which interval the data is exponential, quadratic, and linear. Then find a formula for $f(x)$.

x	y	x	y	x	y
1	2	6	18	11	14
2	4	7	19	12	12
3	8	8	20	13	11
4	16	9	21	14	11
5	17	10	17		

10. The following graph illustrates the four parts of a trip a family took from home. Use the graph to answer the questions.

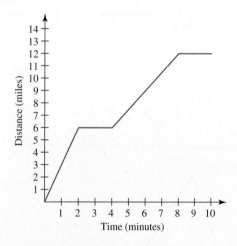

a. When did the family take a break?
b. When was their speed the slowest?
c. When was their speed the fastest?

11. Author Royalties An author is about to publish her first book, which will be sold for $35. The author has negotiated to be paid a set amount of $30,000 for the first 20,000 copies sold, royalties of 13% on the next 7500 copies, and 22% on any additional copies.

a. Write a piecewise function that specifies the total amount the author will be paid, P, based on the number of copies of the book, b, sold.
b. Sketch a graph of the function over a practical domain.

c. Evaluate $P(42{,}556)$ and explain what the numerical value of the solution means in the real-world context of this problem.
d. Solve $P(b) = 42{,}556$ for b and explain what the numerical value of the solution means in the real-world context of this problem.

■ SHOW YOU KNOW

12. What is a piecewise function?

13. Explain how to create the equation for a piecewise function from data in tabular or graphical form.

14. Explain how to evaluate a piecewise function represented in graphical, tabular, or symbolic form.

15. What does it mean for a function to be continuous or discontinuous?

16. What are the different types of discontinuities?

17. What are the advantages and disadvantages of choosing to model data with a piecewise function?

18. Why are some data sets best modeled by piecewise functions?

■ MAKE IT REAL

19. Postage According to the U.S. Postal Service the single-piece rate to send first-class mail (letters, cards, flats, and parcels) in 2007 is displayed in the table below.

Weight Not Over (ounces) w	Single-Piece Charge (dollars) c
1	$0.39
2	$0.63
3	$0.87
4	$1.11
5	$1.35
6	$1.59
7	$1.83
8	$2.07
9	$2.31
10	$2.55
11	$2.79
12	$3.03
13	$3.27
for each additional ounce over 13	+ $0.24

Source: **U.S. Postal Service**

a. Write a function, $c(w)$, where w is the weight of a first-class letter in ounces and c is the required postage fee in dollars.

b. Evaluate $c(7)$ and explain what the numerical value means in the real-world context.

c. Solve $c(w) = 3.99$ for w and explain what the numerical value means in the real-world context.

20. **Ironman Competition** A 2.4-mile swim, 112-mile bike ride, and a 26.2-mile run make up an Ironman Triathlon competition. A certain triathlete averages a swimming speed of 2.6 mph, a cycling speed of 18 mph, and a running speed of 10 mph. Assume there is no transitioning time from one segment of the race to another. (*Source: www.ironmanarizona.com*) (*Hint:* Recall that distance = rate · time.)

a. Develop a piecewise function for the speed, S, of the participant as a function of his time, t, in minutes.

b. Sketch a graph of the piecewise function $S(t)$.

c. Describe how each piece of the function $S(t)$ models the three segments of the race.

21. **Cost of Electricity** Most utility companies charge a higher rate when customers use more than a certain amount of energy. In warm climates, they also charge more during summer months when electric use is higher. As an example, the table shows the residential electric rates for Austin, Texas in 2003. (Note: These figures include a fuel charge of 2.265¢ per kWh.)

Energy Used (kilowatts) k	Energy Charge (cents per kilowatt hour (kWh)) E
$0 < k \leq 500$	$0.058
$k > 500$ May–Oct.	$0.10
$k > 500$ Nov.–Apr.	$0.083

Source: www.michaelbluejay.com

a. Derive a formula for the piecewise function, $E(k)$, that models the energy cost per kilowatt hour used from May to October.

b. Evaluate $E(450)$ for a customer in the month of August.

c. Derive a formula for the piecewise function, $E(k)$, that models the energy cost per kilowatt hour used from November to April.

d. Evaluate $E(450)$ for a customer in the month of December.

22. **Windchill Factor** The windchill factor, W, is a measure of how cold it feels when taking into account both the temperature (degrees Fahrenheit), t, and the wind velocity (miles per hour), v. In the fall of 2001 the National Weather Service began using a revised windchill formula based on new research on how wind and cold air affects people. (*Source: www.weather.gov*) The formula the National Weather Service uses is the multi-variable piecewise function:

$$W(t,v) = \begin{cases} t & \text{if } 0 \leq v < 4 \\ 35.74 + 0.6215t - 35.75v^{0.16} \\ \quad + 0.4275t(v^{0.16}) & \text{if } 4 \leq v \leq 45 \\ 1.60t - 55 & \text{if } v > 45 \end{cases}$$

a. Evaluate $W(32, 25)$ and interpret the meaning of the solution in the real-world context.

b. Sketch a graph of $W(20, v)$ and, using the concept of rate of change, explain what information the shape of the graph gives regarding the windchill factor and the temperature.

c. How fast must the wind blow for the temperature of 20° to feel like −5°?

d. At what wind speeds do the windchill factor and the actual temperature feel the same?

23. **Federal Income Tax Rates** Federal income tax rates depend on the amount of taxable income received. The following tax rate schedule shows how the tax rate was determined for a single filer for 2006.

2006 Federal Tax Rate Schedules		
Note: These tax rate schedules are provided so that you can compute your federal estimated income tax for 2006. To compute your actual income tax, please see the instructions for 2006 Form 1040, 1040A, or 1040EZ as appropriate when they are available.		
Schedule X Single		
If taxable income is over	But not over	The tax is:
$0	$7,550	10% of the amount over $0
$7,550	$30,650	$755 plus 15% of the amount over 7,550
$30,650	$74,200	$4,220.00 plus 25% of the amount over 30,650
$74,200	$154,800	$15,107.50 plus 28% of the amount over 74,200
$154,800	$336,550	$37,675.50 plus 33% of the amount over 154,800
$336,550	no limit	$97,653.00 plus 35% of the amount over 336,550

Source: www.irs.gov

a. Write a piecewise function for income tax, T, as a function of taxable income, i, for single filers.

b. Use the function from part (a) to calculate the tax for the following taxable income levels: $2400; $29,700; $59,300; $129,000; and $345,000.

c. Use the function to estimate your income tax for 2006.

24. **Music Museum** Experience Music Project, an interactive music museum in Seattle, charged adult visitors a $19.95 admission fee in 2005. Groups of 15 or more could enter for $14.50 per person. (*Source: www.emplive.com*)

a. Create a table for group cost as a function of group size. (Show the cost for groups of 1 to 20 people.)

b. Write the total cost of admission, C, as a function of the number of people in a group, n. Confirm that the function and the table from part (a) are in agreement.

c. Calculate the cost of admitting a group of 14 people and the cost of admitting a group of 15 people.

d. Determine the largest group you could bring in and still remain below the cost of a 14-person group.

25. **Buffet Dinner** The following sign shows the admission price for Amazing Jake's entertainment center in 2005.

Amazing All-U-Can-Eat Buffet

Adults
- 11am–3pm Monday–Friday - $5.99
- After 3pm and Weekends - $6.99
- Add Unlimited Fountain Drinks - 99¢
- Ages 65 and over - $1.00 off adult prices

Children
- Ages 3–6 - $3.99
- Ages 7–12 - $4.99
- Under 2 - FREE with paying adult
- Add Unlimited Fountain Drinks - 99¢

Source: www.amazingjakes.com

a. Write an equation for the weekend buffet price, W, as a function of a person's age, a.

b. Use W(a) to calculate the buffet price for a 3-year-old, 6-year-old, 9-year-old, and 24-year-old person.

c. Explain why it is impossible to evaluate W(2) using the table provided.

d. If a family comprised of a husband and wife (under age 65) and children ages infant, 3, 5, 8, and 12 years go to Amazing Jakes on the weekend, what will it cost them to enter?

26. **Disneyland** The following sign shows the cost of Disneyland theme park tickets for 2007.

Ticket Prices

	Ages 3-9	Ages 10+
5-Day Park Hopper® Bonus Ticket Save up to $40 per person!	$199.00 $159.00	$229.00 $189.00
4-Day Park Hopper® Bonus Ticket Save up to $30 per person!	$179.00 $149.00	$209.00 $179.00
3-Day Park Hopper® Bonus Ticket Save up to $20 per person!	$149.00 $129.00	$179.00 $159.00
2-Day Park Hopper® Ticket	$102.00	$122.00
1-Day Park Hopper® Ticket	$73.00	$83.00
Single-Day Theme Park Ticket	$53.00	$63.00

Source: www.disneyland.com

a. Write an equation for the 3-Day Park Hopper admission price, H, as a function of a person's age, a.

b. What is the domain and range of the function from part (a)?

c. Write an equation for the adult admission price for a Park Hopper ticket, A, as a function of the number of days, d, to be spent in the park.

d. What is the domain and range of A(d)?

e. A husband and wife want to take their children to Disneyland on a 3-Day Park Hopper pass next summer. If the children's ages at the time of the trip will be 1, 3, 6, 9, and 13, what will it cost the family to enter?

27. **Taxi Fare** In 2006 in New York City, one company's initial charge for a taxi ride was $2.50 for the first $\frac{1}{5}$ of a mile, and $0.40 for each additional $\frac{1}{5}$ of a mile. (*Source:* www.schallerconsult.com)

a. Create a table showing the cost of a trip as a function of its length. Your table should start at 0 and go up to 1 mile in $\frac{1}{5}$ of a mile intervals.

b. What is the cost for a 10-mile taxi ride?

c. How far can a person ride for $3.00?

d. Graph the taxi cab fare function in part (a).

28. **Museum Admission** The Museum of Science and Industry in Chicago has special rates for school and tour groups with a four-week advance registration. The museum's 2007 prices for tour groups of 20 or more people are shown in the table.

Attendee's Age a	Admission Price P
$a \geq 65$	$8.00
$12 \leq a < 65$	$9.50
$3 \leq a \leq 11$	$6.00
$a < 3$	Free

Source: www.msichicago.org

a. Determine the cost of attending the museum with fifteen 11-year-olds, seven 12-year-olds, two 29-year-olds, and one 42-year-old.

b. Can the table be used to determine how many 16-year-old students can go to the museum for $100.00? Explain.

c. Graph the museum admission price function P(a).

d. Create a piecewise function for the admission price table for groups of 20 or more people. Write the price as a function of a person's age.

29. **Durango Ski Resort** The ski lift rates for 2006–2007 in Durango, CO, are shown in the table, based on the age of the skier.

Skier's Age (years) a	Charge (dollars) c
12 and under	31
13 to 18	43
19 to 61	59
62 and over	43

Source: www.onthesnow.com

a. A family holding their reunion in Durango decides to go skiing. The following table displays the number of people from each age group that want to ski. How much will it cost for them all to go?

Skier's Age (years) a	Number Who Want to Ski
12 and under	11
13 to 18	7
19 to 61	25
62 and over	1

b. Sketch a graph of the function $c(a)$ that models the cost to ski as a function of the age of the skier.

c. Create a piecewise function for $c(a)$ that models the cost to ski as a function of the age of the skier.

30. **The Indianapolis 500** A scatter plot of the qualifying speeds for the Indianapolis 500 from 1911 to 2007 is shown in the figure.

Source: www.indy500.com

a. The function $s(y) = 1.707y + 77.679$ is the line of best fit for these data, where s is the speed and y is the years since 1911. Evaluate $s(100)$ and explain the meaning of the value in this real-world context.

b. Why do you suppose the qualifying speed for 1911 was 0 mph for the Indianapolis 500? Explain what you think would happen to the parameters (slope and y-intercept) of the line of best fit if this point is removed from the scatter plot and a new linear regression is done.

c. Examine the scatter plot of the qualifying speeds from 1911 to 2007. Give a reasonable explanation for the horizontal and vertical gaps in the data that occur for the following years:
- 1916–1919
- 1941–1946
- 1971–1972
- 1996–1997

d. Would a piecewise function be a better mathematical model for the Indianapolis 500 qualifying speeds than one linear function? Why or why not?

31. **Cost of Natural Gas** In many parts of the country, the natural gas bills for customers are calculated on gas meter readings that are taken every month. A *therm* is a unit that measures the amount of natural gas needed to produce 100,000 BTUs of heat. Each month, a customer's gas bill states how many therms were used. For single-family residential customers of Southwest Gas Corporation in Southern Nevada during the 2007 Summer Season (May–October), the gas delivery charge for natural gas was billed in the following manner.

- $8.50 basic service charge + $1.20 per therm for 0 to 15 therms (including all tariffs and adjustments)
- $8.50 basic service charge + $1.02 per therm (including all tariffs and adjustments) for greater than 15 therms

(*Source*: www.swgas.com)

a. Construct a table for this data, using t to represent the total number of therms used and C to represent the total monthly charge for natural gas delivery. Use therms from 0 to 50 in increments of 5 therms.

b. Write a formula for the piecewise function, $C(t)$, that represents the monthly delivery charge to a residential customer based on the number of therms, t, used.

c. Sketch the graph of the function $C(t)$.

d. What is the delivery charge if a customer used 76 therms during the month of June?

STRETCH YOUR MIND

Exercises 32–39 are intended to challenge your understanding of piecewise functions.

Many companies that have mail order catalogs charge their customers shipping and handling fees for purchased products. The fees that JCPenney charged in 2007 for their catalog purchases are shown in the table.

Total Product Cost (dollars) t	Shipping and Handling Charge (dollars) S
Up to $25.00	$5.95
$25.01 to $40.00	$7.50
$40.01 to $50.00	$8.50
$50.01 to $75.00	$11.50
$75.01 to $100.00	$14.95
$100.01 to $150.00	$18.95
$150.01 to $200.00	$22.95
$200.01 to $300.00	$25.95
$300.01 to $500.00	$29.95
Over $500.00	$39.95

Source: www4.jcpenney.com

32. What are the shipping and handling charges a customer incurs if the total product cost is $50.00? $50.01?

33. Notice that there is a $3.00 increase in shipping and handling charges for product costs that differ by only $0.01 from $50.00 to $50.01. Give a possible reason for this increase.

34. What are the shipping and handling charges a customer incurs if the total product cost is $562.00? $1000.00?

35. Sketch a graph of the shipping and handling cost, S, as a function of the total product cost, t, in dollars. Be sure to label and scale the axes appropriately.

36. Write a formula for the shipping and handling cost, S, as a function of the total cost, t, in dollars.

37. Sketch a graph of the total cost of a purchase including the shipping and handling fee, B, as a function of the total product cost, t, in dollars. Be sure to label and scale the axes appropriately.

38. Write a formula for the total cost of a purchase including the shipping and handling fee, B, as a function of the total of the product cost, t, in dollars.

39. Compare and contrast the graphs of $S(t)$ and $B(t)$. Explain in terms of the real-world context the differences that exist in the graphs.

SECTION 7.3 Composition of Functions

LEARNING OBJECTIVES

- Compose two or more functions using tables, equations, or graphs
- Create, use, and interpret function composition notation in a real-world context

GETTING STARTED

In many fields, an employee with an advanced educational degree can move into positions of responsibility and higher pay more rapidly than less-educated coworkers. This may be due in part to the belief that an employee with an advanced degree has a deeper knowledge base to draw from and is more capable of handling responsibilities. In effect, the time a student spends earning a degree can in turn produce a higher salary.

In many real-world situations such as this, it is common for the output of one function to be used as the input of another function, resulting in a composition of functions. In this section we investigate the concepts behind function composition and learn how to compose and decompose functions using tables, graphs, and equations.

■ Understanding Composition of Functions

Consider Table 7.10, which shows the 2004 median annual salaries of American employees, the associated educational degree acquired, and the accumulated full-time academic years typically spent to attain such degrees.

SECTION 7.3 ■ COMPOSITION OF FUNCTIONS

Table 7.10

Accumulated Time in College (academic years) t	Degree Attained d	Median Annual Salary (dollars) s
0	High School Diploma	20,733
2	Associate's	30,026
4	Bachelor's	38,880
6	Master's	50,693
8	Doctorate	72,073

Source: U.S. Census Bureau, www.census.gov/population/
www/socdemo/education/cps2005.html

EXAMPLE 1 ■ Understanding Composition of Functions

Use Table 7.10 to answer the following questions.

a. If an employee spent 4 years in college, what degree did she earn?
b. What is the median annual salary of an employee with a Doctorate?
c. If an employee attended school for 6 years, what degree did he earn? What is the median annual salary of an employee with a Master's degree?
d. What is the median annual salary of an employee who spent 6 years in college?

Solution

a. According to Table 7.10, a person who attends college for 4 years earns a Bachelor's degree. Notice *academic years* is the input of the function and *degree* is the output. The function used does not reveal anything about the person's median annual salary.

b. According to the table, an employee with a Doctorate degree earns a median annual salary of $72,073. Observe *degree attained* is the input of the function and *median annual salary* is the output. This function does not reveal anything about the academic years spent in college.

c. According to the table, a person who attends college for 6 years earns a Master's degree. The median annual salary of a person with a Master's degree is $50,693. For the first question, *academic years* is the input and *degree attained* is the output. However, for the second question, the *degree attained* is the input and the *median annual salary* is the output of the function.

d. According to the table, an employee with 6 years of college earns a median annual salary of $50,693. (This assumes the employee earned a Master's degree.) We are using a function that has *academic years* as the input and *median annual salary* as the output. In effect, we skipped the step of finding the employee's degree and calculated the salary directly.

In parts (c) and (d) of Example 1 we arrived at the same answer for the employee's median annual salary but took two different paths to get there. Table 7.11 compares the processes involved.

Table 7.11

The two-step method used in part (c)		
Input →	Output/Input →	Output
Accumulated Time in College (academic years)	Degree Attained	Median Annual Salary (dollars)
6 →	Master's →	50,693
The function composition method used in part (d)		
Input →		Output
Accumulated Time in College (academic years)		Median Annual Salary (dollars)
6 →		50,693

The process outlined in part (d) demonstrates function composition, which occurs when the output of one function is used as an input to another function. In this case, the output of the *degree attained* function was the input to the *median annual salary* function.

■ **Composition Notation**

Continuing with the salary example, we define the variables as follows:

- t is the accumulated time in college (academic years).
- d is the degree attained.
- s is the median annual salary.

It is important to understand the relationship between these variables.

- d and t

  We know d is a function of t because the degree attained by an employee depends on the accumulated number of academic years. We write the relationship as $d(t)$.

- s and d

  We know s is a function of d because the median annual salary of an employee depends on the degree attained. We write the relationship as $s(d)$.

- s, d, and t

  We have shown that s is a function of d and that d is a function of t. That is, salary (s) depends on the degree earned (d), which depends on the accumulated number of years spent in college (t). However, we have also seen these two functions can be combined in a **composition of functions**: s is a function of d and d is a function of t. This relationship is sometimes represented with the notation $(s \circ d)(t)$; however, we will use the more intuitive notation, $s(d(t))$.

> **COMPOSITION OF FUNCTIONS**
>
> The function $h(x) = f(g(x))$ is the **composition** of the function f with the function g. Function h is called a *composite* function.

■ **Creating Composite Functions**

In the next few examples, we'll look at how to create and evaluate composite functions defined by tables, equations, and graphs.

SECTION 7.3 ■ COMPOSITION OF FUNCTIONS

EXAMPLE 2 ■ Creating a Composite Function Defined by Tables

More than 500 new energy drinks were launched worldwide in 2006, each one with promises of weight loss, increased endurance, and legal highs. Newer products joined top-sellers Red Bull and Monster to make up a $3.4 billion a year industry that grew by 80% in 2005. (*Source:* www.cnn.com) Most of these beverages are caffeine-laden, and nutritionists warn that the drinks can hook people, especially teenagers, on an unhealthy jolt-and-crash cycle. Although energy drinks such as Red Bull, Rock Star, Rush, Monster, and Hype contain some nutritional vitamins and supplements, many contain other ingredients that are not necessarily good for you. One particular brand of energy drink has 80 milligrams of caffeine in an 8.3-ounce can.

The effects of caffeine in the bloodstream on the heart rate of a typical 175-pound male adult are displayed in Table 7.12. It is also known that the amount of caffeine in a person's bloodstream dissipates over time. The amount of caffeine in a 175-pound adult male's bloodstream would typically dissipate as shown in Table 7.13, where time, t, is the number of hours after the caffeine is ingested.

Table 7.12

Caffeine Level (milligrams) c	Heart Rate (beats per minute) r
0	80
50	82
131	84
150	85
198	86
261	88
300	89

Source: Dr. Brent Alvar, Chandler-Gilbert Community College Wellness and Fitness Director

Table 7.13

Time (hours) t	Caffeine Level (milligrams) c
0	300
1	261
2	227
3	198
4	172
5	150
6	131

Source: Dr. Brent Alvar, Chandler-Gilbert Community College Wellness and Fitness Director

Using Tables 7.12 and 7.13, create a table that shows the heart rate, r, as a function of the time, t.

Solution We begin by writing the two tables adjacent to each other, as shown in Tables 7.14 and 7.15.

Table 7.14

Time (hours) t	Caffeine Level (milligrams) c
0	300
1	261
2	227
3	198
4	172
5	150
6	131

Table 7.15

Caffeine Level (milligrams) c	Heart Rate (beats per minute) r
0	80
50	82
131	84
150	85
198	86
261	88
300	89

We see that the values for the caffeine levels appear to be the same in many cases. We create a three-column table (Table 7.16) to capture what we know. We now eliminate the rows of the table with blank spaces and delete the middle column to get Table 7.17, which relates time, t, to heart rate, r.

Table 7.16

Time (hours) t	Caffeine Level (milligrams) c	Heart Rate (beats per minute) r
0	300	89
1	261	88
2	227	
3	198	86
4	172	
5	150	85
6	131	84
	50	82
	0	80

Table 7.17

Time (hours) t	Heart Rate (beats per minute) r
0	89
1	88
3	86
5	85
6	84

EXAMPLE 3 ■ Evaluating a Composite Function from a Table

Use the tables and the solution in Example 2 to answer the following.

a. Evaluate $r(c(3))$ first by using $c(t)$ and $r(c)$ and then by using the composite function $r(c(t))$. Explain the meaning of the answer in its real-world context.

b. Explain the difference between the two ways to arrive at the answer in part (a).

Solution

a. One way to evaluate $r(c(3))$ is to first evaluate $c(3)$. We find from Table 7.14 that $c(3) = 198$. Now we evaluate $r(198)$ by looking at Table 7.15 and find $r(198) = 86$. The second way is to skip the intermediate step and evaluate $r(c(3))$ using Table 7.17. We see that $r(c(3)) = 86$, which means that 3 hours after the caffeine is ingested, the heart rate is 86 beats per minute.

b. The difference is that we can find the answer in three steps by using the two functions $r(c)$ and $c(t)$ separately, or we can skip the middle step by using the composite function $r(c(t))$.

EXAMPLE 4 ■ Creating a Composition of Functions Defined by Equations

The effect of caffeine, c, in milligrams on a person's heart rate, r, in beats per minute can be modeled by the linear function $r(c) = 80 + 0.03c$. The dissipation of caffeine, c, in milligrams from the bloodstream over time since ingestion, t, in hours can be modeled by the exponential function $c(t) = 300(0.87)^t$.

a. Write the equation for the composite function $r(c(t))$. Explain the input and output variables in terms of the real-world context.

b. Evaluate $r(c(3))$ and explain the meaning of the solution in terms of the real-world context.

c. Find the practical domain of $r(c)$, $c(t)$, and $r(c(t))$. Refer to the tables in Example 2 as needed.

Solution

a. To form the composition $r(c(t))$, we begin by placing the equation for $c(t)$ into function $r(c(t))$ as the input.

$$r(c) = 80 + 0.03c$$
$$r(c(t)) = r(300(0.87^t)) \quad \text{since } c(t) = 300(0.87)^t$$
$$= 80 + 0.03(300(0.87^t))$$
$$= 80 + 9(0.87^t)$$

Therefore, $r(c(t)) = 80 + 9(0.87^t)$. The input to the function $r(c(t))$ is time in hours and the output is heart rate in beats per minute.

b. To evaluate $r(c(3))$, we input 3 hours into the composite function $r(c(t))$.
$r(c(3)) = 80 + 9(0.87^3) = 85.9$. This means that 3 hours after the caffeine is ingested the person's heart rate is about 86 beats per minute.

c. Referring to Table 7.15, we see that the practical domain of $r(c)$ is all values from 0 milligrams to 300 milligrams. That is, $0 \leq c \leq 300$. Referring to Table 7.14, we see that $c(t)$ has a practical domain of all values from 0 to 6 hours. That is, $0 \leq t \leq 6$. (Note that the domain should realistically include more hours to allow for the caffeine to be totally removed from the person's system, which could take 24 hours or more.) Finally, the practical domain for the composite function, $r(c(t))$, is again $0 \leq t \leq 6$ because this function is defined in this context for only the first 6 hours after the person ingested the caffeine.

EXAMPLE 5 ■ Creating a Composition of Functions Defined by Graphs

Figure 7.19 shows the functions f and g. Draw the graph of $f(g(x))$.

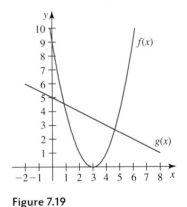

Figure 7.19

Solution We begin by creating a table of values for $g(x)$. Then, using the values of $g(x)$ as inputs, we use the graph of $f(x)$ to find $f(g(x))$. See Table 7.18. For example, we see $g(0) = 5$ and $f(5) = 4$. We plot the points using the first and third columns of the table and connect the points with a smooth curve, as shown in Figure 7.20.

Table 7.18

x	$g(x)$	$f(g(x))$
−2	6	9
0	5	4
2	4	1
4	3	0
6	2	1
8	1	4

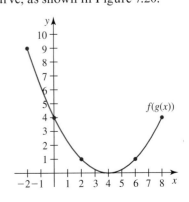

Figure 7.20

EXAMPLE 6 ■ **Creating Composite Functions Symbolically**

Given $f(x) = \sqrt{x+3}$ and $g(x) = \dfrac{2}{x+1}$, find $f(g(x))$ and $g(f(x))$.

Solution We first find $f(g(x))$.

$$f(g(x)) = f\left(\frac{2}{x+1}\right) \quad \text{The function } g(x) \text{ is the input.}$$

$$= \sqrt{\left(\frac{2}{x+1}\right) + 3} \quad \text{Replace the } x \text{ in } f(x) \text{ with } \frac{2}{x+1}.$$

We next find $g(f(x))$.

$$g(f(x)) = g\left(\sqrt{x+3}\right) \quad \text{since } f(x) = \sqrt{x+3}$$

$$= \frac{2}{\sqrt{x+3}+1}$$

EXAMPLE 7 ■ **Decomposing Functions Symbolically**

Each given function is a composite function of the form $f(g(x))$. Decompose the functions into two functions $f(x)$ and $g(x)$.

a. $f(g(x)) = |3x - 2|$

b. $f(g(x)) = \dfrac{2}{x^2 - 1}$

c. $f(g(x)) = 2^{6x-1}$

d. $f(g(x)) = (x-2)^4 - (x-2)^3 + (x-2)^2$

Solution

a. The function $f(g(x))$ takes the absolute value of $3x - 2$. Two functions that can be used for the composition are $f(x) = |x|$ and $g(x) = 3x - 2$. This is not a unique solution; $f(x) = |x - 2|$ and $g(x) = 3x$ also work.

b. The function $f(g(x))$ consists of a rational function and a quadratic function. Two functions that can be used for the composition are $f(x) = \dfrac{2}{x}$ and $g(x) = x^2 - 1$. This is not a unique solution; $f(x) = \dfrac{1}{x-1}$ and $g(x) = x^2$ also work.

c. The function $f(g(x))$ raises 2 to the power $6x - 1$. Two functions that can be used for the composition are $f(x) = 2^x$ and $g(x) = 6x - 1$. Additional solutions exist.

d. The function $f(g(x))$ raises $x - 2$ to the fourth, third, and second power. Two functions that can be used for the composition are $f(x) = x^4 - x^3 + x^2$ and $g(x) = x - 2$. Additional solutions exist.

SUMMARY

In this section you investigated the concepts that underlie function composition. You discovered that in many real-world situations it is common for the output of one function to be the input of another function. You also learned how to create composite functions and to evaluate them using tables, graphs, and equations and then how to decompose such functions symbolically.

7.3 EXERCISES

SKILLS AND CONCEPTS

In Exercises 1–10, rewrite each pair of functions, f(g) and g(t), as one composite function, f(g(t)), if possible. Then evaluate f(g(2)).

1. $f(g) = 3g^2 - g + 4$ $g(t) = 5 - 4t$
2. $f(g) = 3e^g$ $g(t) = 2t^2$
3. $f(g) = \dfrac{5}{g}$ $g(t) = 1 + 4(2^{-0.4t})$
4. $f(g) = \sqrt{8g^2 + 8g - 1}$ $g(t) = |3t^3| - t$
5. $f(g) = 5g - 4$ $g(t) = t^2 - 3t + 6$
6. $f(g) = \sqrt[3]{g}$ $g(t) = t - 5$
7. $f(g) = 6g + 3$ $g(t) = t^2 - 2t - 6$
8. $f(g) = \dfrac{4}{1 - 5g}$ $g(t) = \dfrac{1}{t}$
9. $f(g) = g^3 - 4g^2 + 2g - 3$ $g(t) = t + 1$
10. $f(g) = \dfrac{1 - g}{g}$ $g(t) = \dfrac{2t - 2}{4t + 1}$

In Exercises 11–16, decompose the composite function h(x) = f(g(x)) into f(g) and g(x). (Note: There may be more than one possible correct answer.)

11. $h(x) = (2x - 2)^5$
12. $h(x) = \sqrt[3]{x^2 - 7}$
13. $h(x) = \dfrac{1}{(x - 2)^6}$
14. $h(x) = \left(\dfrac{2 - x^3}{2 + x^3}\right)^2$
15. $h(x) = (\sqrt{x} - 3)^4$
16. $h(x) = e^{(3x - \pi)}$

In Exercises 17–20, decompose the function into two functions. The unfamiliar operational symbols are used to focus on the idea of decomposition rather than the operations involved.

17. $h(x) = (x - 1)^{\heartsuit}$
18. $h(x) = \sqrt[\Uparrow]{x^{\oplus} - x}$
19. $h(x) = \left(\dfrac{4}{(x \Diamond 5)}\right)^{\copyright}$
20. $h(x) = |4x \int 9|$

21. Let $u(x)$ and $v(x)$ be two functions defined by the two graphs in the figure. Estimate the following.
    a. $v(u(-2))$
    b. $u(v(1))$
    c. $v(u(0)) + v(u(3))$
    d. $v^{-1}(u(2))$

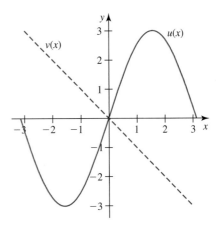

22. Using the two graphs in the figure, estimate the following.
    a. $f(g(2))$
    b. $g(f(1))$
    c. $f(f(-0.5))$
    d. $g(g(3))$
    e. $g^{-1}(f(1.5))$
    f. $g^{-1}(f(3))$

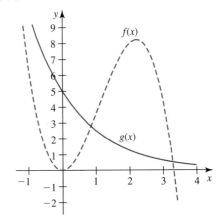

23. Use the first three columns in the table to fill in the values for $k(n(x))$. (Note: Some function values may be undefined.)

x	n(x)	k(x)	k(n(x))
0	2	6	
1	1	3	
2	6	4	
3	3	2	
4	4	5	
5	5	9	

24. Let $k(x)$ and $n(x)$ be the functions from Exercise 23. Construct a table of values for $n(k(x))$. (Note: some function values may be undefined.)

25. Find the equations for the functions in parts (a)–(f). Let $f(x) = x^2 - 2$, $g(x) = \dfrac{2}{x+3}$, and $h(x) = \sqrt{x}$.

   a. $f(g(x))$
   b. $g(f(x))$
   c. $f(h(x))$
   d. $h(f(x))$
   e. $h(h(x))$
   f. $g(f(h(x)))$

26. Use the following graphs of $f(x)$ and $g(x)$ to sketch a graph of functions $f(g(x))$ and $g(f(x))$. Do the two composite functions have the same graphs? Why or why not?

27. Find $f(f(3))$ for
$$f(x) = \begin{cases} 4 & \text{if } x \leq 0 \\ 2x + 1 & \text{if } 0 < x < 4 \\ x^2 - 4 & \text{if } x \geq 4 \end{cases}$$

In Exercises 28–29, select the correct answer from those offered and explain why you chose as you did.

28. **Ripple Spread** A pebble is dropped into a lake, creating a circular ripple that travels outward at a speed of 4 centimeters per second. Which of the following functions express the area, A, of the circle as a function of the number of seconds, s, that have passed since the pebble hits the lake?

A. $A = 16\pi s$
B. $A = 16\pi r^2$
C. $A = 16\pi s^2$
D. $A = 4\pi s^2$
E. $A = \pi r^2$

29. Which of the following functions defines the area, A, of a square as a function of its perimeter, p?

A. $A = \dfrac{p^2}{16}$
B. $A = s^2$
C. $A = \dfrac{p^2}{4}$
D. $A = 16s^2$
E. $p = 4\sqrt{A}$

30. **Quality Control** A baker suspects his oven may be malfunctioning so he begins inspecting muffins at his bakery. He notes the first two batches of muffins he cooks each day are not acceptable because they are not cooked enough in the middle. After those first two batches, the oven generally is warmed up enough and works fine. Since the muffins will be packaged for shipping, the muffins are left to cool in a climate-controlled environment after cooking. The muffins are then properly packaged by a machine in cellophane 95% of the time. Let m be the number of muffin batches cooked in a day.

   a. Write an equation for $n(m)$ that represents the number of muffin batches that are properly cooked.
   b. Write an equation for $p(n)$ that represents the number of muffin batches that are properly packaged.
   c. Now find $p(n(52))$, and interpret its meaning in this real-world context.

31. **Seaside Restaurant** A world-renowned seaside restaurant's motto is "shellfish directly from the sea." The restaurant prides itself in its delectable scallop dish. Management hires local fishermen to harvest scallops in the nearby area, and customers can enjoy their meals while watching the fishermen arrive at the dock with that day's catch. Unfortunately, the number of shellfish in the area has been decreasing due to the increased amount of pollutants that have been found to be in the ocean.

   There are two related functions involved in this scenario. One is the number of scallops per acre, f, as a function of the amount of pollutants found in the ocean, w, in grams per liter $\left(\dfrac{\text{g}}{\text{liter}}\right)$, modeled by $f(w) = 23 - 4w$.

   The second is the price (in dollars) of a scallop dinner, p, as a function of the number of scallops per acre, f, modeled by $p(f) = 39.95 - 1.10f$.

   a. Determine the composite function $p(f(w))$ and explain what it defines.
   b. How much did a scallop dinner cost before there were any measurable pollutants in the ocean? Solve this

problem in two different ways—one using composition and one not.

c. According to the model, what will be the price of a dinner when pollution has killed all of the scallops? (Note: This situation will require scallops to be imported.)

d. Solve $p(f(w)) = 21.50$ and interpret the practical meaning of the results.

32. **College Student Workload** Many college students work to pay tuition. The number of hours students work may affect the number of credits they can take. In turn, study time in preparation for class is dependent on the number of credits taken. The tables show examples of these relationships.

Hours Worked per Week h	Credits Taken c
$0 \leq h < 4$	18
$4 \leq h < 8$	17
$8 \leq h < 12$	16
$12 \leq h < 16$	15
$16 \leq h < 20$	14
$20 \leq h < 24$	13
$24 \leq h < 28$	12

Credits Taken c	Study Hours per Week s
12	12
13	14
14	16
15	20
16	25
17	31
18	39

a. Construct a table showing the relationship between the number of hours worked per week and the number of hours of study per week.

b. Sketch a graph of the new relationship of the number of hours worked per week and the resulting number of hours of study per week.

c. Describe how the new relationship between the number of hours worked per week and the resulting number of hours of study per week is an example of a composition of functions.

d. Considering the information given in the tables, give an example of a situation in which a student reduces the hours worked and thus increases the number of credits and makes more time available for leisure activities.

SHOW YOU KNOW

33. Explain what is meant by function composition.

34. Make up your own situation using a composition of functions and three interrelated variables.

In Exercises 35–39, determine whether it is reasonable for the pairs of functions to be combined by function composition. If so, give function notation for the new function and explain what the input and output values would be.

35. $p(l)$ is the revenue in pesos from the sale of l laptop computers. $D(p)$ is the dollar value of p pesos.

36. $l(s)$ is the profit generated by the sale of s servings of lemonade. $s(h)$ is the number of servings of lemonade a child can make in h hours.

37. $r(y)$ is the home loan interest rate as a function of the year. $h(r)$ is the number of new homes that are built in a city as a function of the home loan interest rate.

38. $c(s)$ is the number of calories consumed as a function of the number of sodas consumed. $s(w)$ is the number of sodas consumed after w weeks.

39. $a(t)$ is the number of accidents caused by drunk drivers in the year t. $c(t)$ is the number of automobiles that have passed by an intersection in t hours.

MAKE IT REAL

40. Table A shows temperatures measured in Celsius, C, and their corresponding temperatures in Fahrenheit, F. Table B shows temperatures in Fahrenheit, F, and their corresponding temperatures in Celsius, C.

Table A

Temperature (°F)	Temperature (°C)
−4	−20
5	−15
14	−10
23	−5
32	0
41	5
50	10
59	15
68	20
77	25
86	30

Table B

Temperature (°C)	Temperature (°F)
−20	−4
−15	5
−10	14
−5	23
0	32
5	41
10	50
15	59
20	68
25	77
30	86

Use the tables to evaluate the following expressions.
a. $C(F(5))$
b. $F(C(23))$
c. $F(0)$

41. **Temperature Conversion** In chemistry it is sometimes necessary to convert temperatures from one unit of measure to another. Specifically, there are three measures of temperature typically used: Fahrenheit, Celsius, and kelvin.

   The formula $C(F) = \frac{5}{9}(F - 32)$ is used to convert Fahrenheit to Celsius and the formula $K(C) = C + 273$ is used to convert Celsius to kelvin.

   a. Find the formula for the composite function $K(C(F))$.
   b. Explain the input and output for $K(C(F))$ in terms of the real-world context.
   c. Evaluate $K(C(81))$ and explain the meaning of the solution in the real-world context.
   d. Solve $K(C(F)) = 572$ for F and explain the meaning of the solution in the real-world context.

42. **Heat Index** The heat index is used to describe the *apparent temperature*, how hot it feels when both temperature and humidity are taken into consideration. For mild temperatures, humidity has little effect on the apparent temperature. However, for high temperatures, humidity dramatically affects the apparent temperature. When the humidity is 50%, the apparent temperature can be modeled by $a(F) = 0.04643F^2 - 6.464F + 299.9$ degrees Fahrenheit, where F is the air temperature in degrees Fahrenheit. This model is valid for air temperatures between 80 and 105 degrees Fahrenheit. (*Source:* www.crh.noaa.gov) Outside of the United States, Celsius is the common unit of measure for temperature, and U.S. citizens traveling abroad sometimes have difficulty translating Celsius temperatures into the more familiar Fahrenheit temperatures. The function $F(C) = 1.8C + 32$ converts degrees Celsius, C, into degrees Fahrenheit, F.

Apparent Temperature at 50% Humidity

Using the graphs of $F(C)$ and $a(F)$,

a. Graph the composite function $a(F(C))$ using axes that are labeled and scaled appropriately.
b. Describe the graphs of functions $F(C)$ and $a(F)$ and compare them with the graph of the composite function $a(F(C))$.
c. Use both $F(C)$ and $a(F)$ as well as $a(f(c))$ to determine what the apparent temperature is when the air temperature is 35 degrees Celsius.

43. **Gasoline Prices** In the following table, $P(y)$ is the purchasing power of a dollar (how much the dollar is "worth") as measured by the consumer price index in year y, using 2000 as the base year for the value of $1.00. $G(P)$ is the average price of a gallon of regular unleaded gasoline when the dollar purchasing power is P dollars. Using function notation, write a composition of functions showing the price of gasoline in year-2000 dollars.

Years Since 2000 y	Dollar Purchasing Power P	Average U.S. Gasoline Price (per gallon) G
0	1.00	1.39
1	1.03	1.66
2	1.04	1.36
3	1.07	1.44
4	1.10	1.81
5	1.13	2.19
6	1.17	2.87

Sources: www.eia.doe.gov and www.measuringworth.com

44. **Take-Home Pay** A student earns $8.75 per hour working as a tutor in her college's learning center. Her weekly take-home pay (after withholdings) is 92.35% of her gross weekly earnings.

   a. Write a function for her gross weekly earnings as a function of the number of hours she works.
   b. Write a function for her weekly take home pay as a function of her gross weekly earnings.

c. Create a composite function that will calculate her weekly take-home pay as a function of the number of hours she works. Then calculate her take-home pay when she works 16 hours.

45. **Dress Sizes** A dress that is size J in Japan is size A in Australia, where $J(A) = A - 1$. A dress that is size A in Australia is size K in the United Kingdom, where $A(K) = K + 2$. A dress that is size K in the United Kingdom is size E in Europe, where $K(E) = E - 28$. A dress that is size E in Europe is size U in the United States, where $E(U) = U + 30$. (*Source:* www.asknumbers.com)

a. Find an equation for the composite function $J(A(K(E(U))))$ and explain what the input and output values to this function represent in the real-world context.

b. Evaluate $J(A(K(E(10))))$ and explain what the numerical solution means in the real-world context.

c. Solve $J(A(K(E(U)))) = 17$ for U and explain the solution in its real-world context.

46. **Money Conversion** In North America, there are three forms of currency: the Mexican peso, the Canadian dollar, and the American dollar. Although exchange rates vary, the conversion rates for these monetary systems can be modeled mathematically.

In May 2007 the function $A(C) = 0.911078C$ converted Canadian dollars, C, to American dollars, A. The function $A(M) = 0.0925834M$ converted Mexican pesos, M, to American dollars, A. (*Source:* www.x-rates.com)

a. Using the information that the two functions $A(C)$ and $A(M)$ provide, find a function to convert Mexican pesos to Canadian dollars.

b. Sketch graphs for $A(C)$, $A(M)$, and $C(A(M))$. Explain what the independent and dependent variables are for each function.

c. What would the composite function $M(A(C))$ have as the input and output values? Create a table of values for $M(A(C))$ for five monetary values of your choice.

47. **Height and Weight** The height and weight of healthy children generally increase as they get older. The function

$$a(h) = 38.05 - \sqrt{-24.39h + 2137}$$

models the age a of a 2- to 18-year-old boy with height h (in inches). The function

$$w(a) = -0.034a^3 + 1.245a^2 - 5.077a + 34.346$$

models the average weight of a 2- to 18-year-old boy, where w is the weight (in pounds) when the boy is a years old.
(*Source:* Modeled from www.cdc.gov data)

a. Find the function composition $w(a(h))$.

b. Determine the independent and dependent variables of $w(a(h))$ and explain their meaning in this real-world context.

c. Calculate $w(a(60))$ and interpret its meaning in this real-world context.

48. **Lumber and Wood Products Industry** Based on data from 1995 to 1999, the average annual earnings of an employee in the lumber and wood products industry can be modeled by $s(n) = -0.7685n^2 + 1295n - 516{,}596$ dollars where n is the number of employees in thousands. (*Source: Statistical Abstract of the United States, 2001,* Table 979)

Based on data from 1995 to 1999, the number of employees in the lumber and wood products industry can be modeled by

$$n(t) = 3.143t^2 + 5.029t + 772.5 \text{ thousand employees}$$

where t is the number of years since 1995. (*Source: Statistical Abstract of the United States, 2001,* Table 979)

a. Calculate $s(n(4))$.

b. Interpret the meaning of the answer in part (a).

c. According to the model, what was the average annual earnings in 1998?

49. **Insurance Premiums** Among other things, car insurance premiums depend on the value of the vehicle, which depends on the age (year) of the vehicle. Suppose you are shopping for a used Cadillac Escalade with approximately 40,000 miles and need to determine how much the insurance premium will be for any particular car. Table C gives the 2007 clean retail value of the Escalade with 40,000 miles for different years.

Table C

Model Year of Vehicle t	2007 Clean Retail Value of Vehicle (with 40,000 miles) v
2000	$17,250
2001	$22,350
2002	$27,225
2003	$28,950
2004	$32,900
2005	$36,525
2006	$40,450

Source: www.nada.com

Table D gives estimated insurance premiums for the Escalade from multiple car insurance companies.

Table D

Vehicle Value v	Insurance Premium (every 6 months) p
$10,000–$20,000	$200.00
$20,001–$25,000	$500.00
$25,001–$30,000	$800.00
$30,001–$40,000	$1000.00
$40,001–$50,000	$1200.00

a. Create a linear model to represent the 2007 value, v, of the vehicle as a function of the model year, t. Let $t = 0$

correspond to the year 2000. Write several sentences explaining exactly what this model represents in this context. Finally, use the model to predict the 2007 value of a 2007 Cadillac Escalade.

b. Create a graph of the insurance premium paid every 6 months, *p*, as a function of the value of the vehicle, *v*. Label and scale the axes appropriately.

c. Explain how the situation of determining the insurance premium paid every 6 months, *p*, given the model year of the vehicle, *t*, is an example of the composition of two functions.

d. What is the 2007 insurance premium charged for a 2003 Cadillac Escalade? Express your answer using function notation.

e. Determine which vehicle you could buy (by indicating the model year) if you could afford insurance premiums of $800 every 6 months. Express your results using function notation.

f. Use *v(t)*, the 2007 value of the Escalade as a function of the model year, to predict how much the insurance premiums would be for a 2005 Escalade.

■ STRETCH YOUR MIND

Exercises 50–52 are intended to challenge your understanding of composite functions.

50. **Craftsmanship** At a factory in Louisville, Kentucky, workers assemble custom chairs using traditional techniques that take more time than modern assembly techniques. However, a worker assembling chairs at the factory slows down throughout the day as he gets tired and bored. The time it takes him to assemble one chair can be modeled by $c(t) = 3 + 0.1t$, where *c* is the number of hours it takes to assemble a chair and *t* is the number of hours he has been working that day. (Note: Since it takes him a range of time to complete the chair, *t* is assumed to be the number of hours he has been working when he begins making the chair.) Assume the worker is paid $15.50 per hour.

a. Find $c(6)$ and explain what this means in the context of the situation.

b. Explain what the rate of change is for the function $c(t)$ and what it represents in the context of the situation. Be as descriptive as possible.

c. What does the "3" represent in the function equation? Be as descriptive as possible.

d. Write a function $w(t)$ representing the worker's total wages after working for *t* hours.

e. What are the labor costs for the factory to have a worker make a chair when that worker starts on the chair 3.5 hours into his shift? Explain how you determined your answer.

f. Write the function $w(c(t))$. What are the input and output variables of this function?

g. Write the function $c(w(t))$. Does this function make sense in the context of the problem? Explain your reasoning.

h. The factory currently has workers completing 12-hour shifts (meaning the workers work for 12 hours a day instead of the standard 8 hours, but the workers work fewer days per week). Should the factory continue to use this schedule or go to a traditional 8-hour shift? Explain your reasoning.

i. If a worker does not complete a chair during his shift, he leaves the chair for a worker on the next shift to complete. How many hours of work is put into a chair if one worker (working a 12-hour shift) starts on the chair 11 hours into his shift and leaves the unfinished chair for the next worker to complete? What are the labor costs to the factory for making this chair?

j. Each chair requires approximately $27 worth of materials to manufacture. Write a function that will give the total costs to the factory for making a single chair.

k. Based on your responses to the preceding questions, what is a reasonable price to charge per chair to ensure that the company earns a profit on each chair? Write a function that shows the factory's profit on a single chair.

51. **Solar Energy** Solar panels convert energy from the sun into usable electricity to power homes and businesses. The larger the solar panel, the more energy from the sun it can absorb. Figure A shows the relationship, $s(A)$, between the size of a solar panel (in square meters) and the amount of energy it can absorb from the sun (in watts) in San Antonio, Texas. However, solar panels are not 100% efficient, meaning they cannot convert 100% of the energy they absorb into usable electricity. Figure B shows the relationship, $p(s)$, between the amount of energy absorbed and the amount of usable electricity created (both in watts).

Figure A

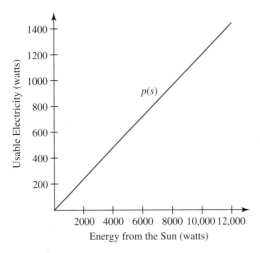

Figure B

a. How much usable electricity can be created from a solar panel that has a surface area of 8 square meters?

b. Your friend in San Antonio is bragging about how his house is hooked up to a solar panel, which saves a bunch of money on his electric bill. He claims he gets 950 watts of usable electricity from his solar panel. About how large is his panel?

c. Create a table of values for the function $p(s(A))$ using at least five coordinate points, then make a graph of this function making sure to label your axes with the appropriate variables.

52. **Life Insurance** Many employers offer supplemental life insurance to their employees to help offset the costs should an employee suddenly die. The cost of the insurance depends on the age of the employee. The following table shows the monthly insurance rates for $20,000 of supplemental insurance.

Rates for $20,000 Supplemental Insurance		
Age	Mid-Interval A	Monthly Rate (dollars) R
25–27	26	0.72
28–30	29	0.80
31–33	32	1.00
34–36	35	1.20
37–39	38	1.44
40–42	41	2.16
43–45	44	3.44
46–48	47	4.16
49–51	50	5.76
52–54	53	7.40
55–57	56	10.28
58–60	59	12.80
61–63	62	17.64
64–66	65	25.40
67–69	68	35.48
70–72	71	54.72
73–75	74	78.32

Source: www.bussvc.wisc.edu

a. Find the *percentage* change in monthly insurance rates between each mid-interval age. Then find the average of these percentage changes and explain what this value means in the context of this situation.

b. We use the mid-interval age to represent each age interval so that we may create a function for this situation. (Note the ages change by 3 years). Write an exponential function that models the insurance rate, R, as a function of the mid-interval age, A.

c. Describe the relationship between the quantities R and A as provided by the function. In other words, verbally describe what information the function provides.

d. Use logarithms to find the value of A if $R(A) = 36$ and explain what the value of A means in the real-world context.

SECTION 7.4 Logistic Functions

LEARNING OBJECTIVES

- Graph logistic functions from equations and tables
- Use logistic models to predict and interpret unknown results

GETTING STARTED

In North America, the influenza season typically starts around the month of October. The "flu" is a contagious respiratory illness caused by influenza viruses that can cause mild to severe illness sometimes leading to death. The number of people infected with the flu grows slowly at first, then more quickly, then slowly again, eventually leveling off. Such behavior (in growth or decay) can be modeled with a logistic function.

450 CHAPTER 7 ■ MODELING WITH OTHER TYPES OF FUNCTIONS

In this section we show how to discriminate between exponential and logistic models, illustrate how to graph logistic functions from tables and equations, and use logistic regression to model real-world data and make predictions. We also use the language of rate of change to describe the behavior of logistic growth and decline.

■ Exponential Growth with Constraints

An important fact about exponential growth functions is that even though growth may seem slow over the short run, it ultimately becomes very rapid. For instance, a population growth rate of 2% per year may seem small; but it means that the population will double every 35 years. A population of 10,000 today that is growing at 2% annually will double three times over the next 105 years, yielding a population of 80,000. However, it is unrealistic to expect exponential growth to continue forever. In most situations, there are factors that ultimately limit the growth (such as available land, natural resources, or food supplies).

■ Logistic Growth

Flu viruses mainly spread from person to person through the coughing or sneezing of people with the virus. Sometimes people become infected by touching something with flu viruses on it and then touching their mouth or nose. Most healthy adults are able to infect others beginning one day before symptoms develop and up to five days after becoming sick. (*Source:* www.cdc.gov) As more and more people catch the flu they, in turn, also begin to infect others. Thus the total number of people who have contracted the flu begins to grow exponentially, increasing at an increasing rate, as in Figure 7.21.

However, this trend cannot continue indefinitely. People with the flu will begin to come in contact with people who already have been infected, have built up immunity, or have been vaccinated. As Figure 7.22 shows, this will cause the rate of increase in the number of people who have had the flu to begin to decline—that is, we see the number of people who have had the flu increasing at a decreasing rate.

Figure 7.21   Figure 7.22

Eventually, the total number of flu cases will level off because there is only a limited number of people who will be able to contract the flu. Therefore, Figure 7.23 represents the total number of people who have had the flu as a combination of very slow exponential growth (labeled *a*), rapid exponential growth (labeled *b*), followed by a slower increase (labeled *c*), and then a leveling off (labeled *d*). At the inflection point the rate at which the flu is spreading is the greatest. The horizontal asymptote (red dashed horizontal line) represents the **limiting value** for the number of people who will contract the flu.

The mathematical model for such behavior is called a **logistic function**.

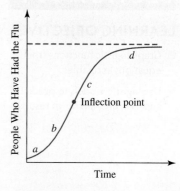

Figure 7.23

PEER INTO THE PAST

LOGISTIC FUNCTIONS

The English economist Thomas Robert Malthus (1766–1834) was a pioneer in population study. He believed that poverty and starvation were unavoidable because the human population tends to grow exponentially while food supplies tend to grow linearly. Malthusian theory laid the groundwork for others to investigate the notion of carrying capacities (limiting values) for populations.

P. F. Verhulst followed up Malthus by introducing the logistic model in 1838 to better describe the growth of human populations. According to the PNAS (Proceedings of the National Academy of Sciences of the United States of America), R. Pearl and L. J. Reed derived the same model in 1920 to describe the growth of the population of the United States since 1790 and to attempt predictions of that population at future times.

It is known that the logistic model for population growth, like the exponential model, has limitations. According to Pearl and Reed, in 1920 the population was to stabilize at about 197 million people. Of course, this level has been far surpassed. At the time of this text's writing there are over 300 million residents of the United States.

Source: www.pubmedcentral.nih.gov

LOGISTIC FUNCTIONS

A **logistic function** is a function of the form

$$N(t) = \frac{L}{1 + Ae^{-Bt}}$$

The number L is the **upper limiting value** of the function $N(t)$. The **lower limiting value** is $y = 0$, the horizontal axis.

EXAMPLE 1 ■ Exploring Logistic Functions in a Real-World Context

The Centers for Disease Control monitor flu infections annually, paying particular attention to flu-like symptoms in children from birth to 4 years old. The cumulative number of children 0–4 years old who visited the CDC's sentinel providers with flu-like symptoms during the 2006–2007 flu season are displayed in Figure 7.24, together with a logistic model. (A sentinel provider is a medical provider who has agreed to report flu data to a national surveillance network from early October through mid-May.) (*Source:* www.cdc.gov)

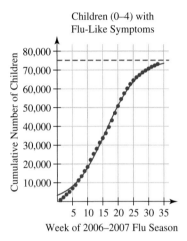

Figure 7.24

a. Explain why a logistic function may better model the 2006–2007 flu infection rate than an exponential function.

b. Using the language of rate of change, describe the behavior of the graph and relate this to what it tells about the real-world context.

c. The formula for the logistic function that models the cumulative number of reported children with flu-like symptoms is $H(w) = \dfrac{75{,}700}{1 + 21.31e^{-0.1897w}}$, where w is the week of the flu season. What does the limiting value mean in the real-world context? How is the limiting value represented in the formula for the function $H(w)$ and its graph?

d. Estimate the coordinates for the point of inflection and explain what each coordinate means in the real-world context.

Solution

a. A logistic function models the growth in the cumulative number of children with flu-like symptoms better than an exponential function because a logistic function grows slowly at first, then more quickly, and finally levels off at a limiting value. An exponential function, on the other hand, may grow slowly at first but then will increase at an ever increasing rate, ultimately exceeding the available number of children who could conceivably become infected with flu-like illnesses.

b. The graph of the function model is increasing throughout the interval; however, the rate at which the graph is increasing varies. Initially, the graph is concave up, indicating that the cumulative number of reported cases is increasing at an ever-increasing rate. Around week 17, the graph changes to concave down, indicating that the cumulative number of reported cases are increasing at a lesser and lesser rate. Around week 32, the graph is increasing at such a slow rate that it appears to level off. This indicates that the number of newly reported cases in weeks 32 and beyond is so small that it has a negligible effect on the cumulative number of cases.

c. The limiting value is approximately 75,700 children. It appears that no more than 75,700 children were reported to have flu-like symptoms during the season. This value is represented in the formula for H(w) by the value 75,700 found in the numerator and on the graph by the horizontal asymptote at 75,700.

   d. From the graph of H(w), an estimate for the coordinates of the inflection point is approximately (17, 40,000). This means at week 17 of the 2006–2007 flu season (late January), the number of children with flu-like symptoms was increasing most rapidly.

Logistic Decay

Many real-world data sets are modeled with decreasing rather than increasing logistic functions. As was the case with logistic growth functions, logistic decay functions have an upper limiting value, L, and a lower limiting value of $y = 0$.

EXAMPLE 2 ■ Recognizing a Logistic Decay Function

As shown in Table 7.19, the infant mortality rate in the United States has been falling since 1950. (An *infant* is a child under 1 year of age.)

a. Using Table 7.19, show how the rate of change records the decline in the infant death mortality rate and discuss why this suggests that a logistic model may fit the data.

b. From the table, predict the future lower limiting value and explain what the numerical value means in the real-world context. Does this value seem reasonable?

c. Create a scatter plot of the data and estimate the upper and lower limiting values.

Table 7.19

Years Since 1950 y	Infant Mortality Rate (deaths per 1000 live births) M
0	29.2
10	26.0
20	20.0
30	12.6
35	10.6
40	9.2
45	7.6
50	6.9
59	6.4

Source: www.cdc.gov

Solution

a. If we calculate the decrease in the infant mortality rate (ΔM) and the yearly rate of change of the infant mortality rate over each interval of time given in Table 7.20, we can determine the behavior of the function $M(y)$. (Note: We must exercise caution because the intervals between the years provided are not the same.)

Table 7.20

Years Since 1950 y	Infant Mortality Rate (deaths per 1000 live births) M	Change in Infant Mortality Rate (deaths per 1000 live births) ΔM	Rate of Change in Infant Mortality Rate per Year $\dfrac{\Delta M}{\Delta y}$
0	29.2		
10	26.0	−3.2	−0.32
20	20.0	−6.0	−0.60
30	12.6	−7.4	−0.74
35	10.6	−2.0	−0.40
40	9.2	−1.4	−0.28
45	7.6	−1.6	−0.32
50	6.9	−0.7	−0.14
59	6.4	−0.5	−0.056

From the rate of change, we can see that the decline in the infant mortality rate tends to drop slowly at first, then more dramatically, and then levels off. The change in the infant mortality rate suggests a logistic model.

b. We estimate the lower limiting value to be approximately 6.0 deaths per 1000 live births because we expect the values to level off near 6.0.

c. The scatter plot is shown in Figure 7.25. From the scatter plot, we predict the upper limiting value will be around 35 and the lower limiting value will be around 6.

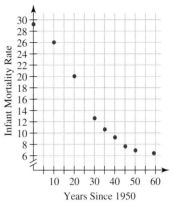

Figure 7.25

We said earlier that all logistic functions have a lower limiting value at $y = 0$. We can use logistic regression to model data sets with this lower limiting value. But when a data set appears to be logistic but has a different lower limiting value, as in Example 2, we need to align the data before using logistic regression to model the function. We demonstrate in Example 3.

EXAMPLE 3 ■ Using Logistic Regression to Model Data

Refer to Table 7.20 and Figure 7.25 in Example 2.

a. Use logistic regression to find the logistic function, $M(y)$, for the data. Comment on the model's fit to the data.

b. Align the data by subtracting the estimated lower limiting value from each output value in the data set. Then redo the logistic regression for the aligned data.

c. Create the aligned logistic model by adding back the lower limiting value to the model equation in part (b). Describe how well this new function fits the data set.

Solution

a. Using the Technology Tip for logistic regression at the end of this section, we find

$$M(y) = \frac{46.96}{1 + 0.5616e^{0.0481y}},$$ graphed in Figure 7.26.

Although the original data looked logistic, this logistic model does not appear to fit the data as well as we might have expected.

b. We align the data by subtracting the estimated lower limiting value of 6.0 from the outputs as shown in Table 7.21. (See also the second Technology Tip at the end of this section.)

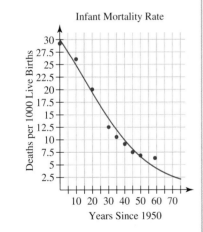

Figure 7.26

Table 7.21

Years Since 1950 y	Infant Mortality Rate (deaths per 1000 live births) M	Aligned Output Data $M - 6.0$
0	29.2	23.2
10	26.0	20.0
20	20.0	14.0
30	12.6	6.6
35	10.6	4.6
40	9.2	3.2
45	7.6	1.6
50	6.9	0.9
59	6.4	0.4

After data alignment, we recalculate the logistic regression to get $M(y) = \dfrac{25.18}{1 + 0.08327e^{0.1144y}}$, graphed in Figure 7.27.

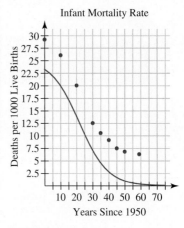

Figure 7.27

c. We notice the logistic regression model for the aligned data lies below the scatter plot. Therefore, we add the limiting value of 6.0 to the equation to shift the data upward. This process gives the function $N(y) = \dfrac{25.18}{1 + 0.08327e^{0.1144y}} + 6.0$. The original scatter plot and the graph of the function N are shown in Figure 7.28. The aligned model fits the data much better than the original logistic model.

Figure 7.28

EXAMPLE 4 ■ Extrapolating Exponential and Logistic Growth

The digital video disc (DVD) player was introduced to the market in the first quarter of 1997. At the end of 2006, Nielsen Media Research reported more U.S. households own DVD players (81.2% of all households) than VCRs (79.2% of households). (*Source:* www.nielsenmedia.com) The sales of DVD hardware from 1997 to 2006 are given in Table 7.22.

Table 7.22

Years Since 1997 y	DVD Hardware Sales ($ millions) D
0	0.305
1	0.946
2	3.550
3	9.877
4	16.662
5	25.113
6	33.734
7	37.125
8	36.737
9	32.660

a. Use logistic regression to determine the function, $D(y)$, of the form $N(t) = \dfrac{L}{1 + Ae^{-Bt}}$ that best models the growth in DVD sales from 1997 to 2004. Determine the value of L and explain what this number means in the real-world context.

b. Based on the data from 2005 and 2006, does it appear the model can be used to accurately forecast future sales? Explain why or why not.

Solution

a. Using the Technology Tip and logistic regression on the data from 1997 to 2004, we find $L(y) = \dfrac{40.69}{(1 + 57.17e^{-0.92y})}$ million dollars is the best-fit logistic function for DVD sales (see Figure 7.29). The limiting value, L, is 40.69, which means DVD sales will level off at \$40,690,000 per year.

b. For the years 2005 and 2006, we evaluate the function at $t = 8$ and $t = 9$.

$$L(8) = \dfrac{40.69}{(1 + 57.17e^{-0.92(8)})} \qquad L(9) = \dfrac{40.69}{(1 + 57.17e^{-0.92(9)})}$$
$$= 39.3 \qquad\qquad\qquad\qquad = 40.1$$

The model forecasts 39.3 million dollars in sales in 2005 and 40.1 million dollars in sales in 2006. In actuality, there were 36.7 and 32.7 million dollars in sales, respectively. The model predicts a leveling off of sales whereas the actual data shows a decline in sales in 2005 and 2006 (see Figure 7.30). Consequently, the model does not appear to accurately model future sales.

Figure 7.29

Figure 7.30

SUMMARY

In this section you learned to graph logistic functions from tables and equations, and use logistic regression to model and make predictions from real-world data. You also learned to use the language of rate of change to describe the behavior of logistic growth and decay.

TECHNOLOGY TIP ■ LOGISTIC REGRESSION

1. Enter the data using the Statistics Menu List Editor.

2. Bring up the Statistics Menu Calculate feature and select item **B:Logistic** and press ENTER.

3. If you want to automatically paste the regression equation into the Y = Editor so that you can easily graph the model, press the key sequence VARS; Y-VARS; Function; Y1 and press ENTER. Otherwise press ENTER.

TECHNOLOGY TIP ■ ALIGNING A DATA SET

1. Enter the data using the Statistics Menu List Editor.

2. Move the cursor to the top of L3. We want the entries in L3 to equal the entries in L2 minus the amount of the vertical shift (in this case, 210). To do this, we must enter the equation **L3=L2−210** as in Step 3.

3. Press 2nd, then type 2 to place L2 on the equation line at the bottom of the viewing window. Then type − and **210** to subtract 210.

4. Press ENTER to display the list of aligned values in L3.

7.4 EXERCISES

■ SKILLS AND CONCEPTS

1. Create a table of values for a logistic function that is increasing.

2. Create a table of values for a logistic function that is decreasing.

3. A logistic growth curve is sketched in the figure. Estimate the limiting value and the inflection point, and explain what each means in the real-world context. Also provide the year you would recommend that the com-

pany release the upgraded product so their profits do not begin to fall.

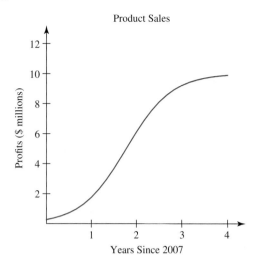

4. Explain why the following table of values can be modeled by a decreasing logistic function.

x	y
0	15
1	14
2	12
3	8
4	2
5	1
6	0.5
7	0.3
8	0.1

5. If $f(x) = \dfrac{12}{1 + 16e^{-0.8x}}$ is a logistic function, what is the limiting value?

6. If $f(x) = \dfrac{113}{1 + 46e^{-4x}}$ is a logistic function, what is the limiting value?

In Exercises 7–12, identify the scatter plots as linear, exponential, logistic, or none of these. If you identify the scatter plot as none of these, explain why.

7.

8.

9.

10.

11.

12.
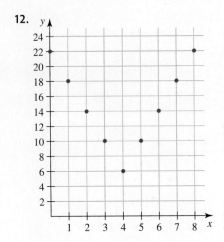

13. **Agriculture** Agriculture plays a significant role in the economy of the midwestern United States. One of the most important crops produced is corn. Much research is done to produce the most hearty, disease- and drought-resistant plants possible. The growth rate of the corn crop is important to farmers as well. With the relatively short growing seasons in some parts of the Midwest, adapting varieties of corn to fit a region's growing season is an important science. (*Source:* www.age.uiuc.edu)

Suppose a new corn crop is planted and measured each day after it first breaks through the soil. The height of the plant, $c(t)$, can be modeled by the logistic function $c(t) = \dfrac{96}{1 + 18(0.955)^t}$ inches, where t is the number of days since the corn plant first emerged from the soil.

  a. Use a graphing calculator to graph the height over 130 days.
  b. What is the practical domain of $c(t)$? What is the range? What does this tell about the height of the corn plant?
  c. Estimate the coordinates of the inflection point and explain what each value means in the real-world context.
  d. Estimate the limiting value and explain what this value means in the real-world context.

14. **Value of Stocks** Suppose the value of a share of one of your stocks increases at an increasing rate from 2004 through 2008 and then increases at a decreasing rate from 2008 to 2012. Fill in the table below with reasonable values that reflect these value changes.

Year	Value of Stock (dollars)	Year	Value of Stock (dollars)
2004	22.00	2009	
2005		2010	
2006		2011	
2007		2012	
2008	50.00		

15. **Pyramid Schemes** A pyramid scheme is a nonsustainable business model that involves the exchange of money primarily for enrolling other people into the scheme, usually without a product or service being rendered. Most pyramid schemes take advantage of the confusion that exists between genuine business models and convincing scams. The essential idea behind each scam is that the individual makes only one payment, but is promised to receive exponential benefits from other people as a reward.

Suppose the scheme initiator promises huge financial gains for investors who join his Profit Club for a $100 fee. He recruits two investors, who in turn each recruit two more investors, and so on. Each new club member pays the $100 joining fee. Half of the fee goes to the person who recruited the club member and the other half is paid to the person who recruited the recruiter as a "bonus." Each recruiter must pay half of all bonuses received to the person who recruited him as a "bonus." This process is to continue indefinitely. Describe the flaw in this business model.

SHOW YOU KNOW

16. How are exponential and logistic functions similar and how are they different?
17. Explain how to graph logistic functions represented by a table or an equation.
18. Describe a logistic function in terms of a rate of change.
19. What is important to keep in mind when modeling data sets with a logistic function?

MAKE IT REAL

20. **Environmental Carrying Capacity** The graph approximates the hypothetical population of roadrunners, R, in thousands of birds within the confines of a designated area within the Tonto National Forest in Arizona as a function of the year y.

Ecologists refer to growth of the type shown here as *logistic population growth*. Coyotes are one of the roadrunners' natural predators and affect the number of roadrunners within the national forest.

a. Explain in general terms how the roadrunner population changes over time.

b. When did the population reach 60,000?

c. Evaluate $R(6)$ and explain what the numerical value means in this context.

d. During which period of years is the graph concave up? Explain what this means about the population growth during this period.

e. During which period is the graph concave down? Explain what this means about the population growth during this period.

f. When does a point of inflection appear to occur on the graph? Explain how this point may be interpreted in terms of the growth rate.

g. What is the *environmental carrying capacity* of this forest for the roadrunner? That is, what is the maximum number of roadrunners that the environment seems to be able to support (the limiting value)?

21. **Population Density** The population of Sydney, Australia, is distributed in suburbs around the city's center. Let S be the number of people (in millions) living within r kilometers of the city's center. Then $S(r)$ can be approximated by

$$S(r) = \frac{1.796}{(1 + 3.05e^{(-0.21r)})}$$

(*Source:* www.environment.gov.au)

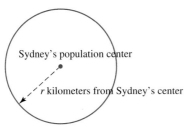

a. Use the graph of $S(r)$ to approximate the number of people living in Sydney including the suburbs.

b. Use the graph of $S(r)$ to estimate the distance in kilometers from Sydney's center to the edge of town where not very many people live.

c. Estimate the distance from Sydney's center to the most densely populated neighborhoods.

22. **Computer Transistors** In 1965 Moore's Law predicted that the number of transistors on a circuit would double every 2 years. In Chapter 6, we modeled the growth of the number of transistors with the exponential function $T(y) = 2300(2)^y$, where y is the number of years since 1971. The following table shows the actual number of transistors for different Intel computers. Do you believe the mathematical model $T(y)$ can be used to extrapolate the future growth in the number of transistors or would a logistic function be a better choice? Justify your answer.

Intel Product Name	Years Since 1971 y	Number of Transistors on Integrated Circuit T
4004 Microprocessor	0	2300
8008 Microprocessor	1	3500
8080 Microprocessor	3	6000
8086 Microprocessor	7	29,000
286 Microprocessor	11	134,000
386 Microprocessor	14	275,000
486 Microprocessor	18	1,200,000
Pentium	22	3,100,000
Pentium II	26	7,500,000
Pentium III	28	9,500,000
Pentium IV	29	42,000,000
Itanium 2	31	220,000,000
Dual Core Itanium 2	35	1,700,000,000

Source: www.intel.com

23. **Social Diffusion** Sociologists use the phrase "social diffusion" to describe the way information spreads through a population. (*Source:* www.jstor.org) The information might be a rumor, a cultural fad, or news about a technical innovation.

Assume the number of people, P, who have received some type of information as a function of the time, t, in days is modeled with the logistic function,

$$P(t) = \frac{145}{1 + 150e^{-0.3t}}$$

a. Based on the graph of $P(t)$, describe what the rate of change of the function tells about how the news of an event spreads over time.

b. Estimate the coordinates of the point of inflection and explain what each value means in this context.

c. Find the average rate of change from $t = 5$ to $t = 25$ for the function and explain the meaning of this rate of change.

d. Is there a limiting value for the function? If so, estimate the value and explain what it means in the context of the spread of information through a population.

24. **Stay-at-Home Moms** In recent years, the number of married mothers who choose to stay at home to care for their

families and bypass an outside career has increased. The function

$$M(t) = \frac{948.2}{1 + 27.23e^{-1.110t}} + 4700$$

models the number of married couple families with stay-at-home mothers (in thousands) as a function of the number of years since 1999, t. (*Source: Statistical Abstract of the United States, 2006*, Table 59)

a. Describe what the rate of change of the function tells about how the number of married, stay-at-home mothers is changing over time.

b. Estimate the coordinates of the inflection point and explain what each value means in this context.

c. Find the average rate of change from $t = 3$ to $t = 7$ for the function and explain the meaning of this rate of change.

d. What are the limiting values for the function and what do they mean in the context of the stay-at-home mothers?

25. **Internet Access** Based on data from 1994 to 2000, the percentage of public school classrooms with Internet access can be modeled by

$$P(y) = \frac{85.88}{1 + 31.40e^{-0.9250y}}$$

where P is the percentage of public schools with Internet access and y is the number of years since the end of 1994.

a. Graph the function over a practical domain.

b. Does it seem reasonable that the function would have a limiting value? Explain.

c. Estimate the rate at which the percentage of public schools with Internet access was changing at the end of 2000.

26. **Women with Ph.D.s** The number of women in the United States who have earned a Ph.D., N, has dramatically increased in the years since 1900, t. The table displays the data for selected years from 1900 to 1999.

Years Since 1900 t	Number of Ph.D.s Awarded to Women N
0	23
10	44
20	88
30	313
40	429
50	620
60	1042
70	3971
80	9408

Years Since 1900 t	Number of Ph.D.s Awarded to Women N
90	13,106
95	16,414
99	17,493

Source: Modeled from U.S. Doctorates in the 20th Century, National Science Foundation, 2006, Figure 3-2

a. Explain in general terms how the number of women earning Ph.D.s changes with time.

b. Estimate the year in which the number of women earning Ph.D.s first surpassed 1000.

c. Looking at the average rate of change between successive years, does it appear there is a limiting value for the number of Ph.D.s earned by women? If so, estimate what this number will be.

d. Create a scatter plot of the data and determine during which period the data is increasing and concave up. Explain what this means about the growth in Ph.D.s for women.

e. During which period is the data increasing and concave down? Explain what this means about the growth in Ph.D.s for women.

f. Estimate where the point of inflection occurs from the graph. What historical events may have been occurring at this time to affect the number of Ph.D.s being earned by women?

27. **VCR Prices vs. DVD Sales** The price of a VCR since 1997 (the year DVD players entered the market) and the sales of DVD players since 1997 in the United States are shown in the following two graphs.

(*continued*)

a. Using the graphs, explain how the two may be interrelated. Make sure to include in your explanation why the price of a VCR is declining while the sales of DVD players is increasing.
b. Explain why each graph behaves in such a way as to follow the pattern of a logistic function.

28. **Telephone Usage** The following table gives the percentages of American households with telephones.

Years Since 1935 t	Percentage of American Households with a Telephone T
0	31.8
5	36.9
10	46.2
15	61.8
20	71.5
25	78.3
30	84.6
35	90.5
45	93.0
50	93.0
55	93.3
60	93.9

Source: Statistical Abstract of the United States, 2007, Table 1111

a. Use the table to estimate the point of inflection. (Be sure to show how the table was used.) What is the significance of the inflection point in this situation?
b. From the table, estimate the instantaneous rate of change in 1970.
c. What is the meaning of the instantaneous rate of change you estimated in part (b)? What are its units?
d. Do you think there should be a limiting value for this data? If so, explain what you think it is. Justify your conclusion.

29. **Cable TV Subscribers** The following table gives the number of basic cable television subscribers since 1975 (in thousands).

Years Since 1975 t	Number of Basic Cable Subscribers (thousands) C
0	9,800
5	17,500
10	35,440
15	50,520
20	60,550
25	66,250
26	66,732
27	66,472
28	66,050
29	65,727
30	65,337

Source: Statistical Abstract of the United States, 2007, Table 1126

a. Use the table to estimate the point of inflection. (Be sure to show how the table was used.) What is the significance of the inflection point in this situation?
b. From the table, estimate the instantaneous rate of change in 2002.
c. What is the meaning of the value you found in part (b)? What are its units?

30. **Pediatric AIDS** Based on data from 1992 to 2001, the estimated number of new pediatric AIDS cases in the United States in year t can be modeled by

$$P(t) = \frac{919.9}{1 + 0.0533e^{0.07758t}} + 90 \text{ cases}$$

where t is the number of years since the end of 1992. (*Source: Centers for Disease Control and Prevention data*)

a. How many new pediatric AIDS cases are estimated to have occurred in 2000?
b. At what rate was the estimated number of new pediatric AIDS cases changing at the end of 2000?
c. Based on the results of part (b) and the graph of $P(t)$, does it look like efforts to reduce pediatric AIDS in the United States are generating positive results? Explain.

31. **Adult Internet Usage** The percentage of adults using the Internet at home, school, or work to access websites or to send and receive email has grown dramatically since 1995. Use the table to suggest whether an exponential or logistic model will be the best choice to extrapolate the percentage of adult Internet users past 2005.

Years Since 1995 t	Percentage of Adult Internet Users P
0	14
5	53
9	59
10	69

32. **Hotel/Motel Occupancy** Based on data from 1990 to 1999, the average hotel/motel room rate can be modeled by

$$R(t) = \frac{27.80}{(1 + 47.23e^{-0.6425t})} + 57 \text{ dollars per day}$$

where t is the number of years since the end of 1990. (*Source: Statistical Abstract of the United States, 2007, Table 1264*)

Approximate the rate at which the hotel/motel room rate was changing at the end of 1997 and at the end of 1999.

33. **Information Technology** Based on data from 1990 to 2000, the percentage of the economy attributed to the information technology sector may be modeled by

$$P(t) = \frac{2.956}{1 + 30.15e^{-0.5219t}} + 7.9$$

where P is the percentage of the economy and t is the number of years since the end of 1990. (*Source: Statistical Abstract of the United States, 2001, Table 1122*) Approximate the rate at which the information technology percentage of the economy was changing at the end of 1995 and at the end of 2000.

34. Chicken Pox

Years Since 1985 T	Reported Cases of Chicken Pox (thousands) C
0	178.2
5	173.1
10	120.6
14	46
15	27.4
16	22.5
17	22.8
18	20.9

Source: Statistical Abstract of the United States, 2006, Table 177

A chicken pox vaccine became widely available in the United States in 1995. According to the model, does the vaccine appear to be having an effect? Does the model project that chicken pox will be eradicated? Explain.

35. Video Game Sales
Blizzard Entertainment released the engaging real-time strategy game "Warcraft III: Reign of Chaos" on July 2, 2002. By July 22, 2002, more than a million copies of the game had been sold.
(*Source:* www.pcgameworld.com)

Assuming a total of 8 million copies of the game will be sold, sketch a graph of a logistic growth model for "Warcraft III" game sales.

In Exercises 36–40, use logistic regression to find the logistic model for the data. Then answer the given questions.

36. Deadly Fights over Money

Years Since 1990 t	Homicides H
1	520
2	483
3	445
4	387
5	338
6	328
7	287
8	241
9	213
10	206

Source: Crime in the United States 1995, 2000, Uniform Crime Report, FBI

Use the logistic model $H(t)$ to estimate the point of inflection and the rate that money-related homicides were decreasing in the year 2001.

37. Cassette Tape Sales

Years Since 1993 t	Percent of Music Market (percentage points) P
0	38.0
1	32.1
2	25.1
3	19.3
4	18.2
5	14.8
6	8.0
7	4.9
8	3.4
9	2.4

Source: Recording Industry of America

According to the model, estimate the year that the cassette market share will drop below 1%. Approximately at what rate will the cassette tape market share be decreasing at that time?

38. *My Big Fat Greek Wedding* Box Office Sales

Weekend	Week Number	Cumulative Gross Box Office Sales (dollars)
Apr. 19–21 (2002)	1	597,362
June 28–30	11	19,340,988
Sept. 6–8	21	95,824,732
Nov. 15–17	31	199,574,370
Jan. 24–26 (2003)	41	236,448,697
Apr. 4–6	51	241,437,427

Source: www.boxofficeguru.com

a. According to the table, were cumulative box office sales for *My Big Fat Greek Wedding* increasing at a higher rate in week 1 or in week 51?

b. IFC Films was the distributor for *My Big Fat Greek Wedding*. If you were a marketing consultant to IFC Films, what would you tell the company about forecasted box office sales beyond week 51?

39. Value of Fabricated Metals Shipments

Years Since 1992 t	Shipment Value ($ millions) V
0	170,403
1	177,967
2	194,113
3	212,444
4	222,995
5	242,812

(*continued*)

Years Since 1992 t	Shipment Value ($ millions) V
6	253,720
7	256,900
8	258,960

Source: Statistical Abstract of the United States, 2001, **Table 982**

(*Hint:* Before creating the model, align the data by subtracting 170,000 from each value of V. After doing logistic regression, add back the 170,000 to the resultant model equation.)

Explain what the point of inflection and limiting value mean in the real-world context.

40. **Calculator Technology** The rapid advancement in calculator technology over the past 30 years has led to a decrease in the price of a basic, four-function calculator.

In August 1972, the four-function Sinclair Executive calculator was introduced for around $150. By the end of the decade, comparable calculators were selling for under $10.

Suppose the table gives the average price of a four-function calculator from 1972 to 1980.

Years Since 1972 t	Calculator Cost (dollars) C
0	150
1	132
2	110
3	85
4	62
5	42
6	27
7	17
8	10

a. Create a scatter plot of the data. Is this scatter plot increasing or decreasing? Does there appear to be an inflection point? What does the inflection point tell about four-function calculator prices?

b. Do you think there is a limiting value as t increases? If so, what is it and what does it represent?

■ **STRETCH YOUR MIND**

Exercises 41–42 are intended to challenge your understanding of logistic functions.

41. **AIDS Deaths in the United States** From 1981 to 1995, the number of adult and adolescent AIDS deaths in the United States increased dramatically. However, from 1995 to 2001 the annual death rate plummeted, as shown in the table.

Years Since 1981 t	Number of Deaths during Year A
0	122
1	453
2	1481
3	3474
4	6877
5	12,016
6	16,194
7	20,922
8	27,680
9	31,436
10	36,708
11	41,424
12	45,187
13	50,071
14	50,876
15	37,646
16	21,630
17	18,028
18	16,648
19	14,433
20	8963

Source: HIV/AIDS Surveillance Report, Dec. 2001; Centers for Disease Control and Prevention, p. 30

a. Create a scatter plot of the data.
b. Find a best-fit function, $d(t)$, for the number of deaths from 1981 to 1995.
c. Find a best-fit function, $c(t)$, for the number of deaths from 1995 to 2001.
d. Explain the trend in the number of AIDS deaths from 1981 to 1995.
e. Explain the trend in the number of AIDS deaths from 1995 to 2001.
f. Write the piecewise function $A(t)$ that models the number of deaths due to AIDS from 1981 to 2001.
g. Evaluate $A(32)$ and $A(35)$. Do these values seem reasonable? Explain.

42. **Canadian Population** The population of Canada has grown at different rates from 1861 to 2001 and is displayed in the table.

Years Since 1861 t	Canadian Population (millions) C
0	3.230
10	3.689
20	4.325
30	4.833
40	5.371
50	7.207
60	8.788
70	10.377
80	11.507

(continued)

Years Since 1861 t	Canadian Population (millions) C
90	13.648
100	18.238
110	21.568
120	24.820
130	28.031
140	31.050

Source: www.sustreport.org

The Sustainability Reporting Program estimates the population of Canada to be 33,369,000 in the year 2011; 35,393,000 in the year 2021; and 42,311,000 in the year 2050. Create a scatter plot of the data in the table and use it to find the mathematical model that those who wrote the Sustainability Report could have used to arrive at their estimates.

SECTION 7.5 Choosing a Mathematical Model

LEARNING OBJECTIVES

- Select the "best" function to model a real-world data set given in equation, graph, table, or word form
- Use the appropriate function to model real-world data sets
- Use appropriate models to predict and interpret unknown results

GETTING STARTED

The U.S. Congress periodically debates the pros and cons of increasing the federal minimum wage. For instance, lawmakers must consider how the increase in the wage will impact business owners' ability to hire employees as well as how the increase in pay will positively impact employees' lives. The process of determining a fair and reasonable minimum wage includes investigating previous wage levels and inflation rates as well as analyzing economic trends. Knowing how to choose a mathematical model is an important skill in situations such as this.

In this section we revisit what we have learned throughout the text by studying how interpolation, extrapolation, function end behavior, and a function's rate of change are used when choosing the "best" mathematical model. We also consider the reasonableness of predictions and determine the practical domain of the models we choose.

■ Choosing a Mathematical Model from a Table of Values

Recall that a *mathematical model* is a graphical, verbal, numerical, and symbolic representation of a problem situation. The model helps us understand the nature of the problem situation and make predictions. Mathematical models are used frequently to represent real-world situations so it is important to master the process of choosing an appropriate model.

EXAMPLE 1 ■ Choosing a Model from a Table of Values

The *Fair Labor Standards Act* was passed by Congress in 1938 to set a minimum hourly wage for American workers. Table 7.23 displays the wages set for the year 1938 and each year in which the wage was raised.

Table 7.23

Years Since 1938 t	Minimum Wage (dollars) M	Years Since 1938 t	Minimum Wage (dollars) M
0	0.25	37	2.10
1	0.30	38	2.30
7	0.40	40	2.65
12	0.75	41	2.90
18	1.00	42	3.10
23	1.15	43	3.35
25	1.25	52	3.80
29	1.40	53	4.25
30	1.60	58	4.75
36	2.00	59	5.15

Source: usgovinfo.about.com

a. Determine whether a linear, a quadratic, or an exponential function is the most appropriate mathematical model for these data. Explain as you go along why you did not choose the two types of functions you reject.

b. Use your model to predict the minimum wage in 2008.

Solution

a. To determine if these data are best modeled by a linear function or a nonlinear function (quadratic or exponential), we first calculate the average rate of change over each time interval, as shown in Table 7.24. (Note that the time intervals are not equal.)

Table 7.24

Years Since 1938 t	Minimum Wage (dollars) M	Average Rate of Change $\frac{\Delta M}{\Delta t}$
0	0.25	
1	0.30	0.05
7	0.40	0.02
12	0.75	0.07
18	1.00	0.04
23	1.15	0.03
25	1.25	0.05
29	1.40	0.04
30	1.60	0.20
36	2.00	0.07
37	2.10	0.10
38	2.30	0.20
40	2.65	0.18
41	2.90	0.25
42	3.10	0.20
43	3.35	0.25
52	3.80	0.05
53	4.25	0.45
58	4.75	0.10
59	5.15	0.40

We know that to be effectively modeled with a linear function, the data must have a relatively constant rate of change. This data set has neither a constant nor nearly constant rate of change so we rule out the linear model. To confirm our conclusion, we draw the scatter plot in Figure 7.31. The overall trend seems to show the data increasing at an increasing rate and, therefore, it is nonlinear.

We next check to see if the data may be best modeled by a quadratic function. Using quadratic regression, we find the quadratic model of best fit, $Q(t) = 0.0012t^2 + 0.01t + 0.29$, and graph it in Figure 7.32.

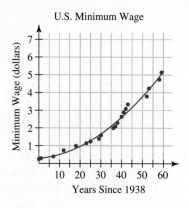

Figure 7.31   **Figure 7.32**

The graph of the quadratic function appears to be a good fit, but exponential functions can also model data that are increasing at an increasing rate. To test the exponential model, we calculate the annual growth factors to see if they are relatively constant, as shown in Table 7.25. Recall that to find the annual growth factor over a time span greater than 1 year, we must find the nth root of the quotient between two consecutive years. For example, for the growth over the years from 1939 and 1945, we find the 6-year growth factor of $\dfrac{0.40}{0.30} \approx 1.33$ and the annual growth factor of $\sqrt[6]{1.33} \approx 1.05$.

Table 7.25

Years Since 1938 t	Minimum Wage (dollars) M	Annual Growth Factor
0	0.25	$\dfrac{0.30}{0.25} \approx 1.20$
1	0.30	$\dfrac{0.40}{0.30} \approx 1.33$ $\sqrt[6]{1.33} \approx 1.05$
7	0.40	$\dfrac{0.75}{0.40} \approx 1.88$ $\sqrt[5]{1.88} \approx 1.13$
12	0.75	1.05
18	1.00	1.03
23	1.15	1.04
25	1.25	1.03
29	1.40	1.14
30	1.60	1.04

Table 7.25 (continued)

Years Since 1938 t	Minimum Wage (dollars) M	Annual Growth Factor
36	2.00	1.05
37	2.10	1.10
38	2.30	1.07
40	2.65	1.09
41	2.90	1.07
42	3.10	1.08
43	3.35	1.01
52	3.80	1.12
53	4.25	1.02
58	4.75	1.08
59	5.15	

Since the growth factors are nearly constant—ranging between 1.01 and 1.20—we use exponential regression to determine the exponential model of best fit: $E(t) = 0.33(1.05^t)$. Its graph is shown (in red) in Figure 7.33, along with the quadratic model (in blue). Like the quadratic model, the exponential model appears to fit the trend in the data relatively well over the domain from 1938 to 1998.

To determine which of the two functions is the best choice, we need to consider which one will give us the most reasonable estimate for the minimum wage in years after 1998. To find out, we expand the graphs to year 75 (or 2013) as shown in Figure 7.34.

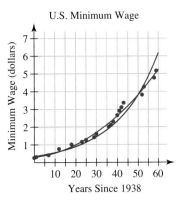

Figure 7.33   Figure 7.34

Now we can see that the exponential function departs from the trend in the data quite markedly while the quadratic function continues to model the trend in the data reasonably well. Therefore, we choose the quadratic function, $Q(t) = 0.0012t^2 + 0.01t + 0.29$ as the most appropriate model.

b. To predict the minimum wage in year 70 (or 2008), we use the model from part (a) to evaluate $Q(70)$.

$$Q(70) = 0.0012(70)^2 + 0.01(70) + 0.29$$
$$= 6.87$$

We estimate the minimum wage in 2008 (year 70) will be $6.87.

The following strategies are useful when you are given a table of data to model.

> **MODEL SELECTION STRATEGIES TO USE WHEN GIVEN A TABLE OF DATA**
>
> 1. Consider the first and second differences for linear and quadratic functions, respectively. Also consider the change factor between outputs for exponential functions.
> 2. Create a scatter plot of the data.
> 3. Determine if the scatter plot looks like one or more of the standard mathematical functions: linear, quadratic, cubic, quartic, exponential, logarithmic, logistic.
> 4. Find a mathematical model for each function type selected in Step 3.
> 5. Use all available information to anticipate the expected behavior of the data outside of the data set. Eliminate models that do not exhibit the expected behavior. (Sometimes it is expedient to switch the order of Steps 4 and 5.)
> 6. Choose the simplest model from among the models that meet the criteria.

In accomplishing Step 3, it is helpful to recognize key graphical features exhibited by the data set, especially the concavity of the scatter plot. Table 7.26 summarizes these features.

Table 7.26

Function Type	Model Equation	Key Graphical Features
Linear	$y = mx + b$	Line
Quadratic	$y = ax^2 + bx + c$	Concave up everywhere or concave down everywhere
Cubic	$y = ax^3 + bx^2 + cx + d$	Changes concavity exactly once, no horizontal asymptotes
Quartic	$y = ax^4 + bx^3 + cx^2 + dx + f$	Changes concavity zero or two times, no horizontal asymptotes
Exponential	$y = ab^x$ with $a > 0$	Concave up, horizontal asymptote at $y = 0$
Logarithmic	$y = a + b \ln(x)$ with $b > 0$	Concave down, vertical asymptote at $x = 0$
Logistic	$y = \dfrac{c}{1 + ae^{-bx}} + k$	Changes concavity exactly once, horizontal asymptotes at $y = k$ and $y = c + k$

SECTION 7.5 ■ CHOOSING A MATHEMATICAL MODEL

EXAMPLE 2 ■ Choosing a Model from a Table of Values

As shown in Table 7.27, the average hotel/motel room rate increased between 1990 and 1999.

Table 7.27

Years Since 1990 t	Room Rate (dollars) R	Years Since 1990 t	Room Rate (dollars) R
0	57.96	5	66.65
1	58.08	6	70.93
2	58.91	7	75.31
3	60.53	8	78.62
4	62.86	9	81.33

Source: *Statistical Abstract of the United States, 2007*, Table 1264

a. Choose a mathematical model for the data.
b. Forecast the average hotel/motel room rate for 2004.
c. Compare your estimated room rate for 2004 with the actual amount of $86.24. What does this tell you about the model you chose?

Solution

a. We first draw the scatter plot shown in Figure 7.35. The graph is initially concave up but changes concavity near $t = 6$. Since the graph changes concavity exactly once, a cubic or a logistic model may fit the data.

Figure 7.35

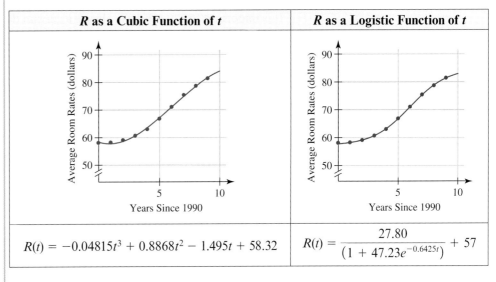

R as a Cubic Function of t	R as a Logistic Function of t
$R(t) = -0.04815t^3 + 0.8868t^2 - 1.495t + 58.32$	$R(t) = \dfrac{27.80}{1 + 47.23e^{-0.6425t}} + 57$

Both the cubic and logistic models appear to fit the data extremely well. To determine which model will be best for forecasting R in 2004 ($t = 14$), we extend the domain for each graph to include the interval $-1 \le t \le 15$.

470 CHAPTER 7 ■ MODELING WITH OTHER TYPES OF FUNCTIONS

R as a Cubic Function of t | **R as a Logistic Function of t**

Now we see that the cubic model decreases at about 2002 ($t = 12$) and continues to decrease. Since room rates have increased every year since 1990, it is unlikely the room rate will drop at any point in the future. On the other hand, the logistic model levels off, which seems to better model room rates.

b. Evaluating the logistic model at $R(14)$, we forecast the 2004 room rate to be $84.64.

c. When we compare the projected average room rate of $84.64 to the actual cost in 2004 of $86.24, we see the logistic model is quite accurate and our choice was correct. However, we need to exercise caution in extrapolating much farther because it is doubtful the average room rate will remain constant.

It is important to note there is always a certain degree of uncertainty when choosing a model from several that seem to fit the data. Two different people may select two different models as their "best" model. For this reason, it is important to always explain the reasoning behind a particular selection.

EXAMPLE 3 ■ Choosing a Model from a Table of Values

As shown in Table 7.28, the average price of a movie ticket increased between 1975 and 2005. Find a mathematical model for the data and forecast the average price of a movie ticket in 2010.

Table 7.28

Years Since 1975 t	Movie Ticket Price (dollars) P
0	2.05
5	2.69
10	3.55
15	4.23
20	4.35
25	5.39
30	6.40

Source: www.boxofficemojo.com

Solution We first draw the scatter plot shown in Figure 7.36. The first four data points appear to be somewhat linear so our initial impression is that a linear model may fit the data well. However, the fifth data point is not aligned with the first four, so we know the data set is not perfectly linear. However, a linear model may still fit the data fairly well.

We could also conclude that the scatter plot is concave down on the interval $0 \leq t \leq 15$ and concave up on $15 \leq t \leq 30$. Since the scatter plot appears to change concavity once and does not have any horizontal asymptotes, a cubic model may work well.

Figure 7.36

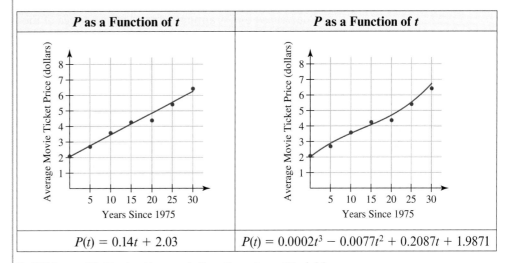

P as a Function of t	P as a Function of t
$P(t) = 0.14t + 2.03$	$P(t) = 0.0002t^3 - 0.0077t^2 + 0.2087t + 1.9871$

In 2010, $t = 35$. Evaluating each function at $t = 35$ yields

$$P(35) = 0.14(35) + 2.03$$
$$\approx \$6.93$$
$$P(35) = 0.0002(35)^3 - 0.0077(35)^2 + 0.2087(35) + 1.9871$$
$$\approx \$8.43$$

The linear model predicts a ticket price of $6.93, and the cubic model predicts a price of $8.43.

It is difficult to know which of these estimates of the average price of a movie ticket is the most accurate since both models seem to fit the data equally well. Additionally, we have no other information that would lead us to believe one model would be better than the other. Because both models seem to fit the data equally well, we choose the simplest one, which is $P(t) = 0.14t + 2.03$.

It is important to note that in this type of problem we are not trying to "hit" each data point with our model. Rather, we are attempting to capture the overall trend.

Choosing between Multiple Best-Fit Models

When two or more models fit a data set equally well and we have no additional information (such as more current data) that would lead us to believe one model would be better than the other, we choose the simplest function as the model.

Choosing a Mathematical Model from a Verbal Description

Certain verbal descriptions often hint at a particular mathematical model to use. By watching for key phrases, we can narrow the model selection process. Table 7.29 presents some typical phrases, their interpretation, and possible models.

Table 7.29

Choosing a Mathematical Model from a Verbal Description		
Phrase	**Interpretation**	**Possible Model**
Salaries are projected to increase by $800 per year for the next several years.	The graph of the salary function will have a constant rate of change, $m = 800$.	Linear $S(t) = 800t + b$
The company's revenue has decreased by 5% annually for the past 6 years.	Since the annual decrease is a *percentage* of the previous year's revenue, the dollar amount by which revenue decreases annually is decreasing. The revenue graph is decreasing and concave up.	Exponential $R(t) = a(1 - 0.05)^t$ $= a(0.95)^t$
Product sales were initially slow when the product was introduced but sales increased rapidly as the popularity of the product increased. Sales are continuing to increase but not as quickly as before.	The graph of the sales function may have a horizontal asymptote at or near $y = 0$ and a horizontal asymptote slightly above the maximum projected sales amount.	Logistic
Enrollments have been dropping for years. Each year we lose more students than we did the year before.	The graph of the enrollment function is decreasing since enrollments are dropping. It will also be concave down since the rate at which students are dropping continues to increase in magnitude.	Quadratic
Company profits increased rapidly in the early 1990s but leveled off in the late 1990s. In the early 2000s, profits again increased rapidly.	The graph of the profit function increases rapidly, then levels off, then increases rapidly again.	Cubic

EXAMPLE 4 ■ Choosing a Model from a Verbal Description

In its 2001 Annual Report, the Coca-Cola Company reported:

> "Our worldwide unit case volume increased 4 percent in 2001, on top of a 4 percent increase in 2000. The increase in unit case volume reflects consistent performance across certain key operations despite difficult global economic conditions. Our business system sold 17.8 billion unit cases in 2001." (*Source:* Coca-Cola Company 2001 Annual Report, p. 46)

Find a mathematical model for the unit case volume of the Coca-Cola Company.

Solution Since the unit case volume is increasing at a constant percentage rate (4%), an exponential model should fit the data well. We have

$$V(t) = ab^t, \text{ where } t \text{ is the number of years since the end of 2001}$$

The report says that the initial number of unit cases sold was 17.8 billion, so

$$V(t) = 17.8(1 + 0.04)^t \text{ billion unit cases}$$
$$= 17.8(1.04)^t \text{ billion unit cases}$$

The mathematical model for the unit case volume is $V(t) = 17.8(1.04)^t$.

Sometimes a data set cannot be effectively modeled by any of the aforementioned functions. In these cases, we look to see if we can model the data with a piecewise function.

EXAMPLE 5 ■ Choosing a Piecewise Function Model

From 1981 to 1995, the annual number of adult and adolescent AIDS deaths in the United States increased dramatically after an initial outbreak of the disease. However, from 1995 to 2001 the annual death rate plummeted as shown in Table 7.30.

Table 7.30

Years Since 1981 t	Number of Deaths during Year D	Years Since 1981 t	Number of Deaths during Year D
0	122	11	41,424
1	453	12	45,187
2	1481	13	50,071
3	3474	14	50,876
4	6877	15	37,646
5	12,016	16	21,630
6	16,194	17	18,028
7	20,922	18	16,648
8	27,680	19	14,433
9	31,436	20	8963
10	36,708		

Source: HIV/AIDS Surveillance Report; Dec. 2001; Centers for Disease Control and Prevention, p. 30

a. Find the mathematical model that best models the data.
b. Use the model to forecast the number of adult and adolescent AIDS deaths in 2004.
c. Then identify factors you think may have led to the dramatic decline in AIDS deaths after 1995.

Solution

a. We first draw the scatter plot shown in Figure 7.37. The data set appears to exhibit logistic behavior up until 1995. After 1995, the graph appears to display cubic behavior. We will use a piecewise function to model the data set. We use a graphing calculator to determine each of the model pieces by using logistic and cubic regression. (The data point associated with $t = 14$ was

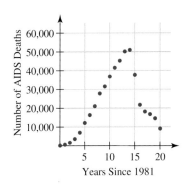

Figure 7.37

used in both pieces of the model.) We obtain the following function and the graph in Figure 7.38.

$$P(t) = \begin{cases} \dfrac{53{,}955}{1 + 38.834e^{-0.45127t}} & \text{if } 0 \le t \le 14 \\ -381.06t^3 + 20{,}770t^2 - 379{,}469t + 2{,}339{,}211 & \text{if } t > 14 \end{cases}$$

Figure 7.38

Our piecewise model appears to fit the data set very well.

b. To forecast the number of AIDS deaths in 2004 ($t = 23$), we use the second part of the piecewise function since $23 > 14$.

$$P(23) = -381.06(23)^3 + 20{,}770(23)^2 - 379{,}469(23) + 2{,}339{,}211$$
$$= -37{,}603 \text{ adult and adolescent AIDS deaths}$$

Despite the fact the model fit the data well, using the model to forecast the 2004 mortality rate yields an unreasonable result: It is impossible to have a negative number of deaths. Even from the data set, we can only estimate that the number of AIDS deaths in years beyond 2001 will be somewhere between 0 and 8963 (the 2001 figure).

c. Public education into the causes of the disease and advances in medical interventions may have been factors that led to the dramatic decline in AIDS deaths.

Some data sets cannot be effectively modeled by any of the standard mathematical functions or even a piecewise function. We look at this in the next example.

EXAMPLE 6 ■ Analyzing Data Not Easily Modeled with a Common Function

The number of firearms detected during airport passenger screening is shown in Table 7.31. After the terrorist attacks on September 11, 2001, airline screening became much more thorough. The increased security initially had a deterrent effect. According to the Bureau of Transportation Statistics, there were 1071 firearms detected in 2001 and 650 firearms detected in 2004. Nevertheless, in 2005 there were 2217 firearms detected, indicating the deterrent effect may have worn off or that detection methods may have improved. (*Source:* www.bts.gov)

Estimate the number of firearms detected by airport screeners in 2008.

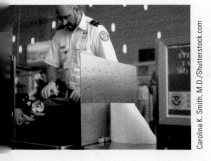

Table 7.31

Years Since 1980 t	Firearms Detected F	Years Since 1980 t	Firearms Detected F
0	1914	14	2994
5	2913	15	2390
10	2549	16	2155
11	1644	17	2067
12	2608	18	1515
13	2798	19	1552

Source: Statistical Abstract of the United States, 2001, Table 1062

Solution The scatter plot in Figure 7.39 does not resemble any of the standard mathematical functions. It also does not appear that a piecewise function will fit the data well. Based on the table data and additional data, the number of firearms detected in a given year appears to be somewhat random, ranging from about 650 firearms to roughly 3000 firearms. Using this additional information, we can estimate that the number of firearms detected in 2008 was somewhere between 650 and 3000.

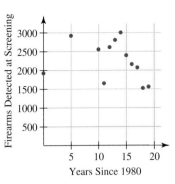

Figure 7.39

SUMMARY

In this section you learned strategies and techniques for selecting mathematical models. You learned to select the "best" function to model based on the rate of change of the data, the scatter plot, and the reasonableness of using the data to extrapolate. You also discovered that sometimes more than one function can be used to model the same data set, and sometimes no function works.

7.5 EXERCISES

SKILLS AND CONCEPTS

In Exercises 1–8, use a graph of the function to determine over what interval(s) the function is increasing or decreasing. Determine over what interval(s) the function is concave up or concave down. Identify the coordinate of the y-intercept, if there is one.

1. $y = 2(0.75)^x$
2. $y = 5 - 15x$
3. $y = 3x^2 - 2x + 5$
4. $y = \sqrt{x} + 7$
5. $y = \dfrac{45}{1 + 7e^{.05x}}$
6. $y = \dfrac{4}{1 - 5x}$
7. $y = -0.3(2.8)^x$
8. $y = -0.5x^2 - 2x$

9. **Candy Bar Prices** Candy bars currently cost $0.60 each. The price of a candy bar is expected to increase by 3% per year in the future. Find a mathematical model for the price of a candy bar.

10. **Calculator Prices** A calculator currently costs $87. The price of the calculator is expected to increase by $3 per year. Find a mathematical model for the price of the calculator.

11. **Mortality Rates** There are presently 95 members of a high school graduating class who are still living. The number of surviving class members is decreasing at a rate of 4% per year. Find a mathematical model for the number of surviving class members.

12. **Product Sales Growth** An electronics company is introducing a new product next year. The company anticipates sales will be initially slow but will increase rapidly once people become aware of the product. It anticipates that monthly sales will start to level off in 18 months at about $200,000. It predicts sales for the first two months will be $12,000 and $19,000, respectively. Develop a mathematical model to forecast monthly product sales.

13. **Club Membership** A business club is concerned about its decreasing number of members. Two years ago, it had 200 members. Last year it had 165 members and this year it has 110 members. If nothing changes, the club expects to lose even more members next year than it lost this year. Develop a mathematical model for the club membership.

14. **Baseball Card Values** A baseball card increases in value according to the function, $b(t) = \dfrac{5}{2}t + 100$, where b is the value of the card (in dollars) and t is the time (in years) since the card was purchased (that is, $t = 1$ represents one year after the card was purchased). Which of the following describe(s) what $\dfrac{5}{2}$ conveys about the situation?

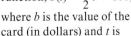

I. The card's value increases $5 every 2 years.

II. The card's value increases by a factor of $\dfrac{5}{2}$ every year.

III. The card's value increases by $\frac{5}{2}$ dollars per year.

A. I only
B. II only
C. III only
D. I and III only
E. I, II, and III

15. Create four data sets with at least eight data points in each: the first linear, the second quadratic, the third cubic, and the fourth exponential. Show why each data set represents the type of function it does by using successive differences or consecutive ratios.

16. Below is a portion of the graph of f. Is $f(x)$ exponential, quadratic, cubic, or is there not enough information to tell? Explain.

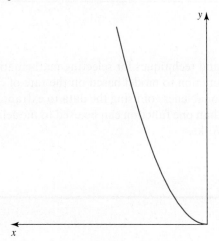

SHOW YOU KNOW

17. What makes a function the "best" to model a data set?
18. How do you use the "best" function to extrapolate effectively?
19. If a cubic model and a linear model both fit a data set with a coefficient of determination of $r^2 = 0.999$, which model should be used to interpolate? Justify your conclusion.

MAKE IT REAL

20. **Population Growth** The town of Queen Creek, Arizona, was founded in 1989. In 1990, there were 2667 people living in the town. The town grew rapidly in the 1990s, due in large part to new home construction in the area. There were 4316 people living in the town in 2000 and 4940 people in 2001. The Arizona Department of Commerce estimated the 2002 population of Queen Creek at 5555 people.

    Find an exponential function to model the population of Queen Creek. Using this model, what is the maximum projected population of the city? (*Hint:* First align the data.)

    The Arizona Department of Commerce estimated the 2005 population of Queen Creek at 15,890. How accurately did the model predict the 2005 population? Explain.

21. **Housing Prices** On March 12, 2004, a builder priced a new home in Queen Creek, Arizona, at $143.9 thousand. The builder's sales rep told the author that the price for that home style would increase on March 16, March 30, and April 13, to $148.9 thousand, $151.9 thousand, and $153.9 thousand, respectively. (*Source:* Fulton Homes) Find a mathematical model for the price of the new home style.

22. **Federal Tax Rates** Federal income tax rates depend on the amount of taxable income received. In 2003, federal income taxes were calculated as follows. For single filers, the first $7000 earned was taxed at 10%. The next $21,400 earned was taxed at 15%. The next $40,400 earned was taxed at 25%.

    For example, the tax of a single filer who earned $25,000 would be calculated as follows:

    10% tax on the first $7000:
    $7000 × 0.10 = $700

    Amount to be taxed at a higher rate:
    $25,000 − $7000 = $18,000

    15% tax on the next $18,000:
    $18,000 × 0.15 = $2700

    The filer's total tax is

    $700 + $2700 = $3400

    Find a mathematical model for income tax as a function of taxable income for single filers.

23. **Measurement of Bacterial Growth** The study of how bacteria grow requires bacterial enumeration (cell counting). If bacteria are held in a closed system like a test tube, after a period of time bacterial growth can occur. Initially this happens when a wall is created that divides the one bacterium into two "daughter cells."

    Two graphs are shown. The first graph shows the number of bacteria in the population over time. Microbiologists typically represent the bacterial growth shown in the first graph with the second graph, which is called the *bacterial growth curve*. On the bacterial growth curve, the vertical axis uses a logarithmic scale. Notice that each tick mark on the vertical axis represents a power of 2. Moving from one tick mark to the next represents a doubling of the population.

    Describe each of the four phases of bacterial growth labeled on the bacterial growth curve (lag phase, exponential phase, stationary phase, and death phase) in terms of the number of viable (living) bacteria and rate of change.

In Exercises 24–25, choose an appropriate model for the graphs. Then explain what the rate of change tells about the relationship between the independent and dependent variables.

24. Confirmed Cases of Chicken Pox

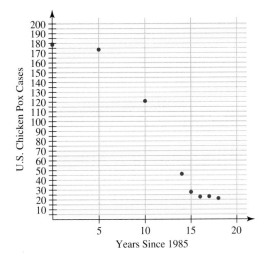

Source: Statistical Abstract of the United States, 2006, Table 177

25. Number of Hospitals

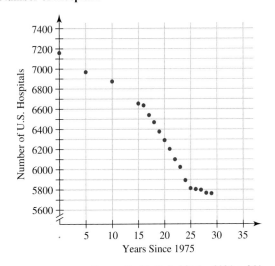

Source: Health, United States, 2006, p. 364

In Exercises 26–36, find the equation of the mathematical model that will most accurately forecast the indicated result (if possible). Then use the model to find the value. Justify your conclusions.

26. Health Care Spending

Years Since 1990 t	Per Capita Spending on Physician and Clinical Services (dollars) p
0	619
5	813
8	914
9	954
10	1010
11	1085
12	1162
13	1249

Source: Statistical Abstract of the United States, 2006, Table 121

Forecast spending on physician and clinical services for 2008.

27. Chapter 11 Bankruptcies

Years Since 1998 t	Nonbusiness Chapter 11 Bankruptcies B
0	981
1	731
2	722
3	745
4	894
5	966
6	935

Source: Statistical Abstract of the United States, 2006, Table 749

Forecast the number of nonbusiness Chapter 11 bankruptcies in 2010.

28. Community College Education Costs

Years Since 1997–1998 t	Maricopa Community College Resident Cost per Credit (dollars) C
0	37
1	38
2	40
3	41
4	43
5	46
6	51
7	55
8	60
9	65
10	65

Source: www.dist.maricopa.edu

Forecast the cost per credit for county resident students attending college in the district in 2010–2011.

29. DVD Players Sold

Average Price of DVD (dollars) p	DVD Players Sold (thousands) D
489.97	349
390.18	1079
270.00	4072
201.55	8499
165.00	12,707
142.00	17,090
123.00	21,994
108.60	19,990

Source: Consumer Electronics Association, www.ce.org

Forecast the DVD players sold when the average price of a DVD player is $100.

30. Prescription Drug Prices

Years Since 1995 t	Average Brand-Name Drug Price (dollars) D
0	40.22
2	49.55
3	53.51
4	60.66
5	65.29
6	69.75
7	77.49
8	85.57
9	95.86

Source: Statistical Abstract of the United States, 2006, Table 126

Forecast the average prescription drug price in 2010.

31. Community College Education Costs

Years Since 1984–1985 t	Washington State Community Colleges Tuition and Fees (dollars) F
0	581
1	699
2	699
3	759
4	780
5	822
6	867
7	945
8	999
9	1125
10	1296

(continued)

Years Since 1984–1985 t	Washington State Community Colleges Tuition and Fees (dollars) F
11	1350
12	1401
13	1458
14	1515
15	1584
16	1641
17	1743

Source: Washington State Higher Education Coordinating Board, Higher Education Statistics, September 2001

Forecast the annual tuition and fees at a Washington state community college in 2008.

32. Golf Course Facilities

Years Since 1980 t	Golf Facilities G	Years Since 1980 t	Golf Facilities G
0	12,005	14	13,683
5	12,346	15	14,074
6	12,384	16	14,341
7	12,407	17	14,602
8	12,582	18	14,900
9	12,658	19	15,195
10	12,846	20	15,489
11	13,004	21	15,689
12	13,210	22	15,827
13	13,439	23	15,899

Source: Statistical Abstract of the United States, 2004–2005, Table 1240

Forecast the number of golf facilities in the United States in 2008.

33. Per Capita Personal Income

Years Since 1993 t	Per Capita Personal Income—Florida (dollars) F
0	21,320
1	21,905
2	22,942
3	23,909
4	24,869
5	26,161
6	26,593
7	27,764

Source: Bureau of Economic Analysis, www.bea.gov

Forecast the per capita personal income in Florida in 2004.

34. Air Carrier Accidents

Year t	Air Carrier Accidents A	Year t	Air Carrier Accidents A
1992	18	1997	49
1993	23	1998	50
1994	23	1999	52
1995	36	2000	54
1996	37		

Source: Statistical Abstract of the United States, 2001, Table 1063

Forecast the number of air carrier accidents in 2002.

35. Chipotle Mexican Grill Revenue

Franchise Royalties and Fees ($1000s) f	Restaurant Sales ($1000s) s
267	131,331
753	203,892
1493	314,027
2142	468,579
2618	625,077

Source: Chipotle Mexican Grill, Inc, 2005 Annual Report

Forecast the amount of restaurant sales (in thousands) when the franchise royalties and fees reach $3,000,000.

36. Yogurt Production

Years Since 1997 t	Yogurt (million pounds) y	Years Since 1997 t	Yogurt (million pounds) y
0	1574	5	2311
1	1639	6	2507
2	1717	7	2707
3	1837	8	2990
4	2003		

Source: Statistical Abstract of the United States, 2007, Table 846

Forecast the amount of yogurt produced in 2010.

In Exercises 37–43, find the model that best fits the data. Then use the model to answer the given questions.

37. Magazine Advertising

Years Since 1990 t	Magazine Advertising Expenditures ($ millions) M
0	6803
1	6524
2	7000
3	7357
4	7916
5	8580
6	9010

(continued)

Years Since 1990 t	Magazine Advertising Expenditures ($ millions) M
7	9821
8	10,518
9	11,433
10	12,348

Source: Statistical Abstract of the United States, 2001, Table 1272

According to the model, how much money was spent on magazine advertising in 2002?

38. Cable TV Advertising

Years Since 1990 t	Cable TV Advertising Expenditures ($ millions)
0	2457
1	2728
2	3201
3	3678
4	4302
5	5108
6	6438
7	7237
8	8301
9	10,429
10	12,364

Source: Statistical Abstract of the United States, 2001, Table 1272

According to the model, how much money was spent on cable television advertising in 2002?

39. Advertising
Using the models from Exercises 37 and 38, determine in what year cable television advertising expenditures are expected to exceed magazine advertising expenditures.

40. Radio Advertising

Years Since 1990 t	Radio Advertising Expenditures ($ millions) R
0	8726
1	8476
2	8654
3	9457
4	10,529
5	11,338
6	12,269
7	13,491
8	15,073
9	17,215
10	19,585

Source: Statistical Abstract of the United States, 2001, Table 1272

According to the model, when will radio advertising exceed $25 billion?

41. Yellow Pages Advertising

Years Since 1990 t	Yellow Pages Advertising Expenditures ($ millions) Y
0	8926
1	9182
2	9320
3	9517
4	9825
5	10,236
6	10,849
7	11,423
8	11,990
9	12,652
10	13,367

Source: Statistical Abstract of the United States, 2001, Table 1272

According to the model, when will yellow page advertising exceed $15 billion?

42. Average Hourly Earnings in Manufacturing Industries

Years Since 1980 t	Michigan Average Hourly Earnings (dollars per hour) M
0	9.52
1	10.53
2	11.18
3	11.62
4	12.18
5	12.64
6	12.80
7	12.97
8	13.31
9	13.51
10	13.86
11	14.52
12	14.81
13	15.36
14	16.13
15	16.31
16	16.67
17	17.18
18	17.61
19	18.38
20	19.20

Source: Statistical Abstract of the United States, 2001, Table 978

According to the model, what will the average hourly wage be in Michigan manufacturing industries in 2003?

43. Average Hourly Earnings in Manufacturing Industries

Years Since 1980 t	Florida Average Hourly Earnings (dollars per hour) F
0	5.98
1	6.53
2	7.02
3	7.33
4	7.62
5	7.86
6	8.02
7	8.16
8	8.39
9	8.67
10	8.98
11	9.30
12	9.59
13	9.76
14	9.97
15	10.18
16	10.55
17	10.95
18	11.43
19	11.83
20	12.28

Source: Statistical Abstract of the United States, 2001, Table 978

According to the model, what will the average hourly wage be in Florida manufacturing industries in 2003?

44. Hourly Earnings Based on the wage data in Exercises 42 and 43, do you think it would be better to start up a manufacturing business in Florida or Michigan? Justify your answer and explain what other issues might impact your decision.

45. Engine Torque Torque is the power a car engine generates. The horsepower reflects the amount of work that the engine is doing based on the gearing and revolutions per minute. Therefore, the torque of an engine depends on the speed of the engine. The table gives the hypothetical torque for different V8 engine speeds (in thousands of revolutions per minute).

Engine Speed (rpm in 1000s) s	Engine Torque (ft-lb) T
1	180
2	390
3	510
4	545

(continued)

Engine Speed (rpm in 1000s) s	Engine Torque (ft-lb) T
5	517
6	410
7	360

Source: **craig.backfire.ca**

a. Draw a scatter plot of the data.

b. Is a linear or quadratic function more appropriate to model this data? Explain.

c. Use regression to find the linear or quadratic model that best fits the data.

d. Use the model in part (c) to find $T(5.5)$. Explain what the solution means in its real-world context.

e. Use the model in part (c) to estimate $T(s) = 100$. Explain what the solution means in its real-world context.

f. Find the coordinates of the vertex (turning point) of the graph of the quadratic function in part (c). Explain what this point means in this context.

g. Find the average rate of change from $s = 2$ to $s = 4$. Explain what the solution means in this context.

h. Find the rate of change at $s = 4$. Explain what the solution means in this context.

■ STRETCH YOUR MIND

Exercises 46–54 are intended to challenge your understanding of mathematical modeling.

46. The graph of a mathematical model passes through all of the points of a data set. A classmate claims that the model is a perfect forecaster of future results. How would you respond?

47. A scatter plot is concave up and increasing on the interval $0 \leq x \leq 5$, concave down and increasing on the interval $5 \leq x \leq 8$, and decreasing at a constant rate on the interval $8 \leq x \leq 15$. Describe two different mathematical models that may fit the data set.

48. Describe how a business owner can benefit from mathematical modeling, despite the imprecision of a model's results.

49. Daily fluctuations in the stock market make the share price of a stock very difficult to model. What approach would you take if you wanted to model the long-term performance of a particular stock?

50. You are asked by your boss to model the data shown in the following scatter plot. How would you respond?

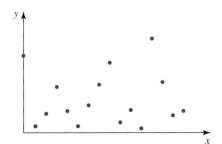

51. **Men's Health** Medical researchers have found that there is a relationship between a person's blood pressure and weight. In males 35 years of age, for every 5-pound increase in the person's weight there is generally an increase in the systolic blood pressure of 2 millimeters of mercury (mmHg). Moreover, for a male 35 years of age and 190 pounds the preferred systolic blood pressure is 125. (*Source:* **Chandler-Gilbert Community College Wellness program, www.cgc.edu**)

a. Construct a formula for a 35-year-old male's systolic blood pressure as a function of weight.

b. What is the slope of the equation? What is the meaning of the slope in terms of the real-world context?

c. What is the vertical intercept of the equation? What is the meaning of the vertical intercept in this context?

52. **Cigarette Consumption** The data for the per capita consumption of cigarettes over the last century is shown in the table.

Years Since 1900 y	Per Capita Cigarette Consumption c
0	54
10	151
20	665
30	1485
40	1976
50	3552
60	4171
70	3985
80	3851
90	2827
100	2092

Source: Tobacco Outlook Report, **Economic Research Service, U.S. Dept. of Agriculture**

a. Determine what type of mathematical model may best fit the data. Choose from linear, quadratic, or exponential. Explain why you think the function you chose is the *best* choice. Be sure to justify your choice by showing differences and ratios, the scatter plot, and what would seem most reasonable for the future. Defend your choice and refute the other two possibilities.

b. Use regression to find the function that best fits the data and name the function $c(y)$.

c. Evaluate and interpret the real-world meaning of $c(109)$.

d. Solve and interpret the real-world meaning of $c(y) = 100$.

e. Using the function $c(y)$ from part (b), approximate the year in which per capita cigarette consumption will reach its maximum.

f. Find the average rate of change from 1940 to 1950 and explain what your answer means in the context of this problem. Considering this decade, why would your answer seem reasonable?

g. Find the average rate of change from 1980 to 1990 and explain what your answer means in the context of this problem. Considering this decade, why would your answer seem reasonable?

h. Estimate the instantaneous rate of change for the year 1995 using the model you chose. Interpret the meaning of your answer in the context of this problem.

53. **Population Explosion** Phoenix, Arizona is located in Maricopa County. Maricopa County's population is expected to nearly double in less than four decades, which could stretch the limits of currently projected water supplies and all other municipal services. The table gives the population of Maricopa County every 10 years from 1900 through 2000.

Years Since 1900 t	Population of Maricopa County P
0	20,457
10	34,488
20	89,567
30	150,970
40	186,193
50	331,770
60	663,510
70	971,228
80	1,509,260
90	2,122,100
100	3,000,000

Source: Maricopa Association of Governments and the U.S. Bureau of Census

a. What kind of function best models the data? Choose from linear, quadratic, or exponential. Explain why the function you chose is the best choice. (Note: None of these three functions fits really well, but you still need to choose the "best.") Be sure to justify your choice using differences and ratios, a scatter plot, and what would seem most reasonable for the future. Defend your choice and refute the other two possibilities.

b. Use regression to find the function that best fits the data and name the function $m(y)$.

c. Evaluate and interpret the real-world meaning of $m(150)$.

d. Solve and interpret the real-world meaning of $m(y) = 4,000,000$.

e. Evaluate $\dfrac{m(88) - m(52)}{88 - 52}$ and interpret what it means in this context.

f. Estimate the rate of change for the year 1995 using the model you chose. Interpret the meaning of your answer in the context of this problem.

g. How far into the future do you feel the model will be effective in predicting the future population? Give the reasons you chose the year you did.

54. **Major League Baseball Players' Strike** During the 2001 Major League Baseball season while the players were contemplating going on strike, one of the authors wrote the following letter to the editor of the Arizona Republic newspaper. Read the letter and then answer the question.

As a mathematics professor at a community college in the Valley, I couldn't help but do some mathematical analysis of the Major League Baseball Players' salaries to determine for myself what the "predicament" is that the players find themselves in that may "force" them to strike. I found the players' **average** *salaries for the last 20 years on the USA Today website. If the trend continues as it has over this time period predicted* **average** *players' salaries for future years are:*

2003	$2,883,458
2004	$3,408,631
2005	$4,055,009
2010	$9,854,997
2015	$22,307,696

Oh, now I know what to tell my 11-year-old son who asked me why the players want to strike. They really do need more money! I would be worried about my income too with low wages like these. Maybe my son's school can start a penny drive for our beloved Diamondbacks!

Signed,
A citizen concerned about our precious baseball players

Year	t	Average Players' Salary
1982	0	241,497
1983	1	289,194
1984	2	329,408
1985	3	371,571
1986	4	412,520
1987	5	412,454
1988	6	438,729
1989	7	497,254
1990	8	597,537
1991	9	851,492
1992	10	1,028,667
1993	11	1,076,089
1994	12	1,168,263
1995 (player's strike year)	13	1,110,766
1996	14	1,119,981
1997	15	1,336,609
1998	16	1,398,831
1999	17	1,611,166
2000	18	1,895,630
2001	19	2,264,403
2002	20	2,383,235

Determine what mathematical model the author used to estimate the average player's salary in the years 2010 and 2015.

CHAPTER 7 Study Sheet

As a result of your work in this chapter, you should be able to answer the following questions, which are focused on the "big ideas" of this chapter.

SECTION 7.1
1. How do you compute the sum, difference, product, and quotient of functions?
2. How do you determine the sum or difference of functions from a graph or table?
3. What is the practical and theoretical domain for a combination of functions?

SECTION 7.2
4. What is a piecewise function?
5. How do you create the formula for a piecewise function given data in tabular or graphical form?
6. How do you evaluate a piecewise function represented in graphical, tabular, or symbolic form?
7. What does it mean for a function to be continuous or discontinuous?
8. What are the different types of discontinuities?
9. What are the advantages and disadvantages of choosing to model data with a piecewise function?
10. Why are some data sets best modeled by piecewise functions?

SECTION 7.3
11. What is a composition of functions?
12. How do you create composite functions represented in tabular, graphical, and formula form?
13. What is function composition notation and how is it interpreted?
14. How do you evaluate a function composition represented in graphical, tabular, or symbolic form?
15. How do you decompose function compositions?

SECTION 7.4
16. How are exponential and logistic functions similar and how are they different?
17. How do you graph logistic functions represented by a table or formula?
18. How can a logistic function be described in terms of a rate of change?
19. What is important to keep in mind when modeling data sets with a logistic function?

SECTION 7.5
20. What makes a function the "best" to model a data set?
21. How do you use the "best" function to extrapolate effectively?

REVIEW EXERCISES

■ SECTION 7.1 ■

In Exercises 1–2, use the given graph or table of functions f and g to evaluate each given function. If it is not possible to evaluate the function, explain why not.

1.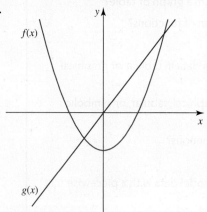

   a. $(f + g)(0)$
   b. $(f - g)(0)$
   c. $\left(\dfrac{f}{g}\right)(0)$
   d. $(f \cdot g)(0)$

2.
x	−8	−4	0	4	8
f(x)	−4	0	4	8	12
g(x)	−4	−2	0	2	4

   a. $(f + g)(0)$
   b. $(f - g)(0)$
   c. $\left(\dfrac{f}{g}\right)(0)$
   d. $(f \cdot g)(0)$

In Exercises 3–6, combine the functions $f(x)$ and $g(x)$ symbolically as indicated. Then state the domain and range of the combination.

3. $f(x) = -2x + 3 \qquad g(x) = x - 4$
   a. $(f + g)(x)$
   b. $(f - g)(x)$

4. $f(x) = x^2 - 1 \qquad g(x) = x - 3$
   a. $(f - g)(x)$
   b. $\left(\dfrac{f}{g}\right)(x)$

5. $f(x) = e^x - 1 \qquad g(x) = 2x$
   a. $(f \cdot g)(x)$
   b. $\left(\dfrac{g}{f}\right)(x)$

6. $f(x) = \ln(x) \qquad g(x) = 2x - 4$
   a. $(f + g)(x)$
   b. $(f \cdot g)(x)$

7. Use the given graphs of the functions f and g to sketch a graph of each combination of functions: $(f + g)(x)$, $(f - g)(x)$, $(f \cdot g)(x)$, and $\left(\dfrac{f}{g}\right)(x)$.

   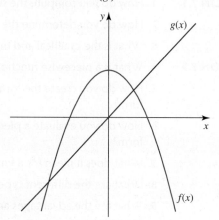

8. **Milk Consumption** The graph shows the amount of milk consumed per person in the United States in selected years.

   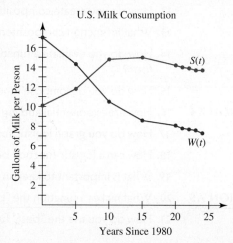

   Source: Statistical Abstract of the United States, 2007, Table 201

   The per person consumption of whole milk as a function of years since 1980 is given by $W(t)$, in gallons. The per person consumption of reduced fat, light, and skim milk as a function of years since 1980 is given by $S(t)$, in gallons.

   a. Interpret the meaning of $(W - S)(t)$.
   b. Estimate and interpret $(W - S)(6)$.
   c. Sketch a graph of $(W - S)(t)$.
   d. Interpret the meaning of $(W + S)(t)$.
   e. Estimate and interpret $(W + S)(6)$.
   f. Sketch a graph of $(W + S)(t)$.

9. **Cheddar Cheese Consumption** The amount of cheddar cheese consumed in the United States, in pounds per person, can be modeled by

$$C(t) = 0.034t^2 - 0.036t + 3.76$$

(*Source: Statistical Abstract of the United States, 2007,* **Table 202**) The cost per pound of cheddar cheese in the United States can be modeled by

$$P(t) = 0.117t^3 - 0.600t^2 + 0.683t + 9.7$$

(*Source: Statistical Abstract of the United States, 2007,* **Table 712**) In both models, t is the number of years since 2000.

   a. Interpret the meaning of $(C \cdot P)(t)$.
   b. Evaluate and interpret $(C \cdot P)(10)$.
   c. Solve $(C \cdot P)(t) = 50$ and explain what the result means in the context of this situation.

10. **Presidential Elections** The total number of votes cast in presidential elections, in millions, and the total number of people in the United States of voter age, in millions, are presented in the table, where t is the number of years since 1980. (*Source: Statistical Abstract of the United States, 2007,* Tables 385 and 405)

Years Since 1980 t	Votes Cast (millions) $C(t)$	Voter-Age Population (millions) $V(t)$
0	86.5	157.1
4	92.7	170.0
8	91.6	178.1
12	104.6	185.7
16	96.4	193.7
20	105.6	202.6
24	122.3	215.7

   a. Interpret the meaning of $\left(\dfrac{C}{V}\right)(t)$.
   b. Evaluate and interpret $\left(\dfrac{C}{V}\right)(24)$.
   c. Create a scatter plot of $\left(\dfrac{C}{V}\right)(t)$. Is there an apparent trend? Explain.

■ **SECTION 7.2** ■

In Exercises 11–12, graph each piecewise function.

11. $f(x) = \begin{cases} -x & \text{if } x \leq 4 \\ x^2 & \text{if } 4 < x < 11 \\ 3\ln(x) & \text{if } x \geq 11 \end{cases}$

12. $f(x) = \begin{cases} \dfrac{2}{x-4} & \text{if } -3 \leq x < 2 \\ 2 & \text{if } 2 < x < 4 \\ 2^x & \text{if } x \geq 6 \end{cases}$

In Exercises 13–14, write the equation for each piecewise function.

13.

14.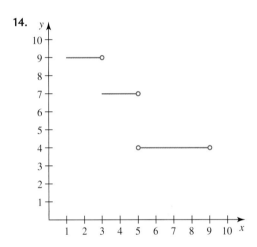

15. **Cesarean Births** The percentages of babies delivered by Cesarean birth are listed in the table for selected years from 1970 to 2005.

Years Since 1970 y	Cesarean Births (percent) C
0	5
5	11
10	17
15	23
20	23
25	21
30	23
35	30

Source: www.census.gov

   a. Create a scatter plot of these data.
   b. Assuming the data from 1970 to 1995 was quadratic and the data from 1995 to present is exponential, write the equation for the piecewise function $C(y)$ that models the data.
   c. Calculate the average rate of change from $y = 0$ to $y = 15$ and explain what this value means in this context.

d. Estimate $C(17)$ and interpret its real-world meaning.

e. Estimate the instantaneous rate of change at $y = 17$ and interpret its real-world meaning.

f. Predict $C(35)$ assuming the trend in the data continues.

16. **Filling a Swimming Pool** The function,

$$V(t) = \begin{cases} 100t & \text{if } 0 \leq t < 20 \\ 2000 & \text{if } 20 \leq t < 23 \\ 1500t - 28{,}000 & \text{if } 23 \leq t < 38 \\ 32{,}000 & \text{if } 40 \leq t \leq 48 \end{cases}$$

models the volume of water, V, (in gallons) in a swimming pool in t hours as it is filled.

a. Graph the function $V(t)$ and explain what information each piece of the function gives about how the pool is filled.

b. Evaluate $V(24)$ and explain what the numerical value represents in the real-world context.

17. **Grading Scales** At Chandler-Gilbert Community College, a professor may assign a student who completes a course a grade of A, B, C, D, or F. An A is worth 4.0 grade points per credit, a B is worth 3.0 grade points per credit, a C is worth 2.0 grade points per credit, a D is worth 1.0 grade point per credit, and an F is worth 0.0 grade points per credit. The table shows a typical grading scale in terms of grade points earned per credit.

Final Course Percentage p	Grade Points per Credit G
$p < 60$	0.0
$60 \leq p < 70$	1.0
$70 \leq p < 80$	2.0
$80 \leq p < 90$	3.0
$90 \leq p < 100$	4.0

a. Sketch a graph of the function $G(p)$ where G is the grade points per credit and p is the final course percentage.

b. Solve $G(p) = 3.0$ for p and explain what the numerical answer(s) mean in the real-world context.

18. **Credit Card Bill** One of the authors received a credit card offer from Capital One® Visa® that offered the following minimum payment terms:

If the minimum balance is less than $10, the full balance must be paid. If the balance is $10 or more, the minimum payment is 3% or $10, whichever is more.
(*Source:* **Capital One**)

a. Create a table of values for the minimum amount due, p, as a function of the unpaid balance, b. Start the table with a $5 unpaid balance and continue to $1000 in increments of $100.

b. Write a piecewise function $p(b)$ for the minimum payment due.

c. Sketch a graph of the function $p(b)$ over the domain $0 \leq b \leq 1000$.

19. **Legoland Ride Policy** The following picture was taken at Legoland in California in 2007 while one of the authors was on vacation.

At the time of the visit, the author was accompanied by children of the following ages and heights.

Child Number	Age (years)	Height (inches)
1	1	33
2	4	42
3	7	49
4	10	59
5	14	65

a. Determine which of the five children could go on the ride and under what conditions.

b. When reading the sign, the author found the phrase "between 34 inches tall and 6 years of age and 48 inches tall" confusing. Describe the ambiguities that are created by this verbiage.

c. Represent the ride restrictions with a piecewise function. (*Hint:* There are two input variables: age and height.)

d. Referring to the piecewise function in part (c), rewrite the language on the sign to be more clear.

■ **SECTION 7.3** ■

In Exercises 20–23, rewrite each pair of functions, $f(g)$ and $g(t)$, as one composite function, $f(g(t))$, if possible. Then evaluate $f(g(2))$.

20. $f(g) = 2g + 3 \qquad g(t) = 3t$

21. $f(g) = 5e^g \qquad g(t) = t^2 + 4$

22. $f(g) = \dfrac{3}{g-1} \qquad g(t) = \dfrac{2}{t}$

23. $f(g) = \sqrt{g} \qquad g(t) = 2t + 3$

In Exercises 24–29, decompose the composite function $h(x) = f(g(x))$ into $f(g)$ and $g(x)$. (Note: There may be more than one possible correct answer.)

24. $h(x) = \sqrt{\ln(x^2 - 2)}$

25. $h(x) = e^{x^2 - 2}$

26. $h(x) = \dfrac{9}{(x+3)^4}$

27. $h(x) = |-x^3 + 4|$

28. $h(x) = (2x+1)^2 - 5$

29. $h(x) = \dfrac{2}{1+\sqrt{x}}$

30. Use the first three columns in the table to fill in the values for $f(g(x))$. (Note: Some function values may be undefined.)

x	$f(x)$	$g(x)$	$f(g(x))$
0	3	1	
1	5	4	
2	4	0	
3	1	2	
4	0	5	
5	2	3	

31. Find $g(g(4))$ for

$$g(x) = \begin{cases} 4 & \text{if } x \le 2 \\ 2x+1 & \text{if } 2 < x < 6 \\ x^2 - 4 & \text{if } x \ge 8 \end{cases}$$

32. **Cigarettes and Heart Disease** The function $D(p) = 14.08p - 53.87$ models the death rate due to heart disease from 1974 to 2003 in deaths per 100,000 people as a function of the percentage of people, p, who smoke. The function $p(t) = -0.52t + 35.7$ models the percentage of people who smoke as a function of the years, t, since 1974. (*Source: Statistical Abstract of the United States, 2006*, **Table 106**) Write a formula for $D(p(t))$ and then evaluate and interpret $D(p(36))$.

33. **Motion Picture Screens and Movie Attendance** The number of motion picture screens can be modeled by

$$S(m) = -0.000917m^3 + 4.21m^2 - 6440m + 3{,}320{,}000$$

screens where m is the movie attendance, in millions of people. Movie attendance may be modeled by

$$m(t) = -30.36t^2 + 153.12t + 1407.29$$

million people where t is the number of years since 2000. (*Source: Statistical Abstract of the United States, 2006*, **Table 1234**) Write a formula for $S(m(t))$ then evaluate and interpret $S(m(9))$.

34. **Gross Pay vs. Net Pay** An employee working as a lifeguard earns $8.40 per hour.

   a. Write a function that gives the employee's gross pay as a function of the number of hours worked. Be sure to identify the meaning of the variables you use.

   b. Gross pay is reduced by Social Security (6.2%) and Medicare (1.45%). Additionally, income tax may be withheld and unemployment insurance may be required. Assuming the deductions total 12% of the employee's gross pay, write the employee's net pay as a function of the gross pay. Be sure to identify the meaning of the variables you use.

   c. Based on the results of part (a) and (b), write an equation for net pay as a function of the number of hours worked.

   d. Starting with the equation in part (c), write an equation for number of hours worked as a function of the net pay. This function is the inverse of the function in part (c). Explain what the input and output variables represent.

35. **Cell Phone Billing** In November 2006, Cingular Wireless advertised the following rate plan on their website.

Nation 1350 w/Rollover Rate Plan Details	
Monthly Cost (for 1350 Anytime minutes)	$79.99
Night & Weekend minutes	Unlimited
Mobile to Mobile minutes	Unlimited
Long Distance	$0.00
Roaming Charges	$0.00
Additional minutes	$0.35/minute

   a. Write a function that will allow you to calculate the monthly cell phone cost as a function of weekday daytime minutes used. Indicate the meaning of the variables you use.

   b. Use the function from part (a) to complete the table. Show all steps leading to your answers.

Minutes Used	Monthly Cost
0	
1000	
1500	
	$364.19

   c. A subscriber to the Nation 1350 plan has observed that when he travels his weekday daytime cell phone usage increases dramatically. When he is at home, he uses roughly 1300 weekday daytime minutes monthly; however, his cell phone usage increases by 100 weekday daytime minutes per day that he travels. (For example, if he travels 2 days in the month, he uses $1300 + 2(100) = 1500$ weekday daytime minutes in the month.)
   Write a function that gives the number of weekday daytime minutes he uses in a month as a function of the number of days he travels in the month. Be sure to state the meaning of the variables you use.

   d. Assuming the Nation 1350 plan subscriber from part (c) travels at least one day a month, write a function that gives his monthly cell phone bill as a function of the number of days he travels in the month. (*Hint:* Refer to your results for parts (a) and (c).)

   e. What will be his monthly cell phone bill if he travels five days in the month?

   f. Explain how this situation of determining the monthly cost of this cell phone plan (refer back to parts (d) and (e)) is an example of the composition of two functions.

■ **SECTION 7.4** ■

36. Create a table of values for a logistic function that is increasing and has a limiting value at 13.

37. Create a table of values for a logistic function that is decreasing and has a limiting value at 38.

38. If $f(x) = \dfrac{15}{1 + 23e^{-1.8x}}$ is a logistic function, what is the limiting value?

In Exercises 39–40, identify the scatter plots as linear, exponential, logistic, or none of these. If you identify the scatter plot as none of these, explain why.

39.

40.

41. Lawn Mowing At the beginning of winter, I do not have to mow the yard. As spring approaches, I have to mow it some. In spring, the number of times I have to mow grows dramatically as the spring rains come. In the summer, this number declines because it is so hot. In the fall I do not have to mow very much as winter approaches. Sketch a graph that models the cumulative number of times I mow the lawn during a year starting with January 1.

42. Audio Cassette Sales The sales of audio cassettes in 1990 were high before the advent of CDs. As CDs began to catch on, the sales of audio cassettes began to drop but not dramatically at first because CD players were not widespread. In the mid-1990s the sales of audio cassettes dropped quickly and then leveled off in the late 1990s. Sketch a graph that models the annual sales of audio cassettes as a function of the years since 1990.

43. Amusement Park Attendance The cumulative number of visitors to an amusement park that is open year round is given in the table.

Month m	Cumulative Number of Visitors by the End of the Month (thousands) V
January	32
February	55
March	122
April	255
May	530
June	910
July	1465
August	1954
September	2187
October	2401
November	2411
December	2415

a. Find a logistic model to fit the data.

b. Estimate the limiting value and explain what it means in the real-world context.

c. In what month does it appear the point of inflection occurs? What does this mean in this context? Does this seem reasonable? Explain.

44. College Graduates According to the U.S. Department of Education, the number of college graduates increased significantly during the 20th century. The following table gives the number of college graduates (in thousands) from 1900 to 2000.

Years Since 1900 y	Number of College Graduates (thousands) g
0	30
10	54
20	73
30	127
40	223
50	432
60	530
70	878
80	935
90	1017
100	1180

Source: U.S. Department of Education

a. Create a scatter plot of these data.

b. Find the equation for the logistic function $g(y)$ that best models the data.

c. Describe how the rate of change of the function $g(y)$ details the growth in the number of college graduates in the 20th century.

d. Estimate the limiting value for the logistic function $g(y)$.

e. Do you think the limiting value you found in part (d) will accurately project the number of college graduates in the beginning years of the 21st century? Why or why not?

45. Male Growth The weight W (in pounds) of the average boy is a function of his age y (in years). Below is a graph of $W(y)$ for males from 2 to 20 years of age.

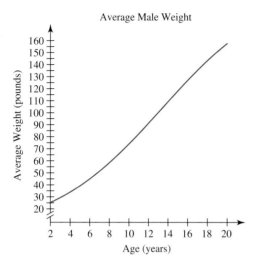

Source: www.cdc.gov

a. Evaluate $W(16)$ and explain what the value means in the real-world context.

b. Solve $W(t) = 140$ and then explain what the numerical value means in the context of the problem.

c. Over what time period is the graph concave down? Explain what information this provides about the average weight of males.

d. Estimate the point of inflection on the graph and explain what the coordinates of this point mean in terms of the average male.

In Exercises 46–47, imagine you are writing the headline for a particular situation and that you are analyzing the following graphs for your story. Write a headline that accurately represents each of the graphs. You may create your own context.

46.

47.

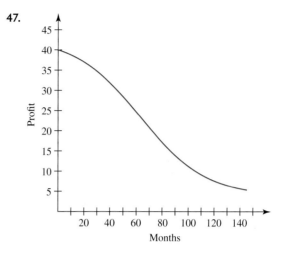

■ SECTION 7.5 ■

48. Create four data sets with at least eight data points each. The first set should be linear; the second, quadratic; the third, exponential; and the fourth, logistic. Explain why each data set represents the type of function it does.

In Exercises 49–60, find the equation of the mathematical model that will most accurately forecast the indicated result (if possible). Then use the model to find the value. Justify your conclusions.

49. Average National Football League Salaries

Years Since 1976 t	Average NFL Player's Earnings (dollars) E
0	47,500
10	244,700
20	787,500
30	1,700,000

Source: Sports Illustrated, July 31, 2006; "Average Joe"

Forecast the average NFL player's salary in 2010.

50. Average Major League Baseball Salaries

Years Since 1976 t	Average MLB Player's Earnings (dollars) E
0	52,802
10	402,579
20	1,101,455
30	2,857,932

Source: Sports Illustrated, July 31, 2006; "Average Joe"

Forecast the average MLB player's salary in 2010.

51. Average National Basketball Association Salaries

Years Since 1976 t	Average NBA Player's Earnings (dollars) E
0	130,000
10	382,000
20	2,000,000
30	5,000,000

Source: Sports Illustrated, July 31, 2006; "Average Joe"

Forecast the average NBA player's salary in 2010.

52. Average National Hockey Association Salaries

Years Since 1976 t	Average NHL Player's Earnings (dollars) E
0	85,000
10	159,000
20	892,000
30	1,500,000

Source: Sports Illustrated, July 31, 2006; "Average Joe"

Forecast the average NHL player's salary in 2010.

53. Average Professional Golf Association Salaries

Years Since 1976 t	Average PGA Golfer's Earnings (dollars) E
0	25,814
10	97,610
20	257,840
30	973,495

Source: Sports Illustrated, July 31, 2006; "Average Joe"

Forecast the average PGA golfer's salary in 2010.

54. Average Ladies' Professional Golf Association Salaries

Years Since 1976 t	Average LPGA Golfer's Earnings (dollars) E
0	7,185
10	37,850
20	98,363
30	162,043

Source: Sports Illustrated, July 31, 2006; "Average Jane"

Forecast the average LPGA golfer's salary in 2010.

55. Average Association of Tennis Professionals Salaries

Years Since 1976 t	Average ATP Player's Earnings (dollars) E
0	18,034
10	62,892
20	231,827
30	260,000

Source: Sports Illustrated, July 31, 2006; "Average Joe"

Forecast the average ATP player's salary in 2010.

56. Average Women's Tennis Association Salaries

Years Since 1976 t	Average WTA Player's Earnings (dollars) E
10	55,965
20	165,477
30	345,000

Source: Sports Illustrated, July 31, 2006; "Average Jane"

Forecast the average WTA player's salary in 2010.

57. Average Engineer Salaries

Years Since 1976 t	Average Engineer's Earnings (dollars) E
0	20,749
10	42,667
20	60,684
30	80,122

Source: American Federation of Teachers

Forecast the average engineer's salary in 2010.

58. Average Attorney Salaries

Years Since 1976 t	Average Attorney's Earnings (dollars) E
0	24,205
10	50,119
20	66,560
30	100,852

Source: American Federation of Teachers

Forecast the average attorney's salary in 2010.

59. Average Accountant Salaries

Years Since 1976 t	Average Accountant's Earnings (dollars) E
0	24,205
10	50,119
20	66,560
30	71,200

Source: American Federation of Teachers

Forecast the average accountant's salary in 2010.

60. Average K–12 Teacher Salaries

Years Since 1976 t	Average K–12 Teacher's Earnings (dollars) E
0	12,591
10	25,260
20	37,594
30	47,602

Source: American Federation of Teachers

Forecast the average teacher's salary in 2010.

Make It Real Project

What to Do

1. Find a salary schedule for one of the local school districts in your area or use the one provided on the next page for the Maricopa Community College District.

2. For each of the vertical salary "lanes," determine how much each *vertical* step on the pay scale increases a teacher's salary.

3. Teachers who participate in professional development activities move *horizontally* on the pay scale. For each professional development credit hour earned, determine by how much the annual pay is increased.

4. From the salary schedule, pick a realistic salary for a newly hired teacher in the district. Write a function that models her salary as a function of years worked given that she expects to move one step vertically and earn six professional development credits each year.

5. As inflation increases prices, the buying power of the dollar decreases. Consequently, many employers offer a *cost of living allowance* (COLA) to their employees. Assuming a COLA of 3% is applied to the salary schedule annually, revise the function in part (4) to address this fact.

6. Use the model in part (5) to forecast the future salary of a teacher 5 years and 8 years into the future.

Make It Real Project — continued

Maricopa Community College District
Residential Faculty Salary Schedule
2007–2008
Effective 7/1/2007

Base Salary
Credit Hour .33% $136.04
Vertical Increment 7% $2886

Step	IP	IP+12	IP+20	IP+24	IP+36	IP+40	IP+48	IP+60	IP+75	Ph.D.
1	$41,225	$42,857	$43,946	$44,490	$46,122	$46,667	$47,755	$49,387	$51,428	$53,469
2	$44,111	$45,743	$46,832	$47,376	$49,008	$49,553	$50,641	$52,273	$54,314	$56,355
3	$46,997	$48,629	$49,718	$50,262	$51,894	$52,439	$53,527	$55,159	$57,200	$59,241
4	$49,883	$51,515	$52,604	$53,148	$54,780	$55,325	$56,413	$58,045	$60,086	$62,127
5	$52,769	$54,401	$55,490	$56,034	$57,666	$58,211	$59,299	$60,931	$62,972	$65,013
6	$55,655	$57,287	$58,376	$58,920	$60,552	$61,097	$62,185	$63,817	$65,858	$67,899
7	$58,541	$60,173	$61,262	$61,806	$63,438	$63,983	$65,071	$66,703	$68,744	$70,785
8	$61,427	$63,059	$64,148	$64,692	$66,324	$66,869	$67,957	$69,589	$71,630	$73,671
9	$64,313	$65,945	$67,034	$67,578	$69,210	$69,755	$70,843	$72,475	$74,516	$76,557
10	$67,199	$68,831	$69,920	$70,464	$72,096	$72,641	$73,729	$75,361	$77,402	$79,443
11	$70,085	$71,717	$72,806	$73,350	$74,982	$75,527	$76,615	$78,247	$80,288	$82,329
12	$72,971	$74,603	$75,692	$76,236	$77,868	$78,413	$79,501	$81,133	$83,174	$85,215
13	$75,857	$77,489	$78,578	$79,122	$80,754	$81,299	$82,387	$84,019	$86,060	$88,101
14				$82,008	$83,640	$84,185	$85,273	$86,905	$88,946	$90,987

Initial Placement (IP) indicates initial placement on the salary schedule for any faculty member with an associates, bachelors, or masters degree. Credit hours are paid for each hour earned.

Wage & Salary for MCCD

CHAPTER 8

Trigonometric Functions

© Chip East/Reuters/Corbis

On August 13, 2003, the largest electrical blackout in North American history struck Connecticut, New Jersey, New York, Massachusetts, Vermont, Michigan, Ohio, Pennsylvania, and the Canadian province of Ontario. The outage impacted the lives of about 50 million people, leaving some without power for days. Financial losses due to the outage were estimated to be between $4 billion and $10 billion. (*Source*: U.S.-Canada Power System Outage Task Force, reports.energy.gov) The photo shows the New York skyline during the outage.

To reduce the frequency of such outages, energy organizations monitor and attempt to predict power usage each day throughout the year. The cyclical nature of daily power consumption can be modeled by periodic functions, making power usage predictions remarkably accurate.

8.1 Periodic Functions
8.2 Angle Measure
8.3 Unit Circle and Trigonometric Functions
8.4 Graphing Cosine and Sine Functions
8.5 Modeling with Trigonometric Functions
8.6 Other Trigonometric Functions
8.7 Inverse Trigonometric Functions
STUDY SHEET
REVIEW EXERCISES
MAKE IT REAL PROJECT

SECTION 8.1 Periodic Functions

LEARNING OBJECTIVES

- Determine if a real-world data table or graph is periodic
- Determine the period, midline, amplitude, and frequency of a periodic function
- Extrapolate function values for a periodic function given in a graph or table

GETTING STARTED

Many people commute to work or school and must plan for travel time to arrive at their destination on time. In addition, delivery truck drivers and others whose work efficiency is affected by freeway travel times can benefit from a detailed analysis of average travel times throughout the day. For example, by scheduling deliveries before and after peak travel times, drivers can reduce the amount of time spent in traffic and increase the number of deliveries that can be made in a day. Typical commute times on a given segment of a road can be modeled by a periodic function.

In this section we explore periodic functions and see how to apply these functions to real-world situations such as analyzing commute times. We define certain aspects of these functions—namely, the midline, amplitude, and frequency. We also show how to recognize the periodicity of a function given a graph or a table.

■ Periodic Functions

A **periodic function** is a function whose output values repeat at regular intervals, or **periods**. Such a function is said to have **periodicity**.

> **PERIODIC FUNCTIONS**
>
> A function f is **periodic** if $f(x + p) = f(x)$ for all x and for a positive constant p. The value p is called the **period** of the function.

From our knowledge of transformations, we know the graph of $f(x + p)$ is the graph of $f(x)$ shifted left p units. A periodic function has the property that when its graph is shifted left or right p units (where p is the period of the function), the resultant graph is the same as the original graph.

Let's explore the concepts related to periodic functions in the context of commuter travel times. The Washington Department of Transportation regularly monitors traffic flows and reports average and current travel times to commuters via its website (www.wsdot.wa.gov). The data is updated every five minutes throughout the day so commuters can estimate their travel times on key segments of freeway. A sample report is shown in Figure 8.1. The Washington State Transportation Department graph in Figure 8.2 shows the average travel time from Bellevue to Seattle on I-90 on Tuesdays in 2003 at various times throughout the day.

If we assume that average travel times for Wednesdays and Thursdays are identical to those for Tuesdays, we can say the travel-time function has a period of 24 hours. That is, every 24 hours, the travel-time pattern will repeat. (Note: The actual weekly data show that travel times for Fridays through Mondays vary significantly from the Tuesday travel times; therefore, we limit the periodic model to Tuesdays through Thursdays.)

When working with periodic functions, we are often interested in the **midline** of the function.

Travel times as of 10:10 A.M. Wednesday, June 13, 2007

State Route/Interstate	Route Description	Distance (miles)	Average Travel Time (minutes)	Current Travel Time (minutes)	Via HOV (min.)
167	Auburn to Renton	9.8	17	12	10
405	Bellevue to Bothell	9.7	11	15	10
405 5	Bellevue to Everett	23.2	25	31	24
405 5	Bellevue to Federal Way	24.9	29	30	26
405 90	Bellevue to Issaquah	9.8	10	11	11
405 520	Bellevue to Redmond	6.8	8	10	8
	Bellevue to Seattle	10.7	14	13	12
405 90 5	Via Westbound Express Lanes	10.7	15	13	12

Figure 8.1

Source: www.wsdot.wa.gov

Figure 8.2

Source: www.wsdot.wa.gov

MIDLINE (CENTERLINE)

The **midline** or **centerline** of a periodic function f is the horizontal line that lies halfway in between the maximum and minimum output values of the function. In other words, the midline is the horizontal line with equation

$$y = \frac{\max \text{ of } f + \min \text{ of } f}{2}$$

For the travel-time function, it appears the maximum travel time is about 24 minutes and the minimum travel time is about 12 minutes. The midline is

$$y = \frac{24 + 12}{2}$$

$$= 18 \text{ minutes}$$

We draw a red dashed line on the graph, as shown in Figure 8.3, to indicate the location of the midline.

Figure 8.3

Source: www.wsdot.wa.gov

If the graph is above the midline, this means the travel time exceeds the average of the maximum and minimum travel times. We see this occurs between about 4:45 P.M. and 7:00 P.M. (indicated by black dashed lines). At all other times, the travel time is less than the average of the maximum and minimum travel times.

Another important characteristic of periodic functions is the **amplitude** of the function.

Amplitude

The **amplitude** of a periodic function f is the distance from the midline to the maximum or minimum output value of the function. In symbolic terms,

$$\text{amplitude} = \text{max of } f - \text{midline}$$

Equivalently, the amplitude is half of the distance between the maximum and minimum output values of the function. That is,

$$\text{amplitude} = \frac{\text{max of } f - \text{min of } f}{2}$$

For the travel-time function, we have

$$\text{amplitude} = \frac{\text{max of } f - \text{min of } f}{2}$$
$$= \frac{24 - 12}{2}$$
$$= 6 \text{ minutes}$$

This tells us the *maximum* commute time is 6 minutes *more* than the average of the maximum and minimum travel times. Similarly, the *minimum* commute time is 6 minutes *less* than the average of the maximum and minimum travel times. We indicate the amplitude with additional red dashed lines on the graph in Figure 8.4.

Figure 8.4

Source: www.wsdot.gov

A third characteristic of periodic functions is **frequency**. The frequency of a periodic function is closely related to its period.

Frequency

The **frequency** of a periodic function is the reciprocal of its period. That is,

$$\text{frequency} = \frac{1}{\text{period}}$$

SECTION 8.1 ■ PERIODIC FUNCTIONS   497

For the travel-time function, the period is 24 hours so the frequency is $\frac{1}{24}$. Because the units of the frequency are not immediately obvious, let's investigate. Recall that there are 24 hours per 1 period. We can represent this as

$$\frac{24 \text{ hours}}{1 \text{ period}}$$

Now we take the reciprocal of this expression to find the units of the frequency.

$$\frac{1 \text{ period}}{24 \text{ hours}}$$

Thus the frequency is $\frac{1}{24}$ period per hour. That is, for each hour that passes, the travel-time function moves through $\frac{1}{24}$ of its period.

A periodic function's frequency is generally measured in the number of cycles (periods) per unit of time such as minutes, seconds, and so on. Figure 8.5 shows periodic functions with different frequencies, where the green graph has the highest frequency and the black graph has the lowest frequency.

Figure 8.5

EXAMPLE 1 ■ Determining the Frequency of a Periodic Function

Two siblings (a boy and a girl) begin to play on the swing set. At rest, both swings are at approximately 2 feet off the ground but, as the children pump their legs, each swing's height off the ground changes. The graphs in Figure 8.6 display a 14-second interval of time (from 16 seconds to 30 seconds) and the heights of each swing seat (in feet). Use the graphs to determine the midline, amplitude, period, and frequency of each graph and explain what they mean in terms of each child's swinging action.

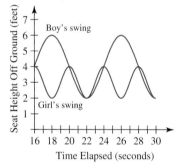

Figure 8.6

Solution We find the midline of each graph by calculating the average between the maximum and minimum swing seat heights for each child and display the midlines with dashed lines in Figure 8.7.

boy's: $\frac{6 + 2}{2} = 4$ feet   girl's: $\frac{4 + 2}{2} = 3$ feet

We calculate the amplitude of each graph by calculating half the distance between the maximum and minimum swing seat heights.

boy's: $\frac{6 - 2}{2} = 2$ feet   girl's: $\frac{4 - 2}{2} = 1$ foot

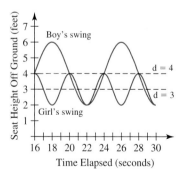

Figure 8.7

The amplitude of the boy's graph is 2 feet and the amplitude of the girl's graph is 1 foot. This means the boy's swing seat height varies from 2 feet above and 2 feet below the midline and the girl's from 1 foot below to 1 foot above the midline.

We determine the period for each of the graphs by finding out how long it takes each child's seat to return to its starting position. At 16 seconds, the boy's seat height is 4 feet. Eight seconds later, the seat returns to the same position. Therefore, the boy's period is 8 seconds. At 16 seconds, the girl's seat height is 4 feet. Four seconds later, her seat is at the same position. Therefore, the girl's period is 4 seconds.

Since frequency is the reciprocal of period, we have

$$\text{boy's frequency: } \frac{1 \text{ cycle}}{8 \text{ seconds}} = 0.125 \text{ cycles per second}$$

and

$$\text{girl's frequency: } \frac{1 \text{ cycle}}{4 \text{ seconds}} = 0.25 \text{ cycles per second}$$

Since the girl's frequency (0.25) is twice as large as the boy's frequency (0.125), the girl is traveling twice as fast as the boy.

Let's now turn our attention to how periodic functions may be transformed vertically or horizontally and what this means in a real-world context.

EXAMPLE 2 ■ Transforming Periodic Functions

The California Independent System Operator (ISO) is a not-for-profit electric transmission organization that predicts and monitors California's energy usage. Through close monitoring, the organization seeks to avoid the crippling electrical blackouts that can shut down a region. To help electricity consumers do their part in conserving energy, the organization publishes and regularly updates a daily power usage forecast on its website (www.caiso.com).

Figure 8.8 shows California ISO's electricity demand forecast for June 12 and June 13, 2007.

Figure 8.8

Source: www.caiso.com

a. Explain in terms of the rate of change why you think the data change as they do. Also, give a reasonable rationale for why the graph appears to be periodic in nature.

b. Estimate the midline and amplitude and explain what each value means in terms of power usage.

c. Suppose that through an aggressive energy conservation campaign, consumers were persuaded to reduce their collective electricity usage by 2000 megawatts each hour. How would the graph of the actual usage compare to the graph of the usage forecast?

d. Let $E(h)$ represent the predicted power usage function with E representing megawatts and h representing the number of hours since midnight June 11, 2007. Use function notation to symbolize the transformed function that approximates the actual amount of electricity used each hour if consumers reduce their collective electricity usage by 2000 megawatts each hour.

e. Over the two-day interval from June 12–June 13, 2007, the data appear to be periodic. Do you think this graphical model would be accurate in predicting usage for January 12–January 13, 2008? Explain.

Solution

a. Between midnight and 5 A.M., power usage decreases with a rate of change that is becoming less negative. Between 5 A.M. and around 4 to 5 P.M., hourly power usage increases rapidly from roughly 23,500 megawatts to about 39,500 megawatts. Between 5 P.M. and midnight, energy usage tends to decline, becoming more and more negative. The units of the hourly power usage rate of change are megawatts per hour.

Since most people sleep at night, it seems reasonable that energy usage at night will be less than that in the day. Furthermore, since air conditioners consume a lot of electricity during hot summer days, we expect the demand in the day to be greater than that during the cooler summer nights. The highest predicted usage on both dates occurred between noon and 8 P.M. It seems reasonable that the daily pattern of power usage should be about the same for most weekdays. The data appears to be periodic, at least in the short term. However, variable temperatures may increase or decrease the demand so using the model to extrapolate values too far away from June 12, 2007, will likely yield unreliable results.

b. Although this data set is not perfectly periodic, the maximum power usage appears to be around 39,750 megawatts and the minimum power usage appears to be around 23,500 megawatts. The midline is given by

$$y = \frac{39{,}750 + 23{,}500}{2}$$
$$= 31{,}625$$

The usage value midway between the maximum and minimum usage values is 31,625 megawatts.

The amplitude is given by

$$\text{amplitude} = 39{,}750 - 31{,}625$$
$$= 8125$$

The maximum usage is 8125 megawatts above the midline and the minimum usage is 8125 megawatts below the midline. We graph both the midline and the amplitude on Figure 8.9.

c. If all of the usage forecast values are decreased by 2000 megawatts, the graph will be shifted downward by 2000, as shown in Figure 8.10.

Figure 8.9   Figure 8.10

d. Since $E(h)$ gives the original energy forecast, the function $N(h) = E(h) - 2000$ gives the energy usage if the consumers collectively use 2000 fewer megawatts of electricity each hour.

e. Although the function may be effective in predicting energy usage for days on or around June 12, 2007, it is unlikely to be accurate in forecasting electricity usage for January 12–13, 2008. There is a substantial difference in daytime temperatures between June 12 and January 12; heaters (many powered by natural gas), not air conditioners, will be more widely used on January 12 and are more likely to be running at night.

EXAMPLE 3 ■ Determining Periodic Behavior from Tabular Data

Table 8.1 shows the number of new houses sold each month for two consecutive years in the United States. Is the data set periodic from year to year? Explain.

Table 8.1

Month (1 = January) m	New Houses Sold in 2005 (in thousands) H	New Houses Sold in 2006 (in thousands) N
1	1203	1185
2	1319	1084
3	1328	1126
4	1260	1097
5	1286	1087
6	1274	1073
7	1389	969
8	1255	1000
9	1244	1004
10	1336	952
11	1214	987
12	1239	1019

Source: www.census.gov

Solution To be perfectly periodic, the number of new homes sold each month in 2006 must be the same as the number sold in 2005 in the same month. Even if the data values do not match up perfectly, we may be able to model the data by a periodic function if the data demonstrates the same increasing and decreasing behavior each year.

We immediately conclude that this data set is not perfectly periodic since the data values differ from year to year. Over the first three months of the year, home sales increased in 2005. However, in 2006, home sales first decreased then increased. In 2006, home sales decreased monthly between March and July. In contrast, in 2005 home sales oscillated between increasing and decreasing over the same time period. These results convince us home sales in these years were not periodic by year.

We now look at how to extrapolate periodic data given in both graphical and tabular forms.

EXAMPLE 4 ■ Extrapolating Periodic Graphical Data

The graph in Figure 8.11 models the K–8 enrollment in American public schools from 1970 to 2000. Assuming that the pattern established here is periodic, sketch a graph of the function from 1970 to 2030. Also, provide an estimate for how many K–8 students there will be in the years 2010, 2020, and 2030.

Figure 8.11

Source: Statistical Abstract of the United States, 2001, Table 232

Solution We have been told to assume the data is periodic so we may use the current graph to generate the data for the subsequent 30-year period. We do so in Figure 8.12, where the graph extends the pattern to the year 60.

We find the output for each year on the graph, as indicated on Figure 8.13, and predict the public school K–8 enrollment for the years 2010, 2020, and 2030 to be 27,900, 29,900, and 33,400, respectively.

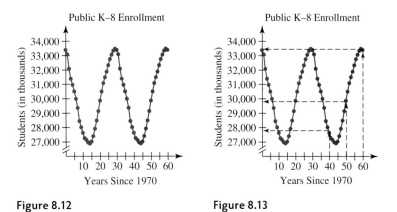

Figure 8.12

Figure 8.13

EXAMPLE 5 ■ Extrapolating Periodic Data from a Table

Table 8.2 shows the numbers of hours of daylight in Minneapolis, Minnesota, on the 21st day of the month for 2007 and part of 2008. Determine if the data appears to be periodic. If so, predict the number of hours of daylight for the remaining days of 2008.

Table 8.2

Day	2007 Hours of Daylight	2008 Hours of Daylight
Jan 21	9:22	9:21
Feb 21	10:44	10:44
Mar 21	12:12	12:14
Apr 21	13:47	13:49
May 21	15:03	15:05
Jun 21	15:37	15:37
Jul 21	15:06	
Aug 21	13:48	
Sep 21	12:14	
Oct 21	10:42	
Nov 21	9:21	
Dec 21	8:46	

Source: aa.usno.navy.mil

Solution Since the hours of daylight for 2008 are very close to the hours of daylight for 2007, it appears the hours of daylight are periodic with a period of 12 months. We estimate the remaining values for 2008 simply by copying the 2007 values, as shown in Table 8.3.

Table 8.3

Day	2007 Hours of Daylight	2008 Hours of Daylight
Jul 21	15:06	15:06
Aug 21	13:48	13:48
Sep 21	12:14	12:14
Oct 21	10:42	10:42
Nov 21	9:21	9:21
Dec 21	8:46	8:46

SUMMARY

In this section you learned to recognize the periodicity of a function given a graph or table. You also learned about the midline and amplitude of periodic functions and learned to distinguish between the period and frequency.

8.1 EXERCISES

SKILLS AND CONCEPTS

In Exercises 1–8, determine if the given function appears to be periodic in nature.

1.

2.

3.

4.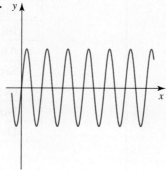

SECTION 8.1 EXERCISES 503

5.
x	y
0	2
1	10
2	14
3	2
4	10
5	14
6	2
7	10
8	14
9	2
10	10

6.
x	y
−5	9
−4	15
−3	13
−2	12
−1	6
0	7
1	98
2	1
3	−9
4	6
5	4

7.
x	y
−1	7
0	8
1	9
2	8
3	7
4	8
5	9
6	8
7	7
8	8
9	9

8.
x	y
60	a
70	b
80	c
90	f
100	r
110	f
120	c
130	b
140	a
150	b
160	c

11.

12.

(graph)

13.
x	y
0	2
1	10
2	14
3	2
4	10
5	14
6	2
7	10

14.
x	y
12	3
24	6
36	9
48	8
60	3
72	6
84	9
96	8

In Exercises 9–16, estimate the period of each function. (Note: Assume each horizontal line of the grid on the graphs is one unit.)

9.

10.

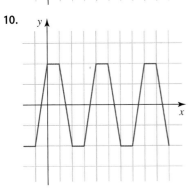

15.
x	y
−38	2
−37	1
−36	2
−35	1
−34	2
−33	1
−32	2

16. Assume the x-values are equally spaced.

x	y
a	1
b	2
c	3
d	4
e	5
f	4
g	3
h	2
i	1
j	2
k	3

504 CHAPTER 8 ■ TRIGONOMETRIC FUNCTIONS

In Exercises 17–20, assume the functions are periodic and that one period is shown. Using the concept of periodicity, complete one more cycle before and after the values given in each graph and table.

17.

18.

19.

x	y
0	
1	
2	
3	
4	10
5	14
6	2
7	8
8	
9	
10	
11	

20.

x	y
$\pi - 4$	
$\pi - 3$	
$\pi - 2$	
$\pi - 1$	
π	∂
$\pi + 1$	Σ
$\pi + 2$	©
$\pi + 3$	®
$\pi + 4$	∂
$\pi + 5$	
$\pi + 6$	
$\pi + 7$	
$\pi + 8$	

x	g(x)
0	3
1	4
2	5
3	3
4	4
5	5
6	3

21. Sketch $f(x) + 3$.

22. Sketch $f(x - 1)$.

23. Add a third column in the table of $g(x)$ for values of $g(x + 2)$.

24. Add a third column in the table of $g(x)$ for values of $2g(x - 1)$.

In Exercises 25–28, indicate which of the graphs A–D may possibly model the statement. The independent and dependent variables are defined for each exercise. (Note: There may be more than one correct answer.) Justify your conclusion.

25. Ferris Wheel A man takes a ride on a Ferris wheel. The independent variable is elapsed time since the man got on the Ferris wheel and the dependent variable is the man's height above the ground.

A.

B.

In Exercises 21–24, two periodic functions, f and g, are given. Transform the function as described.

C.

D.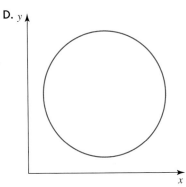

26. **A Child Swinging** A child swings back and forth on a swing. The independent variable is elapsed time from when he got on the swing and the dependent variable is the child's height above the ground.

A.

B.

C.

D.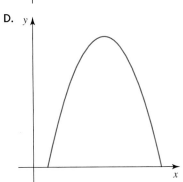

27. **Family Vacation** A family's motor home uses a large quantity of gasoline during an extended family trip. The independent variable is elapsed time since the family began their trip and the dependent variable is the amount of gasoline remaining in their tank. On a particular day, the family does not stop the vehicle other than to gas up the motor home.

A.

B.

C.

D.

28. Subway Commute A subway makes repeated trips transporting passengers back and forth between two destination points within the city. The independent variable is elapsed time since the subway began transporting passengers and the dependent variable is the distance from its original starting point.

A.

B.

C.

D.
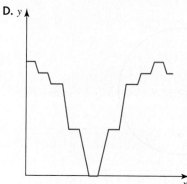

29. Daily Sales The table depicts the hypothetical number of customers that are served each hour at a 24-hour fast-food restaurant on a typical weekday.

Time of Day ($0 = 12{:}00$ A.M.) h	Number of Customers c
0	1
1	0
2	0
3	1
4	3
5	9
6	15
7	23
8	55
9	50
10	45
11	75
12	115
13	114
14	64
15	35
16	36
17	98
18	111
19	89
20	60
21	44
22	31
23	9

a. Estimate the average rate of change between 6 A.M. and 11 A.M. Interpret your answer in terms of the context.

b. Estimate the instantaneous rate of change at 12:00 P.M. (noon). Interpret your answer.

c. Estimate where the highest values are in the morning, mid-day, and evening, and explain what happens before and after each in terms of rate of change. Why would it make sense that these three highest values occur when they do?

d. Do you think this table of data would be periodic over a week of time? Explain.

■ SHOW YOU KNOW

30. In your own words, explain what it means for a function to be "periodic."

31. Explain the difference between the period and the frequency of a periodic function. Be sure to talk about their relative meanings—do not simply state formulas or algorithms for calculating their values.

32. What is the relationship between the maximum value, minimum value, midline, and amplitude of a periodic function?

33. In this section we have seen many examples of periodic functions, including hours of daylight during the year and traffic commute times during the week. Describe a real-world situation not found in this section that you think can be represented by a periodic function and explain why this situation is periodic.

■ MAKE IT REAL

34. **Female Hormones** The cycle of hormone production in the female human body regulates many important bodily functions. Progesterone is one such hormone that in the average woman fluctuates on approximately a 28-day cycle, as shown in the graph.

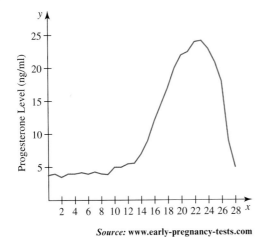

Source: www.early-pregnancy-tests.com

Describe what the graph tells about women's monthly cycle of progesterone production.

35. **Alternating Current** The electricity used in American households is in the form of alternating current (AC). Alternating current is typically 120 volts and 60 hertz in the United States. This means that the voltage cycles from -120 volts to $+120$ volts and back to -120 volts and 60 cycles occur each second. In answering the following questions, assume that at $t = 0$ there is no voltage at a given outlet.

a. Sketch $v(t)$, the voltage as a function of time, for the first 0.1 second.

b. State the period, the amplitude, and the midline of the graph you made in part (a). Explain the real-world meaning of each.

36. **Pneumonia and Influenza Mortality** According to the Centers for Disease Control and Prevention, the percentage of all deaths caused by pneumonia and influenza over the years 2003 to 2006 in 122 American cities is shown in the following graph together with a *seasonal baseline* and an *epidemic threshold*. The input values (which repeat) are measured in weeks since January 1 of the corresponding year.

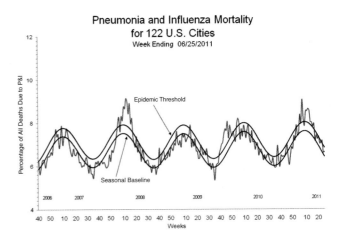

Source: www.cdc.gov/flu/weekly/

a. Explain the purpose of the periodic functions representing the *seasonal baseline* and the *epidemic threshold*.

b. Estimate the period, the amplitude, and the midline for the *seasonal baseline* and the *epidemic threshold*. Explain the real-world meaning of each.

c. Determine from the graph the weeks and years in which there was an epidemic level of deaths from pneumonia and influenza.

37. **Utilities Budget** A utility bill for electricity usage may comprise a large part of a family's yearly budget. To predict the amount of money required for electricity, a family may look at a prior year's costs. The table shows the monthly utility costs for the year 2006 for a single-family residence.

Month (1 = January) m	Cost (dollars) c
1	124.19
2	111.16
3	116.30
4	133.71
5	270.07
6	320.11
7	250.23
8	391.19
9	345.91
10	256.44
11	124.87
12	126.98

Source: Author's data

a. Assume the charge for electricity remained the same for 2007, and create a graph of $c(m)$ that will predict the cost each month in 2007.

b. Now suppose the cost of electricity increased by 4% for 2008 and remained at that level through 2009. Create a graph showing the function that will predict the 2008 and 2009 monthly electric bills.

c. If you chose to add an electrically heated swimming pool to your home at the start of 2007, how would this change affect the graph of this function?

38. **Population Cycles** Populations of some species go through periods of repeated growth (boom) and decline (bust). The graph shows the 10-year cyclical fluctuations in the population of the snowshoe rabbit and its primary predator, the lynx, in the Hudson Bay area of Canada from 1850 to 1910. What claims can one make based on this data, if any? Justify your response.

Source: users.rcn.com

STRETCH YOUR MIND

Exercises 39–40 are intended to challenge your understanding of periodic functions.

39. **Pendulum Swings** The period of a pendulum is defined as the time taken for two swings of the pendulum (left to right and back again). The formula for the period, T (in seconds), is the multivariable function, $T(l, g) = 2\pi\sqrt{\dfrac{l}{g}}$, where

l is the length of the pendulum measured in meters from the pivot point to the bob's center of gravity, and g is the gravitational field strength (or acceleration due to gravity).

a. Sketch the graph of $T(l, 9.8)$ from $0 < l \leq 5$.

b. Explain what information the graph of this function provides regarding the period of a pendulum and its length.

c. Evaluate $T(0.5, 9.8)$, $T(0.7, 9.8)$, and $T(1.0, 9.8)$.

40. **Sunspots** In the mid-1800s, Samuel Heinrich Schwabe (an amateur astronomer) declared that the number of sunspots varied cyclically, reaching a peak about every 10 years. He based his claim on 17 years of daily observations. Advancing Schwabe's work, Swiss astronomer Rudolf Wolf compiled the history of sunspot variations as far back as 1745 and determined that the period for sunspot activity would be more accurately estimated to be 11 years rather than 10. A minimum number of sunspots occurred in 1760 and a maximum occurred about 5.5 years later. (*Source:* www-spof.gsfc.nasa.gov)

Use this information and the 11-year cycle of sunspots to sketch a possible graph of the sunspot activity for the years 2000 to 2050. Then estimate when the minimum and maximum number of sunspots will occur over this 50-year timespan.

SECTION 8.2 Angle Measure

LEARNING OBJECTIVES

- Explain the useful definition of angle measure
- Measure angles in degrees and radians
- Convert angle measure between units of degree and radian

GETTING STARTED

Many of us have an intuitive understanding of *angle* measure. Angles play a big role in architecture, construction, surveying, geometry, backyard pool design, highway construction, and even some sporting activities. For example, to be successful in the game of billiards, a player must bank the balls off of the cushions at specific angles. In construction, surveyors measure distances and angles to map out

locations for the construction project. Surveyors use a tool called a theodolite, which is a telescope mounted on a tripod that allows the surveyor to measure angles with great precision: $\frac{1}{60}$ of 1 degree.

In this section we define the term *angle* and look at the different ways that angles are measured. We also provide a way to think about angle measure that will be useful in our later study of trigonometric functions.

■ Angles

From a geometric point of view, an **angle** is the figure formed by two rays (or segments) sharing a common endpoint. We could think of the angle as the amount of opening between the two rays. While this is a fairly common interpretation of an angle measure, we will formalize this notion to create a method of measuring angles that will be very useful in subsequent sections.

To measure an angle, we imagine drawing an arc, centered at the vertex of the angle, from one ray to the other. The length of this arc, s, is a portion of the circumference of an entire circle of radius, r. That is, the arc length s is directly proportional to the circumference C of the circle. So for some constant k, $s = kC$. Recall the circumference of a circle is $C = 2\pi r$. The constant of proportionality k tells what proportion of the circumference is included in the arc length.

Since $s = kC$ and $C = 2\pi r$, we have

$$s = kC$$
$$= k(2\pi r)$$
$$= (2\pi k)r$$

This result shows us the arc length, s, is directly proportional to the radius, r, of the circle with constant of proportionality $2\pi k$.

For example, if the arc length is equal to half of the circumference of the corresponding circle, $k = \dfrac{1}{2}$ and

$$s = (2\pi k)r$$
$$= (2\pi \tfrac{1}{2})r$$
$$= \pi r$$

When the arc length is half of the circumference of the corresponding circle, the arc length is equal to π times the radius.

If the arc length is equal to one-sixth of the circumference, $k = \frac{1}{6}$ and

$$s = (2\pi k)r$$
$$= (2\pi \tfrac{1}{6})r$$
$$= \frac{\pi}{3}r$$

When the arc length is one-sixth of the circumference of the corresponding circle, the arc length is equal to $\frac{\pi}{3}$ times the radius.

For simplicity, we represent the constant of proportionality in the equation $s = (2\pi k)r$ with K. That is, $K = 2\pi k$. The simplified equation is $s = Kr$, where $K = 2\pi k$.

JUST IN TIME ■ PROPORTIONALITY

A quantity y is said to be **directly proportional** to a quantity x if $y = kx$ for some constant value k, called the **constant of proportionality**. Consider the following examples.

1. The circumference of a circle is directly proportional to its radius with a constant of proportionality 2π. That is, $C = 2\pi r$.
2. The total cost of filling a gas tank is directly proportional to the number of gallons put in the tank. The constant of proportionality is the gasoline price per gallon. At a price of $3.489 per gallon, the cost is $C = 3.489x$.

Relationship between Arc Length and Radius

The length, s, of an arc is directly proportional to the radius of the arc, r, with constant of proportionality K. That is,

$$s = Kr$$

If k is the proportion of the circumference of a circle of radius r that coincides with the arc, then $K = 2\pi k$.

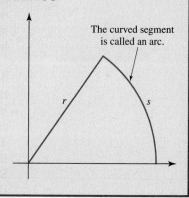

The curved segment is called an arc.

EXAMPLE 1 ■ Relating Radius and Arc Length

An arc's length is equal to $\frac{2}{3}$ of the circumference of the corresponding circle. Write the equation for the arc length as a function of the radius.

Solution Since $K = 2\pi k$,

$$K = 2\pi\left(\frac{2}{3}\right) \quad \text{since } k = \frac{2}{3}$$
$$= \frac{4\pi}{3}$$

So $s = \frac{4\pi}{3}r$ is the equation for arc length as a function of radius for an arc that is equal to $\frac{2}{3}$ of the circumference of the corresponding circle.

EXAMPLE 2 ■ Determining When Arc Length Equals Radius

Under what conditions is the arc length equal to the radius?

Solution When the arc length equals the radius, the equation $s = Kr$ becomes $r = Kr$. Solving this equation for K yields $K = 1$. We have

$$K = 2\pi k$$
$$1 = 2\pi k$$
$$\frac{1}{2\pi} = k$$
$$0.159 \approx k$$

So when the arc length is about 15.9% of the circumference, the arc length and the corresponding radius are equal.

■ Measuring Angles Using Radians

When measuring quantities, we may use any number of different measurement systems. For example, although the numbers 212, 100, and 373.15 are different from each other, they each represent the boiling temperature for water when the appropriate units are used: 212 degrees Fahrenheit = 100 degrees Celsius = 373.15 kelvin. Likewise, we may assign any numeric value we want to a particular angle as long as we have clearly defined the measurement system. For angles, two different measurement systems are commonly used: degrees and radians. Due to its close connection with arc length, we first look at radian measure.

We begin by considering the **unit circle**, a circle of radius 1 centered at the origin. Imagine the radius of the circle is a piece of string. Take the piece of string and lay it on the outer edge of the circle, starting at the point (1, 0). Draw a line from the center of the circle to the uppermost end of the string as indicated in Figure 8.14. The angle formed by this line and the horizontal axis is defined to be 1 **radian**. That is, the measure of the angle θ that corresponds with an arc length of 1 radius is 1 radian.

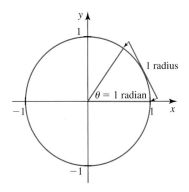

Figure 8.14

RADIAN MEASURE

The measure of an angle θ that corresponds with an arc length of 1 radius in the counterclockwise direction is 1 **radian**.

EXAMPLE 3 ■ Determining the Number of Radians in One Revolution of the Unit Circle

Using the idea that 1 radian is the measure of the angle that corresponds with an arc length of 1 radius, determine how many radians are in one complete revolution of a circle.

Solution We continue to imagine that the radius of the unit circle is a piece of string and that we want to use the piece of string to measure the circumference of the whole circle. We imagine laying the string on the outer edge of the circle, with one end beginning at the point (1, 0), and measuring the number of 1-radius lengths required to span the entire circle. In Figure 8.15, each change in color represents the length of 1 radius.

It takes six complete radius lengths to span the circumference of the whole circle plus a little bit more. In fact, the "little bit more" is approximately 0.28 of one radius. It takes approximately 6.28 radius lengths to span the circumference of the circle. The exact number of lengths is 2π ($2\pi \approx 6.2832$).

Since each radian corresponds with an arc length of 1 radius on the unit circle, it takes 2π radians to generate an arc length equal to the circumference of the circle. This is where we get the formula for the circumference of a circle, $C = 2\pi r$.

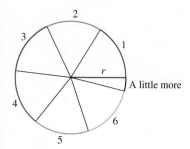

Figure 8.15

Linking the idea of angle measure with the idea of arc length has significant consequences. One radian yields an arc length of 1 radius, 2 radians yields an arc length of 2 radii, and so on. In general, θ radians yields an arc length of θ radii. Symbolically, we write $s = \theta r$, where s is the arc length and r is the length of the radius.

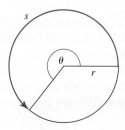

Figure 8.16

ARC LENGTH AS A FUNCTION OF ANGLE MEASURE IN RADIANS

The length, s, of the arc corresponding with an angle measure of θ radians is

$$s = \theta r$$

where r is the length of the radius. See Figure 8.16. For the unit circle, $s = \theta$.

We showed earlier that the length, s, of an arc is directly proportional to the radius of the arc, r, with constant of proportionality K. That is, $s = Kr$. When measuring angles in radians, the constant of proportionality equals the angle measure. That is, $K = \theta$.

EXAMPLE 4 ■ Determining the Radian Measure of an Angle

What is the radian measure of the angle that corresponds with $\dfrac{1}{4}$ of the circumference of a circle of radius 3?

Solution The circumference of a circle corresponds with an angle of 2π. Therefore, $\dfrac{1}{4}$ of the circumference of a circle corresponds with $\dfrac{1}{4}$ of 2π.

$$\theta = \frac{1}{4}(2\pi)$$
$$= \frac{2\pi}{4}$$
$$= \frac{\pi}{2}$$

Observe that the radius of the circle was not used in the calculation of the angle measure. For any circle (regardless of the length of the radius), $\dfrac{1}{4}$ of the circumference corresponds with an angle measure of $\dfrac{\pi}{2}$. The measure of the angle is $\dfrac{\pi}{2}$ radians.

EXAMPLE 5 ■ Determining the Proportion of a Circumference That Corresponds with an Angle

What proportion of a circle's circumference corresponds with an angle of $\dfrac{4\pi}{3}$ radians?

Solution The arc length is

$$s = \theta r$$

$$s = \frac{4\pi}{3} r$$

We calculate the proportion by dividing the arc length by the circumference.

$$\frac{\text{arc length}}{\text{circumference}} = \frac{\frac{4\pi}{3} r}{2\pi r}$$

$$= \left(\frac{4\pi r}{3}\right) \frac{1}{2\pi r}$$

$$= \left(\frac{4}{3}\right)\left(\frac{1}{2}\right)$$

$$= \frac{4}{6}$$

$$= \frac{2}{3}$$

An angle of $\frac{4\pi}{3}$ radians corresponds with $\frac{2}{3}$ of the circumference of the circle.

■ Sketching Angles

An angle is represented graphically by an *initial side* and a *terminal side*, as shown in Figure 8.17. The positive horizontal axis forms the *initial side* of any angle in standard position.

To sketch a *positive* angle, we begin on the positive horizontal axis and sweep out an arc in the *counterclockwise* direction. To sketch a *negative* angle, we begin on the positive horizontal axis and sweep out an arc in the *clockwise* direction. The magnitude of the angle determines the position of the terminal side, as is demonstrated in Example 6.

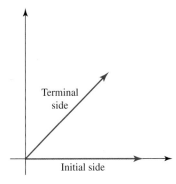

Figure 8.17

EXAMPLE 6 ■ **Sketching Angles Measured in Radians**

Sketch each of the following angles measured in radians.

a. $\theta = \pi$
b. $\theta = -\pi$
c. $\theta = \frac{\pi}{4}$
d. $\theta = -\frac{2\pi}{3}$
e. $\theta = 4$

Solution

a. We know an angle of 2π represents an arc length that corresponds with the circumference of an entire circle. Therefore, to sketch the angle $\theta = \pi$, we represent an arc length that traverses one-half the circumference of a circle in the counterclockwise direction, as shown in Figure 8.18.

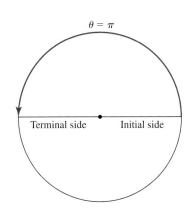

Figure 8.18

b. The magnitude of the angle is π; however, since the angle is negative, the arc length traverses the circumference in the clockwise direction to represent $\theta = -\pi$. See Figure 8.19.

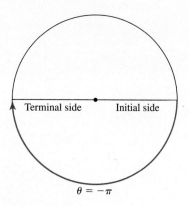

Figure 8.19

c. The angle $\theta = \dfrac{\pi}{4}$ is equivalent to $\theta = \dfrac{1}{4}\pi$. This angle is one-quarter of the angle $\theta = \pi$, as shown in Figure 8.20.

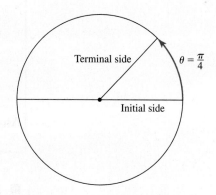

Figure 8.20

d. The angle $\theta = -\dfrac{2\pi}{3}$ is equivalent to $\theta = -\dfrac{2}{3}\pi$. We then see this angle is two-thirds of the angle $\theta = \pi$ and, since it is negative, the arc length traverses the circle in the clockwise direction. See Figure 8.21.

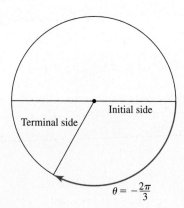

Figure 8.21

e. Recall for the unit circle, the numeric value of the angle is equivalent to the arc length. In this case, the angle and arc length are 4. We first determine the proportion of the circumference that the arc length represents.

$$\frac{\text{arc length}}{\text{circumference}} = \frac{4}{2\pi}$$

$$\approx 0.637$$

The terminal side of the angle will occur close to $\dfrac{2}{3}$ (66.7%) of the way around the circle in the counterclockwise direction, as shown in Figure 8.22.

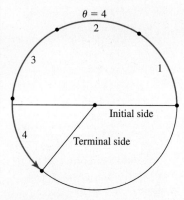

Figure 8.22

Angles That Measure Greater Than 2π Radians

We can measure and sketch angles with $\theta > 2\pi$. Since each measure of 2π corresponds with one complete revolution around the circle, an angle greater than 2π indicates that the angle represents more than one revolution around the circle, as indicated in Figure 8.23. For example, we may think of $\theta = 2.5\pi$ as

$$\begin{aligned} \theta &= 2\pi + 0.5\pi \\ &= 2\pi + \tfrac{1}{2}\pi \\ &= 2\pi + \tfrac{1}{4}(2\pi) \\ &= 1 \text{ revolution} + \tfrac{1}{4} \text{ revolution} \end{aligned}$$

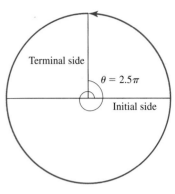

Figure 8.23

The terminal side of $\theta = 2.5\pi$ is in the same position as the angle whose arc length is one-quarter of the circumference of the circle, in this case $\dfrac{\pi}{2}$.

In Section 8.3, we will discuss *reference angles*, which will further help us identify angles such as $\theta = 2.5\pi$ and $\theta = \dfrac{\pi}{2}$ that share a terminal side.

Measuring Angles Using Degrees

Another common way to measure angles is in *degrees*. Because of historical convention established by the ancient Babylonians, an angle whose arc length is $\dfrac{1}{360}$ of the circumference of a circle is known as 1 **degree** (notated 1°). The number 360 was a good choice because 2, 3, 4, 5, 6, 8, 9, 10 and so on are factors of 360. Consequently, determining the angle measure for $\dfrac{1}{2}, \dfrac{1}{3}, \dfrac{1}{4}, \dfrac{1}{5}, \dfrac{1}{6}, \dfrac{1}{8}, \dfrac{1}{9}, \dfrac{1}{10}$ of a circle's circumference results in whole-number angles. For example, the measure of the angle that corresponds with $\dfrac{1}{10}$ of the circumference of the circle is $\dfrac{1}{10}(360°) = 36°$.

PEER INTO THE PAST

WHY ARE THERE 360° IN A CIRCLE?

The reason that we divide a circle into 360° is based on historical convention. The ancient Babylonians (approximately 3000 B.C.), who lived in Mesopotamia (southern Iraq), used a base-60 number system. Therefore, it made sense to them to use 360° as opposed to 100°, which might make more sense to us today. Back then, people believed that the Sun circled Earth and that it took about 360 days for the Sun to complete one circuit. It is believed that the ancient Babylonians invented the 360° sundial as a way to keep track of time.

We should note that the use of 360 is very convenient since 360 is divisible by 2, 3, 4, 5, 6, 8, 9, 10, 12, 15, 18, 20, and so on. This makes it very easy to divide a circle up into smaller parts using nice, "neat" fractions.

Sources: www.wonderquest.com and mathforum.org

DEGREE MEASURE

The measure of an angle θ that corresponds with an arc length of $\dfrac{1}{360}$ of the circumference of a circle is 1 degree.

EXAMPLE 7 ■ Sketching Angles Measured in Degrees

Sketch each of the following angles measured in degrees.

a. $\theta = 180°$
b. $\theta = -180°$
c. $\theta = 120°$
d. $\theta = -225°$
e. $\theta = 72°$

Solution

a. $\theta = 180°$ is the angle whose arc length is $\dfrac{180}{360}$ or $\dfrac{1}{2}$ of the full circumference of the circle, as shown in Figure 8.24.

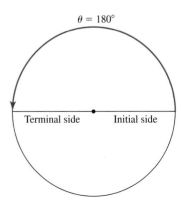

Figure 8.24

b. $\theta = -180°$ is the same as $\theta = 180°$ except the arc length traverses the circumference of the circle in the clockwise direction, as indicated in Figure 8.25.

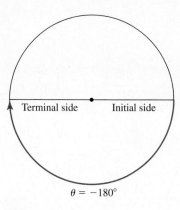

Figure 8.25

c. $\theta = 120°$ is the angle whose arc length is $\frac{120}{360}$ or $\frac{1}{3}$ of the full circumference of the circle. See Figure 8.26.

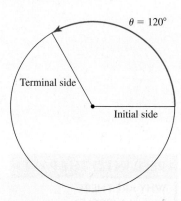

Figure 8.26

d. $\theta = -225°$ is the angle whose arc length is $\frac{225}{360}$ or $\frac{5}{8}$ of the full circumference of the circle traversed in the clockwise direction, as shown in Figure 8.27.

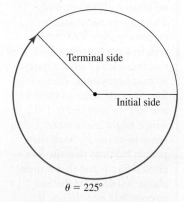

Figure 8.27

e. $\theta = 72°$ is the angle whose arc length is $\frac{72}{360}$ or $\frac{1}{5}$ of the full circumference of the circle. See Figure 8.28.

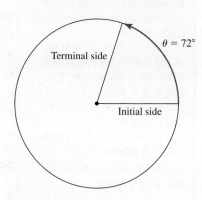

Figure 8.28

Earlier in this section, we established that an arc length, s, is related to its radius, r, by the equation $s = Kr$. Using our knowledge of the circumference of a circle, we showed that $K = 2\pi k$, where k represented the proportion of the circumference included in the arc length. We revisit these equations now to determine the relationship between the arc length and the degree measure of the corresponding angle. We have

$$s = Kr$$
$$= (2\pi k)r \quad \text{since } K = 2\pi k$$

If θ is an angle in degrees, then $k = \dfrac{\theta}{360}$ gives us the proportion of the circumference included in the arc length. Substituting this result, we have

$$s = (2\pi k)r$$
$$= \left(2\pi\left(\dfrac{\theta}{360}\right)\right)r$$
$$= \left(\dfrac{\pi}{180}\theta\right)r$$

> **ARC LENGTH AS A FUNCTION OF ANGLE MEASURE IN DEGREES**
>
> The length, s, of the arc corresponding with θ degrees is
>
> $$s = \left(\dfrac{\pi}{180}\theta\right)r$$
>
> where r is the length of the radius. For the unit circle, $s = \dfrac{\pi}{180}\theta$.

■ Converting from Degree Measure to Radian Measure

Since 360 degrees is equivalent to 2π radians, we can easily convert from degrees to radians. We first observe that

$$\dfrac{2\pi \text{ radians}}{360 \text{ degrees}} = \dfrac{\pi \text{ radians}}{180 \text{ degrees}}$$
$$= 1 \quad \text{since the quantities in the numerator and denominator are equal}$$

We will use this fact as we convert from degrees to radians. For example, to convert 30° to radians we have

$$30 \text{ degrees} = 30 \text{ degrees} \cdot \dfrac{\pi \text{ radians}}{180 \text{ degrees}} \quad \text{since } \dfrac{\pi \text{ radians}}{180 \text{ degrees}} = 1$$
$$= \dfrac{30 \text{ degrees}}{180 \text{ degrees}} \cdot \pi \text{ radians}$$
$$= \dfrac{1}{6}\pi \text{ radians}$$
$$= \dfrac{\pi}{6} \text{ radians}$$

So $30° = \dfrac{\pi}{6}$ radians.

> **HOW TO:** ■ **CONVERT AN ANGLE MEASURE FROM DEGREES TO RADIANS**
>
> To convert degree measure to radian measure, multiply θ degrees by $\dfrac{\pi \text{ radians}}{180 \text{ degrees}}$.

EXAMPLE 8 ■ **Converting Degree Measure to Radian Measure**

Convert each angle from degree to radian measure. Explain why your result makes sense.

a. 90° **b.** 150° **c.** −45° **d.** 1°

Solution

a. To convert from degree to radian measure, we multiply the degree measure by $\frac{\pi \text{ radians}}{180°}$.

$$90° \cdot \frac{\pi \text{ radians}}{180°} = \frac{90° \cdot \pi}{180°} \text{ radians}$$

$$= \frac{\pi}{2} \text{ radians}$$

This makes sense because 90° is $\frac{1}{4}$ of 360° and $\frac{\pi}{2}$ radians is $\frac{1}{4}$ of 2π radians.

b. To convert from degree to radian measure, we multiply the degree measure by $\frac{\pi \text{ radians}}{180°}$.

$$150° \cdot \frac{\pi \text{ radians}}{180°} = \frac{150° \cdot \pi}{180°} \text{ radians}$$

$$= \frac{5\pi}{6} \text{ radians}$$

We see $150° = \frac{5\pi}{6}$ radians.

To check our work, we verify that each angle represents the same proportion of the circumference included in the associated arc length.

$$\frac{150°}{360°} = \frac{5}{12} \qquad \frac{\frac{5\pi}{6} \text{ radians}}{2\pi \text{ radians}} = \frac{5\pi}{6} \cdot \frac{1}{2\pi}$$

$$= \frac{5}{12}$$

Both angles correspond with arc lengths that are $\frac{5}{12}$ of the circumference of the circle.

c.
$$-45° \cdot \frac{\pi \text{ radians}}{180°} = \frac{-45° \cdot \pi}{180°} \text{ radians}$$

$$= -\frac{\pi}{4} \text{ radian}$$

This makes sense because both angles correspond with arc lengths that are $\frac{1}{8}$ of the circumference of the circle traversed in the clockwise direction $\left(\frac{45}{360} = \frac{\pi/4}{2\pi} = \frac{1}{8} \right)$.

d.
$$1° \cdot \frac{\pi \text{ radians}}{180°} = \frac{1° \cdot \pi}{180°} \text{ radians}$$

$$= \frac{\pi}{180} \text{ radian}$$

This makes sense because both angles correspond with arc lengths that are $\frac{1}{360}$ of the circumference of the circle $\left(\frac{1}{360} = \frac{\pi/180}{2\pi} \right)$.

■ Converting from Radian Measure to Degree Measure

Since 360 degrees is equivalent to 2π radians, we can easily convert from radians to degrees. We first observe that

$$\frac{360 \text{ degrees}}{2\pi \text{ radians}} = \frac{180 \text{ degrees}}{\pi \text{ radians}}$$
$$= 1 \qquad \text{since the quantities in the numerator and denominator are equal}$$

We will use this fact as we convert from radians to degrees. For example, to convert $\frac{7\pi}{4}$ radians to degrees we have

$$\frac{7\pi}{4} \text{ radians} = \frac{7\pi}{4} \text{ radians} \cdot \frac{180 \text{ degrees}}{\pi \text{ radians}} \qquad \text{since } \frac{180 \text{ degrees}}{\pi \text{ radians}} = 1$$
$$= \frac{1260\pi \text{ radians}}{4\pi \text{ radians}} \text{ degrees}$$
$$= 315 \text{ degrees}$$

So $\frac{7\pi}{4}$ radians is equal to 315°.

> **HOW TO: ■ CONVERT AN ANGLE MEASURE FROM RADIANS TO DEGREES**
>
> To convert radian measure to degree measure, multiply θ radians by $\frac{180 \text{ degrees}}{\pi \text{ radians}}$.

EXAMPLE 9 ■ Converting Radian Measure to Degree Measure

Convert each angle from radian to degree measure. Then explain why the result makes sense.

a. $\frac{\pi}{6}$ radian

b. $\frac{3\pi}{2}$ radians

c. 3.5 radians

d. $-\frac{3\pi}{4}$ radians

e. 1 radian

Solution

a. To convert from radian to degree measure, we multiply the radian measure by $\frac{180°}{\pi}$.

$$\frac{\pi}{6} \cdot \frac{180°}{\pi} = \frac{180°}{6}$$
$$= 30°$$

This result makes sense because $\frac{\pi}{6}$ represents $\frac{1}{12}$ of the full circumference of a circle and 30° is $\frac{1}{12}$ of 360°.

b.
$$\frac{3\pi}{2} \cdot \frac{180°}{\pi} = \frac{540°}{2}$$
$$= 270°$$

This result makes sense because $\frac{3\pi}{2}$ represents $\frac{3}{4}$ of the full circumference of a circle and 270° is $\frac{3}{4}$ of 360°.

c.
$$3.5 \cdot \frac{180°}{\pi} = \frac{630°}{\pi}$$
$$\approx 200.54°$$

This result makes sense because 3.5 radians is between π radians and $\frac{3\pi}{2}$ and 200.54° is between 180° and 270°.

d.
$$-\frac{3\pi}{4} \cdot \frac{180°}{\pi} = -\frac{540°}{4}$$
$$= -135°$$

This result makes sense because $-\frac{3\pi}{4}$ represents $\frac{3}{8}$ of the full circumference of a circle traversed in a clockwise direction and 135° is $\frac{3}{8}$ of 360°.

e.
$$1 \cdot \frac{180°}{\pi} = \frac{180°}{4}$$
$$\approx 57.3°$$

This result makes sense because 1 radian is the arc length spanned when 1 radius length is wrapped around the circumference of the circle. We have seen in this section that this always corresponds with a length along the circumference that is in the first quadrant and less than 90°.

Ideally, all angles would be appropriately labeled with units (e.g., degrees, radians) all of the time; however, mathematicians frequently drop the unit "radian" off of angles in radian measure. Therefore, we will assume any unitless angle measure is measured in radians. For example, $\theta = 4$ is interpreted as $\theta = 4$ radians.

SUMMARY

In this section you learned that the length of an arc is directly proportional to the radius of the corresponding circle. You discovered that when the angle is measured in radians, the angle measure is the constant of proportionality. You also learned that angles are typically measured using radians or degrees. Additionally, you discovered how to convert from one angle measure to another.

8.2 EXERCISES

SKILLS AND CONCEPTS

In Exercises 1–8, describe what fraction of the circumference of a full circle is spanned by an angle with the given measure.

1. $\theta = \dfrac{\pi}{4}$ radian
2. $\theta = 2.5$ radians
3. $\theta = \dfrac{\pi}{10}$ radian
4. $\theta = \dfrac{3\pi}{5}$ radians
5. $\theta = \dfrac{11\pi}{6}$ radians
6. $\theta = \dfrac{4\pi}{3}$ radians
7. $\theta = \dfrac{7\pi}{8}$ radians
8. $\theta = 8$ radians

In Exercises 9–16, draw a sketch of an angle with the given measure. Clearly indicate what you are looking at when you view the angle you have sketched.

9. $\theta = \dfrac{1}{2}$ radian
10. $\theta = 210°$
11. $\theta = 6\pi$ radians
12. $\theta = -330°$
13. $\theta = -\dfrac{4\pi}{5}$ radians
14. $\theta = 175°$
15. $\theta = 45$ radians
16. $\theta = 12\pi°$

In Exercises 17–21, convert each angle measure from degrees to radians.

17. $\theta = 15°$
18. $\theta = -60°$
19. $\theta = 335°$
20. $\theta = 120°$
21. $\theta = 410°$

In Exercises 22–26, convert each angle measure from radians to degrees.

22. $\theta = \dfrac{\pi}{7}$
23. $\theta = \dfrac{\pi}{18}$
24. $\theta = \dfrac{4\pi}{3}$
25. $\theta = 8.5$
26. $\theta = -10$

27. What is the length of the arc that corresponds with an angle whose measure is 150° in a circle of radius 2 feet?
28. What is the length of the arc that corresponds with an angle whose measure is $\dfrac{2\pi}{3}$ radian in a circle of radius 3 meters?
29. What is the length of the arc that corresponds with an angle whose measure is 315° in a circle of radius 4 yards?
30. What is the length of the arc that corresponds with an angle whose measure is 4 radians in a circle of radius 6 centimeters?
31. What is the length of the arc that corresponds with an angle whose measure is 10° in a circle of radius 3 meters?
32. Without using a calculator, rank the following angle measures in order from most negative to most positive.

    $\dfrac{3\pi}{4}, \dfrac{\pi}{2}, -\dfrac{\pi}{5}, \dfrac{5\pi}{6}, -\dfrac{\pi}{9}$

33. What angle measure, in radians, corresponds to 2.25 rotations around a unit circle?
34. What angle measure, in radians, corresponds to -0.6 rotations around a unit circle?
35. Consider a circle of radius 2π. Find the arc length that corresponds with an angle of 60°.
36. Consider a circle of radius 2π. Find the arc length that corresponds with an angle of $-270°$.
37. Consider a circle of radius 2π. Find the arc length that corresponds with an angle of $A°$.

SHOW YOU KNOW

38. Pretend you are explaining the idea of radian measure of angles to a classmate who missed class. Write out an explanation or describe how you would help your classmate understand this idea.
39. Pretend you are explaining to another student the procedure for converting the measure of an angle from degrees to radians. She asks, "Why do you multiply by $\dfrac{\pi}{180}$?" How do you respond?
40. Explain why, without using computations, an angle measuring 180° is equivalent to one measuring π radians.
41. Explain why the measure of the angle shown is the same regardless of the location in which the angle is measured. Use the definition $\theta = k\dfrac{s}{r}$ in your explanation.

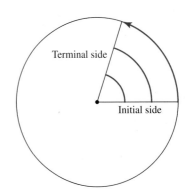

522 CHAPTER 8 ■ TRIGONOMETRIC FUNCTIONS

■ MAKE IT REAL

In Exercises 42–46, match each angle measured in radians with the picture where that angle measure is seen, approximately. Each angle measure matches with one picture.

42. $\dfrac{5\pi}{6}$

43. $\dfrac{\pi}{3}$

44. 0.25

45. π

46. 2π

A.

B.

C.

D.

E.

In Exercises 47–51, describe a real-world object, picture, or situation where you would see approximately each of the following angle measures. Draw a sketch of the angle to accompany your description.

47. $90°$

48. $\dfrac{\pi}{4}$ radian

49. 2π radians

50. $180°$

51. 1 radian

■ STRETCH YOUR MIND

Exercises 52–54 are intended to challenge your understanding of angle measure. Another unit of angle measure is the gradian. Just as there are 2π radians and $360°$ in a circle, there are 400 gradians in a circle.

52. What gradian measure represents the arc length spanning one-quarter of a circle? Half circle? One and one-half circle?

53. Convert 1 radian to gradians.

54. Recall the definition of angle uses $\theta = k\dfrac{s}{r}$. What is the constant of proportionality, k, when using gradians as the angle measure?

SECTION 8.3 Unit Circle and Trigonometric Functions

LEARNING OBJECTIVES

- Describe the relationship between the cosine and sine functions and the length of the arc swept out by the angle
- Describe the relationship between the coordinates of a point on the unit circle and the cosine and sine functions
- Use reference angles to determine cosine and sine values

GETTING STARTED

The Ferris wheel is named after George Ferris, a bridge engineer who designed the attraction for the 1893 World's Columbian Exposition in Chicago, Illinois. Ferris's wheel had a diameter of 250 feet and was an engineering marvel that astonished the world (*Source:* www.hydeparkhistory.org) It is now common to find Ferris wheels all over the world.

In this section we learn about the properties of circles as well as how to describe the coordinates of a point on the circle. These insights will help us understand ideas such as a person's height above the ground while riding a Ferris wheel.

■ Defining the Circle

A circle is defined as the set of points that are all located a fixed distance (the radius) away from another point called the center of the circle (Figure 8.29a).

The *unit circle* is the basic example of a circle from which we will develop our understanding of circle-related ideas (Figure 8.29b).

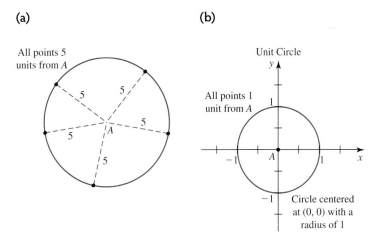

Figure 8.29

UNIT CIRCLE

The **unit circle** is a circle with its center at (0, 0) and a radius of 1 unit.

We can describe points on the unit circle in a number of ways. One is to describe the angle that corresponds with where a point is located on the arc of the circle. For example, Figure 8.30a shows that the line through the center of the circle and point B creates a 30° angle $\left(\dfrac{\pi}{6} \text{ radians}\right)$. Similarly, Figure 8.30b shows that the line through the center of the circle and point C creates a 255° angle $\left(\dfrac{17\pi}{12} \text{ radians}\right)$.

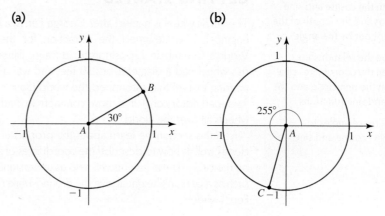

Figure 8.30

When talking about angle measures throughout this section, we assume the angles are in standard position. That is, the initial side of the angle coincides with the positive x-axis. We place the circle on a coordinate plane so that we may define points on the circle using familiar (x, y) coordinates.

EXAMPLE 1 ■ Estimating $x(\theta)$ and $y(\theta)$

As the arc length increases along a unit circle and the corresponding angle changes from 0° to 90° in increments of 10°, estimate the x- and y-values of the endpoint of the arc.

Solution We start by estimating the coordinates of the endpoint of the arc corresponding with 10°, comparing its position to the x- and y-axes. See Figure 8.31.

It appears this point is located at approximately (0.98, 0.17). Similarly, we can find the coordinates for points where the arcs correspond with 20° and 30°, as shown in Figures 8.32a and 8.32b.

Figure 8.31

Figure 8.32

SECTION 8.3 UNIT CIRCLE AND TRIGONOMETRIC FUNCTIONS

We can fill in a table of values for the remaining coordinates around the circle in the first quadrant in the same way (see Table 8.4 and Figure 8.33). We have also estimated the endpoint of the arc corresponding with 45° due to its unique characteristic of being equidistant from both axes.

Table 8.4

Angle (degrees) θ	Angle (radians) θ	Horizontal Position $x(\theta)$	Vertical Position $y(\theta)$	Coordinate Point (x, y)
0°	0	1	0	(1, 0)
10°	$\pi/18$	0.98	0.17	(0.98, 0.17)
20°	$\pi/9$	0.94	0.34	(0.94, 0.34)
30°	$\pi/6$	0.87	0.5	(0.87, 0.5)
40°	$2\pi/9$	0.77	0.64	(0.77, 0.64)
45°	$\pi/4$	0.71	0.71	(0.71, 0.71)
50°	$5\pi/18$	0.64	0.77	(0.64, 0.77)
60°	$\pi/3$	0.5	0.87	(0.5, 0.87)
70°	$7\pi/18$	0.34	0.94	(0.34, 0.94)
80°	$4\pi/9$	0.17	0.98	(0.17, 0.98)
90°	$\pi/2$	0	1	(0, 1)

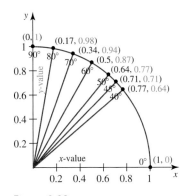

Figure 8.33

Unlike many of the functions we have worked with up to this point, functions describing these coordinates are not functions with x as the input and y as the output. Rather, both the horizontal position (x) and the vertical position (y) are functions of the angle (θ).

The ideas seen in Example 1 may be applied to arcs of any measure. Note that the x- and y-values may become negative depending on the quadrant in which the endpoint of the arc is located. See Figures 8.34a and 8.34b.

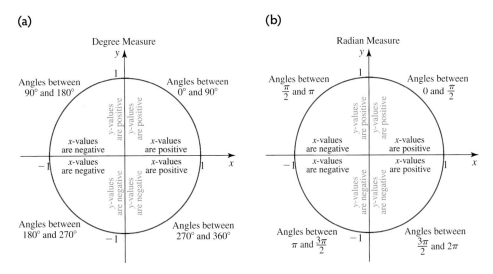

Figure 8.34

EXAMPLE 2 ■ Estimating $x(\theta)$ and $y(\theta)$

Estimate $x(\theta)$ and $y(\theta)$ for the endpoint of the arc corresponding with the given angle.

a. $\theta = 325° \left(\dfrac{65\pi}{36} \text{ radians} \right)$

b. $\theta = 110° \left(\dfrac{11\pi}{18} \text{ radians} \right)$

Figure 8.35

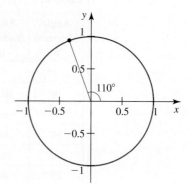

Figure 8.36

Solution

a. By estimation, $x(325°) \approx 0.82$ and $y(325°) \approx -0.57$, so we estimate that the endpoint of the arc corresponding with a 325° angle $\left(\dfrac{65\pi}{36} \text{ radians} \right)$ is located at $(0.82, -0.57)$. See Figure 8.37.

b. By estimation, $x(110°) \approx -0.34$ and $y(110°) \approx 0.94$, so we estimate that the endpoint of the arc corresponding with $110° \left(\dfrac{11\pi}{18} \text{ radians} \right)$ is at $(-0.34, 0.94)$. See Figure 8.38.

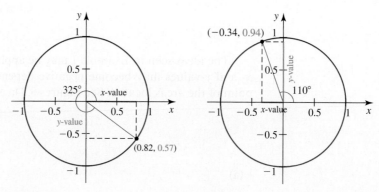

Figure 8.37

Figure 8.38

■ Circles with Radii Other Than 1

When we estimated the horizontal and vertical components of the points on the unit circle (Examples 1 and 2), each value was between -1 and 1 since the endpoint positions may not exceed the radius of the circle. However, if the circle is larger, the horizontal and vertical components will increase in magnitude. For example, if the radius of the circle is 5, the coordinates of the endpoint of an arc will range from −5 to 5, as shown in Figures 8.39a and 8.39b.

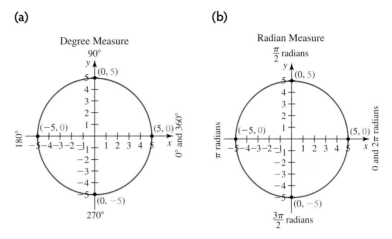

Figure 8.39

Instead of repeating the point estimation process to create a new table, we look at this as a transformation of the original situation. The unit circle has grown by a factor of 5, so the *x*- and *y*-values of the coordinates will all change by a factor of 5, as shown in Table 8.5.

Table 8.5

Angle (degrees) θ	Angle (radians) θ	Horizontal Position on the Unit Circle	Vertical Position on the Unit Circle	Horizontal Position on a Circle with Radius 5 $x(\theta)$	Vertical Position on a Circle with Radius 5 $y(\theta)$	Coordinate Point on a Circle of Radius 5 (x, y)
0°	0	1	0	5(1) = 5	5(0) = 0	(5, 0)
30°	$\pi/6$	0.87	0.5	5(0.87) ≈ 4.4	5(0.50) = 2.5	(4.4, 2.5)
60°	$\pi/3$	0.5	0.87	5(0.50) = 2.5	5(0.87) ≈ 4.4	(2.5, 4.4)
90°	$\pi/2$	0	1	5(0) = 0	5(1) = 5	(0, 5)

A similar conclusion can be made about the points when $\theta > 90°$.

This demonstrates the importance of the unit circle and the estimations of $x(\theta)$ and $y(\theta)$ made earlier. For example, $x(60°) = 0.5$ for the unit circle. This means that the *x*-coordinate of the endpoint of the arc corresponding with a 60° angle on the unit circle is 0.5, which is half the length of the radius. This relationship is the same for a circle of *any* size: If the radius of the circle is 10, then $x(60°)$ is half the radius, or 5; if the radius is 4, then $x(60°)$ is half the radius, or 2.

Thus, if we have a circle with a radius of 12 and want to know the coordinates of the endpoint of the arc corresponding with a 70° angle on the circle, we may look to the unit circle to find the solution. Using Table 8.6, we see the horizontal position is 34% of one radius length and the vertical position is 94% of one radius length.

Table 8.6

Angle (degrees) θ	Angle (radians) θ	Horizontal Position on the Unit Circle $x(\theta)$	Percentage of the Length of One Radius for Any Circle	Vertical Position on the Unit Circle $y(\theta)$	Percentage of the Length of One Radius for Any Circle
0°	0	1	100%	0	0%
10°	$\pi/18$	0.98	98%	0.17	17%
20°	$\pi/9$	0.94	94%	0.34	34%
30°	$\pi/6$	0.87	87%	0.5	50%
40°	$2\pi/9$	0.77	77%	0.64	64%
50°	$5\pi/18$	0.64	64%	0.77	77%
60°	$\pi/3$	0.5	50%	0.87	87%
70°	$7\pi/18$	0.34	34%	0.94	94%
80°	$4\pi/9$	0.17	17%	0.98	98%
90°	$\pi/2$	0	0%	1	100%

This gives us a coordinate point at (4.1, 11) since

$$x(70°) = 0.34(12) \qquad y(70°) = 0.94(12)$$
$$\approx 4.1 \qquad\qquad \approx 11$$

■ Cosine and Sine Functions

The ideas we just examined are the basis for the trigonometric functions *cosine* and *sine*. The **cosine** function (abbreviated as **cos**) is defined as the horizontal position of the endpoint of an arc on the unit circle. This is exactly what the function $x(\theta)$ represents when applied to the unit circle. Thus, when dealing with a unit circle,

$$x(\theta) = \cos(\theta)$$

> **COSINE FUNCTION**
>
> The **cosine** of an angle θ, denoted $\cos(\theta)$, is the horizontal position of the endpoint of the corresponding arc on the unit circle.

We have just discussed how to find the horizontal position of the endpoint of an arc on any circle by multiplying $\cos(\theta)$ by the radius. Therefore, for a circle of radius r, the horizontal position—denoted by $x(\theta)$—is found by

$$x(\theta) = r \cdot \cos(\theta)$$

> **HORIZONTAL POSITION OF A POINT ON A CIRCLE**
>
> The horizontal position of a point on the arc of a circle of radius r is given by $x(\theta) = r \cdot \cos(\theta)$, where θ is the corresponding angle.

PEER INTO THE PAST

WHY DO WE USE THE TERMS *COSINE* AND *SINE*?

The first time the sine value as a function of an angle appears in writing is in a sixth-century Hindu document called the *Aryabhatiya*. In this work, the sine value is identified as half of a chord within the circle, and thus it was called *ardha-jya*, which means "half-chord." This term was eventually shortened to *jiva*. Later, this important document was translated to Arabic, but the term *jiva* was left untranslated. When it was read, however, its sound was very similar to another Arabic word, *jaib*, which means "fold" or "bay" (as in "Hudson Bay"). Thus, when the term was finally translated into Latin—the language of European scholars—it was written as *sinus*, which is the Latin word for "bay" or "curve." *Sinus* was eventually shortened to *sine*.

Cosine earned its name through its connection with the sine values of angles. For angles between 0° and 90°, the cosine of the angle measure is equivalent to the sine of the complementary angle (complementary angles are angles with measures that sum to 90°). Thus, it is possible to find the sine and cosine values of angles from 0° to 45° and have enough information to fill in the values of sine and cosine for the remaining angles up to 90°.

Source: Eli Maor, *Trigonometric Delights*, Princeton University Press, 1998.

Similarly, the **sine** function (abbreviated as **sin**) is defined as the vertical position of the endpoint of an arc on the unit circle. This is what the function $y(\theta)$ represents when applied to the unit circle. Hence, when working with the unit circle,

$$y(\theta) = \sin(\theta)$$

SINE FUNCTION

The **sine** of an angle θ, denoted $\sin(\theta)$, is the vertical position of the endpoint of the corresponding arc on the unit circle.

And for a circle with radius r, the position is

$$y(\theta) = r \cdot \sin(\theta)$$

VERTICAL POSITION OF A POINT ON A CIRCLE

The vertical position of a point on the arc of a circle of radius r is given by $y(\theta) = r \cdot \sin(\theta)$, where θ is the corresponding angle.

Cosine and sine are often written without units, but it can be useful to think of their unit as "radius length" since the numbers they return tell us the horizontal and vertical positions of a point as portions of one radius length. For example, we can think of $\cos(60°) = 0.5$ as "The horizontal position of the endpoint of the arc corresponding with a 60° angle is equal to half of one radius length."

EXAMPLE 3 ■ Using and Interpreting Cosine and Sine

Evaluate each of the following and explain what the answer represents in the context of arcs of circles.

a. $\cos(20°)$
b. $4\sin(10°)$
c. $(\cos(40°), \sin(40°))$

Solution

a. $\cos(20°)$ represents the horizontal (x) position of the endpoint of the arc on the unit circle corresponding with a 20° angle. From Figure 8.40, we estimate $\cos(20°) \approx 0.94$.

b. When a constant is multiplied by either a cosine or sine value, then the circle we are working with has a radius other than 1. In this case, the radius of the circle is 4. From Table 8.6, our estimate of the vertical position of the endpoint of an arc corresponding to a 10° angle was 0.17, or 17% of the radius of the circle (see Figure 8.41a). Thus, the vertical position of the endpoint of the arc on a circle of radius 4 is 17% of 4 (see Figure 8.41b):

$$4\sin(10°) \approx 4(0.17)$$
$$\approx 0.69$$

Figure 8.40

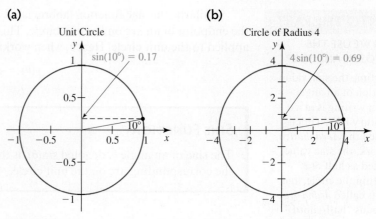

Figure 8.41

c. $(\cos(40°), \sin(40°))$ is the (x, y) coordinate point at the endpoint of the arc on the unit circle corresponding with a 40° angle, as shown in Figure 8.42. The coordinate is (0.77, 0.64) since

$$\cos(40°) \approx 0.77 \qquad \sin(40°) \approx 0.64$$

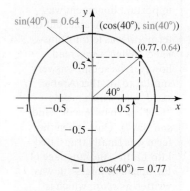

Figure 8.42

Using a Calculator to Evaluate Cosine and Sine

The visual estimates of the coordinate points on a circle we have used so far are limiting because the numbers are not exact values. Furthermore, it is difficult to accurately estimate the difference between positions of the endpoint of an arc when the corresponding angles differ by small amounts, such as between $\cos(13°)$ and $\cos(13.1°)$. Fortunately, calculators evaluate the cosine and sine values with much greater precision and for any angle value. (See the Technology Tip at the end of the section for details.)

EXAMPLE 4 ■ Evaluating Cosine and Sine Using Technology

Evaluate $\cos(213.68°)$ and $\sin(213.68°)$ and interpret their meaning for the unit circle and for a circle with a radius other than 1.

Solution A calculator gives the screens shown in Figure 8.43.

Figure 8.43

Thus the endpoint of the arc on the unit circle corresponding to an angle of 213.68° is located at about $(-0.8321, -0.5546)$.

If the radius of the circle is not 1, then the x-coordinate is 83.21% of one radius length in the negative horizontal direction and the y-coordinate is 55.46% of one radius length in the negative vertical direction. For example, if the radius is 7, then the coordinates of the endpoint of the arc are $(-5.8247, -3.8822)$ since

$$x\text{-coordinate} = (-0.8321)(7) \qquad y\text{-coordinate} = (-0.5546)(7)$$
$$= -5.8247 \qquad\qquad\qquad\qquad = -3.8822$$

$$(-5.8247, -3.8822)$$

We have thus far used $x(\theta)$ and $y(\theta)$ to name the functions finding horizontal and vertical positions, respectively. However, the use of cosine and sine now make this distinction clear, so we can now use function notation such as $f(\theta) = \cos(\theta)$.

EXAMPLE 5 ■ Using the Sine Function

The original Ferris wheel designed by George Ferris had a radius of 125 feet. If a person boards the Ferris wheel from the bottom (ground level, ignoring seat height), how high off the ground is the person after they have traveled $\frac{2}{3}$ of the way around the wheel?

Solution Since a full revolution of the wheel corresponds to an angle measuring 360°, traveling $\frac{2}{3}$ of the way around the Ferris wheel corresponds to an angle of 240° $\left(\frac{2}{3}(360°) = 240°\right)$. However, in this case, the angle is not being measured from its standard position on the positive x-axis. Instead, it is starting from the position corresponding with $-90°$. See Figure 8.44. By subtracting 90° from 240°, we can use the sine function to determine the height. Since

$$125 \sin(240° - 90°) = 125 \sin(150°)$$
$$= 62.5 \text{ feet}$$

The person is 62.5 feet above the horizontal line passing through the center of the Ferris wheel (the x-axis). To find the total height, we will need to add the length of the radius, 125 feet. See Figure 8.45 and the following calculation.

$$\text{height above ground} = 125 + 125 \sin(150°)$$
$$= 125 + 62.5$$
$$= 187.5 \text{ feet}$$

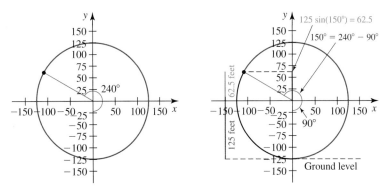

Figure 8.44   Figure 8.45

A person who has traveled $\frac{2}{3}$ around the Ferris wheel is 187.5 feet above the ground.

■ Domain and Range of the Cosine and Sine Functions

Earlier in the chapter, we discussed what it meant for angles to be negative (measured in the clockwise direction). We also indicated that when an angle surpasses 360° (2π radians), it is making more than one revolution around the circle. As the endpoint of the arc moves around the circle more than one time in a clockwise or counterclockwise direction, it will repeat all of the same values for $\cos(\theta)$ and $\sin(\theta)$ for every 360° (or 2π radians) that the angle changes. This occurs because no matter which direction we move on the circle or how many revolutions we make, we are still on the same circle and still passing through the same points. Thus, we may find the cosine or sine of *any* angle measure. Therefore, the domain of $\cos(\theta)$ and $\sin(\theta)$ includes all real numbers.

Since $\cos(\theta)$ and $\sin(\theta)$ represent horizontal and vertical positions of a point on the unit circle, and since their units are portions of a radius length, the values for these functions will always be between -1 and 1. It is impossible for the x- or y-value of a coordinate on the unit circle to be farther from the origin than the one radius. Thus, the ranges for both $x(\theta) = \cos(\theta)$ and $y(\theta) = \sin(\theta)$ are identical and include all values between -1 and 1.

EXAMPLE 6 ■ Finding Values of Cosine and Sine

Find the coordinates at the endpoints of the arc corresponding with a 405° angle and a $\dfrac{19\pi}{6}$ radian angle on the unit circle. Compare these to the coordinates of the endpoints of arcs corresponding with angles between 0° and 360° or 0 and 2π radians on the unit circle.

Solution

$$\cos(405°) \approx 0.7071 \qquad \cos\left(\dfrac{19\pi}{6}\right) \approx -0.8660$$

$$\sin(405°) \approx 0.7071 \qquad \sin\left(\dfrac{19\pi}{6}\right) \approx -0.5$$

$$(0.7071, 0.7071) \qquad (-0.8660, -0.5)$$

Since $405° = 360° + 45°$, the terminal side of a 405° angle is in the same position as the terminal side of a 45° angle. The endpoint of the corresponding arc is approximately $(0.7071, 0.7071)$, as shown in Figure 8.46.

Since $\dfrac{7\pi}{6} + 2\pi = \dfrac{19\pi}{6}$, the terminal side of a $\dfrac{19\pi}{6}$ radian angle is in the same position as the terminal side of a $\dfrac{7\pi}{6}$ radian angle. See Figure 8.47.

Figure 8.46 Figure 8.47

As Example 6 shows, we determine the endpoints of arcs corresponding with angles that are greater than 360° (or 2π radians) by looking at corresponding angles between 0° and 360° (or 0 and 2π radians). If the arc completes between one and two revolutions around the circle, we can find the corresponding angle by subtracting 360° or 2π once. If the arc makes between two and three complete revolutions around the circle, we can find the corresponding angle by subtracting 360° or 2π twice. This pattern will continue for arcs that make more than three complete revolutions. For arcs marked off in the clockwise direction (negative arcs), we instead *add* 360° or 2π radians to the angle the appropriate number of times until we get an angle that is between 0° and 360° or 0 and 2π radians.

■ Reference Angles

You may have noticed that some coordinates for the endpoints of arcs outside of the first quadrant are very similar to those corresponding to angles between 0° and 90° (or 0 and $\frac{\pi}{2}$ radians). For example, the endpoint of an arc corresponding with 225° on the unit circle is located at $(-0.71, -0.71)$ while the endpoint of an arc corresponding with a 45° arc on the unit circle is located at $(0.71, 0.71)$. This is not a coincidence. We observe that circles are symmetric about both the *x*- and *y*-axes, meaning that every point (x, y) on the circle also has corresponding points at $(-x, y)$, $(x, -y)$, and $(-x, -y)$ on the circle, as illustrated in Figure 8.48. These points have the same *reference angle*, as shown in Figures 8.49a–8.49d (on the next page).

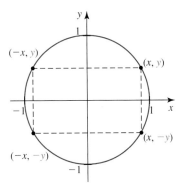

Figure 8.48

> **REFERENCE ANGLE**
>
> A **reference angle** is the smallest positive angle formed by the terminal side of the angle θ and the *x*-axis.

These graphs in Figure 8.49 demonstrate two important points. First, the reference angle will always measure between 0° and 90° (or 0 and $\frac{\pi}{2}$ radians). Second, when the reference angles are equivalent, the *x*- and *y*-values of the points on the circle will have the same magnitudes, although the signs will differ depending on the quadrant in which the point is located.

534   CHAPTER 8   ■   TRIGONOMETRIC FUNCTIONS

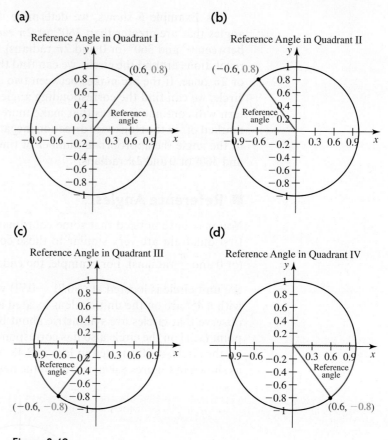

Figure 8.49

EXAMPLE 7 ■ Using Reference Angles

a. The coordinate (0.8480, 0.5299) lies on the endpoint of the arc on the unit circle corresponding with a 32° angle. Find three angles between 0° and 360° with the same reference angle and give the coordinates of the endpoint of each corresponding arc.

b. The coordinate (0.3827, 0.9239) lies on the endpoint of an arc on the unit circle corresponding with $\frac{3\pi}{8}$ radians. Find three angles between 0 and 2π with the same reference angle and give the coordinates of the endpoint of each corresponding arc.

Solution

a. The three remaining arcs will be in the second, third, and fourth quadrants and will all have a 32° reference angle. Thus, we may find these angle measures by adding or subtracting 32° from 180° and 360°. The coordinates of the endpoint of each arc will match the coordinates for the endpoint of the arc corresponding with a 32° angle, but the signs will vary according to the quadrant. See Figure 8.50.

Figure 8.50

Quadrant II	Quadrant III	Quadrant IV
$180° - 32° = 148°$	$180° + 32° = 212°$	$360° - 32° = 328°$
$(-0.8480, 0.5299)$	$(-0.8480, -0.5299)$	$(0.8480, -0.5299)$

b. Each angle will have a reference angle measuring $\frac{3\pi}{8}$ radians. Thus, we may find these angle measures by adding or subtracting $\frac{3\pi}{8}$ radians from π radians and 2π radians. The coordinates of the endpoint of each arc will match the coordinates for the endpoint of the arc corresponding with an angle measuring $\frac{3\pi}{8}$ radians, but the signs will vary according to the quadrant. See Figure 8.51.

Quadrant II	Quadrant III	Quadrant IV
$\pi - \frac{3\pi}{8} = \frac{5\pi}{8}$	$\pi + \frac{3\pi}{8} = \frac{11\pi}{8}$	$2\pi - \frac{3\pi}{8} = \frac{13\pi}{8}$
$(-0.3827, 0.9239)$	$(-0.3827, -0.9239)$	$(0.3827, -0.9239)$

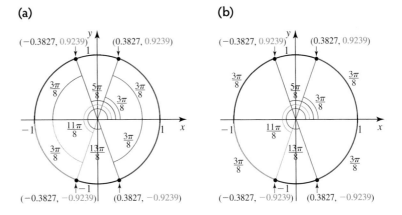

Figure 8.51

■ Exact Values of Cosine and Sine on the Unit Circle

The values we have been using for cosine and sine are rounded versions of the exact values of these functions. Certain angles appear often enough that it is beneficial to find the *exact* cosine and sine values for these angles. We may find these values using basic concepts from geometry. We begin with an equilateral triangle.

Consider an equilateral triangle arranged inside of a unit circle as shown in Figure 8.52a. Since two of the three sides are radii of the circle, and since an equilateral triangle has three sides of the same length, we know all three sides are 1 unit long. In addition, an equilateral triangle has three angles of the same measure. Since the angles in any triangle have a sum of 180°, each of the angles of the equilateral triangle must measure 60°. By using an **altitude** of the triangle (a line drawn from a vertex perpendicular to the opposite side), we can bisect this equilateral triangle, creating the right triangle shown in Figure 8.52b.

PEER INTO THE PAST

PYTHAGORAS AND HIS THEOREM

Pythagoras of Greece was born between 580 and 572 B.C. and died between 500 and 490 B.C. He is respected as a great mathematician and scientist. Because folklore and legend cloud his work, little is known about his life and teachings. He taught his followers that everything is related to mathematics and that numbers are the ultimate reality. Pythagoras believed that through mathematics everything could be predicted and measured through the rhythmic cycles and patterns found in nature.

Pythagoras is best known for the theorem that bears his name—by tradition he is credited with its discovery and proof.

The Pythagorean theorem states that in any right triangle the square of the length of the hypotenuse (side opposite the right angle) is equal to the sum of the squares of the lengths of the legs of the triangle. The theorem can be written symbolically $a^2 + b^2 = c^2$, where c represents the length of the hypotenuse, and a and b represent the lengths of the other two sides.

One of the simplest proofs of the theorem is shown here. (*Source:* www-history.mcs.st-and.ac.uk)

$$c^2 + 4\left(\frac{1}{2}ab\right) = (a+b)^2$$
$$c^2 + 2ab = a^2 + 2ab + b^2$$
$$c^2 = a^2 + b^2$$

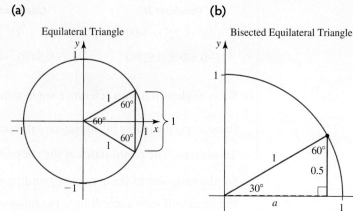

Figure 8.52

The length of the hypotenuse of the triangle is 1 and the length of the shortest leg is 0.5 or $\frac{1}{2}$. To find the length of the other leg (marked a in the figure), we use the Pythagorean theorem.

$$a^2 + b^2 = c^2$$
$$a^2 + \left(\frac{1}{2}\right)^2 = (1)^2$$
$$a^2 + \frac{1}{4} = 1$$
$$a^2 = \frac{3}{4}$$
$$a = \sqrt{\frac{3}{4}}$$
$$a = \frac{\sqrt{3}}{\sqrt{4}}$$
$$a = \frac{\sqrt{3}}{2}$$

The horizontal position of the vertex lying on the circle is $a = \frac{\sqrt{3}}{2}$. The vertical position is $\frac{1}{2}$. Thus $\cos(30°) = \frac{\sqrt{3}}{2}$ and $\sin(30°) = \frac{1}{2}$, as shown in Figure 8.53a. We may

Figure 8.53

find the exact values for cos(60°) and sin(60°) by drawing an equilateral triangle bisected by the y-axis and following the same process. The results are shown in Figure 8.53b, giving us $\cos(60°) = \frac{1}{2}$ and $\sin(60°) = \frac{\sqrt{3}}{2}$.

In radians, $\cos\left(\frac{\pi}{6}\right) = \frac{\sqrt{3}}{2}$, $\sin\left(\frac{\pi}{6}\right) = \frac{1}{2}$, $\cos\left(\frac{\pi}{3}\right) = \frac{1}{2}$, and $\sin\left(\frac{\pi}{3}\right) = \frac{\sqrt{3}}{2}$. Since the values of cosine and sine are reversed for these angles, it is easy to get them confused. However, by remembering that $\frac{\sqrt{3}}{2} \approx 0.8660$ and $\frac{1}{2} = 0.5$, we simply need to ask "to which axis is the coordinate at the endpoint of the arc closer, the x-axis or the y-axis?" If it is closer to the x-axis, the x-coordinate will be the larger value $\left(\frac{\sqrt{3}}{2}\right)$. If it is closer to the y-axis, the y-coordinate will be the larger value $\left(\frac{\sqrt{3}}{2}\right)$.

In addition to 30° and 60°, 45° is considered to be one of the *standard* angles in the first quadrant. This result is derived from an *isosceles right triangle*. An **isosceles right triangle** is a right triangle whose legs are of equal length. In Figure 8.54, we arrange a right triangle in the unit circle, with a hypotenuse of length 1 and each leg of length a. Since in triangles sides of equal length have equal angles opposite the sides, the two non-90° angles must be equal. The sum of the two non-right angles is 90°, so each angle must be $\frac{90°}{2} = 45°$.

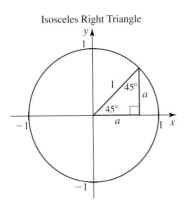

Figure 8.54

The length of the legs of the triangle represent cos(45°) and sin(45°). We again utilize the Pythagorean theorem.

$$a^2 + b^2 = c^2$$
$$a^2 + a^2 = (1)^2$$
$$2a^2 = 1$$
$$a^2 = \frac{1}{2}$$
$$a = \sqrt{\frac{1}{2}}$$
$$a = \frac{\sqrt{1}}{\sqrt{2}}$$
$$a = \frac{1}{\sqrt{2}}$$

Therefore $\cos(45°) = \frac{1}{\sqrt{2}}$ and $\sin(45°) = \frac{1}{\sqrt{2}}$. However, when writing fractions that include radicals, such as $\sqrt{2}$, it is customary to **rationalize the denominator**, which means to write the fraction so that the denominator does not contain a radical.

$$a = \frac{1}{\sqrt{2}}$$
$$= \frac{1}{\sqrt{2}} \cdot \frac{\sqrt{2}}{\sqrt{2}}$$
$$= \frac{1 \cdot \sqrt{2}}{\sqrt{2} \cdot \sqrt{2}}$$
$$= \frac{\sqrt{2}}{(\sqrt{2})^2}$$
$$= \frac{\sqrt{2}}{2}$$

So $\cos(45°) = \dfrac{\sqrt{2}}{2}$ and $\sin(45°) = \dfrac{\sqrt{2}}{2}$. In radians, $\cos\left(\dfrac{\pi}{4}\right) = \dfrac{\sqrt{2}}{2}$ and $\sin\left(\dfrac{\pi}{4}\right) = \dfrac{\sqrt{2}}{2}$. See Figure 8.55.

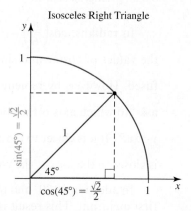

Figure 8.55

JUST IN TIME ■ RATIONALIZING DENOMINATORS

Historically, it has been a common practice to rationalize denominators for a number of reasons. First, rationalizing denominators gives us a standard way to write numbers, which is helpful when comparing two values. For example, it is not clear that $\dfrac{1}{\sqrt{2}}$ and $\dfrac{\sqrt{2}}{2}$ have the same value, but it becomes apparent after rationalizing. Thus, a student who gets $\dfrac{1}{\sqrt{2}}$ as an answer might be very frustrated when their teacher, peers, and textbook say the answer is $\dfrac{\sqrt{2}}{2}$ unless she understands that the two are equivalent. Furthermore, when doing calculations by hand, such as dividing, it is easier to use long division with $\dfrac{\sqrt{2}}{2}$ than with $\dfrac{1}{\sqrt{2}}$.

The importance of rationalizing denominators has decreased in recent years as calculators and computers now perform most calculations. For example, it only takes a few seconds to verify that $\dfrac{1}{\sqrt{2}} = \dfrac{\sqrt{2}}{2}$ using a calculator. On the other hand, many textbooks use $\dfrac{\sqrt{2}}{2}$ as the value for $\cos(45°)$ and $\sin(45°)$ instead of $\dfrac{1}{\sqrt{2}}$. Thus, learners should be aware of the practice of rationalizing denominators and know how to recognize equivalent fractions in rationalized form.

We combine the preceding results into Table 8.7 to show the exact values of cosine and sine for major angles in the first quadrant and on the adjacent axes. Then, based on what we know about reference angles, we extend this table to include angles outside of the first quadrant, as shown in Figure 8.56.

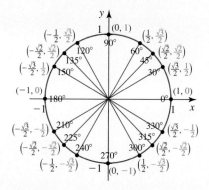

Figure 8.56

Table 8.7

Angle Measure (degrees) θ	Angle Measure (radians) θ	Reference Angle's Measure	$\cos(\theta)$	$\sin(\theta)$
0°	0	0	1	0
30°	$\frac{\pi}{6}$	30° or $\frac{\pi}{6}$	$\frac{\sqrt{3}}{2}$	$\frac{1}{2}$
45°	$\frac{\pi}{4}$	45° or $\frac{\pi}{4}$	$\frac{\sqrt{2}}{2}$	$\frac{\sqrt{2}}{2}$
60°	$\frac{\pi}{3}$	60° or $\frac{\pi}{3}$	$\frac{1}{2}$	$\frac{\sqrt{3}}{2}$
90°	$\frac{\pi}{2}$	90° or $\frac{\pi}{2}$	0	1
120°	$\frac{2\pi}{3}$	60° or $\frac{\pi}{3}$	$-\frac{1}{2}$	$\frac{\sqrt{3}}{2}$
135°	$\frac{3\pi}{4}$	45° or $\frac{\pi}{4}$	$-\frac{\sqrt{2}}{2}$	$\frac{\sqrt{2}}{2}$
150°	$\frac{5\pi}{6}$	30° or $\frac{\pi}{6}$	$-\frac{\sqrt{3}}{2}$	$\frac{1}{2}$
180°	π	0	-1	0
210°	$\frac{7\pi}{6}$	30° or $\frac{\pi}{6}$	$-\frac{\sqrt{3}}{2}$	$-\frac{1}{2}$
225°	$\frac{5\pi}{4}$	45° or $\frac{\pi}{4}$	$-\frac{\sqrt{2}}{2}$	$-\frac{\sqrt{2}}{2}$
240°	$\frac{4\pi}{3}$	60° or $\frac{\pi}{3}$	$-\frac{1}{2}$	$-\frac{\sqrt{3}}{2}$
270°	$\frac{3\pi}{2}$	90° or $\frac{\pi}{2}$	0	-1
300°	$\frac{5\pi}{3}$	60° or $\frac{\pi}{3}$	$\frac{1}{2}$	$-\frac{\sqrt{3}}{2}$
315°	$\frac{7\pi}{4}$	45° or $\frac{\pi}{4}$	$\frac{\sqrt{2}}{2}$	$-\frac{\sqrt{2}}{2}$
330°	$\frac{11\pi}{6}$	30° or $\frac{\pi}{6}$	$\frac{\sqrt{3}}{2}$	$-\frac{1}{2}$
360°	2π	0	1	0

EXAMPLE 8 ■ Finding Exact Values for Cosine and Sine

Find the exact value of each of the following expressions.

a. $\sin\left(\frac{4\pi}{3}\right)$

b. $\cos(855°)$

c. $\cos\left(-\dfrac{11\pi}{6}\right)$

Solution

a. Consulting the table, we find $\sin\left(\dfrac{4\pi}{3}\right) = -\dfrac{\sqrt{3}}{2}$.

b. First, we need to find the corresponding angle between 0° and 360°.

$$855° - 2(360°) = 135°$$
$$\cos(855°) = \cos(135°)$$
$$= -\dfrac{\sqrt{2}}{2}$$

c. We first find the corresponding angle between 0 and 2π radians.

$$-\dfrac{11\pi}{6} + 2\pi = \dfrac{\pi}{6}$$
$$\cos\left(-\dfrac{11\pi}{6}\right) = \cos\left(\dfrac{\pi}{6}\right)$$
$$= \dfrac{\sqrt{3}}{2}$$

SUMMARY

In this section you learned about the unit circle and saw how the coordinate points on a circle are found using cosine and sine. You discovered how to use reference angles to find cosine and sine values of any angle. You also learned about the domain and range of these functions and how to express the exact values of cosine or sine for standard angles.

TECHNOLOGY TIP ■ CALCULATING COSINE AND SINE

1. Press [MODE] to verify that your calculator is in the proper units (Radian or Degree) according to the angle given.

2. Press [COS] or [SIN] on your calculator, then input the angle value and close the parentheses. Rounding will often be necessary when using the calculator. It is customary to round the values of trigonometric functions to 4 decimal places.

$$\sin\left(\dfrac{3\pi}{10}\text{ radians}\right) \approx 0.8090$$

$$\cos(277°) \approx 0.1219$$

8.3 EXERCISES

SKILLS AND CONCEPTS

In Exercises 1–4, use the following figure.

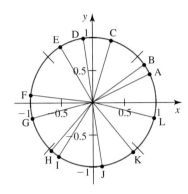

For the indicated point,

  a. Estimate θ, the angle between 0° and 360° where the point lies.
  b. Estimate cos(θ) and sin(θ).

1. A
2. E
3. G
4. K

In Exercises 5–8, use the preceding figure. For the indicated point,

  a. Estimate θ, the angle between 0 and 2π radians where the point lies.
  b. Estimate cos(θ) and sin(θ).

5. B
6. D
7. H
8. L

In Exercises 9–12, use the following figure. When we are working with a circle whose radius is not 1, we use $x = r\cos(\theta)$ and $y = r\sin(\theta)$ to find the coordinates of the endpoint of an arc. We can rewrite these formulas as $\frac{x}{r} = \cos(\theta)$ and $\frac{y}{r} = \sin(\theta)$ by dividing by r.

  a. Estimate the (x, y) coordinates for the indicated point.
  b. Use your answer in part (a) and the fact that the circle has a radius of 6 to find cos(θ) and sin(θ).
  c. Explain what your answer in part (b) represents.

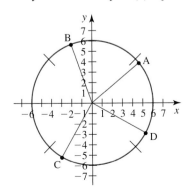

9. A
10. B
11. C
12. D

In Exercises 13–16, use a calculator to evaluate each expression. Then explain what the value means in the context of circles.

13. cos(13.5°)
14. sin(119°)
15. 0.5 sin(537°)
16. (cos(−203.6°), sin(−203.6°))

In Exercises 17–20, use a calculator to evaluate each expression. Then explain what the value means in the context of circles. The angles are all measured in radians.

17. $\cos\left(\dfrac{4\pi}{7}\right)$
18. $\sin\left(-\dfrac{\pi}{9}\right)$
19. $4.2\sin\left(-\dfrac{7\pi}{6}\right)$
20. $\left(8\cos\left(\dfrac{11\pi}{3}\right), 8\sin\left(\dfrac{11\pi}{3}\right)\right)$

In Exercises 21–26, find the exact value of each expression, then draw a picture to represent the meaning behind the value you found.

21. cos(30°)
22. cos(−315°)
23. 2.5 sin(90°)
24. 18 cos(π)
25. $\sin\left(\dfrac{2\pi}{3}\right)$
26. $-\sin\left(\dfrac{5\pi}{4}\right)$

In Exercises 27–30, you are given the cosine value for the endpoint of the arc corresponding with an angle measuring θ°. For each value of cosine,

  a. Draw a unit circle.
  b. Indicate the two possible positions on this circle where the given value of cosine would be true.
  c. Estimate the value of sin(θ) for each of these points.

27. cos(θ) = 0.43
28. cos(θ) = 0.76
29. cos(θ) = −0.6
30. cos(θ) = −0.24
31. Based on your responses to Exercises 27–30, how many positions on a circle generally have the same value for cosine? What is true about the sine values at each of these positions?

In Exercises 32–34, you are given the sine value for the endpoint of the arc corresponding with an angle of θ°. For each value of sine,

   a. Draw a unit circle.
   b. Indicate the two possible positions on this circle where the corresponding angle will have the given value of sine.
   c. Estimate the value of cos(θ) for each of these points.

32. $\sin(\theta) = 0.16$

33. $\sin(\theta) = -0.52$

34. $\sin(\theta) = -0.1$

35. Based on your responses to Exercises 32–34, how many positions on a circle generally have the same value for sine? What is true about the cosine values at each of these positions?

36. For each of the indicated intervals, describe whether the cosine and sine of θ are positive or negative and determine whether they are increasing or decreasing.

   a. $0 < \theta < \dfrac{\pi}{2}$
   b. $\dfrac{\pi}{2} < \theta < \pi$
   c. $\pi < \theta < \dfrac{3\pi}{2}$
   d. $\dfrac{3\pi}{2} < \theta < 2\pi$

In Exercises 37–40, you are given the cosine and sine value for the endpoint of an arc.

   a. Find three angles between 0° and 360° with the same reference angles.
   b. Find the cosine and sine values for each of the angles found in part (a).
   c. Draw a diagram that shows all four angles and their cosine and sine values.

37. $\cos(53°) = 0.6018$ and $\sin(53°) = 0.7986$

38. $\cos(76.2°) = 0.2385$ and $\sin(76.2°) = 0.9711$

39. $\cos(148°) = -0.8480$ and $\sin(148°) = 0.5299$

40. $\cos(262.4°) = -0.1323$ and $\sin(262.4°) = -0.9912$

In Exercises 41–44, you are given the cosine and sine value for the endpoint of an arc.

   a. Find three angles between 0 and 2π radians with the same reference angles.
   b. Find the cosine and sine values for each of the angles found in part (a).
   c. Draw a diagram that shows all four angles and their cosine and sine values.

41. $\cos\left(\dfrac{\pi}{5}\right) = 0.8090$ and $\sin\left(\dfrac{\pi}{5}\right) = 0.5878$

42. $\cos\left(\dfrac{2\pi}{9}\right) = 0.7660$ and $\sin\left(\dfrac{2\pi}{9}\right) = 0.6428$

43. $\cos\left(\dfrac{9\pi}{8}\right) = -0.9239$ and $\sin\left(\dfrac{9\pi}{8}\right) = -0.3827$

44. $\cos\left(\dfrac{12\pi}{7}\right) = 0.6235$ and $\sin\left(\dfrac{12\pi}{7}\right) = -0.7818$

45. Throughout this section you may have noticed that the cosine value of one angle may be the same sine value as another angle. In fact, there is a pattern to when $\cos(\theta) = \sin(\phi)$. In the first quadrant, the rule is that the sine of an angle is the same as the cosine of its complement (complementary angles have a sum of 90° or $\dfrac{\pi}{2}$ radians). Use the following diagram to explain why this is true.

46. As the length of an arc increases, the cosine and sine values of its endpoint pass through all of the points in their range in a specific order. In fact, they pass through the same values *in the same order* as each other. The only difference is that they do not begin at the same value.

   a. Starting with θ = 90°, find sin(θ) in 15° intervals until θ = 180°.
   b. Starting with θ = 0°, find cos(θ) in 15° intervals until θ = 90°.
   c. Compare your results in parts (a) and (b) and explain what you observe.

47. Will the trend observed in Exercise 46 continue all the way around the circle? Defend your reasoning using diagrams to strengthen your argument.

48. Fill in the blanks to formalize the observations you made in Exercises 46 and 47.
   a. cos(θ) = sin(_____)
   b. sin(θ) = cos(_____)

SHOW YOU KNOW

49. When you evaluate cos(342°) with your calculator and get the value 0.9511, does this value represent a length? Justify your answer.

50. Explain why it is impossible for $\cos(\theta) = 1.5$ or $\sin(\theta) = -3$.

51. Cosine and sine functions are periodic, meaning that their values repeat at regular intervals. Explain, with reference to the unit circle, why these functions are periodic and state the period of each function.

52. Explain how the cosine and sine values of angles on the unit circle can be used to find coordinates of points on *any* circle, and explain why this works.

53. A fellow classmate mistakenly thinks that cos(45°) = 0.5 and sin(45°) = 0.5. Explain why he might think this and why it is not possible. Use diagrams as necessary.

■ MAKE IT REAL

In Exercises 54–56, use the following information. The circumference of a circle is $C = 2\pi r$. Positions on Earth's surface are described according to lines of latitude and longitude. Lines of latitude are drawn around the Earth parallel to the equator at regular intervals determined by the arc drawn from the equator (see the figure). Each line of latitude forms a circle around the Earth.

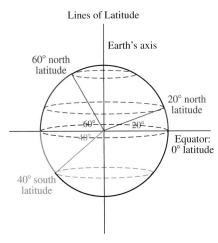

54. **Earth's Circumference** What is the circumference of Earth at the equator, assuming the Earth's radius is 3963 miles?

55. **Earth's Circumference**
    a. What happens to the circumference of the circles formed by the lines of latitude as you move farther away from the equator?
    b. What is the largest possible degree for a line of latitude and what is the circumference of the circle at that line? Where is this located?

56. **Earth's Circumference** The formula $C(\theta) = 2\pi(3963 \cos(\theta))$ models the circumference of the circle formed by a line of latitude θ degrees from the equator.
    a. Explain why $3963 \cos(\theta)$ models the radius of the circle formed by a line of latitude θ degrees from the equator.
    b. Find the radius of the circle formed by the Tropic of Cancer (23.5° north latitude) and the Tropic of Capricorn (23.5° south latitude).
    c. Explain why it is not required to use negative values for θ to get the correct circumference in the Southern Hemisphere.

57. **Ferris Wheels** The London Eye is a Ferris wheel constructed on the banks of the River Thames in London. The London Eye has a radius of about 221 feet and is boarded from the bottom. (*Source:* www.aboutbritain.com)

   Determine the height of a person from the bottom of the London Eye after traveling each of the following portions of a revolution.
    a. 1/3 of the way around
    b. 5/12 of the way around
    c. 9/10 of the way around
    d. after completing 6/5 revolutions

58. **Ferris Wheels** The Singapore Flyer is a Ferris wheel constructed on top of a three-story building. The wheel itself has a diameter of 492 feet, but the bottom of the wheel sits nearly 50 feet off the ground. Determine the height from the ground of a person after boarding from the bottom and traveling each of the following portions of a revolution. (*Source:* www.singaporeflyer.com.sg)
    a. halfway around
    b. 1/10 of the way around
    c. 4/5 of the way around
    d. 4.65 full revolutions

■ STRETCH YOUR MIND

Exercises 59–63 are intended to challenge your understanding of cosine and sine functions.

59. Use the Pythagorean theorem to write an equation involving $\cos(\theta)$, $\sin(\theta)$, and the radius in a unit circle. (*Hint:* Draw a picture of their relationship first.)

60. Explain how you can use your answer to Exercise 59 to check if it is possible for $\cos(\theta) = 0.7642$ and $\sin(\theta) = 0.4391$.

61. The following diagram shows an angle of 40° beginning at the positive y-axis. Explain why the expression $-\sin(40°)$ may be used to find the horizontal position a.

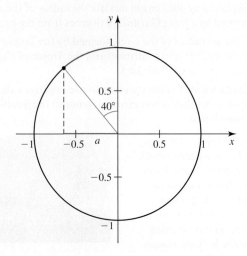

62. Consider the equation $\sin(-\theta) = -\sin(\theta)$. Is this equation true for all values of θ? Defend your answer using an explanation and diagrams.

63. Given that $\cos(\theta) = x$ and $\sin(\theta) = y$, evaluate each expression.

   a. $\cos(-\theta)$
   b. $\sin(\theta + 2\pi n)$
   c. $\sin(\pi - \theta)$
   d. $\cos(\theta + \pi)$

SECTION 8.4 Graphing Cosine and Sine Functions

LEARNING OBJECTIVES

- Use the unit circle to construct the graphs of the cosine and sine functions
- Graph cosine and sine including transformations of each
- Solve trigonometric equations graphically

GETTING STARTED

The number of hours of daylight varies throughout the year, with more hours of sunlight in the summer and less in the winter. Depending on where you live, the change in the number of hours of daylight is more or less dramatic. For example, there are up to 7.5 more hours of sunlight in the summer than in the winter in Seattle, Washington, while the difference is only 4.5 hours in Los Angeles, California. (*Source:* http://aa.usno.navy.mil)

In this section we explain how to graph cosine and sine functions. We also show how to transform them and what these transformations represent with respect to the unit circle. We also apply our understanding of cosine and sine functions to real-world periodic phenomena, such as the number of hours of daylight throughout the year.

■ Graphing $x(\theta) = \cos(\theta)$ and $y(\theta) = \sin(\theta)$

To graph $x(\theta) = \cos(\theta)$ and $y(\theta) = \sin(\theta)$, we need to change from the xy-coordinate system to a system with the angle measure as the independent variable and the trigonometric function value of the angle as the dependent variable. In these systems, the angle measure is the input value and the trigonometric function value of the angle is the output.

The graphs in Figure 8.57 show the transition from the unit circle to the graph of $x(\theta) = \cos(\theta)$. In Figure 8.57a, we see the points on the unit circle labeled with their angle measure (in degrees) and the corresponding horizontal (x) position of the endpoint for the corresponding arcs. However, these points are still on the xy-plane. We take these points and graph them as θx-coordinates in Figure 8.57b. Note the x-value appears on the vertical axis in this situation because it is the output of the function.

SECTION 8.4 ■ GRAPHING COSINE AND SINE FUNCTIONS

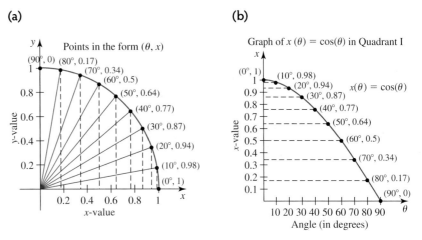

Figure 8.57

Looking closely at the function, we see that as θ changes, the x-value is not changing as much for angles close to 0° as it is for angles closer to 90°. To see why, let's examine what happens as we move around the circle using 10° intervals of θ in the first quadrant, as illustrated in Figure 8.58.

We see that the same change in the angle measurement yields a greater change in the horizontal component of each coordinate point as the angle measure approaches 90° and a greater change in the vertical component of each coordinate point when the angle is closer to 0°.

We follow the same process to graph $y(\theta) = \sin(\theta)$ from the first quadrant of the unit circle. See Figures 8.59a and 8.59b.

Figure 8.58

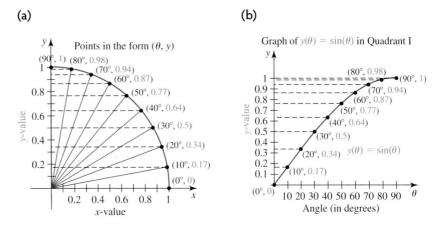

Figure 8.59

To complete the graphs of $x(\theta) = \cos(\theta)$ and $y(\theta) = \sin(\theta)$, we apply the same reasoning as we continue around the unit circle. See Figures 8.60a–8.60d.

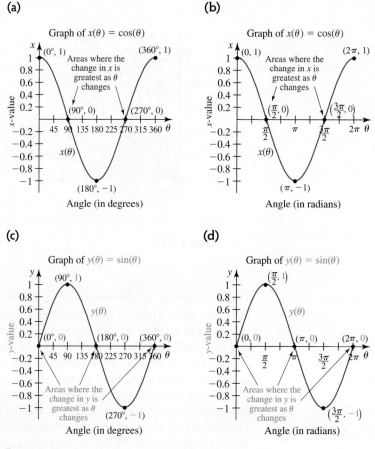

Figure 8.60

JUST IN TIME ■ INTERVAL NOTATION

The inequality of $a \leq x \leq b$ is represented as $[a, b]$ and the inequality $a < x < b$ is represented as (a, b) in interval notation. Similarly, $a \leq x < b$ is represented as $[a, b)$ and $a < x \leq b$ is represented as $(a, b]$. The interval $[a, b]$ is called a **closed** interval because it includes the endpoints. The interval (a, b) is called an **open** interval because it excludes the endpoints. Since interval notation is less cumbersome than inequality notation, it is frequently used.

The graphs of $x(\theta) = \cos(\theta)$ and $y(\theta) = \sin(\theta)$ in Figure 8.60 show the points corresponding with the interval $[0°, 360°]$ and $[0, 2\pi \text{ radians}]$, respectively. But we know we may calculate cosine and sine for *any* angle, although all of the possible range values are shown in the graphs of Figure 8.60. Cosine and sine are *periodic* functions and these graphs show only one period for each. As the angle continues to increase past 360° or 2π radians, the values shown will repeat themselves every 360° or 2π radians. As θ decreases to be less than 0° or 0 radians, the values will also repeat, but in the opposite order as θ continues to decrease (moving left on the graph). This allows us to expand the graphs to include more periods, as shown in Figures 8.61a–8.61d.

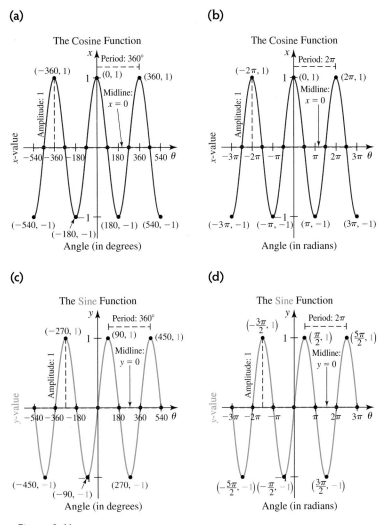

Figure 8.61

■ Characteristics of Cosine and Sine Functions

The characteristics of cos(θ) and sin(θ) are described in Table 8.8 and Figure 8.62.

Table 8.8

Function	Amplitude	Period	Frequency	Midline	Vertical Intercept	Horizontal Intercepts
$x(\theta) = \cos(\theta)$	1	360° or 2π	$\dfrac{1}{360}$ or $\dfrac{1}{2\pi}$	$x = 0$	$x(0) = 1$	$\theta = 90° + 180°n$ or $\theta = \dfrac{\pi}{2} + \pi n$
$y(\theta) = \sin(\theta)$	1	360° or 2π	$\dfrac{1}{360}$ or $\dfrac{1}{2\pi}$	$y = 0$	$y(0) = 0$	$\theta = 180°n$ or $\theta = \pi n$

A careful look at the graphs of cosine and sine shows they have identical patterns. The two functions are separated by a horizontal shift of one-quarter of a period. We will explore this idea in greater depth in the exercises at the end of this section.

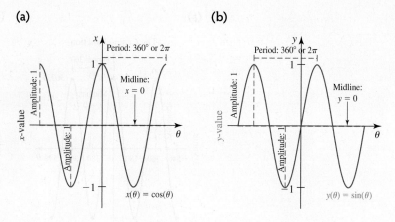

Figure 8.62

Transformations of Cosine and Sine Functions

The transformations of cosine and sine work just like the transformations we studied in Chapter 3.

EXAMPLE 1 ■ Shifting Cosine and Sine Functions

Graph each of the following functions.

a. $x(\theta) = \cos(\theta) + 2$; θ is in degrees.

b. $y(\theta) = \sin\left(\theta + \dfrac{2\pi}{3}\right)$; θ is in radians.

Solution

a. As shown in Figure 8.63, the graph of $x(\theta) = \cos(\theta) + 2$ is the graph of $x(\theta) = \cos(\theta)$ shifted upward 2 units. A vertical shift of the function changes the vertical position of the function but otherwise has no effect on the shape of the graph.

b. As shown in Figure 8.64, the graph of $y(\theta) = \sin\left(\theta + \dfrac{2\pi}{3}\right)$ is the graph of $y(\theta) = \sin(\theta)$ shifted left $\dfrac{2\pi}{3}$ radians.

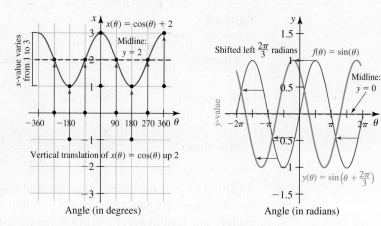

Figure 8.63   Figure 8.64

A horizontal shift of the function changes the horizontal position of the function but otherwise has no effect on the shape of the graph.

EXAMPLE 2 ■ Reflecting the Cosine and Sine Functions

Graph $f(\theta) = \sin(-\theta)$ and $g(\theta) = -\sin(\theta)$, then compare the graphs. Repeat for $h(\theta) = \cos(-\theta)$ and $j(\theta) = -\cos(\theta)$.

Solution Figure 8.65a shows that the graph of $f(\theta) = \sin(-\theta)$ is a horizontal reflection of the sine function. Figure 8.65b shows that the graph of $g(\theta) = -\sin(\theta)$ is a vertical reflection of the sine function.

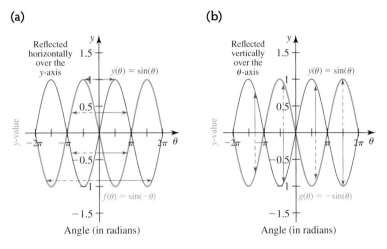

Figure 8.65

The graphs of $f(\theta) = \sin(-\theta)$ and $g(\theta) = -\sin(\theta)$ are identical, meaning that the sine function possesses *odd symmetry* (see Chapter 3).

The graph of $h(\theta) = \cos(-\theta)$ is a horizontal reflection of the cosine function, and the graph of $j(\theta) = -\cos(\theta)$ is a vertical reflection of the cosine function. However, since the cosine function is symmetric about the vertical axis, a horizontal reflection will not change the graph's appearance. Thus, the cosine function possesses *even symmetry* (see Chapter 3). A vertical reflection does change the appearance of the cosine function, as shown in Figure 8.66.

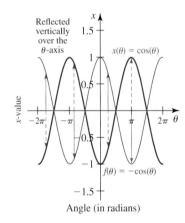

Figure 8.66

EXAMPLE 3 ■ Vertically Stretching and Compressing Cosine and Sine Functions

Graph each of the following functions. The angle θ is in degrees.

a. $x(\theta) = 5\cos(\theta)$

b. $y(\theta) = \dfrac{2}{5}\sin(\theta)$

Solution

a. Figure 8.67 shows that the graph of $x(\theta) = 5\cos(\theta)$ is the cosine function vertically stretched by a factor of 5.

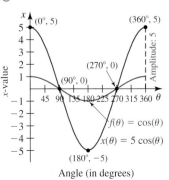

Figure 8.67

b. Figure 8.68 shows that the graph of $y(\theta) = \frac{2}{5}\sin(\theta)$ is the sine function vertically compressed by a factor of $\frac{2}{5}$.

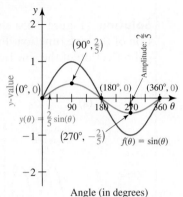

Figure 8.68

Therefore, vertical compressions and stretches affect the amplitude of a trigonometric function. The new amplitude of a vertically stretched cosine or sine function will be identical to the vertical stretch factor.

EXAMPLE 4 ■ Horizontally Stretching and Compressing Cosine and Sine Functions

Graph each of the following functions and interpret them with reference to the unit circle.

a. $x(\theta) = \cos(4\theta)$

b. $y(\theta) = \sin\left(\frac{1}{2}\theta\right)$

Solution

a. The graphs (in degrees and in radians) are shown in Figures 8.69a and 8.69b. Since 4θ changes 4 times as much when compared to θ, $x(\theta) = \cos(4\theta)$ will have a period one-fourth of 360°. Thus, θ must only change from 0° to 90° for 4θ to change from 0° to 360°, making the period of $x(\theta) = \cos(4\theta)$ 90°, or $\frac{360°}{4}$. In radians, the period is $\frac{\pi}{2}\left(\text{or } \frac{2\pi}{4}\right)$. The "4" in $x(\theta) = \cos(4\theta)$ is called the **angular frequency**—the number of periods the function will complete as θ changes by 360° or 2π radians.

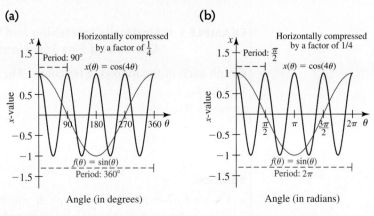

Figure 8.69

b. The graphs (in degrees and in radians) are shown in Figures 8.70a and 8.70b. The measure $\frac{1}{2}\theta$ changes only half as much as θ, meaning that θ will have to change twice as much for $y(\theta) = \sin\left(\frac{1}{2}\theta\right)$ to complete one period of sine than for $f(\theta) = \sin(\theta)$. Specifically, θ must change from 0° to 720° for $\frac{1}{2}\theta$ to change from 0° to 360°, making the period of $y(\theta) = \sin\left(\frac{1}{2}\theta\right)$ equal 720° or $\frac{360°}{1/2}$. In radians, the period is $4\pi\left(\text{or } \frac{2\pi}{1/2}\right)$. Furthermore, the angular frequency is $\frac{1}{2}$, telling us that the function will complete half of a period as θ changes by 360° or 2π radians.

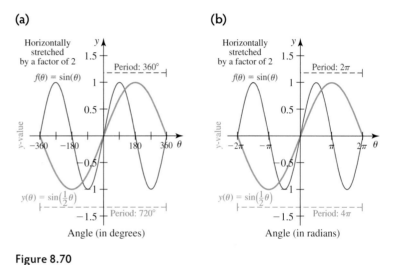

Figure 8.70

Example 4 demonstrates that horizontal stretches and compressions impact the period of a trigonometric function.

We summarize all of the above observations as follows.

Transformed Cosine and Sine Functions

For functions $x(\theta) = A\cos(B(\theta - C)) + D$ or $y(\theta) = A\sin(B(\theta - C)) + D$, we generalize the graphical meaning of their parameters as follows.

- $|A|$, the vertical stretch factor, is the *amplitude* of the graph. If $A < 0$, then the function is also reflected vertically.
- $|B|$ is the *angular frequency*, which is the number of periods that will be completed in the interval [0°, 360°] or [0, 2π radians] for θ. If $B < 0$, then the function is also reflected horizontally.
- $\frac{360°}{|B|}$ or $\frac{2\pi}{|B|}$ radians is the *period* of the function, the length of the interval required to complete one complete cycle.
- $\frac{|B|}{360°}$ or $\frac{|B|}{2\pi}$ radians is the *frequency* of the function, the fraction of a period completed within a one-unit interval.
- $|C|$ is the *horizontal shift* of the function. If C is positive, the shift is to the right. If C is negative, the shift is to the left.
- $|D|$ is the *vertical shift* of the function. If D is positive, the shift is upward. If D is negative, the shift is downward. The midline is located at $x = D$ or $y = D$.

EXAMPLE 5 ■ Multiple Transformations of Cosine and Sine Functions

Graph each of the following functions.

a. $y(\theta) = -\sin\left(2\left(\theta + \dfrac{\pi}{6}\right)\right) - 5$; θ is measured in radians.

b. $x(\theta) = \dfrac{4}{3}\cos\left(\dfrac{1}{4}(\theta - 180°)\right)$; θ is measured in degrees.

Solution

a. We create the graph for $y(\theta) = -\overset{3}{\sin}\left(\overset{1}{2}\left(\theta + \overset{2}{\dfrac{\pi}{6}}\right)\right) \overset{4}{- 5}$ by performing four transformations on $f(\theta) = \sin(\theta)$ in the following order: (1) a horizontal compression by a factor of $\dfrac{1}{2}$, (2) a horizontal shift left $\dfrac{\pi}{6}$, (3) a vertical reflection, and (4) a shift downward 5 units. See Figures 8.71a–8.71d.

(a) Step 1

(b) Step 2

(c) Step 3

(d) Step 4

Figure 8.71

SECTION 8.4 ■ GRAPHING COSINE AND SINE FUNCTIONS

b. We create the graph for $x(\theta) = \overset{3}{\tfrac{4}{3}} \cos\left(\overset{1}{\tfrac{1}{4}}(\overset{2}{\theta - 180°})\right)$ by performing three transformations on $f(\theta) = \cos(\theta)$ in the following order: (1) a horizontal stretch by a factor of 4, (2) a horizontal shift to the right 180°, and (3) a vertical stretch by a factor of $\tfrac{4}{3}$. See Figures 8.72a–8.72d.

(a) Step 1

(b) Step 2

(c) Step 3

(d) Step 4

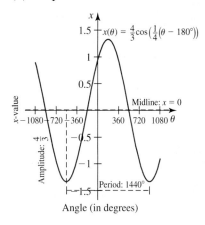

Figure 8.72

■ Phase Shifts

In Example 5, we saw that the horizontal shifts of a function are apparent when the input to the cosine or sine function is written in factored form, such as $2\left(\theta + \tfrac{\pi}{6}\right)$ in $y(\theta) = -\sin\left(2\left(\theta + \tfrac{\pi}{6}\right)\right) - 5$ or $\tfrac{1}{4}(\theta - 180)$ in $x(\theta) = \tfrac{4}{3}\cos\left(\tfrac{1}{4}(\theta - 180)\right)$. Writing these inputs in expanded form "hides" the horizontal shift, but it gives us another piece of information: the **phase shift**.

> **PHASE SHIFT**
>
> The **phase shift** is the *portion of one period* by which the function is shifted horizontally instead of the *amount* it is shifted.

Consider $x(\theta) = \frac{4}{3}\cos\left(\frac{1}{4}(\theta - 180)\right)$. Distributing the $\frac{1}{4}$ yields $x(\theta) = \frac{4}{3}\cos\left(\frac{1}{4}(\theta - 45)\right)$. The phase shift is 45°, which means that shifting the graph right 180° is equivalent to shifting the graph right $\frac{45}{360}$ of one period. Reducing the fraction tells us that this is a translation of $\frac{1}{8}$ of one period. See Figure 8.73.

With $y(\theta) = -\sin\left(2\left(\theta + \frac{\pi}{6}\right)\right) - 5$ in radians the same reasoning applies. Distributing the 2 inside of sine yields $y(\theta) = -\sin\left(2\theta + \frac{\pi}{3}\right) - 5$.

The phase shift is $\frac{\pi}{3}$. This tells us that we need to shift the graph horizontally a portion of one period equal to $\frac{\pi/3}{2\pi}$.

$$\frac{\pi/3}{2\pi} = \frac{\pi}{3} \cdot \frac{1}{2\pi}$$

$$= \frac{\pi}{6\pi}$$

$$= \frac{1}{6}$$

Thus, shifting the graph left by $\frac{\pi}{6}$ means that we are shifting it left $\frac{1}{6}$ of one period.

Figure 8.73

DETERMINING PHASE SHIFT

For the functions $x(\theta) = A\cos(B(\theta - C)) + D$ or $y(\theta) = A\sin(B(\theta - C)) + D$, the **phase shift** is $|BC|$. The value $\frac{|BC|}{360°}$ or $\frac{|BC|}{2\pi}$ tells us the portion of one period that the graph has been shifted horizontally.

■ Solving Trigonometric Equations by Graphing

So far we have worked with calculating cosine and sine values when we are given the angle. With the ability to graph these functions, we may now find the angle given the values of cosine and sine.

EXAMPLE 6 ■ Solving Trigonometric Equations by Graphing

Solve each of the following equations for θ values over the interval $[0, 2\pi]$ by graphing. Then interpret the solution.

a. $\cos(\theta) = 0.33$
b. $3\sin(2\theta) = 2.15$

Solution

a. Using the Technology Tip at the end of this section, we first verify that the calculator is in Radian mode. To solve this equation by graphing, we need to graph two functions, $y = \cos(x)$ and $y = 0.33$, and find their intersection point(s). Since we only want angles between 0 and 2π, we adjust our viewing window accordingly to get the results shown in Figure 8.74.

Figure 8.74

Angles measuring 1.2345 radians and 5.0487 radians have cosine values equal to 0.33.

b. We graph the equations $y = 3\sin(2x)$ and $y = 2.15$ and find their intersection point(s) as shown in Figure 8.75. Notice, however, that the sine function is being horizontally compressed by $\frac{1}{2}$. This means that as θ changes from 0 to 2π we will have two periods of sine, giving us four solutions to this equation on the interval $[0, 2\pi]$.

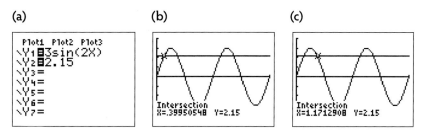

Figure 8.75

We can follow the same process to find the remaining two solutions, or we can simply add π to the first two solutions since the remaining solutions will be in the same positions in the second period. Our results tell us that a vertical position of 2.15 occurs when θ is 0.3995, 1.1713, 3.5411, and 4.3129 radians.

■ **Modeling with Cosine and Sine Functions**

As we saw earlier in this chapter, there are many situations in the real world that display periodic behavior. Cosine and sine functions can be used to model many of these situations.

EXAMPLE 7 ■ Using a Sine Model in a Real-World Context

The function $H(d) = 3.78 \sin\left(\dfrac{2\pi}{365} d\right) + 12.2$ models the number of hours of daylight in Seattle d days after March 21, 2010. Use this information to answer the following questions. (*Hint:* Be sure to use radian mode on your calculator.) (*Source:* Modeled from data at aa.usno.navy.mil)

a. What is the period, the amplitude, and the equation of the midline? What do these represent in the real-world context?

b. How many hours of daylight are there on March 21? March 1?

c. Graph H over the interval $-80 \le t \le 285$ (the data for all of 2010). Describe the rate of change and concavity of the function and explain what this information tells you about this situation.

d. What day(s) of the year will have 15 hours of daylight?

Solution

a. This is a sine function being stretched horizontally by $\frac{365}{2\pi}$. So, while the original period was 2π, the new period will be $\frac{365}{2\pi}$ times as much, or 365 days $\left(\frac{365}{2\pi} \cdot 2\pi = 365\right)$. Alternatively, we could use the fact that $\frac{2\pi}{|B|}$ will calculate the period.

$$\frac{2\pi}{|B|} = \frac{2\pi}{2\pi/365}$$

$$= 2\pi \cdot \frac{365}{2\pi}$$

$$= 365$$

This tells us that the cycle will repeat every 365 days.

The equation of the midline is $H = 12.2$ and the amplitude is 3.78 hours. We use these to find the greatest and least number of hours of daylight.

$$12.2 + 3.78 \qquad 12.2 - 3.78$$
$$15.98 \qquad\qquad 8.42$$

Seattle has at most 15.98 hours and at least 8.42 hours of daylight in a day.

b. On March 21, $d = 0$, and on March 1, $d = -20$.

$H(0) = 3.78 \sin\left(\frac{2\pi}{365}(0)\right) + 12.2 \qquad H(-20) = 3.78 \sin\left(\frac{2\pi}{365}(-20)\right) + 12.2$

$ = 3.78 \sin(0) + 12.2 \qquad\qquad = 3.78 \sin(-0.3443) + 12.2$

$ = 0 + 12.2 \qquad\qquad\qquad = -1.276 + 12$

$ = 12.2 \text{ hours} \qquad\qquad\qquad \approx 10.9 \text{ hours}$

c. The graph and viewing window specifications are shown in Figure 8.76. From $t = -80$ to $t = 0$ the graph appears to be concave up and increasing, meaning that the number of daylight hours is increasing, and increasing by more and more for each day that passes. After $t = 0$ the function becomes concave down. The number of daylight hours is still increasing, but now by smaller and smaller increments each day. After $t = 92$ or 93 (around June 21), the number of daylight hours begins to decrease, and it does so by larger and larger amounts each day until about $t = 176$ (around September 21) when the graph becomes concave up, meaning that the number of daylight hours decreases by smaller amounts each day.

(a)   (b)

Figure 8.76

d. As shown in Figure 8.77, we graph the equation $y = 15$ and find the intersection of this graph and function H. We should expect to find two solutions since we have graphed one period of the function and we know that for one period of cosine or sine there are two domain values that give us the same cosine or sine value.

(a) (b) (c)

Figure 8.77

There are about 15 hours of daylight on two days during the year: $t \approx 48$ (May 8) and $t \approx 134$ (August 2).

SUMMARY

In this section you learned how to graph the cosine and sine functions and transformations of these functions. You also learned how to solve trigonometric equations using a graph. Additionally, you discovered that cosine and sine functions can be used to model real-world situations.

TECHNOLOGY TIP ■ GRAPHING COSINE AND SINE

1. Press MODE to verify that your calculator is in the proper units (Radian or Degree) according to the angle given.

2. Use Y= to enter the sine or cosine function you want to graph using the COS or SIN button, as appropriate. The calculator will open a set of parentheses for you after pressing COS or SIN. Be sure to close the set of parentheses at the appropriate time.

3. Under the ZOOM menu, select 7:ZTrig to create a graph using a window size that fits most basic trigonometric functions.

4. Press WINDOW and adjust the maximum and minimum values to see more or less of the graph.

8.4 EXERCISES

SKILLS AND CONCEPTS

In Exercises 1–7, answer all of the following questions for each function.

a. What is the period, the amplitude, the frequency, and the equation of the midline?

b. What are the maximum and minimum values of f?

c. Graph the function without using a calculator. Make sure to graph at least one complete period.

d. Where are the horizontal intercepts (if they exist) located?

e. Where is the vertical intercept located?

f. Describe the transformation of the function as related to the graph of $f(\theta) = \sin(\theta)$ or $f(\theta) = \cos(\theta)$, as appropriate.

1. $f(\theta) = 4\cos(\theta)$; θ is in degrees.
2. $f(\theta) = \cos(2\theta)$; θ is in radians.
3. $f(\theta) = \sin\left(\frac{1}{4}\theta\right)$; θ is in degrees.
4. $f(\theta) = \cos(\theta - 120°)$; θ is in degrees.
5. $f(\theta) = 3\sin\left(\theta + \frac{\pi}{3}\right)$; θ is in radians.
6. $f(\theta) = -0.75\cos\left(\frac{1}{3}(\theta + 30°)\right)$; θ is in degrees.
7. $f(\theta) = -\sin\left(4\left(\theta - \frac{\pi}{6}\right)\right)$; θ is in radians.

8. Determine the phase shift for each function in Exercises 1–7, then interpret what the phase shift tells you about that function.

In Exercises 9–10,

a. Determine the number of periods the function will complete on the interval $0 \leq \theta \leq 360°$.

b. How many values of θ between $0°$ and $360°$ will make the equation $f(\theta) = 0.5$ true?

c. Graph the function using a calculator. (*Hint:* Set your calculator to Degree mode and your window range to Xmin=0 and Xmax=360.) Find the solutions described in part (b).

9. $f(\theta) = \cos(2\theta)$
10. $f(\theta) = 2\sin(4(\theta + 200°)) - 1$

In Exercises 11–12,

a. Determine the number of periods the function will complete on the interval $0 \leq \theta < 2\pi$.

b. How many values of θ between 0 and 2π radians will make the equation $f(\theta) = 0.25$ true?

c. Graph the function using a calculator. (*Hint:* Set your calculator to Radian mode and your window range to Xmin=0 and Xmax=2π.) Find the solutions described in part (b).

11. $f(\theta) = \sin(3\theta)$
12. $f(\theta) = 4\cos(2(\theta + \pi))$

In Exercises 13–18, state whether or not each of the following equations is true. Justify your conclusion.

13. $-\cos(\theta) = \cos(-\theta)$
14. $-\sin(\theta) = \sin(-\theta)$
15. $\cos(\theta) = \cos(-\theta)$
16. $\sin(\theta) = \sin(-\theta)$
17. $\cos(\theta - \pi) = \sin\left(\theta + \frac{3\pi}{2}\right)$
18. $\cos(\theta) = -\sin(\theta - 90°)$

In Exercises 19–20, calculate the average rate of change of the function from $\theta = \frac{\pi}{8}$ to $\theta = \frac{\pi}{6}$.

19. $f(\theta) = 2\sin(\theta) + 1$
20. $f(\theta) = 3\sin(3\theta)$

In Exercises 21–24, use the graph of $f(\theta) = \cos(\theta)$ and $g(\theta) = \sin(\theta)$ to sketch graphs of the each of the following functions.

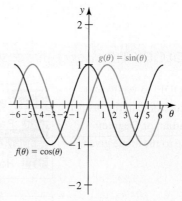

21. $f(\theta) + g(\theta)$
22. $f(\theta) - g(\theta)$
23. $(f \circ g)(\theta)$
24. $(g \circ f)(\theta)$
25. Which of the functions in Exercises 21–24 are periodic?
26. We know $f(\theta) = \cos(\theta)$ has even symmetry and that $g(\theta) = \sin(\theta)$ has odd symmetry. Do any of the functions in Exercises 21–24 possess even or odd symmetry?
27. Consider the function $f(\theta) = \cos(\theta) + \theta$, where θ is in radians.

a. Create a table and graph of f.

b. Does this function have a midline or amplitude?

c. Is this function periodic? Justify your conclusion.

■ SHOW YOU KNOW

28. Explain the similarities and differences between the graphs of the cosine and sine functions.
29. Explain why horizontally or vertically shifting the cosine or sine function does not change the period.
30. Explain why changes to the period of a cosine or sine function will never affect the amplitude.
31. A classmate tells you that it is never necessary to use the cosine function—that *every* cosine function may instead be written as a shifted sine function. Do you agree or disagree? Explain.

■ MAKE IT REAL

In Exercises 32–34, use the following information. Seasonal employment is a major part of the economy in many places. The function

$$L(m) = -550\cos\left(\frac{\pi}{6}(m-1)\right) + 3300$$

models the number of people employed in leisure and hospitality fields in Clatsop County, Oregon, m months after January 2004. (Hint: Be sure to use Radian mode on your calculator.) (**Source:** Modeled using estimates from a graph at www.qualityinfo.org)

32. **Seasonal Employment** Find the amplitude, period, equation of the midline, and phase shift for L. Then explain what each tells about this situation.

33. **Seasonal Employment**
    a. During what month are the fewest number of people employed in these industries in Clatsop County?
    b. During what month are the most people employed?
    c. What are the fewest and most employees working in this field during the year?
    d. Give a possible explanation for why employment in these industries follows this pattern.

34. **Seasonal Employment** Graph L over the interval $0 \le m \le 12$. Describe its rate of change and concavity throughout this interval and explain what this tells about employment in these industries throughout the year.

In Exercises 35–40, use the table showing the average temperature in Indianapolis, Indiana. Average temperatures are calculated by taking the average temperature for each day over a 24-hour period, and then finding the average of the daily temperatures for each month. (Hint: Be sure to use Radian mode on your calculator.)

Month of the Year (1 = January) m	Average Temperature (°F) $T(m)$
1	29.8°
2	33.0°
3	41.9°
4	53.3°
5	62.8°

(continued)

Month of the Year (1 = January) m	Average Temperature (°F) $T(m)$
6	71.3°
7	75.3°
8	74.4°
9	66.9°
10	55.1°
11	43.6°
12	32.8°

Source: www.engr.udayton.edu

35. **Average Temperatures**
    a. Plot the data for Indianapolis's average monthly temperature.
    b. Explain why a cosine or sine function will better model the data than a polynomial function.
    c. Determine the period, amplitude, and equation of the midline from the table or graph.

36. **Average Temperatures**
    a. Graph the function
    $$T(m) = -22.75\cos\left(\frac{\pi}{6}(m-1)\right) + 52.55$$
    together with the points in the table on the same set of axes.
    b. Is this model a good fit for the data?
    c. State the period, amplitude, and equation of the midline of the function. Then compare these with your answers from Exercise 35, part (c).
    d. Describe the rate of change and concavity of the function, then explain what this tells about the average temperatures in Indianapolis throughout the year.
    e. Explain why a negative cosine function was used to model this situation.

37. **Average Temperatures** Using the function $T(m)$ given in Exercise 36,
    a. Evaluate $T(10)$ and explain the meaning of the value in this context.
    b. During what month is the average temperature the greatest? What is the greatest average temperature?
    c. During what month is the average temperature the lowest? What is the lowest average temperature?
    d. What does the value $T(2.5)$ tell you? Find this value.

38. **Average Temperatures** The function
    $$H(m) = -18.20\cos\left(\frac{\pi}{6}(m-1)\right) + 60.46$$
    models the average temperatures in degrees Fahrenheit in Huntsville, Alabama, during month m of the year. Using the function $T(m)$ given in Exercise 37, graph H and T together using a calculator. Describe the similarities and differences between the two functions.

39. Average Temperatures
The function

$$S(m) = 9.40\cos\left(\frac{\pi}{6}(m-1)\right) + 57.8$$

models the average temperatures in degrees Fahrenheit in Sydney, Australia, during month m of the year. (*Source:* Modeled from data at www.engr.udayton.edu)

a. Graph H (from Exercise 38) and S together using a calculator.

b. Describe the similarities and differences between these two functions.

c. Why was a positive cosine function used to model S?

d. What accounts for the differences between the average temperatures in these locations?

e. What is the phase shift in S? Interpret this value.

f. Suppose we wanted to use a negative cosine function to model S instead. Fill in the blank to create a function that models the average temperatures in Sydney.

$$S(m) = -9.40\cos\left(\frac{\pi}{6}(\underline{\quad})\right) + 57.8$$

40. Average Temperatures
Solve the following equations by graphing using the appropriate functions from Exercises 35–39. Then interpret the solutions.

a. $T(m) = 45$
b. $H(m) = 75$
c. $S(m) = 58$
d. $S(m) = 80$

In Exercises 41–47, use the following information. The function

$$M(t) = 50\cos\left(\frac{2\pi}{29.53}(t)\right) + 50$$

models the percentage M of the full moon visible t days after January 1, 2010. (Hint: Be sure to use Radian mode on your calculator.) (Source: Modeled from data at http://aa.usno.navy.mil)

41. Phases of the Moon
What are the smallest and largest percentages of the full moon visible? Explain how these values are represented in the formula.

42. Phases of the Moon
What percentage of the full moon was visible on each of the following dates?

a. January 5
b. January 16
c. February 5

43. Phases of the Moon
Function M is created by transforming the function $f(t) = \cos(t)$. However, it was not necessary to horizontally shift the cosine function to create M. What does this tell you?

44. Phases of the Moon
What is the period of M? Interpret this value in the context of the situation.

45. Phases of the Moon
How many periods will be completed over the interval $0 \le t \le 119$ (the first four months of the year)? How many days during this interval will 75% of the full moon be visible? Explain.

46. Phases of the Moon

a. Graph M over the interval $0 \le t \le 119$.

b. Find the days during this interval where 75% of the full moon will be visible.

c. Describe the rate of change and concavity over the interval $0 \le t \le 15$ and explain what this information means in the context of the situation.

47. Phases of the Moon
In the year 2020, the first full moon will occur on January 10. Write the function $P(d)$, the percentage of the full moon visible d days after January 1, 2020 and explain how this function is a transformation of M.

In Exercises 48–49, use the following information. If you have traveled to a foreign country, you may have discovered that different countries use different standards for the voltage and frequency of their alternating current (AC) power supply. In an AC system, the current changes direction, creating positive and negative voltages according to the sine formula

$$V(t) = P\sin(2\pi f t)$$

where V is the voltage, P is the peak voltage, f is the frequency, and t is time in seconds.

48. Alternating Current
The United States uses a peak voltage of 120 volts with a frequency of 60 cycles per second.

a. Find the function for the voltage of the alternating current in the United States.

b. What is the period of this function? What does this tell you?

c. How many times per second will the voltage be 0.5? Explain how you found this answer.

d. Graph one period of this function, then use your graph to find the time(s) when the voltage will be 50.

49. Alternating Current
The United Kingdom uses a peak voltage of 240 volts with a frequency of 50 cycles per second.

a. Explain how you could create the function for the United Kingdom's voltage by performing transformations on the function created in Exercise 48. Then write the function.

b. What is the period of the United Kingdom's voltage function? What does this tell you?

c. How many times per second will the voltage be 0.5? Explain how you found this answer.

d. Graph one period of this function, then use your graph to find the time(s) when the voltage will be 50.

■ STRETCH YOUR MIND

Exercises 50–55 are intended to challenge your understanding of graphing cosine and sine functions.

50. It can be said that $f(\theta) = \sin(\theta)$ will complete one full cycle (period) whenever θ changes by 360° or 2π radians. Using this idea, answer the following questions.

  a. Describe the behavior of $h(x) = \sin(x^2)$ for $x > 0$. Then make a rough sketch of the graph of h without using a calculator. Is this function periodic? Explain.

  b. Describe the behavior of $j(x) = \sin\left(\dfrac{1}{x}\right)$ for $x > 0$. Then sketch a rough graph of j without using a calculator. Is this function periodic? Explain.

51. Create two functions—one using cosine and the other using sine—that have the following graph if θ is in degrees.

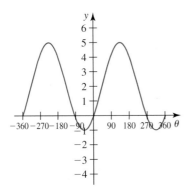

52. Create two functions—one using cosine and the other using sine—that have the following graph if θ is in radians.

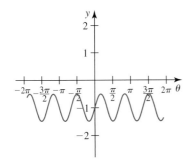

53. The following graphs show how to find $\cos\left(\theta - \dfrac{\pi}{4}\right)$ and $\sin\left(\theta - \dfrac{\pi}{4}\right)$ using the unit circle.

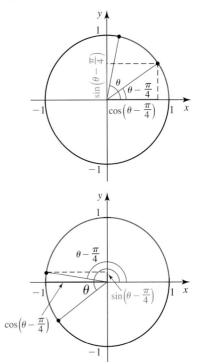

Explain, using these figures, why the graphs of $x(\theta) = \cos\left(\theta - \dfrac{\pi}{4}\right)$ and $y(\theta) = \sin\left(\theta - \dfrac{\pi}{4}\right)$ will be shifts of the original cosine and sine functions and what the new graphs represent with respect to the unit circle.

54. The following figures show the relationship between θ and 4θ.

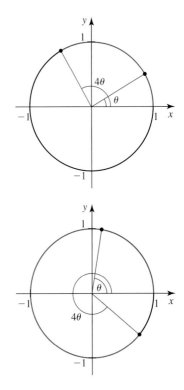

Explain, using these figures, why the graphs of $x(\theta) = \cos(4\theta)$ and $y(\theta) = \sin(4\theta)$ will be horizontal compressions of the original cosine and sine functions and what the new graphs represent with respect to the unit circle.

55. The following diagram shows a circle of radius 1 centered at $(2, -1)$.

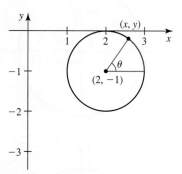

Draw the graphs of the *x*- and *y*-coordinates of the endpoints of arcs on this circle as functions of the angle θ. Explain what transformation(s) of the original cosine and sine functions are necessary to create these graphs.

SECTION 8.5 Modeling with Trigonometric Functions

LEARNING OBJECTIVES

- Generate equations for sine and cosine functions from tables and graphs
- Graph sine and cosine functions from equations and tables
- Use sine and cosine functions to model real-world data sets

GETTING STARTED

The Japanese word *tsunami* translates as "harbor wave" and is used internationally to refer to a series of waves (known as a *wave train*) traveling at hundreds of miles per hour across the ocean with extremely long wavelengths. While in the deep ocean, wave trains may have hundreds of miles between wave crests. As the waves approach shore, their speed decreases due to friction as they begin to "feel" the bottom. As the waves strike shore, they may flood low-lying coastal areas and cause massive destruction and significant loss of life, such as in Japan in March 2011. The behavior of tsunamis can be modeled and analyzed with sinusoidal functions (cosine and sine).

In this section we generate graphs and formulas for cosine and sine functions from verbal descriptions and tables of data. We also show how to use sinusoidal functions to model real-world situations to make predictions. In addition, we show how to solve sinusoidal equations using tables and graphs.

■ Sinusoidal Models

Tsunamis are generally the result of a sudden rise or fall of a section of the earth's crust under or near the ocean (in other words, an earthquake). This type of seismic disturbance can displace the water column, creating a rise or fall in the level of the ocean above. This rise or fall in sea level is the initial formation of a tsunami wave. (*Source:* www.tsunami.org)

On April 1, 1946, an earthquake with a reported magnitude of 7.8 on the Richter scale occurred in the Aleutian Islands off of the coast of Alaska. Almost 5 hours later, the largest and most destructive tsunami waves ever to strike the Hawaiian Islands penetrated more than half a mile inland in some areas. Between wave crests, the drawdown

Laupahoehoe Point Beach Park, today

SECTION 8.5 ■ MODELING WITH TRIGONOMETRIC FUNCTIONS

(pull back of the ocean from the shoreline) exposed some areas of the sea floor 500 feet in the seaward direction.

No warning was given or possible for this tsunami. As a consequence, a total of 159 tsunami-related fatalities resulted from this single destructive event. In quick response to the 1946 tsunami tragedy, scientists developed a warning system that has detected nearly every Pacific-wide tsunami since. (*Source:* www.tsunami.org)

1946 tsunami hitting Hawaiian Islands

EXAMPLE 1 ■ Generating a Sinusoidal Model from a Verbal Description

Tsunami waves generated by an earthquake of magnitude 7.8 on the Richter scale occur 10 to 20 minutes apart and travel at speeds greater than 500 miles per hour. The amount of time between successive wave crests, known as the wave period, varies from only a few minutes to more than an hour. When the April 1946 tsunami hit Laupahoehoe Point, Hawaii, it had a 27.3-foot amplitude (from normal level) and a wave period of 12 minutes. (*Source:* www.wcatwc.gov)

a. Assuming the height of the water surface above the sea floor at the base of the pier can be modeled with a sinusoidal function, generate a graph for the sine function model, given that the height was 12 feet before the tsunami hit Laupahoehoe Point. For the purposes of the graphical model, assume the water level at the pier first went down from its normal level, then rose an equal distance above its normal level, and finally returned to its normal level. Label the dependent and independent axes.

b. Write a function, $d(t)$, that models the situation.

Solution

a. We know that the height of the water surface above the sea floor at the pier is 12 feet initially. We represent this on the graph in Figure 8.78 by making 12 feet the vertical intercept at time 0 minutes. We also know the amplitude of the waves is 27.3 feet. We use this figure to find the maximum and minimum values that the water surface height reaches.

$= 12 + 27.3$ $= 12 - 27.3$

$= 39.3$ feet   *maximum water surface height* $= -15.3$ feet   *minimum water surface height*

We were told the water level went down before returning to its normal level and then rose an equal amount. This tells us the graph decreases first 27.3 feet before increasing to 27.3 feet above normal level.

We also know the wave period is 12 minutes. This is shown on the graph by making the graph return to 12 feet (the initial water surface height) after 12 minutes. Finally, because it is stated that the wave is sinusoidal, we draw a negative sine curve.

b. In the prior section, we learned about transforming the sine functions using the standard equation $y(\theta) = A \sin(B(\theta - C)) + D$. The value of A is -27.3 feet because the amplitude of the tsunami is 27.3 feet. We make it negative because the water receded first from the pier.

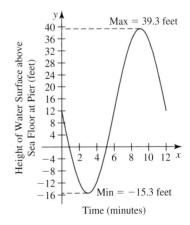

Figure 8.78

We know the tsunami had a wave period of 12 minutes. We use this information to find B.

$$\frac{2\pi}{|B|} = \text{period}$$

$$\frac{2\pi}{|B|} = 12$$

$$\frac{2\pi}{12} = |B|$$

$$|B| = \frac{\pi}{6}$$

So $B = \pm\frac{\pi}{6}$. We choose $B = \frac{\pi}{6}$ since we only need a horizontal stretch, not a horizontal stretch and horizontal reflection. We do not need to shift the graph horizontally, so $C = 0$. The vertical shift is $D = 12$ because the normal water surface height at the pier is 12 feet. The sine function that models this tsunami is

$$d(t) = -27.3 \sin\left(\frac{2\pi}{12}(t - 0)\right) + 12$$

$$= -27.3 \sin\left(\frac{\pi}{6}(t)\right) + 12$$

EXAMPLE 2 ■ Using a Sinusoidal Graph to Model Real-World Data

The graph in Figure 8.79 models the water surface height above the ocean floor at the base of the pier in Laupahoehoe Point, Hawaii, during the April 1946 tsunami.

a. Evaluate $d(4)$ and explain what this means in the real-world context of this exercise.

b. Solve the equation $d(t) = 25$ for t. Find all values of t that occur between time 0 minutes and time 12 minutes.

c. According to the graph of $d(t)$, what will be the minimum water surface height? How do you interpret this answer in terms of what happened in the real world?

Figure 8.79

Solution

a. Using the graph in Figure 8.79 to evaluate $d(4)$, we find the input of 4 minutes on the horizontal axis and then locate the output value on the vertical axis. It appears that $d(4) \approx -12$. This means that at 4 minutes, the water surface height is 12 feet below the level of the ocean floor at the pier. Since the ocean floor slants away from the pier, the ocean floor is exposed between the pier and the ocean for some distance.

b. To solve the equation $d(t) = 25$, we draw the horizontal line at $y = 25$, as shown in Figure 8.80. This line intersects the graph in two places on the interval from 0 minutes to 12 minutes. We move vertically from the points of intersection to the horizontal axis and determine $t \approx 7$ and $t \approx 11$.

Figure 8.80

The water surface height is at 25 feet at approximately 7 minutes and 11 minutes into the first period.

c. From the graph of *d(t)*, we estimate the minimum water surface height is about −15 feet. This means the base of the pier is 15 feet *above* the water surface height. As a result, the ocean floor at the pier and some distance away from the pier is exposed. How much of the ocean floor between the pier and the ocean depends on the slope of the ocean floor, which is unknown to us.

EXAMPLE 3 ■ Generating a Sinusoidal Model from a Table of Data

The Willis Tower (formerly the Sears Tower) in Chicago was designed to be flexible enough to bend rather than break in strong winds, but rigid enough so people are comfortable working and living in it without feeling the movements of the building. (*Source:* www.911research.wtc7.net)

On a windy Chicago day (35-mph winds), the distance, *s*, that the top of the Willis Tower sways varies sinusoidally with the time, *t*, in seconds. For such days, engineers have designed the tower to sway a maximum distance of 12 inches from vertical in each direction. (*Source:* www.dupont.com) Table 8.9 gives hypothetical sway distances for specific times based on a sinusoidal function.

View from top of Willis Tower

Table 8.9

Time (seconds) t	Sway Distance from Vertical (inches) d
0	0.0
6	6.0
12	10.4
18	12.0
24	10.4
30	6.0
36	0.0
42	−6.0
48	−10.4
54	−12.0
60	−10.4
66	−6.0
72	0.0

a. Using the table of values for *d(t)*, give the midline, amplitude, period, and frequency. Explain what each means in the real-world context.
b. Write a sinusoidal equation expressing the distance that the building sways.
c. Using *d(t)*, predict the sway after 9 seconds.

Solution

a. The maximum sway value is 12 and the minimum sway value is −12.

$$\text{amplitude} = \frac{\text{max} - \text{min}}{2} \qquad \text{midline} = \frac{\text{max} + \text{min}}{2}$$

$$= \frac{12 - (-12)}{2} \qquad\qquad = \frac{12 + (-12)}{2}$$

$$= 12 \qquad\qquad\qquad\qquad = 0$$

The amplitude is 12 and the midline is $y = 0$. This means the building sways 12 inches in either direction from the vertical position.

From the data table, we observe the function values begin at 0 inches, increase to 12 inches, decrease to -12 inches, and then return to 0 inches. Since this cycle takes 72 seconds, the period is 72.

The frequency is the reciprocal of the period and equals $\frac{1}{72}$. The portion of a period that occurs over a 1-second interval is $\frac{1}{72}$.

b. Since the function values start at the midline, increase to the maximum value, decrease to the minimum value, and return to the midline, a sine function model should be used for the data. The standard form for a sine function is $y(\theta) = A \sin(B(\theta - C)) + D$. From the description in part (a), we know $A = 12$, $C = 0$, and $D = 0$. To find the value of B, we use the formula $B = \frac{2\pi}{\text{period}}$ to get $B = \frac{2\pi}{72} = \frac{\pi}{36}$. Therefore, the equation for the distance d in inches that the Willis Tower sways in t seconds is $d(t) = 12 \sin\left(\frac{\pi}{36} t\right)$.

c. We evaluate $y(9)$ to find the sway after 9 seconds.

$$d(9) = 12 \sin\left(\frac{\pi}{36}(9)\right)$$
$$\approx 8.5$$

The sway is 8.5 inches after 9 seconds.

EXAMPLE 4 ■ Generating an Equation from the Graph of a Sinusoid

Pure sounds produce single sine waves on an oscilloscope. Find both the sine and cosine formula for the wave in the graph of a pure sound shown in Figure 8.81. On the vertical scale, the voltage is measured with each increment representing 0.5 volts, and on the horizontal scale each increment represents 30 milliseconds.

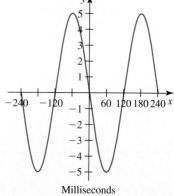

Figure 8.81

Solution To find the sine formula for the wave, we need to determine what A, B, C, and D are in the standard sine function $y(\theta) = A \sin(B(\theta - C)) + D$. The amplitude is $|A|$ and on the graph it is 5 because that is the height of the wave above the midline. Also, we note the graph is the reflection of the standard sine wave because it decreases after 0 milliseconds instead of increasing. Therefore, $A = -5$. The frequency, B, is the constant that determines the period. We know that the period for a sine wave is 360 milliseconds. This wave's period is 240 milliseconds so $B = \frac{360}{240} = 1.5$. The graph of the wave is not shifted horizontally from the vertical axis so C, the horizontal shift, is 0. Neither is the graph of this wave shifted vertically, so D, the vertical shift, is also 0. The equation for this function is therefore $y(\theta) = -5 \sin(1.5\theta)$.

Likewise, to find a cosine equation we need to know what the parameters A, B, C, and D are in the standard cosine function $y(\theta) = A \cos(B(\theta - C)) + D$. The sine and cosine functions will have the same amplitude, period, and vertical shift. We must only determine the horizontal shift. To do this, we observe there is a maximum value for the function at -60 milliseconds instead of at 0 milliseconds. Therefore, we shift the graph left 60 milliseconds from the vertical axis. This gives us the equation $y(\theta) = 5 \cos(1.5(\theta + 60))$.

8.5 EXERCISES

SKILLS AND CONCEPTS

In Exercises 1–4, state the period, amplitude, and midline.

1. $y = 5\sin(4(t-2)) + 3$
2. $y = 7\cos(0.5(t+4)) - 2$
3. $y = 8\cos(2t)$
4. $y = 3\sin(\pi(t-2)) - 0.8$

In Exercises 5–8, determine the vertical and horizontal shifts from a basic sine or cosine function.

5. $y = 3\sin(7(t-1)) + 2$
6. $y = -2\cos(t-2) - 3.5$
7. $y = 6\cos\left(3\left(t+\dfrac{\pi}{2}\right)\right) - 9$
8. $y = \sin(\pi t) - 1$

In Exercises 9–12, sketch one period of the function without using a calculator.

9. $y = 4\cos(3t)$
10. $y = 4\sin\left(t + \dfrac{\pi}{4}\right)$
11. $y = 4\sin(\pi(t+\pi)) - 3$
12. $y = 4\cos\left(2\left(t - \dfrac{\pi}{2}\right)\right) + 4$

In Exercises 13–20, find an equation for the sinusoidal function shown in the graph. You may use either sine or cosine.

13.

14.

15.

16.

17.

18.

19.

20.

In Exercises 21–25, find a possible equation for the trigonometric function whose values are shown in the table.

21.

x	y
0	2
0.1	3.2
0.2	3.9
0.3	3.9
0.4	3.2
0.5	2
0.6	0.8
0.7	0.1
0.8	0.1
0.9	0.8
1.0	2

22.

x	y
−5	4
−4	5.2
−3	6
−2	6
−1	5.2
0	4
1	2.8
2	2
3	2
4	2.8
5	4

23.

x	y
0	1
5	1.3
10	1.5
15	1.5
20	1.3
25	1
30	0.7
35	0.5
40	0.5
45	0.7
50	1

24.

x	y
−2	−4
−1	−5.2
0	−6
1	−6
2	−5.2
3	−4
4	−2.8
5	−2
6	−2
7	−2.8
8	−4

25.

x	y
1	0.5
3	0.7
5	0.5
7	0.3
9	0.5
11	0.7
13	0.5
15	0.3
17	0.5
19	0.7
21	0.5

26. Forearm The photo shows a schematic diagram of a rhythmically moving arm. The upper arm rotates back and forth from the point A; the position of the arm is measured by the angle a (in radians) between the actual position and the downward vertical position as shown in the graph.

a. Find an equation of the form $y(\theta) = A\sin(B(\theta - C)) + D$ for the graph $a(t)$.

b. How long does it take for a complete movement (swing from vertical then back-and-forth back to vertical) of the upper arm?

c. At what times during one cycle is the upper arm in the vertical position?

d. What is the fifth time that the arm is at -0.1 radians?

In Exercises 27–28, find a sine and cosine function for the wave of a pure sound shown in the graph. On the vertical scale, each increment represents 1 volt, and on the horizontal scale each increment represents 30°.

27.

28.

29. **Bouncing Spring** When a weight is attached to the end of a spring, pulled down to extend the spring, and then released, the weight begins to bounce up and down. When this occurs, the distance the weight is from the floor, d, in centimeters varies sinusoidally over a short period of time, t, in seconds. Assume a stopwatch reads 0.5 seconds when the weight reaches its first high point 42 centimeters above the floor. The next low point, 11 centimeters above the floor, occurs at 1.2 seconds.

a. Sketch a graph of this sinusoidal function.

b. Write an equation expressing the distance the weight is from the floor in terms of the number of seconds the stopwatch reads.

c. Predict the distance from the floor when the stopwatch reads 3.4 seconds.

d. What was the distance from the floor when timing started?

e. Predict the first positive value of time at which the weight is 38 centimeters above the floor.

f. Can the height the weight is off the floor be accurately modeled by the function $d(t)$ indefinitely? Explain.

30. **Product Demand** Cyclical behavior is common in the business world as the seasonal fluctuations in the demand for certain equipment are largely dependent on the weather. Common examples would include sales of surfing equipment, swimwear, and snow shovels. Assume the amount of money spent on snow skis, S, in the Colorado Springs, Colorado, vicinity can be modeled by

$$S(m) = 480.02\sin(0.49m + 1.76) + 1026.1$$

thousand dollars where m is the number of months since January 1 ($m = 0$).

a. Calculate the amplitude and explain what the numerical value means in this context.

b. Determine the vertical shift and explain what the numerical value means in this context.

c. Determine the phase shift and explain what the numerical value means in this context.

d. Determine the period and explain what the numerical value means in this context.

31. **Swinging** Two siblings (a boy and a girl) begin to play on the swing set. The graphs in the figure display a 14-second interval of time and the heights of each child in feet. Use the graphs to determine the sinusoidal equations for each child's distance from the ground as a function of the time.

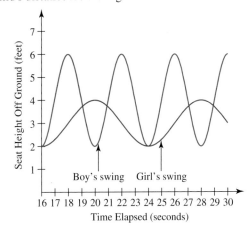

32. Paddlewheel Boat As a paddlewheel turns to drive a boat downriver, each paddle blade moves in such a way that its distance, d, from the water's surface is a sinusoidal function of time. When a stopwatch reads 5 seconds, one of the paddle blades was at its highest, 18 feet above the water's surface. Assume the wheel's diameter is 20 feet, and it completed a revolution every 9 seconds.

a. Sketch a graph of this sinusoidal function.

b. Write a function, $d(t)$, expressing the distance above the river the blade is in terms of t.

c. Find each value and interpret its meaning in the real-world context:

  i. $d(6)$ ii. $d(2.9)$
  iii. $d(-3.5)$ iv. $d(t) = 9.7$
  v. $d(t) = d(6)$ vi. $d(t) = -8$

d. Where was the paddlewheel blade when the stopwatch was started?

e. Find when the paddle entered the surface of the water. Give all answers for the first complete turn of a wheel.

f. Find when the paddle is 7 feet above the water during the second turn of the wheel.

SHOW YOU KNOW

33. How can you tell if a sinusoidal function can model a data set well?

34. Are sinusoidal functions appropriate for modeling all periodic functions? Defend your conclusion.

35. One of your classmates was absent and doesn't know how to construct a sinusoidal model when given periodic data in a table. Write an explanation to help her understand the process.

36. Explain why the period of a function and the horizontal stretch/compression factor, B, are related by the formula period $= \dfrac{360°}{|B|}$ or period $= \dfrac{2\pi}{|B|}$.

MAKE IT REAL

37. The Bay of Fundy In Canada, the Bay of Fundy tides are in constant motion and are the highest in the world. The tides may rise and fall at a rate between 6 and 8 feet per hour and have a difference between low and high water levels of nearly 50 feet. In addition, the period of time between low and high tides is 6 hours and 13 minutes. At a particular location the depth of the water,

f, in feet is given as a function of time, t, in hours since midnight by $f(t) = A \cos(B(t - C)) + D$.

a. What is the physical meaning of D?

b. What is the value of A?

c. What is the value of B?

d. What is the physical meaning of C?

38. Utilities Budget The table shows the monthly electricity costs for the year 2006 of a single-family residence. Explain why a sinusoidal function should not be used to model the electricity cost as a function of month.

Month (1 = January) m	Cost (dollars) c
1	124.19
2	111.16
3	116.30
4	133.71
5	270.07
6	320.11
7	250.23
8	391.19
9	345.91
10	256.44
11	124.87
12	126.98

39. Voltage The voltage V in an electrical circuit is given by the formula $V(t) = 5\cos(120\pi t)$, where t is the time measured in seconds.

a. Find the amplitude and the period.

b. Determine the frequency of the function and describe what this means in the real-world context.

c. Evaluate $V(0)$, $V(0.03)$, $V(0.06)$, $V(0.09)$, and $V(0.12)$.

d. Graph $V(t)$ for $0 \leq t \leq \dfrac{1}{30}$.

40. Voltage The voltage V in an electrical circuit is given by the formula $V(t) = 3.8\cos(40\pi t)$, where t is the time measured in seconds.

a. Find the amplitude and the period.

b. Determine the frequency of the function and describe what this means in the real-world context.

c. Evaluate $V(0.02)$, $V(0.04)$, $V(0.08)$, $V(0.12)$, and $V(0.14)$.

d. Graph one period of $V(t)$.

41. Pneumonia and Influenza Mortality According to the Centers for Disease Control and Prevention, the epidemic threshold and the seasonal baseline over the years 2003 to 2006 for pneumonia and influenza in 122 American cities can be modeled as shown in the graph in weeks since January 1 of each year. The epidemic threshold means that when the number of cases exceeds this number it is considered an epidemic or very serious problem. The seasonal baseline is the expected minimum number of cases. (*Source:* www.cdc.gov/flu/weekly)

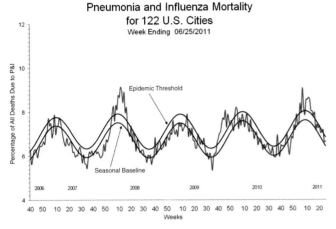

a. Use the graph of the epidemic threshold to determine the sinusoidal equation for the percent of all deaths, *p*, as a function of the week of the year, *w*.

b. Explain what the period, the amplitude, and the midline mean in terms of the real-world context for the epidemic threshold graph.

c. Evaluate *p*(30) for the year 2005 and explain what this number means.

d. Estimate in what weeks and what years the maximum epidemic threshold for the percentage of deaths due to pneumonia and influenza occur from 2003 to 2006.

e. Estimate in what weeks and what years the minimum in the epidemic threshold occurs from 2003 to 2006.

f. Give the intervals over which the function *p*(*w*) is increasing and decreasing for 2005. What information does each interval give us?

g. What does the gap between the epidemic threshold and seasonal baseline graphs mean?

42. **Hours of Daylight** The table displays the number of hours of daylight, *H*, in Seattle, Washington, *d* days after March 21, 2010. Use the table of values to answer the following questions. (*Hint:* Be sure to use Radian mode on your calculator.)

Days after March 21, 2010 d	Number of Daylight Hours H
0	12.20
30	14.07
60	15.45
90	15.98
120	15.53
150	14.21
180	12.36
210	10.48
240	9.04
270	8.43
300	8.80
330	10.06
360	11.88

Source: aa.usno.navy.mil

a. Find an equation for the sinusoidal function, *H*(*d*), to model the number of daylight hours in Seattle.

b. Use *H*(*d*) to estimate when the number of daylight hours will be 13 hours.

c. Find the average rate of change of *H*(*d*) from *d* = 30 to *d* = 120. What does this rate mean?

43. **Full Moon** The table displays the percentage *M* of the full moon visible *t* days after January 1, 2010.

Days after January 1, 2010 t	Percentage of Moon Visible M
0	100
3	91
6	63
9	31
12	8
15	0
18	8
21	29
24	59
27	87
30	100

Source: aa.usno.navy.mila

a. Find an equation for the sinusoidal function, *M*(*t*), to model the percentage of the full moon visible *t* days after January 1, 2010.

b. Use *M*(*t*) to estimate when the moon will be 33% and 50% visible.

c. Find the average rate of change of *M*(*t*) from *t* = 10 to *t* = 25. What does this rate mean?

44. **New York Temperature** The table displays the average high and low temperatures over the course of a year since January.

Month (1 = January) m	New York Average High (degrees Fahrenheit) H	New York Average Low (degrees Fahrenheit) L
1	39	26
2	42	29
3	50	35
4	60	45
5	71	55
6	79	64
7	85	70
8	83	69
9	76	61
10	65	50
11	54	41
12	44	32

Source: weather.com

a. Find an equation for the average high temperature in New York, H(m), as a function of the month of the year.

b. Find an equation for the average low temperature in New York, L(m), as a function of the month of the year.

c. Compare the two functions H(m) and L(m) in terms of the amplitude, period, and horizontal and vertical shifts. Explain any discrepancies in terms of the context of the real world.

d. Use a function transformation to write L(m) in terms of H(m).

45. Death Valley Temperature Air temperature is a cyclic phenomenon with a cycle of 12 months. We derive the function T(t) to estimate the temperature T in degrees at time t in months based on the average taken over several years. Note this function is not an accurate indicator of temperature at a specific time of day since it is normally cooler in the morning, warmer during the day, and cooler again at night. Also, for any particular year and at different times of day the temperature will vary from the average. However, it will allow us to approximate the average temperature for certain dates.

Death Valley in California is in the Mojave Desert and even during the coldest times of the year the average low temperature is quite high. Based on the average temperature data, we make the following assumptions:

- The minimum monthly low temperature of 37°F occurs at the end of January ($t = 1$).
- The maximum monthly low temperature of 86°F occurs at the end of July ($t = 7$).
- Temperature is periodic (cyclic) on an annual basis.

a. Write a function T(t) to describe the average daily low temperature at time t using a sinusoidal function.

b. Graph one cycle of the function T(t) and label the axes appropriately.

c. Label the minimum and maximum values on the graph of T(t) noting both the input and output values.

d. What is the period?

e. Determine the amplitude.

f. Determine the horizontal shift.

g. Determine the midline.

h. What is the temperature at the end of April? In what month does the temperature again reach that level?

i. When will the average low temperature be 82°F? Give the month and approximate day of the month.

j. Indicate on the graph when $T(t) = 78$ and then find the solutions for t over the interval $0 \leq t \leq 12$.

■ STRETCH YOUR MIND

Exercises 46–50 are intended to challenge your understanding of sinusoidal models.

46. Space Shuttle When the Space Shuttle is fired into orbit from a site such as Cape Canaveral, Florida, which is not on the equator, it goes into an orbit that takes it alternatively north and south of the equator. Its distance from the equator is approximately sinusoidal over time. (*Source:* www.cse.ssl.berkeley.edu)

Suppose the Space Shuttle is fired into orbit from Cape Canaveral and 12 minutes later it reaches its farthest distance north of the equator, 2485 miles. Half a cycle later it reaches its farthest distance south of the equator on the other side of Earth, also 2485 miles. Assume the Space Shuttle completes one orbit every 82 minutes.

a. Sketch one complete cycle of the number of miles, m, the Space Shuttle is from the equator at time t minutes since liftoff. (When the Shuttle is north of the equator consider the distance positive and when it is south of the equator consider the distance negative.)

b. Find an equation expressing m(t).

c. Evaluate m(30), m(55), and m(180) and explain what the values mean in this context.

d. What is the smallest positive time t for when the Space Shuttle is 1000 miles south of the equator?

e. Estimate how far Cape Canaveral is from the equator.

47. Satellite Communications Satellites orbit Earth in an elliptical path. The apogee of the orbit is the highest point the satellite is from Earth and the perigee is the lowest point. Suppose on a Low Earth Orbit (ranging from 0 to 1240 miles) a satellite is at its apogee of $d = 900$ miles at $t = 0$ hours and at its perigee of $d = 56$ miles at $t = 1$ hour.

a. Write an equation for the distance, d, the satellite is above the surface of Earth as a function of time, t.

b. Predict the first four positive values of t for which the satellite is 280 miles from the surface.

c. For how many consecutive minutes will the satellite be below 300 miles?

48. Power Utilization The first table displays the estimated power usage for the California ISO service area for a day in 2007.

Time Since Midnight (hours) t	California Power Usage (megawatts) C
0	14,235
1	12,565
2	13,637
3	13,750
4	14,432
5	13,797
6	14,013
7	12,968
8	12,159
9	11,661
10	10,661
11	8928
12	7277
13	5809

(continued)

Time Since Midnight (hours) t	California Power Usage (megawatts) C
14	3880
15	2681
16	2253
17	2904
18	4056
19	5353
20	7693
21	7432
22	7563
23	9992

Source: www.caiso.com

The second table displays the amount of power used by Arizona Power Service (APS) customers per hour for a day in the winter of 2007.

Time Since Midnight (hours) t	Arizona Power Usage (megawatts) A
0	3830
1	3813
2	3859
3	3887
4	4010
5	4273
6	4630
7	4825
8	4731
9	4553
10	4322
11	4096
12	3897
13	3698
14	3547
15	3472
16	3510
17	3822
18	4305
19	4367
20	4318
21	4213
22	3921
23	3665

Source: **Tom Glock, Manager, Power Operations for Arizona Public Service**

a. After analyzing a scatter plot for both the power usage of California, $C(t)$, and Arizona, $A(t)$, explain if it makes sense to use a sinusoidal function to model each.

b. Describe the power requirements of each state in words.

c. Find $P(t) = C(t) + A(t)$ and explain what the input and output values of $P(t)$ are.

d. What is the total power that the two power companies must prepare to provide and why is this important to know? Keep in mind the overall maximum power needed at any particular hour.

49. **Octaves** Musical notes are sounds with specific frequencies that are spaced apart according to a mathematical rule. In most Western music, the notes are organized into a 12-tone scale (see the table). A set of 12 consecutive notes is called an *octave*.

Note	Frequency (Hz)
C (middle C)	262
C sharp	277
D	294
D sharp	311
E	330
F	349
F sharp	370
G	392
G sharp	415
A	440
A sharp	466
B	494
C	523

Source: www.vibrationdata.com

The mathematical models for the frequencies of the notes D sharp and G are

$$F_{\text{D sharp}}(t) = k \sin(2\pi \cdot 311t)$$

and

$$F_G(t) = k \sin(2\pi \cdot 392t)$$

where k is the loudness of the notes at the instant each note is played, and t is the time in seconds. If we assume that the loudness k is the same for both notes, we can graph each note as shown in the figure. (Note the very short time interval graphed: from 0 to 0.02 seconds.)

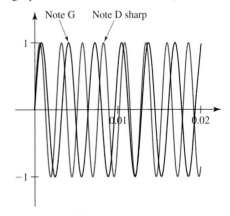

a. Graph the sum of these two notes.

b. At what times do the maximum and minimum points occur on the interval 0 to 0.02 seconds?

c. What is the frequency of the musical chord formed by playing both D sharp and G?

50. **Damped Oscillation** When a weight attached to a spring is released and begins to bob up and down, it does not continue to rise and drop the same distance for very long. As the weight moves toward equilibrium, its up-and-down movements become smaller. Suppose that we raise the weight 7 centimeters above its rest position and release it at time $t = 0$, where t is in seconds. A function that can model the behavior of the weight is $w(t) = 7(0.5^t)\cos(2\pi t)$. Graph the function $w(t)$ and determine how long it takes for the weight to come to a rest. (Assume the object is at rest when $w(t) < 0.005$.)

SECTION 8.6 Other Trigonometric Functions

LEARNING OBJECTIVES

- Define tangent, cotangent, secant, and cosecant in terms of sine and cosine
- Use the language of rate of change to describe the behavior of trigonometric functions

GETTING STARTED

Steep downhill portions of roads are very dangerous for large trucks carrying heavy loads. A driver must use a combination of vehicle brakes and engine braking (using a low gear) to maintain a safe speed and control of the vehicle. If the brakes fail from overheating or if the driver does not maintain an engaged transmission, there is potential for a deadly accident. (*Source:* www.usroads.com) What exactly does it mean when a road sign warns of an 8% downhill grade, such as this sign near Moab, Utah?

In this section we discuss the trigonometric functions tangent, cotangent, secant, and cosecant. We show how each of these functions are connected to cosine and sine, how each behaves, and how to graph each of them. We also see how these trigonometric functions are used to model real-world situations such as understanding road grades.

■ The Tangent Function

The **tangent function** (abbreviated **tan**) has an angle measure as its input, and its output gives the slope of the terminal side of an angle in standard position. Recall that the terminal side of an angle θ passes through the points $(0, 0)$ and $(\cos(\theta), \sin(\theta))$. We define tangent using the basic equation for slope.

$$\tan(\theta) = m = \frac{\text{change in } y}{\text{change in } x} = \frac{y_2 - y_1}{x_2 - x_1} = \frac{\sin(\theta) - 0}{\cos(\theta) - 0} = \frac{\sin(\theta)}{\cos(\theta)}$$

Using the unit circle, we can see how the tangent function uses this equation and how it relates to cosine and sine.

EXAMPLE 1 ■ Finding the Slope of the Terminal Side of an Angle

Find the slope of the terminal side of a 35° angle and explain how cosine and sine are involved in this calculation.

Solution We first locate the endpoint of the arc corresponding with a 35° angle on the unit circle shown in Figure 8.82. Recall that we can express the coordinates of any point on the unit circle in terms of the angle measure. In this case, the coordinate is $(\cos(35°), \sin(35°))$. We need to find the slope of the line going through this point and through the origin.

Figure 8.82

$$\tan(35°) = \frac{\sin(35°) - 0}{\cos(35°) - 0}$$

$$= \frac{\sin(35°)}{\cos(35°)}$$

$$\approx 0.7002$$

Thus we see that the tangent function is the ratio of the sine and cosine functions.

Tangent Function

The **tangent function** gives the slope of the terminal side of an angle in standard position measured in θ degrees or radians. That is,

$$m = \tan(\theta) = \frac{\sin(\theta)}{\cos(\theta)}$$

The Tangent Function and Technology

We can calculate tangent values with a calculator by using the $\boxed{\text{TAN}}$ key or by finding the ratio $\frac{\sin(\theta)}{\cos(\theta)}$. For example, the calculations for the slope of the terminal side of an angle measuring $\frac{\pi}{8}$ radians is shown in Figure 8.83 using both methods.

Figure 8.83

EXAMPLE 2 ■ Using the Tangent Function

Evaluate the expression $\tan\left(\frac{8\pi}{7}\right)$ and explain what this value represents in the context of angles.

Solution Using technology, we find $\tan\left(\frac{8\pi}{7}\right) \approx 0.4816$. This tells us that the slope of the terminal side of an $\frac{8\pi}{7}$ angle in standard position is 0.4816, as shown in Figure 8.84.

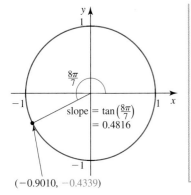

Figure 8.84

Graph of the Tangent Function

To understand what the graph of the tangent function will look like, let's examine Figure 8.85 to see how the rates of change of the terminal sides of angles change as we move around the circle.

In the first quadrant, the slope starts at 0 (horizontal) and increases until it becomes infinitely large near the positive y-axis. When we reach the positive y-axis, the slope is undefined because the line segment is vertical. As we continue into the second quadrant, we see that the slope is very steep in the negative direction and increases toward 0 as we approach the negative x-axis. Thereafter the slope continues to increase until it is once again infinitely large near the

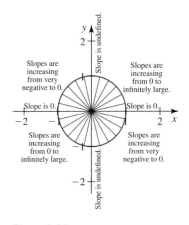

Figure 8.85

negative y-axis. The slope becomes undefined as the line segment becomes vertical, then in the fourth quadrant the slope again starts very steep in the negative direction and increases to 0.

The graph of $f(\theta) = \tan(\theta)$, in degrees and in radians, is shown in Figure 8.86. The tangent function has horizontal intercepts at (0°, 0), (180°, 0), and (360°, 0) and the function is undefined when $\theta = 90°$ and $\theta = 270°$.

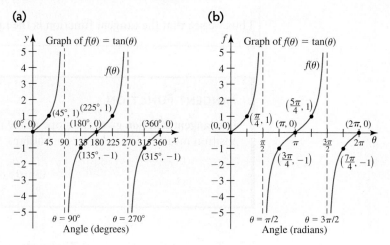

Figure 8.86

Looking closely at the graph of $f(\theta) = \tan(\theta)$, we see there are two angles between 0° and 360° (0 and 2π radians) with the same tangent value, and we observe these occur for θ-values that are separated by 180° (π radians). The tangent function, like cosine and sine, is a periodic function but has a period of 180° (π radians).

Due to the tangent function's periodic nature, its horizontal intercepts and vertical asymptotes occur at regular intervals. Since $\tan(\theta) = \dfrac{\sin(\theta)}{\cos(\theta)}$, the horizontal intercept occurs when $\dfrac{\sin(\theta)}{\cos(\theta)} = 0$, which happens each time $\sin(\theta) = 0$. Similarly, the vertical asymptotes occur when $\dfrac{\sin(\theta)}{\cos(\theta)}$ is undefined, which happens each time $\cos(\theta) = 0$. See Figure 8.87.

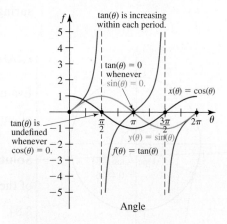

Figure 8.87

Table 8.10

	Period	Vertical Intercept	Horizontal Intercepts Occur When $\sin(\theta) = 0$	Vertical Asymptotes Occur When $\cos(\theta) = 0$
$f(\theta) = \tan(\theta)$ $= \dfrac{\sin(\theta)}{\cos(\theta)}$	180° or π radians	$f(0) = 0$	$\theta = 180°n$ or $\theta = \pi n$ radians for integer n	$\theta = 90° + 180°n$ or $\theta = \dfrac{\pi}{2} + \pi n$ radians for integer n

The domain of the tangent function includes all real numbers except where the function is undefined, which happens when $\cos(\theta) = 0$ (at $\theta = 90° + 180°n$ or $\theta = \dfrac{\pi}{2} + \pi n$ radians for all integer values of n). The range includes all real numbers since the slope of the terminal side of an angle can vary between being infinitely steep in the negative and positive directions.

EXAMPLE 3 ■ Using the Tangent Function

Roads have varying degrees of steepness. A road's steepness is reported as a percent and is calculated by dividing the change in elevation by the change in horizontal distance. Explain what it means to have an 8% downhill grade on a road. Then find the angle of descent for the road in degrees.

Solution An 8% downhill grade is a road that declines in elevation $\frac{8 \text{ vertical feet}}{100 \text{ horizontal feet}}$. In other words, for every 100-foot change in horizontal position, the road elevation decreases 8 feet (or, for every 1-foot horizontal change the elevation decreases by 0.08 feet).

If we denote the angle of descent as θ, then the 8% downhill grade tells us the following.

$$\tan(\theta) = \text{slope of the road}$$
$$= \frac{\text{vertical change}}{\text{horizontal change}}$$
$$= -\frac{8}{100}$$

Since we know the road's slope but not the angle whose terminal side gives us this slope, we solve the equation $\tan(\theta) = -\frac{8}{100}$ by graphing, as shown in Figure 8.88.

Figure 8.88

There are multiple solutions to this equation, but the incline or decline of roads will not vary too far from 0°. Keep in mind that a road that is 0° is completely level and a road that is 90° is completely straight up. The angle of descent for this road is about 4.57° below the horizontal.

■ Reciprocal Trigonometric Functions

Cosine, sine, and tangent are the three trigonometric functions typically given the most attention. The remaining three trigonometric functions are created by finding the *reciprocals* of the three main functions. Recall that the **reciprocal** of a quantity a is $1/a$. The product of a number and its reciprocal is 1. For example, the values in the following number pairs are reciprocals of each other: 2 and $\frac{1}{2}$, $\frac{5}{3}$ and $\frac{3}{5}$, or -5 and $-\frac{1}{5}$.

Although the reciprocal trigonometric functions are not used as often as cosine, sine, and tangent, they are still widely used for simplifying and solving trigonometric equations.

Cotangent Function

Similar to the tangent function, the **cotangent function**, abbreviated **cot**, is a way to describe the slope of a line. The cotangent of an angle is defined to be the reciprocal of the tangent function.

$$\cot(\theta) = \frac{1}{\tan(\theta)}$$

Substituting $\tan(\theta) = \frac{\sin(\theta)}{\cos(\theta)}$ for $\tan(\theta)$ will give us a clearer picture of how the cotangent value relates to cosine and sine.

$$\cot(\theta) = \frac{1}{\tan(\theta)}$$

$$= \frac{1}{\frac{\sin(\theta)}{\cos(\theta)}} \qquad \text{Substitute } \tan(\theta) = \frac{\sin(\theta)}{\cos(\theta)}.$$

$$= 1 \cdot \frac{\cos(\theta)}{\sin(\theta)}$$

$$= \frac{\cos(\theta)}{\sin(\theta)}$$

This shows us that the cotangent function is a ratio of the change in x to the change in y of the terminal side of an angle in standard position, which is the reciprocal of the slope equation.

COTANGENT FUNCTION

The **cotangent function** of an angle, denoted **cot(θ)**, is the reciprocal of the tangent function and represents the reciprocal of the slope of the terminal side of an angle measuring θ degrees or radians.

$$\cot(\theta) = \frac{1}{\tan(\theta)}$$

Also,

$$\cot(\theta) = \frac{\cos(\theta)}{\sin(\theta)}$$

EXAMPLE 4 ■ Finding and Interpreting the Cotangent Value

Some of the most recognizable structures on Earth are the pyramids of Egypt. While building these structures, the Egyptians utilized the cotangent relationship (which they called the *seked*) by keeping careful measurements of the lengths of each horizontal level as they increased the pyramid's height. (*Source:* Eli Maor, *Trigonometric Delights*, Princeton University Press, 1998)

The Great Pyramid at Giza, built for Khufu between 2589 and 2566 B.C. has sides that make a 51.843° angle with the ground. Find cot(51.843°) and explain what this value represents. (*Source:* www.pbs.org)

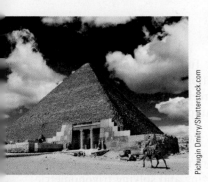

Solution To find cot(51.843°), we use a calculator to evaluate $\frac{1}{\tan(51.843°)}$ and find cot(51.843°) ≈ 0.7857, or about $\frac{78.57}{100}$. This tells us the distance from the horizontal center of the pyramid to the edge gains 0.7857 foot each time the height increases by 1 foot.

Since the width is twice the distance from the horizontal center to the edge, a 1-foot increase in height corresponds with a 1.5714-foot increase in width.

Since the cotangent function is so closely connected to the tangent function, we see many similarities between their graphs. Because they are reciprocals, as $\tan(\theta)$ increases, the fraction $\dfrac{1}{\tan(\theta)}$ decreases. Much like the tangent function, the cotangent function has horizontal intercepts and vertical asymptotes at regular intervals. We find a horizontal intercept whenever $\cos(\theta) = 0$ since this makes the ratio $\dfrac{\cos(\theta)}{\sin(\theta)} = 0$, and a vertical asymptote whenever $\sin(\theta) = 0$, which makes the ratio $\dfrac{\cos(\theta)}{\sin(\theta)}$ undefined. See Figure 8.89 and Table 8.11.

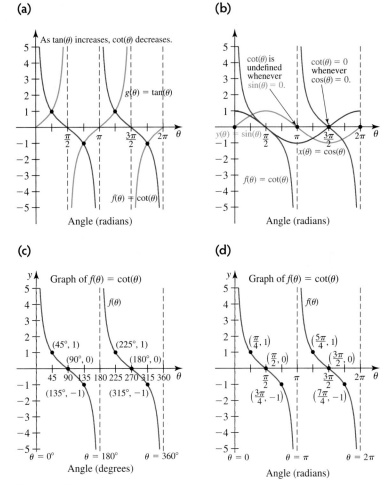

Figure 8.89

Table 8.11

	Period	Vertical Intercept	Horizontal Intercepts Occur When $\cos(\theta) = 0$	Vertical Asymptotes Occur When $\sin(\theta) = 0$
$f(\theta) = \cot(\theta)$ $= \dfrac{\cos(\theta)}{\sin(\theta)}$	180° or π radians	none	$\theta = 90° + 180°n$ or $\theta = \dfrac{\pi}{2} + \pi n$ radians for integer n	$\theta = 180°n$ or $\theta = \pi n$ radians for integer n

The cotangent function has a domain of all real numbers except where it is undefined, which occurs when $\sin(\theta) = 0$ (at $\theta = 180°n$ or $\theta = \pi n$ radians for all integer values of n). The range of the cotangent function includes all real numbers.

EXAMPLE 5 ■ Using the Graph of Cotangent

Pharoah Khafre, believed to be Khufu's son, constructed his pyramid next to his father's in Giza between 2558 and 2532 B.C. His pyramid was 471 feet tall with a square base measuring 704 feet on each side. Use the cotangent function to find the angle of incline for the side of Khafre's pyramid. (*Source:* www.pbs.org)

Solution We begin the problem with a diagram representing the situation, shown in Figure 8.90.

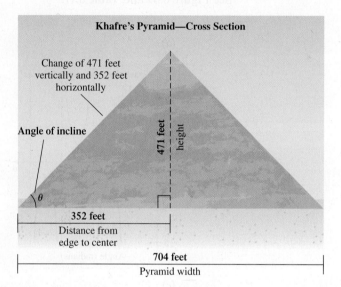

Figure 8.90

This diagram shows that the distance from the center to the edge of the pyramid changes by 352 feet as the height increases by 471 feet, giving a cotangent value of $\frac{352}{471}$, or about 0.7473 (the horizontal changes 0.7473 foot as the height increases by 1 foot). To find the angle, we solve the equation $\cot(\theta) = \frac{352}{471}$ by graphing, as shown in Figure 8.91. See the Technology Tip at the end of the section for details.

Since the function is periodic, there are many solutions. However, only an angle between 0° and 90° makes sense in this situation because 0° is flat and 90° is vertical. Also, we expect the angle of the pyramid's incline to be closer to 45°, which is halfway between flat (0°) and vertical (90°). The side of the pyramid has a 53.223° incline.

Figure 8.91

Secant Function

The **secant function** is the reciprocal of the cosine function.

> ### SECANT FUNCTION
>
> The **secant** of an angle θ, denoted **sec(θ)**, is the reciprocal of the cosine value at θ.
>
> $$\sec(\theta) = \frac{1}{\cos(\theta)}$$

The graph of the secant function, in degrees and in radians, is shown in Figure 8.92. The value of the secant function is the value of the fraction $\frac{1}{\cos(\theta)}$, so the secant varies as the cosine varies. As $\cos(\theta)$ increases, $\sec(\theta)$ decreases and as $\cos(\theta)$ decreases, $\sec(\theta)$ increases. Furthermore, the value of secant is undefined at regular intervals, whenever $\cos(\theta) = 0$. It also has the same period as the cosine function since $\frac{1}{\cos(\theta)}$ repeats its values as $\cos(\theta)$ repeats its values. (See Table 8.12.)

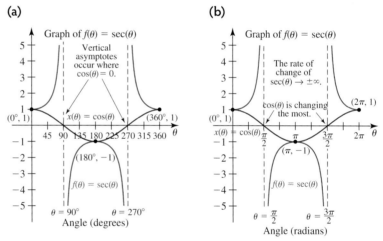

Figure 8.92

Table 8.12

	Period	Vertical Intercept	Horizontal Intercepts	Vertical Asymptotes Occur When $\cos(\theta) = 0$
$f(\theta) = \sec(\theta)$ $= \frac{1}{\cos(\theta)}$	360° or 2π radians	$f(0) = 1$	none	$\theta = 90° + 180°n$ or $\theta = \frac{\pi}{2} + \pi n$ for integer n

To further explore the connectedness of $\cos(\theta)$ and $\sec(\theta)$, consider how the rate of change of $\cos(\theta)$ determines the rate of change of $\sec(\theta)$. As indicated in Figure 8.92b, when the rate of change of $\cos(\theta)$ is close to zero (near its extrema), the rate of change for $\sec(\theta)$ is also close to zero. As the rate of change of $\cos(\theta)$ moves away from 0, the rate of change of $\sec(\theta)$ also moves away from 0.

The value of $\sec(\theta)$ will never fall between -1 and 1 because $|\cos(\theta)|$ is never greater than 1. Instead, $|\cos(\theta)| < 1$ gives us secant values such as $\frac{1}{0.5} = 2$ or $\frac{1}{0.1} = 10$. Therefore, the range of the secant function does not include all real numbers; there is a gap between -1 and 1 that the secant function graph does not cross.

The domain includes all real numbers except where $\cos(\theta) = 0$ (at $\theta = 90° + 180°n$ or $\theta = \frac{\pi}{2} + \pi n$ radians for all integer values of n).

Cosecant Function

The **cosecant function** is the reciprocal of the sine function.

> **COSECANT FUNCTION**
>
> The **cosecant function** of an angle θ, denoted **csc(θ)**, is the reciprocal of the sine value at θ.
>
> $$\csc(\theta) = \frac{1}{\sin(\theta)}$$

The graph of the cosecant function, in degrees and in radians, is shown in Figure 8.93. The summary of its characteristics is shown in Table 8.13. All our observations regarding the secant function apply to the cosecant function as well, with one small difference—the cosecant function is associated with the sine function instead of the cosine function.

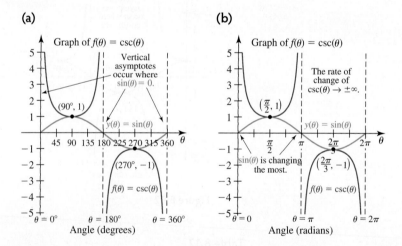

Figure 8.93

Table 8.13

	Period	Vertical Intercept	Horizontal Intercepts	Vertical Asymptotes Occur When $\sin(\theta) = 0$
$f(\theta) = \csc(\theta)$ $= \frac{1}{\sin(\theta)}$	360° or 2π radians	none	none	$\theta = 180°n$ or $\theta = \pi n$ for integer n

Like the secant function, the range includes all real numbers except those between -1 and 1. The domain includes all real numbers except where $\sin(\theta) = 0$ (at $\theta = 180°n$ or $\theta = \pi n$ radians for all integer values of n), which makes the function undefined.

Exact Values with Other Trigonometric Functions

We have already determined the exact cosine and sine values at key angles, and this information will allow us to determine the exact values of the remaining trigonometric functions at the same angles. We demonstrate a few examples of how to find these exact values and then provide a complete table. In the examples, some of these values contain radicals in the denominators. All of the values in Table 8.14 have rationalized denominators.

$$\tan(60°) = \frac{\sin(60°)}{\cos(60°)} \qquad \sec\left(\frac{\pi}{2}\right) = \frac{1}{\cos(\pi/2)} \qquad \csc(30°) = \frac{1}{\sin(30°)} \qquad \cot\left(\frac{\pi}{4}\right) = \left(\frac{\cos(\pi/4)}{\sin(\pi/4)}\right)$$

$$= \frac{\sqrt{3}/2}{1/2} \qquad\qquad = \frac{1}{0} \qquad\qquad = \frac{1}{1/2} \qquad\qquad = \frac{\sqrt{2}/2}{\sqrt{2}/2}$$

$$= \frac{\sqrt{3}}{2} \cdot \frac{2}{1} \qquad\qquad \text{undefined} \qquad\qquad = 2 \qquad\qquad = 1$$

$$= \sqrt{3}$$

Table 8.14

Exact Values of the Six Trigonometric Functions for Major Reference Angles							
Angle (degrees) θ	Angle (radians) θ	Cosine $\cos(\theta)$	Sine $\sin(\theta)$	Tangent $\tan(\theta)$	Cotangent $\cot(\theta)$	Secant $\sec(\theta)$	Cosecant $\csc(\theta)$
0°	0	1	0	0	undefined	1	undefined
30°	$\frac{\pi}{6}$	$\frac{\sqrt{3}}{2}$	$\frac{1}{2}$	$\frac{\sqrt{3}}{3}$	$\sqrt{3}$	$\frac{2\sqrt{3}}{3}$	2
45°	$\frac{\pi}{4}$	$\frac{\sqrt{2}}{2}$	$\frac{\sqrt{2}}{2}$	1	1	$\sqrt{2}$	$\sqrt{2}$
60°	$\frac{\pi}{3}$	$\frac{1}{2}$	$\frac{\sqrt{3}}{2}$	$\sqrt{3}$	$\frac{\sqrt{3}}{3}$	2	$\frac{2\sqrt{3}}{3}$
90°	$\frac{\pi}{2}$	0	1	undefined	0	undefined	1

Other exact values can be determined using reference angles while making necessary adjustments to the sign of each value based on the quadrant. The diagrams in Figure 8.94 show the signs of the six trigonometric functions in each of the four quadrants.

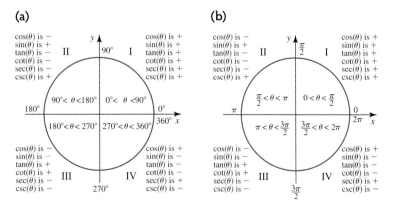

Figure 8.94

EXAMPLE 6 ■ **Evaluating Secant and Cosecant**

Evaluate each of the following expressions.

a. $\sec\left(\dfrac{3\pi}{5}\right)$ using a calculator

b. $\csc\left(\dfrac{8\pi}{3}\right)$ without using a calculator

Solution

a.
$$\sec\left(\frac{3\pi}{5}\right) = \frac{1}{\cos\left(\dfrac{3\pi}{5}\right)}$$
$$\approx -3.2361$$

b. First, we recognize $\dfrac{8\pi}{3}$ indicates an angle greater than 2π. We find the corresponding angle between 0 and 2π.

$$\frac{8\pi}{3} - 2\pi = \frac{8\pi}{3} - \frac{6\pi}{3}$$
$$= \frac{2\pi}{3}$$

So $\csc\left(\dfrac{8\pi}{3}\right) = \csc\left(\dfrac{2\pi}{3}\right)$. The angle $\dfrac{2\pi}{3}$ has a reference angle measuring $\dfrac{\pi}{3}$, and $\csc\left(\dfrac{\pi}{3}\right) = \dfrac{2\sqrt{3}}{3}$ according to Table 8.14. However, $\dfrac{2\pi}{3}$ is in the second quadrant, so we need to double-check the sign value for cosecant in this quadrant. In this case, cosecant is positive in Quadrant II, so $\csc\left(\dfrac{2\pi}{3}\right) = \dfrac{2\sqrt{3}}{3}$, telling us that $\csc\left(\dfrac{8\pi}{3}\right) = \dfrac{2\sqrt{3}}{3}$.

SUMMARY

In this section you learned to define, graph, and use the tangent function. You also learned about the three reciprocal trigonometric functions: cotangent, secant, and cosecant. Additionally, you learned how to graph these four trigonometric functions and how to describe their behaviors in relationship to the cosine and sine functions.

SECTION 8.6 EXERCISES

TECHNOLOGY TIP ■ GRAPHING OTHER TRIGONOMETRIC FUNCTIONS

1. Press MODE to verify that your calculator is in the proper units (Radian or Degree) according to the situation.

2. Use Y= to enter the trigonometric function you wish to graph. Note that only the tangent function is directly programmed into the calculator. For all other functions, you must express them using cosine, sine, or tangent according to the relationships $\cot(x) = \dfrac{1}{\tan(x)}$, $\sec(x) = \dfrac{1}{\cos(x)}$, and $\csc(x) = \dfrac{1}{\sin(x)}$. For example, the correct way to enter $y = 0.5\sec(2(x - 45))$ is shown.

3. Under the ZOOM menu, select 7:ZTrig to create a graph using a window size that fits most basic trigonometric functions. Notice that some technology does not show the entire range of values due to the limitations of pixel density on the screen. Use the WINDOW feature to further adjust the viewing window if necessary.

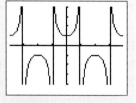

4. In some cases your calculator may create extra lines that are not part of your actual graph because it connects points that should not be connected, as shown. To remove these lines you may change your calculator from Connected mode to Dot mode after pressing the MODE button.

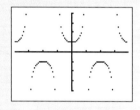

8.6 EXERCISES

■ SKILLS AND CONCEPTS

In Exercises 1–4, use your calculator to evaluate each expression, then explain what your solution means in the context of angles.

1. $\tan(175°)$
2. $\tan\left(-\dfrac{5\pi}{7}\right)$
3. $\cot(80°)$
4. $\cot(8)$

In Exercises 5–10, find the exact value of each expression without using a calculator, then explain what your solution means in the context of angles.

5. $\tan(30°)$
6. $\cot\left(\dfrac{3\pi}{2}\right)$
7. $\cot(810°)$
8. $\tan(-\pi)$
9. $\tan(135°)$
10. $\cot\left(-\dfrac{2\pi}{3}\right)$

In Exercises 11–14, determine whether the expression is equivalent to $\cos(\theta)$, $\sin(\theta)$, $\tan(\theta)$, or $\cot(\theta)$.

11. $\dfrac{\sec(\theta)}{\csc(\theta)}$
12. $\dfrac{\csc(\theta)}{\sec(\theta)}$
13. $\dfrac{\tan(\theta)}{\sec(\theta)}$
14. $\dfrac{\cot(\theta)}{\csc(\theta)}$

In Exercises 15–20, find the exact value of each expression.

15. $\sec(0)$
16. $\csc(180°)$
17. $\sec(-30°)$
18. $\csc\left(\dfrac{5\pi}{6}\right)$
19. $\csc(-855°)$
20. $\sec\left(\dfrac{8\pi}{3}\right)$

In Exercises 21–22, determine the exact values of the indicated trigonometric functions without a calculator.

21. Use the fact that $\cos(60°) = \dfrac{1}{2}$ to determine $\sec(60°)$.

22. Use the facts that $\cos(210°) = -\dfrac{\sqrt{3}}{2}$ and $\sin(210°) = -\dfrac{1}{2}$ to find $\cot(210°)$.

In Exercises 23–26, draw a circle and an angle θ such that the given information is true.

23. $\tan(\theta)$ is positive and $\cos(\theta)$ is negative.
24. $\sin(\theta)$ is positive and $\cot(\theta)$ is negative.
25. $\sec(\theta)$ is positive and $\csc(\theta)$ is positive.
26. $\csc(\theta)$ is negative and $\tan(\theta)$ is negative.
27. Given that $\sec(\theta) = 2$, draw a circle and show the two possible locations for the endpoint of an arc corresponding with an angle θ. Repeat for ϕ given that $\csc(\phi) = -3$.

SHOW YOU KNOW

28. The figure shows the signs of tangent, cotangent, secant, and cosecant values in each of the four quadrants. Explain why each of the functions has the indicated sign in each quadrant.

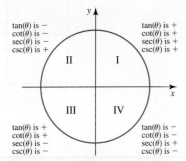

29. Let θ and ϕ be two angles such that $0 \leq \theta \leq 2\pi$, $0 \leq \phi \leq 2\pi$, and $\theta \neq \phi$. What must be true about these angles if $\tan(\theta) = \tan(\phi)$? Explain your reasoning, then explain whether $\cot(\theta) = \cot(\phi)$ will also be true for that relationship.

30. Explain why it is possible for tangent, cotangent, secant, and cosecant values to be greater than 1 even though it is not possible for cosine or sine values to be greater than 1.

31. Explain, in your own words, why the slope of the terminal side of an angle in standard position is always found using $\tan(\theta)$ instead of $r\tan(\theta)$ even if the angle is measured on a circle with a radius other than 1.

32. Fill in the blanks with either $<$, $>$, or $=$ to correctly complete each statement, then explain why the statement must be true. Do not use a calculator.

   a. $\cos\left(\dfrac{\pi}{8}\right)$ _____ $\sin\left(\dfrac{\pi}{8}\right)$

   b. $\tan\left(\dfrac{\pi}{8}\right)$ _____ $\cot\left(\dfrac{\pi}{8}\right)$

   c. $\sec\left(\dfrac{\pi}{8}\right)$ _____ $\csc\left(\dfrac{\pi}{8}\right)$

   d. $\cos\left(\dfrac{\pi}{8}\right)$ _____ $\sec\left(\dfrac{\pi}{8}\right)$

   e. $\sin\left(\dfrac{\pi}{8}\right)$ _____ $\csc\left(\dfrac{\pi}{8}\right)$

MAKE IT REAL

33. **Road Grades** The Clinton-Pavilion Road through the Pavilion Mountains in British Columbia has an 8.53° uphill climb at its steepest point. Find $\tan(8.53°)$ and $\cot(8.53°)$, then explain what these values represent in this situation. (*Source: www.pleadwalk.org*)

34. **Road Grades** The westbound U.S. Highway 74 truck route through North Carolina near the Tuckaseegee River has a downhill grade of 5%. (*Source: www.southeastroads.com*)

   a. Explain what a 5% downhill grade means, then find the angle of descent for the road using the graph of the tangent function.

   b. Explain how you could have used the cotangent function to have found the answer to part (a).

   c. What is the elevation change over 2.5 miles on a road with a 5% downhill grade?

35. **Pyramids** Menkaure ruled Egypt from 2490 to 2472 B.C. and was the third pharaoh to build a pyramid at Giza. His pyramid has an incline of 51.34°. (*Source: www.pbs.org*)

   a. Find $\tan(51.34°)$ and $\cot(51.34°)$, then explain what these values mean in this context.

   b. Menkaure's pyramid was 216 feet tall when constructed. How wide is his pyramid?

   c. Write a function $w(\theta)$ to model the width of a 216-foot-tall pyramid that has an incline of θ degrees.

36. **Pyramids** The Luxor Casino in Las Vegas is built in the shape of a pyramid to match its Egyptian theme. The casino stands 350 feet tall and has a square base with sides that are 600 feet long. (*Source: Luxor Hotel Fact Sheet*)

   a. Find the angle of incline for the Luxor Hotel by graphing using the cotangent function.

   b. Repeat part (a) using the tangent function and compare your results.

37. **Wheelchair Ramps** The federal government has established guidelines for the construction of wheelchair ramps. One of these guidelines says that no single ramp may rise more than 30 inches in height over any horizontal distance. (*Source: www.usdoj.gov*)

   The function $L(\theta) = 30\cot(\theta)$ models the horizontal length of a ramp in inches that rises 30 inches off the ground and makes an angle of θ with the ground.

a. Only changes in height of 1 inch over every 20 inches or less are considered "ramps" by the federal government, and no ramp may increase its vertical distance more than 1 inch for every change of 12 inches of horizontal distance. What is the domain for L? What is its range?

b. Graph L over its domain, then describe the rate of change and concavity and what this information tells you about this situation.

c. The government suggests that ramps increase 1 inch for every 16 to 20 inches of horizontal distance. What is the domain of L for "ideal" ramps?

38. **Wire Tension** When a 12-pound picture is hung as shown, the tension T in pounds on each side of the wire is given by the function $T(\theta) = 12 \sec\left(\dfrac{\theta}{2}\right)$.

a. What is the practical domain for T? What is the practical range?

b. Graph T over the practical domain. Describe the rate of change and concavity of T, then explain what this information tells you about this situation.

c. How much tension is on each side of the wire when $\theta = \dfrac{2\pi}{3}$?

d. If the tension on each side of the wire is 18.6 pounds, what is the value of θ?

39. **Electrical Conduits** Electrical wiring in homes and businesses is run through conduits to help protect and route the wiring throughout a structure. When this conduit must be bent, such as to move around an obstacle or change levels, electricians use a rule involving the cosecant function to know where to bend the conduit to create the right amount of offset.

The function $L(\theta) = 10 \csc(\theta)$ determines the length of conduit between the bends of an offset of 10 inches at an angle of θ degrees. (*Source: www.porcupinepress.com*)

a. What is the reasonable domain of this function? What is the reasonable range?

b. Graph the function over its reasonable domain. Describe the rate of change and concavity of the function, then explain what this information tells you about this situation.

c. If a contractor created a 10-inch offset with a 40° angle, what is the distance between the bends in the offset?

d. At what angle must a contractor bend a conduit to avoid a 10-inch obstruction using 18 inches between the bends in the offset?

e. Electricians try to use a 30° angle for offsets whenever practical. Explain some possible reasons for this.

■ **STRETCH YOUR MIND**

Exercises 40–44 are intended to challenge your understanding of trigonometric functions.

40. Write a possible formula for the graph of $f(\theta)$.

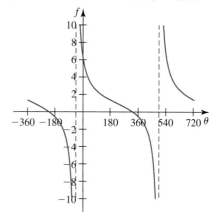

41. Write a possible formula for the graph of $f(\theta)$.

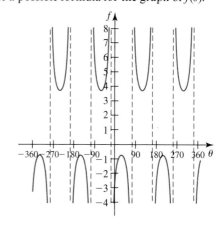

42. a. Create a table of values for $f(\theta) = \tan(\theta)$ over the domain $-\dfrac{\pi}{2} < \theta < \dfrac{\pi}{2}$, then create the graph for f^{-1}.

b. Discuss why the domain had to be restricted to $-\dfrac{\pi}{2} < \theta < \dfrac{\pi}{2}$ to talk about f^{-1}.

c. What does the function f^{-1} do?

43. Using a calculator, graph the function $f(\theta) = (\tan(\theta))^2$. (*Hint:* It may be helpful to look at a table of values in addition to the graph.)

   a. Compare this function to the function $g(\theta) = \tan(\theta)$. Discuss any similarities and differences and the reasons for each.
   b. Graph the function $h(\theta) = (\sec(\theta))^2$ on the same screen. Discuss the relationship between f and h.
   c. Write a formula involving $\tan(\theta)$ and $\sec(\theta)$ based on your observations in part (b). How might this formula be useful when working with trigonometric functions?

44. Using a calculator, graph the function $f(\theta) = (\cot(\theta))^2$. (*Hint:* It may be helpful to look at a table of values in addition to the graph.)

   a. Compare this function to the function $g(\theta) = \cot(\theta)$. Discuss any similarities and differences and the reasons for each.
   b. Graph the function $h(\theta) = (\csc(\theta))^2$ on the same screen. Discuss the relationship between f and h.
   c. Write a formula involving $\cot(\theta)$ and $\csc(\theta)$ based on your observations in part (b). How might this formula be useful when working with trigonometric functions?

SECTION 8.7 Inverse Trigonometric Functions

LEARNING OBJECTIVES

- Use inverse trigonometric functions to solve trigonometric equations algebraically
- Use trigonometric models of real-world data sets for questions whose answers require inverse trigonometric functions

GETTING STARTED

The changing tides are an example of periodic phenomena in our world. Many tidal patterns may be modeled using cosine and sine functions. Monitoring tide heights is important for many reasons including safely docking ships, military coordination during times of war, and making adjustments to satellite elevation readings. Novice sailors who fail to consider tidal patterns may find their boats beached on the shore. (*Source:* www.oceanservice.noaa.gov)

In this section we discuss inverse trigonometric functions and their graphs. We look at how to use inverse trigonometric concepts to solve trigonometric equations, such as finding the time of day when the tide will be at a certain height.

■ Solving for an Angle Measure

Suppose we have three angles whose measures are θ, ϕ, and α, and that $\cos(\theta) = 0.8$, $\cos(\phi) = 0.3$, and $\cos(\alpha) = -0.5$. These three equations show us the *x*-coordinate at the endpoint of the arcs associated with each angle, but in each case we do not know the actual angle measure. Figure 8.95a shows these equations as they relate to the unit circle.

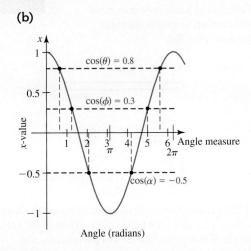

Figure 8.95

We can see there are two points on the circle that make each of the equations true, meaning there are two arcs whose endpoints have each given x-coordinate in one period of the cosine function. We can also draw the cosine function on the interval $[0, 2\pi]$ and solve these equations graphically as we learned earlier in this chapter (see Figure 8.95b.) Using a graphing calculator, we determine the solutions to the equations by finding the points of intersection. The solutions are

$$\cos(\theta) = 0.8 \qquad \cos(\phi) = 0.3 \qquad \cos(\alpha) = -0.5$$
$$\theta \approx 0.6435 \qquad \phi \approx 1.2661 \qquad \alpha \approx 2.0944$$
$$\text{and} \qquad \text{and} \qquad \text{and}$$
$$\theta \approx 5.6397 \qquad \phi \approx 5.0171 \qquad \alpha \approx 4.1888$$

However, recall that the cosine function is periodic, and angles with measures less than 0 and greater than 2π are possible. This means we can find these same x-coordinates for an *infinite* number of angles, as shown in Figure 8.96.

We can find each of the remaining solutions by adding or subtracting 2π from the initial solutions a certain number of times. We write the solutions to the equation $\cos(\theta) = 0.8$ as follows for all integer values of n.

$$\theta \approx 0.6435 + 2\pi n$$

and

$$\theta \approx 5.6397 + 2\pi n$$

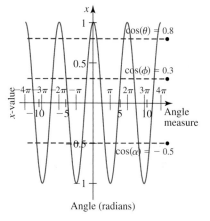

Figure 8.96

For example, let $n = 1$. This gives us angles that have completed one period plus a portion of another period of the cosine graph.

$$\theta \approx 0.6435 + 2\pi(1) \qquad \theta \approx 5.6397 + 2\pi(1)$$
$$\approx 6.9267 \qquad \approx 11.9229$$

We can repeat this process indefinitely, including using negative values for n, to find any of the infinite solutions to these equations.

The process of solving for the input value when the output value is given is a common mathematical procedure. We have learned in previous chapters how to create *inverse functions* to make this process easier. We now demonstrate how to do this with trigonometric functions.

■ Inverse Cosine Function

Consider the equation $\cos(\theta) = 0.8$. This may also be written as $\cos^{-1}(0.8) = \theta$. The notation **cos**$^{-1}$ refers to the **inverse cosine function**, which is the function that shows the same relationship as the cosine function, but whose inputs and outputs are reversed. The inverse cosine function takes the x-coordinate of the endpoint of the arc as its input and outputs the angle measure that corresponds to an arc whose endpoint has this x-coordinate.

cosine function   *inverse cosine function*

angle measure   angle measure
↓   ↓   ↓
$\cos(\theta) = 0.8$ $\cos^{-1}(0.8) = \theta$
↑   ↑   ↑
x-coordinate of the x-coordinate of the
endpoint of the arc   endpoint of the arc

If we define the function $f(\theta) = \cos(\theta)$, then the outputs of f are x-values, meaning $f(\theta) = x$ or $x = \cos(\theta)$. In the inverse form, $\cos^{-1}(x) = \theta$, the x-value is now the input and the θ-value is the output. We write this function as $f^{-1}(x) = \cos^{-1}(x)$.

INVERSE COSINE FUNCTION

The function $f^{-1}(x) = \cos^{-1}(x)$ is the **inverse cosine function**. The inverse cosine function takes the x-coordinate of the endpoint of an arc as its input and outputs an angle measure that corresponds with an arc whose endpoint has this x-coordinate.

Using Technology to Evaluate the Inverse Cosine Function

Once the relationship $\cos(\theta) = 0.8$ is written in its inverse form $\cos^{-1}(0.8) = \theta$, we can use technology to solve the equation without having to graph the function and determine intersection points. See the Technology Tip at the end of the section for details.

Figure 8.97 shows that the endpoint of an arc that corresponds to an angle of 0.6435 radians has an x-coordinate of 0.8, which is the same angle measure we determined previously by graphing. However, the calculator only shows one of the infinite number of solutions to this equation. We find the remaining solutions using reference angles, as indicated in Figure 8.98. For example, the other solution between 0 and 2π is approximately 5.6397 radians, and is found using a reference angle of 0.6435 radians in Quadrant IV.

Figure 8.97

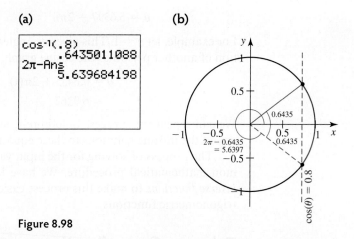

Figure 8.98

We can perform the same process with a calculator in Degree mode if θ is in degrees.

EXAMPLE 1 ■ Using the Inverse Cosine Function

Solve each of the following equations for θ over the interval $0 \leq \theta \leq 2\pi$, then explain what the solution represents with respect to circles.

a. $\cos(\theta) = -0.25$

b. $6\cos(\theta) = 1.04$

Solution

a. We first write the equation $\cos(\theta) = -0.25$ in its inverse form, $\cos^{-1}(-0.25) = \theta$. Using a calculator, we find $\cos^{-1}(-0.25) \approx 1.8235$, telling us that an angle measuring 1.8235 radians corresponds with an arc on the unit circle whose endpoint has an x-coordinate of -0.25. Using reference angles as shown in Figure 8.99, we find another solution in the third quadrant (the other quadrant where negative x-coordinates exist) at approximately 4.4597 radians.

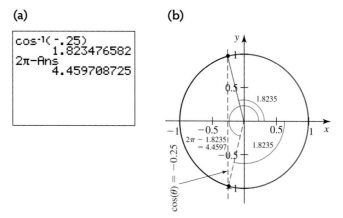

Figure 8.99

b. We recall 6 cos(θ) represents the *x*-coordinate at the endpoint of an arc on a circle with radius 6. We first divide the equation 6 cos(θ) = 1.04 by 6 to get $\cos(\theta) = \dfrac{1.04}{6} \approx 0.1733$. This tells us that, on the unit circle, an arc whose endpoint has an *x*-coordinate of 0.1733 will correspond with an angle θ. We now use the inverse cosine function and reference angles to find our solutions at θ ≈ 1.397 and θ ≈ 4.887. (See Figure 8.100.)

$$\cos(\theta) \approx 0.1733$$
$$\cos^{-1}(0.1733) \approx \theta$$
$$1.397 \approx \theta$$

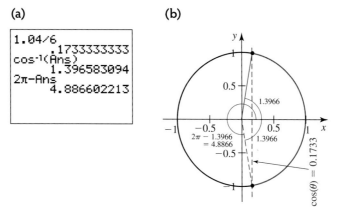

Figure 8.100

This also tells us that the arcs on a circle of radius 6 that correspond with angles of 1.397 and 4.887 radians will have endpoints whose *x*-coordinates are 1.04.

In the preceding examples we note that the calculator only displays one solution and that this solution is the angle measure closest to 0. We will explore this important observation in greater depth later in this section.

Finding Exact Values Using Inverse Cosine

Recall there are some angles whose exact trigonometric values we use whenever possible. Table 8.15, which we developed in Sections 8.3 and 8.6, can help us find the exact value solutions to inverse trigonometric functions as well.

Table 8.15

Exact Values of the Six Trigonometric Functions for Major Reference Angles							
Angle (degrees) θ	Angle (radians) θ	Cosine $\cos(\theta)$	Sine $\sin(\theta)$	Tangent $\tan(\theta)$	Cotangent $\cot(\theta)$	Secant $\sec(\theta)$	Cosecant $\csc(\theta)$
0°	0	1	0	0	undefined	1	undefined
30°	$\frac{\pi}{6}$	$\frac{\sqrt{3}}{2}$	$\frac{1}{2}$	$\frac{\sqrt{3}}{3}$	$\sqrt{3}$	$\frac{2\sqrt{3}}{3}$	2
45°	$\frac{\pi}{4}$	$\frac{\sqrt{2}}{2}$	$\frac{\sqrt{2}}{2}$	1	1	$\sqrt{2}$	$\sqrt{2}$
60°	$\frac{\pi}{3}$	$\frac{1}{2}$	$\frac{\sqrt{3}}{2}$	$\sqrt{3}$	$\frac{\sqrt{3}}{3}$	2	$\frac{2\sqrt{3}}{3}$
90°	$\frac{\pi}{2}$	0	1	undefined	0	undefined	1

EXAMPLE 2 ■ **Finding Exact Values Using Inverse Cosine**

Solve each equation for θ using exact values over the interval $0 \leq \theta \leq 2\pi$, then explain what the solution represents with regard to the unit circle.

a. $\cos(\theta) = \frac{\sqrt{3}}{2}$

b. $9\cos(\theta) + 3 = -1.5$

Solution

a. The cosine value is positive in Quadrants I and IV, and is $\frac{\sqrt{3}}{2}$ for a reference angle of $\frac{\pi}{6}$. For $\cos(\theta) = \frac{\sqrt{3}}{2}$, we write this in inverse form $\cos^{-1}\left(\frac{\sqrt{3}}{2}\right) = \theta$ and say $\cos^{-1}\left(\frac{\sqrt{3}}{2}\right) = \frac{\pi}{6}$ since $\cos\left(\frac{\pi}{6}\right) = \frac{\sqrt{3}}{2}$. The remaining solution will be an angle measuring $2\pi - \frac{\pi}{6}$, or $\frac{11\pi}{6}$ as shown in Figure 8.101. The solutions are $\theta = \frac{\pi}{6}$ and $\frac{11\pi}{6}$.

Figure 8.101

b. We begin by solving the equation for $\cos(\theta)$ and then rewrite in inverse form.

$$9\cos(\theta) + 3 = -1.5$$
$$9\cos(\theta) = -4.5$$
$$\cos(\theta) = -0.5$$
$$\cos^{-1}(-0.5) = \theta$$

We are looking for two angle measures that correspond to arcs on the unit circle whose endpoints have an x-coordinate of -0.5. This occurs first in Quadrant II at a reference angle of $\frac{\pi}{3}$, at $\theta = \frac{2\pi}{3}$. Thus, $\cos^{-1}(-0.5) = \frac{2\pi}{3}$ since $\cos\left(\frac{2\pi}{3}\right) = -0.5$. The other solution occurs in Quadrant III (the other quadrant where negative x-coordinates exist) at $\pi + \frac{\pi}{3}$, or $\frac{4\pi}{3}$ (see Figure 8.102). The solutions are $\theta = \frac{2\pi}{3}$ and $\theta = \frac{4\pi}{3}$.

Figure 8.102

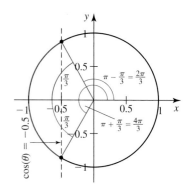

EXAMPLE 3 ■ Using the Inverse Cosine Function to Solve Equations

Tide heights are measured in feet above the mean lower low water (MLLW) height, the average of the lowest height during low tide for a given day over a 19-year period. The function $H(t) = 5\cos\left(\frac{2\pi}{13}(t + 3)\right) + 7$ models the height of the tide in feet above MLLW at Cutler Naval Base in Maine t hours after midnight on May 29, 2007. (*Source:* Modeled from a graph at www.tidesonline.noaa.gov)

a. What is the period of H and what does it tell about this situation?

b. Explain what the solution(s) to the equation $5\cos\left(\frac{2\pi}{13}(t + 3)\right) + 7 = 7.25$ will represent, then determine all of the solutions over one period of the function beginning at midnight on May 29, 2007.

Solution

a. We find the period by using $\frac{2\pi}{|B|}$ for the given function.

$$\frac{2\pi}{|B|} = \frac{2\pi}{2\pi/13} \qquad B = \frac{2\pi}{13}$$

$$= 2\pi\left(\frac{13}{2\pi}\right)$$

$$= 13$$

The tides at Cutler Naval Base at this time during the year have a period of 13 hours, meaning they go from high tide to low tide and back to high tide once every 13 hours.

b. The solution(s) to $5\cos\left(\frac{2\pi}{13}(t + 3)\right) + 7 = 7.25$ give the number of hours after midnight on May 29, 2007 when the tide height was 7.25 feet above MLLW.

$$5\cos\left(\frac{2\pi}{13}(t + 3)\right) + 7 = 7.25$$

$$5\cos\left(\frac{2\pi}{13}(t + 3)\right) = 0.25$$

$$\cos\left(\frac{2\pi}{13}(t + 3)\right) = 0.05$$

$$\cos^{-1}(0.05) = \frac{2\pi}{13}(t + 3)$$

At this step we use a calculator to determine $\cos^{-1}(0.05)$. Remember that for one period of cosine, there will be two values that make $\cos^{-1}(0.05)$ true. We use reference angles to find the second solution.

$$\cos^{-1}(0.05) \approx 1.5208 \quad \text{One solution is between 0 and } \frac{\pi}{2}.$$

$$2\pi - 1.5208 \approx 4.7624 \quad \text{The other solution is between } \frac{3\pi}{2} \text{ and } 2\pi.$$

Since $\cos^{-1}(0.05) = \frac{2\pi}{13}(t + 3)$, these solutions tell us the two values that $\frac{2\pi}{13}(t + 3)$ may equal over one period of cosine. We set up and solve two equations.

$$\frac{2\pi}{13}(t + 3) \approx 1.5208 \quad \cos^{-1}(0.05) \approx 1.5208$$

$$t + 3 \approx 3.1465 \quad \text{Multiply by } \frac{13}{2\pi}.$$

$$t \approx 0.1465$$

$$\frac{2\pi}{13}(t + 3) \approx 4.7624 \quad \cos^{-1}(0.05) \approx 4.7624$$

$$t + 3 \approx 9.8535 \quad \text{Multiply by } \frac{13}{2\pi}.$$

$$t \approx 6.8535$$

Approximately 0.1465 hours after midnight (about 12:09 A.M.) and 6.8535 hours after midnight (about 6:51 A.M.) the tide height will be 7.5 feet above MLLW.

■ Inverse Sine and Inverse Tangent Functions

The sine and tangent functions can also be written in inverse form. We can use these inverse functions to solve equations involving sine and tangent.

EXAMPLE 4 ■ Using the Inverse Sine Function

Given the equation $0.5 \sin(\theta) = 0.35$,

a. Explain what the solution to the equation will represent on the unit circle.

b. Solve the equation for θ on the interval $[0, 2\pi]$ by graphing.

c. Solve the equation for θ on the interval $[0, 2\pi]$ using the inverse sine function.

Solution

a. This equation is used to find angles that correspond with arcs whose endpoints on a circle with a radius of 0.5 have y-coordinates of 0.35. If we multiply each side of the equation by 2, we get $\sin(\theta) = 0.7$. Thus, on the unit circle, the same angle measures will correspond with arcs whose endpoints have a y-coordinate of 0.7. See Figure 8.103.

Figure 8.103

b.

Figure 8.104

Figure 8.104 shows that angles measuring about 0.7754 and 2.3662 radians will correspond with arcs whose endpoints have a y-coordinate of 0.35 on a circle with a radius of 0.5.

c. This equation will have two solutions over the interval $[0, 2\pi]$. We use the inverse sine function to find the first solution and reference angles (Figure 8.105) to find the second.

$$0.5 \sin(\theta) = 0.35$$
$$\sin(\theta) = 0.7$$
$$\sin^{-1}(0.7) = \theta$$
$$0.7754 \approx \theta \quad \text{One solution is between 0 and } \frac{\pi}{2}.$$

$$\pi - 0.7754 \approx \theta \quad \text{The other solution is between } \frac{\pi}{2} \text{ and } \pi.$$
$$2.3662 \approx \theta$$

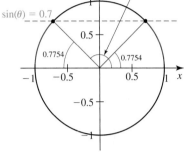

Figure 8.105

EXAMPLE 5 ■ Using the Inverse Tangent Function

Given the equation $\tan(\theta) = -1$,

a. Explain what the solution to the equation will represent.
b. Solve the equation for θ over the interval $[0°, 360°]$ by graphing.
c. Solve the equation for θ over the interval $[0°, 360°]$ using the inverse tangent function.

Solution

a. The equation $\tan(\theta) = -1$ is used to find an angle measure such that the terminal side of the angle has a slope of -1.

b.

Figure 8.106

Figure 8.106 shows that angles in standard position measuring 135° and 315° have terminal sides with a slope of –1. We note that 135° and 315° are separated by 180°, a property consistent with our observations in Section 8.6 about angles with the same tangent values.

c. We begin by writing the equation $\tan(\theta) = -1$ in inverse form $\tan^{-1}(-1) = \theta$. From Table 8.15, we see that the tangent function has a magnitude of 1 when the reference angle is 45°, and it will be −1 in Quadrants II and IV.

$$180° - 45° = 135°$$ One solution will be between 90° and 180°.
$$135° + 180° = 315°$$ The other solution will be 180° away.

Another approach is to use a calculator to evaluate $\tan^{-1}(-1)$. When we do this, we get $\tan^{-1}(-1) = -45°$. In this case, a calculator returned a negative angle, which is not in the interval [0°, 360°]. To find the two solutions in the given interval, we need to add 180° (to find the next angle with the same tangent value) and 360° (to find the corresponding angle to −45° on the circle). See Figure 8.107.

$$-45° + 180° = 135°$$
$$-45° + 360° = 315°$$

Figure 8.107

INVERSE SINE FUNCTION

The function $f^{-1}(y) = \sin^{-1}(y)$ is the **inverse sine function**. The inverse sine function takes the y-coordinate of the endpoint of an arc as its input and outputs an angle measure that corresponds with an arc whose endpoint has this y-coordinate.

INVERSE TANGENT FUNCTION

The function $f^{-1}(x) = \tan^{-1}(x)$ is the **inverse tangent function**. The inverse tangent function takes the slope of the terminal side of an angle as its input and outputs one of the angle measures whose terminal side has this slope.

EXAMPLE 6 ■ Using Inverse Trigonometric Functions

Given $\cos(\theta) = 0.6410$ for some θ in the interval $\left[0, \dfrac{\pi}{2}\right]$, use your knowledge of inverse trigonometric functions to find $\sin(\theta)$.

Solution The equation $\cos(\theta) = 0.6410$ tells us the endpoint of an arc on the unit circle has an x-coordinate of 0.6410. We write this in the inverse form:

$\cos^{-1}(0.6410) = \theta$ for a value of θ between 0 and $\dfrac{\pi}{2}$. Using a calculator, we get

$$\theta = \cos^{-1}(0.6410)$$
$$\approx 0.8750$$

So $\theta \approx 0.8750$ (which falls between 0 and $\dfrac{\pi}{2}$). Thus, we find $\sin(0.8750) \approx 0.7675$. An arc that corresponds with a 0.8750 radian angle has its endpoint at (0.6410, 0.7675).

SECTION 8.7 ■ INVERSE TRIGONOMETRIC FUNCTIONS

EXAMPLE 7 ■ Using Inverse Trigonometric Functions with a Horizontal Compression

Find all of the solutions for the equation $\sin(3\theta) = 0.4690$ over the interval $0 \leq \theta \leq 360°$ without graphing.

Solution Over the interval $0 \leq \theta \leq 360°$ we expect there to be six solutions to the equation $\sin(3\theta) = 0.4690$. As θ varies from $0°$ to $360°$, 3θ varies from $0°$ to $1080°$ and the sine function completes three full periods, yielding six different angles that correspond with arcs whose endpoints have 0.4690 as their y-coordinate. We begin by finding the six angle measures between $0°$ to $1080°$ that meet this criterion.

$$\sin^{-1}(0.4690) \approx 27.97°$$

The first angle that meets the criterion measures about $27.97°$. The next angle is in Quadrant II and can be found using reference angles.

$$180° - 27.97° = 152.03°$$

There are four remaining angles to find over the interval $[0°, 1080°]$, and these will correspond to the first two angles as we make a second and third revolution. See Figure 8.108.

$$27.97° + 360° = 387.97° \qquad 27.97° + 2(360°) = 747.97°$$
$$152.03° + 360° = 512.03° \qquad 152.03° + 2(360°) = 872.03°$$

(a)

(b)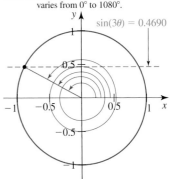

Figure 8.108

As 3θ varies from $0°$ to $1080°$, the angles $27.97°$, $152.03°$, $387.97°$, $512.03°$, $747.97°$, and $872.03°$ correspond with arcs whose endpoints have a y-coordinate of 0.4690. We therefore solve the following equations to find the six values of θ over the interval $[0°, 360°]$ that make this true.

$3\theta \approx 27.97°$ $3\theta \approx 152.03°$ $3\theta \approx 387.97°$
$\theta \approx 9.32°$ $\theta \approx 50.68°$ $\theta \approx 129.32°$

$3\theta \approx 512.03°$ $3\theta \approx 747.97°$ $3\theta \approx 872.03°$
$\theta \approx 170.68°$ $\theta \approx 249.32°$ $\theta \approx 290.68°$

EXAMPLE 8 ■ Solving Equations with Other Trigonometric Functions

Solve the following equations over the interval $[0°, 360°]$ using inverse trigonometric functions.

a. $\csc(\theta) = -1.4532$
b. $\cot(\theta) = 4.2308$

Solution

a. Since calculators are not able to work with the cosecant, we need to convert this into an equation involving sine.

$$\csc(\theta) = -1.4532$$

$$\frac{1}{\sin(\theta)} = \frac{-1.4532}{1} \qquad \csc(\theta) = \frac{1}{\sin(\theta)}$$

$$\frac{\sin(\theta)}{1} = \frac{1}{-1.4532} \qquad \text{Use the reciprocal of both sides.}$$

$$\sin(\theta) \approx -0.6881 \qquad \text{Evaluate } \frac{1}{-1.4532}.$$

$$\sin^{-1}(-0.6881) \approx \theta \qquad \text{Write in inverse form.}$$

$$-43.48° \approx \theta \qquad \text{Evaluate with a calculator.}$$

The solution given by a calculator is not in the interval [0°, 360°], but it does have a reference angle of 43.48°. The solutions will have this reference angle and will be located in Quadrants III and IV.

$$180° + 43.48° = 223.48° \qquad \text{One solution will be between 180° and 270°.}$$

$$360° - 43.48° = 316.52° \qquad \text{One solution will be between 270° and 360°.}$$

Thus, $\csc(\theta) = -1.4532$ is true for $\theta \approx 223.48°$ and $\theta \approx 316.52°$ over the interval $0° \geq \theta \geq 360°$.

b. We first transform this equation to use tangent instead of cotangent.

$$\cot(\theta) = 4.2308$$

$$\frac{1}{\tan(\theta)} = \frac{4.2308}{1} \qquad \cot(\theta) = \frac{1}{\tan(\theta)}$$

$$\tan(\theta) = \frac{1}{4.2308} \qquad \text{Use the reciprocal of both sides.}$$

$$\tan(\theta) \approx 0.2364 \qquad \text{Evaluate } \frac{1}{4.2308}.$$

$$\tan^{-1}(0.2364) \approx \theta \qquad \text{Write in inverse form.}$$

$$13.30° \approx \theta \qquad \text{Evaluate with a calculator.}$$

The terminal sides of angles measuring 13.30° and 193.30° (180° greater than 13.30°) have a slope of 0.2364. Therefore, $\cot(\theta) = 4.2308$ is true on the interval $0° \geq \theta \geq 360°$ for $\theta \approx 13.30°$ and $\theta \approx 193.30°$.

■ Graphing Inverse Trigonometric Functions

We have noticed that a calculator does not automatically give us all of the solutions to an equation when we use inverse trigonometric functions. In fact, sometimes we get positive angle measures and sometimes we get negative measures. Why? There is a very important reason for this involving their definition as *functions*. Recall that a function has only one output for any given input. To create an inverse trigonometric function, we must first restrict the domain of the trigonometric function so that the function is strictly increasing or decreasing. We can then define the corresponding inverse trigonometric function.

Graphing the Inverse Cosine Function

We use some of the exact values for the cosine function to create a partial table of values that has the cosine of an angle as the input and the angle itself as the output. See Tables 8.16 and 8.17.

SECTION 8.7 ■ INVERSE TRIGONOMETRIC FUNCTIONS

Table 8.16

Angle (degrees) θ	Cosine (x-coordinate) $\cos(\theta) = x$
0	1
$\pi/4$	$\sqrt{2}/2$
$\pi/2$	0
$3\pi/4$	$-\sqrt{2}/2$
π	-1
$5\pi/4$	$-\sqrt{2}/2$
$3\pi/2$	0
$7\pi/4$	$\sqrt{2}/2$
2π	1

Table 8.17

Cosine (x-coordinate) x	Angle (degrees) $\cos^{-1}(x) = \theta$
1	0
$\sqrt{2}/2$	$\pi/4$
0	$\pi/2$
$-\sqrt{2}/2$	$3\pi/4$
-1	π
$-\sqrt{2}/2$	$5\pi/4$
0	$3\pi/2$
$\sqrt{2}/2$	$7\pi/4$
1	2π

When we graph the cosine function and the data from Table 8.16, we see it is a function (Figure 8.109a), but when we investigate the graph of Table 8.17, we see it is not a function (Figure 8.109b).

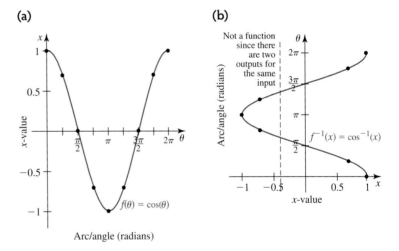

Figure 8.109

In fact, when we graph functions such as $f^{-1}(x) = \cos^{-1}(x)$ using a calculator as shown in Figure 8.110, we see that a calculator graphs the inverse based on a restricted domain for the cosine function.

Figure 8.110

It appears that a calculator only shows inputs (x-values) between -1 and 1 and outputs angles between 0 and π. Let's examine the reasons for this domain and range.

As we saw in Figure 8.95, repeated here in Figure 8.111, there were two solutions to each of the equations $\cos(\theta) = 0.8$, $\cos(\phi) = 0.3$, and $\cos(\alpha) = -0.5$ on the unit circle.

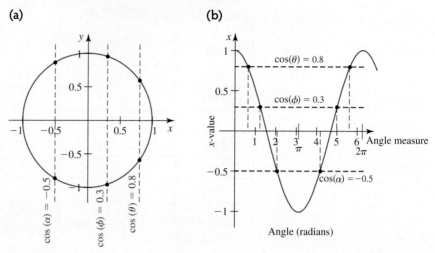

Figure 8.111

If there are two different input values with the same output value, then when we create the inverse relationship we have two outputs for the same input value, which is not a function. Thus, we need to restrict the original domain of the cosine function such that there will be one output associated with a given input.

We notice all of the possible x-coordinates for the endpoint of an arc occur in Quadrants I and II (the top half of the unit circle). That is, for angles between 0° and 180° or 0 and π radians, all of the possible x-coordinates are generated. All of the x-coordinates in Quadrants III and IV match those found in Quadrants I and II. Thus, by restricting the domain of the cosine function to the interval $0° \leq \theta \leq 180°$ or $0 \leq \theta \leq \pi$, we will still be able to represent all of the possible values of the cosine function (values between −1 and 1), but each of these values will only be associated with one angle.

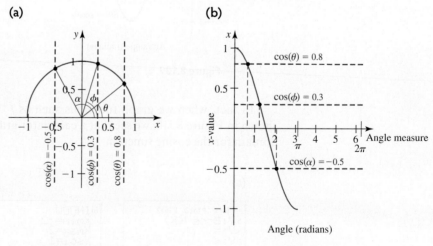

Figure 8.112

When the inverse of the function in Figure 8.112 is graphed (shown in Figure 8.113), we now see the same graph generated by a calculator, showing us that a calculator considers the domain of the inverse cosine function to be [−1, 1] and its range to be [0°, 180°] or [0, π]. This is why a calculator only returns one angle measure when calculating inverse cosine values. Other angles with the same trigonometric values can always be found using reference angles.

Figure 8.113

Graphing the Inverse Sine and Inverse Tangent Functions

As with the inverse cosine function, the inverse sine and inverse tangent relations will not be functions unless we restrict the domain of the original sine and tangent functions.

Let's consider the equation $\sin(\theta) = -0.25$. We can examine the solutions to this equation on the unit circle (Figure 8.114a) and the graph of the sine function (Figure 8.114b).

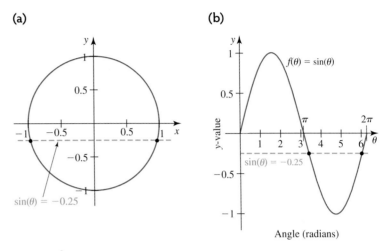

Figure 8.114

As with cosine, we find two solutions to this equation on the circle. We notice each time we get two solutions to an equation such as $\sin(\theta) = -0.25$, one of the solutions is in Quadrant I or IV and the other is in Quadrant II or III. In fact, all of the values of the sine function (y-coordinates from -1 to 1) will occur on either the right side or the left side of the circle (as opposed to either the top or the bottom of the circle with cosine). To create the inverse sine relation and have it be a function, we need to restrict the domain such that we include all of the values of sine and have an interval closest to the origin. We choose the interval $[-90°, 90°]$ or $\left[-\dfrac{\pi}{2}, \dfrac{\pi}{2}\right]$ shown in Figure 8.115a. (Note: There are other intervals that we could have chosen, but the common convention is to choose the intervals closest to the origin.) With this restricted domain for the sine function, the inverse sine relation will now be a function, shown in Figure 8.115b.

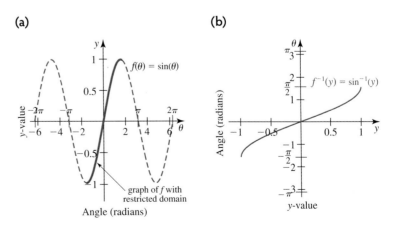

Figure 8.115

The graph shown in Figure 8.115b is identical to what a calculator will display when you graph the inverse sine function. All of the angle measures the calculator returns when we use the inverse sine function will be in the interval $[-90°, 90°]$ or $\left[-\dfrac{\pi}{2}, \dfrac{\pi}{2}\right]$, showing that this is how a calculator defines the range of the inverse sine function.

To restrict the domain of the tangent function, we simply choose the complete period nearest to the origin since the tangent function outputs all of its slope values exactly once over each of its periods. Thus, if we restrict the domain of tangent to the interval $[-90°, 90°]$ or $\left[-\dfrac{\pi}{2}, \dfrac{\pi}{2}\right]$ (Figure 8.116a), its inverse will be a function (Figure 8.116b).

Figure 8.116

The graph we see in Figure 8.116b is identical to the one a calculator creates when we graph the inverse tangent function. Likewise, the calculator restricts the range of the inverse tangent function to the interval $[-90°, 90°]$ or $\left[-\dfrac{\pi}{2}, \dfrac{\pi}{2}\right]$, which is why we only get angle measures within this interval when we use a calculator to find inverse tangent values.

DOMAIN AND RANGE FOR INVERSE TRIGONOMETRIC FUNCTIONS

- The inverse cosine function $f^{-1}(x) = \cos^{-1}(x)$ has a domain of $[-1, 1]$ and a range of $[0°, 180°]$ or $[0, \pi]$.
- The inverse sine function $f^{-1}(y) = \sin^{-1}(y)$ has a domain of $[-1, 1]$ and a range of $[-90°, 90°]$ or $\left[-\dfrac{\pi}{2}, \dfrac{\pi}{2}\right]$.
- The inverse tangent function $f^{-1}(y) = \tan^{-1}(x)$ has a domain of all real numbers and a range of $[-90°, 90°]$ or $\left[-\dfrac{\pi}{2}, \dfrac{\pi}{2}\right]$.

EXAMPLE 9 ■ Graphing Inverse Trigonometric Functions and Solving Equations

The sport of rodeo (including the bull riding event) is popular in Cheyenne, Wyoming. The function

$$T(m) = -20.6 \cos\left(\dfrac{\pi}{6}(m-1)\right) + 50$$

models the average temperature in Cheyenne in degrees Fahrenheit during month m of the year. (For this model, m is in radians.) (*Source:* Modeled using data from aa.usno.navy.mil)

a. Graph the function and determine a restricted domain for T such that the inverse will also be a function.

b. Find the formula for the inverse function of T. Explain what the inputs and outputs of this function are.

c. Graph the inverse function you found in part (b).

d. According to your function in part (b), when is the average temperature 35°F in Cheyenne?

Solution

a. After graphing the function as shown in Figure 8.117, we choose to restrict the domain to [1, 7]. This domain ensures that we have all of the output values of T represented once.

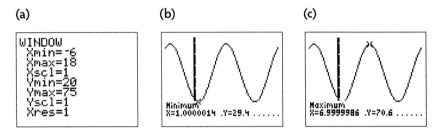

Figure 8.117

b. From part (a) or from using the amplitude and midline of the function we know that the inverse function will have inputs that are average temperatures (T) between 29.4°F and 70.6°F and outputs that are months of the year (m) between 1 and 7 (beginning of January to the beginning of July). We now solve for the input variable in our original function to create the inverse function.

$$-20.6 \cos\left(\frac{\pi}{6}(m-1)\right) + 50 = T$$

$$-20.6 \cos\left(\frac{\pi}{6}(m-1)\right) = T - 50$$

$$\cos\left(\frac{\pi}{6}(m-1)\right) = -\frac{T-50}{20.6}$$

$$\frac{\pi}{6}(m-1) = \cos^{-1}\left(-\frac{T-50}{20.6}\right) \quad \text{Write in inverse form.}$$

$$m - 1 = \frac{6}{\pi}\cos^{-1}\left(-\frac{T-50}{20.6}\right)$$

$$m = \frac{6}{\pi}\cos^{-1}\left(-\frac{T-50}{20.6}\right) + 1$$

The inverse function is $m(T) = \frac{6}{\pi}\cos^{-1}\left(-\frac{T-50}{20.6}\right) + 1$.

c. The graph is shown in Figure 8.118.

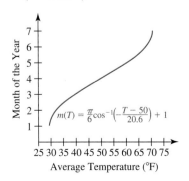

Figure 8.118

d.
$$m(35) = \frac{6}{\pi}\cos^{-1}\left(-\frac{35-50}{20.6}\right) + 1$$
$$= \frac{6}{\pi}\cos^{-1}\left(-\frac{15}{20.6}\right) + 1$$
$$\approx \frac{6}{\pi}\cos^{-1}(-0.7282) + 1$$
$$\approx 2.44 \quad \text{Evaluate with a calculator.}$$

The average temperature will be 35°F in about the middle of February. We note that, due to the restricted domain and range of function m, we only have one solution to this equation. We verify this solution using the graph from part (c).

■ Alternative Notation for Inverse Trigonometric Functions

Since the outputs of the inverse trigonometric functions are angle measures that correspond with endpoints of arcs on the unit circle, they are sometimes called arc functions, as in **arc cosine (arccos)**, **arc sine (arcsin)**, and **arc tangent (arctan)**. We use this notation interchangeably with inverse trigonometric notation: for example, $f^{-1}(x) = \cos^{-1}(x)$ can be written $f^{-1}(x) = \arccos(x)$ without changing its meaning. Arc function notation is sometimes preferred because it eliminates the confusion between an exponent of -1 and the -1 that is part of inverse trigonometric function notation.

SUMMARY

In this section you learned about inverse trigonometric functions and how they are used to solve equations where the value of a trigonometric function is known but not the angle measure. You also learned how to find and graph inverse trigonometric functions and how to restrict the domain of a trigonometric function so that its inverse is also a function.

TECHNOLOGY TIP ■ USING COS⁻¹, SIN⁻¹, AND TAN⁻¹

1. Press **MODE** to verify that your calculator is in the proper units (Radian or Degree) according to the situation.

2. Press **2nd**, followed by **COS**, **SIN**, or **TAN** to use $\cos^{-1}$, $\sin^{-1}$, or $\tan^{-1}$, respectively. This will give you one angle that has the given value for the trigonometric function you selected. You will need to use reference angles to find additional angles with the given trigonometric value if this is not the desired solution.

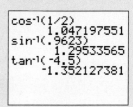

3. To find the values of the inverse secant, cosecant, or cotangent functions, you will need to use the relationships $\sec^{-1}(x) = \cos^{-1}\left(\frac{1}{x}\right)$, $\csc^{-1}(x) = \sin^{-1}\left(\frac{1}{x}\right)$, and $\cot^{-1}(x) = \tan^{-1}\left(\frac{1}{x}\right)$. The method for evaluating $\csc^{-1}(-1.3)$ is shown.

8.7 EXERCISES

SKILLS AND CONCEPTS

In Exercises 1–4, find at least one value for the given expression over the interval $[0, 2\pi]$, then explain what your answer represents.

1. $\sin^{-1}(0.68)$
2. $\tan^{-1}(0.3267)$
3. $\arccos(-0.4682)$
4. $\arctan(8.4)$

In Exercises 5–8, evaluate each expression exactly, finding at least one value over the interval $[0, 2\pi]$ without using a calculator. Explain what your answer represents.

5. $\cos^{-1}\left(\dfrac{1}{2}\right)$
6. $\tan^{-1}\left(\dfrac{\sqrt{3}}{3}\right)$
7. $\sin^{-1}\left(-\dfrac{\sqrt{2}}{2}\right)$
8. $\csc^{-1}(-1)$

In Exercises 9–12, find all solutions to each equation over the interval $[0°, 360°]$ using inverse trigonometric functions. Check your solutions by graphing.

9. $\cos(\theta) = 0.7767$
10. $\tan(\theta) = -11.2960$
11. $6 = 8\cos(\theta)$
12. $-9\cot(\theta) = 3\sqrt{3}$

In Exercises 13–16, find all solutions to each equation over the interval $[0, 2\pi]$ using inverse trigonometric functions. Check your solutions by graphing.

13. $\sin(\theta) = -0.4390$
14. $3\sin(\theta) = 2.6402$
15. $\cos\left(\dfrac{1}{2}\theta\right) = -\dfrac{\sqrt{3}}{2}$
16. $1.4650 = \sec(3\theta)$

17. a. Explain how the tangent and inverse tangent functions can be used to solve the equation $\dfrac{\sin(\theta)}{\cos(\theta)} = 0.8361$, then find at least one solution.
    b. Using the reasoning you gave in part (a), find at least one solution to each of the following equations.
       i. $\sin(\theta) = 2.5\cos(\theta)$
       ii. $1.32\cos(\theta) = -4\sin(\theta)$
       iii. $\cos(\theta) - \sin(\theta) = 0$

18. a. Graph the cotangent, secant, and cosecant functions over restricted domains such that their inverses will be a function.
    b. Graph the inverse cotangent function, inverse secant function, and inverse cosecant function based on the restricted domains in part (a).
    c. Identify the domain and range for these three inverse functions.
    d. Explain how to use your calculator to graph these inverse functions, then graph each using your calculator and compare these graphs to your graphs in part (b).

In Exercises 19–22, find all solutions to the given equation over the interval $[0°, 360°]$ using inverse trigonometric functions.

19. $\cos(\theta + 15°) = 0.8265$
20. $1.5\tan(\theta + 45°) = 3$
21. $-\dfrac{1}{4}\sin(\theta - 60°) = 1$
22. $3\sec(\theta - 40°) = 8.6$

In Exercises 23–30, evaluate each of the expressions exactly. All angles are in radians in the first quadrant.

23. $\tan\left(\cos^{-1}\left(\dfrac{1}{2}\right)\right)$
24. $\sin(\tan^{-1}(1))$
25. $\sec(\cot^{-1}(\sqrt{3}))$
26. $\cot(\sec^{-1}(1))$
27. $\tan^{-1}\left(\cot\left(\dfrac{\pi}{3}\right)\right)$
28. $\cot^{-1}(\sin(0))$
29. $\csc^{-1}\left(\sec\left(\dfrac{\pi}{6}\right)\right)$
30. $\sin(\csc^{-1}(\sqrt{2}))$

In Exercises 31–34, evaluate each expression using a calculator. All angles are in radians in the first quadrant.

31. $\tan(\sin^{-1}(0.7053))$
32. $\cos(\sin^{-1}(0.2879))$
33. $\sec(\tan^{-1}(1.4602))$
34. $\csc(\sec^{-1}(2.3760))$

In Exercises 35–40, evaluate the indicated trigonometric expression using the given information.

35. If $\cos(\theta) = 0.0532$ over the interval $[0°, 90°]$, find $\sin(\theta)$.
36. If $\tan(\theta) = 1.4269$ over the interval $[0°, 90°]$, find $\cos(\theta)$.
37. If $\sin(\theta) = -0.4850$ over the interval $\left[\pi, \dfrac{3\pi}{2}\right]$, find $\tan(\theta)$.
38. If $\tan(\theta) = -4.8035$ over the interval $\left[\dfrac{3\pi}{2}, 2\pi\right]$, find $\csc(\theta)$.
39. If $\sec(\theta) = -3.1333$ over the interval $[90°, 180°]$, find $\cot(\theta)$.
40. If $\csc(\theta) = -4.0023$ over the interval $[270°, 360°]$, find $\sec(\theta)$.

SHOW YOU KNOW

41. One of your classmates says that the equation $\sin\left(\dfrac{1}{2}\theta\right) = 0.5$ will only have one solution over the interval $0° \leq \theta \leq 360°$ and gives the following rationale. "The

equation $\sin(\theta) = 0.5$ has two solutions over the interval $0° \le \theta \le 360°$, but $\sin\left(\dfrac{1}{2}\theta\right)$ means that the sine function only completes half of one period as θ goes from $0°$ to $360°$. Thus, there will only be one solution to the equation $\sin\left(\dfrac{1}{2}\theta\right) = 0.5$."

a. Explain what part of his conclusion is correct.

b. Explain what part of his conclusion is incorrect and why.

42. One of your classmates says that the equation $\sec(\theta) = 3$ has no solution. She justifies this claim by plugging $\dfrac{1}{\cos^{-1}(3)}$ into the calculator and showing you that the calculator outputs a domain error. Is she correct? Explain.

43. Compare and contrast the expressions $\cos(1)$ and $\cos^{-1}(1)$. Assume all angles are in radians.

44. Explain why the six basic inverse trigonometric functions will never have a vertical asymptote despite the fact that four of the original trigonometric functions have vertical asymptotes.

■ MAKE IT REAL

In Exercises 45–46,

a. Using inverse trigonometric functions, find a solution to the given equation that is reasonable in the context of the problem.

b. Explain what your solution means, and explain why it is a reasonable solution.

45. **Lines of Latitude** The formula $C(\theta) = 2\pi(3963\cos(\theta))$ models the circumference of the circle in miles formed by a line of latitude θ degrees from the equator.

Lines of Latitude

Solve $2\pi(3963\cos(\theta)) = 11{,}880$. Explain what your answer represents.

46. **Wheelchair Ramps** The federal government has established guidelines for the construction of wheelchair ramps, one of which says that a single ramp may not increase its height more than 30 inches over any horizontal distance. The function $L(\theta) = 30\cot(\theta)$ models the horizontal length of a ramp in inches that rises 30 inches off the ground and makes an angle of θ with the ground. Solve $30\cot(\theta) = 490$. Explain what your answer represents. (*Source:* www.usdoj.gov)

In Exercises 47–50,

a. Using inverse trigonometric functions, find all solutions to the given equation over the interval provided.

b. Explain what your solution(s) mean in the given context.

c. Use a calculator to verify your solutions.

47. **Seasonal Employment** The function
$$L(m) = -550\cos\left(\dfrac{\pi}{6}(m-1)\right) + 3300$$
models the number of people employed in leisure and hospitality fields in Clatsop County, Oregon, m months after January 2004. Solve the equation
$$-550\cos\left(\dfrac{\pi}{6}(m-1)\right) + 3300 = 4200$$
over the interval [1, 13]. (*Source:* Modeled from a graph at www.qualityinfo.org)

48. **Average Temperatures** The function
$$S(m) = 9.40\cos\left(\dfrac{\pi}{6}(m-1)\right) + 57.8$$
models the average monthly temperatures in degrees Fahrenheit in Sydney, Australia during the mth month of the year. Solve
$$53.1 = 9.40\cos\left(\dfrac{\pi}{6}(m-1)\right) + 57.8$$
on the interval [1, 12]. (*Source:* Modeled using data from www.engr.udayton.edu)

49. **Hours of Daylight** The number of hours of daylight in New York City d days after March 21, 2010 can be modeled by
$$N(d) = 2.925\sin\left(\dfrac{2\pi}{365}d\right) + 12.18$$
Solve
$$2.925\sin\left(\dfrac{2\pi}{365}d\right) + 12.18 = 11.5$$
over the interval [−79, 285]. (*Source:* Modeled using data from aa.usno.navy.mil)

50. **Ferris Wheels** The function
$$H(\theta) = 246\sin(\theta - 90°) + 296$$
models a person's height in feet on the Singapore Flyer Ferris wheel after boarding at the bottom and traveling θ degrees around the wheel. Solve
$$246\sin(\theta - 90°) + 296 = 110$$
over the interval [0°, 720°]. (*Source:* www.singaporeflyer.com.sg)

In Exercises 51–54,

a. Find a possible formula for the curve.

b. Find all of the intersection points of the curve and the line shown using inverse trigonometric functions.

c. Explain what your solutions represent.

51. Average High Temperature

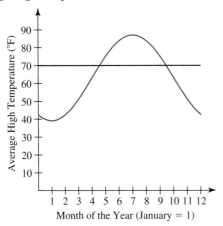

Source: www.weatherbase.com

52. Average Low Temperature

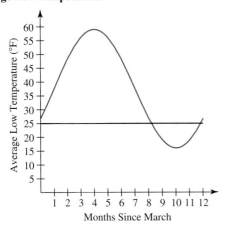

Source: www.weatherbase.com

53. Hours of Daylight

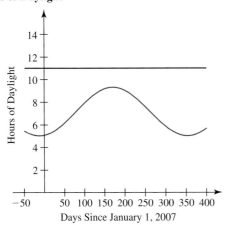

Source: aa.usno.navy.mil

54. Hours of Daylight

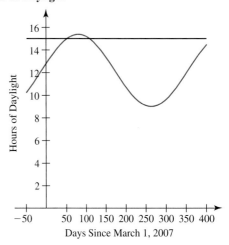

Source: aa.usno.navy.mil

In Exercises 55–56,

a. Graph the function, then identify an appropriate restricted domain such that its inverse will also be a function.

b. Find the inverse function and state its domain and range.

c. Graph the inverse function over the domain and range found in part (b).

d. Use the inverse function found in part (b) to answer the given question.

55. Ferris Wheels The function

$$H(t) = 221 \sin\left(\frac{\pi}{15} t - \frac{\pi}{2}\right) + 221$$

models a person's height in feet from the bottom of the London Eye Ferris wheel after boarding at the bottom and traveling around the wheel for t minutes. When is a rider 155 feet higher than where the rider boarded? (*Source:* www.aboutbritain.com)

56. Seasonal Employment The function

$$A(m) = -18{,}000 \cos\left(\frac{\pi}{6}(m-1)\right) + 59{,}000$$

models the number of workers in the agricultural industry m months after January 2004 in Clatsop County, Oregon. (*Source:* Modeled from a graph at www.qualityinfo.org) When will the number of workers be 41,000?

STRETCH YOUR MIND

Exercises 57–58 are intended to challenge your understanding of inverse trigonometric functions.

57. The graph is the inverse cosine relation for two periods of the cosine function. Write a series of formulas necessary to duplicate this graph on your calculator.

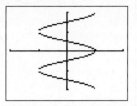

58. Consider the functions $f(x) = \sin^{-1}(\sin(x))$ and $g(x) = \sin(\sin^{-1}(x))$. Assume all angles are in radians.
    a. What are the domain and range of each function?
    b. Graph each function separately and explain why they appear the way they do.
    c. One of your classmates says that both of these functions are identical, and that they can be simplified to $f(x) = x$ and $g(x) = x$, so $f(x) = g(x)$. Explain why he might have come to this conclusion, but why this is not a correct assumption.

In Exercises 59–60, evaluate each expression without using a calculator. All angles are in radians in the first quadrant.

59. $\cos(\tan^{-1}(\cot(\csc^{-1}(2))))$

60. $\csc^{-1}\left(\cot\left(\sec^{-1}\left(2\sin\left(\dfrac{\pi}{4}\right)\right)\right)\right)$

CHAPTER 8 Study Sheet

As a result of your work in this chapter, you should be able to answer the following questions, which are focused on the "big ideas" of this chapter.

SECTION 8.1
1. What does it mean for something to be "periodic"?
2. What is the relationship between maximum value, minimum value, midline, and amplitude?

SECTION 8.2
3. How are arc length and radius related?
4. What is a radian?

SECTION 8.3
5. What is the relationship between the sine and cosine functions and the unit circle?
6. Why will the values of $f(\theta) = r\sin(\theta)$ and $g(\theta) = r\cos(\theta)$ never exceed r?

SECTION 8.4
7. How do you solve a trigonometric equation graphically?
8. How do you solve a trigonometric equation with a data table?

SECTION 8.5
9. How can you tell if a sinusoidal function will model a data set well?
10. How do you construct a sinusoidal model?

SECTION 8.6
11. How are tangent, cotangent, secant, and cosecant related to sine and cosine?
12. Why do tangent, cotangent, cosecant, and secant have restricted domains?

SECTION 8.7
13. What is the relationship between a trigonometric function and its inverse?
14. What are the benefits of using inverse trigonometric functions in solving trigonometric equations?
15. What are the drawbacks of using inverse trigonometric functions in solving trigonometric equations?

REVIEW EXERCISES

■ SECTION 8.1 ■

1. **Merry-Go-Round** A child begins a ride on a merry-go-round. The independent variable is the elapsed time from when he began to ride and the dependent variable is the child's position in relation to where he started. Sketch a graph to model this scenario. (*Note:* There may be more than one correct answer.)

2. **Solar Irradiance** *Solar irradiance* is radiant energy emitted by the sun resulting from a nuclear fusion reaction that creates electromagnetic energy. The graph shows the pattern that solar irradiance measured in watts per square meter (W/m^2) has been found to follow over time.

   Source: http://serc.carleton.edu. Courtesy SERC, the Science Education Resource Center at Carleton College.

   a. Estimate the period, the amplitude, and the midline for the solar irradiance. Explain the real-world meaning of each.
   b. Determine from the graph the years in which solar irradiance was at its lowest.

3. **Surface Water Temperature** A scientific study of the Jack Bay in the lower Patuxent River in Maryland involved measuring the surface temperature of the water. Data were collected from 1998 to 1999.

Time in Months Since Jan. 1, 1998 t	Surface Temperature (degrees Celsius) C
1	5.90
2	6.10
3	6.95
4	15.00
5	19.90
6	23.70
7	27.45
8	28.05
9	25.95
10	19.10

   (*continued*)

Time in Months Since Jan. 1, 1998 t	Surface Temperature (degrees Celsius) C
11	11.70
12	12.90
13	5.30
14	6.90
15	6.25
16	13.45
17	19.20
18	23.70
19	26.25
20	28.00
21	23.60
22	17.00
23	12.90
24	8.30

   Source: www.dnr.state.md.us

   a. Estimate the average rate of change between months 8 and 12. Interpret your answer in terms of the context.
   b. Estimate the instantaneous rate of change at month 20. Interpret your answer.
   c. Estimate where the highest and lowest temperatures are in each year and explain what happens before and after each in terms of rate of change.
   d. Does this table of data appear to be periodic? Explain.

4. **Earthquakes** The number of worldwide earthquakes E of different intensities that occurred each year for 2005 and 2006 is displayed in the table.

Magnitude (Richter scale value) m	Number of Earthquakes in 2005 E	Number of Earthquakes in 2006 N
8.0 to 9.9	1	1
7.0 to 7.9	10	10
6.0 to 6.9	141	132
5.0 to 5.9	1697	1483
4.0 to 4.9	13,918	13,069
3.0 to 3.9	9189	9953
2.0 to 2.9	4636	4016
1.0 to 1.9	26	19
0.1 to 0.9	0	2

   Source: neic.usgs.gov/neis/eqlists/eqstats.html

   a. Do these data appear to be periodic? Explain.
   b. Using these data, predict how many earthquakes there will be of magnitude 3.5 on the Richter scale in 2010.

5. **Freezer Temperature** The graph models the fluctuation in temperature in degrees Fahrenheit inside of a freezer with its door closed.

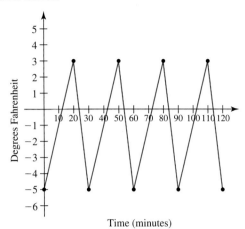

a. Estimate the temperature in the freezer after 50 and 70 minutes.
b. Does this graph represent a periodic function? Explain.
c. Suppose the freezer door is opened after being closed for 120 minutes and is left open for 30 minutes in a 70°F room. Sketch a graph representing the temperature inside the freezer for the 30-minute period that the door is open after the 120-minute period represented in the graph.

6. **Blood Pressure** The graph gives the variation in blood pressure for a typical person. Systolic and diastolic pressures are the upper and lower limits of the periodic changes in pressure that produce the pulse. The length of time between peaks is called the period of the pulse.

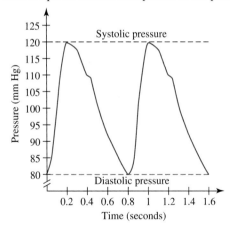

Source: www.vaughns-1-pagers.com/
medicine/blood-pressure.htm

a. Find the amplitude, midline, and period.
b. Find the frequency (also known as the pulse rate—the number of pulse beats in one minute) for the typical person.
c. When a person's blood pressure reaches 160 (systolic pressure) over 100 (diastolic pressure) many doctors require their patients to receive treatment. What aspect(s) of the graph would be affected with these blood pressure readings? Estimate the numerical value(s) of the parameter(s) that has to change.

7. **Pilot-Induced Oscillation** *Pilot-induced oscillation* is caused when a pilot makes a sequence of corrections in opposing directions. For example, a pilot aiming for a 450-foot-per-minute descent may find herself descending too rapidly at 475 feet per minute. She may then try to correct this by slowing the vertical speed to 450 feet per minute. However, the vertical speed indicator generally lags the actual vertical speed. Consequently, the vertical speed may actually drop below 450 feet per minute before the indicator shows a speed of 450. When the indicator is updated with this slower speed she may then choose to increase her vertical speed causing her to travel at 475 feet per minute again. Given that each cycle of corrections is 5 minutes in length, sketch two cycles of the pilot trying to correct her vertical speed. (*Source:* www.answers.com/topic/pilot-induced-oscillation)

8. **Electrocardiogram** An electrocardiogram is a record of the electrical activity of the heart as a function of time in seconds. Typically doctors analyze the various waves on the printout and use the information as part of a diagnostic process to determine the health of a patient. An electrocardiograph machine outputs paper at a constant rate of 25 millimeters per second and the grid lines on the graph paper are 5 millimeters apart. Estimate the period and frequency of the graph. Also estimate how many beats per minute this person's heart is beating.

Note: Image is not drawn to scale.

■ SECTION 8.2 ■

For Exercises 9–10, describe what fraction of the circumference of a full circle is spanned by an angle with the given measure.

9. $\theta = \dfrac{5\pi}{6}$ radians

10. $\theta = 3$ radians

For Exercises 11–12, convert each angle measure in degrees to radian measure.

11. $\theta = 35°$

12. $\theta = -120°$

For Exercises 13–14, convert each angle measure from radians to degrees.

13. $\theta = \dfrac{\pi}{6}$

14. $\theta = -\dfrac{2\pi}{3}$

15. What is the length of the arc that corresponds with an angle with a measure of 75° in a circle of radius 1.5 centimeters?

16. What is the length of the arc that corresponds with an angle of 1.75 radians in a circle of radius 5 inches?

17. In a circle of radius 10 meters, what is the measure of the angle that corresponds with an arc of length 30 meters?

18. List two foundational characteristics of the idea of angle. Explain why you think they are foundational to your understanding of the idea of angle.

19. **Boarding** For boarders, an "inverted 540" is "an inverted aerial where the athlete performs a 540 degree rotational flip. In other words, the athlete approaches the half pipe wall riding forward, becomes airborne, rotates 540 degrees in a backside direction while performing a front flip, and lands riding forward. Invented by skateboarder Mike McGill (www.snowboarding.com)." Describe what is meant by "rotates 540 degrees."

■ SECTION 8.3 ■

In Exercises 20–22, use the following graph, which shows the terminal side of three angles (A, B, and C) associated with an arc in standard position.

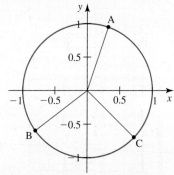

If the angle measures θ, estimate $\cos(\theta)$ and $\sin(\theta)$.

20. A   21. B   22. C

In Exercises 23–24, use a calculator to evaluate each expression, then draw a picture to represent the meaning behind the value you found.

23. $\cos(125°)$

24. $4\sin(213.5°)$

In Exercises 25–26, use a calculator to evaluate each expression, then draw a picture to represent the meaning behind the value you found. The angles are all measured in radians.

25. $\cos\left(\dfrac{\pi}{7}\right)$

26. $\left(\cos\left(\dfrac{9\pi}{5}\right), \sin\left(\dfrac{9\pi}{5}\right)\right)$

In Exercises 27–28, find the exact values of each expression, then draw a picture to represent the meaning behind the value you found.

27. $\cos(45°)$

28. $\sin(240°)$

29. **Ferris Wheels** The London Eye is a Ferris wheel constructed on the banks of the River Thames in London. The London Eye has a radius of about 221 feet and is boarded from the bottom. (*Source:* www.aboutbritain.com)

   Determine the height of a person from the bottom of the London Eye after traveling each of the following portions of a revolution.

   a. 3/4 of the way around
   b. 1/8 of the way around
   c. 7/10 of the way around

30. **Ferris Wheels** The Singapore Flyer is a Ferris wheel constructed on top of a three-story building. The wheel itself has a diameter of 492 feet, but the bottom of the wheel sits nearly 50 feet off the ground. Determine the height from the ground of a person after boarding from the bottom and traveling each of the following portions of a revolution. (*Source:* www.singaporeflyer.com.sg)

   a. 1/5 of the way around
   b. 5/6 of the way around
   c. after 8/5 revolutions

■ SECTION 8.4 ■

In Exercises 31–36, answer all of the following questions for each function.

   a. What are the period, amplitude, frequency, and equation of the midline?
   b. What are the maximum and minimum values of f?
   c. Graph the function without using a calculator. Make sure to graph at least one complete period.
   d. Where are the horizontal intercepts located?
   e. Where is the vertical intercept located?
   f. Describe the transformation of the function as related to the graph of $f(\theta) = \sin(\theta)$ or $f(\theta) = \cos(\theta)$, as appropriate.

31. $f(\theta) = 3\sin(\theta)$; θ is in degrees.

32. $f(\theta) = \sin\left(\dfrac{1}{2}\theta\right)$; θ is in radians.

33. $f(\theta) = \dfrac{2}{3}\cos\left(\theta - \dfrac{\pi}{4}\right)$; θ is in radians.

34. $f(\theta) = \sin\left(1.5\left(\theta + \dfrac{\pi}{3}\right)\right) + 2$; θ is in radians.

35. $f(\theta) = -2\cos\left(\dfrac{1}{5}(\theta + 90°)\right)$; θ is in degrees.

36. $f(\theta) = -4\cos\left(3\left(\theta - \dfrac{\pi}{6}\right)\right) + 1$; θ is in radians.

37. Explain the difference between a phase shift and a horizontal shift.

38. State which transformations have the ability to alter the given characteristic of a trigonometric function.
    a. period
    b. amplitude
    c. midline
    d. frequency
    e. horizontal intercepts
    f. vertical intercept

39. **Seasonal Employment** The function
$$C(m) = -5000 \cos\left(\frac{\pi}{6}(m-3)\right) + 77{,}000$$
models the number of people employed in construction in Oregon m months after January 2002. (*Source:* Modeled using estimates from a graph at www.qualityinfo.org)
    a. Find the amplitude, period, equation of the midline, and phase shift for L, then explain what each tells you about this situation.
    b. During what month are the fewest number of people employed in construction in Oregon? The most people?
    c. Graph C over the interval $0 \le m \le 12$. Describe its rate of change and concavity throughout this interval and explain what this tells you about construction employment in Oregon.

40. **Sunrise** The function
$$S(d) = 1.325 \cos\left(\frac{2\pi}{365}(d+2)\right) + 5.775$$
models the time of sunrise in Washington, Indiana, in Eastern Standard Time (EST) on day d of the year. (*Note:* Daylight savings time is not taken into consideration in this model.)
(*Source:* Modeled from data at aa.usno.navy.mil)
    a. Find the period, the amplitude, and the equation of the midline for S. Explain what these values tell you about the time that the sun rises in Washington, Indiana.
    b. Graph S over the interval [0, 365]. Describe the rate of change and concavity of the function, then explain what this information tells you about this situation.
    c. Solve the equation
$$5 = 1.325 \cos\left(\frac{2\pi}{365}(d+2)\right) + 5.775$$
    over the same interval by graphing, then explain what your solution(s) mean.

41. **Average Temperatures** The function
$$T(m) = 24.8 \sin\left(\frac{\pi}{6}(m-4)\right) + 63.3$$
models the average high temperature in Washington, Indiana, in degrees Fahrenheit during month m of the year.
(*Source:* www.weatherbase.com)
    a. Find the period, the amplitude, and the equation of the midline for T. Explain what these values tell you about the average high temperature in Washington, Indiana.
    b. Solve the equation
$$24.8 \sin\left(\frac{\pi}{6}(m-4)\right) + 63.3 = 75$$
    over the interval [1, 12] by graphing, then explain what your solution(s) mean.

SECTION 8.5

In Exercises 42–45, state the period, amplitude, and midline.
42. $y = 4 \sin(2(t+1)) + 6$
43. $y = 9 \cos(3.1(t+2)) - 4$
44. $y = 6 \cos(0.5t)$
45. $y = \pi \sin(3t) - 0.04$

In Exercises 46–49, determine the vertical and horizontal shifts.
46. $y = 4 \sin(2(t+1)) + 6$
47. $y = 9 \cos(3.1(t+2)) - 4$
48. $y = 6 \cos(0.5t)$
49. $y = \pi \sin(3t) - 0.04$

In Exercises 50–53, sketch one period of the function without using a calculator.
50. $y = 2 \cos(5t)$
51. $y = 11 \sin\left(t + \frac{3\pi}{4}\right)$
52. $y = 3 \sin(\pi(t - \pi)) + 1$
53. $y = \cos\left(t - \frac{5\pi}{2}\right) + 8$

54. Find an equation for the sinusoidal function shown in the graph. You may use either sine or cosine.

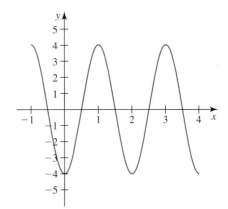

SECTION 8.6

In Exercises 55–56, use your calculator to evaluate each expression, then explain what your solution means in the context of angles.
55. $\tan(113°)$
56. $\cot(230°)$

In Exercises 57–58, find the exact value of each expression without using a calculator, then explain what your solution means in the context of angles.
57. $\cot\left(\frac{2\pi}{3}\right)$
58. $\tan\left(\frac{7\pi}{4}\right)$

59. **Road Grades** The U.S. Department of Agriculture recommends building forest access roads on which water drains away from the road so that the roads may still be used under wet conditions. One method of doing this is to *outslope* the road, or give it a slight slope so that water runs down the road instead of collecting on it. Roads with an outslope have an angle of incline of 2.3°. (*Source:* www.na.fs.fed.us)

   a. What is the minimum road grade for an outsloped road?
   b. What does cot(2.3°) represent in this situation?
   c. What is the elevation change over 0.25 miles on a road with a 2.3° incline?

60. **Pyramids** The ancient Romans built a pyramid in the Egyptian style called the Pyramid of Cestius. However, the Romans built this pyramid much steeper than the famous pyramids at Giza. (Their use of concrete made it easier to do so.) The sides of the Pyramid of Cestius have an incline of about 67.8°. (*Source:* www.romeguide.it)

   a. Find tan(67.8°) and cot(67.8°), then explain what these values mean in this context.
   b. The Pyramid of Cestius is 27 meters tall. How wide is it?

In Exercises 61–62, use a calculator to evaluate each expression, then use your answer to find the cosine or sine value for the same arc or angle.

61. sec(81°)
62. csc(163°)

In Exercises 63–64, find the exact value of each expression. Angles are in radians.

63. $\sec\left(\dfrac{\pi}{6}\right)$
64. $\csc\left(\dfrac{7\pi}{6}\right)$

For Exercises 65–66, use the following information. A person observing a 100-meter sprinter race at a track meet is located at point P. When the sprinter reaches point X, the sprinter and observer are 7 meters apart.

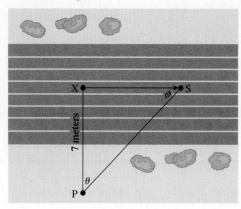

65. The function $d(\theta) = 7\sec(\theta)$ models the distance in meters between the sprinter and the observer (the distance from P to S) as a function of the angle θ. What is the distance between the sprinter and observer when $\theta = 32°$?

66. The function $s(\phi) = 7\csc(\phi)$ will also model the distance in meters between the sprinter and observer (the distance from P to S), but will be a function of ϕ instead of θ. Find a reasonable solution to the equation $20 = 7\csc(\phi)$ using graphing, then explain what your solution means.

■ SECTION 8.7 ■

In Exercises 67–68, find at least one value for the given expression over the interval [0, 2π], then explain what your answer represents.

67. $\cos^{-1}(0.82)$
68. $\sin^{-1}(-0.34)$

In Exercises 69–70, evaluate each expression exactly, finding at least one value over the interval [0°, 360°] without using a calculator.

69. $\csc^{-1}(\sqrt{2})$
70. $\tan^{-1}(-\sqrt{3})$

In Exercises 71–73, find all solutions to each equation over the interval [0°, 360°] using inverse trigonometric functions. Check your solutions by graphing.

71. $\cos(\theta) = -0.3926$
72. $3.5013 = 2.1\tan(\theta)$
73. $6\csc(\theta - 20°) = -13$

In Exercises 74–76, find all solutions to each equation over the interval [0, 2π] using inverse trigonometric functions. Check your solutions by graphing.

74. $\sin(\theta) = 0.0184$
75. $\sec\left(\theta + \dfrac{\pi}{15}\right) = -3.5$
76. $16.8 = 7.6\cot\left(\theta - \dfrac{\pi}{6}\right)$

In Exercises 77–78,

   a. Using inverse trigonometric functions, find a solution to the given equation that is reasonable in the context of the problem.
   b. Explain what your solution means, and explain why it is a reasonable solution.

77. **Road Grades** The function $g(\theta) = 100\tan(\theta)$ models the percent road grade measure as a function of θ, the angle of ascent or descent. Solve $9 = 100\tan(\theta)$.

78. **Lines of Latitude** The function $C(\theta) = 2\pi(3963\cos(\theta))$ models the circumference of the circle (in miles) formed by a line of latitude θ degrees from the equator. Solve $2\pi(3963\cos(\theta)) = 5650$.

Make It Real Project

What to Do

1. Go to www.weatherbase.com and select "United States" as the region. Select the state and city where you live to get weather data for your city.

2. Use the data provided to create a formula and graph for the average high temperature in your city over the course of one year.

3. Go to http://aa.usno.navy.mil/, then select "Data Services," followed by "Duration of Daylight/Darkness Table for One Year." Enter your state and city information to get the number of hours of daylight per day in your city throughout the year.

4. Use the data provided to create a formula and graph for the number of hours of daylight in a day throughout the year. (*Hint:* To turn 8 hours 14 minutes into a number of hours in decimal form, find $8 + \frac{14}{60}$.)

5. Compare and contrast the graphs you created. Explain the possible reasons for any similarities in these graphs.

CHAPTER 9
Triangle Trigonometry and Applications

Tyler Olson/Shutterstock.com

Scaling a cliff is difficult for even the most experienced athletes. Success requires physical strength and agility as well as mental focus. Safety-conscious climbers arm themselves with proper gear and with an in-depth knowledge of the surfaces they climb. They ask, "Are the rocks stable?", "Are there a sufficient number of handholds?", "How high is this cliff?"

Using basic trigonometry skills, a climber can estimate the height of a cliff before she begins her ascent. These skills can be used no matter the height of the cliff or the difficulty of the terrain.

9.1 Right Triangle Trigonometry
9.2 Law of Cosines
9.3 Law of Sines
9.4 Polar Coordinates
9.5 Vectors
STUDY SHEET
REVIEW EXERCISES
MAKE IT REAL PROJECT

SECTION 9.1 Right Triangle Trigonometry

LEARNING OBJECTIVES

- Describe the six trigonometric ratios
- Use the appropriate trigonometric ratio to solve real-world problems involving right triangles
- Describe the relationships among the trigonometric ratios

GETTING STARTED

In September 2006, two amateur California naturalists purportedly discovered the "World's Tallest Tree." (*Source:* Associated Press, September 8, 2006) The duo found three trees that reportedly are taller than the world record holder at that time. Prior to this, the tallest tree as recorded in the Guinness Book of World Records was the Stratosphere Giant, a redwood tree found in Redwood National Park, California, measuring 370 feet tall. Each of the three newly discovered redwood trees, also found in Redwood National Park, are taller than 370 feet. The tallest, named Hyperion, is 378.1 feet tall. The next tallest, the Helios, stands at 376.3 feet; Icarus, the third, reaches 371.2 feet. (*Source:* sfgate.com)

In this section we investigate methods of measuring very tall objects, such as redwood trees. With right triangle trigonometry, we can solve a variety of real-world problems such as this.

■ Measuring with Shadows and Similar Triangles

One way to measure a very tall object such as a redwood tree is to use the geometry of similar triangles. To make sense of this idea, we will consider a simplified situation. Imagine a shadow cast by the tree along the ground as shown in Figure 9.1. Using a tape measure, we determine the shadow is 504 feet in length. Now imagine a 6-foot-tall man positioning himself so that his shadow, measuring 8 feet, reaches the top of the tree's shadow.

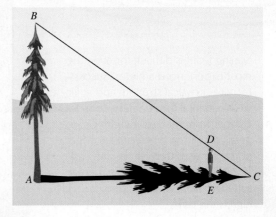

Figure 9.1

The larger triangle made by the tree and its shadow, $\triangle ABC$, is **similar** to the smaller triangle made by the man and his shadow, $\triangle EDC$. This means that the lengths of the corresponding sides of the two triangles are proportional and the corresponding angles have equal measure. That is, the ratio of $\overline{AB}$ to $\overline{ED}$ is the same as the ratio of $\overline{AC}$ to $\overline{EC}$ and the ratio of $\overline{BC}$ to $\overline{DC}$. Because these ratios are the same, we say that the three pairs of sides are **proportional**. Angles with equal measure are said to be **congruent**. We write $\angle A \cong \angle E$ and $\angle B \cong \angle D$. The symbol "$\cong$" means "is congruent to."

SIMILAR TRIANGLES

Two triangles, $\triangle ABC$ and $\triangle DEF$, are similar if
- Corresponding angles are **congruent**.
- Corresponding sides are **proportional**.

Armed with this information, we can now find the height of the tree. We know the following:

Height of person: $\overline{DE} = 6$ feet

Length of person's shadow: $\overline{EC} = 8$ feet

Length of tree's shadow: $\overline{AC} = 504$ feet

Height of tree: $\overline{AB} = H$ feet

Since $\triangle ABC$ is similar to $\triangle EDC$ we know their corresponding sides are proportional. Therefore,

$$\frac{\overline{AB}}{\overline{DE}} = \frac{\overline{AC}}{\overline{EC}}$$

$$\frac{H \text{ feet}}{6 \text{ feet}} = \frac{504 \text{ feet}}{8 \text{ feet}}$$

There are several ways to solve this equation for H, the height of the tree. We choose to cross multiply.

$$\frac{H \text{ feet}}{6 \text{ feet}} = \frac{504 \text{ feet}}{8 \text{ feet}}$$

$$\frac{H}{6} = \frac{504}{8}$$

$(H)(8) = (6)(504)$ **Cross multiply.**

$H = \dfrac{(6)(504)}{8}$ **Divide by 8.**

$H = 378$

The height of this hypothetical redwood tree is 378 feet.

■ Making the Connection between Geometry and Trigonometry

The *hypotenuse* of a right triangle is the side opposite the right angle. The *opposite* side for an angle whose measure is θ is the side directly across from the angle. The *adjacent* side for an angle whose measure is θ is the remaining side (see Figure 9.2).

Now let's scale this right triangle so that it fits inside of the unit circle, as shown in Figure 9.3.

Figure 9.2

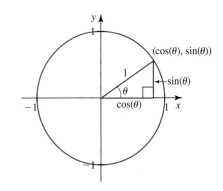

Figure 9.3

Because the original triangle is similar to the resized triangle contained within the unit circle, the side lengths are proportional. We have the following right triangle definitions:

$$\frac{\sin(\theta)}{1} = \frac{b}{h} \qquad \frac{\cos(\theta)}{1} = \frac{a}{h} \qquad \frac{\sin(\theta)}{\cos(\theta)} = \frac{b}{a}$$

$$\sin(\theta) = \frac{\text{opposite}}{\text{hypotenuse}} \qquad \cos(\theta) = \frac{\text{adjacent}}{\text{hypotenuse}} \qquad \tan(\theta) = \frac{\text{opposite}}{\text{adjacent}}$$

We find the right triangle definitions of the remaining trigonometric functions by taking the reciprocal of these three. We get

$$\csc(\theta) = \frac{\text{hypotenuse}}{\text{opposite}} \qquad \sec(\theta) = \frac{\text{hypotenuse}}{\text{adjacent}} \qquad \cot(\theta) = \frac{\text{adjacent}}{\text{opposite}}$$

We summarize the results.

THE SIX TRIGONOMETRIC FUNCTION RATIOS

$$\sin(\theta) = \frac{\text{opposite}}{\text{hypotenuse}} \qquad \cos(\theta) = \frac{\text{adjacent}}{\text{hypotenuse}} \qquad \tan(\theta) = \frac{\text{opposite}}{\text{adjacent}}$$

$$\csc(\theta) = \frac{\text{hypotenuse}}{\text{opposite}} \qquad \sec(\theta) = \frac{\text{hypotenuse}}{\text{adjacent}} \qquad \cot(\theta) = \frac{\text{adjacent}}{\text{opposite}}$$

EXAMPLE 1 ■ Using Trigonometric Functions

While standing 200 feet away from the base of the Hyperion redwood tree, a forestry major measures an angle of 62.12° to the top of the tree. What is the height of the tree?

Solution We draw a picture to represent the situation, as shown in Figure 9.4. We then use trigonometry to solve for H.

$$\tan(62.12°) = \frac{H}{200}$$

$$200 \tan(62.12°) = H$$

$$200(1.890) \approx H$$

$$H \approx 378$$

The height of the tree is 378 feet.

Figure 9.4

EXAMPLE 2 ■ Measuring the Hypotenuse

One of the Seven Wonders of the Ancient World is the Great Pyramid of Giza located near Cairo, Egypt. The pyramid was built by the Egyptian pharaoh Khufu of the Fourth Dynasty around the year 2560 B.C. to serve as a tomb when he died.

(*Source:* ce.eng.usf.edu) The pyramid was 481 feet high when it was built. The angle the side makes with the ground is 51.83°. The square bottom of the pyramid measures roughly 755.3 feet on each side. What is the distance along one of the slanted edges from the ground to the tip of the pyramid?

Solution We begin by drawing a picture of one of the faces of the pyramid and labeling the information we know. See Figure 9.5. The hypotenuse, d, of the right triangle is the length that is unknown; the lengths of the sides opposite and adjacent are known. Therefore, we may use either the cosine or sine ratios to solve for the length, d. To show that either may be used, we will solve the problem both ways.

$$\cos(51.83°) = \frac{377.7}{d} \qquad \sin(51.83°) = \frac{481}{d}$$

$$d\cos(51.83°) = 377.7 \qquad d\sin(51.83°) = 481$$

$$d = \frac{377.7}{\cos(51.83°)} \qquad d = \frac{481}{\sin(51.83°)}$$

$$d \approx \frac{377.7}{0.6180} \qquad d \approx \frac{481}{0.7862}$$

$$d \approx 611.2 \qquad d \approx 611.8$$

Figure 9.5

We attribute the slight difference to rounding and inexact measurements.

■ Angles of Elevation and Depression

Often we are given information about either the **angle of elevation** (Figure 9.6a) or **angle of depression** (Figure 9.6b) when solving problems involving right triangle trigonometry. These angles can be measured using professional tools such as the theodolite or homemade tools such as a transit.

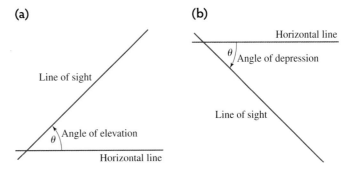

Figure 9.6

> ### Angles of Elevation and Depression
>
> The **angle of elevation** is the angle from an imaginary horizontal line and the observer's line of sight to an object that is *above* the horizontal line.
>
> The **angle of depression** is the angle from an imaginary horizontal line and the observer's line of sight to an object that is *below* the horizontal line.

EXAMPLE 3 ■ Using Angle Measures

Prior to Global Positioning System (GPS) technology, fire lookout towers played a major role in helping people spot forest fires. While some are still in operation, they are being used less and less. Upon spotting a fire, the person in the lookout tower uses a tool called an Osborne Fire Finder (see photo) to determine the distance to the fire. Suppose the fire lookout spotted a fire at an angle of depression of 2.86° and that the height above the ground at the point of spotting the fire was 150 feet. How far from the fire is the base of the fire tower?

Solution We begin with the sketch of the problem situation shown in Figure 9.7. Note that the sketch is not drawn to scale. We consider the right triangle where one of the legs of the triangle is the height of the lookout tower and the other leg is the unknown distance to the fire. The angle of depression, whose measure is $\theta = 2.86°$, is not a part of the triangle that we need to use to solve the problem. However, we use that angle's measure to determine the measure of the angle from the fire tower to the line of sight (hypotenuse): $90° - 2.86° = 87.14°$. Now we use the tangent ratio to solve the problem.

Figure 9.7

$$\tan(87.14°) = \frac{D}{150}$$

$$150\tan(87.14°) = D$$

$$D = 3002.5$$

The distance from the base of the lookout tower to the fire is about 3002.5 feet.

EXAMPLE 4 ■ Expressing the Six Trigonometric Functions

Given the right triangle shown in Figure 9.8, find the value of each of the following: $\sin(\theta)$, $\cos(\theta)$, $\tan(\theta)$, $\sec(\theta)$, $\csc(\theta)$, and $\cot(\theta)$.

Solution We use the definitions of the six trigonometric ratios as follows.

$$\sin(\theta) = \frac{12}{13} \qquad \cos(\theta) = \frac{5}{13} \qquad \tan(\theta) = \frac{12}{5}$$

$$\csc(\theta) = \frac{1}{\sin(\theta)} = \frac{13}{12} \qquad \sec(\theta) = \frac{1}{\cos(\theta)} = \frac{13}{5} \qquad \cot(\theta) = \frac{1}{\tan(\theta)} = \frac{5}{12}$$

Figure 9.8

■ Determining Angle Measure

Recall that we can use inverse trigonometric functions to determine the measure of an angle when the trigonometric ratio is known. In the next example, we do this to determine the measure of an angle of elevation.

EXAMPLE 5 ■ Calculating an Angle of Elevation

In 1999, two of the authors visited San Francisco for a math conference. While walking along a steep San Francisco street, they wondered what the angle of elevation of the street was. To answer their question, they measured a brick in the foundation of a nearby home, as represented in Figure 9.9. Determine the angle of elevation of the hill.

Figure 9.9

Solution We wish to find the measure of the angle the street makes with the hypothetical horizontal line. Since the top and bottom of the brick are parallel lines, this angle has the same measure, θ, as the angle the street makes with the top of the brick. Using this angle as our point of reference, we see that the opposite side length of the right triangle is known (3 inches). Also, the adjacent side length is known (12 inches). Therefore, we use the tangent function to find the required angle measure.

$$\tan(\theta) = \frac{3}{12}$$
$$\theta = \tan^{-1}\left(\frac{3}{12}\right)$$
$$\theta \approx 14.04°$$

The street has an angle of elevation that measures 14.04°.

SUMMARY

In this section you learned how to solve problems that can be represented using right triangles. You learned the six trigonometric ratios and how to use them to find missing lengths and missing angle measures in right triangles.

9.1 EXERCISES

SKILLS AND CONCEPTS

In Exercises 1–6, evaluate the six trigonometric functions of θ. That is, evaluate $\sin(\theta)$, $\cos(\theta)$, $\tan(\theta)$, $\sec(\theta)$, $\csc(\theta)$, $\cot(\theta)$.

1.

2.

3.

4.

5.

6.

In Exercises 7–11, find each of the values.

   a. sin(*A*)   **b.** cos(*B*)
   c. sin(*B*)   **d.** cos(*A*)

7.

8.

9.

10.

11. (symbols are intentional)
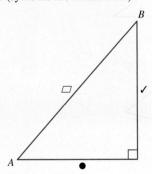

12. Refer back to Exercises 7–11 and express the relationship between the measures of ∠*A* and ∠*B*. Explain this relationship as clearly as you can.

13. Refer back to Exercises 7–11 and express the relationship between sin(*A*) and cos(*B*). Explain this relationship as clearly as you can.

14. Refer back to Exercises 7–11 and express the relationship between sin(*B*) and cos(*A*). Explain this relationship as clearly as you can.

In Exercises 15–23, a trigonometric value is given. Determine if the angle measure, θ, is

   a. Less than 30°.
   b. Between 30° and 60°.
   c. Between 60° and 90°.

Explain how you know.

15. sin(θ) = 0.15   **16.** tan(θ) = 3.8
17. sin(θ) = 0.65   **18.** cos(θ) = 0.95
19. tan(θ) = 1   **20.** cos(θ) = 0.75
21. tan(θ) = 0.05   **22.** sin(θ) = $\frac{1}{4}$
23. cos(θ) = 0.01

In Exercises 24–28, determine the indicated angle measure, θ.

24.

25.

26.   **27.**

28.

29. **Hot Air Balloon** To tether a hot air balloon to the ground, long ropes must be used. Suppose that the rope makes an angle of elevation with the ground measuring 30° and is tethered 100 feet from the bottom of the balloon.

   a. Assuming the rope is attached to the top of the balloon, what is the height of the balloon?

   b. Suppose that when fully inflated, the balloon is 100 feet tall and the angle of elevation of the rope measures 42°. What is the minimum amount of additional rope needed to tether the balloon when fully inflated?

30. **Space Shuttle Launch** Suppose you are 4000 feet away from the launch pad of the Space Shuttle. You are told that 7 seconds after liftoff, the Space Shuttle is 530 feet above the ground.

   a. What is the measure of the angle of elevation from you to the Space Shuttle?

   b. What is the average rate of change of the measure of the angle of elevation between liftoff (0 seconds) and 7 seconds?

   c. Do you think that the rate of change of the measure of the angle of elevation later in the flight of the shuttle will be faster or slower than the result in part (b)? Explain your answer.

31. **Speed Trap** A theodolite is used to measure the distance from a particular point (A) to a car traveling down a flat and level road ($\overline{BC}$). Ten seconds later, the device measures the distance to the car and the angle of rotation (C). Using the diagram, determine if the car was speeding. Assume the speed limit is 45 mph.

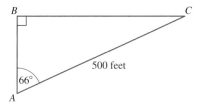

SHOW YOU KNOW

32. You are given a right triangle problem situation to solve. You know the measure of one side and one acute angle. Can you find the measure of every other side and every other angle? If so, explain how. If not, explain why not.

33. You are given a right triangle problem situation to solve. You know the measure of all three sides. Can you find the measure of each of the angles? If so, explain how. If not, explain why not.

34. You are given a right triangle problem situation to solve. You know the measure of all three angles. Can you find the measure of each of the three sides? If so, explain how. If not, explain why not.

35. Is there an angle measure whose sine and cosine are equal? If so, what is it and why are these ratios equal?

36. Which of the three ratios—$\sin(A)$, $\cos(A)$, or $\tan(A)$—is equal to the slope of a line m? Explain.

MAKE IT REAL

37. **Pittsburgh Incline** Downtown Pittsburgh is located at the intersection of three rivers: the Ohio, the Allegheny, and the Monongahela. In the late 1800s workers came down to the city on a cable car system, or incline, from the residential area located at the top of a ridge. The incline is now a historical site. At the top of the ridge, a sign with some of the facts related to the incline are given. The grade (angle of elevation's measure) is listed as 35.58°. The length of the incline track (the hypotenuse) is 635 feet. What is the vertical distance of the incline track from the bottom to the top of the ridge?

38. **Ladder Safety** A report describing a firefighter training accident stated that a 24-foot extension ladder was placed against a wall at a 75° angle with the ground. At the time of the accident, the top of the ladder was at a height of 20 feet 1 inch. (*Source:* www.cdc.gov)

   a. Was the ladder at its full 24-foot length? If so, explain how you know. If not, how long was the ladder?

   b. The report states the ladder "was positioned at an angle of about 75° (4 to 1 ratio)." Use trigonometry to confirm or refute the claim that there was a 4 to 1 ratio. What do you think was meant by this?

39. **Largest Escalator** The CNN Center in Atlanta, Georgia, has the largest freestanding escalator in the country. It ascends 8 stories (85 feet) and is approximately 200 feet in length.

   a. What is the measure of the angle of elevation of the escalator?

   b. If it takes 2 minutes to get to the top of the escalator, how fast is the escalator moving?

40. **Leaning Tower of Pisa** The Leaning Tower of Pisa is leaning because of unstable ground upon which it was built. According to the following statement: *"The inclination is c. $5\frac{1}{2}$ degrees towards the south; this means that the seventh cornice protrudes about 4.5m over the first cornice."* (*Source:* terre.duomo.pisa.it)

   The horizontal protrusions at each level are referred to as cornices. What is the vertical distance between the first and the seventh cornice?

41. **Steep Grade Road Sign** In mountainous regions, it is common to see signs warning drivers of a steep grade. If a sign warns of a 14% grade, that means that for every 100 feet traveled horizontally, the elevation will decrease by 14 feet. Determine the angle of elevation on a portion of roadway with a 14% grade.

42. **Evel Knievel's Rocket Launching** On September 8, 1974, daredevil Evel Knievel attempted to jump the Snake River Canyon in southern Idaho. A special rocket was built for the stunt. It was launched off a ramp shown in the picture. The ramp was 300 feet in length and rose vertically 200 feet. (*Source:* www.canosoarus.com)

    a. Calculate the measure of the angle of elevation that the ramp makes with the ground.

    b. The horizontal crossbeam is located halfway to the top of the ramp. Calculate the length of this crossbeam.

43. **Parasailor** A parasailor is tethered to a boat with a 110-foot cable and is 74 feet above the point of attachment.

    a. What is the measure of the angle of depression from the parasailor to the boat?

    b. How far, horizontally, is the parasailor from the boat?

44. **Hoover Dam** Two trigonometry students wrote the following in a report they created about Hoover Dam.

    > My friend stood at the top of the dam and used a theodolite to measure the angle of depression to my position, 1558 feet from the base of the dam. The measure of the angle of depression was 25°. Using right triangle trigonometry, we calculated the height of Hoover Dam to be 726 feet.

    Discuss the accuracy of the report. Are the values reported by the students reasonably accurate? Why or why not? (*Note:* The true height of Hoover Dam is 726.4 feet.) (*Source:* www.usbr.gov)

45. **Trigonometry in Sports** During the Olympic Games, judges use a theodolite to measure angles and distances in track and field events such as the discus, hammer throw, and shot put. (*Source:* www.leica-geosystems.com)

In the discus event, a referee sticks a target into the ground at the place the discus hits the turf. The theodolite measures the distance to the target using lasers.

Suppose a discus thrower (*A*) tosses the discus and the discus lands at point (*B*). A theodolite is set up 40 meters from the throwing circle (*C*). The theodolite calculates the distance to the target (*B*) is 45 meters. What is the official distance of the throw? What is the measure of angle *A*?

■ **STRETCH YOUR MIND**

Exercises 46–52 are intended to challenge your understanding of right triangle trigonometry.

46. Sketch three right triangles, $\triangle ABC$, such that $\sin(A) = \frac{2}{3}$. In each, label all side and angle measures.

47. Sketch three right triangles, $\triangle ABC$, such that $\cot(A) = 2$. In each, label all side and angle measures.

48. Sketch three right triangles, $\triangle ABC$, such that $\sec(A) = \frac{5}{3}$. In each, label all side and angle measures.

49. Describe how right triangle trigonometry can be used to find the area of a parallelogram.

50. In the diagram, suppose the length w is fixed. Write two equations that relate the length h and the angle measure θ.

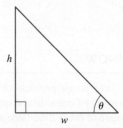

51. In the diagram, suppose the length w is fixed. Write two equations that relate the length h and the angle measure θ.

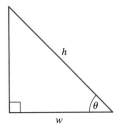

52. In the diagram, suppose the length w is fixed while h is free to vary. Express the function $\theta(h)$ and explain what information it provides.

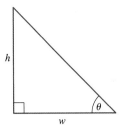

SECTION 9.2 Law of Cosines

LEARNING OBJECTIVES

- Develop the formula for the Law of Cosines
- Demonstrate the relationship between the Pythagorean theorem and the Law of Cosines
- Apply the Law of Cosines to real-world scenarios

GETTING STARTED

The usefulness of trigonometry in surveying, navigation, and astronomy has been seen throughout history. The Egyptians used concepts from trigonometry in land surveying and in building the pyramids. The Greeks and Babylonians, pioneers in the field of astronomy, used trigonometry to develop early theories for planetary position and motion. Today, trigonometry is also used in physics to further our understanding of space. In 2004, NASA's Gravity Probe B satellite was launched into orbit to test a key prediction of Albert Einstein's general theory of relativity: that the fabric that makes up space and time is curved and dragged by the gravity of celestial bodies such as Earth. (*Source:* www.einstein.stanford.edu) To test this hypothesis, scientists employed a law of trigonometry known as the Law of Cosines.

In this section we develop the formula for the Law of Cosines, demonstrate the relationship between the Pythagorean theorem and the Law of Cosines, and apply the law to real-world scenarios.

■ Types of Triangles

If none of the angles in a triangle is a right angle, the triangle is called **oblique**. Each angle of the triangle is either **acute** (greater than 0° but less than 90°) or **obtuse** (greater than 90° but less than 180°). An oblique triangle with three acute angles is called an **acute triangle** (Figure 9.10a). An oblique triangle with two acute angles and one obtuse angle is called an **obtuse triangle** (Figure 9.10b).

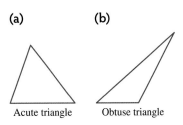

Figure 9.10

As before, we label angles with capital letters such as A, B, and C and the sides directly across from these angles with their corresponding lowercase letters such as a, b, and c. See Figure 9.11.

The phrase **solve a triangle** means to find the lengths of its sides and the measurements of its angles. To solve any triangle, we only need to know the length of one side and two other pieces of information: two more sides, a side and an angle, or two angles. Letting S represent a side length and A represent an angle measure, the three pieces of information may be arranged as shown in Table 9.1 and Figure 9.12.

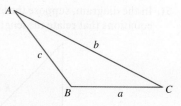

Figure 9.11

Table 9.1

Case	Meaning	Description
SSS	Side–Side–Side	Three sides
SAS	Side–Angle–Side	Two sides and the angle between them
ASA	Angle–Side–Angle	Two angles and the side shared by the angles
SAA	Side–Angle–Angle	Two angles and a side opposite one angle
SSA	Side–Side–Angle	Two sides and an angle opposite one side

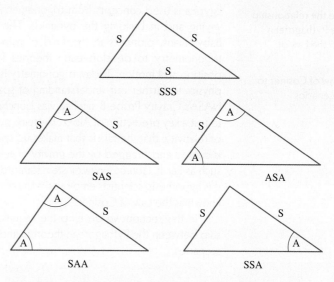

Figure 9.12

■ **Law of Cosines**

The Law of Cosines is used to solve oblique triangles with all three sides known (SSS) and triangles with two sides and the included angle (SAS) given. (We will investigate the remaining cases (ASA, SAA, SSA) when we investigate the Law of Sines in the next section.)

EXAMPLE 1 ■ Solving an Oblique Triangle Using Right Triangle Trigonometry

Solve the oblique triangle shown in Figure 9.13.

Solution All of our trigonometric tools up to this point rely on right triangles, and this is not a right triangle. However, we can turn this oblique triangle into two right triangles by drawing an **altitude**, a line

Figure 9.13

segment drawn from one vertex perpendicular to the opposite side. We also add labels to the unknown sides and angles. See Figure 9.14. Notice how the altitude breaks up the 8-unit side into two pieces. We do not know the exact length of each piece, so we use x and $8 - x$ to represent their lengths, noting that their sum must be 8 units. As shown in Figure 9.15, we begin solving with the right triangle on the left, using a cosine and sine ratio to find the values of x and h, respectively.

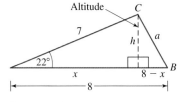

Figure 9.14

$$\cos(22°) = \frac{x}{7} \qquad \sin(22°) = \frac{h}{7}$$
$$x = 7\cos(22°) \qquad h = 7\sin(22°)$$
$$x \approx 6.49 \qquad h \approx 2.62$$

Figure 9.15

Now we solve the triangle on the right. Since $x \approx 6.49$, then $8 - x \approx 1.51$, and since $h \approx 2.62$, then we get the triangle shown in Figure 9.16. The known sides allow us to set up a tangent ratio to solve for the measure of angle B.

$$\tan(B) \approx \frac{2.62}{1.51}$$
$$B \approx \tan^{-1}\left(\frac{2.62}{1.51}\right)$$
$$B \approx 60.0°$$

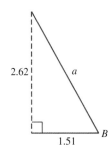

Figure 9.16

We can solve for the length of a using either a trigonometric ratio or the Pythagorean theorem with the smaller right triangle. We use the Pythagorean theorem.

$$1.51^2 + 2.62^2 \approx a^2$$
$$a^2 \approx 9.14$$
$$a \approx \sqrt{9.14}$$
$$a \approx 3.02$$

Since the sum of the angle measures of a triangle is 180°, we complete the solution by finding the measure of angle C, as follows. The solved triangle is illustrated in Figure 9.17.

$$180° - 22° - 60.0° = 98.0°$$

Figure 9.17

Example 1 demonstrates it is possible to solve oblique triangles using basic trigonometric ratios. However, the process can be tedious and time consuming. The Law of Cosines is a shorter process that will achieve the same goal.

■ Developing the Law of Cosines

To develop the Law of Cosines, we begin with oblique triangle ABC. As in Example 1, we begin by drawing an altitude, as in Figure 9.18.

If we apply the Pythagorean theorem to the triangle on the right, we get

$$(a - x)^2 + h^2 = c^2$$
$$a^2 - 2ax + x^2 + h^2 = c^2$$

Likewise, if we apply the Pythagorean theorem to the triangle on the left, we get

$$x^2 + h^2 = b^2$$

Figure 9.18

Since $x^2 + h^2 = b^2$, we may substitute this result into the first equation.

$$a^2 - 2ax + (x^2 + h^2) = c^2$$
$$a^2 - 2ax + b^2 = c^2 \quad \text{Substitute } b^2 \text{ for } x^2 + h^2.$$
$$a^2 + b^2 - 2ax = c^2 \quad \text{Rearrange terms.}$$

Observe from the triangle that $\cos(C) = \dfrac{x}{b}$. Therefore, $x = b \cos(C)$. We use this fact to eliminate the x in the equation.

$$a^2 + b^2 - 2a(x) = c^2$$
$$a^2 + b^2 - 2a(b \cos(C)) = c^2$$

This result is called the **Law of Cosines**.

LAW OF COSINES

For ΔABC, $c^2 = a^2 + b^2 - 2ab \cos(C)$, where c is the length of the side opposite angle C. Similarly, $a^2 = b^2 + c^2 - 2bc \cos(A)$ and $b^2 = a^2 + c^2 - 2ac \cos(B)$.

EXAMPLE 2 ■ **Solving an SSS Triangle**

Solve the triangle ABC shown in Figure 9.19. The measures of the three sides are $a = 4$, $b = 3$, and $c = 6$.

Solution We need to find the measurements of its angles since we know the length of all three sides. We use the Law of Cosines $c^2 = a^2 + b^2 - 2ab \cos(C)$ with $a = 4$, $b = 3$, and $c = 6$, as shown in Figure 9.20. To find angle C, we have

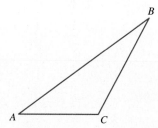

Figure 9.19

$$6^2 = 4^2 + 3^2 - 2(4)(3)\cos(C)$$
$$36 = 16 + 9 - 24\cos(C)$$
$$11 = -24\cos(C)$$
$$\frac{11}{-24} = \cos(C)$$
$$-0.4583 = \cos(C)$$
$$\cos^{-1}(-0.4583) = C$$
$$117.28° = C$$

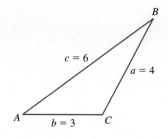

Figure 9.20

We can now find angle B by using a variation of the Law of Cosines.

$$b^2 = a^2 + c^2 - 2ac\cos(B)$$
$$3^2 = 4^2 + 6^2 - 2(4)(6)\cos(B)$$
$$9 = 16 + 36 - 48\cos(B)$$
$$-43 = -48\cos(B)$$
$$\frac{-43}{-48} = \cos(B)$$
$$0.8958 = \cos(B)$$
$$\cos^{-1}(0.8958) = B$$
$$26.39° = B$$

Since the sum of the angle measures of a triangle is 180°, the measure of angle A is given by

$$180° - 117.28° - 26.39° = 36.33°$$

Thus $A = 36.33°$, $B = 26.39°$, and $C = 117.28°$.

The Law of Cosines and the Pythagorean Theorem

The Law of Cosines and the Pythagorean theorem are related. Observe the Law of Cosines ($c^2 = a^2 + b^2 - 2ab\cos(C)$) and the Pythagorean theorem ($c^2 = a^2 + b^2$) differ only by the term $-2ab\cos(C)$. Whenever $\cos(C) = 0$, the two equations are equal. Since $\cos^{-1}(0) = 90°$, the Law of Cosines and Pythagorean theorem are equivalent whenever $C = 90°$. Thus the Pythagorean theorem is a special case of the Law of Cosines that works specifically with right triangles.

EXAMPLE 3 ■ Applying the Law of Cosines

In a follow-up study to the 2004 Gravity Probe B investigation mentioned earlier, a physics experiment on a grand scale has been initiated. At NASA's Jet Propulsion Laboratory, scientists are testing Einstein's theory of relativity, which states that the Sun's gravity causes starlight to bend, shifting the apparent position of stars in the sky. Using the Solar System as a giant laboratory, the scientists plan to test two seemingly competing theories regarding the nature and behavior of space and time: Einstein's theory of relativity and the more recent idea of quantum mechanics. Einstein's theory states that gravity from celestial bodies such as the Sun causes light to bend. Quantum mechanics says that gravity bends light because it causes space to warp, as though space and time were a sheet of flat flexible material that depresses under the weight of celestial objects. The depressions in the sheet cause light to bend or curve slightly as it passes by the large objects. (See illustration at left.) (For more information go to science.nasa.gov/headlines/y2004/26mar_einstein.htm.)

The experiment, known as the Laser Astrometric Test Of Relativity (LATOR), will look at how the sun's gravity deflects beams of laser light emitted by two small satellites. The two satellites will orbit on opposite sides of the Sun at about the same distance from the Sun as Earth. This will form an oblique triangle between Earth and the satellites, with laser beams as its sides, as shown in Figure 9.21. Note that one beam passes very close to the Sun. As is shown in the figure, the scientists

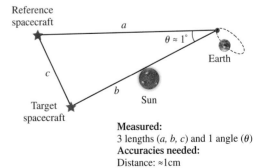

Figure 9.21

Source: NASA

anticipate that the angle between the two satellites as viewed from Earth will be approximately 1°. We want to verify that prediction.

Using Figure 9.21 and assuming the distance from Earth to the target spacecraft is 297 million kilometers, determine the measure of angle θ. Also find the other two angles formed by the spacecrafts and Earth.

Solution   We use the Law of Cosines because we know all three side lengths of the triangle (SSS) and do not know any of the angle measures. We denote the sides as $a = 300$ million kilometers, $b = 297$ million kilometers, and $c = 5$ million kilometers. We want to find θ, which we denote as C in our formula.

$$c^2 = a^2 + b^2 - 2ab\cos(C)$$
$$5^2 = 300^2 + 297^2 - 2(300)(297)\cos(C) \quad \text{$c = 5, a = 300, b = 297$}$$
$$25 = 90{,}000 + 88{,}209 - 178{,}200\cos(C) \quad \text{Solve for angle C.}$$
$$-178{,}184 = -178{,}200\cos(C)$$
$$\frac{-178{,}184}{-178{,}200} = \cos(C)$$
$$0.9999102 = \cos(C)$$
$$\cos^{-1}(0.9999102) = C$$
$$0.768° = C$$
$$C \approx 1°$$

We see angle C rounds to 1°, as the scientists predict.

To find the next angle, we use an equivalent form of the Law of Cosines.

$$a^2 = b^2 + c^2 - 2bc\cos(A)$$
$$300^2 = 297^2 + 5^2 - 2(297)(5)\cos(A) \quad \text{$c = 5, a = 300, b = 297$}$$
$$90{,}000 = 88{,}209 + 25 - 2970\cos(A) \quad \text{Solve for angle A.}$$
$$1766 = -2970\cos(A)$$
$$\frac{1766}{-2970} = \cos(A)$$
$$-0.5946 = \cos(A)$$
$$\cos^{-1}(-0.5946) = A$$
$$126.48° = A$$

To find the measure of angle B we use the fact that the sum of the angles of a triangle is 180°.

$$B = 180° - 0.768° - 126.48°$$
$$B = 52.75°$$

The solved triangle is shown in Figure 9.22.

Figure 9.22

EXAMPLE 4 ■ Applying the Law of Cosines

At the 2000 Summer Olympics in Sydney, Australia, surveying equipment was used to save valuable television transmission time and to avoid long, unnecessary delays. A *total station* (an advanced, electronic theodolite) was set up near all discus, shot-put, and javelin events. The total station measured angles and distances quickly and recorded the thrown distances to millimeter accuracy. (*Source:* www.leica-geosystems.com)

Figure 9.23 displays the discus event. The total station is set up prior to each event a known distance from the thrower. After a contestant throws the discus, a referee sticks a target in the ground at the impact location. The surveyor then points the telescope of the total station to the target and reads the distance using a laser aimed at the landing location. Using the figure, find the distance d the discus was thrown.

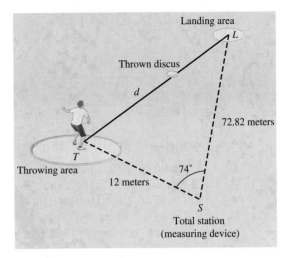

Figure 9.23

Solution We need to find the measure of d, the distance the discus was thrown. Since we are given two sides and the included angle (SAS), we can use the Law of Cosines to find d.

$$d^2 = 72.82^2 + 12^2 - 2(72.82)(12) \cos(74°)$$
$$d^2 = 5302.75 + 144 - 1747.68 \cos(74°)$$
$$d^2 = 5446.75 - 1747.68 \cos(74°)$$
$$d^2 = 5446.75 - 481.73$$
$$d^2 = 4965.02$$
$$d = \sqrt{4965.02}$$
$$d = 70.46$$

The discus was thrown 70.46 meters.

EXAMPLE 5 ■ Choosing between Techniques

Given the triangles in Figures 9.24–9.26, state whether you can use the Pythagorean theorem, a trigonometric function, the Law of Cosines, or a combination of these to solve each triangle.

a.

Figure 9.24

b.

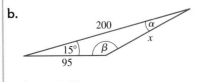

Figure 9.25

Figure 9.26

Solution

a. The triangle is a right triangle and we know the lengths of two sides. Therefore, we can use the Pythagorean theorem to find the length of the third side. However, we could use the trigonometric functions cosine or secant as well, since we are seeking the length of the hypotenuse. To find the measure of the third angle, we can use the fact that the sum of the angle measures of a triangle is 180°.

b. Since we are given SAS and do not know the measure of the remaining two angles, we would use the Law of Cosines to solve the triangle.

c. Even though the triangle looks like a right triangle, we do not have enough information to determine if it is. Therefore, since we are given the length of all three sides (SSS), we would use the Law of Cosines to solve the triangle.

SUMMARY

In this section you learned the Law of Cosines and its relationship to the Pythagorean theorem. You also learned to apply the Law of Cosines to real-world scenarios.

9.2 EXERCISES

SKILLS AND CONCEPTS

In Exercises 1–10, find the length of the side opposite the given angle.

1. In $\triangle ABC$, $b = 4$, $c = 5$, and $A = 53°$
2. In $\triangle ABC$, $a = 7$, $c = 10$, and $B = 37°$
3. In $\triangle PQR$, $p = 3$, $q = 2$, and $R = 131°$
4. In $\triangle JKL$, $j = 8$, $k = 5$, and $L = 175°$
5. In $\triangle EFG$, $e = 36.5$, $g = 46.3$, and $F = 71°$
6. In $\triangle TRE$, $r = 2.875$, $e = 6.013$, and $T = 116°$
7. In $\triangle MNO$, $m = 6$, $n = 2$, and $O = 31°$
8. In $\triangle ABC$, $a = 4$, $b = 2$, and $C = 35°$
9. In $\triangle ABC$, $a = 1$, $b = 3$, and $C = 28°$
10. In $\triangle ABC$, $a = 7$, $b = 8$, and $C = 10°$

In Exercises 11–20, find the measure of the specified angle.

11. $\angle A$ in $\triangle ABC$; if $a = 2$, $b = 3$, and $c = 4$
12. $\angle C$ in $\triangle ABC$; if $a = 6$, $b = 7$, and $c = 9$
13. $\angle T$ in $\triangle PAT$; if $p = 6$, $a = 7$, and $t = 12$
14. $\angle E$ in $\triangle MEG$; if $m = 13$, $e = 21$, and $g = 17$
15. $\angle Y$ in $\triangle XYZ$; if $x = 7$, $y = 5$, and $z = 11$
16. $\angle O$ in $\triangle NOT$; if $n = 1604$, $o = 2143$, and $t = 1448$
17. $\angle Q$ in $\triangle SQR$; if $s = 1643$, $q = 2455$, and $r = 1998$
18. $\angle A$ in $\triangle ABC$; if $a = 3$, $b = 5$, and $c = 6$
19. $\angle C$ in $\triangle ABC$; if $a = 5$, $b = 6$, and $c = 8$
20. $\angle C$ in $\triangle ABC$; if $a = 5$, $b = 5$, and $c = 9$

In Exercises 21–26, solve each triangle using any correct technique.

21.

22.

23.

24.

25.

26.

27. **Height of a Flagpole** A student views the flagpole outside her school and wonders how tall it is. The flagpole has a shadow 66 feet, 5 inches long. She estimates the measure of the angle of elevation to the top of the flagpole to be 36°. How tall is the pole?

28. **Ship at Sea** A ship travels 76 miles due east, then adjusts its course 18° southward. After traveling 78 miles, how far is the ship from where it began?

SHOW YOU KNOW

29. Explain how the Law of Cosines and the Pythagorean theorem are related. Use diagrams, tables, and/or graphs in your explanation.

30. Describe how you know when the Law of Cosines can be used to solve a triangle.

31. Is there such a triangle that allows you to find the third side with either the Pythagorean theorem or the Law of Cosines? Justify your answer.

32. As C changes, describe what happens to the term $-2ab\cos(C)$ in the Law of Cosines. Make sure to explain what happens when $0° < C < 90°$, when $C = 90°$, and when $90° < C < 180°$.

MAKE IT REAL

In Exercises 33–40, draw a diagram when one is not supplied. Then apply the Law of Cosines, the Pythagorean theorem, or a trigonometric function to find the length of the specified side or measure of the specified angle.

33. **Depth of Oil** With oil prices high and demand outstripping supply, oil companies are venturing into rugged areas that were previously ignored. Geologists work with engineers to locate oil-rich areas offshore. Recent advances in technology make it possible to access oil reserves as deep as 21,000 feet below the water's surface. Computers monitor the treacherous currents swirling around floating oil rigs and adjust drills to make sure they remain on course. (*Source:* www.msnbc.msn.com) Use the diagram to answer the following questions.

   a. How far is the oil tanker from the oil source?
   b. At what angle measure is the tanker from the oil source in relation to the oil rig?

34. **Yankee Stadium** Yankee Stadium, home of the New York Yankees, has a distance of 408 feet from home plate

to dead center field. (*Source:* www.ballparks.com) The distance from one base to the next base is 90 feet for all major league fields. How far is it from dead center to first base? Third base?

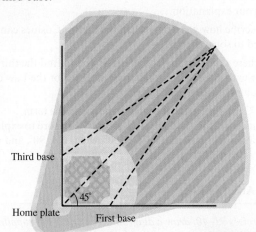

35. Submarine Races The nuclear-powered USS Triton was the first submarine to go completely around the world without surfacing, a feat she accomplished in 1960. Her underwater speed was estimated to be around 21 knots. In contrast, one of the United States' most recent submarines, the USS Virginia, has an underwater speed of 25 knots.
(*Source:* www.navy.mil)

Although built decades apart, suppose these two submarines engaged in an underwater race. The submarines start at the same point and travel for 3 hours in different directions. Given that the angle between their paths measures 28°, how far apart are they at the end of the 3 hours? (*Hint:* 1 knot = 1.15 miles per hour.)

36. Airplane Distances The Blue Angels, the Navy's elite flying squadron, have performed for more than 427 million fans since their inception in 1946. (*Source:* www.blueangels.navy.mil)

Suppose the two Blue Angel jets at points P and A in the diagram have elevations of 1575 feet and 1000 feet, respectively. Both are flying east toward the viewing crowd at T. From T, the measure of the angle of elevation of jet P is 5°, and the measure of the angle of elevation of jet A is 2.8°.

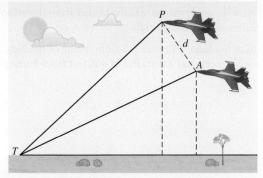

a. How far from the crowd is each of the jets?
b. How far in front of jet A is the nose tip of jet P?
c. What is the ground distance from the nose tip of each jet from the crowd?

37. Sailboats and Sea Lions The National Marine Fisheries Service has issued mandatory regulations requiring people stay at least 100 yards from certain marine mammals in the wild, including sea lions. (*Source:* http://www.nmfs.noaa.gov)

As a sailboat sails through a harbor, the captain sights some sea lions lounging on a buoy at an angle measuring 13° to the path of the boat. The sailboat then goes another 3925 feet and records that the buoy now makes an angle measuring 32°.

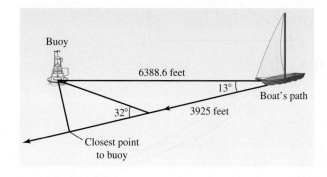

a. How far is the sailboat from the sea lions at the second sighting if at the first sighting the distance to the buoy was 6388.6 feet?
b. The children onboard want a close-up view of the sea lions. What is the closest they will come to the sea lions if the boat stays on it current course? At the closest point, will the boat be a legal distance from the mammals?
c. The ship continues past the buoy on the same path. After going 20,000 feet past the buoy, one child takes one last look at the sea lions. At what angle is the child's line of sight back to the buoy?

38. Airport Positions The distance between the Baltimore, Maryland, airport (BWI) and the Albuquerque, New Mexico, airport (ABQ) is 1448.1 nautical miles. The distance between BWI and the Fresno, California, airport (FAT) is 2023.2 nautical miles. The distance between ABQ and FAT is 645.2 nautical miles. (*Source:* www.cgsnetwork.com) With the three airports being the vertices of a triangle, what is the measure of the angle at BWI?

39. Aircraft Landing The process of landing an aircraft is one of the most difficult challenges a pilot encounters. In the late 1920s airports used rotating lights to mark landing fields after dark. Approach lighting was added in the early 1930s to indicate to pilots the correct angle of descent.
(*Source:* www.centennialofflight.gov)

Suppose an airplane flying at an altitude of 8 miles has an angle of depression measuring 7.78° to a landing strip.

a. What is the plane's ground distance from the landing strip?

b. If the plane descends on a path with a constant angle of depression measuring 7.78°, how far will it travel in air distance before it touches down?

40. **Mountain Height** Radhanath Sikdar, a mathematician and surveyor from India, calculated the height of Mount Qomolangma (Mount Everest) in 1852. While standing a distance of 240 kilometers from the peak, he determined that the height of the mountain was 8839 meters. (*Source:* www.wikipedia.com)

   a. What was the measure of the angle of elevation from his position to the peak?
   b. If he would have made his calculation from 100 kilometers closer to the peak, what would have been the measure of the angle of elevation?

■ **STRETCH YOUR MIND**

Exercises 41–44 are intended to challenge your understanding of the Law of Cosines.

41. For a certain triangle with side lengths a and b of a given length, demonstrate the effect the term $-2ab \cos(C)$ from the Law of Cosines has on the measure of the third side c as the measure of angle C grows larger than 90° and approaches 180°. Use diagrams as a part of your explanation.

42. From the Pythagorean theorem, we know that the square of the length of the hypotenuse of a right triangle is equal to the sum of the squares of the other two side lengths ($c^2 = a^2 + b^2$) with a, b, and c as shown below in $\triangle ABC$.

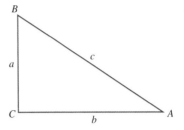

   a. What is the effect on the relationship of $c^2 = a^2 + b^2$ if $\angle C$ becomes acute while a and b remain constant?
   b. What is the effect on the relationship of $c^2 = a^2 + b^2$ if $\angle C$ becomes obtuse while a and b remain constant?
   c. Is there an $\angle C$ that maintains the relationship $c^2 = a^2 + b^2$? Explain.

43. In $\triangle XYZ$, $x = 3$, $y = 7$, and $z = 11$. Explain why you cannot find the measure of angle Z.

44. Solve the following oblique triangle using right triangle trigonometry and not the Law of Cosines.

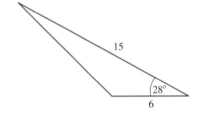

SECTION 9.3 Law of Sines

LEARNING OBJECTIVES

- Develop the formula for the Law of Sines
- Apply the Law of Sines to real-world scenarios

GETTING STARTED

After years of lobbying the federal government and private fundraising, city leaders from St. Louis, Missouri, secured the approval and resources necessary to build a monument to revitalize the city. The Jefferson National Expansion Memorial (commonly called the Gateway Arch) on the shore of the Mississippi River would commemorate the westward expansion of the early American pioneers who faced and overcame daunting odds. In 1948, after the biggest nationwide architectural design competition in American history, a committee chose the plan of a little-known engineer, Eero Saarinen, for its deceptively simple, yet elegant stainless steel arch structure. At 630 feet tall and 630 feet wide, the arch would be more than twice as tall as the Statue of Liberty.

During construction from 1963 to 1965, engineers made numerous important mathematical calculations to ensure the "Gateway to the West" would be the masterpiece they had all envisioned. (*Source:* www.nps.gov/jeff) Specifically, they used the Law of Sines to keep each leg centered correctly. In this section, we will develop the formula for the Law of Sines, apply it to real-world scenarios, and use it in solving triangles.

Law of Sines

When solving oblique triangles for missing angle measures and side lengths, we use the Law of Sines to compute the remaining side lengths of a triangle if two angle measures and a side length are known (AAS or ASA) or when two side lengths and one of the nonincluded angle measures are known (SSA). Before we develop the Law of Sines, let's analyze these types of triangles using basic trigonometric ratios.

EXAMPLE 1 ■ Solving an Oblique Triangle Using Right Triangle Trigonometry

Solve the oblique triangle shown in Figure 9.27.

Solution A logical first step is to find the measure of the missing angle. Since the sum of the angle measures in a triangle is 180°, we have

$$180° - 48° - 50° = 82°$$

We draw an altitude to create two right triangles, then label the diagram as in Figure 9.28. Starting with the triangle on the right (Figure 9.29), we use a sine and cosine ratio to find the lengths of h and x, respectively.

Figure 9.27

Figure 9.28

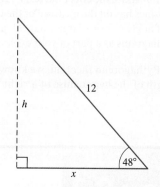

Figure 9.29

$$\sin(48°) = \frac{h}{12} \qquad \cos(48°) = \frac{x}{12}$$

$$h = 12\sin(48°) \qquad x = 12\cos(48°)$$

$$h \approx 8.92 \qquad x \approx 8.03$$

Now that we know the value of h, we use the triangle on the left (Figure 9.30) and a tangent and sine ratio to find the lengths of y and a, respectively.

$$\tan(82°) = \frac{h}{y} \qquad \sin(82°) = \frac{h}{a}$$

$$y\tan(82°) = h \qquad a\sin(82°) = h$$

$$y = \frac{h}{\tan(82°)} \qquad a = \frac{h}{\sin(82°)}$$

$$y \approx \frac{8.92}{\tan(82°)} \qquad a \approx \frac{8.92}{\sin(82°)}$$

$$y \approx 1.25 \qquad a \approx 9.01$$

Figure 9.30

We add the lengths of x and y to get approximately 9.28. The solved triangle is shown in Figure 9.31.

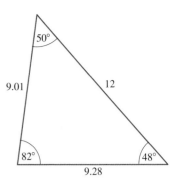

Figure 9.31

Example 1 shows that we can solve an oblique triangle using basic trigonometric ratios. However, with the Law of Sines we can solve them more quickly.

■ Developing the Law of Sines

To develop the formula for the Law of Sines, let's consider the triangle ABC shown in Figure 9.32, with h as the length of the altitude to the side with length a and k as the length of the altitude to the side with length b. We will determine the sine of various angles and compare their ratios. (Recall that the sine function is defined as $\sin(\theta) = \dfrac{\text{opposite}}{\text{hypotenuse}}$.)

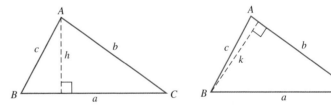

Figure 9.32

We have

$$\sin(B) = \frac{h}{c} \qquad \sin(C) = \frac{h}{b} \qquad \sin(C) = \frac{k}{a} \qquad \sin(A) = \frac{k}{c}$$
$$h = c\sin(B) \qquad h = b\sin(C) \qquad k = a\sin(C) \qquad k = c\sin(A)$$

We now set the two pairs of ratios equal to each other.

$$h = c\sin(B) \quad \text{and} \quad h = b\sin(C) \qquad\qquad k = a\sin(C) \quad \text{and} \quad k = c\sin(A)$$
$c\sin(B) = b\sin(C)$ **Substitute for h.** $\qquad a\sin(C) = c\sin(A)$ **Substitute for k.**

$\dfrac{b}{\sin(B)} = \dfrac{c}{\sin(C)}$ **Divide by sine terms.** $\qquad \dfrac{c}{\sin(C)} = \dfrac{a}{\sin(A)}$ **Divide by sine terms.**

Thus we conclude that in triangle ABC,

$$\frac{b}{\sin(B)} = \frac{c}{\sin(C)} = \frac{a}{\sin(A)}$$

By taking reciprocals of each of these terms, we additionally discover that

$$\frac{\sin(B)}{b} = \frac{\sin(C)}{c} = \frac{\sin(A)}{a}$$

This relationship between sides and angles is known as the **Law of Sines**.

> **LAW OF SINES**
>
> In any ΔABC, the ratio of the sine of an angle to the length of the side opposite the angle is a constant. That is,
>
> $$\frac{\sin(B)}{b} = \frac{\sin(C)}{c} = \frac{\sin(A)}{a} \quad \text{and} \quad \frac{b}{\sin(B)} = \frac{c}{\sin(C)} = \frac{a}{\sin(A)}$$
>
> for angles A, B, and C and side lengths a, b, and c.

■ Triangulation

We now return to the Gateway Arch example to see how the Law of Sines can be applied in a real-world situation. Engineers working on the arch knew it was critical to make precise measurements as the two arch legs were being erected simultaneously. Even a slight deviation in any direction would result in the legs failing to meet at the top. Throughout the entire process, engineers had to maintain accuracy to $\frac{1}{64}$ of an inch. (*Source:* The History Channel, "*The St. Louis Arch*," 2001) To achieve the extreme accuracy that was required, *triangulation* was used. **Triangulation** is the process of using two known points to find the coordinates of an unknown point by constructing a triangle with the three points as vertices and using the Law of Sines to determine the unknown point.

EXAMPLE 2 ■ Using the Law of Sines for an ASA Triangle

During the process of constructing the legs of the Gateway Arch, engineers used four monuments (posts) that formed a rectangle 300 feet by 720 feet centered on the arch legs (see Figure 9.33). The four posts were set, two behind each leg of the arch, and the distance between those two monuments was measured. Surveyors used a theodolite to measure angles. Engineers used the two north monuments for triangulation on the south leg and the two south monuments for triangulation on the north leg. By applying the Law of Sines, the engineers precisely determined any given point along the arch legs. (*Source:* The History Channel, "*The St. Louis Arch*," 2001) Figure 9.33 represents one of the triangles formed by the surveyors as they were aligning the south leg.

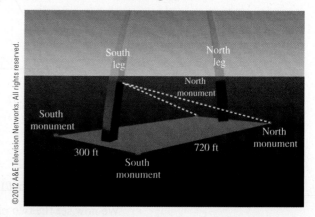

Figure 9.33

The triangle in Figure 9.34 provides the measurements for one such triangle as the section was being moved into place. Determine the values of the missing angle measure A, and the two side lengths b and c.

Solution In Figure 9.34, the unknown angle A measures $180° - B - C$, since the sum of the three angle measures of a triangle is 180°. Therefore, the measure of angle A is

$$180° - 76° - 77° = 27°$$

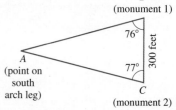

Figure 9.34

To determine the lengths of the remaining two sides of the triangle, we use the Law of Sines since we are given the measure of two angles (*B* and *C*) and the included side (*a*). See Figure 9.35. Using the Law of Sines to find *c*, we have

$$\frac{\sin(77°)}{c} = \frac{\sin(27°)}{300 \text{ feet}}$$

$$c(\sin(27°)) = (300 \text{ feet})(\sin(77°))$$

$$c = \frac{(300 \text{ feet})(\sin(77°))}{\sin(27°)}$$

$$c \approx 643.8 \text{ feet}$$

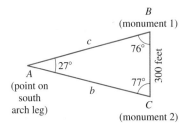

Figure 9.35

Using the Law of Sines to find *b* gives

$$\frac{\sin(76°)}{b} = \frac{\sin(27°)}{300 \text{ feet}}$$

$$b(\sin(27°)) = (300 \text{ feet})(\sin(76°))$$

$$b = \frac{(300 \text{ feet})(\sin(76°))}{\sin(27°)}$$

$$b \approx 641.2 \text{ feet}$$

Therefore, the solved triangle *ABC* has the measures shown in Figure 9.36.

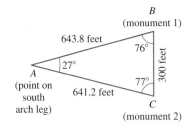

Figure 9.36

EXAMPLE 3 ■ Using the Law of Sines to Solve an ASA Triangle

In triangle *ABC*, $A = 35°$, $B = 15°$, and $c = 5$. Find *a*, *b*, and angle *C*.

Solution We first draw a diagram for the triangle described, as shown in Figure 9.37. The measure of angle *C* is

$$C = 180° - 35° - 15°$$
$$C = 130°$$

Figure 9.37

We use the Law of Sines to find the missing sides because we are given two angle measures and their included side. Although we may use either form of the Law of Sines, $\frac{b}{\sin(B)} = \frac{c}{\sin(C)} = \frac{a}{\sin(A)}$ or $\frac{\sin(B)}{b} = \frac{\sin(C)}{c} = \frac{\sin(A)}{a}$, it is most efficient to have the variable in the numerator. We'll use $\frac{b}{\sin(B)} = \frac{c}{\sin(C)} = \frac{a}{\sin(A)}$ since we are solving for a side. (If we were solving for an angle, we would have used the other form.)

$$\frac{a}{\sin(35°)} = \frac{5}{\sin(130°)}$$

$$a = \frac{5 \sin(35°)}{\sin(130°)}$$

$$a \approx 3.74$$

$$\frac{b}{\sin(15°)} = \frac{5}{\sin(130°)}$$

$$b = \frac{5 \sin(15°)}{\sin(130°)}$$

$$b \approx 1.69$$

Therefore, the solved triangle has the measures shown in Figure 9.38.

Figure 9.38

EXAMPLE 4 ■ Using the Law of Sines to Solve an AAS Triangle

The diagram in Figure 9.39, which is drawn to scale, shows the relative positions of Chicago, New York City, and Washington, DC. The angle at Chicago measures 14.91° and the angle at New York City measures 47.18°. The distance between New York City and Washington, DC, is 209 miles. How far is Chicago from New York City and Washington, DC?

Figure 9.39

Solution We relabel the triangle as in Figure 9.40 for computational ease. We are told that $A = 14.91°$, $B = 47.18°$, and $a = 209$. We need to find b (the distance from Chicago to Washington, DC) and c (the distance from Chicago to New York City). We use the Law of Sines because we are given two angles and a side.

Figure 9.40

$$\frac{209}{\sin(14.91°)} = \frac{b}{\sin(47.18°)}$$

$$b = \frac{209\sin(47.18°)}{\sin(14.91°)}$$

$$b \approx 596$$

The distance between Chicago and Washington, DC, is about 596 miles. To set up the next ratio, we first need to know C.

$$C = 180° - A - B$$
$$C = 180° - 14.91° - 47.18°$$
$$C = 117.91°$$

Using the Law of Sines, we find c.

$$\frac{209}{\sin(14.91°)} = \frac{c}{\sin(117.91°)}$$

$$c = \frac{209\sin(117.91°)}{\sin(14.91°)}$$

$$\approx 718$$

The distance between Chicago and New York City is about 718 miles.

Using the Law of Sines with an SSA triangle often results in what is referred to as the *ambiguous case* because more than one triangle may be generated. In fact, the known information may result in one triangle, two triangles, or no triangle at all. We look at this next.

EXAMPLE 5 ■ Exploring the SSA Triangle

Both of the triangles shown in Figure 9.41 have sides of length 7 and 9 and contain a 43° angle. However, angle A for the first triangle is acute whereas the corresponding angle of the second triangle, B, is obtuse. Find the measure of angles A and B in the triangles.

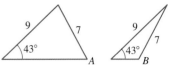

Figure 9.41

Solution As shown in Figure 9.42, we begin with the first triangle. To find A, we use the Law of Sines since we have two sides and a nonincluded angle.

$$\frac{\sin(A)}{9} = \frac{\sin(43°)}{7}$$

$$7\sin(A) = 9\sin(43°)$$

$$\sin(A) = \frac{9\sin(43°)}{7}$$

$$\sin(A) = 0.8769$$

$$\sin^{-1}(\sin(A)) = \sin^{-1}(0.8769)$$

$$A = \sin^{-1}(0.8769)$$

$$A = 61.27°$$

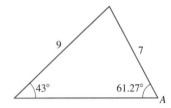

Figure 9.42

Now, as shown in Figure 9.43, we move to the second triangle. To find B, we recall that sine is positive in both the first and second quadrants. Therefore, there are two angles such that $\sin(\theta) = 0.8769$. $A = 61.27°$ is one of the angles. To find the measure of the other angle, B, we find the angle in the second quadrant with a reference angle of 61.27°.

$$B = 180° - 61.27° = 118.73°$$

Thus there are two triangles with sides 7 and 9 and nonincluded angle measuring 43°.

Figure 9.43

EXAMPLE 6 ■ Using the Law of Sines for an SSA Triangle

The design for a new playground calls for a children's slide that is 12.5 feet long inclined 30° with the horizontal. (This is the measure of the angle the slide would create with the ground if it did not bend at the bottom.) To reach the top of the slide the children will climb a ladder from the other side that is 6.4 feet long. For slides to be safe for children to use, the angle that the ladder makes with the horizontal should measure 80° or less. Determine whether or not the design of the playground slide is in compliance with the standard. Draw a sketch that shows the slide, ladder, and ground. (*Note:* There are two correct answers. In your solution show them both and describe why only one will make sense in the real-world context.

Solution We first draw a diagram for the triangle described. See Figure 9.44. To show that the measure of angle B is less than or equal to 80°, we employ the Law of Sines.

Figure 9.44

$$\frac{\sin(B)}{12.5} = \frac{\sin(30°)}{6.4}$$

$$6.4 \sin(B) = 12.5 \sin(30°)$$

$$\sin(B) = \frac{12.5 \sin(30°)}{6.4}$$

$$\sin(B) = 0.9766$$

$$B = \sin^{-1}(0.9766)$$

Because sine is positive in the first and second quadrants, there are two answers for the measure of angle B.

$$B = 77.6° \quad \text{and} \quad \begin{array}{l} B = 180° - 77.6° \\ B = 102.4° \end{array}$$

We show the first angle, $B = 77.6°$, in Figure 9.45. This slide is in compliance since $77.6° < 80°$. Angle $B = 102.4°$ is not in compliance since $102.4° > 80°$. Besides, as shown in Figure 9.46, an angle measuring $102.4°$ is unreasonable since a ladder sloped backward would be impossible for young children (and most adults) to climb.

Figure 9.45   **Figure 9.46**

Summary of Triangle Solving Techniques

To solve any triangle with one known side length and two other pieces of known information (two side lengths, a side length and an angle measure, or two angle measures), use the following techniques.

Right Triangles
- Use the Pythagorean theorem to determine side lengths.
- Use the six trigonometric functions to find side lengths or angle measures.

Oblique Triangles
- Use the Law of Cosines for triangles with three known side lengths or two known side lengths and the measure of the angle between them (SAS or SSS).
- Use the Law of Sines for triangles with two known angle measures and one known side length (ASA, AAS) and for triangles with two known side lengths and the measure of an angle opposite a known side (SSA).

SUMMARY

In this section you learned where the Law of Sines comes from and how to apply it to real-world scenarios. You also learned how to use the Law of Sines to solve triangles.

9.3 EXERCISES

SKILLS AND CONCEPTS

In Exercises 1–20, use triangle ABC to find the missing side lengths and angle measures (if possible). If there are two solutions, find both. If there is no solution, so state.

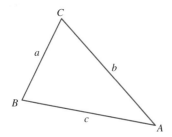

1. $a = 6, B = 77°, C = 14°$
2. $a = 8, B = 19°, C = 75°$
3. $a = 15, B = 98°, C = 15°$
4. $a = 4, B = 48°, C = 114°$
5. $A = 16°, B = 37°, c = 3$
6. $A = 96°, B = 23°, c = 6$
7. $A = 106°, B = 8°, c = 16$
8. $A = 76°, B = 48°, c = 10$
9. $A = 19°, C = 82°, b = 24$
10. $A = 10°, C = 160°, b = 4$
11. $A = 94°, C = 31°, b = 8$
12. $A = 4°, C = 7°, b = 4$
13. $A = 19°, B = 82°, b = 24$
14. $A = 34°, B = 80°, b = 12$
15. $A = 28°, B = 67°, b = 23$
16. $A = 4°, B = 111°, b = 19$
17. $a = 34, A = 80°, b = 12$
18. $a = 26, A = 140°, b = 43$
19. $a = 5, A = 64°, b = 15$
20. $a = 13, A = 136°, b = 3$

In Exercises 21–23, apply the Law of Sines, the Pythagorean theorem, or a trigonometric function to the given situation.

21. **Distance to a Ship** An observer at A measures angle A between the shore and the ship, and an observer at B does likewise for angle B. If the length l is 524 feet, find the distance d from shore to ship.

22. **Storm at Sea** A ship receives notification of a storm to its east. To avoid the storm, it proceeds at a heading of 43° north of east for 167 nautical miles. It then changes to a heading of 59° south of east.

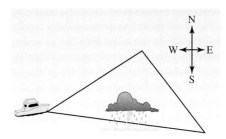

a. How far does the ship travel on a heading of 59° south of east until it intersects its original course?

b. How far out of its way did the ship go to avoid the storm?

23. **U.S. Coast Guard** One U.S. Coast Guard station is located 105 miles due west of a second station. A ship in distress sends a signal received by each station. The location of the ship from the first station is 34° west of south and the location of the ship from the second station is 41° east of south.

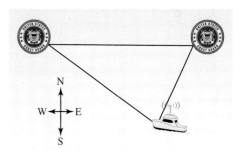

a. How far is each station from the ship?

b. If a search and rescue helicopter capable of flying 175 miles per hour is dispatched from the nearest station to the ship, how long will it take to reach the ship?

In Exercises 24–26, solve each problem for the missing values. Use the Law of Sines or the Law of Cosines as appropriate.

24. **Mountain Height** Two observers view the same mountain peak from two points on level ground and 3 miles apart. The angle of elevation at P to the peak is 29°. For

the other observer, the angle of elevation at O to the peak measures 49°.

a. Find the distance from P to the summit.
b. Find the height of the mountain h.

25. **Radio Tower** A radio tower is located on top of a hill, as shown in the figure. The distance from point T on the ground to the top of the hill is 1964 feet. The angle of elevation from T to the base of the tower measures 23° and the angle of elevation from T to the top of the tower measures 27°.

a. Find the height of the hill.
b. Find the height of the tower.

26. **Height of a Kite** Three children are flying a kite and are located as displayed in the diagram.

a. Estimate the angle of elevation from each child to the kite by looking at the figure.
b. If child C_1 is approximately 3 feet, child C_2 is approximately 5 feet, and child C_3 is approximately 11 feet from being directly under the kite, estimate the distance from each child's feet to the kite using the angle of elevation from part (a).

In Exercises 27–32, two sides and an angle are given. Determine whether the given information results in one triangle, two triangles, or no triangle at all. Find the missing values of all sides and angles for the triangles that result.

27. $a = 3, b = 2, A = 50°$
28. $b = 5, c = 3, B = 100°$
29. $b = 4, c = 6, B = 20°$
30. $a = 4, b = 5, A = 60°$
31. $a = 2, c = 1, C = 25°$
32. $a = 2, c = 1, C = 120°$

■ SHOW YOU KNOW

33. Describe under what conditions it is appropriate to use the Law of Sines.
34. Explain how triangulation is used to solve real-world problems.
35. Explain what is meant by the ambiguous case as it relates to the Law of Sines.
36. Suppose a classmate who missed the development of the Law of Sines is wondering why there are three parts to the Law of Sines formula. That is, how would you explain that

$$\frac{b}{\sin(B)} = \frac{c}{\sin(C)} = \frac{a}{\sin(A)}$$

37. The Law of Sines states that

$$\frac{b}{\sin(B)} = \frac{c}{\sin(C)} = \frac{a}{\sin(A)}$$

and

$$\frac{\sin(B)}{b} = \frac{\sin(C)}{c} = \frac{\sin(A)}{a}$$

Explain why the Law of Sines may be expressed either way.

■ MAKE IT REAL

38. **Height of a Tree** Trees located on a sloping hillside turn to grow vertically by responding to gravity and light rays. (*Source: www.madsci.org*) To measure the height of a tree on a slope it is helpful to know the angle between the tree and the sloped ground.
    If a tree is growing vertically at an angle of 76° with respect to the sloped ground and the angle to the top of the tree from a location uphill 110 feet from the base of the tree is 48°, how tall is the tree?

39. **Building Height** As of 2006, officials in Bellevue, Washington, limited the heights of community business buildings in the city to 45 feet. (*Source: www.ci.bellevue.wa.us*) One of the city officials wants to know if an existing building meets the height limits. At a distance of 100 feet away from the base of the building, the official measures the angle of elevation to the top of the building to be 26°.

a. Does the building meet the city code for height? Explain.
b. What is the maximum angle of elevation from a distance of 100 feet for which a building will meet the city code for height?

40. Forest Fire Observation

In the Adirondack Mountains near Tupper Lake, New York, forest fire observers used fire observation towers for over 70 years. The people who staffed the towers reported forest fires and smoke sightings. As many as 57 towers were used by the observers; however, in the early 1980s, the State of New York determined that the towers were no longer needed since aerial observations had become commonplace. (*Source:* www.tupperlake.net)

Suppose a forest ranger in one of the observation towers in the Adirondacks sights a fire 33° east of north while a ranger in a tower 15 miles due east of the first ranger sights the same fire at 46° west of north. How far is the fire from each ranger?

41. Baseball Diamond

A major league baseball diamond is a square with 90 feet on a side. The pitching rubber is located 60.5 feet from home plate on the diagonal from home plate to second base.

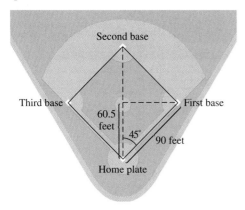

a. How far is a pitcher's throw to first base from the pitching rubber?

b. How far is a pitcher's throw to second base from the pitching rubber?

c. If a pitcher faces home plate, through what angle does he need to turn to face third base?

42. Airplane Distance

While flying from New Orleans, Louisiana, *N*, to San Antonio, Texas, *S*, a distance of approximately 545 miles, a pilot had to go off course by 15° to avoid turbulence.

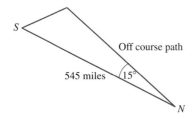

a. The pilot corrects his course after 10 minutes and maintains an average speed of 245 miles per hour. Through what angle should the pilot now turn to head toward San Antonio?

b. For the trip to last a total of 2.5 hours, what new average speed should the pilot achieve?

43. Patriot Missile Launch

A Patriot missile may be launched at a variety of angles. After release, its guidance system turns the missile and adjusts its speed to reach its desired destination. (*Source:* www.army-technology.com)

Suppose a missile is turned to an angle of 87° after its initial launch. An observer 2 miles from the launch pad sees the missile after launch at an angle of elevation measuring 19°. Five seconds later, the missile is observed at an angle of elevation of 37°.

a. How far did the missile travel during the 5-second interval?

b. What was the missile's average speed during this 5-second interval?

c. If the missile continues on the same path (87° from vertical), what will its angle of elevation be 17 seconds after the first sighting?

44. Height of the Tower of Pisa

Using the photograph, estimate the height, *h*, of the Leaning Tower of Pisa. Also, estimate how far from vertical (90°) the tower is leaning.

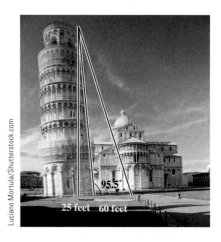

45. The "Phoenix Lights"

Something bizarre appeared in the Arizona sky on the night of March 13, 1997. It was witnessed by thousands of people. Neither researchers nor witnesses have yet figured out what Arizonans saw in the event now dubbed "the Phoenix Lights." (*Source:* www.azcentral.com) That has not stopped many from trying to figure it out, however.

648 CHAPTER 9 ■ TRIANGLE TRIGONOMETRY AND APPLICATIONS

be 152 poles (836 yards). (*Hint:* Rotate your textbook so that North is at the top of the photo.) (*Note:* The notation N 84° W means 84° west of north and S $28\frac{1}{2}°$E means $28\frac{1}{2}°$ east of south.)

The images above include a photograph of the lights and an artist's interpretation of the source of the lights. Assume the lights were first sighted by a family at a point L_1 due east from them at an angle of elevation of 23° from the ground and at an altitude of 225 meters. Then suppose the lights were next sighted at a point L_2 due east at an angle of 54° and an altitude of 375 meters. What is the distance that the lights traveled from L_1 to L_2?

■ **STRETCH YOUR MIND**

Exercises 46–47 are intended to challenge your understanding of triangle trigonometry.

46. Width of the Columbia River While Lewis and Clark were exploring the American west in 1805, William Clark used triangulation to measure the width of the Columbia River near its junction with the Snake River. Lewis placed two markers 148 poles apart on the side of the river he was on. (*Note:* a pole was a 5.5-yard measuring device used by surveyors at the time.)

Using the reproduced photograph of the region from 1945, show how Lewis found the width of the Columbia to

47. Solve the following oblique triangle using right triangle trigonometry and not the Law of Sines.

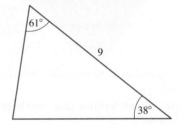

SECTION 9.4 Polar Coordinates

LEARNING OBJECTIVES

- Graph and classify equations in polar coordinates
- Convert between polar and rectangular coordinates
- Describe the relationship between Cartesian (rectangular) and polar coordinates using trigonometry

GETTING STARTED

There are over 8000 artificial objects orbiting Earth. Of those, nearly 2500 are satellites, some operative and some inoperative. (*Source:* liftoff.msfc.nasa.gov) The rest of the objects are space junk—nose cone shrouds, hatch covers, rocket bodies, payloads that have disintegrated or exploded, and even objects that have escaped from manned spacecraft during operations.

Satellites may orbit Earth in several ways. The three most common ways are geostationary orbits, polar orbits, and inclined orbits. According to NASA (*source:* asd-www.larc.nasa.gov), a geostationary orbit is one in which the satellite stays in the same relative position over Earth by orbiting in the same plane as Earth's rotation (equatorial plane). A polar orbit is one where the plane in which the satellite orbits is perpendicular to Earth's equatorial plane. An inclined orbit is one where the orbit of the satellite is somewhere in between a geostationary orbit (angle of inclination is 0 degrees) and a polar orbit (angle of inclination is 90 degrees).

In this section we investigate the polar coordinate system, which allows us to locate an object that moves in a circular path around a fixed center. We also discuss the relationship between the more well-known Cartesian, or rectangular, coordinate system and the polar coordinate system.

■ The Polar Coordinate System

> **PEER INTO THE PAST**
>
> **THE CARTESIAN COORDINATE SYSTEM**
>
> The Cartesian, or rectangular, coordinate system is named after the famous mathematician and philosopher René Descartes of France. Legend has it that Descartes was lying in bed and noticed a fly on the ceiling. This inspired him to think of a way to describe the location of the fly. Choosing a beginning location called the origin, (0, 0), he counted horizontally a certain number of tiles and then vertically as a way to describe the fly's location. The development of this system of graphing led to what we now know as analytic geometry, which bridges algebra and geometry.
>
>
> © Bettmann/CORBIS

Let's consider a satellite orbiting Earth in an inclined orbit. Iridium 19, a communications satellite launched in 1997, orbits Earth every 100 minutes at an inclination of 86.4°. (*Source:* www.n2yo.com) If we visualize this orbit in a two-dimensional plane (as shown in Figure 9.47), we see it traces out a circular path around the globe.

To locate the Iridium 19 satellite in this plane, we superimpose a Cartesian coordinate system on this diagram and locate the satellite by indicating the horizontal and vertical distances from the origin, as shown in Figure 9.48.

Figure 9.47

Because of the circular path of the satellite, we can also describe its location in the plane using **polar coordinates**. We locate the satellite by its distance, r, from the origin (called the **pole**) and the angle, θ, this radius makes with the horizontal axis. We name this point (r, θ), as shown in Figure 9.49.

Figure 9.48

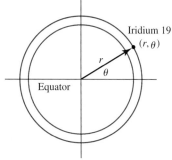

Figure 9.49

> **POLAR COORDINATES**
>
> The **polar coordinates** (r, θ) represent a point in the polar plane where r is the distance from the pole to the point and θ is the measure of the angle formed with the horizontal.

Rather than plotting polar coordinates in the Cartesian coordinate plane, we use the polar coordinate plane. Several concentric circles representing the distances r along with the terminal sides of several different angle measures are shown in the polar coordinate plane in Figure 9.50.

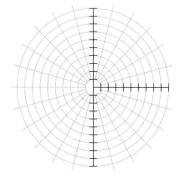

Figure 9.50

EXAMPLE 1 ■ Locating a Point in the Polar Coordinate Plane

The radius of Earth is approximately 6378 kilometers. The Iridium 19 satellite orbits Earth at an altitude of 778 kilometers. At a particular time, the satellite creates an angle measuring 0.5 radian with the horizontal axis. What is its location in terms of polar coordinates? Plot this point using a polar coordinate plane.

Solution If we take the center of Earth to be the pole, then the distance to the satellite is the radius of Earth added to the altitude of the satellite:

$$6378 \text{ km} + 778 \text{ km} = 7156 \text{ km}$$

With a reported angle of 0.5 radian, we write the location of the satellite in polar coordinates as $(r, \theta) = (7156, 0.5)$. To plot this point in the polar coordinate plane, we begin on the positive horizontal axis and measure 0.5 radian or approximately 28.6°. Along the terminal side of that angle, we count out 7156 kilometers, where the scale is 1000 kilometers for each of the concentric circles. See Figure 9.51.

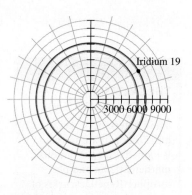

Figure 9.51

EXAMPLE 2 ■ Determining the Coordinates of a Point in the Polar Plane

Give three possible coordinates for the polar point given in Figure 9.52. Give the angle measure in radians.

Solution We first determine the value of θ. Using the positive horizontal axis as the initial side, we see the angle measuring θ has a terminal side of $\frac{4\pi}{12} = \frac{\pi}{3}$. Along the terminal side of this angle, the point is located 4 units from the pole. We label this point $\left(4, \frac{\pi}{3}\right)$, as shown in Figure 9.53.

We could also say the terminal side of the angle is $\frac{\pi}{3}$ plus one complete revolution. That is, $\frac{\pi}{3} + 2\pi = \frac{7\pi}{3}$.

Using this angle, we label the point $\left(4, \frac{7\pi}{3}\right)$, as in Figure 9.54. Extending this idea, we can infer there are infinitely many labels for this point.

Figure 9.52

Figure 9.53

Figure 9.54

Finally, we can express the location of the point using a negative angle measure and negative value for r. Instead of counting out r units along the terminal side of the angle, we count out r units in the opposite direction, labeling the value of r as a negative in $\left(-4, -\dfrac{2\pi}{3}\right)$. See Figure 9.55.

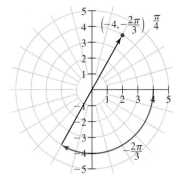

Figure 9.55

■ Using Polar Coordinates to Describe Relationships

Polar coordinates are sometimes preferred for symbolically describing real-world phenomena such as satellite locations. In the two-dimensional plane we examined previously, we saw that a satellite follows a circular path around Earth. To describe this path in polar coordinates, we use equations of the form $r = \textit{something}$. We can also use graphs to express relationships using polar coordinates. The Technology Tip at the end of the section shows how to use a graphing calculator to produce graphs of polar equations.

EXAMPLE 3 ■ Using Polar Coordinates

The DirecTV-4S satellite was launched in November 2001 to provide over 300 television channels to 41 different communities. (*Source:* science.nasa.gov) It is in a geostationary orbit at an altitude of 35,800 kilometers, as depicted in Figure 9.56. Using the center of Earth as the pole, express the geostationary orbit of this satellite using an equation in polar coordinates. (*Hint:* Earth's radius is about 6378 kilometers.) Graph this equation.

Figure 9.56

Solution We can describe the path of the satellite by recognizing that the value of the radius, r, remains constant. That is, for any value of θ, the polar coordinates (r, θ) will have a constant value for r. Using the center of Earth as the pole, we find $r = 6378 + 35{,}800 = 42{,}178$ kilometers. We write the equation $r = 42{,}178$ to represent the circular orbit of the satellite. The graph is shown in Figure 9.57.

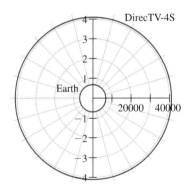

Figure 9.57

■ Natural Applications of Polar Coordinates

The chambered nautilus is a sea creature found in the tropical Pacific Ocean. A newly hatched nautilus wears a shell divided into four small chambers. As a nautilus grows, it gains more living space by building new chambers connected to the old ones. Adult shells have 30 chambers. (*Source:* www.mbayaq.org)

The pattern of the chambered nautilus shell was first described by the Greek mathematician Archimedes and is now known as the "Archimedes spiral." As we follow the curve of the shell from the very center as it curls around to the tip of its shell, we see that as the angle increases, the distance, r, from the pole also increases. We can describe this pattern using polar coordinates as in Example 4.

EXAMPLE 4 ■ Graphing in Polar Coordinates

Consider the equation $r = \theta$. Complete a table of values for the equation and use it to describe the polar curve formed by the equation. Then graph the equation.

Solution We begin by making Table 9.2 to help understand the nature of the polar curve. We use radians for the angle measure and write the values of r as decimals.

Table 9.2

θ	0	$\dfrac{\pi}{4}$	$\dfrac{\pi}{2}$	$\dfrac{3\pi}{4}$	π	$\dfrac{5\pi}{4}$	$\dfrac{3\pi}{2}$	$\dfrac{7\pi}{4}$	2π
r	0	0.785	1.57	2.36	3.14	3.93	4.71	5.50	6.28

We see that as θ increases, r increases as well. That is, the points move farther and farther from the pole as θ increases. We plot these points on the polar plane as in Figure 9.58.

We now plot the equation $r = \theta$ (Figure 9.59a) and see that it creates the Archimedes spiral (Figure 9.59b).

Figure 9.58

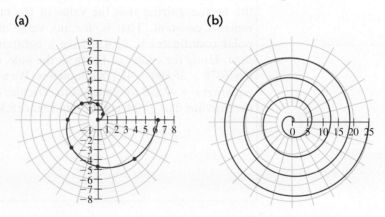

Figure 9.59

Polar equations often create very interesting and beautiful graphs. In the next example, we will look at equations of the form $r = a\cos(n\theta)$.

EXAMPLE 5 ■ Graphing Polar Equations

Graph the equation $r = 2\cos(3\theta)$ for $0 \leq \theta \leq 2\pi$. Write a justification for the behavior of the graph. Then, explore several other equations of the form $r = 2\cos(n\theta)$ and describe the graph and its connection to the value of n.

Solution We begin by graphing $r = 2\cos(3\theta)$ as in Figure 9.60. The graph resembles a flower with three petals. The greatest value of r is $r = 2$ because $-1 \leq \cos(3\theta) \leq 1$ and it follows that $-2 \leq 2\cos(3\theta) \leq 2$. Since $r = 2\cos(3\theta)$, $-2 \leq r \leq 2$. To see why the graph reaches $r = 2$ three times, note that the interval $0 \leq \theta \leq 2\pi$ is equivalent to $0 \leq 3\theta \leq 6\pi$. We know $\cos(3\theta) = \pm 1$ when $3\theta = 0, \pi, 2\pi, 3\pi, 4\pi, 5\pi, 6\pi$—that is, when $\theta = 0, \frac{\pi}{3}, \frac{2\pi}{3}, \pi, \frac{4\pi}{3}, \frac{5\pi}{3}, 2\pi$. At these angle values, $2\cos(3\theta) = \pm 2$. The reason we only see three petals is that the period of this function is $\frac{2\pi}{3}$. After $\frac{2\pi}{3}$, the graph begins repeating itself.

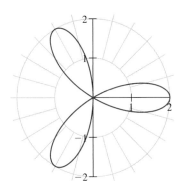

Figure 9.60

Now let's explore other values of n. Consider the graph of $r = 2\cos(5\theta)$ shown in Figure 9.61, a 5-petal flower. In Figure 9.62, we see $r = 2\cos(6\theta)$ produces a 12-petal flower. The interval $0 \leq \theta \leq 2\pi$ is equivalent to $0 \leq 6\theta \leq 12\pi$. Also, $\cos(6\theta) = \pm 1$ when $6\theta = 0, \pi, 2\pi, 3\pi, 4\pi, 5\pi, 6\pi, 7\pi, 8\pi, 9\pi, 10\pi, 11\pi, 12\pi$—that is, when $\theta = 0, \frac{\pi}{6}, \frac{\pi}{3}, \frac{\pi}{2}, \frac{2\pi}{3}, \frac{5\pi}{6}, \pi, \frac{7\pi}{6}, \frac{4\pi}{3}, \frac{3\pi}{2}, \frac{5\pi}{3}, \frac{11\pi}{6}, 2\pi$. Therefore $r = \pm 2$ for each of these 12 values of θ. In general, $r = a\cos(n\theta)$ produces a n-petal rose when n is odd and a $2n$-petal rose when n is even.

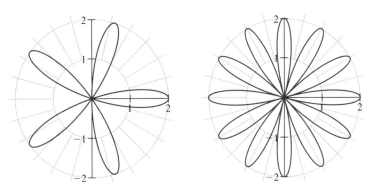

Figure 9.61   Figure 9.62

■ Other Polar Graphs

Polar equations can be used to create a variety of beautiful graphs. We encourage you to experiment! We provide a few graphs here together with their equations to pique your interest. The possibilities are endless. For each, challenge yourself to describe why the graphs appear as they do.

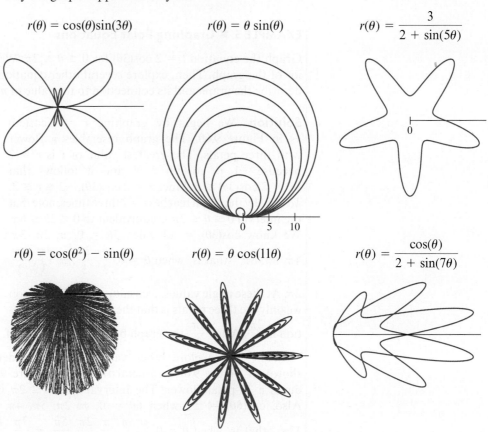

$r(\theta) = \cos(\theta)\sin(3\theta)$

$r(\theta) = \theta \sin(\theta)$

$r(\theta) = \dfrac{3}{2 + \sin(5\theta)}$

$r(\theta) = \cos(\theta^2) - \sin(\theta)$

$r(\theta) = \theta \cos(11\theta)$

$r(\theta) = \dfrac{\cos(\theta)}{2 + \sin(7\theta)}$

EXAMPLE 6 ■ Investigating Polar Graphs

Graph the polar equation $r = \cos(5\theta) + 10$ using a graphing calculator. Choose a viewing window so you obtain a good result. Then write an explanation for why the graph makes sense. For example, explain why the maximum value of r makes sense, why the behavior of the graph makes sense, and so on.

Solution Using the Technology Tip at the end of this chapter, we begin by entering the equation $r = \cos(5\theta) + 10$ into the calculator and producing a graph in the standard viewing window. See Figure 9.63.

We do not consider this a good viewing window since part of the graph is not visible. We know $\cos(5\theta)$ always produces values between -1 and 1, inclusive. Adding 10 will produce values for r such that $9 \leq r \leq 11$. We change the viewing window to include these values for r and obtain the graph shown in Figure 9.64.

The interval $0 \leq \theta \leq 2\pi$ is equivalent to $0 \leq 5\theta \leq 10\pi$. Also, $\cos(5\theta) = 1$ when $5\theta = 0, 2\pi, 4\pi, 6\pi, 8\pi, 10\pi$—that is, when $\theta = 0, \dfrac{2\pi}{5}, \dfrac{4\pi}{5}, \dfrac{6\pi}{5}, \dfrac{8\pi}{5}, \dfrac{10\pi}{5}$. Therefore, $\cos(5\theta) + 10 = 11$ at each of these values of θ.

Figure 9.63

Figure 9.64

Relationship between Cartesian and Polar Coordinates

Using right triangle trigonometry and the Pythagorean theorem, we can determine the relationships between Cartesian and polar coordinates. That is, if we know the coordinates of a point in the Cartesian system, (x, y), we can determine the same location in polar coordinates, (r, θ). Or, if we know the coordinates of a point in the polar system, (r, θ), we can determine the same location in Cartesian coordinates.

We examine the generic graph in Figure 9.65 to determine these relationships. Suppose we know the values of r and θ and wish to determine the corresponding values of x and y.

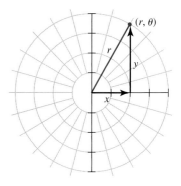

Figure 9.65

We have the following relationships.

$$\cos(\theta) = \frac{x}{r} \qquad \sin(\theta) = \frac{y}{r}$$

$$x = r\cos(\theta) \qquad y = r\sin(\theta)$$

That is, if we know (r, θ), we can compute (x, y).

We can also determine (r, θ) if we know (x, y). Consider the graph in Figure 9.66. Using the Pythagorean theorem and the definition of tangent, we have

$$x^2 + y^2 = r^2 \qquad \tan(\theta) = \frac{y}{x}$$

$$r = \sqrt{x^2 + y^2} \qquad \theta = \tan^{-1}\left(\frac{y}{x}\right), x \neq 0$$

We take θ to be the reference angle and will determine the quadrant in which it lies through the context of the situation.

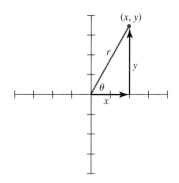

Figure 9.66

Relationships between Polar and Cartesian Coordinates

The polar coordinates (r, θ) and the corresponding Cartesian coordinates (x, y) are related as follows.

$$x = r\cos(\theta) \quad \text{and} \quad y = r\sin(\theta)$$

$$r = \sqrt{x^2 + y^2}$$

$$\theta = \tan^{-1}\left(\frac{y}{x}\right) \text{ for } x \neq 0$$

EXAMPLE 7 ■ **Converting from Polar to Cartesian Coordinates**

Using the polar coordinates for the Iridium 19 satellite, $(r, \theta) = (7156, 0.5)$, determine the equivalent Cartesian coordinates, (x, y).

Solution Since we know the polar coordinates, we find the x- and y-coordinates using $x = r\cos(\theta)$ and $y = r\sin(\theta)$. (Remember that the angle is in radians.)

$$x = 7156\cos(0.5)$$
$$x = 6279.98$$
$$y = 7156\sin(0.5)$$
$$y = 3430.77$$

The polar coordinates $(r, \theta) = (7156, 0.5)$ are equivalent to the Cartesian coordinates $(x, y) = (6279.98, 3430.77)$, as shown in Figure 9.67.

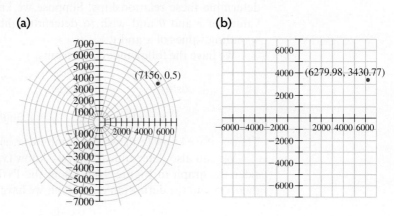

Figure 9.67

EXAMPLE 8 ■ **Converting from Cartesian to Polar Coordinates**

Convert the Cartesian coordinates $(x, y) = (-4, 6)$ to polar coordinates.

Solution We begin by graphing the point in the Cartesian plane, as shown in Figure 9.68. We know $r = \sqrt{x^2 + y^2}$. Thus,

$$r = \sqrt{(-4)^2 + (6)^2}$$
$$r = \sqrt{52}$$
$$r \approx 7.211$$

We also know $\theta = \tan^{-1}\left(\dfrac{y}{x}\right)$, $x \neq 0$. Therefore,

$$\theta = \tan^{-1}\left(\dfrac{6}{-4}\right)$$
$$\theta = \tan^{-1}\left(-\dfrac{3}{2}\right)$$
$$\theta \approx -0.9828$$

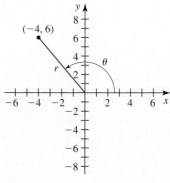

Figure 9.68

The inverse tangent function returns angles in Quadrants I and IV only. Since $\theta \approx -0.9828$ is in Quadrant IV, we need to add π to get to Quadrant II.

$$\theta \approx -0.9828 + \pi$$
$$\theta \approx 2.159$$

The point $(x, y) = (-4, 6)$ is equivalent to $(r, \theta) = (7.211, 2.159)$. See Figure 9.69.

Figure 9.69

Describing Graphs Using Polar Coordinates

In calculus, it is sometimes necessary to describe a region in a plane using polar coordinates. We practice this skill in the next example.

EXAMPLE 9 ■ Describing a Region with Polar Coordinates

Farmers sometimes allow their fields to be shaped so that beautiful geometric designs can be seen from the air. A relatively simple crop circle is shown in the photo on the left. Use polar coordinates to describe the circular, ring region shown in the inset.

Solution Since we do not know the scale used for this ring, we will impose a coordinate system over the picture and generically use "units" as the unit of length. See Figure 9.70.

We know that polar coordinates describe the value of r, the distance from the origin, and the angle θ measured from the positive x-axis. In this crop circle, with the coordinate system shown, the value of r varies from approximately $r = 1.6$ units at the inner circle to approximately $r = 1.9$ units at the outer circle. We express this region using the inequality $1.6 \leq r \leq 1.9$.

Figure 9.70

Since the crop circle ring in question completes one revolution, θ varies from 0 radians to 2π radians. We express this as $0 \leq \theta \leq 2\pi$.

SUMMARY

In this section you learned how to locate a point in a plane using polar coordinates. You also learned how to model real-world phenomena using graphs of polar equations as well as strategies for making sense of the behavior of the polar graph. Additionally, you learned how to convert from Cartesian coordinates to polar coordinates and vice versa.

TECHNOLOGY TIP ■ POLAR GRAPHS

1. Press MODE, scroll to POL and press ENTER. This will change the calculator to polar mode.

2. Press Y= and enter the equation of a polar graph, $r(\theta)$. Press X,T,θ,n to access the variable θ. In this example, we graph the polar equation $r = 2 + 3\sin(\theta)$.

3. Press WINDOW. Enter appropriate values for θmin and θmax. In radians, consider trying θmin=0 and θmax=2π. Enter a value for θstep. This determines the frequency at which pixels will be plotted. The smaller the θstep, the more accurate the graph (but it will plot more slowly). The rest of the window settings can be made as usual.

4. Press GRAPH. You may need to adjust the window to get a good view of the polar graph.

9.4 EXERCISES

SKILLS AND CONCEPTS

In Exercises 1–5, consider each of the points shown in the polar plane. Label each point in two different ways.

1. Point A
2. Point B
3. Point C
4. Point D
5. Point E

In Exercises 6–15, convert the Cartesian coordinates to polar coordinates.

6. (2, 2)
7. (3, 0)
8. (0, −5)
9. (−4, 0)
10. ($\sqrt{3}$, 2)
11. (1, −$\sqrt{3}$)
12. (5, 2)
13. (−1, −5)
14. (4, −3)
15. (−6, 5)

In Exercises 16–25, convert the polar coordinates to Cartesian coordinates.

16. (2, 0)
17. $\left(3, \dfrac{\pi}{2}\right)$
18. (2, π)
19. $\left(-3, \dfrac{\pi}{6}\right)$
20. $\left(5, \dfrac{5\pi}{6}\right)$
21. $\left(2, -\dfrac{\pi}{3}\right)$
22. (0.5, 1)
23. (−2, 4)
24. (−4, 0)
25. (−5, −2)

26. Give inequalities for r and θ to describe the given region in polar coordinates.

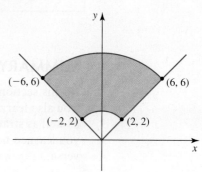

27. Give inequalities for r and θ to describe the given region in polar coordinates. Note the shaded region extends indefinitely in both the x- and y-directions.

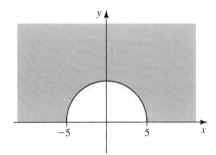

In Exercises 28–32, graph the polar equation using a graphing calculator. Choose a viewing window so that you obtain a good result. Then, write an explanation for why the graph makes sense.

28. $r = 4\cos(5\theta)$
29. $r = 4\sin(2\theta)$
30. $r = 4\sin(3\theta)$
31. $r = 4\sin(5\theta)$
32. $r = 1 + 2\sin(\theta)$

In Exercises 33–42, graph the polar equation using a graphing calculator. Choose a viewing window so that you obtain a good result.

33. $r = 1 + 3\sin(\theta)$
34. $r = 1 + 4\sin(\theta)$
35. $r = 1 + 2\cos(\theta)$
36. $r = 1 + 3\cos(\theta)$
37. $r = 1 + 4\cos(\theta)$
38. $r = 5\sin(\theta)\cos^2(\theta)$
39. $r = 5\cos(\theta)\sin^2(\theta)$
40. $r = \dfrac{2}{\sqrt{\theta}}$
41. $r = \dfrac{4\sin(\theta)}{\theta}$
42. $r = \dfrac{10\sin(\theta)}{\theta}$

SHOW YOU KNOW

43. A classmate missed class and asks you to explain the purpose of polar coordinates. How do you respond?
44. Write a detailed explanation on how to convert from Cartesian to polar coordinates. Then write an explanation for why the formulas work in this process.
45. Write a detailed explanation on how to convert from polar to Cartesian coordinates. Then write an explanation for why the formulas work in this process.
46. Write a detailed explanation of the behavior of the polar graph of $r = C$, where C is a constant.
47. Write a detailed explanation of the behavior of the polar graph of $r = \theta$.

MAKE IT REAL

48. **Compact Disc** An audio compact disc (CD) has a diameter of 12 centimeters. The diameter of the hole in the center of the CD is 1.5 centimeters. Write a description of the CD using polar coordinates. That is, use inequalities to describe the range of values of r and θ that would describe the CD.

49. **Pizza Slice** A 16-inch-diameter pizza is cut into 8 equal pieces. Use inequalities to describe one slice of pizza in polar coordinates.

50. **Pac-Man** The video game Pac-Man was introduced to the world in 1980 by game designer Toru Iwatani. One day during lunch, he took a slice of pizza from the whole pie and was inspired to create Pac-Man. (*Source: www.arcade-history.com*) Use the diagram given to express Pac-Man using inequalities in polar coordinates.

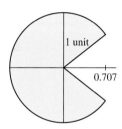

STRETCH YOUR MIND

Exercises 51–55 are designed to challenge your understanding of polar graphs.

51. Graph the equation $r = a + n\sin(\theta)$ for different values of a while holding n constant. Describe the relationship between the behavior of the graph and the value of a.
52. Graph the equation $r = a + n\sin(\theta)$ for different values of n while holding a constant. Describe the relationship between the behavior of the graph and the value of n.
53. Graph the equation $r = a + n\cos(\theta)$ for different values of a while holding n constant. Describe the relationship between the behavior of the graph and the value of a.
54. Graph the equation $r = a + n\cos(\theta)$ for different values of n while holding a constant. Describe the relationship between the behavior of the graph and the value of n.
55. Repeat Exercises 51–54 but for the polar equations $r = a - n\sin(\theta)$ and $r = a - n\cos(\theta)$.

SECTION 9.5 Vectors

LEARNING OBJECTIVES

- Use vectors to model and solve real-world situations
- Determine the magnitude and direction of a vector
- Resolve a vector into components
- Add, subtract, and scale vectors graphically and algebraically

GETTING STARTED

On the ground, the wind has a negligible effect on an airplane's speed, but in the air, airplanes may be significantly impacted by wind. In fact, when pilots calculate the time needed to complete a journey, they distinguish between ground speed and airspeed. *Ground speed* is the speed of the airplane relative to the ground and *airspeed* is the difference between the ground speed and the wind speed. If there is no wind speed, the airspeed is equal to

the ground speed. If there is a tailwind (wind blowing in the direction of the airplane's travel), the airspeed will be greater than the ground speed and if there is a headwind (wind blowing in the opposite direction of the airplane's travel), the airspeed will be less than the ground speed. In addition to having an influence on the speed of the plane, wind may influence the direction the plane is traveling. If the wind is blowing against the side of the plane, the airspeed is not impacted, but the wind may alter the direction the plane is traveling.

In this section we study vector quantities like wind speed where both magnitude (e.g., speed, force, mass, length) and direction are relevant. We learn to use vectors to model and solve real-world situations.

■ Geometric Vectors

The Dubai International Airport in Dubai, United Arab Emirates, is one of the largest and busiest airports in the world. As with most airports, it has a moving walkway system to quickly move people from one part of the airport terminal to another. We may think of the velocity of the walkway as a vector quantity. **Vector** quantities have both a *magnitude* (the speed of the walkway in this case) and *direction* (the direction the walkway is moving). (Scalar quantities have only magnitude.)

> **VECTOR**
>
> A **vector** is any quantity with both magnitude and direction.

Suppose the walkway is moving at a speed of 2 miles per hour and we designate the direction to be in the positive vertical direction on the Cartesian plane. To represent the vector quantity geometrically, we use an arrow connecting two points as in Figure 9.71a, where we connect the points (0, 0) and (0, 2). The length, or **magnitude**, of the vector corresponds to the speed of the walkway. This magnitude, together with the direction of 90°, makes this a vector quantity. The beginning point is known as the **tail** of the vector and the end point is known as the **tip** of the vector. There is no special reason why the tail is located at (0, 0). In fact, as shown in Figure 9.71b, we could place this 2-unit-long vertical vector anywhere on the Cartesian coordinate system. When the tail of the vector is located at the origin, we refer to the vector as a **position vector**.

SECTION 9.5 ■ VECTORS

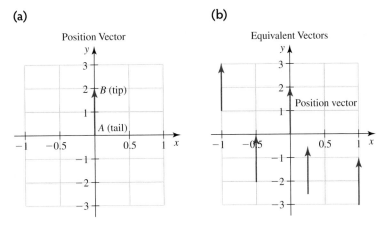

Figure 9.71

Symbolically, we express the position vector by determining how much distance is displaced horizontally and vertically from tail to tip. In this case, the horizontal displacement is 0 units and the vertical displacement is 2 units. Therefore, we express the vector as $\vec{AB} = \langle 0, 2 \rangle$. We call this the **displacement vector in component form**.

> ### DISPLACEMENT VECTOR IN COMPONENT FORM
>
> A **displacement vector** from one point to another is represented by an arrow with the tail located at the first point and the tip located at the second point. If the tail is located at point (a, b) and the tip is located at the point (c, d), then the displacement vector, $\vec{V}$, is written in component form as
>
> $$\vec{V} = \langle c - a, d - b \rangle$$

Suppose we see a person walking on the moving walkway at a pace of 3 miles per hour. Because the walkway is moving at 2 miles per hour, the combined speed of the person is 5 miles per hour. We express this sum geometrically by placing the vectors (the moving walkway and the person walking) tail to tip in the Cartesian plane and connecting the tail of the first vector with the tip of the second vector. See Figure 9.72a.

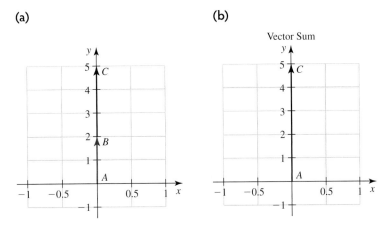

Figure 9.72

We symbolically say $\vec{AB} + \vec{BC} = \vec{AC}$. We can also compute the **vector sum** (Figure 9.72b) by adding the horizontal and then the vertical components of the vectors.

$$\vec{AB} + \vec{BC} = \vec{AC}$$
$$\langle 0, 2 \rangle + \langle 0, 3 \rangle = \langle 0 + 0, 2 + 3 \rangle$$
$$\vec{AC} = \langle 0, 5 \rangle$$

> ### Vector Sum
> Given two vectors $\vec{A} = \langle a, b \rangle$ and $\vec{B} = \langle c, d \rangle$, the sum $\vec{A} + \vec{B}$ is found by adding corresponding components.
>
> $$\vec{A} + \vec{B} = \langle a + c, b + d \rangle$$

■ Measuring Angles in an Aviation Context

In aviation, angles are measured from due north. That is, due north is 0° and angles are described as being east or west of north. For example, 45° east of north (E 45° N) is an angle in the first quadrant, as shown in Figure 9.73.

Figure 9.73

EXAMPLE 1 ■ Using Vectors

Suppose an airplane pilot wishes to file a flight plan from one airport to another. The destination airport is 200 miles due south. The pilot anticipates an average airspeed of 80 miles per hour. Indicate the heading as a vector quantity in component form and then calculate the flight time.

Solution Since the location of the destination airport is due south of the starting point, the airplane should head E 180° N (due south) at 80 miles per hour. Having both magnitude (speed of 80 miles per hour) and direction (due south), this is a vector quantity. We write this velocity vector as $\vec{V} = \langle 0, -80 \rangle$ and graph it as shown in Figure 9.74.

At the average speed of 80 miles per hour, it will take 2.5 hours to travel the 200 miles to the destination airport since $\dfrac{200 \text{ miles}}{80 \text{ miles per hour}} = 2.5$ hours.

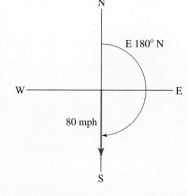

Figure 9.74

■ Computing the Magnitude and Direction of a Vector

Suppose we need to find the magnitude of the vector $\vec{A} = \langle 2, 5 \rangle$. As a position vector, this vector has a horizontal component of 2 and a vertical component of 5. Sketching this as in Figure 9.75, we see that we can use the Pythagorean theorem to compute the magnitude, or length, of the vector (dashed line). We use the symbol $|\vec{A}|$ to represent the magnitude.

$$|\vec{A}|^2 = 2^2 + 5^2$$
$$|\vec{A}|^2 = 4 + 25$$
$$|\vec{A}|^2 = 29$$
$$|\vec{A}| = \sqrt{29}$$

The magnitude of the vector is $|\vec{A}| = \sqrt{29}$ or $|\vec{A}| \approx 5.39$ units.

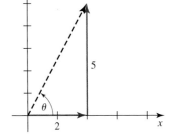

Figure 9.75

Using trigonometric functions, we can also determine the direction of the vector.

$$\tan(\theta) = \frac{5}{2}$$
$$\theta = \tan^{-1}(2.5)$$
$$\approx 68.2°$$

The vector is headed in the direction of 68.2°.

Computing Magnitude and Direction of a Given Vector

The magnitude of a vector $\vec{A} = \langle a, b \rangle$ is given by

$$|\vec{A}| = \sqrt{a^2 + b^2}$$

The direction is determined by

- $\theta = \tan^{-1}\left(\dfrac{b}{a}\right)$ for a vector with a tip in the first or fourth quadrant.
- $\theta = \tan^{-1}\left(\dfrac{b}{a}\right) + 180°$ for a vector with a tip in the second or third quadrant.

■ Adding Vectors Geometrically

It is sometimes helpful to see the sum of two vectors graphically. Suppose we wish to visualize the sum of the vectors $\vec{W} = \langle -2, 5 \rangle$ and $\vec{Z} = \langle 3, 1 \rangle$. We begin by sketching these vectors in the Cartesian plane as position vectors. (See Figure 9.76.) Recall we may move vectors around the plane as long as we maintain their magnitude (length) and direction. To visualize the sum $\vec{W} + \vec{Z}$, we move vector $\vec{Z}$ (red line) and connect it tail to tip with vector $\vec{W}$ (blue line). Then we connect the tail of vector $\vec{W}$ to the tip of the repositioned vector $\vec{Z}$ (light red line). This vector, $\vec{W} + \vec{Z}$, is known as the **resultant vector** (dashed black line). We see that $\vec{W} + \vec{Z} = \langle -2, 5 \rangle + \langle 3, 1 \rangle = \langle 1, 6 \rangle$.

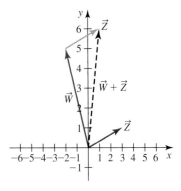

Figure 9.76

EXAMPLE 2 ■ Using Vectors

The airplane in Example 1 was traveling 80 miles per hour due south. Suppose the airplane encounters a 25-mile-per-hour wind blowing from west to east throughout the entire trip. If the pilot keeps the vector heading of due south, how far from the destination will the pilot end up? Express the solution of this problem using geometric vectors. What is the magnitude of the resultant vector?

Solution We first make a diagram of the problem situation, drawing both the vector for the airplane and the wind. See Figure 9.77a. Recall that as long as magnitude and direction are maintained, we can move a vector around the Cartesian plane. Figure 9.77b shows the repositioned wind vector connected tail to tip with the airplane vector.

Figure 9.77

The vector sum of the airplane vector and the wind vector is shown in Figure 9.78. As we might expect, the effect of the wind causes the airplane to miss its target by pushing it to the east of the destination airport. Since the airplane takes 2.5 hours to travel the 200-mile distance to its destination, the wind will cause the airplane to travel 62.5 miles east of the airport.

$$25 \frac{\text{miles}}{\text{hour}} \cdot 2.5 \text{ hours} = 62.5 \text{ miles}$$

The resultant velocity vector is the sum of the velocity of the airplane, $\vec{A} = \langle 0, -80 \rangle$, and the velocity of the wind, $\vec{W} = \langle 25, 0 \rangle$.

Figure 9.78

$$\vec{R} = \vec{A} + \vec{W}$$
$$\vec{R} = \langle 0, -80 \rangle + \langle 25, 0 \rangle$$
$$\vec{R} = \langle 0 + 25, -80 + 0 \rangle$$
$$\vec{R} = \langle 25, -80 \rangle$$

The magnitude of the resultant vector is the airspeed of the plane as a result of the effect of the wind.

$$|\vec{R}| = \sqrt{(25)^2 + (-80)^2}$$
$$|\vec{R}| = \sqrt{625 + 6400}$$
$$|\vec{R}| = \sqrt{7025}$$
$$|\vec{R}| \approx 83.82$$

The airspeed of the airplane is 83.82 miles per hour.

EXAMPLE 3 ■ Using Vectors

Determine the heading necessary so that the airplane in Example 2 will reach the destination airport. The average ground speed is still 80 miles per hour. Sketch a diagram showing the corrected heading and write the heading as a vector in component form.

Solution The pilot will need to set the heading so that the airplane points into the wind in such a way that the horizontal component of its velocity vector compensates for the 25-mile-per-hour wind. We sketch the problem situation in Figure 9.79. We see the hypotenuse of the right triangle is represented by the 80-mile-per-hour average speed and the side opposite the angle is the 25-mile-per-hour wind speed. We can determine the angle needed to correct the heading using trigonometry.

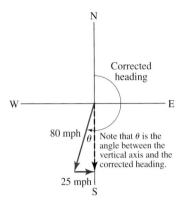

Figure 9.79

$$\sin(\theta) = \frac{25}{80}$$

$$\theta = \sin^{-1}\left(\frac{25}{80}\right)$$

$$\theta = 18.21°$$

The pilot needs to add 18.21° to the heading to correct for the wind. Therefore, the new heading is 180° + 18.21° = 198.21° E of N at 80 miles per hour. We can also refer to this angle as 161.79° W of N. This vector has a horizontal component of −25 to counteract the effects of the wind. We determine the vertical component using the Pythagorean theorem.

$$80^2 = (-25)^2 + y^2$$
$$6400 = 625 + y^2$$
$$y^2 = 6400 - 625$$
$$y^2 = 5775$$
$$y = \sqrt{5775}$$
$$y \approx 75.99$$

The desired heading will be determined by the velocity vector $\vec{V} = \langle -25, -75.99 \rangle$.

■ Scalar Multiplication with Vectors

When we multiply a vector quantity by a scalar, only the magnitude of the vector changes. For example, suppose the velocity of a sprinter in the 100-meter sprint is given by the vector $\vec{S} = \langle 3, 6 \rangle$ measured in meters per second. Suppose the wind is blowing directly at the sprinter's back at a velocity that is $\frac{1}{18}$ the speed of the sprinter. We can compute the velocity of the wind, $\vec{W}$, as a vector quantity.

$$\vec{W} = \frac{1}{18}\vec{S}$$

$$\vec{W} = \frac{1}{18}\langle 3, 6 \rangle$$

$$\vec{W} = \left\langle \frac{1}{18} \cdot 3, \frac{1}{18} \cdot 6 \right\rangle$$

$$\vec{W} = \left\langle \frac{1}{6}, \frac{1}{3} \right\rangle$$

> **SCALAR MULTIPLICATION**
>
> Given a vector $\vec{A} = \langle a, b \rangle$ and a scalar, k, the product $k\vec{A}$ is the vector
>
> $$k \cdot \vec{A} = k \cdot \langle a, b \rangle$$
> $$= \langle k \cdot a, k \cdot b \rangle$$

EXAMPLE 4 ■ Using Scalar Multiplication

Suppose the average velocity vector of a sprinter in the 100-meter sprint is given by $\vec{S} = \langle 3, 6 \rangle$ and that the wind is blowing directly at the sprinter's back at a speed that is $\frac{1}{18}$ as great as the velocity of the sprinter. With the effect of the wind taken into consideration, how long will it take the sprinter to finish the race? According to USA Track and Field rules, if the average speed of the wind blowing in the direction of the runner is greater than 2 meters per second, then a record cannot be accepted. Could a record be accepted in this case?

Solution We saw, using scalar multiplication, that the velocity of the wind is given by the vector $\vec{W} = \left\langle \frac{1}{6}, \frac{1}{3} \right\rangle$. We know the wind velocity vector is in the same direction as the sprinter velocity vector, just smaller in magnitude by $\frac{1}{18}$. We will find the sum of these vectors and use the resultant vector to determine the speed of the sprinter with the effects of the wind considered.

$$\vec{S} + \vec{W} = \langle 3, 6 \rangle + \left\langle \frac{1}{6}, \frac{1}{3} \right\rangle$$

$$\vec{S} + \vec{W} = \left\langle 3 + \frac{1}{6}, 6 + \frac{1}{3} \right\rangle$$

$$\vec{S} + \vec{W} = \left\langle \frac{19}{6}, \frac{19}{3} \right\rangle$$

The magnitude of this resultant vector is the speed of the runner.

$$|\vec{S} + \vec{W}| = \sqrt{\left(\frac{19}{6}\right)^2 + \left(\frac{19}{3}\right)^2}$$

$$|\vec{S} + \vec{W}| = \sqrt{\frac{1805}{36}}$$

$$|\vec{S} + \vec{W}| \approx 7.08 \text{ meters per second}$$

Using this average speed of 7.08 meters per second, we determine the time it will take the sprinter to complete the 100-meter sprint as

$$\frac{100 \text{ meters}}{7.08 \frac{\text{meters}}{\text{second}}} = 14.12 \text{ seconds}$$

While this is not a record time, if the wind speed is less than 2 meters per second, a record time would be accepted.

$$|\vec{W}| = \sqrt{\left(\frac{1}{6}\right)^2 + \left(\frac{1}{3}\right)^2}$$

$$|\vec{W}| = \sqrt{\frac{1}{36} + \frac{1}{9}}$$

$$|\vec{W}| = \sqrt{\frac{5}{36}}$$

$$|\vec{W}| \approx 0.37 \text{ meters per second}$$

The wind speed was 0.37 meters per second, so a record time could be accepted.

■ Vector Subtraction

We subtract the vectors through the process of adding the opposite. That is, $\vec{A} - \vec{B} = \vec{A} + (-\vec{B})$. In other words, we first multiply the second vector by the scalar -1 and then add corresponding components of the vectors.

> **VECTOR SUBTRACTION**
>
> Given two vectors $\vec{A} = \langle a, b \rangle$ and $\vec{B} = \langle c, d \rangle$, the difference $\vec{A} - \vec{B}$ is given by $\vec{A} + (-\vec{B})$, where $-\vec{B} = \langle -c, -d \rangle$. That is,
>
> $$\vec{A} - \vec{B} = \langle a - c, b - d \rangle$$

■ Unit Vectors

Unit vectors are vectors that have a length of 1 unit. We can use two special unit vectors to express displacement vectors in a different form. We define these unit vectors as follows.

$$\vec{i} = \langle 1, 0 \rangle$$

$$\vec{j} = \langle 0, 1 \rangle$$

To show how we can use the unit vectors $\vec{i}$ and $\vec{j}$ to express displacement vectors, consider the following.

$$\vec{V} = \langle 4, 5 \rangle$$
$$\vec{V} = \langle 4, 0 \rangle + \langle 0, 5 \rangle$$
$$\vec{V} = 4 \cdot \langle 1, 0 \rangle + 5 \cdot \langle 0, 1 \rangle$$
$$\vec{V} = 4\vec{i} + 5\vec{j}$$

These equivalent forms of expressing vector quantities are conventionally used in mathematics and science. We will use both forms throughout the section. Within a particular problem situation, it is typical to stay with one particular format.

Vectors Used to Measure Force

Another way that vectors may be used is to represent forces. Simply put, a force is a push or a pull on an object. Using pounds as a unit of measure of force, we might say "20 pounds of force were required to move the box across the room" or "we applied 80 pounds of force to the rope to keep the wagon from rolling down the hill." In the next example, we will express forces as vectors in component form and use these vectors to solve a problem.

EXAMPLE 5 ■ Using Vectors

Suppose a 400-pound force is required to stabilize a hot air balloon. Two ropes, held at the angles shown in Figure 9.80, are used as tether lines to hold the balloon in place. Determine the force needed in each of the tether lines to create the 400-pound force. Express each of these forces as vectors in component form.

Figure 9.80

Solution We determine the measure of the angle between each vector and the *x*-axis. Then we draw the horizontal and vertical components of each vector. See Figure 9.81.

Figure 9.81

We will first determine the vector $\vec{AB}$. We know $\vec{f}_x + \vec{f}_y = \vec{AB}$, so we determine the magnitudes of $\vec{f}_x$ and $\vec{f}_y$ by using the cosine and sine functions.

$$\cos(45°) = \frac{|\vec{f_x}|}{|\vec{AB}|} \qquad \sin(45°) = \frac{|\vec{f_y}|}{|\vec{AB}|}$$

$$|\vec{f_x}| = |\vec{AB}|\cos(45°) \qquad |\vec{f_y}| = |\vec{AB}|\sin(45°)$$

The vector $\vec{AB} = \langle -|\vec{AB}|\cos(45°), -|\vec{AB}|\sin(45°)\rangle$. (We add the negatives to both components because the tip of the vector $\vec{AB}$ in standard position lies in the third quadrant.) We represent the vector in component form as

$$\vec{AB} = -|\vec{AB}|\cos(45°)\vec{i} - |\vec{AB}|\sin(45°)\vec{j}$$

We determine the component form of vector $\vec{AC}$ in a similar way. We know $\vec{g_x} + \vec{g_y} = \vec{AC}$, so we determine the magnitudes of $\vec{g_x}$ and $\vec{g_y}$ by using the cosine and sine functions.

$$\cos(65°) = \frac{|\vec{g_x}|}{|\vec{AC}|} \qquad \sin(65°) = \frac{|\vec{g_y}|}{|\vec{AC}|}$$

$$|\vec{g_x}| = |\vec{AC}|\cos(65°) \qquad |\vec{g_y}| = |\vec{AC}|\sin(65°)$$

The vector $\vec{AC} = \langle |\vec{AC}|\cos(65°), -|\vec{AC}|\sin(65°)\rangle$. (We add the negative to the vertical component because the tip of the vector $\vec{AC}$ in standard position lies in the fourth quadrant.) We represent the vector in component form as

$$\vec{AC} = |\vec{AC}|\cos(65°)\vec{i} - |\vec{AC}|\sin(65°)\vec{j}$$

Next, we apply ideas concerning the balloon situation. For the balloon to be stabilized, $\vec{AB} + \vec{AC} = 0\vec{i} - 400\vec{j}$. That is, the horizontal components should add to 0 since the balloon is not to move horizontally. The vertical components should add to be -400 to overcome the force of the balloon in the upward direction. Thus,

$$-|\vec{AB}|\cos(45°) + |\vec{AC}|\cos(65°) = 0$$

$$-|\vec{AB}|\sin(45°) - |\vec{AC}|\sin(65°) = -400$$

We have a system of two equations with two unknowns, $|\vec{AB}|$ and $|\vec{AC}|$. We will solve the first equation for $|\vec{AC}|$ and then substitute into the second equation.

$$|\vec{AC}|\cos(65°) = |\vec{AB}|\cos(45°)$$

$$|\vec{AC}| = \frac{|\vec{AB}|\cos(45°)}{\cos(65°)}$$

Next, we substitute and solve for $|\vec{AB}|$.

$$-|\vec{AB}|\sin(45°) - |\vec{AC}|\sin(65°) = -400$$

$$-|\vec{AB}|\sin(45°) - \left[\frac{|\vec{AB}|\cos(45°)}{\cos(65°)}\right]\sin(65°) = -400$$

$$|\vec{AB}|\sin(45°) + \left[\frac{|\vec{AB}|\cos(45°)}{\cos(65°)}\right]\sin(65°) = 400$$

$$0.707|\vec{AB}| + 1.52|\vec{AB}| = 400$$

$$2.22|\vec{AB}| = 400$$

$$|\vec{AB}| = \frac{400}{2.22}$$

$$|\vec{AB}| \approx 179.9 \text{ pounds}$$

We now substitute this result into $|\vec{AC}| = \dfrac{|\vec{AB}|\cos(45°)}{\cos(65°)}$ to find the magnitude of $\vec{AC}$.

$$|\vec{AC}| = \frac{179.9\cos(45°)}{\cos(65°)}$$

$$|\vec{AC}| \approx 301 \text{ pounds}$$

The component forms of the vectors are

$$\vec{AB} = -179.9\cos(45°)\vec{i} - 179.9\sin(45°)\vec{j}$$
$$\vec{AB} = -127.2\vec{i} - 127.2\vec{j}$$

$$\vec{AC} = 301\cos(65°)\vec{i} - 301\sin(65°)\vec{j}$$
$$\vec{AC} = 127.2\vec{i} - 272.8\vec{j}$$

SUMMARY

In this section you learned how to use vectors to model and solve real-world situations. You learned to compute the magnitude of a vector and to write a vector in component form. You learned how to add and subtract vectors as well as calculate the scalar multiple of a vector.

9.5 EXERCISES

SKILLS AND CONCEPTS

In Exercises 1–10, determine whether the given quantity is a vector or scalar.

1. Velocity of wind
2. Price of gasoline
3. Air temperature
4. Force needed to lift a box
5. Number of students in a class
6. Tension in a suspension bridge cable
7. Water current in river
8. Velocity of a car on the highway
9. Amount of water in swimming pool
10. Force needed to open a door

In Exercises 11–15, write each vector in component form.

11.

12.

13.

14.

15.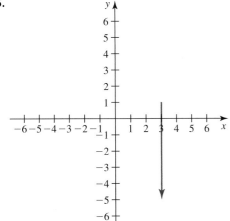

In Exercises 16–25, perform the indicated operation. Then sketch the given vectors and the resultant of the operation.

$$\vec{a} = 3\vec{i} + 5\vec{j} \qquad \vec{b} = -4\vec{i} - \vec{j}$$
$$\vec{c} = 6\vec{j} \qquad \vec{d} = -8i$$

16. $\vec{a} + \vec{b}$
17. $\vec{a} + \vec{c}$
18. $3\vec{a}$
19. $-\vec{b}$
20. $\vec{b} - \vec{a}$
21. $2\vec{c} - \vec{b}$
22. $3\vec{b} - 2\vec{d}$
23. $2\vec{c} - 2\vec{c}$
24. $2\vec{a} + \frac{1}{2}\vec{d}$
25. $2\vec{a} + 3\vec{b}$

In Exercises 26–30, find each of the displacement vectors from one city to another. For these exercises, the origin is the intersection of the equator and the prime meridian. The horizontal and vertical distances are approximated to be the latitude and longitude measurements.

26. From Phoenix (33, −112) to Minneapolis (45, −93)
27. From Minneapolis (45, −93) to Amsterdam (52, 5)
28. From Amsterdam (52, 5) to Dubai (25, 55)
29. From Dubai (25, 55) to Malé (4, 73)
30. From Dubai (25, 55) to Phoenix (33, −112)

31. Find another vector, $\vec{S}$, that is in the same direction as $\vec{T} = 5\vec{i} - 3\vec{j}$ but is only one unit in length.

32. Find another vector, $\vec{P}$, that is twice the length of $\vec{Q} = \langle 2, 6 \rangle$ but is in the opposite direction of $\vec{Q}$.

33. Suppose a displacement vector from (2, 3) to (c, n) is $\vec{W} = \langle 3, -4 \rangle$. Determine the value of c and n.

34. Suppose the sum of three vectors is the vector $\vec{T} = 6\vec{i} + 10\vec{j}$. One of the vectors is $\vec{Z} = \vec{i} - 2\vec{j}$. Determine components of the other two vectors that make the equation true.

35. Determine the point that is the location of the tip of the vector whose tail is located at (4, 5) and is displaced two units in the direction of $\theta = 60°$.

36. Suppose a delivery truck is traveling north at 10 miles per hour and the driver tosses a newspaper out of the window at a speed of 2 miles per hour toward a house located directly to the west relative to the truck.
    a. Find the velocity vector of the paper relative to the ground when the driver releases it.
    b. Find the speed of the paper at that time.

SHOW YOU KNOW

37. Many people think *velocity* and *speed* to mean the same thing. Explain why these words are not synonymous.

672 CHAPTER 9 ■ TRIANGLE TRIGONOMETRY AND APPLICATIONS

38. Read the FoxTrot comic strip. Write an explanation, including graphs or diagrams, confirming that running "10 yards out, then 10 yards to the right" puts you in the same place as going "$10\sqrt{2}$ yards at a 45-degree angle."

39. Suppose a boat is traveling on a river at velocity $\vec{B}$ and speed of 20 miles per hour. Further suppose that the river has a current of $\vec{C}$ and a speed of 5 miles per hour. Under what conditions will the speed of the boat be 25 miles per hour relative to the land alongside the river? Create a geometric argument for a situation where the relative speed of the boat is not 25 miles per hour.

40. The acceleration due to gravity is approximately 9.8 meters per second per second. This means the velocity of a falling object, due to gravity, increases by 9.8 meters per second for each second that the object falls. The force of gravity is a vector that may be expressed as $\vec{G} = -9.8\vec{j}$. Explain why.

41. Vectors express quantities that have magnitude and direction. It is possible to describe the "zero vector." Write an explanation of the zero vector and explain how it is expressed symbolically.

42. Using a geometric representation of vectors, show that a vector sum is commutative. That is, show that $\vec{A} + \vec{B} = \vec{B} + \vec{A}$.

■ **MAKE IT REAL**

43. **Marching Band Drill** Marching band members follow detailed instructions to create formations on the field. For example, one move might instruct a marcher to begin 8 steps (5 yards) in front of one sideline on the 40 yard line and move to the position 8 steps (5 yards) in front of the other sideline on the 30 yard line. The sidelines are 50 yards apart. Using the diagram shown, create a displacement vector to represent this marching move. Answers may vary depending on how you describe the points.

44. **Hanging a Frame** As seen in the photo, a frame is hung on a wall with the center of the hanging cable on the

hook. The angle at the hook is 100° and the frame weighs 8 pounds. Determine the tension in each side of the hanging cable and express the tension as a vector in component form.

45. **Aircraft Heading**   The heading from Minneapolis, Minnesota, to Los Angeles, California, is E 249° N. (*Source:* www.aeroplanner.com) Suppose an airplane is flying from Minneapolis to Los Angeles at a speed of 605 miles per hour at a fixed altitude with no wind factor.

    a. The velocity of the plane may be considered a vector. Explain why it is a vector and describe the two important parts of the vector.

    b. Express the velocity of the plane, $\vec{A}$, as a vector in component form.

    c. Find the magnitude of $\vec{A}$ and describe its significance in the context of the problem situation. Show all work.

46. **Wind Effect**   The heading from Minneapolis, Minnesota, to Los Angeles, California, is E 249° N. (*Source:* www.aeroplanner.com) Suppose an airplane is flying from Minneapolis to Los Angeles at a speed of 605 miles per hour at a fixed altitude with no wind factor. As the plane reaches a certain point, it encounters wind with a velocity of 50 miles per hour in the direction E 30° N.

    a. Express the velocity of the wind, $\vec{W}$, and the velocity of the airplane, $\vec{A}$, as a vector.

    b. Sketch the resultant vector produced by the sum of $\vec{A}$ and $\vec{W}$. What is the resultant velocity of the plane? Round to the nearest tenth.

    c. What is the resultant speed of the plane? Round to the nearest tenth and include units.

    d. What is the resultant direction of the plane? That is, what is the measure of the angle that the resultant vector forms with the positive *x*-axis?

47. **Seed Spitting Contest**   The U.S. Watermelon Seed Spitting Contest is held in Pardeeville, Wisconsin, each September. For men, the current seed spitting record is 61 feet 3 inches. (*Source:* www.uncommondays.com)

Suppose in one particular year, the seed spitting contest was being performed in the direction of $\vec{S} = 4\vec{i} + \vec{j}$. During the contest, a steady wind picked up with a velocity of $\vec{V} = \vec{i} + 0.25\vec{j}$ miles per hour.

Suppose further that the rulebook states that a legal wind in the direction of the seed spit must not exceed 2.5 miles per hour. If a contestant spits a seed 63 feet, will the record stand?

48. **The Wright Brothers**   On December 17, 1903, Orville Wright flew the world's first aircraft. Later that day he sent the following telegram.

    "SUCCESS FOUR FLIGHTS THURSDAY MORNING ALL AGAINST TWENTY ONE MILE WIND STARTED FROM LEVEL WITH ENGINE POWER ALONE AVERAGE SPEED THROUGH AIR THIRTY ONE MILES LONGEST 57 SECONDS INFORM PRESS HOME CHRISTMAS. OREVELLE [*sic*] WRIGHT" (*Source:* www.first-to-fly.com)

    According to Wright's journal, the wind was blowing from the north.

    a. Express the velocity of the airplane as a vector in component form.

    b. Express the velocity of the wind as a vector in component form.

    c. Find the vector sum of the velocity of the airplane and the velocity of the wind.

    d. What is the magnitude of the vector sum found in part (c)? Do you think that Orville's brother Wilbur was able to run alongside the airplane? Why or why not?

■ **STRETCH YOUR MIND**

Exercises 49–53 are intended to challenge your understanding of vectors. Your task is to make an argument for each of the following facts. Your argument may be graphical and/or symbolic.

49. $(c + k)\vec{AB} = c\vec{AB} + k\vec{AB}$

50. $\vec{AB} + \vec{0} = \vec{AB}$

51. $k(\vec{A} + \vec{B}) = k\vec{A} + k\vec{B}$

52. $\vec{AB} + (-\vec{AB}) = 0$

53. $(\vec{A} + \vec{B}) + \vec{C} = \vec{A} + (\vec{B} + \vec{C})$

CHAPTER 9 Study Sheet

As a result of your work in this chapter, you should be able to answer the following questions, which are focused on the "big ideas" of this chapter.

SECTION 9.1
1. What are the six trigonometric ratios and how can you use them to solve problems?
2. How are the six trigonometric ratios connected to the unit circle?
3. What is an angle of elevation? What is an angle of depression?

SECTION 9.2
4. What is the Law of Cosines?
5. Under what conditions is it appropriate to use the Law of Cosines?
6. How is the Law of Cosines connected to the Pythagorean theorem?

SECTION 9.3
7. What is the Law of Sines?
8. Under what conditions is it appropriate to use the Law of Sines?
9. How is triangulation used to solve real-world problems?

SECTION 9.4
10. How are coordinates described in the polar coordinate system?
11. How can you convert coordinates from the Cartesian system to the polar system? How can you convert from polar coordinates to Cartesian coordinates?
12. How can you describe a geometric region using inequalities in polar coordinates?
13. How can you describe the graph of an equation in polar coordinates?
14. In what kinds of real-world situations is it appropriate to use polar coordinates?

SECTION 9.5
15. What is a vector quantity and how is it different from a scalar quantity?
16. How are vectors represented geometrically?
17. How are vectors added and subtracted, and how is a sum or difference represented geometrically?
18. How can you find the magnitude of a vector? What might this magnitude represent?
19. What is a position vector?
20. How are displacement vectors determined?
21. How can vectors be used to represent real-world situations?

REVIEW EXERCISES

■ SECTION 9.1 ■

In Exercises 1–2, evaluate the six trigonometric functions of θ. That is, evaluate sin(θ), cos(θ), tan(θ), sec(θ), csc(θ), cot(θ).

1.

2.
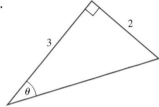

In Exercises 3–4, a trigonometric value is given. Determine if the angle measure, θ, is

  a. Less than 30°.
  b. Between 30° and 60°.
  c. Between 60° and 90°.

Explain how you know.

3. $\sin(\theta) = 0.98$
4. $\cos(\theta) = 0.95$

In Exercises 5–6, determine the angle measure of θ.

5.

6.

7. **Washington Monument** Suppose you are standing 1000 feet away from the base of the Washington Monument, which is 555 feet $\frac{5}{8}$ inches in height. What is the measure of the angle of elevation from your position to the top of the monument?

8. **The Burj Dubai** The Burj Dubai, currently under construction in the city of Dubai in the United Arab Emirates, will be the tallest building in the world when it is completed. While its ultimate height is being kept secret, it is projected to be approximately 800 meters in height upon completion. (*Source:* www.newscientist.com) On June 24, 2007, the height of the building reached 493 meters. (*Source:* www.burjdubaiskyscraper.com)

Suppose you are standing 500 meters from the base of the tower.

  a. Compute the measure of the angle of elevation to the top of the tower (493 meters).
  b. Suppose after the tower is completed, you went back to the same position, 500 meters from the base of the tower. What will be the measure of the angle of elevation to the top of the tower at that time? (Assume the final height will be 800 meters.)

■ SECTION 9.2 ■

In Exercises 9–12, find the length of the side opposite the given angle.

9. In $\triangle ABC$, $b = 15$, $c = 19$, and $A = 24°$
10. In $\triangle ABC$, $a = 71$, $b = 80$, and $C = 7°$
11. In $\triangle ABC$, $b = 5$, $c = 6$, and $A = 59°$
12. In $\triangle ABC$, $a = 5$, $c = 8$, and $B = 31°$

In Exercises 13–14, find the measure of the specified angle.

13. A in $\triangle ABC$; if $a = 4$, $b = 5$, and $c = 6$
14. C in $\triangle ABC$; if $a = 15$, $b = 16$, and $c = 18$

In Exercises 15–16, solve each triangle using any correct technique.

15.

16.
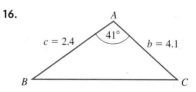

676 CHAPTER 9 ■ TRIANGLE TRIGONOMETRY AND APPLICATIONS

In Exercises 17–18, draw a diagram when one is not supplied. Then apply the Law of Cosines, the Pythagorean theorem, or a trigonometric function to find the length of the specified side or measure of the specified angle.

17. **Distance between Airplanes** Two airplanes leave Reagan National Airport in Washington DC at noon. The air traffic controller notes the two are traveling away from one another at an angle of 107°. One of the airplanes is traveling at 495 miles per hour and the other 502 miles per hour. How far apart are the two planes at 12:50 P.M.?

18. **Distance between Cities** Attempting to fly from Cincinnati, Ohio, to Tallahassee, Florida—a distance of 738 miles—a small plane veers off course by 5°.

   a. If the plane has a constant speed of 210 miles per hour and the pilot notices his error after 30 minutes, how far off course is the plane at that time? In other words, at what angle should he turn to get back on track?

   b. How fast should the pilot fly the plane to get to Tallahassee in four hours?

■ SECTION 9.3 ■

In Exercises 19–22, use triangle ABC to find the missing sides and angles (if possible). If there are two solutions, find both. If there is no solution, so state.

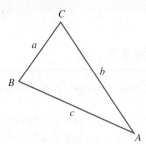

19. $a = 8$, $B = 87°$, $C = 24°$
20. $a = 4$, $B = 15°$, $C = 55°$
21. $A = 26°$, $B = 47°$, $c = 13$
22. $A = 86°$, $B = 13°$, $c = 4.5$

In Exercises 23–26, draw a diagram when one is not supplied. Then apply the Law of Sines, Law of Cosines, the Pythagorean theorem, or a trigonometric function to the situation.

23. **Cloud Height** Two observers wonder how high the cloud is above their heads. The first person can see the cloud at an angle of elevation measuring 35° and the other at 43°. If the two people are 2500 feet apart, how high is the cloud?

24. **Interplanetary Distance** Earth (E) and Mars (M) are approximately 9.3 million and 14.2 million miles from the Sun (S), respectively. (*Source:* www.enchantedlearning.com) Find the distance between the two planets when $\angle SEM$ measures 21°.

25. **Radio Transmission** The figure shows a vehicle moving due west and approaching a 30-watt transmitter on a 30-meter-tall radio tower. The effective range of the transmitter is 15 kilometers. To the nearest kilometer, for what distance x will the vehicle be within range of the transmitter at R if, when the car is at V, the measure of $\angle SVR = 43°$ and the distance from V to R is 21 kilometers?

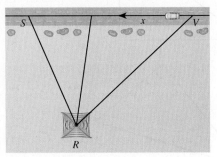

26. **Tree Height** A tree grows vertically on the side of a hill that slopes upward from the horizontal by 6°. The tree casts a shadow 53 feet long down the hill when the angle of elevation of the sun measures 15°. How tall is the tree?

■ SECTION 9.4 ■

In Exercises 27–31, consider each of the points shown in the polar plane. Label each point in two different ways.

27. Point A
28. Point B
29. Point C
30. Point D
31. Point E

In Exercises 32–34, convert the Cartesian coordinates to polar coordinates.

32. $(3, -4)$
33. $(-1, -1)$
34. $(0, 6)$

In Exercises 35–40, convert the polar coordinates to Cartesian coordinates.

35. $(4, 2\pi)$
36. $\left(1, \dfrac{3\pi}{2}\right)$
37. $\left(-2, \dfrac{\pi}{3}\right)$
38. $\left(-1, -\dfrac{5\pi}{6}\right)$
39. $\left(6, \dfrac{2\pi}{3}\right)$

40. Give inequalities for r and θ to describe the shaded region in polar coordinates.

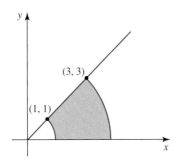

In Exercises 41–42, graph the polar equation using a graphing calculator. Choose a viewing window so you obtain a good result.

41. $r = \cos(5\theta) + 2$
42. $r = \cos(5\theta) + 5$

■ SECTION 9.5 ■

In Exercises 43–44, write each vector in component form.

43.

44.

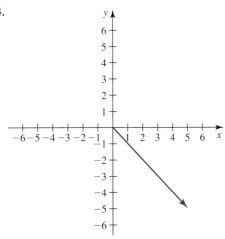

In Exercises 45–46, perform the indicated operation. Then sketch the given vectors and the resultant of the operation.

$$\vec{v} = 6\vec{i} + 2\vec{j} \qquad \vec{w} = -5\vec{i} - 3\vec{j}$$
$$\vec{r} = -4\vec{j} \qquad \vec{u} = 7i$$

45. $\vec{v} + 2\vec{w}$
46. $-\vec{r} + \vec{u}$

47. Find another vector, $\vec{B}$, that is in the opposite direction as $\vec{W} = -6\vec{i} + 3\vec{j}$ and is only one unit in length.

48. Find another vector, $\vec{P}$, that is twice the length of the vector $\vec{K} = \langle 2, -1 \rangle$ but in the opposite direction.

49. Find the value of c such that the magnitude of vector $\vec{Z} = 2\vec{i} + c\vec{j}$ is $|\vec{Z}| = 5$.

50. **Flight Heading** To travel from New York, New York, to Boston, Massachusetts, a pilot needs to fly at a heading of E 52.4° N. (*Source:* www.aeroplanner.com) Suppose an airplane is flying the route at a speed of 540 miles per hour at a fixed altitude with no wind factor.

    a. The velocity of the plane may be considered a vector. Explain why it is a vector and describe the two important parts of the vector.
    b. Express the velocity of the plane, $\vec{A}$, as a vector in component form.
    c. Find the magnitude of $\vec{A}$ and describe its significance in the context of the problem situation. Show all work.

51. **Wind Effects** To travel from New York, New York, to Boston, Massachusetts, a pilot needs to fly at a heading of E 52.4° N. (*Source:* www.aeroplanner.com) Suppose an airplane is flying the route at a speed of 540 miles per hour at a fixed altitude with no wind factor. As the plane reaches a certain point, it encounters wind with a velocity of 35 miles per hour in the direction E 10° N.

    a. Express the velocity of the wind, $\vec{W}$, and the velocity of the airplane, $\vec{A}$, as a vector.
    b. Sketch the resultant vector produced by the sum of $\vec{A}$ and $\vec{W}$. What is the resultant velocity of the plane? Round to the nearest tenth.
    c. What is the resultant speed of the plane? Round to the nearest tenth and include units.
    d. What is the resultant direction of the plane? That is, what is the measure of the angle that the resultant vector forms with the positive x-axis?

Make It Real Project

What to Do: Part I

1. Identify an object that is at least three times your height.
2. Draw a diagram showing the distances and angles you will need to calculate the object's height.
3. Measure the needed angles and distances and use trigonometry to calculate the height of the object.
4. Write a report describing the situation, the measurements, your trigonometric computations, and your results.

What to Do: Part II

1. A polar graph with equation $r = \pm\sqrt{a^2\theta}$ is called Fermat's spiral. Graph several polar graphs for different values of a. Describe the effect that a has on these graphs.

2. A polar graph with equation $r = \dfrac{a}{\theta}$ is called a hyperbolic spiral. Graph several polar graphs for different values of a. Describe the effect that a has on these graphs.

3. A polar graph with equation $r = \dfrac{a\sin(\theta)}{\theta}$ is called a cochleoid. Graph several polar graphs for different values of a. Describe the effect that a has on these graphs.

CHAPTER 10
Trigonometric Identities

As consumers, we are skilled in quickly determining if two amounts of money are equivalent. For example, we easily recognize that two dimes and a nickel are equivalent to a quarter, or that three dimes do not have the same value as three quarters, even though both groups contain the same number of coins.

The skill of recognizing when two different-looking quantities, such as two trigonometric expressions, are mathematically equivalent helps us become more efficient in solving problems.

10.1 Basic Identities
10.2 Verifying Identities
10.3 Other Trigonometric Identities
STUDY SHEET
REVIEW EXERCISES
MAKE IT REAL PROJECT

SECTION 10.1 Basic Identities

LEARNING OBJECTIVES

- Use the fundamental trigonometric identities
- Use the even–odd function identities
- Use the cofunction identities
- Use the Pythagorean identities

GETTING STARTED

In places where roof tiles are used, the tiles are delivered and distributed on the roof prior to being installed. Objects resting on a sloped surface seem to defy the force of gravity, but they actually maintain their position because of friction. Through the use of trigonometric identities, we can quantify how the tile stays on the roof.

In this section we define trigonometric identities and use these identities to help simplify trigonometric expressions and solve trigonometric equations.

■ Definition of an Identity

Consider the following expressions.

$$3 + 1 \qquad 2^2 \qquad 6 - 2 \qquad \frac{48}{12} \qquad \frac{1}{2}(8)$$

Each expression is equivalent to 4. Thus for any of these, we can replace the expression with the number 4.

Even though expressions that contain a variable can take on a number of different values, variable expressions can also be equivalent. For example, each of the following variable expressions simplifies to $2x - 6$.

$$2(x - 3) \qquad x + x - 6 \qquad \frac{4x - 12}{2} \qquad -2x - 6 + 4x$$

Since these expressions are equivalent, evaluating them at the same value of x will yield the same result. For example, when $x = 5$ we have

$$2(5 - 3) = 2(2) \qquad\qquad 5 + 5 - 6 = 10 - 6$$
$$= 4 \qquad\qquad\qquad\qquad\qquad = 4$$

$$\frac{4(5) - 12}{2} = \frac{20 - 12}{2} \qquad -2x - 6 + 4x = -2(5) - 6 + 4(5)$$
$$= \frac{8}{2} \qquad\qquad\qquad\qquad = -10 - 6 + 20$$
$$= 4 \qquad\qquad\qquad\qquad\qquad = 4$$

Likewise, evaluating the simplified expression $2x - 6$ at $x = 5$ also yields the same result.

$$2(5) - 6 = 10 - 6$$
$$= 4$$

In general, if two variable expressions are always equivalent, we say they form an **identity**.

> **IDENTITY**
>
> Two variable expressions that are always equivalent are said to form a mathematical **identity**.

EXAMPLE 1 ■ **Determining If an Identity Exists**

Determine if each of the following equations is an identity.

a. $(x - 6)(x + 2) = x^2 - 4x - 12$
b. $4(x - 2) = 2(2x + 6)$

Solution To be an identity, the expressions on either side of the equal sign must be equivalent for all values of x. To verify if the equation is an identity, we simplify each expression as much as possible.

a. We have

$$(x - 6)(x + 2) = x^2 - 4x - 12$$
$$x^2 + 2x - 6x - 12 = x^2 - 4x - 12$$
$$x^2 - 4x - 12 = x^2 - 4x - 12$$

Since the simplified expression on the left side of the equal sign is the same as the expression on the right side, the equation $(x - 6)(x + 2) = x^2 - 4x - 12$ is an identity.

b. We have

$$4(x - 2) = 2(2x + 6)$$
$$4x - 8 = 4x + 12$$

Multiplying a number by 4 and then subtracting 8 is not the same as multiplying a number by 4 and adding 12. Therefore, $4(x - 2) = 2(2x + 6)$ is *not* an identity.

■ Fundamental Trigonometric Identities

In Chapters 8 and 9 we worked with some identities in trigonometry. For example, the definitions of tangent, cotangent, secant, and cosecant are all identities.

> **FUNDAMENTAL TRIGONOMETRIC IDENTITIES**
>
> $$\tan(\theta) = \frac{\sin(\theta)}{\cos(\theta)} \qquad \cot(\theta) = \frac{\cos(\theta)}{\sin(\theta)}$$
>
> $$\sec(\theta) = \frac{1}{\cos(\theta)} \qquad \csc(\theta) = \frac{1}{\sin(\theta)}$$

Trigonometric identities often make it easier to evaluate a trigonometric expression or solve a trigonometric equation.

EXAMPLE 2 ■ Using Identities to Solve a Trigonometric Equation

Solve $x = \tan(\theta)\csc(\theta)\cos(\theta)$ for x and write the result in simplified form.

Solution Since x is already isolated, we only need to write the result in simplified form. To do this, we rewrite $\tan(\theta)$ and $\csc(\theta)$ using the fundamental trigonometric identities.

$$x = \tan(\theta)\csc(\theta)\cos(\theta)$$
$$= \left[\frac{\sin(\theta)}{\cos(\theta)}\right]\left[\frac{1}{\sin(\theta)}\right][\cos(\theta)]$$
$$= \frac{\sin(\theta)\cos(\theta)}{\sin(\theta)\cos(\theta)}$$
$$= 1$$

Thus $x = \tan(\theta)\csc(\theta)\cos(\theta)$ is equivalent to $x = 1$.

Trigonometric identities allow us to rewrite trigonometric expressions in a simpler form to use when working with expressions or equations.

EXAMPLE 3 ■ Using Fundamental Identities

When an object is at rest, all of the forces acting on it are in balance. The forces acting on a roof tile resting on a sloped roof are related in the formula $w \cdot \sin(\theta) = \mu_s \cdot w \cdot \cos(\theta)$, where w is the weight of the roof tile, θ is the measure of the roof's angle of incline, and μ_s is the *frictional coefficient* for two surfaces. See Figure 10.1. Use a fundamental trigonometric identity to explain how the frictional coefficient is related to the angle of incline for the roof.

Figure 10.1

Solution We begin by solving the formula $w \cdot \sin(\theta) = \mu_s \cdot w \cdot \cos(\theta)$ for μ_s.

$$w \cdot \sin(\theta) = \mu_s \cdot w \cdot \cos(\theta)$$
$$\frac{w \cdot \sin(\theta)}{w \cdot \cos(\theta)} = \frac{\mu_s \cdot w \cdot \cos(\theta)}{w \cdot \cos(\theta)}$$
$$\frac{\sin(\theta)}{\cos(\theta)} = \mu_s$$

We use the fundamental identity $\tan(\theta) = \dfrac{\sin(\theta)}{\cos(\theta)}$ to write μ_3 in terms of a single trigonometric function: $\tan(\theta) = \mu_s$.

Thus, since $\tan(\theta)$ always calculates a slope, the frictional coefficient is equivalent to the slope of the roof with an angle of incline measuring θ. If the frictional coefficient is known for two surfaces, such as wood and tile, we can use this relationship to determine the angle at which an object will begin to slide down the slope. We will explore this concept in the exercise set.

■ Even–Odd Function Identities

In Chapter 3, we explained that even functions are symmetric with respect to the vertical axis. That is, their graphs remain unchanged after being reflected horizontally about the vertical axis. Functions with even symmetry have the property that $f(-x) = f(x)$.

As we consider the graph of the cosine function shown in Figure 10.2, we see that reflecting this graph about the vertical axis will result in the same graph. Therefore, $f(\theta) = \cos(\theta)$ is an even function and $\cos(-\theta) = \cos(\theta)$.

Odd functions are symmetric to the origin. This means the function's graph reflected horizontally will be identical to the function's graph reflected vertically. Functions with odd symmetry thus have the property $f(-x) = -f(x)$.

As we consider the graph of the sine function (blue curve) and the negative sine function (red dashed curve) shown in Figure 10.3, we see that reflecting the sine graph about the vertical axis will result in the same negative sine graph. Thus sine is an odd function and $\sin(-\theta) = -\sin(\theta)$.

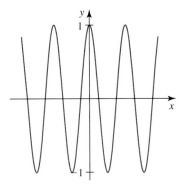

Figure 10.2   Figure 10.3

We use the fundamental trigonometric identities and our knowledge about the even nature of the cosine function and the odd nature of the sine function to determine which of the four remaining trigonometric functions are even and odd.

$$\tan(-\theta) = \frac{\sin(-\theta)}{\cos(-\theta)}$$

$$= \frac{-\sin(\theta)}{\cos(\theta)} \quad \text{since } \cos(-\theta) = \cos(\theta) \text{ and } \sin(-\theta) = -\sin(\theta)$$

$$= -\tan(\theta)$$

Since $\tan(-\theta) = -\tan(\theta)$, tangent is an odd function. It follows that the cotangent function is odd since

$$\cot(-\theta) = \frac{1}{\tan(-\theta)}$$

$$= \frac{1}{-\tan\theta}$$

$$= -\cot(\theta)$$

Similarly, $\sec(\theta)$ is even and $\csc(\theta)$ is odd because

$$\sec(-\theta) = \frac{1}{\cos(-\theta)} \qquad \csc(-\theta) = \frac{1}{\sin(-\theta)}$$

$$= \frac{1}{\cos(\theta)} \qquad\qquad = \frac{1}{-\sin(\theta)}$$

$$= \sec(\theta) \qquad\qquad = -\csc(\theta)$$

We summarize these **even–odd identities** as follows.

EVEN–ODD IDENTITIES

Cosine and secant functions are **even** functions, giving us the following identities.

$$\cos(-\theta) = \cos(\theta) \qquad \sec(-\theta) = \sec(\theta)$$

Sine, cosecant, tangent, and cotangent are **odd** functions, giving us the following identities.

$$\sin(-\theta) = -\sin(\theta) \qquad \tan(-\theta) = -\tan(\theta)$$
$$\csc(-\theta) = -\csc(\theta) \qquad \cot(-\theta) = -\cot(\theta)$$

EXAMPLE 4 ■ Using Even–Odd Identities to Evaluate Expressions

Evaluate each of the following expressions using even–odd identities.

a. $\cos\left(-\dfrac{\pi}{4}\right)$

b. $\csc\left(-\dfrac{7\pi}{3}\right)$

Solution

a. $\cos\left(-\dfrac{\pi}{4}\right) = \cos\left(\dfrac{\pi}{4}\right)$

$= \dfrac{\sqrt{2}}{2}$

b. $\csc\left(-\dfrac{7\pi}{3}\right) = -\csc\left(\dfrac{7\pi}{3}\right)$

A radian measure of $\dfrac{7\pi}{3}$ indicates more than one revolution around the circle. Thus, we subtract 2π to find its corresponding measure between 0 and 2π.

$$\dfrac{7\pi}{3} - 2\pi = \dfrac{7\pi}{3} - \dfrac{6\pi}{3}$$
$$= \dfrac{\pi}{3}$$

We now use $\dfrac{\pi}{3}$ in place of $\dfrac{7\pi}{3}$ in the equation.

$$-\csc\left(\dfrac{7\pi}{3}\right) = -\csc\left(\dfrac{\pi}{3}\right)$$
$$= -\dfrac{2\sqrt{3}}{3}$$

EXAMPLE 5 ■ Using Even–Odd Identities to Solve Equations

Find all solutions to each of the following equations over the interval $[0, 2\pi]$.

a. $\tan(-\theta)\csc(\theta) = 1.4$

b. $\dfrac{3\sin(-\theta)}{\sin(\theta)} = 5\cot(\theta)$

Solution

a. $\tan(-\theta)\csc(\theta) = 1.4$

$-\tan(\theta)\csc(\theta) = 1.4$ $\tan(-\theta) = -\tan(\theta)$

$-\dfrac{\sin(\theta)}{\cos(\theta)} \cdot \dfrac{1}{\sin(\theta)} = 1.4$ $\tan(\theta) = \dfrac{\sin(\theta)}{\cos(\theta)}$ and $\csc(\theta) = \dfrac{1}{\sin(\theta)}$

$-\dfrac{\sin(\theta)}{\cos(\theta)\sin(\theta)} = 1.4$

$-\dfrac{\cancel{\sin(\theta)}}{\cos(\theta)\cancel{\sin(\theta)}} = 1.4$

$-\dfrac{1}{\cos(\theta)} = 1.4$

$-1 = 1.4\cos(\theta)$

$-\dfrac{1}{1.4} = \cos(\theta)$

$\theta = \cos^{-1}\left(-\dfrac{1}{1.4}\right)$ Write in inverse form.

$\theta \approx 2.3664$ One solution is between $\dfrac{\pi}{2}$ and π.

$\theta \approx 3.9168$ One solution is between π and $\dfrac{3\pi}{2}$ (use reference angles).

b. $\dfrac{3\sin(-\theta)}{\sin(\theta)} = 5\cot(\theta)$

$\dfrac{-3\sin(\theta)}{\sin(\theta)} = 5\cot(\theta)$ $\sin(-\theta) = -\sin(\theta)$

$\dfrac{-3\cancel{\sin(\theta)}}{\cancel{\sin(\theta)}} = 5\cot(\theta)$

$-3 = 5\cot(\theta)$

$-\dfrac{3}{5} = \cot(\theta)$

$-\dfrac{3}{5} = \dfrac{1}{\tan(\theta)}$ $\cot(\theta) = \dfrac{1}{\tan(\theta)}$

$-\dfrac{5}{3} = \tan(\theta)$ Use the reciprocal of each side.

$\theta = \tan^{-1}\left(-\dfrac{5}{3}\right)$ Write in the inverse form.

$\theta \approx -1.0304$ Evaluate with a calculator.

$\theta \approx -1.0304 + \pi$ One solution is between $\dfrac{\pi}{2}$ and π.

≈ 2.1112

$\theta \approx -1.0304 + 2\pi$ One solution is between $\dfrac{3\pi}{2}$ and 2π.

≈ 5.2528

Cofunction Identities

Cofunction identities show relationships between pairs of functions, such as the cosine and sine functions and the tangent and cotangent functions. We develop the cofunction identities for cosine and sine here; the remaining four are left to the exercises.

First, we consider a right triangle with acute angles measuring θ and ϕ (read "fee" or "fy"), shown in Figure 10.4. Using the definitions of cosine and sine in right triangle trigonometry, we find the following relationships.

Figure 10.4

$$\cos(\theta) = \frac{x}{r} \qquad \sin(\phi) = \frac{x}{r}$$

$$\cos(\phi) = \frac{y}{r} \qquad \sin(\theta) = \frac{y}{r}$$

This tells us the cosine and sine values of the two acute angles are related as follows.

$$\cos(\theta) = \sin(\phi) \qquad \text{since } \cos(\theta) = \frac{x}{r} \text{ and } \sin(\phi) = \frac{x}{r}$$

$$\sin(\theta) = \cos(\phi) \qquad \text{since } \cos(\phi) = \frac{y}{r} \text{ and } \sin(\theta) = \frac{y}{r}$$

Since the two acute angles in a right triangle add up to 90°, we can rewrite ϕ in terms of θ:

$$\theta + \phi = 90°$$
$$\phi = 90° - \theta$$

Thus

$$\cos(\theta) = \sin(\phi) \qquad \sin(\theta) = \cos(\phi)$$
$$\qquad = \sin(90° - \theta) \qquad \qquad = \cos(90° - \theta)$$

The six **cofunction identities** are summarized in the following box.

COFUNCTION IDENTITIES

θ in degrees	θ in radians
$\cos(\theta) = \sin(90° - \theta)$	$\cos(\theta) = \sin\left(\dfrac{\pi}{2} - \theta\right)$
$\sin(\theta) = \cos(90° - \theta)$	$\sin(\theta) = \cos\left(\dfrac{\pi}{2} - \theta\right)$
$\tan(\theta) = \cot(90° - \theta)$	$\tan(\theta) = \cot\left(\dfrac{\pi}{2} - \theta\right)$
$\sec(\theta) = \csc(90° - \theta)$	$\sec(\theta) = \csc\left(\dfrac{\pi}{2} - \theta\right)$
$\csc(\theta) = \sec(90° - \theta)$	$\csc(\theta) = \sec\left(\dfrac{\pi}{2} - \theta\right)$
$\cot(\theta) = \tan(90° - \theta)$	$\cot(\theta) = \tan\left(\dfrac{\pi}{2} - \theta\right)$

SECTION 10.1 ■ BASIC IDENTITIES

EXAMPLE 6 ■ Using Cofunction Identities to Solve an Equation

Find all solutions to the equation $\cos\left(\dfrac{\pi}{2} - \theta\right) = 0.8$ over the interval $[0, 2\pi]$ using a cofunction identity.

Solution We substitute $\sin(\theta)$ for $\cos\left(\dfrac{\pi}{2} - \theta\right)$ to simplify the problem.

$$\cos\left(\frac{\pi}{2} - \theta\right) = 0.8$$

$$\sin(\theta) = 0.8 \qquad \cos\left(\frac{\pi}{2} - \theta\right) = \sin(\theta)$$

$$\theta = \sin^{-1}(0.8) \qquad \text{Write in inverse form.}$$

$$\theta \approx 0.9273 \qquad \text{Evaluate with a calculator.}$$

We use reference angles with sine to find the second solution in Quadrant II.

$$\theta \approx \pi - 0.9273$$
$$\approx 2.2143$$

Therefore, $\theta \approx 0.9273$ and $\theta \approx 2.2143$ are solutions to $\cos\left(\dfrac{\pi}{2} - \theta\right) = 0.8$.

EXAMPLE 7 ■ Using Cofunction Identities to Simplify an Expression

Use cofunction identities to write the expression $\cot\left(\dfrac{\pi}{2} - \theta\right)\sec\left(\dfrac{\pi}{2} - \theta\right)$ as a single trigonometric function.

Solution

$$\cot\left(\frac{\pi}{2} - \theta\right)\sec\left(\frac{\pi}{2} - \theta\right) = \tan(\theta)\csc(\theta) \qquad \text{since } \cot\left(\frac{\pi}{2} - \theta\right) = \tan(\theta) \text{ and } \sec\left(\frac{\pi}{2} - \theta\right) = \csc(\theta)$$

$$= \frac{\sin(\theta)}{\cos(\theta)} \cdot \frac{1}{\sin(\theta)} \qquad \text{since } \tan(\theta) = \frac{\sin(\theta)}{\cos(\theta)} \text{ and } \csc(\theta) = \frac{1}{\sin(\theta)}$$

$$= \frac{\cancel{\sin(\theta)}}{\cos(\theta)\cancel{\sin(\theta)}}$$

$$= \frac{1}{\cos(\theta)}$$

$$= \sec(\theta)$$

JUST IN TIME ■ EXPONENTS AND TRIGONOMETRIC FUNCTIONS

When a trigonometric function is raised to an exponent, such as $[\cos(\theta)]^2$ and $[\tan(\theta)]^4$, we typically use the notation $\cos^2(\theta)$ and $\tan^4(\theta)$ to avoid confusion with $\cos(\theta)^2$ and $\tan(\theta)^4$, which are equivalent to $\cos(\theta^2)$ and $\tan(\theta^4)$.

Pythagorean Identities

We know that the (x, y) coordinates of the endpoint of an arc on the unit circle are given by $(\cos(\theta), \sin(\theta))$. If we draw a right triangle using the radius of the unit circle as the hypotenuse as shown in Figure 10.5a, we see that the lengths of the legs are also given by $\cos(\theta)$ and $\sin(\theta)$.

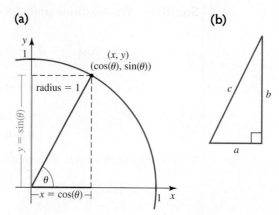

Figure 10.5

Comparing Figures 10.5a and 10.5b and using the Pythagorean theorem, we see the following relationship between cosine and sine.

$$a^2 + b^2 = c^2 \quad \text{Pythagorean theorem}$$
$$\cos^2(\theta) + \sin^2(\theta) = 1 \quad a = \cos(\theta), b = \sin(\theta), \text{ and } c = 1$$

The equation $\cos^2(\theta) + \sin^2(\theta) = 1$ is called a **Pythagorean identity**.

EXAMPLE 8 ■ Using a Pythagorean Identity to Solve an Equation

Find all solutions to the equation $\cos^2(\theta) + \cos^2(\theta)\tan^2(\theta) = 2.6 \cos(\theta)$ over the interval $[0, 2\pi]$.

Solution One problem-solving technique is to write all trigonometric functions in terms of cosine and sine. This will often simplify the problem.

$$\cos^2(\theta) + \cos^2(\theta)\tan^2(\theta) = 2.6 \cos(\theta)$$

$$\cos^2(\theta) + \frac{\cos^2(\theta)}{1} \cdot \frac{\sin^2(\theta)}{\cos^2(\theta)} = 2.6 \cos(\theta) \quad \text{since } \tan^2(\theta) = \frac{\sin^2(\theta)}{\cos^2(\theta)}$$

$$\cos^2(\theta) + \frac{\cancel{\cos^2(\theta)}\sin^2(\theta)}{\cancel{\cos^2(\theta)}} = 2.6 \cos(\theta)$$

$$\cos^2(\theta) + \sin^2(\theta) = 2.6 \cos(\theta)$$

$$1 = 2.6 \cos(\theta) \quad \text{since } \cos^2(\theta) + \sin^2(\theta) = 1$$

$$0.3846 \approx \cos(\theta)$$

$$\theta \approx \cos^{-1}(0.3846)$$

$$\theta \approx 1.1760 \quad \text{Evaluate using a calculator.}$$

$$\theta \approx 2\pi - 1.1760 \quad \text{Use reference angles.}$$

$$\approx 5.1072$$

Using the basic Pythagorean identity $\cos^2(\theta) + \sin^2(\theta) = 1$, we establish the other two Pythagorean identities. First, we see what happens when we divide the Pythagorean identity $\cos^2(\theta) + \sin^2(\theta) = 1$ by $\cos^2(\theta)$.

$$\cos^2(\theta) + \sin^2(\theta) = 1$$

$$\frac{\cos^2(\theta) + \sin^2(\theta)}{\cos^2(\theta)} = \frac{1}{\cos^2(\theta)} \quad \text{Divide each side by } \cos^2(\theta).$$

$$\frac{\cos^2(\theta)}{\cos^2(\theta)} + \frac{\sin^2(\theta)}{\cos^2(\theta)} = \frac{1}{\cos^2(\theta)} \quad \text{Separate the fraction.}$$

$$1 + \tan^2(\theta) = \sec^2(\theta) \quad \frac{\sin^2(\theta)}{\cos^2(\theta)} = \tan^2(\theta) \text{ and } \frac{1}{\cos^2(\theta)} = \sec^2(\theta)$$

This gives us the identity $1 + \tan^2(\theta) = \sec^2(\theta)$. This identity is true for all values of θ except those that make $\cos(\theta) = 0$, in which case both sides of the identity are undefined.

The third Pythagorean identity occurs when we divide the Pythagorean identity $\cos^2(\theta) + \sin^2(\theta) = 1$ by $\sin^2(\theta)$.

$$\cos^2(\theta) + \sin^2(\theta) = 1$$

$$\frac{\cos^2(\theta) + \sin^2(\theta)}{\sin^2(\theta)} = \frac{1}{\sin^2(\theta)} \quad \text{Divide each side by } \sin^2(\theta).$$

$$\frac{\cos^2(\theta)}{\sin^2(\theta)} + \frac{\sin^2(\theta)}{\sin^2(\theta)} = \frac{1}{\sin^2(\theta)} \quad \text{Separate the fraction.}$$

$$\cot^2(\theta) + 1 = \csc^2(\theta) \quad \frac{\cos^2(\theta)}{\sin^2(\theta)} = \cot^2(\theta) \text{ and } \frac{1}{\sin^2(\theta)} = \csc^2(\theta)$$

The identity $\cot^2(\theta) + 1 = \csc^2(\theta)$ will be true for all values of θ except those that make $\sin(\theta) = 0$, in which case both sides of the identity are undefined.

We summarize the **Pythagorean identities** in the following box.

PYTHAGOREAN IDENTITIES

$$\cos^2(\theta) + \sin^2(\theta) = 1 \qquad 1 + \tan^2(\theta) = \sec^2(\theta) \qquad \cot^2(\theta) + 1 = \csc^2(\theta)$$

These identities can appear in various forms. For example, $\cos^2(\theta) + \sin^2(\theta) = 1$ may be written as $\sin^2(\theta) = 1 - \cos^2(\theta)$ or $\cos^2(\theta) = 1 - \sin^2(\theta)$, while $\cot^2(\theta) + 1 = \csc^2(\theta)$ may be written as $\cot^2(\theta) = \csc^2(\theta) - 1$ or $1 - \csc^2(\theta) = -\cot^2(\theta)$. Each of these alternative forms is valid.

EXAMPLE 9 ■ **Using Pythagorean Identities to Simplify Expressions**

Use trigonometric identities to write each of the following as a single trigonometric function or as a whole number.

a. $[1 - \cos^2(\theta)][\cot^2(\theta)]$

b. $[\sec(\theta) + \tan(\theta)][\sec(\theta) - \tan(\theta)]$

Solution

a. We notice $1 - \cos^2(\theta)$ is present in this expression. We also note that solving the identity $\cos^2(\theta) + \sin^2(\theta) = 1$ for $\sin^2(\theta)$ yields $\sin^2(\theta) = 1 - \cos^2(\theta)$. Therefore,

$$[1 - \cos^2(\theta)][\cot^2(\theta)] = \sin^2(\theta)\cot^2(\theta) \quad \text{since } \sin^2(\theta) = 1 - \cos^2(\theta)$$

$$= \frac{\sin^2(\theta)}{1} \cdot \frac{\cos^2(\theta)}{\sin^2(\theta)} \quad \text{since } \cot^2(\theta) = \frac{\cos^2(\theta)}{\sin^2(\theta)}$$

$$= \frac{\sin^2(\theta)\cos^2(\theta)}{\sin^2(\theta)}$$

$$= \cos^2(\theta)$$

b.
$$[\sec(\theta) + \tan(\theta)][\sec(\theta) - \tan(\theta)]$$
$$\sec(\theta) \cdot \sec(\theta) - \sec(\theta) \cdot \tan(\theta) + \tan(\theta) \cdot \sec(\theta) - \tan(\theta) \cdot \tan(\theta)$$
$$\sec^2(\theta) - \sec(\theta)\tan(\theta) + \sec(\theta)\tan(\theta) - \tan^2(\theta)$$
$$\sec^2(\theta) - \tan^2(\theta)$$
$$1 \quad \sec^2(\theta) - \tan^2(\theta) = 1 \text{ since } 1 + \tan^2(\theta) = \sec^2(\theta)$$

So $[\sec(\theta) + \tan(\theta)][\sec(\theta) - \tan(\theta)] = 1$.

SUMMARY

In this section you learned the definition of a mathematical identity. You also learned the fundamental trigonometric identities, cofunction identities, even–odd identities, and Pythagorean identities and saw how to use them to solve trigonometric equations or simplify trigonometric expressions.

10.1 EXERCISES

SKILLS AND CONCEPTS

1. Make a list of all of the trigonometric identities covered in this section. This will be a helpful reference throughout the chapter that you can refer to easily while working on other exercises. You may also want to include alternative versions of the Pythagorean identities. For example, the identity $\cos^2(\theta) + \sin^2(\theta) = 1$ can be written as $\sin^2(\theta) = 1 - \cos^2(\theta)$, $\cos^2(\theta) = 1 - \sin^2(\theta)$, $\sin^2(\theta) - 1 = -\cos^2(\theta)$, and $\cos^2(\theta) - 1 = -\sin^2(\theta)$.

In Exercises 2–7, use fundamental trigonometric identities to simplify each expression by writing it as a single trigonometric function or as a whole number.

2. $\tan(\theta)\cos(\theta)$

3. $\tan(x)\csc(x)\sec(x)$

4. $\dfrac{\sec(x)}{\tan(x)}$

5. $\dfrac{\sec(\beta)}{\csc(\beta)}$

6. $\dfrac{\cos(\theta)}{\tan(\theta)} \cdot \sin(\theta)$

7. $\dfrac{\sin(\theta)\sec(\theta)}{\tan(\theta)}$

In Exercises 8–11, use fundamental trigonometric identities to solve each equation. Find all solutions over the interval $[0, 2\pi]$. Verify your solutions by graphing on a graphing calculator.

8. $\cot(\theta)\sin(\theta) = 0.75$

9. $\csc(x)\sin^2(x) = 0.19$

10. $-2\sec(\alpha) = \csc(\alpha)$

11. $\tan^2(\theta) = 0.33 \sec^2(\theta)$

In Exercises 12–17, use even–odd identities to simplify each expression by writing it in terms of a single trigonometric function with a positive angle or as a whole number.

12. $\cos(-\theta)$

13. $\dfrac{\cos(-\theta)}{\sin(-\theta)}$

14. $\tan(-\beta)\cos(-\beta)$

15. $\cot(\theta)\csc(-\theta)\sec(-\theta)$

16. $\csc^2(-x)$

17. $\dfrac{\sin^2(-\phi)}{\tan^2(\phi)}$

In Exercises 18–21, use even–odd identities to solve each equation. Find all solutions over the interval $[0, 2\pi]$. Verify your solutions by graphing on a graphing calculator.

18. $\sin(-\theta) = 0.73$
19. $-3.95 = \sec(-x) - 2$
20. $-3\left(\dfrac{\sin(-\theta)}{\cos(-\theta)}\right) = 4.12$
21. $-9.1 = \dfrac{1}{\sin(-\phi)} - 3.5$

In Exercises 22–27, use cofunction identities to simplify each expression by writing it in terms of a single trigonometric function with an angle measuring θ or as a whole number.

22. $\sin\left(\dfrac{\pi}{2} - \theta\right)$
23. $\csc\left(\dfrac{\pi}{2} - z\right)$
24. $\cos(90° - \theta)\sec(\theta)$
25. $\sin\left(\dfrac{\pi}{2} - \theta\right)\tan(\theta)$
26. $\dfrac{\sin\left(\dfrac{\pi}{2} - \theta\right)}{\cos\left(\dfrac{\pi}{2} - \theta\right)}$
27. $\dfrac{\tan\left(\dfrac{\pi}{2} - m\right)}{\sin\left(\dfrac{\pi}{2} - m\right)}$

In Exercises 28–31, use cofunction identities to solve each equation. Find all solutions over the interval $[0, 2\pi]$. Verify your solutions by graphing on a graphing calculator.

28. $\cos\left(\dfrac{\pi}{2} - \theta\right) = -0.61$
29. $4 = \csc\left(\dfrac{\pi}{2} - \theta\right) + 1$
30. $-1 = \cos(\theta)\cot\left(\dfrac{\pi}{2} - \theta\right)$
31. $-5\cot\left(\dfrac{\pi}{2} - x\right)\sec\left(\dfrac{\pi}{2} - x\right) - 3.5 = -10$

In Exercises 32–35, evaluate each pair of expressions. (Hint: Refer to the Just In Time box in this section.)

32. $\cos^2\left(\dfrac{\pi}{5}\right)$ and $\cos\left(\dfrac{\pi}{5}\right)^2$
33. $\sin\left(\dfrac{11\pi}{7}\right)^2$ and $\sin^2\left(\dfrac{11\pi}{7}\right)$
34. $\tan^3\left(\dfrac{\pi}{3}\right)$ and $\tan\left(\dfrac{\pi}{3}\right)^3$
35. $\cot^4\left(\dfrac{2\pi}{3}\right)$ and $\cot\left(\dfrac{2\pi}{3}\right)^4$

In Exercises 36–41, use Pythagorean identities to simplify each expression.

36. $\sec^2(\theta) - \tan^2(\theta)$
37. $\sec^2(\alpha)[1 - \sin^2(\alpha)]$
38. $\cos^2(\theta) + \tan^2(\theta)\cos^2(\theta)$
39. $\dfrac{\sec^2(\phi) - 1}{\sin^2(\phi)}$
40. $[\csc(\theta) + \cot(\theta)][\csc(\theta) - \cot(\theta)]$
41. $[\sec(\theta) + 1][\sec(\theta) - 1]$

In Exercises 42–45, use Pythagorean identities to solve each equation. Find all solutions over the interval $[0, 2\pi]$. Verify your solutions by graphing on a graphing calculator.

42. $[1 - \sin^2(x)][\sec(x)] = -0.1$
43. $\dfrac{\cos(\theta)}{1 - \sin^2(\theta)} + 2 = -1$
44. $\sin(w) = 2.7[1 - \cos^2(w)]$
45. $1 - \sin^2(\theta)\cot^2(\theta) = 0.8$

In Exercises 46–48, complete the indicated tasks and answer any given questions.

46. The figure shows the graphs of cosine and sine.

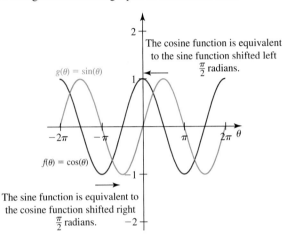

Use the information in the figure to write two cofunction identities that relate the cosine and sine functions.

47. Use graphs and/or diagrams to verify that the secant function has even symmetry while the cosecant and cotangent functions have odd symmetry.

48. In calculus, Pythagorean identities are very useful for rewriting certain expressions to make calculations easier. For example, consider the expressions $\sqrt{1 + x^2}$, $\sqrt{1 - x^2}$, and $\sqrt{x^2 - 1}$. A *trigonometric substitution* is when the x-value is replaced by $\sin(\theta)$, $\tan(\theta)$, or $\sec(\theta)$, and the expression is then simplified.

   a. Which of the three trigonometric values should be substituted for x in each of the three expressions $\sqrt{1 + x^2}$, $\sqrt{1 - x^2}$, and $\sqrt{x^2 - 1}$ to simplify them? Make the substitutions, then simplify each expression.

   b. What would you substitute for x in the expression $\sqrt{16 - x^2}$? Perform the substitution and simplify the expression.

In Exercises 49–52, solve each trigonometric equation by factoring. Find all solutions over the interval $[0, 2\pi]$. (Hint: Start by performing a substitution, such as letting $x = \cos(\theta)$ or $x = \sin(\theta)$.) Verify your solutions by graphing on a graphing calculator.

49. $\cos^2(\theta) + 4\cos(\theta) - 5 = 0$
50. $2\sin^2(\theta) - \sin(\theta) = 1$
51. $3\sin^2(\theta) - 10\sin(\theta) = 6\sin(\theta) - 12$
52. $3\sin^2(\theta) - 2\cos(\theta) = 3\cos(\theta) + 1$

SHOW YOU KNOW

53. What are the differences between an expression, an equation, and an identity?
54. How do you determine if an equation is an identity?
55. Why are trigonometric identities useful in solving equations?
56. Explain why the cofunction identities $\tan\left(\frac{\pi}{2} - \theta\right) = \cot(\theta)$ and $\cot\left(\frac{\pi}{2} - \theta\right) = \tan(\theta)$ are true. Use graphs and/or diagrams in your explanation.
57. Explain why the cofunction identities $\sec\left(\frac{\pi}{2} - \theta\right) = \csc(\theta)$ and $\csc\left(\frac{\pi}{2} - \theta\right) = \sec(\theta)$ are true. Use graphs and/or diagrams in your explanation.

MAKE IT REAL

In Exercises 58–61, refer to Example 3.

58. **Friction** A brick resting on wood has a frictional coefficient of 0.6. What is the minimum angle measure at which the brick will slide down a wooden surface? (*Source:* www.roymech.co.uk)
59. **Friction** A copper object resting on a steel surface has a frictional coefficient of 0.53. What is the minimum angle measure at which the object will slide down a steel surface? (*Source:* frictioncenter.siu.edu)
60. **Friction** A rubber object resting on concrete has a frictional coefficient that varies from 1.0 to 4.0 depending on conditions such as moisture. (*Source:* frictioncenter.siu.edu)
    a. Find the range of angles for which a rubber object would begin to slide down a concrete slope.
    b. Based on your results to part (a), explain at least one reason that rubber is used for the soles of running shoes.
61. **Friction** Following are the frictional coefficients for a number of materials and surfaces. Which of these objects would remain stationary on a 20° incline? (*Source:* www.roymech.co.uk)
    A. A Teflon-coated object on a steel surface: $\mu_s = 0.04$
    B. A leather object resting on an oak surface: $\mu_s = 0.61$
    C. A Plexiglas object resting on a steel surface: $\mu_s = 0.4$
    D. A steel object on a polystyrene surface: $\mu_s = 0.35$

STRETCH YOUR MIND

Exercises 62–66 are intended to challenge your understanding of simplifying trigonometric expressions and solving trigonometric equations.

62. Show that $\dfrac{1}{\sin(\theta) + 1} + \dfrac{1}{\csc(\theta) + 1}$ simplifies to a whole number.
63. Rewrite the expression $\dfrac{\sin(\pi - \theta)}{\cos(\theta - \pi)}$ using a single trigonometric function.
64. Rewrite the expression $\dfrac{1 - 2\sin^2(\theta)}{\sin(\theta)\cos(\theta)} + \tan(\theta)$ using a single trigonometric function.
65. Find all solutions to the equation $\cos^4(\theta) + 2\cos^2(\theta)\sin^2(\theta) + \sin^4(\theta) = 1$ over the interval $[-2\pi, 2\pi]$.
66. Find all solutions to the equation $\dfrac{1 + \sin(\theta)}{\cos(\theta)} + \dfrac{\cos(\theta)}{1 + \sin(\theta)} = -4$ over the interval $[-4\pi, 0]$.

SECTION 10.2 Verifying Identities

LEARNING OBJECTIVES

- Explain how trigonometric identities are determined
- Use algebraic manipulation to verify the equivalency of two identities
- Use trigonometric identities to describe equivalent relationships

GETTING STARTED

In Chapter 8 we defined cosine and sine as "lengths" associated with the unit circle. In Chapter 9 we discovered how cosine and sine are related to right triangles. We used this knowledge to solve for missing angles and side lengths in triangles. In addition, we learned about four other trigonometric functions: tangent, secant, cosecant, and cotangent.

In this section we use the identities introduced in Section 10.1 to determine if two trigonometric expressions are equivalent. We also see that all trigonometric functions may be represented as lengths associated with a unit circle, as shown in Figure 10.6, and that these relationships can be proved using trigonometric identities.

Figure 10.6

Verifying Identities

The process of verifying the equivalence of two expressions is different from the process of solving equations. When solving equations, we move quantities back and forth across the equal sign using mathematical operations. When verifying identities, we seek to make one side of the equation look like the other side of the equation, but we may not move any quantities across the equal sign because we do not *know* the two sides are equivalent—this is what we want to show. Instead, we rewrite the expression on one side of the equal sign using known identities until we succeed in making it look like the other side of the equal sign. This process is difficult for many people because a certain path to the solution is often unclear. The key to success is to recognize that if a chosen strategy begins to make an expression more complex instead of simpler, a restart using a different strategy may be needed. In fact, it is common to restart the simplification process multiple times in a single problem. This should not be viewed as a failure. It is part of the learning process. With practice, anyone can become skilled in verifying identities.

EXAMPLE 1 ■ Verifying Identities

Verify each identity.

a. $\cot(\theta)\tan(\theta) = 1$

b. $\csc(\theta)\tan(\theta) = \cos(\theta)\sec^2(\theta)$

Solution

a. To verify an identity means to show that the expressions on either side of the equal sign are equivalent. Thus, we want to show that, for *any* input for which the expressions are defined, the equation will always be true.

$$\cot(\theta)\tan(\theta) \stackrel{?}{=} 1$$

$$\frac{\cos(\theta)}{\sin(\theta)} \cdot \frac{\sin(\theta)}{\cos(\theta)} \stackrel{?}{=} 1 \qquad \text{Rewrite using cosine and sine.}$$

$$\frac{\cancel{\cos(\theta)} \cdot \cancel{\sin(\theta)}}{\cancel{\cos(\theta)} \cdot \cancel{\sin(\theta)}} \stackrel{?}{=} 1$$

$$1 = 1$$

The expression $\cot(\theta)\tan(\theta)$ will always be 1 for any angle where it is defined. Thus $\cot(\theta)\tan(\theta) = 1$ is an identity.

b. We use identities from Section 10.1 to simplify the right side of $\csc(\theta)\tan(\theta) = \cos(\theta)\sec^2(\theta)$.

$$\csc(\theta)\tan(\theta) \stackrel{?}{=} \cos(\theta)\sec^2(\theta)$$

$$\csc(\theta)\tan(\theta) \stackrel{?}{=} \cos(\theta) \cdot \frac{1}{\cos^2(\theta)} \qquad \text{since } \sec^2(\theta) = \frac{1}{\cos^2(\theta)}$$

$$\csc(\theta)\tan(\theta) \stackrel{?}{=} \frac{\cos(\theta)}{\cos^2(\theta)}$$

$$\csc(\theta)\tan(\theta) \stackrel{?}{=} \frac{\cancel{\cos(\theta)}}{\cos(\theta)\cancel{\cos(\theta)}} \qquad \text{since } \cos^2(\theta) = \cos(\theta)\cos(\theta)$$

$$\csc(\theta)\tan(\theta) \stackrel{?}{=} \frac{1}{\cos(\theta)}$$

$$\csc(\theta)\tan(\theta) \stackrel{?}{=} \sec(\theta) \qquad \text{since } \sec(\theta) = \frac{1}{\cos(\theta)}$$

We have simplified the right side of the potential identity by writing it in terms of a single trigonometric function, but we have yet to show that the two sides are equivalent. We now simplify the left side.

$$\csc(\theta)\tan(\theta) \stackrel{?}{=} \sec(\theta)$$

$$\frac{1}{\sin(\theta)} \cdot \frac{\sin(\theta)}{\cos(\theta)} \stackrel{?}{=} \sec(\theta) \qquad \text{Use fundamental trigonometric identities.}$$

$$\frac{\cancel{\sin(\theta)}}{\cancel{\sin(\theta)} \cdot \cos(\theta)} \stackrel{?}{=} \sec(\theta)$$

$$\frac{1}{\cos(\theta)} \stackrel{?}{=} \sec(\theta)$$

$$\sec(\theta) = \sec(\theta) \qquad \text{since } \sec(\theta) = \frac{1}{\cos(\theta)}$$

Thus $\csc(\theta)\tan(\theta) = \cos(\theta)\sec^2(\theta)$ is an identity.

Notice we did not move any quantities across the equal sign as we verified the identity. Instead, we simplified each side as far as possible. There are no definitive steps to take each time we are verifying an identity. However, some general guidelines will help us be efficient.

HOW TO: ■ VERIFY TRIGONOMETRIC IDENTITIES

To verify trigonometric identities,

1. Simplify one side at a time. It is often best to work with the side that appears more complex. Always stay on one side of the equation at a time.
2. Combine fractions on the same side using addition, subtraction, multiplication, or division as necessary.
3. If any of the trigonometric functions are being squared, you may be able to use a Pythagorean identity to simplify the expression. If there are no exponents, try writing all trigonometric functions in terms of cosine and sine.
4. Keep your goal in mind. Sometimes looking at the less complex side of the potential identity will give you a clue about what identities will help you get there.
5. If one approach does not work, try another. Often, it only takes one substitution or one key step to unlock the entire identity.

EXAMPLE 2 ■ Verifying an Identity

Verify the identity $\sec(\theta) - \tan(\theta)\sin(\theta) = \cos(\theta)$.

Solution We begin by rewriting the trigonometric functions on the left side using cosine and sine.

$$\sec(\theta) - \tan(\theta)\sin(\theta) \stackrel{?}{=} \cos(\theta)$$

$$\frac{1}{\cos(\theta)} - \frac{\sin(\theta)}{\cos(\theta)} \cdot \frac{\sin(\theta)}{1} \stackrel{?}{=} \cos(\theta) \quad \text{Rewrite using cosine and sine.}$$

$$\frac{1}{\cos(\theta)} - \frac{\sin^2(\theta)}{\cos(\theta)} \stackrel{?}{=} \cos(\theta) \quad \text{Multiply } \frac{\sin(\theta)}{\cos(\theta)} \cdot \frac{\sin(\theta)}{1}.$$

$$\frac{1 - \sin^2(\theta)}{\cos(\theta)} \stackrel{?}{=} \cos(\theta) \quad \text{Combine fractions.}$$

$$\frac{\cos^2(\theta)}{\cos(\theta)} \stackrel{?}{=} \cos(\theta) \quad \cos^2(\theta) = 1 - \sin^2(\theta), \text{ since } \cos^2(\theta) + \sin^2(\theta) = 1$$

$$\frac{\cancel{\cos(\theta)}\cos(\theta)}{\cancel{\cos(\theta)}} \stackrel{?}{=} \cos(\theta) \quad \text{since } \cos^2(\theta) = \cos(\theta)\cos(\theta)$$

$$\cos(\theta) = \cos(\theta)$$

We may also use graphing to help verify possible identities. However, as previously mentioned, graphs can sometimes be misleading. The only way to verify identities without any doubt is to use algebra.

EXAMPLE 3 ■ Verifying an Identity

Use a graph to determine if it is possible for each of the following to be an identity. If it is possible, verify the identity. If it is not, provide a counterexample to show the statement is not true.

a. $\dfrac{\sec(-\theta)}{\tan(-\theta)} = -\csc(\theta)$

b. $\cot^2(\theta)\cos^2(\theta) = \sin^2(\theta)$

Solution

a. We begin by graphing $y = \dfrac{\sec(-\theta)}{\tan(-\theta)}$ and $y = -\csc(\theta)$, as shown in Figure 10.7. It appears the two graphs coincide.

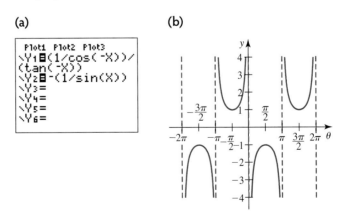

Figure 10.7

Thus, this may be an identity. We verify the identity using algebra, beginning with the more complex left side.

$$\frac{\sec(-\theta)}{\tan(-\theta)} \stackrel{?}{=} -\csc(\theta)$$

$$\frac{\sec(\theta)}{-\tan(\theta)} \stackrel{?}{=} -\csc(\theta) \quad \sec(-\theta) = \sec(\theta) \text{ and } \tan(-\theta) = -\tan(\theta)$$

$$\frac{\sec(\theta)}{-\tan(\theta)} \stackrel{?}{=} -\csc(\theta)$$

$$\frac{\dfrac{1}{\cos(\theta)}}{-\dfrac{\sin(\theta)}{\cos(\theta)}} \stackrel{?}{=} -\csc(\theta) \quad \text{Rewrite the left side using cosine and sine.}$$

$$\frac{1}{\cos(\theta)} \cdot -\frac{\cos(\theta)}{\sin(\theta)} \stackrel{?}{=} -\csc(\theta) \quad \text{To divide, multiply by the reciprocal.}$$

$$-\frac{\cancel{\cos(\theta)}}{\cancel{\cos(\theta)}\sin(\theta)} \stackrel{?}{=} -\csc(\theta)$$

$$-\frac{1}{\sin(\theta)} \stackrel{?}{=} -\csc(\theta)$$

$$-\csc(\theta) = -\csc(\theta)$$

b. We begin by graphing $y = \cot^2(\theta)\cos^2(\theta)$ and $y = \sin^2(\theta)$, as shown in Figure 10.8.

(a)   (b)

Figure 10.8

It is clear the expressions are not equivalent. We will provide a counterexample to prove the point. Let $\theta = \dfrac{\pi}{2}$.

$$\cot^2(\theta)\cos^2(\theta) \stackrel{?}{=} \sin^2(\theta)$$

$$\cot^2\left(\frac{\pi}{2}\right)\cos^2\left(\frac{\pi}{2}\right) \stackrel{?}{=} \sin^2\left(\frac{\pi}{2}\right)$$

$$(0)^2(0)^2 \stackrel{?}{=} (1)^2$$

$$0 \neq 1$$

SECTION 10.2 ■ VERIFYING IDENTITIES

EXAMPLE 4 ■ Verifying an Identity

Verify the identity $\cot^2(x) = \csc^2(x) - \cot(x)\tan(x)$.

Solution We begin with the right side since it is more complex. We will use a Pythagorean identity since we have some squared terms.

$$\cot^2(x) \stackrel{?}{=} \csc^2(x) - \cot(x)\tan(x)$$

$$\cot^2(x) \stackrel{?}{=} \csc^2(x) - \frac{\cos(x)}{\sin(x)} \cdot \frac{\sin(x)}{\cos(x)} \qquad \text{Rewrite using cosine and sine.}$$

$$\cot^2(x) \stackrel{?}{=} \csc^2(x) - \frac{\cancel{\cos(x)} \cdot \cancel{\sin(x)}}{\cancel{\cos(x)} \cdot \cancel{\sin(x)}}$$

$$\cot^2(x) \stackrel{?}{=} \csc^2(x) - 1$$

$$\cot^2(x) = \cot^2(x) \qquad \csc^2(x) - 1 = \cot^2(x) \text{ since } \cot^2(x) + 1 = \csc^2(x)$$

EXAMPLE 5 ■ Verifying an Identity

Verify the identity $\dfrac{\tan\left(\dfrac{\pi}{2} - \theta\right)}{\csc(\theta)} = \cos(\theta)$.

Solution We notice that the input of one of the trigonometric functions is $\dfrac{\pi}{2} - \theta$, indicating that we may use a cofunction identity. In this case, we use the identity $\tan\left(\dfrac{\pi}{2} - \theta\right) = \cot(\theta)$ to rewrite the expression on the left side.

$$\frac{\tan\left(\dfrac{\pi}{2} - \theta\right)}{\csc(\theta)} \stackrel{?}{=} \cos(\theta)$$

$$\frac{\cot(\theta)}{\csc(\theta)} \stackrel{?}{=} \cos(\theta) \qquad \tan\left(\dfrac{\pi}{2} - \theta\right) = \cot(\theta)$$

$$\frac{\dfrac{\cos(\theta)}{\sin(\theta)}}{\dfrac{1}{\sin(\theta)}} \stackrel{?}{=} \cos(\theta) \qquad \text{Rewrite in terms of cosine and sine.}$$

$$\frac{\cos(\theta)}{\sin(\theta)} \cdot \frac{\sin(\theta)}{1} \stackrel{?}{=} \cos(\theta) \qquad \text{To divide, multiply by the reciprocal.}$$

$$\frac{\cos(\theta)\cancel{\sin(\theta)}}{\cancel{\sin(\theta)}} \stackrel{?}{=} \cos(\theta)$$

$$\cos(\theta) = \cos(\theta)$$

EXAMPLE 6 ■ Verifying an Identity

Verify the identity $\csc(\beta) = \dfrac{\cot(\beta) + 1}{\cos(\beta) + \sin(\beta)}$.

Solution Our first step is to write the numerator in terms of cosine and sine.

$$\csc(\beta) \stackrel{?}{=} \frac{\cot(\beta) + 1}{\cos(\beta) + \sin(\beta)}$$

$$\csc(\beta) \stackrel{?}{=} \frac{\dfrac{\cos(\beta)}{\sin(\beta)} + 1}{\cos(\beta) + \sin(\beta)} \qquad \text{Write in terms of cosine and sine.}$$

$$\csc(\beta) \stackrel{?}{=} \frac{\dfrac{\cos(\beta)}{\sin(\beta)} + 1}{\cos(\beta) + \sin(\beta)}$$

$$\csc(\beta) \stackrel{?}{=} \frac{\dfrac{\cos(\beta)}{\sin(\beta)} + \dfrac{\sin(\beta)}{\sin(\beta)}}{\cos(\beta) + \sin(\beta)} \qquad \text{Get common denominators in the numerator.}$$

$$\csc(\beta) \stackrel{?}{=} \frac{\dfrac{\cos(\beta) + \sin(\beta)}{\sin(\beta)}}{\cos(\beta) + \sin(\beta)} \qquad \text{Combine the fractions in the numerator.}$$

$$\csc(\beta) \stackrel{?}{=} \frac{\cos(\beta) + \sin(\beta)}{\sin(\beta)} \cdot \frac{1}{\cos(\beta) + \sin(\beta)} \qquad \text{To divide, multiply by the reciprocal.}$$

$$\csc(\beta) \stackrel{?}{=} \frac{\cancel{\cos(\beta) + \sin(\beta)}}{\sin(\beta)(\cancel{\cos(\beta) + \sin(\beta)})}$$

$$\csc(\beta) \stackrel{?}{=} \frac{1}{\sin(\beta)}$$

$$\csc(\beta) = \csc(\beta)$$

EXAMPLE 7 ■ Verifying an Identity

Verify the identity $\dfrac{1 + \tan(\theta)}{1 + \cot(\theta)} = \tan(\theta)$.

Solution We will begin simplifying the left side since it is more complex than the right side. Since we want the right side to simplify to tangent, we will try to write the left side of the identity using tangent.

$$\frac{1 + \tan(\theta)}{1 + \cot(\theta)} \stackrel{?}{=} \tan(\theta)$$

$$\frac{1 + \tan(\theta)}{1 + \dfrac{1}{\tan(\theta)}} \stackrel{?}{=} \tan(\theta) \qquad \cot(\theta) = \frac{1}{\tan(\theta)}$$

$$\frac{1 + \tan(\theta)}{\dfrac{\tan(\theta)}{\tan(\theta)} + \dfrac{1}{\tan(\theta)}} \stackrel{?}{=} \tan(\theta) \qquad \text{Use common denominators to combine terms.}$$

$$\frac{1 + \tan(\theta)}{\dfrac{\tan(\theta) + 1}{\tan(\theta)}} \stackrel{?}{=} \tan(\theta)$$

$$\frac{1+\tan(\theta)}{1} \cdot \frac{\tan(\theta)}{\tan(\theta)+1} \stackrel{?}{=} \tan(\theta) \quad \text{To divide, multiply by the reciprocal.}$$

$$\frac{\tan(\theta)(1 + \tan(\theta))}{(\tan(\theta) + 1)} \stackrel{?}{=} \tan(\theta) \quad 1 + \tan(\theta) = \tan(\theta) + 1$$

$$\tan(\theta) = \tan(\theta)$$

There are other methods for verifying this identity, such as rewriting the trigonometric functions in terms of cosine and sine. We will return to this example in Exercise 59 at the end of this section and verify the identity using different approaches.

It is sometimes necessary to multiply by an expression equal to the number 1 to help simplify identities. Look for expressions such as $1 - \sin(\theta)$ or $\cos(\theta) + 1$ in a fraction on either side of the potential identity. Our next example will show how we can use a Pythagorean identity to help verify identities of this form.

EXAMPLE 8 ■ Verifying an Identity

Verify the identity $\dfrac{\sin(\theta)}{1 - \cos(\theta)} = \dfrac{1 + \cos(\theta)}{\sin(\theta)}$.

Solution We start on the left side. We recall that the product of two binomials of the form $(a + b)(a - b)$ will expand to become $a^2 - b^2$. Also, if $a = 1$ and $b = \cos(\theta)$ or $b = \sin(\theta)$, we will get the expression $1 - \cos^2(\theta)$ or $1 - \sin^2(\theta)$, which are components of a Pythagorean identity.

$$\frac{\sin(\theta)}{1 - \cos(\theta)} \stackrel{?}{=} \frac{1 + \cos(\theta)}{\sin(\theta)}$$

$$\frac{\sin(\theta)}{1 - \cos(\theta)} \cdot \frac{1 + \cos(\theta)}{1 + \cos(\theta)} \stackrel{?}{=} \frac{1 + \cos(\theta)}{\sin(\theta)} \quad \text{Multiply the left side by } \frac{1 + \cos(\theta)}{1 + \cos(\theta)}.$$

$$\frac{[\sin(\theta)][1 + \cos(\theta)]}{[1 - \cos(\theta)][1 + \cos(\theta)]} \stackrel{?}{=} \frac{1 + \cos(\theta)}{\sin(\theta)}$$

$$\frac{[\sin(\theta)][1 + \cos(\theta)]}{1 - \cos^2(\theta)} \stackrel{?}{=} \frac{1 + \cos(\theta)}{\sin(\theta)} \quad [1 - \cos(\theta)][1 + \cos(\theta)] = 1 - \cos^2(\theta)$$

$$\frac{[\sin(\theta)][1 + \cos(\theta)]}{\sin^2(\theta)} \stackrel{?}{=} \frac{1 + \cos(\theta)}{\sin(\theta)} \quad \sin^2(\theta) = 1 - \cos^2(\theta) \text{ since } \cos^2(\theta) + \sin^2(\theta) = 1$$

$$\frac{[\sin(\theta)][1 + \cos(\theta)]}{\sin(\theta) \cdot \sin(\theta)} \stackrel{?}{=} \frac{1 + \cos(\theta)}{\sin(\theta)} \quad \text{since } \sin^2(\theta) = \sin(\theta)\sin(\theta)$$

$$\frac{1 + \cos(\theta)}{\sin(\theta)} = \frac{1 + \cos(\theta)}{\sin(\theta)}$$

■ A Construction Approach to Analyzing the Six Trigonometric Functions

One approach to understanding the values of the six trigonometric functions for a given angle measure is to use the unit circle. When we first discussed cosine and sine, we located the endpoint of the arc on the unit circle associated with an angle measuring θ and then determined their values by determining the "length" of the horizontal and vertical components of each coordinate. Similarly, the values of all the trigonometric functions can be represented as lengths associated with the unit circle according to Figure 10.9.

The proofs to verify that these lengths are accurate representations for the value of each trigonometric function rely on Pythagorean identities. The three right triangles shown in Figure 10.10 match up exactly with the three Pythagorean identities.

In the exercise set, we will use Pythagorean identities to verify other relationships given in Figure 10.9.

Figure 10.9

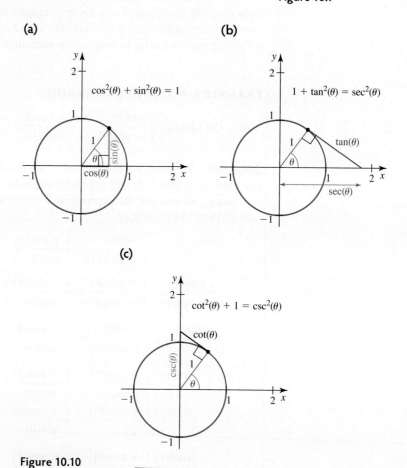

Figure 10.10

SUMMARY

In this section you learned how to verify that two trigonometric expressions are equivalent. You also learned how the six trigonometric functions can be represented as lengths associated with the unit circle.

10.2 EXERCISES

SKILLS AND CONCEPTS

In Exercises 1–7, use fundamental identities to verify each identity. (See Examples 1, 6, and 7.)

1. $\cot(\theta)\cos(\theta) = \sin(\theta)$
2. $\cos(\theta)\sec(\theta) = 1$
3. $\tan^2(\theta) = \sin^2(\theta)\sec^2(\theta)$
4. $\cos(\beta) + [\cot(\beta)][1 - \sin(\beta)] = \cot(\beta)$
5. $\dfrac{\cot(\theta)}{\cos(\theta)} = \csc(\theta)$
6. $\dfrac{\sin(\theta) + \cos(\theta)}{\sin(\theta)\cos(\theta)} = \csc(\theta) + \sec(\theta)$
7. $\dfrac{\sin(\theta)}{\sin(\theta) - \cos(\theta)} = \dfrac{1}{1 - \cot(\theta)}$

In Exercises 8–13, use Pythagorean identities to verify each identity. (See Examples 2 and 4.)

8. $[\sin^2(x)][1 + \cot^2(x)] = 1$
9. $\cot(\beta)\tan(\beta) - \cos^2(\beta) = \sin^2(\beta)$
10. $\cos^2(\theta) = \dfrac{\cos^2(\theta) + \sin^2(\theta)}{1 + \tan^2(\theta)}$
11. $[\cos(x) + \sin(x)]^2 = 2\cos(x)\sin(x) + 1$
12. $2\cos^2(\theta) + \sin^2(\theta) = 1 + \cos^2(\theta)$
13. $[1 + \tan^2(x)][\cos^2(x) - 1] = -\tan^2(x)$

In Exercises 14–19, use even–odd identities to verify each identity. (See Example 3.)

14. $-\csc(\theta) = \dfrac{1}{\sin(-\theta)}$
15. $\dfrac{\csc(-x)}{\sec(-x)} = -\cot(x)$
16. $\sec^2(\theta) = 1 + \tan^2(-\theta)$
17. $\cos(-\theta)\cos(\theta) - \sin(\theta)\sin(-\theta) = 1$
18. $1 = \csc^2(-y)\sin^2(-y)$
19. $\dfrac{\sin^2(-\theta)}{\tan^2(\theta)} = \cos^2(\theta)$

In Exercises 20–25, use cofunction identities to verify each identity. (See Example 5.)

20. $\sin\left(\dfrac{\pi}{2} - \theta\right)\sec(\theta) = 1$
21. $\cot(x) = \dfrac{\sin\left(\dfrac{\pi}{2} - x\right)}{\cos\left(\dfrac{\pi}{2} - x\right)}$
22. $\sin(\beta) = \cos(\beta)\cot\left(\dfrac{\pi}{2} - \beta\right)$
23. $\cos^2\left(\dfrac{\pi}{2} - \alpha\right) + \cos^2(-\alpha) = 1$
24. $-\cot^2(x) = \dfrac{\cot(x)\cos(x)}{\sin\left(\dfrac{\pi}{2} - x\right)\tan(-x)}$
25. $\dfrac{\sec(\theta)\tan\left(\dfrac{\pi}{2} - \theta\right)}{1 - \csc^2(\theta)} = -\sec(\theta)\tan(\theta)$

In Exercises 26–29, use a graph to determine if it is likely that each trigonometric equation is an identity. If it appears to be an identity, use algebra to verify. If it does not appear to be an identity, provide a counterexample to demonstrate that the two expressions are not equivalent. (Hint: To input $\tan^2(\theta)$ into your calculator, use $(\tan(\theta))^2$.)

26. $\dfrac{\sin(\theta)}{\tan(\theta)} = \sec(\theta)$
27. $\cos(\theta)\sec(\theta) + \tan^2(\theta) = \sec^2(\theta)$
28. $[\cos(\theta)][\tan(\theta) + \cot(\theta)] = \csc(\theta)$
29. $\csc(\theta)\tan(\theta) + \sec(\theta) = 2\cos(\theta)$

In Exercises 30–39, verify each identity.

30. $\cot(\theta)\tan(-\theta) = -1$
31. $1 = \tan(\phi)\tan\left(\dfrac{\pi}{2} - \phi\right)$
32. $\sec^2\left(\dfrac{\pi}{2} - \theta\right) - 1 = \cot^2(\theta)$
33. $\tan^2(\theta)\cos^2(\theta) + \cot^2(\theta)\sin^2(\theta) = 1$
34. $\sec(\theta)\cos(-\theta) - \sin^2(\theta) = \cos^2(\theta)$
35. $1 = \sec(\alpha) \cdot \dfrac{\sin(\alpha)}{\tan(\alpha)}$
36. $\sin(\theta)\tan(\theta) = \dfrac{\cos^2\left(\dfrac{\pi}{2} - \theta\right)}{\cos(\theta)}$
37. $\dfrac{\sec(\theta)}{\cos(\theta)} = \sec^2(\theta)$
38. $\tan^2(y) - \sin^2(y)\tan^2(y) = \sin^2(y)$
39. $\tan(\theta) - \dfrac{\sec^2(\theta)}{\tan(\theta)} = -\cot(\theta)$

In Exercises 40 and 41,

a. Trace the given figure.
b. Draw the construction showing the six trigonometric functions as lengths.
c. Using a ruler, measure the length of the radius and each of the six lengths that represent the values of the trigonometric functions for the given angle measure.
d. Divide each of the six lengths for the trigonometric values you measured by the length of the radius. These are the values for the six functions of the given angle measure on the unit circle.
e. Using a calculator, find the values of the six trigonometric functions for the given angle measure and compare these values to the values you found in part (d).

40.

41.

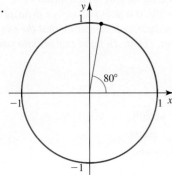

42. Explain how the values of each of the six trigonometric functions change as the angle measure increases in the first quadrant. Compare this behavior to the graphs of each of the six functions on the interval (0°, 90°).

In Exercises 43–46, you are shown identities implied by the diagram of the six trigonometric functions as they relate to the unit circle. Verify each identity.

43. Use similar triangles.

$$\frac{\sin(\theta)}{\csc(\theta)} = \frac{\tan(\theta)}{\tan(\theta) + \cot(\theta)}$$

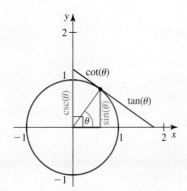

44. Use similar triangles.

$$\frac{\sec(\theta) - \cos(\theta)}{\sec(\theta)} = \frac{\sin(\theta)}{\csc(\theta)}$$

45. Use the Pythagorean theorem.

$$\sin^2(\theta) + [\sec(\theta) - \cos(\theta)]^2 = \tan^2(\theta)$$

46. Use the Pythagorean theorem.

$$\sec^2(\theta) + \csc^2(\theta) = [\cot(\theta) + \tan(\theta)]^2$$

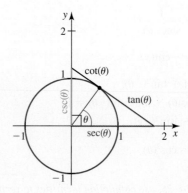

In Exercises 47–50, verify each identity. In each case, you will be multiplying binomials of the form $(a + b)(a - b) = a^2 - b^2$. This skill will be necessary to complete Exercises 51–58.

47. $[1 + \sin(\theta)][1 - \sin(\theta)] = \cos^2(\theta)$

48. $[1 + \cos(\theta)][1 - \cos(\theta)] = \sin^2(\theta)$

49. $[\csc(\theta) + \cot(\theta)][\csc(\theta) - \cot(\theta)] = 1$

50. $\dfrac{1}{1 + \cos(\theta)} + \dfrac{1}{1 - \cos(\theta)} = 2\csc^2(\theta)$

In Exercises 51–54, verify each identity. (See Example 8.)

51. $\dfrac{\sec(\theta) - 1}{1 - \cos(\theta)} = \sec(\theta)$

52. $\sec(\theta) - \tan(\theta) = \dfrac{\cos(\theta)}{1 + \sin(\theta)}$

53. $\dfrac{\sec(x) + 1}{\tan(x)} = \dfrac{\tan(x)}{\sec(x) - 1}$

54. $\dfrac{\cos(-\theta)}{1 + \sin(-\theta)} = \sec(\theta) + \tan(\theta)$

In Exercises 55–58, use factoring to verify each identity. (See Example 9.)

55. $\dfrac{\sin^2(\theta)}{1 - \cos(\theta)} = 1 + \cos(\theta)$

56. $1 - \sin(x) = \dfrac{\cos^2(x)}{1 + \sin(x)}$

57. $\dfrac{\csc^2(\phi) - 1}{\csc(\phi) - 1} = \csc(\phi) + 1$

58. $\dfrac{\sin^2(\beta) - 9}{\sin(\beta) - 3} = \sin(\beta) + 3$

59. In Example 7, we verified the identity $\dfrac{1 + \tan(\theta)}{1 + \cot(\theta)} = \tan(\theta)$ using only one of several different options available. Verify this identity using two methods different from the one used in Example 7.

▦ SHOW YOU KNOW

60. This section provided you with some tips on verifying identities. Provide a list of suggestions (in your own words) on how to verify identities and share it with a classmate who is struggling.

61. Explain why rewriting trigonometric functions using cosine and sine is a good technique to use when verifying an identity.

62. A classmate says $\sin(\theta)\cot(\theta) = 1 - \sin(\theta)$ is an identity. His justification is that if $\theta = \dfrac{\pi}{2}$, then

$$\sin\left(\dfrac{\pi}{2}\right)\cot\left(\dfrac{\pi}{2}\right) = 1 - \sin\left(\dfrac{\pi}{2}\right)$$

$$1 \cdot 0 = 1 - 1$$

$$0 = 0$$

Is your classmate correct? Explain.

63. A classmate says since $\cos^2(\theta) + \sin^2(\theta) = 1$ is an identity, then solving for cosine yields $\cos(\theta) = \sqrt{1 - \sin^2(\theta)}$, which must also be an identity. Do you agree the expressions $\cos(\theta)$ and $\sqrt{1 - \sin^2(\theta)}$ are equivalent? Explain.

▪ STRETCH YOUR MIND

Exercises 64–69 are intended to challenge your abilities to verify identities. For each, verify the identity.

64. $\sin^4(\theta) = 1 - 2\cos^2(\theta) + \cos^4(\theta)$

65. $\dfrac{\cos^4(\theta) + \sin^2(\theta)\cos^2(\theta) + \sin^2(\theta)}{\cos^2(\theta)} = \sec^2(\theta)$

66. $\dfrac{\tan(\theta) + \tan(\phi)}{1 - \tan(\theta)\tan(\phi)} = \dfrac{\cot(\theta) + \cot(\phi)}{\cot(\theta)\cot(\phi) - 1}$

67. $\dfrac{\sin(\theta)\cos(\phi) + \cos(\theta)\sin(\phi)}{\cos(\theta)\cos(\phi) - \sin(\theta)\sin(\phi)} = \dfrac{\tan(\theta) + \tan(\phi)}{1 - \tan(\theta)\tan(\phi)}$

68. $\sec^4(\theta) - \tan^4(\theta) = \dfrac{1 + \sin^2(\theta)}{\cos^2(\theta)}$

69. $x^2 + y^2 = [x\sin(\theta) + y\cos(\theta)]^2 + [x\cos(\theta) - y\sin(\theta)]^2$

SECTION 10.3 Other Trigonometric Identities

LEARNING OBJECTIVES

▪ Use the sum and difference identities

▪ Use the sum-to-product and product-to-sum identities

▪ Use double-angle and half-angle identities

GETTING STARTED

When two or more musical notes of different frequencies are played together, the combination of their intensities and frequencies create acoustic *beats*. The manner in which the sounds combine can either be *consonant* or *dissonant*. Consonant sounds, such as common chords used by most musicians, are pleasant. Dissonant sounds sound as if they conflict. To tune an instrument, a person compares the notes the instrument plays with a standard note played by an electronic tuning machine or a tuning fork. The person adjusts the instrument until the two notes, when played together, produce no beat. Trigonometric identities can help us work with the formulas used to model the combination of musical notes.

In this section we discuss trigonometric identities involving the sum or difference of angles, the sum or

Plamens art./Shutterstock.com

difference of two trigonometric functions, the product of two trigonometric functions, and trigonometric functions involving double angles and half angles. We focus on stating and using common identities rather than verifying them.

■ Sum and Difference Identities

In Chapter 8, we demonstrated how to find the exact values of trigonometric functions with reference angles measuring $0, \frac{\pi}{6}, \frac{\pi}{4}, \frac{\pi}{3}$, or $\frac{\pi}{2}$. For all other measures, we used a calculator to generate rounded estimates of the function values. With the sum and difference identities, we can begin to find exact values of trigonometric functions for many more angles. Specifically, the sine, cosine, or tangent of the sum of two angle measures can be determined using the **sum and difference identities**.

SUM AND DIFFERENCE IDENTITIES

$$\cos(\theta + \phi) = \cos(\theta)\cos(\phi) - \sin(\theta)\sin(\phi)$$
$$\cos(\theta - \phi) = \cos(\theta)\cos(\phi) + \sin(\theta)\sin(\phi)$$

$$\sin(\theta + \phi) = \sin(\theta)\cos(\phi) + \cos(\theta)\sin(\phi)$$
$$\sin(\theta - \phi) = \sin(\theta)\cos(\phi) - \cos(\theta)\sin(\phi)$$

$$\tan(\theta + \phi) = \frac{\tan(\theta) + \tan(\phi)}{1 - \tan(\theta)\tan(\phi)}$$

$$\tan(\theta - \phi) = \frac{\tan(\theta) - \tan(\phi)}{1 + \tan(\theta)\tan(\phi)}$$

EXAMPLE 1 ■ **Using the Difference Identity for Cosine**

Find the exact value for $\cos(15°)$.

Solution To find $\cos(15°)$, we need to write 15° as the difference of two angle measures whose exact trigonometric values we know. In this case, we use 45° − 30°.

$$\cos(\theta - \phi) = \cos(\theta)\cos(\phi) + \sin(\theta)\sin(\phi)$$

$$\cos(45° - 30°) = \cos(45°)\cos(30°) + \sin(45°)\sin(30°) \quad \theta = 45° \text{ and } \phi = 30°$$

$$\cos(15°) = \frac{\sqrt{2}}{2} \cdot \frac{\sqrt{3}}{2} + \frac{\sqrt{2}}{2} \cdot \frac{1}{2} \quad \text{Use exact values.}$$

$$\cos(15°) = \frac{\sqrt{6}}{4} + \frac{\sqrt{2}}{4}$$

$$\cos(15°) = \frac{\sqrt{6} + \sqrt{2}}{4}$$

We use a calculator to check our work: $\dfrac{\sqrt{6} + \sqrt{2}}{4} \approx 0.9659$ and $\cos(15°) \approx 0.9659$.

EXAMPLE 2 ■ Using Sum and Difference Identities

Find the exact value for each of the following.

a. $\sin\left(-\dfrac{\pi}{12}\right)$

b. $\sin(285°)$

Solution

a. Our first step is to find two angle measures whose sum or difference is $-\dfrac{\pi}{12}$.

$$\dfrac{2\pi}{12} - \dfrac{3\pi}{12} = -\dfrac{\pi}{12}$$

$$\dfrac{\pi}{6} - \dfrac{\pi}{4} = -\dfrac{\pi}{12}$$

We find $\dfrac{\pi}{6} - \dfrac{\pi}{4}$ meets these requirements. We now use the difference identity for sine.

$$\sin(\theta - \phi) = \sin(\theta)\cos(\phi) - \cos(\theta)\sin(\phi)$$

$$\sin\left(\dfrac{\pi}{6} - \dfrac{\pi}{4}\right) = \sin\left(\dfrac{\pi}{6}\right)\cos\left(\dfrac{\pi}{4}\right) - \cos\left(\dfrac{\pi}{6}\right)\sin\left(\dfrac{\pi}{4}\right) \qquad \theta = \dfrac{\pi}{6} \text{ and } \phi = \dfrac{\pi}{4}$$

$$\sin\left(-\dfrac{\pi}{12}\right) = \dfrac{1}{2} \cdot \dfrac{\sqrt{2}}{2} - \dfrac{\sqrt{3}}{2} \cdot \dfrac{\sqrt{2}}{2} \qquad \text{Use exact values.}$$

$$= \dfrac{\sqrt{2}}{4} - \dfrac{\sqrt{6}}{4} \qquad \sqrt{a} \cdot \sqrt{b} = \sqrt{a \cdot b}$$

$$= \dfrac{\sqrt{2} - \sqrt{6}}{4}$$

We check this answer with a calculator: $\dfrac{\sqrt{2} - \sqrt{6}}{4} \approx -0.2588$ and $\sin\left(-\dfrac{\pi}{12}\right) \approx -0.2588$. We also note $-\dfrac{\pi}{12}$ is an arc whose endpoint lies in Quadrant IV, and that sine values are negative in Quadrant IV since the endpoint lies below the x-axis.

b. We begin by finding two angles whose measures have a sum or difference of 285° and for which we know the exact values of the trigonometric functions. We use 225° + 60°. We now use the sum identity for sine. We expect the final value to be negative since 285° terminates in Quadrant IV, where the sine values are negative.

$$\sin(\theta + \phi) = \sin(\theta)\cos(\phi) + \cos(\theta)\sin(\phi)$$

$$\sin(225° + 60°) = \sin(225°)\cos(60°) + \cos(225°)\sin(60°) \qquad \theta = 225° \text{ and } \phi = 60°$$

$$\sin(285°) = \left(-\dfrac{\sqrt{2}}{2}\right) \cdot \dfrac{1}{2} + \left(-\dfrac{\sqrt{2}}{2}\right) \cdot \dfrac{\sqrt{3}}{2} \qquad \text{Use exact values.}$$

$$= -\dfrac{\sqrt{2}}{4} + \left(-\dfrac{\sqrt{6}}{4}\right)$$

$$= \dfrac{-\sqrt{2} - \sqrt{6}}{4}$$

Again, we check our answer using a calculator: $\dfrac{-\sqrt{2} - \sqrt{6}}{4} \approx -0.9659$ and $\sin(285°) \approx -0.9659$.

706 CHAPTER 10 ■ TRIGONOMETRIC IDENTITIES

EXAMPLE 3 ■ Using a Sum or Difference Identity

The Ferris wheel at Odaiba in Tokyo, Japan, has a diameter of 100 meters. Riders reach a maximum height of 115 meters off the ground during a 16-minute revolution. Rider A rides the Odaiba Ferris wheel and her height above the ground in meters can be modeled by the formula $H_A(t) = -50\cos\left(\dfrac{\pi}{8}t\right) + 65$, where t is in minutes since boarding. After Rider A travels for 2 minutes, Rider B boards the Ferris wheel. Rider B's height above the ground in meters can be modeled by the formula $H_B(t) = -50\cos\left[\dfrac{\pi}{8}(t-2)\right] + 65$, where t is in minutes since Rider A boarded. When was the first time both riders were the same height above the ground? (*Source:* Tokyo-i.tripod.com)

Solution The riders will be the same height above the ground when $H_A(t) = H_B(t)$. Thus, we set their formulas equal to each other and simplify.

$$-50\cos\left(\dfrac{\pi}{8}t\right) + 65 = -50\cos\left[\dfrac{\pi}{8}(t-2)\right] + 65$$

$$-50\cos\left(\dfrac{\pi}{8}t\right) = -50\cos\left[\dfrac{\pi}{8}(t-2)\right]$$

$$\cos\left(\dfrac{\pi}{8}t\right) = \cos\left[\dfrac{\pi}{8}(t-2)\right]$$

$$\cos\left(\dfrac{\pi}{8}t\right) = \cos\left(\dfrac{\pi}{8}t - \dfrac{\pi}{4}\right)$$

At this point, we can use the difference identity for cosine on the right side.

$$\cos\left(\dfrac{\pi}{8}t\right) = \cos\left(\dfrac{\pi}{8}t\right)\cos\left(\dfrac{\pi}{4}\right) + \sin\left(\dfrac{\pi}{8}t\right)\sin\left(\dfrac{\pi}{4}\right) \quad \text{Use the difference identity.}$$

$$\cos\left(\dfrac{\pi}{8}t\right) = \cos\left(\dfrac{\pi}{8}t\right)\cdot\dfrac{\sqrt{2}}{2} + \sin\left(\dfrac{\pi}{8}t\right)\cdot\dfrac{\sqrt{2}}{2} \quad \text{Use exact values.}$$

$$\dfrac{\cancel{\cos\left(\dfrac{\pi}{8}t\right)}}{\cancel{\cos\left(\dfrac{\pi}{8}t\right)}} = \dfrac{\cancel{\cos\left(\dfrac{\pi}{8}t\right)}\cdot\dfrac{\sqrt{2}}{2}}{\cancel{\cos\left(\dfrac{\pi}{8}t\right)}} + \dfrac{\sin\left(\dfrac{\pi}{8}t\right)\cdot\dfrac{\sqrt{2}}{2}}{\cos\left(\dfrac{\pi}{8}t\right)}$$

$$1 = \dfrac{\sqrt{2}}{2} + \tan\left(\dfrac{\pi}{8}t\right)\cdot\dfrac{\sqrt{2}}{2} \quad \text{since } \dfrac{\sin\left(\dfrac{\pi}{8}t\right)}{\cos\left(\dfrac{\pi}{8}t\right)} = \tan\left(\dfrac{\pi}{8}t\right)$$

$$1 = \dfrac{\sqrt{2}}{2}\left[1 + \tan\left(\dfrac{\pi}{8}t\right)\right] \quad \text{Factor out } \dfrac{\sqrt{2}}{2} \text{ from the right side.}$$

$$1\cdot\dfrac{2}{\sqrt{2}} = 1 + \tan\left(\dfrac{\pi}{8}t\right) \quad \text{Multiply both sides by } \dfrac{2}{\sqrt{2}}.$$

$$\dfrac{2}{\sqrt{2}} - 1 = \tan\left(\dfrac{\pi}{8}t\right)$$

$$\dfrac{\pi}{8}t = \tan^{-1}\left(\dfrac{2}{\sqrt{2}} - 1\right) \quad \text{Write in inverse form.}$$

We evaluate $\tan^{-1}\left(\dfrac{2}{\sqrt{2}} - 1\right)$ using a calculator to get about 0.3927 radians. The other angle with the same tangent value during one rotation will be $0.3927 + \pi \approx 3.5343$ radians. We therefore set up two equations and solve.

$$\dfrac{\pi}{8} t \approx 0.3927 \qquad \dfrac{\pi}{8} t \approx 3.5343$$

$$t \approx \dfrac{8}{\pi} \cdot 0.3927 \qquad t \approx \dfrac{8}{\pi} \cdot 3.5343$$

$$t = 1 \qquad\qquad t = 9$$

We find they will be at the same height 1 minute and 9 minutes after Rider A boards the Ferris wheel. However, only the answer $t = 9$ makes sense in this context because Rider B has not yet boarded the Ferris wheel at $t = 1$.

EXAMPLE 4 ■ Using Sum and Difference Identities

Find the exact value for $\csc(195°)$.

Solution We begin by rewriting the expression using sine.

$$\csc(195°) = \dfrac{1}{\sin(195°)}$$

We rewrite 195° as 240° − 45° and use the difference identity for sine. (Note: There are other correct choices, such as using 135° + 60°.)

$$\csc(195°) = \dfrac{1}{\sin(195°)}$$

$$= \dfrac{1}{\sin(240° - 45°)}$$

$$= \dfrac{1}{\sin(240°)\cos(45°) - \cos(240°)\sin(45°)} \qquad \text{Use the difference identity.}$$

$$= \dfrac{1}{\left(-\dfrac{\sqrt{3}}{2}\right)\dfrac{\sqrt{2}}{2} - \left(-\dfrac{1}{2}\right)\dfrac{\sqrt{2}}{2}} \qquad \text{Use exact values.}$$

$$= \dfrac{1}{-\dfrac{\sqrt{6}}{4} + \dfrac{\sqrt{2}}{4}}$$

$$= \dfrac{1}{\dfrac{-\sqrt{6} + \sqrt{2}}{4}}$$

$$= \dfrac{4}{-\sqrt{6} + \sqrt{2}}$$

If we rationalize the denominator by multiplying by $\dfrac{-\sqrt{6} - \sqrt{2}}{-\sqrt{6} - \sqrt{2}}$, we get the simplified solution $-\sqrt{6} - \sqrt{2}$.

Common Student Errors

It is important to avoid common errors when working with sum and difference identities. The following box shows some of these common errors.

COMMON ERRORS

$$\cos(\theta + \phi) \neq \cos(\theta) + \cos(\phi) \qquad \cos(\theta - \phi) \neq \cos(\theta) - \cos(\phi)$$
$$\sin(\theta + \phi) \neq \sin(\theta) + \sin(\phi) \qquad \sin(\theta - \phi) \neq \sin(\theta) - \sin(\phi)$$
$$\tan(\theta + \phi) \neq \tan(\theta) + \tan(\phi) \qquad \tan(\theta - \phi) \neq \tan(\theta) - \tan(\phi)$$

As these errors indicate, we must be careful to use the appropriate identities and not try to apply the distributive property to these expressions. The parentheses in, say, the expression $\cos(\theta + \phi)$ do not indicate multiplication because "cos" does not represent a number we can use to multiply. Rather "cos" is an operation itself indicating that we are to find the x-coordinate of the endpoint of an arc with angle measuring $\theta + \phi$.

■ Sum-to-Product Identities

When calculating the sum or difference of the sine or cosine of two angles, it is useful to rewrite the sum or difference as a product. The **sum-to-product identities** are used for this purpose.

SUM-TO-PRODUCT IDENTITIES

$$\cos(\theta) + \cos(\phi) = 2\cos\left(\frac{\theta + \phi}{2}\right)\cos\left(\frac{\theta - \phi}{2}\right)$$

$$\cos(\theta) - \cos(\phi) = -2\sin\left(\frac{\theta + \phi}{2}\right)\sin\left(\frac{\theta - \phi}{2}\right)$$

$$\sin(\theta) + \sin(\phi) = 2\sin\left(\frac{\theta + \phi}{2}\right)\cos\left(\frac{\theta - \phi}{2}\right)$$

$$\sin(\theta) - \sin(\phi) = 2\sin\left(\frac{\theta - \phi}{2}\right)\cos\left(\frac{\theta + \phi}{2}\right)$$

EXAMPLE 5 ■ Using Sum-to-Product Identities to Find Exact Values

Find the exact value for $\sin\left(\dfrac{\pi}{12}\right) - \sin\left(\dfrac{17\pi}{12}\right)$.

Solution

$$\sin(\theta) - \sin(\phi) = 2\sin\left(\frac{\theta - \phi}{2}\right)\cos\left(\frac{\theta + \phi}{2}\right)$$

$$\sin\left(\frac{\pi}{12}\right) - \sin\left(\frac{17\pi}{12}\right) = 2\sin\left(\frac{\frac{\pi}{12} - \frac{17\pi}{12}}{2}\right)\cos\left(\frac{\frac{\pi}{12} + \frac{17\pi}{12}}{2}\right)$$

$$= 2\sin\left(\frac{-\frac{16\pi}{12}}{2}\right)\cos\left(\frac{\frac{18\pi}{12}}{2}\right)$$

$$= 2\sin\left(-\frac{16\pi}{24}\right)\cos\left(\frac{18\pi}{24}\right)$$

$$\sin\left(\frac{\pi}{12}\right) - \sin\left(\frac{17\pi}{12}\right) = 2\sin\left(-\frac{2\pi}{3}\right)\cos\left(\frac{3\pi}{4}\right)$$

$$= -2\sin\left(\frac{2\pi}{3}\right)\cos\left(\frac{3\pi}{4}\right) \qquad \text{since } \sin(-\theta) = -\sin(\theta)$$

$$= -2\left(\frac{\sqrt{3}}{2}\right)\left(-\frac{\sqrt{2}}{2}\right)$$

$$= \frac{\sqrt{6}}{2}$$

Sinusoidal functions can be used to model the intensity of sounds, such as those produced when a note is played on a piano. The function $I(t) = \cos(2\pi f t)$ models the intensity, I, of sound, where f is the frequency of the note played in Hz (cycles per second) and t is the time in seconds. When two sound waves interact, they produce either *constructive interference* (the sounds combine to become louder), *destructive interference* (the sounds combine to become softer), or some combination of the two that creates *beats* in the noise. A beat makes the combined sound seem to "vibrate" as it alternates between sounding louder and sounding softer. (*Note:* In a graph of the model, louder sounds have intensities farther from the horizontal axis and softer sounds have intensities closer to the horizontal axis.)

EXAMPLE 6 ■ Using a Sum-to-Product Identity

A piano tuner strikes a tuning fork that produces a sound with a frequency of 440 Hz (the frequency of key A4 on the piano. This is called "Concert A" and is the A above Middle C that is used to tune instruments for concert performances). At the same time, the note A4 is played on a perfectly tuned piano.

a. Model the intensity of both the tuning fork and the note A4.

b. Add the equations for the intensities of these two sounds, then rewrite the result as a single trigonometric function.

c. Graph the equations in part (a) and the result from part (b) and explain what the graph shows. Use the interval [0, 0.01] for your graph.

d. Suppose the note A4 is played on a piano that is out of tune such that the key associated with A4 produces a sound with a 420 Hz frequency instead of 440 Hz. Repeat parts (a) and (b), then graph the equation for the combined intensity of the two notes on the interval [0, 0.05]. Explain how the piano tuner knows the piano is out of tune.

Solution

a. We use the function $I(t) = \cos(2\pi ft)$ for two notes with the frequency 440 Hz, denoting them I_T for the tuning fork and I_A for the key A4 on the piano.

$$I_T(t) = \cos(2\pi 440 t) \qquad I_A(t) = \cos(880\pi t)$$
$$= \cos(880\pi t)$$

b.
$$I_T(t) + I_A(t) = \cos(880\pi t) + \cos(880\pi t)$$
$$I(t) = 2\cos(880\pi t)$$

c. We graph $I(t) = \cos(880\pi t)$ and $I(t) = 2\cos(880\pi t)$ on the same axes, as shown in Figure 10.11. (*Note:* We do not need to graph $I_A(t)$ since it is identical to $I_T(t)$.)

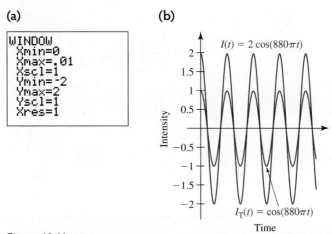

Figure 10.11

The graph shows us that the tuning fork and note played on the piano are creating constructive interference. At all points in time the intensity of the sound is exactly double what it was previously, which causes the combined sound to be perceived as being louder. (This same phenomenon happens with more complex sounds, such as music from a stereo, when two or more properly synchronized speakers are used.)

d. With the piano out of tune, $I_A(t)$ will be altered to reflect a frequency of 420 Hz.

$$I_T(t) = \cos(880\pi t) \qquad I_A(t) = \cos(2\pi \cdot 420 t)$$
$$= \cos(840\pi t)$$

We now add the intensities of the notes to create a function for their combined intensities.

$$I_T(t) + I_A(t) = \cos(880\pi t) + \cos(840\pi t)$$

$$I(t) = 2\cos\left(\frac{880\pi t + 840\pi t}{2}\right)\cos\left(\frac{880\pi t - 840\pi t}{2}\right) \quad \text{Use a sum-to-product identity.}$$

$$= 2\cos\left(\frac{1720\pi t}{2}\right)\cos\left(\frac{40\pi t}{2}\right)$$

$$= 2\cos(860\pi t)\cos(20\pi t)$$

Figure 10.12

The graph of $I(t)$ in Figure 10.12 shows that two notes with frequencies of 440 Hz and 420 Hz create an interesting pattern that is repeated at regular intervals. This pattern shows intervals where the intensity of the sound is greater than either of the original notes and intervals where the intensity of the sound is less than either of the original notes. These alternating periods of high and low intensity that make the combined sound seem to vibrate are what tell the tuner the piano is out of tune. The tuner then tightens or loosens the piano strings until the notes combine perfectly.

■ Product-to-Sum Identities

Just as we can use identities to convert sums to products, we can use identities to convert products to sums through the use of the **product-to-sum identities**.

PRODUCT-TO-SUM IDENTITIES

$$\cos(\theta)\cos(\phi) = \frac{\cos(\theta + \phi) + \cos(\theta - \phi)}{2}$$

$$\sin(\theta)\sin(\phi) = \frac{\cos(\theta - \phi) - \cos(\theta + \phi)}{2}$$

$$\sin(\theta)\cos(\phi) = \frac{\sin(\theta + \phi) + \sin(\theta - \phi)}{2}$$

EXAMPLE 7 ■ Using a Product-to-Sum Identity

Find the exact value of $\cos\left(\dfrac{5\pi}{4}\right)\cos\left(\dfrac{5\pi}{12}\right)$.

Solution

$$\cos(\theta)\cos(\phi) = \frac{\cos(\theta + \phi) + \cos(\theta - \phi)}{2}$$

$$\cos\left(\frac{5\pi}{4}\right)\cos\left(\frac{5\pi}{12}\right) = \frac{\cos\left(\frac{5\pi}{4} + \frac{5\pi}{12}\right) + \cos\left(\frac{5\pi}{4} - \frac{5\pi}{12}\right)}{2}$$

$$= \frac{1}{2}\left[\cos\left(\frac{15\pi}{12} + \frac{5\pi}{12}\right) + \cos\left(\frac{15\pi}{12} - \frac{5\pi}{12}\right)\right]$$

$$= \frac{1}{2}\left[\cos\left(\frac{20\pi}{12}\right) + \cos\left(\frac{10\pi}{12}\right)\right]$$

$$= \frac{1}{2}\left[\cos\left(\frac{5\pi}{3}\right) + \cos\left(\frac{5\pi}{6}\right)\right] \quad \text{Reduce fractions.}$$

$$= \frac{1}{2}\left[\frac{1}{2} + \left(-\frac{\sqrt{3}}{2}\right)\right] \quad \text{Use exact values.}$$

$$= \frac{1}{4} - \frac{\sqrt{3}}{4}$$

■ Double-Angle Identities

The **double-angle identities** are developed by applying the sum and difference identities using $\phi = \theta$.

DOUBLE-ANGLE IDENTITIES

$$\cos(2\theta) = \cos^2(\theta) - \sin^2(\theta)$$
$$= 2\cos^2\theta - 1$$
$$= 1 - 2\sin^2\theta$$

$$\sin(2\theta) = 2\sin(\theta)\cos(\theta)$$

$$\tan(2\theta) = \frac{2\tan(\theta)}{1 - \tan^2(\theta)}$$

EXAMPLE 8 ■ **Using Double-Angle Identities to Solve Equations**

Find all of the solutions to the equation $\sin(2\theta) + \cos(\theta) = 0$ over the interval $[0, 2\pi]$.

Solution We use the double-angle identity for sine and then factor to solve the equation.

$$\sin(2\theta) + \cos(\theta) = 0$$
$$2\sin(\theta)\cos(\theta) + \cos(\theta) = 0 \quad \text{since } \sin(2\theta) = 2\sin(\theta)\cos(\theta)$$
$$\cos(\theta)(2\sin(\theta) + 1) = 0 \quad \text{Factor out } \cos(\theta).$$
$$\cos(\theta) = 0 \quad \text{or} \quad 2\sin(\theta) + 1 = 0$$

The solutions will occur when $\cos(\theta) = 0$ and when $2\sin(\theta) + 1 = 0$.

$$\cos(\theta) = 0 \qquad\qquad 2\sin(\theta) + 1 = 0$$
$$\theta = \cos^{-1}(0) \qquad\qquad 2\sin(\theta) = -1$$
$$\theta = \frac{\pi}{2} \quad \text{and} \quad \theta = \frac{3\pi}{2} \qquad \sin(\theta) = -\frac{1}{2}$$
$$\theta = \sin^{-1}\left(-\frac{1}{2}\right)$$
$$\theta = \frac{7\pi}{6} \quad \text{and} \quad \theta = \frac{11\pi}{6}$$

There are four solutions: $\theta = \frac{\pi}{2}, \theta = \frac{7\pi}{6}, \theta = \frac{3\pi}{2},$ and $\theta = \frac{11\pi}{6}$.

■ **Half-Angle Identities**

To find the sine, cosine, or tangent of half of an angle, we use the **half-angle identities**.

HALF-ANGLE IDENTITIES

$$\cos\left(\frac{\theta}{2}\right) = \pm\sqrt{\frac{1 + \cos(\theta)}{2}} \qquad \tan\left(\frac{\theta}{2}\right) = \frac{1 - \cos(\theta)}{\sin(\theta)}$$

or

$$\sin\left(\frac{\theta}{2}\right) = \pm\sqrt{\frac{1 - \cos(\theta)}{2}} \qquad \tan\left(\frac{\theta}{2}\right) = \frac{\sin(\theta)}{1 + \cos(\theta)}$$

The quadrant where $\left(\frac{\theta}{2}\right)$ falls determines whether the positive or negative square root is used.

EXAMPLE 9 ■ **Using Half-Angle Identities**

Find the exact value of sin(15°) using a half-angle identity.

Solution We use a half-angle identity since 15° is half of 30°, and we know the exact values for trigonometric functions at 30°.

$$\sin\left(\frac{\theta}{2}\right) = \pm\sqrt{\frac{1 - \cos(\theta)}{2}}$$

$$\sin\left(\frac{30°}{2}\right) = \pm\sqrt{\frac{1 - \cos(30°)}{2}}$$

$$\sin(15°) = \pm\sqrt{\frac{1 - \frac{\sqrt{3}}{2}}{2}} \qquad \cos(30°) = \frac{\sqrt{3}}{2}$$

$$= \pm\sqrt{\frac{1}{2} - \frac{\sqrt{3}}{4}} \qquad \text{Dividing by 2 is the same as multiplying by } \frac{1}{2}.$$

$$= \pm\sqrt{\frac{2 - \sqrt{3}}{4}} \qquad \text{since } \frac{1}{2} = \frac{2}{4}$$

$$= \frac{\sqrt{2 - \sqrt{3}}}{2} \qquad \text{15° is in Quadrant I, so sine is positive.}$$

SUMMARY

In this section you learned many additional identities involving trigonometric functions. You also learned how to use these identities to rewrite trigonometric expressions, solve equations, and find the exact values of trigonometric expressions. The identities covered in this section are summarized on the next page.

Sum and Difference Identities

$$\cos(\theta + \phi) = \cos(\theta)\cos(\phi) - \sin(\theta)\sin(\phi)$$

$$\cos(\theta - \phi) = \cos(\theta)\cos(\phi) + \sin(\theta)\sin(\phi)$$

$$\sin(\theta + \phi) = \sin(\theta)\cos(\phi) + \cos(\theta)\sin(\phi)$$

$$\sin(\theta - \phi) = \sin(\theta)\cos(\phi) - \cos(\theta)\sin(\phi)$$

$$\tan(\theta + \phi) = \frac{\tan(\theta) + \tan(\phi)}{1 - \tan(\theta)\tan(\phi)}$$

$$\tan(\theta - \phi) = \frac{\tan(\theta) - \tan(\phi)}{1 + \tan(\theta)\tan(\phi)}$$

Sum-to-Product Identities

$$\cos(\theta) + \cos(\phi) = 2\cos\left(\frac{\theta + \phi}{2}\right)\cos\left(\frac{\theta - \phi}{2}\right)$$

$$\cos(\theta) - \cos(\phi) = -2\sin\left(\frac{\theta + \phi}{2}\right)\sin\left(\frac{\theta - \phi}{2}\right)$$

$$\sin(\theta) + \sin(\phi) = 2\sin\left(\frac{\theta + \phi}{2}\right)\cos\left(\frac{\theta - \phi}{2}\right)$$

$$\sin(\theta) - \sin(\phi) = 2\sin\left(\frac{\theta - \phi}{2}\right)\cos\left(\frac{\theta + \phi}{2}\right)$$

Product-to-Sum Identities

$$\cos(\theta)\cos(\phi) = \frac{\cos(\theta + \phi) + \cos(\theta - \phi)}{2}$$

$$\sin(\theta)\sin(\phi) = \frac{\cos(\theta - \phi) - \cos(\theta + \phi)}{2}$$

$$\sin(\theta)\cos(\phi) = \frac{\sin(\theta + \phi) + \sin(\theta - \phi)}{2}$$

Double-Angle Identities

$$\cos(2\theta) = \cos^2(\theta) - \sin^2(\theta) \qquad \sin(2\theta) = 2\sin(\theta)\cos(\theta)$$

$$= 2\cos^2(\theta) - 1 \qquad \tan(2\theta) = \frac{2\tan(\theta)}{1 - \tan^2(\theta)}$$

$$= 1 - 2\sin^2(\theta)$$

Half-Angle Identities

$$\cos\left(\frac{\theta}{2}\right) = \pm\sqrt{\frac{1 + \cos(\theta)}{2}} \qquad \tan\left(\frac{\theta}{2}\right) = \frac{1 - \cos(\theta)}{\sin(\theta)}$$

or

$$\sin\left(\frac{\theta}{2}\right) = \pm\sqrt{\frac{1 - \cos(\theta)}{2}} \qquad \tan\left(\frac{\theta}{2}\right) = \frac{\sin(\theta)}{1 + \cos(\theta)}$$

10.3 EXERCISES

SKILLS AND CONCEPTS

1. In Section 10.1 you made a list of the basic trigonometric identities for easy reference. Add the identities from this section to that reference sheet.

In Exercises 2–5,

a. For the given angle measure, find a pair of angle measures that meets the following criteria: They have a sum or difference equivalent to the given angle measure and both angles have reference angles measuring 0°, 30°, 45°, 60°, or 90°.

b. Find the exact value of each expression.

2. $\cos(15°)$
3. $\sin(-165°)$
4. $\tan(75°)$
5. $\sec(285°)$

In Exercises 6–9,

a. For the given angle measure, find a pair of angle measures that meets the following criteria: They have a sum or difference equivalent to the given angle measure and both angles have reference angles measuring $0, \frac{\pi}{6}, \frac{\pi}{4}, \frac{\pi}{3}$, or $\frac{\pi}{2}$. For example, the angle measure $\frac{\pi}{12}$ can be written as $\frac{3\pi}{12} - \frac{2\pi}{12}$ (simplified to $\frac{\pi}{4} - \frac{\pi}{6}$).

b. Find the exact value of each expression.

6. $\tan\left(-\frac{7\pi}{12}\right)$
7. $\cos\left(\frac{11\pi}{12}\right)$
8. $\sin\left(\frac{17\pi}{12}\right)$
9. $\tan\left(\frac{61\pi}{12}\right)$

In Exercises 10–15, find the exact value of each expression.

10. $\sin(43°)\cos(17°) + \cos(43°)\sin(17°)$
11. $\dfrac{\tan(302°) - \tan(167°)}{1 + \tan(302°)\tan(167°)}$
12. $\dfrac{1}{\cos(216°)\cos(546°) + \sin(216°)\sin(546°)}$
13. $\cos\left(\frac{13\pi}{15}\right)\cos\left(\frac{23\pi}{15}\right) + \sin\left(\frac{13\pi}{15}\right)\sin\left(\frac{23\pi}{15}\right)$
14. $\dfrac{\tan\left(\frac{8\pi}{9}\right) + \tan\left(\frac{7\pi}{9}\right)}{1 - \tan\left(\frac{8\pi}{9}\right)\tan\left(\frac{7\pi}{9}\right)}$
15. $\dfrac{1}{\sin\left(\frac{11\pi}{6}\right)\cos\left(\frac{7\pi}{12}\right) - \cos\left(\frac{11\pi}{6}\right)\sin\left(\frac{7\pi}{12}\right)}$
16. Show that using $\cos(315° - 60°)$ or $\cos(225° + 30°)$ give the same result for $\cos(255°)$.

In Exercises 17–22, use the product-to-sum identities to find the exact value of each expression.

17. $\sin(45°)\sin(15°)$
18. $\cos(75°)\cos(165°)$
19. $\sin(15°)\cos(75°)$
20. $\sin\left(\frac{\pi}{12}\right)\cos\left(\frac{\pi}{12}\right)$
21. $\sin\left(\frac{3\pi}{4}\right)\sin\left(\frac{11\pi}{12}\right)$
22. $\cos\left(\frac{13\pi}{24}\right)\sin\left(\frac{31\pi}{24}\right)$

In Exercises 23–30,

a. Rewrite the given angle measure to express it as half of the measure of one of the special angles for which we know the exact trigonometric values. For example, 15° may be rewritten as $\frac{1}{2} \cdot 30°$ or $\frac{30°}{2}$ since we know the exact trigonometric values at 30°. Also, $\frac{5\pi}{8}$ may be written as $\frac{\frac{5\pi}{4}}{2}$ since $\frac{5\pi}{8} = \frac{1}{2} \cdot \frac{5\pi}{4}$ and since we know the exact trigonometric values at $\frac{5\pi}{4}$.

b. Find the exact value of the expression.

23. $\sin(75°)$
24. $\cos(15°)$
25. $\tan(-22.5°)$
26. $\sin(-105°)$
27. $\tan\left(\frac{\pi}{8}\right)$
28. $\sin\left(\frac{\pi}{12}\right)$
29. $\tan\left(\frac{7\pi}{12}\right)$
30. $\cos\left(\frac{13\pi}{12}\right)$

In Exercises 31–36, use the sum-to-product identities to find the exact value of each expression.

31. $\sin(75°) + \sin(15°)$
32. $\cos(75°) - \cos(345°)$
33. $\sin(-15°) - \sin(285°)$
34. $\cos\left(\frac{7\pi}{12}\right) + \cos\left(\frac{\pi}{12}\right)$
35. $\sin\left(\frac{11\pi}{12}\right) - \sin\left(-\frac{5\pi}{12}\right)$
36. $\cos\left(\frac{\pi}{12}\right) + \cos\left(\frac{19\pi}{12}\right)$

In Exercises 37–42, use double-angle identities to solve each equation. Find all solutions over the interval $[0, 2\pi]$.

37. $\sin(2\theta) + \sin(\theta) = 0$
38. $\sin(2\theta) - \cos(\theta) = 0$
39. $2\sin(x)\cos(x) = 0.7$
40. $-6\sin(\theta)\cos(\theta) = 0.5$
41. $2\cos^2(\theta) - 1 = -\dfrac{2}{3}$
42. $4 - \sin^2(x) = 3.5$

■ SHOW YOU KNOW

43. Discuss the relationship between θ and ϕ if $\sin(\theta + \phi) = 0$.
44. Explain the relationship between θ and ϕ if $\cos(\theta) + \cos(\phi) = 0$.
45. State the range for the function $f(\theta) = \sin(2\theta)$. Use your answer to explain why the following *cannot* be an identity: $\sin(2\theta) = 2\sin(\theta)$.
46. Choose a value of θ, then use a diagram of the unit circle to explain what $\cos(2\theta)$ and $\tan\left(\dfrac{\theta}{2}\right)$ represent.
47. The identity formulas for $\cos\left(\dfrac{\theta}{2}\right)$ and $\sin\left(\dfrac{\theta}{2}\right)$ contain the symbol $\pm$. Explain how you know whether to use the positive or negative value obtained using the formula.

■ MAKE IT REAL

48. **Guy Wires** Guy wires are wires attached to tall towers to provide stability and support. An 80-foot microwave tower is constructed with guy wires attached at the top of the tower and halfway to the top of the tower s as shown in the diagram.

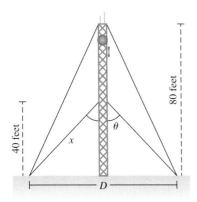

The distance between where guy wires are attached to the ground on opposite sides of the tower is given by the equation $D = 2x\sin\left(\dfrac{\theta}{2}\right)$. Use a trigonometric identity to rewrite this equation using trigonometric functions of θ instead of $\dfrac{\theta}{2}$.

49. **Range of a Launched Object** If we neglect air resistance and other environmental factors, the total horizontal distance (in feet) an object will travel when it is launched at an angle of θ with an initial velocity of v_0 feet per second is given by the equation

$$R = \dfrac{1}{16}v_0^2\sin(\theta)\cos(\theta).$$

a. Use an identity to rewrite this formula using only one trigonometric function.
b. The U.S. Golf Association regulates the manufacture of golf balls. Under very specific testing conditions, the balls to be sold cannot exceed an initial velocity of 250 feet per second when hit. Suppose a golfer hits a ball at a 19° angle (considered optimal) with an initial velocity of 250 feet per second. According to our formula, how far will the ball travel? (*Source:* www.golfdigest.com)
c. Graph R with an initial velocity of 250 feet per second and a variable trajectory angle. What angle allows a golf ball to travel the farthest?
d. Explain some possible reasons for the difference between your answer in part (c) and the 19° optimal angle described in part (b). Research the Internet to better address this question.

In Exercises 50–51, answer the questions that relate to sound. The function $I(t) = \cos(2\pi f t)$ models the intensity of a sound where f is the frequency in Hz (cycles per second) and t is the time elapsed in seconds. For the model, use radian mode on the calculator.

50. **Musical Notes** A piano player strikes the note A4 (frequency of 440 Hz) and the white key to its left, G4 (frequency of 392 Hz), simultaneously on a piano.

a. Write an equation to model the intensity of each note.
b. Add the intensities of the notes to create a function for the combined intensity of the notes. Use a sum-to-product identity to write your final answer as the product of trigonometric functions.
c. Graph the result to part (b) over the interval $[0, 0.05]$ using a calculator. Explain how the combination of these two notes will sound, especially to a trained ear.
d. Graph the formulas from part (a) together over the interval $[0, 0.02]$. Use this graph to explain the behavior of the function graphed in part (c).

51. **Noise Cancellation** Noise cancellation technology, such as that being used in high-end headphones and car mufflers, utilize sounds with the same frequency to cancel out unwanted noise. Sounds with the same frequency that are half of a period out of phase with each other will cancel each other with total destructive interference.

a. Compare the graphs of the functions modeling the intensity of two different sounds, A and B, given by $I_A(t) = \cos(120\pi t)$ and $I_B(t) = \cos(120\pi t - \pi)$.
b. Find the sum of the intensities of the sounds from part (a). Use a sum-to-product identity to simplify the result. Explain what you found in terms of noise cancellation technology.

52. Ferris Wheels The function

$$H_A(t) = -67.5 \cos\left(\frac{\pi}{15} t\right) + 67.5$$

models the height from the bottom of the London Eye Ferris wheel, in meters, for Rider A t minutes after she boards.

a. Write a formula to model the height from the bottom of the Ferris wheel for Rider B—who boards the London Eye 5 minutes after Rider A—in terms of t, the number of minutes since Rider A boarded.

b. Each rider only makes one revolution on the London Eye before getting off. When were both riders at the same height?

STRETCH YOUR MIND

Exercises 53–59 are intended to challenge your understanding of using the trigonometric identities from this section.

53. Find an identity for $\cos(\theta + \phi + \beta)$. (*Hint:* you will need to group angles.)

54. Verify the identity.
$$\sin(6\theta) - \sin(2\theta) + \sin(12\theta) - \sin(8\theta)$$
$$= 4 \sin(2\theta)\cos(7\theta)\cos(3\theta)$$

55. Verify the identity.
$$\cos(5\theta) + \cos(3\theta) + \cos(8\theta) + \cos(10\theta)$$
$$= 4 \cos(\theta)\cos(6.5\theta)\cos(2.5\theta)$$

56. Find the exact value of $\cos(7.5°)$.

57. Find the exact value of $\tan\left(\frac{\pi}{16}\right)$.

58. Rewrite $\sin(3\theta)$ as an expression in terms of $\cos(\theta)$ and $\sin(\theta)$.

59. Rewrite $\cos(3\theta)$ as an expression in terms of $\cos(\theta)$ and $\sin(\theta)$.

CHAPTER 10 Study Sheet

As a result of your work in this chapter, you should be able to answer the following questions, which are focused on the "big ideas" of this chapter.

SECTION 10.1
1. What are the differences between an expression, an equation, and an identity?
2. How do you determine if an equation is an identity?

SECTION 10.2
3. What are some guidelines to follow when attempting to verify trigonometric identities?
4. How can graphing be used to help in verifying identities?

SECTIONS 10.1 AND 10.3
5. Why are trigonometric identities useful in solving equations?

REVIEW EXERCISES

■ SECTION 10.1 ■

In Exercises 1–8, use trigonometric identities to simplify each expression by writing it in terms of a single trigonometric function or as a whole number.

1. $\cos^2(\alpha)\sec(\alpha)$
2. $[\cos^2(\theta)][1 + \tan^2(\theta)]$
3. $\cos\left(\dfrac{\pi}{2} - \theta\right)\sec(\theta)$
4. $\dfrac{\cos(-\theta)}{\sin(-\theta)}$
5. $\dfrac{\sec(\theta) - \cos(\theta)}{\tan(\theta)}$
6. $\sec^2(\theta) - \cot^2\left(\dfrac{\pi}{2} - \theta\right)$
7. $\cos^2(\theta) - \sin^2(\theta)$
8. $[\sin(\theta) + \cos(\theta)]^2 + [\sin(\theta) - \cos(\theta)]^2$

In Exercises 9–16, use trigonometric identities to solve each equation. Find all solutions over the interval $[0, 2\pi]$. Verify your solutions by graphing on a graphing calculator.

9. $-3.1 = \dfrac{\csc(\theta)}{\cot(\theta)}$
10. $\cos(-\theta)\tan(\theta) = 0.98$
11. $\cos\left(\dfrac{\pi}{2} - \theta\right)\cot(\theta) = 0.43$
12. $4\sin(\theta)\csc\left(\dfrac{\pi}{2} - \theta\right) = 4.35$
13. $\dfrac{3\sin(\theta)}{1 - \cos^2(\theta)} = 7$
14. $\dfrac{1 + \cot^2(-\theta)}{\csc(\theta)} = 2.6$
15. $\sin(\theta)\cos(\theta)\tan(\theta) + \cos^2(\theta) = -2\tan(\theta)$
16. $\dfrac{1}{\cot^2(\theta) + 1} = 0.15$

■ SECTION 10.2 ■

In Exercises 17–25, verify each identity.

17. $\sin(\theta)\csc(\theta) - \sin^2(\theta) = \cos^2(\theta)$
18. $[\sec(x)][\sec(x) - \cos(x)] = \tan^2(x)$
19. $\cos^2(\beta) + \cos^2\left(\dfrac{\pi}{2} - \beta\right) = 1$
20. $\csc^2(\theta) = \cot^2(\theta) + \sin^2(\theta) + \cos^2(-\theta)$
21. $\sin^2(y) = \dfrac{1 - \sin^2(y)}{\csc^2(y) - 1}$
22. $\tan(x) = \dfrac{\cos\left(\dfrac{\pi}{2} - x\right)}{\sin\left(\dfrac{\pi}{2} - x\right)}$
23. $\dfrac{1}{\tan(\theta) + \sec(\theta)} = \dfrac{\cos(\theta)}{\sin(\theta) + 1}$
24. $\sec(\theta) - \cos(\theta) = \tan(\theta)\sin(\theta)$
25. $\dfrac{\cos^2(\theta)}{1 + \sin(\theta)} = 1 - \sin(\theta)$

In Exercises 26–29, use a graph to determine if it is likely each trigonometric equation is an identity. If it appears to be an identity, use algebra to verify. If it does not appear to be an identity, provide a counterexample to demonstrate that the two expressions are not equivalent.

26. $\sec^2(x) = \dfrac{\sec^2(x) - 1}{\sin^2(x)}$
27. $[\csc(x) - 1][\csc(x) + 1] = \tan^2(x)$
28. $[\csc(x)]\left[\dfrac{\sin(x)}{\tan(x)}\right] = 1$
29. $\sec^2(\theta) + 2 = \tan^2(\theta) + 3$

■ SECTION 10.3 ■

In Exercises 30–35, find the exact value of each expression using sum and difference identities.

30. $\tan(15°)$
31. $\cos(165°)$
32. $\dfrac{\tan(140°) - \tan(5°)}{1 + \tan(140°)\tan(5°)}$
33. $\sin\left(\dfrac{23\pi}{12}\right)$
34. $\cos\left(-\dfrac{13\pi}{12}\right)$
35. $\cos\left(\dfrac{13\pi}{8}\right)\cos\left(\dfrac{3\pi}{8}\right) + \sin\left(\dfrac{13\pi}{8}\right)\sin\left(\dfrac{3\pi}{8}\right)$

In Exercises 36–37, use the sum-to-product identities to find the exact value of each expression.

36. $\sin(255°) - \sin(15°)$
37. $\cos\left(\dfrac{5\pi}{12}\right) + \cos\left(\dfrac{13\pi}{12}\right)$

In Exercises 38–39, use the product-to-sum identities to find the exact value of each expression.

38. $\cos(165°)\cos(135°)$
39. $\cos\left(-\dfrac{\pi}{4}\right)\sin\left(\dfrac{\pi}{12}\right)$

In Exercises 40–41, use double-angle identities to solve each equation. Find all solutions over the interval $[0, 2\pi]$.

40. $\sin(2\theta) - \sin(\theta) = 0$
41. $\cos^2(\theta) - \sin^2(\theta) = 0.75$

In Exercises 42–45, use half-angle identities to find the exact value of each expression.

42. $\sin(67.5°)$
43. $\tan(-105°)$
44. $\cos\left(\dfrac{\pi}{12}\right)$
45. $\sin\left(-\dfrac{7\pi}{8}\right)$

In Exercises 46–55, verify each identity using the trigonometric identities from this chapter.

46. $\cos(\pi - \theta) = -\cos(\theta)$
47. $\cos\left(\dfrac{3\pi}{2} + \theta\right) = \sin(\theta)$
48. $\cot(\pi - \theta) = -\cot(\theta)$
49. $\dfrac{\sin(4\theta)}{\cos(2\theta) + \cos^2(2\theta)} = 2\tan(\theta)$
50. $\dfrac{\cos(8x) - \cos(2x)}{\cos(8x) + \cos(2x)} = \tan(5x)\tan(3x)$
51. $10\sin(\theta)\cos(\theta) = 5\sin(2\theta)$
52. $3\cos(2\theta) = 3\cos^2(\theta) - 3\sin^2(\theta)$
53. $\csc(2\theta) = \dfrac{\csc(\theta)}{2\cos(\theta)}$
54. $\cos(\theta) + 2\sin^2\left(\dfrac{\theta}{2}\right) = 1$
55. $2 - \cos(\theta) = \cos^2\left(\dfrac{\theta}{2}\right) + 3\sin^2\left(\dfrac{\theta}{2}\right)$

Make It Real Project

What to Do

1. Search the Internet to find a list of the frequencies for different notes on a musical instrument (such as a piano). Choose two different notes, and write down their names and frequencies. Also, give the name of the website where you found the information.

2. Use the equation $I(t) = \cos(2\pi f t)$ to model the intensity of each note, where f is the frequency in Hz (cycles per second).

3. If the two notes are played simultaneously their intensities will combine. Add the equations for the intensities of both notes, then use a trigonometric identity to write the combined equation as the product of trigonometric functions.

4. Graph the combined intensity equation found in Question 3. You may need to adjust your window range to see the graph properly. A domain from 0 to 0.05 is a good starting point.

5. Describe the pattern of the combined intensities of the two notes, then explain how the combined notes will sound.

6. Using the Internet, research the term *longitudinal wave*. Write a few paragraphs explaining how sound travels and what sinusoidal curves are modeling when they represent longitudinal waves.

CHAPTER 11

Conic Sections

Planets, asteroids, dwarf planets, and space junk orbit the Sun in elliptical orbits. However, because they are not gravitationally bound to the Sun, some comets have parabolic or hyperbolic trajectories. These comets enter the solar system and then escape back into space. The paths of motion for many celestial objects can be modeled by circles, ellipses, parabolas, or hyperbolas. These four shapes are together called the conic sections.

Giovanni Benintende/Shutterstock.com

11.1 Slices of a Cone
11.2 Ellipses and Circles
11.3 Parabolas
11.4 Hyperbolas
STUDY SHEET
REVIEW EXERCISES
MAKE IT REAL PROJECT

SECTION 11.1 Slices of a Cone

LEARNING OBJECTIVE

- Explain how slices of a cone form the different conic sections: circle, ellipse, parabola, and hyperbola

GETTING STARTED

The oldest publication containing a description of the conic sections was given by Appollonius of Perga in about 200 B.C. It is believed Appollonius was the first to give the names ellipse, hyperbola, and parabola for the conic sections. Many famous mathematicians, including Archimedes, Pascal, Fermat, Descartes, Newton, and Kepler, have contributed to the development of conic sections. As a result of the work of these mathematical giants, we are able to apply the ideas of conic sections to destroy kidney stones (using elliptical tanks), to build headlights for cars (using parabolic reflectors), and for long range navigation (using the intersection of two hyperbolas). (*Sources: Great Books of the Western World: Euclid, Archimedes, Appollonius of Perga, Nicomachus;* www.usma.edu; www.loc.gov)

In this section we give a brief introduction to the conic sections. We investigate each conic section in greater detail throughout the rest of the chapter.

■ Slices of a Cone

Consider a double cone (two cones joined at their vertices) like the one shown in Figure 11.1. When a plane intersects these cones at different angles, the **conic sections** are formed.

Figure 11.1

The Circle as a Conic Section

A circle is formed when a horizontal plane intersects either of the cones (not at the vertex), as shown in Figure 11.2. The radius of the circle depends on the site of the intersection.

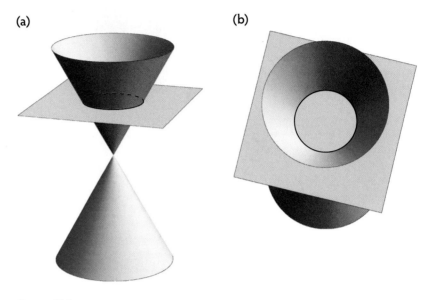

Figure 11.2

The Ellipse as a Conic Section

An ellipse is formed when a nonhorizontal plane intersects both sides of either cone, as shown in Figure 11.3.

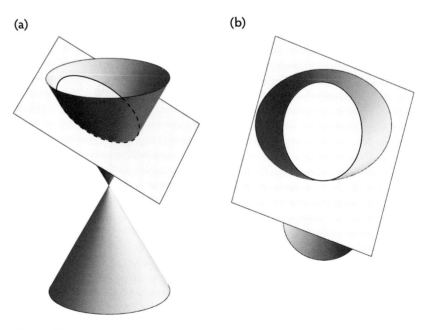

Figure 11.3

The Parabola as a Conic Section

Figure 11.4 shows a parabola, formed when a plane intersects the top (or bottom) of a cone and a corresponding side.

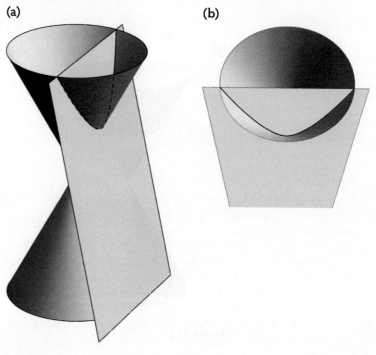

Figure 11.4

The Hyperbola as a Conic Section

A hyperbola, which consists of two parts, is formed when a vertical plane intersects both cones, as shown in Figure 11.5.

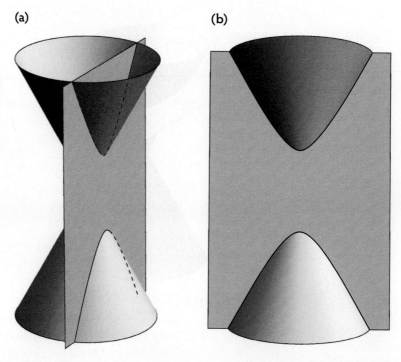

Figure 11.5

SUMMARY

In this section you learned how a double cone is used to create the four conic sections: circle, ellipse, parabola, and hyperbola.

SECTION 11.2 Ellipses and Circles

LEARNING OBJECTIVES

- Generate equations for ellipses and circles given a graph, a set of points, or words
- Apply ellipses and circles to real-world scenarios

GETTING STARTED

St. Louis Union Station opened in 1894 and was the largest and busiest passenger rail terminal in the world at the time. Its semi-elliptical arch located near the main entrance to the station has some intriguing acoustics. Two people located at specific locations beneath the arch can whisper and hear each other, even though they are some distance apart. Remarkably, people nearby the whisperers cannot hear them speaking. Tradition has it that some of the country's most famous train robbers (including Jesse James) may have used the arch in this way to carry out their heists. (*Source:* www.stlouisunionstation.com)

In this section we look at ellipses and the special properties they possess including those that make the whispering arch possible. We also generate the standard equation for an ellipse and apply it to real-world scenarios.

■ The Ellipse

An **ellipse** is a geometric shape defined as follows.

ELLIPSE

An **ellipse** is the set of all points in a plane, the sum of whose distances d_1 and d_2 from two fixed points in the plane (the foci) is constant.

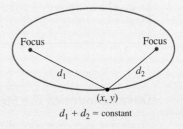

We can construct an ellipse by placing two thumbtacks into a piece of paper at points labeled F_1 and F_2. Each of these points is called a **focus** and together they are called **foci**. We attach the end of a piece of string to the base of each thumbtack. With a pencil holding the string taut, we move the pencil around the foci. The resultant shape is an ellipse, as shown in Figure 11.6. Since the string length remains constant, the sum $d_1 + d_2$ remains constant for all points on the ellipse.

If we move the foci farther apart, we create a long, slim ellipse like the one in Figure 11.7a. Moving the foci closer together creates an ellipse that is closer to being circular, as in Figure 11.7b. In fact, if the two foci overlap to become a single focus, the ellipse becomes a circle. See Figure 11.7c.

Figure 11.6

Figure 11.7

ANATOMY OF AN ELLIPSE

- The fixed points F_1 and F_2 on the interior of an ellipse are called **foci** (or, individually, a **focus**).
- The midpoint of the segment between F_1 and F_2 is called the **center** C.
- The segment between V_1 and V_2 is called the **major axis**.
- The endpoints V_1 and V_2 of the major axis are referred to as the **vertices**.
- The segment between N_1 and N_2 is called the **minor axis**.

Development of the Standard Equation

To derive the equation of the ellipse, we consider an ellipse centered at the origin. The foci are c units from the center and have coordinates $F_1(-c, 0)$ and $F_2(c, 0)$. See Figure 11.8.

We know the sum of the distances from P to F_1 (d_1) and from P to F_2 (d_2) is constant. We let this sum be $2a$ and use the distance formula and algebraic techniques to derive the standard formula for the ellipse.

Figure 11.8

$$d_1 = \sqrt{[x-(-c)]^2 + (y-0)^2} \quad \text{and} \quad d_2 = \sqrt{(x-c)^2 + (y-0)^2}$$

$$\sqrt{[x-(-c)]^2 + y^2} + \sqrt{(x-c)^2 + y^2} = 2a \qquad \text{Sum of two distances equals } 2a.$$

$$\sqrt{(x+c)^2 + y^2} = 2a - \sqrt{(x-c)^2 + y^2}$$

$$\left[\sqrt{(x+c)^2 + y^2}\right]^2 = \left[2a - \sqrt{(x-c)^2 + y^2}\right]^2 \qquad \text{Square both sides and expand.}$$

$$(x+c)^2 + y^2 = 4a^2 - 4a\sqrt{(x-c)^2 + y^2} + (x-c)^2 + y^2$$

$$x^2 + 2cx + c^2 + y^2 = 4a^2 - 4a\sqrt{(x-c)^2 + y^2} + x^2 - 2cx + c^2 + y^2$$

$$2cx = 4a^2 - 4a\sqrt{(x-c)^2 + y^2} - 2cx \qquad \text{Simplify.}$$

$$2cx + 2cx + 4a\sqrt{(x-c)^2 + y^2} - 4a^2 = 0 \qquad \text{Rearrange terms.}$$

$$4cx + 4a\sqrt{(x-c)^2 + y^2} - 4a^2 = 0 \qquad \text{Combine like terms.}$$

$$4a\sqrt{(x-c)^2 + y^2} = 4a^2 - 4cx \qquad \text{Rearrange terms.}$$

$$a\sqrt{(x-c)^2 + y^2} = a^2 - cx \qquad \text{Divide by 4.}$$

$$\left[a\sqrt{(x-c)^2 + y^2}\right]^2 = (a^2 - cx)^2 \qquad \text{Square both sides.}$$

$$a^2(x^2 - 2cx + c^2 + y^2) = a^4 - 2a^2cx + c^2x^2 \qquad \text{Expand the expression.}$$

$$a^2x^2 - 2a^2cx + a^2c^2 + a^2y^2 = a^4 - 2a^2cx + c^2x^2 \qquad \text{distributive property}$$

$$a^2x^2 - c^2x^2 - 2a^2cx + 2a^2cx + a^2y^2 = a^4 - a^2c^2 \qquad \text{Rearrange terms.}$$

$$x^2(a^2 - c^2) + a^2y^2 = a^2(a^2 - c^2) \qquad \text{Factor.}$$

$$\frac{x^2(a^2 - c^2)}{a^2(a^2 - c^2)} + \frac{a^2y^2}{a^2(a^2 - c^2)} = \frac{a^2(a^2 - c^2)}{a^2(a^2 - c^2)} \qquad \text{Divide by } a^2(a^2 - c^2).$$

$$\frac{x^2}{a^2} + \frac{y^2}{a^2 - c^2} = 1 \qquad \text{Simplify.}$$

$$\frac{x^2}{a^2} + \frac{y^2}{b^2} = 1 \qquad \text{Letting } a^2 - c^2 = b^2 \text{ we get the standard equation for an ellipse.}$$

The coordinates of every point (x, y) on the ellipse satisfy this equation.
We find the x-intercepts by setting $y = 0$.

$$\frac{x^2}{a^2} + \frac{y^2}{b^2} = 1$$

$$\frac{x^2}{a^2} + \frac{0^2}{b^2} = 1$$

$$\frac{x^2}{a^2} = 1$$

$$x^2 = a^2$$

$$x = \pm\sqrt{a^2}$$

$$x = a \quad \text{or} \quad x = -a$$

Therefore, the coordinates of the vertices V_1 and V_2 are $(a, 0)$ and $(-a, 0)$. The major axis has length $2a$.

We find the y-intercepts by setting $x = 0$.

$$\frac{x^2}{a^2} + \frac{y^2}{b^2} = 1$$

$$\frac{0^2}{a^2} + \frac{y^2}{b^2} = 1$$

$$\frac{y^2}{b^2} = 1$$

$$y^2 = b^2$$

$$y = \pm \sqrt{b^2}$$

$$y = b \quad \text{or} \quad y = -b$$

Therefore, the coordinates of N_1 and N_2 are $(0, b)$ and $(0, -b)$. The minor axis has length $2b$.

THE ELLIPSE: MAJOR AXIS HORIZONTAL, CENTER AT (0, 0)

The standard equation of an ellipse with center at the origin and major axis on the x-axis is $\frac{x^2}{a^2} + \frac{y^2}{b^2} = 1$, where $a > b > 0$ and $b^2 = a^2 - c^2$.

- The vertices of the ellipse are $(a, 0)$ and $(-a, 0)$.
- The major axis has length $2a$.
- The endpoints of the minor axis are $(0, b)$ and $(0, -b)$.
- The minor axis has length $2b$.

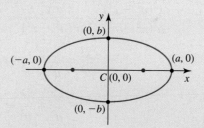

The formulas for certain transformed ellipses follow. The development of each is given as a challenge exercise at the end of this section.

THE ELLIPSE: MAJOR AXIS VERTICAL, CENTER AT (0, 0)

The standard equation of an ellipse centered at the origin with the major axis vertical is $\dfrac{x^2}{b^2} + \dfrac{y^2}{a^2} = 1$, where $a > b > 0$ and $b^2 = a^2 - c^2$.

- The vertices of the ellipse are $(0, a)$ and $(0, -a)$.
- The major axis has length $2a$.
- The endpoints of the minor axis are $(b, 0)$ and $(-b, 0)$.
- The minor axis has length $2b$.

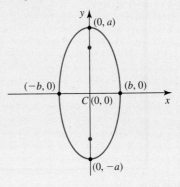

Shifting an ellipse horizontally or vertically changes the location of the center and vertices; however, it does not affect the length of the major and minor axes.

THE ELLIPSE: MAJOR AXIS HORIZONTAL, CENTER AT (h, k)

The standard equation of an ellipse centered at (h, k) with major axis horizontal is $\dfrac{(x-h)^2}{a^2} + \dfrac{(y-k)^2}{b^2} = 1$, where $a > b > 0$ and $b^2 = a^2 - c^2$.

- The vertices of the ellipse are $(h + a, k)$ and $(h - a, k)$.
- The endpoints of the minor axis are $(h, k + b)$ and $(h, k - b)$.

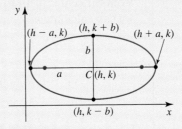

THE ELLIPSE: MAJOR AXIS VERTICAL, CENTER AT (h, k)

The standard equation of an ellipse centered at (h, k) with major axis vertical is $\dfrac{(x - h)^2}{b^2} + \dfrac{(y - k)^2}{a^2} = 1$, where $a > b > 0$ and $b^2 = a^2 - c^2$.

- The vertices of the ellipse are $(h, k + a)$ and $(h, k - a)$.
- The endpoints of the minor axis are $(h + b, k)$ and $(h - b, k)$.

EXAMPLE 1 ■ Finding the Equation of an Ellipse

Find the equation of each ellipse and graph it.

a. Centered at the origin, one focus at $(2, 0)$, and a vertex at $(-3, 0)$
b. One focus at $(0, 4)$ and vertices at $(0, 6)$ and $(0, -6)$
c. Center at $(1, -5)$, one focus at $(2, -5)$, and one vertex at $(4, -5)$

Solution

a. The ellipse has its center at the origin and its focus and vertex are on the x-axis.

Therefore, the major axis is the x-axis. Consequently, we use the formula $\dfrac{x^2}{a^2} + \dfrac{y^2}{b^2} = 1$, where $a > b > 0$ and $b^2 = a^2 - c^2$.

The distance from the center $(0, 0)$ to one of the foci $(2, 0)$ is $c = 2$. The distance from the center $(0, 0)$ to one of the vertices $(-3, 0)$ is $a = 3$. With this information, we can calculate b.

$$b^2 = a^2 - c^2$$
$$b^2 = 3^2 - 2^2$$
$$b^2 = 9 - 4$$
$$b^2 = 5$$
$$b = \sqrt{5}$$

This gives us the equation $\dfrac{x^2}{9} + \dfrac{y^2}{5} = 1$ and the graph shown in Figure 11.9.

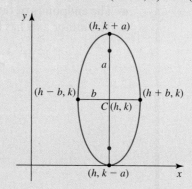

Figure 11.9

b. One of the foci and the vertices are on the y-axis, so the major axis is the y-axis. The center must be halfway between the vertices, so the center is at the origin, $(0, 0)$. We use the formula $\dfrac{x^2}{b^2} + \dfrac{y^2}{a^2} = 1$, where $a > b > 0$ and $b^2 = a^2 - c^2$. We have $a = 6$ and $c = 4$ and use these to find b.

$$b^2 = a^2 - c^2$$
$$b^2 = 6^2 - 4^2$$
$$b^2 = 36 - 16$$
$$b^2 = 20$$
$$b = \sqrt{20}$$

This gives us the equation $\dfrac{x^2}{20} + \dfrac{y^2}{36} = 1$ and the graph shown in Figure 11.10.

c. The ellipse has its center at $(1, -5)$, one focus at $(2, -5)$, and one vertex at $(4, -5)$. Since the center, focus, and vertex have y-values equal to -5, the major axis is the horizontal line $y = -5$. To create the equation, we use $\dfrac{(x - h)^2}{a^2} + \dfrac{(y - k)^2}{b^2} = 1$. We plot points to help us determine the values of a and c, as shown in Figure 11.11. We observe $c = 1$ and $a = 3$ and use these to find b.

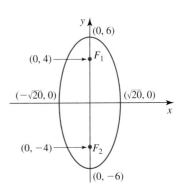

Figure 11.10

$$b^2 = a^2 - c^2$$
$$b^2 = 3^2 - 1^2$$
$$b^2 = 9 - 1$$
$$b^2 = 8$$
$$b = \sqrt{8}$$

This gives us the equation $\dfrac{(x - 1)^2}{9} + \dfrac{(y + 5)^2}{8} = 1$ and the graph shown in Figure 11.12.

Figure 11.11

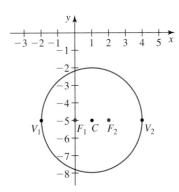

Figure 11.12

EXAMPLE 2 ■ Describing an Ellipse from Its Equation

Determine the center, foci, vertices, and the endpoints of the minor axis of each ellipse. Then graph the ellipse.

a. $\dfrac{x^2}{36} + \dfrac{y^2}{16} = 1$

b. $\dfrac{(x - 4)^2}{4} + \dfrac{(y + 3)^2}{25} = 1$

c. $2x^2 + 3y^2 - 8x + 6y + 5 = 0$

Solution

a. We use the formula $\dfrac{x^2}{a^2} + \dfrac{y^2}{b^2} = 1$, where $a > b > 0$ and $b^2 = a^2 - c^2$. We know that the center is $(0, 0)$, $a^2 = 36$, and $b^2 = 16$. Therefore, we know that $a = 6$ and $b = 4$, and we use these to find c.

$$b^2 = a^2 - c^2$$
$$16 = 36 - c^2$$
$$16 - 36 = -c^2$$
$$-20 = -c^2$$
$$20 = c^2$$
$$\sqrt{20} = c$$

Therefore, the foci are $(-\sqrt{20}, 0)$ and $(\sqrt{20}, 0)$ and the vertices are $-6, 0)$ and $(6, 0)$. The endpoints of the minor axis are $(0, -4)$ and $(0, 4)$. We sketch the ellipse as in Figure 11.13.

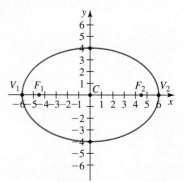

Figure 11.13

b. We use the formula $\dfrac{(x-h)^2}{b^2} + \dfrac{(y-k)^2}{a^2} = 1$, where $a > b > 0$ and $b^2 = a^2 - c^2$. We know the center is $(4, -3)$, $a^2 = 25$, and $b^2 = 4$. Therefore, we know that $a = 5$ and $b = 2$, and we use these to find c.

$$b^2 = a^2 - c^2$$
$$4 = 25 - c^2$$
$$4 - 25 = -c^2$$
$$-21 = -c^2$$
$$21 = c^2$$
$$\sqrt{21} = c$$

Therefore, the foci are $(4, -3 + \sqrt{21})$ and $(4, -3 - \sqrt{21})$ and the vertices are $(4, 2)$ and $(4, -8)$. The endpoints of the minor axis are $(2, -3)$ and $(6, -3)$. We sketch the ellipse as in Figure 11.14.

Figure 11.14

c. We need to complete the square to get the equation in the standard form.

$$2x^2 - 8x + 3y^2 + 6y = -5$$
$$2(x^2 - 4x) + 3(y^2 + 2y) = -5 \quad \text{common factor}$$
$$2(x^2 - 4x + 4) + 3(y^2 + 2y + 1) = -5 + 8 + 3 \quad \text{Complete the square.}$$
$$2(x - 2)^2 + 3(y + 1)^2 = 6 \quad \text{Factor.}$$
$$\dfrac{2(x-2)^2}{6} + \dfrac{3(y+1)^2}{6} = \dfrac{6}{6}$$
$$\dfrac{(x-2)^2}{3} + \dfrac{(y+1)^2}{2} = 1 \quad \text{Simplify.}$$

SECTION 11.2 ELLIPSES AND CIRCLES

We know the center is $(2, -1)$, $a^2 = 3$, and $b^2 = 2$. Therefore, we know that $a = \sqrt{3}$ and $b = \sqrt{2}$, and we use these to find c.

$$b^2 = a^2 - c^2$$
$$2 = 3 - c^2$$
$$2 - 3 = -c^2$$
$$-1 = -c^2$$
$$1 = c^2$$
$$1 = c$$

Therefore, the foci are $(1, -1)$ and $(3, -1)$ and the vertices are $(2 + \sqrt{3}, -1)$ and $(2 - \sqrt{3}, -1)$. The endpoints of the minor axis are $(2, -1 + \sqrt{2})$ and $(2, -1 - \sqrt{2})$. We sketch the ellipse as in Figure 11.15.

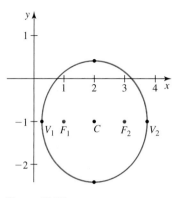

Figure 11.15

PEER INTO THE PAST
WHISPERING GALLERIES: NATIONAL STATUARY HALL

Statuary Hall is an area in the United States Capitol devoted to statues of people and symbols of American history. It was originally the meeting place for the House of Representatives. After his presidency, John Quincy Adams was elected to the House. Many people were amazed at his ability to predict the plans made by the opposite party. Like today, the House's two main parties were separated by an aisle. Adams's desk was situated at one of the foci of the elliptical hall opposite from the opposing political party's members. The sound waves bounced off the walls and ceiling and came to his seat. He could and would listen in on what members of the other party would talk about and used the information as he saw fit. The location of his seat is marked today in the National Statuary Hall with a name plate on the floor.

(*Source:* www.aoc.gov/cc/capitol/nat_stat_hall.cfm)

EXAMPLE 3 ■ Using an Ellipse in a Real-World Context

If rays of light or sound waves are emitted from one focus of an ellipse, they will be reflected off an elliptical ceiling and be directed to the other focus. The elliptical chamber in the United States Capitol Building in Washington, D.C., has this property. A person whispering at one focus is easily heard by a person standing at the other focus, while persons at other positions cannot hear the whispers. Let's look at this further by returning to the archway at St. Louis's Union Station.

Figure 11.16 shows the dimensions of the elliptical ceiling in the entrance of Union Station in St. Louis. (*Source:* www.stlouisunionstation.com) Assuming the center of the ellipse is at $(0, 5.83)$, find the equation for the ellipse to model the archway. Where should two individuals 5.83 feet tall stand so that they can whisper back and forth using the acoustics of the elliptical arch?

Figure 11.16

Solution We create an ellipse with a horizontal major axis along the horizontal line at 5.83 feet (5 feet 10 inches) above the floor. See Figure 11.17.

Figure 11.17

We now know $a = 25$ and $b = 20$. Since the ellipse is horizontal, we generate the equation for the ellipse using $\dfrac{(x-h)^2}{a^2} + \dfrac{(y-k)^2}{b^2} = 1$. We know $b^2 = a^2 - c^2$ and the major axis is horizontal. The center is at (0, 5.83).

$$\frac{(x-h)^2}{a^2} + \frac{(y-k)^2}{b^2} = 1$$

$$\frac{(x-0)^2}{25^2} + \frac{(y-5.83)^2}{20^2} = 1$$

$$\frac{x^2}{625} + \frac{(y-5.83)^2}{400} = 1$$

To find the coordinates of foci, we use the equation $b^2 = a^2 - c^2$ to determine the value of c.

$$b^2 = a^2 - c^2$$
$$20^2 = 25^2 - c^2$$
$$400 = 625 - c^2$$
$$-225 = -c^2$$
$$225 = c^2$$
$$15 = c$$

The distance from the center of the ellipse to its foci is 15 feet. Therefore, the coordinates of the foci are (−15, 5.83) and (15, 5.83). Two individuals 5.83 feet tall should stand on opposite sides beneath the arch 15 feet away from the center so they can whisper and hear each other.

Eccentricity of the Ellipse

Recall that the shape of an ellipse can vary depending on the length of the major axis and where the foci are located. We use the **eccentricity** of an ellipse to quantify its shape.

SECTION 11.2 ■ ELLIPSES AND CIRCLES

ECCENTRICITY

The **eccentricity** (*e*) of an ellipse is the ratio of the distance between the foci and the length of the major axis or

$$e = \frac{2c}{2a} = \frac{c}{a}$$

If the eccentricity is close to 1, the ellipse is relatively narrow.

If the eccentricity is close to 0, the ellipse is close to circular.

Figure 11.18 shows a group of ellipses centered at (0, 0) with $a = 10$ together with their corresponding eccentricities.

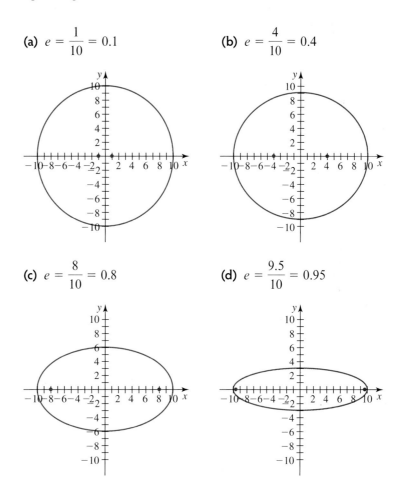

(a) $e = \dfrac{1}{10} = 0.1$

(b) $e = \dfrac{4}{10} = 0.4$

(c) $e = \dfrac{8}{10} = 0.8$

(d) $e = \dfrac{9.5}{10} = 0.95$

Figure 11.18

One of the many amazing aspects of the cosmos is the intricate placement of the planets in our Solar System. Earth is extraordinary because it is the only planet known to be able to support life. It has an elliptical orbit and an axial tilt of 23.45°. Its four seasons (spring, summer, winter, fall) are a result of the tilt of the planet's axis relative to its orbit.

PEER INTO THE PAST
ELLIPTICAL ORBITS

Johannes Kepler (1571–1630), a German scientist and mathematician, was motivated by his deep faith in God to try and discover God's design in the universe. His goal was to "read the mind of God" because he believed that mathematics was the language of God. At the time of his study, there were two commonly held beliefs. The *geocentric model* was the belief that Earth was the center of the universe and all of the bodies in the heavens go around it. A competing theory known as the *heliocentric model* was that the planets orbited the Sun. Both models accepted that the paths of the planets were circular because circles were believed to be the perfect shape.

Kepler began his work based on Copernicus's heliocentric model. Kepler sought to refine the model but after years of work, Kepler became convinced that Earth's orbit around the Sun was not circular. This idea was contrary to the prevailing belief that had been held for 2000 years. His contemporaries (including Galileo) openly disagreed with the conclusion now known as Kepler's **First Law of Planetary Motion**, which states that the planetary orbits are ellipses with the Sun positioned at one focus.

(*Source:* http://kepler.nasa.gov/johannes/)

EXAMPLE 4 ■ Calculating Orbital Eccentricity

The orbits of the planets around the Sun are elliptical with the Sun positioned at one of the foci. (See the Peer into the Past.) The **orbital eccentricity** of a planet is a measure of how far a planet's orbit deviates from the shape of a circle. The amount that Earth varies in its distance from the Sun throughout the year is relatively small. Earth is closest to the Sun (91.4 million miles) around January 1 and is the farthest away from the Sun (94.8 million miles) around July 1. Earth's mean distance of 93 million miles from the Sun is referred to as 1 **astronomical unit** or **AU**.

Use the information provided in Table 11.1 and the equation $\frac{x^2}{a^2} + \frac{y^2}{b^2} = 1$ where $a > b > 0$, $b^2 = a^2 - c^2$, and $e = \frac{c}{a}$ to answer the following.

Table 11.1

Planet	Distance from Center of Orbit to Vertex (AU) a	Distance from Center of Orbit to Focus (AU) c	Eccentricity e
Mercury	0.3870		0.2056
Venus	0.7233		0.0068
Earth	1.0000		0.0167
Mars	1.5237		0.0934
Jupiter	5.2034		0.0484
Saturn	9.5371	0.5150	
Uranus	19.1913	0.9058	
Neptune	30.0690	0.2586	
Pluto*	39.4817	9.8309	

Source: www.windows.ucar.edu

* Although Pluto is no longer officially classified as a planet, we include it here for the educational value of its orbit.

a. Fill in the missing cells in the table.
b. What do we learn from the eccentricity of the planets in our Solar System?
c. Find the equation and sketch the graph for Earth's elliptical orbit.
d. How far is the Sun from the center of Earth's orbit?

Solution

a. Since $e = \frac{c}{a}$, we have $c = ae$. We use this to find the values of c in the table. For example, for Mercury we get $c = (0.3870)(0.2056) \approx 0.0796$.

To find the missing eccentricity values, we use $e = \frac{c}{a}$. For example, Saturn's eccentricity is given by

$$e = \frac{0.5150}{9.5371} = 0.0540$$

The same process is followed for the other planets, and the results are shown in Table 11.2.

SECTION 11.2 ELLIPSES AND CIRCLES 739

Table 11.2

Planet	Distance from Center of Orbit to Vertex (AU) a	Distance from Center of Orbit to Focus (AU) c	Eccentricity e
Mercury	0.3870	0.0796	0.2056
Venus	0.7233	0.0049	0.0068
Earth	1.0000	0.0167	0.0167
Mars	1.5237	0.1423	0.0934
Jupiter	5.2034	0.2518	0.0484
Saturn	9.5371	0.5150	0.0540
Uranus	19.1913	0.9058	0.0472
Neptune	30.0690	0.2586	0.0086
Pluto	39.4817	9.8309	0.2490

b. Within the inner solar system, Mercury has the highest orbital eccentricity (0.2056). In fact, only Pluto has a higher eccentricity. All other orbital eccentricities are less than 0.100. This means that the orbits of the planets are nearly circular. This may explain why it was hard for early scientists to detect that the planetary orbits were elliptical.

c. For Earth, $a = 1.000$ and $c = 0.0167$. We calculate b^2 as follows.

$$c^2 = a^2 - b^2$$
$$(0.0167)^2 = (1.000)^2 - b^2$$
$$b^2 = 1 - 0.0002789$$
$$b^2 \approx 0.9997$$

We know $a^2 = (1.000)^2 = 1$. The equation for the elliptical orbit of Earth is given by

$$\frac{x^2}{a^2} + \frac{y^2}{b^2} = 1$$

$$\frac{x^2}{1} + \frac{y^2}{0.9997} = 1$$

Notice from the graph in Figure 11.19 that the ellipse is nearly circular. The Sun is at one focus and the other focus is near the Sun.

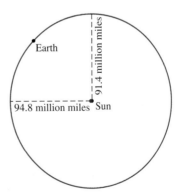

Figure 11.19

d. In part (a) we found that $c = 0.0167$. For ellipses, c is the distance from the center to the foci. Therefore the Sun, being at one focus, is 0.0167 AU from the center of Earth's orbit. In miles this would be 0.0167 AU × 93,000,000 miles/1 AU = 1,553,100 miles.

The Circle

As explained earlier, when the two foci of an ellipse overlap to become a single focus, a **circle** results.

> ### CIRCLE
>
> A **circle** is the set of all points in a plane at a fixed distance r (called the **radius**) from a given point C (called the **center**).
>
>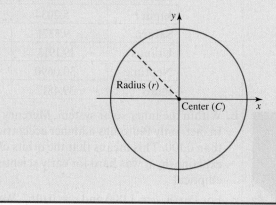

Development of the Standard Equation

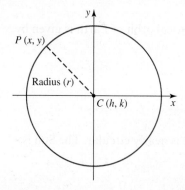

Figure 11.20

We use the distance formula $\sqrt{(x_2 - x_1)^2 + (y_2 - y_1)^2} = d$ to develop an equation for the graph of a circle. If we let (h, k) be the center C, r be the radius, and (x, y) be a point P on the circle, we have the graph in Figure 11.20. Using the distance formula we get

$$\sqrt{(x - h)^2 + (y - k)^2} = r$$

$$\left[\sqrt{(x - h)^2 + (y - k)^2}\right]^2 = (r)^2$$

$$(x - h)^2 + (y - k)^2 = r^2 \quad \text{standard equation of a circle}$$

The coordinates of every point (x, y) on the circle satisfy this equation.

> ### STANDARD EQUATION OF THE CIRCLE
>
> The standard equation of a circle with center at (h, k) and radius r is
>
> $$(x - h)^2 + (y - k)^2 = r^2$$
>
>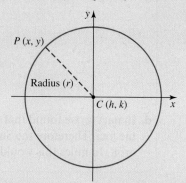

EXAMPLE 5 ■ Using the Circle in a Real-World Context

Meteorologists track the path of a storm using Doppler radar. Circles are used to locate points on the radar scope. If a storm is 5 miles from the radar station, find an equation to monitor the distance the storm is away. Let the radar station be (0, 0).

Solution We are given the center and the radius so we use the standard equation for a circle to get

$$(x - 0)^2 + (y - 0)^2 = 5^2$$
$$x^2 + y^2 = 25$$

The graph of the circle is shown in Figure 11.21.

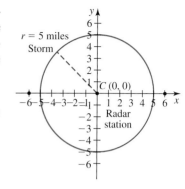

Figure 11.21

EXAMPLE 6 ■ Finding the Equation of a Circle by Completing the Square

Find the center and radius of the circle with the equation $x^2 + y^2 - 4x + 6y - 3 = 0$.

Solution We begin by arranging the terms of the equation as follows.

$$(x^2 - 4x) + (y^2 + 6y) = 3$$

Next we complete the square by adding appropriate numbers within the parentheses and the equivalent amount to the other side of the equation.

$$(x^2 - 4x + 4) + (y^2 + 6y + 9) = 3 + 4 + 9$$
$$(x - 2)^2 + (y + 3)^2 = 16$$

The circle has the center (2, −3) and radius 4.

SUMMARY

In this section you learned to create the standard equation for various ellipses and circles. You also saw how to apply ellipses and circles to real-world scenarios.

11.2 EXERCISES

SKILLS AND CONCEPTS

In Exercises 1–4, the graph of an ellipse is given. Match each graph with its equation.

A.

B.

C.

D.

1. $\dfrac{x^2}{9} + \dfrac{y^2}{36} = 1$

2. $\dfrac{x^2}{4} + y^2 = 1$

3. $\dfrac{x^2}{16} + \dfrac{y^2}{25} = 1$

4. $\dfrac{x^2}{25} + y^2 = 1$

In Exercises 5–12, write the equation of the ellipse. Then graph the equation.

5. Center at the origin; focus at (1, 0); 4 is the length of the major axis.
6. Center at the origin; focus at (0, 2); 8 is the length of the major axis.
7. Center at the origin; focus at (1.5, 0); 3 is the length of the minor axis.
8. Center at the origin; focus at (0, 6); 1 is the length of the minor axis.
9. Center at (2, −2); vertex at (7, −2); focus at (4, −2)
10. Center at (−4, 1); vertex at (−4, 3); focus at (−4, 1.5)
11. Vertices at (3, 3) and (3, 9); focus at (3, 8)
12. Vertices at (2, 5) and (2, −1); $c = 2$

In Exercises 13–16, find the vertices and foci of each ellipse. Then graph each equation.

13. $\dfrac{x^2}{25} + \dfrac{y^2}{9} = 1$

14. $\dfrac{x^2}{4} + \dfrac{y^2}{16} = 1$

15. $16x^2 + 4y^2 = 16$
16. $x^2 + 9y^2 = 18$

In Exercises 17–20, write the equation for each ellipse.

17.

18.

19.

20.

In Exercises 21–32, find the center, foci, and vertices of each ellipse. Then graph the equation.

21. $4x^2 + 18y^2 = 36$

22. $\dfrac{x^2}{16} + \dfrac{y^2}{12} = 1$

23. $9x^2 + 4y^2 = 36$

24. $4x^2 + y^2 = 16$

25. $\dfrac{(x-4)^2}{4} + \dfrac{(y+2)^2}{9} = 1$

26. $\dfrac{(x+5)^2}{25} + \dfrac{(y+1)^2}{8} = 1$

27. $x^2 + 4x + 4y^2 - 8y + 4 = 0$

28. $2x^2 - 8x + 3y^2 + 6y + 5 = 0$

29. $4x^2 + 16x + y^2 + 2y + 13 = 0$

30. $9x^2 + 9x + 4y^2 - 4y = \dfrac{131}{4}$

31. $\dfrac{x^2 - 2x}{4} + y^2 + 4y + \dfrac{13}{4} = 0$

32. $\dfrac{x^2 + 4x}{9} + \dfrac{y^2 + 10y}{25} = -\dfrac{4}{9}$

In Exercises 33–38, calculate the eccentricity of each ellipse.

33. $\dfrac{x^2}{9} + \dfrac{y^2}{4} = 1$

34. $\dfrac{x^2}{16} + \dfrac{y^2}{25} = 1$

35. Vertices at (3, 3) and (3, 9); focus at (3, 8)
36. Vertices at (2, 5) and (2, −1); $c = 3$
37. Minor axis length 4 and major axis length 8
38. Minor axis length 2 and major axis length 16

In Exercises 39–42, give the coordinates of the circle's center and its radius.

39. $(x-4)^2 + (y+5)^2 = 4$
40. $x^2 + y^2 = 49$
41. $x^2 + y^2 = 3$
42. $2(x-5)^2 + 2y^2 = 11$

In Exercises 43–48, write the equation of each circle.

43.

44.

45.

46.

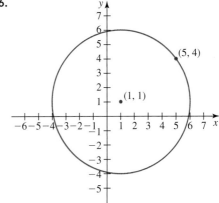

47. Radius of 5; center at (−9, 4)
48. Radius of 6; center at (3, 6)

In Exercises 49–52, graph each equation.

49. $x^2 + y^2 = 9$
50. $x^2 - 2x + y^2 = 15$
51. $4x^2 + 4y^2 - 24x + 32y = 4$
52. $3x^2 + 3y^2 - 12x - 6y = 12$

53. **Television Signal** A television tower broadcasts a signal with a circular range represented by the equation $x^2 + y^2 = 10{,}000$. Can a city 60 miles east and 80 miles north of the tower receive the signal?

SHOW YOU KNOW

54. Explain what *eccentricity* means in your own words.

55. Explain the difference between an ellipse and a circle in terms of eccentricity.

56. The eccentricity of an ellipse is defined as $e = \dfrac{c}{a}$. If a is fixed and c is allowed to vary, describe the shape of the ellipse when the eccentricity nears 1 and when it is close to 0.

57. You are given the task of drawing an ellipse that is 28 inches wide and 18 inches tall. Describe how to use a piece of string, two tacks, and a pencil to accomplish the task. Be specific.

MAKE IT REAL

58. **London Bridge** One of the most famous landmarks in England is the London Bridge that spans the River Thames. Its history can be traced back to the first century and the Roman occupation of England. Since then, multiple structures have spanned the river's banks. The most recent bridge was built in 1960. The previous bridge, built in 1831, was transported, stone by stone, to Lake Havasu, Arizona.

One of the earliest bridges on the site and the first made of stone had twenty arches that were 60 feet high and 30 feet wide.

Write an equation for the ellipse in which the x-axis coincides with the water level and the y-axis passes through the center of the arch.

59. **Pont du Gard** Built by the Romans in the early first century, the Pont du Gard ("Bridge of the River") in France is constructed of multiple semi-elliptical arches. Each of the arches on the first level span a width of approximately 70 feet and are about 65 feet high. On the second level, the arches span 65 feet and are 60 feet high. On the top level, the arches span almost 24 feet and are 21 feet high.
(*Source:* www.pontdugard.fr)

a. Find the eccentricity of each of the arches.

b. Based on the photo, do the eccentricity values in part (a) seem to be reasonable?

60. **Radio Broadcast** The effective broadcast range in any one direction of the radio station KYLV 88.5 FM centered in Oklahoma City is 30 miles. The radio station KRMG 740 AM broadcast range in any one direction from Tulsa is 85 miles. (*Source:* www.ontheradio.net) Tulsa's broadcast station is 91 miles east of Oklahoma City and 52 miles north. Use the diagram, which shows Oklahoma City's broadcast station at (0, 0), to answer the following.

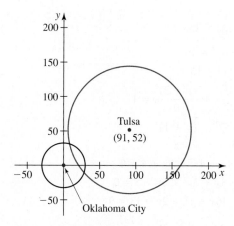

a. Generate the equation for each circle that models the broadcast area for the two stations.

b. Is there any location that can receive both stations' signals?

61. **Lithotripsy** Lithotripsy is a medical procedure used to break up kidney stones. One technique is to place a patient in an elliptical tank of water in such a way as to locate the person's kidney at one focus. Ultrasound shock waves are then emitted from the other focus, thereby concentrating them on the stone and pulverizing it. (*Source:* www.nlm.nih.gov/medlineplus)

Suppose the tank that a patient is placed in is 8 feet long and 5 feet wide.

a. Referring to the figure, find the equation of the ellipse that models the tank.

b. Where should the person's kidney be placed along the x-axis so that the sound waves are focused on the right location?

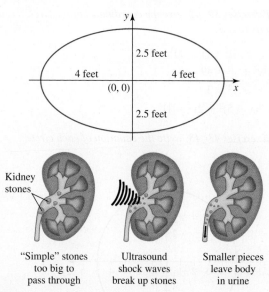

62. **The Orbit of Halley's Comet** Compared to the planets in our Solar System, the orbit of Halley's comet is highly elliptical with a small eccentricity. Its aphelion (farthest distance the comet ever gets from the Sun/farthest foci) is 35 AU, the semi-major axis is 17.8 AU, and eccentricity is 0.967. (Recall that AU is an astronomical unit equal to the average distance the Earth is from the Sun or 93 million miles.)

a. Find the equation for the ellipse that models the orbit of Halley's comet around the Sun.
b. How far from the center of the elliptical path is the Sun?
c. Sketch a graph of the ellipse modeling the comet's path.

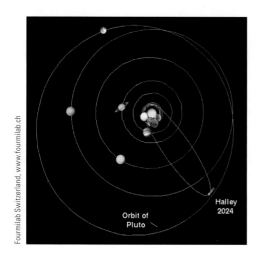

Halley's Comet 2024—Top View

■ STRETCH YOUR MIND

Exercises 63–66 are intended to challenge your understanding of ellipses.

63. Derive the equation $\dfrac{x^2}{b^2} + \dfrac{y^2}{a^2} = 1$ for the ellipse with its major axis on the y-axis and its center at $(0, 0)$.

64. Find the points of intersection of the graphs of the given equations. Sketch both graphs on the same axes, showing the points of intersection.

$$\begin{cases} x^2 + 4y^2 = 36 \\ x^2 + y^2 = 12 \end{cases}$$

65. The eccentricity of an ellipse is defined to be $e = \dfrac{c}{a}$, where c is the distance from the foci to the center and a is the length of half of the major axis. Show that it follows from $a > c$ that $e < 1$.

66. In the ellipse shown, segment $\overline{W_1 W_2}$ (called the focal width) passes through the focus F and is perpendicular to the major axis.

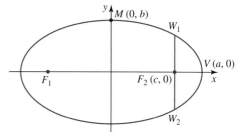

Show that the length of $\overline{W_1 W_2}$ is $\dfrac{2b^2}{a}$.

SECTION 11.3 Parabolas

LEARNING OBJECTIVES

- Generate equations for parabolas given a graph, a set of points, or words
- Apply parabolas to real-world scenarios

GETTING STARTED

With over 6 billion people living on Earth, the need for accessible and sustainable energy solutions is becoming increasingly important. In impoverished parts of the world, seemingly simple energy tasks such as cooking food often become extremely difficult. One relatively low-cost solution, a parabolic solar oven, relies heavily on mathematical principles.

In this section we investigate the anatomy and equations of parabolas and look at real-world applications of parabolas.

■ The Parabola

A **parabola** is a geometric shape formally defined as follows.

Parabola

A **parabola** is the set of all points in a plane that are equidistant from a line called the **directrix** and a fixed point not on the line called the **focus**.

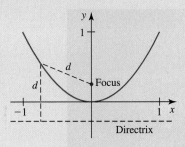

The directrix may be parallel to the x-axis, as shown above, or parallel to any other line. However, for simplicity, we focus on parabolas that open vertically (as above) or horizontally.

Anatomy of a Parabola

- The line perpendicular to the directrix that passes through the focus is called the **axis of symmetry**.
- The point on the axis of symmetry that is equidistant from the focus and the directrix is the **vertex**.
- The line segment that crosses the interior of the parabola parallel to the directrix and through the focus is the **latus rectum**.

■ Development of the Standard Equation

To develop the standard equation for the parabola, we position the vertex of the parabola at the origin $(0, 0)$ for simplicity's sake. See Figure 11.22. Then we label the distance between the vertex and the focus p. Consequently, the coordinate of the focus is $(0, p)$. Since the vertex lies halfway between the focus and directrix, the directrix is the line $y = -p$.

We pick an arbitrary point (x, y) on the graph of the parabola. We know the distance between this point and the focus is equal to the shortest distance between this point and the directrix. That is, the distance between $(0, p)$ and (x, y) is the same as the distance between (x, y) and $(x, -p)$. We will use the distance formula $d = \sqrt{(x_2 - x_1)^2 + (y_2 - y_1)^2}$ to determine each of these distances.

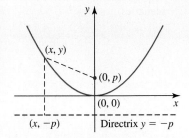

Figure 11.22

- distance between $(0, p)$ and (x, y):
$$d = \sqrt{(x-0)^2 + (y-p)^2}$$
$$= \sqrt{x^2 + y^2 - 2py + p^2}$$

- distance between (x, y) and $(x, -p)$:
$$d = \sqrt{(x-x)^2 + [y-(-p)]^2}$$
$$= \sqrt{y^2 + 2py + p^2}$$

Since these distances are equal, we have

$$\sqrt{x^2 + y^2 - 2py + p^2} = \sqrt{y^2 + 2py + p^2}$$
$$x^2 + y^2 - 2py + p^2 = y^2 + 2py + p^2$$
$$x^2 \cancel{+ y^2} - 2py \cancel{+ p^2} = \cancel{y^2} + 2py \cancel{+ p^2}$$
$$x^2 - 2py = 2py$$
$$x^2 = 4py \quad \text{standard equation of a parabola}$$

The coordinates of every point (x, y) on the parabola satisfy this equation.

THE PARABOLA: AXIS OF SYMMETRY VERTICAL, VERTEX AT (0, 0)

The standard equation of a parabola with vertex at the origin and axis of symmetry *vertical* is

$$x^2 = 4py$$

The equations for certain transformed parabolas follow.

THE PARABOLA: AXIS OF SYMMETRY HORIZONTAL, VERTEX AT (0, 0)

The standard equation of a parabola with vertex at the origin and axis of symmetry *horizontal* is

$$y^2 = 4px$$

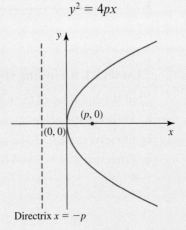

As shown in the next two boxes, shifting a parabola horizontally or vertically changes the location of the vertex but does not affect the overall shape of the parabola.

THE PARABOLA: AXIS OF SYMMETRY VERTICAL, VERTEX AT (h, k)

The standard equation of a parabola with vertex at (h, k) and axis of symmetry *vertical* is

$$(x - h)^2 = 4p(y - k)$$

THE PARABOLA: AXIS OF SYMMETRY HORIZONTAL, VERTEX AT (h, k)

The standard equation of a parabola with vertex at (h, k) and axis of symmetry *horizontal* is

$$(y - k)^2 = 4p(x - h)$$

EXAMPLE 1 ■ Finding the Equation of a Parabola

Find the equation of the parabola. Then graph the parabola.
a. Vertex at the origin and focus at $(0, 2)$
b. Directrix at $x = 3$ and vertex at $(5, 4)$
c. Directrix at $y = 3$ and focus at $(4, -1)$

Solution

a. Since the vertex is at the origin and the focus is *above* the vertex, this parabola will have the form $x^2 = 4yp$. Since the distance between the vertex $(0, 0)$ and the focus $(0, 2)$ is 2, $p = 2$. So

$$x^2 = 4yp$$
$$x^2 = 4y(2)$$
$$x^2 = 8y$$

This graph is shown in Figure 11.23.

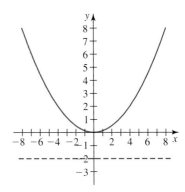

Figure 11.23

b. Since the directrix is vertical, the axis of symmetry will be horizontal. Furthermore, since the vertex is not the origin, the parabola will have an equation of the form $(y - k)^2 = 4p(x - h)$. Since the vertex is $(5, 4)$, we have

$$(y - 4)^2 = 4p(x - 5)$$

Since the vertex is two units away from the directrix, $p = 2$. Thus

$$(y - 4)^2 = 4(2)(x - 5)$$
$$(y - 4)^2 = 8(x - 5)$$

The graph is shown in Figure 11.24.

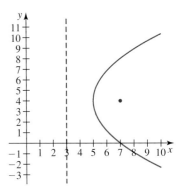

Figure 11.24

c. Since the directrix is horizontal, the axis of symmetry is vertical. The vertex is on the axis of symmetry halfway between the directrix and the focus. Since $\dfrac{3 + (-1)}{2} = 1$, the vertex is at $(4, 1)$. The distance from the vertex to the focus is 2 units, so $p = 2$. Notice that the vertex lies *below* the directrix, so this parabola is concave down. In developing the standard equation, we assumed the focus was *above* the directrix, so we must change the sign on the value of p in the equation for this parabola. We have

$$(x - h)^2 = 4p(y - k)$$
$$(x - 4)^2 = 4(-2)(y - 1)$$
$$(x - 4)^2 = -8(y - 1)$$

The graph is shown in Figure 11.25.

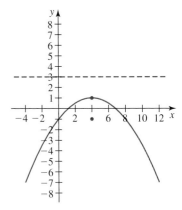

Figure 11.25

EXAMPLE 2 ■ Describing a Parabola from Its Equation

Determine the vertex, focus, directrix, and axis of symmetry of each parabola. Then graph the parabola.

a. $x^2 = 16y$
b. $y^2 = 2x$
c. $(y - 4)^2 = 16(x + 1)$
d. $x^2 + 6x - y = 0$

Solution

a. The vertex is (0, 0) and the parabola is vertical. Therefore, the axis of symmetry is $x = 0$. We know $4p = 16$, so $p = 4$. The directrix is 4 units below the vertex at $y = -4$ and the focus is 4 units above the vertex at (0, 4). The graph is shown in Figure 11.26.

Figure 11.26

b. The vertex is (0, 0) and the parabola is horizontal. Therefore, the axis of symmetry is $x = 0$. We know $4p = 2$, so $p = \frac{1}{2}$. The directrix is $\frac{1}{2}$ unit to the left of the vertex at $x = -\frac{1}{2}$ and the focus is $\frac{1}{2}$ unit to the right of the vertex at $\left(0, \frac{1}{2}\right)$. The graph is shown in Figure 11.27.

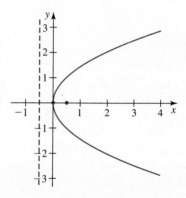

Figure 11.27

c. The vertex is $(-1, 4)$ and the parabola is horizontal. Therefore, the axis of symmetry is $y = 4$. We know $4p = 16$, so $p = 4$. The directrix is 4 units to the left of the vertex at $x = -5$ and the focus is 4 units to the right of the vertex at (3, 4). The graph is shown in Figure 11.28.

Figure 11.28

d. We need to complete the square to get the equation in the standard form.

$$x^2 + 6x - y = 0$$
$$x^2 + 6x = y$$
$$x^2 + 6x + \left(\frac{6}{2}\right)^2 = y + \left(\frac{6}{2}\right)^2 \quad \text{Add the square of } \frac{b}{2a}.$$
$$(x + 3)^2 = y + 9 \quad \text{Simplify.}$$

The vertex is $(-3, -9)$ and the parabola is vertical. Therefore, the axis of symmetry is $x = -3$. We know $4p = 1$, so $p = \frac{1}{4}$. The directrix is $\frac{1}{4}$ unit below the vertex at $y = -9\frac{1}{4}$ and the focus is $\frac{1}{4}$ unit above the vertex at $\left(-3, -8\frac{3}{4}\right)$. The graph is shown in Figure 11.29.

Figure 11.29

It is important to note that the equation $(x + 3)^2 = y + 9$ from part (d) can be written as $y = (x + 3)^2 - 9$, which is the vertex form of a quadratic function we learned in Chapter 4. In fact, the concepts from Chapter 4 can be applied to any parabolic graphs that are functions.

■ Applications of Parabolas

A parabola has one truly amazing property: Various types of rays (e.g., light, sound) striking the interior of the parabola parallel to the axis of symmetry of the parabola are concentrated into a single point (*the focus*), as shown in Figure 11.30. Thus an intense amount of energy is concentrated into a single point in the interior of the parabola. This property is key to the design of a number of parabolic devices such as solar panels, contacts, eyeglasses, satellite dishes, and flashlights.

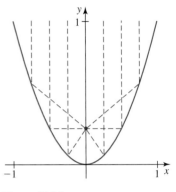

Figure 11.30

Parabolic Troughs

Mr. Matt Orosz, founder of the MIT Solar Turbine Group, discovered firsthand how important solar parabolas can be for cooking while serving as a Peace Corps volunteer in Lesotho. Upon returning to MIT, Mr. Orosz organized a team of colleagues to streamline the manufacturing process of parabolic solar troughs and to investigate additional ways to use the troughs to harness energy. The team invented a solar thermal microgenerator that produces electricity, pumped water, and hot water supply. In May 2006, the team won a prestigious World Bank Development Marketplace grant, which provided a way for them to implement two solar thermal microgenerators in Lesotho. The solar troughs have increased the standard of living for numerous families in that region of Africa.*

The solar troughs use the reflective nature of a parabolic surface to concentrate the sun's rays into a single area of the interior of the trough, as illustrated in Figure 11.31.

Trough—front view

Figure 11.31

Source: http://serc.carleton.edu. Courtesy SERC, the Science Education Resource Center at Carleton College.

EXAMPLE 3 ■ **Determining the Equation of a Parabola with a Known Focal Length**

In constructing the solar troughs, the Solar Turbine Group typically uses a focal length of 40 centimeters. Assuming the vertex is at the origin and the trough is in a vertical position, determine the equation of the parabola that shapes the trough.

---

* Mr. Orosz set up the non-profit Solar Turbine Group to sustain the humanitarian work of the team. Individuals interested in learning more about the projects or interested in contributing to the work may learn more at www.solarturbinegroup.org.

Solution Since the focal length is 40 centimeters, we substitute $p = 40$ into the standard equation.

$$x^2 = 4py$$
$$x^2 = 4(40)y$$
$$x^2 = 160y$$

EXAMPLE 4 ■ Comparing Parabolas with Different Focal Lengths

Figure 11.32 shows four parabolas that could be used in the design of a solar trough together with their foci. The vertex of each parabola is at the origin and the foci are at (0, 10), (0, 25), (0, 40), and (0, 50), respectively.

a. Describe what effect moving the focus away from the vertex has on the shape of the graph.

b. Determine the length of the latus rectum for each graph.

Figure 11.32

Solution

a. As the focus moves away from the vertex, the sides of the parabola become less steep.

b. The equation of each parabola is of the form $x^2 = 4py$. Since the latus rectum is a horizontal line segment passing through the focus $(0, p)$, the endpoints of the latus rectum will be of the form $(-a, p)$ and (a, p). Since both of these points lie on the parabola, they will satisfy the equation $x^2 = 4py$.

$$x^2 = 4py$$
$$(a)^2 = 4p(p)$$
$$a^2 = 4p^2$$
$$a = \pm\sqrt{4p^2}$$
$$a = \pm 2p$$

So the endpoints of the latus rectum are $(-2p, p)$ and $(2p, p)$. Since this segment is horizontal, the length of the latus rectum is the difference in the x-values, which is $4p$. So the length of the latus rectum is four times the focal length. The results for the four parabolas are shown in Table 11.3.

Table 11.3

Focal length p	Latus Rectum Length $4p$
10	40
25	100
40	160
50	200

Parabolic Antennas

Parabolic antennas are designed by rotating a parabola about its axis of symmetry. The resultant shape is called a **paraboloid**. Parabolic antennas are commonly used for radio, television, data communication, and RADAR. The feed antenna is positioned at the focus of the parabola. The radio rays striking the surface of the reflector are concentrated on the feed antenna at the focus.

EXAMPLE 5 ■ Estimating the Equation of a Parabola from a Physical Device

Many consumers have a small satellite dish mounted on an exterior wall of their homes. For one brand of dish, the focal length is roughly 17 inches and the diameter of the circular cross section of the dish is roughly 22 inches. What is the distance between the focus and the plane described by the circular edge of the dish? (See Figure 11.33.)

Solution Since we know the focal length $p = 17$, the equation of the parabola is given by

$$x^2 = 4py$$
$$x^2 = 4(17)y$$
$$x^2 = 68y$$

We evaluate this function at $x = 11$.

$$(11)^2 = 68y$$
$$y = \frac{121}{68}$$
$$y \approx 1.8 \text{ inches}$$

Figure 11.33

So the vertex is about 1.8 inches from the plane described by the circular edge of the dish. Since the distance from the vertex to the focus is 17 inches, the distance from the plane to the focus is $17 - 1.8 = 15.2$ inches.

Parabolic Paths

When water under pressure is released upward into the air, it describes a parabolic path. If we know the equation of the parabola, we can predict the path of the water.

EXAMPLE 6 ■ Determining the Equation of a Parabola from a Picture

A photo of a drinking fountain is overlaid with a square grid in Figure 11.34. We define the point (0, 0) to be the point where the water emerges from the fountain. Determine the equation of the parabola.

Solution The water passes through (0, 0). The vertex appears to be at (−6, 9). We know the equation will be of the form $(x - h)^2 = -4p(y - k)$ since the graph is concave down. We substitute the vertex into the equation.

Figure 11.34

$$(x - h)^2 = -4p(y - k)$$
$$[x - (-6)]^2 = -4p(y - 9)$$
$$(x + 6)^2 = -4p(y - 9)$$

We determine the value of p by substituting in the point (0, 0).

$$(0 + 6)^2 = -4p(0 - 9)$$
$$36 = 36p$$
$$p = 1$$

The equation of the parabola is $(x + 6)^2 = -4(y - 9)$. As shown in Figure 11.35, the graph fits the image.

Figure 11.35

SUMMARY

In this section you learned the anatomy of the parabola and how to generate its standard equations. You also discovered various applications of parabolas.

11.3 EXERCISES

SKILLS AND CONCEPTS

In Exercises 1–4, the graph of a parabola is given. Match each graph with its equation.

A.

B.

C.

D.

1. $(y - 1)^2 = 16x$
2. $x^2 = -20(y - 1)$
3. $y^2 = -16x$
4. $x^2 = 20y$

In Exercises 5–16, write the equation of each parabola. Then graph the equation.

5. Focus at (4, 0); vertex at (6, 0)
6. Focus at (0, 3); vertex at (0, 5)
7. Focus at (1, 0); directrix at $y = 4$
8. Focus at (0, 2); directrix at $x = 3$
9. Focus at (1.5, 0); directrix at the y-axis
10. Focus at (0, 6); focal length 2; concave down
11. Vertex at (7, −2); focal length 3; concave up
12. Vertex at (−4, 3); focus at (−4, 1.5)
13. Vertex at (3, 3); focus at (3, 8)
14. Focal length 6; vertex at (4, 2); opens to the left
15. Focus at (1, 5); latus rectum of length 20; opens to the right
16. Focus at (1, 3); latus rectum of length 12; concave down

In Exercises 17–24, find the vertex and focus of each parabola. Then graph the equation.

17. $x^2 = 12y$
18. $y^2 = 24x$
19. $(x + 2)^2 = -y$
20. $(x - 9)^2 = 36y$
21. $y^2 = 18(x - 5)$
22. $(y - 7)^2 = -2(x + 3)$
23. $x^2 + 8x = -4y$
24. $y^2 + 4y = -12x$

In Exercises 25–28, write the equation for each parabola.

25.

26.

27.

28.
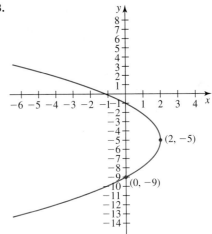

In Exercises 29–46, find the directrix, focus, vertex, and axis of symmetry for each parabola. Then graph the equation.

29. $x^2 = 28y$
30. $y^2 = 14x$
31. $y^2 = x$
32. $x^2 = y$
33. $(x - 4)^2 = 16y$
34. $(x + 2)^2 = 4(y - 1)$
35. $y^2 = 6(x + 5)$
36. $(y - 7)^2 = 8x$
37. $y^2 = -6(x + 9)$
38. $(x + 5)^2 = 2(y - 8)$
39. $(x + 12)^2 = 3(y + 7)$
40. $(y + 2)^2 = -10x$
41. $(y + 1)^2 = -16(x + 4)$
42. $(x + 8)^2 = -22(y - 1)$
43. $(x - 9)^2 = -48(y + 2)$
44. $\left(x + \frac{1}{2}\right)^2 = 2\left(y - \frac{3}{2}\right)$
45. $\left(y - \frac{1}{2}\right)^2 = 2\left(x + \frac{3}{2}\right)$
46. $\left(y - \frac{3}{4}\right)^2 = x - 3$

SHOW YOU KNOW

47. What is the relationship between the vertex, the focus, and the directrix of a parabola?
48. From the equation of a parabola, how can you tell if it is concave down?
49. From the equation of a parabola, how can you tell if it opens horizontally or vertically?

50. A classmate claims that a particular horizontal parabola has focus (2, 4) and vertex (3, 5). How do you know that he's wrong?

MAKE IT REAL

51. Water Fountain Trajectory Using the square grid overlain on the picture, determine the equation of the graph of the parabola. Assume the origin is the point on the grid where the water first emerges.

52. Architectural Design This beautiful planetarium is located in City of Arts and Sciences in Valencia, Spain. In this two-dimensional photograph, is the top arch spanning the photo parabolic? (*Hint:* If we designate the vertex at the top of the arch as (0, 0), the following points lie on the arch: (2, −0.3), (4, −1.2), (−2, −0.3), and (−4, −1.2).)

53. Architectural Design The Sheffield Winter Gardens in Sheffield, England, are well known for their beautiful parabolic arches. The 21 arches span a space 230 feet long and 72 feet wide. The arches are of varying heights; however, the tallest arch is 72 feet tall.

Assuming the tallest arch is 72 feet tall and 72 feet wide at its base, determine the equation of the parabola. Assume the origin is the point on the floor of the garden directly below the vertex of the parabola.

54. Architectural Design Referring to the specifications given in Exercise 53, determine the equation of the parabola whose height is 12 feet shorter than the tallest parabola.

55. Telescope Mirror Manufacturing The Steward Observatory Mirror Lab at the University of Arizona creates telescope mirrors by spinning molten glass into the shape of a paraboloid. The focal ratio, commonly called the f/number, is the ratio of the focal length to the diameter of the mirror. The f/number of a mirror with 6.5-meter diameter is typically 1.125. (*Source:* mirrorlab.as.arizona.edu)

Determine the equation of the parabola created by the surface of the mirror. Assume the vertex is at the origin.

56. Telescope Mirror Manufacturing (For background on mirror manufacturing, refer to Exercise 55.) A telescope mirror with 8.4-meter diameter typically has an f/number of 1.14. (*Source:* mirrorlab.as.arizona.edu) Determine the equation of the parabola formed by the surface of the mirror. Assume the vertex is at the origin.

57. Parabolic Solar Troughs Solar power plants often use parabolic solar troughs to collect energy from the sun (see photo). The troughs track the sun as it moves across the sky.

Assume the origin is on the surface of the ground and the vertex is at (0, 3). When the trough is vertical, the right edge of the trough is at (14, 6). Determine the equation of the parabola used to generate the trough.

58. **Parabolic Solar Troughs** (For background information on solar troughs, see Exercise 57.) Assume the origin is on the surface of the ground and the vertex is at (0, 3). When the trough is vertical, the right edge of the trough is at (15, 8). Determine the equation of the parabola used to generate the trough.

59. **Solar Wind Concentrator** A cross section of a mirror in the Genesis Solar Wind Concentrator is a parabola with the focus about 20 centimeters from the vertex of the parabola. (*Source*: www.nasa.gov)

    Assuming the vertex is at the origin, determine the equation of the parabola formed by the surface of the mirror.

60. **Catenary** When a flexible cable is supported on both ends, the shape it takes on is referred to as a **catenary**. Galileo claimed a hanging cable would take on the shape of a parabola but later mathematicians disproved that theory. The image here shows a catenary.

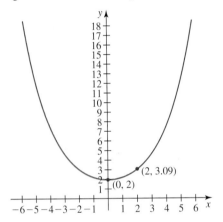

Prove that the figure is not a parabola. (*Hint:* Find the equation of the parabola with vertex (0, 2) that passes through the point (2, 3.09). Then show a point on the catenary that does not satisfy the parabolic equation.)

■ **STRETCH YOUR MIND**

Exercises 61–65 are intended to challenge your understanding of parabolas.

61. What is the relationship between the standard form of a quadratic equation $y = ax^2 + bx + c$ and the standard form of a parabola $x^2 = 4py$?

62. How is the function $y = 2\sqrt{x}$ related to the horizontal parabola $y^2 = 4x$?

63. A vertical parabola with vertex at the origin has focus $(0, p)$. A horizontal parabola with vertex at the origin has focus $(p, 0)$. What is the focus of a parabola angled at 45° (halfway between horizontal and vertical)?

64. In polar coordinates, the equation $r = 4p \sin(\theta)\sec^2(\theta)$ describes a parabola. Determine the corresponding equation in rectangular coordinates.

65. In polar coordinates, the equation

    $$r = 4p \sin\left(\theta + \frac{\pi}{4}\right)\sec^2\left(\theta + \frac{\pi}{4}\right)$$

    is a parabola with vertex at the origin and angled to the right halfway between horizontal and vertical. Determine the corresponding equation in rectangular coordinates.

SECTION 11.4 Hyperbolas

LEARNING OBJECTIVES

- Generate equations for hyperbolas given a graph, a set of points, or words
- Apply hyperbolas to real-world scenarios

GETTING STARTED

During World War II, the LORAN (LOng RAnge Navigation) was used to locate ships in the ocean. Although not as popular today as the Global Positioning System (GPS), the hyperbolic LORAN system is still used and is considered a backup to the GPS. The LORAN system uses low-frequency transmitters positioned all over the world. It sends signals to a receiver on the ship that is being tracked. The amount of time it takes for the signal to reach the receiver and return is recorded and the distance from the ship to the transmitter is calculated. This process is repeated using a different transmitter and the difference in the distances is calculated. Armed with these two calculations, trackers determine that the ship lies somewhere on a hyperbola. The process is repeated with a second pair of transmitters and a second hyperbola is determined. The location of the ship is determined by calculating the intersection of the two hyperbolas. The entire process is very rapid, taking less time than it took for you to read this paragraph.

In this section, we investigate hyperbolas and their properties, including that which makes the LORAN system possible. We also generate the standard equation for hyperbolas and apply it to real-world scenarios.

■ The Hyperbola

The definition of a hyperbola is similar to that of an ellipse. However, for the hyperbola the difference (rather than the sum) of the distances from the foci is constant. We define **hyperbola** as follows.

> **HYPERBOLA**
>
> A **hyperbola** is the set of all points in a plane with the property that the difference in the distances d_1 and d_2 from two fixed points in the plane (the foci) is constant.

We represent the two distances as line segments and align their endpoints on the right, as shown in Figure 11.36. The distance $d_1 - d_2$ (represented by the dashed line segment) must remain constant although d_1 and d_2 may change in length. If we increase d_1 by 1 unit we must also increase d_2 by 1 unit in order to keep their difference constant:

Figure 11.36

$$(d_1 + 1) - (d_2 + 1) = d_1 - d_2$$

Graphically speaking, if we move the right endpoint of the d_1 segment any number of units, we must also move the right endpoint of the d_2 segment the same number of units.

Consider Table 11.4, a table of values showing pairs of distances that differ by exactly four units.

Table 11.4

d_1	d_2	$d_1 - d_2$
6	2	4
7	3	4
9	5	4
11	7	4

To help us visualize this concept, we draw a line segment between a focus at $(-3, 0)$ and a focus at $(3, 0)$. As shown in Figure 11.37, we construct a series of triangles that share this focal line segment as a common side. The other two sides of the triangle have lengths d_1 and d_2, respectively. By construction, the vertex of each triangle that is formed by legs of length d_1 and d_2 has the property that $d_1 - d_2 = 4$. Therefore, each of these vertices lies on a single hyperbola. We add a few additional triangles in Figure 11.38 and then connect the vertices.

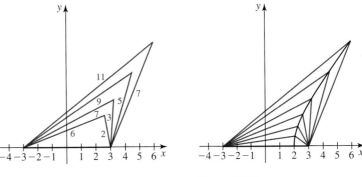

Figure 11.37   **Figure 11.38**

Although we placed these triangles above the *x*-axis, a corresponding set of similar triangles lies below the *x*-axis. We draw those triangles in Figure 11.39 and continue to graph the hyperbola.

We construct two more sets of triangles to generate the second branch (piece) of the hyperbola shown in Figure 11.40.

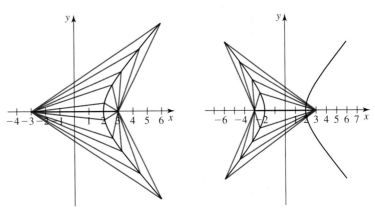

Figure 11.39   **Figure 11.40**

ANATOMY OF A HYPERBOLA

- The fixed points F_1 and F_2 are called **foci** (or, individually, a **focus**).
- The midpoint of the segment between F_1 and F_2 is called the **center**, C.
- The segment between V_1 and V_2 is called the **transverse axis**.
- The endpoints V_1 and V_2 of the transverse axis are referred to as the **vertices**.
- When the transverse axis is vertical, the hyperbola opens up and down. When the transverse axis is horizontal, the hyperbola opens left and right.

Development of the Standard Equation

To derive the equation of the hyperbola, we consider a hyperbola centered at the origin. The foci are c units away from the center and have the coordinates $F_1(-c, 0)$ and $F_2(c, 0)$. See Figure 11.41.

We define $2a$ to be the constant difference. That is, $d_1 - d_2 = 2a$. Using the distance formula and algebraic techniques, we derive the standard formula for the hyperbola.

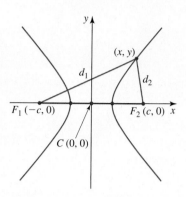

Figure 11.41

$$d_1 = \sqrt{[x-(-c)]^2 + (y-0)^2} \quad \text{and} \quad d_2 = \sqrt{(x-c)^2 + (y-0)^2}$$

$$\sqrt{[x-(-c)]^2 + y^2} - \sqrt{(x-c)^2 + y^2} = 2a \quad \text{Difference of two distances equals 2a.}$$

$$\sqrt{(x+c)^2 + y^2} = 2a + \sqrt{(x-c)^2 + y^2}$$

$$\left[\sqrt{(x+c)^2 + y^2}\right]^2 = \left[2a + \sqrt{(x-c)^2 + y^2}\right]^2 \quad \text{Square both sides and expand.}$$

$$(x+c)^2 + y^2 = (2a)^2 + 2(2a)\left(\sqrt{(x-c)^2 + y^2}\right) + \left(\sqrt{(x-c)^2 + y^2}\right)^2$$

$$(x+c)^2 + y^2 = 4a^2 + 4a\sqrt{(x-c)^2 + y^2} + x^2 - 2cx + c^2 + y^2$$

$$x^2 + 2cx + c^2 + y^2 = 4a^2 + 4a\sqrt{(x-c)^2 + y^2} + x^2 - 2cx + c^2 + y^2$$

$$2cx = 4a^2 + 4a\sqrt{(x-c)^2 + y^2} - 2cx \quad \text{Simplify.}$$

$$4cx - 4a\sqrt{(x-c)^2 + y^2} - 4a^2 = 0 \quad \text{Combine like terms.}$$

$$-4a\sqrt{(x-c)^2 + y^2} = 4a^2 - 4cx \quad \text{Rearrange terms.}$$

$$-a\sqrt{(x-c)^2 + y^2} = a^2 - cx \quad \text{Divide by 4.}$$

$$\left[-a\sqrt{(x-c)^2 + y^2}\right]^2 = (a^2 - cx)^2 \quad \text{Square both sides.}$$

$$a^2(x^2 - 2cx + c^2 + y^2) = a^4 - 2a^2cx + c^2x^2 \quad \text{Expand the expression.}$$

$$a^2x^2 - 2a^2cx + a^2c^2 + a^2y^2 = a^4 - 2a^2cx + c^2x^2 \quad \text{distributive property}$$

$$a^2x^2 - c^2x^2 - 2a^2cx + 2a^2cx + a^2y^2 = a^4 - a^2c^2 \quad \text{Rearrange terms.}$$

$$x^2(a^2 - c^2) + a^2y^2 = a^2(a^2 - c^2) \quad \text{Factor.}$$

$$-x^2(c^2 - a^2) + a^2y^2 = -a^2(c^2 - a^2) \quad a^2 - c^2 = -(c^2 - a^2)$$

$$\frac{a^2y^2}{-a^2(c^2 - a^2)} - \frac{x^2(c^2 - a^2)}{-a^2(c^2 - a^2)} = \frac{-a^2(c^2 - a^2)}{-a^2(c^2 - a^2)} \quad \text{Divide by } -a^2(c^2 - a^2).$$

$$\frac{x^2}{a^2} - \frac{y^2}{c^2 - a^2} = 1 \quad \text{Simplify.}$$

$$\frac{x^2}{a^2} - \frac{y^2}{b^2} = 1 \quad \text{Let } b^2 = c^2 - a^2.$$
standard equation of a hyperbola

The coordinates of every point (x, y) on the hyperbola satisfy this equation. The x-intercepts can be found by setting $y = 0$.

$$\frac{x^2}{a^2} - \frac{y^2}{b^2} = 1$$

$$\frac{x^2}{a^2} - \frac{0^2}{b^2} = 1$$

$$\frac{x^2}{a^2} = 1$$

$$x^2 = a^2$$

$$x = \pm\sqrt{a^2}$$

$$x = a \text{ or } x = -a$$

Therefore, the coordinates of the vertices V_1 and V_2 are $(a, 0)$ and $(-a, 0)$.

THE HYPERBOLA: HORIZONTAL TRANSVERSE AXIS, CENTER AT (0, 0)

The standard equation of a hyperbola with center at the origin and with a horizontal transverse axis is $\frac{x^2}{a^2} - \frac{y^2}{b^2} = 1$, where $a^2 + b^2 = c^2$.

- The vertices of the hyperbola are $(a, 0)$ and $(-a, 0)$.
- The transverse axis has length $2a$.

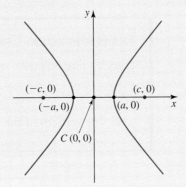

The equations for certain transformed hyperbolas are as follows. The developments of these are left as challenge exercises.

THE HYPERBOLA: VERTICAL TRANSVERSE AXIS, CENTER AT (0, 0)

The standard equation of a hyperbola with center at (0, 0) and with a vertical transverse axis is $\dfrac{y^2}{a^2} - \dfrac{x^2}{b^2} = 1$, where $a^2 + b^2 = c^2$.

- The vertices of the hyperbola are $(0, a)$ and $(0, -a)$.
- The transverse axis has length $2a$.

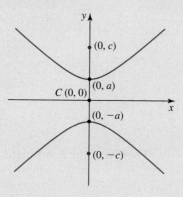

THE HYPERBOLA: CENTER AT (h, k), HORIZONTAL TRANSVERSE AXIS

The standard equation of a hyperbola with center at (h, k) and with a horizontal transverse axis is $\dfrac{(x - h)^2}{a^2} - \dfrac{(y - k)^2}{b^2} = 1$, where $a^2 + b^2 = c^2$.

- The vertices of the hyperbola are $(h + a, k)$ and $(h - a, k)$.

THE HYPERBOLA: CENTER AT (h, k), VERTICAL TRANSVERSE AXIS

The standard equation of a hyperbola with center at (h, k) and with a vertical transverse axis is $\dfrac{(y - k)^2}{a^2} - \dfrac{(x - h)^2}{b^2} = 1$, where $a^2 + b^2 = c^2$.

- The vertices of the hyperbola are $(h, k + a)$ and $(h, k - a)$.

EXAMPLE 1 ■ Finding the Equation of a Hyperbola

Find an equation of the hyperbola centered at the origin and with vertex $(2, 0)$ and one focus at $(\sqrt{13}, 0)$.

Solution With a vertex of $(2, 0)$ and focus of $(\sqrt{13}, 0)$, we know the transverse axis is horizontal. Thus the equation is of the standard form $\dfrac{(x - h^2)}{a^2} - \dfrac{(y - k)^2}{b^2} = 1$. The center is $(0, 0)$ so we know the equation will be $\dfrac{x^2}{a^2} - \dfrac{y^2}{b^2} = 1$ since $h = 0$ and $k = 0$. Since one vertex is at $(2, 0)$, the other vertex is at $(-2, 0)$. This means $a = 2$. Finally, since one focus is at $(\sqrt{13}, 0)$, we know $c = \sqrt{13}$. Using the relationship $a^2 + b^2 = c^2$, we determine the value of b^2 as follows.

$$(2)^2 + b^2 = (\sqrt{13})^2$$
$$4 + b^2 = 13$$
$$b^2 = 13 - 4$$
$$b^2 = 9$$

The equation of the hyperbola is

$$\frac{x^2}{a^2} - \frac{y^2}{b^2} = 1$$

$$\frac{x^2}{4} - \frac{y^2}{9} = 1$$

Analyzing Asymptotes for a Hyperbola

One of the unique features of the hyperbola is that the graph becomes nearly linear as $x \to \pm\infty$. Consider the standard equation of a hyperbola, $\frac{x^2}{a^2} - \frac{y^2}{b^2} = 1$. Observe $x \neq 0$ because when $x = 0$ the resultant equation has no solution. When we solve the standard equation for y, we get

$$y = \pm\sqrt{b^2\left(\frac{x^2}{a^2} - 1\right)}$$

$$= \pm b\sqrt{\frac{x^2}{a^2} - 1} \qquad \text{since } \sqrt{b^2} = \pm b$$

$$= \pm b\sqrt{\frac{x^2}{a^2}\left(1 - \frac{a^2}{x^2}\right)} \qquad \text{Factor out } \frac{x^2}{a^2}.$$

$$= \pm\frac{bx}{a}\sqrt{1 - \frac{a^2}{x^2}} \qquad \text{since } \sqrt{\frac{x^2}{a^2}} = \pm\frac{x}{a}$$

Notice as $x \to \pm\infty$, $x^2 \to \pm\infty$ and $\frac{a^2}{x^2} \to 0$. So for values of x with large magnitude,

$$y \approx \pm\frac{bx}{a}\sqrt{1 - 0}$$

$$= \pm\frac{b}{a}x$$

We use this result to define the **asymptotes** of a hyperbola.

ASYMPTOTES OF A HYPERBOLA

- A hyperbola with a horizontal transverse axis; centered at the origin, has **asymptotes** with equations $y = \frac{b}{a}x$ and $y = -\frac{b}{a}x$.
- A hyperbola with a vertical transverse axis, centered at the origin, has **asymptotes** with equations $y = \frac{a}{b}x$ and $y = -\frac{a}{b}x$.

EXAMPLE 2 ■ Determining the Equations of the Asymptotes for a Hyperbola

Find the equations of the asymptotes for the hyperbola $\dfrac{x^2}{4} - \dfrac{y^2}{9} = 1$. Then graph the hyperbola along with the asymptotes.

Solution From the given equation, we know $a = 2$ and $b = 3$. We substitute these values in the general asymptote equation, $y = \pm \dfrac{b}{a} x$.

$$y = \pm \dfrac{3}{2} x$$

The equations of the asymptotes are $y = \dfrac{3}{2} x$ and $y = -\dfrac{3}{2} x$. We graph all three equations in Figure 11.42 to confirm our work.

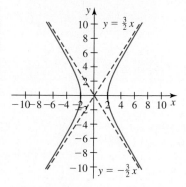

Figure 11.42

EXAMPLE 3 ■ Describing a Hyperbola from Its Equation

Determine the center, foci, vertices, asymptotes, and direction of each hyperbola. Then graph the equation.

a. $\dfrac{x^2}{36} - \dfrac{y^2}{16} = 1$

b. $\dfrac{y^2}{1} - \dfrac{x^2}{9} = 1$

c. $\dfrac{(x-4)^2}{4} - \dfrac{(y+3)^2}{25} = 1$

d. $2y^2 - 3x^2 - 12x + 4y - 16 = 0$

Solution

a. Since the equation follows the general form $\dfrac{x^2}{a^2} - \dfrac{y^2}{b^2} = 1$ where $a^2 + b^2 = c^2$, we know the center is $(0, 0)$, the transverse axis is horizontal, $a^2 = 36$, and $b^2 = 16$. Therefore, we use $a = 6$ and $b = 4$ to find c.

$$a^2 + b^2 = c^2$$
$$36 + 16 = c^2$$
$$52 = c^2$$
$$\sqrt{52} = c$$

The foci are $(-\sqrt{52}, 0)$ and $(\sqrt{52}, 0)$ and the vertices are $(-6, 0)$ and $(6, 0)$. The hyperbola opens left and right. We determine the equations of the asymptotes using $y = \pm \dfrac{b}{a} x$.

$$y = \pm \dfrac{4}{6} x$$
$$= \pm \dfrac{2}{3} x$$

We draw the asymptotes and vertices and then sketch the graph of the hyperbola, as shown in Figure 11.43.

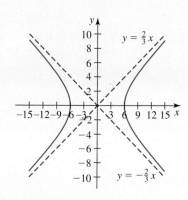

Figure 11.43

b. The equation follows the form $\frac{y^2}{a^2} - \frac{x^2}{b^2} = 1$ where $a^2 + b^2 = c^2$. We know the center is (0, 0), the transverse axis is vertical, $a^2 = 1$, and $b^2 = 9$. Therefore, we use $a = 1$ and $b = 3$ to find c.

$$a^2 + b^2 = c^2$$
$$1 + 9 = c^2$$
$$10 = c^2$$
$$\sqrt{10} = c$$

Therefore the foci are $(0, \sqrt{10})$ and $(0, -\sqrt{10})$ and the vertices are (0, 1) and (0, −1). The hyperbola opens up and down. We determine the equations of the asymptotes using $y = \pm\frac{a}{b}x$.

$$y = \pm\frac{1}{3}x$$

We plot the vertices, draw the asymptotes, and then sketch the graph of the hyperbola, as shown in Figure 11.44.

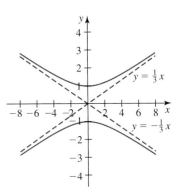

Figure 11.44

c. The equation follows the form $\frac{(x-h)^2}{a^2} - \frac{(y-k)^2}{b^2} = 1$ where $a^2 + b^2 = c^2$. We know the center is (4, −3), the transverse axis is horizontal, $a^2 = 4$, and $b^2 = 25$. Therefore, we use $a = 2$ and $b = 5$ to find c.

$$a^2 + b^2 = c^2$$
$$4 + 25 = c^2$$
$$29 = c^2$$
$$\sqrt{29} = c$$

Therefore the foci are $(\sqrt{29}, 0)$ and $(-\sqrt{29}, 0)$ and the vertices are (2, 0) and (−2, 0). The hyperbola opens left and right.

Since the center of the hyperbola is (4, −3), $\frac{(x-4)^2}{4} - \frac{(y+3)^2}{25} = 1$ is the graph of $\frac{x^2}{4} - \frac{y^2}{25} = 1$ shifted right 4 units and down 3 units. The asymptotes of $\frac{x^2}{4} - \frac{y^2}{25} = 1$ are $y = \pm\frac{5}{2}x$. Shifting the asymptotes right 4 units and down 3 units results in the equations $y = \pm\frac{5}{2}(x-4) - 3$ or, in slope-intercept form, $y = \frac{5}{2}x - 13$ and $y = -\frac{5}{2}x + 7$.

We plot the vertices, draw the asymptotes, and sketch the graph of the hyperbola, as shown in Figure 11.45.

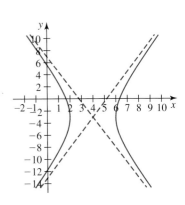

Figure 11.45

d. Since the equation is not in standard form, we complete the square to get it in standard form.

$$2y^2 + 4y - 3x^2 - 12x = 16$$
$$2(y^2 + 2y) - 3(x^2 + 4x) = 16 \quad \text{common factor}$$
$$2(y^2 + 2y + 1) - 3(x^2 + 4x + 4) = 16 + 2 - 12 \quad \text{Complete the square.}$$
$$2(y + 1)^2 - 3(x + 2)^2 = 6 \quad \text{Factor.}$$
$$\frac{2(y+1)^2}{6} - \frac{3(x+2)^2}{6} = \frac{6}{6}$$
$$\frac{(y+1)^2}{3} - \frac{(x+2)^2}{2} = 1 \quad \text{Simplify.}$$

Now we see that the equation $\frac{(y+1)^2}{3} - \frac{(x+2)^2}{2} = 1$ follows the form $\frac{(y-k)^2}{a^2} - \frac{(x-h)^2}{b^2} = 1$ where $a^2 + b^2 = c^2$. We know the center is $(-2, -1)$, the transverse axis is vertical, $a^2 = 3$, and $b^2 = 2$. Therefore, $a = \sqrt{3}$ and $b = \sqrt{2}$. We find c.

$$a^2 + b^2 = c^2$$
$$3 + 2 = c^2$$
$$5 = c^2$$
$$\sqrt{5} = c$$

Therefore the foci are $(0, \sqrt{5})$ and $(0, -\sqrt{5})$ and the vertices are $(0, \sqrt{3})$ and $(0, -\sqrt{3})$. The hyperbola opens up and down. Since the center of the hyperbola is $(-2, -1)$, $\frac{(y+1)^2}{3} - \frac{(x+2)^2}{2} = 1$ is the graph of $\frac{y^2}{3} - \frac{x^2}{2} = 1$ shifted left 2 units and down 1 unit. The asymptotes of $\frac{y^2}{3} - \frac{x^2}{2} = 1$ are $y = \pm\frac{\sqrt{3}}{\sqrt{2}}x$. Shifting the asymptotes left 2 units and down 1 unit results in the equations $y = \pm\frac{\sqrt{3}}{\sqrt{2}}(x+2) - 1$. In slope-intercept form, we have $y = \frac{\sqrt{6}}{2}x + \sqrt{6} - 1$ and $y = -\frac{\sqrt{6}}{2}x - \sqrt{6} - 1$.

We plot the vertices, draw the asymptotes, and then sketch the hyperbola, as shown in Figure 11.46.

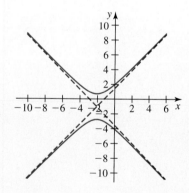

Figure 11.46

Applications of Hyperbolas

In early 2000, the Environmental Protection Agency (EPA) completed a report in response to a petition to track and control discharges from cruise ships. (*Source:* www.epa.gov) As a part of the data collection process, the EPA used the LORAN system (discussed in the Getting Started of this section) to monitor the location of several cruise ships as they traveled near the coast of Florida.

Visualize a ship in the ocean as being on a flat plane with a coordinate system superimposed over the region. Now suppose, as shown in Figure 11.47, that three

LORAN transmitting stations are located at (200, 0), (−200, 0), and (−200, 500), where all units are in miles. We will name these stations T_1, T_2, and T_3, respectively. By measuring the time it takes for the signal to travel from the transmitting station to the receiver on the ship and back, the LORAN system determines that the distance from the ship to T_1 is 120 miles, to T_2 is 307 miles, and to T_3 is 528 miles.

Figure 11.47

Using the transmitting stations T_1 and T_2 as the foci of a hyperbola, we can determine the equation of a hyperbola on which the ship lies. From the definition of a hyperbola, the difference in distance from the two fixed points (the transmitting stations acting as the foci) is defined to be $2a$. In this case, the difference in the distance from the ship to the two transmitting stations is $|307 - 120| = 187$ miles. So we have

$$2a = 187$$
$$a = 93.5$$

With this value of a and the known value of c, we can compute b.

$$a^2 + b^2 = c^2$$
$$93.5^2 + b^2 = 200^2$$
$$b^2 = 200^2 - 93.5^2$$
$$b^2 = 31257.75$$
$$b \approx 176.8$$

We write the equation of the hyperbola with horizontal transverse axis.

$$\frac{(x-0)^2}{93.5^2} - \frac{(y-0)^2}{176.8^2} = 1$$
$$\frac{x^2}{93.5^2} - \frac{y^2}{176.8^2} = 1$$

As shown in Figure 11.48, the ship must lie somewhere on this hyperbola. We continue the process in Example 4.

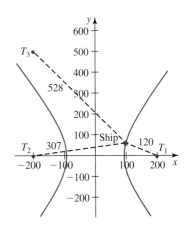

Figure 11.48

EXAMPLE 4 ■ Using Hyperbolas in a Real-World Context

Find the equation of the hyperbola shown in Figure 11.48 with foci at transmitting stations T_2 and T_3.

Solution Due to the location of the transmitting stations, we know the center of the hyperbola is (−200, 250). The value of c is 250 and the transverse axis is vertical. The difference in the distance from the ship to the two transmitting stations is |528 − 307| = 221 miles. We use this information to compute the value of a.

$$2a = 221$$
$$a = 110.5$$

Using these values of a and c, we can compute b.

$$a^2 + b^2 = c^2$$
$$110.5^2 + b^2 = 250^2$$
$$b^2 = 250^2 - 110.5^2$$
$$b^2 = 50289.75$$
$$b \approx 224.3$$

We write the equation of this hyperbola, which opens up and down.

$$\frac{(y-250)^2}{110.5^2} - \frac{(x+200)^2}{224.3^2} = 1$$

The ship must lie somewhere on this hyperbola, as shown in Figure 11.49.

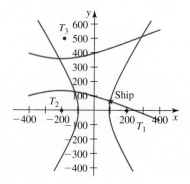

Figure 11.49

EXAMPLE 5 ■ Using Hyperbolas in a Real-World Context

Determine the exact location of the cruise ship that lies on the hyperbola $\frac{x^2}{93.5^2} - \frac{y^2}{176.8^2} = 1$ and the hyperbola $\frac{(y-250)^2}{110.5^2} - \frac{(x+200)^2}{224.3^2} = 1$. These hyperbolas intersect in four places. Determine at which intersection point the ship is located.

Solution From Figures 11.47–11.49, the solution is obvious because we marked the location of the ship on the coordinate system at the start of the problem. However, even without this visual clue, we can determine which intersection point is the right one.

First of all, we know the ship is closest to T_1 (120 miles versus 327 miles and 508 miles). Consequently, we may throw out the two intersection points in the second quadrant since they are closest to T_2 and T_3, respectively.

The distance from the ship to T_3 (528 miles) is more than four times larger than the distance from the ship to T_1 (120 miles). However, the intersection point in the uppermost part of the first quadrant appears to be about equidistant from T_1 and T_3. Therefore, we throw it out. The remaining point of intersection is where the ship is located.

We zoom in on the graph as in Figure 11.50 to visually estimate the coordinates of the point of intersection. From the graph, we estimate the ship is located at (100, 65). The algebraic method to find a solution involves solving a complicated system of quadratic equations. Using a computer algebra system, we find the solution is $x = 99.7$ and $y = 65.6$.

Figure 11.50

SUMMARY

In this section you learned the anatomy of a hyperbola and how to generate its equations. You also learned how to apply the equation of the hyperbola to real-world scenarios.

11.4 EXERCISES

SKILLS AND CONCEPTS

In Exercises 1–5, the graph of a hyperbola is given. Match each graph with its equation.

A.

B.

C.

D.

E.

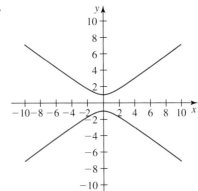

1. $\dfrac{y^2}{1} - \dfrac{x^2}{2} = 1$

2. $\dfrac{x^2}{1} - \dfrac{y^2}{2} = 1$

3. $\dfrac{x^2}{4} - \dfrac{y^2}{64} = 1$

4. $\dfrac{(y-3)^2}{4} - \dfrac{(x-3)^2}{64} = 1$

5. $\dfrac{y^2}{4} - \dfrac{x^2}{64} = 1$

In Exercises 6–20, write the equation of the hyperbola in standard form. Then graph the equation.

6. Center at the origin; focus at (6, 0); vertex at (4, 0)
7. Center at the origin; focus at (0, 5); vertex at (0, 3)

8. Center at the origin; focus at (2, 0); vertex at (1, 0)
9. Center at the origin; focus at (0, 7.5); vertex at (0, 5)
10. Center at the origin; focus at (10, 0); vertex at (−5, 0)
11. Center at the origin; focus at (0, 8); vertex at (0, 6)
12. Center at (2, −2); focus at (7, −2); vertex at (4, −2)
13. Center at (−4, 1); focus at (−4, 3); vertex at (−4, 1.5)
14. Vertices at (3, 3) and (3, 9); focus at (3, 10)
15. Vertices at (2, 5) and (2, −1); $c = 4$
16. Center at (1, 3); focus at (1, 5); vertex at (1, 2)
17. Center at (1, 3); vertex at (1, 5); focus at (1, −2)
18. Center at (−1, 2); focus at (−1, 5); vertex at (−1, 3)
19. Center at (−1, 2); focus at (−1, 5); vertex at (−1, 1)
20. Center at (−1, −1); vertex at (−1, −5); focus at (−1, −8)

In Exercises 21–30, find the vertices and foci and state whether the hyperbola opens up and down or left and right.

21. $\dfrac{x^2}{25} - \dfrac{y^2}{9} = 1$

22. $\dfrac{x^2}{4} - \dfrac{y^2}{16} = 1$

23. $\dfrac{y^2}{36} - \dfrac{x^2}{49} = 1$

24. $(y - 1)^2 - \dfrac{(x + 2)^2}{2} = 1$

25. $16x^2 - 4y^2 = 16$
26. $x^2 - 9y^2 = 18$
27. $4y^2 - x^2 = 12$
28. $9y^2 - 4x^2 = 36$
29. $2x^2 + 4x - 3y^2 - 6y - 25 = 0$
30. $3x^2 - 12x - 4y^2 + 40y - 124 = 0$

31. For a hyperbola with equation in standard form $\dfrac{x^2}{a^2} - \dfrac{y^2}{b^2} = 1$, solve for y.

32. For a hyperbola with equation in standard form $\dfrac{y^2}{a^2} - \dfrac{x^2}{b^2} = 1$, solve for y.

33. For a hyperbola with equation in standard form $\dfrac{(x - h)^2}{a^2} - \dfrac{(y - k)^2}{b^2} = 1$, solve for y.

34. For a hyperbola with equation in standard form $\dfrac{(y - k)^2}{a^2} - \dfrac{(x - h)^2}{b^2} = 1$, solve for y.

35. You must have completed one or more of Exercises 31–34 to complete this exercise. Explain why it is useful to solve the standard equation for a hyperbola for y.

In Exercises 36–40, write an equation for each hyperbola and the equations for the asymptotes.

36. $b = 4$

37. $b = 3$

38. $b = 2$

39. $b = \sqrt{13}$

40. $b = \sqrt{21}$

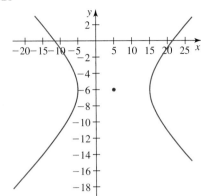

41. Lightning Strike Suppose you are on the phone with a friend that lives 2 miles directly east of you. You hear a loud thunder bang and about 9 seconds later, your friend hears the same bang.

  a. Using the fact that the speed of sound is 1100 feet per second, express the location of the lightning strike using a hyperbolic equation.

  b. Suppose you have a third friend that lives 1 mile north of you. Is it possible that the lightning strike hit her house? Explain your reasoning.

42. Concentric Circles and Hyperbolas Imagine dropping two stones into a quiet lake and observing the ripple effect from the two stones. In this exercise, you will focus on the points of intersection of the circles caused by the ripples from each rock. The center of these circles is the fixed point where the rock entered the water. The following image is intended to represent this ripple effect.

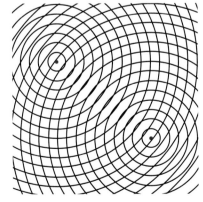

  a. Choose any point of intersection of the circles caused by the ripple effect. Find the distance (by counting the number of "rings") from each focus to this point. Compute the difference in these distances.

  b. Repeat part (a) for several points of intersection such that the difference in the distances is constant. What do you notice?

  c. Sketch the hyperbola that contains the points of intersection such that the difference in the distances from each focus is constant.

SHOW YOU KNOW

43. Suppose a classmate missed class the day that the hyperbola was introduced. Write an explanation for your classmate explaining how the difference between the distances from two fixed points creates the two branches of a hyperbola.

44. Explain the role that asymptotes play in the graph of a hyperbola.

45. Using the idea of transformation of functions, explain the relationship between the hyperbola with equation $\dfrac{x^2}{a^2} - \dfrac{y^2}{b^2} = 1$ and the hyperbola with equation $\dfrac{(x-h)^2}{a^2} - \dfrac{(y-k)^2}{b^2} = 1$.

46. When the distance between a ship and two different transmitting stations is known, the LORAN system is able to determine a hyperbola on which the ship is located. Explain why the known information generates a hyperbola instead of an ellipse or parabola.

47. For the LORAN system to work, a minimum of three transmitting stations must be used. Explain why.

MAKE IT REAL

48. Nuclear Cooling Towers Three Mile Island is a nuclear power plant near Middletown, Pennsylvania. On March 28, 1979, a serious nuclear accident occurred there that later resulted in sweeping changes in the way nuclear power plants are operated. (*Source:* www.nrc.gov) As shown in the photo, the nuclear cooling towers are hyperbolic in shape.

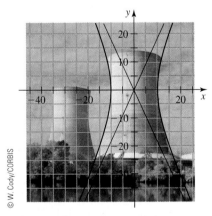

  a. Using the coordinate system and the asymptotes that are superimposed on the photo, write an equation for the hyperbola that describes the shape of the towers.

  b. The height of the cooling tower is approximately 300 feet. Using the ratio 1 unit = 75 feet, convert the equation in part (a) so that it represents the actual size of the cooling tower.

49. Lamp Light on Wall A table lamp casts a hyperbolic pattern on the wall as shown in the picture. The vertices of the hyperbola are (0, 4) and (0, −4).

Using the coordinate system and the asymptotes that are superimposed on the photo, write an equation for the hyperbolic shape of the light pattern.

50. LORAN Navigation System Suppose three LORAN transmitting stations are located at (0, 400), (0, −400), and (300, 400), where all units are in miles. Name these stations T_1, T_2, and T_3, respectively. By measuring the time it takes for the signal to travel from the transmitting station to the receiver on a cruise ship and back, the LORAN system determines that the distance from the ship to T_1 is 224 miles. Similarly, it determines that the distance from the ship to T_2 is 608 miles and the distance from the ship to T_3 is 447 miles. Use the transmitting stations as the foci of two different hyperbolas, determine the location of the ship in the coordinate plane.

51. Sonic Boom When an aircraft breaks the sound barrier (741 miles per hour) in perfect conditions—just the right amount of water vapor, density, and temperature—shock waves can be momentarily seen by the naked eye. In this photo of an F-4F Phantom II, the plane appears to fly through the vertex of a cone whose base is perpendicular to the ground. The radius of the conic shock waves will continue to increase until they intersect the plane of the earth. Using a diagram, explain why the conic shock waves intersect the plane of the earth in the shape of one branch of a hyperbola (see photo). (You may find it helpful to refer back to Section 11.1.)

■ STRETCH YOUR MIND

Exercises 52–56 are designed to challenge your understanding of hyperbolas.

52. Geocaching Geocaching is a worldwide treasure hunt game where participants post clues on a website (www.geocaching.com) for finding their "cache." Using a GPS device, participants search for the cache given the clues. Once found, geocachers are asked to take an item (typically some small item like a coin, trinket, or handmade object), leave an item, or sign a guestbook. One clue reads as follows:

Hyperbolas are the most complex and mysterious of the four conic sections, and to find this cache you will have to find the intersection of the two hyperbolas described, as if you were using LORAN. There are actually four intersections, but only one will complete the coordinates in a way that makes sense.

The coordinates are N 41° 39.XXX and W 72° 38.YYY.

Hyperbola 1 is centered at (−1, 829). Its foci are located at (−5.123, 829) and (3.123, 829). Everywhere along the hyperbola, the difference between the distances to the two foci is always 8.

Hyperbola 2 is centered at (99, −1021). Its foci are located at (99, −1007.252) and (99, −1034.748). Everywhere along the hyperbola, the difference between the distances to the two foci is always 26.832. When you find the cache area, you'll see that there are conic sections in practical use there. Perhaps some clever cachers will be able to find this cache based on that clue alone!

Determine two possible locations for the cache.

53. Derive the equation $\dfrac{(y-k)^2}{a^2} - \dfrac{(x-h)^2}{b^2} = 1$ for the hyperbola with a vertical transverse axis and its center at (h, k).

54. Solve the standard equation for a hyperbola, $\dfrac{x^2}{a^2} - \dfrac{y^2}{b^2} = 1$, for y. By investigating the behavior of this graph when $x \to \infty$, derive the equation for the asymptotes for the hyperbola.

55. A rectangular hyperbola is one whose asymptotes intersect at a right angle. Show that hyperbola $\dfrac{x^2}{1} - \dfrac{y^2}{1} = 1$ is a rectangular hyperbola. Create the equation of another rectangular hyperbola and prove that it is rectangular.

56. Consider the equation $xy = 9$. By investigating the graph of this equation,

  a. Explain why it may be labeled as a hyperbola.

  b. Find the vertices.

  c. Find the equations of its asymptotes.

CHAPTER 11 Study Sheet

As a result of your work in this chapter, you should be able to answer the following questions, which are focused on the "big ideas" of this chapter.

SECTION 11.1 1. What are the four conic sections and how are they formed?

SECTION 11.2 2. What is the geometric definition of an ellipse?

3. How do each of the key elements of the standard equation of an ellipse relate to the graph of an ellipse?

4. What is meant by eccentricity and what does it tell us about the ellipse?

SECTION 11.3 5. What is the geometric definition of a parabola?

6. How do each of the key elements of the standard equation of a parabola relate to the graph of a parabola?

SECTION 11.4 7. What is the geometric definition of a hyperbola?

8. How do each of the key elements of the standard equation of a hyperbola relate to the graph of a hyperbola?

9. What is the role of the asymptotes when graphing and analyzing a hyperbola?

SECTIONS 11.2, 11.3, 11.4

10. How can the conic sections be used to model real-world situations?

REVIEW EXERCISES

■ SECTION 11.1 ■

You have seen how to slice the cone so that the conic sections can be formed. In these exercises, you will investigate what is known as the degenerate conics. In Exercises 1–3, describe how to slice the cone to obtain the given degenerate conic.

1. Describe how to slice the given double cone so that the shape of the cross section is a single point.

2. Describe how to slice the given double cone so that the shape of the cross section is a single line.

3. Describe how to slice the given double cone so that the shape of the cross section is two intersecting lines.

4. **Sonic Boom Cloud**
The speed of sound is about 741 miles per hour and when an object exceeds this speed it breaks the sound barrier.

Under the right weather conditions, a "sonic boom cloud" is formed when a jet breaks the sound barrier. The cloud forms the shape of a cone whose base is perpendicular to the ground. If we sliced through the cloud shown in the picture with a vertical plane (geometric shape not flying object), which of the conic sections would the cross section most resemble?

■ SECTION 11.2 ■

In Exercises 5–14, write the equation of the ellipse. Then graph the equation.

5. Center at the origin; focus at (6, 0); vertex at (7, 0)
6. Center at the origin; focus at (0, 2); vertex at (0, 4)
7. Center at the origin; focus at (3, 0); 10 is the length of the major axis.
8. Center at the origin; focus at (0, 3.5); 9.6 is the length of the major axis.
9. Center at the origin; focus at (2.3, 0); 3.4 is the length of the semi-minor axis.
10. Center at the origin; focus at (0, 5); $\frac{3}{2}$ is the length of the minor axis.
11. Vertices at (5, 3) and (5, 9); length of minor axis is 2.
12. Vertices at (−2, 10) and (12, 10); length of minor axis is 4.
13. Vertices at (2, −3) and (2, −9); focus at (2, −8)
14. Vertices at (6, 6) and (6, −2); $c = 3$

In Exercises 15–18, find the vertices and foci of each ellipse. Then graph each equation.

15. $\dfrac{x^2}{15} + \dfrac{y^2}{3} = 1$
16. $\dfrac{x^2}{4} + \dfrac{y^2}{64} = 1$
17. $5x^2 + 2y^2 = 10$
18. $\dfrac{1}{2}x^2 + 2y^2 = 8$

In Exercises 19–22, write an equation for each ellipse.

19.

20.

21.

22.
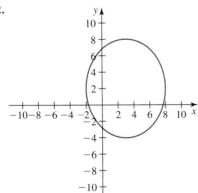

In Exercises 23–30, determine the center, foci, and vertices of each ellipse. Then graph each equation.

23. $4x^2 + 9y^2 = 36$

24. $\dfrac{x^2}{12} + \dfrac{y^2}{36} = 1$

25. $\dfrac{x^2}{16} + \dfrac{y^2}{25} = 1$

26. $9x^2 + 25y^2 = 225$

27. $x^2 + 2x + 3y^2 = 8$

28. $9x^2 - 54x + 4y^2 + 24y + 81 = 0$

29. $4x^2 - 48x + 9y^2 + 72y + 144 = 0$

30. $3x^2 - 6x + 2y^2 + 8y + 5 = 0$

In Exercises 31–34, calculate the eccentricity of each ellipse.

31. $\dfrac{x^2}{16} + \dfrac{y^2}{36} = 1$

32. Vertices at (0, 6) and (0, −6); foci at (0, 5) and (0, −5)

33. Vertices at (3, 6) and (3, 0); $c = 2$

34. Minor axis length 3 and major axis length 5

35. **Architecture** Architectural structures often include elliptical and semi-elliptical windows to enhance the appeal of the building. The window in the photo is an example of a 19th century Louis Seize style gate lodge with an oval window.

Suppose that the axes of the ellipse in the photograph are 18 inches and 24 inches long. Write the equation for the ellipse and find its eccentricity.

36. **Sunburst Window Heading** A sunburst window heading is a covering for a window or door opening. The configuration is sun-like in that it displays radiating rays (see photo). This particular semi-elliptical sunburst is 74.625 inches wide and 22.875 inches high.

Determine the eccentricity of the sunburst window heading.

37. **Astronomy—Jupiter** The planet Jupiter has an elliptical orbit with the Sun at one focus. Jupiter's major axis is 967,360,000 miles and its minor axis is 966,220,000 miles.

   a. Find the equation for the orbit of Jupiter.

   b. Calculate the eccentricity of Jupiter's orbit.

38. **Astronomy—The Moon** The moon has an elliptical orbit with Earth at one focus. The moon's orbit is nearly circular with an eccentricity of $e = \dfrac{c}{a} = \dfrac{11{,}000 \text{ miles}}{200{,}000 \text{ miles}}$.

   a. Find the equation for the orbit of the moon around Earth with the center of the orbit located at (0, 0).

   b. Determine how many miles Earth is away from the center of the moon's orbit.

39. **Racetrack Design** Many racetracks around the world are either elliptical or have portions of the track that are elliptical. (*Source:* www.oval.race-cars.com) The Dover International Speedway in Dover, Delaware, is a racetrack designed roughly in the shape of an ellipse that is approximately 1276 feet long and 638 feet wide. (*Source:* www.nascar.com)

Write the equation of the ellipse and find its **focal width** (the length of the line segment that is perpendicular to the major axis and passes through either focus of the ellipse).

40. **The Rose Bowl** The Rose Bowl in Pasadena, California, is in the shape of an ellipse. Use the following dimensions of the field to find an equation for the ellipse. Locate the foci and calculate the eccentricity.

   The stadium measures 880 feet from north to south rims and 695 feet from east to west rims. (*Source:* www.rosebowlstadium.com/RoseBowl_general-info.htm)

41. Describe how to use a string and thumbtacks to construct an ellipse that is 10 inches wide and 6 inches high.

42. Describe how to use a string and two thumbtacks to construct an ellipse that has an eccentricity of $e = \dfrac{c}{a} = \dfrac{3}{7}$ and a minor axis of length 4 inches.

■ SECTION 11.3 ■

In Exercises 43–45, the graph of a parabola is given. Match each graph with its equation.

A.

B.

C.
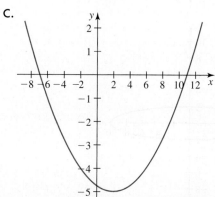

43. $(y - 5)^2 = -16(x - 2)$
44. $(x - 5)^2 = -16(y - 2)$
45. $(x - 2)^2 = 16(y + 5)$

In Exercises 46–47, write the equation of each parabola. Then graph the equation.

46. Vertex (2, 4); passes through (3, 5); vertical parabola
47. Focus (3, 4); directrix $y = 6$

In Exercises 48–49, find the vertex and focus of each parabola. Then graph the equation.

48. $(x - 3)^2 = 12(y + 2)$
49. $(x + 5)^2 = -6(y + 1)$

50. Write the equation for the parabola shown in the figure.

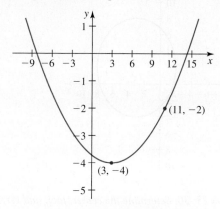

In Exercises 51–52, find the directrix, focus, vertex, and axis of symmetry for each parabola. Then graph the equation.

51. $(x - 3)^2 = y + 2$
52. $(y + 1)^2 = -x$

SECTION 11.4

In Exercises 53–55, write the equation of the hyperbola in standard form. Then graph the equation.

53. Center at the origin; vertex at (−3, 0); and focus at (−5, 0)
54. Center at (1, 1); vertex at (3, 1); and focus at (6, 1)
55. Center at the origin; vertex at (0, −5); and focus at (0, 7)

In Exercises 56–57, find the vertices, foci, and asymptotes of each hyperbola. Then indicate whether the hyperbola opens up and down or left and right.

56. $\dfrac{x^2}{9} - \dfrac{y^2}{4} = 1$

57. $\dfrac{y^2}{25} - \dfrac{x^2}{9} = 1$

Make It Real Project

What to Do

1. Find a photograph, physical object, or real-world situation that can be modeled with an ellipse, parabola, or hyperbola.
2. Determine the model equation.
3. Graph the model equation and verify that it accurately represents the thing being modeled.
4. Write a report explaining why the conic section you chose was the right one to use. Provide evidence such as the equation, graph, or other results to back up your conclusion.

CHAPTER 12
Sequences and Series

Valeev/Shutterstock.com

As humans, we constantly look for patterns. By recognizing patterns, we can more easily find order in the randomness of life. For example, people who are paid every two weeks know to expect a paycheck every fourteen days. Drivers who travel the same route to work every day soon memorize the route and find themselves making the trip with little mental effort.

Number patterns are intriguing because they frequently occur in real-world phenomena. Consider one of the most famous sequences of numbers: the Fibonacci sequence. The first few numbers of the sequence are 1, 1, 2, 3, 5, 8, 13, 21, 34, ... Each new number of the sequence is created by adding the two numbers before it: for example, $13 + 21 = 34$. This sequence is evident in the sequence of leaves on many plant stems. The spiral arrangement of sunflower seeds is also closely linked to the Fibonacci sequence.

12.1 Sequences
12.2 Arithmetic and Geometric Sequences
12.3 Series
STUDY SHEET
REVIEW EXERCISES
MAKE IT REAL PROJECT

SECTION 12.1 Sequences

LEARNING OBJECTIVES

- Determine whether a given sequence converges or diverges using graphs, tables, and symbols
- Find explicit and recursive formulas for sequences
- Use sequences to analyze real-world situations

GETTING STARTED

As many as one in five adults has high cholesterol. (*Source:* www.forcholessterol.com) People with high cholesterol are at a higher risk of heart disease (including heart attack and stroke) than those with healthier levels of cholesterol. Many people take prescription drugs to help lower the amount of cholesterol in their bloodstream. In most cases, this type of medicine is considered a maintenance medicine—users take the medicine every day for the rest of their lives. This brings up many interesting questions: "If I take the medicine every day, will the concentration of medicine in my blood continue to increase?" "Should I take a double dose if I forget to take the medicine at the right time?" "Is it possible for the amount of medicine in my body to reach unsafe levels?"

In this section we introduce the mathematical concept of sequences. With this concept we can address a number of real-world applications including the use of cholesterol-lowering medicine.

■ Sequences

Lipitor® is a common cholesterol-lowering medicine taken by about 18 million people in the United States. On average, people take a 20-milligram (mg) dosage once per day. Lipitor has a half-life of 14 hours. This is the amount of time it takes for half of the medicine to be eliminated from the body. (*Source:* www.medguides.medicines.org.uk) Will the amount of Lipitor in a person's body increase continuously if he continues to take the medicine everyday?

We can determine the amount of Lipitor left in the body by using the following formula.

$$\text{percent left} = \left(\frac{1}{2}\right)^{\text{time/half life}}$$
$$= \left(\frac{1}{2}\right)^{24/14}$$
$$\approx 0.305$$

Since the half-life is 14 hours, roughly 30.5% of the dose is left in the body after 24 hours. We assume when the medicine is taken, it is instantly absorbed into the bloodstream. With these assumptions, we can compute how much medicine is in the body on the second day, assuming the medicine is taken at the same time each day. Initially, a 20-mg dose is taken. The second day 30.5% of the first 20-mg dose of medicine remains in the body when the second 20-mg dose is taken. Thus

$$(20)(0.305) + 20 = 26.1$$

After taking the second dose, there will be 26.1 mg of medicine in the body. By the time the third dose is taken, 30.5% of the 26.1 mg of medicine remains in the body, so

$$(26.1)(0.305) + 20 \approx 27.96$$

After the third dose, there is 27.96 mg of medicine in the body. If we keep track of the amount of medicine in the body each day, we generate a **sequence** of numbers.

$$20, 26.1, 27.96, \ldots$$

SEQUENCE

A **sequence** a_n is a function whose domain is the set of natural numbers $\{1, 2, 3, \ldots\}$. The function values are called **terms** of the sequence. The terms are represented by $a_1, a_2, \ldots, a_n, \ldots$.

For the cholesterol drug example, the domain values represent the cumulative number of doses taken. The range values represent the amount of medicine remaining in the body at that time.

EXAMPLE 1 ■ Using Sequence Notation

Given the sequence 20, 26.1, 27.96, . . . , use proper notation to write the first three terms of the sequence. Then use sequence notation to define the function represented verbally by the instruction "multiply the previous number in the sequence by 0.305 (30.5% of the medicine remains in the body) and add 20 (the new dose is 20 mg of medicine)."

Solution The first term of the sequence corresponds with $n = 1$ and is written a_1. The second term of the sequence corresponds with $n = 2$ and is written a_2. In general, the ith term of the sequence corresponds with $n = i$ and is written a_i. Each of these terms represents the amount of medicine in the body after n doses.

$$a_1 = 20 \text{ mg}$$
$$a_2 = 26.1 \text{ mg}$$
$$a_3 = 27.96 \text{ mg}$$

We write the verbal instructions in sequence notation as

$$a_n = 0.305 a_{n-1} + 20$$

In this formula, a_n represents any term of the sequence. The term a_{n-1} is the term right before a_n.

Even after analyzing the amount of medicine in the body after three doses, we see the amount of medicine is increasing by less and less each dose. The second dose adds 6.1 mg of medicine to the total amount in the body while the third dose only adds another 1.86 mg of medicine. This suggests that, over time, the amount of medicine in the body must reach a limit. This is called the **maintenance level**. In mathematical terms, we say a sequence that reaches a limit is a *converging* sequence. A sequence **converges** if the terms of the sequence get closer and closer to a particular finite value as the number of terms increases; it **diverges** otherwise.

CONVERGENCE OF SEQUENCES

A sequence $a_1, a_2, a_3, \ldots, a_n, \ldots$ **converges** (or reaches a **maintenance level**) if $a_n \to L$ as $n \to \infty$, where L is a finite number. Otherwise, we say the sequence **diverges**.

To determine the maintenance level in this case, we analyze the formula and graph for the sequence representing the amount of medicine in the body each day.

EXAMPLE 2 ■ Representing a Sequence Using a Graph

Lipitor is taken in 20-mg doses once per day. Each day, 69.5% of the medicine is eliminated from the body. Use a graph to find the maintenance level and explain what this value represents.

Solution Since 69.5% of the medicine is eliminated, 30.5% remains. The formula representing the situation is $a_n = 0.305a_{n-1} + 20$. We calculate the first nine terms of the sequence and graph them as shown in Figure 12.1.

$$20.00, 26.10, 27.96, 28.53, 28.70,$$
$$28.75, 28.77, 28.775, 28.776, \ldots$$

The graph suggests that the maintenance level is about 28.8 mg of medicine. This means that if 20 mg of Lipitor is taken each day, the amount of medicine in the body will reach 28.8 mg and is maintained at that amount. A Technology Tip detailing the process for creating graphs of sequences is given at the end of the section.

Figure 12.1

EXAMPLE 3 ■ Finding the Maintenance Level of a Sequence

An initial dose of 40 mg of Lipitor is taken. After this initial dose, a single 20-mg tablet is to be taken each day. Every 24 hours, 69.5% of the medicine is eliminated from the body. Find the maintenance level and explain what it means.

Solution We begin by writing the first four terms of the sequence. The first term is 40 mg. The following day, only 30.5% of the medicine remains in the body and a new dose of 20 mg is taken.

$$a_1 = 40$$
$$a_2 = (0.305)(40) + 20 = 32.2$$

This pattern continues for subsequent days.

$$a_3 = (0.305)(32.2) + 20 \approx 29.8$$
$$a_4 = (0.305)(29.8) + 20 = 29.1$$

We now have the first four terms of the sequence.

$$40, 32.2, 29.8, 29.1, \ldots$$

The sequence suggests that, over time, the amount of medicine in the body will become stable. A graph of the sequence will tell us more. In general, the nth term of the sequence may be expressed as a formula.

$$a_n = 0.305a_{n-1} + 20$$

Using $a_1 = 40$, we calculate additional terms and analyze a graph of the sequence shown in Figure 12.2. The graph shows that the amount of medicine in the body drops very quickly after the initial 40-mg dose and suggests that the maintenance level is about 28.8 mg. Additionally, since 69.5% of 28.8 mg is about 20 mg, the amount of medicine the body eliminates is the same amount as the next dose taken. In other

Figure 12.2

words, once the body has 28.8 mg of medicine, the amount eliminated (20 mg) and the new dose taken (20 mg) are the same. Therefore, the body is able to maintain 28.8 mg of medicine and unsafe levels will not build up.

An alternative way to find the maintenance level is to tackle the problem symbolically. The maintenance level k will have the property that

$$0.305k + 20 = k$$

Solving this equation for k yields

$$0.305k + 20 = k$$
$$20 = 0.695k$$
$$k = \frac{20}{0.695}$$
$$\approx 28.8$$

Phase4Photography/
Shutterstock.com

EXAMPLE 4 ■ Using Sequences

Atacand® (candesartan cilexetil) is a medicine for high blood pressure. According to the technical pamphlet for the drug, the half-life of candesartan is about 9 hours. The typical recommended starting dose is 16 mg once daily. A doctor may recommend the dose be taken twice daily; however, total daily doses should not exceed 32 mg. (*Source:* www.astrazeneca-us.com).

Suppose a doctor prescribes a patient to take a 16-mg dose once every 12 hours.

a. What is the maintenance level for the drug?

b. Suppose after reaching the maintenance level, the patient accidentally misses a dose and takes a double dose the next time he is scheduled to take the medicine. (We do not recommend doing this.) Assuming the drug is instantly absorbed and distributed evenly throughout the bloodstream, how much of the drug will be in the bloodstream immediately after the double dose is taken?

Solution

a. We first determine how much of the drug is left after 12 hours.

$$\text{percent left} = \left(\frac{1}{2}\right)^{\text{time/half life}}$$
$$= \left(\frac{1}{2}\right)^{12/9}$$
$$\approx 0.397$$

We calculate the first eight terms of the sequence representing the amount of medicine in the body. Since 39.7% remains after 12 hours, we have

$$a_1 = 16$$
$$a_2 = (0.397)(16) + 16 = 22.4$$
$$a_3 = (0.397)(22.4) + 16 = 24.9$$
$$a_4 = (0.397)(24.9) + 16 = 25.9$$
$$a_5 = (0.397)(25.9) + 16 = 26.3$$
$$a_6 = (0.397)(26.3) + 16 = 26.4$$
$$a_7 = (0.397)(26.4) + 16 = 26.5$$
$$a_8 = (0.397)(26.5) + 16 = 26.5$$

The maintenance level seems to be approaching 26.5 mg and the graph shown in Figure 12.3 confirms this observation. To check our conclusion algebraically, we note that the maintenance level k will have the property that

$$0.397k + 16 = k$$

Solving this equation for k yields

$$0.397k + 16 = k$$
$$16 = 0.603k$$
$$k = \frac{16}{0.603}$$
$$\approx 26.5$$

Figure 12.3

b. After reaching the maintenance level, 26.5 mg of medicine will remain in the body. If a new dose is not taken at the scheduled time, by the next scheduled time 10.5 mg [(0.397)(26.5) = 10.5] of medicine will remain in the body. If the patient takes two doses, there will be 10.5 mg + 16 mg + 16 mg = 42.5 mg of medicine in the body. This is noticeably higher than the maintenance level and may result in adverse side effects. However, if the patient does not have an adverse reaction, the amount of medicine in the body will return to the maintenance level over time. See Figure 12.4.

Figure 12.4

Not all sequences reach a maintenance level or a stable limit. Example 5 illustrates a sequence that diverges.

EXAMPLE 5 ■ Identifying a Divergent Sequence

An employee asks his employer to withhold $1200 for retirement savings during the first year of employment and to increase the amount withheld by 2% in each subsequent year. If this continues for 20 years, how much money is withheld in the 20th year?

Solution We calculate the first four terms of the sequence.

$$a_1 = 1200$$
$$a_2 = 1200 + (0.02)(1200) = 1224$$
$$a_3 = 1224 + (0.02)(1224) = 1248.48$$
$$a_4 = 1248.48 + (0.02)(1248.48) = 1273.45$$

These terms represent the amount of money withheld each of the first four years.

$$1200, 1224, 1248.48, 1273.45, \ldots$$

So far the amount withheld does not appear to be stabilizing. Using the Technology Tip at the end of this section to graph the sequence $a_n = 0.02a_{n-1} + a_{n-1} = 1.02a_{n-1}$ as shown in Figure 12.5, we determine the amount withheld in the 20th year will be $1748.17. As we can see from the graph, this sequence does not have a limiting value. Therefore, this sequence diverges.

Figure 12.5

Each of the situations we have seen so far have been situations where the formula for the sequence was expressed using a *recursive formula*. A **recursive formula** defines the *n*th term of the sequence as a function of the previous term, a_{n-1}. Sometimes it is possible to express the formula for the *n*th term of a sequence *explicitly* in terms of the value of *n*. An **explicit formula**, which allows us to compute any term of the sequence easily, defines the *n*th term of the sequence as a function of *n*.

EXAMPLE 6 ■ Finding an Explicit Formula for a Sequence

In Example 5 we found a recursive formula for the situation where an employer withheld $1200 for retirement savings during the first year of employment and increased the amount withheld by 2% each year. The recursive formula was

$$a_n = 0.02a_{n-1} + a_{n-1} = 1.02a_{n-1}$$

Write an explicit formula for the amount of money withheld in the *n*th year of employment.

Solution We write the first four terms of the sequence.

Value of term	1200	1224	1248.48	1273.45
Number of term	$n = 1$	$n = 2$	$n = 3$	$n = 4$

Notice we start with the term 1200, multiply it by 1.02, then multiply that answer by 1.02, and finally multiply that answer by 1.02.

	×1.02	×1.02	×1.02
Value of term	1200	1224	1248.48	1273.45
Number of term	$n = 1$	$n = 2$	$n = 3$	$n = 4$

We express the idea of starting with a value (1200) and repeatedly multiplying by another value (1.02) using exponential notation:

$$a_n = 1200(1.02)^{n-1}$$

Using this explicit formula generates the same results as before.

$$a_1 = 1200(1.02)^{1-1} = 1200(1.02)^0 = 1200$$
$$a_2 = 1200(1.02)^{2-1} = 1200(1.02)^1 = 1224$$
$$a_3 = 1200(1.02)^{3-1} = 1200(1.02)^2 = 1248.48$$
$$a_4 = 1200(1.02)^{4-1} = 1200(1.02)^3 = 1273.45$$

EXAMPLE 7 ■ Writing Recursive and Explicit Formulas for a Sequence

A $400 car payment is made each month. Write a recursive formula and an explicit formula for the total amount of money paid out in car payments over time. After making 60 payments (5 years), how much money has been paid out?

Solution We begin by writing the first four terms of the sequence, representing a running total of how much money is spent on car payments.

$$400, 800, 1200, 1600, \ldots$$

To get a particular term in the sequence, we add 400 to the previous term in the sequence. We express this as a recursive formula for the sequence.

$$a_n = a_{n-1} + 400$$

To get an explicit formula, we organize the sequence as follows.

Number of term	$n = 1$	$n = 2$	$n = 3$	$n = 4$
Function value	400	800	1200	1600

We see if we multiply the number of the term ($n = 1$, $n = 2$, and so on) by 400, we obtain the value of the term. Therefore, the explicit formula for this sequence is

$$a_n = 400n$$

We use the explicit formula to determine how much money has been paid out in car payments after 60 payments have been made. That is, we compute

$$a_{60} = (400)(60) = 24{,}000$$

This means $24,000 has been paid through monthly car payments after 60 payments have been made.

SUMMARY

In this section you learned how to create a sequence of numbers to represent a particular situation as well as to determine whether the sequence converges or diverges. You also discovered how to represent sequences with graphs. Additionally, you learned how to express sequences using recursive and explicit formulas.

TECHNOLOGY TIP ■ GRAPHING A SEQUENCE

1. Press MODE. Change the mode to sequence mode by using the arrows to highlight **Seq** and pressing ENTER.

2. Edit the parameters using the Y = Editor. The parameter *n*Min refers to the minimum term number you choose to display. The parameter u(*n*) represents the formula. Access "u" by pressing 2nd and the number 7 (note the blue "u" above the 7 button). Access "*n*" by pressing the X,T,Θ,n button. The parameter u(*n*Min) represents the first term of the sequence.

3. Edit the WINDOW parameters. The parameters *n*Min and *n*Max refer to the number terms in the sequence to be evaluated. **PlotStart** and **PlotStep** determine which term to plot first and the incremental value of *n*. The rest of the parameters are as usual, where *x* represents the term number and *y* represents the value in the sequence.

4. Press GRAPH.

5. Use TRACE as usual. You will see the term number, *n*, as well as the sequence value.

12.1 EXERCISES

SKILLS AND CONCEPTS

In Exercises 1–5,
  a. List the first six terms of the given sequence.
  b. Predict whether the sequence will converge or diverge.
  c. Check the accuracy of your prediction by using a graphing calculator to graph the first ten terms of the sequence.

1. $a_n = \dfrac{n}{3n+1}$
2. $a_n = \dfrac{n^2}{2^n - 1}$
3. $a_n = \dfrac{1}{2^n}$
4. $a_n = 4 + (-1)^n$
5. $a_n = \left(1 + \dfrac{1}{n}\right)^n$

In Exercises 6–14, find a formula for the general term of the sequence, a_n, and state whether the sequence converges. Assume the pattern in the given terms continues.

6. 3, 8, 13, 18, 23, ...
7. 10, 5, 2.5, 1.25, ...
8. $\dfrac{1}{2}, \dfrac{2}{5}, \dfrac{3}{10}, \dfrac{4}{17}, \ldots$
9. 2, 4, 8, 16, 32, ...
10. 0, 5, 0, 5, 0, ...
11. $\dfrac{1}{2}, -\dfrac{1}{4}, \dfrac{1}{8}, -\dfrac{1}{16}, \ldots$
12. 0, 50, 87.5, 115.625, ...
13. 100, 102, 104, 106, ...
14. 300, 600, 900, 1200, ...
15. Consider the general sequence $1, k, k^2, k^3, k^4, \ldots$. Determine for which values of k the sequence will converge and for which values of k the sequence will diverge.
16. **Geometry** A regular polygon is a geometric shape where each of the sides has the same length. For example, a square is a regular polygon since all four sides are the same length. A diagonal is a line segment drawn from one corner of a polygon to another. For example, a square has two diagonals, as shown in the figure.

   Beginning with a triangle, we can express the number of diagonals as a sequence.
   Triangle: $a_1 = 0$
   Square: $a_2 = 2$
   Write an explicit formula for the sequence giving the number of diagonals, a_n, for a regular polygon with n sides.

17. **Fibonacci Sequence** The Fibonacci sequence, named for Italian mathematician Leonardo Fibonacci, is the sequence 1, 1, 2, 3, 5, 8, 13, ...
   a. Find the next five terms of the sequence. (*Hint:* For this sequence, a_n depends on a_{n-1} and a_{n-2}.)
   b. Write a recursive formula for the sequence.

18. Using a calculator, repeatedly take the square root of 5. Express the decimal approximations of this as a sequence. That is, the sequence will be $\sqrt{5}, \sqrt{\sqrt{5}}, \sqrt{\sqrt{\sqrt{5}}}$, and so on. Does this sequence converge? If so, to what does it converge?

19. A sequence is considered **strictly monotone** if it is either always increasing or always decreasing from term to term. Explain whether each of the following sequences is strictly monotonic or not. Then, determine if the sequence converges or diverges.
   a. $a_n = \dfrac{1}{2}n$
   b. $a_n = \dfrac{1}{n^2}$
   c. $a_n = \dfrac{(-1)^n}{n^2}$

SHOW YOU KNOW

20. Explain what it means for a sequence to converge. Use graphs, numbers, symbols, and words in your explanation.
21. Create two sequences, both of which diverge. Explain why the sequences diverge and include graphs to support your explanation.
22. Create two sequences, both of which converge. Explain why the sequences converge and include graphs to support your explanation.
23. Using the idea of rate of change, describe the shape of a graph of a sequence that is ever increasing (strictly monotonic) but converges. Then, describe the shape of a graph of a sequence that is ever decreasing but converges.

MAKE IT REAL

24. **Inflation** Inflation is an increase in the price of a particular product. For example, suppose the value of a MP3 player increases by the May 2007 inflation rate of 2.69% per year for the next 5 years. (*Source:* www.inflationdata.com) If the MP3 player costs $90, write its predicted price for the next 5 years as a sequence. Does this sequence converge? Explain.

25. **Chain Letter** One chain letter scam, called the "Make Money Fast Pyramid," claims that by mailing $1 to five people, you can make thousands of dollars. (*Source:* hoaxbusters.ciac.org) It allegedly works like this: You receive a letter with five names. You mail a copy of the letter to five friends and send $1 to the top name on the list. Before you mail the letter to your friends, you remove the name at the top of the list and add your name to the bottom of the list.

a. If the chain is not broken, how much money will you receive before your name is removed from the list?

b. What if there were only three names on the list instead of five?

c. What if there are n names on the list? Write your answer as a sequence where C_n is the name of the sequence.

26. **Dog Years** Depending on the type of breed, 1 year for a human is equivalent to 7 years for a dog. A baby human and a puppy are born on the same day. Write a sequence showing the age of the dog (in dog years) as a function of the baby's age for the first 10 years of the baby's life. Then write an explicit formula for this sequence.

27. **Purchasing a Car** As a rule of thumb, a car depreciates in value by 15% every year. (*Source:* www.bankrate.com) The price for a 2007 Toyota Camry Hybrid 4-door Sedan was $25,900 on June 19, 2006. (*Source:* www.nada.com) If this car was purchased and kept indefinitely, does the value of the car converge or diverge? If it converges, to what value does it converge?

28. **Pharmaceuticals—Aspirin** Aspirin washes out of the bloodstream rapidly. The amount of effective aspirin left in the bloodstream is reduced by 50% approximately every 3 hours. (*Source:* www.rxlist.com)

    Suppose a patient takes 650 mg of aspirin every 4 hours, as recommended. (Assume the aspirin is absorbed into the bloodstream immediately.)

    a. Generate the first five terms of the sequence that represent this situation. Explain why this situation can be represented with sequences.

    b. Write an implicit formula giving the medicine level in the patient's bloodstream after each dose is taken and absorbed.

    c. Create a graph that shows units of medicine as a function of number of doses.

    d. How many doses can the patient safely take? Support your answer graphically, numerically, symbolically, and verbally.

    e. Discuss the consequences of missing a dose. Does taking a double dose make up for the missed dose? Explain.

29. **Forestry Management** In its 5-year Urban Forestry Management Plan, the city of Muskego, Wisconsin, planned to increase the number of trees in city parks. According to the plan, there were 500 trees in city parks and the city planned to increase the number of trees by 50% over 5 years. The plan also indicated that the city expected 10% of the newly planted trees to die during the 5-year period. Additionally, the city planned to remove 85 trees over the same time period. (*Source:* City of Muskego Urban Forestry Management Plan at www.ci.muskego.wi.us) Assuming the city will implement the same 5-year plan indefinitely,

    a. Write a sequence of numbers that represents the number of trees in city parks for the first four 5-year periods.

    b. Based on the sequence from part (a), write a rule for the number of trees in city parks after n years.

    c. Represent this situation graphically.

    d. Does repeated application of the 5-year plan make sense in the long run? Justify your answer.

    e. Will the forest size stabilize? If so, in how many years and with how many trees? Justify your answer.

30. **Semiconductor Manufacturing** In the manufacturing of semiconductors, highly toxic chemicals are used. When the need arises, environmental engineers wear protective suits designed to neutralize specific toxic chemicals. However, the ability of a suit to neutralize a chemical is usually reduced with prolonged exposure. Suppose an engineer has put on a new, clean protective suit and begins to clean up a chemical spill in the factory. During each hour of cleanup, the suit neutralizes 30% of the toxic chemical on it and accumulates an additional 12 micrograms of the chemical.

    a. Make a table of the amounts of unneutralized chemical on the suit during each hour of the cleanup. Describe the growth of these values.

    b. Write the first five terms of the sequence related to this situation. Then write a formula and create a graph for the sequence.

    c. The suit is safe to use until it has accumulated 35 micrograms or more of the toxic chemical. How long can the engineer safely continue to clean up the spill before the suit must be changed?

    d. Reanalysis of the toxicity of the chemicals indicates that the suit can be used safely with up to 40 micrograms of accumulated chemical. Does this make a significant difference in the amount of time the engineer can continue to work on the chemical spill? Explain.

31. **Pharmaceuticals—Lamisil®** Lamisil tablets are taken by people with fingernail and toenail fungus problems. Typically, 250 mg tablets are taken once daily for six weeks for people with fingernail fungus and for 12 weeks for people with toenail fungus. With a half-life of 36 hours, 37% of the medicine is removed from the bloodstream each day. (*Source:* www.drugs.com)

    How much medicine will be in the bloodstream at the end of each treatment period for each of the two treatment groups? How close to the maintenance level did each group get?

32. **Pharmaceuticals—Singulair®** Singulair is medicine taken by people for the long-term management of asthma. People aged 15 and over are typically prescribed a single dose of 10 mg of Singulair each day. Children aged 2 to 5 years of age are typically prescribed a single dose of 4 mg of Singulair each day. With a half-life of 4 hours, about 98.4% of the medicine is removed from the bloodstream every day. (*Source:* www.drugs.com)

    Use what you have learned in this section to analyze the effects of the two different dosage levels. Include graphs and formulas and indicate what the maintenance levels are in each case.

33. **Pharmaceuticals—Prilosec®** Prilosec is used to eliminate acid reflux disease in people. Typically, people take a 20-mg dose once daily. Studies have shown that in healthy people, Prilosec has a half-life of about one-half hour. This is equivalent to having 99.9% of the medicine eliminated from the bloodstream every day. The study also showed that in people with unhealthy livers, the half-life is 3 hours. This is equivalent to having 99.51% of the medicine elimi-

nated every day. (*Source:* www.drugs.com) Analyze the effects of the two different half-lives. Include graphs and formulas and indicate what the maintenance levels are in each case.

34. **Pharmaceuticals—Percocet®** Percocet is a pain relief medicine that contains acetaminophen and oxycodone, which is highly addictive. Research has shown it is extremely unhealthy for a person to maintain a level of 15 mg of Percocet in the bloodstream. Dosing depends on the severity of the pain that a person is experiencing. Percocet has a half-life of about 2 hours. (*Source:* en.wikipedia.org)

    Determine the maximum amount of Percocet a doctor should prescribe to a patient who will take the medicine every 4 hours. What will be the maximum dosage in this situation? Use graphs, formulas, and tables to justify your answers.

35. **Pharmaceuticals—Claritin®** Claritin tablets relieve nasal congestion, runny nose, and watering eyes due to allergies. The half-life of Claritin depends on urinary pH: The half-life is approximately 4 hours when the urinary pH is 5 and 12 hours when the urinary pH is 8. Typically, people take a 240-mg dose of Claritin once daily. (*Source:* www.drugs.com)

    With a half-life of 4 hours, the body eliminates approximately 98.4% of the medicine every day. With a half-life of 12 hours, the body eliminates approximately 75% of the medicine every day. Analyze the effects of the two different half-lives. Include graphs and formulas and indicate what the maintenance levels are in each case.

■ STRETCH YOUR MIND

Exercises 36–41 are intended to challenge your understanding of sequences.

36. Using a calculator, a student starts with a number greater than 1 and repeatedly presses the square root button. If each successive result represents the term of a sequence of numbers, write a formula for the nth term of this sequence. Does this sequence converge? If so, to what value does it converge?

37. Using a calculator, a student starts with a positive number less than 1 and repeatedly presses the square root button. If each successive result represents the term of a sequence of numbers, write a formula for the nth term of this sequence. Does this sequence converge? If so, to what value does it converge?

38. The Fibonacci sequence (1, 1, 2, 3, 5, 8, 13, ...) can be defined recursively by the formula $a_n = a_{n-1} + a_{n-2}$. How does the ratio $\dfrac{a_{n+2}}{a_{n+1}}$ compare to $\dfrac{a_n}{a_{n+1}}$ for $n \geq 1$?

39. Investigate the sequence of numbers defined by the formula $a_n = \dfrac{4\sin(n)}{n}$. Does the sequence converge? If so, to what value does it converge?

40. A sequence is considered monotonic if the terms always increase (that is, $a_{n+1} > a_n$) or always decrease (that is, $a_{n+1} < a_n$) for all $n \geq 1$. Determine whether the sequence $a_n = \dfrac{n}{n^2 + 1}$ is monotonic. Justify your answer.

41. The following sequence is known as the harmonic sequence:
$$a_n = 1, \frac{1}{2}, \frac{1}{3}, \frac{1}{4}, \frac{1}{5}, \ldots$$

    Determine whether or not the sequence defined by $b_n = \dfrac{a_n}{a_{n-1}}$ converges or diverges. Justify your answer.

SECTION 12.2 Arithmetic and Geometric Sequences

LEARNING OBJECTIVES

- Determine whether a sequence is arithmetic or geometric
- Use arithmetic and geometric sequences to analyze real-world problems

GETTING STARTED

The cost of tuition is a financial concern for many students. At the University of Minnesota–Duluth, undergraduate students paid $253.50 per credit during the 2006–2007 academic year. (*Source:* www.d.umn.edu) To determine the total cost of tuition, students add an additional $253.50 for every additional credit hour for which they enroll.

Students who graduate from college expect that the investment in tuition costs will pay off. In many cases, it does. A recent survey showed that new college graduates with a degree in chemical engineering earned an average of $59,361. (*Source:* money.cnn.com) This represents a 5.4% increase from the previous year. Assuming a 5.4% increase each year, the starting salary for chemical engineering majors will increase at an ever-increasing rate with a growth factor of 1.054.

In this section we investigate the types of sequences that can be used to represent these two situations: arithmetic and geometric sequences.

■ Arithmetic Sequences

Suppose we plan to take classes at the University of Minnesota–Duluth. We can express the total cost for tuition as a sequence where the value of n represents the number of credits taken and a_n is the total tuition cost.

$$a_1 = 253.50$$
$$a_2 = 507.00$$
$$a_3 = 760.50$$
$$a_4 = 1014.00$$

Each subsequent term in the sequence is obtained by adding the constant amount of $253.50 to the previous term. This is an example of an **arithmetic sequence**.

ARITHMETIC SEQUENCE

An **arithmetic sequence** is a sequence in which each subsequent term of the sequence is found by adding a common difference, d, to the previous term. The formula for the nth term of an arithmetic sequence is

$$a_n = a_1 + (n-1)d$$

EXAMPLE 1 ■ Determining the Explicit Formula of an Arithmetic Sequence

The cost of tuition at the University of Minnesota–Duluth is $253.50 per credit. Write an explicit formula for the cost of enrolling for n credits. What is the total tuition cost if a student enrolls for 12 credits?

Solution The cost of enrolling for 1 credit is $253.50, therefore, $a_1 = 253.50$. With a cost of $253.50 per credit, the common difference is $d = 253.50$. The explicit formula for the total cost is

$$a_n = 253.50 + (n-1)(253.50)$$
$$a_n = 253.50 + 253.50n - 253.50$$
$$a_n = 253.50n$$

We want to know the cost for 12 credits, so we need to compute the value of a_{12}.

$$a_{12} = (253.50)(12)$$
$$a_{12} = 3042.00$$

The total tuition cost for 12 credits is $3042.

Arithmetic Sequences and Linear Functions

There is a strong connection between arithmetic sequences and linear functions. The common difference, d, corresponds to the constant rate of change, or slope. The value of a_1 of the sequence corresponds to the initial value of the linear function. However, unlike linear functions, the domain of an arithmetic sequence is limited to the natural numbers. (In the tuition example, we assume students may not enroll for a fraction of a credit.)

EXAMPLE 2 ■ Graphing an Arithmetic Sequence

Graph the sequence where $a_1 = 10$ and $d = 1.5$. Explain why this is an arithmetic sequence and how it is similar to a linear function.

Solution We first write a formula for this sequence.

$$a_n = a_1 + (n - 1)d$$
$$a_n = 10 + (n - 1)(1.5)$$
$$a_n = 10 + 1.5n - 1.5$$
$$a_n = 1.5n + 8.5$$

We then plot the sequence shown in Figure 12.6 by creating several values of a_n.

$$a_1 = (1.5)(1) + 8.5 = 10$$
$$a_2 = (1.5)(2) + 8.5 = 11.5$$
$$a_3 = (1.5)(3) + 8.5 = 13$$

Since each subsequent term is computed by adding the constant 1.5, the sequence follows a linear trend, but is discrete in nature.

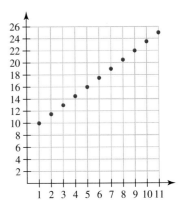

Figure 12.6

■ Geometric Sequences

Suppose the starting salary for a chemical engineer is \$59,361 and is increasing by 5.4% per year. This salary schedule can be expressed as a sequence where the value of n represents the number of years into the future and a_n is the starting salary. We begin with \$59,361 and multiply by 1.054 (because increasing by 5.4% is equivalent to multiplying by 1.054).

$$a_1 = 59,361.00$$
$$a_2 = 62,566.49$$
$$a_3 = 65,945.08$$
$$a_4 = 69,506.12$$

Each subsequent term in the sequence is obtained by multiplying the previous term by the constant factor 1.054. The ratio of a term to the previous term is also equal to 1.054. In this example of a **geometric sequence**, observe that

$$a_1 = 59,361.00$$
$$a_2 = 59,361.00 \cdot 1.054$$
$$a_3 = (59,361.00 \cdot 1.054) \cdot 1.054 = 59,361.00 \cdot 1.054^2$$
$$a_4 = [(59,361.00 \cdot 1.054) \cdot 1.054] \cdot 1.054 = 59,361.00 \cdot 1.054^3$$

GEOMETRIC SEQUENCE

A **geometric sequence** is a sequence in which each subsequent term of the sequence is found by multiplying the previous term by a *common ratio, r*. The formula for the nth term of a geometric sequence is

$$a_n = a_1 r^{n-1}$$

The ratio is called a common ratio because if we take any term in the sequence and divide it by the previous term, we get a constant.

EXAMPLE 3 ■ Using a Geometric Sequence

The starting salary for a new chemical engineer is $59,361. Assuming this salary is increasing by 5.4% annually, write an explicit formula for the salary amount after n years. What will the salary be in 5 years?

Solution We know $a_1 = \$59,361$ and $r = 1.054$. The salary schedule is a geometric sequence with the explicit formula

$$a_n = a_1 r^{n-1}$$
$$a_n = 59,361(1.054)^{n-1}$$

To find the starting salary when $n = 5$, we need to compute the value of a_5.

$$a_5 = 59,361(1.054)^{5-1}$$
$$a_5 = 59,361(1.054)^4$$
$$a_5 = 73,259.45$$

The starting salary for a chemical engineer in 5 years is predicted to be $73,259.45.

Geometric Sequences and Exponential Functions

There is a strong connection between geometric sequences and exponential functions. The common ratio, r, corresponds to the constant change factor. The value of a_1 of the sequence corresponds to the initial value of the exponential function. As was the case with arithmetic sequences, the domain of a geometric sequence is limited to the positive integers.

EXAMPLE 4 ■ Graphing a Geometric Sequence

Graph the sequence where $a_1 = 50$ and $r = 0.75$. Explain why this is a geometric sequence and how it is similar to an exponential function.

Solution We first write a formula for this sequence.

$$a_n = a_1 r^{n-1}$$
$$a_n = 50(0.75)^{n-1}$$

We then plot the sequence shown in Figure 12.7 by creating several values of a_n.

$$a_1 = 50(0.75)^{1-1} = 50$$
$$a_2 = 50(0.75)^{2-1} = 37.5$$
$$a_3 = 50(0.75)^{3-1} = 28.125$$
$$a_4 = 50(0.75)^{4-1} = 21.09375$$

The graph of the exponential function $y = 50(0.75)^x$ has the same basic shape as $a_n = 50(0.75)^{n-1}$; however, the exponential function is defined for all real numbers while the geometric sequence is only defined for positive integers.

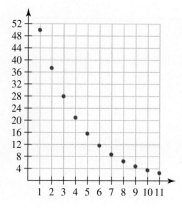

Figure 12.7

EXAMPLE 5 ■ **Distinguishing between Arithmetic and Geometric Sequences**

For each scenario, decide whether an arithmetic or geometric sequence will best model the situation. Explain how you know, then write an explicit formula for the sequence.

a. Each time a picture is taken with a digital camera, 1.5 megabytes of memory is used.

b. One gallon of milk cost $0.23 in 1950. Milk has increased in price by an average of 2.2% annually. (*Sources:* data.bls.gov, www.cbo.gov)

c. Each time you fold a piece of paper with a thickness of 0.0125 inches, the thickness of the stack doubles.

d. Each time a field-goal kicker makes a field goal, 3 points are added to his total points scored for the season.

Solution

a. Since the amount of memory or storage space used increases by a constant amount of 1.5 megabytes per picture, this situation is best modeled by an arithmetic sequence with a constant difference, d, of 1.5. Assuming there were no pictures stored on the camera initially, we have the sequence 0, 1.5, 3, 4.5, ... This is represented by the following equation.

$$a_n = a_1 + (n - 1)d$$
$$a_n = 0 + (n - 1)(1.5)$$
$$a_n = 1.5n - 1.5$$

b. Since the cost of a gallon of milk is increasing by a constant factor of 2.2%, the situation is best modeled by a geometric sequence. The first term is 0.23 and the common ratio, r, of consecutive terms is 1.022 so

$$a_n = 0.23(1.022)^{n-1}$$

c. Doubling a small amount of thickness will not add much. But, doubling larger and larger amounts will add larger and larger amounts to the thickness. Therefore, this situation is best represented by a geometric sequence that has a constant ratio of 2.

$$a_n = 0.0125(2)^{n-1}$$

d. Each time a field goal is successfully attempted, three points are added to the total points scored by the kicker. An arithmetic sequence with a common difference of 3 best represents this situation. We assume the total points scored starts at the beginning of the season, so $a_1 = 0$. We have the sequence 0, 3, 6, 9, . . . , which is represented by the following equation.

$$a_n = a_1 + (n - 1)d$$
$$a_n = 0 + (n - 1)(3)$$
$$a_n = 3n - 3$$

SUMMARY

In this section you learned the difference between arithmetic and geometric sequences and how these are related to linear and exponential functions, respectively. You also learned how to use arithmetic and geometric sequences to represent and analyze real-world situations.

12.2 EXERCISES

SKILLS AND CONCEPTS

In Exercises 1–10, determine if the sequence is arithmetic, geometric, or neither. Then write a formula for the nth term of the sequence.

1. 3, 5, 7, 9, ...
2. 2, 3, 4.5, 6.75, ...
3. 40, 10, 2.5, 0.625, ...
4. 105, 125, 145, 165, ...
5. 1, −1, 1, −1, ...
6. 10, 5, 0, −5, ...
7. $\frac{1}{2}, \frac{1}{3}, \frac{2}{9}, \frac{4}{27}, \ldots$
8. $\frac{1}{3}, \frac{2}{3}, 1, \frac{4}{3}, \ldots$
9. 5, 2.5, 0, −2.5, ...
10. $27, -3, 1, -\frac{1}{3}, \ldots$

In Exercises 11–15, determine if the graph represents an arithmetic or a geometric sequence. If the sequence is arithmetic, determine whether the value of d is positive or negative. If it is geometric, determine whether the value of r is greater than 1 or between 0 and 1. For all the graphs, assume both the horizontal and vertical axes use the same scale.

11.

12.

13.

14.

15.

In Exercises 16–25, two terms of an arithmetic sequence are given. Write an explicit formula for the sequence and find the indicated term.

16. $a_1 = 2, a_5 = 10, a_{10} = ?$
17. $a_1 = 5, a_6 = 12, a_9 = ?$
18. $a_1 = -2, a_7 = 19, a_{15} = ?$
19. $a_1 = \frac{1}{2}, a_4 = -\frac{1}{2}, a_8 = ?$
20. $a_2 = 6, a_3 = 9, a_9 = ?$
21. $a_3 = 5, a_5 = 15, a_{15} = ?$
22. $a_2 = -1, a_6 = 7, a_{20} = ?$
23. $a_5 = 12, a_9 = 22, a_2 = ?$
24. $a_8 = \frac{25}{4}, a_{10} = \frac{83}{12}, a_{11} = ?$
25. $a_3 = \frac{9}{4}, a_6 = \frac{3}{2}, a_{16} = ?$

In Exercises 26–35, two terms of a geometric sequence are given. Write an explicit formula for the sequence and find the indicated term.

26. $a_1 = 2, a_3 = 8, a_6 = ?$
27. $a_1 = 60, a_4 = 7.5, a_7 = ?$
28. $a_1 = 5, a_2 = -5, a_{10} = ?$
29. $a_1 = 81, a_5 = 16, a_6 = ?$
30. $a_2 = -2, a_3 = 4, a_7 = ?$
31. $a_3 = 50, a_5 = 0.5, a_7 = ?$
32. $a_2 = 10, a_4 = 12.1, a_8 = ?$
33. $a_5 = 81, a_9 = 6561, a_2 = ?$
34. $a_8 = \frac{49}{2}, a_{10} = \frac{1}{2}, a_{11} = ?$
35. $a_3 = 100, a_6 = -0.1, a_2 = ?$

36. Two parents decided to put $20 per week into a coffee can beginning on the day their baby boy was born. They continued this practice until the baby turned 18. At that time, they gave the money to their son to help pay for college.
    a. Is this situation best modeled by a geometric or an arithmetic sequence? Why?
    b. Write an explicit formula for the nth term of the sequence.
    c. Use the formula from part (b) to determine how much money the child was given on his 18th birthday.
    d. Suppose the son earns a scholarship and does not need the money that his parents had saved. Instead, he takes the money he receives on his 18th birthday and deposits it into an account that earns 4% annual interest. Is this situation best expressed using a geometric or an arithmetic sequence? Why?
    e. Write an explicit formula for the nth term of the sequence.
    f. Use the formula from part (e) to determine how much money the son will have on his 25th birthday.

■ SHOW YOU KNOW

37. What are the similarities and differences between arithmetic and geometric sequences?
38. Explain the relationship between an arithmetic sequence and a linear function.
39. Explain the relationship between a geometric sequence and an exponential function.

■ MAKE IT REAL

40. **Starting Salaries** According to a poll, the starting salary of a newly hired accountant in 2007 was $46,718. (*Source: money.cnn.com*) This was a 2.3% increase over the previous year.
    a. Explain why this situation can be represented by a geometric sequence. What information do the terms of the sequence provide?
    b. Write an explicit formula for the nth term of the sequence.
    c. Compute a_5 and explain what it means.

41. **Tuition Payment** Tuition at the University of Hawaii in 2007 was $214 per credit for the first 11 credit hours. (*Source: www.hawaii.edu*)
    a. Explain why this situation is best represented by an arithmetic sequence. What information do the terms of the sequence provide?
    b. Write an explicit formula for the nth term of the sequence.
    c. Compute a_{11} and explain what it means.

42. **Rebounding Ball** In a classroom experiment conducted by students in one of the authors' classrooms, it was discovered that when a racquetball is dropped on the classroom floor, it rebounds about $\frac{2}{3}$ of the height from which it was dropped. Imagine that we drop a ball and it always rebounds to $\frac{2}{3}$ of the previous height.
    a. Explain why this situation is best represented by a geometric sequence. What information do the terms of the sequence provide?
    b. Write an explicit formula for the nth term of the sequence.
    c. Compute a_5 and explain what it means.
    d. Will $a_n = 0$ for some value of n? Explain why or why not.

43. **World's Tallest Champagne Fountain** The tallest champagne fountain in the world was built by Luuk Broos in December 1999. (*Source: timesofindia.indiatimes.com*) The 56-story tower of champagne glasses was on display at the Steigenberger Kurhaus Hotel, Scheveningen, Netherlands. The tower was quite complicated with many sections. For one section, there was one champagne glass at the top of the section with one additional champagne glass added to each subsequent row.
    a. Is this situation best expressed using a geometric or an arithmetic sequence? Why?
    b. Write an explicit formula for the nth term of the sequence.

■ STRETCH YOUR MIND

Exercises 44–47 are intended to challenge your understanding of arithmetic and geometric sequences.

44. A classmate tells you that she knows of an arithmetic sequence where $a_5 = 50$, $a_{10} = 55$, and $a_{20} = 65$. Write a convincing argument to support or refute her claim.
45. A classmate tells you that he knows of a geometric sequence where $a_1 = \frac{1}{2}$, $a_2 = \frac{1}{4}$, and $a_{16} = 55$. Write a convincing argument to support or refute his claim.
46. Construct an arithmetic sequence of the form $a_n = a_1 + (n-1)d$ and a geometric sequence of the form $b_n = b_1 r^{n-1}$ such that $a_n = b_n$ for at least one value of n. If you think it is not possible, clearly explain why you think so.
47. Create a geometric sequence of the form $a_n = a_1 r^{n-1}$ that oscillates between positive and negative values as $n \to \infty$. That is, for some arbitrarily large values of n, the values of a_n should be positive, then negative, then positive, and so on. If you think it is not possible, clearly explain why you think so.

SECTION 12.3 Series

LEARNING OBJECTIVES

- Determine whether a series is arithmetic or geometric
- Determine whether a series converges or diverges using graphs, tables, and symbols
- Compute the partial sums of arithmetic and geometric series
- Use series to solve real-world problems

GETTING STARTED

The Port Theatre in Nanaimo, British Columbia, Canada, was built to "stimulate and enhance artistic, cultural and economic activity of central Vancouver Island." (*Source:* www.porttheatre.nisa .com) As shown in the seating chart, the number of seats in each row of the front center section increases by one seat per row from the first to the eighth row. That is, there are 8 seats in the front row, 9 seats in the second row, 10 seats in the third row, and so on. To find how many seats are in this particular section, we could count the individual chairs, but if we use a mathematical *series* we obtain the answer more quickly.

In this section we investigate mathematical series, that is, the sums of the terms in a sequence, and see how to use them in real-world scenarios.

■ The Sum of the Terms in an Arithmetic Sequence

We can express the number of seats in the specified section of the Port Theatre as a sequence, where each term represents the number of chairs in the given row.

$$8, 9, 10, 11, 12, 13, 14, 15$$

To determine the total number of chairs, we sum the terms in the sequence. Since the sequence has eight terms, we write this sum using the notation S_8.

$$S_8 = 8 + 9 + 10 + 11 + 12 + 13 + 14 + 15$$

In the Port Theatre situation, it is relatively easy to sum the eight terms in the sequence.

$$S_8 = 8 + 9 + 10 + 11 + 12 + 13 + 14 + 15 = 92$$

However, if we were adding hundreds or thousands of terms, we would want a more efficient method.

Notice that if we pair up the first and last terms, the second and seventh terms, the third and sixth terms, and so on, each pair adds to 23.

$$S_8 = 8 + 9 + 10 + 11 + 12 + 13 + 14 + 15$$

Since there are four pairs of terms that sum to 23, the sum of the eight terms is $23 \times 4 = 92$. We use this observation to develop a formula for the sum of an arithmetic sequence.

If we take an arbitrary arithmetic sequence with n terms, the sum of the first and last terms is $a_1 + a_n$. The sum of the second and second-to-last terms is also $a_1 + a_n$ since $a_2 + a_{n-1} = (a_1 + d) + (a_n - d) = a_1 + a_n$. In fact, working inward from the ends, each pair of terms will sum to $a_1 + a_n$. How many of these pairs are there? Since there are n terms, there are $\dfrac{n}{2}$ pairs.

SUM OF A FINITE ARITHMETIC SEQUENCE

The sum of the first n terms of a finite arithmetic sequence with first term a_1 and common difference d is given by

$$S_n = \frac{n}{2}(a_1 + a_n)$$

EXAMPLE 1 ■ Using a Finite Arithmetic Sequence

The Port Theatre has a section of seats in the front and center that has 8 seats in the first row, 9 seats in the second row, 10 seats in the third row, and so on. Find the number of seats in the eight rows in this section.

Solution We need to find the sum of the first eight terms of the sequence whose first term is $a = 8$ and whose common difference is $d = 1$. We first write the formula for the nth term of the arithmetic sequence as

$$a_n = a_1 + (n - 1)d$$
$$a_n = 8 + (n - 1)(1)$$
$$a_n = 8 + n - 1$$
$$a_n = 7 + n$$

We now find the eighth term of the sequence:

$$a_8 = 7 + 8$$
$$a_8 = 15$$

Now we can find the sum of the first eight terms of the sequence.

$$S_8 = \frac{8}{2}(8 + 15)$$
$$= 4(23)$$
$$= 92$$

There are 92 seats in this section of the Port Theatre.

EXAMPLE 2 ■ Computing the Sum of an Arithmetic Sequence

In the Nintendo Wii bowling game, the player tries to roll a series of strikes (knock down all pins in one roll of the ball) with more and more pins in the alley at each level. The final level challenges the player to knock down pins that are arranged so that there is one pin in the first row, two pins in the second row, and so on. Given there are 13 rows of pins, how many pins will the player need to knock down to complete the level?

Solution We know the first term of the arithmetic series is $a_1 = 1$, the last term is $a_{13} = 13$, and the common difference is $d = 1$. We use the formula for the sum of a finite arithmetic sequence to determine the sum of the 13 terms:

$$S_n = \frac{n}{2}(a_1 + a_n)$$

$$S_{13} = \frac{13}{2}(1 + 13)$$

$$= \frac{13}{2}(14)$$

$$= 91$$

There are 91 pins to be knocked down.

■ Geometric Series

Azithromycin is an antibiotic used to fight diseases such as strep throat. A typical prescription requires taking two 250-mg tablets initially followed by a single 250-mg tablet each day for four consecutive days. The half-life of azithromycin is 68 hours. (*Source:* www.drugs.com) This means that 50% of the medicine is eliminated from the body every 68 hours. (This is equivalent to eliminating about 22% of the medicine from the body daily.) We will investigate this situation to determine how much medicine will be in the body when the final dose is taken.

We assume the medicine is absorbed into the bloodstream immediately after taking a dose. Initially, 500 mg of medicine is taken. We represent this as $M_1 = 500$ mg. On day 2, a second dose of 250 mg is taken. At that time, the amount of medicine that will be in the body is about

$$M_2 = 250 + 500(0.78) = 640 \text{ mg}$$

This includes the 250-mg second dose and the 78% of the 500-mg first dose that remains. On day 3, a third dose of 250 mg is taken. At that time, the amount of medicine that will be in the body is

$$M_3 = 250 + 640(0.78) \approx 749 \text{ mg}$$

This includes the 250-mg third dose and 78% of the 640 mg that remains from the first two doses.

To better see a pattern develop, consider the following nonsimplified version of the computations:

$$M_3 = 250 + (0.78)(M_2)$$
$$= 250 + (0.78)[250 + (500)(0.78)]$$
$$= 250 + (250)(0.78) + (500)(0.78)^2$$
$$\approx 749 \text{ mg}$$

On day 4, a fourth dose of 250 mg is taken. At that time, the amount of medicine that will be in the body is

$$M_4 = 250 + (0.78)(M_3)$$
$$= 250 + (0.78)[250 + (250)(0.78) + (500)(0.78)^2]$$
$$= 250 + (250)(0.78) + (250)(0.78)^2 + (500)(0.78)^3$$
$$\approx 834 \text{ mg}$$

With the exception of the final term, we see a sum in the form

$$a + ar + ar^2 + ar^3 + \cdots + ar^{n-1}$$

Expressions of this form are known as a finite geometric series. A **finite geometric series** is a sum in which each term in the sum is found by multiplying the previous term by a constant. In other words, a geometric series is the sum of the terms of a geometric sequence.

FINITE GEOMETRIC SERIES

A finite geometric series with n terms can be expressed as

$$a + ar + ar^2 + ar^3 + \cdots + ar^{n-1}$$

where a is the first term of the series and r is the *common ratio* or *constant multiple*.

In our example, each term (except for the last) is multiplied by 0.78 since 78% of the medicine remains in the body upon taking the next dose. We can express this sum as an explicit formula once we recognize the pattern.

250	250(0.78)	250(0.78)^2	250(0.78)^3	500(0.78)^4
$n = 1$	$n = 2$	$n = 3$	$n = 4$	$n = 5$

Expressing each term in the series as a_n, we have

$$a_1 = 250$$
$$a_2 = 250(0.78)$$
$$a_3 = 250(0.78)^2$$
$$a_4 = 250(0.78)^3$$
$$a_5 = 500(0.78)^4$$

We use sigma notation to express the first four terms of this sum. The Greek letter sigma (Σ) means to create a sum.

$$\sum_{n=1}^{4} 250(0.78)^{n-1}$$

In this case, the terms of the sum are formed by letting $n = 1$, $n = 2$, $n = 3$, $n = 4$. Of course, in this case we also have to add the final term.

$$a_5 = 500(0.78)^4$$

Therefore, the amount of azithromycin in the body after taking the fifth dose is

$$\sum_{n=1}^{4} 250(0.78)^{n-1} + a_5 = 250 + (250)(0.78) + (250)(0.78)^2 + (250)(0.78)^3 + (500)(0.78)^4$$
$$= 900.8 \text{ mg}$$

EXAMPLE 3 ■ Using a Finite Geometric Series

A person deposits $2000 every year into an account that earns 8% compounded annually. Using sigma notation, express the amount of money that will be in the account after 8 years, then determine the amount.

Solution Let A_n represent the amount of money in the account immediately after the $2000 is deposited into the account for the nth time. Thus,

$$A_1 = 2000$$
$$A_2 = 2000 + (2000)(1.08) = 4160$$

We find A_2, the amount of money in the account after the second year's deposit is made, by adding the new $2000 deposit to the original $2000 plus the interest the original deposit earned. We continue this pattern. After the third year's deposit, we will have $2000, plus the previous years' deposits and the interest they have earned.

$$A_3 = 2000 + (4160)(1.08)$$
$$= 2000 + (2000 + (2000)(1.08))(1.08)$$
$$= 2000 + (2000)(1.08) + (2000)(1.08)^2$$
$$= 6492.80$$

We investigate what happens after the deposit is made for year 4 and then look for a pattern to help us write the sum in sigma notation.

$$A_4 = 2000 + (6492.80)(1.08)$$
$$= 2000 + [2000 + (2000)(1.08) + (2000)(1.08)^2](1.08)$$
$$= 2000 + (2000)(1.08) + (2000)(1.08)^2 + (2000)(1.08)^3$$
$$= 9012.22$$

We see we have a geometric series that begins with $2000. Each subsequent term is multiplied by the common multiple 1.08. Using sigma notation, we have

$$\sum_{n=1}^{8} 2000(1.08)^{n-1}$$
$$= 2000 + (2000)(1.08) + (2000)(1.08)^2 + (2000)(1.08)^3 + \cdots + (2000)(1.08)^7$$
$$= 21{,}273.26$$

After the eighth deposit, the account will have $21,273.26.

Finding the Formula for a Finite Geometric Series

In the previous examples, we needed to add a finite number of terms of the particular geometric sequence. In the first case, we added a total of five terms. In the second case, we added a total of eight terms. If we needed to add 100 terms, our current method would be tedious and inefficient.

To develop a more efficient method, we begin with a general finite geometric sequence and represent the sum of the first n terms using S_n.

$$S_n = a + ar + ar^2 + \cdots + ar^{n-1}$$

Next, we multiply each side of this equation by r.

$$rS_n = ar + ar^2 + ar^3 + \cdots + ar^n$$

Now we subtract rS_n from S_n.

$$S_n - rS_n = (a + ar + ar^2 + \cdots + ar^{n-1}) - (ar + ar^2 + ar^3 + \cdots + ar^{n-1} + ar^n)$$
$$= (a + ar + ar^2 + \cdots + ar^{n-1}) - ar - ar^2 - ar^3 - \cdots - ar^{n-1} - ar^n$$
$$= a + \cancel{ar} + \cancel{ar^2} + \cdots + \cancel{ar^{n-1}} - \cancel{ar} - \cancel{ar^2} - \cancel{ar^3} - \cdots - \cancel{ar^{n-1}} - ar^n$$
$$= a - ar^n$$

We see that all of the terms except the first and the last ones cancel out. Since our goal is to find a simplified way to express S_n, we solve this equation for S_n.

$$S_n - rS_n = a - ar^n$$
$$S_n(1 - r) = a - ar^n$$
$$S_n = \frac{a - ar^n}{1 - r}$$
$$= \frac{a(1 - r^n)}{1 - r}$$

SUM OF A FINITE GEOMETRIC SEQUENCE

The sum of the first n terms of a finite geometric sequence with first term a and common ratio r is

$$S_n = \frac{a(1 - r^n)}{1 - r}$$

provided that $r \neq 1$.

EXAMPLE 4 ■ **Finding the Sum of a Finite Geometric Sequence**

Use the formula for the sum of a finite geometric sequence to confirm that the amount of azithromycin in the body will be about 901 mg after the fifth dose. Recall that 22% of the medicine is eliminated from the body each day and that the prescription calls for an initial dose of 500 mg followed by daily doses of 250 mg.

Solution We first need to find the sum of the first four terms of the finite geometric sequence where $a = 250$ mg and $r = 0.78$ (since 22% is eliminated, 78% remains).

$$S_4 = \frac{250 - 250(0.78)^4}{1 - 0.78}$$

$$= 715.7$$

Next, we determine how much of the original dose is remaining in the body.

$$500(0.78)^4 = 185.1$$

We add these results to get the final amount of medicine in the body after the fifth dose is taken.

$$715.7 + 185.1 = 900.8 \text{ mg}$$

EXAMPLE 5 ■ **Finding the Sum of a Finite Geometric Sequence**

How much azithromycin will be in the body if the medicine is taken for 10 days? Recall that the prescription calls for an initial dose of 500 mg followed by daily doses of 250 mg and that 22% of the medicine is eliminated from the body each day.

Solution We need to find the sum of the first nine terms of the geometric sequence where $a = 250$ and $r = 0.78$.

$$S_9 = \frac{250 - 250(0.78)^9}{1 - 0.78}$$

$$= 1014.9 \text{ mg}$$

We also need to consider the amount of the initial 500-mg dose that is left in the body.

$$500(0.78)^9 = 53.4 \text{ mg}$$

We add these results to get the final amount of medicine in the body after the tenth dose is taken.

$$1014.9 + 53.4 = 1068.3 \text{ mg}$$

EXAMPLE 6 ■ Using the Sum of a Finite Geometric Sequence

A person deposits $1500 every year into an account that earns 12% compounded annually. How much money will be in this account after 20 years? After 40 years?

Solution We have a finite geometric series where $a = \$1500$ and $r = 1.12$.

$$S_{20} = \frac{1500 - 1500(1.12)^{20}}{1 - 1.12}$$
$$= 108{,}078.66$$

After 20 years, the account will contain over $108,000.

We can use the same formula to find the amount of money in the account after 40 years.

$$S_{40} = \frac{1500 - 1500(1.12)^{40}}{1 - 1.12}$$
$$= 1{,}150{,}637.13$$

After 40 years, the account will contain over $1 million.

Convergence and Infinite Geometric Series

When we look at the examples involving medicine and saving money, we see that their end behavior differs. The amount of medicine in the bloodstream stabilized over time, but the amount of money in the savings account continued to grow at an increasing rate. If we were to let the number of terms summed in each series go to infinity, we would see two very different results. A series that stabilizes (reaches a maintenance level) is said to **converge**. A series that does not stabilize is said to **diverge**.

We now consider the sum of an infinite geometric series for different cases of r to determine under what circumstances the infinite sum will converge or diverge.

Case 1: $r = 1$

Recall that we are working with the geometric series of the form $S_n = a + ar + ar^2 + \cdots + ar^{n-1}$. If $r = 1$, then

$$S_n = a + a + a + \cdots + a$$
$$= na$$

If $n \to \infty$, then $S_n \to \infty$ and the series diverges.

Case 2: $r > 1$

If $r > 1$, then subsequent terms in the sequence will get larger and larger (since we are multiplying a by values larger than 1). As $n \to \infty$, the sum gets larger and larger and thus diverges. This is what happens in the case of the money account.

Case 3: $r < -1$

If $r < -1$, then we multiply a by positive values when the power of r is even and negative values when the power of r is odd. Since $|r| > 1$, $|r^n| > |r^{n-1}|$ for all values of n. Therefore, the sum will oscillate and become larger and larger in magnitude. Again, the infinite sum diverges.

Case 4: $-1 < r < 1$

The only remaining case is a series with values of r that are between -1 and 1 as in the case of the azithromycin, where $r = 0.22$. If we analyze the formula for a finite geometric sum, we recognize that the series will stabilize.

$$S_n = \frac{a - ar^n}{1 - r}$$

$$= \frac{a(1 - r^n)}{1 - r}$$

As $n \to \infty$, $r^n \to 0$ since $-1 < r < 1$. Therefore,

$$\text{as } n \to \infty, S_n \to \frac{a(1 - 0)}{1 - r} = \frac{a}{1 - r}$$

In general, we have the following.

SUM OF AN INFINITE GEOMETRIC SERIES

The sum of the first n terms of an infinite geometric series

$$S_n = a + ar + ar^2 + \cdots + ar^{n-1}$$

converges to

$$S = \frac{a}{1 - r}$$

as $n \to \infty$ provided that $-1 < r < 1$. Otherwise, the series diverges.

We can verify the formula for the sum of an infinite geometric series if we consider the sum for very large values of n.

EXAMPLE 7 ■ Using an Infinite Series

A person is given a prescription that requires a single 500-mg dose of medicine every day. Each day, 25% of the medicine is eliminated from the body. How much medicine is in the body after 10 days? 50 days? 100 days? What if the person takes the medicine indefinitely?

Solution We recognize this to be a situation involving a finite geometric series with $a = 500$ and $r = 0.75$ (since 25% of the medicine is eliminated, 75% remains). We use the formula for the sum of a finite geometric series to find the amount of medicine in the body after 10, 50, and then 100 days.

$$S_{10} = \frac{500 - 500(0.75)^{10}}{1 - 0.75}$$

$$= 1887.37 \text{ mg}$$

After the 10th dose, there will be 1887.37 mg in the body.

$$S_{50} = \frac{500 - 500(0.75)^{50}}{1 - 0.75}$$

$$= 2000.00 \text{ mg}$$

After the 50th dose, there will be 2000 mg of medicine in the body. It appears as though we are reaching the maintenance level. Let's verify by finding the amount of medicine in the body after the 100th dose.

$$S_{100} = \frac{500 - 500(0.75)^{100}}{1 - 0.75}$$
$$= 2000 \text{ mg}$$

The amount of medicine is being maintained at 2000 mg. If the medicine was to be taken indefinitely, then we could say that $n \to \infty$. We could then find the sum of the infinite geometric series where $a = 500$ mg and $r = 0.75$.

$$S = \frac{500}{1 - 0.75}$$
$$= 2000 \text{ mg}$$

As predicted, the amount of medicine in the body will stabilize at 2000 mg.

EXAMPLE 8 ■ Using an Infinite Series

Attracting large events such as the Super Bowl to a city can boost the local economy. The boost comes in the form of money spent by visitors who pay for lodging, food, transportation, souvenirs, and so on. Economists estimate that about 70% of every dollar introduced into the local economy is spent again locally. Then, 70% of that money is spent again locally. The process continues with 70% of the money spent in the community being spent again locally. It is estimated that during the 1998 Super Bowl in San Diego, visitors contributed well over $125 million in direct spending to the region while they were there. (*Source:* www.nfl.com) What was the total impact, in dollars, on the local San Diego economy considering how the money was re-spent locally?

Solution We begin with the $125 million that was spent by visitors to San Diego. Of that amount, 70% was spent again locally, and this pattern continues indefinitely. We can express the economic impact as the infinite geometric series, S.

$$S = 125 + (125)(0.70) + (125)(0.70)^2 + (125)(0.70)^3 + \cdots$$
$$= \sum_{n=1}^{\infty} 125(0.70)^{n-1}$$

Thus we can find the sum of this infinite geometric series where $a = 125$ and $r = 0.70$.

$$S = \frac{125}{1 - 0.70}$$
$$= 416.67$$

We estimate the total economic impact of the Super Bowl on the San Diego area to be $416,670,000.

Finding a Rational Number Given a Repeating Decimal

Many people know the repeating decimal $0.3333\ldots$ represents the fraction $\frac{1}{3}$. But what fraction does the repeating decimal $0.82222\ldots$ represent? We can use the ideas from this section to find a rational number that is equivalent to a given repeating decimal.

EXAMPLE 9 ■ **Writing a Repeating Decimal as a Rational Number**

Express the repeating decimal 0.82222 . . . as a rational number.

Solution We begin by rewriting the repeating decimal as a sum.

$$0.8 + 0.02 + 0.002 + 0.0002 + \cdots$$

After the first term of 0.8, we see a geometric series where the first term is 0.02 and each subsequent term is found by multiplying by $\frac{1}{10} = 0.1$.

$$0.8 + \underbrace{0.02 + 0.002 + 0.0002 + \cdots}_{\text{Geometric series}}$$

We can find the sum of the infinite geometric series.

$$S = \frac{0.02}{1 - 0.1}$$

$$= \frac{0.02}{0.9}$$

$$= \frac{2}{90} \quad \text{Multiply by } \frac{100}{100}.$$

$$= \frac{1}{45}$$

We keep the calculations in fraction form since we ultimately want to express the repeating decimal in fraction form. We now add the initial 0.8 to the fraction $\frac{1}{45}$ to obtain the final rational number.

$$0.82222\ldots = 0.8 + \frac{1}{45}$$

$$= \frac{8}{10} + \frac{1}{45}$$

$$= \frac{(8)(9)}{90} + \frac{(1)(2)}{90}$$

$$= \frac{72 + 2}{90}$$

$$= \frac{74}{90}$$

$$= \frac{37}{45}$$

SUMMARY

In this section you learned how to find the sum of finite arithmetic and geometric sequences. You also learned how to determine whether a series converges or diverges. Additionally, you saw how to calculate the sum of a convergent infinite geometric series and use these series in real-world situations.

12.3 EXERCISES

SKILLS AND CONCEPTS

In Exercises 1–5, determine whether the series is arithmetic, geometric, or neither. Explain how you know.

1. $\sum_{n=1}^{10} 500\left(\dfrac{1}{2}\right)^n$
2. $\sum_{n=1}^{10} 500 + \dfrac{1}{2}n$
3. $\sum_{n=1}^{20} 2n^2$
4. $\sum_{n=1}^{5} (4 + 2n)$
5. $\sum_{n=1}^{5} 4(2)^n$

In Exercises 6–10, find the sum of the first 20 terms in the arithmetic series.

6. $3 + 6 + 9 + 12 + \cdots + 3n + \cdots$
7. $-5 + -1 + 3 + 7 + \cdots + (4n - 9) + \cdots$
8. $10 + 9.5 + 9 + 8.5 + \cdots + (9.5 + 0.5k) + \cdots$
9. $\dfrac{1}{4} + \dfrac{3}{8} + \dfrac{1}{2} + \dfrac{5}{8} + \cdots + \left(\dfrac{1}{8} + \dfrac{1}{8}k\right) + \cdots$
10. $7 + 9 + 11 + 13 + \cdots + (5 + 2n) + \cdots$

11. A pyramid of logs has 2 logs in the top row, 4 logs in the second row from the top, 6 logs in the third row from the top, and so on, until there are 200 logs in the bottom row.

    a. Write and interpret the first 10 terms of this sequence of numbers.

    b. Provide the formula for the nth term of the sequence and use it to find the number of logs in the 76th row.

    c. Use an arithmetic series to determine the total number of logs in the pyramid.

In Exercises 12–16, find the sum of the first four terms in the geometric series. Then determine a formula for the sum of the first n terms in the series. Determine whether the series converges or diverges. If the series converges, state its sum.

12. $\dfrac{3}{4^0} + \dfrac{3}{4^1} + \dfrac{3}{4^2} + \dfrac{3}{4^3} + \cdots + \dfrac{3}{4^{n-1}} + \cdots$
13. $\dfrac{1}{3} + \dfrac{1}{6} + \dfrac{1}{12} + \dfrac{1}{24} + \cdots + \dfrac{1}{3 \cdot 2^{n-1}} + \cdots$
14. $2 + \dfrac{2}{2^{1/2}} + \dfrac{2}{3^{1/2}} + \dfrac{2}{4^{1/2}} + \cdots + \dfrac{2}{k^{1/2}} + \cdots$
15. $10 + 1 + \dfrac{1}{10} + \dfrac{1}{100} + \cdots + \dfrac{10}{10^{k-1}} + \cdots$
16. $2 + 4 + 6 + 8 + \cdots + 2k + \cdots$

In Exercises 17–21, match the given series with the appropriate graph (A–E) of the partial sums.

A.

B.

C.

D.

E.
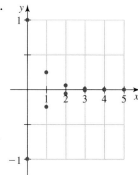

17. $\sum_{n=1}^{\infty} 4\left(\frac{1}{2}\right)^n$
18. $\sum_{n=1}^{\infty} \frac{1}{2}(2^n)$
19. $\sum_{n=1}^{\infty} \left(-\frac{1}{4}\right)^n$
20. $\sum_{n=1}^{\infty} (2 + 3n)$
21. $\sum_{n=1}^{\infty} \frac{10}{(n+1)(n+2)}$

In Exercises 22–26, an infinite geometric series is given. Explain how you can tell if the infinite series will converge or diverge simply by looking at the expression. Do not determine the sum of the infinite series.

22. $\sum_{n=1}^{\infty} \frac{1}{5}(5^n)$
23. $\sum_{n=1}^{\infty} 5\left(\frac{1}{5}\right)^n$
24. $\sum_{n=1}^{\infty} 5\left(-\frac{1}{5}\right)^n$
25. $\sum_{n=1}^{\infty} 1\left(\frac{2}{3}\right)^n$
26. $\sum_{n=1}^{\infty} \frac{2}{3}(1^n)$

In Exercises 27–31, express each repeating decimal as a rational number.

27. 0.2222…
28. 0.9999…
29. 2.242424…
30. 1.7373737…
31. 0.181818…

In Exercises 32–41, find the missing value.

32. $a = 6$, $d = 5$, $S_{20} = 1070$, $a_{20} = ?$
33. $a = 250$, $d = -2$, $S_{50} = 10{,}050$, $a_{50} = ?$
34. $a_{100} = 304$, $d = 3$, $S_{100} = 15{,}550$, $a = ?$
35. $a_{200} = 89.5$, $d = 0.5$, $S_{200} = 15{,}900$, $a = ?$
36. $r = 1.5$, $a = 5$, $S_5 = ?$
37. $r = 0.25$, $a = 40$, $S_{10} = ?$
38. $r = 3$, $S_6 = 1164.8$, $a = ?$
39. $r = 2$, $S_5 = 310$, $a = ?$
40. $r = 4$, $S_7 = 2730.5$, $a = ?$
41. $r = 1.5$, $S_{12} = 100$, $a = ?$

■ SHOW YOU KNOW

42. Write a finite geometric series where the common ratio is between -1 and 1. Explain why the series you created is a geometric series. Find the total sum of your finite geometric series.

43. Explain why the formula for the sum of a finite geometric series includes the stipulation that $r \neq 1$.

44. Explain why the formula for the sum of an infinite geometric series requires that $-1 < r < 1$ in order for the infinite series to converge.

45. Write an infinite geometric series that will converge. Explain how you know that the series you created will converge. Find the total sum of your infinite series.

46. Explain why the idea of an infinite geometric series can be used to find a rational number that is equivalent to a repeating decimal. Then, write your own repeating decimal and then express your number as a rational number.

■ MAKE IT REAL

47. **Economics** According to the Federal Reserve Bank of Boston, $180 million of counterfeit money was put into the economy in 1999. (*Source:* www.bos.frb.org) Suppose each time this counterfeit money is used, 26% of it is seized. What is the total dollar impact of the counterfeit money from 1999 on the economy?

48. **Economic Impact of an Event** Officials from the City of Indianapolis predicted that up to $45 million was going to be spent by visitors who were in town for the NCAA Final Four basketball tournament in 2006. It was expected that much of this money would be spent on national chain restaurants and hotels and therefore may not be spent again locally. (*Source:* www.courier-journal.com) Suppose only 40% of the money is spent again locally. What is the total economic impact of the Final Four basketball tournament in Indianapolis?

49. **Pharmaceuticals—Azithromycin** Azithromycin is an antibiotic often prescribed to treat strep throat. The half-life of azithromycin is 68 hours. Suppose 753.3 mg of azithromycin is in the bloodstream after the fifth dose is taken. If the person stops taking the medicine at this time, how long will it take before the medicine is eliminated from the bloodstream? (*Source:* www.drugs.com)

■ STRETCH YOUR MIND

Exercises 50–56 are intended to challenge your understanding of series.

50. A ball is dropped from a height of 15 meters. It rebounds one-half of the vertical height from which it just fell. It continues to rebound in this way indefinitely. Find the total distance that the ball traveled.

51. The transcendental number e can be expressed as an infinite series. The number e can be approximated using the infinite series

$$e \approx 1 + \frac{1}{1} + \frac{1}{2} + \frac{1}{6} + \frac{1}{24} + \frac{1}{120} + \frac{1}{720} + \cdots$$

Write this infinite series using sigma notation and verify that the partial sums approach e as the number of terms approaches ∞.

52. The transcendental number π can be expressed as an infinite series. The number π may be approximated using the infinite series

$$\pi \approx \frac{4}{1} - \frac{4}{3} + \frac{4}{5} - \frac{4}{7} + \frac{4}{9} + \frac{4(-1)^{n+1}}{2n-1} + \cdots$$

Write this infinite series using sigma notation and verify that the partial sums approach π as the number of terms approaches ∞.

53. Verify, using multiple representations, that the following series converges to 1.

$$\sum_{n=1}^{\infty} \frac{1}{n(n+1)} = 1$$

54. What is the value(s) of p if $\sum_{n=1}^{\infty} (2+p)^{-2n} = \frac{1}{8}$?

55. Write a discussion revealing the errors in reasoning in the following "proof."

$0 = 0 + 0 + 0 + \cdots$

$0 = (1-1) + (1-1) + (1-1) + \cdots$

$0 = 1 - 1 + 1 - 1 + 1 - 1 + \cdots$

$0 = 1 + (-1+1) + (-1+1) + \cdots$

$0 = 1 + 0 + 0 + \cdots$

$0 = 1$

56. Construct an argument that explains that if the series $\sum_{n=1}^{\infty} \frac{1}{n}$ diverges, then the series $\sum_{n=1}^{\infty} \frac{1}{n-5}$ must also diverge.

CHAPTER 12 Study Sheet

As a result of your work in this chapter, you should be able to answer the following questions, which are focused on the "big ideas" of this chapter.

SECTION 12.1
1. What is a sequence and what does it mean to say that a sequence converges?
2. What is the difference between recursive and explicit formulas for sequences?

SECTION 12.2
3. What is the difference between a geometric sequence and an arithmetic sequence?
4. What is the relationship between an arithmetic sequence and a linear function? What is the relationship between a geometric sequence and an exponential function?

SECTION 12.3
5. What is the difference between a sequence and a series?
6. How do you find the sum of a finite arithmetic series and why does this procedure make sense?
7. What does it mean to say that we can find the sum of an infinite series?
8. How do you find the sum of an infinite geometric series?

REVIEW EXERCISES

■ SECTION 12.1 ■

In Exercises 1–4,
- a. List the first six terms of the given sequence.
- b. Use a graphing calculator to graph the sequence.
- c. State whether the sequence appears to converge or diverge.

1. $a_n = \dfrac{n^2 + 1}{n}$
2. $a_n = \dfrac{10n}{3^n}$
3. $a_n = 0.5a_{n-1} + 2$
4. $a_n = 1.5a_{n-1} + 1$

In Exercises 5–8, find a formula for the general term of the sequence, a_n, and state whether the sequence converges. Assume the pattern in the given terms continues.

5. $-2, 1, 4, 7, \ldots$
6. $18, 6, 2, \dfrac{1}{3}, \ldots$
7. $\dfrac{2}{3}, \dfrac{4}{5}, \dfrac{6}{7}, \dfrac{8}{9}, \ldots$
8. $2, 3, \dfrac{9}{2}, \dfrac{27}{4}, \dfrac{81}{8}, \ldots$

9. **Heartworm Medicine** Dog owners give their dogs heartworm medicine once every month to prevent heartworm disease. One type of heartworm medicine, Interceptor Flavor Tabs, contains 11.5 mg of medicine and has a half-life of about 36 hours. (*Source:* www.beaconforhealth.org)
   - a. Generate the first five terms of the sequence that would represent the amount of heartworm medicine in the bloodstream every 36 hours.
   - b. Create an explicit formula representing the amount of medicine in the bloodstream.
   - c. Estimate the amount of medicine in the bloodstream at the end of 2 weeks.
   - d. How much medicine will be in the bloodstream at the time the next dose is administered one month later? Use a graph to confirm your answer.

10. **Pharmaceuticals—Naproxen** Naproxen is an anti-inflammatory medicine often used by patients with arthritis to reduce swelling in joints. With a half-life of 15 hours, about 42.6% of the medicine is eliminated from the bloodstream every 12 hours. A typical prescription calls for the patient to take one 500-mg tablet every 12 hours. (*Source:* www.drugs.com)
    - a. Generate the first five terms of the sequence that would represent the amount of medicine in the bloodstream every 12 hours.
    - b. What is the maintenance level in this situation?
    - c. How long does it take to reach the maintenance level? Use a graph to confirm your answer.

■ SECTION 12.2 ■

In Exercises 11–15, two terms of an arithmetic sequence are given. Write an explicit formula for the sequence and find the indicated term.

11. $a_1 = -5, a_5 = 3, a_{10} = ?$
12. $a_1 = 10, a_8 = 45, a_{18} = ?$
13. $a_5 = 1, a_8 = -14, a_2 = ?$
14. $a_{20} = 100, a_{25} = 102.5, a_{65} = ?$
15. $a_8 = 2, a_{11} = \dfrac{11}{4}, a_1 = ?$

In Exercises 16–20, two terms of a geometric sequence are given. Write an explicit formula for the sequence and find the indicated term.

16. $a_1 = 10, a_3 = 2.5, a_6 = ?$
17. $a_1 = -1, a_5 = -16, a_{10} = ?$
18. $a_2 = 1, a_6 = 81, a_1 = ?$
19. $a_5 = 8, a_7 = \dfrac{1}{8}, a_3 = ?$
20. $a_{10} = \dfrac{128}{25}, a_{15} = \dfrac{4096}{25}, a_{38} = ?$

21. **Starting Salaries** According to a poll, the starting salary of a newly hired college English major in 2007 was $32,553. (*Source:* money.cnn.com) This was a 5.3% increase over the previous year.
    - a. Explain why this situation is best represented by a geometric sequence. What information do the terms of the sequence provide?
    - b. Write an explicit formula for the nth term of the sequence.
    - c. Compute a_5 and explain what it means.

22. **Skydiving** In November 2003, a group of 70 skydivers performed one of the largest canopy formation jumps ever. (*Source:* www.boeing.com) The canopy formation is created, after the parachutes are opened, so that there is one person at the top, two people below the top person, then three people below them, and so on. There are nine rows following this pattern before the pattern is inverted. For this exercise, only consider the first nine rows of the formation.

    Photo by Bryan Scott

    - a. Is this situation better expressed using a geometric or an arithmetic sequence? Why?
    - b. Write an explicit formula for the nth term of the sequence.

■ SECTION 12.3 ■

In Exercises 23–26, determine whether the series is arithmetic, geometric, or neither. Justify your answer.

23. $\displaystyle\sum_{n=1}^{10} 2 + 3n$
24. $\displaystyle\sum_{n=1}^{5} 10\left(\dfrac{1}{14}\right)^{n-1}$
25. $\displaystyle\sum_{n=1}^{15} n$
26. $\displaystyle\sum_{n=1}^{5} 2n^2 + 1$

In Exercises 27–31, find the sum of the first 20 terms of the arithmetic series.

27. $5 + 10 + 15 + 20 + \cdots + 5n + \cdots$
28. $-2 + 0 + 2 + 4 + \cdots + (2n - 4) + \cdots$
29. $10 + \dfrac{41}{4} + \dfrac{21}{2} + \dfrac{43}{4} + \cdots + \left(\dfrac{1}{4}n + \dfrac{39}{4}\right) + \cdots$
30. $100 + 103 + 106 + 109 + \cdots + (3n + 97) + \cdots$
31. $(-5) + (-9) + (-13) + (-17) + \cdots + (-4n - 1) + \cdots$

In Exercises 32–36, find the sum of the first four terms of the geometric series. Determine whether the series converges or diverges. If the series converges, state its sum.

32. $\dfrac{1}{5^0} + \dfrac{1}{5^1} + \dfrac{1}{5^2} + \dfrac{1}{5^3} + \cdots + \dfrac{1}{5^{n-1}} + \cdots$
33. $\dfrac{1}{7} + \dfrac{2}{7} + \dfrac{4}{7} + \dfrac{8}{7} + \cdots + \dfrac{2^{n-1}}{7} + \cdots$
34. $3 + \dfrac{3}{\sqrt{2}} + \dfrac{3}{2} + \dfrac{3}{2\sqrt{2}} + \dfrac{3}{4} + \cdots + \dfrac{3}{(\sqrt{2})^{k-1}} + \cdots$
35. $5 + 1 + \dfrac{1}{5} + \dfrac{1}{25} + \cdots + \dfrac{5}{5^{k-1}} + \cdots$
36. $3 + 6 + 12 + 24 + \cdots + 3(2)^{k-1} + \cdots$

In Exercises 37–40, express each repeating decimal as a rational number.

37. $0.1111\ldots$
38. $0.123123\ldots$
39. $0.1818\ldots$
40. $0.23\overline{1}\ldots$

41. **Pharmaceuticals—Naproxen** Suppose a maintenance level of 871.1 mg of naproxen is in the bloodstream. If the person stops taking the medicine at this time, how long before the medicine is eliminated from the bloodstream? The half-life of naproxen is 15 hours. (*Source:* www.drugs.com)

42. **Pharmaceuticals—Singulair®** Suppose a maintenance level of 10 mg of Singulair is in the bloodstream. If the person stops taking the medicine at this time, how long before the medicine is eliminated from the bloodstream? The half-life of Singulair is 4 hours. (*Source:* www.drugs.com)

Make It Real Project

What to Do

1. Choose a prescription drug that you are interested in researching. Find the half-life for the drug and a typical dose.
2. Write the general term of the sequence describing the amount of the drug in the bloodstream for a person who takes the typical dose.
3. Write the general term of the geometric series describing the amount of the drug in the bloodstream.
4. Provide a full analysis for the drug including its maintenance level using graphs, tables, and symbols.
5. Describe the effects of missing a dose. Determine the effect of taking a double dose the next time the medication is to be taken.
6. Write a report summarizing your findings.

Where to Find Data

Data for this project can be found at the following sites, among others. Look under "Metabolism," "Pharmacokinetics," or "Elimination" in the technical information section for half-life data.

- www.drugs.com
- www.rxmed.com
- www.astrazeneca-us.com

At these sites, you will also see the type of information that is included concerning information to the consumer.

ANSWERS

CHAPTER 1 ■ SECTION 1.1

1. The decision-factor equation suggests an economy car with high gas mileage and low miles.

3. The style would be type 1: midsize sedan.

5. The maximum mileage it can have is 3298 miles.

7. Answers vary. decision factor = −500(cost) + 10(style) + 0(color) − 0.1(mileage) + 100(year)

9. Answers vary. decision factor = 0(cost) + 15,000(style) + 5000(color) − 0(mileage) − 2(year)

11. A mathematical model is a table, graph, or formula that represents a real-world situation and is used to make predictions and/or answer questions about that situation. Mathematical models are similar to other types of models in that they represent something else. They are dissimilar in their use for predictions and problem solving.

13. Mathematical models can be in table, graph, formula, or verbal form.

15. a. The total revenue (in millions) of McDonald's increases each year.
    b. The total revenue (in millions) of McDonald's increases each year at a variable rate.

17. a. $8624.95
    b. The differences were: from 2006 to 2007, −$2916; from 2007 to 2008, −$2478; from 2008 to 2009, −$2107; and from 2009 to 2010, −$1791. The dollar amount of depreciation decreases each year by a lesser amount.
    c. No, the car will technically always have some value left.

19. a. Female non-teachers earned less than female teachers from 1940 to some time in the 1980s and then they began to earn more. It also appears that the percentage of female non-teachers who earn a higher salary is increasing.
    b. Answers vary.

21. a. The ticket face value for the Super Bowl is increasing at an increasing rate.
    b. Answers vary.

23. a. As the price of a gallon of gasoline increases, the fuel consumption in billions of gallons varies and does not have a clear trend for the costs between $1.06 to $1.59 per gallon.
    b. Answers vary.

25. a. The time period that hurricanes decreased the longest was 1940–1979.
    b. The decade that the greatest number of hurricanes struck was 1940–1949.
    c. The decade that the least number of hurricanes struck was 1970–1979.

27. a. As the years increase, the number of doctoral degrees awarded in mathematics has decreased.
    b. ≈ 1000 − 900 ≈ 100
    c. ≈ 1000 + 100(3) ≈ 1300

29. a. As the number of women in the workforce (in 1000s) increases, the number of children enrolled in the Head Start program increases.
    b. Answers vary.

31. a. No trend can be seen because the scatter plot does not show an increasing or a decreasing pattern.
    b. Answers vary.

33. a. As the years increase, the average SAT math scores increased and the average SAT verbal scores both increased and decreased.
    b. Answers vary.

35. The fact that a trend is "perfectly" captured by a mathematical model does not necessarily mean that the trend will continue into the future or even between particular data points. Also, if the data is not collected properly due to human error or incorrect assumptions, the predictions can also be wrong.

CHAPTER 1 ■ SECTION 1.2

1. $T(d)$ where d = number of dollars your home is assessed for, T = amount of property tax. T is the dependent variable and d is the independent variable.

3. $c(w)$. Independent variable = weight of a package in ounces. Dependent variable = cost to mail the package.

5. $B(a)$ where B = a person's blood alcohol level, a = number of alcoholic drinks consumed in a 2-hour period. B is the dependent variable and a is the independent variable.

7. a function

9. not a function

11. not a function

13. a function
15. not a function
17. a function
19. not a function
21. not a function
23. a function
25. not a function

27. $v(t) = \dfrac{\sqrt{0.3t}}{10}$; $v(3) = \dfrac{\sqrt{0.3(3)}}{10} \approx 0.095$

29. $t(v) = -v^2 + 3v - \dfrac{4}{v}$;
    $t(-4) = -(-4)^2 + 3(-4) - \dfrac{4}{(-4)} = -27$

31. $m(x) = \sqrt{x^2 - 4x}$;
    $m(\# + 3) = \left[\sqrt{(\# + 3)^2 - 4(\# + 3)}\right]$
    $= (\sqrt{\#^2 + 6\# + 9 - 4\# - 12})$
    $= (\sqrt{\#^2 + 2\# - 3})$

33. $r(s) = |9s^3 - 2s + 18|$;
    $r(\Theta + \Delta) = |9(\Theta + \Delta)^3 - 2(\Theta + \Delta) + 18|$

35. $h(x) = 3^x - 17x + x^2$;
    $h(b^2 + 7) = 3^{b^2+7} - 17(b^2 + 7) + (b^2 + 7)^2$

37. $C(2.76, 310)$
    $C(2.76, 310) = \dfrac{(2.76)(310)}{25} \approx \$34.22$

39. Answers vary. A process is a function in which the input undergoes some sort of a process to produce an output while a correspondence is a function in which there is simply a pairing up of one input with only one output.

41. In many real-world problems to have multiple outputs for one input does not make sense. For example, to have more than one high temperature in Milwaukee, Wisconsin on March 17, 2015, is impossible.

43. Answers vary.

45. $E(t) = 1019.65t + 27861.97$;
    $E(20) = 1019.65(20) + 27,861.97 = 48,254.97$. In 2010 the average annual expenditures of all U.S. consumers is $\$48,254.97$. $E(20) = 48,254.97$

47. $300,000 = 5844.95t + 56,589.91$; $243,410 = 5844.95t$; $41.64 = t$. The median sales price of new homes will be $\$300,000$ in 2021.

49. a. $v(s) = 630$. $s =$ Arizona. In 2004 the state with 630,000 overseas visitors was Arizona.
    b. $v(Hawaii) = 2215$. In 2004 Hawaii had 2,215,000 overseas visitors.

51. $f(x) = 4x - 3$ • okay
    $f(x) - 4x = -3$ • okay
    $x(f - 4) = -3$ • incorrect
    The error is thinking that $f(x)$ means f times x. $f(x)$ is function notation and **not** multiplication.

b. For a sphere with radius 100, the volume is 4,188,790.21 cubic centimeters.

5. a. $A(9, 14, 24) = \dfrac{9(14 + 24)}{2} = 171$
   b. For a trapezoid with height 9 and bases 14 and 24, the area is 171 square meters.

7. The "average rate of change" means that if some sort of change were constant over a particular interval then it would be whatever the average rate of change is. For example, if the average rate of change were a speed of 25 miles per hour for a time period of 4 hours, then if the speed was the same for those 4 hours, that speed would be 25 miles per hour (even though the speed in actuality may very quite significantly).

9. Car insurance premiums are based on age, driving record, style of car, value of car, sex of driver, and so on. For these, the independent variables would be age, driving record, style of car, value of car, and sex of driver; the dependent variable would be the premium.

11. a. The data does represent a function: There is only one output for each input value.
    b.

$\dfrac{25 - 21}{40 - 35}$	$= \dfrac{4}{5}$	$= \$0.80$
$\dfrac{41 - 25}{45 - 40}$	$= \dfrac{16}{5}$	$= \$3.20$
$\dfrac{57 - 41}{50 - 45}$	$= \dfrac{16}{5}$	$= \$3.20$
$\dfrac{88 - 57}{55 - 50}$	$= \dfrac{31}{5}$	$= \$6.20$
$\dfrac{130 - 88}{60 - 55}$	$= \dfrac{42}{5}$	$= \$8.40$
$\dfrac{209 - 130}{65 - 60}$	$= \dfrac{79}{5}$	$= \$15.80$
$\dfrac{361 - 209}{70 - 65}$	$= \dfrac{152}{5}$	$= \$30.40$

c. (sample answer for each 5-year interval): The monthly premium for $\$1,000,000$ of coverage increases at a constant rate of $\$0.80$ for each year in the female's age from 35 to 40 years old.

13. a. The data does represent a function: There is only one output for each input value.
    b.

$\dfrac{16.0 - 7.5}{6 - 0}$	$= \dfrac{8.5}{6}$	$= 1.42$
$\dfrac{21.0 - 16.0}{12 - 6}$	$= \dfrac{5}{6}$	$= 0.83$
$\dfrac{24.0 - 21.0}{18 - 12}$	$= \dfrac{3}{6}$	$= 0.50$
$\dfrac{26.5 - 24.0}{24 - 18}$	$= \dfrac{2.5}{6}$	$= 0.417$
$\dfrac{28.5 - 26.5}{30 - 24}$	$= \dfrac{2}{6}$	$= 0.33$
$\dfrac{30.5 - 28.5}{36 - 30}$	$= \dfrac{2}{6}$	$= 0.33$

CHAPTER 1 ■ SECTION 1.3

1. a. $P(5, 8) = 2(5) + 2(8) = 26$
   b. For a 5×8 rectangle, the perimeter is 26 inches.

3. a. $V(100) = \dfrac{4}{3}\pi(100)^3 \approx 4,188,790.21$

c. (sample answer for each 6-month interval): The weight of girls in pounds increases at a constant rate of 1.42 pounds for each year for girls from birth to 6 months.

15. a. The data does represent a function: There is only one output for each input value.

b.

$$\frac{10.1 - 9.2}{1965 - 1960} = \frac{0.9}{5} = 0.18$$

$$\frac{15.0 - 10.1}{1970 - 1965} = \frac{4.9}{5} = 0.98$$

$$\frac{20.1 - 15.0}{1975 - 1970} = \frac{5.1}{5} = 1.02$$

c. (sample answer for each 5-year interval): The divorce rate increased at a constant rate of 0.18 divorces per 1000 married women for each year from 1960 to 1965.

17. The average rate of change in China's demand for oil (in millions of barrels per day) between 2004 and 2005 is $\frac{6.3 - 6.1}{1} = 0.2$ millions of barrels per day per year. Assuming the change in oil demand remains constant from 2005 to 2006, we can estimate $D(2006)$ by finding $6.3 + 0.2(1) = 6.5$. Thus the estimated number of barrels in China's oil demand is 6.5 million barrels per day in 2006.

19. From February to March the number of private trips given to U.S. lawmakers in 2006 dropped by 32. It would not be reasonable to assume the same decrease from March to April because there would be a negative number of private trips. Therefore a reasonable estimate would be any number less than or equal to 29.

21. The average rate of change in Amazon's net income (in billions of dollars) between 2006 and 2007 is $\frac{14.8 - 10.7}{1} = 4.1$ billion dollars per year. Assuming the change in Amazon's net income remained constant from 2007 to 2008, we can estimate $R(2008)$ by finding $14.8 + 4.1 = 18.9$. Thus the estimated amount of Amazon's net income in 2008 is 18.9 billion dollars.

23. a. $F(2003) = 3426.0$, which means that in 2003 there were 3426 fines levied on passengers by the TSA.
  b. Since the rate of change from 2003 to 2004 is $\frac{9741.0 - 3426.0}{2004 - 2003} = \frac{6315}{1}$, we can estimate the number of fines levied in 2005 as $9741.0 + 6315 = 16,056$.

25. a. $N(2) = 6.9$, which means that in the 2nd quarter of 2005 the net income of Exxon Mobil was 6.9 billion dollars.
  b. Since the rate of change from the 3rd to 4th quarter is $\frac{10.5 - 10.0}{4 - 3} = \frac{0.5}{1}$, we can estimate the amount of net income in the 1st quarter of 2006 as $10.5 - 0.5 = 11.0$ billion dollars.

27. a. The largest average rate of change from 2000 to 2001, 2003 to 2004, and 2004 to 2005 was Toyota, and from 2001 to 2002 and 2002 to 2003 was Honda.
  b. Since Toyota had the largest average rate of change from 2003 to 2004 and 2004 to 2005, it is possible that Toyota also had the largest average rate of change in 2006.

29. a. From February to March, April to July, September to October, and from November to December $N(m)$ is increasing. From February to March the average rate of change was $\frac{0.06}{1}$, from April to July the average rate of change was $\frac{0.09}{3} = 0.03$, from September to October the average rate of change was $\frac{0.10}{1}$, and from November to December the average rate of change was $\frac{0.04}{1}$.

  b. There are three intervals of time when $N(m)$ is decreasing. They are from March to April $\left(\frac{-0.03}{1}\right)$, July to September $\left(\frac{-0.12}{2} = -0.06\right)$, and October to November $\left(\frac{-0.11}{1}\right)$.

  c. There are three intervals of time when $E(m)$ is decreasing. They are from April to May $\left(\frac{-0.01}{1}\right)$, June to July $\left(\frac{-0.16}{1}\right)$, and August to December $\left(\frac{-0.52}{4} = -0.13\right)$.

31. $B(350, 0.06, 12, 5) = 350\left(1 + \frac{0.06}{12}\right)^{12 \cdot 5} \approx \$472.10$. The balance would be \$472.10 for an investment of \$350 invested at an interest rate of 6% with 12 compoundings per year for 5 years.

33. $S(155, 0.2, 0.9) = \sqrt{30(155)(0.2)(0.9)} \approx 28.93$. The minimum speed of a car at the beginning of a skid would be 28.93 miles per hour, if the skid distance is 155 feet, if 0.2 is the drag factor for the road surface, and if 0.9% is the braking efficiency.

35. a. $800,000 = s$
  b. When the total salary of people in sales is \$60,000 the amount of sales is \$800,000.

37. $294.26 \approx K$

39. $F = 158$

41. $50 \approx F$

43. $10.91 \approx h$

45. $4644.80 \approx p$

47. a. $T(m) = 172.55m + 3000$
  b. $T(72) = \$15,423.60$
  c. $15,423.60 - 13,210 = \$2213.60$

49. a. $M(600, 0.0082, 12) \approx 52.45$
  b. The monthly payment on a 12-month loan of \$600 at 0.82% is \$52.45.

51. $FV(100, 0.00425, 60) \approx \$6817.99$. The future value of an investment with a constant monthly payment of \$100 after 60 months and monthly interest rate of 0.00425 is \$6817.99.

53. a. $FV(75, 0.00387, 60) = 75\left[\frac{(1 + 0.00387)^{60} - 1}{0.00387}\right]$

b. $FV(75, 0.00387, 60) \approx \$5054.40$. The future value of an investment with a constant monthly payment of $75 after 60 months and monthly interest rate of 0.00387 is $5054.40.

55. The average rate of change in the price per month is $0.0125.

CHAPTER 1 ■ SECTION 1.4

1. $f(-3) \approx 8.5$
3. $x \approx 0.6$
5. $f(4) \approx -2$
7. no solution
9. $g(0) = 0$
11. $f(g) = e$
13. $f(x) = e$, $x = g$ and $x = n$
15. $f(j) = b$
17. Graph E
19. Graph B
21. The vertical intercept represents the initial amount of air in the balloon. The horizontal intercepts represent the two times when there was no air in the balloon.
23. The vertical intercept represents the initial number of times the lawn was mowed (which would be 0). The horizontal intercept represents the time when there were no cumulative numbers of lawn mowing.
25. Answers vary.
27. **a.** Answers vary.
    b. As the years increase after 1900, the cigarette consumption increases until 1960 and then decreases until 2000.
    c. the cigarette consumption in the year 1900
29. **a.** When the tram is 100 horizontal feet from its original position it would be approximately 400 vertical feet off the ground.
    b. When the tram is 500 vertical feet off the ground it is either 150 or 450 horizontal feet from its original position.
    c. No, the arch is only 600 feet wide.
    d. At 200 feet above the ground, the width of the arch is approximately 500 horizontal feet. If the wingspan of the plane is 46 feet then there should be $\dfrac{(500 - 46) \text{ feet}}{2}$ on each side. So there would be 227 feet on either side of the plane to the inside edge of the arch.
31. **a.** Hiker A is ahead after 1 hour because he has been hiking faster the whole time.
    b. Hiker A is hiking faster than Hiker B because his speed (the vertical coordinates) is greater at that time.
    c. Hiker A is pulling away from Hiker B because his speed (vertical coordinates) is greater during the interval of time from 45 minutes to 1 hour.

CHAPTER 1 ■ SECTION 1.5

1. Let C = total cost of gasoline and g = gallons of gasoline. Then
   $$C = f(g)$$
   $$C(g) = 4.109g$$

3.
Bushels/Acre	Production (billion bushels)
42.4	1.32
42.4 + 2 = 44.4	1.32 + 0.18 = 1.50

5.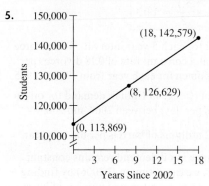

   $m = 1595$ students/year

   $113{,}869 + 1595(8) = 126{,}629$

   In 2010, Arizona universities will have approximately 126,629 students.

7.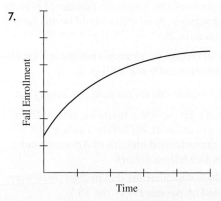

9. The variable to be assigned the horizontal axis is the independent variable and the variable to be assigned the vertical axis is the dependent variable.
11. Choosing letters that can more easily be tied to the variable they represent can help one better recall what the letter stands for and understand the context of the problem.
13. When there were 14,000,000 public college students, there were 4,303,000 private college students.
15. $281.38 \approx c$. According to this mathematical model, net sales of Apple Computer Corporation will be $0 when there is $281,380,000 in sales cost.
17. When there are 9 pounds of combined per capita consumption of apricots, avocados, cherries, cranberries, kiwis, mangoes, papayas, and honeydew, there are 4.4 pounds of per capita pineapple consumed.
19. When there are 3 pounds of combined per capita consumption of apricots, avocados, cherries, cranberries, kiwis, mangoes, papayas, and honeydew, there are −188 pounds of per capita pineapple consumed. This

does not make sense in the real-world context because it is impossible to eat negative pounds of pineapple.

21. As the years increased from 1990 to 2003, the per capita spending on physicians and clinical services increased.

23. As the number of tenpin bowling establishments decreased, the number of bowling memberships in thousands decreased.

25. As the average price of a DVD player increases, the number of DVD players sold in thousands decreases at a decreasing rate.

27. As the number of music cassette tapes shipped in millions increases, the value of the cassette tape shipments in millions of dollars increases at a decreasing rate.

29. As the years increase from 1990 to 2004, cassette tape sales decreases at an increasing rate and then decreases at a decreasing rate.

31. In 1980 there were 44 cases of autism per 100,000 live births and in 1994 there were 208 cases per 100,000 live births. This is an increase of 164 cases per 100,000 live births compared to the initial number of 44 cases in 1980.

Calculating the percentage increase we get $\frac{164}{44} \approx 3.73$ or 373% increase in autism compared to only a 10% increase in immunizations. So most likely MMR vaccinations are not responsible for the increase.

33. $I(R) = -40138.18 + 0.124R$

CHAPTER 1 ■ SECTION 1.6

1. **a.** independent variable: time water has been running; dependent variable: height of bathtub water
   b. independent variable: height of bathtub water; dependent variable: time water has been running
   c. The inverse relationship is a function.

3. **a.** independent variable: time after getting a haircut; dependent variable: length of hair
   b. independent variable: length of hair; dependent variable: time after getting a haircut
   c. The inverse relationship is a function.

5. **a.** independent variable: temperature at which thermostat is set; dependent variable: electric bill
   b. independent variable: electric bill; dependent variable: temperature at which thermostat is set
   c. The inverse relationship may not be a function due to multiple possible outputs for a specific input value.

7. **a.** independent variable: time since entering freeway; dependent variable: speed of car
   b. independent variable: speed of car; dependent variable: time since entering freeway
   c. The inverse relationship may not be a function due to multiple possible outputs for a specific input value.

9. **a.** independent variable: age of adult; dependent variable: weight
   b. independent variable: weight; dependent variable: age of adult

   c. The inverse relationship may not be a function due to multiple possible outputs for a specific input value.

11. $S(t)$ is the function relationship that links the dependent variable, speed, to the independent variable, time.

13.

15.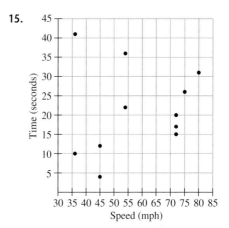

The inverse relationship does not represent a function because there are multiple times that the car was going 72 miles per hour.

17.

x	y
1	19
2	21
3	23
4	25

y	x
19	1
21	2
23	3
25	4

The input and output values are interchanged in each table.

19. **a.** $f^{-1}(C) = t$ is the inverse function relationship of $C(t)$ where $C = \%$ of people cremated is the independent variable and $t =$ the number of years since 1980 is the dependent variable.
    b. $37.69 = t$. Sometime between the year 2017 and 2018 the percentage of people that are cremated will be 40%.
    c. $f^{-1}(40) = 37.69$. When the percentage of people choosing cremation is 40% the year will be between 2017 and 2018.
    d. The results to (b) and (c) represent the same data point and the relationship between the year and % cremated. In part (b), 40% is the output so we had to solve for the independent variable t. In part (c), the inverse uses the percentage as the input so we just had to evaluate.

818 ANSWERS ■ CHAPTER 1

21. a. $C(B)$ means the claw weight is a function of the body weight.

b.

$C(300.2) = 78.1$. When the body weight of a fiddler crab is 300.2 grams its large claw weight is 78.2 grams.

c.

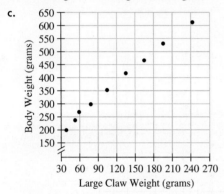

This is a function because each body weight input has only one claw weight output.

d. $f^{-1}(52.5) = 238.3$. When the large claw weight is 52.5 grams, the body weight is 238.3 grams.

23. Where there is $5,597,000,000 international net sales of Apple Computer Corporation there is $8,338,000,000 domestic net sales.

25. For the function $g(y) = 2y + 4$ the input value y is multiplied by 2 and added to 4 whereas for the inverse function $f(x)$ the input is subtracted from 4 and divided by 2.

27. $f^{-1}(3) = 10$

$g(3) = 2(3) + 4 = 10$

CHAPTER 1 ■ REVIEW

1. Answers vary.

3. a. As people's age increases to the teenage years participation in tennis grows then in the early 20s participation drops. In the late 20s participation in tennis grows again then drops from the age of 34 through old age.

b. Answers vary.

5. a. The average number of persons per vehicle decreased for 15 years and then leveled off.

b. The phrase "1.6 persons per vehicle" means that if we hypothetically take all of the people in all the vehicles and divide them up equally there would be 1.6 people in each.

c. Answers vary.

7. $f(-1) = -5$

9. a. $r =$ southwest. In 2004 the average cost of tuition was $4569 in the American Southwest.

b. $6839. In 2004 the average cost of tuition was $6839 in New England.

11. a. $E(210, 74, 42) = 3144.175$

b. Answers vary.

13. a.

$\dfrac{26.5 - 19.6}{6 - 0} = 1.15$
$\dfrac{29.8 - 26.5}{12 - 6} = 0.55$
$\dfrac{32.3 - 29.8}{18 - 12} \approx 0.42$
$\dfrac{34.5 - 32.3}{24 - 18} \approx 0.37$
$\dfrac{36.2 - 34.5}{30 - 24} \approx 0.28$
$\dfrac{37.8 - 36.2}{36 - 30} \approx 0.27$

b. Answers vary. For the average rate of change from 12 to 18 months as the baby boys' age increases and assuming his height increases at a constant rate, that rate would be 0.42 inches per month.

c. $f(27) = 35.35$. A 27-month-old is estimated to be 35.35 inches tall.

15. Answers vary. If we assume the same decrease in receipts from weekend 6 to 7 we estimate $1.343 - 1.678 = -0.335$, which is unreasonable. Therefore, we estimate 0.6.

17. $B(5000, 0.067, 12, 3) \approx 6109.71$

19. a. $T(m) = 49.99 + 0.2(m)$

b. $T(100) = \$69.99$

21. $S(64.3, 0.7, 1.00) \approx 36.75$ miles per hour

23. a. Answers vary. domain: $0 \leq m \leq 5{,}000{,}000$; range: $0 \leq g \leq 1.6$

b. As an American's income increases from 0 to $100,000 the percentage of assets given to charity increases from 0.6 to 1.5% then as income increases from $100,000 to $1,000,000 the percentage decreases. As income increases from $1,000,000 the percentage given to charity slowly increases from 0.7 to 1.4%.

25. a. Graph D; vertical intercept: the initial (time = 0) temperature of the car; horizontal intercept: there will not be one. This would mean the time the temperature in the car is 0°.

Graph E; vertical intercept: the person's initial (time = 0) heart rate; horizontal intercept: there will not be one. This would mean the time when the heart rate is 0 beats per minute.

Graph F; vertical intercept: the initial (time = 0) value of the car; horizontal intercept: there will not be one. This would mean the time when the car is worth $0.

b. Graph D; (answers vary) Let $m =$ the time in car in minutes and $T =$ temperature of the car in degrees Fahrenheit. practical domain: $0 \leq m \leq 10$, practical range: $78 \leq T \leq 92$.

Graph E; (answers vary) Let m = months working out and H = heart rate in beats per minute. practical domain: $0 \leq m \leq 12$, practical range: $100 \leq H \leq 130$.

Graph F; (answers vary) Let t = time car is owned in years since 1964 and V = value of the car in dollars. practical domain: $0 \leq t \leq 45$, practical range: $5000 \leq V \leq 40{,}000$.

27. At 415 feet horizontally the vertical distance of Gunsight Butte is 290 feet.

29. a. $p = f(t)$, where p = price of a new electronics item and t = time
    b. Given the price of a new electronics item, we can determine the time it was purchased.
    c. It may not be a function because the input price can possibly produce multiple times that the price was incurred.

31. a. The notation describes the inverse function with the number of Head Start children (in 1000s) as the input and the number of women in the workforce (in 1000s) as the ouptut.
    b. $60{,}960.13 \approx N$. When there are 700,000 children in Head Start the number of women in the workforce is 60,960,130.
    c. $f^{-1}(700) = 60{,}958$
    d. The answers are the same because of the inverse relationship between the functions.

33. a. The inverse function whose input is the weekly gross income of a minimum wage worker and output is the number of hours worked.
    b. $40 = h$. For 40 hours of work, the weekly gross income is $206.
    c. $f^{-1}(206) = 40$. For the weekly gross income of $206, the time worked is 40 hours.
    d. Parts (b) and (c) should be equivalent.

CHAPTER 2 ■ SECTION 2.1

1. slope: -4; v. intercept: $(0, 10)$; h. intercept: $\left(\dfrac{5}{2}, 0\right)$

3. slope: $\dfrac{-4}{5}$; v. intercept: $\left(0, \dfrac{3}{10}\right)$; h. intercept: $\left(\dfrac{3}{8}, 0\right)$

5. slope: -2; v. intercept: $(0, 8)$ [given]; h. intercept $(4, 0)$ [given]

7. slope: $\dfrac{-5}{3}$; v. intercept: $\left(0, \dfrac{37}{3}\right)$; h. intercept $\left(\dfrac{37}{5}, 0\right)$

9. slope: -2; v. intercept: $(0, -5)$; h. intercept: $\left(\dfrac{-5}{2}, 0\right)$

11. $5 per year. The change in annual fees (in dollars) will be 5 times as great as the number of years elapsed.

13. $112 per credit hour. The change in tuition costs are 112 times as great as the change in the number of enrolled credit hours.

15. 0 dollars per year. There is no change in my salary as time elapses.

17. a.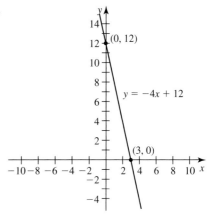

    b. v. intercept: $(0, 12)$ [given]; h. intercept: $(3, 0)$

19. a.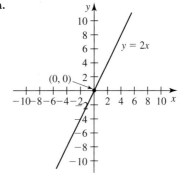

    b. v. intercept and h. intercept both occur at $(0, 0)$

21. a.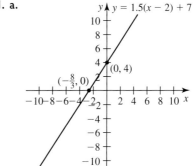

    b. v. intercept: $(0, 4)$; h. intercept: $\left(\dfrac{-8}{3}, 0\right)$

23. a.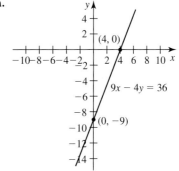

    b. v. intercept: $(0, -9)$; h. intercept: $(4, 0)$

25. a.

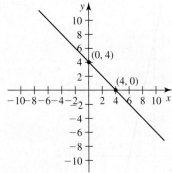

b. v. intercept: (0, 4); h. intercept: (4, 0)

27. $y = x + 3$

29. $y = 27$

31. $y = \dfrac{1}{4}x + \dfrac{1}{2}$

33. $s(t) = t + 50$

35. $a = \dfrac{-56}{3}$

37. $a = -12$

39. The numerator represents the vertical change or "rise" between any two points on the line and the denominator represents the horizontal change or "run."

41. We know that the slope-intercept form of a line is $y = mx + b$ and the point-slope form of a line is $y - y_1 = m(x - x_1)$. If we solve for y we get $y = m(x - x_1) + y_1$ and the vertical y intercept will be the point $(0, b)$. We can substitute this point for (x_1, y_1) and get $y = m(x - 0) + b = mx + b$.

43. *No slope* indicates that no relationship exists between x and y. *Zero slope* means that there is no change in y as x changes, which indicates that the graph of the function is horizontal. *Undefined slope* indicates that a relationship exists that doesn't conform to our meaning for slope. It is used to describe a change in y without a corresponding change in x.

45. Let d be the death rate due to heart disease (measured in deaths per 100,000 people) and t be the number of years since 1980.

a. The vertical intercept, (0, 412.1), is given in the problem. The horizontal intercept is about (52.66, 0).

b.

c. The vertical intercept (0, 412.1) tells us that in 1980 the death rate due to heart disease was 412.1 deaths per 100,000 people. The horizontal intercept (52.66, 0) tells us that, assuming this trend continues, the death rate due to heart disease will be zero deaths per 100,000 people about 52.66 years after 1980 (in 2033). In addition to the fact that 0 deaths is not probable, we are using extrapolation far into the future, so we are skeptical of this conclusion.

47. Let v represent the road elevation in feet above sea level and h represent the horizontal distance from the road sign in feet in the downhill direction.

a. The vertical intercept, (0, 8240), is given in the problem. The horizontal intercept is (103,000, 0).

b.

Horizontal Distance from Road Sign (in feet in the downhill direction)

c. The vertical intercept (0, 8240) tells us that the elevation at the sign is 8240 feet above sea level. The horizontal intercept (103,000, 0) tells us that, if the road continued at this rate of decline indefinitely, the road would reach sea level 103,000 horizontal feet from the road sign.

49. Let w represent the cups of Total Whole Grain cereal eaten per day and p represent the scoops of ProFiber fiber supplement eaten per day.

a. The vertical intercept is about $\left(0, \dfrac{5}{3}\right)$; the horizontal intercept is $\left(\dfrac{5}{2}, 0\right)$.

b.

Cups of Cereal Eaten per Day

c. The vertical intercept $\left(0, \dfrac{5}{3}\right)$ tells us that the person can reach her fiber intake goal of 10 grams per day by eating $\dfrac{5}{3}$ scoops of ProFiber and no Total Whole Grain cereal. The horizontal intercept $\left(\frac{5}{2}, 0\right)$ tells us that she can skip taking the ProFiber fiber supplement and reach her fiber intake goal by eating $\frac{5}{2}$ cups of Total Whole Grain cereal.

51. Let P be the number of participants in the Federal Civil Servant Retirement System (in millions) and t be years since 1980. Then $P(t) \approx -0.0925t + 5.12$.

53. Let I be the number of income tax returns filed and t be years since 1990. Then $I(t) = 1{,}555{,}000t + 109{,}868{,}000$.

55. Let b be the number of beneficiaries in millions and c be the amount of employee contributions in billions of dollars. Then $b(c) = 0.17(c - 13.9) + 4.03$.

57. $m = n$ but $b \neq c$ since the lines must have the same slope but are not the same line. (If $b = c$ the lines would intersect at every point on each line and essentially be the same line.)

59. Since the relationship between the changes in x and y for a situation modeled by a vertical line cannot be expressed as a constant rate of change (there is no value m for which (change in y) = m(change in x) is true along a vertical line), and since the formula for a linear function is developed from the meaning of constant rate of change, a vertical line cannot be expressed in this form.

61. **a.** $32.5 + 72 = 104.5 \approx 105$
   b. $20 + 62 = 82$
   c. $136 \approx s$. A golfer would expect to score 30 strokes over par on a course with a slope rating of 136.
   d. Slope on a golf course is a method of comparing the level of difficulty of a course against the average level of course difficulty, which is set as a slope rating of 113. This can be used to help a golfer estimate how a golfer will do on a course he is about to play. For a course with slope of 110 this would compare to 113 by the ratio of $\frac{110}{113} \approx \frac{0.97}{1}$. This means that he could expect to shoot almost the same score as he would on an average level of course but will do slightly better because a single stroke on an average course will only take 0.97 strokes on a course with a 110 slope rating.

CHAPTER 2 ■ SECTION 2.2

1. $f^{-1}(y) = \dfrac{y - 10}{-4}$

3. $f^{-1}(y) = \dfrac{y + 18}{9}$

5. $f^{-1}(y) = \dfrac{y - 4}{\frac{1}{3}}$

7. *Directly proportional* describes a relationship between two quantities such that as the independent variable changes by a certain amount the dependent variable changes by a factor as much.

9. For one function to be the inverse of another each of the input and output values are interchanged. In other words, the domain and range values of the original function become the range and domain values of the inverse.

input x → $\boxed{f}$ → y output
output x ← $\boxed{f^{-1}}$ ← y input

11. Linear because the cost per gallon is constant. $C(g) = 1.889g$, where C = the total cost of gasoline purchased and g = number of gallons purchased.

13. Linear because the cost per pair of pants is constant. R = total retail revenue; p = number of pairs of pants; $R(p) = 34.5p$

15. The restaurant expenses, e, as a function of the year, y, is nearly linear because the rate of change between consecutive years is nearly constant. $e(y) = 104.5y + 1087$

17. The children under 5 in Madagascar, C, as a function of the year, t, is nearly linear because the rate of change between consecutive years is nearly constant. $C(t) = 75.6t + 2110.88$

19. We are to model the number of people (in millions) that are age 65 and older, S, as a function of the years since 1950, t, with a linear function. $S(t) = 12 + 0.45t$

21. $\dfrac{P - 6}{16/75} = f^{-1}(P); f^{-1}(14) = 37.5$; When the price of a 40-pound box of oranges was \$14.00, there was 37.5% of the citrus crop frozen.

23. $\dfrac{D + 53.87}{14.08} = f^{-1}(D); f^{-1}(101) = 10.99$; When the death rate is 101 per 100,000, the percentage of smokers is about 11%.

25. $\dfrac{L - 146.8}{10.47} = f^{-1}(L); f^{-1}(251.5) = 10$; When the marketing labor cost is 251.5 billion dollars, it is the year 2000.

27. Function modeling is valuable when we have enough data provided to make reasonable assumptions about patterns and trends. Even with an accurate model we must use caution in terms of how far into the future we predict (extrapolate) due to potential changes.

29. Answers vary. You can calculate the average rate of change between some of the data points to see if they are constant. If the input values are changing at a constant rate, you can simply look at the change in the output values to see if it is constant.

CHAPTER 2 ■ SECTION 2.3

1. $y_1 = \dfrac{7}{4}x + 1$; sum of squares is 10

   $y_2 = \dfrac{3}{2}x + 2$; sum of squares is 27

   y_1 fits the data better because the smaller sum of squares indicates that there is less error with this model.

3. $y_1 = 17.5x + 12.5$; sum of squares is 10,487.5

   $y_2 = 30.5x - 180$; sum of squares is 35,618.75

   y_1 fits the data better.

5. $y_1 = -3x + 15$; sum of squares is 9

   $y_2 = -3x + 13.9$; sum of squares is 4.05

   y_2 fits the data better.

7. The coefficient of determination is a way of determining how well a set of data is modeled by the linear regression line. (More specifically, it is the ratio of the explained error to the total error.)

9. Both values help us to determine the accuracy of the model for making predictions through interpolation and extrapolation.

11. a. $T(I) = -2149.48 + 188.19I$
    b. The slope means that for every increase in $1000 in taxable income an estimated additional $188.19 in taxes is due. The vertical intercept means that when a married couple has an income of $0 they owe −$2149.48 in taxes. This is unreasonable so they either owe $0 in taxes or may receive a refund.
    c. $T(72) = \$11,400.24$; interpolation
    d. $T(110) = \$18,551.48$; extrapolation
    e. Since the coefficient of determination is 0.98 we could say that the values found should be reasonably accurate.

13. a. $P(t) = -0.248t + 5.986$
    b. The slope means that the carbon monoxide pollutant concentration is decreasing by 0.248 parts per million each year on average. The vertical intercept means that in the year 0 (1990) the carbon monoxide pollutant concentration is 5.986 parts per million.
    c. $P(2) = 5.49$; interpolation
    d. $P(10) = 1.026$; extrapolation
    e. The coefficient of determination is $r^2 = 0.9949$, which means any predictions would be very accurate.

15. a. $h(n) = 838.72 + 5.99n$
    b. The slope means that each story added to a building increases the height by 5.99 feet. The vertical intercept means that a building with 0 stories would be 838.72 feet, which is of course unreasonable.
    c. $h(50) = 1138.22$ feet; extrapolation
    d. $h(120) = 1557.52$ feet; extrapolation
    e. The coefficient of determination is $r^2 = 0.30$, which means that any predictions would not be very accurate.

17. a.

    b. $M(b) = -599.71 + 1.08b$
    c. The slope means that for every additional 1000 registered boats there are 1.08 more manatee deaths. The vertical intercept means that when there are 0 registered boats there are −599.71 manatee deaths, which is of course unreasonable.
    d. The coefficient of determination is $r^2 = 0.32$, which means that any predictions would not be very accurate.
    e. $M(3000) = 2640.29$. Even though we get a numerical answer (2640.29) for the number of manatee deaths this model should not be used to extrapolate so far out.

19. a.

    b. $P(t) = 35.71 - 0.52t$
    c. Since $r^2 = 0.97$ and $r = -0.986$, the model fits the data well and the predictions may be quite accurate.
    d. $68.67 \approx t$. This would be year 2043 but this would be extrapolating too far out and it is very unlikely that there will ever be no people smoking.

21. a.

    b. $G(m) = 0.66m + 8.10$
    c. The slope means that the average number of golf balls found per month is increasing by 0.66 balls per month.
    d. $G(26) = 25.26$

23. No, the linear model does not fit the original data set well because the residuals are quite spread out and form a pattern over the horizontal line, showing that there is a lot of difference between the actual and estimated values of the function.

25. Yes, the linear model fits the original data set well because the residuals are spread out over the horizontal line, showing that there is some difference between the actual and estimated values of the function but there is no pattern and there are as many data values over as under the line of best fit.

CHAPTER 2 ■ SECTION 2.4

1. a. (2, 1)
  b. Graphing the system confirms our solution.

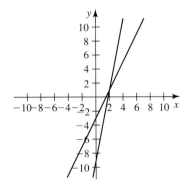

3. a. about (0.212, 5)
  b. Graphing the system confirms our solution.

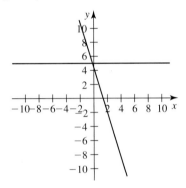

5. a. no solution
  b. Graphing the system confirms that the lines are parallel, and thus there is no point of intersection.

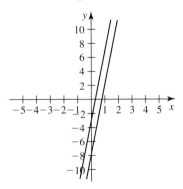

7. a. (0, −9)
  b. Graphing the system confirms our solution.

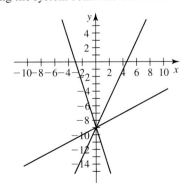

9. a. no solution
  b. Graphing the system confirms that there is no point of intersection and no solution exists.

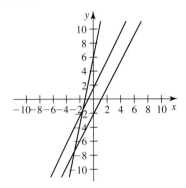

11. (7, 3)
13. (−10, −10)
15. (−1.5, 2.4)
17. a. $m_g = -2$ and $m_f = 3$

x	f(x)	g(x)
−2	1	6
−1	4	4
0	7	2
1	10	0
2	13	−2

  b. $f(x) = 3x + 7$ and $g(x) = -2x + 2$
  c. (−1, 4). The table of values verifies that when $x = -1$, both functions' output values are 4.
  d.

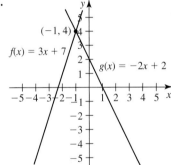

19. a. $m_f = -1$ and $m_g = 5$

x	f(x)	g(x)
−5	4	−38
0	−1	−13
1	−2	−8
4	−5	7
5	−6	12

  b. $f(x) = -x - 1$ and $g(x) = 5x - 13$
  c. (2, −3). From the table of values, we see that $f(1) > g(1)$ and $f(4) < g(4)$, so it makes sense that $f(2) = g(2)$. Also, $f(4) < -3 < f(1)$ and $g(1) < -3 < g(4)$, so an output value of −3 also makes sense.

d.

21. a. $m_f = 0.4$ and $m_g = -1.5$

x	f(x)	g(x)
−2	3.36	11.15
−0.5	3.96	8.9
0	4.16	8.15
2.5	5.16	4.4
3	5.36	3.65

b. $f(x) = 0.4x + 4.16$ and $g(x) = -1.5x + 8.15$

c. (2.1, 5). From the table of values, we see that $f(0) < g(0)$ and $f(2.5) > g(2.5)$, so it makes sense that $f(2.1) = g(2.1)$. Also, $f(0) < 5 < f(2.5)$ and $g(2.5) < 5 < g(0)$, so an output value of 5 also makes sense.

d.

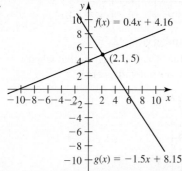

23. The solution to a linear system of equations in two variables is an ordered pair (x, y) where, when the independent value x is used as the input for each function in the system, all functions have the same output value. This common output value is the dependent value y. If we plot the solution as a coordinate point, the graphs of all functions in the system will pass through this point.

25. No. An inconsistent system is defined as a system with no solution. A dependent system has infinitely many solutions and therefore cannot be inconsistent.

27. (answers vary) In the substitution method, we use one of the equations to solve for one of the variables, thus getting an expression equivalent to this variable's value. We then substitute this equivalent expression for the variable in the second equation. After simplifying, we have an equation of one variable, which is easy to solve. Once the value of one variable is determined, we plug this value into one of the original equations to find the value of the second variable.

29. a. about (1.35, 136.08). We expect that about 1.35 years after 1994 the incidence of diabetes in men and women aged 45–64 years was the same, both being about 136.08 people per thousand people in the population.

b. We check our solution:

$F(t) = 3.75t + 131$ $\qquad$ $M(t) = 20t + 109$

$F(1.35) = 3.75(1.35) + 131$ $\qquad$ $M(1.35) = 20(1.35) + 109$

$136.08 \stackrel{?}{\approx} 3.75(1.35) + 131$ $\qquad$ $136.08 \stackrel{?}{\approx} 20(1.35) + 109$

$136.08 \stackrel{?}{\approx} 5.06 + 131$ $\qquad$ $136.08 \stackrel{?}{\approx} 27 + 109$

$136.08 \approx 136.06$ $\qquad$ $136.08 \approx 136$

Our solution appears correct with a slight discrepancy due to rounding.

31. No, according to the models the food costs will never be the same. We can see from the graph that there is no ordered pair in common for all three functions. Thus, at no time is the weekly food cost for a 15–19-year-old male in a family of four expected to be the same for all three meal plans.

33. the year 2022

35. a. (using linear regression) Let y represent the number of years since 1990. Then the number of men and women (in millions) participating in high school sports in the United States can be modeled by $M(y) = 0.05y + 3.36$ and $W(y) = 0.09y + 1.89$, respectively.

b. 0.05 million men per year and 0.09 million women per year. According to our models, 0.05 million more men (50,000 more men) participated in high school athletics each year from 1990 to 2002 while 0.09 million more women (90,000 more women) participated in high school athletics each year from 1990 to 2002.

c. about (36.75, 5.20). About 37 years after 1990 (the year 2027), the number of men and women participating in high school sports would be equal at 5.2 million athletes.

d. We are assuming that the trends observed and modeled between 1990 and 2002 will continue for the next few decades since we have to extrapolate over 20 years beyond the domain of our models to locate a solution.

37. 15 cups of peanuts and 15 cups of raisins

39. a. Let m represent the number of miles traveled beyond the flag drop. Then the total costs in dollars for hiring a taxi in Baltimore and Seattle are $B(m) = 2.20m + 2.60$ and $S(m) = 2.00m + 3.50$, respectively.

b. (4.5, 12.50)

c. Our solution tells us that the taxi fare in both Baltimore and Seattle is $12.50 for a trip 4.5 miles beyond the flag drop (assuming two minutes of waiting time).

41. a. Let m represent the number of miles traveled beyond the flag drop. Then the total costs in dollars for hiring a taxi in Seattle and Atlanta are $S(m) = 2.00m + 2.50$ and $A(m) = 2.00m + 2.50$, respectively.

   b. infinitely many solutions (dependent system). We observe that these are equivalent functions, and thus every input value will yield the same output value in both functions.

   c. No matter how much distance beyond the flag drop a rider travels the fares will be identical in Seattle and Atlanta (assuming no waiting time). Makes sense.

43. a. Let h represent the price in dollars of an HDX 16t Premium Series laptop and p represent the price in dollars of a Pavilion dv7t Series laptop. Then $7h + 6p = 11,449.87$ represents the combinations possible for the $11,449.87 order and $6h + 5p = 9699.89$ represents the combinations possible for the $9699.89 order.

   b. An HDX 16t Premium Series laptop costs $949.99 and a Pavilion dv7t Series laptop costs $799.99.

45. a. the number of ounces of Planters Honey Nut Medley Trail Mix in the mixture and the number of ounces of Planters Dry Roasted Peanuts in the mixture

   b. Let t represent the number of ounces of Honey Nut Medley Trail Mix and p represent the number of ounces of Dry Roasted Peanuts. Then $t + p = 8$ represents the combinations possible to meet the 8-ounce requirement for the total size of the mixture and $4t + 8p = 42$ represents the combinations possible to meet the goal of exactly 42 grams of protein in the entire mixture.

   c. (5.5, 2.5). The mixture should contain 5.5 ounces of Honey Nut Medley Trail Mix and 2.5 ounces of Dry Roasted Peanuts to meet both requirements.

   d. We check our solution:

   $$t + p = 8 \qquad\qquad 4t + 8p = 42$$
   $$5.5 + 2.5 \stackrel{?}{=} 8 \qquad 4(5.5) + 8(2.5) \stackrel{?}{=} 42$$
   $$8 = 8 \qquad\qquad 22 + 20 \stackrel{?}{=} 42$$
   $$42 = 42$$

47. a. Let m represent the additional miles traveled beyond the flag drop and w represent the number of minutes spent waiting. Then $1.70m + 0.37w = 4.03$ models the combinations in St. Louis that result in a total fare of $6.53 ($4.03 plus the flag drop) and $1.80m + 0.33w = 4.02$ models the combinations in Phoenix that result in a total fare of $6.52.

   b. $m = 1.5$, $w = 4$

   c. Both passengers traveled 1.5 miles beyond the flag drop and experienced 4 minutes of waiting time in their respective trips.

49. a. Let m represent the additional miles traveled beyond the flag drop and w represent the number of minutes spent waiting. Then $1.80m + 0.33w = 8.19$ models the combinations in Chicago that result in a total fare of $10.44 ($8.19 plus the flag drop) and $1.80m + 0.33w = 8.19$ models the combinations in Phoenix that result in a total fare of $10.69.

   b. infinitely many solutions (dependent system). We observe that these are equivalent functions, and thus every input value will yield the same output value in both functions.

   c. Any combination of distance traveled beyond the flag drop and wait time experienced that generates a total fare of $10.44 in Chicago will generate a total fare of $10.69 in Phoenix.

51. The system is inconsistent if $m = 1$ and $b \neq 0$. The system will be dependent when the two functions are actually the same, or when $m = \dfrac{1}{m}$ and $b = -\dfrac{b}{m}$. This occurs when $m = -1$ (regardless of the value of b) or when $m = 1$ and $b = 0$.

53. No. For example, consider a system that has two equations and three variables. We could use the two equations and the elimination method to yield an equation with two variables, but then there would be no way to determine a single set of values for these variables that must represent the only solution. Alternatively, we could use one of the two equations, solve for one of the variables to get an equivalent expression, then use the substitution method to replace the variable with its equivalent expression in the other equation. This will also yield an equation with only two variables, but again there is no way to then find a single set of values that must represent the only solution. Instead, we get an entire range of possible solutions described by the final two-variable equation. Similar reasoning can be used for systems with more equations and variables.

CHAPTER 2 ■ SECTION 2.5

1. 2×3

3. 1×3

5. 1×1

7. $x + 3y = 12$
   $4x + y = 7$
   $-x + 6y = 9$

9. $3x - y + 9z = 13$
   $-3x + y = -31$

11. a. $\begin{bmatrix} 2 & 5 & | & 2 \\ 3 & -5 & | & 3 \end{bmatrix}$

    b.

    c. The solution to the system is $x = 1$, $y = 0$.

13. a. $\begin{bmatrix} -2 & 6 & 4 & | & 10 \\ 4 & -12 & 2 & | & -20 \\ 3 & 4 & -1 & | & 11 \end{bmatrix}$

b. Using technology we reduce the augmented matrix to:

```
rref([A])
    [[1 0 0 1]
     [0 1 0 2]
     [0 0 1 0]]
```

c. The solution to the system is $x = 1$, $y = 2$, $z = 0$.

15. a. $\begin{bmatrix} 3 & -2 & 1 & | & 6 \\ 11 & -20 & -1 & | & 0 \\ 0 & 1 & 1 & | & 3 \end{bmatrix}$

b. Using technology we reduce the augmented matrix to:

```
rref([A])
    [[1 0 0 2]
     [0 1 0 1]
     [0 0 1 2]]
```

c. The solution to the system is $x = 2$, $y = 1$, $z = 2$.

17. a.

b.
```
rref([A])
    [[1 0 2]
     [0 1 7]]
```

c. The solution to the system is $x = 2$, $y = 7$.

19. a.

b. Using technology we reduce the augmented matrix to:

```
rref([A])
    [[1 0 0 0]
     [0 1 0 -4]]
```

c. The solution to the system is $x = 0$, $y = -4$.

21. The solution to a system of equations is the point or points that make the equations in the system all true at the same time.

23. multiplying or dividing an equation in the system by a constant and then adding or subtracting rows together

25. Reduced row echelon form is a matrix with the coefficients and solutions of a system of equations input into a table that has been simplified through row operations (adding, subtracting, multiplying, and dividing) to show the solutions to the system.

27. $287.32 should be invested in Global Equities, $1436.61 in Money Markets, and $3276.06 in Social Choice.

29. Three possible solutions would be:

C	L	T	Associated Cost
1	8	6	1(7.99) + 8(7.99) + 6(8.49) = $122.85
5	10	0	5(7.99) + 10(7.99) = $119.85
3	9	3	3(7.99) + 9(7.99) + 3(8.49) = $121.35

31. The chef would use up all of the ingredients making 4 batches of custard and 3 batches of bread pudding (it is not necessary to make any batches of vanilla pudding).

33. Three possible solutions would be:

S	J	B
110	1065	25
120	1030	50
200	750	250

35. 15,000 of the $49.50 tickets, 6375 of the $94.50 tickets, 2125 of the $154.50 tickets, and 1500 of the $304.50 tickets would have to be sold to reach $2,130,000 in ticket revenue.

37. Mathematically, there are infinitely many solutions to this system of equations. However, in this context, we know that all values must be whole numbers. The whole number solutions are 14 pennies, 22 nickels, 22 dimes, and 6 quarters.

39. Answers vary. Two possible solutions would be:

x_1	x_2	x_3	x_4
100	50	0	150
500	450	400	550

CHAPTER 2 ■ SECTION 2.6

1. $\begin{bmatrix} -4 & 6 \\ 10 & 12 \\ -4 & 3 \end{bmatrix}$

3. Matrices B and C cannot be added since their dimensions are not the same.

5. $\begin{bmatrix} 3 & 0 & 4 \\ -1 & 9 & 8 \\ 14 & 6 & 2 \end{bmatrix}$

7. $\begin{bmatrix} -15 & 0 \\ 21 & 24 \\ -27 & 3 \end{bmatrix}$

9. $\begin{bmatrix} -2 & -12 & 8 \\ 2 & 2 & 12 \\ 8 & 20 & 4 \end{bmatrix}$

11. $\begin{bmatrix} 1.44 & 7.56 & 0.48 \\ -10.92 & 5.04 & 2.04 \\ 1.08 & -2.4 & 0.36 \end{bmatrix}$

13. $\begin{bmatrix} -1.78 & 8.25 & -0.44 \\ -4.48 & -7.61 & -14.29 \\ -20.08 & -22.18 & -4.24 \end{bmatrix}$

15. $\begin{bmatrix} 2.74 & -7.80 & 0.72 \\ 1.89 & 11.33 & 18.72 \\ 25.69 & 27.14 & 5.47 \end{bmatrix}$

17. $\begin{bmatrix} 1.74 & -57.42 & 0 \\ 54.81 & 11.31 & 46.98 \\ 72.21 & 92.22 & 14.79 \end{bmatrix}$

19. $\begin{bmatrix} 23.22 & 46.17 & 7.08 \\ -98.70 & 87.21 & 83.55 \\ 98.10 & 69.54 & 22.14 \end{bmatrix}$

21. $\begin{bmatrix} 2700 \\ 27{,}700 \\ 8400 \\ 2950 \\ 82{,}500 \end{bmatrix}$

23. Scalar multiplication is when a matrix is multiplied by a factor to alter all of the entries by the factor.

25. If the entries in the two matrices are additive inverses of each other the result of the sum of the two matrices would be a matrix consisting only of zeros.

27. The relationship between $A - B$ and $B - A$ will be that the resulting matrices will be inverses of each other.

29. $\begin{bmatrix} 40{,}207.79 & 44{,}340.62 & 48{,}474.59 \\ 42{,}365.19 & 46{,}498.02 & 50{,}630.84 \\ 44{,}521.45 & 48{,}654.28 & 52{,}787.10 \\ 46{,}677.71 & 50{,}810.53 & 54{,}943.36 \end{bmatrix}$

31. $\begin{bmatrix} 1.79 & 1.88 & 5.06 \\ 3.58 & 3.76 & 10.13 \\ 1.11 & 0 & 0 \end{bmatrix}$

33. $\begin{bmatrix} 1.16 & 1.21 \\ 1.14 & 1.20 \end{bmatrix}$

The average markup price will be between 14% and 21%.

35. The amount of natural gas produced is not meeting the amount consumed so we need to produce more natural gas. This will also most likely reduce our dependence on using coal for energy production and lower greenhouse gases.

37. Nuclear electric power best supports the environmentalist's position because the United States consumed all of the electricity produced by nuclear energy.

39. $\begin{bmatrix} 3.64 \\ 0.33 \\ 3.16 \\ 0.07 \\ 0.03 \end{bmatrix}$

41. No, you cannot determine the percentage of ninth graders who exercise 20 or more minutes per class because we don't know the actual number of males and females enrolled in a P.E. class. There might be a large discrepancy between the number of males and females enrolled and this would not allow us to simply add the percentages exercising for 20 or more minutes together.

43. $\begin{bmatrix} 5\frac{1}{3} & 10\frac{1}{3} & 1\frac{1}{3} \\ 5 & -2\frac{2}{3} & -4\frac{1}{3} \\ -6 & \frac{5}{6} & \frac{1}{3} \end{bmatrix}$

45. $\begin{bmatrix} -\frac{104}{5} & -\frac{183}{20} & -\frac{33}{5} \\ -\frac{159}{20} & 2 & -\frac{15}{4} \\ \frac{73}{10} & -\frac{109}{40} & \frac{13}{60} \end{bmatrix}$

CHAPTER 2 ■ REVIEW

1. slope: -1; v. intercept: $(0, 8)$; h. intercept: $(8, 0)$

3. slope: $\frac{2}{3}$; v. intercept: $(0, -2)$; h. intercept: $(3, 0)$

5. slope: -3; v. intercept: $(0, 18)$ [given]; h. intercept: $(6, 0)$

7. a. slope: 5; v. intercept: $(0, 4)$ [given]; h. intercept: $\left(-\frac{4}{5}, 0\right)$

b.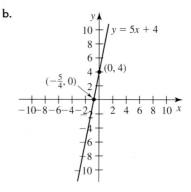

9. a. slope: $\frac{15}{8}$; v. intercept: $\left(0, -\frac{27}{8}\right)$; h. intercept: $\left(\frac{9}{5}, 0\right)$

b.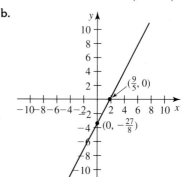

828 ANSWERS ■ CHAPTER 2

11. a. slope: $\frac{2}{5}$ mmHg per pound; v. intercept: (0, 49); h. intercept: (−122.5, 0)

b.

c. The slope is $\frac{2}{5}$ mmHg per pound. The change in systolic blood pressure (in mmHg) is $\frac{2}{5}$ times as great as the change in weight (in pounds). Neither intercept makes sense in this context. The vertical intercept says a man whose weight is zero pounds has a blood pressure of 49 mmHg. The horizontal intercept says that blood pressure of 0 mmHg requires a weight less than 0 pounds.

13. a. $c(a) = \frac{1}{250}a + 300$ or $c(a) = \frac{1}{250}(a - 50{,}000) + 500$

b. $\frac{1}{250}$ car per dollar. The number of cars sold changes by $\frac{1}{250}$ times as much as the change in advertising spending (in dollars).

c. The vertical intercept is (0, 300). We expect the car company to sell 300 cars if it does not spend any money on advertising.

15. Let t = production year and P = average price of a Toyota Prius. Then $P(t) = 14{,}800 + 2383.33t$.

17. The average rate of change over the 3-year interval is $1.3 billion per year. Therefore in 2004 the estimated admissions to spectator amusements is $37.3 billion and in 2005 it is $38.6 billion.

19. The correlation coefficient would be negative because as the baby gets older the number of hours of sleep per day would decrease.

21. The negative sign tells us that as the number of hours per week an individual watches television increased the grade point average tended to decrease. The numerical value of 0.95 means that the relationship is pretty "strong," meaning that there is not much spread around the linear regression line.

23. a.

Total Sleep Hours per Day vs Age (months)

b. $H(a) = -0.11a + 15.38$

c. The linear model fits the data reasonably well—the correlation coefficient is −0.92 and the coefficient of determination is 0.84.

d. $H(36) = 11.42$. A 3-year-old child is recommended to get 11.42 hours of sleep per day.

25. a. $w(h) = 6.15h - 263.66$

b. The slope of the linear model means that for every additional inch the weight increases by 6.15 pounds for Phoenix Suns players. The vertical intercept hypothetically means that a player that is 0 inches tall would weigh −263.66 pounds, which of course is impossible and therefore outside the practical domain.

c. (70) = 166.84. This means a player that is 70 inches tall would weigh 166.84 pounds.

d. $75.39 \approx h$. This means a player that weighs 200 pounds would be 75.39 inches tall (6' 3").

27. a. about (2.33, 20.71)

b.

29. a. (4, 9)

b.

31. a. $P(t) = -0.26t + 54.25$ and $B(t) = -0.61t + 77.21$, where t represents the number of years since 1980

b. The constant rate of change for our model of pork consumption is -0.26 pounds per person per year, meaning that, according to the model, each person in the United States consumed about 0.26 fewer pounds of pork per year during this period. The constant rate of change for our model of beef consumption is -0.61 pounds per person per year, meaning that, according to the model, each person in the United States consumed about 0.61 fewer pounds of beef per year during this period.

c. $(65.6, 37.19)$. According to our models, 65.6 years after 1980 the pork and beef consumption in the United States will both be about 37.19 pounds per person.

d. We assume that each equation accurately models the corresponding data set. That is, we are assuming a constant rate of change and vertical intercept for each of the two models.

33. $(-1, 13)$

35. a. the number of ounces of trail mix and the number of ounces of dry roasted peanuts in the final mixture

b. $t + p = 8$ and $4t + 8p = 42$

c. The mixture should include 5.5 ounces of trail mix and 2.5 ounces of peanuts to meet our requirements of an 8-ounce mixture with 42 total grams of protein.

d. We check our solution:

$$5.5 + 2.5 \stackrel{?}{=} 8 \qquad 4(5.5) + 8(2.5) \stackrel{?}{=} 42$$
$$8 = 8 \qquad 22 + 20 \stackrel{?}{=} 42$$
$$42 = 42$$

37. $\begin{bmatrix} 5.2 & -1.3 & | & 12 \\ -10.4 & 2.6 & | & 24 \end{bmatrix} = \begin{bmatrix} 1 & -0.25 & | & 0 \\ 0 & 0 & | & 1 \end{bmatrix}$; no solution

39. $\begin{bmatrix} 2.9 & -8.1 & | & 4 \\ 3.7 & 16.2 & | & 1.5 \end{bmatrix} = \begin{bmatrix} 1 & 0 & | & 1 \\ 0 & 1 & | & -11/81 \end{bmatrix}$; $x = 1, y = -\dfrac{11}{81}$

41. $\begin{bmatrix} 8.1 & 4.3 & 6.2 & | & 3.1 \\ 2.7 & -2.8 & 7.5 & | & 7.2 \\ 8.2 & -5.3 & -1.1 & | & 5.9 \end{bmatrix} = \begin{bmatrix} 1 & 0 & 0 & | & 0.33 \\ 0 & 1 & 0 & | & -0.72 \\ 0 & 0 & 1 & | & 0.57 \end{bmatrix}$

43. $\begin{bmatrix} 1 & 5 & -1 & | & -11 \\ 0 & 0 & 3 & | & 12 \\ 2 & 4 & -2 & | & 8 \end{bmatrix} = \begin{bmatrix} 1 & 0 & 0 & | & 18 \\ 0 & 1 & 0 & | & -5 \\ 0 & 0 & 1 & | & 4 \end{bmatrix}$

45. $\begin{bmatrix} 7 & 9 \\ 0 & -8 \\ 9 & 9 \end{bmatrix}$

47. $\begin{bmatrix} 4 & 6 \\ -6 & -8 \\ 0 & 14 \end{bmatrix}$

49. $\begin{bmatrix} 19 & 24 \\ 3 & -20 \\ 27 & 20 \end{bmatrix}$

51. $\begin{bmatrix} 18.02 & -1.59 & -2.12 \\ -20.14 & 29.68 & 38.16 \\ 11.66 & 13.78 & 10.60 \end{bmatrix}$

53. $\begin{bmatrix} 5.26 & 24.98 & -1.68 \\ -37.69 & 17.78 & 7.69 \\ 3.91 & 12.56 & 13.73 \end{bmatrix}$

55. $\begin{bmatrix} 1.17 & 1.14 \\ 1.16 & 1.13 \end{bmatrix}$ Therefore the markup is between 13% and 17%.

CHAPTER 3 ■ SECTION 3.1

1. $y = 4x - 8$

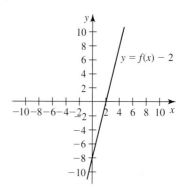

3. $y = 4x + 0.5$

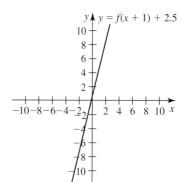

5. a. Answers vary.

x	$f(x) = x^2$
-3	9
-2	4
-1	1
0	0
1	1
2	4
3	9

x	$g(x) = f(x + 2) + 7$
-5	16
-4	11
-3	8
-2	7
-1	8
0	11
1	16

b.

 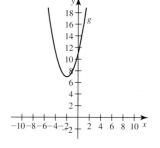

c. The graph of g can be created by shifting the graph of f left 2 units and upward 7 units.

7. a. Answers vary.

x	f(x) = |x| + 5
−3	8
−2	7
−1	6
0	5
1	6
2	7
3	8

x	g(x) = f(x − 1) − 3
−2	5
−1	4
0	3
1	2
2	3
3	4
4	5

b.

c. The graph of g can be created by shifting the graph of f right 1 unit and downward 3 units.

9. 12

11. −7

13. 7

15. $x = 1$

17. $x = 3$

19.

x	g(x)
0	12
1	10
2	7
3	2
4	−1
5	−3
6	4
7	5
8	6

21. A

23. D

25.

27.

29.

31. a. A can be created by shifting f upward 4 units.
 b. $A(x) = f(x) + 4$
 c. $A(x) = \sqrt{x} + 4$

33. a. C can be created by shifting f left 4 units and downward 5 units.
 b. $C(x) = f(x + 4) − 5$
 c. $C(x) = \sqrt{x + 4} − 5$

35. g can be created by shifting f left 2 units. $g(x) = f(x + 2)$

37. j can be created by shifting f right 1 unit and shifting it upward 1 unit. $j(x) = f(x − 1) + 1$

39. We graph $f(x) = x^2$, and calculate its average rate of change over the interval from $x = −3$ to $x = 1$.

$$\frac{f(1) − f(−3)}{1 − (−3)} = −2$$

The average rate of change from $x = −3$ to $x = 1$ is $−2$. Suppose we shift the graph downward 5 units and recalculate the average rate of change.

$$\frac{(f(1) − 5) − (f(−3) − 5)}{1 − (−3)} = −2$$

The average rate of change from $x = -3$ to $x = 1$ was not affected by the vertical shift. To generalize this observation, the average rate of change of a function f from $x = a$ to $x = b$ can be found by

$$\frac{f(b) - f(a)}{b - a}$$

If f were shifted vertically k units, then the points $(a, f(a))$ and $(b, f(b))$ become $(a, f(a) + k)$ and $(b, f(b) + k)$, and the average rate of change from $x = a$ to $x = b$ is

$$\frac{f(b) + k - (f(a) + k)}{b - a} = \frac{f(b) - f(a)}{b - a}$$

41. Define the function h by $h(x) = f(x - 4)$. This statement means

$$\underbrace{h}_{\text{the output of function } h} \underbrace{(x)}_{\text{at an input of } x} \underbrace{=}_{\text{is the same as}} \underbrace{f}_{\text{the output of function } f} \underbrace{(x - 4)}_{\text{when the input is 4 units less than } x}$$

So, since the same outputs occur for $h(x) = f(x - 4)$ when the input is 4 units larger in h (or 4 units less in f), the graph of h can be created by shifting the graph of f right 4 units. Similar reasoning can justify why, if $g(x) = f(x + 4)$, then the graph of g can be created by shifting the graph of f left 4 units.

43. $P(t) = W(t) + 13$ or $P(t) = 225(1.00465)^t + 13$

45. a.

Years Since 1980	Number of Public Airports in the U.S.
0	4814
5	5858
10	5589
15	5415
20	5317
22	5286

b. The data was shifted left 1980 units.

c. $g(x) = f(x + 1980)$

47. a.

Years Since 1998	Number of Doctorate Degrees Awarded in Astronomy
-3	173
0	206
1	159
2	185
3	186
4	144
5	167

b. The data was shifted left 1998 units.

c. $g(x) = f(x + 1998)$

49. a. The new function, which we shall call L, would model the number golf facilities in the United States as a function of the number of years since 1975. If y represents the number of years since 1975, then $L(y) = G(y - 5)$.

b. The new function, which we shall call F, would model the number of golf facilities in the United States as a function of the number of years since 1990. If n represents the number of years since 1990, then $F(n) = G(n + 10)$.

c. The new function, which we shall call K, would model the increase in the number of golf facilities in the United States since 1980. If t represents the number of years since 1980, then $K(t) = G(t) - 12,000$.

51. a.

b. $P(a) = -0.0012a + 14.09$. A linear model is reasonable. When calculating the regression model, $|r|$ is very close to 1.

c. The new function can be created by shifting P left 4000 units. This table represents the atmospheric pressure for a given change of altitude from 4000 meters. We should label the left column "change in altitude from 4000 meters" and give it a variable name, say c, and we should label the right column "atmospheric pressure (psi)" and give the function a new name, say D. $D(c) = -0.0012c + 9.29$

d. The new function can be created by shifting P downward 14.70 units. This table represents the change in atmospheric pressure from the pressure at sea level (in psi) for a given altitude (in meters). We should label the left column "altitude in meters, a," and label the right column "change in atmospheric pressure from the pressure at sea level (in psi)" and give it a new function name, say N. $N(a) = -0.0012a - 0.61$

53. We could interpret the expressions as representing the average summit temperatures in degrees Fahrenheit for the same month in the following year, $f(m + 12)$, or in the previous year, $f(m - 12)$.

55. No. Shifts require that either the outputs of two functions be the same for input values with a constant difference, or that for any input value the output values always differ by the same amount (or both). However, when you spend more money, a discount of 15% translates to a larger amount of savings. Therefore, for the same input values, we want a function that outputs sale prices that are not a constant amount different from the original price.

57. Answers vary.

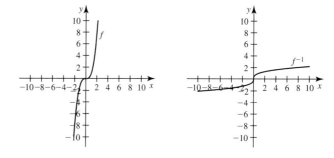

A horizontal shift in the original function causes a vertical shift in the inverse function. Recall that an inverse function is a function that reverses the input-output process of another function. Thus, if a function has the ordered pair (x, y) as a solution, its inverse will have the ordered pair (y, x) as a solution. Suppose that the ordered pairs $(-4, 7)$ and $(1, 10)$ were solutions of the original function. Then the inverse would have the ordered pairs $(7, -4)$ and $(10, 1)$ as solutions. If the original function undergoes a horizontal shift of 5 units left, say, then the ordered pairs $(-4, 7)$ and $(1, 10)$ become $(-9, 7)$ and $(-4, 10)$, meaning that for the inverse function the ordered pairs $(7, -4)$ and $(10, 1)$ become $(7, -9)$ and $(10, -4)$, a vertical shift downward 5 units. (Horizontal shifts of the original function to the right will shift the inverse function upward.) Similar reasoning can verify that a vertical shift of the original function results in a horizontal shift in the inverse function.

15.

x	j(x)
−8	50
−6	37
−4	25
−2	14
0	4
2	−5
4	−13
6	−20
8	−26

17. B

19. A

21.

CHAPTER 3 ■ SECTION 3.2

1. $y = -\dfrac{2}{3}x - 1$

3. $y = -\dfrac{2}{3}x + 1$

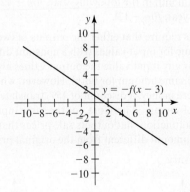

5. −20

7. 21

9. −42

11. $x = 4$

13. $x = -2$

23.

25. a. The graph of A can be created by reflecting f vertically about the horizontal axis and shifting the resulting graph upward 4 units.
    b. $A(x) = -f(x) + 4$
    c. $A(x) = -\sqrt{x} + 4$

27. a. The graph of C can be created by reflecting f horizontally about the vertical axis and shifting the resulting graph downward 2 units.
    b. $C(x) = f(-x) - 2$
    c. $C(x) = \sqrt{-x} - 2$

29. Function g can be created by shifting function f right 2 units and then reflecting it vertically about the horizontal axis.

31. Function j can be created by horizontally reflecting f about the vertical axis and shifting it upward 3 units.

33. a. Not possible. The function has a vertical intercept at about (0, 2.5). To have odd symmetry, the graph of the function must also pass through the point (−0, −2.5) or (0, −2.5), but this is not possible since the graph would then fail the vertical line test and would not represent a function.

b.

c. Answers vary.

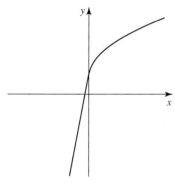

35. a. Not possible. The function has a vertical intercept at (0, 5). To have odd symmetry, the graph of the function must also pass through the point (−0, −5) or (0, −5), but this is not possible since the graph would then fail the vertical line test and would not represent a function.

b.

c. Answers vary.

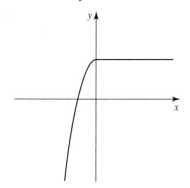

37. odd

39. neither

41. The function must pass through the origin. Since odd functions have the property that $f(-x) = -f(x)$, then for every point (x, y) on the graph there exists the point $(-x, -y)$ on the graph as well. Therefore, if the function has the vertical intercept $(0, b)$, then the point $(0, -b)$ must also be on the graph, and the function would have two vertical intercepts unless $b = 0$. Since a graph with two vertical intercepts would fail the vertical line test and thus not be a function, it must be that an odd function has the vertical intercept $(0, 0)$. Thus, if the function does not pass through the origin, it cannot possess odd symmetry.

43. a. m is any real number, $b = 0$

b. $m = 0$, b is any real number

c. $m = 0$, $b = 0$

d. m and b are both nonzero real numbers

45. The statement "$g(x) = f(-x)$" is interpreted as saying "the output of g at x is the same as the output of f at the opposite of x." Therefore, for every point (x, y) on the graph of g, we would have the point $(-x, y)$ on the graph of f (and vice versa).

47. a. Yes. Calculating the rate of change between any two ordered pairs in the table yields a constant $6.653 payment per $1000 borrowed.

b. $P(b) = 6.653b$

c. The horizontally reflected function can be interpreted as a function that inputs the impact of the loan on a person's net worth in thousands of dollars and outputs the monthly payment in dollars.

Effect on Personal Net Worth ($ thousands)	Monthly Payment (dollars)
−100	665.30
−200	1330.60
−225	1496.93
−280	1862.85
−330	2195.50

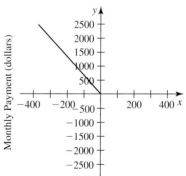

Impact on Net Worth ($ thousands)

d. The vertically reflected function can be interpreted as a function that inputs the amount borrowed in thousands of dollars and outputs the impact of the loan on a person's monthly budget in dollars (a negative number meaning there is a decrease in funds available for other expenses and purchases).

Amount Borrowed ($ thousands)	Impact on Monthly Budget (dollars)
100	−665.30
200	−1330.60
225	−1496.93
280	−1862.85
330	−2195.50

e. A horizontal shift would mean that the same monthly payments in dollars would correlate with amounts borrowed that were all adjusted by some constant amount. However, this is not a reasonable transformation since, all other things being equal, increasing or decreasing the amount borrowed by fixed amounts would require that the monthly payments change as well. Therefore, a horizontal shift does not make sense in this context.

f. A vertical shift would mean that for the same amount borrowed, the monthly payment would increase or decrease by some fixed amount. It is possible that banks might add fixed amounts to the monthly payment of a loan to cover processing fees or other costs, so it is reasonable that a vertical shift could have a practical meaning in this context.

49. a.

Feet Less Than the Maximum Recommended Safe Diving Depth z	Increased Pressure on the Human Body (psi) $B(z)$
0	45
20	36
40	27
60	18
80	9
100	0

b. B can be created by horizontally reflecting P about the vertical axis and shifting it right 100 units.

c. $z = 100 - d$ or $d = 100 - z$

d. $B(z) = P(100 - z)$, $B(z) = P(-z + 100)$, or $B(z) = P(-(z - 100))$

51. a.

Altitude above Sea Level (kilometers) A	Difference between Room Temperature and Temperature Outside (in °C) $R(A)$
−0.25	5.36
0	6.99
0.25	8.62
0.5	10.24
0.75	11.87
1	13.50
1.25	15.12
1.5	16.75

b. R can be created by reflecting T vertically about the horizontal axis and shifting it upward 22 units.

c. $R(A) = 22 - T(A)$, $R(A) = 6.505A + 6.99$; $R(A) = 22 - T(A) = 6.505A + 6.99$

53. It is possible to have translational symmetry under horizontal shifts, but not under vertical shifts. An example of a function with horizontal translational symmetry is given below (assume that the domain includes all real numbers). Any graph with vertical translational symmetry would fail the vertical line test and would not represent a function.

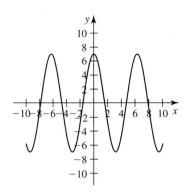

55. None of the transformations we have studied can change a function relationship into a relationship that no longer represents a function. Each transformation operates on an ordered pair in some way, and so the correspondence of one input to its output does not change (even if the values themselves change).

CHAPTER 3 ■ SECTION 3.3

1. $y = 4x + 1$

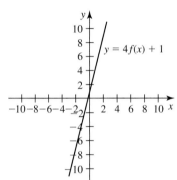

3. $y = -1.5x - 12$

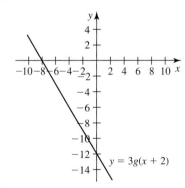

5. $y = -0.5x + 2.5$

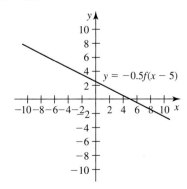

7. g can be created by shifting f right 2 units and vertically stretching it by a factor of 8.

9. g can be created by reflecting f horizontally about the vertical axis.

11. $\dfrac{-26}{3}$

13. 19

15. 55

17. $x = -1$

19. $x = 5$

21. Answers vary.

x	$h(x)$
-4	-3.5
-3	-6.5
-2	-14.5
-1	-10
0	-3
1	0.5
2	2
3	6.5
4	14

23. $g(x) = -x^2 + 1$

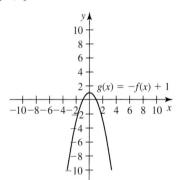

25. $j(x) = \dfrac{1}{3}(x - 2)^2 + 5$

27.

29.

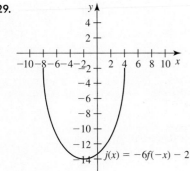

31. Function g can be created by vertically stretching function f by a factor of 4 (the output of g is 4 times as large as the output of f when the inputs are the same). $g(x) = 4f(x)$

33. Function j can be created by vertically stretching function f by a factor of 3 and vertically reflecting it about the horizontal axis (the output of j is -3 times as large as the output of f when the inputs are the same). $j(x) = -3f(x)$

35. a. Answers vary. $(-5, 48)$, $(-3, 0)$, and $(-2, 12)$
   b. Answers vary. $(-2, -86)$, $(0, -2)$, and $(1, -23)$
   c. Answers vary. $(2, 102)$, $(0, 0)$, and $(-1, 25.5)$

37. We could model the cost of electricity after a 3% increase in cost with a vertical stretch of C by a factor of 1.03 (a 3% increase means the new cost of electricity will be 103% of the previous cost). In other words, if $E(m)$ models the average cost in dollars of electricity for a single-family home in Chicago during month m of the year after a 3% increase in the cost of electricity, $E(m) = 1.03 \cdot C(m)$.

39. The horizontal intercepts do not change. A vertical stretch or compression multiplies the output of a function by the stretch or compression factor for a given input. Since horizontal intercepts are of the form $(a, 0)$, the output of the function is 0, and the product of any number and 0 remains 0. Thus, if a horizontal intercept is located at $(a, 0)$, then it will remain at $(a, 0)$ after a vertical stretch or compression.

41. Yes. A linear function can always be written in slope-intercept form ($y = mx + b$). Therefore, if a linear function is defined by $g(x) = mx + b$, then g can be created by performing a vertical stretch/compression of f by a factor of m and vertical shift upward/downward b units. In other words, g can be created from f by $g(x) = m \cdot f(x) + b$.

43. a. $\dfrac{1}{2}$ dollars per orange or $0.50 per orange
   b. $F(x) = 1.50x$; $F(x) = 3C(x)$

c. $1.50 per orange; the cost of an orange after the freeze was expected to reach $1.50. The slope of F is 3 times as large as the slope of C.

45. a. Let f represent the per-day cost of a BritRail 15-day Flexipass in dollars for d days of train travel. Then $F(d) = \dfrac{644}{d}$. Note that the domain of the function contains discrete values, so the graph should not be drawn as a smooth curve.

b. If C represents the per-day cost of a BritRail 15-day Consecutive Pass in dollars for d days of travel, then $C(d) = \dfrac{499}{d}$. Since both F and C are multiples of $\dfrac{1}{d}$, we write C in terms of F as $C(d) = 0.775F(d)$. Therefore, C can be created by vertically compressing F by a factor of 0.775.

c. The 15-day Consecutive Pass is always a less expensive option for the same number of days traveling on the trains. BritRail charges more for the flexibility offered by the Flexipass, and it is more likely that a customer will use all 15 days of travel with the Flexipass than the Consecutive Pass.

47. a. M can be created by vertically compressing N by a factor of $\dfrac{1}{1{,}000{,}000}$, or 0.000001. $M(t) = \dfrac{1}{1{,}000{,}000} N(t)$ or $M(t) = 0.000001 N(t)$

b. L models the number of registered vehicles in the United States (in millions) x years since 1970. $L(x) = 0.000001 N(x + 1970)$

49. a. R can be created by vertically compressing N by a factor of $\dfrac{1}{1000}$, or 0.001.

$$R(t) = \frac{1}{1000} N(t) \text{ or } M(t) = 0.001 N(t)$$

b.
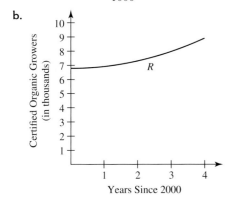

51. a. $L(t) = 1000 \cdot S(t);\ L(t) = 1899t + 4175$

  b. $Y(t)$ models the median salary in millions of dollars for individuals in the science and engineering field as a function of the number of years since 1990.
  $Y(t) = 0.001899t + 0.04175$

53. a. $T(c) = \left(\dfrac{1}{19}\right) 2.19c$ or $T(c) \approx 0.115c$

  b. We could model the cost in Amarillo of driving c miles in the city by vertically stretching T by a factor of $\dfrac{2.44}{2.19}$, or about 1.114. If we call our new function A, then $A(c) = \dfrac{2.44}{2.19} T(c)$ or $A(c) \approx 1.114 T(c)$.

  c. A new function B can be created by vertically compressing T by a factor of $\dfrac{19}{26}$ or about 0.731.
  $B(h) = \dfrac{19}{26} T(h)$ or $B(h) \approx 0.731 \cdot T(h)$

  d. A new function C can be created by vertically compressing T by a factor of $\left(\dfrac{19}{26}\right)\left(\dfrac{2.44}{2.19}\right)$ or about 0.814.
  $C(h) = \left(\dfrac{19}{26} \cdot \dfrac{2.44}{2.19}\right) T(h)$ or $C(h) \approx 0.814 \cdot T(h)$

55. A vertical compression preserves the horizontal intercepts of a function, but f and g have different horizontal intercepts. Therefore, g cannot be created by only performing a vertical compression on f. It would require at least one additional transformation.

57. The phrase "per capita" means that the number being reported is a proportion. For example, if the per capita spending on drugs in the United States was $320 for some year, that means that during the given year the average spending per person in the country on prescription drugs was $320. My classmate has the right idea that multiplying the per capita value by the U.S. population will give the total amount spent on prescription drugs in the United States. However, the population of the United States is not a constant value, and a different population number would be required for each year. Since the population of the U.S. is not constant, we could not use it as a vertical stretch factor for a function that models spending over multiple years to achieve what my classmate desires.

CHAPTER 3 ■ SECTION 3.4

1.

3.
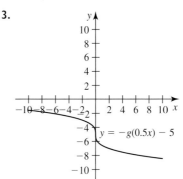

5. 24

7. -3

9. 87.5

11. $x = 9$

13. $x = -1.2$

15. Answers vary.

x	$g(x)$
-1	3
$-3/4$	6
$-1/2$	1
$-1/4$	10
0	-2
1/4	15
1/2	-5
3/4	24
1	-9

17.

x	$j(x)$
$-4/3$	-6
-1	-12
$-2/3$	-2
$-1/3$	-20
0	4
1/3	-30
2/3	10
1	-48
4/3	18

19. $F(x)$

21. $D(x)$

23. $E(x)$

25.

27.

29.

31.

33. a. g can be created by horizontally stretching f by a factor of 4.
 b. $g(x) = f\left(\dfrac{1}{4}x\right)$
 c. $g(x) = \left(\dfrac{1}{4}x\right)^2$ or $g(x) = \dfrac{1}{16}x^2$

35. a. j can be created by horizontally stretching f by a factor of 2 and shifting it right 3 units.

 b. $j(x) = f\left(\dfrac{1}{2}(x-3)\right)$

 c. $j(x) = \left(\dfrac{1}{2}(x-3)\right)^2$ or $j(x) = \dfrac{1}{4}(x-3)^2$

37. a. Answers vary. $(-2, -0.25)$, $(0.5, 1)$, and $(5, 0.1)$
 b. Answers vary. $(16, 6.75)$, $(-4, 8)$, and $(-40, 7.1)$
 c. Answers vary. $\left(-\dfrac{16}{7}, 0.5\right)$, $\left(\dfrac{4}{7}, -2\right)$, and $\left(\dfrac{40}{7}, -0.2\right)$

39. a. g can be created by vertically stretching f by a factor of 2. $g(x) = 2f(x)$
 b. g can be created by horizontally compressing f by a factor of $\dfrac{1}{4}$. $g(x) = f(4x)$

41. There is no change to the vertical intercept of a function under a horizontal stretch or compression.

43. They are both correct. For this particular function f, g can be created with either a vertical compression by a factor of $\dfrac{1}{4}$ or a horizontal stretch by a factor of 2.

45. The horizontal stretch/compression should be done first, then the vertical stretch/compression should follow. This is because when working with functions we would first evaluate the input to the function before using this to determine the corresponding output. However, in this particular case, the final result of performing the transformations is the same regardless of the order.

47. Answers vary. We graph $f(x) = x^2$ and calculate its average rate of change over the interval from $x = -3$ to $x = 1$.

$$\dfrac{f(1) - f(-3)}{1 - (-3)} = -2$$

The average rate of change from $x = -3$ to $x = 1$ is -2. Suppose we vertically stretch the graph by a factor of 2 and recalculate the average rate of change.

$$\dfrac{2f(1) - 2f(-3)}{1 - (-3)} = -4$$

The average rate of change doubled on the interval $x = -3$ to $x = 1$ after a vertical stretch by a factor of 2. To generalize this observation, the average rate of change of a function f from $x = a$ to $x = b$ can be found by

$$\dfrac{f(b) - f(a)}{b - a}$$

If f were stretched/compressed vertically by a factor of k, then the points $(a, f(a))$ and $(b, f(b))$ become $(a, k \cdot f(a))$ and

$(b, k \cdot f(b))$, and the average rate of change from $x = a$ to $x = b$ is

$$\frac{k \cdot f(b) - k \cdot f(a)}{b - a} = k\left(\frac{f(b) - f(a)}{b - a}\right)$$

Therefore, if the vertical stretch or compression factor is k, the average rate of change will be k times as large over a given interval after the vertical stretch.

49. a.

Time Since Taking the 150-mg Dose (hours)	Amount of Gentamicin Remaining in the Person's System (mg)
0	150
3	75
6	37.5
9	18.75
12	9.375

b. The new situation can be modeled by horizontally compressing the function in part (a) by a factor of $\frac{2}{3}$. If we let $G(t)$ model the amount of Gentamicin remaining in the system (in mg) t hours after taking a 150-mg dose with a 4-hour half-life and $M(h)$ model the amount of Gentamicin remaining in the system (in mg) h hours after taking a 150-mg dose with a 3-hour half-life, then $M(2) = G(3)$, $M(4) = G(6)$, $M(6) = G(9)$, and so on.

c. The new situation can be modeled by horizontally stretching the function in part (a) by a factor of $\frac{70}{3}$.

51. a. The 1/600th scale model will be about $\left(\frac{1}{600}\right)630$, or about 1.05, feet tall. The 1/1000th scale model will be about $\left(\frac{1}{1000}\right)630$, or about 0.63, feet tall.

b. For a 1/600th scale model, we would need to horizontally compress the function by a factor of $\frac{1}{600}$ as well as vertically compress the function by a factor of $\frac{1}{600}$. For a 1/1000th scale model, we would need to horizontally compress the function by a factor of $\frac{1}{1000}$ as well as vertically compress the function by a factor of $\frac{1}{1000}$.

c. Let $S(d)$ model the height of the 1/600th scale model in feet as a function of the distance (in feet) from the base of the model and $T(n)$ model the height of the 1/1000th scale model in feet as a function of the distance (in feet) from the base of the model. Then $S(d) = \frac{1}{600}A(600d)$ and $T(n) = \frac{1}{1000}A(1000n)$.

d. $M(u) = \frac{12}{600}A\left(\frac{600}{12}u\right)$ and $N(v) = \frac{12}{1000}A\left(\frac{1000}{12}v\right)$,

where $M(u)$ is the height of the 1/600th scale model in inches as a function of the distance (in inches) from the base of the model and $N(v)$ is the height of the 1/1000th scale model as a function of the distance (in inches) from the base of the model.

53. a. The new table represents a horizontal compression of W by a factor of $\frac{1}{10}$.

b. If the new function is named M, and its input named d, then $M(d) = W(10d)$.

c. The left column should be labeled "Decades Since 1950" and the right column "Federal Minimum Wage (dollars)."

d. No. Although the inputs of both functions represent amounts of time, their units are different so it would be inappropriate to graph them on the same set of axes.

55. a. R can be created by horizontally compressing P by a factor of $\frac{1}{1000}$. $R(k) = P(1000k)$

b. $R(k) = 14.96(0.9998)^{1000k}$

c. S can be created by horizontally stretching P by a factor of 100. $S(c) = P\left(\frac{1}{100}c\right)$

d. $S(c) = 14.96(0.9998)^{\frac{1}{100}c}$

57. a. Answers vary.

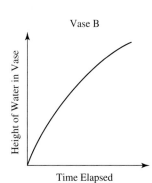

b. The graphs would be stretched horizontally since it would take a proportionally longer amount of time for the height of the water to reach the same levels than for the original vases. We could also justify the transformation as a vertical compression since for the same amount of elapsed time the water would reach a proportionally smaller height in the vases.

c. The graphs would be compressed horizontally since it would take a proportionally smaller amount of time to reach the same height of water when compared to the original vases. We could also justify the transformation as a vertical stretch since for the same amount of time elapsed the height of the water will be proportionally greater.

59. Answers vary.

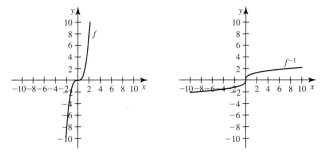

A vertical stretch/compression of the original function causes a horizontal stretch/compression (respectively) in the inverse function. Recall that an inverse function is a function that reverses the input–output process of another function. Thus, if a function has the ordered pair (x, y) as a solution, its inverse will have the ordered pair (y, x) as a solution. Suppose that the ordered pairs $(-3, -27)$ and $(2, 8)$ were solutions of the original function. Then the inverse would have the ordered pairs $(-27, -3)$ and $(8, 2)$ as solutions. If the original function undergoes a vertical stretch by a factor of 3, then the ordered pairs $(-3, -27)$ and $(2, 8)$ become $(-3, -81)$ and $(2, 24)$, meaning that for the inverse function the ordered pairs $(-27, -3)$ and $(8, 2)$ become $(-81, -3)$ and $(24, 2)$, a horizontal stretch by a factor of 3. Similar reasoning can verify that a vertical compression of the original function will horizontally compress the inverse function by the same factor.

CHAPTER 3 ■ REVIEW

1. a.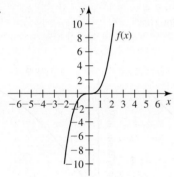

b. g can be created by shifting f right 1 unit and upward 6 units.

c.

d. Answers vary. An example is given. $(-1, -2)$, $(1, 6)$, and $(2, 7)$

e. $g(x) = (x - 1)^3 + 6$

3. a.

b. g can be created by shifting f right 4 units and downward 2 units.

c.

d. Answers vary. An example is given. $(0, -1)$, $(6, 2)$, and $(10, 4)$

e. $g(x) = \left[\dfrac{1}{2}(x - 4) + 3\right] - 2$ or $g(x) = \dfrac{1}{2}x - 1$

5. The outputs of h match the outputs of f when the inputs of f are 3 units greater than the inputs of h (or the inputs of h are 3 units less than the inputs of f). Thus, h is created by shifting f left 3 units. $h(x) = f(x + 3)$

7. g can be created by shifting f left 2 units and downward 5 units. $g(x) = f(x + 2) - 5$

9. a.

Years Since 1992	Median Family Income in Constant (2003) Dollars
-12	44,452
-7	45,223
-2	48,248
0	46,992
2	47,615
4	49,378
6	52,675
8	54,191

b. The new function can be created by shifting the given function to the left 1992 units.

c. $g(x) = f(x + 1992)$

11. a.

Year	Change in Median Family Income Since 1980 (in constant 2003 dollars)
1980	0
1985	771
1990	3796
1992	2540
1994	3163
1996	4926
1998	8223
2000	9739

b. The new function can be created by shifting the given function downward 44,452 units.

c. $g(x) = f(x) - 44,452$

13. -2

15. -8

17. $x = 1$

19. Answers vary.

x	$g(x)$
-5	-11
-4	-8
-3	-1
-2	3
-1	5
0	4
1	2

21. a.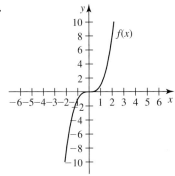

b. g can be created by horizontally reflecting f about the vertical axis and shifting it upward 2 units.

c.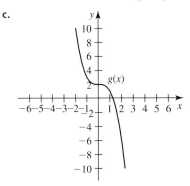

d. Answers vary. An example is given. $(1, 1), (0, 2),$ and $(-2, 10)$

e. $g(x) = (-x)^3 + 2$ or $g(x) = -x^3 + 2$

23. a.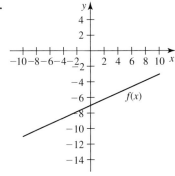

b. g can be created by reflecting f both horizontally and vertically and shifting it downward 6 units.

c.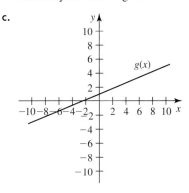

d. Answers vary. An example is given. $(5, 3), (0, 1),$ and $(-5, -1)$

e. $g(x) = -\left[\dfrac{2}{5}(-x) - 7\right] - 6$ or $g(x) = \dfrac{2}{5}x + 1$

25. a. Labor costs are considered negative numbers when determining their impact on profit because labor costs (as well as any other costs) are removed from a company's revenues to determine profit. This can be represented by vertically reflecting L about the horizontal axis.

b. $N(t) = -L(t)$

c. S represents the impact Chipotle's labor costs have on Chipotle's profits as a function of the year, y.

d. Perhaps. Graphing the points shows that a linear model might closely approximate the relationship. However, since we have equal changes in the year, the changes in the labor costs would need to be roughly equal over each interval for a linear model to be a great fit. In the table we create, we see that the labor costs change by about $20,000,000 from 2001 to 2002, about $28,000,000 from 2002 to 2003, about $45,000,000 from 2003 to 2004, and about $39,000,000 from 2004 to 2005. Therefore, while a linear model might be reasonable, it may not be the best choice for this context.

Year y	Labor Costs ($ thousand) $S(y)$
2001	$-46,048$
2002	$-66,515$
2003	$-94,023$
2004	$-139,494$
2005	$-178,721$

27. a. A function has even symmetry if $f(-x) = f(x)$ for any x in the domain of f. We can also say that the graph of the function would be identical after performing a horizontal reflection about the vertical axis.

b. A function has odd symmetry if $f(-x) = -f(x)$ for any x in the domain of f. We can also say that the graph of the function would be identical after performing either a horizontal reflection about the vertical axis or a vertical reflection about the horizontal axis.

29. even

31. a.

b. g can be created by shifting f right 4 units and vertically stretching it by a factor of 5.

c.

d. Answers vary. An example is given. $(2, -20)$, $(5, 10)$, and $(6, 20)$

e. $g(x) = 5[2(x - 4)]$ or $g(x) = 10x - 40$

33. a.

b. g can be created by horizontally reflecting f about the vertical axis, vertically stretching it by a factor of 1.5, and shifting it upward 8 units.

c.

d. Answers vary. An example is given. $(0, 6.5)$, $(-1, 8)$, and $(-4, 9.5)$

e. $g(x) = 1.5(\sqrt{-x} - 1) + 8$ or $g(x) = 1.5\sqrt{-x} + 6.5$

35. 40

37. $\dfrac{47}{3}$

39. 105

41. $x = 1$

43. Answers vary.

x	$g(x)$
-3	-31
-2	-22
-1	-16
0	-12
1	-10
2	-11
3	-14

45. a. $A(x) = 0.80R(x)$

b. $G(x) = 0.80R(x) - 50$

c. The vertical compression (the percent discount) must occur first, followed by the shift downward (applying the gift card). This makes sense in this context because when you purchase an item with a gift card, it pays down the balance owed just like cash and is always applied after coupons and sales.

d. The total cost of the books, after the 20% discount was applied, was less than $50. This results in the gift card still having some of its original value remaining after the books are purchased.

47. a. S can be created by vertically stretching P by a factor of 1000. $S(y) = 1000 \cdot P(y)$

b. M can be created by vertically compressing P by a factor of $\dfrac{1}{1000}$ or 0.001. $M(y) = 0.001 \cdot P(y)$

49. a.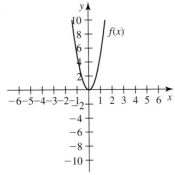

b. g can be created by horizontally stretching f by a factor of 4 and shifting it downward 10 units.

c.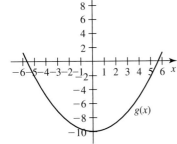

d. Answers vary. An example is given. $(-4, -5)$, $(0, -10)$, and $(4, -5)$

e. $g(x) = 5(0.25x)^2 - 10$ or $g(x) = 0.3125x^2 - 10$

51. a.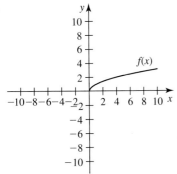

b. g can be created by horizontally compressing f by a factor of $\dfrac{1}{3}$ and horizontally reflecting it about the vertical axis.

c.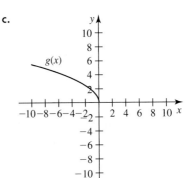

d. Answers vary. An example is given. $\left(-\dfrac{1}{3}, 1\right)$, $\left(-\dfrac{4}{3}, 2\right)$, and $(-3, 3)$

e. $g(x) = \sqrt{-3x}$

53.

55.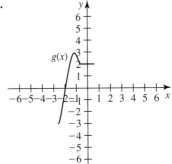

57. a. f models the approximate number of vacant housing units in a state that contains x total housing units.

b. $f(x) = 1000[0.096345(0.001x) + 15.099]$ or $f(x) = 0.096345x + 15{,}099$

59. h can be created by horizontally compressing f by a factor of $\dfrac{1}{3}$. $h(x) = f(3x)$

61. It can be thought of either way, as $g(x) = \dfrac{1}{3}(x)$, a vertical compression, or $g(x) = \left(\dfrac{1}{3}x\right)$, a horizontal stretch. In the case of $f(x) = x$, the two transformations create identical functions, $g(x) = \dfrac{1}{3}x$.

63. a. E can be created by horizontally compressing C by a factor of $\dfrac{1}{10}$. $E(d) = C(10d)$

b. $E(d) = 122.81(1.122)^{10d}$

c. $P(d)$ represents the value of U.S. goods (in dollars) exported to Colombia d decades since 1990.

65. a. The vertical intercept will move upward or downward the same distance as the vertical shift.
   b. The vertical intercept will likely change, but the distance it moves depends on the function.
   c. The vertical intercept will reflect vertically over the horizontal axis. If the vertical intercept was located at $(0, b)$, it will be located at $(0, -b)$ after the vertical reflection.
   d. no effect
   e. The vertical intercept will change (unless the vertical intercept is $(0, 0)$). If $(0, b)$ is the vertical intercept of the original function and the vertical stretch or compression factor is n, then it will be located at $(0, nb)$ after the vertical stretch/compression.
   f. no effect

67. a. no effect
   b. no effect
   c. no effect
   d. no effect
   e. The magnitude of the slope will change. The magnitude of the new slope will be the magnitude of the product of the original slope and the stretch or compression factor.
   f. The magnitude of the slope will change. The magnitude of the new slope will be the magnitude of the product of the original slope and the reciprocal of the stretch or compression factor.

CHAPTER 4 ■ SECTION 4.1

1. Answers vary.

x	y
0	0
1	1
2	4
3	9
4	16
5	25

3. Answers vary.

x	y
1	1
4	2
16	4
25	5
36	6

5. Answers vary.

x	y
0	5
1	7
2	9
3	11
4	13

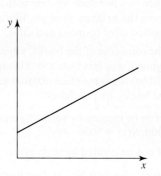

7. Answers vary.

x	y
3	12
4	12
5	12
6	12
7	12
8	12

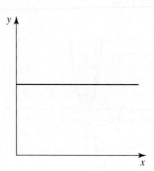

9. Answers vary.

x	y
2	4
3	5
4	8
5	13
6	15
7	16
8	15
9	12
10	7
11	4
12	3
13	4
14	8

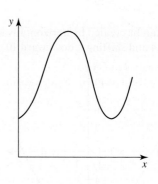

11. a. and $h(x)$ b. and $f(x)$
    c. and $g(x)$ d. does not fit any of the functions

13. $f(x)$ and $g(x)$ both have variable rates of change, $h(x)$ has a constant rate of change.

15. Answers vary.

17. $-\dfrac{4}{3}$

19. Since the intervals for the change in x are 1 unit apart, the best we can estimate the instantaneous rate of change at $x = 0$ is $\dfrac{4-3}{0--1} = \dfrac{1}{1} = 1$. (We could also use $x = 0$ and $x = 1$ to estimate.)

21. a. Answers vary. Increasing from $0 \le f \le 35$
    Answers vary. Decreasing from $35 \le f \le 74$

    b. The function is concave down, which means that initially even though the apple production increases with additional fertilizer, the increase in apple production becomes less and less until the increase eventually levels off. Eventually apple production decreases with additional fertilizer applied.

23.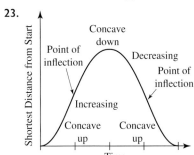

25. Variable rate of change; for equal changes in x, there are different corresponding changes in y.

x	y	$\dfrac{\Delta y}{\Delta x}$
0	1	$\dfrac{4}{2}$
2	5	$\dfrac{1}{2}$
4	6	$\dfrac{6}{2}$
6	12	$\dfrac{4}{2}$
8	16	$\dfrac{3}{2}$
10	19	$\dfrac{1}{2}$
12	20	$\dfrac{5}{2}$
14	25	$\dfrac{4}{2}$
16	29	

27. Constant rate of change; for equal changes in x there are equal changes of y by a factor of -2.

29. The classmate is only looking at the change in the outputs when considering the average rate of change and not paying attention to the fact that the change in the input values is over an interval of 2 units. To get the correct response, we must divide the change of 18 in the outputs over an interval of 2 to determine the average rate of change of 9.

31. a. The function is increasing for posted speed limits from 20 to about 57 miles per hour and it is decreasing for posted speed limits from about 57 to 70 miles per hour.

    b. $F(s)$ is primarily concave down. This tells us that as the posted speed limit increases from 20 to 57 miles per hour the fatality rate increases but at a decreasing rate. After posted speed limits of 57 miles per hour the fatality rate decreases at an increasing rate.

33. a.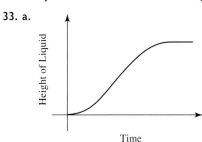

    b. Yes, it's important; if it were not being poured at a constant rate then we could not graph height as a function of time. We must consider small intervals of time and how the height is changing in order to sketch the graph. If the rate that the liquid was poured changed then it would effect the height at different times.

35. a.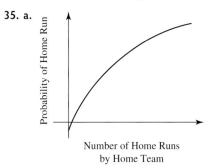

    b. Going into a baseball game there already exists some factors that benefit the home team such as playing on a field the players are familiar with, being at home and at a close proximity to where the game is played, and fans cheering for the home team. As the home team scores additional runs the team, fans, and players become more and more excited about the prospects of winning, which increases the probability of winning. The manner in which the probability goes up could not be constant or at an increasing rate because the maximum probability of winning is 1. So it would increase less and less with each run scored, reaching a limit of winning probability at 1.

    c. It could be true because as the home team scores more runs, the excitement of the crowd and players is contagious and often leads to more runs. On the other hand, the visiting team, without the support of fans on their side, may feel more and more defeated with each additional run scored by the home team.

37. a. $\dfrac{4334 + 1832}{2} = 3083$. This gives us the "middle" per capita income (in dollars) between the years 1960 to 1970.

    b. $\dfrac{4334 - 1832}{10 - 0} = \dfrac{\$250.20}{\text{year}}$. $P(5) = 1832 + 250.2(5) = \$3083$. If we assume that U.S. residents all received the same increase in their income they would have received $250.2 each year for 5 years, resulting in an income of $3083 in 1965.

c. No; since $P(13)$ is not exactly halfway between the years 10 and 20 the average per capita income would not be an accurate estimate. Therefore we would need to use the average rate of change.

d. $P(50) = 44,630$; In 2010, the estimated per capita income is $44,630 in the United States.

39. a. From the table it appears that the point of inflection would occur near the year 15 (1985). This means that the number of basic cable TV subscribers was increasing at an increasing rate from 1970 to 1985 and then the number continued to increase but at a decreasing rate.

b. An estimate for the instantaneous rate of change in 2002 would be -422. This means that in 2002 the number of basic cable TV subscribers decreased by 422,000 people.

41. a. The graph is decreasing. This means that as time passes from the year 1900 the world record time in the 100-meter Men's Freestyle Swim has been decreasing.

b. The graph is concave up. This means that the world record times in the 100-meter Men's Freestyle Swim has been decreasing at a decreasing rate.

43. a. The average rate of change of the world record times for the mile from 1913 to 1999 would be $\dfrac{3.72 - 4.24}{1999 - 1913} \approx -0.006$. This means that from 1913 to 1999 if the world record time decreased the same amount each year the times would have decreased by 0.006 minutes each year.

b.

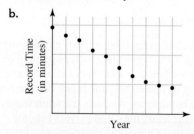

The scatter plot is concave down and then concave up. This means that for over the years of about 1913 to 1954 the world record times for the mile were decreasing at an increasing rate and then from 1954 to 1999 the times were decreasing at a decreasing rate.

45. a. The percentage of the total U.S. population that were immigrants increased concave down from 1850 to 1870, decreased from 1870 to 1880, increased from 1880 to 1890, decreased from 1890 to 1900, increased from 1900 to 1910, decreased dramatically from 1910 to 1970, and then increased from 1970 to 2005.

b. From 1850 to 1870 the country was still relatively young and growing primarily due to European immigration. With the many wars from 1910 through 1970, including two World Wars, the percentage of immigrants decreased. After the wars we see that the percentage began to increase again.

47.

49.

CHAPTER 4 ■ SECTION 4.2

1. quadratic because the second difference is constant

x	y	Δy	$\Delta(\Delta y)$
0	1		
1	3	2	
2	9	6	
3	19	10	4
4	33	14	4
			4

3. linear because the first difference is constant

x	y	Δy	$\Delta(\Delta y)$
5	2		
10	4	2	
15	6	2	0
20	8	2	0
25	10	2	0

5. quadratic because the second difference is constant

x	y	Δy	$\Delta(\Delta y)$
-3	-1		
0	-10	-9	
3	-1	9	18
6	26	27	18
9	71	45	18

7. In a data set with equally spaced inputs if the second consecutive differences are constant, the set represents a quadratic function.

9. the change in the rate of change

11. The change in the rate of change is negative/decreasing.

13. We know that $c = 3$ and a is a positive number.

15. The graphical significance of the value of c is the initial value (or vertical intercept).

17. We are talking about the rate of change at the initial value.

19. If $a > 0$, then the rate of change is increasing.

21. The parameter a is -0.00472 and represents that the rate at which the million enrollees per year is changing by is -0.00944 million Medicare enrollees per year. The parameter b is 0.663 and represents the increase in the number of Medicare enrollees per year in the initial year of 1980. The parameter c is 28.4 and represents that the number of Medicare enrollees in 1980 is 28,400,000.

23. The parameter a is 1.046 and represents that the rate at which the number of children under 5 years old per year is changing by is 2.092 children per year. The parameter b is 60.82 and represents the rate at which the number of children under 5 years old is changing per year. The parameter c is 2152 and represents the number of children under 5 years old in 1990.

25. The parameter a is 0.05 and represents that the rate at which the change in the number of members of the USAA per year is increasing 0.10 million people per year. The parameter b is 0.15 and represents the increase in the number of members of the USAA per year in the initial year of 2003. The parameter c is 5.0 million and represents the number of members of the USAA in the initial year of 2003.

27. The parameter a is 141.25 and represents that the rate at which the change in the number of students enrolled in the Arizona Virtual Academy is increasing is 282.5 students per year each year. The parameter b is 358.75 and represents the increase in the number of students enrolled in the Arizona Virtual Academy in the initial year of 2003. The parameter c is 318.75 and represents the number of students enrolled in the Arizona Virtual Academy in the year 2003.

29. The parameter a is 4.29 and represents that the rate at which the change in the difference between U.S. oil field production and net oil imports is increasing is 8.58 million barrels per year each year. The parameter b is -278.3 and represents the decrease in the difference between U.S. oil field production and net oil imports in millions of barrels per year in 1985. The parameter c is 2251 and represents the difference between U.S. oil field production and net oil imports in millions of barrels in 1985.

31.
$w(b) = -0.0009b^2 + 3.54b - 2101.29$
$w(1500) \approx 1049$ live *white* births when there are 1500 live births

33.
$e(x) = 23.19x^2 - 218.07x + 1825.11$
$e(9) \approx 1741$ thousand computer and electronic products industry employees in 2009

35.
$s(x) = -20.32x^2 + 221.01x + 790.93$
$s(8) \approx 1258.41$ thousand dollars is the NFL player's average salary in 2008

37. No; the constant second difference has to be determined when the increments between the x values are all the same.

x	y	$\frac{\Delta y}{\Delta x}$	$\Delta\left(\frac{\Delta y}{\Delta x}\right)$
0	-1		
2	9	$\frac{9 - -1}{2 - 0} = 5$	
4	27	$\frac{27 - 9}{4 - 2} = 9$	4
8	53	$\frac{53 - 27}{8 - 4} = 6.5$	-2.5
12	87	$\frac{87 - 53}{12 - 8} = 8.5$	2

CHAPTER 4 ■ SECTION 4.3

1. vertex (3, 2); no real horizontal intercepts (since the solution is a complex number)

3. vertex (2, -1); horizontal intercepts at $x = 0$ and $x = 4$

5. vertex (0.5, -2); horizontal intercepts at $x = 0$ and $x = 1$

7. vertex (-2, -1.6); horizontal intercepts at $x = 2$ and $x = -6$

9. vertex (0, -4); horizontal intercepts $\left(\frac{-2}{3}, 0\right)$ and $\left(\frac{2}{3}, 0\right)$

11. vertex (1, -54); horizontal intercepts (4, 0) and (-2, 0)

13. vertex (-9, 12); horizontal intercepts (-7, 0) and (-11, 0)

15. vertex (0, 7); horizontal intercepts (-1, 0) and (1, 0)

17. The vertex is found with $x = \dfrac{-b}{2a} = \dfrac{-4}{2(-2)} = 1$ and $y = -6$. Thus $(1, -6)$ is the vertex and the parabola is concave down.

19. The vertex is found with $x = \dfrac{-b}{2a} = \dfrac{-50}{2(-5)} = 5$ and $y = 125$. Thus $(5, 125)$ is the vertex and the parabola is concave down.

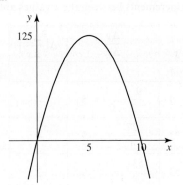

21. $y = 2(x + 4)(x - 6)$
23. $y = \dfrac{2}{3}(x - 6)^2 + 0$
25. $y = -(x - 0)^2 + 0$
27. $y = (x + 1)(x + 1)$
29. $y = (x - 6)(x + 4)$
31. For a 12-inch by 12-inch picture the mat frame area is 224 square inches and for a 15-inch by 15-inch picture the mat frame area is 155 square inches.
33. The dimensions that maximize the area for the pens are 125 feet by $\dfrac{250}{3}$ feet.
35. The standard form gives the vertical stretch/compression and reflection over the x-axis as well as the vertical intercept. Its disadvantage is the work that must be done to find the vertex and horizontal intercepts. The vertex form gives the vertex and the vertical stretch/compression and reflection over the x-axis. The disadvantage is that additional work is needed to find the horizontal intercepts and the vertical intercept. While the factored form lists the horizontal intercepts and vertical stretch/compression and reflection over the x-axis, more work is required to find the vertex and the vertical intercept.
37. The factored form of a quadratic function can be changed into the standard form by expanding the expression and combining like terms.
39. The horizontal intercepts are equal horizontal distance to the left and right from the line of symmetry (x-coordinate of the vertex). Find the difference in the horizontal distance between the known horizontal intercept and the vertex. Use that distance to determine the other horizontal intercept.
41. Kostadinova was airborne for approximately 1.14 seconds.
43. The dimensions that maximize the area of a swimming pool with a perimeter of 110 feet are 27.5 feet by 27.5 feet.
45. Make the pool 36.2 feet wide and 13.8 feet in length (according to the picture). The area of the deck will be 322.8 square feet.
47. The rocket's height at any time t is $s(t) = -16(t - 11.42)^2 + 2088$ and the coordinates of the vertex are $(11.42, 2088)$. This means it took approximately 11.42 seconds to reach its maximum and (assuming a quadratic function models the height) it will take another 11.42 seconds for the rocket to return to the ground for a total flight time of 22.84 seconds.
49. Since the factors of the quadratic function are derived from the horizontal intercepts of the function, we will not be able to put into factored form a quadratic function that does not intersect the x-axis, which is a quadratic function that has only positive function values.

CHAPTER 4 ■ REVIEW

1. Answers vary.

x	y
0	5
1	4
2	1
3	-4
4	-11
5	-20
6	-31

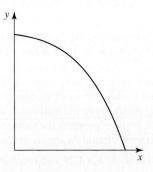

3. Answers vary.

x	y
-1	0.5
0	1
1	0.5
2	-0.4
3	-1
4	-0.7
5	0.3
6	1
7	0.8
8	-0.2

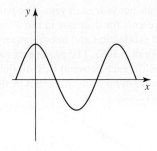

5. In function $h(x)$ there is an inflection point near $x = 4$ because from $x = -6$ to $x = 4$ the function values decrease at a decreasing rate and then from $x = 4$ to $x = 8$ the function values decrease at an increasing rate. We can find inflection points on a graph where the concavity changes.

7. **Germany**
   a. The function for Germany's aircraft production between 1940 and 1944 is concave up and therefore increased at an increasing rate.
   b.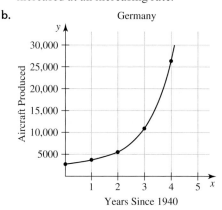

   USSR
   a. The function for the USSR's aircraft production between 1940 and 1944 increased at a nearly constant rate and therefore did not have any concavity.
   b.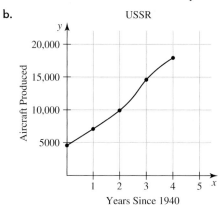

   United Kingdom
   a. The function for the United Kingdom's aircraft production between 1940 and 1944 is concave down and therefore increased at a decreasing rate.
   b.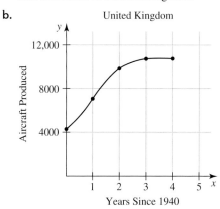

9. average rate of change for USSR ≈ 3338.7

   For the years 1940 to 1943, USSR's aircraft production increased approximately 3339 aircraft per year.

11. instantaneous rate of change for Germany = 15,428

    For the year 1943, Germany's aircraft production increased approximately 15,428 aircraft per year.

13. a. Graphs vary.

    b. The function is increasing from hour 10 to hour 16, decreasing from hour 0 to hour 10, and is constant from hour 4 to 6. The temperature of Seattle cooled off on March 13, 2007 from 43°F to 41°F at hour 4 and then stayed the same temperature for 2 hours before beginning to cool off until hour 10 at which time the temperatures started to increase.

    c. The function is concave down from hour 0 to about hour 4 and concave up from hour 6 to hour 16. From hour 0 to hour 4 the temperatures were cooling off at an increasing rate and from hour 6 to hour 10 the temperatures were cooling off at a decreasing rate. From hour 10 to hour 16 the temperatures were increasing at an increasing rate.

    d. $F(3) \approx 41.5$
       $F(9) \approx 38.45$
       $F(13) \approx 40.5$

    e. $F(3) \approx 42 - 0.5 = 41.5$
       $F(9) \approx 39.9 - 1.45 = 38.45$
       $F(13) \approx 39 + 1.5 = 40.5$

    f. $t = 6.75$ would represent 6.75 hours after 11 P.M. on March 13, 2007. You could use the average rate of change to estimate $F(6.75)$ by taking the temperature at 6 hours and then adding 0.75 times the average rate of change onto it.

15. The coach may have been experiencing an increase in team wins at an increasing rate (concave up) and then continued to win more games than the year before but at a decreasing rate (concave down).

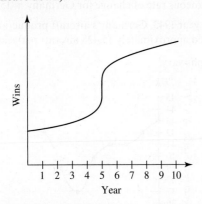

17. from a table: if the second differences are constant; from a graph: if it has one type of concavity; from a formula: if it is of the form $y = ax^2 + bx + c$

19. One method is to create a scatter plot and do a quadratic regression. Another method is to find a, b, and c for $y = ax^2 + bx + c$ by taking one-half of the second difference for a, estimate b by calculating the rate of change at $x = 0$, and find c by determining the value of the function at $x = 0$.

21. $a = 1.008$; rate of change in consumer expenditure = $2(1.008) = 2.016$; Based on data from 1990 to 2004, the rate of change in consumer expenditure is increasing at a rate of 2.016 billion dollars per year each year.

    $b = 9.951$; Based on data from 1990 to 2004, the rate of change in consumer expenditure for 1990 is increasing 9.951 billion dollars per year.

    $c = 450.5$; Based on data from 1990 to 2004, the consumer expenditure for 1990 is 450.5 billion dollars.

23. $C(t) = 15.838t^2 + 181.401t + 4218.779$ million dollars, where $t = 0$ represents 1990

25. The vertex is $(-2, 14)$ and there are no real horizontal intercepts. Therefore this function does not cross the horizontal axis.

27.

29.

31. $C(n) = -0.12n^2 + 1.73n$, where n is the number of pounds purchased and C is the cost per pound. This means the price the store should charge for a 10-pound bag of oranges is $5.22 and the price per pound is $0.52.

33. The equation in vertex form is $y = -x^2 + 8$.

CHAPTER 5 ■ SECTION 5.1

1. $f(x)$ is quadratic because the second differences are constant. $g(x)$ is cubic because the third differences are constant. $h(x)$ is linear because the first differences are constant.

3. $f(x)$ is cubic because the third differences are constant. $g(x)$ is linear because the first differences are constant. $h(x)$ is none of these because neither the first, second, nor third differences are constant.

5. The concavity changes one time so the type of polynomial function represented by the graph is a cubic function.

7. The concavity changes two times so the lowest-degree polynomial function that could represent the graph is a fourth-degree polynomial.

9. a. As $x \to \infty$, $y \to -\infty$ and as $x \to -\infty$, $y \to -\infty$.
   b. Answers vary.

x	y
$-1{,}000{,}000$	-2.0×10^{12}
$-100{,}000$	-2.0×10^{10}
$-10{,}000$	-2.0×10^{8}
0	-10
$10{,}000$	-2.0×10^{8}
$100{,}000$	-2.0×10^{10}
$1{,}000{,}000$	-2.0×10^{12}

11. a. As $x \to \infty$, $f(x) \to -\infty$ and as $x \to -\infty$, $f(x) \to \infty$.
    b. Answers vary.

x	$f(x)$
$-1{,}000{,}000$	3×10^{18}
$-100{,}000$	3×10^{15}
$-10{,}000$	3×10^{12}
0	-3
$10{,}000$	-3×10^{12}
$100{,}000$	-3×10^{15}
$1{,}000{,}000$	-3×10^{18}

13. a. As $x \to \infty, f(x) \to \infty$ and as $x \to -\infty, f(x) \to -\infty$.
b. Answers vary.

x	$f(x)$
$-1{,}000{,}000$	-1×10^{30}
$-100{,}000$	-1×10^{25}
$-10{,}000$	-1×10^{20}
0	2
$10{,}000$	1×10^{20}
$100{,}000$	1×10^{25}
$1{,}000{,}000$	1×10^{30}

15. a. As $x \to \infty, f(x) \to -\infty$ and as $x \to -\infty, f(x) \to \infty$.
b. Answers vary.

x	$f(x)$
$-1{,}000{,}000$	1×10^{30}
$-100{,}000$	1×10^{25}
$-10{,}000$	1×10^{20}
0	0
$10{,}000$	-1×10^{20}
$100{,}000$	-1×10^{25}
$1{,}000{,}000$	-1×10^{30}

17.

19.

21.

23.

25.

27.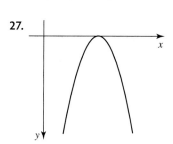

29. If a scatter plot demonstrates a constant rate of change then the best model is linear, if a variable rate of change and one type of concavity then the best model is quadratic, if a variable rate of change and two types of concavity then the best model is quadratic, if a variable rate of change and three types of concavity then the best model is cubic, etc.

31. It is not polynomial because for a polynomial function each term must have a whole-number exponent, not a fraction; $\sqrt{x} = x^{1/2}$.

33. As we consider $x \to \infty$ and $x \to -\infty$ and the leading term we can see if the function values $f(x) \to \infty$ or $f(x) \to -\infty$ by thinking about what would happen as we input larger and larger numbers in. This helps us to know what type of function to choose as the mathematical model.

35. a.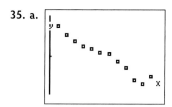

There are three different types of concavity (concave down, concave up, and concave up) so a quartic function would fit well.

b. $P(t) = 0.000809t^4 - 0.0208t^3 + 0.177t^2 - 0.671t + 7.233$

37. a.

A cubic model best models the data because the concavity changes from concave up to concave down.

b. $R(t) = -0.048t^3 + 0.887t^2 - 1.495t + 58.324$

39. a.

With two different types of concavity, a cubic function will fit best.

b. $B(t) = -12.58t^3 + 134.116t^2 - 358.81t + 979.21$

41. When $a > 0$, as $x \to -\infty$, $y \to -\infty$ and as $x \to \infty$, $y \to \infty$.
When $a < 0$, as $x \to -\infty$, $y \to \infty$ and as $x \to \infty$, $y \to -\infty$.

43. When $a > 0$, as $x \to -\infty$, $y \to -\infty$ and as $x \to \infty$, $y \to \infty$.
When $a < 0$, as $x \to -\infty$, $y \to \infty$ and as $x \to \infty$, $y \to -\infty$.

45. When $a > 0$, as $x \to -\infty$, $y \to -\infty$ and as $x \to \infty$, $y \to \infty$.
When $a < 0$, as $x \to -\infty$, $y \to \infty$ and as $x \to \infty$, $y \to -\infty$.

47. If n is odd and $a > 0$, as $x \to -\infty$, $y \to -\infty$ and as $x \to \infty$, $y \to \infty$.

If n is odd and $a < 0$, as $x \to -\infty$, $y \to \infty$ and as $x \to \infty$, $y \to -\infty$.

If n is even and $a > 0$, as $x \to -\infty$, $y \to \infty$ and as $x \to \infty$, $y \to \infty$.

If n is even and $a < 0$, as $x \to -\infty$, $y \to -\infty$ and as $x \to \infty$, $y \to -\infty$.

CHAPTER 5 ■ SECTION 5.2

1. direct variation; as the inputs increase, the outputs increase

3. inverse variation; as the inputs increase, the outputs decrease

5. direct variation; as the inputs increase, the outputs increase

7. inverse variation; as the inputs increase, the outputs decrease

9. direct variation; as the inputs increase, the outputs increase

11. $A(s) = s^2$ **13.** $W(h) = 7.95h$

15. $t(s) = \dfrac{400}{s}$ **17.** $g = \pm\sqrt{\dfrac{1}{5}}$

19. $m \approx 75.82$ **21.** $t \approx 4.23$

23. $v \approx 27.626$ **25.** $x = 64$

27. $y = \dfrac{1}{\sqrt{x}}$

x	1	4	16	25
y	1	$\dfrac{1}{2}$	$\dfrac{1}{4}$	$\dfrac{1}{5}$

29. $y = 5x^{0.25}$

x	1	4	16	25
y	5	7.07	10	11.18

31. $y = ax^n$ is the function $g(x)$ because as $n \to \infty$ the function $g(x)$ increases at an increasing rate and $y = ax^{1/n}$ is the function $f(x)$ because as $n \to \infty$ the function $f(x)$ increases at a decreasing rate. The point of intersection can be found by setting the two functions equal to one another and solving for x. Therefore, the points of intersection are $(0, 0)$ and $(1, a)$.

33. Answers vary. $y = x^{0.50}$ or $y = \dfrac{1}{8}x^2$ are possible answers.

35. a. Graph B
b. Graph A
c. Graph D
d. Graph C

37. a. direct variation; as the thickness of the board increases, the length of the nail also increases
b. 3 is the constant of proportionality
c. $N(b) = 3b$

39. a. As the radius of the blood vessel increases, the velocity of the blood decreases at a decreasing rate.
b. $b < 0$
c. $V(r) = 23.4r^{-1.14}$
d. As the blood vessel becomes increasingly larger, the velocity of the blood decreases at a decreasing rate.
e. As the blood vessel becomes increasingly smaller, the velocity increases at an increasing rate.

41. a. $V(m) = k\sqrt{m}$
b. The terminal velocity will be $\sqrt{2}$ times as much.
c. The terminal velocity will be twice as much.
d. If the mass of the skydiver is cut in half, the velocity will be $\sqrt{\dfrac{1}{2}}$ as much.

43. a. The period of the pendulum increases at a decreasing rate so a power function is appropriate.
b. $P(L) = 2L^{0.5}$
c. $P(9.1) \approx 6.03$
d. $L = 1$

45. a.

Distance (meters) D	Light Intensity (mw/cm²) L	Rate of Change
0.5	0.462	$\dfrac{-0.121}{0.1} = -1.21$
0.6	0.341	$\dfrac{-0.061}{0.1} = -0.61$
0.7	0.280	$\dfrac{-0.077}{0.1} = -0.77$
0.8	0.203	$\dfrac{-0.042}{0.1} = -0.42$
0.9	0.161	$\dfrac{0.003}{0.1} = 0.03$
1.0	0.164	$\dfrac{-0.053}{0.1} = -0.53$

(continued)

Distance (meters) D	Light Intensity (mw/cm²) L	Rate of Change
1.1	0.111	$\frac{-0.014}{0.1} = -0.14$
1.2	0.097	$\frac{-0.006}{0.1} = -0.06$
1.3	0.091	$\frac{-0.011}{0.1} = -0.11$
1.4	0.091	
1.5	0.080	

Yes, the function is decreasing at a decreasing rate.

b. $L(d) = 0.145d^{-1.65}$

c. At 2 meters the intensity is 0.046 mw/cm².

d. The light must be 0.72 meters away.

47. Answers vary.

49. Answers vary.

51. Answers vary.

CHAPTER 5 ■ SECTION 5.3

1. horizontal asymptote: $f(x) = 2$; vertical asymptote: $x = -4$

3. horizontal asymptote: $g(x) = \frac{-3}{2}$; no vertical asymptotes

5. horizontal intercept: $(3, 0)$; vertical intercept: $(0, 0.6)$; horizontal asymptote: $y = 1$; vertical asymptote: $x = 5$

7. horizontal intercepts are $(-3, 0)$ and $(3, 0)$; no vertical intercepts; horizontal asymptote: $g(x) = 0$; vertical asymptotes: $x = -9$ and $x = 0$

9. a. as $x \to 2$ from the right, $f(x) \to \infty$; as $x \to 2$ from the left, $f(x) \to -\infty$

x	f(x)
1.00	-1
1.90	-10
1.99	-100
2.00	undefined
2.01	100
2.10	10
3.00	1

b. as $x \to \infty$, $f(x) \to 0$; as $x \to -\infty$, $f(x) \to 0$

x	f(x)
5	0.35
50	0.021
500	0.002
5000	0.0002
50,000	0.00002

x	f(x)
-5	-0.143
-50	-0.019
-500	-0.002
-5000	-0.0002
$-50,000$	-0.00002

c.

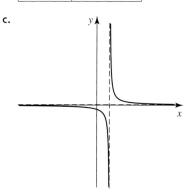

vertical asymptote: $x = 2$; horizontal asymptote: $y = 0$

11. a. The graph of $h(x)$ will be shifted downward 3 units and reflected vertically.

b. The graph $h(x)$ will be shifted upward 4 units and shifted right 2 units.

c. The graph of $h(x)$ will be shifted downward 5 units, shifted left 3 units, stretched vertically by a factor of 2, and reflected vertically.

13. a. $f(x) = \frac{1}{x - 6} - 2$

b. horizontal intercept: $(6.5, 0)$; vertical intercept: $(0, -2.17)$

c. horizontal asymptote: $y = -2$; vertical asymptote: $x = 6$

15. a. $f(x) = -\frac{1}{x^2} - 2$

b. no intercepts

c. horizontal asymptote: $y = -2$; vertical asymptote: $x = 0$

17. y-intercept: $(0, 0)$; horizontal asymptote: $y = 1$

19.

21. Answers vary.

x	f(x)
−13	10
−12	500
−11	10,000
−10	undefined
−9	10,000
−8	0
−7	−5000
−6	−7000
−5	−8000
−4	−8500

23. Answers vary. $f(x) = \dfrac{-1}{(x-6)^2}$

25.

27.

29. B

31. A table will show function values approaching positive and/or negative infinity at input values near the asymptote. For example, for $x = -\dfrac{1}{2}$ we get the following table.

x	h(x)
−1	−9
−0.9	−15
−0.8	−55
−0.7	−125

(*continued*)

x	h(x)
−0.6	−1300
−0.5	undefined
−0.4	1300
−0.3	125
−0.2	55

In terms of a graph, we would see the following.

For an asymptote at $x = -\dfrac{1}{2}$, we could say in symbols "as $x \to -\dfrac{1}{2}$ from the left, $f(x) \to -\infty$ and as $x \to -\dfrac{1}{2}$ from the right, $f(x) \to \infty$." In words, as the input values approach $-\dfrac{1}{2}$ from the left we see the function values approaching negative infinity and as the input values approach $-\dfrac{1}{2}$ from the right the function values approach infinity.

33. Vertical asymptotes show where the function is undefined and also where the function approaches positive and/or negative infinity. Horizontal asymptotes show what the function values do as the input values approach positive and/or negative infinity. We can determine where vertical asymptotes occur by determining at what x-values the denominator will be equal to zero (but will not simultaneously make the numerator equal to zero). Horizontal asymptotes can be determined by considering what the function values approach as $x \to \infty$ and $x \to -\infty$.

35. Answers vary. A real-world interpretation of a vertical asymptote could be when there is a pollution cleanup and 100% of the cleanup cannot be achieved (see example in this section). A real-world interpretation of a horizontal asymptote could be the cost of 1 ounce of a drug on the market as the production amounts continue for extremely large amounts (see example in this section).

37. a. We know that distance equals the rate times the time. Therefore, 150 miles = 75 miles/hour times the time. Time would equal 2 hours.

  b. $t(s) = \dfrac{150}{s}$

  c. As the average speed increases, the travel time decreases. This means that if one's average speed increases then the time it takes to complete the journey decreases.

d.

s	t
20	7.5
15	10
10	15
5	30
4	37.5
3	50
2	75
1	150

e.

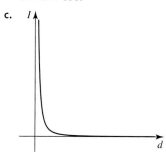

f. The horizontal asymptote is $t = 0$, which means that as the person's speed increases the time it takes to complete the 150-mile journey becomes shorter and shorter until it can almost reach 0 hours for *very* high speeds. The vertical asymptote is $s = 0$, which means that as the person's speed approaches very small values (s approaches 0 miles per hour) the time it takes to complete the 150-mile journey becomes very large.

39. a.

d	I
0.1	149,500
0.5	5980
1	1495
2	373.75
5	59.8
10	14.95
20	3.7375
30	1.6611

b. Answers vary. A practical domain would be $0 \leq d \leq 100$.

c.

d. As you move closer to the person speaking, the sound intensity increases at an increasing rate.

41. a. A 200-pound person can have about 4.5 beers.

A 110-pound person can have about 2.5 beers.

b. Yes, the lighter a person is, the fewer beers can be consumed.

43. a. 1 share: $T = 532.95$ dollars
10 shares: $T = 5150$ dollars
100 shares: $T = 51{,}320$ dollars
1000 shares: $T = 513{,}020$ dollars

b. 1 share: $B = 532.92$ dollars
10 shares: $B = 515$ dollars
100 shares: $B = 513.20$ dollars
1000 shares: $B = 513.02$ dollars

c. As the number of shares purchased increases, the break-even share price decreases, approaching the initial price per share.

45. a. $0 \leq t \leq 5$

b. 0.907 hours

c. 3.23 hours

d. 5 hours

47. a. False; asymptotes can be functions other than linear such as quadratic or other functions.

b. False; this would be a true statement if it only included vertical asymptotes but horizontal asymptotes can be crossed by the function.

49. The table shows how the values fluctuate between highs and lows, but as $x \to \infty$, $f(x) \to 3$ because the highs and lows get closer and closer to 3.

x	f(x)
0	10
5	−4
10	8
15	−1
20	6
25	1
30	5
35	2
40	4

CHAPTER 5 ■ REVIEW

1. $f(x)$ is quadratic because the second differences are constant. $g(x)$ is linear because the first differences are constant. $h(x)$ is cubic because the third differences are constant.

3. a. As $x \to \infty$, $y \to \infty$ and as $x \to -\infty$, $y \to \infty$.

b. Answers vary.

x	y
−1000	997,992
−100	9792
−10	72
0	−8
10	112
100	10,192
1000	1,001,992

5. a. As $x \to \infty$, $y \to -\infty$ and as $x \to -\infty$, $y \to \infty$.
   b. Answers vary.

x	y
-1000	$\approx 1 \times 10^{15}$
-100	$10,000,029,793$
-10	$100,273$
0	-7
10	$-99,687$
100	$-9,999,969,807$
1000	$\approx -1 \times 10^{15}$

7. a. As $x \to \infty$, $y \to \infty$ and as $x \to -\infty$, $y \to \infty$.
   b. Answers vary.

x	y
-1000	$1,000,000,000,000,000,000$
-100	$1,000,000,000,000$
-10	$1,000,000$
0	0
10	$1,000,000$
100	$1,000,000,000,000$
1000	$1,000,000,000,000,000,000$

9. Degree 4

11. a.

   b. There are three different types of concavity, so a fourth-degree polynomial would fit the scatter plot well.
   c. $C(t) = 0.00787t^4 - 0.21016t^3 + 1.5041t^2 - 2.849t + 45.7598$

   d. Using the graph and increasing the domain we see that this model shows that the percentage of eighth graders who use cigarettes will never be 0. Yes, this is reasonable—there will most likely always be teenagers who smoke.

13. direct; as the x-values increase, the y-values also increase

15. inverse; as the x-values increase, the y-values decrease

17. $L(t) = kT^4$

19. $F(m) = -9.8m$

21. $S(t) = \dfrac{1200}{t}$

23. a. As rank increases, the public debt decreases at a decreasing rate.
   b. $P(c) = 221.6c^{-0.38}$
   c. $P(11) \approx 89.1\%$ of GDP

25. horizontal asymptote: $y = 0$; vertical asymptotes: $x = 6$ and $x = 1$

27. horizontal intercept: $(0, 0)$; vertical intercept: $(0, 0)$; horizontal asymptote: $y = 1$; vertical asymptote: $x = 2$; hole: $(4, 0)$

29. The vertical asymptote is at $x = 7$ because as the input values approach 7 the output values increase from one side and decrease from the other.

31.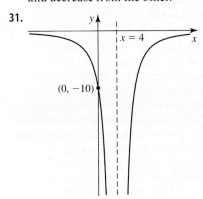

CHAPTER 6 ■ SECTION 6.1

1. exponential; there is a common factor of 0.5

3. none of these; the first and second differences are not constant and there is no change factor

5. a. B: $f(x)$
   b. C: $h(x)$
   c. A: $k(x)$

7. D

9. No, without any points labeled or scales given it is impossible to determine if $f(x)$ is quadratic or exponential because we cannot determine how the function is changing (either a constant change in the rate of change or percentage change).

11. Answers vary.

x	$f(x)$	Change Factor
-2	$\dfrac{1}{9}$	3
-1	$\dfrac{1}{3}$	3
0	1	3
1	3	3
2	9	

13. Percentage growth or decay means that as the input values increase the output values either increase or decrease by a certain percentage.

15. C

17. If a person's salary is $30,000 and decreased by 5%, the salary would be 0.95(30,000) = $28,500. If the person was then given a 5% increase the following year, the salary would be 1.05(28,500) = $29,925, not the original salary.

19. a. 80-year growth factor: 2641
    b. 80-year percentage growth rate: 264,000%
    c. average annual growth factor: $\sqrt[80]{2641} = 1.103$
    d. average annual percentage change: 10.3%
    e. $I(y) = 1(1.103)^y$
    f. $I(90) = \$6788.85$. After 90 years, the initial investment would have risen to a value of $6788.85 with a 10.3% percentage change.

21. Either quadratic or exponential

23. a. (i) 0.4214; (ii) 0.8564; (iii) 0.8812
    b. (i) 58% decrease; (ii) 14% decrease; (iii) 12% decrease. From 1980 to 2005 the value of the dollar decreased by 58%, from 1990 to 1995 the value of the dollar decreased by 14%, and from 2000 to 2005 the value of the dollar decreased by 12%.
    c. average rate of change: $\frac{-\$0.0208}{1 \text{ year}}$; percentage change: about 0.9707
    d. If the percentage change is negative consumers' purchasing power is decreasing (their money is worth less); if the percentage change is positive consumers' purchasing power is increasing (their money is worth more); and if the percentage change is zero consumers' purchasing power is unchanging.

25. a. Neither a linear function nor an exponential function fits perfectly but both can reasonably model the data due to the fact that the rate of change is reasonably constant and the change factor is too.
    b. The slope is 415.49 and means that the average tuition for American public universities was estimated to be increasing by $415.49 per year.
    c. Comparing the rate of increase of private to public universities we can see that the private universities ($991.69 per year) tuition is increasing at a faster rate than the public universities ($415.49 per year).

27. annual percentage growth rate for Atlanta: 3.2%; annual percentage decay rate for New Orleans: 4.1%

29. a. $118 - 10w$ b. $118(0.90)^w$

31. Answers vary. In an *L.A. Times* article the term *exponential growth* refers simply to a rapidly increasing relationship. The term is used in a data-less context about the fact that China and India are growing "by leaps and bounds." The article goes on to discuss why recycling water is essential.

33. about $\frac{7}{100}$

35. a. 2 b. 4
    c. 8 d. 16
    e. $A = 2^g$

37. 19%

39. $n(x) = 1(2)^x$

41. a. On the 14th day the bottom of the pool will be covered with algae.
    b. one day

CHAPTER 6 ■ SECTION 6.2

1. A
3. C
5. E
7. decreasing; concave down; vertical intercept $(0, -4)$; horizontal asymptote $y = 0$
9. increasing; concave up; vertical intercept $(0, 21)$; horizontal asymptote $y = 17$
11. There is a constant change factor.
13. Doubling, tripling, etc., half-life, percentage change. Any of these mean that the output values change by a constant factor.
15. the amount of time for something that is growing to double
17. B; approximately 23,000,000 in 2005
19. E; approximately 19,250,000 in 2005
21. $P(t) = 298,213(1.009)^t$
23. An exponential function is not appropriate because there is a constant rate of change (0.27 quadrillion BTUs per year), meaning it is linear.
25. $S(r) = 7779(0.96)^r$; $S(5) = 7779(0.96)^5 \approx 6342.78$
27. $P(v) = 14.696(0.9999)^v$; $P(6000) \approx 8.065$
29. An exponential function is not appropriate. Over the first 40 years the percentage of adults with only an elementary education had a decay factor of 0.466 and then over the next 40 years the decay factor was only 0.2162.
31. $E(t) = 4747.1(1.04)^t$; $E(20) = \$10,401.48$
33. $H(t) = 989.27(1.07)^t$; Health expenditures are increasing at 7%.
35. $p(t) = 49.39(0.97)^t$; The percentage of highway accidents resulting in injuries is decreasing by 3% per year.
37. $E(y) = 1863.2(1.05)^y$; $E(20) = \$4943.62$
39. $b(t) = 161.43(1.169)^t$; The average NBA salary will be much more than $3 million in 2010.
41. a. average rate of change $= \dfrac{ab^c(b^h - 1)}{h}$
    b. Consider that h is a specific value and for any value c that x takes on, the average rate of change is also an exponential function. We see that $\dfrac{b^h - 1}{h}$ will be a specific value if we know the value of h. Therefore, the average rate of change is also an exponential function of the form

    average rate of change $= kab^c$, where $k = \dfrac{b^h - 1}{h}$

43. The reason why $y = ab^x$ does not have a horizontal intercept is because the function will always have a horizontal

asymptote at $y = 0$ (the x-axis) regardless of what a and b are.

45. $f(x) = \dfrac{-1}{x}$

CHAPTER 6 ■ SECTION 6.3

1. $e \approx 2.7$
3. $e + 3 \approx 5.7$
5. $\dfrac{e}{2} \approx 1.4$
7. $e^3 < 27$
9. $e^{-2} < 0.25$
11. a. C b. A
    c. B d. D
13. a. 0.904% b. 0.901%
15. a. 5.71% b. 5.59%
17. 5.49%
19. 3.59%
21. Yes; although the future value of the account increases as the account is compounded more frequently, its value increases at a decreasing rate and has a limit that is equal to the future value of the account calculated using the continuous compound interest formula $A(t) = Pe^{rt}$.
23. Gained money; since the values of the stocks are changing by a constant percentage each year, they are not changing by a constant amount each year. 3% of a larger number is greater than 3% of a smaller number, and therefore the stock that is increasing in value is changing more each year than the stock that is decreasing in value. Thus, the amount the first stock increases in value each year after the first is more than the second stock loses each year.
25. $1199.28
27. $3062.03
29. $2549.50
31. a. 4.75% b. $548.63
    c. $24.32 per year d. 4.75%
    e. $4556.82 f. $1.20
33. a. 1.045% b. $1132.86
    c. $44.29 per year d. 4.25%
    e. $4413.59 f. $0.74
35. ING Direct has a higher APY by 0.01%.
37. They have effectively the same APY.
39. NBC Bank has a higher APY by 0.01%.
41. a. $b = e^k = e^{0.0198} = 1.02$
    b. $C(t) = 9.038(1.02)^t$
    c. The population is increasing: $k > 0$ in the form ae^{kt}. The population is increasing by about 2% annually.
43. a. $e^k = 1.05$ so $k = 0.0488$; the population is increasing at a continuous rate of 4.88%.
    b. $R(t) = 9.038e^{0.0488t}$
    c. The population is increasing: $b > 1$ in the form ab^t.

45. $k = -0.0173$; the amount of medicine will decrease to 98.28% of the amount 1 hour before.
47. a. 0.215%
    b. $d \approx 22.6$ meters for 50% of available surface light.
       $d \approx 150$ meters for 1% of available surface light.
    c. If the water is not clear, we should assume that the percent of available surface light is less for any given depth (meaning the percent of available surface light will decay more rapidly). This will require $k < -0.0307$.
    d. $P(d) \approx 100(0.9698)^d$
49. The second account has a nominal rate about 3.906 times greater than that of the first account.
51. $P(t) \approx 1.853e^{0.028t}$

CHAPTER 6 ■ SECTION 6.4

1. $7^3 = 343$ 3. $10^{2.778} \approx 600$
5. $\log_4(16{,}384) = 7$ 7. $\ln(24.53) \approx 3.2$
9. 3 11. 6
13. 4 15. -2
17. 0.2 19. -1.3
21. between 4 and 5; $\dfrac{\log(20)}{\log(2)} \approx 4.322$
23. between 2 and 3; $\dfrac{\log(10)}{\log(e)} \approx 2.303$
25. $\log(m + n) \stackrel{?}{=} \log(m) + \log(n)$
    $\log(10 + 100) \stackrel{?}{=} \log(10) + \log(100)$
    $\log(110) \stackrel{?}{=} \log(10) + \log(10^2)$
    $2.04 \stackrel{?}{\approx} 1 + 2$
    $2.04 \ne 3$
27. $\log(m \cdot n) \stackrel{?}{=} [\log(m)][\log(n)]$
    $\log(10 \cdot 100) \stackrel{?}{=} [\log(10)][\log(100)]$
    $\log(1000) \stackrel{?}{=} [\log(10)][\log(10^2)]$
    $\log(10^3) \stackrel{?}{=} [\log(10)][\log(10^2)]$
    $3 \stackrel{?}{=} (1)(2)$
    $3 \ne 2$
29. $x = 4$ 31. $x \approx 2.33$
33. $x \approx 1.0986$ 35. $x = 243$
37. $x = 1000$ 39. $x = 9$
41. $x = \dfrac{\ln\left(\dfrac{y}{a}\right)}{\ln(b)}$ or $x = \dfrac{\ln(y) - \ln(a)}{\ln(b)}$
43. Consider $10^{\log(x)}$. Look first at $\log(x)$. Its value is "the exponent we place on 10 to get x." If we call this value t, then we know $t = \log(x)$, and that $10^t = x$. So
    $$10^{\log(x)} = 10^t$$
    $$= x$$
45. We use the rules of logarithms to rewrite $\log_3(400)$ in terms of $\log_3(10)$ and $\log_3(4)$ and substitute their values.

$$\log_3(400) = \log_3(100 \cdot 4)$$
$$= \log_3(100) + \log_3(4) \quad \text{Rule 3}$$
$$= \log_3(10)^2 + \log_3(4)$$
$$= 2 \cdot \log_3(10) + \log_3(4) \quad \text{Rule 5}$$
$$\approx 2(2.1) + 1.26 \quad \log_3(10) \approx 2.1, \log_3(4) \approx 1.26$$
$$\approx 4.2 + 1.26$$
$$\approx 5.46$$

47. $t(P) = \dfrac{\ln\left(\dfrac{P}{1{,}103.4}\right)}{0.0149}$. This function models the number of years since 2005 when the population of India will reach a given number (in millions).

49. $t(A) = -\dfrac{\ln\left(\dfrac{A}{500}\right)}{0.1386}$. This function models the time it takes (in hours) for the amount of medicine in the body (in milligrams) to reach a given amount.

51. $t \approx 0.57$, during 2005

53. $t \approx 46.21$ years

55. *battery acid*

$10^{-0.3}$, or about 0.5 moles per liter

orange juice

$10^{-4.3}$, or about 5.0×10^{-5} moles per liter

sea water

1.0×10^{-8} moles per liter

bleach

$10^{-12.6}$, or about 2.5×10^{-13} moles per liter

57. **a.** $x = 4$
 b. $x = 2$
 c. $x = \dfrac{5 + \sqrt{145}}{6}$

59. $t \approx 35.6$; during 1995, the annual costs reached $300 billion

CHAPTER 6 ■ SECTION 6.5

1. $\log(p) = 0.301t - 0.301$

3. $\log(p) = 0.18t + 3$

5. $\log(p) = -t + 4.301$

7.
a	Average Rate of Change
1	0.693
10	0.095
100	0.00995
1000	0.000995

9. The initial value of an exponential function will be the zero (horizontal intercept) of its logarithmic function inverse.

11. The graph is increasing, concave down with vertical asymptote at $x = 0$.

13. As $x \to \infty$ the rate of change approaches 0.

15. The instantaneous rate of change is very high near $x = 0$ and then the instantaneous rate of change approaches 0 as $x \to \infty$.

17. The data demonstrate that insurance expenditures are increasing by larger amounts as time continues, so this data could be effectively modeled as a logarithmic function.

19. Yes, a logarithmic function would be a good fit with $0 < b < 1$.

21. $m(Y) = 17.659 + 0.491 \ln(Y)$ thousand local municipal governments, where Y is years since 1965

23. $N(Y) = -11{,}392.136 + 5527.173 \ln(Y)$ black elected officials, where Y is years since 1960

25. $v(t) = 1.838 - 0.392 \ln(t)$ value of the dollar for years since 1975

27. $T(d) = 80.797 + 68.515 \ln(d)$ Park Hopper Bonus Adult Ticket cost, where d is the days spent in the park

29. $m(Y) = 17.659 + 0.491 \ln(Y)$ thousand local municipal governments, where Y is years since 1965

31. $N(Y) = -11{,}392.136 + 5527.173 \ln(Y)$ black elected officials, where Y is years since 1960

33. $v(t) = 1.838 - 0.392 \ln(t)$ value of the dollar for years since 1975

35. $T(d) = 80.797 + 68.515 \ln(d)$ Park Hopper Bonus Adult Ticket cost, where d is the days spent in the park

37. **a.**

School Year Since 1990–1991 t	Expenditure per Pupil (dollars) E	$\log(E)$
0	4902	3.6904
2	5160	3.7126
4	5529	3.7426
6	5923	3.7725
8	6508	3.8134
10	7380	3.8681
12	8044	3.9055

b. $\log(E) = 0.018t + 3.68$

c. $(1.04)^t(4786.30) \approx E$

39. **a.**

Years Since 1995 t	Health Expenditures ($ billions) H	$\log(H)$
0	1020	3.0086
1	1073	3.0306
2	1130	3.0531
3	1196	3.0777
4	1270	3.1038
5	1359	3.1332
6	1474	3.1685
7	1608	3.2063
8	1741	3.2408

b. $\log(H) = 0.030t + 2.995$
c. $(1.07)^t(988.55) \approx H$

41. a.

Years Since 2000 t	Accidents Resulting in Injuries (percent) p	$\log(p)$
0	49.9	1.6981
1	48	1.6812
2	46.3	1.6656
3	45.6	1.659
4	45.1	1.6542

b. $\log(p) = -0.011t + 1.69$
c. $(0.97)^t(48.977) \approx p$

43. Disagree. A logarithm can be transformed by shifting it left of the y-axis; for example, consider the function $f(x) = \log(x + 6)$.

45. Vertical asymptotes; for example, $f(x) = \log(x)$ has a vertical asymptote at $x = 0$.

CHAPTER 6 ■ REVIEW

1. linear; the first differences are constant.

3. quadratic; the second differences are constant.

5. exponential; the ratios are nearly constant.

7. 2008: $119,574.36; 2009: $123,819.25

9. a. annual growth factor: $\sqrt[50]{2941} \approx 1.17$; annual percentage increase: 17%
b. $25,662,152.84

11. $34,504.16

13. The graph has an initial value of 2. As x increases, y will increase at an increasing rate, and the graph will be concave up.

15. The graph has an initial value of 17. As x increases, y will decrease at a decreasing rate, and the graph will be concave up.

17. The exponential regression model is $I(t) = 106.12(0.97)^t$ and the Consumer Price Index for the year 2005 is predicted to be $49.56 for television sets.

19. a. 0.367% **b.** $3113.94
c. $122.79 per year **d.** 4.49%
e. $9634.10 \approx P$ **f.** $1.25

21. a. 0.0169% **b.** $10,155.18
c. $620.72 per year **d.** 6.35%
e. $4,923.60 \approx P$ **f.** $0.01

23. a. 1.85% **b.** 1.834%

25. 3.247%

27. a. The population is decreasing by about 0.9% annually.
b. $B(t) \approx 7.45(0.991)^t$

29. a. The population is decreasing at a continuous rate of about 0.3%.
b. $L(t) \approx 3.6e^{-0.003t}$

31. The half-life is about 11 hours. The formula $A(t) = 200(0.9389)^t$ assumes an initial dosage of 200 milligrams, so one can find the half-life by determining the time it takes for the amount of medicine to reach 100 milligrams.

33. $\log\left(\dfrac{1}{10,000}\right) = -4$

35. $e^{2.944} \approx 19$

37. -5

39. 16

41. $3 < x < 4$; $\dfrac{\log(600)}{\log(5)} \approx 3.97$

43. a. $17.71 \approx t$; about 17.71 years after the end of 2006, so during 2023
b. $t(G) = \dfrac{\ln(G) - \ln(12.18)}{0.02205}$

This function models the number of years after the end of 2006 for Guatemala's population to reach a given number of people (in millions).

45. $t(A) \approx \dfrac{\ln(A) - \ln(8000)}{\ln(1.0485)}$

This function models the amount of time (in years) it will take for the account's value to reach a given value (in dollars).

47. $t \approx 9.95$ years

49. If an object's apparent magnitude is 6, but its magnitude would be 2.3 if it was located 10 parsecs from Earth, then the object must be about 54.95 parsecs from Earth.

51. The graph of $y = \log_2(x)$ will be increasing and concave down due to the base being 2.

53. The graph of $y = 3\log_4(x)$ will be increasing and concave down due to the base being 4 and it will be stretched vertically by a factor of 3.

55. $t(i) = -166.63 + 37.17 \ln(i)$ and $t(i) \approx 17.05$. This means that when the Consumer Price Index for wine consumed at home was $140 it was the year 1997.

CHAPTER 7 ■ SECTION 7.1

1. a. $f(2) = 2$
$g(2) = 6$
$(f + g)(2) = 8$
b. $(f \cdot g)(2) = 12$

3. a. $f(-2) = -1$
$g(-2) = 0$
$(g \cdot f)(-2) = 0$

b. $\left(\dfrac{f}{g}\right)(-2) =$ undefined

5. a. $g(1) = 3$
$f(1) = -4$
$(g + f)(1) = -1$

b. $(g \cdot f)(2) = -6$

7. a. $f(4) = 5$
$g(4) = 0$
$(f + g)(4) = 5$

b. $(f - g)(4) = 5$

c. $\left(\dfrac{f}{g}\right)(4) = \dfrac{5}{0} =$ undefined

d. $(f \cdot g)(4) = 0$

9. a. $f(4) = 0$
$g(4) = -5$
$(f + g)(4) = -5$

b. $(f - g)(4) = 5$

c. $\left(\dfrac{f}{g}\right)(4) = 0$

d. $(f \cdot g)(4) = 0$

11. a. $(f + g)(x) = -x + 2$
The domain is all real numbers.

b. $(f - g)(x) = 7x - 12$
The domain is all real numbers.

13. a. $(f \cdot g)(x) = x^2 + 2x - 3$
The domain is all real numbers.

b. $\left(\dfrac{g}{f}\right)(x) = \dfrac{x + 3}{x - 1}$
The domain is all real numbers except for $x = 1$.

15. a. $(g - f)(x) = -x^2 + 2x + 12$
The domain is all real numbers.

b. $\left(\dfrac{f}{g}\right)(x) = \dfrac{x^2 - 4}{2x + 8}$
The domain is all real numbers except for $x = -4$.

17. a. $(g \cdot f)(x) = x \cdot 2^x$
The domain is all real numbers.

b. $\left(\dfrac{g}{f}\right)(x) = \dfrac{x}{2^x}$
The domain is all real numbers.

19. a. $(f + g)(x) = 2x^3 + 3x + 5$
The domain is all real numbers.

b. $(g \cdot f)(x) = -2x^5 + 13x^4 + 7x^3 - 37x^2 + 57x - 14$
The domain is all real numbers.

21. a. In Month 6 Store 2 historically has 700 customers per day at that store.

b. $(f + g)(m)$ represents the total number of customers per day at both Stores 1 and 2.

c. A rough sketch using only integer years as estimates of output values will give approximately this graph.

23.

25.

27.

29.

31.

33. $h(a) = f(a) + f(a) = 0 + b = b$

35. $h(x)$ will have vertical asymptotes for all values of x such that $g(x) = 0$.

37. We note that the ratio of units will be
$\dfrac{\text{million \$}}{\text{\$/hr}} = \dfrac{\text{million \$/1}}{\text{\$/hr}} = \dfrac{\text{million \$}}{1} \times \dfrac{\text{hr}}{\$} = \dfrac{\text{million hr}}{1}$.
Therefore, the resulting units will be millions of hours.

39. $j(x)$ will be concave down and will be below the x-axis.

41. In multiplying f and g, the units are also multiplied:
$\dfrac{\text{dollars}}{\text{day}} \times \dfrac{\text{days}}{\text{week}} = \dfrac{\text{dollars}}{\text{week}}$, or dollars per week.

43. a. $R(S) = 1.1S$
  b. $B(S) = 0.056S^2 - 0.001S + 0.021$
  c. $(R + B)(S) = 0.056S^2 + 1.099S + 0.021$

  The function $(R + B)(S)$ gives the total stopping distance including both the reaction distance and the braking distance.

  d. $(R + B)(75) \approx 397.4$; A car traveling 75 mph will need 397.4 feet to come to a complete stop once the brakes are applied.

45. a. $\left(\dfrac{V}{P}\right)(t) = \dfrac{3.383t + 154.8}{2.821t + 225.5}$ represents the percentage of the U.S. population that has a registered vehicle.

  b. $t \approx 126$; In 126 years from 1980 (which would be the year 2106) every person in the United States would own a registered vehicle.

47. a. $(S \cdot P)(t) = (-0.22t + 53.18)(2870.9t + 282{,}426)$ and represents the total number of 1000s of gallons of soda consumed in the United States in year t. Note: $P(t)$ was in 1000s of gallons.

  b. The function is increasing as t increases.

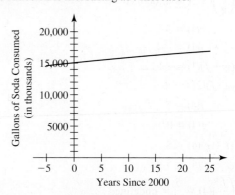

  c. $(S \cdot P)(10) = 15{,}861{,}662$ thousand gallons

  According to this model, in the year 2010 there were 15,861,662,000 gallons of soda consumed in the United States.

49. a. $A(18) = 23{,}103.4$; In 2008 there were 23,103,400 fans attending NFL games.

  b. $S(18) = 951.1048$; In 2008 the average salary of an NFL player was $951,104.

  c. $\left(\dfrac{S}{A}\right)(t) = \dfrac{-0.1812t^4 + 5.75t^3 - 54.05t^2 + 199.6t + 311.5}{323.3t + 17{,}284}$ and represents the amount of money in dollars that each fan in attendance paid toward the average salary of each player in year t.

  d. Answers vary. The practical domain could be the interval $0 \leq t \leq 15$ (after year 15 the per fan amount of the salary begins to drop, which would seem unlikely) and the practical range could be $0 \leq \dfrac{S}{A} \leq 0.07$.

51. It is impossible to tell if $(f + g)(x)$ is always increasing or always decreasing because it depends on the difference between the two functions' output values, their relative rates of change, and if the function values are positive or negative.

53. It is impossible to tell if $(f \cdot g)(x)$ is always increasing or always decreasing because it depends on the difference between the two functions' output values, their relative rates of change, and if the function values are positive or negative.

55. It is impossible to tell if $f^{-1}(x) + g^{-1}(x)$ is always increasing or always decreasing because it depends on the difference between the two functions' output values, their relative rates of change, and if the function values are positive or negative.

CHAPTER 7 ■ SECTION 7.2

1.

3.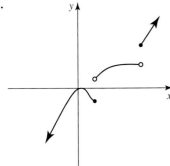

5. $f(x) = \begin{cases} 5 - x & \text{if } 0 \leq x < 3 \\ 6.5 - \dfrac{1}{2}x & \text{if } 3 \leq x \leq 9 \end{cases}$

7. $f(x) = \begin{cases} 1 + \dfrac{5}{2}x & \text{if } 0 \leq x \leq 3 \\ 13.5 - \dfrac{3}{2}x & \text{if } 3 < x \leq 9 \end{cases}$

9. $f(x) = \begin{cases} 2^x & \text{if } 2 \leq x \leq 4 \quad \text{exponential} \\ x + 14 & \text{if } 4 \leq x < 9 \quad \text{linear} \\ 0.5x^2 - 13.5x + 102 & \text{if } 9 \leq x \leq 14 \quad \text{quadratic} \end{cases}$

11. a. $P(b) = \begin{cases} 30{,}000 & \text{if } c \leq 20{,}000 \\ 30{,}000 + 0.13(35(b - 20{,}000)) & \text{if } 20{,}000 < b \leq 27{,}500 \\ 30{,}000 + 34{,}125 \\ \quad + 0.22(35(b - 27{,}500)) & \text{if } b > 27{,}500 \end{cases}$

b.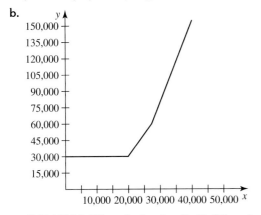

c. $180,056.20; When the book sells 42,556 copies the total amount the author will be paid is $180,056.20.

d. $b = 22{,}759$; The author will be paid $42,556 if 22,759 books are sold.

13. We consider what type of function might best model each "piece" of the overall function over different intervals of the domain.

15. Continuous: For every value of x over an interval there is an output value and for input values "nearby" the output values are also "near" each other so there are no sudden "jumps" or changes in output values. (A more formal definition is given in calculus.) A function is discontinuous if there is a hole, jump, or break.

17. The advantage is that a piecewise model most likely will more closely model the data over specified intervals. The disadvantage is that piecewise functions are generally more complicated than simple functions.

19. a. $c(w) = \begin{cases} 0.39 & \text{if } 0 < w \leq 1 \\ 0.63 & \text{if } 1 < w \leq 2 \\ 0.87 & \text{if } 2 < w \leq 3 \\ 1.11 & \text{if } 3 < w \leq 4 \\ 1.35 & \text{if } 4 < w \leq 5 \\ 1.59 & \text{if } 5 < w \leq 6 \\ 1.83 & \text{if } 6 < w \leq 7 \\ 2.07 & \text{if } 7 < w \leq 8 \\ 2.31 & \text{if } 8 < w \leq 9 \\ 2.55 & \text{if } 9 < w \leq 10 \\ 2.79 & \text{if } 10 < w \leq 11 \\ 3.03 & \text{if } 11 < w \leq 12 \\ 3.27 & \text{if } 12 < w \leq 13 \\ 3.27 + 0.24(w - 13) & \text{if } w > 13 \end{cases}$

b. $c(7) = \$1.83$; If you are sending a letter that weighs 7 ounces, it will cost $1.83.

c. $c(w) = 3.99$, $w = 16$ ounces; If the cost of sending a package is $3.99, then the package weighed 16 ounces.

21. a. $E(k) = \begin{cases} 0.058k & \text{if } 0 < k \leq 500 \\ 0.10k & \text{if } k > 500 \end{cases}$

b. $E(450) = \$26.10$

c. $E(k) = \begin{cases} 0.058k & \text{if } 0 < k \leq 500 \\ 0.083k & \text{if } k > 500 \end{cases}$

d. $E(450) = \$26.10$

23. a. $T(i) = \begin{cases} 0.10i & \text{if } 0 \leq i \leq 7550 \\ 755 \\ \quad + 0.15(i - 7550) & \text{if } 7550 < i \leq 30{,}650 \\ 4220 \\ \quad + 0.25(i - 30{,}650) & \text{if } 30{,}650 < i \leq 74{,}200 \\ 15{,}107.50 \\ \quad + 0.28(i - 74{,}200) & \text{if } 74{,}200 < i \leq 154{,}800 \\ 37{,}675.50 \\ \quad + 0.33(i - 154{,}800) & \text{if } 154{,}800 < i \leq 336{,}550 \\ 97{,}653 \\ \quad + 0.35(i - 336{,}550) & \text{if } i > 336{,}550 \end{cases}$

b. 240; 4077.50; 11,382.50; 30,451.50; 100,610.50

c. Answers vary.

25. a. $W(a) = \begin{cases} 0 & \text{if } a < 2 \\ 3.99 & \text{if } 3 \leq a \leq 6 \\ 4.99 & \text{if } 7 \leq a \leq 12 \\ 6.99 & \text{if } 13 \leq a \leq 64 \\ 5.99 & \text{if } a \geq 65 \end{cases}$

b. $W(3) = 3.99$
$W(6) = 3.99$
$W(9) = 4.99$
$W(24) = 6.99$

c. Evaluating $W(2)$ is impossible since the sign tells us that "children *under* 2" are free. What about children that *are* 2? (We assume that they meant to say that children 2 and under are free.)

d. $6.99 + $6.99 + $0 + $3.99 + $3.99 + $4.99 + $4.99 = $34.94

27. a.

Length L	Cost C
0	0
0.2	2.5
0.4	2.9
0.6	3.3
0.8	3.7
1.0	4.1

b. $22.10

c. A person can ride for 0.45 miles for $3.00.

d.

29. a. $2160

b.

c. $c(a) = \begin{cases} 31 & \text{if } 0 < a \leq 12 \\ 43 & \text{if } 13 \leq a \leq 18 \\ 59 & \text{if } 19 \leq a \leq 61 \\ 43 & \text{if } a \geq 62 \end{cases}$

31. a.

t	C
0	8.50
5	14.50
10	20.50
15	26.50
20	28.90
25	34.00
30	39.10
35	44.20
40	49.30
45	54.40
50	59.50

b. $C(t) = \begin{cases} 8.50 + 1.20t & \text{if } 0 \leq t \leq 15 \\ 8.50 + 1.02t & \text{if } t > 15 \end{cases}$

c.

d. $86.02

33. Packages may be anticipated to be larger at that cost and therefore the company anticipates greater charges being necessary.

35.

37.

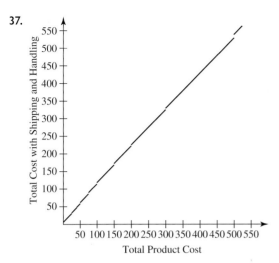

39. The graphs of S(t) and B(t) are both piecewise discontinuous functions. There are jump discontinuities in each. S(t) is a constant function over each specified interval; B(t) has a constant rate of change over each specified interval.

CHAPTER 7 ■ SECTION 7.3

1. $f(g(t)) = 3(5 - 4t)^2 - (5 - 4t) + 4$
$f(g(2)) = 34$

3. $f(g(t)) = \dfrac{5}{1 + 4(2^{-0.4t})}$
$f(g(2)) \approx 1.52$

5. $f(g(t)) = 5t^2 - 15t + 26$
$f(g(2)) = 16$

7. $f(g(t)) = 6(t^2 - 2t - 6) + 3$
$f(g(2)) = -33$

9. $f(g(t)) = (t + 1)^3 - 4(t + 1)^2 + 2(t + 1) - 3$
$f(g(2)) = -6$

11. $h(x) = (2x - 2)^5$
$g(x) = 2x - 2$
$f(g) = g^5$

13. $h(x) = \dfrac{1}{(x - 2)^6}$
$g(x) = x - 2$
$f(g) = \dfrac{1}{g^6}$

15. $h(x) = (\sqrt{x} - 3)^4$
$g(x) = \sqrt{x} - 3$
$f(g) = g^4$

17. $h(x) = (x - 1)^\heartsuit$
$g(x) = x - 1$
$f(g) = g^\heartsuit$

19. $h(x) = \left(\dfrac{4}{x \diamond 5}\right)^\copyright$
$g(x) = \dfrac{4}{x \diamond 5}$
$f(g) = (g)^\copyright$

21. a. ≈ 2.8
b. ≈ -2.4
c. ≈ 0.2
d. ≈ -2.8

23.

x	k(n(x))
0	4
1	3
2	undefined
3	2
4	5
5	9

25. a. $f(g(x)) = \left(\dfrac{2}{x + 3}\right)^2 - 2$
b. $g(f(x)) = \dfrac{2}{(x^2 - 2) + 3}$
c. $f(h(x)) = x - 2$
d. $h(f(x)) = \sqrt{x^2 - 2}$
e. $h(h(x)) = \sqrt{\sqrt{x}}$
f. $g(f(h(x))) = g(f(\sqrt{x})) = \dfrac{2}{x + 1}$

27. $f(f(3)) = 45$

29. A

31. a. $p(f(w)) = 14.65 + 4.4w$; The input of the function $p(f(w))$ is the amount of pollution found in the river, w, and the output the price of a scallop dinner, p.
b. $p(f(0)) = 14.65$
c. $p(0) = 39.95$
d. When 1.56 grams per liter of pollutants are found in the ocean, the price is $21.50.

33. Function composition occurs when the output of one function is used as an input to another function.

35. $D(p(l))$; With an input of l computers, the output is the revenue in dollars for l computers.

37. $h(r(y))$; input: year; output: number of new homes

39. This composition is not reasonable because the output of neither function can be the input to the other function.

41. a. $K(C(F)) = \dfrac{5}{9}(F - 32) + 273$
b. Input: Fahrenheit Output: kelvin
c. A temperature of 81 degrees Fahrenheit is equal to 300.2° kelvin.
d. A kelvin temperature of 572° is equal to 570.2 degrees Fahrenheit.

43. $G(P(y))$

45. a. $J(A(K(E(U)))) = U + 3$; if we know the dress size, U, in the United States, we can determine the dress size, J, in Japan by adding 3 to the dress size in the United States.
b. $J(A(K(E(10)))) = 13$; a dress size of 10 in the United States would be a dress size of 13 in Japan.
c. $U = 14$; a dress size of 17 in Japan is equivalent to a dress size of 14 in the United States.

47. a. $w(a(h)) = -0.034(38.05 - \sqrt{-24.39h + 2137})^3 + 1.245(38.05 - \sqrt{-24.39h + 2137})^2 - 5.077(38.05 - \sqrt{-24.39h + 2137}) + 34.346$

b. The independent variable is h (height) and the dependent variable is w (weight). The meaning of $w(a(h))$ is the average weight (in pounds) as a function of the height (in inches).

c. $w(a(60)) \approx 94.42$; For a healthy child that is 60 inches tall the weight would be 94.92 pounds.

49. a. $v(t) = 18{,}275.89 + 3700.89t$; The retail value of a clean 2000 Cadillac Escalade with 40,000 miles in 2007 is estimated to be \$18,275.89 and the value increases by \$3700.89 each year if the retail value is assumed to increase at a constant rate each year.

$$v(7) = \$44{,}182.12$$

b.

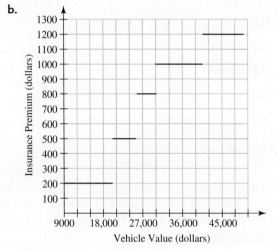

c. If we take the function $p(v)$ with the function $v(t)$ we would get the composition function $p(v(t))$. This is a composition function because we take the output value of the first function, "retail value," and make it the input value to the next function, which produces the output of insurance premium.

d. The insurance premium charged for a 2003 Cadillac Escalade is \$800. In function notation this is written: $p(v(3)) = \$800.00$.

e. If you could afford insurance premiums of \$800 every 6 months you could buy either a 2002 or 2003 Cadillac Escalade (or earlier models if you wanted to pay less than \$800). In function notation this would be written as $p(v(2)) = \$800.00$ or $p(v(3)) = \$800.00$.

f. $v(5) = \$36{,}525$ and $p(36{,}525) = \$1000.00$

51. a. About 990 watts of usable electricity can be created from a solar panel with 8 square meters of surface area.

b. a solar panel of about 8 square meters of surface area for 950 watts of usable electricity created

c.

Solar Panel Surface Area (square meters) A	Electricity Produced (watts) p
1	100
3	370
5	600
7	840
9	1100
11	1300

CHAPTER 7 ■ SECTION 7.4

1. Answers vary.

x	y
0	5
1	6.2
2	7.3
3	8.2
4	8.8
5	9.2
6	9.5
7	9.7
8	9.8
9	9.9
10	9.9

3. The limiting value is $P \approx 10$, which means that profits are leveling off as the years increase toward year 4. The inflection point is approximately (2, 6). The profits are beginning to increase at a decreasing rate at year 2 (2009). The company may want to release the upgraded product in approximately year 3 because profits are beginning to level off then.

5. 12

7. exponential; The scatter plot is increasing at an increasing rate.

9. logistic; It is decreasing at an increasing rate then decreasing at a decreasing rate.

11. logistic; It is increasing at an increasing rate then increasing at a decreasing rate.

13. a.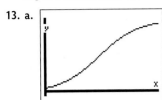

b. domain: (0, 130); range: (0, 96); Even with an unlimited number of growing days, the corn plant will not get taller than 96 inches.

c. (60, 45); After 60 days of growing, the corn plant is 45 inches tall and will continue to grow but at a slower pace.

d. 96; Even if the number of days since the corn broke the soil increased past 130, the corn wouldn't grow any taller than 96 inches.

15. Since this is an example of exponential growth (each investor recruits two more investors) the model is unsustainable because very quickly the number of required investors grows enormous. One would run out of people to recruit, causing the whole model to crumble.

17. The best way to graph a logistic function given a table is to create a scatter plot and then find the logistic regression model and graph it using graphing technology. For a logistic function given as an equation it is best to use graphing technology as well.

19. It is important to remember that logistic functions will have two types of concavity, a point of inflection, and a limiting value.

21. a. The graph shows that the number of people living r kilometers from the center of the city reaches a limiting value of 1.796 million people. We can say that the approximate number of people living in Sydney is therefore 1.796 million.

b. Answers vary. $r \approx 24$ kilometers

c. $r \approx 5$ kilometers

23. a. Initially, the rate of change is relatively small, which means that the news is spreading relatively slowly. As time increases, the rate of change increases. This means that the news is spreading at a faster rate. The rate then decreases and eventually gets close to zero. This tells us that the news spreads at a slower and slower rate until it nearly stops spreading.

b. The inflection point occurs at approximately (15, 53.8). This means that 15 days after the news began about 54 people have heard the news.

c. $\dfrac{P(25) - P(5)}{25 - 5} \approx 6.5$; Between the 5th and 25th days, the rate at which the news spread was as if it spread at a constant rate of 6.5 people per day.

d. The limiting value appears to be 145 people. This means that there are 145 people in the population who can potentially hear the news. No more than 145 people can hear the news.

25. a. Using the practical domain of 10 years:

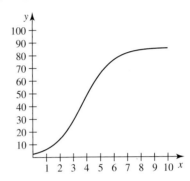

b. Yes. There will be some point when the percentage of schools with Internet access will become stable.

c. approximately 8% per year

27. a. From 1997 to 2000 the price of VCRs decreased at an increasing rate as the sales of DVD players increased at an increasing rate. VCRs were becoming cheaper because fewer people were buying them because they were buying DVD players instead.

b. VCR price decreases at an increasing rate as DVD players become more and more popular and then it decreases at a decreasing rate until it reaches a limiting value of about $60 because the price will never be zero. Sales of DVD players increase at an increasing rate when they first come out and then increase at a decreasing rate as the market becomes saturated and then reaches a limiting value of about 35,000,000 sales.

29. a.

Years Since 1975 t	Number of Basic Cable Subscribers (thousands) C	Average Rate of Change
0	9,800	1540
5	17,500	3588
10	35,440	3016
15	50,520	2006
20	60,550	1140
25	66,250	482
26	66,732	−260
27	66,472	−422
28	66,050	−323
29	65,727	−390
30	65,337	

The point of inflection will occur in a year between 5 and 10. This is when the increase in the number of basic cable subscribers began to decrease.

b. $\dfrac{66472 - 66732}{1} = -260$

c. The instantaneous rate of change means that in 2002 the number of homes with basic cable was decreasing by 260,000 per year.

31. It is hard to tell which to choose but knowing that there cannot be more than 100% of the adults using the Internet we know that there has to be a limiting value. So a logistic function would most likely be the best model.

33. approximately 0.3% per year in 1995 and approximately 0.22% per year in 2000

35.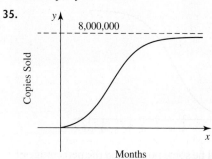

37. 2005; $\dfrac{-0.451\%}{\text{year}}$

39. The point of inflection is approximately (3.39, 215,932.63) and means that after the year 1995 the value of fabricated metal shipments was increasing at a decreasing rate. The limiting value is 260,721.10 and represents the value that the shipments will approach into the future.

41. a.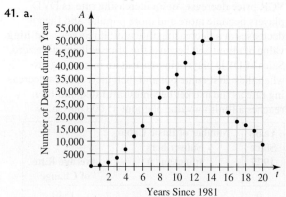

b. $d(t) = \dfrac{53{,}955.02}{1 + 38.8339e^{-0.4513t}}$

c. $c(t) = -810.33t^3 + 7118.18t^2 - 21{,}058.85t + 37{,}342.64$

d. From 1981 to 1995 the number of deaths due to AIDS was increasing at an increasing rate until about year 1989 and then continued to increase but at a decreasing rate until year 1995.

e. From 1995 to 1997 the number of deaths due to AIDS decreased at a decreasing rate and then from 1997 to 2001 the number of deaths due to AIDS decreased at an increasing rate.

f. $A(t) = \begin{cases} \dfrac{53{,}955.02}{1 + 38.8339e^{-0.4513t}} & \text{if } 0 \le t \le 14 \\ -810.33t^3 + 7118.18t^2 \\ \quad - 21{,}058.85t + 37{,}342.64 & \text{if } t > 14 \end{cases}$

g. $A(32) = -90{,}455$ and $A(35) = -4{,}770{,}273$; No, these are not reasonable because there cannot be negative deaths. The practical domain for $A(t)$ does not go beyond the year 5.

CHAPTER 7 ■ SECTION 7.5

1. decreasing over the interval $(-\infty, \infty)$; concave up; y-intercept: (0, 2)

3. decreasing on $\left(-\infty, \dfrac{1}{3}\right)$ and increasing on $\left(\dfrac{1}{3}, \infty\right)$; concave up on $(-\infty, \infty)$; y-intercept: (0, 5)

5. decreasing over the interval $(-\infty, \infty)$; concave down approximately from $(-\infty, -50)$ and concave up approximately from $(-50, \infty)$; y-intercept: (0, 5.625)

7. decreasing and concave down over the interval $(-\infty, \infty)$; y-intercept: (0, −0.3)

9. $c(t) = 0.60(1.03)^t$

11. $M(t) = 95(0.96)^t$

13. $m(t) = -10t^2 - 25t + 200$

15. Linear; for every increase in x of 2, the first differences increase by 4.

x	y	First Differences
0	0	
2	4	4
4	8	4
6	12	4
8	16	4
10	20	4
12	24	4
14	28	4

Quadratic; for every increase in x of 2, the first differences increase by 8.

x	y	First Differences	Second Differences
0	0		
2	4	4	
4	16	12	8
6	36	20	8
8	64	28	8
10	100	36	8
12	144	44	8
14	196	52	8

Cubic; for every increase in x of 2, the second differences increase by 48.

x	y	First Differences	Second Differences	Third Differences
0	0			
2	−16	−16		
4	8	24	40	
6	120	112	88	48
8	368	248	136	48
10	800	432	184	48
12	1464	664	232	48
14	2408	944	280	48

Exponential; for every increase in x of 1, the consecutive ratio is 2.

x	y	Consecutive Ratio
0	1	
1	2	2
2	4	2
3	8	2
4	16	2
5	32	2
6	64	2
7	128	2

17. A function is "best" if it represents the data set well by passing through or near the points in the scatter plot.

19. It doesn't really matter because both models will give accurate predictions. We generally choose the function of the lesser degree (linear before quadratic, quadratic before cubic, etc.) because it is easier to interpret the real-world meaning from "simpler" models.

21. $p(d) = 144.05 + 2.78(\ln d)$

23. During the *lag phase*, bacteria begin to adapt themselves to the environmental conditions. It is the period when the individual bacteria are maturing and not yet able to divide. The rate of change during the lag phase is 0 as no increase in the number of viable bacteria occurs.

    The *exponential phase* is a period characterized by cells doubling at a constant rate. If growth is not limited, doubling will continue at a constant rate so both the number of cells and the rate of population increase doubles with each consecutive time period.

    During the *stationary phase*, the growth rate slows to zero as a result of the depletion of nutrients and the accumulation of waste products. During this phase, the rate of bacterial growth is equal to the rate of bacterial death.

    At the *death phase*, the viable cell population declines as bacteria run out of nutrients and die. During the death phase, the number of viable cells decreases exponentially, essentially reversing the growth during the exponential phase.

25. Logistic; the rate of change indicates that as the years increase the number of hospitals decreases at an increasing rate and then decreases at a decreasing rate.

27. We choose a linear model using the final two data points.
    $B(t) = -31t + 1121$
    $B(12) = 749$

29. The number of DVD players sold is decreasing as the price increases and most likely the number of DVD players sold will not begin to increase as the price of a DVD player increases. This means that exponential is probably the best choice for a mathematical model.
    $D(p) = 77{,}113.769(0.989^p)$
    $D(100) = 25{,}512{,}974$

31. Although the function values (tuition and fees) increase at a varying rate, they don't vary much. The increase overall is relatively constant so a linear function can be used to model the data.
    $F(t) = 69.610t + 534.08$
    $F(24) = \$2204.72$

33. This function is linear because the per capita personal income appears to be increasing at close to a constant rate.
    $F(t) = 942.44t + 21{,}134.33$
    $F(11) \approx \$31{,}501$

35. Quadratic because the data is increasing at an increasing rate and it fits better than an exponential function does.
    $s(f) = 0.05f^2 + 54.89f + 120{,}026.48$
    $s(3000) \approx 734{,}696$

37. $M(t) = 42.29t^2 + 158.32t + 6574.52$
    $M(12) \approx 14{,}564$

39. in year 10 or 2010

41. $Y(t) = 34.99t^2 + 94.91t + 8963$; When $t \approx 12$, advertising expenditures are about $15 billion.

43. $F(t) = 0.00087t^3 - 0.0238t^2 + 0.438t + 6.1098$
    $F(23) \approx \$14.20$

45. a.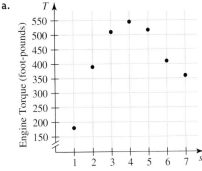

b. A quadratic function is most appropriate since the function increases to a maximum and then decreases. There is a changing rate of change so a linear function is not appropriate.

c. $T(s) = -30.49s^2 + 264.87s - 33.71$

d. $T(5.5) \approx 500$; An engine speed of 5500 rpm will produce engine torque of about 500 foot-pounds.

e. There are two engine speeds that produce a torque of 100 foot-pounds. From the graph, we estimate them to be $s = 0.5$ and $s = 8.1$. That is, an engine speed of

500 rpm and an engine speed of 8100 rpm both produce an engine torque of 100 foot-pounds.

f. (4.343, 541.53); This vertex tells us the engine speed that produces the maximum engine torque. This occurs at an engine speed of 4343 rpm, which produces a maximum engine torque of about 541 foot-pounds.

g. The average rate of change on the interval [2, 4] is 81.9 foot-pounds per rpm. That is, for every increase in engine speed of 1000 rpm, the engine torque will increase by 81.9 foot-pounds.

h. $\dfrac{T(4.001) - T(4)}{4.001 - 4} = 21$; We estimate that when the engine speed is 4000 rpm, the engine torque is changing at a rate of 21 foot-pounds per rpm. That is, when engine speed is 4000 rpm, an increase of 1 (in thousands) rpm will cause an increase in engine torque of 21 foot-pounds.

47. Two different models that may fit the data would be a piecewise function (quadratic, quadratic, and linear) and perhaps a quartic function although it would not be as accurate as the piecewise function.

49. A linear model would balance out the highs and lows of the market to help minimize the risk and the temptation to sell when the market dips and hold on too long when it is rising.

51. a. $BP(w) = \dfrac{2}{5}w + 49$

b. The slope indicates that if the person's blood pressure increases by 2 millimeters of mercury, his weight increases by 5 pounds.

c. The vertical intercept is (0, 49). In the context of the situation, this says that a person who weighs 0 pounds would have a systolic blood pressure of 49 millimeters of mercury. This vertical intercept is not part of the practical domain since there is no such thing as a 35-year-old male who weighs 0 pounds.

53. a. After considering the first and second differences as well as the ratios, we can see that none of the three functions fits really well, as was noted in the exercise directions. However, as we consider the scatter plot we can narrow our decision down to either quadratic or exponential because the data is increasing at an increasing rate. When we consider extrapolating into the future it would seem most reasonable to choose a quadratic function because an exponential model allows the population to grow too quickly. (In fact a logistic model may be the best choice but that is not an option provided in this exercise.)

b. $m(y) = 453.955y^2 - 18{,}463.616y + 159{,}751.217$

c. $m(150) \approx 7{,}604{,}196$ is the estimated population in the year 2050.

d. $m(115) \approx 4{,}000{,}000$; In the year 2015 the estimated population would be 4,000,000.

e. $\dfrac{m(88) - m(52)}{88 - 52} \approx 45{,}090.08$; If the population of Maricopa County had increased at the same rate from 1952 to 1988 it would have increased by 45,090 people each year.

f. $\dfrac{m(95) - m(94)}{95 - 94} \approx \dfrac{67{,}333.9}{1}$; In 1995 the population of Maricopa County is estimated to have increased at the rate of 67,334 people per year.

g. Answers vary. Predicting about 10 to 15 years into the future would probably be the maximum due to changing economic conditions. The growth rate will most likely slow.

CHAPTER 7 ■ REVIEW

1. a. $(f + g)(0) = -4$
b. $(f - g)(0) = -4$
c. $\left(\dfrac{f}{g}\right)(0) =$ undefined
d. $(f \cdot g)(0) = 0$

3. a. $(f + g)(x) = -x - 1$;
D (domain): all real numbers
b. $(f - g)(x) = -3x + 7$
D: all real numbers

5. a. $(f \cdot g)(x) = (e^x - 1)(2x)$
D: all real numbers
b. $\left(\dfrac{g}{f}\right)(x) = \dfrac{2x}{e^x - 1}$
D: all real numbers

7.

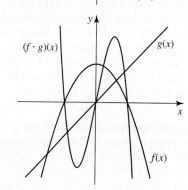

9. a. It represents how much was spent on cheddar cheese in dollars per person in years since 2000.
  b. $(C \cdot P)(10) = 500.004$; In 2010 each person in the United States spent $500 on cheddar cheese.
  c. $50 per person was spent on cheddar cheese in the United States in the year 2005.

11.

13. $f(x) = \begin{cases} 4 + \dfrac{5}{6}x & \text{if } 0 \le x \le 6 \\ 72 - 8x & \text{if } 8 \le x \le 9 \end{cases}$

15. a.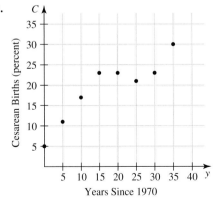

  b. $C(y) = \begin{cases} -0.05y^2 + 1.84y + 4.14 & \text{if } 0 \le y \le 25 \\ 8.36(1.036)^y & \text{if } 25 < y \le 35 \end{cases}$

  c. $\dfrac{C(15) - C(0)}{15 - 0} = 1.09$; From 1970 to 1985, if the percentage of babies delivered by Cesarean birth increased at a constant rate they would have increased by 1.09% per year.

  d. $C(17) \approx 20.97$; In 1987 an estimated 20.97% of babies were delivered by Cesarean birth.
  e. $\dfrac{C(17) - C(16.9)}{17 - 16.9} \approx 0.1$; In 1987 the percentage of babies delivered by Cesarean birth was increasing at the rate of 0.1% per year.
  f. $C(35) \approx 28.8$; In 2005 an estimated 28.8% of babies were delivered by Cesarean birth.

17. a.

  b. A final semester grade of B is earned with an overall average exam grade of $80 \le p < 90$.

19. a. Child Number 3, 4, and 5 will be able to go on the ride without a chaperone.
  b. It is hard to tell what is meant by the phrase "34 inches tall and 6 years of age and 48 inches tall." With the "6 years of age" being placed between 34 inches and 48 inches it is very confusing. Does this mean that you have to be 34 inches tall, 6 years of age, and 48 inches tall at the same time?
  c. $A(a, h) = \begin{cases} \text{no admittance} & \text{if } h < 34 \text{ and } a < 6 \\ \text{admitted with a chaperone} & \text{if } 34 \le h \le 48 \text{ and } a \ge 6 \\ \text{admitted} & \text{if } h \ge 48 \text{ and } a \ge 6 \end{cases}$
  d. All children 6 years of age and older and between 34 inches tall and 48 inches tall must be accompanied by a responsible chaperone.

21. $f(g(t)) = 5e^{t^2 + 4}$
  $f(g(2)) \approx 14{,}904.8$

23. $f(g(t)) = \sqrt{2t + 3}$
  $f(g(2)) = \sqrt{7}$

25. $f(g) = e^g$
  $g(x) = x^2 - 2$

27. $f(g) = |g|$
  $g(x) = -x^3 + 4$

29. $f(g) = \dfrac{2}{g}$
  $g(x) = 1 + \sqrt{x}$

31. $g(g(4)) = 77$

33. $S(m(t)) = -0.000917(-30.36t^2 + 153.12t + 1407.29)^3 + 4.21(-30.36t^2 + 153.12t + 1407.29)^2 - 6440(-30.36t^2 + 153.12t + 1407.29) + 3{,}320{,}000$
  $S(m(9)) \approx 1{,}635{,}374$
  In 2009, the estimated number of movie screens is 1,635,374.

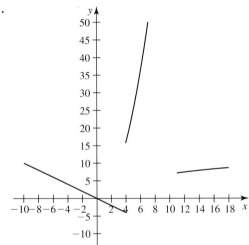

35. a. If we let C = monthly cell phone cost and m = weekday, daytime minutes used, then

$$C(m) = \begin{cases} 79.99 & \text{if } 0 \leq m \leq 1350 \\ 79.99 + 0.35(m - 1350) & \text{if } m > 1350 \end{cases}.$$

b.

Minutes Used	Monthly Cost
0	$79.99
1000	$79.99
1500	$132.49
2162	$364.19

c. If we let m = weekday, daytime minutes used and d = number of days he travels for business in the month, then $m(d) = 1300 + 100d$.

d. $C(m(d)) = 62.49 + 35d$

e. $C(m(5)) = \$237.49$

f. The function $m(d)$ has an output of the number of minutes used, which is the input to the function $c(m)$.

37. Answers vary. We need to create a table that decreases at an increasing rate and then decreases at a decreasing rate and approaches 38.

x	y
0	100
1	95
2	80
3	55
4	43
5	40
6	39
7	38.2
8	38.1

39. logistic

41.

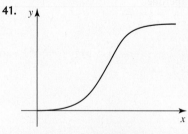

43. a. $V(m) = \dfrac{2468.62}{1 + 299.11e^{-0.871m}}$

b. The limiting value is 2,468,620 and represents the total number of visitors by the end of the year at the amusement park.

c. June; In June the rate of visitors going to the amusement park is the greatest. Yes, this is midsummer when families take vacations, and before the weather is too hot.

45. a. $W(16) \approx 132$; An average 16-year-old male would weigh approximately 132 pounds.

b. $t \approx 17$; An average male weighing 140 pounds would be about 17 years old.

c. (12, 20); From age 12 to 20 an average male continues to increase his weight but at a decreasing rate.

d. (12, 90); At age 12 when the average male weighs 90 pounds his weight is increasing at the fastest rate it ever will.

47. Answers vary. "Company Profits Decrease But Are Leveling Off"

49. A quadratic function is the best choice because the data are increasing and concave up and an exponential model would grow too quickly.

$E(t) = 1788.25t^2 + 1355.5t + 48{,}705$

$E(34) \approx \$2{,}162{,}009$

51. A quadratic function is the best choice because the data are increasing and concave up and an exponential model would grow too quickly.

$E(t) = 6870t^2 - 43{,}820t + 130{,}800$

$E(34) \approx \$6{,}582{,}640$

53. A quadratic function is the best choice because the data are increasing and concave up and an exponential model would grow too quickly.

$E(t) = 1609.65t^2 - 18{,}256.7t + 49{,}163.55$

$E(34) \approx \$1{,}289{,}188$

55. A logistic function is the best choice because the data are increasing at an increasing rate and then increasing at a decreasing rate.

$E(t) = \dfrac{263{,}327.604}{1 + 63e^{-0.303t}}$

$E(34) \approx \$262{,}772$

57. A linear function is the best choice because the data are increasing at nearly a constant rate.

$E(t) = 1961.36t + 21{,}635.1$

$E(34) \approx \$88{,}321$

59. If we let 1970 be year $t = 0$ and not 1976 then a logarithmic function is the best choice because the data are increasing and concave down.

$E(t) = -24{,}213.26 + 27{,}085.57 \ln(t)$

$E(34) \approx \$71{,}300$

CHAPTER 8 ■ SECTION 8.1

1. no **3.** no
5. yes **7.** yes
9. 2 units **11.** 4 units
13. 3 units **15.** 2 units

17.

19.

x	y
0	10
1	14
2	2
3	8
4	10
5	14
6	2
7	8
8	10
9	14
10	2
11	8

21.

23.

x	g(x)	g(x + 2)
0	3	5
1	4	3
2	5	4
3	3	5
4	4	3
5	5	undefined
6	3	undefined

25. C

27. D

29. a. $12 \frac{\text{customers}}{\text{hour}}$; From 6 A.M. to 11 A.M. the number of customers changed at a rate of 12 per hour.

b. $40 \frac{\text{customers}}{\text{hour}}$; At noon the number of customers changed at a rate of 40 per hour.

c. At 8 A.M., noon, and 6 P.M. The number of customers increases and then decreases right before and after each of these times. These times would make sense as they are typically when people eat breakfast, lunch, and dinner.

d. Yes, over a week of time the number of customers would vary in a similar fashion.

31. The period of a periodic function describes the interval of input values over which one period occurs and the frequency of a periodic function describes how many periods occur over one unit.

33. Answers vary. The time of day a rooster begins to crow throughout the year.

35. a.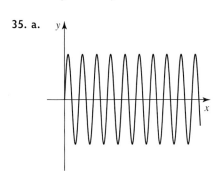

b. The period is 1/60 of a second, the amplitude is 120 volts, and the midline of the graph is at $v = 0$. The voltage will complete one cycle from −120 volts to +120 volts in 1/60 or 0.0166 seconds.

37. a.

b.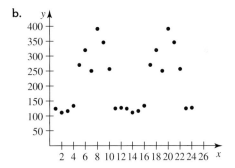

c. It would increase the monthly costs and thereby shift all of the output values upward.

39. a.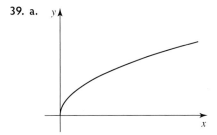

b. As the length of the pendulum increases, the period of the pendulum increases at a decreasing rate.

c. $T(0.5, 9.8) = 1.4$
$T(0.7, 9.8) = 1.7$
$T(1.0, 9.8) = 2.0$

CHAPTER 8 ■ SECTION 8.2

1. $\dfrac{1}{8}$

3. $\dfrac{1}{20}$

5. $\dfrac{11}{12}$

7. $\dfrac{7}{16}$

9.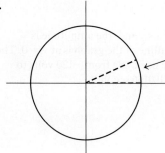
A portion of a complete rotation that is formed when the arc length is ½ the length of the radius

11.
3 complete rotations

13.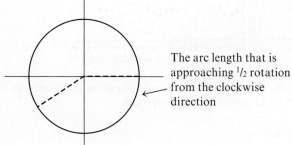
The arc length that is approaching ½ rotation from the clockwise direction

15.
7 complete rotations (43.96 radians) and a little more (1.04 radians)

17. $\dfrac{\pi}{12}$

19. ≈ 5.84

21. $\dfrac{41\pi}{18}$

23. $10°$

25. $\approx 487.3°$

27. ≈ 5.23598 feet

29. ≈ 21.98 yards

31. ≈ 0.52 meters

33. ≈ 14.14

35. $\approx \dfrac{2\pi^2}{3}$

37. $\approx \dfrac{A\pi^2}{90}$

39. We multiply by $\dfrac{\pi}{180°}$ because it is a unit multiplier that is equal to 1. There are 180 degrees in π radians.

41. When using radians to measure an angle, the radius is used as a measuring stick. As you choose larger and larger radii from the three given, they mark off the same number of radians.

43. D

45. C

47. Answers vary. wall framing for a home

49. Answers vary. the pitcher's mound on a baseball field

51. Answers vary. road intersection

53. 1 radian ≈ 63.66 gradians

dence to support this conclusion. In each case we see that the cosine and sine values of angles separated by 90° are equivalent when the input to the sine function is the larger of the two.

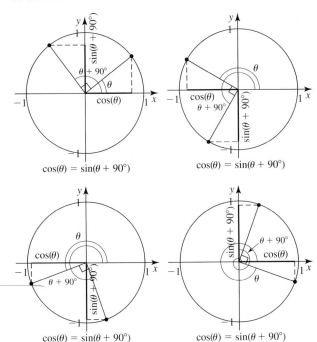

49. When working with the unit circle, we can interpret 0.9511 as representing a length, specifically the horizontal distance from the origin to the endpoint of an arc corresponding with 342°. However, in the general case of circles with any radius, 0.9511 is interpreted as the horizontal position of this endpoint in terms of the radius of the circle, and this is not a length but a constant of proportionality.

51. Since cosine and sine functions refer to the horizontal and vertical position of the endpoints of arcs on a unit circle, these endpoints repeat for each revolution around the circle. Therefore, cosine and sine functions are periodic, and their periods are 360° or 2π radians.

53. In Quadrant I, angles measure between 0° and 90° and the values of cosine and sine measure between 0 and 1. My classmate likely thinks that, since 45° is halfway between 0° and 90°, then cos(45°) and sin(45°) should be halfway between 0 and 1, or 0.5. Looking at a diagram of a circle, however, it is clear that this is not the case as both cos(45°) > 0.5 and sin(45°) > 0.5.

55. a. The farther we move away from the equator (the closer to the poles we get), the smaller the circumferences become of the circles formed by the lines of latitude.
    b. 90° and a circumference of 0 miles; 90° north latitude is located at the North Pole and 90° south latitude is located at the South Pole.

57. a. about 231.5 feet above the bottom of the London Eye
    b. about 412.4 feet above the bottom of the London Eye
    c. about 42.2 feet above the bottom of the London Eye
    d. about 152.7 feet above the bottom of the London Eye

59. $\cos^2(\theta) + \sin^2(\theta) = 1$

61. Imagine the diagram rotated one-quarter turn clockwise. In this view, we see that the magnitude of a will equal the value of sin(40°), that is, $|a| = \sin(40°)$. Since $\sin(40°) > 0$ while $a < 0$, we conclude that $a = -\sin(40°)$.

63. a. $\cos(-\theta) = x$
    b. $\sin(\theta + 2\pi n) = y$
    c. $\sin(\pi - \theta) = y$
    d. $\cos(\theta + \pi) = -x$

CHAPTER 8 ■ SECTION 8.4

1. a. period: 360°; amplitude: 4; frequency: $\dfrac{1}{360}$; midline: $y = 0$
   b. maximum: 4; minimum: -4
   c.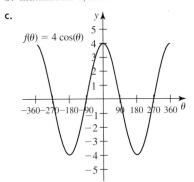
   d. The horizontal intercepts will be located at $(x, 0)$ for every x such that $x = 90° + 180°n$ for some integer n.
   e. (0, 4)
   f. vertical stretch by a factor of 4

3. a. period: 1440°; amplitude: 1; frequency: $\dfrac{1}{1440}$; midline: $y = 0$
   b. maximum: -4; minimum: -6
   c.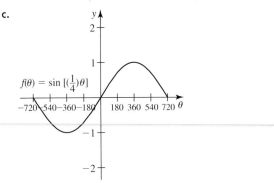
   d. The horizontal intercepts will be located at $(x, 0)$ for every x such that $x = 720°n$ for some integer n.
   e. (0, 0)
   f. horizontal stretch by a factor of 4

5. a. period: 2π; amplitude: 3; frequency: $\dfrac{1}{2\pi}$; midline: $y = 0$
   b. maximum: 3; minimum: -3

c.

d. The horizontal intercepts will be located at $(x, 0)$ for every x such that $x = \frac{2\pi}{3} + \pi n$ for some integer n.

e. $\left(0, \frac{3\sqrt{3}}{2}\right)$

f. shift left $\frac{\pi}{3}$ and vertically stretched by a factor of 3

7. a. period: $\frac{\pi}{2}$ radians; amplitude: 1; frequency: $\frac{2}{\pi}$; midline: $y = 0$

b. maximum: 1; minimum: -1

c.

d. The horizontal intercepts will be located at $(x, 0)$ for every x such that $x = \frac{\pi}{6} + \frac{\pi}{4} n$ for some integer n.

e. $\left(0, -\frac{\sqrt{3}}{2}\right)$

f. horizontal compression by a factor of $\frac{1}{4}$, shift right $\frac{\pi}{6}$, reflected about the horizontal axis

9. a. 2 periods

b. four (two for each period)

c. $\theta = 30°, 150°, 210°, 330°$

11. a. 3 periods

b. six (two for each period)

c. $\theta \approx 0.08, 0.96, 2.18, 3.06, 4.27, 5.15$ radians

13. false

15. true

17. true

19. about 1.79

21. For various input values θ, find the sum of the outputs $f(\theta)$ and $g(\theta)$.

23. For various input values θ, find the output value $g(\theta)$, then use this as the input for $f(\theta)$. This output is plotted against θ.

25. They are all periodic.

27. a. A table is shown for the first two quadrants.

θ	$f(\theta) = \cos(\theta) + \theta$
0	1
$\pi/12$	1.2277
$2\pi/12$	1.3896
$3\pi/12$	1.4925
$4\pi/12$	1.5472
$5\pi/12$	1.5678
$6\pi/12$	1.5708
$7\pi/12$	1.5738
$8\pi/12$	1.5944
$9\pi/12$	1.6491
$10\pi/12$	1.7520
$11\pi/12$	1.9139
π	3.1416

b. no

c. No. The outputs do not repeat at regular intervals, and therefore the function is not periodic.

29. The period is the fixed interval length over which the output values of the function repeat. Horizontal shifts only affect which inputs yield specific outputs; they do not impact the length of the interval over which these outputs repeat.

31. My classmate is correct. The outputs of the cosine and sine functions follow identical patterns, but for different inputs. The outputs of the sine function match the outputs of the cosine function when the inputs of the sine function are one-quarter of 1 period greater. Therefore, every cosine function could be written as a sine function shifted one-quarter of 1 period.

33. a. January ($m = 1$)

b. July ($m = 7$)

c. 2750 and 3850 people

d. Tourism and leisure pursuits in northern areas of the United States tend to peak in the summer months when the weather is best, and thus more employees are needed in leisure and hospitality fields, and decline in winter months when temperatures are colder, and thus fewer employees are needed.

35. a.

[Scatter plot: Average Temperature (T) vs. Months of the Year (m), showing data points rising from ~30 at m=1 to ~75 at m=7, then falling back to ~33 at m=12]

b. It appears that a polynomial function could model this data well, but when we consider the end behavior of the function we realize that a trigonometric function is a better fit. The end behavior of the polynomial function will tend to positive or negative infinity, which will be well outside the domain of possible temperatures in Indiana. This would restrict our model to only being effective in the given domain. Temperatures tend to be periodic, with, for example, temperatures in January being similar from year to year. Thus, if we use a trigonometric function, we can use the model outside the given domain to examine temperatures over several years. So using a trigonometric model we could consider $m = 8$, $m = 15$, and $m = 31$ as representing temperatures in months over several years.

c. period: 12 months; amplitude: 22.75; midline: $T = 52.55$

37. a. $T(10) = 52.55$. In October, the average temperature in Indianapolis is 52.55°.

b. The greatest average monthly temperature occurs in July and is 75.3°.

c. The lowest average monthly temperature occurs in January and is 29.8°.

d. The function outputs average monthly temperatures, which is the average of the average daily temperatures over a 1-month period. For example, $T(3)$ is the average temperature over the entire month of March. Therefore, $T(2.5)$ is interpreted as the average temperature in Indianapolis for the 1-month period beginning in the middle of February until the middle of March. $T(2.5) \approx 36.46°$.

39. a.

[Graph showing two sinusoidal curves, one peaking where the other troughs]

b. Both functions have the same period (12 months). However, the maximum average monthly temperatures occur 6 months apart. The maximum average monthly temperature in Huntsville occurs in July while Sydney's occurs in January. We also note that Huntsville's graph has a larger amplitude, which means that the average monthly temperature varies more throughout the year in Huntsville than it does in Sydney.

c. A positive cosine function was used to model Sydney's average monthly temperatures since the maximum average monthly temperature occurs in January ($m = 1$), which is close to an input of zero, the location of the maximum in a positive cosine function.

d. Sydney lies in the Southern Hemisphere while Huntsville, Alabama lies in the Northern Hemisphere. The seasons are reversed in the two hemispheres, so warmer temperatures in the Northern Hemisphere tend to correlate with colder temperatures in the Southern Hemisphere, and vice versa.

e. $\frac{1}{12}$. The horizontal shift of this function is equivalent to $\frac{1}{12}$ of the function's period.

f. $m - 7$

41. 0% and 100%. This is represented with a midline of $M = 50$ and an amplitude of 50.

43. In 2010 there was a full moon (100% of the full moon was visible) on January 1.

45. Just over four complete periods will occur on the interval $0 \le t \le 119$. Eight times during this interval 75% of the full moon will be visible, which we can verify by graphing the function $M(t) = 50 \cos\left(\frac{2\pi}{29.53}(t)\right) + 50$ along with $M = 75$ on the interval $0 \le t \le 119$ and counting the number of intersections.

47. $P(d) = 50 \cos\left(\frac{2\pi}{29.53}(d - 9)\right) + 50$. We need the maximum to occur for an input $d = 9$ (January 10 is 9 days after January 1). Also, the pattern of the percent of the full moon visible will be the same in 2020 as it was in 2010, only with a different starting day. Therefore, we need only perform a horizontal shift of the function $M(t)$ right 9 units to create the function $P(d)$.

49. **a.** The UK's voltage function can be generated by performing a vertical stretch of the U.S. voltage function by a factor of 2 (since the peak voltage is twice as large in the UK as in the U.S.) and by performing a horizontal stretch of the U.S. voltage function by a factor of $\frac{6}{5}$ (since the period of the UK voltage function, 1/50 of a second, is 6/5 times as large as the period of the U.S. voltage function, 1/60 of a second). $V_{UK}(t) = 240 \sin(100\pi t)$

b. $\frac{1}{50}$ of a second. The current changes direction and then changes back to its original direction in $\frac{1}{50}$ of a second.

c. The voltage will be -240 exactly 250 times in 5 seconds. The current will reach peak voltage in the alternate direction once per period, and since a period is $\frac{1}{50}$ of a second, this will occur 50 times per second for 5 seconds, or 250 total times.

d.

$t \approx 0.015$

51. Answers vary. cosine function: $f(\theta) = 3\cos(\theta + 225°) + 2$ or $f(\theta) = 3\cos(\theta - 135°) + 2$; sine function: $g(\theta) = 3\sin(\theta + 315°) + 2$ or $g(\theta) = 3\sin(\theta - 45°) + 2$. We first identify the midline ($y = 0$), the amplitude (3), and the period (360°). This information is needed for both a cosine and a sine function. Next we identify the horizontal shift from the basic cosine function to the given function. The basic cosine function has a maximum when $\theta = 0°$, and as θ increases, the function values decrease. This occurs for the given function at $\theta = -225°$ and $\theta = 135°$, so we have a horizontal shift either left 360° or right 360° for the function involving cosine. The basic sine function's output is equal to the midline when $\theta = 0°$, and as θ increases the function values increase. This occurs at $\theta = -315°$ and $\theta = 45°$ in the given function, so we have a horizontal shift left 315° or right 45° for the function involving sine.

53. The graphs of $x(\theta) = \cos\left(\theta - \frac{\pi}{4}\right)$ and $y(\theta) = \sin\left(\theta - \frac{\pi}{4}\right)$ may be interpreted as showing the horizontal and vertical coordinates, respectively, of the endpoint of the arc corresponding to $\theta - \frac{\pi}{4}$ radians on the unit circle, that is, an angle measuring $\frac{\pi}{4}$ radians less than θ radians. Therefore, the position for the endpoint of a given arc will occur for inputs that are $\frac{\pi}{4}$ radians larger in the functions $x(\theta) = \cos\left(\theta - \frac{\pi}{4}\right)$ and $y(\theta) = \sin\left(\theta - \frac{\pi}{4}\right)$ than in the functions $x_1(\theta) = \cos(\theta)$ and $y_1(\theta) = \sin(\theta)$. So the graphs of $x(\theta) = \cos\left(\theta - \frac{\pi}{4}\right)$ and $y(\theta) = \sin\left(\theta - \frac{\pi}{4}\right)$ will be the graphs of $x_1(\theta) = \cos(\theta)$ and $y_1(\theta) = \sin(\theta)$ shifted right $\frac{\pi}{4}$ radians.

55.

The horizontal coordinate of the endpoint varies between 1 and 3 with a period of 2π radians, so we can model it with a shift of the cosine function 2 units upward ($x(\theta) = \cos(\theta) + 2$). The vertical position of the endpoint varies between -2 and 0 with a period of 2π radians, so we can model it with a shift of the sine function downward 1 unit ($y(\theta) = \sin(\theta) - 1$).

CHAPTER 8 ■ SECTION 8.5

1. period $= \frac{\pi}{2}$; amplitude $= 5$; midline: $y = 3$

3. period $= \pi$; amplitude $= 8$; midline: $y = 0$

5. upward 2 units and right 1 unit

7. downward 9 units and left $\frac{\pi}{2}$ units

9.

11.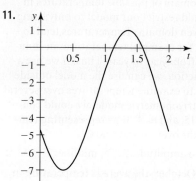

13. Answers vary.

$y = 3\cos\left[\frac{\pi}{2}(x)\right]$

15. Answers vary.
$$y = -1 + 3\cos\left(\frac{\pi}{2}x\right)$$

17. Answers vary.
$$y = 3 + 5\cos(\pi x)$$

19. Answers vary.
$$y = 1.5 + 3\cos\left[\frac{\pi}{4}(x+1)\right]$$

21. Answers vary. It appears that the amplitude is 2, the midline is $y = 2$, and that the period is 1 unit so $y = 2 + 2\sin(2\pi x)$.

23. Answers vary. It appears that the amplitude is 0.5, the midline is $y = 1$, and that the period is 50 units so
$$y = 1 + 0.5\sin\left(\frac{2\pi}{50}x\right).$$

25. Answers vary. It appears that the amplitude is 0.2, the midline is $y = 0.5$, the phase shift is right 1, and that the period is 8 units so $y = 0.5 + 0.2\sin\left[\frac{2\pi}{8}(x-1)\right]$.

27. $V_1(d) = 5\sin(3d)$
    $V_2(d) = 5\cos[3(d - 30°)]$

29. a.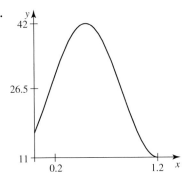

  b. $d(t) = 26.5 + 15.5\cos[\pi(t - 0.2)]$
  c. $d(3.4) = 31.29$ centimeters
  d. $d(0) = 39.04$ centimeters
  e. approximately 0.434 seconds
  f. No, because the spring will begin to not fluctuate as much over time. Eventually it will stop bouncing.

31. $B(t) = 4 + 2\sin\left[\frac{\pi}{2}(t - 17)\right]$
    $G(t) = 3 + \sin\left[\frac{\pi}{4}(t - 18)\right]$

33. if the data is periodic and takes on the shape of a sine or cosine curve

35. We know that sinusoidal models can be represented by the function $y = C + A\sin(k(x - P))$ so we need to determine the midline (C), amplitude (A), period $\left(\frac{2\pi}{\text{period}} = k\right)$, and phase shift ($P$). We can determine the period by inspecting the interval over which the output values repeat and then calculate k using $\frac{2\pi}{\text{period}} = k$ for radians and $\frac{360°}{\text{period}} = k$ for angles in degrees. We determine the midline by averaging the maximum and minimum values. We determine the amplitude by calculating the distance above or below the midline. For the cosine function, we can determine the phase shift by calculating when the output values start at a maximum and then decrease to a minimum and then return to a maximum. For the sine function, we can determine the phase shift by calculating when the output values start at the midline and then increase to a maximum, then to a minimum, and then back to the midline.

37. a. D is the "typical" depth of the water and most likely sea level at this location.
  b. A is the amplitude and would be 25 feet.
  c. B is the number of minutes for a complete cycle between low tides or 373 minutes (6 hours and 13 minutes) times 2 or 746 minutes (12 hours and 26 minutes).
  d. C is the number of hours (or minutes) we would need to shift the graph in order to "hit" a maximum or minimum since we have been told to use a cosine function to model the tides.

39. a. amplitude: 5; period: $\frac{1}{60}$
  b. The frequency of the function is 60 cycles per second and means that the voltage will complete 60 cycles from +5 to −5 in 1 second.
  c. $V(0) = 5$; $V(0.03) = 1.5$; $V(0.06) = -4.05$; $V(0.09) = -4.05$; $V(0.12) = 1.5$
  d.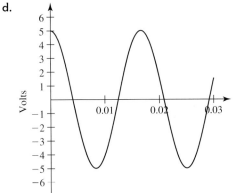

41. a. Answers vary. $p(w) = 7 - 1\cos\left[\frac{2\pi}{52}(w - 10)\right]$
  b. The period is 52 weeks and represents the yearly cycle of pneumonia and influenza deaths. The amplitude is 1% and shows the amount above and below the deaths go from the midline or average of 7% of all deaths.
  c. $p(30) \approx 7.7$. Thirty weeks into the year 2005, 7.7% of all deaths were due to pneumonia and influenza.
  d. approximately in each year from 2003 to 2008 from week 48 in the prior year to week 10 in the subsequent year
  e. approximately in each year from 2003 to 2008 in week 32
  f. The function is increasing from week 0 to 10 and from week 32 to 52, which tells us over what interval of time the percentage of deaths due to pneumonia and

882 ANSWERS ■ CHAPTER 8

influenza is rising. The function is decreasing from week 10 to 32, which tells us over what interval of time the percentage of deaths due to pneumonia and influenza is falling.

g. The gap tells us when the percentage of deaths due to pneumonia and influenza is higher than normal but not considered to be an epidemic yet.

43. a. $M(t) = 50 + 50 \cos\left[\left(\frac{2\pi}{30}\right)t\right]$

b. 33%

$t \approx 9$ $t \approx 21$
January 10, 2010   January 22, 2010

50%

$t \approx 8$ $t \approx 23$
January 9, 2010   January 24, 2010

c. From day 10 to 25 the percentage of the moon visible increases by 3.33% each day on average.

45. a. $T(t) = 61.5 - 24.5 \cos\left[\frac{\pi}{6}(t-1)\right]$

b.

c.

d. period: 12 months
e. amplitude: 24.5°
f. horizontal shift: 1 month
g. midline: 61.5°
h. $T(4) = 61.5 - 24.5 \cos\left[\frac{\pi}{6}(4-1)\right]$
$= 61.5°$

The temperature reaches that temperature again in the 10th month (October).

i. The average low temperature will occur twice during the year. The first will be at 5.89 (about June 26) and 8.1 (about September 3).

j.

47. a. $d(t) = 422 + 366 \cos(\pi t)$
b. 0.656, 1.345, 2.655, and 3.345 hours
c. approximately from 0.64 hours to 1.36 hours

49. a. Assuming $k = 1$ we get the following graph.

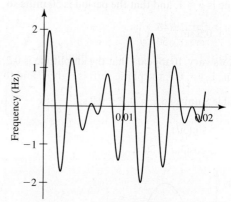

b. Assuming that $k = 1$, at 0.0006 seconds the sum of the two notes reaches a maximum frequency of about 1.95 Hz and after 0.012 seconds the sum of the notes reaches a minimum frequency of −2 Hz.

c. $S_{\text{D sharp + G}}(t) = k \sin(2\pi \cdot 311t) + k \sin(2\pi \cdot 392t)$

CHAPTER 8 ■ SECTION 8.6

1. −0.0875; the slope of the terminal angle at 175°
3. 0.1763; the inverse of the slope of the terminal angle at 80°
5. $\frac{\sqrt{3}}{3}$; the slope of the terminal angle at 30°
7. undefined
9. −1; the slope of the terminal angle at 135°
11. $\tan(\theta)$
13. $\sin(\theta)$
15. 1
17. $\frac{2}{\sqrt{3}}$ or $\frac{2\sqrt{3}}{3}$
19. $\frac{2}{\sqrt{2}}$ or $\sqrt{2}$
21. $\sec(60°) = 2$

23. We know tan(θ) is positive in Quadrants I and III and cos(θ) is negative in Quadrants II and III. Therefore, θ must be an angle in Quadrant III.

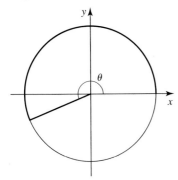

25. We know sec(θ) is positive when cos(θ) is positive, which occurs in Quadrants I and IV, and csc(θ) is positive when sin(θ) is positive, which occurs in Quadrants I and II. Therefore, θ must be an angle in Quadrant I.

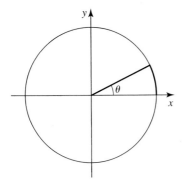

27. If sec(θ) = 2, then cos(θ) = $\frac{1}{2}$.

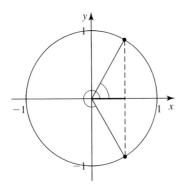

If csc(θ) = −3, then sin(θ) = $-\frac{1}{3}$.

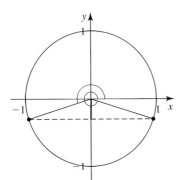

29. Angles that differ by π radians have the same tangent value, that is, the slopes of their terminal sides are the same. Therefore, it must be true that either θ = φ + π or θ = φ − π. In this case, it would also be true that cot(θ) = cot(φ) since

$$\tan(\theta) = \tan(\phi)$$
$$\frac{1}{\tan(\theta)} = \frac{1}{\tan(\phi)}$$
$$\cot(\theta) = \cot(\phi)$$

31. Answers vary. The tangent value is the slope of a line passing through the origin. Regardless of the radius of the circle used to measure the angle, the result of the slope calculation will be the same since the slope of a line is constant, meaning we can use any two points on the line to calculate its value.

33. tan(8.53°) ≈ 0.1500; cot(8.53°) ≈ 6.6673

35. a. tan(51.34°) ≈ 1.2500; cot(51.34°) ≈ 0.8000
    b. about 345.6 feet
    c. w(θ) = 432 cot(θ)

37. a. The domain is approximately the interval 2.86° ≤ θ ≤ 4.76°.
    b. The function has a negative rate of change with a positive concavity, meaning that as the angle of incline for the ramp increases, the horizontal length of the ramp decreases, but that the horizontal length decreases by smaller and smaller amounts for equal increases in the angle of incline.
    c. The ideal domain is approximately the interval 2.86° ≤ θ ≤ 3.58°.

39. a. The reasonable domain is the interval 0° < θ < 90°. The corresponding range is L > 10, or all values greater than 10 inches.
    b. The rate of change is negative and the concavity is positive, so as the angle of the offset increases, the distance between the bends decreases, but the distance between bends decreases by less and less for equal increases in the angle of the offset.
    c. about 15.56 inches
    d. about 33.75°
    e. Answers vary. We know csc(30°) = 2, which is an easy value to remember, so the electrician can determine the necessary distance between bends by simply doubling the length of the obstruction.

41. Answers vary. f(θ) = −2.25 csc[(2(θ + 15°)] + 1.5

43. a. The functions have the same period, as well as the same vertical asymptotes. However, since the outputs of f are the result of squaring a number, it will never be negative, while g has both positive and negative output values. In addition, the output value for f will always be the square of the output of g for the same input.
    b. The output of h is always 1 greater than the output of f for the same input.
    c. [tan(θ)]² + 1 = [sec(θ)]². This formula may be useful in rewriting expressions involving trigonometric functions

to make them simpler. For example, $5[\tan(\theta)]^2 + 5$ could be rewritten as $5[\sec(\theta)]^2$ using this relationship.

CHAPTER 8 ■ SECTION 8.7

1. 0.7478. The vertical position of the endpoint of the arc corresponding with 0.7478 radians is 0.68 on the unit circle, or 0.68 of the radius length above the horizontal axis on a circle of any radius.

3. 2.0580. The horizontal position of the endpoint of the arc corresponding with 2.0580 radians is -0.4682 on the unit circle, or 0.4682 of the radius length to the left of the vertical axis on a circle of any radius.

5. $\theta = \dfrac{\pi}{3}$. The horizontal position of the endpoint of the arc corresponding with $\dfrac{\pi}{3}$ radians is $\dfrac{1}{2}$ on the unit circle, or half the radius length to the right of the vertical axis on a circle of any radius.

7. $\theta = \dfrac{7\pi}{4}$. The vertical position of the endpoint of the arc corresponding with $\dfrac{7\pi}{4}$ radians is $-\dfrac{\sqrt{2}}{2}$, or $\dfrac{\sqrt{2}}{2}$ times as far as one radius length below the horizontal axis for a circle with any radius.

9. $\theta \approx 39.04°, 320.96°$

11. $\theta \approx 41.41°, 318.59°$

13. $\theta \approx 3.60, 5.83$ radians

15. $\theta = \dfrac{5\pi}{3}$ radians

17. a. The ratio $\dfrac{\sin(\theta)}{\cos(\theta)}$ can be rewritten as $\tan(\theta)$, so the equation becomes $\tan(\theta) = 0.8361$. We can then use the inverse tangent function to rewrite the equation as $\theta = \tan^{-1}(0.8361)$, which we can determine using a calculator. One solution is about 39.90°, or about 0.70 radians.

  b. i. 68.20° or 248.20°, or 1.19 or 4.33 radians
  ii. $-18.26°$ or 161.74°, or -0.32 or 2.40 radians
  iii. 45° or 225° or $\dfrac{\pi}{4}$ or $\dfrac{5\pi}{4}$ radians

19. $\theta \approx 19.26°, 310.74°$ 21. no solution

23. $\sqrt{3}$ 25. $\dfrac{2\sqrt{3}}{3}$

27. $\dfrac{\pi}{6}$ 29. $\dfrac{\pi}{3}$

31. 0.9949 33. 1.7698

35. $\sin(\theta) \approx 0.9986$ 37. 0.5546

39. -0.3368

41. a. My classmate is correct when he says that $\sin(\theta) = 0.5$ has two solutions on the interval $0° \leq \theta \leq 360°$ and that $\sin\left(\dfrac{1}{2}\theta\right)$ will complete only the first half of its period on this interval.

  b. My classmate is incorrect when he says that $\sin\left(\dfrac{1}{2}\theta\right) = 0.5$ will only have one solution on the interval $0° \leq \theta \leq 360°$. As θ increases from 0° to 360°, $\dfrac{1}{2}\theta$ increases from 0° to 180°, passing completely through Quadrant I and Quadrant II, and sine is positive in both of these quadrants. So $\sin\left(\dfrac{1}{2}\theta\right) = 0.5$ has two solutions on the interval $0° \leq \theta \leq 360°$ even though $\sin\left(\dfrac{1}{2}\theta\right)$ does not complete a full period on this interval.

43. The inputs of both functions are "1," but these inputs stand for different quantities, and the values of $\cos(1)$ and $\cos^{-1}(1)$ represent different quantities. The input "1" in $\cos(1)$ represents an angle of 1 radian, and $\cos(1)$ represents the horizontal coordinate of the endpoint of the corresponding arc on the unit circle. The input "1" in $\cos^{-1}(1)$ represents a cosine value of 1 (a horizontal coordinate of 1 on the unit circle), and $\cos^{-1}(1)$ is the angle measure in radians that corresponds to an arc whose endpoint has this horizontal coordinate.

45. a. $\theta \approx 61.50°$
  b. 61.50° North or South latitude is a circle with a circumference of 11,880 miles. This is reasonable because latitudes must be between 0° and 90°, and we expect to find a solution since 11,880 miles is less than the circumference at the equator, which is about 24,900 miles.

47. a. no solution
  b. According to our model, there were never 4200 people employed in leisure and hospitality fields in Clatsop County between 1 and 13 months after January 2004.
  c.

49. a. $d \approx -13.63, 196.13$ (*Note:* Use radians.)
  b. There were 11.5 hours of daylight in New York City about 14 days before and about 196 days after March 21, 2010.
  c.

51. a. Answers vary. $T(m) = -24\cos\left[\dfrac{2\pi}{12}(m-1)\right] + 63$ (using radians)
  b. $m \approx 4.57, 9.43$
  c. The average high temperature in Philadelphia is 70°F about halfway through April and halfway through August.

53. a. Answers vary. $C(d) = 2.15 \cos\left[\dfrac{2\pi}{365}(d - 170)\right] + 12.15$ (using radians)

b. $d \approx 45.96, 294.04$

c. There were 11 hours of daylight in Charleston about 46 and 294 days after January 1, 2007 (February 16 and October 22).

55. a. We restrict the domain to the interval $0 \le t \le 15$.

b. $t(H) = \dfrac{15}{\pi}\left[\sin^{-1}\left(\dfrac{H - 221}{221}\right) + \dfrac{\pi}{2}\right]$; domain: $0 \le H \le 442$; range: $0 \le t \le 15$

c.

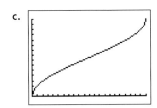

d. about 6.05 minutes after boarding; Evaluate $t(155)$. Note that there are other solutions as well, but that this is the only solution between 0 and 15 minutes after the rider boards the London Eye.

57. $f_1(x) = -\cos^{-1}(x) + 2\pi$
$f_2(x) = \cos^{-1}(x)$
$f_3(x) = -\cos^{-1}(x)$
$f_4(x) = \cos^{-1}(x) - 2\pi$

59. $\dfrac{1}{2}$

CHAPTER 8 ■ REVIEW

1.

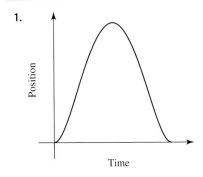

3. a. $\dfrac{-3.79°C}{1 \text{ month}}$

b. $\dfrac{1.75°C}{1 \text{ month}}$

c. The highest temperature occurs at 7.6 and the lowest at 1.6. Near the month 7.6, the surface temperature increases at a decreasing rate, reaches the maximum temperature of 27.50 then decreases at an increasing rate. Near the month 1.6, the surface temperature decreases at a decreasing rate, then increases at an increasing rate.

d. Yes, it would make sense that the surface temperature would follow a cyclic pattern with the yearly weather patterns.

5. a. After 50 minutes it would be 3°F and after 70 minutes it would be −1°F.

b. Yes, the temperature repeats output values every 30 minutes.

c. The graph would increase at a relatively constant rate until the door is closed again.

7.

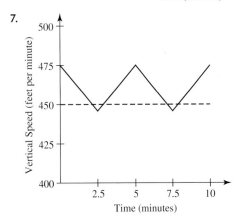

9. $\dfrac{5}{12}$

11. ≈ 0.61

13. $30°$

15. 1.96 centimeters

17. 3 radians

19. A 540° rotation is the same as one complete rotation and 1/2 of another. In other words if you rotate 540° you spin completely around and then 1/2 again to where you would be facing the opposite direction from where you started.

21. $\cos(\theta) \approx -0.8$; $\sin(\theta) \approx -0.6$

23. −0.5736

25. 0.9010

27. $\frac{\sqrt{2}}{2}$

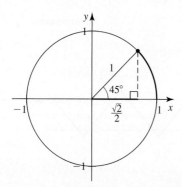

29. a. 221 feet above the bottom of the London Eye
  b. about 64.73 feet above the bottom of the London Eye
  c. about 289.29 feet above the bottom of the London Eye

31. a. period: 360°; amplitude: 3; frequency: $\frac{1}{360}$; midline: $y = 0$
  b. maximum: 3; minimum: −3
  c.

  d. The horizontal intercepts are located at $(x, 0)$ for every x such that $x = 180°n$ for some integer n.
  e. (0, 0)
  f. vertical stretch by a factor of 3

33. a. period: 2π radians; amplitude: $\frac{2}{3}$; frequency: $\frac{1}{2\pi}$; midline: $y = 0$
  b. maximum: $\frac{2}{3}$; minimum: $-\frac{2}{3}$
  c.

  d. The horizontal intercepts are located at $(x, 0)$ for every x such that $x = \frac{3\pi}{4} + \pi n$ for some integer n.
  e. about (0, 0.47)
  f. shift right $\frac{\pi}{4}$ radians; vertical compression by a factor of $\frac{2}{3}$

35. a. period: 1800°; amplitude: 2; frequency: $\frac{1}{1800}$; midline: $y = 0$
  b. maximum: 2; minimum: −2
  c.

  d. The horizontal intercepts are located at $(x, 0)$ for every x such that $x = 360° + 900°n$ for some integer n.
  e. about (0, −1.90)
  f. horizontal stretch by a factor of 5; shift left 90°; vertical stretch by a factor of 2; reflected vertically about the horizontal axis

37. The horizontal shift describes *how much* the graph shifts horizontally from the parent function to create the graph of the given function. The phase shift describes the horizontal shift in terms of the portion of one period the graph is shifted.

39. a. amplitude: 5000; midline: $y = 77,000$. The construction employment in Oregon varies between 72,000 and 82,000 people during the year. Period: 12 months. The pattern for construction employment repeats every year. Phase shift: $\frac{\pi}{2}$. When modeling with a negative cosine function, the minimum value occurs for an input of 0 (and then periodically from there). In this context, the minimum occurs for an input of $m = 3$, showing that a horizontal shift of $\frac{1}{4}$ of one period is involved in using a negative cosine function to model this context.

b. The fewest number of workers are employed in construction in Oregon during April ($m = 3$), the most during October ($m = 9$).

c. On the intervals $0 < m < 3$ and $9 < m < 12$, the rate of change is negative and increasing toward 0 (the concavity is positive). For equal changes in m, fewer and fewer construction jobs are lost in Oregon. On the interval $3 < m < 6$, the rate of change is positive and increasing (the concavity is positive). For equal changes in m, more and more people are employed in construction jobs in Oregon. On the interval $6 < m < 9$, the rate of change is positive and decreasing (the concavity is negative). The number of people employed in construction in Oregon is still increasing, but for equal changes in m the increase in construction employees is less and less.

41. a. amplitude: 24.8, midline: $y = 63.3$. The average high temperature in Washington, Indiana, varies between 38.5°F and 88.1°F during the year. Period: 12 months. The pattern for the average high temperature in Washington, Indiana, repeats every year.

b. $m \approx 4.94, 9.06$. The average high temperature in Washington, Indiana, is 75°F at the end of April and the beginning of September.

43. period: $\frac{2\pi}{3} \approx 2.03$; amplitude: 9; midline: $y = -4$

45. period: $\frac{2\pi}{3}$; amplitude: π; midline: $y = -0.04$

47. The vertical shift is downward 4 and the horizontal shift is left 2.

49. The vertical shift is downward 0.04 and there is no horizontal shift.

51.

53.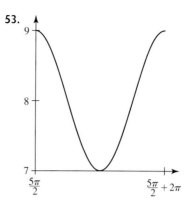

55. -2.3559. The slope of the terminal side of an angle measuring 113° is about -2.3559.

57. $-\frac{\sqrt{3}}{3}$. The slope of the terminal side of an angle measuring $\frac{2\pi}{3}$ radians as a ratio of the horizontal change to the vertical change is $-\frac{\sqrt{3}}{3}$.

59. a. about 4%

b. It represents the horizontal distance in feet the road extends for each 1-foot change in the elevation of the road.

c. about 0.01 miles or about 53.02 feet

61. $\sec(81°) \approx 6.3925$; $\cos(81°) \approx 0.1564$

63. $\frac{2\sqrt{3}}{3}$

65. about 8.25 meters

67. 0.6094 or 5.6738. The horizontal position of the endpoint of the arc corresponding with about 0.6094 or 5.6738 radians is 0.82 on the unit circle, or 0.82 times the radius length to the right of the vertical axis on a circle of any radius.

69. $\theta = 45°$ or $135°$

71. $\theta \approx 113.12°, 246.88°$

73. $\theta \approx 227.49°, 352.51°$

75. $\theta \approx 1.65, 4.21$ radians

77. a. $\theta \approx 5.14°$

b. A road with a 9% grade makes an angle of about 5.14° with the horizontal (either above or below). This solution is reasonable because the angles roads form with the horizontal must be relatively close to 0° in order to be usable by vehicles.

CHAPTER 9 ■ SECTION 9.1

1. $\sin(\theta) = \dfrac{8}{10} = \dfrac{4}{5}$, $\cos(\theta) = \dfrac{6}{10} = \dfrac{3}{5}$, $\tan(\theta) = \dfrac{8}{6} = \dfrac{4}{3}$,

$\sec(\theta) = \dfrac{10}{6} = \dfrac{5}{3}$, $\csc(\theta) = \dfrac{10}{8} = \dfrac{5}{4}$, $\cot(\theta) = \dfrac{6}{8} = \dfrac{3}{4}$

3. $\sin(\theta) = \dfrac{2.5}{6}$, $\cos(\theta) = \dfrac{\sqrt{29.75}}{6}$, $\tan(\theta) = \dfrac{2.5}{\sqrt{29.75}}$, $\sec(\theta) = \dfrac{6}{\sqrt{29.75}}$, $\csc(\theta) = \dfrac{6}{2.5}$, $\cot(\theta) = \dfrac{\sqrt{29.75}}{2.5}$

5. $\sin(\theta) = \dfrac{3}{\sqrt{13}}$, $\cos(\theta) = \dfrac{2}{\sqrt{13}}$, $\tan(\theta) = \dfrac{3}{2}$, $\sec(\theta) = \dfrac{\sqrt{13}}{2}$, $\csc(\theta) = \dfrac{\sqrt{13}}{3}$, $\cot(\theta) = \dfrac{2}{3}$

7. **a.** $\sin(A) = \dfrac{4}{5}$ **b.** $\cos(B) = \dfrac{4}{5}$ **c.** $\sin(B) = \dfrac{3}{5}$ **d.** $\cos(A) = \dfrac{3}{5}$

9. **a.** $\sin(A) = \dfrac{12}{13}$ **b.** $\cos(B) = \dfrac{12}{13}$ **c.** $\sin(B) = \dfrac{5}{13}$ **d.** $\cos(A) = \dfrac{5}{13}$

11. **a.** $\sin(A) = \dfrac{\checkmark}{\square}$ **b.** $\cos(B) = \dfrac{\checkmark}{\square}$ **c.** $\sin(B) = \dfrac{\bullet}{\square}$ **d.** $\cos(A) = \dfrac{\bullet}{\square}$

13. $\sin(A)$ and $\cos(B)$ are equivalent because $\sin(A) = \dfrac{\text{opp}}{\text{hyp}} = \dfrac{a}{c}$ and $\cos(B) = \dfrac{\text{adj}}{\text{hyp}} = \dfrac{a}{c}$.

15. The ratio of the opposite side to the hypotenuse is close to 0 so the angle has to be relatively small and therefore less than 30°.

17. The ratio of the opposite side to the hypotenuse is greater than $\dfrac{1}{2}$, but not very close to 1, so the angle is likely between 30° and 60°.

19. The tangent function is equivalent to the slope of the hypotenuse. If the slope is 1 then the angle would have to be halfway between 0 and 90° (or 45°) so between 30° and 60°.

21. The slope is close to 0 so the angle has to be less than 30°.

23. The ratio of the adjacent side to the hypotenuse is fairly close to 0 so the angle has to be close to 90° and therefore between 60° and 90°.

25. $\theta \approx 38.66°$

27. $\theta \approx 55.15°$

29. **a.** 57.74 feet **b.** 33.98 feet

31. The car's speed was 31.14 mph, so it was not speeding.

33. Yes. First find the angle measure of one of the acute angles using one of the trig functions such as $\cos(\theta) = \dfrac{3}{5}$. Then find the third angle by subtracting the two angles you now know from 180°.

35. Yes. The angle that passes through the first and third quadrants will have the same sine and cosine values. This would be the 45° or $\dfrac{\pi}{4}$ and 225° or $\dfrac{5\pi}{4}$ because the x- and y-coordinates will be the same.

37. 369.47 feet

39. **a.** 25.15° **b.** 100 feet per minute

41. $\theta \approx 7.96°$

43. **a.** $\theta \approx 42.28°$ **b.** 81.39 feet

45. The throw was 60.21 feet; the angle measures 48.37°.

47. Answers vary.

49.

For the rectangle that has the same area as the parallelogram,
$$A_\square = l \cdot w$$
$$b = l$$

The measure of both angles labeled θ are the same because they are on alternate sides of the transversal so
$$\sin(\theta) = \dfrac{w}{a}$$
$$a\sin(\theta) = w$$
$$A_\square = b \cdot a \sin(\theta)$$

51. $\cos(\theta) = \dfrac{w}{h}$ and $\sec(\theta) = \dfrac{h}{w}$

CHAPTER 9 ■ SECTION 9.2

1. $a \approx 4.11$
3. $r \approx 4.57$
5. $f \approx 48.74$
7. $o \approx 4.408$
9. $c \approx 2.168$
11. $A \approx 28.96°$
13. $T \approx 134.62°$
15. $Y \approx 19.69°$
17. $Q \approx 84.19°$
19. $C \approx 92.87°$
21. $C \approx 46.7°$, $a \approx 7.55$
23. $A \approx 70.97°$, $B \approx 32.04°$, $C \approx 76.99°$
25. $a \approx 26.098$, $B \approx 171.94°$, $C \approx 3.06°$
27. 48.26 feet
29. The Law of Cosines is based on the Pythagorean theorem. The Law of Cosines will add length to side c if angle C is greater than 90° and subtract length from side c if angle C is less than 90°.

31. Yes, any right triangle in which we know two sides
$$c^2 = 3^2 + 4^2$$
$$c^2 = 25$$
$$c = 5$$
or
$$c^2 = 3^2 + 4^2 - 2(3)(4)\cos(90°)$$
$$c^2 = 25$$
$$c = 5$$

33. a. 22,010.41 feet
  b. 72.48°

35. 40.66 miles

37. a. 2711.67 feet
  b. 1437.13 feet; Yes, the minimum distance required is 100 yards (300 feet).
  c. 4.11°

39. a. 58.55 miles
  b. 59.10 miles

41.

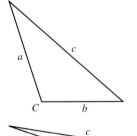

The term $-2ab\cos(C)$ will become larger and larger (and positive), which in turn makes c larger and larger.

43.

$$11^2 = 3^2 + 7^2 - 2(3)(7)\cos(Z)$$
$$-1.5 \approx \cos(Z)$$

The cosine of an angle can never be greater than 1 or less than -1. Therefore, there is no such triangle that has sides 3, 7, and 11.

CHAPTER 9 ■ SECTION 9.3

1. $A = 89°$, $b \approx 5.85$, $c \approx 1.45$
3. $A = 67°$, $b \approx 16.14$, $c \approx 4.22$
5. $C = 127°$, $b \approx 2.26$, $a \approx 1.04$
7. $C = 66°$, $a \approx 16.84$, $b \approx 2.44$

9. $B = 79°$, $a \approx 7.96$, $c \approx 24.21$
11. $B = 55°$, $a \approx 9.74$, $c \approx 5.03$
13. $C = 79°$, $a \approx 7.89$, $c \approx 23.79$
15. $C = 85°$, $a \approx 11.73$, $c \approx 24.89$
17. $B \approx 20.34°$, $C \approx 79.66°$, $c \approx 33.96$
19. no solution
21. 242.88 feet
23. a. 90.12 miles and 82.64 miles
   b. 0.47 hours
25. a. 767.4 feet
   b. 148.83 feet
27. one triangle; $B \approx 30.71°$, $C \approx 99.29°$, $c \approx 3.86$
29. two triangles; For the first possible triangle, $A \approx 129.14°$, $C \approx 30.86°$, $a \approx 9.07$. For the second possible triangle, $A \approx 10.86°$, $C \approx 149.14°$, $a \approx 2.20$.
31. two triangles; For the first possible triangle, $A \approx 57.7°$, $B \approx 97.3°$, $b \approx 2.35$. For the second possible triangle, $A \approx 122.3°$, $B \approx 32.7°$, $b \approx 1.28$.
33. It is appropriate to use the Law of Sines if we are given two angles and a side (AAS or ASA) or two sides and one of the non-included angles (SSA).
35. When we are given two sides and a non-included angle (SSA), sometimes more than one triangle may be generated. In fact, the known information may result in one triangle, two triangles, or no triangle at all.
37. We know from algebra that if a proportion is true, then the reciprocal of the same proportion will also be true. For example, if $\dfrac{1}{2} = \dfrac{4}{8}$, then $\dfrac{2}{1} = \dfrac{8}{4}$. In terms of the Law of Sines, if $\dfrac{b}{\sin(B)} = \dfrac{c}{\sin(C)} = \dfrac{a}{\sin(A)}$, then $\dfrac{\sin(B)}{b} = \dfrac{\sin(C)}{c} = \dfrac{\sin(A)}{a}$.
39. a. No, it is 48.77 feet.
   b. 24.23°
41. a. 63.72 feet
   b. 66.78 feet
   c. 92.81°
43. a. 0.78 miles
   b. 561 mph
   c. 61.17°
45. 298.1 meters
47. 81°, 5.61, 7.97

CHAPTER 9 ■ SECTION 9.4

1. Answers vary. $(5, 0)$, $(5, 2\pi)$
3. Answers vary. $\left(3, \dfrac{2\pi}{3}\right)$, $\left(-3, -\dfrac{\pi}{3}\right)$
5. Answers vary. $\left(4, \dfrac{3\pi}{2}\right)$, $\left(-4, \dfrac{\pi}{2}\right)$
7. Answers vary. $(3, 0)$

9. Answers vary. $(4, \pi)$

11. Answers vary. $\left(2, \dfrac{5\pi}{3}\right)$

13. Answers vary. $(\sqrt{26}, 4.51)$

15. Answers vary. $(\sqrt{61}, 2.45)$

17. Answers vary. $(0, 3)$

19. Answers vary. $(-2.60, -1.5)$

21. Answers vary. $(1, -1.73)$

23. Answers vary. $(1.31, 1.51)$

25. Answers vary. $(2.08, 4.55)$

27. $0 \leq \theta \leq \pi, r \geq 5$

29. Explanations vary. The value of r is the greatest in magnitude when $\sin(2\theta) = 1$. Since $\sin\left(\dfrac{\pi}{2}\right) = 1$, we find values of θ such that $\sin(2\theta) = 1$.

$$2\theta = \dfrac{\pi}{2}$$

$$\theta = \dfrac{\pi}{4}$$

We extend this to find values of θ on $(0, 2\pi)$. We say that r has the greatest magnitude when $\theta = \dfrac{1}{4}\pi, \dfrac{3}{4}\pi, \dfrac{5}{4}\pi, \dfrac{7}{4}\pi$. The value of r is maximized at 4 values of θ thus revealing why we see 4 petals.

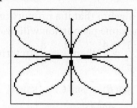

31. Explanations vary. The value of r is the greatest in magnitude when $\sin(5\theta) = 1$. Since $\sin\left(\dfrac{\pi}{2}\right) = 1$, we find values of θ such that $\sin(5\theta) = 1$.

$$5\theta = \dfrac{n\pi}{2}, n \text{ is a positive odd integer}$$

$$\theta = \dfrac{n\pi}{10}, n \text{ is a positive odd integer}$$

We extend this to find values of θ on $(0, 2\pi)$. We say that r has the greatest magnitude when $\theta = \dfrac{1}{10}\pi, \dfrac{3}{10}\pi, \dfrac{5}{10}\pi, \dfrac{7}{10}\pi, \dfrac{9}{10}\pi$. Note that after $\theta = \dfrac{9}{10}\pi$, we get the same points as before so there are only 5 distinct points where the value of r is maximized (in magnitude), thus revealing why we see 5 petals.

33.

35.

37.

39.

41.

43. The polar coordinate system is especially useful in situations where the relationship between two points is most easily expressed in terms of angles and distance.

45. If we know the coordinates of a point in the polar system, (r, θ), we can determine the same location in Cartesian coordinates. We examine a generic graph to determine these relationships. Suppose we know the values of r and θ and wish to determine the corresponding values of x and y.

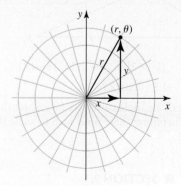

We have the following relationships.

$$\cos(\theta) = \dfrac{x}{r} \qquad \sin(\theta) = \dfrac{y}{r}$$

$$x = r\cos(\theta) \qquad y = r\sin(\theta)$$

47. Since r stands for the radius and θ is the angle measure, we know that the radius and angle measure are always equal when $r = \theta$. Therefore as the radius grows from 0 to 2π the angle will be the same, creating the graph shown here.

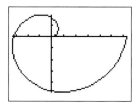

49. $0 < r \le 8$ for all $0 \le \theta \le \dfrac{\pi}{4}$

51. The graph has an elliptical shape and as *a* increases the graph's width and height grow. See the graphs below for when $n = 1$ and *a* varies.

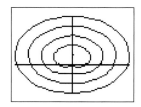

53. The graph has an elliptical shape and as *a* increases the graph's width and height grow. See the graphs below for when $n = 1$ and *a* varies.

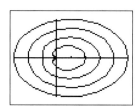

55. Same as Exercise 51 but with $r = a - n\sin(\theta)$. We get the same graph with an elliptical shape and as *a* increases the graph's width and height grow except it is reflected vertically.

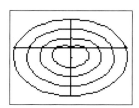

Same as Exercise 52 but with $r = a - n\sin(\theta)$. We get the same graph where the inner and outer loops grow as the size of *n* increases and *a* is held constant except it is vertically flipped over.

Same as Exercise 53 but with $r = a - n\cos(\theta)$. We get the same graph with an elliptical shape and as *a* increases the graph's width and height grow except that it is flipped horizontally.

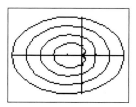

Same as Exercise 54 but with $r = a - n\cos(\theta)$. We get the same graph whose inner and outer loops grow as the size of *n* increases and *a* is held constant except that it is flipped horizontally.

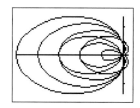

CHAPTER 9 ■ SECTION 9.5

1. vector **3.** scalar
5. scalar **7.** vector
9. scalar **11.** ⟨4, 3⟩
13. ⟨−7, 0⟩ **15.** ⟨0, −6⟩
17. $3\vec{i} + 11\vec{j}$

19. $4\vec{i} + \vec{j}$

21. $4\vec{i} + 13\vec{j}$

23. 0

25. $-6\vec{i} + 7\vec{j}$

27. $\langle 7, 98 \rangle$

29. $\langle -21, 18 \rangle$

31. $\vec{S} = \left(\dfrac{5}{\sqrt{34}}\right)\vec{i} - \left(\dfrac{3}{\sqrt{34}}\right)\vec{j}$

33. $c = 5, n = -1$

35. $(5, 5 + \sqrt{3})$

37. Velocity and speed are not the same thing because velocity is a vector (magnitude and direction) while speed is a scalar (magnitude only).

39. The speed of the boat will be 25 mph when the current is heading in the same direction as the boat.

A situation where the relative speed is not 25 mph would be when the current is heading in any direction other than the same direction as the boat. The figure shows a geometric example of when the current is heading the opposite direction as the boat for a resultant speed of 15 mph.

41. The zero vector would be a force with no magnitude or direction and could be expressed

$$\vec{z} = 0\vec{i} + 0\vec{j}$$
$$\langle 0, 0 \rangle$$

43. Answers vary. $\langle 10, -40 \rangle$

45. a. The magnitude is 605 mph and the direction is E 249° N.
   b. $\langle -564.82, -216.81 \rangle$
   c. 605 mph; represents the actual speed of the airplane

47. Yes, the magnitude of the wind is only 1.03 mph.

49. Let $c = 3, k = 2, \vec{AB} = |-1, 2|$, then
$$(c + k)\vec{AB} = c\vec{AB} + k\vec{AB}.$$
$$(3 + 2)\vec{AB} = 3\vec{AB} + 2\vec{AB}$$
$$5\vec{AB} = 3\vec{AB} + 2\vec{AB}$$
$$5|-1, 2| \stackrel{?}{=} 3|-1, 2| + 2|-1, 2|$$
$$|-5, 10| \stackrel{?}{=} |-3, 6| + |-2, 4|$$
$$|-5, 10| = |-5, 10|$$

51. Let $k = 5, \vec{A} = |6, 9|$ and $\vec{B} = |2, 4|$, then
$$k(\vec{A} + \vec{B}) = k\vec{A} + k\vec{B}.$$
$$5(|6, 9| + |2, 4|) \stackrel{?}{=} 5|6, 9| + 5|2, 4|$$
$$5(|6 + 2, 9 + 4|) \stackrel{?}{=} |30, 45| + |10, 20|$$
$$5|8, 13| \stackrel{?}{=} |40, 65|$$
$$\langle 40, 65 \rangle = \langle 40, 65 \rangle$$

53. Let $\vec{A} = |-2, 4|, \vec{B} = |0, 6|$, and $\vec{C} = |3, 5|$, then
$$(\vec{A} + \vec{B}) + \vec{C} = \vec{A} + (\vec{B} + \vec{C}).$$
$$(|-2, 4| + |0, 6|) + |3, 5| \stackrel{?}{=} |-2, 4| + (|0, 6| + |3, 5|)$$
$$|-2, 10| + |3, 5| \stackrel{?}{=} |-2, 4| + |3, 11|$$
$$\langle 1, 15 \rangle = \langle 1, 15 \rangle$$

CHAPTER 9 ■ REVIEW

1. $\sin(\theta) = \dfrac{18}{30} = \dfrac{3}{5}, \cos(\theta) = \dfrac{24}{30} = \dfrac{4}{5}, \tan(\theta) = \dfrac{18}{24} = \dfrac{3}{4}$,

$\sec(\theta) = \dfrac{30}{24} = \dfrac{5}{4}, \csc(\theta) = \dfrac{30}{18} = \dfrac{5}{3}, \cot(\theta) = \dfrac{24}{18} = \dfrac{4}{3}$

3. θ is between 60° and 90° because the sine of an angle is equivalent to the y-coordinate of a point on the unit circle and since the sine is 0.98 it is close to 1 which would be at 90°.

5. 36.87° **7.** 29.06°
9. $a \approx 8.08$ **11.** $a \approx 5.49$
13. $A \approx 41.41°$ **15.** $a \approx 11.82$
17. 667.88 miles
19. $A = 69°, b \approx 8.56, c \approx 3.49$
21. $C = 107°, a \approx 5.96, b \approx 9.94$
23. 999.79 feet
25. 10.9 kilometers or 8.92 kilometers
27. Answers vary. (4, 180°), (4, −180°)
29. Answers vary. (3, 60°), (3, −300°)
31. Answers vary. (6, 90°), (−6, 270°)
33. $(\sqrt{2}, 45°)$
35. (4, 0)
37. $(-1, -\sqrt{3})$
39. $(-3, 3\sqrt{3})$
41.
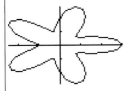
43. $\langle -5, 5 \rangle$
45. $-4\vec{i} - 4\vec{j}$

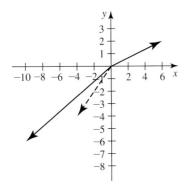

47. $\vec{B} = \dfrac{6}{3\sqrt{5}}\vec{i} - \dfrac{3}{3\sqrt{5}}\vec{j} = \dfrac{2}{\sqrt{5}}\vec{i} - \dfrac{1}{\sqrt{5}}\vec{j}$
49. $c = \sqrt{21}$
51. a. $\vec{W} = 6.08\vec{i} + 34.47\vec{j}, \vec{A} = 427.84\vec{i} + 329.48\vec{j}$
 b. The resultant velocity is 566.3 mph at E 50° N.

 c. 566.3 mph
 d. E 50° N; 40°

CHAPTER 10 ■ SECTION 10.1

3. $\sec^2(x)$ **5.** $\tan(\beta)$
7. 1 **9.** $x \approx 0.1912, 2.9504$
11. $\theta \approx 0.6119, 2.5297, 3.7535, 5.6712$
13. $-\cot(\theta)$ **15.** $-\csc^2(\theta)$
17. $\cos^2(\theta)$ **19.** $\theta \approx 2.1093, 4.1739$
21. $\phi \approx 0.1795, 2.9621$ **23.** $\sec(z)$
25. $\sin(\theta)$ **27.** $\csc(m)$
29. $\theta \approx 1.2310, 5.0522$ **31.** $\theta \approx 0.6932, 5.5900$
33. $\sin\left(\dfrac{11\pi}{7}\right)^2 \approx -0.6895, \sin^2\left(\dfrac{11\pi}{7}\right) \approx 0.9505$
35. $\cot^4\left(\dfrac{2\pi}{3}\right) \approx 0.1111, \cot\left(\dfrac{2\pi}{3}\right)^4 \approx 2.4207$
37. 1
39. $\sec^2(\theta)$
41. $\tan^2(\theta)$
43. $\theta \approx 1.9106, 4.3726$
45. $\theta \approx 1.1071, 2.0344, 4.2487, 5.1760$
47. $\sec(-\theta) = \sec(\theta)$, so the secant function has even symmetry; $\csc(-\theta) = -\csc(\theta)$, so the cosecant function has odd symmetry.

$$\text{Since } \sec(\theta) = \dfrac{1}{\cos(\theta)}, \text{ then}$$
$$\sec(-\theta) = \dfrac{1}{\cos(-\theta)}$$
$$= \dfrac{1}{\cos(\theta)}$$
$$= \sec(\theta)$$

Thus, $\sec(-\theta) = \sec(\theta)$, meaning that the secant function has even symmetry.

$$\text{Since } \csc(\theta) = \dfrac{1}{\sin(\theta)}, \text{ then}$$
$$\csc(-\theta) = \dfrac{1}{\sin(-\theta)}$$
$$= \dfrac{1}{-\sin(\theta)}$$
$$= -\dfrac{1}{\sin(\theta)}$$
$$= -\csc(\theta)$$

Thus, $\csc(-\theta) = -\csc(\theta)$, meaning that the cosecant function has odd symmetry. The demonstration for the tangent function's odd symmetry is identical.

49. $\theta = 0, 2\pi$
51. $\theta = \dfrac{\pi}{2}$

53. An expression is a representation of a number, whether it is a known quantity or variable quantity, such as −5, 6^2, or $2x + 1$. An equation exists when we describe one expression as having the same value as another expression, such as stating that $2x = 5$. We are stating that the expressions $2x$ and 5 have the same value. This allows us to, for example, find the value(s) of x that satisfy the equality. An identity is an equation that is true for all values of the variable(s) involved in the expressions. For example, $2(3x) = 6x$ is an identity because the expressions $2(3x)$ and $6x$ have the same value for *all* values of x.

55. Trigonometric identities can be useful in solving equations because they can sometimes simplify complicated equations. Identities may help us by reducing the number of trigonometric functions present or by simplifying the inputs to the trigonometric functions being used.

57.

Using the diagram above,

$$\sec\left(\frac{\pi}{2} - \theta\right) = \frac{1}{\cos\left(\frac{\pi}{2} - \theta\right)} \quad \text{and} \quad \csc(\theta) = \frac{1}{\sin(\theta)}$$

$$= \frac{1}{\frac{y}{r}} \qquad\qquad\qquad = \frac{1}{\frac{y}{r}}$$

$$= \frac{r}{y} \qquad\qquad\qquad = \frac{r}{y}$$

Thus, $\sec\left(\frac{\pi}{2} - \theta\right) = \csc(\theta)$.

59. $\theta \approx 27.92°$

61. B and C; Any two objects with a frictional coefficient greater than 0.3640 would remain stationary on a 20° incline.

63. $-\tan(\theta)$

65. The equation is true for all real numbers, so any value of θ on the interval $[-2\pi, 2\pi]$ is a solution.

CHAPTER 10 ■ SECTION 10.2

For many exercises, answers will vary. Consult the solutions manual for answers not shown here.

27. From the graph, it appears that it is an identity. (Proof not shown.)

29. From the graph, it is clear this is not an identity. Counterexample: Let $\theta = \frac{\pi}{4}$. Then

$$\csc\left(\frac{\pi}{4}\right)\tan\left(\frac{\pi}{4}\right) + \sec\left(\frac{\pi}{4}\right) = 2\sqrt{2} \text{ and } 2\cos\left(\frac{\pi}{4}\right) = \sqrt{2}.$$

41. b.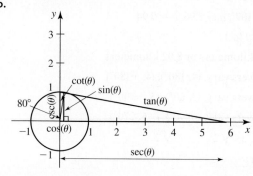

e. $\cos(80°) \approx 0.1736$, $\sin(80°) \approx 0.9848$, $\tan(80°) \approx 5.6713$, $\cot(80°) \approx 0.1763$, $\sec(80°) \approx 5.7588$, $\csc(80°) \approx 1.0154$

61. Answers vary. Since all trigonometric functions can be rewritten using cosine and sine, using this technique can help to combine like terms, create common denominators to add or subtract fractions, or assist in seeing how to simplify expressions.

63. This is not accurate. Since the value of $\cos(\theta)$ could be either positive or negative, and since $\sqrt{1 - \sin^2(\theta)}$ indicates only the positive root of $1 - \sin^2(\theta)$, the correct statement is $\cos(\theta) = \pm\sqrt{1 - \sin^2(\theta)}$.

CHAPTER 10 ■ SECTION 10.3

3. Answers vary.
   a. $45° − 210°$
   b. $\dfrac{-\sqrt{6} + \sqrt{2}}{4}$

5. Answers vary.
   a. $330° − 45°$
   b. $\dfrac{4}{\sqrt{6} - \sqrt{2}}$ or $\sqrt{6} + \sqrt{2}$

7. Answers vary.
   a. $\dfrac{2\pi}{3} + \dfrac{\pi}{4}\left(\text{or } \dfrac{8\pi}{12} + \dfrac{3\pi}{12}\right)$
   b. $\dfrac{-\sqrt{2} - \sqrt{6}}{4}$ or $-\dfrac{\sqrt{2} + \sqrt{6}}{4}$

9. Answers vary.
   a. $\dfrac{13\pi}{3} + \dfrac{3\pi}{4}\left(\text{or } \dfrac{52\pi}{12} + \dfrac{9\pi}{12}\right)$
   b. $\dfrac{\sqrt{3} - 1}{1 + \sqrt{3}}$ or $2 - \sqrt{3}$

11. -1

13. $-\dfrac{1}{2}$

15. $-\sqrt{2}$

17. $\dfrac{\sqrt{3}-1}{4}$

19. $\dfrac{2-\sqrt{3}}{4}$

21. $\dfrac{\sqrt{3}-1}{4}$

23. a. $\dfrac{150°}{2}$ b. $\dfrac{\sqrt{2+\sqrt{3}}}{2}$

25. a. $\dfrac{-45°}{2}$

   b. $-\dfrac{2-\sqrt{2}}{\sqrt{2}}$ or $\dfrac{\sqrt{2}-2}{\sqrt{2}}$ or $1-\sqrt{2}$

27. a. $\dfrac{\pi/4}{2}$ b. $\dfrac{2-\sqrt{2}}{\sqrt{2}}$ or $\sqrt{2}-1$

29. a. $\dfrac{7\pi/6}{2}$ b. $-(2+\sqrt{3})$ or $-2-\sqrt{3}$

31. $\dfrac{\sqrt{6}}{2}$

33. $\dfrac{\sqrt{2}}{2}$

35. $\dfrac{\sqrt{6}}{2}$

37. $\theta = 0, \dfrac{2\pi}{3}, \pi, \dfrac{4\pi}{3}, 2\pi$

39. $x \approx 0.3877, 1.1831, 3.5293, 4.3247$

41. $\theta \approx 1.1503, 1.9913, 4.2919, 5.1329$

43. The equation $\sin(\alpha) = 0$ will only be true when $\alpha = 0$, $\alpha = \pi$, or $\alpha = 2\pi$ on the interval $[0, 2\pi]$. Therefore, $\sin(\theta + \phi) = 0$ will only be true if one of the following relationships is true (taking into account the periodic nature of the sine function): $\theta + \phi = 0 + 2\pi n$, $\theta + \phi = \pi + 2\pi n$, or $\theta + \phi = 2\pi + 2\pi n$ [alternatively stated as $\theta + \phi = 2\pi n$, $\theta + \phi = \pi + 2\pi n$, or $\theta + \phi = 2\pi(n+1)$].

45. The range of $f(\theta) = \sin(2\theta)$ is $[-1, 1]$. However, the range of $g(\theta) = 2\sin(\theta)$ is $[-2, 2]$. Therefore, $2\sin(\theta)$ takes on values that $\sin(2\theta)$ does not, so $\sin(2\theta) = 2\sin(\theta)$ cannot be an identity.

47. We must determine in which quadrant the angle $\dfrac{\theta}{2}$ terminates, which tells us the sign of $\cos\left(\dfrac{\theta}{2}\right)$ and $\sin\left(\dfrac{\theta}{2}\right)$. The value of $\sin\left(\dfrac{\theta}{2}\right)$ will be positive if $\dfrac{\theta}{2}$ terminates in Quadrants I or II and the value of $\cos\left(\dfrac{\theta}{2}\right)$ will be positive if $\dfrac{\theta}{2}$ terminates in Quadrants I or IV.

49. a. $R = \dfrac{1}{32} v_0^2 \sin(2\theta)$

   b. $R \approx 1202.5$ feet

   c. hitting the ball at an angle of 45° will allow the ball to travel the farthest

   d. Our model discounts air resistance and other environmental factors. As a golf ball travels through the air, the movement of air over the dimple of the balls and the spin of the ball alters air pressure around the ball, creates drag, and changes the theoretical trajectory we were using in this problem. For more detailed information, you can perform research on impetus trajectories, Magnus lift, and drag.

51. a. The intensity of sounds A and B always have the same magnitude, but their signs are always opposite.

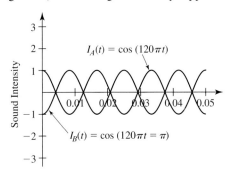

   b. The combined intensity of these two sounds is always "zero" meaning that the two sounds completely cancel each other out. Thus, sound B could be generated by a noise cancellation device to block sound A.

53. $\cos(\theta)\cos(\phi)\cos(\beta) - \cos(\theta)\sin(\phi)\sin(\beta) - \sin(\theta)\sin(\phi)\cos(\beta) - \sin(\theta)\cos(\phi)\sin(\beta)$

57. $\dfrac{2-\sqrt{2+\sqrt{2}}}{\sqrt{2-\sqrt{2}}}$

59. $\cos(\theta) - 4\sin^2(\theta)\cos(\theta)$

CHAPTER 10 ■ REVIEW

For many exercises, answers will vary. Consult the student solutions manual for answers not shown here.

1. $\cos(\alpha)$

3. $\tan(\theta)$

5. $\sin(\theta)$

7. $2\cos^2(\theta) - 1$ or $1 - 2\sin^2(\theta)$

9. $\theta \approx 1.8993, 4.3839$

11. $\theta \approx 1.1263, 5.1569$

13. $\theta \approx 0.4429, 2.6987$

15. $\theta \approx 2.6779, 5.8195$

27. From the graph, it is clear that this is not an identity. Let $x = \dfrac{\pi}{6}$. Then $\left[\csc\left(\dfrac{\pi}{6}\right) - 1\right]\left[\csc\left(\dfrac{\pi}{6}\right) + 1\right] = 3$ and $\tan^2\left(\dfrac{\pi}{6}\right) = \dfrac{1}{3}$.

29. From the graph, it appears that this is an identity.

31. $\dfrac{-\sqrt{2}-\sqrt{6}}{4}$ or $-\dfrac{\sqrt{2}+\sqrt{6}}{4}$

33. $\dfrac{\sqrt{2} - \sqrt{6}}{4}$

35. $-\dfrac{\sqrt{2}}{2}$

37. $-\dfrac{\sqrt{2}}{2}$

39. $\dfrac{\sqrt{3} - 1}{4}$

41. $\theta \approx 0.3614, 2.7802, 3.5030, 5.9218$

43. $2 + \sqrt{3}$

45. $-\dfrac{\sqrt{2 - \sqrt{2}}}{2}$

CHAPTER 11 ■ SECTION 11.1

NOTE: There are no exercises in Section 11.1.

CHAPTER 11 ■ SECTION 11.2

1. C

3. B

5. $\dfrac{x^2}{4} + \dfrac{y^2}{3} = 1$

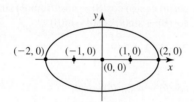

7. $\dfrac{x^2}{4.5} + \dfrac{y^2}{2.25} = 1$

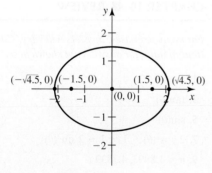

9. $\dfrac{(x-2)^2}{25} + \dfrac{(y+2)^2}{21} = 1$

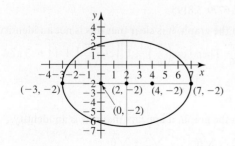

11. $\dfrac{(x-3)^2}{5} + \dfrac{(y-6)^2}{9} = 1$

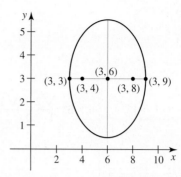

13. vertices: $(5,0)$ and $(-5, 0)$; foci: $(4,0)$ and $(-4, 0)$

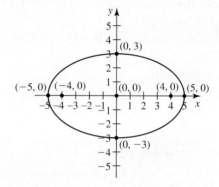

15. vertices: $(0, 2)$ and $(0, -2)$; foci: $(0, \sqrt{3})$ and $(0, -\sqrt{3})$

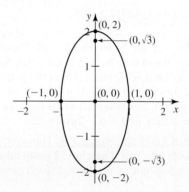

17. $\dfrac{(x-3)^2}{16} + \dfrac{y^2}{1} = 1$

19. $\dfrac{(x-6)^2}{36} + \dfrac{(y+3)^2}{9} = 1$

21. vertices: (3, 0) and (−3, 0);
foci: ($\sqrt{7}$, 0) and (−$\sqrt{7}$, 0);
center: (0, 0)

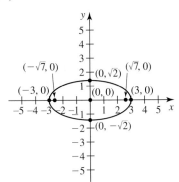

23. vertices: (0, 3) and (0, −3);
foci: (0, $\sqrt{5}$) and (0, −$\sqrt{5}$);
center: (0, 0)

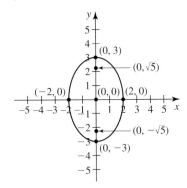

25. vertices: (4, 1) and (4, −5);
foci: (4, −2 + $\sqrt{5}$) and (4, −2 − $\sqrt{5}$);
center: (4, −2)

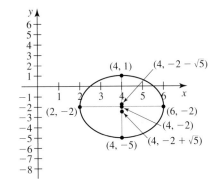

27. vertices: (−4, 1) and (0, 1);
foci: (−2 − $\sqrt{3}$, 1) and (−2 + $\sqrt{3}$, 1);
center: (−2, 1)

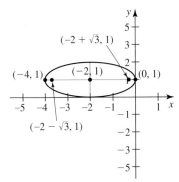

29. vertices: (−2, 1) and (−2, −3);
foci: (−2, −1 + $\sqrt{3}$) and (−2, −1 − $\sqrt{3}$);
center: (−2, −1)

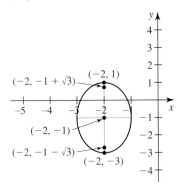

31. vertices: (−1, −2) and (3, −2);
foci: (1 + $\sqrt{3}$, −2) and (1 − $\sqrt{3}$, −2);
center: (1, −2)

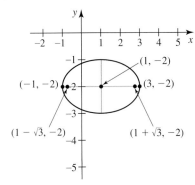

33. $e = \dfrac{\sqrt{5}}{3}$

35. $e = \dfrac{2}{3}$

37. $e = \dfrac{\sqrt{3}}{2}$

39. $C = (4, -5)$; $r = 2$

41. $C = (0, 0)$; $r = \sqrt{3}$

43. $x^2 + y^2 = 25$

45. $(x + 2)^2 + (y - 2)^2 = 25$

47. $(x + 9)^2 + (y - 4)^2 = 25$

49.

51.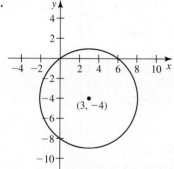

53. Yes, the city would be 100 miles away from the tower, which is within its broadcast range.

55. A circle has eccentricity of $e = 0$ and an ellipse has eccentricity $0 < e < 1$.

57. We need to know where to place the two tacks so we can draw the ellipse. The tacks need to be placed at the foci. If we place the center of the ellipse at $(0, 0)$ on a rectangular coordinate system, the foci are located at $(c, 0)$ and $(-c, 0)$. We know that $b^2 = a^2 - c^2$ where $a > b > 0$ and $9^2 = 14^2 - c^2$, so $c \approx 10.72$. If we place a tack at $(10.72, 0)$ and $(-10.72, 0)$ and use a string that is 28 inches long $(10.72 + 14 + 3.28 = 28)$, we can draw the specified ellipse.

59. a. bottom (1st level): $e \approx 0.843$
       middle (2nd level): $e \approx 0.841$
       top (3rd level): $e \approx 0.821$
    b. Yes. Based on the photo, it appears that the ellipses get more circular as you move up levels (to the top).

61. a. $\dfrac{x^2}{16^2} + \dfrac{y^2}{6.25^2} = 1$
    b. 3.12 feet from the center of the tank

63. $d_1 = \sqrt{(x - 0)^2 + (y - c)^2}$ and
    $d_2 = \sqrt{(x - 0)^2 + [y - (-c)]^2}$
    $\sqrt{x^2 + (y - c)^2} + \sqrt{x^2 + (y + c)^2} = 2a$
    $\sqrt{x^2 + (y - c)^2} = 2a - \sqrt{x^2 + (y + c)^2}$
    $[\sqrt{x^2 + (y - c)^2}]^2 = [2a - \sqrt{x^2 + (y + c)^2}]^2$
    $x^2 + (y - c)^2 = 4a^2 - 4a\sqrt{x^2 + (y + c)^2} + x^2 + (y + c)^2$
    $x^2 + y^2 - 2cy + c^2 = 4a^2 - 4a\sqrt{x^2 + (y + c)^2} + x^2 + y^2 + 2cy + c^2$
    $-4cy = 4a^2 - 4a\sqrt{x^2 + (y^2 + c)^2}$
    $0 = 4cy - 4a\sqrt{x^2 + (y + c)^2} + 4a^2$
    $4a\sqrt{x^2 + (y + c)^2} = 4cy + 4a^2$
    $a\sqrt{x^2 + (y + c)^2} = cy + a^2$
    $[a\sqrt{x^2 + (y + c)^2}]^2 = [cy + a^2]^2$
    $a^2(x^2 + y^2 + 2cy + c^2) = c^2y^2 + 2a^2cy + a^4$
    $a^2x^2 + a^2y^2 + 2a^2cy + a^2c^2 = c^2y^2 + 2a^2cy + a^4$
    $a^2x^2 + a^2y^2 - c^2y^2 + 2a^2cy - 2a^2cy = a^4 - a^2c^2$
    $a^2x^2 + y^2(a^2 - c^2) = a^2(a^2 - c^2)$
    $\dfrac{a^2x^2}{a^2(a^2 - c^2)} + \dfrac{y^2(a^2 - c^2)}{a^2(a^2 - c^2)} = \dfrac{a^2(a^2 - c^2)}{a^2(a^2 - c^2)}$
    $\dfrac{x^2}{a^2 - c^2} + \dfrac{y^2}{a^2} = 1$
    $\dfrac{x^2}{b^2} + \dfrac{y^2}{a^2} = 1$

65. If a is always greater than c then the numerator and denominator will never be equal and therefore the fraction cannot equal 1. Also since the denominator will always be greater than the numerator then the value of the fraction has to be less than 1.

CHAPTER 11 ■ SECTION 11.3

1. C

3. D

5. $y^2 = -8(x - 6)$

7. $(x-1)^2 = -8(y-2)$

9. $(y-0)^2 = 3(x-0.75)$

11. $(x-7)^2 = 4(3)(y+2)$

13. $(x-3)^2 = 4(5)(y-3)$

15. $(y-5)^2 = 4(5)(x-1)$

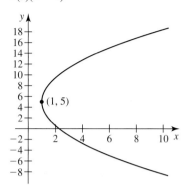

17. vertex: $(0, 0)$;
focus: $(0, 3)$

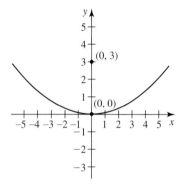

19. vertex: $(-2, 0)$;

focus: $\left(-2, -\dfrac{1}{4}\right)$

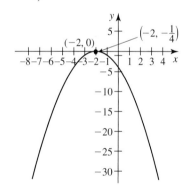

21. vertex: $(5, 0)$;
focus: $(9.5, 0)$

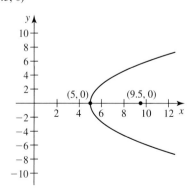

900 ANSWERS ■ CHAPTER 11

23. vertex: (−4, 4);
focus: (−4, 3)

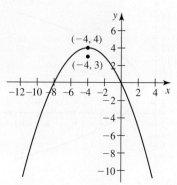

25. $(x - 3)^2 = 4(y - 2)$
27. $(y + 3)^2 = 16(x + 4)$
29. vertex: (0, 0); focus: (0, 7); axis of symmetry: $x = 0$;
directrix: $y = -7$

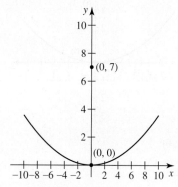

31. vertex: (0, 0); focus: $\left(\dfrac{1}{4}, 0\right)$; axis of symmetry: $y = 0$;
directrix: $x = \dfrac{1}{4}$

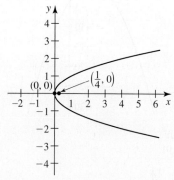

33. vertex: (4, 0); focus: (4, 4); axis of symmetry: $x = 4$;
directrix: $y = -4$

35. vertex: (−5, 0); focus: (−3.5, 0); axis of symmetry: $y = 0$;
directrix: $x = -6.5$

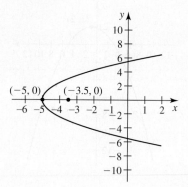

37. vertex: (−9, 0); focus: (−10.5, 0); axis of symmetry: $y = 0$;
directrix: $x = -7.5$

39. vertex: (−12, −7); focus: (−12, −6.25); axis of symmetry:
$x = -12$; directrix: $y = -7.75$

41. vertex: (−4, −1); focus: (−8, −1); axis of symmetry: $y = -1$;
directrix: $x = 0$

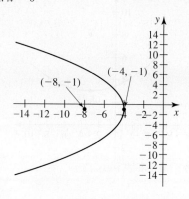

43. vertex: (9, −2); focus: (9, −14); axis of symmetry: $x = 9$; directrix: $y = 10$

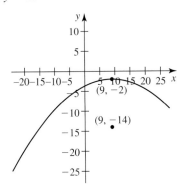

45. vertex: $\left(-\frac{3}{2}, \frac{1}{2}\right)$; focus: $\left(-1, \frac{1}{2}\right)$; axis of symmetry: $y = \frac{1}{2}$; directrix: $x = -2$

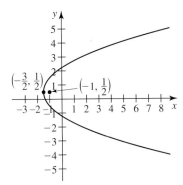

47. The relationship is that the focus and the directrix are both p units from the vertex.

49. If the squared term is x^2 the parabola opens vertically, if the squared term is y^2 the parabola opens horizontally.

51. $(x − 4)^2 = −2(y − 7)$

53. $x^2 = −18(y − 72)$

55. $x^2 = 29.25y$

57. $x^2 = 65.3(y − 3)$

59. $y^2 = 80x$

61. In $x^2 = 4py$, we assume the vertex is at (0, 0) and the directrix is at $y = −p$. With these assumptions, we can say $y = ax^2$ and $x^2 = 4py \Rightarrow y = \frac{x^2}{4p} = \frac{1}{4p}x^2$. Therefore, $a = \frac{1}{4p}$ so the relationship is that both forms give the same graph and we know that $a = \frac{1}{4p}$ or $p = \frac{1}{4a}$.

63. $\left(\frac{p}{2}, \frac{p}{2}\right)$

65. $\sqrt{2}(x − y)^2 = 8p(y + x)$ or $y = \pm 2\sqrt{2}\sqrt{p^2 + \sqrt{2}px} + 2\sqrt{2}p + x$

CHAPTER 11 ■ SECTION 11.4

1. E

3. A

5. B

7. $\dfrac{y^2}{9} - \dfrac{x^2}{16} = 1$

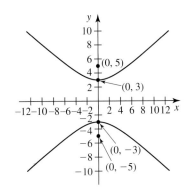

9. $\dfrac{y^2}{25} - \dfrac{x^2}{31.25} = 1$

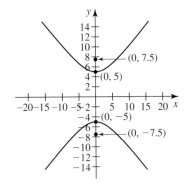

11. $\dfrac{y^2}{36} - \dfrac{x^2}{28} = 1$

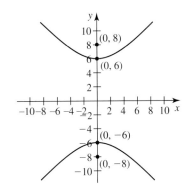

13. $\dfrac{(y − 1)^2}{0.25} - \dfrac{(x + 4)^2}{3.75} = 1$

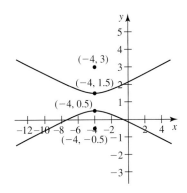

15. $\dfrac{(y-2)^2}{9} - \dfrac{(x-2)^2}{7} = 1$

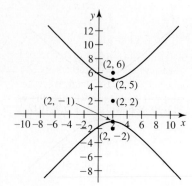

17. $\dfrac{(y-3)^2}{4} - \dfrac{(x-1)^2}{21} = 1$

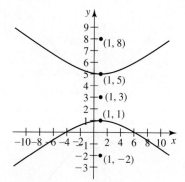

19. $\dfrac{(y-2)^2}{1} - \dfrac{(x+1)^2}{8} = 1$

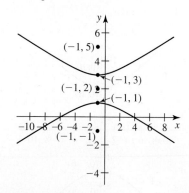

21. vertices: $(5, 0), (-5, 0)$;
foci: $(\sqrt{34}, 0), (-\sqrt{34}, 0)$;
opens left and right

23. vertices: $(0, 6), (0, -6)$;
foci: $(0, \sqrt{85}), (0, -\sqrt{85})$;
opens up and down

25. vertices: $(-1, 0), (1, 0)$;
foci: $(-\sqrt{5}, 0), (\sqrt{5}, 0)$;
opens left and right

27. vertices: $(0, \sqrt{3}), (0, -\sqrt{3})$;
foci: $(0, \sqrt{15}), (0, -\sqrt{15})$;
opens up and down

29. vertices: $(-1 + \sqrt{12}, -1), (-1 - \sqrt{12}, -1)$;
foci: $(-1 + \sqrt{20}, -1), (-1 - \sqrt{20}, -1)$;
opens left and right

31. $y = \pm\sqrt{\dfrac{x^2 b^2}{a^2} - b^2}$

33. $y = k \pm \sqrt{\dfrac{b^2(x-h)^2}{a^2} - b^2}$

35. One reason it is useful to solve the standard equation of a hyperbola for y is so we can graph it using technology.

37. $\dfrac{(x-1)^2}{4} - \dfrac{(y+2)^2}{9} = 1$

$y = \dfrac{3}{2}x - 3.5$

$y = -\dfrac{3}{2}x - 0.5$

39. $\dfrac{y^2}{16} - \dfrac{x}{13} = 1$

$y = \pm\dfrac{\sqrt{13}}{4}x$

41. a. $\dfrac{256x^2}{225} - \dfrac{256y^2}{31} = 1$

b. No. The location of the lightning strike would not lie on the hyperbola defined in part (a).

43. Although the difference in the distances from the fixed foci remains constant, the points on the hyperbola become farther away from each foci as d_1 and d_2 increase. This can best be seen from a table and graph. (See pages in this section that demonstrate the graphical development of a hyperbola.)

45. The hyperbola $\dfrac{(x-h)^2}{a^2} - \dfrac{(y-k)^2}{b^2} = 1$ is shifted h units right or left and k units up or down.

47. The third transmitting station allows us to get two hyperbolas from which we can consider the points (up to four) of intersection for the appropriate location of the ship.

49. $\dfrac{y^2}{16} - \dfrac{x^2}{9} = 1$

51. When the conic shock wave intersects the plane of Earth (even though Earth is a sphere we can consider it a plane over such a small distance), it will form one branch of a hyperbola because hyperbolas are formed by intersecting one branch of a cone with a plane.

53. To derive the equation of the hyperbola, we consider a hyperbola centered at the origin. The foci are c units away from the center and have the coordinates $F_1(0, -c)$ and $F_2(0, c)$.

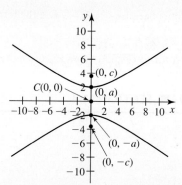

We define $2a$ to be the constant difference. That is, $d_1 - d_2 = 2a$. Using the distance formula and algebraic techniques, we derive the standard formula for the hyperbola.

$$d_1 = \sqrt{(x-0)^2 + (y-c)^2} \text{ and}$$
$$d_2 = \sqrt{(x-0)^2 + (y-(-c))^2}$$
$$\sqrt{x^2 + (y-c)^2} - \sqrt{x^2 + (y+c)^2} = 2a$$
$$\sqrt{x^2 + (y-c)^2} = 2a + \sqrt{x^2 + (y+c)^2}$$
$$\left[\sqrt{x^2 + (y-c)^2}\right]^2 = \left[2a + \sqrt{x^2 + (y+c)^2}\right]^2$$
$$x^2 + (y-c)^2 = 4a^2 + 4a\sqrt{x^2 + (y+c)^2} + x^2 + y^2 + 2cy + c^2$$
$$x^2 + y^2 - 2cy + c^2 = 4a^2 + 4a\sqrt{x^2 + (y+c)^2}$$
$$+ x^2 + y^2 + 2cy + c^2$$
$$-2cy = 4a^2 + 4a\sqrt{x^2 + (y+c)^2} + 2cy$$
$$-4cy - 4a\sqrt{x^2 + (y+c)^2} - 4a^2 = 0$$
$$-4a\sqrt{x^2 + (y+c)^2} = 4a^2 + 4cy$$
$$-a\sqrt{x^2 + (y+c)^2} = a^2 + cy$$
$$\left[-a\sqrt{x^2 + (y+c)^2}\right]^2 = (a^2 + cy)^2$$
$$a^2(x^2 + y^2 + 2cy + c^2) = a^4 + 2a^2cy + c^2y^2$$
$$a^2x^2 + a^2y^2 + 2a^2cy + a^2c^2 = a^4 + 2a^2cy + c^2y^2$$
$$a^2x^2 + a^2y^2 + 2a^2cy - 2a^2cy - c^2y^2 = a^4 - a^2c^2$$
$$a^2x^2 + y^2(a^2 - c^2) = a^2(a^2 - c^2)$$
$$a^2x^2 - y^2(c^2 - a^2) = -a^2(c^2 - a^2)$$
$$\frac{a^2x^2}{-a^2(c^2-a^2)} - \frac{y^2(c^2-a^2)}{-a^2(c^2-a^2)} = \frac{-a^2(c^2-a^2)}{-a^2(c^2-a^2)}$$
$$-\frac{x^2}{c^2-a^2} + \frac{y^2}{a^2} = 1$$
$$\frac{y^2}{a^2} - \frac{x^2}{b^2} = 1$$

The coordinates of every point (x, y) on the hyperbola satisfy this equation. This is known as the standard form for the equation of a hyperbola with foci on the y-axis and center at the origin.

55. Answers vary. We know that the asymptotes are $y = \pm\frac{b}{a}x$ and $a = 1 = b$. We also know that the center is $(0, 0)$. Therefore the formulas are $y = \pm\frac{1}{1}x$. The slopes are negative reciprocals of each other so the asymptotes are perpendicular. Similarly, the formula $\frac{y^2}{1} - \frac{x^2}{1} = 1$ is a rectangular hyperbola because $a = 1 = b$ and the formulas for the asymptotes are $y = \pm\frac{1}{1}x$. The slopes are negative reciprocals of each other so the asymptotes are perpendicular.

CHAPTER 11 ■ REVIEW

1. The plane must slice the cone where the two cones meet (at the vertices). We say that the circular conic section degenerates to a point.

3. When a vertical plane intersects the cone and goes through the vertices of the cones, two intersecting lines are formed. This is the same vertical plane used to create the hyperbola. But when this plane goes through the vertices of the cones, we say that the hyperbolic conic section degenerates to two intersecting lines.

5. $\dfrac{x^2}{49} + \dfrac{y^2}{13} = 1$

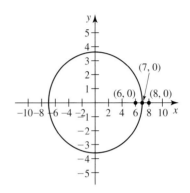

7. $\dfrac{x^2}{25} + \dfrac{y^2}{16} = 1$

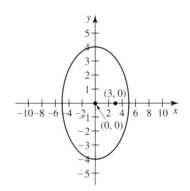

9. $\dfrac{x^2}{16.85} + \dfrac{y^2}{11.56} = 1$

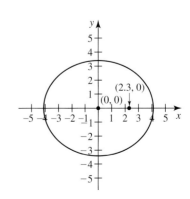

11. $\dfrac{(x-5)^2}{1} + \dfrac{(y-6)^2}{9} = 1$

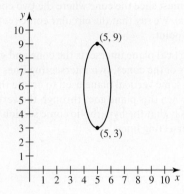

13. $\dfrac{(y+6)^2}{9} + \dfrac{(x-2)^2}{5} = 1$

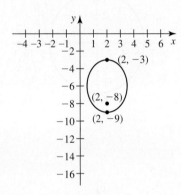

15. vertices: $(\sqrt{15}, 0)$ and $(-\sqrt{15}, 0)$;
foci: $(\sqrt{12}, 0)$ and $(-\sqrt{12}, 0)$

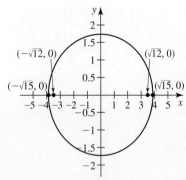

17. vertices: $(0, \sqrt{5})$ and $(0, -\sqrt{5})$;
foci: $(0, \sqrt{3})$ and $(0, -\sqrt{3})$

19. $\dfrac{x^2}{4} + \dfrac{y^2}{81} = 1$

21. $\dfrac{(x-2)^2}{64} + \dfrac{(y+2)^2}{16} = 1$

23. vertices: $(-3, 0)$ and $(3, 0)$;
foci: $(\sqrt{5}, 0)$ and $(-\sqrt{5}, 0)$;
center: $(0, 0)$

25. vertices: $(0, 5)$ and $(0, -5)$;
foci: $(0, 3)$ and $(0, -3)$;
center: $(0, 0)$

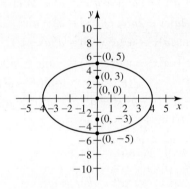

27. vertices: $(2, 0)$ and $(-4, 0)$;
foci: $(-1 - \sqrt{6}, 0)$ and $(-1 + \sqrt{6}, 0)$;
center: $(-1, 0)$

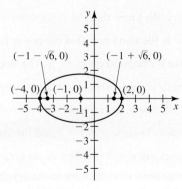

29. vertices: $(12, -4)$ and $(0, -4)$;
foci: $(6 - \sqrt{20}, -4)$ and $(6 + \sqrt{20}, -4)$;
center: $(6, -4)$

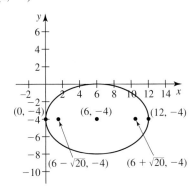

31. $e = \dfrac{\sqrt{20}}{6}$

33. $e = \dfrac{2}{3}$

35. $\dfrac{x^2}{81} + \dfrac{y^2}{144} = 1$

$e = \dfrac{\sqrt{63}}{12}$

37. a. $\dfrac{x^2}{483{,}680{,}000^2} + \dfrac{y^2}{483{,}110{,}000^2} = 1$

b. $e \approx 0.05$

39. $\dfrac{x^2}{638^2} + \dfrac{y^2}{319^2} = 1$

The focal width is 17.86 feet.

41. An ellipse that is 10 inches wide and 6 inches tall can be constructed by placing two thumbtacks into a piece of paper at two focal points that are 4 inches from a center point labeled C. We know that the foci have to be 4 inches apart because

$$b^2 = a^2 - c^2$$
$$9 = 25 - c^2$$
$$c^2 = 16$$
$$c = 4$$

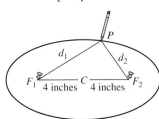

We will need a string that is 10 inches long because if we use the Pythagorean theorem we see that d_1 and d_2 in the diagram above must each be 5 inches long.

$$3^2 + 4^2 = d_2^2$$
$$25 = d_2^2$$
$$5 = d_2$$

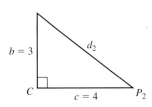

We now attach the end of the 10-inch piece of string to the base of each thumbtack. With a pencil holding the string taut, we move the pencil around the foci. The resultant shape is an ellipse that is 10 inches wide and 6 inches high.

43. B

45. C

47. $(x - 3)^2 = -4(y - 5)$

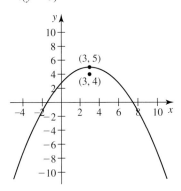

49. vertex: $(-5, -1)$;
focus: $(-5, -2.5)$

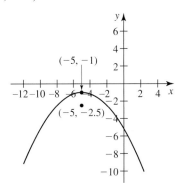

51. vertex: $(3, -2)$;
focus: $(3, -1.75)$;
axis of symmetry: $x = 3$;
directrix: $y = -2.25$

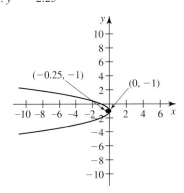

53. $\dfrac{x^2}{9} - \dfrac{y^2}{16} = 1$

55. $\dfrac{y^2}{25} - \dfrac{x^2}{24} = 1$

57. vertices: $(0, -5)$, $(0, 5)$;

foci: $(0, \sqrt{34})$, $(0, -\sqrt{34})$;

asymptotes: $y = \pm \dfrac{5}{3} x$;

opens up and down

CHAPTER 12 ■ SECTION 12.1

1. a. $a_1 = \dfrac{1}{4}, a_2 = \dfrac{2}{7}, a_3 = \dfrac{3}{10}, a_4 = \dfrac{4}{13}, a_5 = \dfrac{5}{16}, a_6 = \dfrac{6}{19}$

   b. converges to approximately 0.328

   c.

3. a. $a_1 = \dfrac{1}{2}, a_2 = \dfrac{1}{4}, a_3 = \dfrac{1}{8}, a_4 = \dfrac{1}{16}, a_5 = \dfrac{1}{32}, a_6 = \dfrac{1}{64}$

   b. converges to 0

   c.

5. a. $a_1 = 2, a_2 = 2.25, a_3 \approx 2.37, a_4 \approx 2.44, a_5 \approx 2.49, a_6 \approx 2.52$

   b. converges to approximately 2.71

   c.

7. $a_n = 10\left(\dfrac{1}{2}\right)^{n-1}$ where $a_1 = 10$

   converges to 0

9. $a_n = 2^n$

   does not converge

11. $a_n = \left(\dfrac{1}{2}\right)^n (-1)^{n-1}$ where $a_1 = \dfrac{1}{2}$

    converges to 0

13. $a_n = 100 + 2(n-1)$ where $a_1 = 100$

    does not converge

15. converges when $-1 < k \le 1$; otherwise diverges

17. a. 21, 34, 55, 89, 144

    b. $a_n = a_{n-1} + a_{n-2}$

19. a. The sequence $a_n = \dfrac{1}{2} n$ is strictly monotonic. The larger n becomes, the larger a_n becomes since multiplying larger and larger values of n results in larger and larger values of a_n. The sequence diverges.

b. The sequence $a_n = \dfrac{1}{n^2}$ is strictly monotonic. The larger n becomes, the smaller a_n becomes since squaring larger and larger values of n results in larger and larger values in the denominator, making a_n smaller and smaller. The sequence converges.

c. The sequence $a_n = \dfrac{(-1)^n}{n^2}$ is not strictly monotonic.

The numerator $(-1)^n$ has the effect of causing the terms of the sequence to oscillate. The sequence converges to 0.

21. Answers vary. The sequence $a_n = 1.25 a_{n-1}$ diverges. As the number of terms increase, subsequent terms get larger and larger since they are being multiplied by 1.25. The graph of this sequence supports this fact.

A second example is $a_n = 1.5n + 1$. Again, as the number of terms increase, the value of the term increases. This is because, in this explicitly defined formula, the term number n is multiplied by 1.5 then increased by 1. This will result in larger and larger terms of the sequence meaning it diverges. The graph confirms this fact.

23. For a strictly increasing sequence to converge, it will have to increase at a decreasing rate. The graph of such a sequence would resemble a function that is concave down. If increasing and concave down, the sequence will continue to increase but at a lesser and lesser rate, causing the sequence to converge.

   For a strictly decreasing function to converge, it will have to decrease at a smaller and smaller rate. The graph of this sequence would resemble a function that is concave up. If decreasing and concave up, the sequence will continue to decrease, but by less and less, causing the sequence to converge.

25. **a.** $3125
    b. $27
    c. $C_n = n^n$

27. The value of the car converges to 0.

29. **a.** 500, 700, 895, 1202
    b. $a_n = 500(1.4)^n - 85(n-1)$ where $n = 0$ represents the initial number of trees and n represents the number of 5-year periods
    c.

    d. No. As the 5-year plan is repeated, the number of trees increases without bound.
    e. The forest size does not stabilize.

31. Fingernail fungus: The maintenance level is about 675.5 mg and is reached after about 18 days (2.5 weeks), well before the 6-week treatment period is over.

    Toenail fungus: Those with toenail fungus issues will maintain this level for a longer period of time since the treatment time is 12 weeks.

33. Half-hour half-life
$$a_n = 0.001a_{n-1} + 20 \text{ where } a_1 = 20$$

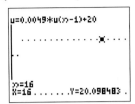

With such a short half-life, we see that Prilosec has a maintenance level of 20.02 mg. This is nearly the daily dosage.

3-hour half-life
$$a_n = 0.0049a_{n-1} + 20 \text{ where } a_1 = 20$$

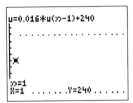

Even with a longer half-life, it is still short enough to have a maintenance level of 20.1 mg. This is slightly greater than the normal situation, but still nearly the daily dosage.

35. 4-hour half-life
$$a_n = 0.016a_{n-1} + 240 \text{ where } a_1 = 240$$

With such a short half-life, the maintenance level of 243.9 mg is quickly reached.

12-hour half-life
$$a_n = 0.25a_{n-1} + 240 \text{ where } a_1 = 240$$

In this case, a maintenance level of 320 mg is reached more slowly than in the other case.

37. converges to 1
39. converges to 0
41. converges to 1

CHAPTER 12 ■ SECTION 12.2

1. arithmetic; $a_n = 3 + 2(n - 1)$
3. geometric; $a_n = 40\left(\dfrac{1}{4}\right)^{n-1}$
5. geometric; $a_n = -1(-1)^{n-1} = (-1)^n$
7. geometric; $a_n = \dfrac{1}{2}\left(\dfrac{2}{3}\right)^{n-1}$
9. arithmetic; $a_n = 5 - 2.5(n - 1)$
11. arithmetic; $d > 0$
13. geometric; $r > 1$
15. geometric; $r > 1$
17. $a_n = 5 + (n-1)\left(\dfrac{7}{5}\right)$; $a_9 = 16.2$

19. $a_n = \frac{1}{2} + (n-1)\left(-\frac{1}{3}\right); a_8 = -\frac{11}{6}$

21. $a_n = -5 + (n-1)5; a_{15} = 65$

23. $a_n = 2 + (n-1)2.5; a_2 = 4.5$

25. $a_n = 2.75 + (n-1)\left(-\frac{1}{4}\right); a_{16} = -1$

27. $a_n = 60\left(\frac{1}{2}\right)^{n-1}; a_7 = 0.9375$

29. $a_n = 81\left(\frac{2}{3}\right)^{n-1}; a_6 = \frac{32}{3}$

31. $a_n = 5000(0.1)^{n-1}; a_7 = 0.005$

33. $a_n = 1(3)^{n-1}; a_2 = 3$

35. $a_n = 10{,}000(-0.1)^{n-1}; a_2 = -1000$

37. Similarity: there is a list of numbers that are added together; difference: in an arithmetic sequence the terms are generated by a common difference whereas in a geometric sequence the terms are generated by a common ratio.

39. The relationship between a geometric sequence and an exponential function is that both have a common ratio or change factor.

41. a. This is an arithmetic sequence because the amount increases by a common difference of $217 per credit. The terms of the sequence tell us what the tuition amount is that a student would owe for a given number of credits.
    b. $a_n = 214 + (n-1)214$
    c. The tuition for 11 credits will be $2354.

43. a. This is an arithmetic sequence because the number of champagne glasses increases by a common difference of 1 per row.
    b. $a_n = 1 + (n-1)1$

45. I refute the classmate's claim because of the following argument:

$$r = \frac{\frac{1}{4}}{\frac{1}{2}} = \frac{1}{2}$$

$$a_n = \frac{1}{2}\left(\frac{1}{2}\right)^{n-1}$$

$$a_{16} = \frac{1}{2}\left(\frac{1}{2}\right)^{16-1}$$

$$\neq 65$$

47. If we select $r = -1$, then any choice of a_n will allow the sequence to fluctuate from positive to negative for arbitrarily large values of n.

CHAPTER 12 ■ SECTION 12.3

1. geometric; the series has a constant ratio of $\frac{1}{2}$. That is, each subsequent term in the series is found by multiplying the previous term by $\frac{1}{2}$.

3. neither; the series does not have a constant difference or a constant ratio.

5. geometric; the series has a constant ratio of 2. That is, each subsequent term in the series is found by multiplying the previous term by 2.

7. 660

9. $\frac{115}{4}$

11. a. 2, 4, 6, 8, 10, 12, 14, 16, 18, and 20 are the numbers of logs in the pyramid in the first 10 rows.
    b. $a_n = 2 + (n-1)2$; In the 76th row there will be $a_{76} = 2 + (76-1)2 = 152$ logs.
    c. 10,100 logs

13. $S_4 = \frac{5}{8}, S_n = \left(\frac{2}{3}\right)\left(\frac{1}{2}\right)^n$; The series converges since $r = \frac{1}{2}$ and converges to $\frac{4}{3}$.

15. $S_4 = 11.11, S_n = \frac{100}{9} - \frac{100}{9}\left(\frac{1}{10}\right)^n$; The series converges to $\frac{100}{9}$.

17. B

19. E

21. D

23. This series will converge since $r = \frac{1}{5}$, which is between -1 and 1.

25. This series will converge since $r = \frac{2}{3}$, which is between -1 and 1.

27. $\frac{2}{9}$

29. $\frac{222}{99}$

31. $\frac{2}{11}$

33. 152

35. 69.5

37. 53.333

39. 10

41. 0.3884

43. If the common ratio, r, was equal to 1, each term of the sequence would be the same. If we were to repeatedly add a number to itself for a finite number of times, it certainly would not converge.

45. Answers vary. An infinite geometric series will only converge if $-1 < r < 1$. For example, $\sum_{n=1}^{\infty}\left(\frac{1}{5}\right)^n$ converges to a sum of $\frac{5}{4}$. This infinite series converges to a sum of $\frac{5}{4}$.

47. $692.3 million

49. Answers vary. We will assume that when $a_n < 0.1$ mg, the medicine is essentially removed, which occurs after 12.87 sixty-eight-hour periods, or about 37 days.

51. $e = \sum_{n=0}^{\infty} \frac{1}{n!} \approx 2.708$

53.

$$\sum_{n=1}^{\infty} \frac{1}{n(n+1)} = 1$$

$$\frac{1}{1(1+1)} + \frac{1}{2(2+1)} + \frac{1}{3(3+1)} + \frac{1}{4(4+1)} + \cdots$$

$$= \frac{1}{2} + \frac{1}{6} + \frac{1}{12} + \frac{1}{20} + \frac{1}{30} + \frac{1}{42} + \frac{1}{56} + \frac{1}{72}$$

$$+ \frac{1}{90} + \frac{1}{110} \cdots \approx 1$$

55. Answers vary. To group a finite number of terms of the infinite series in the way given is problematic.

CHAPTER 12 ■ REVIEW

1. a. $a_1 = 2, a_2 = \frac{5}{2}, a_3 = \frac{10}{3}, a_4 = \frac{17}{4}, a_5 = \frac{26}{5}, a_6 = \frac{37}{6}$

b.

c. diverges

3. a. $a_1 = 50, a_2 = 27, a_3 = 15.5, a_4 = 9.75, a_5 = 6.875, a_6 = 5.4375$

b.

c. converges to 4

5. $a_n = 3n - 5$; diverges

7. $a_n = \frac{2n}{2n+1}$; converges to 1

9. a. 11.5, 5.75, 2.875, 1.4375, 0.71875

b. $a_n = 11.5\left(\frac{1}{2}\right)^{n-1}$

c. about 0.045 mg

d. about 0.0000219 mg, or almost all of the medicine will be removed

11. $a_n = -5 + (n-1)2; a_{10} = 13$

13. $a_n = 21 + (n-1)(-5); a_2 = 16$

15. $a_n = \frac{1}{4} + (n-1)\left(\frac{1}{4}\right); a_1 = \frac{1}{4}$

17. $a_n = (-1)(2)^{n-1}; a_{10} = -512$

19. $a_n = 32{,}768\left(\frac{1}{8}\right)^{n-1}; a_3 = 512$

21. a. geometric; there is a common ratio of 1.053. The terms provide the starting salary of English majors in future years with a constant 5.3% increase in starting salary.

b. $a_n = 32{,}553(1.053)^{n-1}$

c. $40,022.53; This is the estimated starting salary for English majors in the year 2011.

23. arithmetic; each subsequent term in the series is found by adding 3 to the previous term.

25. arithmetic; each subsequent term in the series is found by adding 1 to the previous term.

27. 1050

29. 247.5

31. -860

33. $S_4 = 2.143$; diverges

35. $S_4 = 6.24$; converges to $\frac{25}{4}$

37. $\frac{1}{9}$

39. $\frac{2}{11}$

41. If we assume that the medicine is essentially gone from the bloodstream when the amount is less than 1 mg, it will take about 147 hours or a little over 6 days.

INDEX

A

Absolute value function, 430–431
Acute angle, 627
Acute triangle, 627
Adams, John Quincy, 735
Addition
  matrix, 141–143, 145
  vector, geometric, 663–665
Alignment, of data, 169
Altitude, of triangle, 535
Amplitude, 496
Angle(s)
  acute, 627
  in aviation context, 662
  congruent, 618
  definition of, 509
  in degrees, 515–520
  of depression, 621–622
  of elevation, 621–622
  greater than 2π radians, 515
  measure, 509–520, 622–623
  obtuse, 627
  positive, 513
  proportion of circumference corresponding with, 512–513
  in radians, 511–513
  reference, 533–535
  sketching, 513–514
  terminal side
    definition of, 513
    slope of, 574–575
Annual percentage yield (APY), 372
Antenna, parabolic, 752–753
Appollonius of Perga, 724
Arc cosine, 604
Archimedes, 86
Arc length
  degree measure and, 517
  radius and, 510
Arc sine, 604
Arc tangent, 604
Arithmetic sequence, 790–791, 793, 796–798
Astronomical unit (AU), 738
Asymptote
  horizontal
    definition of, 319, 321
    from equation, 323
    in exponential functions, 361
    finding, 322–325
    in real-world context, 319–321
  of hyperbola, 763–766
  vertical
    definition of, 315
    finding, from function equation, 317
    in real-world context, 315–317
Augmented matrix, 131
Average rate of change, 23–25
  calculating, 23–24, 25
  identifying limitations in, 24
  interpreting, 23–24, 25
  in nonlinear functions, 230–231
Axis of symmetry, of parabola, 746

B

Babylonians, 86
Base
  of logarithm, change of, 388–389
  of logarithmic function, 381
Best fit, line of, 96
Break-even point, 114

C

Calculator
  cosine on, 530–531, 540
  logarithms on, 387–389
  sine on, 530–531, 540
  tangent on, 575
Cartesian coordinate system, 35
  converting from, to polar, 656–657
  converting to, from polar, 655–656
  polar coordinate system and, 655–657
Center
  of circle, 740
  of ellipse, 728
  of hyperbola, 759
Centerline, 494–495
Change
  factor, 342, 343–345, 346, 359
  percentage, 340–349, 355–356
  rate of
    average, 23–25
      calculating, 23–24
      identifying limitations in, 24
      interpreting, 23–24
    constant
      calculating, in context, 70–72
      definition of, 69
      functions with, 68–80
      interpreting, 69–70
      using, 72–73
    inflection points and, 285
    percentage, 346
    percentage change rate *vs.*, 356
    units of, 71
China, 86
Circle
  center of, 740
  as conic section, 725
  definition of, 740
  equation for, 740–741
  radius of, 740
  unit
    cosine values on, 535–540
    defining, 523–526
    radians in, 511
    with radius other than 1, 526–528
    reference angle and, 533–535
    sine values on, 535–540
Coefficient
  correlation, 100–101
  of determination, 99
Cofunction identities, 686–687
Column matrix, 131
Combinations of functions, 410–417
  division, 412–414
  graphical method for, 410–411
  multiplication, 414–417
  numerical method for, 410–411
  predictions with, 412
  symbolic method for, 411–412
  understanding, 416–417
Common logarithms, 386–387
Common ratio, 791
Composite functions, 436–442
  creating, 438–442
  decomposition of, 442
  definition of, 438
  from equations, 440–441
  evaluating, 440
  from graphs, 441–442
  notation for, 438
  with tables, 439–440
  understanding, 436–438
Compounding frequency, 368
Compound interest, 368–377
  annual percentage yield and, 372
  compounding frequency and, 368
  continuous, 374–375
  different interest rates and, 371
  as exponential function, 371–372
  formula, 369–375

Compound interest (*continued*)
  future values with, 370–371
  modeling, 372
  periodic rate and, 368
Compression
  from equation, 209
  horizontal
    to change units, 209
    of cosine function, 550–551
    definition of, 207
    of function, 205–207
    inverse trigonometric functions with, 597
    of sine function, 550–551
  vertical
    to change units, 197
    of cosine function, 549–550
    definition of, 194
    of function, 192–194
    graph of, 195
    rate of change and, 195–197
    of sine function, 549–550
Concave down, 235
Concave up, 235
Concavity, 234–235
Condensed function notation, 16
Conic section(s)
  circle as, 725
  definition of, 724
  ellipse as, 725
  hyperbola as, 726
  parabola as, 726
Constant, definition of, 11
Constant of proportionality, 86, 304
Constant rate of change
  calculating, in context, 70–72
  definition of, 69
  functions with, 68–80
  interpreting, 69–70
  using, 72–73
Constant speed, 68–69
Continuous compounding, 374–375
Continuous decay rate, 376
Continuous functions, 197
Continuous growth, 368–377
Continuous growth factor, 373
Continuous growth rate, 376
Convergence
  of sequence, 781
  of series, 802
Coordinate system
  Cartesian, 35
    converting from, to polar, 656–657
    converting to, from polar, 655–656
    polar and, 655–657
  polar, 649–651
    Cartesian and, 655–657
    converting from, to Cartesian, 655–656
    converting to, from Cartesian, 656–657
    definition of, 649
    to describe relationships, 651

describing graphs with, 657
    graphing in, 652–654
    locating point in plane of, 650
    natural applications of, 651–653
    of point in polar plane, 650–651
    system of, 649–651
    using, 651
Correlation coefficient, 100–101
Correspondence, function as, 12
Cosecant function
  definition of, 582
  evaluating, 584
  right triangle and, 620
  values of, 583
Cosine function, 528–530
  calculator for, 530–531, 540
  characteristics of, 547–548
  difference identity for, 704
  domain of, 532–533
  exact values of, on unit circle, 535–540
  graph of, 544–557
  horizontal compression of, 550–551
  horizontal stretch of, 550–551
  interpreting, 529–530
  inverse
    definition of, 589, 590
    domain of, 602
    finding exact values with, 591–593
    graph of, 598–600
    solving equations with, 593–594
    using, 590–591
  modeling with, 555–557
  multiple transformations of, 552
  origin of term, 529
  reflection of, 549
  right triangle and, 620
  shifting of, 548
  transformations of, 548–553
  using, 529–530
  values of, 532, 539–540, 583
  vertical compression of, 549–550
  vertical stretch of, 549–550
Cosines, law of, 627–632
  applying, 632–634
  developing, 629–631
  Pythagorean theorem and, 631–634
Costs
  fixed, 114
  variable, 114
Cotangent function
  definition of, 578
  finding value of, 578–580
  graph of, 580
  interpreting value of, 578–580
  right triangle and, 620
  values of, 583
Cubic functions, 282–287
  definition of, 283
  equation of, algebraic method for finding, 283
  graph of, 284–285, 288
  inflection point of, 284
  modeling with, 285–287

in real-world context, 285–287
  standard form of, 283
Cubic regression model, 291–292

D

Data
  alignment of, 169
  linearization of, 395–396
  predicting unknown, with table, 26
  selecting linear function to model, 95
  trends in, 5
Decay
  exponential, 347–348
  logistic, 452–455
Decay factor, 342, 346, 347–348
Decay rate
  continuous, 376
  percentage, 346
Decay rate, percentage, 347–348, 357
Decimal, rational number from repeating, 804–805
Decision-factor equation, 2–3
Decreasing functions, 229–230
Degrees
  angle measure in, 515–520
  converting from, to radians, 517–518
  converting to, from radians, 519
Denominator, rationalize, 537
Dependent system of equations, 122, 134–135
Dependent variable, 14
Depression, angle of, 621–622
Descartes, René, 35, 649
Determination, coefficient of, 99
Difference identities, 704–708, 715
Differences
  first, 232–233
  second, 232–233, 234–235
  successive, 232–233
Direction, of vector, 662–663
Direct proportionality
  angles and, 510
  for conversion of units of measure, 86–87
  definition of, 86
  linear model for, 86
Directrix, 746
Direct variation
  definition of, 304
  examples of, 304
Discontinuities, removable, 318
Discrete functions, 197
Displacement vector in component form, 661
Divergence, of sequence, 781, 784
Division of functions, 412–414
Domain
  of cosine function, 532–533
  definition of, 36
  of inverse trigonometric functions, 602
  practical, 36–37, 413–414
  of rational function, 324–325
  of sine function, 532–533

Double-angle identities, 712–713, 715
Doubling time, 356–358

E

e
   as continuous growth factor, 373
   exponential functions using, general form of, 376
   history of, 375
   value of, 373

Eccentricity
   of ellipse, 736–739
   orbital, 738

Egypt, 86
Einstein, Albert, 627
Elevation, angle of, 621–622
Elimination method, for system of equations, 117–122

Ellipse
   anatomy of, 728
   center of, 728
   as conic section, 725
   definition of, 727
   eccentricity of, 736–739
   focus of, 727, 728
   major axis of, 728
   minor axis of, 728
   in real-world context, 735–736
   standard equation for, 728–735
   vertices of, 728

End behavior, of polynomial functions, 290–293
Entries, of matrix, 130

Equation(s)
   of circle, 740–741
   composite functions from, 440–441
   of cubic functions, 283
   determining compression from, 209
   for ellipse, 728–735
   exponential, solving, 379–390
   for exponential functions, 342
   horizontal asymptotes from, 323
   for horizontal lines, 76
   of hyperbola, 760–763
   interpreting in words, 44–46
   logarithmic, solving, 379–390
   of parabola, 746–751
   piecewise functions with, 425–431
   power, solving, 301
   quadratic
      a in, 249
      b in, 249
      c in, 249
      parameters in, 249
      in standard form, 246–249
      in vertex form, 257–260
   from sinusoidal graph, 566
   solving from graph, 17
   solving from table, 16–17
   solving function, 15–16
   solving with inverse cosine, 593–594
   solving with trigonometric functions, 597–598
   system of
      definition of, 109
      dependent, 122, 134–135
      elimination method for, 117–122
      equivalent, 118–119
      graph in solving, 111–112
      inconsistent, 122, 136–137
      with infinitely many solutions, 120–122
      linear, 109–123
      matrix for representation of, 130
      modeling with, 128–129
      with no solution, 120
      in real-world context, 136–137
      solving with matrix, 133–134, 134–135
      substitution method for, 112–116
      table with, 110–111
      three or more, 122–123
   trigonometric
      solving by graphing, 554–555
      solving with cofunction identities, 686–687
      solving with even-odd identities, 684–685
      solving with identities, 682
      solving with Pythagorean identities, 688–690
   for vertical lines, 76

Euler, Leonhard, 14
Evaluation, of function, 14–16
Even function identities, 682–684
Even functions, 184
Exponential decay, 347–348
Exponential equations, solving, 379–390
Exponential functions, 341–349
   change factors and, 343–345
   compound interest and, 371–372
   definition of, 342
   with e, 376
   equation for, 342
   geometric sequences and, 792
   graphs of, 355–364
   horizontal asymptote in, 361
   identifying based on graph, 359–360
   initial value of, 360
   inverse of, 393
   inverse relationship with logarithmic functions, 381
   modeling with, 355–364
   percentage change and, 345–349
   from percentage rate, 346–347

Exponential growth, 347, 450
Exponential model, 348–349, 355–356, 355–364, 394–397
Exponents, properties of, 381
Extrapolation, 97–98
Extrema, relative, 293

F

Factor
   change, 342, 343–345, 346, 359
   continuous growth, 373
   decay, 342, 346, 347–348
   growth, 342, 346

Factored form
   definition of, 265
   from graph, 267
   working with, 266

Factoring
   algebraic, 267
   quadratic functions, 266–270

Ferris, George, 523
Fibonacci, 86
Finite geometric series, 799–800
First differences, 232–233
Fixed costs, 114

Focus
   of ellipse, 727, 728
   of hyperbola, 759
   of parabola, 746

Force, vectors used to measure, 668–670

Form
   factored
      definition of, 265
      from graph, 267
      working with, 266
   point-slope
      definition of, 77
      graph of, 78
      of linear function, 77
      writing linear function in, 77
   reduced row echelon, 131
   slope-intercept
      definition of, 74
      graph of, 74–75
      of linear function, 74–77
   standard
      of cubic function, 283
      definition of, 78
      graph of function in, 79–80
      of linear function, 78–80
      quadratic equation in, 246–249
      of quadratic function, 262–265
      writing formula for linear function in, 78–79
   vertex, of quadratic function, 257–260

Formula
   change of base, 388
   compound interest, 369–375
   definition of, 26
   evaluating function from, 14–15
   functions represented by, 22–29
   multivariable, 28–29
   quadratic, 264–265
   for sequence, 785–786
   for transformed functions, 182–183

Frequency
   of periodic function, 496–497, 497–498

Frequency, compounding, 368

Function(s)
   absolute value, 430–431
   combinations of, 410–417
      division, 412–414
      graphical method for, 410–411

INDEX

Function(s)
  combinations of (*continued*)
    multiplication of, 414–417
    numerical method for, 410–411
    predictions with, 412
    symbolic method for, 411–412
    understanding, 416–417
  composite, 436–442
    creating, 438–442
    decomposition of, 442
    definition of, 438
    from equations, 440–441
    evaluating, 440
    from graphs, 441–442
    notation for, 438
    with tables, 439–440
    understanding, 436–438
  concavity of, 235
  condensed notation, 16
  with constant rate of change, 68–80
  constants in, 11
  continuous *vs.* discrete, 197
  as correspondence, 12
  cosecant
    definition of, 581
    evaluation of, 584
    right triangle and, 620
    values of, 583
  cosine, 528–530
    calculator for, 530–531, 540
    difference identity for, 704
    domain of, 532–533
    exact values of, on unit circle, 535–540
    graph of, 544–557
    interpreting, 529–530
    inverse
      definition of, 589, 590
      domain of, 602
      finding exact values with, 591–593
      graph of, 598–600
      range of, 602
      using, 590–591
    modeling with, 555–557
    origin of term, 529
    right triangle and, 620
    solving equations with, 593–594
    using, 529–530
    values of, 532, 539–540, 583
  cosine function, 528–530
    calculator for, 530–531, 540
    characteristics of, 547–548
    domain of, 532–533
    exact values of, on unit circle, 535–540
    graph of, 544–557
    horizontal compression of, 550–551
    horizontal stretch of, 550–551
    interpreting, 529–530
    multiple transformations of, 552
    origin of term, 529
    reflection of, 549
    shifting of, 548
    transformations of, 548–553
    using, 529–530
    values of, 532, 539–540
    vertical compression of, 549–550
    vertical stretch of, 549–550
  cotangent
    definition of, 578
    finding value of, 578–580
    graph of, 580
    interpreting value of, 578–580
    right triangle and, 620
    values of, 583
  cubic, 282–287
    definition of, 283
    equation of, algebraic method for finding, 283
    graph of, 284–285, 288
    inflection point of, 284
    modeling with, 285–287
    in real-world context, 285–287
    standard form of, 283
  decreasing, 229–230
  determining if inverse relationship is, 55–56
  determining if relationship is, 12–13
  determining if table represents, 23
  differences of, 232–235
  discrete *vs.* continuous, 197
  division of, 412–414
  domain of, 36–37
  estimating unknown value of, at specified input value, 26
  estimating values, from graph, 35
  evaluation of, 14–16, 17
  even, 184
  exponential, 341–349
    change factors and, 343–345
    compound interest and, 371–372
    definition of, 342
    with e, 376
    equation for, 342
    geometric sequences and, 792
    graphs of, 355–364
    horizontal asymptote in, 361
    identifying based on graph, 359–360
    initial value of, 360
    inverse of, 393
    inverse relationship with logarithmic functions, 381
    modeling with, 355–364
    percentage change and, 345–349
    from percentage rate, 346–347
  formulas for transformed, 182–183
  graphs of, 35–36
  horizontal reflection of, 178–180
  horizontal stretching of, 203–205
  image, 164
  increasing, 229–230
  interpreting meaning of, using words, 44–46
  inverse
    definition of, 51
    exponential, 393
    graph of, 53–54
    linear, 89–92
    modeling with, 52–55
    power, 306–307
    quadratic, 270
    rational, 325
  linear
    arithmetic sequence and, 790
    definition of, 73
    graph of, in slope-intercept form, 74–75
    horizontal intercept of, 75
    inverses of, 89–92
    modeling with, 85–92
    point-slope form of, 77–78
    selection of, to model data set, 95
    slope-intercept form of, 74–77
    slope of, 73
    standard form of, 78–80
    from verbal description, 73–74
    vertical intercept of, 75
  logarithmic, 381–382
    graphing of, 392–394
    modeling with, 392–397
  logistic, 449–455
    decay with, 452–453
    definition of, 451
    modeling with, 453–454
    in real-world context, 451–452
  multiplication of, 414–417
  multivariable, 18
  name, 14
  nonlinear, average rate of change in, 230–231
  notation, 13–14
  odd, 184
  parent, 164
  periodic
    amplitude of, 496
    definition of, 494
    frequency of, 496–497, 497–498
    graph of, 500–501
    midline of, 494–495
    from tabular data, 499–500
    transformation of, 498–499
  period of, 494
  piecewise, 424–431
    creating, 426–427, 428–429
    definition of, 424–425
    with equations, 425–431
    evaluating, 425–426
    graphing, 429–430
    modeling with, 473–474
    regression for, 428–429
    solving, 426
    using, 426–427
  polynomial
    definition of, 287
    end behavior of, 290–293
    graphs of, 288–289

higher-order, 282–294
inverse of, 293–294
modeling with, 289–290
relative extrema of, 293
power, 298–307
  decreasing, 303
  definition of, 299
  increasing, 303
  in real-world context, 300, 301–303
  with $x > 0$ and $b < 0$, 303
  with $x > 0$ and $0 < b < 1$, 301
quadratic
  of best fit, 252–254
  determining if data set represents, 250–251
  determining if table represents, 251
  difference properties of, 251
  factored form of, 265–266
  factoring, 266–270
  forms of, 257–270
  graphs of, 257–270, 288
    in vertex form by hand, 259–260
  horizontal intercept of, 260–262, 265
  inverse of, 270
  maximum value of, 259
  minimum value of, 259
  modeling with, 246–254, 261–262
  standard form of, 262–265
  vertex form of, 257–260
    standard *vs.*, 262–265
  x-coordinate of vertex of, 263
quartic, graph of, 288
quintic, graph of, 289
range of, 36–37
rational, 314–325
  definition of, 314
  domain of, 324–325
  graph of, 323–324
  horizontal asymptotes and, 319–325
  inverse of, 325
  removable discontinuities and, 318
  vertical asymptotes and, 315–317
recognizing, in words, 42–44
represented by formulas, 22–29
represented by graphs, 34–39
represented by tables, 22–29
represented by words, 42–46
secant
  definition of, 581
  right triangle and, 620
  value of, 584
  values of, 583
sine, 528–530
  calculator for, 530–531, 540
  domain of, 532–533
  exact values of, on unit circle, 535–540
  graph of, 544–557
  interpreting, 529–530
  inverse
    definition of, 596
    domain of, 602

    graph of, 601–602
    using, 594–595
  modeling with, 555–557
  origin of term, 529
  right triangle and, 620
  using, 529–530
  values of, 532, 539–540, 583
sine function, 528–530
  calculator for, 530–531, 540
  characteristics of, 547–548
  domain of, 532–533
  exact values of, on unit circle, 535–540
  graph of, 544–557
  horizontal compression of, 550–551
  horizontal stretch of, 550–551
  interpreting, 529–530
  multiple transformations of, 552
  origin of term, 529
  reflection of, 549
  shifting of, 548
  transformations of, 548–553
  using, 529–530
  values of, 532, 539–540
  vertical compression of, 549–550
  vertical stretch of, 549–550
single-variable, 11
tangent
  on calculator, 575
  definition of, 574
  graph of, 575–577
  inverse
    definition of, 596
    domain of, 602
    graph of, 601–602
    using, 595–596
  right triangle and, 620
  using, 577
  values of, 583
translating, in words, to graph, 43–44
translating, in words, to symbolic notation, 43
trigonometric
  construction approach for, 699–700
  inverse, 588–604
  modeling with, 562–567
  reciprocal, 577–583
variables in, 11
vertical compression of, 192–194
vertical line test and, 13
vertical reflection of, 175–177
vertical stretch of, 192–194
Future values, 370–371

G

Galton, Sir Francis, 98
Geometric sequence, 791–793
Geometric series, 798–805
Geometric vectors, 660–662
Geometry, trigonometry and, 619–621
Graph(s)
  of absolute value function, 430–431
  change factor and, 359

  combining functions with, 410–411
  composite functions from, 441–442
  of cosine function, 544–557
  of cotangent function, 580
  of cubic function, 284–285, 288
  estimating function values from, 35
  of exponential function, 355–364
  exponential function identification based on, 359–360
  factored form from, 267
  of functions, 35–36
  functions represented by, 34–39
  of geometric sequence, 792
  horizontal shift of, 164–166
  initial value and, 360
  input values from, 36
  interpretation of
    in context, 37–39
    in words, 45–46
  of inverse cosine function, 598–600
  of inverse function, 53–54
  of inverse sine function, 601–602
  of inverse tangent function, 601–602
  of inverse trigonometric functions, 598–608
  of logarithmic functions, 392–394
  models presented by, 5–6
  output values from, 36
  of periodic functions, 500–501
  of piecewise functions, 429–430
  of point-slope form, 78
  in polar coordinates, 652–654
  of polynomial functions, 288–289
  of quadratic functions, 257–270, 288
    in real-world context, 268–269
    in vertex form by hand, 259–260
  of quartic functions, 288
  of quintic functions, 289
  of rational function, 323–324
  real-world meaning of, 37–39
  of sequence, 782, 792
  of sine function, 544–557
  sinusoidal, 564–565
  slope and, 77
  of slope-intercept form, 74–75
  solving function equation from, 17
  solving system with, 111–112
  of standard form, 79–80
  of tangent function, 575–577
  of transformations, 167
  translating function in words into, 43–44
  trigonometric equations with, solving, 554–555
  vertex form of quadratic function from, 262
  of vertical compression, 195
  vertical shift of, 162–163
Growth
  continuous, 368–377
  exponential, 347, 450
  factor, 342, 346, 373

Growth (*continued*)
  logistic, 450–452
  patterns of, 340–341
Growth rate
  continuous, 376
  percentage, 346
Growth rate, percentage, 357

H
Half-angle identities, 713–714, 715
Half-life, 356–358
Higher-order polynomial functions, 282–294
Holes, 318
Horizontal alignment of data, 169
Horizontal asymptote
  definition of, 319, 321
  from equation, 323
  in exponential functions, 361
  finding, 322–325
  in real-world context, 319–321
Horizontal compression
  to change units, 209
  of cosine function, 550–551
  definition of, 207
  of function, 205–207
  inverse trigonometric functions with, 597
  of sine function, 550–551
Horizontal intercept
  definition of, 37
  of linear function, 75
  of quadratic function, 260–262, 265
Horizontal lines, 76
Horizontal reflections, 177–180
Horizontal shift
  definition of, 164
  on graph, 164–166
Horizontal stretching
  to change units, 209
  of cosine function, 550–551
  definition of, 207
  of function, 203–205
  of sine function, 550–551
Hyperbola
  anatomy of, 759
  applications of, 766–769
  asymptotes of, 763–766
  center of, 759
  as conic section, 726
  definition of, 758
  equation for, 760–763
  focus of, 759
  transverse axis of, 759
  vertex of, 759
Hypotenuse
  definition of, 619
  measuring, 620–621

I
Identity(ies), trigonometric
  cofunction, 686–687
  definition of, 680–681
  determining existence of, 681
  difference, 704–708, 715
  double-angle, 712–713, 715
  even-odd, 682–684
  fundamental, 681–682
  half-angle, 713–714, 715
  product-to-sum, 711–712, 715
  Pythagorean, 688–690
  for solving trigonometric equations, 682
  sum, 704–708, 715
  sum-to-product, 708–711, 715
  verification of, 692–700
Image function, 164
Inconsistent system of equations, 122, 136–137
Increasing functions, 229–230
Independent variable, 14
Infinite geometric series, 802
Inflection point
  of cubic function, 284
  of nonlinear function, 235–236
  rate of change and, 285
Initial side, 513
Initial value, 37, 74, 360
Input, in function notation, 14
Instant, rates of change at, 231–232
Intercept
  horizontal
    definition of, 37
    of linear function, 75
    of quadratic function, 260–262, 265
  vertical, 37, 75
Interest, compound, 368–377
  annual percentage yield and, 372
  compounding frequency and, 368
  continuous, 374–375
  different interest rates and, 371
  as exponential function, 371–372
  formula, 369–375
  future values with, 370–371
  modeling, 372
  periodic rate and, 368
Interest rates, 371
Interpolation, 97–98
Interval notation, 546
Inverse cosine function
  definition of, 589, 590
  domain of, 602
  finding exact values with, 591–593
  graph of, 598–600
  range of, 602
  solving equations with, 593–594
  using, 590–591
Inverse function
  definition of, 51
  exponential, 393
  graph of, 53–54
  linear, 89–92
  modeling with, 52–55
  polynomial, 293–294
  power, 306–307
  quadratic, 270
  rational, 325
  trigonometric, 588–604
Inverse relationship
  definition of, 53
  determining if function, 55–56
  that are not inverse functions, 55–56
Inverse sine function
  definition of, 596
  domain of, 602
  graph of, 601–602
  using, 594–595
Inverse tangent function
  definition of, 596
  domain of, 602
  graph of, 601–602
  use of, 595–596
Inverse trigonometric functions, 588–604
Inverse variation
  definition of, 304
  examples of, 304
  modeling of, 305–306
Isosceles right triangle, 537
Israelites, 86

K
Kepler, Johannes, 738

L
Latus rectum, 746
Law of cosines, 627–632
  applying, 632–634
  developing, 629–631
  Pythagorean theorem and, 631–634
Law of sines
  definition of, 640
  developing, 639–640
  for oblique triangle, 638–639
  triangulation and, 640–644
Least squares, 96–97
Least squares regression line, 98
Leibniz, Gottfried Wilhelm, 12, 375
Limiting value, 450
  lower, 451
  upper, 451
Line(s)
  of best fit, 96
  horizontal, 76
  least squares regression, 98
  of symmetry, 184
  vertical, 76
Linear equations, system of, 109–123
Linear function
  arithmetic sequence and, 790
  definition of, 73
  graph of, in slope-intercept form, 74–75
  horizontal intercept of, 75
  inverses of, 89–92
  modeling with, 85–92
  point-slope form of, 77–78
  selection of, to model data set, 95

slope-intercept form of, 74–77
slope of, 73
standard form of, 78–80
from verbal description, 73–74
vertical intercept of, 75
Linearization, of data, 395–396
Linear models
   for directly proportional qualities, 86
   knowing when to use, 85–89
   predicting values with, 97–98
   from table, 87–89
   from verbal description, 87
Linear regression, 95
Linear regression model
   computing, 99–100
   correlation coefficient and, 100–101
   definition of, 98
   determining, 98–99
Line of best fit, 96
Logarithmic equations, solving, 379–390
Logarithmic functions, 381–382
   graphing, 392–394
   modeling with, 392–397
Logarithms, 380–381
   with calculator, 387–389
   changing base of, 388–389
   common, 386–387
   errors with, 390
   evaluating, 382
   exact, calculation of, 387–388
   exponential model with, 394–397
   in functions, 381–382
   natural, 386–387
   rules of, 382–386
   solving exponential equations with, 385–386
Logistic decay, 452–455
Logistic functions, 449–455
   decay with, 452–453
   definition of, 451
   modeling with, 453–454
   in real-world context, 451–452
Logistic growth, 450–452
Logistic regression, 453–454
Lower limiting value, 451

M

Machin, John, 86
Magnitude, 660
Maintenance level, 781, 782–783
Major axis, of ellipse, 728
Malthus, Thomas Robert, 451
Matrix(ices)
   addition, 142–144, 146
   applications, 142–148
   augmented, 131
   column, 130
   determining dimensions of, 130
   determining entry values of, 130
   entries of, 130
   multiplication, 144–146
   notation, 129–130
   operations, 142–148

reduced, 132
reduced row echelon form for, 131
row, 130
row operations in, 131–134
solving dependent system of equations with, 134–135
solving system of equations with, 133–134
square, 130
Maximum, relative, 293
Maximum value, of quadratic function, 259
Measure
   angle, 509–520, 622–623
      in degrees, 515–520
      in radians, 511–513
   units of, 86–87
Midline, 494–495
Minimum, relative, 293
Minimum value, of quadratic function, 259
Minor axis, of ellipse, 728
Model(s)
   analysis of
      in graphical form, 6
      in numerical form, 4–5
   of best fit, 252–253
   choosing, 464–475
      from table, 464–471
      from verbal description, 472–475
   compound interest, 372
   with cubic functions, 285–287
   cubic regression, 291–292
   data trends and, 5
   decision-factor equation and, 2–3
   definition of, 2
   of doubling time, 356–358
   exponential, 348–349, 355–356, 355–364, 394–397
   graphical presentation of, 5–6
   of half-life, 356–358
   higher-order polynomial, 289–290
   with inverse functions, 52–55
   of inverse variation, 305–306
   linear
      for directly proportional qualities, 86
      knowing when to use, 85–89
      predicting values with, 97–98
      from table of data, 87–89
      from verbal description, 87
   with linear functions, 85–92
   linear regression
      computing, 99–100
      correlation coefficient and, 100–101
      definition of, 98
      determining, 98–99
   logarithmic, 394
   with logarithmic functions, 392–397
   with logistic regression, 453–454
   numerical presentation of, 3–5
   with percentage rates of change, 355–356

with quadratic functions, 246–254, 261–262
quadratic regression and, 252–254
selecting linear functions for, 95
sinusoidal, 562–567
   from real-world data, 564–565
   from table, 565–566
   from verbal description, 563–564
with system of three equations, 128–129
with trigonometric functions, 562–567
Multiplication
   of functions, 414–417
   matrix, 144–146
   scalar, 144–146
   of vectors, 665–667
Multivariable formula, 28–29
Multivariable functions, 18

N

Natural logarithms, 386–387
Nonlinear functions, average rate of change in, 230–231
Notation
   composite, 438
   condensed function, 16
   function, 13–14
   interval, 546
   matrix, 129–130
   sequence, 781
   symbolic, translating function in words into, 43
Numbers
   mathematical model presentation with, 3–5
   rational, from repeating decimal, 804–805

O

Oblique triangle, 627, 628–629, 638–639
Obtuse angle, 627
Obtuse triangle, 627
Odd function identities, 682–684
Odd functions, 184
Orbital eccentricity, 738
Orosz, Matt, 751
Output, in function notation, 14

P

Parabola, 248
   anatomy of, 746
   applications of, 751–754
   axis of symmetry of, 746
   as conic section, 726
   definition of, 745–746
   equation for, 746–751
   focus of, 746
   latus rectum of, 746
   vertex of, 258, 263–265, 746
Parabolic antenna, 752–753
Parabolic paths, 753–754
Parabolic trough, 751–752
Paraboloid, 752

Parent function, 164
Patterns of growth, 340–341
Pearl, R., 451
Percentage change, 340–349, 355–356
Percentage decay rate, 346, 347–348, 357
Percentage growth rate, 346, 357
Period, of function, 494
Periodic functions
  amplitude of, 496
  definition of, 494
  frequency of, 496–497, 497–498
  graph of, 500–501
  midline of, 494–495
  from tabular data, 499–500
  transformation of, 498–499
Periodicity, 494
Periodic rate, 368
Phase shift
  definition of, 553
  determining, 554
π, 86
Piecewise functions, 424–431
  creating, 426–427, 428–429
  definition of, 424–425
  with equations, 425–431
  evaluating, 425–426
  graphing, 429–430
  modeling with, 473–474
  regression for, 428–429
  solving, 426
  using, 426–427
Point(s)
  inflection
    of cubic function, 284
    of nonlinear function, 235–236
    rate of change and, 285
Point-slope form
  definition of, 77
  graph of, 78
  of linear function, 77
  writing linear function in, 77
Polar coordinates
  Cartesian and, 655–657
  converting from, to Cartesian, 655–656
  converting to, from Cartesian, 656–657
  definition of, 649
  to describe relationships, 651
  describing graphs with, 657
  graphing in, 652–654
  locating point in plane of, 650
  natural applications of, 651–653
  of point in polar plane, 650–651
  system of, 649–651
  using, 651
Pole, 649
Polynomial functions
  definition of, 287
  end behavior of, 290–293
  graphs of, 288–289
  higher-order, 282–294

  inverses of, 293–294
  modeling with, 289–290
  relative extrema of, 293
Position vector, 660
Power equations, solving, 301
Power functions, 298–307
  decreasing, 303
  definition of, 299
  increasing, 303
  inverses of, 306–307
  in real-world context, 300, 301–303
  with $x > 0$ and $b < 0$, 303
  with $x \geq 0$ and $0 < b < 1$, 301
Practical domain, 36–37, 413–414
Practical range, 36–37, 413–414
Product-to-sum identities, 711–712, 715
Proportionality
  constant of, 86, 304
  direct
    angles and, 510
    for conversion of units of measure, 86–87
    definition of, 86
    linear model for, 86
  of sides of triangle, 618
Pythagoras, 536
Pythagorean identities, 688–690
Pythagorean theorem, 536, 631–634

Q

Quadratic equation
  a in, 249
  b in, 249
  c in, 249
  parameters in, 249
  in standard form, 246–249
  in vertex form, 257–260
Quadratic formula, 264–265
Quadratic function(s)
  of best fit, 252–254
  determining if data set represents, 250–251
  determining if table represents, 251
  difference properties of, 251
  factored form of, 265–266
  factoring, 266–270
  forms of, 257–270
  graphs of, 257–270, 288
    in real-world context, 268–269
    in vertex form by hand, 259–260
  horizontal intercept of, 260–262, 265
  inverse of, 270
  maximum value of, 259
  minimum value of, 259
  modeling with, 246–254, 261–262
  standard form of, 262–265
  vertex form of, 257–260
    standard vs., 262–265
  x-coordinate of vertex of, 263
Quadratic regression, 52, 252–254
Quartic functions, graph of, 288
Quintic functions, graph of, 289

R

Radians, 511–513
  converting from degrees, 517–518
  converting to degrees, 519
Radius
  arc length and, 510
  of circle, 740
Range
  definition of, 36
  of inverse trigonometric functions, 602
  practical, 36–37, 413–414
Rate
  continuous decay, 376
  continuous growth, 376
  interest, 371
  periodic, 368
Rate of change
  average, 23–25
    calculating, 23–24, 25
    identifying limitations in, 24
    interpreting, 23–24, 25
    in nonlinear functions, 230–231
  constant
    calculating, in context, 70–72
    definition of, 69
    functions with, 68–80
    interpreting, 69–70
    using, 72–73
  inflection points and, 285
  at instant, 231–232
  percentage, 346, 355–356
  percentage change rate vs., 356
  units of, 71
  variable, 228–236
  vertical compression and, 195–197
  vertical stretch and, 195–197
Ratio, common, 791
Rational functions, 314–325
  definition of, 314
  domain of, 324–325
  graph of, 323–324
  horizontal asymptotes and, 319–325
  inverse of, 325
  removable discontinuities and, 318
  vertical asymptotes and, 315–317
Rationalize denominator, 537
Rational number, from repeating decimal, 804–805
Reciprocal trigonometric functions, 577–583
Rectangular coordinate system, 35
Reduced matrix, 132
Reduced row echelon form, 131
Reed, L. J., 451
Reference angle, 533–535
Reflections
  of cosine function, 549
  horizontal, 177–180
  shifts and, 180–183
  of sine function, 549
  vertical, 175–177

Regression
  cubic, 291–292
  for exponential model, 358, 362–363
  linear, 95
  for logarithmic model, 394
  logistic, 453–454
  for piecewise functions, 428–429
  quadratic, 252–254
Relationship, inverse
  definition of, 53
  determining if function, 55–56
  that are not inverse functions, 55–56
Relative extrema, 293
Relative maximum, 293
Relative minimum, 293
Removable discontinuities, 318
Repeating decimal, rational number from, 804–805
Residual, 96
Right triangle
  adjacent side of, 619
  hypotenuse of, 619
  opposite side of, 619
  trigonometric functions and, 620
Row matrix, 130
Row operations, 131–134

S
Scalar, definition of, 144
Scalar multiplication, 144–146
Secant function
  definition of, 581
  evaluating, 584
  right triangle and, 620
  values of, 583
Second differences, 232–233, 234–235
Sequence(s)
  arithmetic, 790–791, 793, 796–798
  convergence of, 781
  definition of, 780–781
  divergence of, 781, 784
  formula for, 785–786
  geometric, 791–793
  on graph, 782, 792
  maintenance level of, 781, 782–783
  notation, 781
  using, 783–784
Series
  arithmetic sequence and, 796–798
  computing, 797–798
  convergence of, 802
  definition of, 797
  formula for, 800–801
  geometric, 798–805
  infinite, 803–804
Shift
  analyzing, 168–169
  of cosine function, 548
  horizontal
    definition of, 164
    on graph, 164–166
    reflections and, 180–183

of sine function, 548
on table, 168
vertical
  definition of, 163
  on graph, 162–163
  in real-world context, 164
  relationship between two functions differing by, 163
Similar triangle
  definition of, 618
  measuring with, 618–619
Sine function, 528–530
  calculator for, 530–531, 540
  characteristics of, 547–548
  domain of, 532–533
  exact values of, on unit circle, 535–540
  graph of, 544–557
  horizontal compression of, 550–551
  horizontal stretch of, 550–551
  interpreting, 529–530
  inverse
    definition of, 596
    domain of, 602
    graph of, 601–602
    using, 594–595
  modeling with, 555–557
  multiple transformations of, 552
  origin of term, 529
  reflection of, 549
  right triangle and, 620
  shifting of, 548
  transformations of, 548–553
  using, 529–530
  values of, 532, 539–540, 583
  vertical compression of, 549–550
  vertical stretch of, 549–550
Sines, law of
  definition of, 640
  developing, 639–640
  for oblique triangle, 638–639
  triangulation and, 640–644
Single-variable function, 11
Sinusoidal graph, 564–565, 566
Sinusoidal models, 562–567
  from real-world data, 564–565
  from table, 565–566
  from verbal description, 563–564
Sketching, angles, 513–514
Slope
  definition of, 73
  graphical meaning of, 77
  of linear function, 73
  of terminal side of angle, 574–575
Slope-intercept form
  definition of, 74
  graph of, 74–75
  of linear function, 74–77
Speed, constant, 68–69
Square matrix, 130
Squares
  least, 96
  sum of, 96

Standard form
  of cubic function, 283
  definition of, 78
  graph of function in, 79–80
  of linear function, 78–80
  quadratic equation in, 246–249
  of quadratic function, 262–265
  writing formula for linear function in, 78–79
Stretch
  horizontal
    to change units, 209
    of cosine function, 550–551
    definition of, 207
    of function, 203–205
    of sine function, 550–551
  vertical
    to change units, 197
    of cosine function, 549–550
    definition of, 194
    of function, 192–194
    of sine function, 549–550
Substitution method, for system of equations, 112–116
Subtraction, of vectors, 667
Successive differences, 232–233
Sum
  of arithmetic sequence, 797
  of geometric sequence, 801
  of infinite geometric series, 803
Sum identities, 704–708, 714, 715
Sum of squares, 96
Sum-to-product identities, 708–711, 715
Symbolic notation, translating function in words into, 43
Symmetry
  definition of, 184
  even functions and, 184
  line of, 184
  odd functions and, 184
  testing for, 185–186
System
  of equations
    definition of, 109
    dependent, 122, 134–135
    elimination method for, 117–122
    equivalent, 118–119
    graph in solving, 111–112
    inconsistent, 122, 136–137
    with infinitely many solutions, 120–122
    linear, 109–123
    matrix for representation of, 130
    modeling with, 128–129
    with no solution, 120
    in real-world context, 136–137
    solving with matrix, 133–134, 134–135
    substitution method for, 112–116
    with table, 110–111
    three or more, 122–123

T

Table
  composite functions defined by, 439–440
  determining, if represents function, 23
  evaluating function from, 16–17
  functions represented by, 22–29
  interpreting in words meaning of, 45
  linear model from, 87–89
  model from, choosing, 464–471
  periodic behavior from, 499–500
  periodic data from, 501–502
  predicting unknown data values using, 26
  shifts on, 168
  sinusoidal model from, 565–566
  solving equation from, 16–17
  system of equations with, 110–111
Tail, of vector, 660
Tangent function
  on calculator, 575
  definition of, 574
  graph of, 575–577
  inverse
    definition of, 596
    domain of, 602
    graph of, 601–602
    using, 595–596
  right triangle and, 620
  using, 577
  values of, 583
Technology Tips
  adding two matrices, 148
  aligning a data set, 456
  creating lists of values for a data set, 102
  drawing a scatter plot, 102
  entering a matrix, 138
  exponential regression, 364
  finding a point of intersection, 123
  finding the reduced row echelon form of a matrix, 138
  graphing a function, 81
  graphing a sequence, 786
  linear regression, 103
  logarithmic regression, 397
  logistic regression, 456
  polar graphs, 658
  power regression, 307
  quadratic regression, 254
  scalar multiplication, 149
Terminal side
  definition of, 513
  slope of, 574–575
Test
  for symmetry, 185–186
  vertical line, 13
Time, doubling, 356–358
Tip, of vector, 660
Transformations
  combining, 197–198, 209–212
  compression
    from equation, 209
    horizontal
      to change units, 209
      of cosine function, 550–551
      definition of, 207
      of function, 205–207
      horizontal compression of, 597
      of sine function, 550–551
    vertical
      to change units, 197
      of cosine function, 549–550
      definition of, 194
      of function, 192–194
      graph of, 195
      rate of change and, 195–197
      of sine function, 549–550
  of cosine functions, 548–553
  multiple, 181–182
    of cosine function, 552
    of sine function, 552
  of periodic functions, 498–499
  reflections
    of cosine function, 549
    horizontal, 177–180
    shifts and, 180–183
    of sine function, 549
    vertical, 175–177
  shift
    analyzing, 168–169
    of cosine function, 548
    horizontal
      definition of, 164
      on graph, 164–166
      reflections and, 180–183
      of sine function, 548
      on table, 168
    vertical
      definition of, 163
      on graph, 162–163
      in real-world context, 164
      relationship between two functions differing by, 163
  of sine functions, 548–553
  stretch
    horizontal
      to change units, 209
      of cosine function, 550–551
      definition of, 207
      of function, 203–205
      of sine function, 550–551
    vertical
      to change units, 197
      of cosine function, 549–550
      definition of, 194
      of function, 192–194
      of sine function, 549–550
Transverse axis, of hyperbola, 759
Trends, in data, 5
Triangle
  acute, 627
  altitude of, 535
  isosceles right, 537
  oblique, 627, 628–629, 638–639
  obtuse, 627
  right
    adjacent side of, 619
    hypotenuse of, 619
    opposite side of, 619
    trigonometric functions and, 620
  similar
    definition of, 618
    measuring with, 618–619
  solving, 628
  types of, 627–628
Triangulation, 640–644
Trigonometric equations
  solving by graphing, 554–555
  solving with cofunction identities, 686–687
  solving with even-odd identities, 684–685
  solving with identities, 682
  solving with Pythagorean identities, 688–690
Trigonometric functions
  construction approach for, 699–700
  inverse, 588–604
  modeling with, 562–567
  reciprocal, 577–583
Trigonometric identity(ies)
  cofunction, 686–687
  definition of, 680–681
  determining existence of, 681
  difference, 704–708, 715
  double-angle, 712–713, 715
  even-odd, 682–684
  fundamental, 681–682
  half-angle, 713–714, 715
  product-to-sum, 711–712, 715
  Pythagorean, 688–690
  for solving trigonometric equations, 682
  sum, 704–708, 715
  sum-to-product, 708–711, 715
  verification of, 692–700
Trigonometry, geometry and, 619–621
Tsunami, 562

U

Unit circle
  cosine values on, 535–540
  defining, 523–526
  radians in, 511
  with radius other than 1, 526–528
  reference angle and, 533–535
  sine values on, 535–540
Units of measure, 86–87
Unit vectors, 667
Upper limiting value, 451

V

Value(s)
  future, 370–371
  initial, 37, 74, 360
  limiting, 450
    lower, 451
    upper, 451

maximum, 259
minimum, 259
Van Ceulen, Ludolph, 86
Variable
    definition of, 11
    dependent, 14
    independent, 14
Variable costs, 114
Variable rates of change, 228–236
Variation, 303–306
    direct
        definition of, 304
        examples of, 304
    inverse
        definition of, 304
        examples of, 304
        modeling of, 305–306
Vector(s)
    addition of, geometric, 663–665
    in aviation, 662
    definition of, 660
    direction of, 662–663
    displacement, in component form, 661
    for force, 668–670
    geometric, 660–662
    magnitude of, 660, 662–663
    position, 660
    scalar multiplication with, 665–667
    subtraction of, 667
    tail of, 660
    tip of, 660
    unit, 667
Verhulst, P. F., 451
Vertex
    of hyperbola, 759
    of parabola, 258, 263–265, 746
Vertex form, of quadratic functions, 257–260
Vertical asymptote
    definition of, 315
    finding, from function equation, 317
    in real-world context, 315–317
Vertical compression
    to change units, 197
    of cosine function, 549–550
    definition of, 194
    of function, 192–194
    graph of, 195
    rate of change and, 195–197
    of sine function, 549–550
Vertical intercept, 37, 75
Vertical lines, 76
Vertical line test, 13
Vertical reflections, 175–177
Vertical shift
    definition of, 163
    on graph, 162–163
    in real-world context, 164
    relationship between two functions differing by, 163
Vertical stretch
    to change units, 197
    of cosine function, 549–550
    definition of, 194
    of function, 192–194
    of sine function, 549–550
Vertices, of ellipse, 728

W

Words
    functions represented by, 42–46
    indicative of linear model use, 85
    interpreting meaning of function table in, 45
    interpreting meaning of function using, 44–46
    linear function from, 73–74
    linear model from, 87
    model from, 472–475
    recognizing function in, 42–44
    sinusoidal model from, 563–564
    translating function in, to graph, 43–44
    translating function in, to symbolic notation, 43

GENERAL FORMULA FOR COSINE AND SINE FUNCTIONS

For functions of the form

$x(\theta) = A\cos(B(\theta - C)) + D$ or

$y(\theta) = A\sin(B(\theta - C)) + D$

- $|A|$, the vertical stretch factor, is the *amplitude* of the graph. If $A < 0$, then the function is also reflected vertically.
- $|B|$ is the *angular frequency,* which is the number of periods that will be completed in the interval $[0°, 360°]$ or $[0, 2\pi$ radians$]$ for θ. If $B < 0$, then the function is also reflected horizontally.
- $\dfrac{360°}{|B|}$ or $\dfrac{2\pi}{|B|}$ radians is the *period* of the function, the length of the interval required to complete one complete cycle.
- $\dfrac{|B|}{360°}$ or $\dfrac{|B|}{2\pi}$ radians is the *frequency* of the function, the fraction of a period completed within a one-unit interval.
- $|C|$ is the *horizontal shift* of the function. If C is positive, the shift is to the right. If C is negative, the shift is to the left.
- $|D|$ is the *vertical shift* of the function. If D is positive, the shift is up. If D is negative, the shift is down. The midline is located at $x = D$ or $y = D$.

CONIC SECTIONS

The Ellipse: Major Axis Horizontal, Center at (h, k)

The standard equation of an ellipse with center at (h, k) and major axis horizontal is
$$\dfrac{(x-h)^2}{a^2} + \dfrac{(y-k)^2}{b^2} = 1 \text{ where } a > b > 0 \text{ and } b^2 = a^2 - c^2.$$

The vertices of the ellipse are $(h + a, k)$ and $(h - a, k)$.

The endpoints of the minor axis are $(h, k + b)$ and $(h, k - b)$.

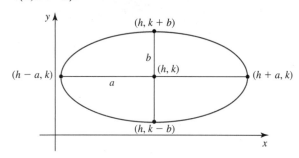

The Ellipse: Major Axis Vertical, Center at *(h, k)*

The standard equation of an ellipse with center at (h, k) and major axis vertical is
$$\dfrac{(x-h)^2}{b^2} + \dfrac{(y-k)^2}{a^2} = 1 \text{ where } a > b > 0 \text{ and } b^2 = a^2 - c^2.$$

The vertices of the ellipse are $(h, k + a)$ and $(h, k - a)$.

The endpoints of the minor axis are $(h + b, k)$ and $(h - b, k)$.

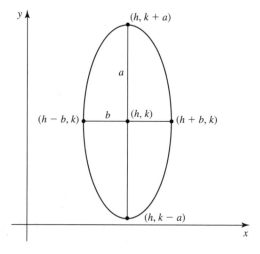

Standard Equation of the Circle

The standard equation of a circle with center at (h, k) and radius r is $(x - h)^2 + (y - k)^2 = r^2$.

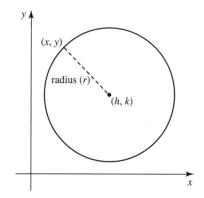